HANDBOOK OF

FOOD PROCESSING

Food Preservation

Contemporary Food Engineering

Series Editor

Professor Da-Wen Sun, Director
Food Refrigeration & Computerized Food Technology
National University of Ireland, Dublin
(University College Dublin)
Dublin, Ireland
http://www.ucd.ie/sun/

Handbook of Food Processing: Food Preservation, *edited by Theodoros Varzakas and Constantina Tzia* (2015)

Handbook of Food Processing: Food Safety, Quality, and Manufacturing Processes, *edited by Theodoros Varzakas and Constantina Tzia* (2015)

Edible Food Packaging: Materials and Processing Technologies, *edited by Miquel Angelo Parente Ribeiro Cerqueira, Ricardo Nuno Correia Pereira, Oscar Leandro da Silva Ramos, Jose Antonio Couto Teixeira, and Antonio Augusto Vicente* (2015)

Advances in Postharvest Fruit and Vegetable Technology, *edited by Ron B.H. Wills and John Golding* (2015)

Engineering Aspects of Food Emulsification and Homogenization, *edited by Marilyn Rayner and Petr Dejmek* (2015)

Handbook of Food Processing and Engineering, Volume II: Food Process Engineering, *edited by Theodoros Varzakas and Constantina Tzia* (2014)

Handbook of Food Processing and Engineering, Volume I: Food Engineering Fundamentals, *edited by Theodoros Varzakas and Constantina Tzia* (2014)

Juice Processing: Quality, Safety and Value-Added Opportunities, *edited by Víctor Falguera and Albert Ibarz* (2014)

Engineering Aspects of Food Biotechnology, *edited by José A. Teixeira and António A. Vicente* (2013)

Engineering Aspects of Cereal and Cereal-Based Products, *edited by Raquel de Pinho Ferreira Guiné and Paula Maria dos Reis Correia* (2013)

Fermentation Processes Engineering in the Food Industry, *edited by Carlos Ricardo Soccol, Ashok Pandey, and Christian Larroche* (2013)

Modified Atmosphere and Active Packaging Technologies, *edited by Ioannis Arvanitoyannis* (2012)

Advances in Fruit Processing Technologies, *edited by Sueli Rodrigues and Fabiano Andre Narciso Fernandes* (2012)

Biopolymer Engineering in Food Processing, *edited by Vânia Regina Nicoletti Telis* (2012)

Operations in Food Refrigeration, *edited by Rodolfo H. Mascheroni* (2012)

Thermal Food Processing: New Technologies and Quality Issues, Second Edition, *edited by Da-Wen Sun* (2012)

Physical Properties of Foods: Novel Measurement Techniques and Applications, *edited by Ignacio Arana* (2012)

Handbook of Frozen Food Processing and Packaging, Second Edition, *edited by Da-Wen Sun* (2011)

Advances in Food Extrusion Technology, *edited by Medeni Maskan and Aylin Altan* (2011)

Enhancing Extraction Processes in the Food Industry, *edited by Nikolai Lebovka, Eugene Vorobiev, and Farid Chemat* (2011)

Emerging Technologies for Food Quality and Food Safety Evaluation, *edited by Yong-Jin Cho and Sukwon Kang* (2011)

Food Process Engineering Operations, *edited by George D. Saravacos and Zacharias B. Maroulis* (2011)

Biosensors in Food Processing, Safety, and Quality Control, *edited by Mehmet Mutlu* (2011)

Physicochemical Aspects of Food Engineering and Processing, *edited by Sakamon Devahastin* (2010)

Infrared Heating for Food and Agricultural Processing, *edited by Zhongli Pan and Griffiths Gregory Atungulu* (2010)

Mathematical Modeling of Food Processing, *edited by Mohammed M. Farid* (2009)

Engineering Aspects of Milk and Dairy Products, *edited by Jane Sélia dos Reis Coimbra and José A. Teixeira* (2009)

Innovation in Food Engineering: New Techniques and Products, *edited by Maria Laura Passos and Claudio P. Ribeiro* (2009)

Processing Effects on Safety and Quality of Foods, *edited by Enrique Ortega-Rivas* (2009)

Engineering Aspects of Thermal Food Processing, *edited by Ricardo Simpson* (2009)

Ultraviolet Light in Food Technology: Principles and Applications, *Tatiana N. Koutchma, Larry J. Forney, and Carmen I. Moraru* (2009)

Advances in Deep-Fat Frying of Foods, *edited by Serpil Sahin and Servet Gülüm Sumnu* (2009)

Extracting Bioactive Compounds for Food Products: Theory and Applications, *edited by M. Angela A. Meireles* (2009)

Advances in Food Dehydration, *edited by Cristina Ratti* (2009)

Optimization in Food Engineering, *edited by Ferruh Erdoğdu* (2009)

Optical Monitoring of Fresh and Processed Agricultural Crops, *edited by Manuela Zude* (2009)

Food Engineering Aspects of Baking Sweet Goods, *edited by Servet Gülüm Sumnu and Serpil Sahin* (2008)

Computational Fluid Dynamics in Food Processing, *edited by Da-Wen Sun* (2007)

HANDBOOK OF FOOD PROCESSING

Food Preservation

EDITED BY

Theodoros Varzakas • Constantina Tzia

CRC Press
Taylor & Francis Group
Boca Raton London New York

CRC Press is an imprint of the
Taylor & Francis Group, an **informa** business

CRC Press
Taylor & Francis Group
6000 Broken Sound Parkway NW, Suite 300
Boca Raton, FL 33487-2742

Printed on acid-free paper
Version Date: 20150819

International Standard Book Number-13: 978-1-4987-2175-2 (Hardback)

Visit the Taylor & Francis Web site at
http://www.taylorandfrancis.com

and the CRC Press Web site at
http://www.crcpress.com

Dedicated to my wife, Elia, and my daughter, Fotini,
for their endless support and love.

To my mother for her love and understanding and to the memory of my father.

Theodoros Varzakas

Dedicated to the memory of my parents.

Constantina Tzia

Contents

Series Preface

CONTEMPORARY FOOD ENGINEERING

Food engineering is a multidisciplinary field of applied physical sciences combined with the knowledge of product properties. Food engineers provide the technological knowledge transfer essential for the cost-effective production and commercialization of food products and services. In particular, food engineers develop and design processes and equipment to convert raw agricultural materials and ingredients into safe, convenient, and nutritious consumer food products. However, food engineering topics are continuously undergoing changes to meet diverse consumer demands, and the subject is being rapidly developed to reflect market needs.

In the development of food engineering, one of the many challenges is to employ modern tools and knowledge, such as computational materials science and nanotechnology, to develop new products and processes. Simultaneously, improving food quality, safety, and security continues to be a critical issue in food engineering studies. New packaging materials and techniques are being developed to provide more protection to foods, and novel preservation technologies are emerging to enhance food security and defense. Additionally, process control and automation regularly appear among the top priorities identified in food engineering. Advanced monitoring and control systems are developed to facilitate automation and flexible food manufacturing. Furthermore, energy saving and minimization of environmental problems continue to be important food engineering issues, and significant progress is being made in waste management, efficient utilization of energy, and reduction of effluents and emissions in food production.

The Contemporary Food Engineering Series, consisting of edited books, attempts to address some of the recent developments in food engineering. The series covers advances in classical unit operations in engineering applied to food manufacturing as well as such topics as progress in the transport and storage of liquid and solid foods; heating, chilling, and freezing of foods; mass transfer of foods; chemical and biochemical aspects of food engineering and the use of kinetic analysis; dehydration, thermal processing, nonthermal processing, extrusion, liquid food concentration, membrane processes, and applications of membranes in food processing; shelf life and electronic indicators in inventory management; sustainable technologies in food processing; and packaging, cleaning, and sanitation. These books are aimed at professional food scientists, academics researching food engineering problems, and graduate-level students.

The editors of these books are leading engineers and scientists from different parts of the world. All the editors were asked to present their books to address the market's needs and pinpoint cutting-edge technologies in food engineering.

All contributions are written by internationally renowned experts who have both academic and professional credentials. All authors have attempted to provide critical, comprehensive, and readily accessible information on the art and science of a relevant topic in each chapter, with reference lists for further information. Therefore, each book can serve as an essential reference source to students and researchers in universities and research institutions.

Da-Wen Sun
Series Editor

Series Editor

Born in Southern China, Professor Da-Wen Sun is a global authority in food engineering research and education; he is a member of the Royal Irish Academy (RIA), which is the highest academic honor in Ireland; he is also a member of Academia Europaea (The Academy of Europe) and a fellow of the International Academy of Food Science and Technology. He has significantly contributed to the field of food engineering as a researcher, as an academic authority and as an educator.

His main research activities include cooling, drying, and refrigeration processes and systems; quality and safety of food products; bioprocess simulation and optimization; and computer vision/image processing and hyperspectral imaging technologies. Especially, his many scholarly works have become standard reference materials for researchers in the areas of computer vision, computational fluid dynamics modeling, vacuum cooling, and so on. Results of his work have been published in over 800 papers, including more than 400 peer-reviewed journal papers (Web of Science h-index = 64). He has also edited 14 authoritative books. According to Thomson Reuters's Essential Science Indicators SM, based on data derived over a period of ten years from Web of Science, there are about 4,500 scientists who are among the top one percent of the most cited scientists in the category of Agriculture Sciences, and in the past many years, Professor Sun has consistently been ranked among the very top 50 scientists in the world (he was at the 25th position in March 2015, and in 2nd position if ranking was based on "Highly Cited Papers").

He received a first class BSc Honors and MSc in mechanical engineering and a PhD in chemical engineering in China before working in various universities in Europe. He became the first Chinese national to be permanently employed in an Irish university when he was appointed college lecturer at the National University of Ireland, Dublin (University College Dublin [UCD]), in 1995, and was then continuously promoted in the shortest possible time to senior lecturer, associate professor, and full professor. Dr. Sun is now a professor of Food and Biosystems Engineering and the director of the Food Refrigeration and Computerised Food Technology Research Group at the UCD.

As a leading educator in food engineering, Professor Sun has significantly contributed to the field of food engineering. He has trained many PhD students, who have made their own contributions to the industry and academia. He has also delivered lectures on advances in food engineering on a regular basis in academic institutions internationally and delivered keynote speeches at international conferences. As a recognized authority in food engineering, he has been conferred adjunct/visiting/consulting professorships from 10 top universities in China, including Zhejiang University, Shanghai Jiaotong University, Harbin Institute of Technology, China Agricultural University, South China University of Technology, and Jiangnan University. In recognition of his significant contribution to food engineering worldwide and for his outstanding leadership in the field, the International Commission of Agricultural and Biosystems Engineering (CIGR) awarded him the "CIGR Merit Award" in 2000, and again in 2006, the Institution of Mechanical Engineers based in the United Kingdom named him "Food Engineer of the Year 2004." In 2008, he was awarded the "CIGR Recognition Award" in honor of his distinguished achievements as the top 1% of agricultural engineering scientists in the world. In 2007, he was presented with the only "AFST(I) Fellow Award" in that year by the Association of Food Scientists and Technologists (India), and in 2010, he was presented with the "CIGR Fellow Award"; the title of fellow is the highest honor in CIGR and is conferred to individuals who have made sustained, outstanding contributions worldwide.

In March 2013, he was presented with the "You Bring Charm to the World" Award by Hong Kong–based Phoenix Satellite Television with other award recipients including the 2012 Nobel Laureate in Literature and the Chinese Astronaut Team for Shenzhou IX Spaceship. In July 2013, he received the "Frozen Food Foundation Freezing Research Award" from the International Association for Food Protection (IAFP) for his significant contributions to enhancing the field of food freezing technologies. This is the first time that this prestigious award was presented to a scientist outside the United States. In June 2015, he was presented with the "IAEF Lifetime Achievement Award". This International Association of Engineering and Food (IAEF) award highlights the lifetime contribution of a prominent engineer in the field of food.

He is a fellow of the Institution of Agricultural Engineers and a fellow of Engineers Ireland (the Institution of Engineers of Ireland). He also serves as the editor in chief of *Food and Bioprocess Technology—An International Journal* (2012 Impact Factor = 4.115), former editor of *Journal of Food Engineering* (Elsevier), and editorial board member for a number of international journals, including the *Journal of Food Process Engineering, Journal of Food Measurement and Characterization,* and *Polish Journal of Food and Nutritional Sciences.* He is also a chartered engineer.

On May 28, 2010, he was awarded membership to the RIA, which is the highest honor that can be attained by scholars and scientists working in Ireland; at the 51st CIGR General Assembly held during the CIGR World Congress in Québec City, Canada, on June 13–17, 2010, he was elected. Incoming President of CIGR, became CIGR President in 2013–2014, and is now CIGR Past President.

On September 20, 2011, he was elected to Academia Europaea (The Academy of Europe), which is functioning as the European Academy of Humanities, Letters and Sciences and is one of the most prestigious academies in the world; election to the Academia Europaea represents the highest academic distinction.

Preface

This book presents the necessary information to design food processing operations and methods. It deals with food preservation and describes the equipment needed to carry them out in detail. For every step in the sequence of converting the raw material to the final product, the book covers the most common food preservation processes required.

Chapter 1 describes blanching. Blanching is an important unit operation before processing fruits and vegetables for freezing, pureeing, or dehydration. A case study on the effect of blanching conditions on sulforaphane content in purple and roman cauliflower (*Brassica oleracea l. Var. Botrytis*) is presented.

Chapter 2 deals with thermal processing of foods referring to the application of heat in order to preserve product quality and extend its shelf life. Principles of thermal processing are well described along with thermal process calculations.

Canning of fishery products is described in detail in Chapter 3.

Chapter 4 refers to extrusion cooking with applications in the production of ready-to-eat cereals, pasta, snacks, pet food, fish foods, and confectionery products.

Drying or dehydration of foods is an extremely important food processing operation used to preserve foods for extended periods of time and is described in Chapter 5.

The most popular method for the preservation of fresh foods, especially meat, fish, dairy products, fruit, vegetables, and ready-made meals is chilling and is explored in Chapter 6.

Freezing is continued in Chapter 7, where freezing equipment used, novel methods proposed for freezing, new approaches for the control and optimization of the current cold chain in frozen food distribution, as well as the latest trends, are presented.

Some more recent thermal technologies, for example, microwave energy heating technology, are explored in Chapter 8 in an attempt to find alternatives to conventional heating methods. A case study on microwave preservation of fruit-based products, application to kiwifruit puree, is shown.

Advances in food additives and contaminants are described in Chapter 9. The use of food additives as agents for the improvement of food quality and preservation is well documented.

Ohmic heating, which is comparable to microwave heating, along with its principles and applications are described in Chapter 10.

Chapters 11 and 12 deal with high pressure (HP) processing and especially HP pasteurization which is one of the most interesting nonthermal processes of foods. They cover process design issues, evaluation, technology, and applications.

Pulsed electric field (PEF) processing is one of the promising novel technologies used to process liquid or low-viscosity foods and is described in Chapter 13.

Chapter 14 deals with the basics of magnetic fields technology for food processing and preservation along with some equipment and devices. The use of magnetic fields for microbial inactivation is briefly discussed, and several cases are presented.

Other nonthermal technologies such as ultrasound in food disinfection are described in Chapter 15. The important issues addressed include mechanism of ultrasound disinfection, parameters affecting the effectiveness of ultrasound in disinfection, effects of ultrasound on food quality, and effects of combining ultrasound with other techniques.

The use of edible films and coatings in fresh fruits and vegetables preservation is described in Chapter 16.

Chapters 17 and 18 deal with food packaging—aseptic packaging and modified-atmosphere packaging in fruits and vegetables.

Finally, Chapter 19 describes biosensor technology in food, and Chapter 20 deals with ozone applications presenting a general overview of the use of ozone in the food industry, along with a discussion on the chemical properties of this chemical.

Editors

Theodoros Varzakas earned a bachelor's (honors) degree in microbiology and biochemistry (1992), a PhD in food biotechnology, and an MBA in food from Reading University, United Kingdom (1998). Dr. Varzakas was a postdoctoral research staff member at the same university. He has worked for large pharmaceutical and multinational food companies in Greece for five years and has also for at least 14 years experience in the public sector. Since 2005, he has served as assistant and associate professor in the Department of Food Technology, Technological Educational Institute of Peloponnese (ex Kalamata), Greece, specializing in the issues of food technology, food processing, food quality, and safety. Dr. Varzakas has been a reviewer in many international journals such as *International Journal of Food Science & Technology*, *Journal of Food Engineering*, *Waste Management*, *Critical Reviews in Food Science and Nutrition*, *Italian Journal of Food Science*, *Journal of Food Processing and Preservation*, *Journal of Culinary Science and Technology*, *Journal of Agricultural and Food Chemistry*, *Journal of Food Quality*, *Food Chemistry*, and *Journal of Food Science*. He has written more than 90 research papers and reviews and has presented more than 90 papers and posters in national and international conferences. He has written two books in Greek; one on genetically modified food and the other on quality control in food. He edited a book on sweeteners that was published by CRC Press in 2012 and another book on biosensors published by CRC Press in 2013. Dr. Varzakas has participated in many European and national research programs as coordinator or scientific member. He is a fellow of the Institute of Food Science & Technology (2007).

Constantina Tzia earned a diploma in chemical engineering (1977) and a PhD in food engineering (1987) from the National Technical University of Athens, Greece. Her current research interests include quality and safety (HACCP) of foods, sensory evaluation, fats and oils, dairy and bakery technology, and utilization of food by-products. Professor Tzia's work has been widely published and presented, appearing in prestigious publications such as the *Journal of Food Science*, *LWT–Food Science and Technology*, *Innovative Food Science and Emerging Technologies*, *Food and Bioprocess Technology*, and *Journal of the American Oil Chemists' Society*.

Contributors

Z. Alexandrakis
Laboratory of Food Chemistry and Technology
School of Chemical Engineering
National Technical University of Athens
Athens, Greece

Gustavo V. Barbosa-Cánovas
Center for Nonthermal Processing of Food
Washington State University
Pullman, Washington

Herna Barrientos
Department of Chemical Engineering
Faculty of Engineering
Universidad de Santiago de Chile
Santiago, Chile

M. Benlloch-Tinoco
Department of Food Technology
Polytechnic University of Valencia
Valencia, Spain

Daniela Bermúdez-Aguirre
Center for Nonthermal Processing of Food
Washington State University
Pullman, Washington

Charikleia Chranioti
Laboratory of Food Chemistry and Technology
School of Chemical Engineering
National Technical University of Athens
Athens, Greece

E. Dermesonlouoglou
Laboratory of Food Chemistry and Technology
School of Chemical Engineering
National Technical University of Athens
Athens, Greece

Gulsun Akdemir Evrendilek
Faculty of Engineering and Architecture
Department of Food Engineering
Abant İzzet Baysal University
Bolu, Turkey

Manuel Garcia-Perez
Center for Nonthermal Processing of Food
Washington State University
Pullman, Washington

Maria Giannakourou
Department of Food Technology
Technological Educational Institute of Athens
Athens, Greece

Virginia Giannou
Laboratory of Food Chemistry and Technology
School of Chemical Engineering
National Technical University of Athens
Athens, Greece

Eleni Gogou
Laboratory of Food Chemistry and Technology
School of Chemical Engineering
National Technical University of Athens
Athens, Greece

T.K. Srinivasa Gopal
Fish Processing Division
Central Institute of Fisheries Technology
Indian Council of Agricultural Research
Kerala, India

Athanasia M. Goula
Department of Food Science and Technology
Aristotle University of Thessaloniki
Thessaloniki, Greece

S. Grabowski
Food Research and Development Centre
Agriculture and Agri-Food Canada
Saint-Hyacinthe, Québec, Canada

Victor Haber-Perez
Center for Nonthermal Processing of Food
Washington State University
Pullman, Washington

Chinnadurai Karunanithy
Department of Food and Nutrition
University of Wisconsin-Stout
Menomonie, Wisconsin

George Katsaros
Laboratory of Food Chemistry and Technology
School of Chemical Engineering
National Technical University of Athens
Athens, Greece

Yuh Xiu Liew
Faculty of Engineering
Manufacturing and Industrial Processes Research Division
The University of Nottingham Malaysia Campus
Selangor, Malaysia

Andrea Mahn
Department of Chemical Engineering
Faculty of Engineering
Universidad de Santiago de Chile
Santiago, Chile

Sivakumar Manickam
Faculty of Engineering
Manufacturing and Industrial Processes Research Division
The University of Nottingham Malaysia Campus
Selangor, Malaysia

E. Manolopoulou
Department of Food Technology
Technological Educational Institute of Peloponnese
Kalamata, Greece

N. Martínez-Navarrete
Department of Food Technology
Polytechnic University of Valencia
Valencia, Spain

Mariela Miranda
Department of Chemical Engineering
Universidad de Santiago de Chile
Santiago, Chile

C.O. Mohan
Fish Processing Division
Central Institute of Fisheries Technology
Indian Council of Agricultural Research
Kerala, India

M. Mondor
Food Research and Development Centre
Agriculture and Agri-Food Canada
Saint-Hyacinthe, Québec, Canada

Kasiviswanathan Muthukumarappan
Department of Agricultural and Biosystems Engineering
South Dakota State University
Brookings, South Dakota

Georgia-Paraskevi Nikoleli
Laboratory of Inorganic and Analytical Chemistry
School of Chemical Engineering
National Technical University of Athens
Athens, Greece

Dimitrios P. Nikolelis
Laboratory of Environmental Chemistry
Department of Chemistry
University of Athens
Athens, Greece

Spyridon E. Papadakis
Department of Food Technology
Technological Educational Institute of Athens
Athens, Greece

Carmen Pérez
Department of Chemical Engineering
Faculty of Engineering
Universidad de Santiago de Chile
Santiago, Chile

C.N. Ravishankar
Fish Processing Division
Central Institute of Fisheries Technology
Indian Council of Agricultural Research
Kerala, India

D. Rodrigo
Department of Food Preservation and Food Quality
Institute of Agrochemistry and Food Technology
Valencia, Spain

Oselys Rodriguez-Justo
Center for Nonthermal Processing of Food
Washington State University
Pullman, Washington

A. Salvador
Department of Food Preservation and Food Quality
Institute of Agrochemistry and Food Technology (IATA-CSIC)
Valencia, Spain

Theodora Spiliotaki
Laboratory of Food Chemistry and Technology
School of Chemical Engineering
National Technical University of Athens
Athens, Greece

Nikolaos G. Stoforos
Department of Food Science and Human Nutrition
Agricultural University of Athens
Athens, Greece

Petros Taoukis
Laboratory of Food Chemistry and Technology
School of Chemical Engineering
National Technical University of Athens
Athens, Greece

Loucas Tasios
Laboratory of Food Chemistry and Technology
School of Chemical Engineering
National Technical University of Athens
Athens, Greece

Constantina Tzia
Laboratory of Food Chemistry and Technology
School of Chemical Engineering
National Technical University of Athens
Athens, Greece

Theodoros Varzakas
Department of Food Technology
Technological Educational Institute of Peloponnese
Kalamata, Greece

S. Villeneuve
Food Research and Development Centre
Agriculture and Agri-Food Canada
Saint-Hyacinthe, Québec, Canada

M. Reza Zareifard
Food Research and Development Centre
Agriculture and Agri-Food Canada
Saint-Hyacinthe, Québec, Canada

1 Blanching

Theodoros Varzakas, Andrea Mahn, Carmen Pérez, Mariela Miranda, and Herna Barrientos

CONTENTS

1.1 INTRODUCTION

Blanching is an important unit operation before processing fruits and vegetables for freezing, pureeing, or dehydration.

Blanching also lowers the mass of vegetables; so process profitability can be affected by overtreatment. Commercial blanchers used in the vegetable canning industry are relatively intensive with energy and water consumption. Energy utilization is affected by the equipment used and also by the configuration of the following freezing step.

Furthermore, conventional blanching produces wastewater that can reduce the nutritional value of vegetables by leaching of soluble compounds and subsequently increasing the pollutant discharge (Poulsen, 1986; Williams et al., 1986).

1.2 BLANCHING AND CARROTS

Carrots are well known for their sweetening, antianemic, healing, diuretic, and sedative properties. The enzymes commonly found to have deteriorative effects in carrots are peroxidases (PODs) and catalase. In order to minimize deteriorative reactions, fruits and vegetables are heat treated or blanched to inactivate the enzymes. Blanching of fruits and vegetables is done either in hot water, steam, or selected chemical solutions (Luna-Guzmán and Barret, 2000; Severini et al., 2004a,b).

Blanching in a hot calcium chloride solution is used to increase the firmness of fruits and vegetables because of the activation of pectin methylesterase (PME) (Quintero-Ramos et al., 2002).

The inactivation of POD is usually used to indicate blanching sufficiency as POD is ubiquitous.

Moreover, optimization of the blanching process with respect to nutrient retention (β-carotene, vitamin C loss) and product yield should be considered along with enzyme inactivation (Shivhare et al., 2009). They determined the optimum blanching conditions for carrots in terms of nutrient (vitamin C and β-carotene) retention and studied the kinetics of the inactivation of POD in carrot juice. Various enzyme inactivation models were tested on the basis of statistical and physical parameters to ascertain a suitable model capable of explaining POD inactivation kinetics.

Steam blanching resulted in nonuniformity of enzyme inactivation, and the inactivation times of catalase and POD during steam blanching were consistently higher than that of hot water, acetic acid, or calcium chloride solution blanching.

The best blanching treatment for carrots based on these process parameters was 95°C for 5 min in water. At this time–temperature combination, both POD and catalase were inactivated and 8.192 mg/100 g vitamin C, 55% yield of carrot juice, and 3.18 mg/100 g β-carotene content were observed.

Blanching treatment of carrots prior to juice extraction has been found to be an important step in the production of carrot juice, which improves color and cloud stability (Martin et al., 2003; Zhou et al., 2009).

The effect of three processing steps (blanching, enzyme liquefaction, and pasteurization) on polyphenol and the antioxidant activity of carrot juices was investigated by Ma et al. (2013).

Water blanching was carried out at 86°C for 10 min.

Polyphenols and antioxidant activity of carrot juices varied with different processes. Five polyphenolic acids were identified in fresh carrot juice, and the predominant compound was chlorogenic acid. Compared with fresh carrot juice, blanching and enzyme liquefaction could result in the increase of total polyphenol content (TPC) and antioxidant activity in scavenging DPPH free radicals (DPPH) and Fe^{2+}-chelating capacity (FC), whereas pasteurization could result in the decrease of TPC and antioxidant activity in DPPH and FC. Meanwhile blanching, enzyme liquefaction, and pasteurization showed little influence on the antioxidant activity in lipid peroxidation protection. The antioxidant activities in DPPH and FC increased with increasing concentration while no correlation between lipid peroxidation protection and polyphenols concentration was evident. Polyphenols still retained high antioxidant activity after the processes, which have potential health benefits for consumers.

Blanching and enzyme liquefaction helped the dissolution of polyphenols into the juice.

Freezing of vegetables is generally accompanied by other processing operations such as blanching, which is applied to inactivate enzymes implicated in color change, flavor deterioration, and tissue softening during frozen storage. This thermal treatment (i.e. blanching), when conducted at temperatures higher than 80°C, catalyzes the degradation of pectins due to β-elimination reaction (Sila et al., 2008) and their solubilization from the cell wall and the middle lamella between adjacent cell walls.

Blanching, freezing, and frozen storage, depending on the process conditions, can cause dramatic effects on the textural properties of frozen products (Prestamo et al., 1998; Roy et al., 2001).

Numerous investigations have been carried out on the texture of carrots effected by different blanching treatments (Lee et al., 1979; Bourne, 1987; Verlinden and De Baerdemaeker, 1997; Vu et al., 2004) and different rates of freezing (Rahaman et al., 1971; Fuchigami et al., 1994). In particular, low-temperature (60°C–75°C) blanching has been acknowledged to increase the cell wall strength in carrots (Fuchigami et al., 1995; Sanjuan et al., 2005) due to PME activation. In fact, this enzyme is able to demethylate cell wall pectins, producing cross-linking of pectin molecules in the presence of calcium ions, and this results in the strengthening of the cell walls (Quintero-Ramos et al., 2002; van Buggenhout et al., 2006; Rastogi et al., 2008).

Typically, blanching is carried out by treating the vegetable with steam or hot water for 1–10 min at 75°C–95°C; the time/temperature combination selected is dependent on the type of vegetable. In the case of carrots, low-temperature/long-time and high-temperature/short-time blanching methods have been applied (Sanjuan et al., 2005; Shivhare et al., 2009).

The effect of previous ultrasound and conventional blanching treatments on drying and quality parameters (2-furoylmethyl amino acids—as indicators of lysine and arginine participation in the Maillard reaction—carbohydrates, total polyphenols, protein profile, rehydration ratio, microstructure changes) of convective dehydrated carrots has been assessed by Gamboa-Santos et al. (2013). The most striking feature was the influence of blanching on the subsequent 2-furoylmethyl-amino acid formation during drying, probably due to changes in the protein structure. The highest values of 2-furoylmethyl amino acids were found in carrots conventionally blanched with water at 95°C for 5 min. However, samples previously treated by ultrasound presented intermediate values of 2-furoylmethyl amino acids and carbohydrates as compared to the conventionally blanched samples. Dried carrots previously subjected to ultrasound blanching preserved their total polyphenol content and showed rehydration properties, which were even better than those of the freeze-dried control sample. The results obtained here underline the usefulness of 2-furoylmethyl amino acids as indicators of the damage suffered by carrots during their blanching and subsequent drying.

1.3 BLANCHING AND ACIDIFIED VEGETABLES

In the case of acidified vegetables, it is important to understand the effect of different pretreatments, such as blanching and equilibration of the product in a solution containing acid and salt, on the dielectric properties of food materials. Within these treatments, factors, such as acid and salt concentrations as well as the equilibration time, may affect dielectric properties, and in turn influence microwave heating.

Sarang et al. (2007) reduced electrical conductivity variation, thereby improving heating uniformity of chicken chow mein through selective blanching treatments of food components in a highly conductive, salt-containing sauce prior to ohmic heating. This finding is highly relevant to dielectric heating since electrical conductivity is a major component of the dielectric loss factor.

Koskiniemi et al. (2013) examined the effects of acid and salt concentration on the dielectric properties of acidified vegetables. Broccoli florets and sweet potato cubes (1.2 cm) were blanched to facilitate acid and salt equilibration by heating for 15 s in boiling deionized water. Red bell pepper cubes were not blanched. The vegetable samples were then acidified in solutions of 1%–2% sodium chloride with 0.5%–2% citric acid. Dielectric properties were measured at 915 MHz from 25°C to 100°C after 0, 4, and 24 h soaking periods in the solutions using an open-ended coaxial probe connected to a network analyzer. Equilibration occurred within 4 h of salting and acidification. Acid and salt concentration had no significant effect on the dielectric constant (ε'). However, ε' was significantly different among vegetables ($p < 0.05$). Dielectric loss factor (ε'') was not affected by the acid, but significantly increased with salt concentration. These results provide the necessary dielectric property information to apply microwave heating technology in the processing of acidified vegetables.

1.4 BLANCHING AND SUGARS

The nutritive, physicochemical, and technological characteristics of several intermediate food products (IFPs) from Spanish Confitera fresh date coproducts were investigated by Martin-Sanchez et al. (2014). Three IFPs were obtained, two from unblanched dates in different ripening stages (Khalal and Rutab) and a third one from blanched Khalal fruits. The IFPs were rich in dietary fiber (13%–16%, dry matter), phenolics (0.56–4.26 g GAE/100 g, dry matter), and sugars (55%–82%, dry matter), with glucose and fructose as the predominant sugars.

Malic acid was the major organic acid, and potassium was the main mineral. Blanching Khalal dates helped prevent browning in the IFP, but the thermal treatment modified the sugars profile. The results indicated that both maturity stages yield IFPs with potential in the food industry; and according to their sugar and phenolic content, they could be suitable for the elaboration of new ingredients with different industrial applications. In addition, it would be recommendable to blanch unripe fruits.

Both IFPs from unblanched fruits presented similar total dietary fiber (TDF) values (p >0.05), although some decrease during maturation has been reported related to the enzymatic activity responsible for the softening of dates by Ashraf and Hamidi-Esfahani (2011). Khalal-blanched dates presented the highest TDF, insoluble dietary fiber (IDF), and soluble dietary fiber (SDF) (p <0.05) effect due to the loss of other components into the boiling water, which has affected the proportion of the components.

Blanching did not affect (p >0.05) the total sugar content; however, it caused a decrease of glucose and fructose but an accumulation of sucrose. The same effect was observed by Perkins-Veazie et al. (1994) and Barrett et al. (2000), who found higher sucrose content in blanched sweet corn and more reducing sugars in unblanched samples.

Scalded dates duplicated their TPC (p <0.05), possibly due to the polyphenoloxidase inactivation during blanching, protecting the phenolics against oxidation, particularly during homogenization.

Wen et al. (2010) explained this increase after blanching in vegetables by a different hypothesis: as a possible breakdown of tannins due to the high temperatures favoring their extractability; as a disruption of the cell membranes, which could provoke that phenolics, usually bonded to dietary fiber, proteins, or sugars in complex structures in the plants, become more available; and, also as the possible formation of phenolic compounds during the thermal process due to a higher availability of precursors.

The potential thermo-protective effect of sugars on the microstructure and the mechanical properties of the carrot tissue during blanching were investigated in a recent study (Neri et al., 2011). The protective effect of trehalose and maltose on the microstructural properties of the carrot tissue was highlighted by cryo-scanning electron microscopy (SEM) analysis.

However, when slices of the vegetable underwent heat treatments at 90°C for 3 and 10 min, no meaningful effects were noticed at the textural level.

Raw carrots and carrots blanched in water and in 4% trehalose and maltose solutions at 75°C for 3 (A) and 10 min (C) and at 90°C for 3 (B) and 10 min (D) were frozen and stored at −18°C for 8 months. The effects of heating conditions and exogenous added sugars on the mechanical properties and microstructure of the vegetable after blanching and during frozen storage were studied by Neri et al. (2014).

By the SEM analysis, no significant differences were observed among samples A and B water-blanched and raw carrots, while a thermo-protective effect due to the addition of sugars was evidenced in sample D, which had undergone the most severe thermal treatment. Freezing and frozen storage determined several fractures on both raw and blanched carrots due to ice crystals formation and recrystallization.

The cryoprotective effect of the sugars on the vegetable microstructure was observed only in the "over-blanched" sample D.

The mechanical properties of carrots were affected by blanching, which caused a decrease in hardness, but after freezing and 1 month of frozen storage, all samples showed a further dramatic reduction of hardness.

Only samples characterized by a pectinesterase residual activity showed softening after 1 month of frozen storage likely due to a competitive effect of the thermo-protective ability of trehalose on this enzyme. The exogenous trehalose was able to limit the loss of hardness of carrots that had undergone B, C, and D blanching pretreatments.

1.5 WATER BLANCHING

Water blanching is employed to extend the shelf life of certain vegetable-based foods such as ready meals and frozen vegetables, since blanching inactivates enzymes responsible for food deterioration, and it also reduces microbial count (Bahceci et al., 2005; Olivera et al., 2008; Volden et al., 2009). However, thermal processing such as blanching can induce losses of important compounds due to thermal degradation and leaching into cooking water (Rungapamestry et al., 2007; Olivera et al., 2008; Volden et al., 2008, 2009).

Alvarez-Jubete et al. (2014) investigated the effect of combined pressure/temperature treatments (200, 400, and 600 MPa, at 20°C and 40°C) on key physical and chemical characteristics of white cabbage (*Brassica oleracea* L. var. *capitata alba*). Thermal treatment (blanching) was also investigated and compared with high-pressure processing (HPP). HPP at 400 MPa and 20°C–40°C caused significantly larger color changes compared to any other pressure or thermal treatment.

All pressure treatments induced a softening effect, whereas blanching did not significantly alter the texture. Both blanching and pressure treatments resulted in a reduction in the levels of ascorbic acid, an effect that was less pronounced for blanching and HPP at 600 MPa and 20°C–40°C. HPP at 600 MPa resulted in significantly higher total phenol content, total antioxidant capacity, and total isothiocyanate content compared to blanching. To conclude, the color and texture of white cabbage were better preserved by blanching. However, HPP at 600 MPa resulted in significantly higher levels of phytochemical compounds. The results of this study suggest that HPP may represent an attractive technology to process vegetable-based food products that better maintains important aspects related to the content of health-promoting compounds. This may be of particular relevance to the food industry sector involved in the development of convenient, novel food products with excellent functional properties.

Conventional blanching and pre-drying are two separate processes and have the drawbacks of having low energy efficiency and long processing time (Tajner-Czopek et al., 2008). In a typical water blanching operation, first the water needs to be procured and heated and second, after a certain amount of blanching operations, this water needs to be replaced since it becomes saturated with sugars leaching from the potato strips. This results in not only excessive energy consumption due to the reheating of the water to the blanching temperatures but also the consumption of high amounts of water.

Among all the pretreatment methods for drying fruits and vegetables, hot-water blanching is one of the most frequently used methods as it can accelerate the drying rate and prevent quality deterioration by expelling intercellular air from the tissues, softening the texture, denaturing the enzymes, and destroying microorganisms (Jayaraman and Gupta, 2007; Neves et al., 2012; Xiao et al., 2012). However, grape drying with hot-water pretreatment has not been reported in the literature due to the special structure of grapes.

1.6 VACUUM PULSE OSMOTIC DEHYDRATION AND BLANCHING

Pulsed vacuum osmotic dehydration (PVOD) is an efficient process for obtaining semi-dehydrated food.

Osmotic dehydration (OD) is an alternative pretreatment for processes such as drying and freezing (Correa et al., 2011; Reno et al., 2011). It consists of immersing the food in a hypertonic solution with the consequent water loss (WL) from the food to the osmotic solution and the solid gain (SG) of osmotic solution by the food. The use of vacuum pulse at the beginning of the process, called PVOD, causes the expansion and subsequent compression of occluded gas in the product pores due to the action of hydrodynamic mechanisms (HDM), enhanced by pressure changes, and promotes the exchange of the pore gas/liquid for the external liquid with higher mass transfers than standard OD (Fito, 1994; Moraga et al., 2009; Correa et al., 2010; Fante et al., 2011; Moreno et al., 2011; Viana et al., 2014).

The effects of temperature (30°C–50°C), solute concentration (NaCl, 0–15 kg per 100 kg solution, sucrose, 15–35 kg per 100 kg solution), and vacuum pulse application (50–150 mbar and 5–15 min) on WL, SG, water activity (aw), and total color difference (ΔE) of previously blanched pumpkin slices were assessed through the Plackett–Burman experimental design by Correa et al. (2014). Temperature was not statistically significant in the process. Later, with the aid of a central composite design (CCD), it was found that the concentration of sucrose and NaCl was influent on the WL, SG, aw, and ΔE, and the pressure and time of application of vacuum were influent on the WL and SG. The optimal conditions of the process were stabilized with the desirable function, and the simulated data were similar to the experimental ones.

Sliced pumpkin samples underwent blanching by immersion in boiling water for 3 min. The blanching was stopped by immersing the samples for 2 min in mineral water. The slices had their surface carefully dried with a paper towel to remove the bath water. The blanching conditions were based on Tunde-Akintunde and Ogunlakin (2011) and Falade and Shogaolu (2010). Blanching is indicated for peroxidase inactivation (Pinheiro et al., 2007), color, and texture improvement (Silva et al., 2011) and higher water loss and solid gain (Kowalska et al., 2008).

1.7 MICROWAVE BLANCHING

Mild blanching for a short period retains the freshness and results in texture softening and the liberation of flavor compounds.

Moreover, blanched fish meat can be used for the preparation of value-added products, as it retains the taste and textural profiles closer to that of fresh fish meat.

Microwave heating is being investigated to improve, replace, or complement conventional processing technology for pasteurizing or sterilizing food products as well as to meet the demands of on-the-go consumers who want quick food preparation and superior taste and texture (Ahmed and Ramaswamy, 2007). Domestic microwave ovens are conveniently used to heat foods as they do it faster than conventional methods. The sensory properties of muscle foods, such as texture and color, primarily depend on the time–temperature history of the product.

On the other hand, mild heating or blanching for a short period using a microwave oven improves the texture of fresh fish as it softens the connective tissue proteins, while maintaining the functionalities of myofibrillar proteins.

The effect of microwave blanching on quality characteristics of vacuum and conventional polyethylene-packed sutchi catfish fillets was evaluated under chilled conditions by Binsi et al. (2014). Emphasis has been given to retain the sensory characteristics such as color and textural properties, which is a major problem in sutchi catfish fillets during extended chill storage. In general, microwave blanching imposed minimum changes on fatty acid and mineral composition of fish meat. A marginal increase in fat content was recorded after microwave heating of fish fillets. The microwave-blanched fillets showed minimum cooking loss of 3.2 mL per 100 g meat. A slower increase in spoilage parameters was obtained with microwave-blanched samples compared with unblanched samples, demonstrating the higher storage stability of the sample under chilled conditions. Microwave heating of fish fillets coupled with quick chilling and packing under vacuum improved the color and texture stability of sutchi catfish fillets to a considerable extent. Microwave blanching increased the hardness and chewiness values and decreased the stiffness values of fish fillets. The biochemical and sensory evaluation of microwave-blanched and vacuum-packed sutchi catfish fillets showed an extended storage life of 21 days, compared with 12 days for unblanched vacuum-packed samples.

Since microwave blanching is considered as a dry technique, the volume of wastewater generated could be diminished and therefore losses of water-soluble nutrients could be minimized (Quenzer and Burns, 1981; Günes and Bayindirli, 1993). Several studies on microwave blanching of vegetables and fruits have been reported. Brewer et al. (1994) considered the effect of different blanching methods on the ascorbic acid content and the peroxidase activity in 225 g batches of green beans, and they concluded that a 3 min microwave treatment at 700 W resulted in a product similar to that obtained by steam blanching. Muftugil (1986) observed that the time to complete the peroxidase inactivation in green beans was less with microwave blanching than with water and steam treatment, whereas a higher greenness remained with the two latter methods. Brewer and Begum (2003) studied the effects of power and irradiation time on ascorbic acid, color, and peroxidase activity in microwave blanching of green beans, among other vegetables.

Microwave blanching of green beans (*Phaseolus vulgaris* L.) was explored as an alternative to conventional hot-water blanching by Ruiz-Ojeda and Peñas (2013). Batches of raw pods were treated similarly to an industrial process employing a hot-water treatment but using a microwave oven for blanching. The effects of microwave processing time and nominal output power on physical properties (shrinkage, weight loss, texture, and color), enzyme activities (guaiacol peroxidase, L-ascorbate peroxidase, and catalase), and the ascorbic acid content of pods were measured and modeled by first-order kinetics. Inactivation of POD was the best indicator to assess the efficiency of microwave blanching of green beans. No significant differences in product quality were found between hot-water blanched and microwaved pods at optimal processing conditions. Furthermore, since shorter processing times and higher ascorbic acid retention were found, microwave processing of green beans can be a good alternative to conventional blanching methods.

Microwave blanching of green bean pods has been proved as a reliable alternative method to the conventional heating process used in the vegetable canning industry. The overall quality of the product processed by microwave heating under optimal conditions was comparable to that of the current industry process.

The microwave treatment of pods, in addition to an effective enzyme inactivation in less processing time, led to a better retention of ascorbic acid.

1.8 INFRARED BLANCHING

In industrial production, the potato strips are generally blanched with water (60°C–85°C) for more than 10 min mainly to inactivate enzymes and to obtain a uniform color (Nonaka et al., 1977; Tajner-Czopek et al., 2008), and then predried with warm air to improve texture (Andersson et al., 1994). The blanched potato strips are par-fried in hot oil (170°C–190°C), cooled at room temperature, frozen, packaged, and distributed.

Infrared (IR) heating, which delivers energy by electromagnetic waves, has been shown to be an effective heating technology with advantages of versatility and simplicity in terms of the equipment required (Sandu, 1986; Chou and Chou, 2003). IR heating was used to dry various agricultural and food materials such as onion slices, carrots, apple slices, and almonds (Hebbar et al., 2004; Sharma et al., 2005; Zhu and Pan, 2009; Yang et al., 2010). In a previous research reported by Bingol et al. (2012), they have observed that by using IR heat, complete inactivation of polyphenol oxidase (PPO) enzyme could be achieved in 3 min with 4.7% moisture loss for 9.43 mm regular cut french fries. Furthermore, for fresh-finish-fried french fries, at the end of 7 min frying, compared to unblanched samples, IR-blanched samples had 37.5%, 32%, and 30% less total oil at frying temperatures of 146°C, 160°C, and 174°C, respectively.

Given the successful application of IR blanching for fresh-finish-fried french fries (Bingol et al., 2012) and due to the widespread use of water blanching in industry for par-finish-fried french fries, Bingol et al. (2014) compared IR blanching (IRB) with water blanching (WB) for par-finish-fried french fries in terms of (1) oil uptake, (2) color formation, and (3) the cost of blanching.

Bingol et al. (2014) compared IRB with WB as a pretreatment method for producing lower calorie french fries. It was observed that complete inactivation of polyphenol oxidase enzyme for 9.43 mm potato strips could be achieved in 200 s and 16 min by using IRB and WB, respectively. Following the blanching, the samples were deep-fat par-fried at 174°C for 1 min and were then deep-fat finish-fried at 146°C, 160°C, and 174°C for 2, 3, 4, and 5 min. At all frying times and temperatures, IR-blanched samples had less oil content than water-blanched ones. The energy analysis of both blanching operations showed that energy expenditure-wise operation cost for pretreating french fries with IRB would be head-to-head with WB. The final moisture contents of IR and water-blanched samples were between 40% and 50% after 5 min of finish-frying. The chromatic color components of IR and water-blanched samples were significantly (p <0.05) affected by finish-frying time and temperature, and a^* and b^* values for IR-blanched samples developed faster than water-blanched samples during deep-fat finish frying.

To prevent the swelling of the strip surfaces which were exposed to IR heat, they applied a three-stage blanching process. In the first stage only IR heat was applied for 120 s, and then in the second stage, IR heat was coupled with an air flow of 2.49 ± 0.24 m/s for 45 s, and finally in the last stage, the air velocity was increased to 5.14 ± 0.27 m/s. The total blanching time was 200 s.

Generally, exposure of PPO to temperatures of 70°C–90°C destroys their catalytic activity (Queiroz et al., 2008), and low-temperature blanching (55°C–70°C) reduces the porosity of potato strips, which thereby will reduce oil absorption (Aguilar et al., 1997).

Therefore, for WB, the potato strips were immersed in a 2 L beaker, containing 1 L of water at 70°C, which was held in a water bath for 16 min. Following water blanching, the strips were dried in a convective dryer for 15 min at 60°C.

The surface and center temperatures of potato strips during IRB and WB were measured using type T thermocouples (response time <0.15 s) and were recorded every 1 s with a data logger thermometer.

1.9 BLANCHING AND LEAFY VEGETABLES

Rai et al. (2014) studied the effect of different food processing techniques like blanching, microwave processing, boiling, frying, and different drying methods on the depletion of minerals especially magnesium in green leafy vegetables (leaves of *Trigonella foenum*, common name methi, and *Spinacia oleracea*, common name spinach) using laser-induced breakdown spectroscopy (LIBS). These processing techniques are frequently used at home as well as in food processing industries. The LIBS spectra of the fresh leaves of methi and spinach and their pellets (made by drying, grinding, and pressing the leaf) were recorded in a spectral range from 200 to 500 nm. After applying the aforementioned processing techniques, different pellets of these leaves were made in the same way. The LIBS spectra of these processed leaf samples were also recorded using the same experimental parameters as used for the fresh samples. Their results showed that among the aforementioned processing techniques, frying most significantly reduces the content of magnesium, whereas the least loss of Mg is observed in the case of boiling. They have verified this result by recording the LIBS spectra of the intact fresh leaves and of those processed with different techniques. The same results were also obtained from the LIBS spectra of the intact leaves and their pellets. The LIBS spectra of methi and spinach leaves were also recorded after drying them using two different techniques—drying in vacuum and in a hot air oven. The results show that vacuum drying is more suitable in terms of minimizing the loss of Mg content in leaves.

The loss of magnesium is attributed to the leaching of Mg in water and degradation of the pigment during processing. In blanching, magnesium is leached in two stages, that is, a substantial amount of it is lost in hot water and a small fraction in cold water.

Blanching treatment involved blanching of 100 g of the sample in 1 L of hot water (~85°C) in a stainless steel vessel for 3 min. Following hot water blanching, the samples were plunged into icy water (1 L, temperature of 3°C–4°C) and analyzed both immediately and after forming the pellets.

1.10 BLANCHING AND HIGH-PRESSURE PROCESSING

HPP is known as an alternative nonthermal food preservation method that can be applied to a wide variety of products. HPP is based on the application of pressures between 200 and 900 MPa to food that inactivates foodborne microorganisms and certain enzymes implied in food spoilage (Bayindirli et al., 2006).

HPP is particularly useful for acid foods such as fruit pieces, purees, and juices (Jordan et al., 2001; Bull et al., 2004; Bayindirli et al., 2006; Garcia-Parra et al., 2014).

The application of HPP to some fruit purees at industrial level is limited due to the resistance of browning-related enzymes, such as the PPO, to the treatment. This enzyme reduces the shelf life of the product due to the formation of brown compounds, which could modify the original color of the puree during storage (Gonzalez-Cebrino et al., 2012).

Other practice frequently applied during the manufacture process of purees is the application of thermal blanching. It consists in a short heating of the puree at the beginning of the processing line to maintain the original color of the puree by the inactivation of enzymes such as the PPO and also to reduce initial microbial levels. However, in some cases, this heating could also increase the oxidation of nutritive compounds and reduce the original quality of the processed fruit products. The application of pretreatments would be a necessary step before HPP as suggested by Contador et al. (2012) in pumpkin puree, Landl et al. (2010) in apple puree, and Gonzalez-Cebrino et al. (2012) in plum puree.

A nectarine puree was manufactured with different pretreatments (thermal blanching or ascorbic acid—AA—addition), and then, the puree was processed by high-pressure treatment to evaluate the effect of the initial manufacture conditions in the stability of the processed purees as described by Garcia-Parra et al. (2014). A thermal treatment was also carried out to compare the effect with the HPP. All applied processes were effective to ensure the microbiological safety of the purees. However, the pretreatment (thermal blanching or AA addition) applied during manufacturing affected the final quality of the processed purees. Initially, the AA addition had a protective effect on color degradation during the manufacture of the purees; however, when these purees were treated by HPP, they showed less color stability during storage, lower bioactive compounds content, and antioxidant activity. In contrast, purees with an initial thermal blanching maintained better quality after HPP and during storage.

Puree with thermal blanching was manufactured at 80°C during the last 40 s of the blending process to simulate industrial preheating.

1.11 STEAM BLANCHING

Freeze drying can be combined with heat treatments to promote good quality of the final product and a simultaneous improvement on color and nutritional value.

When tomatoes are submitted to treatments of drying, depending on the parameters and methods used, the concentration or degradation of nutrients can occur. The changes in the composition and color were verified when different drying processes were used. Freeze drying, oven drying, the combination of both, and also the effect of the pretreatment (blanching) using steam were studied by Jorge et al. (2014).

Tomato quarters were placed on a sieve and exposed to boiling water steam for 5 min.

The fresh tomato composition was compared with the composition of dehydrated tomato powder. After dehydration, the moisture content reduced 78% from the total initial moisture. In addition, a nutrient concentration was observed with an increase of about 57% of citric acid content and 3% in the pH. The ash content also increased from 0.53% to 8% (15 times) and 60%, the carbohydrates from 3.94% to 60% (15 times) and the proteins were increased from 1% to 11% (10 times). The blanching resulted in different types of changes, such as greater stability for the proteins, carbohydrates, fat, lycopene, and β-carotene.

There was a significant fat content difference in all treatments when exposed to steam blanching. A slight increase in the fat content was noted in the powder obtained by all drying treatments when the blanching treatment was used. The enzymatic action of lipase was inactivated with the steam application, reducing the degradation reactions of this nutrient in the product (Anese and Sovrano, 2006).

The increase in lycopene and β-carotene content after the blanching was evidenced because heating promotes the change from *cis* to *trans* conformation form, intensifying the detection of these components. Furthermore, it is seen that the heating time can cause the degradation of these pigments. However, the degradation was lower in the drying treatments with previous blanching when compared with those that did not undergo steam blanching (Jorge et al., 2014).

Steam and water blanching seem to be suitable initial operations when processing parsley into paste-like products. The parsley products obtained were characterized by bright green colors and enhanced antioxidant capacity; however, the total phenolic contents were lowered due to leaching (Kaiser et al., 2012). Furthermore, steam and water blanching at various temperature–time regimes of parsley and marjoram had different effects on polyphenol stabilities. Both increases and decreases of individual phenolic compounds were observed (Kaiser et al., 2013a). In a previous study, coriander leaves and fruits were blanched and subsequently processed into a powder. Blanching resulted in reduced microbial loads and retention of bright green color (Schweiggert et al., 2005).

Fresh coriander leaves were steam and water blanched at 100°C and at 90°C and 100°C, respectively, for 1–10 min, and subsequently comminuted to form a paste as reported by Kaiser et al. (2013b). Pasty products obtained from coriander fruits were processed after water blanching, applying the same time–temperature regimes. Among the 11 phenolics characterized in leaves by high-performance liquid chromatography coupled with mass spectrometric detection, several caffeic acid derivatives, 5-feruloylquinic, and 5-*p*-coumaroylquinic acids were tentatively identified for the first time. In fruits, 10 phenolics were detected, whereas rutin, a dicaffeic acid derivative and 2 feruloylquinic and caffeoylquinic acid isomers were newly detected. Upon steam blanching for 1 min, phenolic contents and antioxidant capacities remained virtually unchanged. In contrast, water blanching and extended steam blanching even yielded increased levels compared to the unheated control, whereas short-time water blanching resulted in higher values than prolonged heat treatment. Thus, short-time water blanching is recommended as the initial unit in the processing of coriander leaves and fruits into novel pasty products.

1.12 BLANCHING AND FOLATE REDUCTION

Blanching is commonly used to reduce enzyme activity, which can cause undesirable changes in color, flavor, odor, or nutritive value during frozen storage of vegetables (Selman, 1994). It has also been shown to reduce the folate content in vegetables (McKillop et al., 2002; Stea et al., 2006). Blanching is still used in the Egyptian food industry prior to canning of dried legumes as reported by Hefni and Witthöft (2014). They reported that blanching according to common practice reduced the folate content in faba beans and chickpeas by only 10% and 20%, respectively, probably by leaching into the blanching water as reported by others (Hoppner and Lampi, 1993; Dang et al., 2000).

Industrial food processing and household cooking are reported to affect folate content. This study by Hefni and Witthöft (2014) examined the effects of industrial and household processing methods on folate content in traditional Egyptian foods from faba beans (*Vicia faba*) and chickpeas (*Cicer arietinum*). Overnight soaking increased folate content by ~40%–60%. Industrial canning including soaking, blanching, and retorting did not affect folate content ($p = 0.11$) in faba beans but resulted in losses of ~24% ($p = 0.0005$) in chickpeas.

Germination increased folate content 0.4–2.4-fold. Household preparation increased the folate content in germinated faba bean soup (nabet soup) onefold and in bean stew (foul) by 20% ($p < 0.0001$). After deep-frying of falafel balls made from soaked faba bean paste, losses of 10% ($p = 0.2932$) compared with the raw faba beans were observed. The folate content (fresh weight) in the traditional Egyptian foods, foul and falafel, and in the beans in nabet soup was 30 ± 2, 45 ± 2, and 56 ± 6 µg/100 g, respectively. The traditional Egyptian foods foul, falafel, and nabet soup are good folate sources and techniques like germination and soaking, which increase the folate content, can therefore be recommended.

There are some studies on folate losses in vegetables during cooking, blanching, or freezing. McKillop in 2002 determined that spinach blanching for 3.5 min involves a folate loss of 51% (McKillop et al., 2002). Holasova et al. (2008) obtained similar results with a percentage of retention of around 40% after 12 min boiling. DeSouza and Eitenmiller (1986), studied the impact of different treatments such as blanching and freezing on folate loss in spinach, showed 17% retention after blanching at 100°C for 4 min.

Folates are described to be sensitive to different physical parameters such as heat, light, pH, and leaching. Most studies on folates degradation during processing or cooking treatments were carried out on model solutions, or vegetables only with thermal treatments.

Delchier et al. (2013) identified the steps involved in folates loss in industrial processing chains and the mechanisms underlying these losses. For this, the folates contents were monitored along an industrial canning chain of green beans and along an industrial freezing chain of spinach.

Folates contents decreased significantly by 25% during the washing step for spinach in the freezing process, and by 30% in the green beans canning process after sterilization, with 20% of the initial amount being transferred into the covering liquid. The main mechanism involved in folate loss during both canning green beans and freezing spinach was leaching.

Limiting the contact between vegetables and water or using steaming seems to be an adequate measure to limit folates losses during processing.

1.13 BLANCHING AND FROZEN VEGETABLES

Commercially frozen vegetables undergo blanching prior to freezing, a process utilizing hot water or steam to inactivate enzymes that otherwise cause degradative changes, limiting shelf life severely (Andress and Harrison, 2006). Destruction of the thermally stable enzyme peroxidase is most frequently the endpoint used in determining the choice of temperature and time for the blanching process (USDA 2013). However, the use of peroxidase as an indicator enzyme is controversial, due to the fact that it often has no role in causing or enhancing degradation during storage of frozen product (Barrett and Theerakulkait, 1995). With an increase in blanching time and temperature, not only does the cost increase, but there is also a greater loss of nutrient content (Lim et al., 1989).

Yet typical blanching protocols for processing broccoli prior to freezing often exceed the limit of myrosinase stability (Lund, 1977). It was previously determined that commercially frozen broccoli lacks the ability to form sulforaphane pre- and post-cooking (Dosz and Jeffery, 2013a).

Frozen broccoli can provide a cheaper product, with a longer shelf life and less preparation time than fresh broccoli. Dosz and Jeffery (2013b) previously showed that several commercially available frozen broccoli products do not retain the ability to generate the cancer-preventative agent sulforaphane. They hypothesized that this was because the necessary hydrolyzing enzyme myrosinase was destroyed during blanching, as part of the processing that frozen broccoli undergoes. This study was carried out to determine a way to overcome loss of hydrolyzing activity. Industrial blanching usually aims to inactivate peroxidase, although lipoxygenase plays a greater role in product degradation during frozen storage of broccoli.

Blanching at a temperature of 86°C or higher inactivated peroxidase, lipoxygenase, and myrosinase. Blanching at 76°C inactivated 92% of lipoxygenase activity, whereas there was only an 18% loss in myrosinase-dependent sulforaphane formation. They considered that thawing frozen broccoli might disrupt membrane integrity, allowing myrosinase and glucoraphanin to come into contact. Thawing frozen broccoli for 9 h did not support sulforaphane formation unless an exogenous source of myrosinase was added. Thermal stability studies showed that broccoli root, as a source of myrosinase, was not more heat stable than broccoli floret. Daikon radish root supported some sulforaphane formation even when heated at 125°C for 10 min, a time and temperature comparable to or greater than microwave cooking. Daikon radish (0.25%) added to frozen broccoli that was then allowed to thaw supported sulforaphane formation without any visual alteration to that of untreated broccoli.

1.14 HIGH-HUMIDITY HOT AIR IMPINGEMENT BLANCHING

High-humidity hot air impingement blanching (HHAIB) is a new and effective thermal treatment technology which combines the advantages of steam blanching and impingement technologies, resulting in minimum solids loss, a uniform, rapid and energy-efficient blanching process.

In HHAIB jets of high-humidity hot air impinge on the product surface at high velocity to achieve a high rate of heat transfer. It has been observed that the heat transfer coefficient of HHAIB at the initial stage is about 1400 W/(m^2 K) at 14.4 m/s, 135°C, and 35% as its velocity, temperature, and relative humidity, respectively, which is about 12 times that of pure hot air impingement at the same temperature and velocity (Du et al., 2006). Furthermore, the materials are heated by steam or high humidity hot air, not dipped in water, which avoids loss of water-soluble nutrients during blanching. Xiao et al. (2012) found that appropriately HHAIB pretreatment can accelerate drying and improve the whiteness index of yam slices probably due to the absence of oxygen. Bai et al. (2013a) reported that HHAIB pretreatment is an effective pretreatment for Fuji apple quarters to inactivate PPO and, meanwhile, to maintain produce quality.

Seedless grapes blanched by HHAIB at different temperatures (90°C, 100°C, 110°C, and 120°C) and several durations (30, 60, 90, and 120 s) were air-dried at temperatures ranging from 55°C to 70°C. The PPO activity, drying kinetics, and the product color parameters were investigated to evaluate the effect of HHAIB on drying kinetics and color of seedless grapes. The results clearly show that HHAIB not only extensively decreases the drying time but also effectively inhibits enzymatic browning and results in desirable green–yellow or green raisins (Bai et al., 2013b). In view of the PPO residual activity, drying kinetics and color attributes, HHAIB at 110°C for 90 s followed by air drying at 60°C are proposed as the most favorable conditions for drying grapes. These findings indicate a new pretreatment method to try to enhance both the drying kinetics and quality of seedless grapes.

Drying grapes is more difficult than some other biological materials, since a thin layer of wax covers on its surface peel. Currently, chemical pretreatment methods are used frequently to dissolve the wax layer and accelerate dry rate. However, the chemical additive residue in the raisins may cause food safety problems and how to deal with larger quantities of corrosive chemicals is a serious problem. HHAIB is a new and effective thermal treatment technology with advantages such as minimum solids loss, uniform, rapid and energy-efficient blanching process. The current work indicates that HHAIB may be a useful nonchemical pretreatment technology for seedless grape drying, which can not only accelerate drying kinetics but also improve color parameters of seedless grape.

1.15 BLANCHING AND ANTIOXIDANT CAPACITY OF FOODS

Processing often results in either a depletion of or increase of the antioxidant properties of foods. Processing can induce the formation of compounds with novel antioxidant properties, which can maintain or even enhance the overall antioxidant potential of foods (Ioannou et al., 2012). However, during processing, loss of antioxidants or formation of compounds with prooxidant action may lower the antioxidant capacity.

In previous studies, leek extracts lost 20% of their total phenolic content when subjected to a thermal treatment (100°C, 60 min; this mimicked typical soup preparation).

The degree to which antioxidants change during processing depends on the sensitivity of the compound to modification or degradation and the length of exposure to a processing technique. But losses or gains of antioxidants can also vary with cooking or processing method (Ewald et al., 1999; Ioku et al., 2001; Lee et al., 2008).

Evaluating the effect of domestic cooking on the health benefits of vegetables has great practical importance. However, only a limited number of reports provide information on the effect of these treatments on the antioxidant capacity, polyphenol and *S*-alk(en)yl-L-cysteine sulfoxide

(ACSO, e.g., isoalliin and methiin) content of the white shaft and green leaves of leek (*Allium ampeloprasum* var. *porrum*).

Bernaert et al. (2013) studied the antioxidant capacity of leek and reported that it was highly influenced by cooking (blanching, boiling, and steaming). Boiling had a negative effect on total phenolic content in the white shaft and green leaves. An obvious increase could be observed in the antioxidant capacity of the steamed green leaves, while steaming did not influence the polyphenolic content. Remarkably, blanching resulted in a slight increase in the ACSO content. Subjecting leek samples to a longer thermal treatment appeared to have a negative influence on the ACSO content in leek. Steaming was also responsible for a decrease in ACSOs. Methiin was less susceptible to heat treatment than isoalliin.

Blanching and boiling did not influence the antioxidant capacity of the white shaft of leek, as measured using the oxygen radical absorbance capacity (ORAC assay). Blanching of the green leaves resulted in a 19% higher antioxidant capacity compared with the raw samples.

In general, steaming appeared to be responsible for better retention of the bioactive compounds present in leek compared with boiling.

Incorporation of ground peanut skins (PS) into peanut butter at 1.25%, 2.5%, 3.75%, and 5.0% (w/w) resulted in a marked concentration-dependent increase in both the TPC and antioxidant activity as reported by Ma et al. (2014).

PS, as the other edible part of peanuts, have attracted attention because they are a rich, inexpensive source of potentially health-promoting phenolics and dietary fiber (DF).

Using dry-blanched PS to illustrate, the TPC increased by 86%, 357%, 533%, and 714%, respectively, compared to the peanut butter control devoid of PS; the total proanthocyanidins content (TPACs) rose by 633%, 1933%, 3500%, and 5033%, respectively.

PACs are complex flavonoid polymers; their phenolic nature makes them excellent candidates as food antioxidants.

Normal phase high-performance liquid chromatography (NP-HPLC) detection confirmed that the increase in the phenolics content was attributed to the endogenous proanthocyanidins of the PS, which were characterized as dimers to nonamers by NP-HPLC electrospray ionization mass spectrometry (NP-HPLC/ESI-MS).

Ferric reducing antioxidant power assay (FRAP) values increased correspondingly by 62%, 387%, 747%, and 829%, while hydrophilic-oxygen radical absorbance capacity-fluorescein (H-$ORAC_{FL}$) values grew by 53%, 247%, 382%, and 415%, respectively.

Dry blanching of raw peanuts, to remove the seed coat (i.e., testa) from the kernel, is achieved by transporting peanuts on a belt through a low-temperature heating zone (at a maximum of 96°C) for ~45 min.

The dietary fiber content of dry-blanched PS was ~55%, with 89%–93% being insoluble fiber. Data revealed that PS addition enhances the antioxidant capacity of the peanut butter, permits a "good source of fiber" claim, and offers diversification in the market's product line.

1.16 LOW-TEMPERATURE BLANCHING

Low-temperature blanching (LTB), in the temperature range of 55°C–75°C, had been shown to improve the firmness of cooked vegetables and fruits, reducing physical breakdown and sloughing during further processing and providing an excellent and safe way of texture preserving (Verlinden et al., 2000; Dominguez et al., 2001; Ni et al., 2005; Perez-Aleman et al., 2005). Pectin methylesterase (PME), naturally present in many fruits and vegetables including sweet potato, had the potential to play a major role in cell wall strengthening at LTB (Ni et al., 2005; Abu-Ghannam and Crowley, 2006).

Free starch rate has been one of the most important criteria to evaluate the quality of sweet potato flour. LTB of sweet potatoes before steam cooking has shown significant increase in tissue firmness and cell wall strengthening by He et al. (2013). This research indicated that pectin methylesterase (PME) activity decreased by 87.8% after 30 min of blanching in water at 60°C, while polygalacturonase (PG) and β-amylase activity decreased 69.4% and 7.44%, respectively, under the same condition. Both PME and β-amylase played important roles in tissue firmness. Further studies of tissue firmness and methyl esterification showed that the combination of LTB and Ca^{2+} could increase the activity of PME and significantly enhance the pectin gel hardness to strengthen the cell walls and decrease free starch rate from 12.83% to 7.28%.

1.17 BLANCHING AND SORPTION ISOTHERMS

Several studies concern the influence of blanching on the progress of drying and the quality attributes (Severini et al., 2005; Prajapati et al., 2011).

Jin et al. (2014) used the Flory Huggins free volume (FHFV) theory to interpret the sorption isotherms of broccoli from its composition and using physical properties of the components.

This theory considers the mixing properties of water, biopolymers, and solutes and has the potential to describe the sorption isotherms for varying product moisture content, composition, and temperature. The required physical properties of the pure components in food became available in recent years and allow now the prediction of the sorption isotherms with this theory. Sorption isotherm experiments have been performed for broccoli florets and stalks, at two temperatures. Experimental data shows that the FHFV theory represents the sorption isotherm of fresh and blanched broccoli samples accurately. The results also show that blanching affects the sorption isotherm due to the change of composition via leaching solutes and the change of interaction parameter due to protein denaturation.

Blanching changes the cellular structure (Gómez et al., 2004; Galindo et al., 2005), and consequently changes the organization of the cell structure.

1.18 BLANCHING AND REHYDRATION

In order to tackle the problem of rehydration of freeze-dried vegetables van der Sman et al. (2013a) distinguished three length scales: (1) the microscale of molecules, (2) the mesoscale of pores, and (3) the macroscale of the product.

At the microscale water interacts with the molecules, which constitute the food. This interaction determines the driving force and kinetics for the moisture transport. Recently, they have developed predictive theories for them (van der Sman and Meinders, 2011, 2013; Jin et al., 2011; van der Sman, 2012, 2013; van der Sman et al., 2013b).

At the mesoscale they describe the simultaneous transport of water and solutes via the food matrix and the pore space. While moisture transport in the solid food matrix is governed by diffusion and swelling, the moisture transport in the pore space is governed by capillary transport. Experimental data are obtained via an experimental multiscale approach, combining NMR, Magnetic resonance (MRI), and XRT, targeting the molecular scale, the pore scale, and the macroscopic product scale (Vergeldt et al., 2012; Voda et al., 2013).

They presented a pore-scale model describing the multiphysics occurring during the rehydration of freeze-dried vegetables. This pore-scale model is part of a multiscale simulation model, which should explain the effect of microstructure and pretreatments on the rehydration rate. Simulation results are compared to experimental data, obtained by MRI and XRT. Time scale estimates based on the pore-scale model formulation agree with the experimental observations. Furthermore, the pore-scale simulation model provides a plausible explanation for the strongly increased rehydration rate, induced by the blanching pretreatment.

The increased insight in the physical processes governing the rehydration of porous or freeze-dried food gives more rationale for optimizing all processing steps. Industry is seeking for means to give dried fruits and vegetables more conveniently, but also higher quality concerning health and texture. This study shows that blanching pretreatment prior to freeze-drying strongly enhances the rehydration, while the loss of nutrients is hardly affected.

1.19 CASE STUDY: EFFECT OF BLANCHING CONDITIONS ON SULFORAPHANE CONTENT IN PURPLE AND ROMAN CAULIFLOWER (*BRASSICA OLERACEA L. VAR. BOTRYTIS*)

ABSTRACT

Brassicaceae offer many health-promoting properties due a high content of glucosinolates (glucoraphanin), whose hydrolysis through myrosinase yields sulforaphane, the most powerful anti-cancer compound derived from foodstuffs. Depending on the chemical conditions, a competition reaction occurs catalyzed the epithiospecifier protein, yielding sulforaphane-nitrile, a non-bioactive and potentially toxic compound. Epithiospecifier protein is more thermo-labile than myrosinase, then its inactivation through an adequate blanching step should be possible, thus favoring sulforaphane synthesis. The effect of blanching conditions on sulforaphane content in roman and purple cauliflower was investigated. A factorial 2^2 design in two blocks was used; whose factors were temperature (50°C and 70°C) and immersion time (5 and 15 min). Both factors affected significantly sulforaphane content. The maximum sulforaphane content was achieved after blanching at 70°C. Our results demonstrate that it is possible to favor, and even optimize, sulforaphane synthesis by blanching using an adequate combination of temperature and immersion time.

Keywords: cauliflower, sulforaphane, blanching

INTRODUCTION

The *Brassicaceae* plants, such as broccoli, white cauliflower, roman cauliflower, Brussels sprouts, and radish, offer many health-promoting properties owing to their high content of ascorbic acid, vitamin K, dietary fiber, and carotenoids. Besides, they exhibit a high content of glucosinolates (GSL), a group of secondary metabolites that share a common basic structure comprising a β-D thioglucose group, a sulphonated oxime moiety and a variable side chain derived from amino acids (Kushad et al., 1999; Fahey et al., 2001; Jia et al., 2009). In recent years GSL have become popular due to the chemoprotective properties that offer some of their hydrolysis products: isothiocyanates. Epidemiological studies have shown that consumption of a *Brassicaceae* rich diet significantly reduces the risk of developing some types of cancer, such as lung, colorectal, prostate, and breast cancer. This anticancer effect has been related to the GSL glucoraphanin, whose hydrolysis results in sulforaphane [4-(methylsulfinyl) butyl isothiocyanate] (Giovannucci et al., 2003; Ambrosone et al., 2004; Manchali et al., 2012). Sulforaphane has been recognized as the most powerful anticancer compound derived from foodstuff (Matusheski et al., 2004). Sulforaphane exerts its anticancer effect by inducing phase II enzymes (quinone reductase and glutathione S-transferase), as shown by *in vitro* and *in vivo* studies (Zhang et al., 1992; Maheo et al., 1997; Chung et al., 2000; Matusheski and Jeffery, 2001). Additionally, sulforaphane has been associated with the prevention of cardio vascular diseases (Wu et al., 2004) and inflammatory illnesses such as arteriosclerosis (Kim et al., 2012).

In intact vegetal tissues, sulforaphane is absent, since its synthesis proceeds through the hydrolysis of glucoraphanin by the action of the enzyme myrosinase (thioglucoside glucohydrolase, EC 3.2.1.147), and this enzyme is differently compartmentalized in specific myrosin cells. When the vegetal tissue is broken by mastication, harvesting, or processing, myrosinase enters in contact with glucoraphanin and then the hydrolysis proceeds (Latté et al., 2011). However, depending on the chemical conditions, a competition reaction occurs through the action of the epithiospecifier protein (ESP), which results in sulforaphane nitrile, a non-bioactive and potentially toxic compound (Mithen et al., 2000, 2003). At neutral pH the spontaneous conversion to sulforaphane proceeds, while as at acidic pH or in the presence of Fe^{2+}, the production of sulforaphane nitrile by the action of ESP is favored (Williams et al., 2008; Mahn and Reyes, 2012). Figure 1.1 shows a scheme of

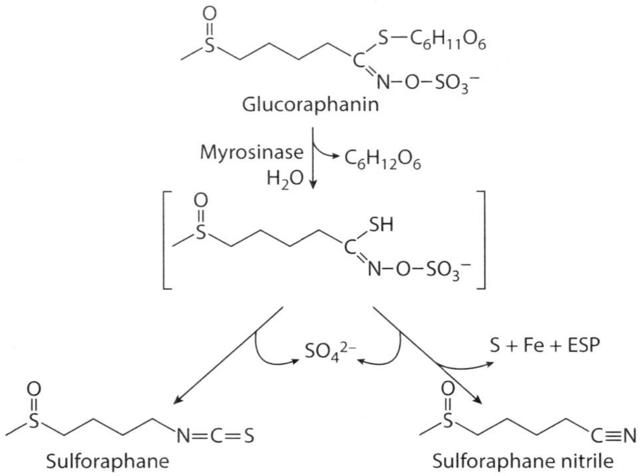

FIGURE 1.1 Mechanism of glucoraphanin hydrolysis. (Adapted from Matusheski, N.V. et al., *J. Agric. Food Chem.*, 54, 2069, 2006. With permission.)

glucoraphanin hydrolysis. ESP is more thermolabile than myrosinase, and accordingly it should be possible to inactivate it through an adequate blanching step. The effect of blanching and cooking conditions of *Brassicaceae* vegetables on GSL content has been studied by several authors (Matusheski et al., 2006; Cieślik et al., 2007; Jones et al., 2010; Pellegrini et al., 2010; Sarvan et al., 2012). However, no study about the effect of blanching conditions on sulforaphane synthesis or content in cauliflower has been reported so far.

This work presents a study of the effect of different blanching conditions on the sulforaphane content in roman and purple cauliflower (*Brassica oleracea L.* var. *Botrytis*).

EXPERIMENTAL

Chemicals

Sulforaphane standard, Acetonitrile (HPLC grade), anhydrous sodium sulfate were purchased form Sigma-Aldrich (Schnelldorf, Germany) and methylene chloride was purchased from J.T. Baker (USA). HPLC grade water was produced in the laboratory using an ultrapure water system (Barnstead, Thermo Scientific, Waltham, Massachusetts).

Vegetable Material

Purple and roman cauliflower were purchased at the local market (Santiago, Chile). All vegetables had 3 days from harvesting. Leaves and stems were discarded and the inflorescences were cut in 5–7 cm pieces (vertical).

Experimental Design

A factorial 2^2 design in two blocks was used to study the effect of blanching conditions on sulforaphane synthesis. The experimental factors were temperature (50°C and 70°C) and immersion time (5 and 15 min) (see Table 1.1). A total of 300 g of vegetable were used in each run. Blanching was performed by immersion in distilled water using a thermostatic bath (RE300, Stuart). After blanching, samples were immediately put in an ice bath, and then they were stored at −20°C until analyses.

TABLE 1.1
Blanching Conditions

Treatments	Temperature (°C)	Time (min)
1	50	5
2	50	15
3	70	5
4	70	15

Sulforaphane Content

Sulforaphane content was assessed by reverse phase HPLC, using the method proposed by Liang et al. (2006) with some minor modifications. Fresh and blanched vegetable samples were pulverized with liquid nitrogen in a mortar, until obtaining a homogeneous meal. A total of 5 g of the meal were left to autolyze at room temperature for 30 min. After that, the meal was extracted two times with 50 mL methylene chloride. Extracts were combined and salted with 2.5 g sodium sulfate anhydrous. The methylene chloride fractions were dried at 30°C under vacuum on a rotary evaporator (RE300, Stuart). The residue was dissolved in acetonitrile and was then filtered through a 0.22 μm membrane filter prior to injection into HPLC. The equipment was a HPLC-DAD (Agilent mod. 1110), and a reversed-phase C_{18} column (15.5 × 4.6 mm, i.d., 5 μm) was used. The solvent system consisted of 20% acetonitrile in water; this solution was then changed linearly over 10 min to 60% acetonitrile, and maintained at 100% acetonitrile for 2 min to purge the column. The column oven temperature was set at 30°C. The flow rate was 1 mL/min, and 10 mL portions were injected into the column. Sulforaphane was detected by absorbance at 254 nm. Quantification was carried out by comparison with a sulforaphane standard curve.

Statistical Analysis

Statistical analysis was performed by ANOVA, Fisher's protected LSD and Dunett's test for comparison with the control samples. A 95% confidence interval was considered ($p \leq 0.05$). The analyses were made using JMP 9.0.1 software (SAS Institute Inc.)

RESULTS AND DISCUSSION

Figure 1.2 shows the sulforaphane content in purple (Figure 1.2a) and in roman cauliflower (Figure 1.2b) after blanching under the different conditions given in Table 1.1. In all runs, blanching increased significantly the sulforaphane content in purple cauliflower with respect to fresh vegetable, agreeing with the results informed by Matusheski et al. (2004) for broccoli (*Brassica oleracea* var. *Italica*) subjected to blanching at 60°C during 5 min. This demonstrates that it is possible to favor sulforaphane synthesis by blanching using an adequate combination of temperature and immersion time. The highest sulforaphane content was obtained at 70°C and 5 min of immersion (run 3) in purple cauliflower, resulting in 17 mmol/g dw, representing an increase of 170% with respect to the fresh vegetable.

In roman cauliflower, only run 4 (70°C and 15 min immersion) resulted in significantly higher sulforaphane content, achieving a concentration equal to 31 mmol/g dw. This represents an increase of 500% with respect to the untreated vegetable. The increase of sulforaphane synthesis under some blanching conditions can be attributed to the inactivation of ESP at temperatures lower than 70°C. In this temperature range (50°C–70°C), myrosinase remained active (Matusheski et al., 2004), and therefore sulforaphane synthesis was favored in detriment to nitrile formation. This observation agrees with the results reported by Jones et al. (2010), who found that sulforaphane content in

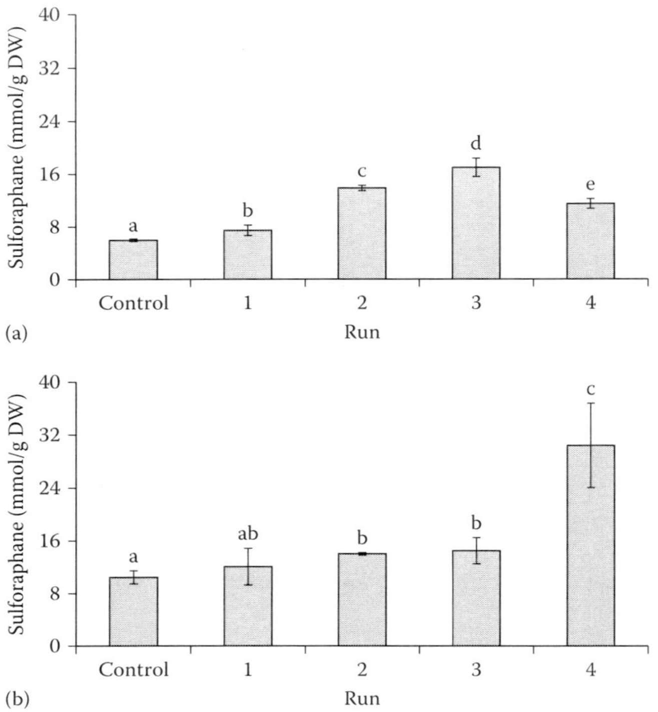

FIGURE 1.2 Sulforaphane content following the different treatments for (a) purple cauliflower and (b) roman cauliflower. Different letters indicate statistically significant differences between treatments.

broccoli diminished after blanching at temperatures higher than 70°C during 2–5 min. The authors attributed this behavior to myrosinase inactivation and also to partial leaching of glucoraphanin in the blanching water.

Matusheski et al. (2004) found that at temperatures higher than 70°C sulforaphane synthesis was disfavored in broccoli, probably due to thermal inactivation of myrosinase. Then, it can be speculated that at 70°C the maximum sulforaphane synthesis would occur. Besides, since the competition reaction that yields sulforaphane nitrile is disfavored, the ESP was probably inactivated at a temperature lower than 70°C.

Figure 1.3 shows the Pareto charts for purple (a) and roman (b) cauliflower. Here, the standardized effects of the experimental factors on sulforaphane content are presented. The experimental factors had significant effects (p-value <0.05) on sulforaphane content in both cauliflower varieties. In purple cauliflower, temperature had a significant positive effect, indicating that an increase in temperature form 50°C to 70°C produced a significant increase in sulforaphane content. Besides, the binary interaction between temperature and immersion time had a significant negative effect, suggesting the existence of an optimum combination of temperature and time that maximizes sulforaphane synthesis. Immersion time had no significant effect on the response. In roman cauliflower both factors, as well as their interaction, had significant positive effects on sulforaphane content, leading to conclude that if there is an optimum, it should be outside the experimental region examined in this work. Our results agree with results obtained in broccoli by Matusheski et al. (2004), van Eylen et al. (2008) and Jones et al. (2010), and can be attributed to thermal inactivation of myrosinase.

Sulforaphane content behavior differed between both cauliflower varieties considered in this study, most likely due to different kinetic and physicochemical properties of myrosinase in both vegetables (Yen and Wei, 1993), or to different glucoraphanin contents in the fresh vegetables.

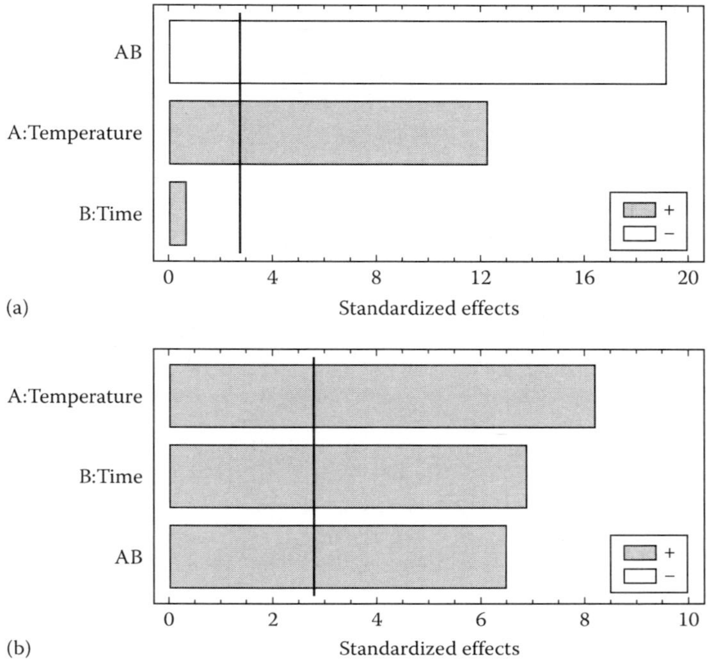

FIGURE 1.3 Pareto chart for the effects of the blanching conditions on sulforaphane content in (a) purple cauliflower and (b) roman cauliflower.

This is supported by the results of Kushad et al. (1999) and Branca et al. (2002), who found that the glucoraphanin content differed considerably between different varieties of cauliflower, ranging from 0.2 to 0.9 μmol/g dw.

The maximum sulforaphane content achieved in both cauliflower varieties in this study were considerably lower than the values reported by Jones et al. (2010) for broccoli subjected to steam blanching during 2 min (145 mmol/g dw). Additionally, Matusheski et al. (2004) reported a maximum sulforaphane content in blanched broccoli (immersion in water at 60°C during 5 min) equal to 10,000 mmol/g dw. The lower sulforaphane contents obtained in cauliflower are attributed to the lower content of glucoraphanin in *Brassica oleracea L.* var. *Botrytis* (up to 900 mmol/g dw), while in broccoli (*Brassica oleracea* var. *Italica*) glucoraphanin content is up to 21,700 mmol/g dw (Kushad et al., 1999). Despite this, cauliflower is still a source of sulforaphane, as evidenced by our results. Besides, purple cauliflower also has a high content of anthocyanins, which show high antioxidant activity, and then this vegetable has great potential as functional food.

CONCLUSION

This work is the first attempt of studying the effect of blanching conditions on sulforaphane content in cauliflower. Blanching temperature and immersion time had statistically significant effect ($p < 0.05$) on sulforaphane synthesis in purple and in roman cauliflower. The maximum sulforaphane content was achieved after blanching at 70°C in both varieties, agreeing with the hypothesis of different inactivation temperatures of ESP and myrosinase. Roman cauliflower showed the highest sulforaphane content after blanching (31 mmol/g dw). Besides, even though purple cauliflower showed a lower sulforaphane content, both cauliflower varieties present high potential as functional food intended to prevent some types of cancer and other diseases. Finally, this study demonstrates that it is possible to favor sulforaphane synthesis by blanching using an adequate combination of temperature and immersion time.

REFERENCES

Abu-Ghannam, N. and Crowley, H. 2006. The effect of low temperature blanching on the texture of whole processed new potatoes. *Journal of Food Engineering*, 74, 335–344.

Aguilar, C.N., Anzaldua-Morales, A., Talamas, R., and Gastelum, G. 1997. Low temperature blanch improves textural quality of French-fries. *Journal of Food Science*, 62(3), 568–571.

Ahmed, J. and Ramaswamy, H.S. 2007. Microwave pasteurization and sterilization of foods. In: M.S. Rahman (ed.), *Handbook of Food Preservation*, 2nd edn., pp. 691–771. Boca Raton, FL: CRC Publication.

Alvarez-Jubete, L., Valverde, J., Patras, A., Mullen, A.M., and Marcos, B. 2014. Assessing the impact of high-pressure processing on selected physical and biochemical attributes of white cabbage (*Brassica oleracea* L. var. *capitata alba*). *Food and Bioprocess Technology*, 7, 682–692.

Ambrosone, C.B., McCann, S.E., Freudenheim, J.L., Marshall, J.R., Zhang, Y., and Shields, P.G. 2004. Breast cancer risk in premenopausal women is inversely associated with consumption of broccoli, a source of isothiocyanates, but is not modified by GST genotype. *Journal of Nutrition*, 134, 1134–1138.

Andersson, A., Gekas, V., Lind, I., Oliveira, F., and Oste, R. 1994. Effect of preheating on potato texture. *Critical Reviews in Food Science and Nutrition*, 34, 229–251.

Andress, E.L. and Harrison, J.A. 2006. *So Easy to Preserve*, 5th edn. Bulletin 989. Cooperative Extension Service. Athens, GA: University of Georgia.

Anese, M. and Sovrano, S. 2006. Kinetics of thermal inactivation of tomato lipoxygenase. *Food Chemistry*, 95, 131–137.

Ashraf, Z. and Hamidi-Esfahani, Z. 2011. Date and date processing: A review. *Food Reviews International*, 27, 101–133.

Bahceci, K.S., Serpen, A., Gokmen, V., and Acar, J. 2005. Study of lipoxygenase and peroxidase as indicator enzymes in green beans: Change of enzyme activity, ascorbic acid and chlorophylls during frozen storage. *Journal of Food Engineering*, 66(2), 187–192.

Bai, J.W., Gao, Z.J., Xiao, H.W., Wang, X.T., and Zhang, Q. 2013a. Polyphenol oxidase inactivation and vitamin C degradation kinetics of Fuji apple quarters by high humidity air impingement blanching. *International Journal of Food Science and Technology*, 48, 1135–1141.

Bai, J.-W., Sun, D.-W., Xiao, H.-W., Mujumdar, A.S., and Gao, Z.-J. 2013b. Novel high-humidity hot air impingement blanching (HHAIB) pretreatment enhances drying kinetics and color attributes of seedless grapes. *Innovative Food Science and Emerging Technologies*, 20, 230–237.

Barrett, D.M., Garcia, E.L., Russell, G.F., Ramirez, E., and Shirazi, A. 2000. Blanch time and cultivar effects on quality of frozen and stored corn and broccoli. *Journal of Food Science*, 65, 534–540.

Barrett, D.M. and Theerakulkait, C. 1995. Quality indicators in blanched, frozen, stored vegetables: Lipoxygenase, rather than peroxidase. *Journal of the American Dietetic Association*, 95, 823–825.

Bayindirli, A., Alpas, H., Bozoglu, F. and Hızal, M. 2006. Efficiency of high pressure treatment on inactivation of pathogenic microorganisms and enzymes in apple orange apricot and sour cherry juices. *Food Control*, 17, 52–58.

Bernaert, N., De Loose, M., Van Bockstaele, E., and Van Droogenbroeck, B. 2013. Antioxidant changes during domestic food processing of the white shaft and green leaves of leek (*Allium ampeloprasum* var. *porrum*). *Journal of the Science of Food and Agriculture* (wileyonlinelibrary.com), 94(6), 1168–1174.

Bingol, G., Wang, B., Zhang, A., Pan, Z., and McHugh, T.H. 2014. Comparison of water and infrared blanching methods for processing performance and final product quality of French fries. *Journal of Food Engineering*, 121, 135–142.

Bingol, G., Zhang, A., Pan, Z., and McHugh, T.H. 2012. Producing lower-calorie deep fat fried French fries using infrared dry-blanching as pretreatment. *Food Chemistry*, 132, 686–692.

Binsi, P.K., Ninan, G., Zynudheen, A.A., Neethu, R., Ronda, V., and Ravishankar, C.N. 2014. Compositional and chill storage characteristics of microwave blanched sutchi catfish (*Pangasianodon hypophthalmus*) fillets. *International Journal of Food Science and Technology*, 49, 364–372.

Bourne, M.C. 1987. Effect of temperature and kinetics of thermal softening of carrots and green beans. *Journal of Food Science*, 52, 667–669.

Branca, F., Li, G., Goyal, S., and Quiros, C.F. 2002. Survey of aliphatic glucosinolates in Sicilian wild and cultivated Brassicaceae. *Phytochemistry*, 59, 717–724.

Brewer, M.S. and Begum, S. 2003. Effect of microwave power level and time on ascorbic acid content, peroxidase activity and color of selected vegetables. *Journal of Food Processing and Preservation*, 27, 411–426.

Brewer, M.S., Klein, B.P., Rastogi, B.K., and Perry, A.K. 1994. Microwave blanching effects on chemical, sensory and color characteristics of frozen green beans. *Journal of Food Quality*, 17, 245–259.

Chou, K.J. and Chou, S.K. 2003. Low-cost drying methods for developing countries. *Trends in Food Science & Technology*, 14, 519–528.

Chung, F.L., Rao, R.V., and Reddy, B.S. 2000. Chemoprevention of colonic aberrant crypt foci in Fischer rats by sulforaphane and phenethyl isothiocyanate. *Carcinogenesis*, 21, 2287–2291.

Cieślik, E., Leszczyńska, T., Filipiak-Florkiewicz, A., Sikora, E., and Pisulewski, P.M. 2007. Effects of some technological processes on glucosinolate contents in cruciferous vegetables. *Food Chemistry*, 105, 976–981.

Contador, R., Gonzalez-Cebrino, F., Garcia-Parra, J., Lozano, M., and Ramırez, R. 2012. Effect of hydrostatic high pressure and thermal treatment on two types of pumpkin puree and changes during refrigerated storage. *Journal of Food Processing and Preservation*, 38(2), 704–712.

Correa, J.L.G., Dev, S.R.S., Gariepy, Y., and Raghavan, G.S.V. 2011. Drying of pineapple by microwave-vacuum with osmotic pretreatment. *Drying Technology*, 29, 1556–1561.

Correa, J.L.G., Ernesto, D.B., Alves, J.G.L.F., and Andrade, R.S. 2014. Optimisation of vacuum pulse osmotic dehydration of blanched pumpkin. *International Journal of Food Science and Technology*, 49(9), 2008–2014.

Correa, J.L.G., Pereira, L.M., Vieira, G., and Hubinger, M.D. 2010. Mass transfer kinetics of pulsed vacuum osmotic dehydration of guavas. *Journal of Food Engineering*, 96, 498–504.

Dang, J., Arcot, J., and Shrestha, A. 2000. Folate retention in selected processed legumes. *Food Chemistry*, 68, 295–298.

Delchier, N., Ringling, C., Le Grandois, J., Aoude-Werner, D., Galland, R., George, S., Rychlik, M., and Renard, C.M.G.C. 2013. Effects of industrial processing on folate content in green vegetables. *Food Chemistry*, 139, 815–824.

DeSouza, S.C. and Eitenmiller, R.R. 1986. Effects of processing and storage on the folate content of spinach and broccoli. *Journal of Food Science*, 51(3), 626–628.

Dominguez, R., Quintero-Ramos, A., Bourne, M. et al. 2001. Texture of rehydrated dried bell peppers modified by low-temperature blanching and calcium addition. *International Journal of Food Science and Technology*, 36, 523–527.

Dosz, E.B. and Jeffery, E.H. 2013a. Commercially produced frozen broccoli lacks the ability to form sulforaphane. *Journal of Functional Foods*, 5(2), 987–990.

Dosz, E.B. and Jeffery, E.H. 2013b. Modifying the processing and handling of frozen broccoli for increased sulforaphane formation. *Journal of Food Science*, 78(9), H1459–H14163.

Du, Z.L., Gao, Z.J., and Zhang, S.X. 2006. Research on convective heat transfer coefficient with air jet impinging. *Transactions of the Chinese Society of Agricultural Engineering*, 22, 1–4 (in Chinese with English abstract).

Ewald, C., Fjelkner-Modig, S., Johansson, K., Sjoholm, I., and Akesson, B. 1999. Effect of processing on major flavonoids in processed onions, green beans, and peas. *Food Chemistry*, 64, 231–235.

Fahey, J.W., Zalcmann, A.T., and Talalay, P. 2001. The chemical diversity and distribution of glucosinolates and isothiocyanates among plants. *Phytochemistry*, 56, 5–51.

Falade, K.O. and Shogaolu, O.T. 2010. Effect of pretreatments on air-drying pattern and color of dried pumpkin (*Cucurbita maxima*) slices. *Journal of Food Processing Engineering*, 33, 1129–1147.

Fante, C., Correa, J., Natividade, M., Lima, J., and Lima, L. 2011. Drying of plums (*Prunus* sp., c.v Gulfblaze) treated with KCl in the field and subjected to pulsed vacuum osmotic dehydration. *International Journal of Food Science and Technology*, 46, 1080–1085.

Fito, P. 1994. Modelling of vacuum osmotic dehydration of food. *Journal of Food Engineering*, 22, 313–328.

Fuchigami, M., Hyakumoto, N., and Miyazaki, K. 1995. Programmed freezing affects texture, pectin composition and electron microscopic structures of carrots. *Journal Food Science*, 60, 137–141.

Fuchigami, M., Hyakumoto, N., Miyazaki, K., Nomura, T., and Sasaki, J. 1994. Texture and histological structure of carrots frozen at a programmed rate and thawed in an electrostatic field. *Journal Food Science*, 59, 1162–1167.

Galindo, F.G., Toledo, R.T., and Sjöholm, I. 2005. *Journal of Food Engineering*, 67(4), 381–385.

Gamboa-Santos, J., Soria, A.C., Villamiel, M., and Montilla, A. 2013. Quality parameters in convective dehydrated carrots blanched by ultrasound and conventional treatment. *Food Chemistry*, 141, 616–624.

Garcia-Parra, J., Contador, R., Delgado-Adamez, J., Gonzalez-Cebrino, F., and Ramirez, R. 2014. The applied pretreatment (blanching, ascorbic acid) at the manufacture process affects the quality of nectarine puree processed by hydrostatic high pressure. *International Journal of Food Science and Technology*, 49, 1203–1214.

Giovannucci, E., Rimm, E.B., Liu, Y., Stampfer, M.J., and Willett, W.C. 2003. A prospective study of cruciferous vegetables and prostate cancer. *Cancer Epidemiology, Biomarkers & Prevention*, 12, 1403–1409.

Gómez, F., Toledo, R.T., Wadsö, L., Gekas, V., and Sjöholm, I. 2004. *Journal of Food Engineering*, 65(2), 165–173.

Gonzalez-Cebrino, F., Garcıa-Parra, J., Contador, R., Tabla, R., and Ramırez, R. 2012. Effect of high-pressure processing and thermal treatment on quality attributes and nutritional compounds of "Songold" plum puree. *Journal of Food Science*, 77, C866–C873.

Günes, B. and Bayindirli, A. 1993. Peroxidase and lipoxygenase inactivation during blanching of green beans, green peas and carrots. *LWT—Food Science and Technology*, 26, 406–410.

He, J., Cheng, L., Gu, Z., Hong, Y., and Li, Z. 2013. Effects of low-temperature blanching on tissue firmness and cell wall strengthening during sweet potato flour processing. *International Journal of Food Science and Technology*, 49(5), 1360–1366.

Hebbar, H.U., Vishwanathan, K.H., and Ramesh, M.N. 2004. Development of combined infrared and hot air dryer for vegetables. *Journal of Food Engineering*, 65, 557–563.

Hefni, M. and Witthöft, C.M. 2014. Folate content in processed legume foods commonly consumed in Egypt. *LWT—Food Science and Technology*, 57, 337–343.

Holasova, M., Vlasta, F., and Slavomira, V. 2008. Determination of folates in vegetables and their retention during boiling. *Czechoslovak Journal of Food Science*, 26(1), 31–37.

Hoppner, K. and Lampi, B. 1993. Folate retention in dried legumes after different methods of meal preparation. *Food Research International*, 26, 45–48.

Ioannou, I., Hafsa, I., Hamdi, S., Charbonnel, C. and Ghoul, M. 2012. Review of the effects of food processing and formulation on flavonol and anthocyanin behaviour. *Journal of Food Engineering*, 111:208–217

Ioku, K., Aoyama, Y., Tokuno, A., Terao, J., Nakatani, N., and Takei, Y. 2001. Various cooking methods and the flavonoid content in onion. *Journal of Nutritional Science and Vitaminology*, 47, 78–83.

Jayaraman, K.S. and Gupta, D.K.D. 2007. Drying of fruits and vegetables. In: A.S. Mujumdar (ed.), *Handbook of Industrial Drying*, 3rd edn., pp. 606–634. U.K.: Taylor & Francis.

Jia, C.G., Xu, C.J., Wei, J., Yuan, J., Yuan, G.F., Wang, B.L., and Wang, Q.M. 2009. Effect of modified atmosphere packaging on visual quality and glucosinolates of broccoli florets. *Food Chemistry*, 114, 28–37.

Jin, X., van der Sman, R.G.M., and van Boxtel, A.J.B. 2011. Evaluation of the free volume theory to predict moisture transport and quality changes during broccoli drying. *Drying Technology*, 29(16), 1963–1971.

Jin, X., van der Sman, R.G.M., van Maanen, J.F.C., van Deventer, H.C., van Straten, G., Boom, R.M., and van Boxtel, A.J.B. 2014. Moisture sorption isotherms of broccoli interpreted with the flory-huggins free volume theory. *Food Biophysics*, 9, 1–9.

Jones, R.B., Frisina, C.L., Winkler, S., Imsic, M., and Tomkins, R.B. 2010. Cooking method significantly effects glucosinolate content and sulforaphane production in broccoli florets. *Food Chemistry*, 123, 237–242.

Jordan, S.L., Pascual, C., Bracey, E. and Mackey, B.M. 2001. Inactivation and injury of pressure-resistant strains of *Escherichia coli* O157:H7 and *Listeria monocytogenes* in fruit juices. *Journal of Applied Microbiology*, 91, 463–469.

Jorge, A., Milleo Almeida, D., Giovanetti Canteri, M.H., Sequinel, T., Toniolo Kubaski, E., and Mazurek Tebcherani, S. 2014. Evaluation of the chemical composition and colour in long-life tomatoes (*Lycopersicon esculentum* Mill) dehydrated by combined drying methods. *International Journal of Food Science and Technology*, 49(9), 2001–2007.

Kaiser, A., Brinkmann, M., Carle, R., and Kammerer, D.R. 2012. Influence of thermal treatment on color, enzyme activities, and antioxidant capacity of innovative pastelike parsley products. *Journal of Agricultural and Food Chemistry*, 60, 3291–3301.

Kaiser, A., Carle, R., and Kammerer, D.R. 2013a. Effects of blanching on polyphenol stability of innovative paste-like parsley (*Petroselinum crispum* (Mill.) Nym ex A. W. Hill) and marjoram (*Origanum majorana* L.) products. *Food Chemistry*, 138, 1648–1656.

Kaiser, A., Kammerer, D.R., and Carle, R. 2013b. Impact of blanching on polyphenol stability and antioxidant capacity of innovative coriander (*Coriandrum sativum* L.) pastes. *Food Chemistry*, 140, 332–339.

Kim, J.Y., Park, H.J., Um, S.H., Sohn, E.H., Kim, B.O., Moon, E.Y., Rhee, D.K., and Pyo, S. 2012. Sulforaphane suppresses vascular adhesion molecule-1 expression in TNF-alpha-stimulated mouse vascular smooth muscle cells: Involvement of the MAPK, *NF-kappa* B and AP-1 signaling pathways. *Vascular Pharmacology*, 56, 131–141.

Koskiniemi, C.B., Truong, V.-D., McFeeters, R.F., and Simunovic, J. 2013. Effects of acid, salt, and soaking time on the dielectric properties of acidified vegetables. *International Journal of Food Properties*, 16, 917–927.

Kowalska, H., Lenart, A., and Leszczyk, D. 2008. The effect of blanching and freezing on osmotic dehydration of pumpkin. *Journal of Food Engineering*, 86, 30–38.

Kushad, M.M., Brown, A.F., Kurilich, A.C., Juvik, J.A., Klein, B.P., Wallig, M.A., and Jeffery, E.H. 1999. Variation of glucosinolates in vegetable crops of *Brassica oleracea*. *Journal of Agricultural and Food* Chemistry, 47, 1541–1548.

Landl, A., Abadias, M., Sárraga, C., Viñas, I., and Picouet, P.A. 2010. Effect of high pressure processing on the quality of acidified Granny Smith apple puree product. *Innovative Food Science & Emerging Technologies*, 11, 557–564.

Latté, K.P., Appel, K.E., and Lampen, A. 2011. Health benefits and possible risks of broccoli—An overview. *Food and Chemical Toxicology*, 49, 3287–3309.

Lee, C.Y., Bourne, M.C., and Van Buren, J.P. 1979. Effect of blanching treatments on the firmness of carrots. *Journal of Food Science*, 44, 615–616.

Lee, S.U., Lee, J.H., Choi, S.H. et al. 2008. Flavonoid content in fresh, home-processed, and light-exposed onions and in dehydrated commercial onion products. *Journal of Agricultural and Food Chemistry*, 56, 8541–8548.

Liang, H., Yuan, Q.P., Dong, H.R., and Liu, Y.M. 2006. Determination of sulforaphane in broccoli and cabbage by high-performance liquid chromatography. *Journal of Food Composition and Analysis*, 19, 473–476.

Lim, M.H., Velasco, P.J., Pangborn, R.M., and Whitaker, J.R. 1989. Enzymes involved in off-aroma formation in broccoli. *ACS Symposium Series*, 405, 72–83.

Luna-Guzmán, I. and Barret, D.M. 2000. Comparison of calcium chloride and calcium lactate effectiveness in maintaining shelf stability and quality of fresh-cut cantaloupes. *Postharvest Biology and Technology*, 19, 61–72.

Lund, D.B. 1977. Design of thermal processing for maximizing nutrient retention. *Food Technology*, 31, 71–78.

Ma, T., Tian, C., Luo, J., Zhou, R., Sun, X., and Ma, J. 2013. Influence of technical processing units on polyphenols and antioxidant capacity of carrot (*Daucus carrot* L.) juice. *Food Chemistry*, 141, 1637–1644.

Ma, Y., Kerr, W.L., Swanson, R.B., Hargrove, J.L., and Pegg, R.B. 2014. Peanut skins-fortified peanut butters: Effect of processing on the phenolics content, fibre content and antioxidant activity. *Food Chemistry*, 145, 883–891.

Maheo, K., Morel, F., Langouët, S., Kramer, H., Le Ferrec, E., Ketterer, B., and Guillouzo, A. 1997. Inhibition of cytochromes P-450 and induction of glutathione S-transferases by sulforaphane in primary human and rat hepatocytes. *Cancer Research*, 57, 3649–3652.

Mahn, A. and Reyes, A. 2012. An overview of health-promoting compounds of broccoli (*Brassica oleracea* var. *italica*) and the effect of processing. *Food Science and Technology International*, 18, 503–514.

Manchali, S., Chidambara Murthy, K.N., and Patil, B.S. 2012. Crucial facts about health benefits of popular cruciferous vegetables. *Journal of Functional Foods*, 4, 94–106.

Martin, R., Monika, S., Sybille, N., and Reinhold, C. 2003. The role of process technology in carrot juice cloud stability. *Swiss Society of Food Science and Technology*, 36, 165–172.

Martin-Sanchez, A.M., Cherif, S., Vilella-Espla, J., Ben-Abda, J., Kuri, V., Perez-Alvarez, J.A., and Sayas-Barbera, E. 2014. Characterization of novel intermediate food products from Spanish date palm (*Phoenix dactylifera* L., cv. *Confitera*) co-products for industrial use. *Food Chemistry*, 154, 269–275.

Matusheski, N.V. and Jeffery, E.H. 2001. Comparison of the bioactivity of two glucoraphanin hydrolysis products found in broccoli, sulforaphane and sulforaphane nitrile. *Journal of Agricultural and Food Chemistry*, 49, 5743–5749.

Matusheski, N.V., Juvik, J.A., and Jeffery, E.H. 2004. Heating decreases epithiospecifier protein activity and increases sulforaphane formation in broccoli. *Phytochemistry*, 65, 1273–1281.

Matusheski, N.V., Swarup, R., Juvik, J.A., Mithen, R., Bennett, M., and Jeffery, E.H. 2006. Epithiospecifier protein from broccoli (*Brassica oleracea* L. ssp *italica*) inhibits formation of the anticancer agent sulforaphane. *Journal of Agricultural and Food Chemistry*, 54, 2069–2076.

McKillop, D.J., Pentieva, K., Daly, D. et al. 2002. The effect of different cooking methods on folate retention in various foods that are amongst the major contributors to folate intake in the UK diet. *British Journal of Nutrition*, 88(06), 681–688.

Mithen, R., Faulkner, K., Magrath, R., Rose, P., Williamson, G., and Marquez, J. 2003. Development of isothiocyanate-enriched broccoli and its enhanced ability to induce phase 2 detoxification enzymes in mammalian cells. *Theoretical and Applied Genetics*, 106, 727–734.

Mithen, R.F., Dekker, M., Verkerk, R., Rabot, S., and Johnson, I.T. 2000. The nutritional significance, biosynthesis and bioavailability of glucosinolates in human foods. *Journal of the Science of Food and Agriculture*, 80, 967–984.

Moraga, M.J., Moraga, G., Fito, P.J., and Martınez-Navarrete, N. 2009. Effect of vacuum impregnation with calcium lactate on the osmotic dehydration kinetics and quality of osmodehydrated grapefruit. *Journal of Food Engineering*, 90, 372–379.

Moreno, J., Simpson, R., Estrada, D., Lorenzen, S., Moraga, D., and Almonacid, S. 2011. Effect of pulsed-vacuum and ohmic heating on the osmodehydration kinetics, physical properties and microstructure of apples (cv. Granny Smith). *Innovative Food Science Emerging Technology*, 12, 562–568.

Muftugil, N. 1986. Effect of different types of blanching on the color and the ascorbic acid and chlorophyll contents of green beans. *Journal of Food Processing and Preservation*, 10, 69–76.

Neri, L., Hernando, I., Perez-Munuera, I., Sacchetti, G., Mastrocola, D., and Pittia, P. 2014. Mechanical properties and microstructure of frozen carrots during storage as affected by blanching in water and sugar solutions. *Food Chemistry*, 144, 65–73.

Neri, L., Hernando, I., Pırez-Munuera, I., Sacchetti, G., and Pittia, P. 2011. Effect of blanching in water and sugar solutions on Texture and Microstructure of Sliced Carrots. *Journal of Food Science*, 76, E23–E30.

Neves, F.I.G., Vieira, M., and Silva, C.L.M. 2012. Inactivation kinetics of peroxidase in zucchini (*Cucurbita pepo* L.) by heat and UV-C radiation. *Innovative Food Science and Emerging Technologies*, 13, 158–162.

Ni, L., Lin, D., and Barrett, D.M. 2005. Pectin methylesterase catalyzed firming effects on low temperature blanched vegetables. *Journal of Food Engineering*, 70, 546–556.

Nonaka, M., Sayre, R.N., and Weaver, M.L. 1977. Oil content of French fries as affected by blanch temperatures, fry temperatures and melting point of frying oils. *American Journal of Potato Research*, 54, 151–159.

Olivera, D.F., Vina, S.Z., Marani, C.M. et al. 2008. Effect of blanching on the quality of brussels sprouts (*Brassica oleracea* L. *Gemmifera* dc) after frozen storage. *Journal of Food Engineering*, 84(1), 148–155.

Pellegrini, N., Chiavaro, E., Gardana, C., Mazzeo, T., Contino, D., Gallo, M., Riso, P., Fogliano, V., and Porrini, M. 2010. Effect of different cooking methods on color, phytochemical concentration, and antioxidant capacity of raw and frozen *Brassica* vegetables. *Journal of Agricultural and Food Chemistry*, 58, 4310–4321.

Perez-Aleman, R., Marquez-Melendez, R., Mendoza-Guzman, V. et al. 2005. Improving textural quality in frozen jalapeno pepper by low temperature blanching in calcium chloride solution. *International Journal of Food Science and Technology*, 40, 401–410.

Perkins-Veazie, P., Collins, J., Wann, E., and Maness, N. 1994. Flavor qualities of minimally processed supersweet corn. *Proceedings of the Florida State Horticultural Society*, 107, 302–305.

Pinheiro, J., Abreu, M., Branda, T.R.S., and Silva, C.L.M. 2007. Modelling the kinetics of peroxidase inactivation, colour and texture changes of pumpkin (*Cucurbita maxima* L.) during blanching. *Journal of Food Engineering*, 81, 693–701.

Poulsen, K.P. 1986. Optimization of vegetable blanching. *Food Technology*, 40, 122–129.

Prajapati, V.K., Nema, P., and Rathore, S.S. 2011. Effect of pretreatment and drying methods on quality of value-added dried aonla (Emblica officinalis Gaertn) shreds. *Journal of Food Science and Technology*, 48(1), 45–52.

Prestamo, G., Fuster, C., and Risueρo, M.C. 1998. Effects of blanching and freezing on the structure of carrots cells and their implications for food processing. *Journal of the Science of Food and Agriculture*, 77, 223–229.

Queiroz, C., Mendes Lopes, M.L., Fialho, E., and Valente-Mesquita, V.L. 2008. Polyphenol oxidase: Characteristics and mechanisms of browning control. *Food Reviews International*, 24(4), 361–375.

Quenzer, N.M. and Burns, E.E. 1981. Effects of microwave, steam and water blanching on freeze-dried spinach. *Journal of Food Science*, 46, 410–413.

Quintero-Ramos, A., Bourne, M., Barnard, J., Gonzalezlaredo, R., Anzaldua-Morales, A., Pensaben-Esquivel, M., and Maquez-Melendez, R. 2002. Low temperature blanching of frozen carrots with calcium chloride solutions at different holding times on texture of frozen carrots. *Journal of Food Processing and Preservation*, 26, 361–374.

Rahaman, A.R., Henning, W.L., and Westcott, D.E. 1971. Histological and physical changes in carrots as affected by blanching, cooking, freezing, freeze drying and compression. *Journal of Food Science*, 36, 500–502.

Rai, D., Agrawal, R., Kumar, R., Rai, A.K., and Rai, G.K. 2014. Effect of processing on magnesium content of green leafy vegetables. *Journal of Applied Spectroscopy*, 80(6), 878–883 (Russian Original Vol. 80, No. 6, November–December, 2013).

Rastogi, N.K., Nguyen, L.T., and Balasubramaniam, V.M. 2008. Effect of pretreatments on carrot texture after thermal and pressure-assisted thermal processing. *Journal of Food Engineering*, 88, 541–547.

Reno, M.J., Prado, E.T., and de Resende, J.V. 2011. Microstructural changes of frozen strawberries submitted to pre-treatments with additives and vacuum impregnation. *Ciencia e Tecnologia de Alimentos*, 31, 247–256.

Roy, S.S., Taylor, T.A., and Kramer, H.L. 2001. Textural and ultrastructural changes in carrot tissue as affected by blanching and freezing. *Journal of Food Science*, 66, 176–180.

Ruiz-Ojeda, L.M. and Peñas, F.J. 2013. Comparison study of conventional hot-water and microwave blanching on quality of green beans. *Innovative Food Science and Emerging Technologies*, 20, 191–197.

Rungapamestry, V., Duncan, A.J., Fuller, Z., and Ratcliffe, B. 2007. Effect of cooking brassica vegetables on the subsequent hydrolysis and metabolic fate of glucosinolates. *Proceedings of the Nutrition Society*, 66(1), 69–81.

Sandu, C. 1986. Infrared radiative drying in food engineering: a process analysis. *Biotechnology Progress*, 2, 109–119.

Sanjuan, N., Hernando, I., Lluch, M.A., and Mulet, A. 2005. Effects of low temperature blanching on texture, microstructure and rehydration capacity of carrots. *Journal of the Science of Food and Agriculture*, 85, 2071–2076.

Sarang, S., Sastry, S.K., Gaines, J., Yang, T.C.S., and Dunne, P. 2007. Product formulation for ohmic heating: Blanching as a pretreatment method to improve uniformity in heating of solid–liquid food mixtures. *Journal of Food Science*, 72(5), 227–234.

Sarvan, I., Verkerk, R., and Dekker, M. 2012. Modelling the fate of glucosinolates during thermal processing of *Brassica* vegetables. *LWT—Food Science and Technology*, 49, 178.

Schweiggert, U., Mix, K., Schieber, A., and Carle, R. 2005. An innovative process for the production of spices through immediate thermal treatment of the plant material. *Innovative Food Science & Emerging Technologies*, 6, 143–153.

Selman, J. 1994. Vitamin retention during blanching of vegetables. *Food Chemistry*, 49, 137–147.

Severini, C., Baiano, A., De Pilli, T., Carbone, B.F., and Derossi, A. 2005. *Journal of Food Engineering*, 68(3), 289–296.

Severini, C., Baiano, A., De Pilli, T., Romanielo, R., and Derossi, A. 2004a. Microwave blanching of sliced potatoes dipped in saline solutions to prevent enzymatic browning. *Journal of Food Biochemistry*, 28, 75–89.

Severini, C., Derossi, A., De Pilli, T., and Baiano, A. 2004b. Acidifying-blanching of "Cicorino" leaves: Effects of recycling of processing solution on product pH. *International Journal of Food Science and Technology*, 39, 811–815.

Sharma, G.P., Verma, R.C., and Pathare, P.B. 2005. Thin-layer infrared radiation drying of onion slices. *Journal of Food Engineering*, 67, 361–366.

Shivhare, U.S., Gupta, M., Basu, S., and Raghavan, G.S.V. 2009. Optimization of blanching process for carrots. *Journal of Food Process Engineering*, 32, 587–605.

Sila, D.N., Duvetter, T., De Roeck, A. et al. 2008. Texture changes of processed fruit and vegetables: Potential use of high pressure processing. *Trends in Food Science and Technology*, 19, 309–319.

Silva, K., Caetano de Souza, L.C., Garcia, C.C., Romero, J.T., Santos, A.B., and Mauro, M.A. 2011. Osmotic dehydration process for low temperature blanched pumpkin. *Journal of Food Engineering*, 105, 56–64.

Stea, T.H., Johansson, M., Jägerstad, M., and Frølich, W. 2006. Retention of folates in cooked, stored and reheated peas, broccoli and potatoes for use in modern large-scale service systems. *Food Chemistry*, 101, 1095–1107.

Tajner-Czopek, A., Figiel, A., and Carbonell-Barrachina, A.A. 2008. Effects of potato strip size and pre-drying method on French fries quality. *European Food Research and Technology*, 227, 757–766.

Tunde-Akintunde, T.Y. and Ogunlakin, G.O. 2011. Influence of drying conditions on the effective moisture diffusivity and energy requirements during the drying of pretreated and untreated pumpkin. *Energy Conversion and Management*, 52, 1107–1113.

USDA. 2013. Technical Procedures Manual—U.S. Department of Agriculture, Agricultural Marketing Service, Washington, DC. Available from: www.ams.usda.gov/AMSv1.0/ getfile?dDocName=stelprdc5091778. Accessed May 11, 2013.

van Buggenhout, S., Lille, M., Messagie, I., van Loey, A., Autio, K., and Hendrickx, M. 2006. Impact of pretreatment and freezing conditions on the microstructure of frozen carrots: Quantification and relation to texture loss. *European Food Research and Technology*, 222, 543–553.

van der Sman, R. G. M. 2012. Thermodynamics of meat proteins. Food Hydrocolloids, 27(2), 529–535.

van der Sman, R.G.M. 2013. Moisture sorption in mixtures of biopolymer, disaccharides and water. *Food Hydrocolloids*, 32(1), 186–194.

van der Sman, R.G.M. and Meinders, M.B.J. 2011. Prediction of the state diagram of starch water mixtures using the Flory–Huggins free volume theory. *Soft Matter*, 7(2), 429–442.

van der Sman, R.G.M. and Meinders, M.B.J. 2013. Moisture diffusivity in food materials. *Food Chemistry*, 138(2–3), 1265.

van der Sman, R.G.M., Vergeldt, F.J., van As, H., van Dalen, G., Voda, A., and van Duynhoven, J.P.M. 2013a. Multiphysics pore-scale model for the rehydration of porous foods. *Innovative Food Science and Emerging Technologies*, 24, 69–79.

van der Sman, R.G.M., Voda, A., Khalloufi, S., and Paudel, E. 2013b. Hydration properties of dietary-fiber-rich foods explained by Flory–Rehner theory. *Food Research International*, 54(1), 804–811.

van Eylen, D., Oey, I., Hendrickx, M., and Loey, A.V. 2008. Effects of pressure/temperature treatments on stability and activity of endogenous broccoli (*Brassica oleracea* L. cv. *Italica*) myrosinase and on cell permeability. *Journal of Food Engineering*, 89, 178–186.

Vergeldt, F.J., Duijster, A.J., van der Sman, R.G.M. et al. 2012. The effect of structure and imbibition mode on the rehydration kinetics of freeze-dried carrots. In: P. Belton, and G. Webb (eds.), *Magnetic Resonance in Food Science: Food for Thought, 11th International Conference on the applications of Magnetic Resonance in Food 2012*. London: RSC Books.

Verlinden, B.E. and De Baerdemaeker, J. 1997. Modelling low temperature blanched carrot firmness based on heat induced processes and enzyme activity. *Journal of Food Science*, 62, 213–218.

Verlinden, B.E., Yuksel, D., Baheri, M., De Baerdemaeker, J., and van Dijk, C. 2000. Low temperature blanching effect on the changes in mechanical properties during subsequent cooking of three potato cultivars. *International Journal of Food Science and Technology*, 35, 331–340.

Viana, A.D., Correa, J.L.G., and Justus, A. 2014. Optimisation of the pulsed vacuum osmotic dehydration of cladodes of fodder palm. *International Journal of Food Science and Technology*, doi:10.1111/ijfs.12357 [Epub ahead of print].

Voda, A., Homan, N., Witek, M., Duijster, A., van Dalen, G., van der Sman, R.G.M., Nijsse, J., van Vliet, L.J., van As, H., and van Duynhoven, J.P.M. 2013. The impact of freeze-drying on microstructure and rehydration properties of carrot. In: J. van Duynhoven, H. Van As, P. Belton, and G. Webb (eds.), *Magnetic Resonance in Food Science: Food for Thought*. U.K.: RSC Books.

Volden, J., Borge, G.I.A., Bengtsson, G.B., Hansen, M., Thygesen, I.E., and Wicklund, T. 2008. Effect of thermal treatment on glucosinolates and antioxidant-related parameters in red cabbage (*Brassica oleracea* L. ssp *capitata* f. *Rubra*). *Food Chemistry*, 109(3), 595–605.

Volden, J., Borge, G.I.A., Hansen, M., Wicklund, T., and Bengtsson, G.B. 2009. Processing (blanching, boiling, steaming) effects on the content of glucosinolates and antioxidant-related parameters in cauliflower (*Brassica oleracea* L. ssp *botrytis*). *Lebensmittel-Wissenschaft und Technologie*, 42(1), 63–73.

Vu, T.S., Smout, C., Sila, D.N., LyNguyen, B., Van Loey, A.M.L., and Hendrickx, M. 2004. Effect of preheating on thermal degradation kinetics of carrot texture. *Innovative Food Science and Emerging Technologies*, 5, 37–44.

Wen, T.N., Prasad, K.N., Yang, B., and Ismail, A. 2010. Bioactive substance contents and antioxidant capacity of raw and blanched vegetables. *Innovative Food Science and Emerging Technologies*, 11, 464–469.

Williams, D.C., Lim, M.H., Chen, A.O., Pangborn, R.M., and Whitaker, J.R. 1986. Blanching of vegetables for freezing: which indicator enzyme to choose. *Food Technology*, 40, 130–140.

Williams, D.J., Critchley, C., Pun, S., Nottingham, S., and O'Hare, T.J. 2008. Epithiospecifier protein activity in broccoli: The link between terminal alkenyl glucosinolates and sulphoraphane nitrile. *Phytochemistry*, 69, 2765–2773.

Wu, L.Y., Ashraf, M.H.N., Facci, M., Wang, R., Paterson, P.G., Ferrie, A., and Juurlink, B.H.J. 2004. Dietary approach to attenuate oxidative stress, hypertension, and inflammation in the cardiovascular system. *Proceedings of the National Academy of Sciences of the United States America*, 101, 7094–7099.

Xiao, H.W., Yao, X.D., Lin, H., Yang, W.X., Meng, J.S., and Gao, Z.J. 2012. Effect of SSB (superheated steam blanching) time and drying temperatures on hot air impingement drying kinetics and quality attributes of yam slices. *Journal of Food Process Engineering*, 35, 370–390.

Yang, J., Bingol, G., Pan, Z., Brandl, M.T., McHugh, T.H., and Wang, H. 2010. Infrared heating for dry-roasting and pasteurization of almonds. *Journal of Food Engineering*, 101(3), 273–280.

Yen, G.C. and Wei, Q.K. 1993. Myrosinase activity and total glucosinolate content of cruciferous vegetables, and some properties of cabbage myrosinase in Taiwan. *Journal of the Science of Food and Agriculture*, 61, 471–475.

Zhang, Y., Talalay, P., Cho, C.G., and Posner, G.H. 1992. A major inducer of anticarcinogenic protective enzymes from broccoli—Isolation and elucidation of structure. *Proceedings of the National Academy of Sciences of the United States America*, 89, 2399–2403.

Zhou, L., Wang, Y., Hu, X., Wu, J., and Liao, X. 2009. Effect of high pressure carbon dioxide on the quality of carrot juice. *Innovative Food Science and Emerging Technologies*, 10, 321–327.

Zhu, Y. and Pan, Z. 2009. Processing and quality characteristics of apple slices under simultaneous infrared dry-blanching and dehydration with continuous heating. *Journal of Food Engineering*, 90, 441–452.

2 Thermal Processing

Nikolaos G. Stoforos

CONTENTS

2.1 INTRODUCTION

While the application of heat in the food industry can serve a variety of purposes, thermal processing of foods refers to the application of heat in order to preserve product quality and extend its shelf life. Thermal processing represents a major food preservation technique. It consists of heating a product at a rather high temperature for a relatively short time in an accurately designed and well-executed process. The product either before (traditional canning) or after (aseptic processing) the thermal treatment is enclosed into hermetically sealed containers. Unlike a number of food preservation methods, such as drying, freezing, or cold storage, which rely on altering product or environmental conditions in order to diminish product degradation reactions, thermal processing acts by destroying the undesirable agents, including pathogenic or spoilage microorganisms, enzymes, and toxins that could limit product shelf life. High-pressure processing, food irradiation, and a number of novel food preservation methods aiming also at destroying unwanted and quality- and shelf-life-reducing parameters are, in fact, following the thermal process design principles introduced almost a century ago (Bigelow et al., 1920).

Thermal processing is the first man-devised preservation procedure not having a homologous prototype in nature, in contrast to some other preservation methods (such as drying, freezing, and cooling), which were developed and improved based on nature's paradigm and systematic observations. Frenchman Nicolas François Appert (1749–1841), a Parisian confectioner and distiller, is considered as the inventor of the process, for which he won the 12,000-franc prize offered by the

French government under Napoleon, to anyone who could present a method for food preservation, in order to supply safe and nutritious food to the French troops (Valigra, 2011a). However, it seems that there was some form of "canned" food for at least 30 years before the publication of Appert's work (Atherton, 1984). The efforts of Appert in developing his technique are extremely well documented in his original work on "the art of maintaining of animal and vegetable substances for several years" (Appert, 1810).

In his book, Appert (1810) presented details on the preparation of the product, the method of filling and sealing the containers (glass bottles and jars with cork stoppers), and the time in boiling water required for the process. Finally, he suggested how to use the processed foods before consumption. Particular emphasis was placed on the quality of processed foods. For example, while describing the production and heat treatment of meat cooked with vegetables (pot-au-feu), he concludes that "after a year or 18 months, the meat and the soup were as good as having been prepared on the same day."

The basic steps followed today during thermal processing practically do not differ from those proposed by Appert (1810), a time at which the reason why thermal processing ensures the stability of the food was not known. The scientific basis of the process was found much later, in 1860, by Luis Pasteur, who explained that during heat treatment the microorganisms responsible for food spoilage were "killed," and by Prescott and Underwood (1897 and 1898), who showed that microorganisms surviving thermal processing were responsible for the spoilage of canned foods. Details of the historical developments of thermal processes are given, inter alia, by Lopez (1987), Holdsworth and Simpson (2008), and Tucker and Featherstone (2011), while a series of some early important works on thermal processing and the microbiology of thermally processed foods were presented by Goldblith et al. (1961).

The intensity of a thermal process depends on its objectives. A thermal process might be applied to a food product for blanching, pasteurization, or commercial sterilization. While sterilization refers to any process, chemical or physical, resulting in complete destruction of all living organisms, in view of the logarithmic nature of thermal destruction of microorganisms, the term "commercial sterilization" is used. The following definitions are adopted here through a compilation of descriptions given in different sources (Lopez, 1987; Potter and Hotchkiss, 1995; Tucker and Featherstone, 2011):

Blanching: A mild heat treatment by direct contact with hot water or live steam applied to fruits and vegetables primarily to inactivate indigenous food enzymes. Depending on its intensity, blanching destroys some microorganisms reducing the initial microbial load of the product. Additionally, blanching softens the tissues, eliminates air from the tissues, washes away raw flavors, expels respiratory gases, and sets the natural color of certain products.

Pasteurization: A relatively mild heat treatment of food, usually lower than 100°C, aimed at destroying the vegetative cells of all pathogenic as well as most nonpathogenic microorganisms. Pasteurization is usually combined with another means of preservation, such as acidity, low water activity, and low-temperature storage.

Commercial sterilization: Application of heat (or other appropriate treatment) to render food free from any viable form of pathogenic and toxin-forming microorganisms, as well as of non-health significant microorganisms, which, if present, could grow in the food under normal conditions of storage and distribution of the product.

We must notice that the same definitions have been adopted for any other appropriate treatment applied to a food product with the same objective. Thus, we talk about "cold" pasteurization when high hydrostatic pressure processing is used to achieve the goals of pasteurization described earlier.

There is no clear borderline between pasteurization and commercial sterilization, especially when the terms are used with acidic or acidified foods. According to Stumbo (1973), "Whether the term sterilization or pasteurization is used to label a heat treatment designed to reduce the microbial population of a food, the basic purpose of the heat treatment is the same—that is, to free the food of microorganisms that may endanger the health of food consumers or cause economically important spoilage of the food in storage and distribution." Moreover, the principles governing the design and evaluation of either one of the earlier described processes (blanching, pasteurization, and commercial sterilization) remain the same. In fact, *cooking* (boiling and frying), a way used to control food texture and palatability, is a process that can also be described with similar mathematics.

Apart from the positive results, as far as the destruction of undesirable agents is concerned, degradation of quality characteristics of the product inevitably occurs during a thermal process. Thus, precise design, implementation, and monitoring of the effects of thermal processes are required. Characteristically, Appert (1810) stated that for some products, even heating for 1 min longer (than planned and needed) would be harmful to the product. In the remaining part of the chapter, the basic principles that govern the design and control of thermal processes will be presented. The production of safe-to-eat products with the highest possible quality is the basis of the presentation that follows.

2.2 PRINCIPLES OF THERMAL PROCESSING

Knowledge of the kinetics of thermal destruction of the heat-labile substance under consideration is the first requirement in analyzing a thermal process. Unless otherwise stated, in the remaining of the chapter, bacterial spores or microbial destruction, in general, will be used as an example when referring to any undesirable agent that can be targeted in a thermal process. Thus, for example, the substitution of spore concentration by enzyme activity into the same equations that will be presented in the following paragraphs will enable enzyme inactivation calculations governing a blanching process.

2.2.1 Kinetics of Microbial Destruction

We assume that thermal destruction of a heat labile substance (e.g., microbial spores, Esty and Meyer, 1922) follows first-order kinetics, that is, at constant temperature:

$$-\frac{dN}{dt} = k_T N \tag{2.1}$$

where

N represents the microbial load, that is, the number of spores/mL (or microorganisms per container, or any other appropriate unit)

t is the processing time in min

k_T is the thermal destruction rate constant in min^{-1} (or s^{-1})

Subscript T in k_T indicates the dependence of the rate constant on temperature.

For N_o being the initial (at $t = 0$) microbial load, the solution of Equation 2.1 is given by

$$\ln(N) = \ln(N_o) - k_T t \tag{2.2}$$

and switching from natural logarithms (ln) to common logarithms (base 10 logarithms, log) we obtain

$$\log(N) = \log(N_o) - \frac{k_T}{\ln(10)} t \tag{2.3}$$

Equation 2.3 indicates the logarithmic nature of microbial thermal destruction. N approaches zero at infinite time. A negative $\log(N)$, that is, an N value less than 1, must be statistically interpreted. Thus, for example, $N = 10^{-4}$ spores/can indicates the surviving of 1 spore in 10^4 cans. Interpreting this reversely, we can say that 9,999 cans (out of 10,000) are spore-free. Thus, in a given experimental

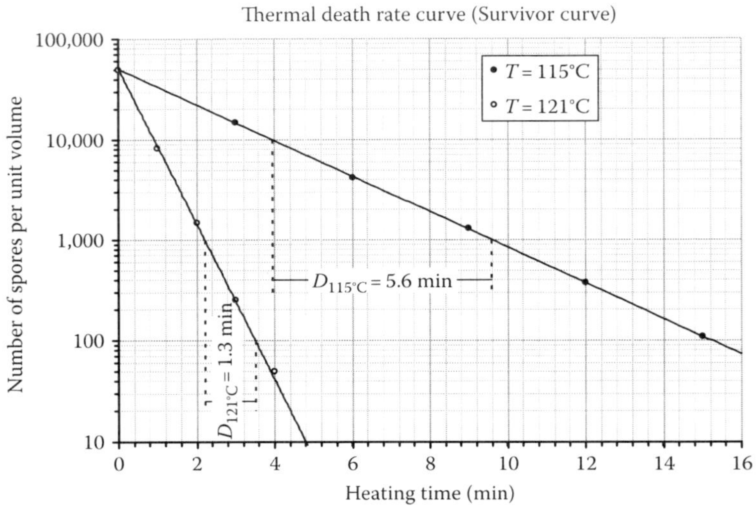

FIGURE 2.1 Log-linear thermal destruction kinetics of heat resistant bacterial spores. Graphical determination of decimal reduction times (D_T) at two temperatures.

observation of, for example, six cans, we will most probably find and report "complete" spore destruction.

As suggested by Equation 2.3, a plot of log(N) versus time is linear with the slope being equal to $-k_T/\ln(10)$. In the classical thermobacteriological literature (Ball and Olson, 1957; NCA, 1968; Stumbo, 1973), experimental N values (in logarithmic scale) were plotted versus heating time (in linear scale) in semi logarithmic paper (as depicted in Figure 2.1); the slope of the "best fit" straight line, the Thermal Death Rate (or Survivor) Curve, was described through the decimal reduction time D_T. The decimal reduction time is defined as the time, at a constant temperature T, required to reduce the microbial population by a factor of 10. In reference to Figure 2.1, D_T can be graphically determined as the time required for the Thermal Death Rate Curve to traverse a logarithmic cycle.

The slope of the straight line (Figure 2.1) is given as

$$slope = \frac{d(\log(N))}{dt} = \frac{\log(N_2) - \log(N_1)}{t_2 - t_1} \tag{2.4}$$

where indices 1 and 2 represent any two points on the straight line. For $N_2 = 0.1 \cdot N$, by definition, $t_2 - t_1 = D_T$. Thus, Equation 2.4 reduces to

$$slope = \frac{\log(0.1N_1) - \log(N_1)}{D_T} \quad \text{or} \quad D_T = -\frac{1}{slope} \tag{2.5}$$

Equation 2.5 can be used to calculate D_T from the slope of the Thermal Death Rate Curve, the slope being estimated from a computerized linear least square regression analysis applied to the experimental log(N) versus t data. Furthermore, recalling the slope of the same line defined by Equation 2.3, an expression between decimal reduction time and thermal destruction rate constant can be obtained

$$D_T = \frac{\ln(10)}{k_T} \tag{2.6}$$

In view of Equation 2.6, Equation 2.3 can be explicitly written in terms of N as

$$N = N_o 10^{-(t/D_T)} \tag{2.7}$$

From Equation 2.7, one can easily evaluate the survivors, N, from a microbial population characterized by a decimal reduction time D_T, for a given processing time t at a given temperature, T, if the initial microbial population, N_o is known. Alternatively, one can calculate the required processing time at a given constant temperature in order to achieve a specific microbial destruction. Rearranging Equation 2.7, the required processing time is given by

$$t_T = D_T(\log(N_o) - \log(N)) \tag{2.8}$$

In Equation 2.8, the symbol t_T was used for the required time in order to emphasize its temperature dependence.

In Figure 2.1, the Thermal Death Rate Curves for bacterial spores at two different temperatures are plotted. As the temperature increases, the time required to reduce the microbial population decreases. For the particular data shown in Figure 2.1, the decimal reduction time at 115°C is equal to 5.6 min, and it reduces to 1.3 min when the temperature rises to 121°C. A secondary model is needed to fully describe the effect of temperature on decimal reduction time or the thermal destruction rate constant. In the classical thermobacteriological analysis, a linear relationship between $\log(D_T)$ and temperature is assumed. In analogy to the decimal reduction time, the D_T values are plotted versus temperature in semilogarithmic paper to form the Phantom Thermal Death Time (TDT) Curve, as depicted in Figure 2.2. The slope of the "best fit" straight line is described through the z value. The z value is defined as the temperature difference required for changing the decimal reduction time by a factor of 10. In reference to Figure 2.2, z can be graphically determined as the temperature required for the Phantom TDT Curve to traverse a logarithmic cycle. As previously shown when analyzing the Thermal Death Rate Curve, similarly, the slope of the Phantom TDT Curve is equal to $-1/z$. Each point on the Phantom TDT Curve represents

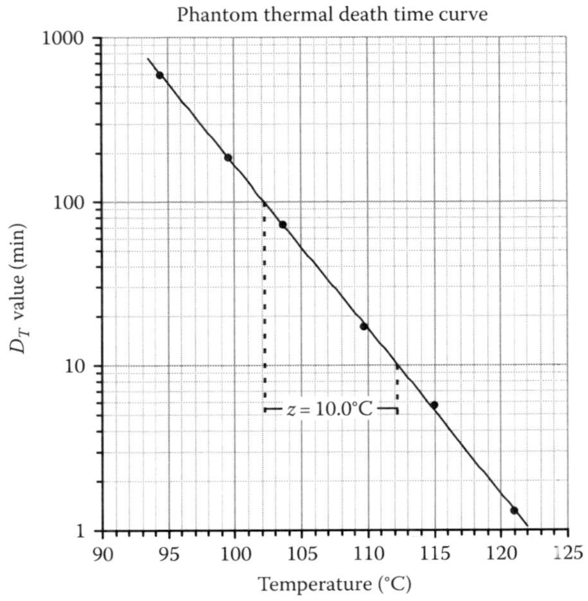

FIGURE 2.2 Effect of temperature on decimal reduction time. Graphical determination of the z value.

time–temperature combinations capable of reducing the microbial population by 90%. On the contrary, a TDT Curve refers to time–temperature combinations needed for "complete" destruction of a given microbial population.

In analogy to Equation 2.7, an expression defining the z value can be written as

$$D_T = D_{T_o} 10^{(-T/z)} \tag{2.9}$$

where D_{To}, a meaningless value, represents the D_T value at zero temperature. For a reference temperature, T_{ref}, within the lethal temperature range of data collection, Equation 2.9 gives

$$D_{T_{ref}} = D_{T_o} 10^{(-T_{ref}/z)} \tag{2.10}$$

and by combining Equations 2.9 and 2.10, we obtain the following expression, the formal mathematical definition of the z value:

$$D_T = D_{T_{ref}} 10^{((T_{ref}-T)/z)} \tag{2.11}$$

In converting z values to different temperature units, we must remember that z represents a temperature difference. Thus, a z value of 10°C is equivalent to 10 K or 18°F.

D_T is used to describe the heat resistance of microorganisms. Indeed, between two different microorganisms, the one characterized by a longer decimal reduction time at a particular temperature is the most heat resistant. However, one must remember the effect of temperature on D_T. If the two microorganisms differ in their sensitivity to temperature changes, that is, if they are characterized by different z values, then, the choice of the most heat-resistant microorganism can be temperature dependent. This is illustrated in Figure 2.3, where below 96°C the microorganism characterized by $D_{100°C}$ = 8 min and z = 5°C is the most heat resistant, while above 96°C the microorganism characterized by $D_{110°C}$ = 2 min and z = 10°C becomes the most heat resistant. In a typical thermal process, where product temperature is not constant, the selection of the target microorganism, among

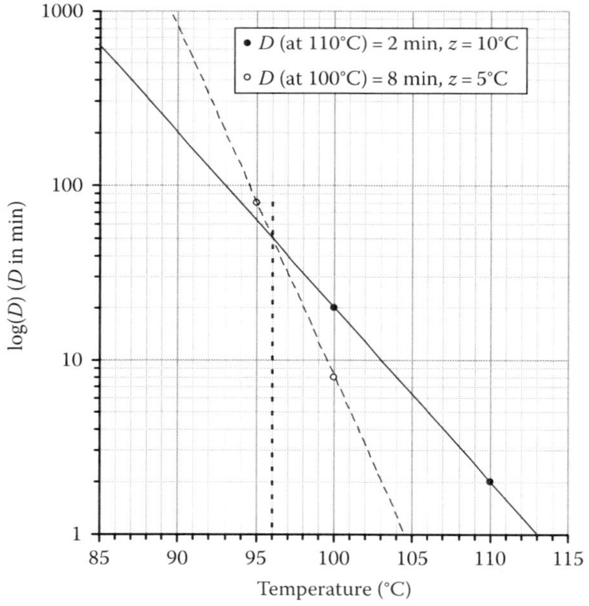

FIGURE 2.3 Selection of the target microorganism based on the Phantom TDT Curves of the microorganisms.

those characterized by different z values, requires complete calculations, the choice being based on the remaining microbial populations at the end of the entire thermal process.

While the secondary model used to describe the effect of temperature on decimal reduction time is given in terms of the z value, the Arrhenius equation is typically used to express the effect of temperature on the reaction rate constant, as given here by (Lund, 1975)

$$k_T = k_{T_{ref}} 10^{-\frac{E_a}{\ln(10) \cdot R_g} \frac{(T_{ref} - T)}{T_{ref} T}} \tag{2.12}$$

Note that T and T_{ref} in Equation 2.12 represent absolute temperatures (i.e., expressed in Kelvin).

The discrepancy between Equations 2.11 and 2.12 is given. Some critical discussion on this matter can be found in the literature (Jonsson et al., 1977; Ramaswamy et al., 1989; David and Merson, 1990; Datta, 1993). Nevertheless, both equations have been used to analyze the thermal destruction of a number of heat-labile substances during heat processes. Equation 2.11 is the choice when dealing with microbial destruction, while Equation 2.12 is preferred when the degradation of chemical substances is considered. Both equations are considered appropriate, as long as they are used within the temperature range where the inactivation data have been collected. However, one should not overestimate their empirical nature (Barsa et al., 2012).

At this point, we must also mention concerns about the appropriateness of first-order kinetics to describe microbial thermal inactivation. Deviations from first-order kinetics do exist. A number of alternative models have been proposed (Valdramidis et al., 2012). However, given the simplicity of the first-order kinetics and its successful application in the thermal processing industry for about a century, the first-order model is considered as an appropriate engineering tool for thermal process design calculations (Pflug, 1987).

2.2.2 *F* Value

If two thermal processes, applied to identical products, are applied for the same processing time, then the one using higher processing temperature is more intense, that is, resulting in higher destruction of a given microbial population. Similarly, if the temperature of two processes is the same, the one lasting longer is more severe. However, if both the processing time and temperature of two thermal processes are different, for example, one is for 30 min at 115°C and the other for 15 min at 120°C, then it is not obvious which one is more destructive.

Among a number of choices, the comparison of different thermal processes in terms of their destructive effect on a heat-labile substance is traditionally made by means of the equivalent processing time, that is, the F value. The F value of a process is formally defined as the equivalent processing time of a hypothetical thermal process at a constant, reference temperature, T_{ref}, that produces the same destructive effect as the actual thermal process. Selection of a different hypothetical thermal process at a different reference temperature results in a different F process value. Thus, when reporting an F value, the reference temperature used must be always explicitly stated. Alternatively, a subscript with the T_{ref} always accompanies the symbol F, that is, F_{Tref}. Furthermore, in view of the discussion when introducing the z value, the F value of a process depends on the z value of the particular microbial population. Thus, the subscript z is also placed on the symbol F when the z value is not explicitly stated; that is, the complete symbol for the F value is $F^z_{T_{ref}}$. If we use the constant product temperature T (when applicable) as the reference temperature, then following the F value definition, the time, t_T, defined by Equation 2.8 for constant temperature processes, coincides with the process F_T value. With the preceding comment, if we write Equation 2.8 for two equivalent thermal processes with constant but different product temperatures, T_1 and T_2, we obtain

$$F^z_{T_1}(=t_{T_1}) = D_{T_1}(\log(N_o) - \log(N)) \tag{2.13}$$

and

$$F_{T_2}^{z}(=t_{T_2}) = D_{T_2}(\log(N_o) - \log(N)) \tag{2.14}$$

On Equations 2.13 and 2.14, the logarithmic microbial reduction, $\log(N_o/N)$, as well as the z value were kept the same since we are referring to equivalent thermal processes (Stoforos and Taoukis, 2006). Expressing D_{T_2} in terms of D_{T_1}, through Equation 2.11, that is,

$$D_{T_2} = D_{T_1} 10^{((T_1 - T_2)/z)} \tag{2.15}$$

and substituting it to Equation 2.14, we end up with

$$F_{T_2}^{z} = 10^{((T_1 - T_2)/z)} F_{T_1}^{z} \tag{2.16}$$

The aforementioned equation suggests how to convert an F value from one reference temperature to another. It also explains mathematically the z dependency of the F value.

The preceding information would be sufficient for thermal process calculations, requiring only minor additions and clarifications, if product temperature remained constant during a thermal process. However, in typical thermal processes, product temperature varies with time, as illustrated in Figure 2.4 for the geometric center of a conduction-heated product processed in metal cans. In reference to Figure 2.4, we must notice that a cooling cycle must always follow the heating cycle of a thermal process in order to cease the detrimental effects of heat on the quality characteristics of the product. In defining the F value for such processes, that is, with product temperature varying with time, we can proceed as follows:

- Total processing time (including heating and cooling time) is divided to n equal-spaced time steps (through $i = 0$ to n nodes); each time step is equal to Δt, while the ith time step, Δt_i, is confined between the $(i - 1)$th and the ith node.
- Within each time step, we approximate product temperature with a constant value, for example, the temperature value at the end of the time step, as illustrated in Figure 2.4. Thus, product temperature at the end of the ith time step is equal to T_i.

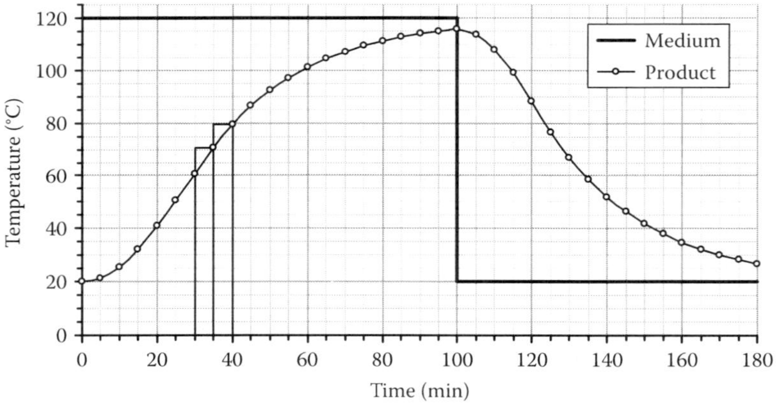

FIGURE 2.4 Typical temperature profile at the geometric center of a conduction-heated canned product during the heating and cooling cycles of a thermal process. The approximation of product temperature, used for the F value derivation, is illustrated at two time steps with the rectangular drawings.

- Next, we calculate the F value for each time step, based on Equation 2.16, due to the constant temperature assumption within the time step, and Equation 2.8:

$$F_{T_{ref}}^{z}\Big|_{i} = 10^{((T_i - T_{ref})/z)}\Delta t_i = D_{T_{ref}}(\log(N_{i-1}) - \log(N_i)) \tag{2.17}$$

where N_i is the microbial population at the end of the ith time interval.

- Doing this for each time step, we progressively obtain

$$\begin{aligned}
F_{T_{ref}}^{z}\Big|_{1} &= 10^{((T_1 - T_{ref})/z)}\Delta t_1 = D_{T_{ref}}(\log(N_o) - \log(N_1)) \\
F_{T_{ref}}^{z}\Big|_{2} &= 10^{((T_2 - T_{ref})/z)}\Delta t_2 = D_{T_{ref}}(\log(N_1) - \log(N_2)) \\
&\vdots \\
F_{T_{ref}}^{z}\Big|_{i-1} &= 10^{((T_{i-1} - T_{ref})/z)}\Delta t_{i-1} = D_{T_{ref}}(\log(N_{i-2}) - \log(N_{i-1})) \\
&\vdots \\
F_{T_{ref}}^{z}\Big|_{n-1} &= 10^{((T_{n-1} - T_{ref})/z)}\Delta t_{n-1} = D_{T_{ref}}(\log(N_{n-2}) - \log(N_{n-1})) \\
F_{T_{ref}}^{z}\Big|_{n} &= 10^{((T_{n-1} - T_{ref})/z)}\Delta t_{n} = D_{T_{ref}}(\log(N_{n-1}) - \log(N_{n}))
\end{aligned} \tag{2.18}$$

- Adding the equations, for equal-spaced time intervals, we end up with

$$\sum_{i=1}^{n} F_{T_{ref}}^{z}\Big|_{i} = \sum_{i=1}^{n} 10^{((T_i - T_{ref})/z)}\Delta t = D_{Tref}(\log(N_o) - \log(N_n)) \tag{2.19}$$

- The first summation is equal to the F value of the entire process. Taking the limit as i goes to infinity, the second summation in Equation 2.19 becomes a definite integral:

$$\lim_{n\to\infty} \sum_{i=1}^{n} 10^{((T_i - T_{ref})/z)}\Delta t = \int_{t=0}^{t=t_p} 10^{((T(t) - T_{ref})/z)}\,dt \tag{2.20}$$

for t_p being the total processing time. Using N instead of N_n, as we have done so far in the text for the microbial population at the end of the process, Equation 2.19 reduces to its final form:

$$F_{T_{ref}}^{z} = \int_{t=0}^{t=t_p} 10^{((T(t) - T_{ref})/z)}\,dt = D_{T_{ref}}(\log(N_o) - \log(N)) \tag{2.21}$$

An alternative derivation of Equation 2.21 has also been presented in the literature (Stoforos and Taoukis, 2006). Analogous expressions can be obtained in the k_T and E_A approach, that is, when Equations 2.2 and 2.12 are used (Stoforos and Taoukis, 2006). One can also derive comparable expressions when other than first-order kinetics are used, for example, when an *n*th order model is used to describe microbial destruction. However, description of microbial inactivation in terms of log reduction in such cases is not meaningful.

2.3 THERMAL PROCESS CALCULATIONS

Equation 2.21 represents the fundamental equation used for thermal process design and evaluation. It enables calculation of the F value of a process, either through microbiological destruction data (right-hand side of Equation 2.21) or through time–temperature data (left-hand side of Equation 2.21). Comparison of the destructive effects of two thermal processes can be easily done through calculation of their corresponding F values. Nevertheless, the F value of a process (noted as $F_{process}$) cannot tell if the goal of the thermal process has been achieved, that is, for example, if the thermal process has achieved commercial sterilization, unless we compare the F value of the process with a target, a required F value.

2.3.1 Required *F* Value

A required F value (noted as $F_{required}$) is defined as the time (at a constant reference temperature) required for destroying a given percentage of microorganisms whose thermal resistance is characterized by a particular z value. It sets the target of a thermal process. If

$$F_{T_{ref}}^{z}\Big|_{process} \geq F_{T_{ref}}^{z}\Big|_{required} \tag{2.22}$$

then the thermal process has accomplished its objectives. In view of the negative effects of heat on product quality, the equality in Equation 2.22 is sought.

The right-hand side of Equation 2.21 is commonly used to estimate the required F value. For example, for a given target microorganism (and thus, for known z and D values, let us assume z = 10°C and $D_{121.11°C}$ = 1 min) based on an estimated maximum initial load of microorganisms (e.g., $N_o = 1.4 \times 10^6$ spores/container) and aiming, after the thermal treatment, for no more than one spore surviving per 50,000 containers (i.e., $N = 2 \times 10^{-5}$ spores/container), the required F value can be calculated as

$$F_{121.11°C}^{10°C}\Big|_{required} = 1 \times (\log(1.4 \times 10^6) - \log(2 \times 10^{-5})) \Rightarrow F_{121.11°C}^{10°C}\Big|_{required} \approx 11 \text{ min}$$

For low-acid products in which there is the potential for *Clostridium botulinum* growth, a minimum $F_{121.11°C}^{10°C}$ required value is taken equal to 3 min, which corresponds to a 12 logarithmic cycle reduction of the microbial load of *C. botulinum* (Pflug and Odlaugh, 1978; Tucker and Featherstone, 2011). Note that the symbol F_o is used for an F value at T_{ref} = 121.11°C (actually, 250°F—a typical processing temperature for commercial sterilization of low-acid foods) and for a z value of 10°C (18°F) (the z value characterizing the thermal inactivation of *C. botulinum* spores). Note that according to the FDA (2014), *low-acid foods mean any foods, other than alcoholic beverages, with a finished equilibrium pH greater than 4.6 and a water activity greater than 0.85. Tomatoes and tomato products having a finished equilibrium pH less than 4.7 are not classed as low-acid foods.* Usually, target F_o values of 6 min or longer are applied to low-acid foods in order to control heat-resistant spoilage microorganisms. However, the rate of quality degradation during thermal processing and storage are key factors in shaping the objectives of the process and selecting the required F value. Some indicative $F_{required}$ values for low-acid foods are given by Holdsworth and Simpson (2008).

Product pH, water activity, the addition of nitrites, or any other hurdle, in addition to heat, used to preserve the product affects the target F value. Thus, for example, according to the International

Olive Council (IOC, 2004), fermented table olives packed in brine require a pasteurization process with an $F_{62.4°C}^{5.25°C}$ value of 15 min, while for commercial sterilization of unfermented olives darkened by oxidation, a required $F_{121.11°C}^{10°C}$ value of 15 min is recommended. Note the different reference temperatures and the different z values used to describe the requirements of the two processes. The difference in z values is due to the different target microorganisms: propionic acid bacteria characterized by a z value of 5.25°C and *C. botulinum* spores with a z value of 10°C. Different reference temperatures are used to provide a "meaningful" F value. Equation 2.16 can be used to convert one reference temperature to another. So, an $F_{62.4°C}^{5.25°C}$ of 15 min corresponds to an $F_{121.11°C}^{5.25°C}$ value of 6.56×10^{-12} min, a number difficult to comprehend. Usually, reference temperatures close to processing temperatures (which in turn are compatible with the severity of the thermal process needed) are used.

For acidic or acidified foods, target $F_{93.33°C}^{8.89°C}$ values suggested by Tucker and Featherstone (2011) are as follows: for products with pH < 3.9, 0.1 min; for 3.9 ≤ pH < 4.1, 1.0 min; for 4.1 ≤ pH < 4.2, 2.5 min; for 4.2 ≤ pH < 4.3, 5.0 min; for 4.3 ≤ pH < 4.4, 10.0 min; and for 4.4 ≤ pH < 4.5, 20.0 min. Note that the thermal processing literature uses temperature in degrees Fahrenheit. When feasible, we convert them to degrees Celsius. This sometimes gives rise to odd decimal numbers (we use two decimal digits). Thus, the reference temperature of 93.33°C and the z value of 8.89°C are temperature conversions from 200°F and 16°F, respectively.

A guide to required F for acidic products is also given by NCA (1968) and Holdsworth and Simpson (2008). In general, a 6-log reduction of the target microbial population is recommended for acidic foods, where processes are based on spoilage, non-health significant microorganisms (Tucker and Featherstone, 2011). In relation to the right-hand side of Equation 2.21, the F value required for a 6-log reduction is equal to

$$F_{T_{ref}}^{z} = 6 \times D_{T_{ref}} \tag{2.23}$$

A collection of relative $D_{T_{ref}}$ and z values for designing pasteurization processes for shelf-stable, high-acid fruit products has been presented by Silva and Gibbs (2004).

In the absence of appropriate data, the determination of the $F_{required}$ value is done through the TDT Curve. The methodology is described in detail in the literature (NCA, 1968). Briefly, it includes the following steps:

- *Preparation of spores of the target microorganism*: As the left-hand side of Equation 2.21 suggests, the F values of different microorganisms characterized by the same z value and experiencing the same time–temperature history are the same. Thus, a surrogate microorganism can be used in place of the target microorganism as long as it is characterized by the same z value; spores of PA 3679, a putrefactive anaerobe, are often used instead of *C. botulinum* spores.
- *Inoculation of the product under study with the spores*: Typically, an initial load of 10^5–10^6 spores per TDT can (a specially designed 208 × 006 metal can for such experiments) is used.
- *Selection of experimental conditions*: Typically, six temperatures and six processing times are selected based on the information of the thermal sensitivity of the target microorganism. If all times and/or all temperatures would result in the total destruction, or no destruction of the tested microorganism, no conclusions can be drawn. Processing times and temperatures should be chosen so that, in each temperature in some samples, microorganisms should be totally destroyed, and in some other samples they should survive after processing.
- *Heating six samples (e.g., TDT cans) for each time and treatment temperature*: Periodically remove samples and check for survivors.
- *Incubate the samples and check for survivors*: For example, for gas-producing microorganisms check for deformation of the lid of the TDT can.

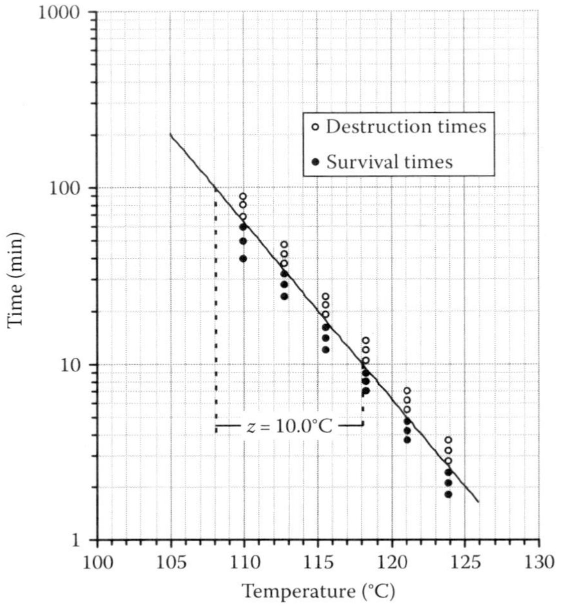

FIGURE 2.5 Thermal death time curve for target $F_{required}$ values determination.

- *Record the results*: Heating times at which no growth of microorganisms is observed in any of the six samples used are recorded as "*destruction times*"; otherwise, heating times which enabled the survival and growth of microorganisms are recorded as "*survival times.*"
- *Plot the results*: On a semilogarithmic paper, plot destruction and survival times (on the logarithmic scale) against processing temperature (on the linear scale) as depicted in Figure 2.5. The straight "best fit" line that is above all survival times and at the same time below as many destruction times as possible defines the TDT Curve.

For a given microorganism, the TDT Curve is parallel to the Phantom TDT Curve, its slope being equal to $-1/z$. Each point on the TDT Curve defines required time–temperature combinations, that is, the required F value at a chosen temperature. In doing such experiments, a number of issues must be resolved, the main one probably being the assurance of constant sample temperature during processing. Again, the reader is referred to the NCA (1968) publication for further details.

2.3.2 *F* Value of a Process

The calculation of the F value of a process is based on Equation 2.21. The right-hand side of Equation 2.21 enables the calculation of $F_{process}$ by measuring the initial and final, at the end of the process, population of the target microorganism (*in situ* method) and knowing the kinetics of its thermal inactivation (D and z values). Any other substance or agent can be alternatively used in place of the target microorganism, as long as the corresponding z values are identical (time–temperature integration—*TTI* method). The left-hand side of Equation 2.21 enables the calculation of the $F_{process}$ value by measuring the product temperature evolution throughout the process (*physical–mathematical* procedures). The latter is the method of choice if such temperature data can become available. Relative discussion can be found in Stoforos and Taoukis (2006).

The calculation of the $F_{process}$ value through the evaluation of the integral of Equation 2.21 is rather straightforward. If experimental product temperature data during the thermal process are available, a numerical integration can be performed. Its accuracy can be controlled by the time step and the method of integration used. If an analytical equation for product temperature predictions is used, the complexity of the equation will direct toward an analytical or a numerical evaluation of the integral. The first published procedure for thermal process calculations used experimental product temperature data and graphical integration for the evaluation of the integral (Bigelow et al., 1920). Since no assumptions about product temperature were made, the method was termed general method.

The ability of calculating the process F value serves a dual purpose. First, it enables the evaluation of the thermal process in terms of its objectives by comparing $F_{process}$ *with* $F_{required}$ (Equation 2.22). In thermal process calculation literature, this is sometimes called *first-type problem*. Second, it enables the design of a thermal process. That is, it allows for the calculation of the heating time required in order for a target $F_{required}$ value to be achieved. This is the *second-type problem*. The mathematical definition of the *second-type problem* is given as follows:

Splitting the integral of Equation 2.21 in two parts, one corresponding to the heating and the other to the cooling cycle of a thermal process, we obtain

$$F_{T_{ref}}^{z} = \int_{t=0}^{t=t_g} 10^{((T(t)-T_{ref})/z)}\,dt + \int_{t=t_g}^{t=t_{end}} 10^{((T(t)-T_{ref})/z)}\,dt \tag{2.24}$$

where

t_g is the time at the end of the heating cycle
t_{end} is the time at the end of the entire process

What we are looking for in designing a thermal process is to determine the heating time, t_g, so that the F value given by Equation 2.24 will be equal to the target $F_{required}$ value. Note that if we totally rely on experimental product temperature data for a single heating time, we will have no means of knowing the product temperature evolution if heating time is to be changed. In fact, this lack of predictive ability associated with the general method is the main drawback of the method. More specifically, in order to attack the second-type problem, general method suggests a "geometric similarity," that is, it assumes that the shape of the integrand (of the second integral of Equation 2.24) versus time curve during cooling is geometrically the same, irrespective of the total heating time. This is not true in general. An example of using the general method for heating time calculations is given by Stoforos and Taoukis (2006). The process F value for the product time–temperature data presented in Figure 2.4, calculated by the general method, is equal to 6.9 min.

A question yet to be addressed in relation to Equation 2.21 (or Equation 2.24) is, at which point inside the product are temperatures to be taken. When product temperature is not uniform, as with conduction heating products, each point will possess a unique $F_{process}$ value, which means that each point will be characterized by a different microbial (log) reduction. The point that receives the least effects of the heat treatment, in terms of microbial destruction, is termed "critical point." It is located at the geometric center of conduction-heated foods in cylindrical cans, or more precisely, at two doughnut-shaped regions located symmetrically along the vertical axis, slightly away from the geometric center (Flambert and Deltour, 1972) and toward the bottom of the container for natural convection-heated foods (Stoforos, 1995). Using computational fluid dynamics, the location of the slowest heating zone and the critical point within the product were assessed for several cases during thermal processing of table olives in still cans (Dimou et al., 2013). Some further discussion on this issue can be found, among others, in the NCA (1968), Holdsworth and Simpson (2008), and Tucker and Featherstone (2011) books.

If the critical point achieves the target microbial reduction, then the whole product will meet process requirements. Thus, an $F_{process}$ value can be based on time–temperature data at the critical point of the product. Such F values are single-point F values and sometimes are symbolized with F_c. On the other hand, design criteria can be based on the whole product. For this purpose, one must evaluate the surviving microbial population at each point in the product and by appropriately integrating such results, assess the microbial reduction over the whole product volume. Such F values are integrated F values, sometimes symbolized with F_s. While there is a preference for the single-point F_c value approach when microbial reduction (safety) is concerned, integrated F_s values are the only choice for most quality calculations cases. For example, the remaining vitamin content at the geometric center of a conduction-heated product is not representative of the vitamin content of the whole product; vitamin retention away from the geometric center, where the product is exposed to higher temperatures, will be much less compared to the center of the product. More discussion and a number of references on this issue are given by Stoforos (1995). Calculations based on F_c values constitute the most common approach for thermal process calculations and will be further discussed.

2.3.3 Ball's Formula Method

Due to the limitations of the general method, a number of thermal process calculation methodologies appeared in the literature (Stoforos et al., 1997). The common characteristic of these methodologies, termed *formula methods*, was the incorporation of an equation, *a formula*, to relate product temperature with time, so that one could transform time–temperature data to different conditions. Charles Olin Ball was a pioneer in this approach (1893–1970); while in graduate school at George Washington University (1919–1922), he did research on sterilization of canned foods for the National Canners Association (Valigra, 2011b). Soon after the introduction of the general method, Ball developed his method (Ball, 1923), which became the most widely used method in the United States for establishing thermal processes and the basis for all subsequently developed formula methods. Figure 2.6 is a scanned copy of a cover of the original publication being in the possession of the present author since 1996, courtesy of Kan-Ichi Hayakawa (1931–2009), a food engineering professor and a major contributor to the thermal processing literature. In fact, the product temperature profile depicted in Figure 2.4 was created using the empirical equations developed by Hayakawa (1970), based on experimental time–temperature data, to describe the product center temperature during the heating and the cooling cycles of a thermal process. A number of publications dealt with Ball's formula method (Merson et al., 1978; Stoforos, 1991, 2010). A brief working presentation of the method will be given in the next paragraphs.

Ball (1923) used the empirical parameters f and j to describe the time–temperature curves of any product during thermal processing. A rather extensive discussion on these parameters as well as their theoretical correlation to product properties and process conditions for a number of practical cases was made in Chapter 4 of the present Handbook (Kookos and Stoforos, 2014). The parameters f and j are related to the slope and the intercept of the heat penetration curves. When the difference between medium and product temperature is plotted, in the logarithmic scale of a semilogarithmic paper, versus time, both heating and cooling curves could be approximated by a straight line. Conventionally, medium temperature during the heating cycle is called retort temperature, symbolized by T_{RT}, whereas during cooling it is called cooling water temperature and is symbolized with T_{CW}. Traditionally, the heating curve ($T_{RT} - T$ vs. heating time) is plotted on a reversed logarithmic paper, with two y-axis scales, as shown in Figure 2.7. A similar plot is used for the cooling curve, where $T - T_{CW}$ versus cooling time data are used (Figure 2.8). The parameter f (f_h for heating and f_c for cooling) is defined as the time needed for the straight-line heating or cooling curve to traverse through a logarithmic cycle. The parameter j, a dimensionless correction factor, is related to the intercept of the straight-line heating or cooling curve. The j_h value is defined as

$$j_h = \frac{T_{RT} - T_A}{T_{RT} - T_{IT}} \qquad (2.25)$$

To Kanichi Hayakawa
with compliments and appreciation
Mar. 27, 1961. C. Olin Ball

Vol. 7. Part 1 OCTOBER, 1923 Number 37

BULLETIN
OF THE
NATIONAL RESEARCH
COUNCIL

THERMAL PROCESS TIME FOR CANNED FOOD

By

CHARLES OLIN BALL
Research Laboratory, National Canners Association
Washington, D.C.

PUBLISHED BY THE NATIONAL RESEARCH COUNCIL
OF
THE NATIONAL ACADEMY OF SCIENCES
WASHINGTON, D.C.
1923

FIGURE 2.6 Copy of a cover of the original publication of Ball's method (Ball, 1923) with Ball's hand writing.

with T_A being an extrapolated pseudo-initial product temperature at the beginning of heating defined as the intercept, with the temperature axis at time zero, of the straight-line heating curve, and T_{IT} is the initial product temperature (Figure 2.7). Similarly, j_c is defined as

$$j_c = \frac{T_B - T_{CW}}{T_g - T_{CW}} \tag{2.26}$$

with T_B being the extrapolated pseudo-initial product temperature at the beginning of cooling, and T_g is the actual product temperature at the beginning of cooling (equal to the product temperature at the end of heating) (Figure 2.8).

Based on the definitions given earlier, the equation for the straight-line portion of the heating curve becomes

$$\frac{T_{RT} - T}{T_{RT} - T_{IT}} = j_h 10^{-t/f_h} \tag{2.27}$$

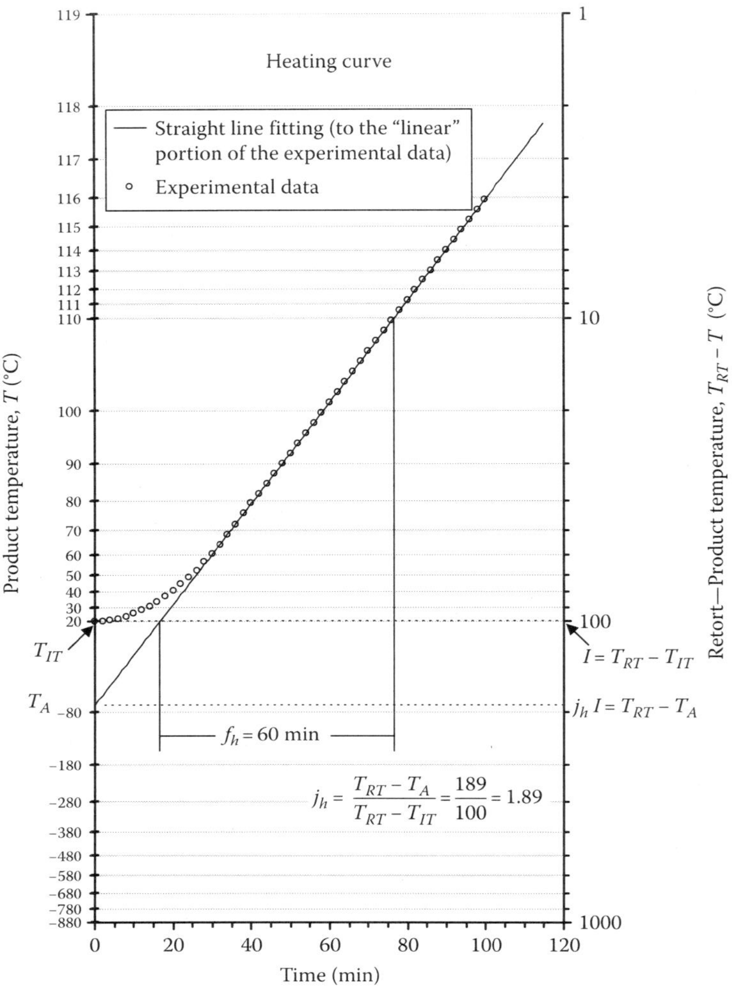

FIGURE 2.7 Typical heating curve, as traditionally plotted, based on heat penetration data presented in Figure 2.4 ($T_{RT} = 120°C$).

The reader should recognize Equation 2.27 based on similar equations presented in Chapter 4 of the present Handbook (Kookos and Stoforos, 2014). Equation 2.27 can be explicitly solved for product temperature to give

$$T = T_{RT} - j_h(T_{RT} - T_{IT})10^{-t/f_h} \tag{2.28}$$

Equation 2.27 (or 2.28) describes only the linear portion of the heating curve, that is, it approximates the heating curve only after some initial time lag. However, Ball (as well as all other investigators that proposed formula methods) used this equation to describe the entire heating curve since the thermal destruction taking place at the beginning of heating, where, under common commercial practices the temperature of the product has not yet reached lethal temperatures, is negligible.

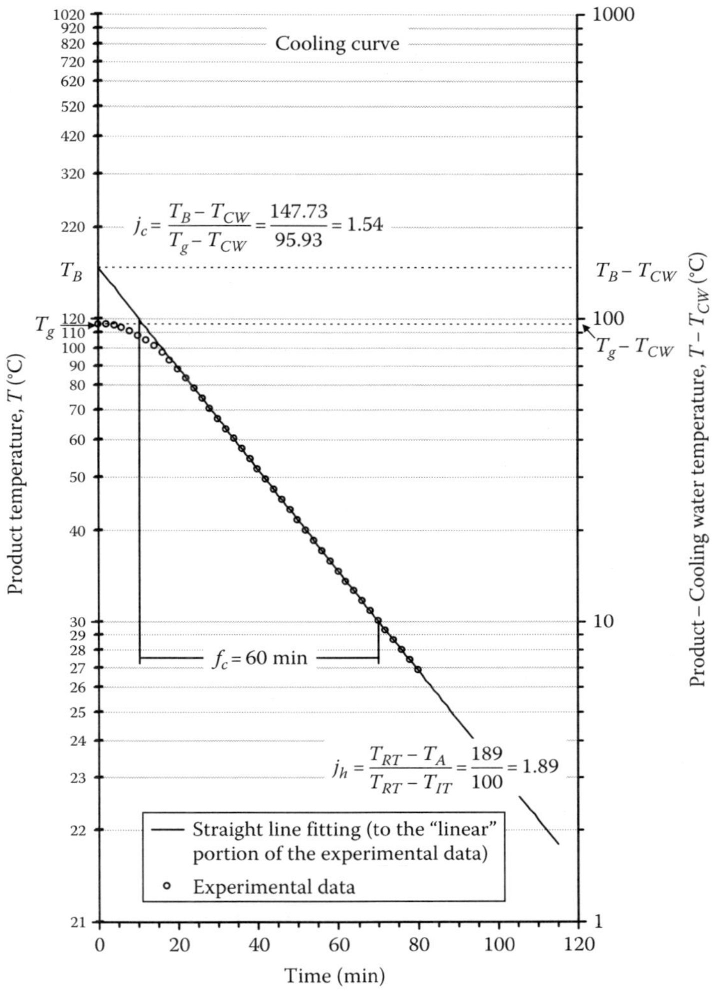

FIGURE 2.8 Typical cooling curve, as traditionally plotted, based on heat penetration data presented in Figure 2.4 ($T_{CW} = 20°C$).

In fact, Ball assumed that no microbial destruction takes place if the product temperature is more than 80°F lower than the retort temperature.

Solving Equation 2.28 for heating time, we obtain

$$t = f_h(\log(j_h(T_{RT} - T_{IT})) - \log(T_{RT} - T)) \tag{2.29}$$

Ball used the symbol B for the time at the end of heating, the symbol I for the initial temperature difference, that is,

$$I = T_{RT} - T_{IT} \tag{2.30}$$

and the symbol g for the temperature difference at the end of heating, that is,

$$g = T_{RT} - T_g \tag{2.31}$$

Under the definitions given earlier, at the end of heating, Equation 2.29 gives

$$B = f_h(\log(j_h I) - \log(g)) \tag{2.32}$$

Thus, from Equation 2.32, one can calculate the (required) heating time if the temperature difference (retort – product temperature) at the end of heating is known.

A similar, to Equation 2.28, equation was used to describe the linear portion of the cooling curve:

$$T = T_{CW} + j_c(T - T_{CW})10^{-t_c/f_c} \tag{2.33}$$

where cooling time, t_c, is used instead of processing time, t. Note that the cooling time is zero at the beginning of cooling, and with the definitions used throughout the chapter:

$$t_c = t - t_g \quad \text{or} \quad t_c = t - B \tag{2.34}$$

Contrary to the heating curve, the initial curvilinear part of the cooling curve, where the product is in the lethal temperature range, cannot be approximated by Equation 2.33, but it needs to be precisely evaluated. In fact, the main difference of the various proposed formula methods for thermal process calculations is based on the way they approach the initial part of the cooling curve (Stoforos et al., 1997). Ball used a hyperbola for the initial curvilinear portion of the cooling curve. For $0 \leq t_c \leq f_c \times \log(j_c/0.657)$, he used the following expression:

$$T = T_g + 0.3\,(T_g - T_{CW})\left[1 - \sqrt{1 + \left(\frac{1}{0.5275\log(j_c/0.657)}\right)^2\left(\frac{t_c}{f_c}\right)^2}\,\right] \tag{2.35}$$

Ball proceeded by substituting Equations 2.28, 2.33, and 2.35 into 2.24 and obtained a relationship between the F value and the heating time. Given the complexity of Equation 2.35, he evaluated the resulting integrals numerically. Before proceeding to the solution of the resulting equation, he did some conservative simplifying assumptions, the most important being the use of a fixed value of 1.41 for j_c and the use of $f_c = f_h$. He presented his results in a tabulated or graphical form (Ball, 1923, 1957; NCA, 1968). Instead of giving F versus B values, in order to significantly reduce the number of figures or tables involved, Ball gave f_h/U versus g or $\log(g)$ data, where U was defined as

$$U = F_{T_{ref}}^z F_i \tag{2.36}$$

for

$$F_i = 10^{((T_{ref} - T_{RT})/z)} \tag{2.37}$$

that is, $U = F_{T_{RT}}^z$, which means that Ball used T_{RT} as T_{ref}, in order to reduce the number of parameters involved. The only additional parameters used were the z value and the $T_{RT} - T_{CW}$ temperature difference, or $m + g$ according to Ball's nomenclature. For m defined as

$$m = T_g - T_{CW} \tag{2.38}$$

then

$$m + g = T_{RT} - T_{CW} \tag{2.39}$$

The final working f_h/U versus log(g) figures associated with Ball's method (NCA, 1968) are three figures for $m + g$ equal to 130°F, 160°F, and 180°F, each one having eight curves for eight different z values: 10°F, 12°F, 14°F, 16°F, 18°F, 20°F, 22°F, and 24°F. Note that since safety (instead of quality) was Ball's intention, the z values used were adequately covering the range of z values involved in microbial destruction calculations.

Corrections for cases where the retort temperature does not instantaneously reach the processing temperature, or when $f_c \neq f_h$, or when a straight line does not adequately describe the heating curve have been addressed by Ball (1923), but they are not considered within the scope of this chapter, and therefore, they will not be presented here. However, they have been discussed elsewhere (Stoforos, 2010).

The steps for solving *first-type problems* (i.e., when looking for the F value of a given process) with Ball's method (or, in fact, any formula method) are

1. Given the heating time, B, calculate log(g) from Equation 2.32.
2. From Ball's figures for the log(g) value calculated in step 1, find f_h/U (for the appropriate z and $m + g$ values).
3. From the f_h/U value calculated in step 2, calculate the $F_{process}$ from the following equation (based on Equation 2.36):

$$F_{T_{ref}}^{z} = \frac{f_h}{(f_h/U)F_i} \tag{2.40}$$

The reverse procedure is followed for solving *second-type problems* (i.e., when looking for the required heating time, B, for a given $F_{required}$ value). Giving explicitly, the steps involved

1. Based on the given required F value, calculate f_h/U from the following equation:

$$\frac{f_h}{U} = \frac{f_h}{F_{T_{ref}}^{z} F_i} \tag{2.41}$$

2. From Ball's figures for the f_h/U value calculated in step 1, find log(g) (for the appropriate z and $m + g$ values).
3. From the log(g) value calculated in step 2, calculate the heating time, B, from Equation 2.32.

To facilitate calculations and make Ball's method easier for computer implementation, regression equations were developed to substitute for Ball's figures (Stoforos, 2010). Thus, for thermal processes with $g \geq 0.1$°F, the following equation was proposed to relate f_h/U with log(g):

$$\log\left(\frac{f_h}{U}\right) = \frac{a_1}{1 + a_2 e^{-a_3(\log(g/z) - z/z_c)}} + \frac{a_4}{1 + a_5 e^{-a_6(\log(g/z) - z/z_c)}} + a_7 \tag{2.42}$$

for the values of the regression parameters a_1 to a_7 and z_c given in Table 2.1.

TABLE 2.1
Values for the Regression Coefficients of Equation 2.42

$m + g$ (°F)	a_1	a_2	a_3	a_4	a_5	a_6	a_7	z_c (°F)
130	40.122199	38.533071	2.3715954	5.305832	2.8885491	0.63534158	−0.63814873	405.49832
180	22.01651	21.598294	2.4586869	38.202986	23.706331	0.49435142	−0.74859566	463.11021

From data by Stoforos (2010).

TABLE 2.2
Values for the Regression Coefficients of Equation 2.43

$m + g$ (°F)	b_1	b_2	b_3	b_4	b_5	b_6	z'_c (°F)
130	−0.088335831	−0.96375429	0.028257272	1.0711536	0.19518983	4.5699218	389.106
180	−3.3545727	−0.34453049	0.42100067	4.005721	0.13211471	3.2971998	389.48491

From data by Stoforos (2010).

Since Equation 2.42 is nonlinear, the following equation was proposed for the inverse problem, that is, to relate log(g) with f_h/U:

$$\log\left(\frac{g}{z}\right) = \frac{b_1}{1 + b_2 e^{-b_3 \log(f_h/U)}} + \frac{b_4}{1 + b_5 e^{-b_6 \log(f_h/U)}} + \frac{z}{z'_c} \tag{2.43}$$

for the values of the regression parameters b_1 to b_6 and z'_c given in Table 2.2. More details and additional information are given by Stoforos (2010).

The ratio of g/z is used in both Equations 2.42 and 2.43 following Hayakawa's (1970) approach Note that the g and z values associated with Equations 2.42 and 2.43 were initially given in degrees Fahrenheit (see Tables 2.1 and 2.2 for the z_c and z'_c values, which are reported as initially calculated in °F), but since temperature difference ratios are used, degrees Celsius can be equally employed, as long as the z_c and z'_c values are converted to degrees Celsius. Regression coefficients for two $m + g$ values, namely 130°F (72.22°C) and 180°F (100°C), are given. For intermediate values, one can do a linear interpolation. For $m + g$ values outside the given range, one can use the closest $m + g$ value. Stumbo (1973) stated that a difference of 10°F in $m + g$ values introduces an error of the order of 1% in the F value.

2.3.3.1 Example Calculation

We will calculate the $F_{process}$ value for the data presented in Figure 2.4 and the associated f_h and j_h values from Figure 2.9, that is, for f_h = 60 min and j_h = 1.89. Further information include T_{IT} = 20°C = 68°F, T_{RT} = 120°C = 248°F, T_{CW} = 20°C = 68°F, and B = 100 min.

1. From Equation 2.32 we calculate

$$B = f_h(\log(j_h I) - log(g)) \Rightarrow \log(g) = \log(jI) - \frac{B}{f_h} \tag{2.44}$$

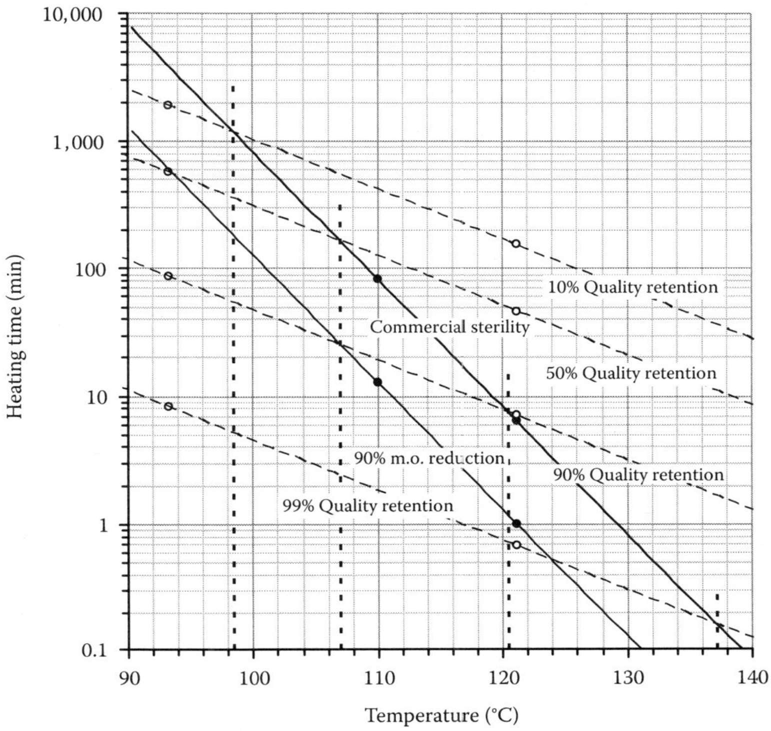

FIGURE 2.9 Time–temperature effects on quality and microbial (m.o.) destruction.

thus,

$$\log(g) = \log(1.89 \times (248 - 68)) - \frac{100}{60} \Rightarrow \log(g) = 0.8651 \Rightarrow g = 7.3294°\text{F}$$

2. From Equation 2.42, for z = 18°F and Table 2.1 coefficients for $m + g$ = 180°F (=$T_{RT} - T_{CW}$ = 248 – 68), we calculate

$$\log\left(\frac{g}{z}\right) - \frac{z}{z_c} = \log\left(\frac{7.3294}{18}\right) - \frac{18}{468.11021} \Rightarrow \log\left(\frac{g}{z}\right) - \frac{z}{z_c} = -0.4287$$

and

$$\log\left(\frac{f_h}{U}\right) = 0.8618 \Rightarrow \frac{f_h}{U} = 7.2748$$

3. From Equation 2.40, for T_{ref} = 250°F and for $F_i = 10^{(250-248)/18} = 1.2915$ (Equation 2.37), we finally calculate

$$F_{T_{250}}^{18} = \frac{f_h}{(f_h/U)F_i} \Rightarrow F_o = \frac{60}{7.2748 \times 1.2915} \Rightarrow F_o = F_{process} = 6.4 \text{ min}$$

We performed the aforementioned calculations in degrees Fahrenheit, as traditionally done in thermal process calculations literature. However, the use of the proposed Equations 2.42 and 2.43 allows direct calculations in degrees Celsius. Repeating the procedure in degrees Celsius, we have

$$\log(g) = \log(1.89 \times (120 - 20)) - \frac{100}{60} \Rightarrow \log(g) = 0.6098 \Rightarrow g = 4.0719°C$$

and for $z = 10°C$, and $z_c = 468.11021/1.8 = 260.06123°C$ we obtain

$$\log\left(\frac{g}{z}\right) - \frac{z}{z_c} = \log\left(\frac{4.0719}{10}\right) - \frac{10}{260.06123} \Rightarrow \log\left(\frac{g}{z}\right) - \frac{z}{z_c} = -0.4287$$

a value exactly the same as previously calculated, which leads to the same f_h/U and F_o values. It must be noted here that an attempt to introduce degrees Celsius for direct thermal process calculations by Shapton et al. (1971) did not receive wide acceptance.

Recall that calculations with the general method gave an $F_{process}$ value equal to 6.9 min. Thus, Ball's method calculations were close to those made by the general method (which can be considered as exact) and slightly conservative. This could be somewhat expected, given the assumptions associated with Ball's method and Hayakawa's (1970) empirical formulas. However, this conclusion should not be generalized.

2.4 THERMAL PROCESS OPTIMIZATION

So far, we presented the principles for establishing *process schedules*, that is, time–temperature combinations for thermal processing of foods. Classical publications such as the NFPA's Bulletins 26-L for metal containers (NFPA, 1982) and 30-L for glass containers (NFPA, 1984) as well as Lopez's (1987) three-volume work entitled *A Complete Course in Canning and Related Processes* suggest process schedules for a variety of products.

It is somewhat intuitively understood that what can be achieved in terms, for example, of microbial destruction, by heating a product at a certain temperature for a certain time, can also be accomplished by heating the product at a higher temperature for a shorter time. Due to differences in temperature sensitivity of quality factors compared to undesirable agents, selection of an optimum process schedule, that is, a time–temperature combination that will provide the desirable degree of safety with minimal quality degradation, is possible.

It is generally accepted that quality degradation follows first-order kinetics, and it can be described by the same kinetic parameters ($D - z$ values or $k - E_a$ parameters) as microbial destruction. Table 2.3 gives practical ranges for z, E_a, and $D_{121°C}$ values for several quality and safety factors (Lund, 1977). An extensive, comprehensive list with kinetic parameters for microbial

TABLE 2.3
Kinetics Parameters for Several Quality and Safety Factors

Constituent	z (°C)	E_a (kcal/mol)	$D_{121°C}$ (min)
Vitamins	25–31	20–30	100–1000
Color, texture, and flavor	25–45	10–30	5–500
Enzymes	7–56	12–100	1–10
Vegetative cells	5–7	100–120	0.002–0.02
Spores	7–12	53–83	0.1–5.0

From data by Lund (1977).

inactivation in a variety of products as well as for a number of quality attributes' degradation is given by Holdsworth and Simpson (2008). The high D values that characterize the majority of the quality parameters compared to the corresponding values for microorganisms (Table 2.3) lead to a high degree of quality retention for most thermally treated products. Differences in z (or, equivalently, E_a) values allow for process optimization, as it will be illustrated in the following examples.

2.4.1 Constant Product Temperature

Let us, at the moment, restrict our discussion to cases where the product temperature is constant throughout the entire thermal process, a situation that can be anticipated during aseptic processing of homogeneous liquid products. Let us further assume that the target microorganism is characterized by a z value equal to 10°C and a $D_{121.11°C}$ equal to 1 min, and commercial sterility is achieved for a required F_o value of 6.4 min. Based on these values one can plot the Phantom TDT Curve (labeled as the "90% m.o. reduction" line in Figure 2.9) and the actual TDT Curve (labeled as the "commercial sterility" line in Figure 2.9).

On the same plot, we can superimpose the retention of a quality parameter characterized by the following parameters: z = 25.56°C (46°F) and a $D_{121.11°C}$ = 154 min. Dotted (regular) lines showing different time–temperature combinations for 10%, 50%, 90%, and 99% quality retention are plotted in Figure 2.9. The intersection of the "commercial sterility" line with the quality retention lines (vertical bold dotted lines in Figure 2.9) gives time–temperature combinations that ensure safety (being on the "commercial sterility" line), but result in different percent retention of the quality parameter (Table 2.4). All calculations were performed through Equations 2.7 and 2.11 and their combination.

For these example data, it is evident (Table 2.4) that high-temperature short-time (HTST) processes preserve better product quality as compared to low-temperature long-time (LTLT) processes. Nevertheless, before generalizing one must remember that we analyzed a constant product temperature case. As a matter of fact, if product temperature is uniform (that is, there is no spatial temperature distribution within the product, although there can be time varying product temperature), the above conclusion still holds. Furthermore, we used a z value for the quality index higher than the z value of the safety parameter. HTST processes have less detrimental effect on parameters characterized with high z values, compared to LTST ones. Thus, the inactivation of a particular enzyme, characterized by a high z value, might be the aim of a HTST process.

2.4.2 Conduction-Heated Foods

If the product temperature is not uniform, one has to integrate the effects of temperature on each portion of the food in order to calculate volume average quality retention values. Mathematically

TABLE 2.4
Time–Temperature Combinations for Constant Product Temperature Thermal Processes Targeting F_o = 6.4 min

Time (min)	Temperature (°C)	Quality Retention (%)
1189.94	98.42	10
165.56	106.98	50
7.50	120.42	90
0.16	137.19	99

Quality retention is based on z = 25.56°C (46°F) and $D_{121.11°C}$ = 154 min.

speaking, the volume average remaining concentration, $\bar{C}_b$, of the parameter under study is given by (Stoforos et al., 1997)

$$\bar{C}_b = \frac{1}{V_p} \int_0^{V_p} C_a(V) \cdot 10^{-F_{T_{ref}}^{z}(V)/D_{T_{ref}}} dV \tag{2.45}$$

where

V_p represents the volume of the product

C_a is the initial concentration of the parameter under study, which can be position dependent, that is, $C_a(V)$

Similar to the right-hand side of Equation 2.21, where we defined point F values, one can define integrated F values, usually noted as F_s, based on average initial and final concentrations, as

$$F_{T_{ref}}^{z}\Big|_s = F_s = D_{T_{ref}}\left(\log(\bar{C}_a) - \log(\bar{C}_b)\right) \tag{2.46}$$

with $\bar{C}_b$ given by Equation 2.45, and $\bar{C}_a$ representing the average initial concentration. Usually, the concentration of the parameter of interest is considered uniformly distributed at the beginning of the process; thus, $\bar{C}_a = C_a$.

2.4.2.1 Stumbo's Method

The significance of using average concentrations and F_s values, when quality calculations are of interest, has been already indicated. The calculations involved for such estimations are rather complicated. Stumbo and his colleagues (Jen et al., 1971; Stumbo, 1973) developed a procedure to facilitate such calculations for conduction-heated products. In the following paragraphs, we will illustrate the use of Stumbo's method through an example, utilizing the product temperature data presented in Figure 2.4, and the kinetic parameters used in the constant product temperature example.

When performing quality retention calculations, the symbol C ($C_{T_{ref}}^{z}$) is usually used interchangeably with the F value, while a prime on the z and D values, that is, z' and D', indicates that the kinetic parameters refer to quality indices. Jen et al. (1971) proposed the following equation for calculating integrated F (or C) values:

$$C_s = C_c + D'_{T_{ref}} \cdot \log\left(\frac{D'_{T_{ref}} + 10.93 \cdot (C_\lambda - C_c)}{D'_{T_{ref}}}\right) \tag{2.47}$$

Subscript "c" in Equation 2.47, as well as everywhere in Stumbo's method, indicates values calculated at the geometric center (the critical point) of the product, while subscript "λ" refers to a point in the product for which

$$g_\lambda = \frac{1}{2} g_c \tag{2.48}$$

and therefore,

$$j_\lambda = \frac{1}{2} j_c \tag{2.49}$$

The C_c and C_λ values, in accordance with Equation 2.40, are given by

$$C_c\left(=C_{T_{ref}}^{z}\Big|_c\right)=\frac{f_h}{(f_h/U)|_c C_i} \tag{2.50}$$

and

$$C_\lambda\left(=C_{T_{ref}}^{z}\Big|_\lambda\right)=\frac{f_h}{(f_h/U)|_\lambda C_i} \tag{2.51}$$

respectively, while the parameter C_i is absolutely equivalent to F_i given by Equation 2.37.

$$C_i=10^{((T_{ref}-T_{RT})/z)} \tag{2.52}$$

The f_h/U values associated with Equations 2.50 and 2.51 are obtained from the appropriate tables. Following Ball's approach, Stumbo (1973) synopsized his method by correlating f_h/U with g, in a tabular form, for a number of z values (ranging from 8°F to 200°F, thus, contrary to Ball's method, allowing also for quality retention calculations), using the j value as the only parameter. Stumbo's method uses temperature in degrees Fahrenheit and restrictively assumes that the f and j parameters during heating and cooling are equal. In relation to Equation 2.47, the 10.93 factor is valid for cylindrical geometries. For spherical objects this factor assumes a 10.73 value, while for cube-shaped objects it is equal to 11.74 (Holdsworth and Simpson, 2008).

2.4.2.1.1 Example Calculation

Utilizing the product temperature data presented in Figure 2.4, the associated values calculated in the example of Ball's method and the kinetic parameters used in the constant product temperature optimization example, we get the following.

We already have found from Equation 2.44 that, $g=g_c=7.3294$°F.

For the case under study, we obtained $j_h=1.89$ and $j_c=1.54$. Since Stumbo assumes $j_h=j_c$, we have to choose a single value, let us assume $j=j_h=j_c=1.89$.

By successive linear interpolations between Stumbo's tabulated values (for $z=46$°F) for $j=1.80$, we have $f_h/U|_c=1.9055$ (by interpolating between values for $f_h/U=1.00$, $g=2.15$°F and for $f_h/U=2.00$, $g=7.87$°F); for $j=2.00$ we have $f_h/U|_c=1.9961$ (by interpolating between $f_h/U=1.00$, $g=2.01$°F and $f_h/U=2.00$, $g=7.35$°F); and for $j=1.89$ we find: $f_h/U|_c=1.9463$.

$$\text{For Equation 2.49:}\quad j=j_\lambda=\frac{1}{2}j_c=\frac{1}{2}1.89\Rightarrow j=0.945$$

$$\text{For Equation 2.48:}\quad g=g_\lambda=\frac{1}{2}g_c=\frac{1}{2}7.3294\Rightarrow g=3.6647\text{°F}$$

By successive linear interpolations between Stumbo's tabulated values (for $z=46$°F) for $j=0.80$, we have $f_h/U|_\lambda=1.6805$ (by interpolating between values for $f_h/U=1.00$, $g=1.31$°F and for $f_h/U=2.00$, $g=4.77$°F); for $j=1.00$, we have $f_h/U|_\lambda=1.5767$ (by interpolating between $f_h/U=1.00$, $g=1.45$°F and $f_h/U=2.00$, $g=75.29$°F); and for $j=0.945$, we find: $f_h/U|_\lambda=1.6053$.

From Equation 2.52:

$$C_i=10^{((T_{ref}-T_{RT})/z)}\Rightarrow C_i=10^{((250-248)/46)}\Rightarrow C_i=1.1053$$

and from Equations 2.50 and 2.51:

$$C_c = \frac{f_h}{f_h/U|_c C_i} \Rightarrow C_c = \frac{60}{1.9463 \times 1.1053} \Rightarrow C_c = 27.8912 \text{ min}$$

$$C_\lambda = \frac{f_h}{f_h/U|_\lambda C_i} \Rightarrow C_\lambda = \frac{60}{1.6053 \times 1.1053} \Rightarrow C_\lambda = 33.8158 \text{ min}$$

Therefore, the C_s value can be calculated from Equation 2.47 as

$$C_s = 27.8912 + 154 \times \log\left(\frac{154 + 10.93 \cdot (33.8158 - 27.8912)}{154}\right) \Rightarrow C_s = 51.3667 \text{ min}$$

From Equation 2.46, the average remaining concentration of the quality parameter can be calculated as

$$\frac{\bar{C}_b}{\bar{C}_a} = 10^{-\left(C_{250\text{ F}}^{46\text{ F}}\Big|_s / D'_{250°\text{F}}\right)} \Rightarrow \frac{\bar{C}_b}{\bar{C}_a} = 10^{-51.3667/154} \Rightarrow \frac{\bar{C}_b}{\bar{C}_a} = 0.4639$$

That is, the average percent remaining concentration is equal to *46.39%*. (Note that if one uses $j = j_h = j_c = 1.54$ and proceeds as described earlier, the resulting average percent remaining concentration can be found equal to *46.51%*.)

In the aforementioned example, an F_o value of 6.4 min was achieved by heating the product for 100 min at a retort temperature of 248°F, and the resulting average percent remaining concentration of the quality factor under study was found equal to 46.39%. With the procedures presented in this chapter, we can perform calculations to find equivalent processes (i.e., processes that result in the same F_o value of 6.4 min) at different retort temperatures and estimate average quality retention for these processes. In doing so, optimum processing conditions can be identified as originally shown by Teixeira et al. (1969) and illustrated here in Figure 2.10 for the aforementioned

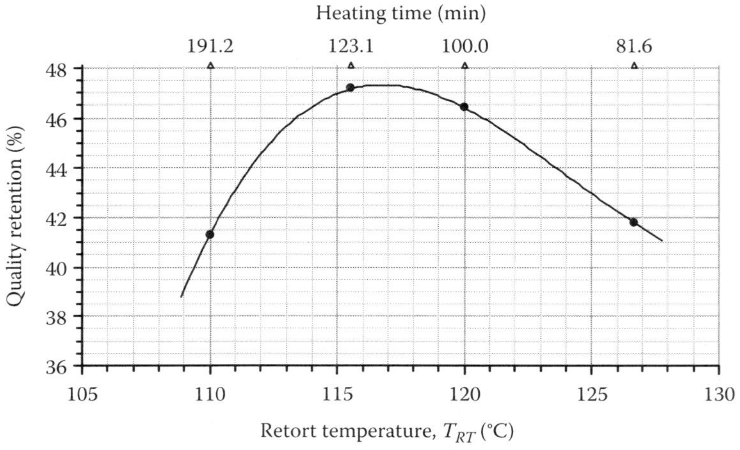

FIGURE 2.10 Thermal process optimization of a conduction-heated product based on the retention of a quality parameter characterized by z = 25.56°C (46°F) and $D_{121.11°C}$ = 154 min. The heat penetration data presented in Figure 2.4 together with a target F_o value of 6.4 min were used.

example data. Compared to the constant product temperature case analyzed earlier, average quality retention is much lower in the case of conduction heating foods where optimum processing conditions exist and should be sought.

2.5 CONCLUDING REMARKS

Kinetic parameters of microbial thermal destruction coupled with product temperature history and the kinetics of thermal degradation of quality factors are absolutely necessary information for designing optimum thermal processes. From the vast thermal processing literature, only key references were given, and only the basic principles in designing thermal processes for food preservation were presented. Determination of the kinetic parameters of heat inactivation through constant or dynamic temperature treatments represents an interesting and challenging issue per se. The same is true for heat transfer studies during thermal processing of foods. A number of issues associated with thermal processing of foods were not covered. For example, no reference was made to process validation, process deviations handling, processing equipment, or to the computational efforts of a number of researches to enlighten the process. Nevertheless, we believe that the information presented here is sufficient to introduce the reader to the subject and guide him for further learning and deeper understanding.

NOMENCLATURE

a_1–a_7 regression coefficients appearing in Equation 2.42, dimensionless

b_1–b_6 regression coefficients appearing in Equation 2.43, dimensionless

B time at the end of heating (steam-off time), min

C concentration of a heat-labile substance, number of microorganisms/mL, spores/container, g/mL, or any other appropriate unit

C_i the F_i factor in reference to quality indices, defined by Equation 2.52, dimensionless

$C^z_{T_{ref}}$ (or simply C) the F value in reference to quality indices, min

D_T (noted also as D) decimal reduction time or death rate constant—time at a constant temperature required to reduce by 90% the initial spore load (or, in general, time required for 90% reduction of a heat-labile substance), min

E_a activation energy, J/mol

F^z_T (or simply F) time at a constant temperature, T, required to destroy a given percentage of microorganisms whose thermal resistance is characterized by z, or, the equivalent processing time of a hypothetical thermal process at a constant temperature that produces the same effect (in terms of spore destruction) as the actual thermal process, min

F_i factor defined by Equation 2.37, which when multiplied by $F_{T_{ref}}$ gives the F value at the retort temperature, dimensionless

f time required for the difference between the medium and the product temperature to change by a factor of 10, min

g difference between retort and product temperature (at the critical point) at steam-off time, °C or °F

I difference between retort and initial product temperature, °C or °F

j a correction factor defined by Equations 2.25 and 2.26 for the heating and cooling curve, respectively, based on the intercept, with the temperature axis at time zero, of the straight line that describes the late, straight portion of the experimental heating or cooling curve plotted in a semilogarithmic temperature difference scale as shown in Figures 2.7 and 2.8, respectively, dimensionless

k_T thermal destruction rate constant, min^{-1}

m difference between product temperature at steam-off and retort temperature during cooling, defined by Equation 2.38, °C or °F

N	number of spores/mL (or microorganisms per container, or any other appropriate unit)
R_g	universal gas constant, 8.314 J/(mol K)
T	(product) temperature, °C or °F
T_A	extrapolated pseudo-initial product temperature at the beginning of heating defined as the intercept, with the temperature axis at time zero, of the straight line that describes the late, straight portion of the experimental heating curve plotted as shown in Figure 2.7, °C or °F
T_B	extrapolated pseudo-initial product temperature at the beginning of cooling defined as the intercept, with the temperature axis at zero cooling time, of the straight line that describes the late, straight portion of the experimental cooling curve plotted as shown in Figure 2.8, °C or °F
t	time, min (or s)
U	the F value at T_{RT}, defined by Equation 2.36, min
V_p	product volume, m^3
z	temperature difference required to achieve a decimal change of the D_T value, °C or °F
z_c, z'_c	correction temperature difference factors appearing in Equations 2.42 and 2.43, respectively, °F (or °C)

Subscripts

1, 2, i, n	referring to a particular condition
a	initial condition
b	final condition
CW	(water) cooling medium
c	cooling phase, or critical point value
h	heating phase
IT	initial condition (for product temperature only)
end	end of cooling cycle
g	condition at steam-off time
p	process
$process$	referring to process values
RT	(retort) heating medium
ref	reference value
$required$	referring to required values
o	initial condition
s	integrated value
T	temperature
λ	refers to a point in the product defined by Equations 2.48 and 2.49

Symbols

Δ	difference
$\bar{}$	volume average value
$'$	a prime on the z and D values indicates that the parameters refer to a quality index

ACKNOWLEDGMENT

This chapter is dedicated to Dr. R.L. Merson, Emeritus Professor, Department of Food Science and Technology and Department of Biological and Agricultural Engineering at U.C. Davis, who was the first to introduce me to the principles of thermal processing.

REFERENCES

Appert, N. 1810. L'art de conserver pendant plusieurs années toutes les substances animales et végétables, pp. 1–147. (Translated by K.G. Bitting, Chicago, IL, 1920.). In: Goldblith, S.A., Joslyn, M.A., and Nickerson, J.T.R. (eds.), (1961) *Introduction to Thermal Processing of Foods*. The AVI Pub. Co., Inc., Westport, CT, 1128pp.

Atherton, D. 1984. *The Heat Processing of Uncured Canned Meat Products*. Guideline No. 6. CFPRA, Chipping Campden, U.K., 62pp.

Ball, C.O. 1923. *Thermal Process Time for Canned Food*, Bulletin of the National Research Council No. 37, Vol. 7, Part 1. National Research Council, Washington, DC, 76pp.

Ball, C.O. and Olson, F.C.W. 1957. *Sterilization in Food Technology. Theory, Practice and Calculations*. McGraw-Hill Book Co., New York, 654pp.

Barsa, C.S., Normand, M.D., and Peleg, M. 2012. On models of the temperature effect on the rate of chemical reactions and biological processes in foods. *Food Engineering Reviews*, 4(4):191–202.

Bigelow, W.D., Bohart, G.S., Richardson, A.C., and Ball, C.O. 1920. *Heat Penetration in Processing Canned Foods*. Bulletin 16-L. National Canners Association Research Laboratory, Washington, DC, 128pp.

Datta, A.K. 1993. Error estimates for approximate kinetic parameters used in food literature. *Journal of Food Engineering*, 18(2):181–199.

David, J.R.D. and Merson, R.L. 1990. Kinetic parameters for inactivation of *Bacillus stearothermophilus* at high temperatures. *Journal of Food Science*, 55(2):488–493, 515.

Dimou, A., Panagou, E., Stoforos, N.G., and Yanniotis, S. 2013. Analysis of thermal processing of table olives using computational fluid dynamics. *Journal of Food Science*, 78(11):E1695–E1703.

Esty, J.R. and Meyer, K.F. 1922. The heat resistance of the spores of *B. botulinus* and allied anaerobes. *Journal of Infectious Diseases*, 34(1):650–663.

FDA. 2014. CFR—Code of Federal Regulations, Title 21, Part 113—Thermally processed low-acid foods packaged in hermetically sealed containers. http://www.accessdata.fda.gov/scripts/cdrh/cfdocs/cfcfr/CFRSearch.cfm?CFRPart=113&showFR=1&subpartNode=21:2.0.1.1.12.1, accessed on September 30, 2014.

Flambert, F. and Deltour, J. 1972. Localization of the critical area in thermally processed conduction heated canned foods. *Lebensmittel-Wissenschaft und -Technologie*, 5(1):7–13.

Goldblith, S.A., Joslyn, M.A., and Nickerson, J.T.R. 1961. *Introduction to Thermal Processing of Foods*. The AVI Publishing Company, Inc., Westport, CT, 1128pp.

Hayakawa, K. 1970. Experimental formulas for accurate estimation of transient temperature of food and their application to thermal process evaluation. *Food Technology*, 24(12):1407–1418.

Holdsworth, D. and Simpson, R. 2008. *Thermal Processing of Packaged Foods*, 2nd edn. Springer Science+Business Media, LLC, New York, 407pp.

IOC (International Olive Council). 2004. Trade standard applying to table olives, resolution no. RES-2/91-IV/04, COI/OT/NC no. 1. Available from: http://www.internationaloliveoil.org, accessed October 14, 2012.

Jen, Y., Manson, J.E., Stumbo, C.R., and Zahradnik, J.W. 1971. A procedure for estimating sterilization of and quality factor degradation in thermally processed foods. *Journal of Food Science*, 36:692–698.

Jonsson, U., Snygg, B.G., Härnulv, B.G., and Zachrisson, T. 1977. Testing two models for the temperature dependence of the heat inactivation rate of *Bacillus stearothermophilus* spores. *Journal of Food Science*, 42(5):1251–1252, 1263.

Kookos, I.K. and Stoforos, N.G. 2014. Heat transfer. In: Tzia, C. and Varzakas, T. (ed.), *Food Engineering Handbook: Food Engineering Fundamentals*. Taylor & Francis Group, Boca Raton, FL, pp. 75–111.

Lopez, A. 1987. *A Complete Course in Canning and Related Processes, Book III, Processing Procedures for canned Food Products*, 12th edn. The Canning Trade Inc., Baltimore, MD, 516pp.

Lund, D.B. 1975. Heat processing. In: Karel, M., Fennema, O.R., and Lund D.B. (eds.), *Principles of Food Science. Part II. Physical Principles of Food Preservation*. Marcel Dekker, Inc., New York, pp. 31–92.

Lund, D.B. 1977. Design of thermal processes for maximizing nutrient retention. *Food Technology*, 31(2):71–78.

Merson, R.L., Singh, R.P., and Carroad, P.A. 1978. An evaluation of Ball's formula method of thermal process calculations. *Food Technology*, 32(3):66–72, 75.

NCA, 1968. *Laboratory Manual for Food Canners and Processors*, Vol. 1, *Microbiology and Processing*. National Canners Association Research Laboratory, The Avi Publishing Company, Inc., Westport, CT.

NFPA, 1982. *Thermal Processes for Low-Acid Foods in Metal Containers*. Bulletin 26-L, 12th edn. National Food Processors Association Research Laboratories, Washington, DC, 68pp.

NFPA, 1984. *Thermal Processes for Low-Acid Foods in Glass Containers*. Bulletin 30-L, 5th edn. National Food Processors Association Research Laboratories, Washington, DC, 26pp.

Pflug, I.J. 1987. Using the straight-line semilogarithmic microbial destruction model as an engineering design model for determining the F-value for heat processes. *Journal of Food Protection*, 50(4):342–346.

Pflug, I.J. and Odlaugh, T.E. 1978. A review of z and F values used to ensure the safety of low-acid canned foods. *Food Technology*, 32:63–70.

Potter, N.N. and Hotchkiss, J.H. 1995. *Food Science*, 5th edn. Springer Science+Business Media, Inc., New York, 608pp.

Prescott, S.C. and Underwood, W.L. 1897. Micro-organisms and sterilizing processes in the canning industries. *Technology Quarterly*, 10(1):183–199.

Prescott, S.C. and Underwood, W.L. 1898. Micro-organisms and sterilizing processes in the canning industries. II. The souring of canned sweet corn. *Technology Quarterly*, 11(1):6–30.

Ramaswamy, H.S., van de Voort, F.R., and Ghazala, S. 1989. An analysis of TDT and Arrhenius methods for handling process and kinetic data. *Journal of Food Science*, 54(5):1322–1326.

Shapton, D.A., Lovelock, D.W., and Laurita-Longo, R. 1971. The evaluation of sterilization and pasteurization processes from temperature measurements in degrees Celsius (°C). *Journal of Applied Bacteriology*, 34(2):491–500.

Silva, F.V. and Gibbs, P. 2004. Target selection in designing pasteurization processes for shelf-stable high-acid fruit products. *Critical Reviews in Food Science and Nutrition*, 44:353–360.

Stoforos, N.G. 1991. On Ball's formula method for thermal process calculations. *Journal of Food Process Engineering*, 13(4):255–268.

Stoforos, N.G. 1995. Thermal process design. *Food Control*, 6(2):81–94.

Stoforos, N.G. 2010. Thermal process calculations through Ball's original formula method: A critical presentation of the method and simplification of its use through regression equations. *Food Engineering Reviews*, 2(1):1–16.

Stoforos, N.G., Noronha, J., Hendrickx, M., and Tobback, P. 1997. A critical analysis of mathematical procedures for the evaluation and design of in-container thermal processes of foods. *Critical Reviews in Food Science and Nutrition*, 37(5):411–441.

Stoforos, N.G. and Taoukis, P.S. 2006. Heat processing: Temperature–time combinations. In: Hui, Y.H. (ed.), *Handbook of Food Science, Technology, and Engineering*, Vol. 3, *Food Engineering and Food Processing, Part L, Thermal Processing*. CRC Press, Taylor & Francis Group, Boca Raton, FL, pp. 109-1–109-16, Chapter 109.

Stumbo, C.R. 1973. *Thermobacteriology in Food Processing*, 2nd edn. Academic Press, Inc., New York, 329pp.

Teixeira, A.A., Dixon, J.R., Zahradnik, J.W., and Zinsmeister, G.E. 1969. Computer optimization of nutrient retention in the thermal processing of conduction-heated foods. *Food Technology*, 23(6):845–850.

Tucker, G. and Featherstone, S. 2011. *Essentials of Thermal Processing*. John Wiley & Sons Ltd., Chichester, U.K., 264pp.

Valdramidis, V.P., Taoukis, P.S., Stoforos, N.G., and Van Impe, J.F.M. 2012. Modeling the kinetics of microbial and quality attributes of fluid food during novel thermal and non-thermal processes. In: Cullen, P.J., Tiwari, B.K., and Valdramidis, V.P. (eds.), *Novel Thermal and Non-Thermal Technologies for Fluid Foods*. Elsevier, Inc., London, U.K., pp. 433–471.

Valigra, L. 2011a. New series: Innovators in food safety and science I. Nicholas Appert: The father of food preservation. *Food Quality*, 18(1):22–24.

Valigra, L. 2011b. New series: Innovators in food safety and science III. C. Olin Ball: A pioneer in thermal death-time standards. *Food Quality*, 18(3):39–41.

3 Canning of Fishery Products

C.O. Mohan, C.N. Ravishankar, and T.K. Srinivasa Gopal

CONTENTS

Food processing was practiced from the prehistoric age mainly to fulfill the requirements of military and sailing persons. The major food processing techniques followed include drying, salted and drying, fermenting, and smoking until the advent of heat-processed products in glass containers by Nicolas Appert. Preservation aims to process foods for storing for longer duration. Human beings are dependent on products of plant and animal origin for food. As most of these products are readily available only during certain seasons of the year and fresh food spoils quickly, methods have been developed to preserve foods. Preservation must be seen as a way of storing excess foods that are abundantly available at certain times of the year, so that they can be consumed in times when food is scarce. Apart from preserving for a longer duration, processing also helps in converting raw foods into edible and palatable form, increases organoleptic quality, makes foods safe for consumption, helps in bringing nutritional and food security, increases product diversification, minimizes wastage, generates employment, and helps in earning foreign exchange by exporting processed food products.

Fish and shellfishes pass through a number of processes immediately after catch before it is consumed or sold for consumption. These processes can be divided into primary processing and secondary processing. Primary processing includes the steps that enable fish to be stored or sold for further processing, packaging, and distribution. Examples include washing, cleaning, heading, gilling, scaling, gutting, grading, filleting, deboning, skinning, chilling, and freezing. Whereas secondary processing includes the production of "value-added products." Examples are salting, drying, smoking, canning, marinating, and packaging ready-to-eat foods. Fresh fish spoils very quickly due to various internal and external factors. Once the fish has been caught, spoilage progresses rapidly. In the high ambient temperatures of the tropics, fish will spoil within 12 h. By adopting good fishing techniques (to minimize fish damage) and cooling the fish immediately, the storage life can be increased. It is well established that spoilage of fish and shellfish is mainly due to three destructive processes. These are

1. Enzymatic decomposition
2. Bacterial action
3. Oxidation process

3.1 ENZYMATIC DECOMPOSITION

Enzymes are powerful biological chemicals that occur in the tissue of all living organisms. They perform important functions, either by breaking down large food compounds into smaller ones in the stomach and gut, as in digestion, or by helping to make new compounds for building new body tissue or for producing energy. In the living animal, the body keeps a close control on the activity of enzymes. However, when the animal dies this control is lost. The enzymes will start attacking the flesh of the body, breaking large compounds down to smaller ones, just like the process of digestion (autolysis). Enzyme activity depends on various factors like temperature, pH, and water activity.

3.2 BACTERIAL ACTION

Bacteria or germs are living organisms that are found everywhere in nature. They are classified mainly as psychrophile, mesophile, and thermophile bacteria. Psychrophilic or psychrotrophic bacteria are microorganisms with optimum growth temperatures in the region of 10°C–15°C and 20°C, respectively, capable of growth down to 0°C. The bacteria growing on fish spoiling in ice are predominately psychrophilic. Mesophiles have an optimum growth temperature, the temperature at which they multiply most rapidly, in the region of 35°C–40°C. Food poisoning organisms are adapted to grow in the body of warm-blooded animals and are mesophiles, though some can grow at chilled temperatures. Thermophilic bacteria grow best at elevated temperatures in the range of 55°C–75°C. *Bacillus stearothermophilus* is an extremely heat-resistant thermophilic spore-forming bacteria, which has been found to be responsible for the flat-sour spoilage of canned foods. Although there are many useful bacteria, they pose problems in the handling of food either by spoiling the food or by causing food poisoning. Fish carry millions of bacteria on their external surfaces (skin and gills) and in their intestines. A healthy, living fish uses its natural defense mechanism to protect it against the harmful effects of bacteria. However, when a fish dies, the defense mechanism stops working. This allows the bacteria the opportunity to feed on the flesh, multiply, and eventually spoil the fish. Conditions which allow bacteria to multiply are suitable temperatures, the presence of water, and a source of food. Bacteria will enter the flesh easily if the fish has been damaged through improper handling and storage.

3.3 OXIDATION

Rancidity is a more widely used term for oxidation. It occurs when oxygen in the air reacts with oil or fat in the flesh of the fish. This leads to a sour or stale, unpleasant smell or taste. Fatty pelagic fishes like skipjack, seer, mackerel, herring, scads, and sardines store fat in their flesh and can turn rancid quickly if not handled and stored properly. White-fleshed demersal fish store fat in their livers, so these must be removed during gutting. Frozen-stored fatty fish can spoil through oxidation if stored improperly, even though the temperature is too low for bacteria to grow or enzymes to work effectively.

3.4 MINIMIZING FISH SPOILAGE

Fish spoilage can be effectively minimized if the effects of enzymes, bacteria, and oxidation are controlled properly. This can be achieved by understanding the optimum conditions that enzymes, bacteria, and oxidation processes prefer and modifying these conditions which helps to preserve food. Many processing procedures aim to alter these conditions to achieve preservation. Some of the approaches are given in Table 3.1.

3.4.1 Lowering the Temperature

Chilling food in the refrigerator or with ice slows down the destructive processes of enzymes and bacteria. The shelf life of food can therefore be extended by many days. If the temperature is lowered further, as in freezing, much longer storage times are possible because all bacterial activity and virtually all enzymatic action is stopped if the temperature is maintained properly.

3.4.2 Raising the Temperature

High temperatures kill bacteria and destroy enzymes. Processes such as cooking (boiling, frying, and baking), hot smoking, canning, and pasteurization extend the keeping time of the food.

3.4.3 Drying or Dehydration

Removing water from the food by drying is a very old and effective method of controlling bacterial and enzymatic spoilage of food. Drying can be achieved under the sun and wind (natural drying) or

TABLE 3.1
Food Preservation Methods

Approaches	Examples of Process
Low temperature	Chilling, refrigeration, freezing
High temperature	Pasteurization, thermal processing, smoking
Reduced water availability	Drying, salt curing, spray drying, freeze drying
Chemical-based preservation	Organic acids, natural extracts from plants
Microbial product based	Bacteriocins
Radiation	Ionizing (Gamma rays) and nonionizing (UV rays) radiation
Hurdle technology	Altered atmosphere (vacuum and modified atmosphere with CO_2, O_2, N_2, and other gases); active packaging; high-pressure treatment; and smoking

in a mechanical drier. Salting helps the drying process too, as it binds the water, making it unavailable to bacteria. Some high-temperature processing such as hot smoking uses a combination of drying and high temperatures to control bacteria and enzymes.

Oxidation problems of fatty fish can be controlled by preventing the contact of oxygen to the product. This is achieved by packing the fish in high barrier plastic bags or by packing under vacuum or packing with oxygen absorbing sachet.

The food preservation methods are aimed at preventing undesirable changes in the wholesomeness, nutritive value, and sensory quality of food by controlling the growth of microorganisms and obviating contamination by adopting economic methods. Thermal processing is one such method, by which food is given sufficient heat treatment in a hermetically sealed container to destroy pathogenic and/or spoilage causing microorganisms and their spores, antinutrients, and enzymes that cause degradation in the food.

An Italian naturalist Spallanzani concluded from his experiments that organisms causing spoilage in a number of food products were carried in the air, and by heating the contaminated infusions in airtight container, the development of the organism was prevented. Although Spallanzani's work was the key to the preservation of food by heat, little use appears to have been made of it until the early part of the nineteenth century when Nicholas Appert first succeeded in preserving food in airtight glass containers. So, the Frenchman Nicholas Appert is credited as the inventor of this noble technology.

3.5 THERMAL PROCESSING

Thermal processing of foods constitutes a significant part of the world's food preservation technique. Thermally processed foods include heat-processed foods in bottles, jars, metal cans, pouches, tubes, and plastic-coated cartons. The heat treatment is applied with the objective of destroying specific, usually pathogenic organisms and also spoilage causing microorganisms. The first publisher on canning of food was French confectioner Nicholas Appert, who in 1810 edited "L art de conserver pendant plusiers ann'ees toutes les substances animals et vegetables," which was later translated to English "Of preserving all kinds of animal and vegetable substances for several years."

Demand for better quality processed food is increasing as civilization developed. This led to the development of a large food preservation industry, aiming to supply food that is sterile, nutritious, and economical. Thermal sterilization of foods is the most significant part of this industry and is one of the most effective means of preserving our food supply (Karel et al., 1975). The objective of sterilization is to extend the shelf life of food products and make them safe for human consumption by destroying the pathogenic microorganisms.

Thermal processing is the most common sterilization method, which employs steam as the heating medium, although other types of heating medium like steam–air mixture, pressurized hot water, and direct flame are available. Saturated steam is the most commonly and highly desirable heating medium used for commercial sterilization of canned foods. Other methods of sterilization such as pulsed electric field (Barbosa-Canovas et al., 1998, 2001), ultrahigh hydrostatic pressure (Barbosa-Canovas et al., 1998; Palou et al., 1999; Furukawa and Hayakawa, 2000), and ultraviolet treatment (Farid et al., 2000) have been widely studied. However, these methods fail to replace the common thermal processing mainly due to their inability to inactivate enzymes, which can cause various negative effects such as discoloration, bitter flavor, and softening (Clark, 2002).

3.5.1 Principles of Thermal Processing

Thermal processing which is also commonly referred as heat processing or canning is a means of achieving long-term microbiological stability for non-dried foods without the use of refrigeration, by prolonged heating in hermetically sealed containers, such as cans or retortable pouches, to render the contents of the container sterile. The concept of thermal processing has come a long way since Nicholas Appert's time. Later, Bigelow and Ball developed the scientific basis for calculating the sterilization process for producing safe foods. Today, thermal processing forms one of the most widely used methods of preserving and extending the shelf life of food products including seafoods. Thermal processing involves the application of high-temperature treatment for sufficient time to destroy all the microorganisms of public health and spoilage concerns. The important factors to be considered in thermal processing are the seal integrity, sufficient process lethality, and post-process hygiene. The seal integrity is achieved by a hermetic seal, which helps in preventing recontamination and creates an environment inside the container that prevents the growth of other microorganisms of higher heat resistance. It also helps in preventing toxin production from pathogens. The time–temperature schedule for the required process lethality should be effective to eliminate the most perilous and heat-resistant mesophilic anaerobic spore-forming pathogen *Clostridium botulinum*. Post-process hygiene is of utmost importance for heat-processed foods as the heat-processed containers, which are still warm and wet, may lead to inward leakage through the seal. Hence, the use of chlorinated water is mandatory for can washing and cooling.

Normally, thermal processing is not designed to destroy all microorganisms in a packaged product. Such a process may destroy all the important nutrients and results in low product quality. Instead of this, the pathogenic microorganisms in a hermetically sealed container are destroyed by heating, and a suitable environment is created inside the container which does not support the growth of spoilage-type microorganisms. Several factors must be considered for deciding the extent of the heat processing (Fellows, 1988). These are

1. Type and heat resistance of the target microorganism, spore, or enzyme present in the food
2. pH of the food
3. Heating conditions
4. Thermo-physical properties of the food and the container shape and size
5. Storage conditions

Food contain different microorganisms and/or enzymes that the thermal process is designed to destroy. In order to determine the type of microorganism on which the process should be based, several factors must be considered. In foods that are vacuum packed in hermetically sealed containers, low oxygen levels are intentionally achieved. Therefore, the prevailing conditions are not conducive to the growth of microorganisms that require oxygen (obligate aerobes) to create food spoilage or public-health problems. Further, the spores of obligate aerobes are less heat resistant than the microbial spores that grow under anaerobic conditions (facultative or obligate anaerobes). The heat resistance of food spoilage microorganisms has been studied extensively, and thermal resistance data are available for the more resistant organisms in a variety of products (Esty and Meyer, 1922). The heat tolerance of microorganisms is greatly influenced by pH or acidity. From a thermal-processing standpoint, foods are divided into three pH groups: high-acid foods (pH < 3.7), acid or medium-acid foods (pH 3.7–4.5), and low-acid foods (pH > 4.5). With reference to thermal processing, the most important distinction in the pH classification is the dividing line between acid and low-acid foods.

Sterilization or its commercial equivalent for the reduction of viable microbes to some predetermined level forms the basis of a substantial class of food preservation operations and is particularly important in canning (Kumar et al., 2001). The main purpose of sterilization is the destruction of microorganisms by heating, which causes spoilage of food during preservation. The usually targeted microorganism in the sterilization of foods is the *C. botulinum*. However, the use of nonpathogenic and more resistant species is preferred, particularly *Clostridium sporogenes* (PA 3679) (Ranganna, 1986). The argument is that once these have been destroyed, all other less heat-resistant spores can be safely assumed to be destroyed.

Most of the research dealing with thermal processing devotes special attention to *C. botulinum*, which is a highly heat-resistant, rod-shaped, spore-forming, anaerobic pathogen that produces an extremely potent exotoxin under favorable conditions, which leads to "*botulism*." It has been generally accepted that *C. botulinum* do not grow and produce toxins below a pH of 4.5 and is a potential health hazard only in foods with a pH above 4.5. Therefore, all low-acid foods should receive a process that is adequate to destroy *C. botulinum*. Generally, canned foods receive a heat treatment that is more severe than that required to destroy *C. botulinum* since several other species of microorganisms have a greater heat resistance. An order-of-the-process factor of 12D is used in the commercial heat processing of low-acid foods that do not contain preservation levels of salt or other bacterio labile or bacteriostatic chemicals (Gillespy, 1951).

Thermal processing is designed to destroy different microorganisms and enzymes present in the food. Normally in thermal processing, an exhausting step is carried out before sealing the containers. In some cases, food is vacuum packed in hermetically sealed containers. In such cases very low levels of oxygen is intentionally achieved. Hence, the prevailing conditions are not favorable for the growth of microorganisms that require oxygen (obligate aerobes) to create food spoilage or public-health problems. Further, the spores of obligate aerobes are less heat resistant than the microbial spores that grow under anaerobic conditions (facultative or obligate anaerobes). The growth and activity of these anaerobic microorganisms are largely pH dependent. From a thermal-processing standpoint, foods are divided into three distinct pH groups which are given below. Changes in the intrinsic properties of food, mainly salt, water activity, and pH are known to affect the ability of microorganisms to survive the thermal processes in addition to their genotype. Due to health-related concerns on the use of salt, there is an increased demand to reduce salt levels in foods. The heat tolerance of microorganisms is greatly influenced by pH or acidity. The United States Food and Drug Administration (FDA) has classified foods in the federal register (21 CFR Part 114) as follows:

1. High-acid foods (pH < 3.7; e.g., apple, apple juice, apple cider, apple sauce, berries, cherry (red sour), cranberry juice, cranberry sauce, fruit jellies, grapefruit juice, grapefruit pulp, lemon juice, lime juice, orange juice, pineapple juice, sour pickles, and vinegar)
2. Acid or medium-acid foods (pH 3.7–4.5; e.g., fruit jams, fruit cocktail, grapes, tomato, tomato juice, peaches, piento, pineapple slices, potato salad, prune juice, and vegetable juice)
3. Low-acid foods (pH > 4.5; e.g., all meats, fish and shellfishes, vegetables, mixed entries, and most soups).

The acidity of the substrate or medium in which microorganisms are present is an important factor in determining the degree of heating required. With reference to thermal processing of food products, special attention should be devoted to *C. botulinum* which is a highly heat-resistant, rod-shaped, spore-forming, anaerobic pathogen that produces the *botulism* toxin. It has been generally accepted that *C. botulinum* and other spore-forming human pathogens does not grow and produce toxins below a pH of 4.6. The organisms that can grow in such acid conditions are destroyed by relatively mild heat treatments. Some spore formers may cause spoilage of this category of foods, for example, *Bacillus coagulans*, *Clostridium butyricum*, and *Bacillus licheniformis*, as well as ascospores of *Byssoclamys fulva* and *Byssoclamys nivea*, which are often present in soft fruits such as strawberries. For food with pH values greater than 4.5, so-called low-acid products which include fishery products, it is necessary to apply a time–temperature regime sufficient to inactivate spores of *C. botulinum* which is commonly referred to as a *botulinum cook* in the industry. Thermal processes are calibrated in terms of the equivalent time the thermal center of the product, that is, the point of the product in the container most distant from the heat source or cold spot, spends at 121.1°C, and this thermal process lethality time is termed F_0 value. Although there are other microorganisms, for example, *B. stearothermophilus*, *Bacillus thermoacidurans*, and *Clostridium thermosaccolyaticum*, which are *thermophilic* in nature (optimal growth temperature ~50°C–55°C) and are more heat resistant than *C. botulinum*, a compromise is drawn on the practical impossibility of achieving full sterility in the contents of a hermetically sealed container during commercial heat processing, whereby the initial bacterial load is destroyed through sufficient decimal reductions to reduce the possibility of a single organism surviving to an acceptably low level. This level depends on the organism, usually *C. botulinum*, which the process is designed to destroy. The time required to reduce the number of spores of this organism (or any other microorganism) by a factor of 10 at a specific reference temperature (121.1°C) is the decimal reduction time, or the D value, denoted by D_0. The D_0 value for *C. botulinum* spores can be taken as 0.25 min. A reduction by a factor of 10^{12}, regarded as an acceptably low level, requires 3 min at 121.1°C, and is known as the process value, or F value, designated F_0 so, in this case, $F_0 = 3$, which is known as a "*botulinum cook*" which is the basis of commercial sterility.

3.5.2 Thermal Resistance of Microorganisms

For establishing safe thermal processing, knowledge on the target microorganism or enzyme, its thermal resistance, microbiological history of the product, composition of the product, and storage conditions are essential. After identifying the target microorganism, the thermal resistance of the microorganism must be determined under conditions similar to the container. Thermal destruction of microorganism generally follow a first-order reaction indicating a logarithmic order of death, that is, the logarithm of the number of microorganisms surviving a given heat treatment at a particular temperature plotted against the heating time (survivor curve) will give a straight line (Figure 3.1). The microbial destruction rate is generally defined in terms of a decimal reduction time (D value) which represents a heating time that results in 90% destruction of the existing microbial population or one decimal reduction in the surviving microbial population. Graphically, this represents the time during which the survival curve passes through one logarithmic cycle (Figure 3.1). Mathematically,

$$D = \frac{(t_2 - t_1)}{(\log a - \log b)}$$

where a and b are the survivor counts following heating for t_1 and t_2 min, respectively. As the survivor or destruction curve follows the logarithmic nature, the complete destruction of the

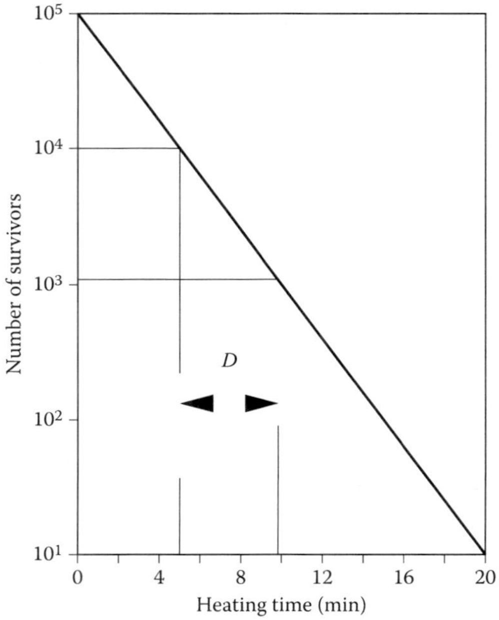

FIGURE 3.1 Survivor curve.

TABLE 3.2
***D* Value (at 121.1°C) of Some Bacterial Spores**

Microorganism	Optimum Growth Temperature (°C)	*D* Value (min)
B. stearothermophilus	55	4–5
Clostridium thermosaccharolyticum	55	3–4
Clostridium nigrificans	55	2–3
C. botulinum types A and B	37	0.1–0.25
C. sporogenes (PA 3679)	37	0.1–1.5
B. coagulans	37	0.01–0.07
Nonspore-forming mesophilic bacterial yeasts and moulds	30–35	0.5–1.0

microorganisms is theoretically not possible. From the survivor curve, as the graph is known, it can be seen that the time interval required to bring about one decimal reduction, that is, 90% reduction, in the number of survivors is constant. This means that the time to reduce the spore population from 10,000 to 1,000 is the same as the time required to reduce the spore population from 1,000 to 100. This time interval is known as the decimal reduction time or the "*D*" value.

The *D* value for bacterial spores is independent of initial numbers, but it is affected by the temperature of the heating medium. The higher the temperature, the faster the rate of thermal destruction and lower the *D* value. This is why thermal sterilization of canned fishery products relies on pressure cooking at elevated temperatures (>100°C) rather than cooking in steam or water, which is open to the atmosphere. The unit of measurement for *D* is "minute." An important feature of the survivor curve is that no matter how many decimal reductions in spore numbers are brought about by a thermal process, there will always be some probability of spore survival. So in practice, the fish canners are satisfied if the probability of pathogenic spore survival is sufficiently remote, and the associated public-health risk is also not significant. In addition to this they also accept, as a commercial risk, a slightly higher probability of non-pathogenic spoilage. Different microorganisms and their spores have different *D* values as shown in Table 3.2.

Since bacterial spores possess different *D* values, a thermal process designed to reduce spore population of one species by a factor of 10^9 (i.e., 9 decimal reductions or 9*D* process) will bring about a different order of destruction for spores of another species. Thermophilic spores are more heat resistant and therefore have higher *D* values than other mesophilic spores.

3.5.3 Thermal Processing Requirements for Canned Fishery Products

Thermal processing is carried out for achieving two objectives; the first is consumer safety from botulism and the second is nonpathogenic spoilage which is deemed commercially acceptable to a certain extent. If heat processing is inadequate, the possibility of spoilage due to *C. botulinum* is more and will endanger the health of the consumer. Safety from botulism is made possible by making the probability of *C. botulinum* spores surviving the heat process sufficiently remote and presents no significant health risk to the consumer. An acceptable low level in the context of this dangerously pathogenic organism means less than one in a billion (10^{-12}) chance of survival. Such a low probability of spore survival is commercially acceptable as it does not represent a significant health risk. The excellent safety record of the canning industry with respect to the incidence of botulism through under processing confirms the validity of this judgment. An acceptable low level in the case of thermophilic nonpathogenic organisms should be arrived at judiciously considering the factors like very high *D* value, risk of flat-sour spoilage, commercial viability, and profitability. Since nonpathogenic organisms do not endanger the health of the consumer, process adequacy is generally assessed in terms of the probability of spore survival which is judged commercially acceptable. Considering all these facts, it is generally found acceptable if thermophilic spore levels are reduced to around 10^{-2} to 10^{-3} per gram. Another reason for this acceptance is that the survivors will not germinate if the storage temperature is kept below the thermophilic optimum growth temperature, that is, below 35°C.

3.5.4 Commercial Sterility

If the thermal process is sufficient to fulfill the criteria of safety and prevention of nonpathogenic spoilage under normal conditions of transport and storage, the product is said to be "commercially sterile." In relation to canned foods, the FAO/WHO Codex Alimentarius Commission defines commercial sterility as the condition achieved by the application of heat, sufficient alone or in combination with other appropriate treatments, to render the food free from microorganisms capable of growing in the food at normal non-refrigerated conditions at which the food is likely to be held during distribution and storage. Apart from this concept, there are circumstances where a canner will select a process which is more severe than that required for commercial sterility as in the case of mackerel and sardine where bone softening is considered desirable.

3.5.5 Thermal Process Evaluation

The primary objective of thermal processing of canned foods is to destroy microorganisms capable of causing deterioration of the foods or endangering the health of consumer (Hersom and Hulland, 1980). The basic ideas of thermal process calculation are well presented in several published articles and textbooks (Ball and Olson, 1957; Lopez, 1969; Stumbo, 1973; Hayakawa, 1978; Merson et al., 1978; Ramaswamy et al., 1992; Teixiera, 1992). The first systematic approach to the problem of applying bacteriology and physical data to the determination of thermal process time for canned foods was that of Bigelow et al. (1920). They derived the "*General method*" of process calculation, which integrates the lethal effects by a graphical or numerical integration procedure based on the time–temperature data obtained from test containers processed under actual commercial processing conditions. Ball (1923) developed a mathematical procedure for determining the heat sterilization required for the safe processing of canned foods, which is known as the "*Formula method*." This makes use of heat penetration (HP) parameters with several mathematical procedures to integrate the lethal effects of heat (Stumbo, 1973; Hayakawa, 1978; Ramaswamy et al., 1992). Over the years, Ball's method has undergone rigorous evaluations, simplifications, and improvements (Ball and Olson, 1957). Despite its limitations, Ball's method is widely used in the canning industry all around the world (Merson et al., 1978). Stumbo's method eliminates some of the oversimplifications and assumptions used in the development of Ball's formula method (Stumbo, 1973). Additional formula methods were developed by Hayakawa (1978) and Steele and Board (1979). These methods involving the solution of formulae for thermal process calculations have been widely used for many years in the food industry. The purpose of these is to estimate the process lethality of a given process, or alternatively to arrive at an appropriate process time under a given set of heating conditions to result in a given process lethality. Smith and Tung (1982) assessed the accuracy of the Ball, Stumbo, Steel and Board, and Hayakawa's formula methods for determining process lethality in conduction-heated foods. They reported that Stumbo's (1973) method gave the best estimates of process lethality under various conditions.

Determination of the time–temperature history of processed food has practical and safety implications. Navankasattusas and Lund (1978), have discussed methods for time–temperature profile evaluation and measurement of lethality in processed foods. Data on thermal process schedules, which indicate the F_0 value, time and temperature of the processing, are available in the literature (Lopez, 1996). Time–temperature histories may be derived by using direct measurements or by mathematical modeling (Tucker and Holdsworth, 1991).

FIGURE 3.2 8 oz and A 2½ aluminum cans, TFS cans, and retortable pouches.

3.5.6 Packaging Materials for Thermal Processing

Packaging material forms the most important component of thermal processed foods. It should be able to withstand the severe process conditions and should prevent recontamination of the product. Various packaging materials have been used historically starting from glass container to metal container, flexible retoratble pouches, and rigid plastic containers (Figure 3.2).

3.5.6.1 Glass Containers

Glass is a natural solution of suitable silicates formed by heat and fusion followed by immediate cooling to prevent crystallization. It is an amorphous transparent or translucent super-cooled liquid. Modern glass container is made of a mixture of oxides namely, silica (S_1O_2), lime (CaO), soda (Na_2O), alumina (Al_2O_3), magnesia (MgO), and potash in definite proportions. Coloring matter and strength improvers are added to this mixture and fused at 1350°C–1400°C and cooled sufficiently quickly to solidify into a vitreous or noncrystalline condition.

Glass jars for food packing have the advantages of very low interaction with the contents and product visibility. However, they require more careful processing and handling. Glass containers used in canning should be able to withstand heat processing at a high temperature and pressure. Breakage occurring due to "thermal shock" is of greater significance in canning than other reasons of breakage. Thermal shock is due to the difference in the temperature between the inside and the outside walls of the container giving rise to different rates of expansion in the glass wall producing an internal stress. This stress can open up microscopic cracks or "clucks" leading to large cracks and container failure. Thermal shock will be greater if the wall thickness is high. Therefore, glass containers in canning should have relatively thin and uniform walls. Similarly, the bottom and the wall should have a thickness as uniform as possible. More defects occur at sharp containers and flat surface and hence these should be avoided. Chemical surface coatings are often applied to make the glass more resistant to "bruising" and to resist thermal shock. Various types of seals are available, including venting and non-venting types, in sizes from 30 to 110 mm in diameter, and made of either tin or tin-free steel. It is essential to use the correct overpressure during retorting to prevent the lid being distorted. It is also essential to preheat the jars prior to processing to prevent shock breakage.

3.5.6.2 Metal Containers

Metal cans are the most widely used containers for thermal processed products. Metal containers are normally made of tin, aluminum, or tin-free steel.

3.5.6.3 Tinplate Cans

Tinplate is low-metalloid steel plate of can making quality (CMQ) coated on both sides with tin, giving a final composition of 98% steel and 2% tin. The thickness varies from 0.19 to 0.3 mm depending on the size of the can. Specifications with respect to the content of other elements are: Carbon (0.04%–0.12%), manganese (0.25%–0.6%), sulfur (0.05% max), phosphorus (0.02% max), silicon (0.01% max), and copper (0.08% max). The corrosive nature of tinplate depends principally on the contents of copper and phosphorous. The higher the contents of these metals, the greater the corrosiveness of steel. However, higher phosphorous content imparts greater stiffness to steel plate which is advantageous in certain applications where higher pressure develops in the container, for example, beer can.

Base plate for can making is manufactured using the cold reduction (CR) process. CR plates are more advantageous over hot reduced plates because of the following characteristics:

1. Superior mechanical properties—possible to use thinner plates without loss of strength
2. More uniform gauge thickness
3. Better resistance to corrosion
4. Better appearance

3.5.6.3.1 Tin Coating

The base plate is coated with tin either by the hot dipping or electrolytic process. The latter is more common because:

- It consumes less tin.
- It gives uniform coating.

In the electrolytic process pure tin is used as the anode, and base steel plate serves as the cathode. Depending upon the number of anodes used and the speed of the plate in the electrolytic cell, the pickup of the can be regulated. Depending upon the different usages, the plate is given even or differential coating. In differential coating, the higher content of tin will always be on the surface coming into contact with the food.

Tinplate for can making has a steel base with four identical layers on either side.

- Steel base—provides strength and formability
- Tin–iron alloy $FeSn_2$ formed at the interface of steel and free tin layer acts as a bond between the two
- Passivating film of chromium oxide that prevents corrosion of tin
- Oil film, usually dioctyl sebacate or acetyl tributyl citrate, for easy handling and prevention of abrasion

3.5.6.3.2 Coating Weights of Tin

The coating weight of tin used to be expressed as units in lbs per basis box. A basis box consists of 112 sheets of size 20 in. × 14 in. The coating of 1 lb per basis box is known as E 100. When the weight of tin is 0.75 lbs per basis box, the plate is described as E 75. However, nowadays this is expressed as grams per square meter (GSM). The recommended differential coating for fish cans is *D* 11.2/5.6 which indicates a coating 11.2 GSM tin on the food contact side and 5.6 GSM on the external surface.

3.5.6.3.3 Lacquering

Tinplate for fish cans is given a coating on the food contact surface with a sulfur-resistant lacquer. The lacquers most commonly employed in fish cans are oleoresinous C-enamels. C-enamels contain zinc oxide that will react with the sulfur compounds released during heat processing producing zinc sulfide (white in color) preventing the formation of tin or iron sulfide.

Open top sanitary cans are generally of "three-piece" or "two-piece" construction. Three-piece cans consist of a cylindrical body with a soldered side seam and two ends attached to the cylinder by double seaming. Circumferential beads are made on certain type of can bodies, especially bodies of larger cans to strengthen it by making it into columns of shorter can bodies. This is also a check against handling abuse and paneling pressures.

Cans ends are stamped out from the tinplate sheet simultaneously producing concentric expansion rings on them. The edges are curled, and the inside of the curl is lined with a sealing compound, which is generally a rubber solution, and is dried quickly. A film of rubber compound is left in the groove that will act as a gasket to ensure airtight sealing of the can.

3.5.6.4 Improvements in Can Making

Beaded cans permit the use of thinner plates for can making. To compensate for the loss in strength, mechanical strength is introduced into the can body by beading.

3.5.6.4.1 Cemented Side Seam

The cemented side seam imparts a better appearance to the can. The end seam also will be better formed. Thermoplastic cement, usually a nylon material, is the bonding material. This is applied to the pre-lacquered body beads, and a lap joint is formed which is bonded by application of heat followed by rapid cooling. The interior of the formed can is sprayed with lacquer.

3.5.6.4.2 Welded Side Seam

This is an alternative to soldering and has several advantages:

- Increased output
- Provides a lead-free side seam
- Improved flanging and double seaming, especially at the junction of the side seam
- Saves can material because of narrow overlap

3.5.6.4.3 Two-Piece Cans

Two-piece cans have a seamless body, and the ends are secured on top by double seaming. Advantages are

- Elimination of side seam and the seam on one end reduces the possibility of leakage
- Eliminates a potential source of lead contamination
- Aesthetic appeal because of the smooth profile and streamlined appearance
- Permits uninterrupted print decoration
- Bottom of the can may be designed and formed for better stackability
- Elimination of overlap at the two seams permits use of less metal

Two-piece cans are "drawn" cans made by cutting, drawing, and flanging from a plate in one operation. Aluminum or tinplate used shall be thicker and more ductile than that used for round cans. The two types are "Drawn and wall-ironed" (DWI) and "Drawn and redrawn" (DRD) cans.

3.5.6.5 Drawn and Wall-Ironed Can

This was first developed with aluminum in the early 1960s and the tinplate version became available by early 1970s. The process involves

- Blanking out a disc of metal from the sheet
- Drawing the disc into a cup
- Forcing it by means of a punch through a series of dies, each slightly smaller than the preceeding one, thus elongating the wall by a stretching or ironing action. During this ironing process, the wall thickness is reduced with a corresponding increase in height
- Trimming the top to the standard height

- Cleaning to remove lubricant and dust of metal
- Flanging
- Lacquering (by spraying)

3.5.6.6 Drawn and Redrawn Can

A blank disc is first drawn into a cup. A can of desired size can be made by a series of press operations, the number of operations being determined by the material used and the depth/diameter ratio of the can. A shallow can could be made by a first draw followed by a redraw; a deeper can by a draw followed by two redraws. In the DRD process

- The plate must be pre-lacquered to assist fabrication.
- The can wall remains substantially equal to the original plate thickness.
- A clear process which avoids the need for washing the cans after fabrication.

Two-piece cans are of different shapes—round, oval, and rectangular. Ordinary seamers for round cans will not be suitable for irregularly shaped cans. In the common seamers used for such cans, the can will remain stationary and two pairs of seaming rolls, the pair of opposite rolls in succession, will move round the can to make a seal. There is a possibility for uneven pressure on the seams and hence of leakage. This is partly overcome by the use of a good quality sealing compound in the cans' ends.

3.5.6.7 Necked-In Can

This is another modification where the can ends have the same diameter as the body. Such cans can be packed compactly.

3.5.6.8 Easy Open Ends

Easy open ends (EOE) are generally made of aluminum, though tinplate versions also became popular later. EOE consists of two parts, the lid and the tab. The tab is secured to the lid by means of an integral rivet. By pulling the tab, the central panel can be removed. Tinplate or aluminum ends are pre-punctured to provide a full aperture sealed by an adhesive strip of metalized polyester film or aluminum foil. Use of aluminum EOE on tinplate cans is limited because of the probability of bimetallic corrosion.

3.5.6.8.1 Aluminum Cans

The standard aluminum of 99.5%–99.7% purity is obtained by addition of one or more elements like magnesium, silicon, manganese, zinc, and copper (Mahadeviah and Gowramma, 1996). Lahiri (1992) described the suitability of different aluminum alloys for various food products. He also reported on the corrosion behavior of aluminum cans. Naresh et al. (1988) studied the corrosion behavior of aluminum cans by electrochemical studies and found that corrosion reaction is faster in plain aluminum cans compared to lacquered ones. Griffin and Sacharow (1972), suggested a suitable food grade lacquer coating for interior corrosion resistance. Balachandran et al. (1998) reported that the best promising alternative to tinplate has been considered as aluminum alloyed with manganese and magnesium. Lopez and Jimenez (1969) reviewed the use of aluminum cans for canning fruits and vegetable products. Srivatsa et al. (1993) studied the suitability of indigenously prepared aluminum cans for canning different food products. The advantages and disadvantages of aluminum alloys have been described by Balachandran (2001). Lakhsminarayan (1992) reported that aluminum containers are 100% recyclable and biodegradable. Gargoming and Astier-Dumas (1995) studied the canning of vegetables and reported that the acidity of tomatoes resulted in greater migration of aluminum into the products. Ranau et al. (2001) studied the aluminum content in fish and fishery products and concluded that aluminum content of seafood does not present a significant health hazard. Ranau et al. (2001) studied the changes in aluminum concentration of canned herring fillets in tomato sauce and curry sauce. Figure 3.4 represents the aluminum can fitted with thermocouples for monitoring time–temperature history of the product.

Advantages of aluminum cans are

- Light weight, slightly more than 1/3 of the weight of a similar tinplate can
- Nonreactive to many food products
- Clear, bright, and aesthetic image
- Not stained by sulfur-bearing compounds
- Nontoxic, does not impart metallic taste or smell to the produce
- Easy to fabricate; easy to open
- Excellent printability
- Recyclability of the metal

However, aluminum cans are not free from some disadvantages like

- Thick gauge sheet needed for strength
- Not highly resistant to corrosion, acid fruits and vegetables need protection by lacquering or other means
- Special protection needed during heat processing to avoid permanent distortion
- Aluminum has a great tendency to bleach some pigmented products
- Service life is less than that of tinplate for most aqueous products

3.5.6.8.2 Tin-Free Steel Containers

Tin-free steel (TFS), apart from aluminum, is a tested and proven alternate to tinplate in food can making. It has the same steel substitute as tinplate. It is provided with a preventive coating of chromium, chromium oxide, and chromate–phosphate. TFS is manufactured by electroplating cold-rolled base plate with chromium in chromic acid. This process does not leave toxin substrate such as chromates or dichromates on the steel, and it can be formed or drawn in the same way as tinplate.

Advantages

- The base chromium layer provides corrosion barrier.
- The superimposed layer of chromium oxide prevents rusting and pickup of iron taste.
- Provides an excellent base for lacquer adhesion.
- Good chemical and thermal resistance.
- Tolerance to high processing temperature and greater internal pressure.
- Improved and more reliable double seam.

Disadvantages

- Low abrasion resistance; hence, compulsory lacquering.
- Difficulty in machine soldering.
- The oxide layer needs removal even for welding.
- Limitations in use for acid foods.

An important problem associated with TFS can ends is scuffing of lacquer on the double seam. This may occur at the seamer or downstream at different stages of lacquering. TFS cans have been found quite suitable for canning different fish in various media. Thus, it holds good scope as an important alternate to tinplate cans.

TABLE 3.3
Can Sizes Used in India

Trade Name	Trade Dimensions	Over Seam Dimensions (mm)
4½ oz prawn	301 × 203	77 × 56
8 oz prawn	301 × 206	77 × 60
1 lb jam	301 × 309	77 × 90
No. 1 Tall	301 × 409	77 × 116
8 oz Tuna	307 × 113	87 × 43

3.5.6.9 Can Sizes

Can sizes are generally denoted by a trade name followed by their dimensions, diameter, and height, in that order, by their digit symbols. The first digit represents integral inches and the next two digits indicate measurements in sixteenth of an inch. Thus, a 301 × 309 can has a diameter of $3^{1/16''}$ and height of $3^{9/10''}$. Presently the dimensions are specified in millimeters. Trade names and dimensions of some cans popular for canning fish in India are given in Table 3.3.

3.5.6.9.1 Rigid Plastic Containers

The rigid plastic material used for thermal processing of food should withstand the rigors of the heating and cooling processes. It is also necessary to control the overpressure correctly to maintain a balance between the internal pressure developed during processing and the pressure of the heating system. The main plastic materials used for heat-processed foods are polypropylene and polyethylene tetraphthalate. These are usually fabricated with an oxygen barrier layer such as ethylvinylalcohol, polyvinylidene chloride, and polyamide. These multilayer materials are used to manufacture flexible pouches and semirigid containers. The rigid containers have the advantage when packing microwavable products.

3.5.6.9.2 Retortable Pouches

The retort pouch is a relatively new type of container for packing foods, which enjoyed a rapid growth in demand since its introduction in the late 1950s. The U.S. Army promoted the concept of flexible retortable pouches for use in combat rations in the 1950s, which replaced the metal containers and glass jars. The choice of materials for the manufacture of retort pouches is very important. Packaging materials for retort pouches should protect against light, degradation, moisture changes, microbial invasion, oxygen ingress and package interactions, toughness and puncture resistance. It is very difficult to get a single material with all the desirable properties. Hence, laminates or co-extruded films are used (Rangarao, 1992). The most common form of pouch consists of 3-ply laminated material, with an outer polyester, middle aluminum foil and inner polypropylene layer. A retort pouch can be defined as a container produced using 2-, 3-, or 4-ply material that, when fully sealed, will serve as a hermetically sealed container that can be sterilized in steam at pressure and temperature similar to those used for metal containers in food canning. The materials used should be tough with good barrier property and heat sealability. Rubinate (1964) and Schulz (1973) suggested the material requirements for the retortable pouch. The development of the retort pouch has been considered as the most significant advance in food packaging since the metal can and has the potential to become a feasible alternative to the metal can and glass jars (Mermeistein, 1978). One of the early studies on the use of the pouch was by Hu et al. (1955). He reported the feasibility of using plastic film packages for heat-processed foods. Ishitani et al. (1980) reported the effect of light and oxygen on the quality changes of retortable pouch-packed foods. Lampi (1967) reported the microbiological problems faced in foods packed in retortable pouch. The review on the flexible packaging material including the history has been documented (Mahadeviah, 1976; Lampi, 1977; Leung, 1984; Gopakumar and Gopal, 1987;

Griffin, 1987; Yamaguchi, 1990; Rangarao, 1992; Gopakumar, 1993). Retort pouches have the advantages of metal cans and boil-in plastic bags. Configuration of some typical pouches are

2 ply	12 μ nylon or polyester/70 μ polyolefin
3 ply	12 μ polyester/9–12 μ aluminum foil/70 μ polyolefin
4 ply	12 μ polyester/9–12 μ aluminum foil/12 μ polyester/70 μ polyolefin

The 3-ply pouch is most commonly used in commercial canning operations. This is a three-layer structure where a thin aluminum foil is sandwiched between two thermoplastic films. The outer polyester layer provides barrier properties as well as mechanical strength. The middle aluminum foil provides protection from gas, light, and water. This also ensures adequate shelf life of the product contained within. The inner film which is generally polyproplyline provides the best heat sealing medium.

The normal design of a pouch is a flat rectangle with rounded corners with four fin seals around 1 cm wide. A tear notch in the fin allows easy opening of the pouch. The rounded corners allow safe handling and help to avoid damage to the adjacent packs. The size of the pouch is determined by the thickness that can be tolerated at the normal fill weight. The size ranges (mm) available are

A_1	130 × 160
A_2	130 × 200
A_3	130 × 240
B_1	150 × 160
B_2	150 × 250
B_3	150 × 240
C_1	170 × 160
C_2	170 × 200
C_3	170 × 240
D_1	250 × 320 (Catering pack)
D_2	250 × 1100
D_3	250 × 480

Advantages

- Thin cross-sectional profile—hence rapid heat transfer—30%–40% saving in processing times—no overheating of the product near the walls
- Better retention of color, flavor, and nutrients
- Shelf life equal to that of the same product in metal can
- Very little storage space for empty pouches—15% of that for cans
- Easy to open

Disadvantages

- Pouches, seals more vulnerable to damage, can be easily damaged by any sharp material, hence necessitates individual coverage.
- With an over wrap cost that may go up above that of cans.
- Slow rate of production, 30 pouches in place of 300–400 cans per minute.
- Needs special equipment.
- Higher packaging cost and low output push up the cost of production.

3.5.7 Unit Operations in Thermal Processing of Fishery Products

Although a very wide variety of canned fishery products are available, there are very few operations that are unique to a certain variety of products. The initial handling steps, retort operation, and post-process handling steps are similar to all the products in a particular container. The major unit operations in canned fishery products are

1. Raw material handling
2. Pretreatment
3. Precooking or blanching
4. Filling into the container
5. Exhausting
6. Sealing
7. Retorting
8. Cooling
9. Post-process handling
10. Storing

Raw material handling is similar among the different products. The fish has to be maintained at a lower temperature from the time of its harvest till it is used for the preparation and maintained in a proper hygienic quality. The quality of the canned product is affected whenever the raw material is not maintained at a proper temperature and/or damaged physically between harvesting and thermal processing. It is usual practice to maintain fish in iced condition or in chilled seawater system or freeze the fishes onboard the fishing vessel to maintain the quality and in the mainland, it is stored in cold storage maintained at −18°C or lower temperature, before it is used for canning purpose.

The pretreatment steps include washing, beheading, gilling, gutting, washing, and cutting into desired sizes. The main purpose of this step is to bring the raw material closer to the usable form. These pretreatment steps have to be carried out under strict hygienic conditions to prevent contamination from handling surfaces and from viscera.

Precooking, sometimes referred to as blanching is normally carried out in steam, water, oil, smoke, microwave, or a combination of these. Precooking reduces enzyme activity and improves sensory quality. The purpose of precooking is to remove excessive water content from the raw material which otherwise will be collected in the container after retorting. Precooking coagulates protein and helps in improving firm texture, imparting desirable flavor, and removing undesirable flavor. As precooking affects yield and sensory quality, the precooking conditions have to be optimized for different fishes.

Filling is the addition of products using filling media, and it can be either manual or automatic. It is a very important step as the variation in filling weight and fill temperature for hot fill products affects the rate of heat transfer. Apart from maintaining proper fill weight, maintaining proper head space in the container is necessary to maintain the seal integrity as it can be affected by the thermal expansion of the product upon retorting.

Exhausting is an important step to remove the air present in the product as well as in the container. The air entrapped inside the container adversely affects the heat transfer capacity. It is also carried out to overcome the loss of oxygen-sensitive nutrients. Different methods are being followed in the industry for exhausting. These are hot filling, thermal exhausting, steam injection, and creating vacuum. In the hot filling method, the filling medium is heated to a specified temperature and filled in a hot condition, and the lids are closed immediately. In thermal exhausting, the filled containers are exposed to a specified temperature maintained either by steam as in an exhaust box (Figure 3.3) or exposed to boiling water. Exhausting is also done by injecting steam directly into the containers till it condenses, which ensures the removal of air. Exhausting is also achieved by creating a vacuum in the containers mechanically.

FIGURE 3.3 Exhausting by steam injection and using exhaust box.

The success of the canning process is mainly assured with the formation of hermetic sealing of the containers whether glass, metal cans, or flexible retortable pouches. Failure in this critical step indicates the compromise of product safety and shelf stability. In cans, sealing is normally done using double seaming machines, either semiautomatic or fully automatic seaming machines (Figure 3.4). Once the sealing is over, the containers are arranged in the perforated trays or cages and loaded into the retort for thermal processing. Heat processing is normally carried out in the temperature range of 110°C–135°C for a specified time to achieve desired lethality and quality product. Retorting can be carried out either in steam, steam air (Figure 3.5), or water immersion retorts (Figure 3.6). After retorting for a specified time, cooling is done inside the retort by pumping cool potable water into the retort. Special attention is to be given while retorting in flexible retortable pouches to overcome the bursting of pouches due to pressure differences in and out of the pouches during cooling as there

FIGURE 3.4 Semi-automatic and automatic double seaming machines for cans.

FIGURE 3.5 Steam retorts used in industry.

FIGURE 3.6 Water immersion retort.

is a sudden drop in the pressure of the retort during the cooling process. This can be overcome by applying an overpressure of around 14 psi using air during the cooling process. Special care is also needed while performing heat processing in glass containers as this requires overpressure to overcome the damage to container. The cooling process is continued till the product temperature reduces to a minimum of 40°C. Upon cooling, post-process handling assumes importance as there is a chance of recontamination of the processed product if sufficient care is not taken. Chlorinated water has to be used to wash the containers after the cooling process maintaining proper hygiene and sanitation. Mechanical damage to containers has to be reduced during handling and storing. The washed and dried containers are stored in an ambient storage condition in a dust-free area.

3.5.8 Thermal Process Validation

Measuring the product temperature at cold point is the major activity of establishing a thermal process. Normally two stages of temperature measurement are employed for establishing a process, whether by general method or mathematical method, for process value calculation. These are *temperature distribution tests* (*TD*) to identify the location of the zone of slowest heating in the retort, and *HP tests* to measure the temperature response at the product cold point.

3.5.8.1 Temperature Distribution Test

TD is the first step in HP studies. Generally, any system for thermal processing, whether retort, autoclave, or sterilizers, will contain regions in which the temperature of the heating medium is lower than that measured by the master temperature indicator. For example, with steam–air processes this can be caused by improper mixture and proportion, and in water immersion processes this can be caused by poor circulation of hot water. The location of these cold spots should be determined by performing "temperature (or heat) distribution" tests throughout the system. The concepts for TD testing are simple; however, the practicalities of making the relevant measurements are loaded with difficulty. A uniform TD throughout the retort does not necessarily indicate uniform lethalities since uniform temperature does not guarantee uniform heat transfer. Therefore, the uniformity in temperature is the minimum that has to be studied and an additional heat distribution study is advisable if there are concerns about air entrapment or heat transfer coefficient reductions throughout a container load. For a steam retort, if the TD is unsatisfactory, it can normally be resolved by increasing the length of the period of air removal at the start of the process (venting). This contrasts with non-steam retorts where a large temperature range may be attributed to the design/loading of the retort, and simple corrective action is not possible.

3.5.8.1.1 When is TD Testing Required?

The TD within a retort should be tested on its installation, with intermittent retesting being required as factors change that could affect the retort performance. Retorts require, at a minimum, retesting in the event of any engineering work likely to affect the TD of the retort, such as

- Relocation of the retort or installation of another retort that uses the same services
- Modification to the steam, water, or air supply
- Failure of the key components (e.g., pumps and valves)
- Repair or modification to water or steam circulation systems within the retort
- If there are any doubts about the performance of the circulation system

In addition, if the load to be processed in a retort changes, retesting is required. Such circumstances include the use of

- New container sizes and shapes
- New container loading patterns
- New crate or layer pad design, or mode of use

It is also necessary to ensure that a retort's performance does not deteriorate over a period of time, as corrosion or fouling in the steam, water, or air supply pipes builds up. Retort instrumentation and process records should be inspected regularly to identify when a TD problem has arisen. Regular retesting of a retort's TD is a good practice to ensure that these faults are not overlooked.

3.5.8.1.2 Objectives of TD Tests

Although TD tests basically sample the performance of retort systems, in the industry they are frequently used as an opportunity to "audit" the installation to ensure long-term compliance (Figure 3.7). Good manufacturing practices for TD tests in batch retorts should be such that in steady-state operation, the temperature spread across the sterilizing vessel should ideally be 1°C

FIGURE 3.7 Arranging thermocouple fittings in overpressure autoclave for heat penetration studies.

or less. However, when this degree of control is not achievable due to design or characteristics of the equipment, any deviation from the limit should be allowed for the scheduled process.

If the retort uses condensing steam as the media, it is necessary to establish the time of vent in order that the distribution of temperatures across the retort is reduced to an acceptable limit. For venting trials, the following guidance are normally adopted:

- Note precisely the time at which the retort reaches 100°C.
- Do not close the main vent until all thermocouples reach the same temperature within 0.5°C.
- Close the vent and record when the master temperature indicator and chart recorder reach the process temperature.
- All thermocouples should indicate the same temperature within 1 min of the first thermocouple indicating that (process) temperature.
- Record the venting time as the number of minutes for which the main vent was left open after 100°C was reached on the thermocouple in the thermometer pocket.

The vent test is specific to condensing steam retorts and is required in addition to TD testing. For retort systems utilizing water or mixtures of steam and air, the TD tests are unlikely to result in a 1°C distribution in temperatures across the crates at the start of the hold phase. This is because of less favorable heat transfer coefficients with these heating media when compared with condensing steam and also the reduced quantity of heat available in, for example, a raining water system. It is common practice to quote a time into the hold phase by which the temperature distribution has stabilized to within 1°C, and to take this into account when establishing the hold time at constant temperature.

3.5.8.2 Heat Penetration Tests

HP is further subdivided into two stages when conducting the tests, first to locate the product cold point in the container and second to establish the process conditions that will lead to the scheduled process.

3.5.8.3 Locating the Product Cold Point

Within each food container there will be a point or region that heats up more slowly than the rest. This is referred to as the "slowest heating point" or "thermal center" and should be located using thermocouples (Figure 3.8) or some other sensing method positioned at different places in a food container (Figures 3.9 and 3.10). For foods that heat mainly by conduction, the slowest heating point

FIGURE 3.8 Different types of thermocouples used in the study.

FIGURE 3.9 Aluminum and TFS cans fitted with thermocouple.

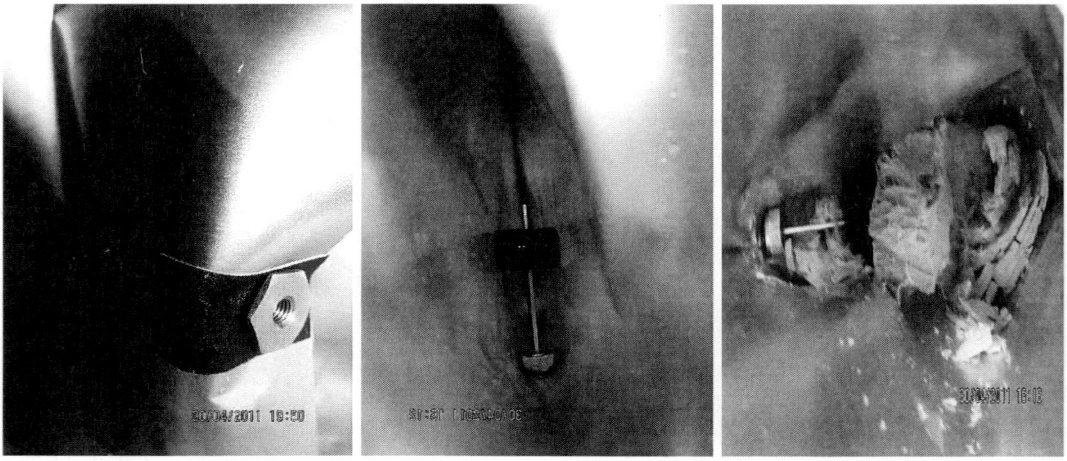

FIGURE 3.10 Retortable pouches fitted with spacer and thermocouple.

will be at the container geometric center. However, for foods that permit movement and can thus convect heat, this point is between the geometric center and approximately one tenth up from the base (in a static process). During a thermal process the food viscosity will decrease in response to increasing temperature, and as a result the slowest heating point will move downwards from the container geometric center. The critical point is when the lethal effect on the target microbiological species is at its most significant, which will be toward the end of the constant temperature hold phase. If the process utilizes rotation or agitation, the slowest heating point will be at the container geometric center.

3.5.8.4 Establishing the Scheduled Process Time and Temperature

The thermal process is finally established by measuring the temperature at the container slowest heating point for a number of replicates that are placed in the cold spot(s) of the thermal processing system. The data obtained are usually referred to as "heat penetration" data. Campden and Chorelywood Food Research Association (CCFRA) (1997) recommended three sensors from each of three replicate runs, and National Food Processors Association (NFPA), USA, suggests at least ten working sensors from a run, with replicate runs required where variability is found. The more common situation now is to take up to ten samples in two replicate runs, providing that the variability between runs is within acceptable limits. However, there can be limitations on the number of probes that can be inserted through a packaging gland or through the central shaft of a rotating system, and in these situations at least two replicate runs should be completed. Various methods can be employed to collect accurate HP data. The aim of an HP study is to determine the heating and cooling behavior of a specific product in order to establish a safe thermal process regime and to provide the data to analyze future process deviations. The design of the study must ensure that all of the critical factors are considered to deliver the thermal process to the product slowest heating point.

3.5.8.4.1 When is HP Testing Required?

The HP study should be carried out before commencing production of a new product, process, or package. Changes to any of the criteria that may change the time–temperature response at the product slowest heating point will require a new HP study to be conducted. The conditions determined in the study are referred to as the scheduled heat process and must be followed for every production batch, with appropriate records taken to confirm that this was followed. No further temperature measurement within containers is required in production, although some companies do measure temperatures in single containers at defined frequencies. However, the conditions used in single container testing will not represent the worst case, and it would be expected that the instrumented container would show a process value in excess of that measured by the HP study. Such data are intended to show carefulness and are more of a comfort factor. Figures 3.11 through 3.14 represent a variety of canned food products like sardine in oil and curry medium, smoked rainbow trout in oil medium, seafood mix in brine and tomato sauce medium, and tuna processed along with vegetables in TFS cans. Figures 3.15 through 3.18 represent the HP curves, and F_0 values of a value-added

FIGURE 3.11 Sardine in oil and curry medium in metal cans and retortable pouches.

FIGURE 3.12 Smoked and canned Rainbow trout in oil in TFS cans.

FIGURE 3.13 Seafood mix in brine and tomato sauce medium in TFS cans.

fish product "Shrimp Kuruma" in aluminum cans and retortable pouches of different sizes. Factors taken into consideration during HP tests are listed below:

1. *Process-related factors*
   a. Retort temperature history, heating, holding, and cooling temperatures and times: The accuracy of temperature measurements has a direct effect on lethality values, and temperature-measuring devices need to be calibrated to a traceable standard.
   b. Heat transfer coefficients: For processes heated by steam and vigorously agitated in boiling water, these are usually so high that no effect is observed; however, with other methods of heating, which have much lower values, it is important to estimate the effective values as accurately as possible.
   c. Come up time in steam, steam–air retorts, and filling stages for raining water or water immersion systems.
   d. Type of retort—stationary (batch or continuous), reel or spiral retort, hydrostatic retort.
2. *Product-related factors*
   a. Size of the individual pieces, composition of the food, product preparation method, ration of solid to liquid content.

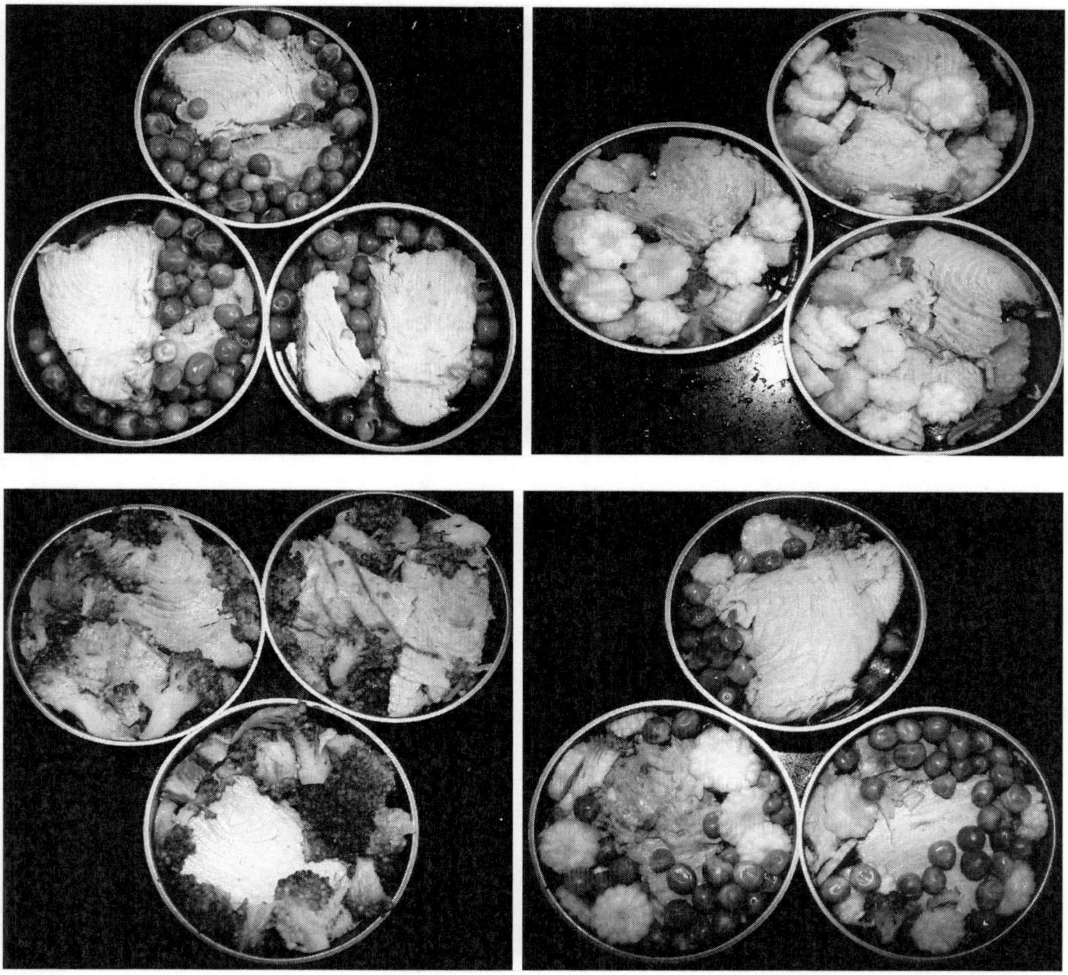

FIGURE 3.14 Tuna with green pea, baby corn, broccoli, and mixed vegetables in TFS cans.

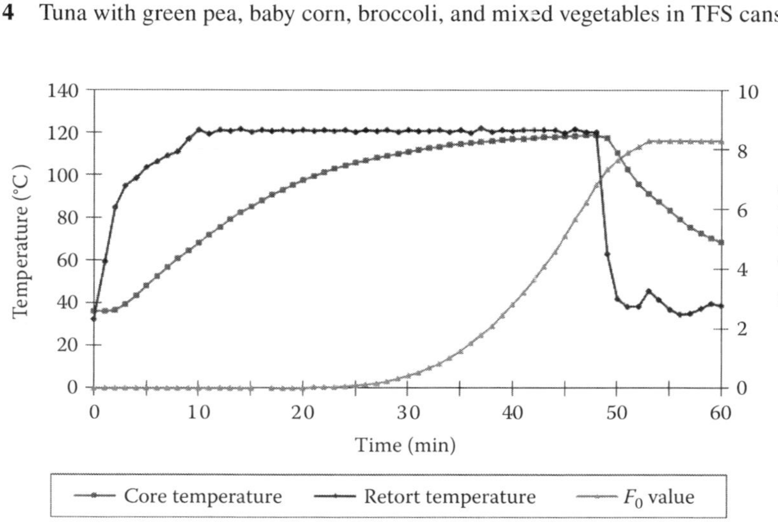

FIGURE 3.15 Heat penetration characteristics and F_0 value for 16 × 20 cm retort pouch.

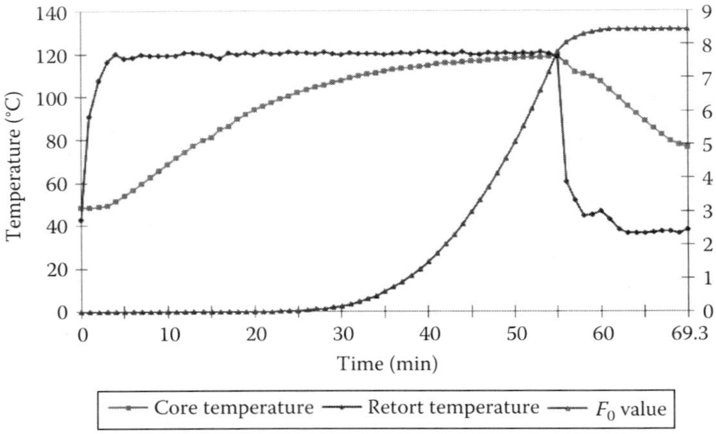

FIGURE 3.16 Heat penetration characteristics and F_0 value for 301 × 206 can.

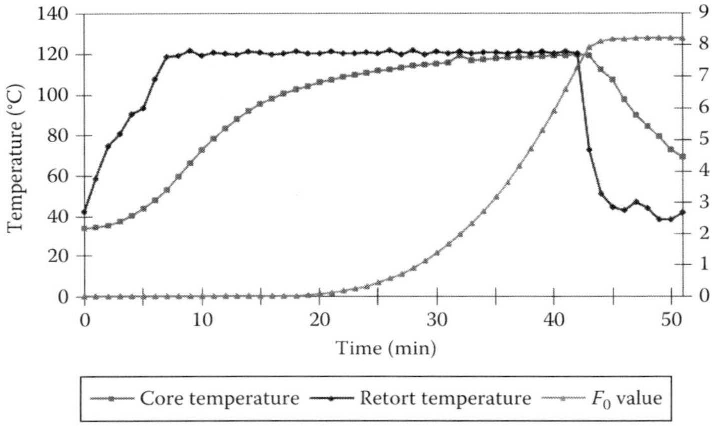

FIGURE 3.17 Heat penetration characteristics and F_0 value for 17 × 30 cm retort pouch.

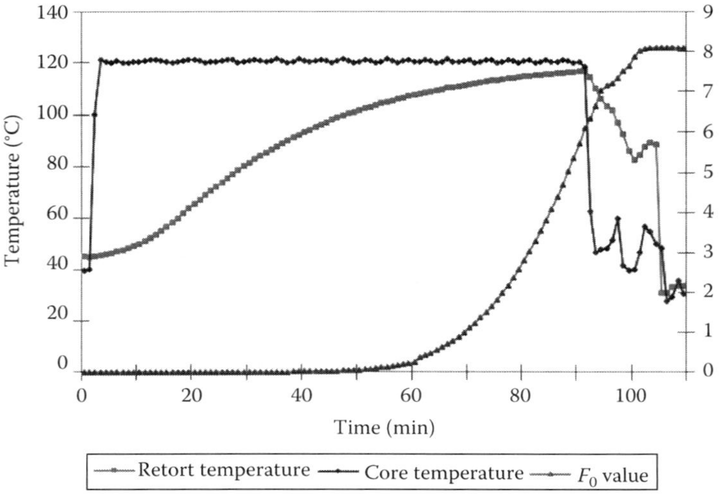

FIGURE 3.18 Heat penetration characteristics and F_0 value for 401 × 411 can.

b. Initial temperature of can contents: The uniformity of the initial filling temperature should be carefully controlled. The higher the initial temperature, the shorter the required process time.
c. Pre-retorting delay temperature and time: This is related to the filling temperatures and results from malfunctioning of the process, and delays affect the initial temperature of the can contents.
d. Thermal diffusivity of product: Most models are very sensitive to changes in the value of this property.
e. Z- and D-value of the target microbial species.

3. *Container-related factors*
   a. Container materials: Apart from tinplate and other metallic containers, all other materials, for example, plastics and glass, impede heat transfer into the container.
   b. Container shape: The most rapidly heating containers have the largest surface area and the thinnest cross section.
   c. Container thickness: The thicker the container wall, the slower the heating rate.
   d. Headspace: This is of particular importance to agitated and rotary processes. The headspace and the rate of rotation need to be carefully controlled.
   e. Container stacking: The position of the containers inside the retort and the type of stacking also affect the heat transfer to individual containers.

3.5.8.5 Effect of Thermal Processing on the Nutrition of Food

The primary objective of a thermal process is to produce a product that is free of pathogenic microorganisms. The target criterion for thermal processing of low-acid foods is to destroy spores of *C. botulinum*, which is the most heat resistant among the common food poisoning bacteria. Although thermal processing makes microorganisms and spores inactive, it may cause destruction of essential nutrients that leads to deterioration of product quality. Much attention has been given to maximizing quality retention for a specified reduction in undesirable microorganism. However, since the degradation of heat-sensitive vitamins and other quality factors such as color and texture will take place along with the microbial destruction, the optimum processing time and temperature must be utilized. Because of these safety and quality factors, care must be taken to avoid either overprocessing or underprocessing. Broek (1965) reported that moderate heating of fish had not affected the nutritive quality, while overheating led to loss of nutrients. In recent years, growing consumer awareness increased the interest in food quality during processing and storage (Lubuza, 1982). Researches have studied the effect of thermal processing on the nutritive value of food. Bender (1972) and Ford (1973) reported the effect of thermal processing on the protein of fish. A reduction of 25% lysine during thermal processing was reported by Tooley and Lowrie (1974). Taguchi et al. (1982), and Taneka and Taguchi (1985) reported an increase in peroxide value. A decrease in TBA value, TMA, and vitamin B1 (thiamin) for shrimp, rainbow trout, and Alaska Pollock (Chia et al., 1983) and a decrease in the TMA-O content of squids (Kolodziejska et al., 1994) after heat processing have been reported. Ma et al. (1983) reported a toughening during the initial stages of heating and softening during the later stage of processing for shrimps and mussels. Tanaka and Taguchi (1985) reported the changes in nutritional and sensory characteristics in canned fishery products. Fellows (1990) reported a reduction of about 10%–20% of the amino acids in canned products. Lou (1997) reported a decrease in purine content of shrimp especially adenine during thermal processing. A combination of tuna and vegetables like green pea and baby corn gave a better product than broccoli (Mohan et al., 2014). Thermal processing of tuna with vegetables like green pea, broccoli, and baby corn to an F_0 value of 8.0 min resulted in the reduction of 4.35%–15.22% in process time compared with tuna without vegetables for the same size cans. Retortable pouches had an advantage over metal cans in many ways. Processing in 16 × 20 cm retortable pouches resulted in 35.67% reduction in process time for equivalent lethality when compared with 301 × 206 cans (Mohan et al., 2006, 2008). For 17 × 30 cm retortable pouches, a reduction of 56.56% process time was obtained than the 401 × 411 cans. In the canned samples, the reduction of SH content was 50.54% more when compared with pouches. Although thermal processing affects some of the nutritional quality, it is by far the most successful preservation technology as it offers various advantages.

REFERENCES

Balachandran, K. K. (2001). Canning. In: *Post-harvest Technology of Fish and Fish Products*. Daya Publishing House, New Delhi, India, pp. 158–220.

Balachandran, K. K., Gopal, T. K. S., and Vijayan, P. K. (1998). Aluminium container for fish canning. In: *Aluminium in Packaging* (ed. Cunha, J. F. D.). Indian Institute of Packaging, Mumbai, India, pp. 570–576.

Ball, C. O. (1923). Thermal process times for canned foods. Bull. No. 37. National Research Council, Washington, DC.

Ball, C. O. and Olson, F. C. W. (1957). *Sterilization in Food Technology*. McGraw-Hill Book Co., New York, 654pp.

Barbosa-Canovas, G. V., Pothakamury, U. R., Palou, E., and Swanson, B. G. (1998). *Non-thermal Preservation of Foods*. Marcel Dekker, New York, 276pp.

Barbosa-Canovas, G. V., Zhang, Q. H., and Tabilo-Munizaga, G. (2001). *Pulsed Electric Field in Food Processing: Fundamental Aspects and Applications*. Technomic Publishing Company Inc., Lancaster, PA, 268pp.

Bender, A. E. (1972). Processing damage to protein food. A review. *J. Food Technol.*, 7: 239–250.

Bigelow, W. D., Bohart, G. S., Richardson, A. C. and Ball, C. O. (1920). Heat penetration in processing of canned foods. National Canner's Association, Bull. 16-L, Washington, DC, p. 128.

Broek, J. C. H. V. (1965). Fish canning. In: *Fish as Food*, Vol. IV (ed. Borgstrom, G.), Academic Press, New York, pp. 127–194.

CCFRA. (1997). In: *HACCP: A Practical Guide* (2nd edn.). CCFRA Technical Manual No. 38 (ed. Leaper, S.). Campden and Chorelywood Food Research Association, Chipping Campden, U.K.

Chia, S. S., Baker, R. C., and Hotchkiss, J. H. (1983). Quality comparison of thermoprocessed fishery products in cans and retortable pouches. *J. Food. Sci.*, 48: 1521–1525, 1531.

Clark, J. P. (2002). Thermal and Non-thermal processing. *Food Technol.*, 56(12): 63–64.

Esty, T. R. and Meyer, F. (1922). The heat resistance of spores of *Clostridium botulinum* and allied anaerobes. *J. Infect. Dis.*, 31: 650–653.

Farid, M. M., Chen, X. D., and Dost, Z. (2000). Ultraviolet sterilization of orange juice. *Eighth International Congress on Engineering and Food*, April 2000, Pueblo, Mexico.

Fellows, P. (1988). *Food Processing Technology*. Ellis Horwood Ltd., Chichester, U.K.

Fellows, P. (1990). *Food Processing Technology, Principle and Practice*. Ellis Horwood Ltd., Cambridge, U.K., 492pp.

Ford, H. E. (1973). Some effects of processing on nutritive value. In: *Protein in Human Nutrition* (ed. Porter, W. G. and Rolls, B. A.). Academic Press, London, U.K., pp. 515–529.

Furukawa, S. and Hayakawa, I. (2000). Investigation of desirable hydrostatic pressure required to sterilize Bacillus stearothermophilus IFO 12550 spores and its sterilization properties in glucose, sodium chloride and ethanol solutions. *J. Food Res. Int.*, 33: 901–905.

Gargoming, N. and Astier-Dumas, M. (1995). Aluminium content of foods: Raw food, canned foods in steel or aluminium cans. *Medecine-et-Nutrition*, 31(5): 253–256.

Gillespy, T. G. (1951). Estimating the sterilizing value of process as applied to canned foods. 1: Packs heating by conduction. *J. Sci. Food Agric.*, 2: 107–125.

Gopakumar, K. and Gopal, T. K. S. (1987). Retort pouch packaging. *Packaging India*, April–June: 3–4.

Gopakumar, K. (1993). Retortable pouch processing. In: *Fish Packaging Technology*, Concept Publishing Company, New Delhi, India, pp. 113–131.

Griffin Jr, R. C. (1987). Retortable plastic packaging. In: *Modern Processing, Packaging and Distribution Systems of Food* (ed. Paine, F.A.), AVI Publishing Co. Inc., West Port, CT, pp. 1–19.

Griffin Jr, R. C. and Sacharow, S. (1972). Packaging materials. In: *Principle of Package Development*. AVI Publishing Co. Inc., West Port, CT, pp. 23–58.

Hayakawa, K. I. (1978). A critical review of mathematical procedures for determining proper heat sterilization processes. *Food Technol.*, 32(3): 59–65.

Hersom, A. C. and Hulland, E. D. (1980). *Canned Foods: An Introduction to their Microbiology*. Chemical Publishing Co. Inc., New York, 180pp.

Hu, K. H., Nelson, A., Legault, R. R., and Steinberg, M. P. (1955). Feasibility of using plastic film packages for heat processed foods. *Food Technol.*, 19(9): 236–240.

Ishitani, T., Hirata, T., Matsushita, K., Hirose, K., Kodani, N., Ueda, K., Yanai, S., and Kumura, S. (1980). The effects of oxygen permeability of pouch, storage temperature and light on the quality change of a retortable pouched food during storage. *J. Food Sci. Technol.*, 27(3): 118–124.

Karel, M., Fennema, O. R., and Lund, D. B. (1975). *Principles of Food Science. Part II*. Marcel Dekker, New York.

Kolodziejska, I., Niecikowska, C., and Sikorski, Z. E. (1994). Dimethylamine and formaldehyde in cooked squid (*Illex argenticus*) muscle extract and mantle. *Food Chem.*, 50(3): 281–283.

Kumar, M. A., Ramesh, M. N., and Nagaraja Rao, S. (2001). Retrofitting of a vertical retort for on-line control of the sterilization process. *J. Food Eng.*, 47: 89–96.

Lahiri, A. (1992). Aluminium Rigid Containers for Processed Foods. *Packaging India*, 3: 19–27.

Lakhsminarayan, S. (1992). Aluminium—The packaging tomorrow. *Packaging India*, 24(5): 33–34.

Lampi, R. A. (1967). Microbial recontamination in flexible films. *Act. Rep. Res. Dev. Assoc. Mil. Food Packag. Syst.*, 19(1): 51–58.

Lampi, R. A. (1977). Flexible packaging for thermoprocessed foods. *Adv. Food Res.*, 23: 306–426.

Leung, H. K. (1984). The retort pouch. In: *Economics and Management of Food Processing* (ed. Greig, W. S.), AVI Publishing Co. Inc., Westport, CT, pp. 219–222.

Lopez, A. (1969). *A Complete Course in Canning* (9th edn.). The Canning Trade, Baltimore, MD, 552pp.

Lopez, A. (1996). *A Complete Course in Canning, Part III* (13th edn.). The Canning Trade, Baltimore, MD, 229pp.

Lopez, A. and Jimenez, M. A. (1969). Canning fruits and vegetable products in aluminium container. *Food Technol.*, 23(10): 1200–1206.

Lou, S. N. (1997). Effects of thermal processing on the purine content of grass shrimp (*P. monodon*). *Food Sci.*, 24(4): 438–447.

Lubuza, T. P. (1982). *Shelf-life Dating of Foods*. Food and Nutrition Press Inc., Westport, CT, 334pp.

Ma, L.Y., Deng, J. C., Ahed, E. M., and Adams, J. P. (1983). Canned shrimp texture as function of its heat history. *J. Food Sci.*, 48: 983.

Mahadeviah, M. (1976). Plastic containers for processed food products. *Indian Food Packer*, 30(4): 35–46.

Mahadeviah, M. and Gowramma, R. V. (1996). Aluminium container. In: *Food Packaging Materials*. Tata McGraw-Hill Publishing Company Ltd., New Delhi, India, pp. 73–83.

Mermeistein, N. H. (1978). Retort pouch earns 1978 IFT food technology industrial achievement award. *Food Technol.*, 32: 22–23, 26, 30, 32–33.

Merson, R. L., Singh, R. P., and Carroad, P. A. (1978). An evaluation of Ball's formula method of thermal process calculations. *Food Technol.*, 32(3): 62–72, 75.

Mohan, C. O., Ravishankar, C. N., Bindu, J., Geethalakshmi, V., and Srinivasa Gopal, T. K. (2006). Effect of thermal process time on quality of shrimp kuruma in retortable pouches and aluminum cans. *J. Food Sci.*, 71: S496–S500.

Mohan, C. O., Ravishankar, C. N., Bindu, J., and Srinivasa Gopal, T. K. (2008). Thermal processing of prawn 'kuruma' in retortable pouches and aluminium cans. *Int. J. Food Sci. Technol.*, 43: 200–207.

Mohan, C. O., Remya, S., Ravishankar, C. N., Vijayan, P. K., and Srinivasa Gopal, T. K. (2014). Effect of filling ingredient on the quality of canned yellowfin tuna (*Thunnus albacares*). *Int. J. Food Sci. Technol.*, 49: 1557–1564.

Naresh, R., Mahadeviah, M., and Gowramma, R. V. (1988). Electrochemical studies of aluminium with model solutions and vegetables. *J. Food. Sci. Technol.*, 25(3): 121–124.

Navankasattusas, S. and Lund, D. B. (1978). Monitoring and controlling thermal process by online measurement of accomplished lethality. *Food Technol.*, 32(3): 79–83.

Palou, E., Lopez-Malo, A., Barbosa-Canovas, and Swanson, B. G. (1999). High-pressure treatment in food preservation. In: *Handbook of Food Preservation* (ed. Rahman, M.S.). Marcel Dekker, New York.

Ramaswamy, H. S., Abdelrahim, K., and Smith, J. (1992). Thermal processing and computer modeling. In: *Encyclopedia of Food Science and Technology* (ed. Hui, H.). Wiley, New York, pp. 2538–2552.

Ranau, R., Oehlenschlaeger, J., and Steinhart, H. (2001). Aluminium content of stored industrially manufactured canned herring products. *Archiv-fuer-lebensmittelhygiene.*, 52(6): 135–139.

Ranganna, S. (1986). *Handbook of Analysis and Quality Control for Fruit and Vegetable Products*. Tata McGraw Hill, New Delhi, India, pp. 718, 754.

Rangarao, G. C. P. (1992). Retortable plastic packaging for thermo-processed foods. *J. Indian Food Ind.*, 11(6): 25–32.

Rubinate, F. J. (1964). Army's obstacle course yields a new look in food packaging. *Food Technol.*, 18 (11): 71–74.

Schulz, G. L. (1973). Test procedures and performance values required to assure reliability. In: *Proceeding of the Symposium on Flexible Packaging for Heat Processed Foods*. Nat. Acad. Sci—Natl. Res. Counc., Hillside III, Illinois, Washington, DC, pp. 71–82.

Smith, T. and Tung, M. A. (1982). Comparison of formula method for calculating thermal process lethality. *J. Food Sci.*, 47: 626–630.

Srivatsa, A. N., Ramakrishna, A., Gopinathan, V. K., Nataraju, S., Leela, R. K., Jayaraman, K. S., and Sankaran, R. (1993). Suitability of indigenously fabricated aluminium cans for canning of Indian foods. *J. Food Sci. Technol.*, 30(6): 429–434.

Steele, R. J. and Board, P. W. (1979). Thermal process calculations using sterilizing ratios. *J. Food Technol.*, 14: 227–235.

Stumbo, C. R. (1973). *Thermobacteriology in Food Processing* (2nd edn.). Academic Press Inc., New York, 329pp.

Taguchi, T., Taneka, M., Okubo, S., and Suzuki, K. (1982). Changes in quality of canned mackerel during long term storage. *Bull. Jpn. Soc. Sci. Fish.*, 48(12): 1765–1769.

Taneka, M. and Taguchi, T. (1985). Non-enzymatic browning during thermal processing of canned sardine. *Bull. Jpn. Soc. Sci. Fish.*, 51(7): 1169–1173.

Teixiera, A. A. (1992). Thermal process calculations, Chapter 11. In: *Handbook of Food Engineering* (eds. Heldman, D. R. and Lund, D. B.). Marcel Dekker, New York.

Tooley, P. J. and Lowrie, R. H. (1974). Effects of deep fat frying on availability of lysine in fish fillets. *J. Food Technol.*, 9: 247–253.

Tucker, G. S. and Holdsworth, S. D. (1991). Mathematical modeling of sterilization and cooking processed for heat preserved foods—Application of a new heat transfer model. *Transl. Inst. Chem. Eng.*, 69(3): 5.

Yamaguchi, K. (1990). Retortable packaging. In: *Food Packaging* (eds. Kadoya, T.). Academic Press, Inc., New York, pp. 185–211.

4 Extrusion
Cooking

Kasiviswanathan Muthukumarappan and Chinnadurai Karunanithy

CONTENTS

4.1 INTRODUCTION

Consumers are looking for foods that provide health, convenience, and taste at an affordable price. Ever-increasing population and urbanization mean high demand for staple foods; moreover, changing eating habits means more convenient foods. To address consumer demand, a key research priority in the food industry is the development of foods that promote health and wellness. Snack foods have become a significant part of modern life and represent a distinct and consistently widening and changing group of food items (Riaz 2005). Extrusion is one of the most versatile operations available to the food industry for transforming ingredients into intermediate or finished products. Extrusion technology provides a method to process raw ingredients on a large scale at a remarkably low cost. Extrusion finds ever-increasing application in the food industry, such as the production of ready-to-eat cereals, pasta, snacks, pet food, fish foods, and confectionary products, apart from its obvious applications in the plastics industry (Muthukumarappan and Karunanithy 2012).

Extrusion is a high-temperature, short-time process that has the ability to perform many operations such as mixing, kneading, cooking, shearing, shaping, and forming simultaneously. Literally, extrusion (*extrudere* in latin) means the action of pushing or forcing material through a narrow gap; thus it describes only a small portion of the industrial applications which are primarily forming or shaping without affecting the properties of the material (e.g., pasta processing). Now, extrusion is not limited to initial application; expanded to cooking, expansion, co-extrusion, supercritical fluid extrusion, destructing antinutritional factors, and some microorganisms. First extrusion was applied to food when chopped meat was stuffed into casings using a piston-type extruder for sausage manufacturing in 1870.

The single-screw extruder (SSE) was introduced in 1930 for mixing semoltheina and water and shaping the dough into macaroni in one continuous operation. High-shear extruders were introduced during the late 1930s and 1940s for making directly expanded products (e.g., corn curls). A literature survey indicates that the twin screw was introduced in the late 1950s. The extrusion process has grown several fold due to continuous research-related efforts as evident from a variety of extruded food products found in the market. The leading food extruder manufacturers around the world are Wenger, United States; American Extrusion, United States; Extru-Tech, United States; Buhler, Switzerland; Pavan Mapimianti, Italy; Krupp Werner & Pfleiderer, Germany; C.W. Brabender, Germany; Clextral, France; APV Baker, England; and Lalasse, Holland. The advantages of extruders can be found in many extrusion text books. This chapter presents the raw materials of an extruder, components of an extruder, types of an extruder, factors influencing expansion and nutritional qualities, heat transfer in an extruder, and process control.

4.2 RAW MATERIALS FOR EXTRUSION COOKING

According to Moley (2013), ingredients are the building blocks of foods—they combine together to form structures at the micro and macro level. In general, most ingredients do not act alone; in order to reap the benefits of ingredients' interactions, the ingredients need to be in the right state; they must be soluble. Different ingredients play different roles in any food product development; therefore, choosing the right ingredients is a critical and challenging task for the food scientists. The properties of raw materials play a vital role on texture, color, and quality of the final product. The type of feed, composition of feed, its moisture content, and particle size have a greater impact on the viscosity of melt within the extruder; ultimately, they dictate the quality of product besides extruder variables. In general, any blend consists of starch, protein, lipid/fat, and fiber, and each one of them have a significant influence on the physical, chemical, nutritional, and sensory qualities. It is well known that starch content up to 60% is required for producing directly expanded products with good expansion. Table 4.1 summarizes the composition of raw materials reported in the literature and can be used as a guide for selecting ingredients to formulate a blend. In general, the blend can be formulated in various combinations of ingredients such as cereal, grains, millets, fruits, vegetables, tubers, legumes, oil seeds, food industry by-products, animal fat, and proteins as presented in Table 4.2. The reason for the selection of a particular ingredient is outlined in the second column of Table 4.2 and significant findings of each study are also given in the fourth column. Starch is the one that provides mechanical strength to the extrudates; the most commonly used starch sources are corn, wheat, rice, and potato. Other less commonly used starch sources such as barley, tapioca, peas, beans, sweet potato, amaranthus, and millets are also explored as evident from Table 4.2. Table 4.3 shows a summary of characteristics of various starches in terms of starch granule size, flavor, color, expansion, and energy requirement that can serve as a guide to select an appropriate starch depending upon the application.

4.3 COMPONENTS OF EXTRUDER

Feed hopper/live bottom bin, preconditioner, screw, barrel, die, cutter, and drive are the major components in addition to sensors/controlling devices such as thermocouples and pressure transducers. The food contact surface should be noncorrosive and nontoxic; therefore, most of the screws and liners are constructed with high-quality wear-resistant stainless steel alloys (heat treatable 400 series).

A feed hopper is a place where the raw material enters into the extruder. The design of a feed hopper depends on the properties of raw materials such as the angle of repose, abrasiveness, ability to free flow/bridging, and the size depends on the capacity. In some cases, the raw material would not flow freely; therefore, a scraper/rotating blade is placed inside the feed hopper and is termed as a live bottom bin where continuous free flow is ensured. It costs more than $20,000 depending upon the capacity.

Preconditioner, as the name implies, is where the raw material is preconditioned before actual extrusion. Addition of moist steam not only hydrates the raw material but also heats

TABLE 4.1
Composition of Raw Materials Reported in the Literature

Raw Material	Moisture	Starch	Protein	Lipid	Fiber	Ash	References
Corn	12.6	76.1	9.0	2.3	NR	NR	Onwulata et al. (1998)
Corn	9.1	78.5	10.23	2.18	7.15	1.46	Pastor-Cavada et al. (2011)
Corn flour	12.0	78.6	6.0	2.2	0.6	0.6	Haley and Mulvaney (2000)
Corn flour	12.0	Diff	6.0	2.2	0.6	0.6	Liu et al. (2000)
Corn flour	9.3	80.4	9.00	2.9	NR	2.1	Veronica et al. (2006)
Cornmeal	10.15	82.21	6.5	1.14	NR	NR	Konstance et al. (1998)
Corn grits	11.77	78.31	6.74	2.33	1.1	0.57	Dehghan-Shoar et al. (2010)
Corn	—	90.54	6.48	0.44	1.98	0.55	Gimenez et al. (2013)
Corn	9.7	48.9	12	3.5	19.3	2.8	Semasaka et al. (2010)
Corn grits	11	77.9	9.2	0.7	0.7	0.5	Nascimento et al. (2012)
Corn grits	—	86.11	11.11	1.01	0.45	1.32	Sobota et al. (2010)
Maize flour	10.61	37.28	9.42	4.47	2.89	1.34	Rodríguez-Miranda et al. (2011)
Nixtamalized maize flour	11.69	59.19	10.81	6.95	1.78	1.91	Rodríguez-Miranda et al. (2011)
Rice	12.0	78.3	8.5	1.2	NR	NR	Onwulata et al. (1998)
Rice flour	9.95	81.43	5.85	0.56	1.74	0.47	Choudhury and Gautam (2003)
Rice grit	11.45	81.43	5.92	0.79	NR	0.42	Yagci and Gögus (2009)
Rice flour	9.95	81.43	5.85	0.56	1.74	0.47	Choudhury and Gautam (1998a)
Rice flour	12.11	79.78	6.75	0.14	0.80	0.28	Dehghan-Shoar et al. (2010)
Rice	8.2	86.9	7.38	0.53	3.81	1.19	Pastor-Cavada et al. (2011)
Durum clear flour	7.39	74.37	13.38	2.93	NR	1.93	Yagci and Gögus (2009)
Wheat semolina	12.48	74.09	10.72	2.15	2.5	0.56	Dehghan-Shoar et al. (2010)
Whole bran wheat flour	—	70	13.8	2.0	NR	NR	Zasypkin and Lee (1998)
Whole wheat bran	4.4	75.2	14.9	1.5	28.5	4.1	Dansby and Bovell-Benjamin (2003)
Wheat bran	—	70.82	17.46	2.45	5.34	3.92	Sobota et al. (2010)
Cassava flour	5.2	89.45	1.55	0.45	1.75	1.60	Rampersad et al. (2003)
Potato	9.0	80.5	10.0	0.5	NR	NR	Onwulata et al. (1998)
Sweet potato flour	2.8	84.3	8.9	0.96	9.9	3.0	Dansby and Bovell-Benjamin (2003)
Oat flour	12.0	Diff	18.0	7.0	9.0	2.0	Liu et al. (2000)
Barely flour	—	85.7	10.3	2.3	16.2	1.7	Altan et al. (2009)

(Continued)

TABLE 4.1 (*Continued*)
Composition of Raw Materials Reported in the Literature

Raw Material	Moisture	Starch	Protein	Lipid	Fiber	Ash	References
Taro flour	6.22	57.55	5.37	0.74	1.47	3.78	Rodríguez-Miranda et al. (2011)
Acha	12.5	87.6	7.98	NR	0.4	2.73	Olapade and Aworh (2012)
Amaranthus flour	12.32	64.17	13.58	7.56	NR	2.37	Chávez-Jáuregui et al. (2000)
Defatted amranthus flour	13.44	69.98	13.70	0.16	NR	2.71	Chávez-Jáuregui et al. (2000)
Glandless cotton seed flour	NR	NR	55.3	12.6	1.0	7.83	Hsieh et al. (1990)
Pulse	5.2	6.03	45	19	20	4.77	Semasaka et al. (2010)
Hard-to-cook common beans	—	75.0	19.3	1.0	NR	4.4	Batista et al. (2010)
Broad bean	—	42.26	31.07	2.66	20.01	3.99	Gimenez et al. (2013)
Pigeon pea flour	8.2	59.68	18.00	1.50	6.10	6.52	Rampersad et al. (2003)
Cowpea	11	67.5	26	NR	1.62	3.15	Olapade and Aworh (2012)
Cowpea	6.95	63.75	23.0	3.75	1.05	1.45	Oduro-Yeboah et al. (2014)
Cowpea	12.0	60.20	22.10	1.70	2.10	1.90	Filli et al. (2011)
Plantain	8.45	85.0	2.15	1.9	0.6	1.85	Oduro-Yeboah et al. (2014)
Full-fat soy flakes	7.67	33.23	39.1	20.0	NR	NR	Konstance et al. (1998)
Partially defatted soy flour	9.4	34.5	44.2	7.7	NR	4.6	Veronica et al. (2006)
Defatted soy flour	—	36.6	55.6	1.3	NR	NR	Zasypkin and Lee (1998)
Semi defatted sesame cake	8.1	14.4	35	11.2	22.7	8.6	Nascimento et al. (2012)
Partially defatted hazelnut flour	2.90	39.72	33.51	17.26	NR	6.61	Yagci and Gögus (2009)
Millet	4.39	70.3	15	2.5	7	0.81	Semasaka et al. (2010)
Pearl millet	11.2	70.80	10.80	3.80	1.80	1.60	Filli et al. (2011)
Tomato pomace	—	82.7	9.7	4.8	52.8	2.8	Altan et al. (2009)
Grape pomace	—	86.6	6.9	2.4	12.3	4.1	Altan et al. (2009)
Brewer's spent grain	6.33	60.64	24.39	6.18	9.19	2.48	Sobukola et al. (2013)
Sweet whey solids	5.0	75.0	12.1	1.5	NR	NR	Onwulata et al. (1998)
WPC	3.0	60.0	34.0	3.0	NR	NR	Onwulata et al. (1998)

NR, not reported.

TABLE 4.2
Selection and Justification of a Variety of Raw Materials Based in Different Extrusion Studies

Raw Materials	Reasons for Selection	Extrusion Conditions	Results	References
Cornmeal, potato, rice flour, WPC, and sweet whey solids	It is usual practice to fortify the low nutrient snacks with protein. However, incorporation of whey protein reduces the expansion. Textural characteristics can be enhanced if TSE is employed due to changing screw configuration/introducing reverse screw elements, enhancing mechanical energy transfer, and lowering moisture content.	TSE with two circular die of 3.18 mm, temperature of 100°C/110°C/125°C, screw speed of 300 rpm, water feed rate of 1.02 L/h, and feed rate of cornmeal, potato, and rice are 5.42, 6.54, and 3.45 kg/h, respectively	Inclusions of whey products require considerable modification for producing expanded crunchy products. Whey products more than 10% in the blend reduces expansion and increases water-holding capacity.	Onwulata et al. (1998)
Corn grits, full-fat and defatted soy flour	Cereals have low protein, high methionine, and cysteine but limited lysine, whereas soy protein is high in lysine but low in methionine and cysteine. Optimal inclusion of defatted/full-fat soy flour into cereal-based extruded snack can increase the protein quality.	Corotating TSE with l/d 35:1, temperature profile of 30°C/35°C/45°C/95°C/135°C/155°C/130°C/125°C, screw speed of 300 rpm, feed moisture of 16.5% wb, and feed rate of 22 kg/h	According to the authors, the extruded snack had higher protein content than that of market snacks. The essential amino acids such as lysine, methionine, and cystine and they are 80% of FAO recommendation.	Boonyasirikool and Charunuch (2000)
Taro flour and maize flour (0%–100%)	Taro is an edible starchy tuber that has high nutritional and health values. Due to high moisture content, sustained metabolism, and microbial attack, taro should be converted from perishable to nonperishable form.	Laboratory-scale SSE with l/d of 20:1 and screw compression ratio of 1:1, a die of 3 mm diameter, temperature of 140°C–180°C, screw speed of 60 rpm, moisture content of 18% wb, and feed rate of 1.68 kg/h	An increase in taro flour proportion has positively enhanced the functional properties. A blend of taro flour and nixtamalized maize flour with 85.4°C–14.6°C extruded at 174.14°C was similar in chemical composition, hardness and overall acceptability, resistant starch content, pH, water activity, apparent density to extrudates prepared taro flour and non-nixtamalized maize flour at the same proportion.	Rodríguez-Miranda et al. (2011)

(Continued)

TABLE 4.2 (*Continued*)
Selection and Justification of a Variety of Raw Materials Based in Different Extrusion Studies

Raw Materials	Reasons for Selection	Extrusion Conditions	Results	References
Oat flour (55%–100%) and corn flour	Soluble fiber (β-glucan) from oats can lower blood cholesterol and prevent coronary heart disease. Production of expanded product from whole grain oat is difficult due to high level of fat and soluble gum. Corn flour can act an expansion agent and blending it with oat flour can produce expanded product.	Corotating and intermeshing TSE with l/d of 15:1 and circular die of 3.18 mm screw speed of 200–400 rpm, feed moisture of 18%–21% wb, and feed rate of 45.4 kg/h	Among the variables studied, oat flour percentage and feed moisture significantly affected its physical and sensory properties, whereas the effect of screw speed was not significant.	Liu et al. (2000)
Pinto bean meal	Though legumes have apparent benefits of soluble fiber in preventing heart disease, they need long cooking time and presence of anti-nutritional substance such as trypsin inhibitors. Legume proteins comes with low biological value because of their low digestibility; therefore their use is limited. Extrusion can address the above issues and make better products.	SSE with l/d of 20:1, screw compression ratio of 2:1 and circular die of 2.4 mm, temperature of 140°C–180°C, screw speeds of 150–250 rpm, and feed moisture of 18%–22% wb	Among the variables studied, temperature and feed moisture significantly influenced expansion index, bulk density, water absorption index, and protein digestibility, whereas temperature affected water solubility index. Authors have not found trypsin inhibitors in extrudate obtained in any of the conditions. Screw speed had no effect on any dependent variable. Feed moisture 22% extruded at 160°C produced the best product.	Balandrán-Quintana et al. (1998)
Whole bran wheat flour and defatted soy flour	A blend of soybean and wheat flour can complement proteins with increased nutritional value and better textural or other functional properties.	SSE with l/d of 20:1 with screw compression ratio of 3:1 and a circular die of 3 mm, temperature of 60°C/160°C/165°C, screw speed of 200 rpm, and feed moisture 16%–18% wb	Expansion ratio of extrudate from wheat flour or soy flour increases with decreasing moisture content, whereas the composite blend has no effect or lowers it.	Zasypkin and Lee (1998)
Degermed cornmeal, full-fat soy flakes, and soy isolates and concentrates	Soy concentrates and soy isolates can provide highly concentrated protein sources, high lysine, bland flavor, reduction of flatulence factors, and reducing sugars. Corn-soy blend eliminates prolonged cooking, thereby reducing nutrient degradation.	TSE with two circular die of 3.18 mm, temperature of 100°C–130°C, screw sped of 300 rpm, and feed moisture of 8.5%–18% wb	Extrusion of these blends at temperature of 100°C–115°C with a moisture of 12%–15% followed by size reduction to a powder form can match the color of current corn-soy blends. Slight alteration in the processing conditions would lead to develop consistencies that would enable products ranging from porridges or gruels to beverages.	Konstance et al. (1998)

(*Continued*)

TABLE 4.2 (*Continued*)
Selection and Justification of a Variety of Raw Materials Based in Different Extrusion Studies

Raw Materials	Reasons for Selection	Extrusion Conditions	Results	References
Yellow peas, texturized soy flour, and soy protein concentrate	Air classified peas are rich in protein, lysine, digestible carbohydrate, and DF. However, it is not widely utilized as a food ingredient, due to flatulence factors and indigestible substances and problems related to flavor and functionality. Extrusion-based texturization would address the above issues; thereby, it enables its use as an ingredient in food products.	Corotating TSE with l/d of 26:1 and circular die of 5 mm, temperature of 130°C–170°C, screw speed of 135–245 rpm, feed moisture of 24%–30.5%, and federate of 27 kg/h	Authors reported that extrusion improved functional (water hydration capacity) as well as nutritional (protein digestibility, trypsin inhibitor activity) properties of texturized pea protein.	Wang et al. (1999)
Pork trimmings and soy protein isolate	TSE has the potential to separate fat from meat products that would alter functional properties, decrease meaty flavor, and increase undesirable flavors. In general, soy proteins are used as meat binders for improving physical and chemical properties of processed meat products.	Intermeshing, corotating TSE with l/d ratio of 15:1, temperature of 25°C–65°C, and soy protein isolate 0%–3%	An increase in barrel temperature separates more fat, reduces lipid oxidation, and darkens the resultant product with more springiness and cohesiveness. Inclusion of soy protein isolate reduces lipid oxidation.	Ahn et al. (1999)
Channel catfish, partially defatted peanut flour, tapioca starch	Snack products mainly contains carbohydrate and fat, along with high quality protein from legumes as well as fish, pork, beef, and chicken for improving nutritional quality. However, the products must retain satisfactory sensory acceptability.	Corotating TSE with l/d of 25:1 and slit die of 1 × 20 mm, temperature of 90°C–100°C, screw speed of 100–400 rpm, feed moisture of 40% wb, and feed rate of 1.6 kg/h	An increase in temperature and screw speed increases expansion and decreases bulk density and shear strength. Authors reported optimum conditions for fish half-products, 94°C–100°C and 220–400 rpm and for peanut half-products, 95°C–100°C and 230–400 rpm.	Suknark et al. (1998)
Defatted amaranthus flour	Amaranth is a potential source of dietary nutrients with protein quality superior than cereal and has high levels of lysine and sulfur amino acids. Development of new products from amaranth through extrusion would enhance the nutritional quality and expand its utilization.	SSE with l/d of 20:1, screw compression ratio of 3.55:1 and a circular die of 3 mm, temperature of 130°C–170°C, screw speed of 200 rpm, feed moisture of 10%–20% wb, and feed rate of 4.2 kg/h	Authors' optimized process conditions such as 15% of moisture and 150°C for maximum expansion coincided with shearing force of the product and sensory texture acceptance of the product.	Chávez-Jáuregui et al. (2000)

(*Continued*)

TABLE 4.2 (*Continued*)
Selection and Justification of a Variety of Raw Materials Based in Different Extrusion Studies

Raw Materials	Reasons for Selection	Extrusion Conditions	Results	References
White cornmeal, corn syrup, blueberry concentrate, and brewer's spent grain (69°B)	Many products found in the market are artificially colored, and a few cereals are now colored with fruit juices and other natural sources. Water-soluble pigments, anthocyanins, are responsible for the red, blue, and purple colors in many food crops, and they are antioxidants. Product development scientists are taking advantage of anthocyanin by including it into new food product development. However, their retention during extrusion has not been studied.	Corotating TSE with l/d of 32:1 and circular die of 4 mm, temperature profile of 38°C/49°C/116°C/138°C/113°C, screw speed of 300 rpm, feed rate of 13.6 kg/h, and liquid feed rate of 4.4 kg/h	Though extrusion reduces anthocyanin content, it retains purple color. There is a research opportunity to minimize the pigment loss through optimizing extrusion parameters.	Camire et al. (2002)
Cornmeal, sucrose (15%), 17% blueberry concentrate (71°B), ascorbic acid (0%–1% w/w)	Anthocyanin can provide color and health benefits; however, extrusion reduces it. Ascorbic acid can function as an acidulant and vitamin in foods; thus it could improve pigment retention, thereby marketability would also improve.	Corotating TSE with l/d of 32:1, circular die of 4 mm, screw speed of 225 rpm, and blueberry concentrate feed rate of 3.6 kg/h	This study revealed that ascorbic acid did not protect anthocyanins nor inhibit browning reactions.	Chaovanalikit et al. (2003)
Sweet potato, whole wheat bran	Sweet potato is one of the crops considered by NASA for long-term space mission. Moreover, its inclusion would result in varieties to breakfast cereals, that too gluten-free.	SSE with l/d of 20:1 and screw compression ratio of 3:1 and rod-shaped die of 0.3 mm, temperature profile of 100°C/135°C/140°C, and screw speed of 165 rpm.	Extrudates with 100% sweet potato, 75% sweet potato, and 25% whole wheat bran were found acceptable by sixth graders. Authors concluded that results are promising to explore for long-term space mission; however, chewiness and hardness of the product should be reduced, and shelf life should be evaluated and there are research opportunities.	Dansby and Bovell-Benjamin (2003)

(*Continued*)

TABLE 4.2 (*Continued*)
Selection and Justification of a Variety of Raw Materials Based in Different Extrusion Studies

Raw Materials	Reasons for Selection	Extrusion Conditions	Results	References
Chickpea flour, corn flour, oat flour, corn starch, carrot powder, and ground raw hazelnut	Investigate the effect of extrusion conditions on phenolic compounds and total antioxidant capacity for developing new snack products.	Corotating TSE with l/d ratio of 27:1 and circular die of 4 mm, screw speed of 230–340 rpm, feed moisture content of 11%–15% wb, and feed rate of 22–26 kg/h	The total antioxidant capacity decreases with an increase in screw speed and decrease in moisture content, whereas extrusion conditions did not influence phenolic compounds.	Ozer et al. (2006)
Rice flour and orange-yellow or red cactus pear pulp concentrate with 40°B at different ratios 100:0–80:20	Cereal flours are poor in sugars, fibers, vitamins, and minerals. Cactus pear pulp is a major source of nutraceuticals and functional components such as betalains (antioxidant), phenolic compounds, pectin (source of DF), vitamin C, calcium, and magnesium. It is rich in amino acid, proline, and taurine (common ingredients in energy drinks) that can meet consumers expectation.	Lab-scale SSE equipped with a screw compression ratio of 4:1 and circular die of 3 mm, temperature profile of 100°C/140°C/160°C, screw speed of 250 rpm, and feed moisture of 16%	Addition of cactus pear pulp increases bulk density, ash content, and color attributes (a* and b*), whereas expansion ratio (ER), water absorption index (WAI), and water soluble index (WSI) decreases, breaking strength (BS) decreases up to 10% of the added ratio, then increases. Addition of concentrated cactus pear pulps to rice flour remarkably enhances the sensory characteristics of final extruded products.	El-Samahy et al. (2007)
Corn flour, rice flour, and egg albumin/cheese powder in a proportion of 35–50:35–50:5–30, respectively	Egg protein is a standard against other proteins due to complete digestion and the highest nutritional quality protein since it provides all the essential amino acids in amounts that closely match human requirements. Egg white/albumin have very little flavor that do not negatively impact the organoleptic properties of foods; hence, it eliminates/reduces the requirements for masking flavors. In general, cheese powder is a good source of protein. It is only used as a flavor coating for extruded snacks but has not been used as an ingredient for extrusion. Cereal-based extruded snacks tend to be low in protein and have low biological value; therefore, incorporation of protein sources would be a basis for a range of highly nutritious snacks.	Corotating, self-wiping TSE with l/d ratio of 27:1 and circular die of 3 mm, temperature of 75°C–100°C, screw speed of 300–350 rpm, feed moisture of 17%–20% wb, and feed rate of 15.2 kg/h	The protein content of the RTE extruded snack with egg albumin powder and cheese powder increased protein content by 20%–50% good physical and sensory properties. A combination of packaging materials with better barrier properties or modified atmospheric packaging can enhance shelf life by retaining the texture of extruded snacks.	Priyanka et al. (2012)

(*Continued*)

TABLE 4.2 (*Continued*)
Selection and Justification of a Variety of Raw Materials Based in Different Extrusion Studies

Raw Materials	Reasons for Selection	Extrusion Conditions	Results	References
Corn/wheat flour	Corn and wheat are the most widely used flour for extrusion-based snacks. Authors studied how these flours influence rheological properties during extrusion and expansion of the product.	Corotating TSE with l/d of 9:1 and circular die of 5 mm, barrel temperature of 160°C, constant SME of 500 kJ/kg by adjusting screw speed (120–190 rpm), and feed rate (25–43 kg/h)	The viscous characteristics, starch, and protein contents of corn and wheat flours affect differently the radial and axial expansion velocities when the global expansion velocity is same; corn flour had the maximum expansion.	Arhaliass et al. (2009)
Barnyard millet and pigeon pea (12%–28%) with moisture content of 12–24% wb	Millets are rich in fibers, and legumes are rich in proteins and vitamins; the resultant product would have high nutritive value.	SSE with l/d of 20:1 and screw compression ratio of 3:1, temperature profile of 100°C–140°C/160°C–200°C, and screw speeds of 100–140 rpm	Authors optimized processing conditions such as moisture content 24% wb; blend ratio 18.7% legume; die head temperature 171.2°C; barrel temperature 140°C; and screw speed 103.8 rpm for product with maximum crispness and with minimum hardness and cutting strength.	Chakraborty et al. (2009)
Corn, millet, and soybean	Cereals are usually fortified with lysine or legume proteins. Soy protein reduces the risk of coronary heart disease due to low saturated fat and cholesterol. Corn, millet, and soybean blend can give a product with high energy value and proteins with high biological value.	Lab-scale corotating fully-intermeshing TSE with 5 mm die, temperature profile of 40°C–80°C/70°C–110°C/100°C–140°C/130°C–170°C; screw speed of 110–150 rpm, moisture of 15%–30% wb, and feed rate 1.2–3.3 kg/h	Temperature 80°C, 110°C, 140°C, and 170°C, screw speed of 110 rpm, feeding speed of 37 g/min and moisture content of 25%–30% are the best for the extrusion of their formulation based on extrudate properties.	Semasaka et al. (2010)
Pearl millet and cowpea	Amino acid profiles of legumes complement those of cereal proteins. It provides a good opportunity to restructure starch and protein-based materials to manufacture a variety of textured convenience foods.	SSE l/d of 20:1 and with circular die of 3 mm, screw speed of 150–250 rpm, and feed moisture of 20%–30% wb	The content of amino acids such as tyrosine, phenylalanine, isoleucine, valine, and leucine increases with increase in the level of cowpea flour.	Filli et al. (2011)

(*Continued*)

TABLE 4.2 (*Continued*)
Selection and Justification of a Variety of Raw Materials Based in Different Extrusion Studies

Raw Materials	Reasons for Selection	Extrusion Conditions	Results	References
Wheat flour and ginseng	Consumers are interested in nutraceutical foods. Ginseng is an herbal medicine known to exhibit various pharmaceutical effects, including antioxidation, antistress, immunostimulation, and anticancer effects. The addition of ginseng to wheat, corn, and rice flour can yield nutraceutical extruded products (snacks or instant powder) that can meet consumer expectations.	TSE with l/d ratio of 25:1, temperature of 110°C–140°C, screw speed of 200–300 rpm, and feed moisture of 25%–35% wb	The ER increased with decreasing feed moisture, decreasing screw speed, and increasing barrel temperature.	Chang and Ng (2011)
Wheat flour, cornstarch, potato starch, skim milk powder with spray-dried apple, banana, strawberry, and tangerine powder	It is used for incorporation of different fruit powders into extruded snacks to improve the nutritional profile with acceptable sensory qualities through understanding the interaction between different fruit powders and extrusion process.	Corotating TSE with circular die of 4 mm and twin-screw volumetric feeder, temperatures of 80°C/120°C, screw speed of 250 rpm, moisture content of 13% wb, and feed rate of 20 kg/h	The incorporation of fruit powder has a negative influence on expansion and antioxidants and a positive influence on density and soluble and insoluble fiber of the extrudates. The extrudates with fruit powder improves nutritional profile compared with other extruded snack products being low in fat and sugar and a good source of fiber.	Potter et al. (2013)
Rice, sorghum (10%–20%), soy (5%–15%), and finger millet (40%–50%).	Millets are rich sources of DF, phytochemicals, and micronutrients. Millets are superior to major cereals due to their protein, energy, vitamins, and minerals and nutritionally comparable to them. Sorghum and millets contain significant amounts of a wide range of phenolic compounds, particularly antioxidant activity.	TSE with l/d of 10:1 and circular die of 2.5 mm, temperature of 184°C, screw speed of 285 rpm, feed moisture content of 18% wb, and a blend of rice to rest of the ingredients 1:2.33	Authors optimized the formulation based on desirable sensory and physical qualities with 42.03% finger millet, 14.95% sorghum, 12.97% soy, and 30% rice.	Seth and Rajamanickam (2012)

(*Continued*)

TABLE 4.2 (*Continued*)
Selection and Justification of a Variety of Raw Materials Based in Different Extrusion Studies

Raw Materials	Reasons for Selection	Extrusion Conditions	Results	References
Degermed white cornmeal (84.3%), sucrose (14.3%), citric acid (0.4%), and dehydrated fruit powder-blueberry, cranberry, concord grape, and raspberry-1%	Consumer demand for food with natural colorants is growing due to their perception as healthy and natural. Dried fruit powders not only can provide natural colors but also provide antioxidants/anthocyanin.	TSE with l/d of 32:1 and circular die of 4 mm, temperature profile of 35°C/45°C/60°C/95°C/113°C/163°C, screw speed of 175 rpm, and feed rate of 15.3 kg/h	Authors found that anthocyanin survived in the extrusion process. Investigations on methods to increase retention of these pigments during processing are necessary. Determination of optimal levels of fruit powders for increasing antioxidant activity with acceptable sensory quality of extruded snacks is a research to explore.	Camire et al. (2007)
Wheat flour, corn grits, oatmeal, and fibers (wheat bran, fine guar gum, inulin, hi-maize 1043, and swede)	Extruded cereal products are considered as high glycemic impact foods equal to, and in most cases exceeding, that of bread. The utilization of DFs as nutritional aids due to their potential in reducing the glycemic index from carbohydrate foods is the need of the hour.	Corotating, self-wiping TSE with a circular die of 3 mm, temperature of 40°C–180°C, screw speed of 315 rpm, feed rate of 6.75 kg/h, and water rate 0.29 L/h	The addition of fiber reduces the amount of readily digestible starch components and increases the amount of slowly digestible starch. This has potential benefits in terms of attenuating the glucose response post-ingestion, and may also lead to an increased feeling of satiety.	Brennan et al. (2008)
Corn grit and wheat bran at various ratios	Dietary fiber is essential in daily diet due to obvious reasons. Wheat bran and meal can be good dietary fiber sources in the production of extrudates.	Counter-rotating TSE with l/d 12:1 and die diameter 4.2 mm and 3.2 mm, temperature of 80°C–200°C, screw speed 72 rpm, and feed moisture of 14% wb	Wheat bran is a source of dietary fiber with only insoluble fraction; therefore, it should be applied together with other sources of soluble fiber with probiotic action.	Brennan et al. (2008)
Corn and cowpea	Cowpea is a good, low-cost source of protein, carbohydrates, vitamins, and minerals; inclusion in diverse products can improve nutritional quality. Lipoxygenase enzyme present in legumes can have negative effects on the color, flavor, texture, and nutritional properties of foods including off-flavor development. Accordingly, the authors investigated the effect of extrusion conditions and lipoxygenase inactivation on the physical and nutritional properties of corn–cowpea blends.	SSE with 4:1 screw compression ratio and circular die of 3.5 mm, temperature of 150°C–180°C, screw speed of 150 rpm, moisture content of 15%–19%, and feed rate of 12 kg/h	The extrusion of corn–cowpea flour blends at 85:15 ratio with 15% moisture content at temperatures of 150°C and 165°C produces extrudates with good physical and nutritional characteristics (40% higher protein).	Sosa-Moguel et al. (2009)

(*Continued*)

TABLE 4.2 (*Continued*)
Selection and Justification of a Variety of Raw Materials Based in Different Extrusion Studies

Raw Materials	Reasons for Selection	Extrusion Conditions	Results	References
Wheat flour, starch, extruded orange pulp (5%–25%), sugar, hydrogenated vegetable fat. leavening agent, powdered milk	Dietary fiber increases the nutritional value and simultaneously alters the rheological properties of the dough, thereby the quality and sensory properties of the final product. Therefore modification of the functional and structural properties of orange pulp as the fiber is essential.	SSE with l/d of 20:1 and a screw compression ratio of 3:1, a die with 20 die-nozzle orifices, each 4 mm diameter, temperature of 83°C–167°C, screw speed of 126–194 rpm, moisture content of 22%–38% wb, and feed rate of 4.2 kg/h	Biscuits of good quality with a good level of acceptance by replacing wheat flour up to 15% of with extruded orange pulp that reduces the energy value.	Larrea et al. (2005)
Rice flour with oleic acid (96% and 4%); rice flour and pistachio nut flour (75% and 25%)	Starch–lipid complex formation is an important reaction during extrusion of flour blends containing fatty meal that affects structure and texture of the extruded products; therefore, understanding their interaction is very important in new extrusion-based products.	Corotating TSE with l/d of 28:1, temperature of 35°C/65°C/72°C–128°C, screw speed of 140 rpm, feed moisture content 16% and 21%, feed rate of 4 kg/h (flour with oleic acid), and 2.9 kg/h (flour with pistachio nut flour)	Lipid content influence the torque and die pressure of the extruder. Because of pistachio's high fat content, energy required for extrusion gradually decreases with increase in barrel temperature due to lubrication effect. Barrel temperature and moisture content play a vital role in the formation of starch–lipid complexes. The processing conditions that favor the maximum formation of starch–lipid complexes leads to highest fat loss and the hardest texture of extrudates made up of pistachio nut flour.	De Pilli et al. (2011)
Cornstarch, navy and small red bean flours (15%–45%)	Cereal-based starch is the main ingredient in extruded snacks; however, its nutritional value is far from expectations of health-conscious consumers. Fortifying cereal-based starch with bean flours and fractions for the production of extruded snacks appears to be promising. In general, beans are rich in fiber and protein, and low in fat, which reduces risk of coronary diseases and some types of cancer. Further, colored beans possess strong antioxidant activity that provides protective benefits on development of degenerative diseases.	Lab-scale TSE with l/d of 25:1 and die diameter of 4.5 mm, temperature profile of 30°C/80°C/120°C/160°C/160°C, screw speed of 150 rpm, feed moisture content of 22% wb, and feed rate of 1.8 kg/h	Authors reported that cornstarch with 30% small red bean flour yielded extrudates with high nutritional functionality, especially an increase in crude protein by 12-fold.	Anton et al. (2009)

(*Continued*)

TABLE 4.2 (*Continued*)
Selection and Justification of a Variety of Raw Materials Based in Different Extrusion Studies

Raw Materials	Reasons for Selection	Extrusion Conditions	Results	References
Rice flour, milk powder, soy flour, cornstarch, and potato starch. Replace rice flour with dry cranberry, beetroots, apple, carrot, and teff flour at a level of 30%.	Gluten-free products generally have low total DF, and they are not enriched and frequently are made from refined flour and/or starch. Extrusion has the ability to increase total DF in gluten-free products through incorporating a number of different fruits and vegetables. Cranberry has high essential nutrients, such as anthocyanins, proanthocyanins, organic acids, vitamins, minerals, and strong antioxidant properties; it compliments other fruits in terms of flavor. Apple and beetroot have a high level of total DF; carrot has high vitamin and fiber contents; teff flour cereal is gluten-free and contains more iron, calcium, and zinc than do other cereal grains.	Corotating TSE with l/d of 27:1 and circular die of 4 mm, temperature of 80°C/80°C–150°C, screw speed of 200 and 350 rpm, water feed of 12%, and feed rate of 15–25 kg/h	An increase in DF of extruded product depends on the types of sources. Authors found a greater increase in DF with teff, followed by apple, cranberry, carrot, and beetroot. However, extrudates containing carrots are the most stable during extrusion in terms of crude protein and texture.	Stojceska et al. (2010)
Corn and lintel flour	All legume containing proteins are low in sulfur-containing amino acids, methionine, cysteine and tryptophan, however, they are much higher in other essential amino acids such as lysine, compared to cereal grains. Legume and cereal proteins are nutritionally complementary with respect to lysine and sulfur amino acid contents.	Corotating TSE with die diameter of 3 mm, temperature of 170°C–230°C, screw speed of 500 rpm, feed moisture content of 13%–19% wb, feed rate of 2.5–6.8 kg/h, and corn–lintel ratio of 10%–50%	Extrusion conditions and corn–lintel properties influenced the functional properties such as WAI, WSI, and oil absorption index.	Lazou and Krokida (2010)
Pigeonpea flour to cassava flour at 0%–15% (db)	Pigeonpea protein is a rich source of lysine, and cassava contains low protein. Extrudates with complementary proteins can increase nutritional value and create better textural and other functional properties.	SSE with l/d ratio 15:1, screw compression ratio of 2:1 with die diameter of 5 mm, temperature of 120°C–125°C, screw speed of 520 rpm, feed moisture of 12% db, and feed rate of 18 kg/h	Extrudates with 95% cassava flour and 5% pigeonpea flour are crisp, more yellow, higher protein, bulk density, and water absorption index with low expansion and water absorption index.	Rampersad et al. (2003)

(*Continued*)

TABLE 4.2 (*Continued*)
Selection and Justification of a Variety of Raw Materials Based in Different Extrusion Studies

Raw Materials	Reasons for Selection	Extrusion Conditions	Results	References
Corn and partially defatted soy flour	Soy is a major source of protein and soy foods, especially isoflavones, reduce blood cholesterol levels, risk of cardiovascular diseases, and certain cancer in humans. Fortification of starch-based snack foods with suitable protein food can improve their nutritional quality.	SSE with die diameter of 3.5 mm, temperature of 200°C, screw speed of 300 rpm, and moisture content of 20% wb.	Incorporating partially defatted soy flour in a corn-based snack has positive effects on chemical properties and negative effects on the physical and sensory characteristics; however, up to 20% replacement is acceptable.	Veronica et al. (2006)
Corn and glandless cotton seed flour (5%–25%)	Glandless cottonseed flour (free from gossypol) can be used as for the production of texturized protein products.	Lab-scale SSE with l/d of 20:1 and screw compression ratio of 3:1 and circular die of 3 mm, temperature of 100°C/110°C/105°C–165°C, screw speed of 90–210 rpm, and feed moisture of 12%–20%	Glandless cottonseed meal (10%)-corn flour extruded snacks increase protein and reduces fat. Authors identified optimal extrusion conditions such as temperature of 120°C, screw speed of 179.9 rpm, cottonseed meal of 10%, and moisture of 16.8%.	Reyes-Jáquez et al. (2012)
Foxtail millet flour, rice flour, chickpea, amaranth seed flour, Bengal gram flour, and cowpea	Foxtail millet is rich in DF; amaranth addresses celiac disease; cowpea has low glycemic responses. These can be blended with rice flour for the production of extruded snacks.	Corotating, self-wiping TSE with l/d of 25:1 and circular die of 2.5 mm, temperature of 90°C/110°C/80°C, screw speed 130–600 rpm, and feed moisture content of 21%–22% wb	This study revealed that composite flour (foxtail millet:amaranth:rice:Bengal gram:cowpea at 60:05:05:20:10) can be used to produce quality extrudates with acceptable sensory properties.	Deshpande and Poshadri (2011)
Glutinous rice, vital wheat gluten, and toasted soy grit	Rice is an attractive ingredient due to its bland taste, attractive white color, hypoallergenicity, and ease of digestion. Further glutinous rice is suitable for producing expanded products such as ready-to-eat snacks and breakfast cereals with low bulk density, high expansion, and low shear stress. Since rice has low protein and limited lysine, it has to be complemented with proteinaceous additives to satisfy nutrient requirement.	Corotating intermeshing TSE with l/d of 28:1 and circular die of 3 mm, temperature 150°C–180°C, screw speed of 400 rpm, feed moisture content of 20%–30% wb, feed rate of 12 kg/h, and feed protein content of 20% and 30%	Increasing feed moisture and reducing barrel temperature reduced non-protein nitrogen and enhanced lysine retention. Any of the variables studied in this study had no significant influence on cysteine and methionine content. A feed moisture content of 20% at a barrel temperature of 180°C resulted in high expansion, low bulk density, and low shear strength of extruded snacks.	Chaiyakul et al. (2009)

(Continued)

TABLE 4.2 (*Continued*)
Selection and Justification of a Variety of Raw Materials Based in Different Extrusion Studies

Raw Materials	Reasons for Selection	Extrusion Conditions	Results	References
Salmon, sucrose, pregelatinized starch, modified tapioca starch, salt, teriyaki flavoring, and oil binding agents such as tapioca starch, high amylose corn starch, and oat fiber	In order to reduce moisture content and improve flow within the extruder, animal meat or fish is minced and combined with binders. Salmon is a good source of protein and omega-three fatty acids. Therefore, development of salmon-based extruded snacks would be an appropriate approach.	Corotating TSE with l/d of 32:1, temperature profile of 65°C/155°C/155°C/80°C, screw speed of 250 rpm, and feed rate of 13.2 kg/h	Extrusion had no influence on the retention of omega-3 fatty acids. Careful selection of barrel temperature and binder would address the adverse conditions resulting from high content of moisture and fat in extrusion.	Kong et al. (2008)
Arrowroot starch	Arrowroot starch is used as a thickener in all kinds of foods, dressings, soups, sauces, candies, cookies, and desserts. It has high digestibility and medicinal properties; further it is gluten-free and underexploited, thus makes a good candidate for developing extrusion-based snacks.	SSE with screw compression ratio of 1:1 and die diameter of 2 mm, temperature profiles of 70°C/80°C/90°C/140°C–190°C, screw speed of 80 rpm, and feed moisture content of 12%–16% wb	The extrusion of arrowroot starch with 12% moisture content at 160°C and 170°C can produce highly expanded, good textured products with appreciable color and appearance.	Jyothi et al. (2008)
Barely, barley-pomace (tomato/grape: 0%–12.7%)	Slowly digestible and absorbable carbohydrates would facilitate dietary management of metabolic disorders such as diabetes. The incorporation of DF in snacks due to its role in health and nutrition has received increased attention in recent years. By-products of fruits and vegetables as fiber supplementation has a great potential; however, how fiber affects functional and digestive properties of snacks should be studied.	Corotating with l/d of 13:1, slit die of 1.47 × 20 × 150 mm, temperature profile of 30°C/60°C/100°C/130°C/136°C–164°C, screw speed of 140–210 rpm, blend moisture content of 22% wb, and feed rate of 2.1 kg/h	Though extrusion increases digestibility of extrudates, the incorporation of pomace decreases the digestibility of samples. Therefore, further studies are required to address the digestibility issues.	Altan et al. (2009)

(*Continued*)

TABLE 4.2 (*Continued*)
Selection and Justification of a Variety of Raw Materials Based in Different Extrusion Studies

Raw Materials	Reasons for Selection	Extrusion Conditions	Results	References
Grape seed or pomace and decorticated white sorghum of 30:70	The presence of procyanidins bears importance as it is partially responsible for organoleptic characteristics of foods and antioxidant capacity and possible protective effects in human health. Procyanidins are present in fruits in a form not readily available for absorption. Thus, investigation is required how much of the procyanidins present in its seed and pomace is in monomoric and polymeric forms and whether extrusion brings about any change in the distribution of these polyphenols.	TSE with rod-shaped die diameter of 3 mm, temperature of 160°C–190°C, screw speed of 100–200 rpm, feed moisture content of 45%, and feed rate of 6 kg/h	Extrusion at 170°C and 200 rpm resulted in the highest increase in monomer contents with 120% increase in grape pomace and 80% increase in grape seed over the control. Extrusion reduces total anthocyanins in pomace.	Khanal et al. (2009a)
Blueberry pomace and decorticated white sorghum of 30:70	Majority of the procyanidins present in fruits are not readily available for absorption, thus it is better to investigate what changes extrusion brings about. Earlier studies revealed that extrusion has the potential to enhance the monomeric and lower oligomeric procyanidin contents, and hence, improve the nutritional value and potential health benefits.	TSE with rod-shaped die diameter of 3 mm, temperature of 160°C–200°C, screw speed of 150–200 rpm, feed moisture content of 45%, and feed rate of 6 kg/h	Extrusion of blueberry pomace increases the monomer, dimer, and trimer contents considerably at both temperature and screw speeds.	Khanal et al. (2009b)
Pea–rice blend of 70:30 with 5%–20% w/w of the locust bean, guar gum, and fenugreek gum	Starch is responsible for mechanical properties of expanded products. At the same time gums can also affect the mechanical, physico-chemical, and micro-structural properties, and may also lower the glycemic index of the food.	TSE with temperature profile of 50°C/50°C/70°C/110°C/150°C/150°C/160°C/118°C, screw speed of 306 rpm, feed rate of 11.5 kg/h, and water feed rate 0.39 L/h	All these gums (15%) resulted in nutritious, organoleptically acceptable, good expanded products with low glycemic index.	Ravindran and Hardacre (2010)

(*Continued*)

TABLE 4.2 (*Continued*)
Selection and Justification of a Variety of Raw Materials Based in Different Extrusion Studies

Raw Materials	Reasons for Selection	Extrusion Conditions	Results	References
Hard-to-cook BRS pontal (carioca) and BRS graphite (black)	High content of protein, carbohydrates, fibers, some minerals, and vitamins make the beans a good source of nutrients; however, the hard-to-cook phenomenon hampers its utilization. Extrusion could be a viable method to utilize common beans through enhancing the functional properties and reducing the antinutritional factors.	SSE with a compression ratio of 3:1 and die of 5 mm, temperature of 150°C, screw speed of 150 rpm, and moisture content of 20% db	Extrusion decreases the anti-nutritional factors such as phytic acid, lectin, α-amylase, and trypsin inhibitors and increases the digestibility of starch and protein.	Batista et al. (2010)
Fresh purple majesty potato flour, split yellow pea with 35/65, 50/50, 65/35 (w/w)	Colored potatoes are rich in anthocyanins, provide natural color to fruits and vegetables, and exhibit antioxidant properties. Thus, producing extruded snacks using colored potato and yellow pea would be a desirable choice.	Corotating TSE, temperature of 120°C–140°C, screw speed of 200–300 rpm, feed moisture of 17%–25% wb, and feed rate of 2.7 kg/h	This study demonstrated the possibility of producing natural-colored, extruded, puffed food products rich in antioxidants. Authors also mentioned that further research on the use of extrusion parameters and their effect on the kinetics of anthocyanins are necessary to study the stability of natural color in the final extrudates.	Nayak et al. (2011)
Sorghum and cowpea (feed ratio 100%, 70%, 50%, 30%, and 0% cowpeas)	Sorghum is an important cereal food for developing countries. Sorghum protein lacks lysine but that can be supplemented with cowpea.	Corotating TSE with die diameter of 5 mm, temperature of 130°C or 165°C, screw speed of 200 rpm, and moisture content of 20% wb	Extrusion of a blend with 50% sorghum and 50% cowpeas at 130°C resulted in a product most similar to a commercial, instant maize–soya composite porridge in terms of composition and functional properties. A 100 g extrudate can provide 28% of the recommended dietary allowance for protein that represents an increase of 110% protein compared to sorghum only.	Pelembe et al. (2002)
Oat bran, defatted soy flour, inulin, and cornstarch	Though DFs have health benefits, their inclusion in extrusion negatively affects the product texture and expansion. This can be minimized if some additives, such as monoglycerides, modified starches, modified gelatin, oligofructose, or inulin, are used.	Lab-scale SSE with l/d of 26:1 and screw compression ratio of 3:1, a die with six 2 mm diameter holes, temperature profile of 80°C/100°C/120°C/120°C, screw speed of 70 rpm, and moisture content of 25% wb	Acceptable extruded products can be produced at 160°C with a moisture content of 25% and inulin content of 45 g/kg of blend, to facilitate the flow of the mixture during extrusion.	Lobato et al. (2011)

(*Continued*)

TABLE 4.2 (*Continued*)
Selection and Justification of a Variety of Raw Materials Based in Different Extrusion Studies

Raw Materials	Reasons for Selection	Extrusion Conditions	Results	References
Rice flour, wheat semolina, corn grits with or without tomato paste or skin powder at a level of 20%	Lycopene is a pigment and nutrient associated with health benefits such as free radical scavenger and anticarcinogen properties. Further DF from tomato skin has beneficial physiological effects such as laxation and modulation of blood glucose. Thus, adding tomato derivatives such as lycopene and fiber to extruded snacks can improve their nutritional value.	Lab-scale TSE with l/d of 28.3:1 and die diameter of 3 mm, temperature profile of 80°C/80°C/80°C/80°C/140°C/160°C/180°C, screw speed of 350 rpm, dry ingredients feed rate of 11.5 kg/h, and water feed rate of 0.5 L/h	High lycopene retention can be achieved with products containing tomato skin powder whereas addition of wheat flour significantly lowers the retention. High processing temperature improves the physicochemical characteristics of the snacks with no significant effect on lycopene retention.	Dehghan-Shoar et al. (2010)
Wheat flour, cauliflower, cornstarch, egg white, oat flour, onion powder, tomato powder, carrot powder, paprika, and salt. Cauliflower powder of 0%–20% replacing wheat flour	Food industries by-products have high DF content and contrasting DF properties that can be used to change physicochemical properties of diets. Cauliflower trimming is high in DF and possesses both antioxidant and anticarcinogenic properties.	Corotating TSE with l/d of 27:1 and circular die of 4 mm, temperature profile of 80°C/120°C, screw speeds of 250–350 rpm, water feed rate 9%–11%, and feed rate of 20–25 kg/h	Inclusion of cauliflower increases dietary fiber, protein, phenolic and antioxidant content whereas it decreases expansion ratios. Product with 10% is acceptable and beyond 10% should be addressed for flavor issues.	Stojceska et al. (2008)
Fenugreek flour (2%–10%) or debittered fenugreek polysaccharides (5%–20%), cowpea, and basmati rice	Fenugreek has antioxidant, restorative, and nutritive properties, can lower blood glucose, and is shown to stimulate digestive processes; however, its distinct aroma and bitter taste limits wide application. Chickpea contains about 50% starches that are slow digestible and can play an important role as a low-glycemic functional ingredient in a healthy diet. Rice flour has a bland taste, low protein content that can limit the non-enzymatic browning reaction rate and contribute to expansion.	Corotating, self-wiping TSE with l/d of 25:1, circular die of 2.5 mm diameter, temperature profile of 50°C/50°C/70°C/110°C/150°C/150°C/160°C/118°C, screw speed of 300 rpm, feed rate of 25 kg/h, and water feed rate of 0.42 kg/h	A blend of 70:30 chickpea and rice supplemented with 15% fenugreek polysaccharide can lower glycemic index with acceptable sensory and physical attributes, whereas fenugreek flour has to be within 2% due to bitter flavor. The bitter flavor can be masked using preprocessed or encapsulated form.	Shirani et al. (2009)

(*Continued*)

TABLE 4.2 (*Continued*)
Selection and Justification of a Variety of Raw Materials Based in Different Extrusion Studies

Raw Materials	Reasons for Selection	Extrusion Conditions	Results	References
Rice flour, potato starch, cornstarch, milk powder, and soya flour and 30% apple, beetroot, carrot, cranberry, and gluten-free teff flour cereal in place of rice flour	The blend is based on the following advantages: apple and beetroot have a high level of DF; carrot has high vitamin and fiber; cranberry has strong antioxidant properties, high essential nutrients, such as anthocyanins, proanthocyanins, organic acids, vitamins, and minerals, and the interesting flavor profile (tart-sweet character) can compliments other fruits, such as orange, lemon, and apple; and teff flour cereal is gluten free and has more iron, calcium, and zinc than do other cereal grains.	Corotating TSE with l/d of 27:1, temperature profile of 80°C/80°C–150°C, screw speed 200–350 rpm, water feed rate 12%, and solid feed rate 15–25 kg/h	Increase in DF depends on the fiber and in this study the order is teff followed by apple, cranberry, carrot, and beetroot. This study demonstrated that formation of a gluten-free expanded product is possible through proper selection of ingredients and controlling operating parameters.	Stojceska et al. (2010)
Peas, rice (70:30), 5%–20% guar gum, locust bean gum, and fenugreek gum	Peas have low glycemic index due to the presence of complex carbohydrates (dietary fiber and water-soluble polysaccharides), further they are rich in protein. Mechanical properties are important for expanded snacks. Starch, a primary contributor, can be altered via inclusion of galactomannan gums (guar gum, locust bean gum, fenugreek gum); further, they alter physico-chemical and micro-structural properties and may lower the glycemic index.	Corotating, self-wiping TSE with l/d of 25:1 and circular die of 2.5 mm diameter, temperature profile of 50°C/50°C/70°C/110°C/150°C/150°C/160°C, screw speed of 306 rpm, feed rate was 11.5 kg/h, and water feed rate of 0.39 L/h	Inclusion of gums up to 15% in a pea–rice blend can result in nutritious snack with low glycemic index that is organoleptically acceptable.	Ravindran et al. (2011)
Corn or rice: wild legume (*Lathyrus annuus* and *Lathyrus clymenum*) of 85:15	Interest in utilizing whole grains in food formulations is increasing due to their rich nutritive, functional, and phytochemical compounds. Nutritional value of corn and rice-based snacks can be improved through addition of legumes in terms of both the amount and quality of the protein mix. Inclusion of wild legumes in formulation is interesting and addresses modern consumers demand such as variety and novelty.	SSE with screw compression ratio of 4:1 and cylindrical die of 3 diameter and 20 mm length, temperature of 175°C, screw speed of 150 rpm, and feed moisture content of 14% wb	Inclusion of legumes has negative effects on expansion of extrudates. Addition of 15% legumes would be appropriate to increase protein content and quality, fiber, and mineral content for acceptable extruded snack production.	Pastor-Cavada et al. (2011)

(*Continued*)

TABLE 4.2 (*Continued*)
Selection and Justification of a Variety of Raw Materials Based in Different Extrusion Studies

Raw Materials	Reasons for Selection	Extrusion Conditions	Results	References
Carrot pomace and rice flour (10%–30%)	Carrot pomace is a by-product from the carrot juice industry that is rich in fiber, β-carotene, and ascorbic acid. Extrusion is one of the ways to utilize dried carrot pomace in a product due to earlier-mentioned facts.	Corotating TSE with l/d of 8:1 and circular die of 4 mm diameter, temperature of 110°C–130°C, screw speed of 270–310 rpm, feed moisture of 17%–21% wb, and feed rate of 4 kg/h	Sensory analysis showed a good product with carrot pomace up to 8.5% and rice flour of 16.5%, moisture content 19.23%, screw speed 310 rpm and die temperature 110°C.	Kumar et al. (2010)
Corn, millet, and soybean flours	Cereals are staple food, legumes are a source of protein and nutrients, and soybean is known for health benefits such as reducing the risk of coronary heart disease. A composite blend of corn, millet, and soybean can give a product, which has a high energy value and proteins, high biological value.	Lab-scale, corotating, fully-intermeshing TSE with 5 mm die, temperature of 130°C–170°C, screw speed of 110–150 rpm, moisture content of 25%–30% wb, and feeding rate of 1.2–3.3 kg/h	Physical and functional properties of extrudates were better at a temperature of 170°C, screw speed of 110 rpm, feeding rate of 2.22 kg/h and moisture content varing from 25% to 30%.	Semasaka et al. (2010)
Yam starch and brewer's spent grain at different ratios 85%–95%:5%–15%	Yams are high in carbohydrates, DF, vitamin C, vitamin B6, potassium, and manganese, with low saturated fat and sodium. Brewer's spent grain is the main by-product of the brewing industry and rich in protein (20%) and fiber (70%).	Lab-scale SSE with l/d of 16.43:1, temperatures of 100°C–110°C, screw speed of 100–140 rpm, and moisture content of 12%	Barrel temperature had a pronounced influence on the physical and functional properties of the extrudates. Authors found an optimum condition such as barrel temperature of 110°C, screw speed of 121.47 rpm, and BSG level of 9.58% based on desirability concept.	Sobukola et al. (2013)
Corn flour and spirulina powder at different ratios of 85%–95%:5%–15%	Cereal-based extruded products are low in protein and biological value due to limited essential amino acid contents. Spirulina—cyanobacteria—is a protein-rich food source.	Corotating TSE with l/d ratio of 8:1 and die of 4 mm diameter, feed rate of 4 kg/h, temperature of 80°C–120°C, screw speed of 250–350 rpm, and feed moisture content of 13%–17% db	Based on the functional and sensory attributes, the following conditions are identified as optimum: corn flour and spirulina of 92.5:7.5, barrel temperature of 109.2°C, screw speed of 280 rpm, and feed moisture of 16% db.	Joshi et al. (2014)

(*Continued*)

TABLE 4.2 (*Continued*)
Selection and Justification of a Variety of Raw Materials Based in Different Extrusion Studies

Raw Materials	Reasons for Selection	Extrusion Conditions	Results	References
Plantain and cowpea flours at different ratios of 50%–100%:0%–50%	Highly perishable plantains are rich in iron and other nutrients used in bakery and confectionery products due to functional benefits such as treating intestinal disorders in infants. Cowpeas are rich in protein (20%–23%), starch (50%–67%), excellent source of niacin, thiamine, riboflavin, and other water-soluble vitamins, and essential minerals such as calcium, magnesium, potassium, and phosphorus. In general, addition of proteins to starches increases sites for cross-linking, thereby affecting textural quality.	TSE with single slit die of 3 × 30 mm and temperatures of 65°C–90°C for half-products; 2 die holes of 5 mm diameter, and temperature of 90°C–140°C for expanded products	Addition of cowpea has enhanced the protein content of the resultant product.	Oduro-Yeboah et al. (2014)
Corn and broad bean flour of 70:30	Broad beans are rich in lysine that can adequately complement the protein of cereals.	SSE with a screw compression ratio of 3:1 and three circular die of 1.5 mm, temperature of 80°C–100°C, screw speed of 60 rpm, and feed moisture content of 28%–34% wb	Authors found that a temperature of 100°C with a feed moisture content of 28%wb is appropriate to obtain corn-broad bean spaghetti-type pasta with high protein and DF content and adequate quality.	Gimenez et al. (2013)

TABLE 4.3
Comparison of Merits and Demerits of Different Starch Sources

Starch Sources	Starch Granule Size	Flavor	Color	Expansion	Energy Requirement for Cooking	Notes
Rice	Smallest	Bland	White	Good	Highest	Easy to flavor, most digestible when cooked
Wheat	Fairly large	Mild	White to off-white	Good	Medium to low	
Corn	Medium	Definite	Yellow	Good	Medium to high	
Barley	Medium to large	Definite	Light brown to gold	Fair	Low	
Oats	Small	Strong	Light brown	Poor	Though starch requires low energy because of high fat it increases	
Potato	Very large	Definite	Gold to light brown	Good	Low	Granules break easily, excellent binder, develops very high viscosity
Tapioca	Medium	Bland	White	Good	Low to moderate	

Source: Riaz, M.N., eds., *Extruders in Food Applications*, Technomic Publishing, Lancaster, PA, 2000. With permission.

and facilitates mixing, thus it increases the capacity and thermal energy input and reduces screw wear and mechanical energy input. Preconditioning augments flavor development and also helps in the final product texture especially in corn- and oat-based products (Huber 2000). Preconditioning is not necessary for all extrusion processes; if the process can benefit from high moisture and long residence time, then preconditioner should be included. As a rule of thumb, preconditioning can be considered to any process running at greater than 18% in-barrel moisture content.

Screws for an SSE are made up of a one piece/solid rod (Figure 4.1) and the type of screw (conical core, standard metering, short metering, venting, and mixing) selection is based on the application. Standard metering screw is the most commonly used screw in a SSE. Screw geometry influences several operations such as mixing, kneading, heat, and pressure development in addition to the capacity of a SSE. Screw compression ratio (channel depth at feed:discharge) plays a vital role in shear development and temperature profile. A compression ratio of 1:1–5:1 is used in food extrusion depending upon the product. Miller and Mulvaney (2000) suggested a compression ratio of less than 3:1 in order to exclude air from cereal products and improve heat transfer efficiency. In a twin screw, hollow screw elements/segments such as forward, kneading, reverse, and conveying can be interchangeably arranged on the screw shaft (Figure 4.1) depending upon the application.

A barrel is a housing where screw(s) is placed, and it can be either smooth or grooved. Friction is the primary mode of conveyance in a SSE. In order to facilitate the conveyance, most of the SSE barrels are grooved. Because most of the food materials are sticky in nature during cooking, twin-screw extruder (TSE) barrels are also grooved for positive movement though the conveyance is primarily due to rotational screws. It is a common practice to relate the SSE capacity with its barrel length to diameter (l/d) ratio (Giles et al. 2005). Food extruders typically have l/d of 1:1–20:1 (Harper 1981). According to Miller and Mulvaney (2000), screws with l/d between 6:1 and 9:1 with a diameter of 120–200 mm are suitable for macaroni extrusion. They also mentioned that l/d of 30 is required in order to accomplish both cooking and forming of cereals in a single extruder. Typically, SSEs have a longer barrel length than that of TSE (Martelli 1983). In general, l/d ratio has no real meaning for TSE because the feeding is controlled by other devices (Martelli 1983); however, l/d has been reported for most of the twin-screw extrusion studies as reported later (Table 4.2).

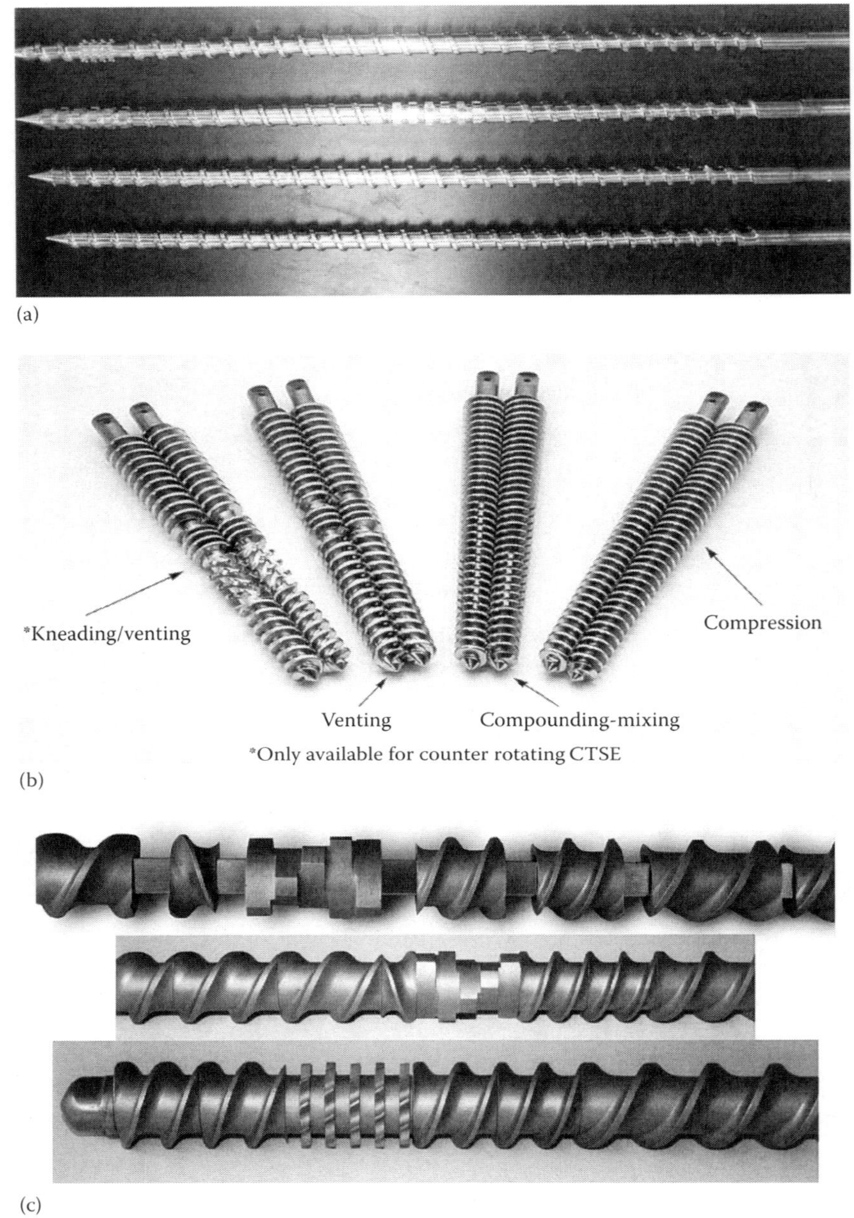

FIGURE 4.1 Screws for (a) single screw extruder, (b) screw elements, and (c) twin screw extruder. (Courtesy of C.W. Brabender, Germany.)

The die may be a single hole or multiple holes depending on the type of extruders and capacity. In general, most of the small-scale extruder/SSE may have only one opening, while large-scale TSEs have multiple openings, for example, breakfast cereal processing. The die plays an important role in deciding a product's physical properties such as density, expansion, surface texture, and final shape based on the die design, extrusion configuration, processing conditions, and blend (Senanayake and Clarke 1999, Riaz 2000) (Figure 4.2).

Most commonly used knife types in food extrusion are ridged-thin knife, flex-knife blades, and turbo knife. A ridged-thin knife is good for an expanded product (second generation) with any

FIGURE 4.2 Die inserts (with single and multiple openings) and die assembly. (Courtesy of C.W. Brabender, Germany.)

shape of 2–10 mm. Flex-knife blades and turbo knives are suitable for third-generation product/pellets (needs additional preparation before consumption). Flex-knife blades are suitable for sticky and difficult-to-cut products and can be cut 0.2–3 mm thick, whereas turbo knife is suitable for ultra-small pellets typically less than 1 mm and is achieved by having more blades operating at low speed.

According to Rokey (2000), an electric motor, a reduction gear, a torque transfer system, and a bearing support mechanism are components of a drive that provides power to rotate the extruder screw(s). The capacity and type of application dictate the size of the motor and may be as large as 400 hp (Harper 1981). The extruder capacity varies from 25 to 25,000 kg/h, correspondingly, the cost also varies from $75,000 to $750,000 (Muthukumrappan and Karunanithy 2012).

4.4 TYPES OF EXTRUDERS

An extruder can be classified based on the number of screws as shown in Figure 4.3, and further TSE is classified based on rotational direction, degree of meshing, and cleaning. Based on the method of operation/moisture content, an extruder can be termed as a wet or dry extruder. In general, most of the food extruders operate with low to intermediate moisture, usually below 40%. High-moisture extrusion is known as wet extrusion. A wet extruder typically has a deep-flighted screw that operates at low speed in a smooth barrel wherein no or little friction develops. Hot dogs, pasta, pastry dough, and certain type of confectionary are produced through wet extrusion. If the food is heated above 100°C, the process is known as *extrusion cooking* (or *hot extrusion*). Here, frictional heat and any additional heating that is used cause the temperature to rise rapidly. Cold extrusion is used to

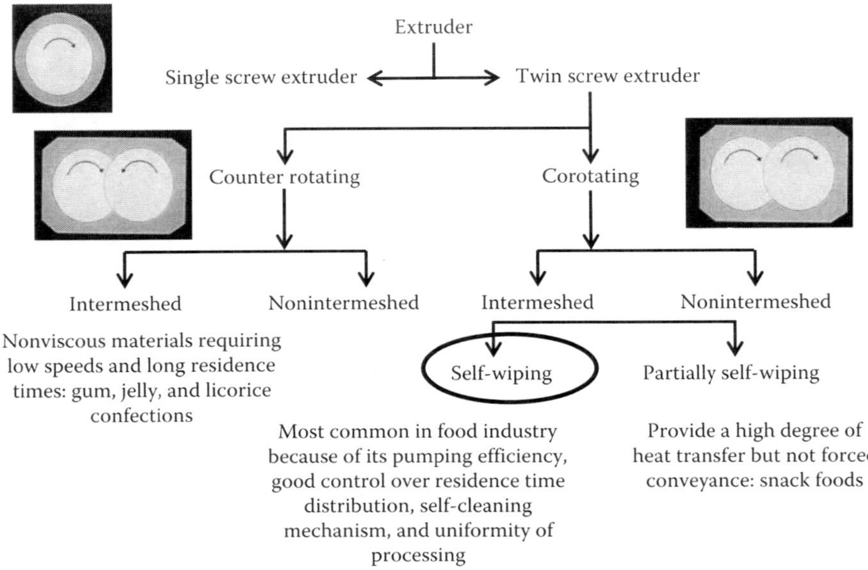

FIGURE 4.3 Classification of extruders.

mix and shape foods such as pasta and meat products wherein the temperature of the food remains at ambient. Similarly, low-pressure extrusion is used to produce, for example, liquorice, fish pastes, surimi, and pet foods wherein temperatures usually below 100°C (van Zuilichem et al. 1990). In hot extrusion, heat is supplied through external means either by steam or electrical heating in addition to internal heating generated due to the friction between the rotating screw and (grooved) barrel for raising the temperature rapidly above 100°C. Snack foods (corn chips, pretzels, and Cheetos), crisp breads, and breakfast cereals are typical products produced through hot extrusion.

In dry extrusion, all the heating is accomplished by mechanical friction without jacket heating by steam or electric. They can handle ingredients with moisture content of 10%–40% depending upon the blend and have provisions to add water during extrusion if needed. Dry extruders are SSEs with screw segments and steam locks (choke plates) on the shaft for increasing shear and creating heat. The basic difference is that more shear occurs in dry extruders to create heat than in wet extruders.

According to Riaz (2000), SSEs are mechanically very simple, and the cost is half the price of similar-sized TSEs; thus, SSEs are used wherever possible in the industry and in academic research (Figure 4.4). One problem with SSE is its poor mixing ability, therefore, premixing of ingredients is necessary before extrusion. The TSE is more flexible in operation than the SSE but it comes with additional cost (50%–150% than a SSE). TSE (Figure 4.5) can handle a wide range of particle size, viscous, oily, sticky, or very wet material that would slip in the SSE. The transport and flow mechanisms differ between co- and counter-rotating TSE that also affect the mixing. The formation of closed "C-shaped" chambers between the screws in counter-rotating intermeshing TSE acts as positive displacement pumps that minimize the mixing and the backflow due to pressure buildup. The material is transported steadily from one screw to the other in corotating extruders where the flow mechanism is a combination of dragflow and positive displacement. According to van Zuilichem et al. (1990), the corotating TSE works at a higher screw speed than a counter-rotating TSE.

Although many sophisticated screw configurations can be made with different screw elements, one can basically identify three different screw sections such as feed section, compression section, and metering section. Intermeshing means one screw penetrates the channel of the second screw that facilitates the pumping action, mixing, and self-cleaning, whereas non-intermeshing is essentially two screws that are placed parallel to each other without engaging, and they work like SSE depending on the friction for conveyance. SSE is used in the food industry since the 1960s, whereas TSE are in use since the 1980s.

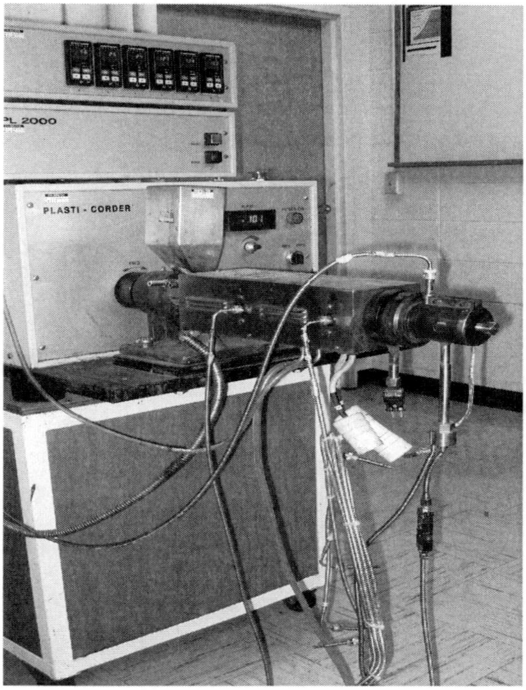

FIGURE 4.4 Lab-scale SSE.

(a) (b)

FIGURE 4.5 Industrial-scale TSE-Poly-twin BCTG-62/20D with a capacity range of 150–800 kg/h for direct expanded cereals. It has 200 kW motor with available screw speed of 1000 rpm. (a) TSE and (b) preconditioner. (Courtesy of Buhler, Switzerland.)

Based on shear stress, extruders can also be classified into low, medium, and high shear. Low-shear extruders are with smooth barrel, high l/d ratio, and deep flights, typically operate at low speed imparting a low level of mechanical energy to densify and shape the high moisture material, for example, pasta, cookies, and certain candies. Medium-shear extruders with grooved barrel and high-compression screw have the ability to handle low moisture materials and require high mechanical energy such as extruding texturized vegetable protein and breadings. High-shear extruders are with grooved barrel and screw with an increase in compression, low l/d ratio, and shallow flights, operate at high speed and mechanical energy input for producing second-generation snacks (highly expanded) with low moisture and bulk density (Riaz 2000).

Interrupted-flight extruders are also called expanders and are similar to screw press wherein revolving interrupted flight pushes the materials. An expander consists of rotating worm shaft with interrupted flights, stationary pins that are intermeshing with interrupted flights that provide high shear and turbulent mixing action. In general, heat generated by mechanical shear of ingredients supplemented with direct steam injection facilitates the process. Anderson International Company (Cleveland, OH) developed and introduced it in the late 1950s in the United States for processing pet foods and other cereal products, and later expanders were found in Brazil, Mexico, Ecuador, Switzerland, Germany, and India. Expanders now play a major role in preparing oil seeds for solvent extraction, for producing pet foods, and floating aquatic feeds.

4.5 TYPES OF EXTRUSION

Conventional extrusion is used for producing a variety of food products that can be seen in the supermarket alleys. Coextrusion and supercritical fluid extrusion are improvements over conventional extrusion with intended purposes. In coextrusion, as the name implies, at least two distinct color, texture, or flavors are extruded separately and formed together; they can be produced in a wide variety of shapes such as pillows, tubes, bars, triangles, trapeze, and stripped products. The outer layer is usually made of wheat but rice, corn, barley, rye, and oats can also be used, whereas the inner layer is filled with cream, jelly, fruit paste, and date filling. Supercritical fluid extrusion is a low-temperature and low-shear extrusion technology wherein CO_2 is introduced into the extruder at supercritical state for enhancing the product expansion (Alavi and Rizvi 2010), further discussion is beyond the scope of this chapter.

4.6 EFFECT OF EXTRUSION ON PRODUCT EXPANSION/QUALITY

Expansion is the most desirable product quality reflecting upon several factors such as feed variables (composition, feed moisture content, and particle size), extruder variables (type of extruder, screw configuration, screw compression ratio, barrel temperature profile, screw speed, die geometry, and feed rate), and their interactions as shown in Figure 4.6. Recent research results are summarized in Table 4.3 showing factors, and their interactions affect the expansion of the final product.

The maximum degree of expansion not only depends on the starch content but also the amylose and amylopectin content. In general, the formulation with a starch level of 60% or less would result in slightly expanded hard and crunchy product after frying or popping (e.g., corn chip snack). When the starch level exceeds 60% of the formulation, the product would expand well with light and crispy texture. Starch content of pure starches, whole grains, seeds, and germs are 100%, 65%–78%, 40%–50%, and 0%–10%, could expand a maximum of 5-fold, 4-fold, and from 1.5- to 2-fold, respectively. According to Riaz (2000), the minimum starch content for expansion is 60%–70%; he also reported that formulation containing 40%–50% starch would result in a normal expansion of two to three folds (Riaz 2005). Some cereals have soft and floury endosperm wherein starch granules and protein layers are loosely bound, while others have hard endosperm (hard wheat, hard durum, vitreous flint maize, and some varieties of barley) wherein the starch and protein layers are strongly bound to form hard particle flour. In milling, soft flour would easily break down into a mixture of starch and protein, whereas hard flour requires more energy; similarly soft and hard flour require low and high mechanical energy, respectively, in the extruder. According to Guy (2001), finely ground hard endosperm with low moisture would yield high expansion; some of the hard endosperm material may be replaced if low to medium expansion is required. Waxy and pregelatinized waxy starches would be better for expanded product, whereas sweet potato starch can be included for extruded pellets (Pszczola 2010). Starch is made of linear amylose and branched amylopectin, and they contribute to mechanical strength and expansion. The ratio of amylose to amylopectin influences both

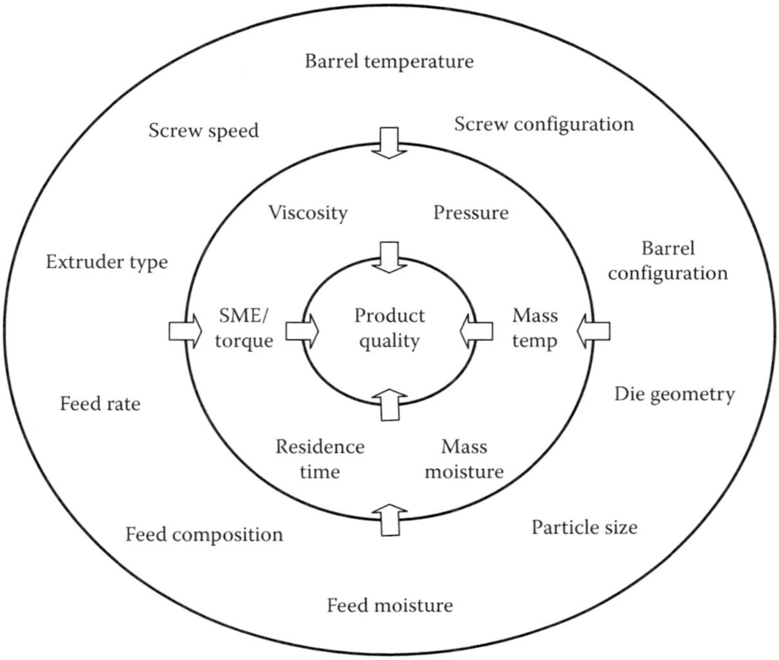

FIGURE 4.6 Interrelationship among extruder and ingredient variables. (From Muthukumarappan, K. and Karunanithy, C., Extrusion process design, in: Ahmed, J. and Rahman, S., eds., *Handbook of Food Process Design*, Wiley-Blackwell, pp. 710–742, 2012, ISBN: 978-1-4443-3011-3. With permission.)

radial (amylopectin) and longitudinal (amylose) expansion. High amylose needs more energy for gelling/gelatinization and retrogradation, which results in hard and dense texture (Agama et al. 2004). High amylopectin starch contributes to a high degree of gelatinization and provides a light and airy expanded texture (Kokini et al. 1992, Guy 2001). Sometimes starch is modified using acidic or alkaline conditions that decrease molecular weight and reduce expansion (Kervinen et al. 1984). The addition of sodium bicarbonate and urea to starches has a negative impact on expansion (Chinnaswamy and Hanna 1988, Lai et al. 1989, Pan et al. 1998).

In general, starch plays a major role in the expansion as mentioned earlier, while protein, sugar, lipid/fat, and fiber act as diluents (Jain 2007). According to Berrios (2006), legume as a protein source can be used to develop products of high nutrition and low calorie with good expansion. The addition of protein to starches competes for water thereby reduces the extensibility of starch polymer (Derby et al. 1975) and increases sites for cross-linking that affects the texture (Aboagye and Stanley 1987); further the effect of protein depends on the type of protein and their concentration (Faubion and Hoseney 1982, Aboagye and Stanley 1987). Soy protein and whey protein are most widely reported in the literature, that too with conflicting results. Faubion and Hoseney (1982) obtained a more expanded product from wheat starch with 10–80 g/kg of soy protein isolate than that of pure starch; however, further increase of soy protein isolate to 100 g/kg decreased expansion. When soy flour is added to wheat flour, the expansion depends upon the moisture content (Zasypkin and Lee 1998). Fernandes et al. (2002) showed an increase of soy flour from 200 to 300 g/kg of mixture with corn grit decreasing the expansion in half. Recently, Li et al. (2005) found that the addition of soy flour (0–400 g/kg) to cornmeal increases expansion. Whey protein concentrate or isolates can be added up to 15% to starch for producing expanded snacks with enhanced nutritional values. Texturization is a process that reduces the water binding capacity of proteins and enables production of texturally firmer and crispier products and is easy to break; increases the addition of protein between 25% and 35% (Onwulata et al. 2001a,b). Recently, Onwulata (2010) developed prototype products such as pretzels, corn chips, and tortilla chips using WPC 80 and WPI at 30 g/100 g of cornmeal.

Type and concentration of sugar influences expansion and shrinkage. Addition of monosaccharide sugars with maize grit lowers expansion and increases shrink than that of disaccharides (Fan et al. 1996). Inclusion of sugar reduces starch in the formulation, reduces specific mechanical energy (SME), die pressure, and melt temperature, and increases viscosity and shrinkage thereby lowering the degree of starch gelatinization (Sopade and Le Grys 1991, Fan et al. 1996, Carvalho and Mitchell 2000). According to El-Samahy et al. (2007), increasing sugar content in formulation increases viscosity, reduces the availability of water for starch gelatinization since sugar competes for water, and thus decreases the available steam for expansion.

The effect of lipid on expansion depends on the concentration. In general, lipid acts as lubricant thereby reduces friction between dough and screw/barrel (Colonna and Mercier 1983, Lin et al. 1997, Guy 2001), decreases mechanical energy input/shear stress (Ilo et al. 2000), prevents break down of starch (Lin et al. 1997), affects dough rheology in the barrel (Schweizer et al. 1986), and retards the degree of gelatinization (Schweizer et al. 1986, Ilo et al. 2000); all of these affect expansion. According to Colonna et al. (1998), the conditioning moisture content can be reduced while increasing the lipid content in order to maintain the expansion of second-generation products. Lipid content up to 3% does not affect the expansion; further increase of lipid above 5% considerably reduces expansion (Harper 1994).

The health benefits of dietary fiber (DF) are well documented; however, its inclusion into the formulation of extruded products have several adverse effects such as reduced expansion and crispiness (Lobato et al. 2011), increased hardness, decreased consumer acceptance (Hsieh et al. 1989, Lue et al. 1990, Jin et al. 1995), compact/dense products that are tough, not crisp, and have an undesirable texture (Lue et al. 1991). The effect of fiber depends on the concentration. Several mechanisms can explain the reduction in expansion due to the addition of fiber: most of the fiber would remain firm and stable without change in size that tends to rupture air cell walls (Riaz 2000, Steel et al. 2012); fiber interferes with bubble formation reducing extensibility of cell wall and causes premature rupture at critical thickness of steam wall facilitating easy escape of steam (Guy 1985); reduces starch content in the formulation (Colonna et al. 1998); competes for water with protein and starch, further the water binding capacity of fiber prevents the water loss at die exit (Camire and King 1991); and the presence of fiber causes incomplete gelatinization of starch (Camire and King 1991). The effect of inulin addition depends upon the concentration and starch type. According to Niness (1999), inulin addition up to 300 g/kg of breakfast cereal contributes to expansion as well bowl life. Later, Ascheri et al. (2006) found that addition of inulin up to 100 g/kg of rice flour had a positive contribution on expansion; however, further inulin increase had a negative effect on expansion.

The effect of feed moisture on expansion depends upon the feed composition such as starch, protein, sugar/salt, fat, and fiber. Feed moisture helps not only in starch gelatinization but also viscosity of the melt/dough; however, low moisture may result in undesirable nutritional quality (Huber 2000). In general, low feed moisture increases the viscosity of the melt (Chang and Ng 2011) that confines the flow and increases residence time, thus enhancing the degree of gelatinization (Chinnaswamy and Hanna 1990); further pressure differential would be greater resulting in more expansion (Singh et al. 2007). Addition of moisture acts as lubricant/plasticizer (Steel et al. 2012) and reduces the friction, decreases shear rate, torque/SME, viscosity (Guy and Horne 1988, Hsieh et al. 1990, Steel et al. 2012), increases elasticity due to plasticization of the melt (Ding et al. 2006, Mahasukhonthachat et al. 2010), reduces starch gelatinization (Kokini et al. 1992), inhibits (Steel et al. 2012), and retards the bubble growth (Ding et al. 2005), thus expansion decreases. Increase in feed moisture would address moisture distribution among protein, sugar, and fiber content in the formulation in addition to starch that would facilitate expansion, otherwise low moisture is preferred for expansion. Particle size also influences moisture distribution, heat transfer, viscosity, and the final product quality.

Temperature is another critical factor that affects product expansion. In general, temperature should be higher than 100°C for the formation of superheated steam and moisture flash-off or increase in water vapor pressure at the die exit to obtain an expanded/puffed product (Panswat 2007,

Steel et al. 2012). An increase in barrel temperature lowers the viscosity/rheology of the melt that favors the bubble growth (Chaiyakul et al. 2009). Expansion increases with an increase in temperature. When the feed moisture is close to 20% (wb,) it lowers viscosity and permits rapid expansion due to greater water vapor pressure (Steel et al. 2012). Beyond a critical temperature (depends on type of starch and moisture content), expansion decreases with an increase in temperature facilitating dextrinization, excessive softening, weakening of starch structure/structural degradation of starch melt to a level where it is not able to withstand the high vapor pressure, resulting in collapse (Colonna et al. 1998, Moraru and Kokini 2003). Temperature is a critical factor for protein cross-linking reactions in the melt; lower than 90°C would impede layer formation as well as expansion (Cheftel et al. 1992); increasing temperature from 140°C to 180°C correspondingly decreases disulfide linkages of soy protein isolates (Areas 1992). In gist, the effect of temperature on expansion depends on composition, moisture, protein, fiber, and so on.

Screw speed not only relates to the rate of shear and shear history of the materials (Smith 1992, Ozer et al. 2004), residence time (van Zuilichem et al. 1988, Colonna et al. 1998), and rheology/viscosity (Blanche and Sun 2004), but also torque and SME that affect expansion (Guy and Horne 1988, Anderson and Ng 2003). Screw configuration (mixing elements-kneading elements, reverse screw elements) influences torque, SME (Yam et al. 1994), die temperature (Choudhary and Guatam 1998), residence time (Altomare and Ghossi 1986), degree of fill (Yam et al. 1994), rheology, physiochemical changes/product transformation, or starch gelatinization (Gogoi 1994, Gogoi et al. 1996, Choudhary and Guatam 1998) and ultimately expansion (Sokhey et al. 1994, Ozer et al. 2004).

Die geometry and diameter has a great effect on expansion. A decrease in die diameter increases the die temperature, and SME results in high degree of gelatinization and expansion (Molina et al. 1978, Janes and Guy 1995).

4.7 EFFECT OF EXTRUSION ON NUTRITION

In general, *protein* undergoes structural unfolding and/or aggregation during extrusion due to moist heat or shear, which leads to insolubilization and inactivation. A single-screw extrusion of dehulled full-fat soybean, defatted soy flour, or corn–soy blend decreases nitrogen solubility and inactivates trypsin inhibitors, thereby improving protein efficiency ratio on rats, pet animal, and infants (Cheftel 1986). Protein quality and nutritional status is usually assessed through protein digestibility (PVID). According to Ainsworth et al. (2007), an increase in screw speed from 100 to 200 rpm increases PIVD values from 76.2% to 89.6% due to increase in shear that opens up protein structure, exposing new sites for enzyme action; however, the addition of brewer's spent grain does not increase the PIVD values probably as antinutritional factors/trypsin inhibitors might block or reduce digestion during analysis. Further increase in screw speed does not have any effect of PIVD since protein structure is already opened up. Twin-screw extrusion of wheat flour and cauliflower (up to 10%) blend decreases PIVD values as a result of a combination of shearing, heat, and pressure during extrusion; the formation of enzyme-resistant, disulfide-bonded oligomers (Stojceska et al. 2008).

As indicated in Figure 4.6, lysine loss depends on raw material, feed moisture, pH, screw compression ratio, screw speed (energy input), temperature, die design, feed rate, torque, and pressure (Noguchi et al. 1982, Bjorck and Asp 1983, Bjirck et al. 1983, Bjorck et al. 1984b, Cheftel 1986, Asp and Bjorck 1989, Camire et al. 1990, Prudencio-Ferreira and Areas 1993, Iwe et al. 2001, Ilo and Berghofer 2003, Chaiyakul et al. 2009, Filli et al. 2011). In general, cereal is a major component in the feed formulation that limits lysine among the essential *amino acids*, thus its retention during extrusion becomes important. According to Filli et al. (2011), lysine content increases with cowpea in the formulation and decreases with feed moisture and screw speed. They also observed tyrosine, phenylalanine, isoleucine, valine, and leucine increases with the amount of cowpea in the formulation. Earlier, Iwe et al. (2004) found that the addition of sweet potato with soy flour increases lysine retention (due to lowering lysine content in the formulation), which increases with screw speed (80–140 rpm) and a reduction of die diameter (10–6 mm) due to reduction in residence time that limits the heat treatment duration. However, Bjorck et al. (1984b) reported that an increase in screw speed decreases lysine content and biological value, while feed rate has opposite effects. It is well known that an increase in feed rate reduces mean residence time, thereby reducing heat transfer to the melt. It is well established that screw speed increases not only the rate of shear/shear force but also the temperature of the melt. Their observation indicates that though increase in screw speed reduces the mean residence time, rate of shear and increase in temperature negatively affect lysine content. Iwe et al. (2001) observed that increase in energy input to extruder (screw speed) significantly reduces the availability of arginine (21%), histidine (15%), aspartic acid (14%), and serine (13%) when maize grit is extruded at 135°C–160°C.

A single-screw extrusion (130°C and 80 rpm, 15% wb, screw compression ratio of 3:1 with die diameter of 3 and 5 mm) of maize significantly reduced contents of the essential amino acids isoleucine, leucine, lysine, threonine, and valine when compared to their original flours; however, the contents of histidine, methionine, phenylalanine, and tryptophan were not reduced (Paes and Maga 2004). Among the essential amino acids, die diameter had a significant effect on leucine. A larger die (5 mm) played a protective role not only for leucine but also for other amino acids due to low shear rate and low residence time (Paes and Maga 2004). Literature survey reveals that moisture content effects on lysine retention are conflicting. For high retention of lysine, Cheftel (1986) suggests moisture content above 15%, others suggest a moisture content of 15%–25% for improving lysine retention (Noguchi et al. 1982, Bjorck and Asp 1983, Asp and Bjorck 1989). These authors attributed the addition of moisture lowers the shearing, dissipation of mechanical energy, viscosity, and product temperature, thus causing high retention of amino acids. A twin-screw extrusion at 150°C–180°C and screw speed of 57–81 rpm with 13%–17% (wb) of maize grits (Ilo and Berghofer 2003) showed a range of loss of lysine (11%–49%), arginine (7%–16%), cysteine (8%–25%), methionine and tryptophan (0%–14%); whereas 82%, 86%, 71%, and 88% arginine, cystine, methionine, and tryptophan, respectively, was reported for extrusion of wheat-based biscuit (Björck et al. 1983). Though increase in screw speed increases shear stress, and product temperature correspondingly reduces the mean residence time; there was no effect on amino acids (Noguchi et al. 1982, Ilo and Berghofer 2003, Iwe et al. 2004). In addition, Asp and Bjorck (1981) found a correlation between lysine loss and screw speed, and they attributed it to the indirect effect of starch hydrolysis at high shear. Prudencio-Ferreira and Areas (1993) and Chaiyakul et al. (2009) reported extrusion conditions such as feed moisture (30%–40% and 20%–30%) and temperature (140°C–180°C and 150°C–180°C) had no influence on the loss of cysteine and methionine from soy grit and a blend of glutinous rice and vital gluten wheat and soy grit, respectively. Due to the complex nature of extruder conditions, nutritional changes including amino acids loss and product quality might not be related to a single factor. Ilo and Berghofer (2003) have developed first order kinetics models for predicting the loss of most unstable amino acids such as lysine, cysteine, and arginine. They found that reaction rate constant is highly dependent on product temperature; lysine is more sensitive to temperature compared to other amino acids.

Maillard reaction is common in foods that occurs between free amino groups of protein and carbonyl groups of reducing sugars leading to browning, flavor production; reducing the availability of amino acids and protein digestibility. Free amino acids are much more sensitive to damage during extrusion cooking than those in proteins. Lysine is the most reactive amino acid since it has two free amino groups (O'Brien and Morrissey 1989). Several researchers have attempted to relate the Maillard reaction and browning/discoloration to loss of lysine (Noguchi et al. 1982, Bjorck and Asp 1983, Cheftel 1986, Asp and Bjorck 1989, O'Brien and Morrissey 1989). Browning of soy–sweet potato extrudate increases with increase of sweet potato in the formulation, screw speed and decrease in die diameter (Iwe et al. 2004). Increase in screw speed increases the browning of soy–sweet potato extrudates because only degraded polysaccharides can take part in the browning process (Areas 1992, Mitchell and Areas 1992). Loss of amino acids can be minimized through utilization of less reactive sugars, and it is a potential area for future research (Singh et al. 2007).

In general, severe extrusion conditions such as temperature above 180°C, screw speed more than 100 rpm, feed with reducing sugars (>3%), and feed moisture less than 15% result in extensive lysine loss and nutritional damage. Therefore, Cheftel (1986) suggested avoiding extrusion above 180°C with a feed moisture below 15% (even though subsequent oven-drying is necessary) and avoiding the presence of reducing sugars during extrusion for keeping lysine losses within the 10%–15% limit accepted in bread baking or in drum cooking and drying of instant flours.

According to American Association of Cereal Chemists (2001), "*Dietary fiber* is the edible parts of plants or analogous carbohydrates that are resistant to digestion and absorption in the human small intestine with complete or partial fermentation in the large intestine." According to the Codex Alimentarius, DF is defined as carbohydrate polymers with 10 or more monomeric units, which are not hydrolyzed by the endogenous enzymes in the small intestine of humans (ALINORM 09/32/26 2009). Based on the solubility in water, DF can be divided into soluble dietary fiber (SDF) and insoluble dietary fiber (IDF). According to Elleuch et al. (2011) and Gajula et al. (2008), IDF such as cellulose, lignin, and hemicelluloses are characterized by their porosity, low density, and ability to increase fecal bulk, whereas SDF includes oligosaccharides, pectins, β-glucans, and galactomannan gums, which is characterized by its capacity to increase viscosity, reduce glycemic response, plasma cholesterol (Tosh and Yada 2010), promoting satiety (Brennan et al. 2008), has a higher capacity to form gels and acts as an emulsifier, further it is easy to incorporate into foods. DFs decrease intestinal transit time, decrease postprandial blood glucose and insulin level, reduce total and low-density lipoprotein cholesterol level in blood, buffer excessive acid in the stomach, and increase stool bulk (Méndez-García et al. 2011). According to Martinez-Flores et al. (2008), consumption DF prevents health problems such as diverticular disease, cardiovascular disease, and colorectal cancer. DFs have been linked to control of weight and obesity (Gordon 1989, Pokitt and Morgan 2005), constipation, diabetes (Hampl et al. 1998, Tapola et al. 2005), and blood pressure (He et al. 2004). β-glucan found in oat has several benefits such as reducing serum cholesterol levels/LDL associated with coronary heart disease (FDA 1997, Brown et al. 1999), serum glucose and insulin after a meal (Wood et al. 1994, Tappy et al. 1996), promotes satiety (Beck et al. 2009), and modulates immune function (Volman et al. 2008). Various jurisdictions of the United States were the first to approve oat's health claim (FDA 1997); recently the European Food Safety Authority endorsed a claim that the consumption of oat β-glucan can help to maintain normal blood cholesterol (EFSA panel 2009). Scientists identified DF as "the seventh basic nutrient," and its importance has attracted a wide range of research and led to the development of a large and potential market for fiber-rich products and ingredients. Because of several benefits, DF becomes the third most sought-after health information in supermarkets in countries like India, Australia, Western Europe, and North America (Mehta 2005).

Whole wheat flour and wheat bran are the most commonly used source of DF for the cereal-based food industry; in order to enhance the fiber content of products, corn bran (Artz et al. 1990); oat bran (Seiz 2006, Zhang et al. 2011); rice bran (Lima et al. 2002); soy hull; citrus fruits (Seiz 2006); fruit powders such as apple and cranberry (Stojceska et al. 2010); vegetables such as carrot, beetroot, and cauliflower (Stojceska et al. 2008, 2010); gums such as fenugreek gum, guar gum, and locust bean gum (Chang et al. 2011, Ravindran et al. 2011); and food industry residues such as okara (Li et al. 2012) are used, and nowadays there is a trend to find new sources of DF. In order to reap the benefits, an ideal DF is expected to have up to 20–30 g SDF/100 g (Li et al. 2012). In 2001, the Food and Nutrition Board of the Institute of Medicine established a recommendation for TDF daily intake of 38 and 25 g for adult men and women, respectively (IOM 2002); Slavin (2003) reported it is 30 and 21 g of fiber per day for men and women over 50 years.

Conflicting findings have been reported about the effect of extrusion on DF. Camire and Flint (1991) and Camire et al. (1997) found extrusion of potato peels increases TDF and SDF, whereas Artz et al. (1990) observed extrusion of a blend of cornstarch and corn fiber had no significant changes in SDF and IDF; Fornal et al. (1987) reported a reduction in fiber content when mixtures of cereal starches were extruded. The increase in SDF is usually at the expense of IDF due to fragmentation or other type of thermo-mechanical decomposition of cellulose and lignin that are major components of insoluble fiber. Twin-screw extrusion of wheat flour and bran redistributes part of the IDF fraction to SDF, thereby SDF increases; however, the decrease in IDF is higher than the accompanying increase in SDF as a portion of the IDF fragments to lower molecular weight fragments and possibly converts to sugars (Gajula et al. 2008).

Extrusion precooking of the flours did not improve the consumer acceptability of cookies and tortillas; however, it did improve their DF profile by increasing the SDF significantly. For example, although extrusion has been shown to increase the TDF content of wheat flour (Theander and Westerlund 1987), it negatively affects the nutritional value of proteins due to Maillard reaction and also leads to loss of heat-labile vitamins (Singh et al. 2007). Apart from improvements in functionality and DF profile due to thermal/mechanical treatment of fiber-based ingredients, the sensory characteristics of the final product could also be positively affected. Bernnan et al. (2008) found that addition of DF such as wheat bran, fine guar gum, inulin, high maize 1043, and swede fiber (5%–15%) into white wheat flour, maize grit, and oatmeal formulation reduces the amount of readily digestible starch components of breakfast products, increases the amount of slowly digestible carbohydrates, and may lead to an increased feeling of satiety.

Single-screw extrusion of orange pulp (83°C–167°C, 126–190 rpm, 22%–38% wb, l/d of 20:1, and screw compression ratio of 3:1) lowered IDF by 39% and increased SDF by 80% due to redistribution of IDF as SDF (Larrea et al. 2005b). They identified the barrel temperature as the most influencing variable followed by the interaction of temperature and screw speed that affects IDF and SDF. Larrea et al. (2005b) showed that the TDF in orange pulps decreases with higher barrel temperatures and lower moisture when screw speed is fixed at 160 rpm. Single screw extrusion of lemon residues (50°C–110°C, 3–37 rpm, 33%–67% wb) increased SDF efficiently and decreased IDF, particularly at a temperature of 110°C, screw speeds of 10 and 20 rpm, with a moisture content of 40% and 50% (Méndez-García et al. 2011).

Twin-screw extrusion (80°C–120°C, 250–350 rpm, solid feed rate of 20–25 kg/h, and water feed rate 9%–11%) of a formulation containing cauliflower (5%–20%) decreases the level of DF due to change in IDF to SDF; heat and moisture/moist heat solubilizes and degrades pectic substances thereby DF decreases (Stojceska et al. 2008). The same group have demonstrated that extrusion increases the level of DF in non-gluten-free, ready-to-eat expanded snacks made from cereal and vegetable coproducts (Stojceska et al. 2008, 2009). Twin-screw extrusion (80°C and 120°C, 200 rpm, feed rate of 25 kg/h, and water feed rate 12%–17%) of a blend containing wheat flour or cornstarch substituting 10% brewer's spent grain from barley or red cabbage shows different trend for DF (Stojceska et al. 2009). Extrusion increases TDF for most of the blends except wheat flour with red cabbage where it decreases due to loss of soluble fiber components into processing water. Authors found that amylose content influences resistant starch formation, thereby TDF. They also observed an increase in moisture up to 15% increases TDF, further increase to 17% decreases TDF. Again the effect of moisture content on TDF depends on the source of starch; TDF increases with moisture for wheat flour and brewer's spent grain, whereas it decreases with moisture for cornstarch and brewer's spent grain.

Twin-screw extrusion (80°C/80°C–150°C, 80 rpm, solid feed rate of 15–25 kg/h, and water feed rate of 12%) of a formulation consisting of rice, milk powder, cornstarch, potato starch, soy flour including fruits (apple and cranberry), vegetables (carrot and beetroot), and teff flour increased TDF depending upon the source of DF (Stojceska et al. 2010). The increase in TDF in teff formulation due to high starch of teff and formation of resistant starch increases IDF. The increase in TDF of fruit and vegetable formulations attributed to the complexes formation between polysaccharides and other components such as proteins and phenolic compounds, which are measured as fiber. The effect of extrusion parameters such as barrel temperature, screw speed, and feed rate on TDF depends upon the type of dietary sources (Stojceska et al. 2008, 2010).

Twin-screw extrusion (80°C/120°C, 250 rpm, 13% wb, 20 kg/h) of a formulation containing wheat flour, corn, potato starch, milk powder, and fruit powder increases IDF and SDF, thus total

DF (Potter et al. 2013). Increase in IDF is attributed to the formation of "resistant starch" whereas increase in SDF is attributed to soluble fiber changing into insoluble fiber. According to Esposito et al. (2005), the break down of glycosidic bonds in polysaccharides due to mechanical stress during extrusion releases oligosaccharides thereby IDF increases. Twin-screw extrusion of different ratios of corn grit and wheat bran formulation with varying moisture and temperature lowered the TDF and IDF and enhanced SDF (Sobota et al. 2010).

The addition of gums such as guar gum, locust bean gum, and fenugreek gum (5%–15%) to a rice and pea formulation increases DF; however, twin-screw extrusion decreases IDF and increases SDF (Ravindran et al. 2011) due to the break down of insoluble fibers into soluble fibers, formation of resistant starch, and enzyme-resistant glucans through transglycosidation (Vasanthan et al. 2002).

Twin-screw extrusion of oat bran (100°C–160°C, 150 rpm, 10%–30% wb, 18 kg/h) increases SDF; feed moisture has negative influence on SDF, whereas temperature has a positive effect (Zhang et al. 2011). Earlier Vasanthan et al. (2002) attributed the changes of DF profile during extrusion of barley flour to a shift from IDF to SDF, the formation of resistant starch and "enzyme-resistant indigestible glucans" formed by transglycosidation.

Okara is a solid residue from processing soy into protein isolate, soymilk, and tofu and is rich in DF (50–60 g/100 g db) that too mainly IDF, which makes it difficult for it to be used as fiber-fortified food products (Li et al. 2012). Twin-screw extrusion of okara with CO_2 producing agents (0–45 g/100 g) such as citric acid and sodium bicarbonate powder (50°C/70°C/110°C/150°C–170°C, 180–200 rpm, 30%–40% db, 12 kg/h) increases SDF. The extent of SDF increase depends on temperature and pressure in the extruder barrel; high temperature and pressure facilitates the break down of polysaccharides' glucosidic bonds. Recently, Robin et al. (2012) published an excellent review on "dietary fiber in extruded cereals: limitations and opportunities" in *Trends in Food Science & Technology*, and it is worth to have a look at it.

Vitamins and minerals are essential nutrients to perform numerous roles in the body including healing wounds, bolstering immune system, repairing cellular damage, and so on. Vitamins differ in their chemical structure, thereby their stability too. Extrusion conditions such as type of extruder, type of food, temperature, screw design, screw speed, moisture content, feed rate, die design, throughput, and storage conditions namely oxygen, light, temperature, pH, moisture, temperature, and storage time influence vitamin losses of extruded products (Killeit 1994). In general, vitamin losses are minimal in cold extrusion; short residence time and rapid cooling of extrudate in hot extrusion would minimize vitamin losses (Fellows 2000). According to Ilo and Berghofer (1998), temperature and screw speed have a positive influence on thiamin losses due to high thermal sensitivity and high mechanical action of shear stress. They attributed the negative influence of feed rate and feed moisture on thiamin destruction to reduction in residence time, and reduction in viscosity leads to less mechanical energy dissipation that affects the temperature profile. Low feed moisture hastens the loss of vitamins; therefore, vitamins and heat-sensitive nutrients are usually added post extrusion especially when moisture is low (Huber 2001).

Thiamin is a small and heat-sensitive molecule that could be destroyed easily during extrusion since it is a high-temperature and short-time process. Several researchers have demonstrated that thiamin loss depends on both thermal and mechanical effects. A school of thought in the late 1980s was that shear-induced local temperature raises increases loss at higher screw speeds (Cheftel 1986, Guzman-Tello and Cheftel 1987, Camire et al. 1990). When this increase in temperature was removed by cooling, the barrel shows that mechanical effects must play a role in destroying thiamin (Harper 1988, Ilo and Berghofer 1998). Earlier Ilo and Berghofer (1998) evaluated three different models for the reaction-rate constant as a function of product temperature just before the die, moisture content, and screw speed or shear stress when maize grits extruded in a counter-rotating TSE (140°C–200°C, 65–81 rpm, 11.8%–14.2% wb, 31–57 kg/h). They found that models that had the effect of screw speed or shear stress fit the data better (r^2 = 0.965, 0.979, respectively) than the model that considered only time and temperature effects (r^2 = 0.809), indicating that shear effects on thiamin retention are significant. Later, Cha et al. (2003) extruded wheat flour and thiamin blend with 25% moisture content in a corotating and intermeshing TSE at 50°C/85°C/115°C/135°C/145°C and 100–300 rpm and found that the mechanical effects had a predominant effect on thiamin loss (90%–94%) than thermal effects. It is important to take into account the average shear rate and residence time for predicting thiamin loss; accordingly, the authors proposed a model where the linear relationship between thiamin retention and shear history shows that shear history can be a more accurate indicator of thiamin retention than screw speed or SME since neither screw speed nor SME account for the duration of shearing (Cha et al. 2003). Recently, Emin et al. (2012) extruded native maize starch (135°C–170°C, 300–800 rpm, and 10% wb) in a TSE wherein β-carotene added after plasticization of starch increases β-carotene by 10% due to a shorter exposure time to thermal and mechanical stress; they found that the β-carotene loss is mainly due to mechanical stress rather than thermal stress.

Vitamin E, tocopheral is an essential nutrient that functions as an antioxidant; extrusion significantly reduced it in fish and peanut containing half-products (Suknark et al. 2001). The addition of α-tocopheral to rolled oatmeal before extrusion inhibits the formation of lipid oxidation products such as pentanal and hexanal during 12 months of storage at room temperature (Guth and Grosch 1994). Vitamin C, ascorbic acid, is an organic acid that can function as acidulant and could improve the pigment retention; thus, fortification with ascorbic acid may improve marketability. Corotating twin-screw extrusion of cornmeal, ascorbic acid, sugar, and blueberry concentrate either does not protect anthocyanin or inhibits browning reactions (Chaovanalikit et al. 2003).

In general, *minerals* are heat stable; meaning they can withstand the extrusion temperature. Minerals absorption can be improved in extrusion through reducing other factors inhibiting absorption such as phytates and condensed tannins. A school of thought is that iron content of the extrudate increases through wearing of screw and barrel. Earlier, Camire et al. (1993) reported extrusion of potato peel increased iron content to the tune of 130%–180%. Though iron can act as a catalyst in favoring lipid oxidation; excess iron (over the hydroperoxide content) can react with free radicals and convert them to ionic form, inhibiting free radical propagation, thus reducing iron's catalytic effect. Accordingly, Camire and Dougherty (1998) extruded cornmeal with synthetic phenolic compounds in a single-screw extrusion at 90°C/105°C/125°C/140°C/150°C, 70 rpm, 23% db, and 30 kg/h; they reported extrusion did not increase iron, and iron sequestration did not contribute to the antioxidant effects of phenolic; however, they haven't reported the color values of the extrudates. Fortification of minerals prior to extrusion results in a dark color due to the formation of iron complex with phenolic compounds and reduces expansion and increases lightness when calcium hydroxide is added (Singh et al. 2007).

Phytic acid/phytates, trypsin inhibitor, tannin, lectin, and hemagglutinins are several *antinutritional factors* found in several food ingredients that can be partially or completely inactivated using extrusion since most of them are thermolabile, thus nutritional value of the extrudates is enhanced. A single-screw extrusion of pinto bean flour at 140°C–180°C, 150–250 rpm, and 18%–22% (wb) moisture completely destroyed trypsin inhibitors (Balandrán-Quintana et al. 1998). Similarly, Alonso et al. (2000) demonstrated the complete destruction of trypsin inhibitors during extrusion of faba and kidney beans. Later, Anton et al. (2009) conducted twin-screw extrusion of cornstarch with varying levels (15%–45%) of navy/small red bean flour (30°C/80°C/120°C/160°C/160°C, 150 rpm, 22% wb, and 1.8 kg/h) and observed a significant reduction in phytic acid and trypsin inhibitors. The effect of extrusion on phytic acid depends on the type of beans, that is, 48%–68% loss for navy bean and 35%–55% for small red beans; the authors attributed hydrolysis of inositol hexaphosphate into lower molecular weight forms or formation of insoluble complexes involving phytate, whereas the type of beans had no influence on the destruction of trypsin inhibitors since it is completely destroyed. Recently, Olapade and Aworh (2012) conducted a single-screw extrusion of a blend consists of acha and cowpea (60–70:30–40 with moisture of 18%–25% wb) at a barrel temperature of 140°C–160°C, and he found an increase in temperature reduces trypsin inhibitor, while increase in feed moisture lowers the trypsin inhibitor to the range of 76.0%–92.1%. Earlier, extrusion of sweet potato and soybean mixture revealed that feed composition and screw speed had a significant impact on trypsin inhibitor destruction (Iwe 2000). According to Harper (1993) as reported by Balandrán-Quintana et al. (1998), a reduction of 70% or more in trypsin inhibition activity in beans is adequate for nutritive quality.

Tannin is a main antinutritional factor present in sesame meal through formation of complexes with the available protein thereby affecting the digestibility. A single-screw extrusion of sesame meal (63°C–97°C, 63–97 rpm, and 31%–48% wb) reduced tannin by 47%–61% depending upon the operating conditions (Mukhopadhyay and Bandyopadhyay 2003). Lectin is another antinutritional factor in beans; its presence with trypsin inhibitors in the diet would increase fecal nitrogen excretion in order to increase nutritional values, thus they have to be reduced. According to Delgado et al. (2012), extrusion of corn (70%) and bean (30%), flour at 150°C, 25 rpm and 20% moisture completely destroyed trypsin inhibitor (from 10,857 to 0 TIU/g) and lectin activity (from 640 HU/mg protein). These studies indicate that high temperature, low moisture, and intense mechanical stress would completely destroy antinutritional factors.

Preconditioning of peas, chickpeas, and faba beans at 70°C–100°C with moisture more than 20% followed by twin-screw extrusion at 70°C/95°C–110°C, 380 rpm, 6.6–7.0 kg/h, and water feed rate of 12.5–15 kg/h greatly reduces phytate, tannin, and trypsin inhibitors (Adamidou et al. 2011). The authors suggested preconditioning would be a valuable tool to partially eliminate antinutritional factor especially trypsin inhibitors from legumes.

Antioxidants technically apply to molecules reacting with oxygen/compounds that can counteract the damaging effects of oxygen in tissues; it is often applied to molecules that protect from any free radical (Velioglu et al. 1998). The consumption of free radicals and oxidation products is considered as a risk factor for cancer and cardiovascular disease (Namiki 1990). Antioxidants can be added before or during extrusion, the combination of high temperature and pressure drop at the extruder die can cause them to volatilize, resulting in loss of effectiveness (Artz et al. 1992). According to Velioglu et al. (1998), possible mechanisms involve direct reaction with and quenching free radicals, chelating transition metals, reducing peroxides, and stimulating the antioxidative defense enzyme activities through which antioxidant contributes to the beneficial effects of grain, colored beans, fruits, and vegetables. Phenolic compounds such as proanthocyanidins/procyanidin and flavonols/kemferol are responsible for antioxidant activity (Skerget et al. 2005). Procyanidins can be found in fruits and vegetables, abundantly in cocoa, grape, and berries. According to Joshi et al. (2001), procyanidins not only possess antioxidant activity but also anti-inflammatory, antibacterial, and anti-arthritic activities that play a major role in preventing heart disease, skin aging, and various cancers; improves insulin sensitivity, ameliorates free radical damage associated with chronic age-related disorders, and inhibits fat-mobilizing enzymes (Al-Awwadi et al. 2005). A substantial amount of procyanidins present in fruits is not readily available for absorption; therefore, the effect of extrusion on the distribution of monomer, dimer, or trimer is essential.

The conflicting results on the impact of extrusion on antioxidant activities indicate that it depends upon the antioxidant sources and extruder conditions. Earlier, Özer et al. (2006) noted that twin-screw extrusion of nutritionally balanced mix (corn flour, cornstarch, enzymatically stabilized whole oat flour, raw chickpea flour, carrot powder, and ground raw hazelnut) at higher screw speeds, lower moisture content of the feed, and lower feed rates reduces only total antioxidant activity, not total phenolic compounds, and they attributed to the destruction of antioxidant compounds other than the phenolic compounds. Later, Camire et al. (2007) examined the role of blueberry, cranberry, raspberry, and grape fruit powders as antioxidants in breakfast cereals and found 62%–64% antioxidant inhibition; the differences is attributed to differences in composition of fruit powders with some fruits being more stable to heat and shear than others. Anton et al. (2009) employed two methods namely oxygen radical absorbance capacity (ORAC) and 2,2-diphenyl-1-picrylhydrazyl (DPPH) for measuring antioxidant activity while cornstarch with navy/small red bean (15%–45%) extruded using twin-screw extrusion. They found that navy and small red bean extrudates had reduction in total phenols, antioxidant activity through DPPH, and ORAC by 10%, 17%, and 10%, and by 70%, 62%, and 17% after extrusion, respectively. This indicates that the reduction in antioxidant activity depends on the source of antioxidant.

Recently, Potter et al. (2013) conducted twin-screw extrusion with a formulation consisting of wheat flour, cornstarch, potato starch, milk powder, and fruit powder (banana, apple, strawberry, and tangerine) and found an increase in antioxidant levels due to Maillard reaction occurring during extrusion or inactive antioxidants being transformed into active antioxidants as a result of nonenzymatic browning (Pokorny and Schmidt 2006). The antioxidant inhibition level depends on the types of fruit and varies between 15%–50%. Sharma et al. (2012) investigated the effect of extrusion moisture (15% and 20%) and temperature (150°C and 180°C) on antioxidant activity of extrudates from eight different barley cultivars using TSE with a screw speed of 400 rpm and a feed rate of 20 kg/h. Irrespective of extrusion conditions, total phenolic content decreased since phenolic compounds are heat labile (Sharma and Gujral 2011), heating over 80°C might destroy or alter their nature (Zielinski et al. 2001); therefore, the authors attributed their decomposition to the high temperature or alteration of molecular structure that reduces the chemical reactivity of them. Extrusion remarkably decreases the total flavonoid content of barley extrudates due to thermal destruction of flavonoids since they are heat sensitive (Sharma and Gujral 2011).

Antioxidant can be incorporated into the blend through either natural ingredients such as grains, legumes, fruits, and vegetables or the addition of synthetic phenolic antioxidants such as butylated hydroxyanisole (BHA), butylated hydroxytoluene (BHT), propyl gallate (PG), and *tert*-butylhydroquinone (THBQ) but are limited to 200 ppm (FDA) based on lipid composition. Accordingly, Camire and Dougherty (1998) evaluated the potential of BHT, cinnamic acid, and vanillin as antioxidants and found cinnamic acid and vanillin better protects the extrudates against lipid oxidation than BHT. A possible research area is the evaluation of other common phenolic compounds with similar antioxidant properties, which should be evaluated for their potential in extruded foods.

Anthocyanins are antioxidants and water-soluble pigments that provide attractive colors such as red, blue, and purple in many food crops. Recent researches have been focused on utilizing fruits with natural colors and antioxidant capacity because of their health benefits such as inhibition of lipid oxidation, hydroperoxide, and hexanal formations. Camire et al. (2002) extruded cornmeal with corn syrup, blueberry concentrate, and grape-juice concentrate using TSE (38°C/49°C/116°C/138°C/113°C, 300 rpm, cornmeal with sugar 13.6 kg/h, and liquid (corn syrup, blueberry concentrate 69°B, or grape-juice concentrate 69°B) for evaluating the stability and acceptability of blueberry and grape anthocyanins in extruded cereals. They found that extrusion decreased anthocyanin content and color by 90% and 78%, 74% and 70%, respectively, in extrudate with blueberry and grape juice concentrate. When Khanal et al. (2009b) extruded a mixture of blueberry pomace and white sorghum with 45% moisture content using TSE at 160°C–200°C, 150–200 rpm reduced total anthocyanin content between 33%–42%. In another study, Khanal et al. (2009a) extruded a mixture of grape pomace and white sorghum with 45% moisture content at 160°C–190°C, and 100–200 rpm reduced total anthocyanin content by 18%–53%, which is lower than blueberry concentrate (Camire et al. 2002). Twin-screw extrusion of colored potatoes which are rich in anthocyanin (80°C/90°C/100°C/110°C/120°C–140°C, 200–300 rpm, 17%–25% moisture, and 2.7 kg/h) reduced total anthocyanin content and increased browning index (Nayak et al. 2011). These aforementioned studies indicated that anthocyanins are unstable during extrusion since they are heat labile; earlier studies showed anthyocyanins degradation and brown pigment formation results in color loss, whereas Camire et al. (2002) reported that not only was there no color change when the products were stored in tri-laminate bags at room temperature for 3 months but also in anthocyanin contents.

Camire et al. (2007) pointed out the necessity of method development for increasing retention or improving stability of anthocyanins/pigments/natural colorants during extrusion for commercially practical. The same group hypothesized that fortification of the blend with ascorbic acid can make it act as an acidulant and the addition of vitamins would improve pigment retention. Accordingly, they extruded cornmeal with ascorbic acid (0%–1%, wt/wt) and sucrose (15%) or blueberry concentrate (17%) for producing ready-to-eat (RTE) breakfast cereals. Contrary to expectation, ascorbic acid fortification accelerated anthocyanin degradation during extrusion through direct condensation mechanism, enhanced polymeric pigment formation, increased browning in the oxygen and nitrogen environment, and interaction of ascorbic acid with oxidation products with anthocyanins (Chaovanalikit et al. 2003). The effects of heat and shear at low moisture conditions on ascorbic acid–anthocyanin is not fully known; therefore, it is better to evaluate extruder operating conditions for the development of ascorbic acid-fortified, natural colored (anthocyanin) products. Chaovanalikit et al. (2003) suggested that other forms of processed blueberry may be a potential/suitable source of anthocyanin in extruded products. Thus, Camire et al. (2007) later examined the effects of dehydrated fruit powders (1% blueberry, cranberry, concord grape, and raspberry) as colorants and antioxidants in extruded (35°C/45°C/60°C/95°C/113°C/163°C, 175 rpm, 15.3 kg/h dry feed, 0.75 kg/h water) white cornmeal breakfast cereals. They observed that anthocyanins from fruit powders survived extrusion and retained some antioxidant activity and concluded that the levels used in their study were too low (1% w/w). Finally, they concluded stating that additional research is necessary to determine optimal levels of fruit powders for extruded cereals that increase antioxidant activity while providing acceptable sensory quality. Therefore, fruit powder levels should be increased and the product should be evaluated in all aspects; increase in production cost due to higher levels of fruit powder can be easily offset by the more attractive and functional cereals that result. There is research potential looking into extrusion parameters and their effect on the kinetics of anthocyanin, their stability, mechanisms of antioxidant destruction, and retention from different sources and forms.

4.7.1 Case Studies: Breakfast Cereals, Snacks, Baby Foods

According to Global Industry Analysts, the world breakfast cereal industry, which was worth $28 billion in 2010 with an annual growth at 4% between 2010 and 2015, would exceed $34 billion because of changing lifestyles, consumer awareness regarding the importance of healthy eating habits, and evolving food consumption patterns. The United States, United Kingdom, France, Spain, Italy, and Russia are the leaders in global breakfast cereal market, and they produce as well as consume the largest volume of cereals; around 35% of the global population buys cereal products, mostly in the United States and the European Union. Cereal has a penetration rate of about 90% in US homes (97% in homes with kids), per capita consumption of breakfast cereals was 4.5 kg in 2010 and is projected to decline in 2016 to 4.2 kg due to a maturing market, the unhealthy image of children's cereals, and competition from foodservice for the hot cereal market (Euromonitor International 2012); however, US retail sales of breakfast cereals are expected to increase from US $9.9 billion in 2010 to US $10.4 billion by 2016 due to an increase in sales volume. According to Thomas et al. (2013), Kellogg (33%) and General Mills (29%) are the major players representing 62% of RTE cereals market in the United States, and the rest is shared among other major national brands and private labels. RTE cereals comprise flaked, puffed, shredded, and granulated products generally made from wheat, corn, or rice, although oats and barley are also used, and they may be enriched with sugar, honey, or malt extract.

The steps in extruded flake production are preprocessing, mixing, extruding, drying, flaking, toasting, and packing. First the required flour blends with other ingredients such as sugar, salt, and color are mixed in a ribbon blender. A bucket elevator/screw conveyor feeds the mixture to the extruder where actual cooking or gelatinization takes place; it depends on the extruder temperature (140°C–180°C), residence time (0.5–1.5 min), moisture (16%–20%), and shear stress through screw speed (200–450 rpm). The cutter placed at the end of the extruder cuts the strand that comes out of the die(s) into small balls that are transported to flaking rollers. Counter-rotating smooth rolls convert the ball into flakes that are conveyed to a belt drier (140°C, 3–8 min, single or multi-pass). The dried flakes are transferred to cylindrical rotating-drum and exposed to spray coating of flavored sugar syrup. It is wise to add heat-sensitive flavors and vitamins after extrusion. The addition of sugar after extrusion would present excessive browning. Finally, the sugar-coated flakes are again dried and cooled in a conveyor belt drier (150°C–200°C for facilitating browning) before packaging. Complete cereal flaking systems are shown in Figures 4.7 and 4.8. Initial moisture content of the

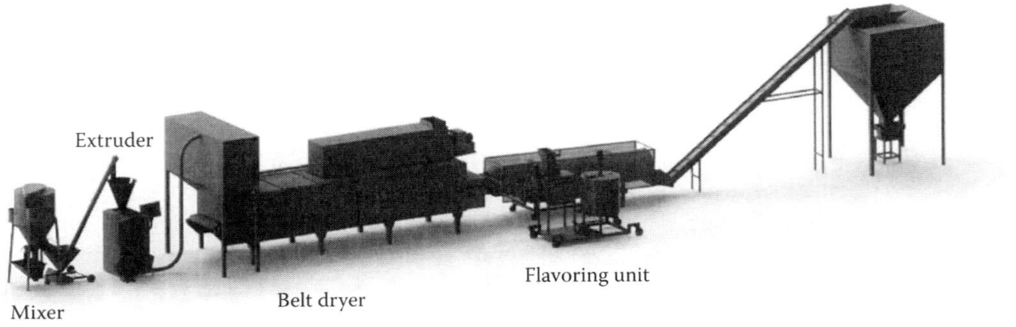

FIGURE 4.7 Breakfast cereal production system. (Courtesy of Lalesse, France.)

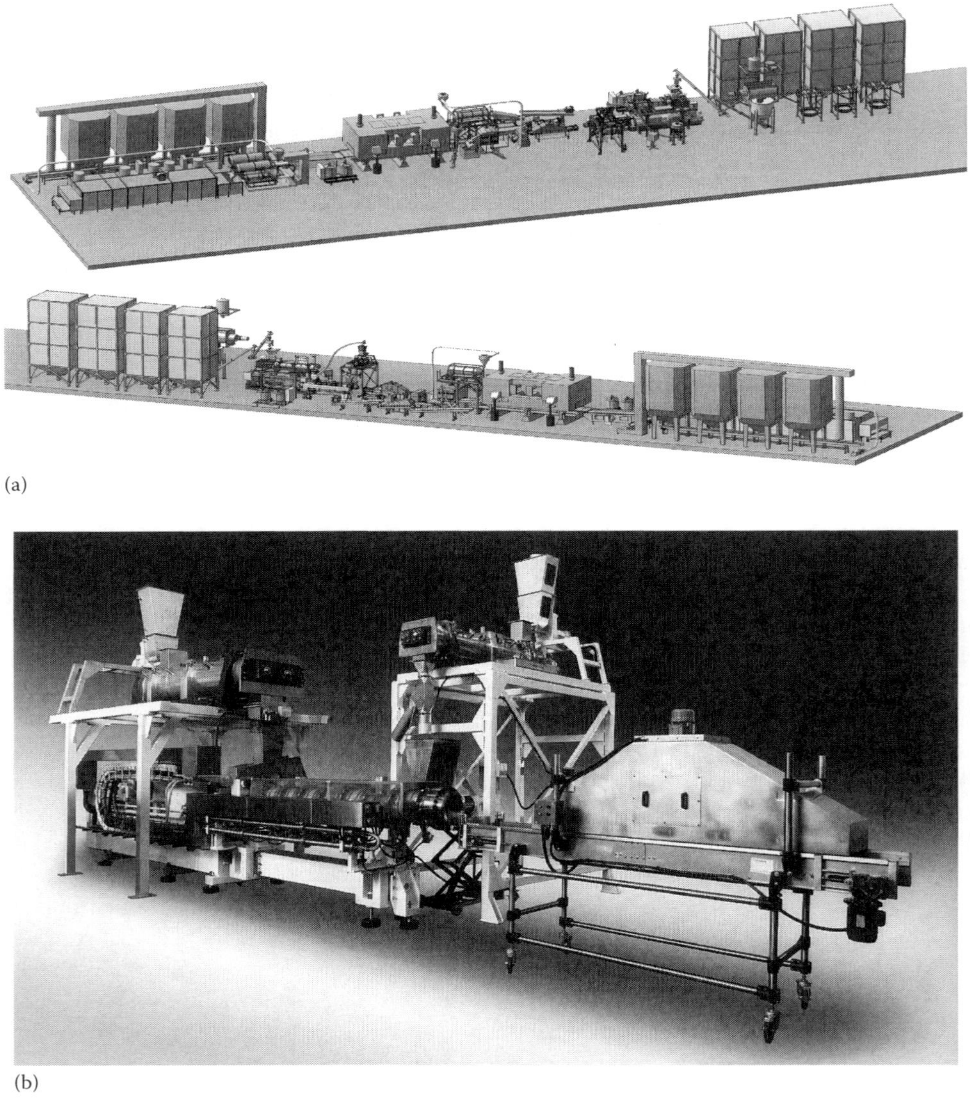

FIGURE 4.8 Cereal flaking system. (a) Schematic diagram and (b) complete actual system. (Courtesy of Clextral, France.)

blend or in-barrel moisture range is 16%–20%, after extrusion moisture may be between 7%–12%, and it should be reduced to 2%–3% for getting the right crispiness when they leave the drier.

For expanded products, flaking rolls are not required; SSE or TSE with an l/d range of 10–20 can be used with proper screw configuration and operating conditions in order to increase the mechanical cooking (Riaz 2000, Guy 2001). According to Riaz (2000), preconditioning is required for producing satisfactory expanded as well flake products if formulation contains bran. Pellets are usually produced using TSE with l/d between 20 and 30 with an appropriate screw configuration and operating conditions for low mechanical cooking (high thermal cooking). Expansion/puffing is avoided by releasing the pressure and cooling it before the melt exit die. Die shape and cutting speed defines the size and shape of the pellets. The feed may have moisture of 22%–26%, and it should be reduced to 2%–3% for final pellets.

4.7.1.1 Snacks

Snacks are an important part of healthy eating plans for many consumers, with 32% consuming snacks as mini-meals throughout the day. Chips, extruded snacks, and tortilla/corn chips accounted for almost 56% of retail sales in 2010. In order to stay in the market, manufacturers have to continue to launch healthier alternatives, such as baked and multigrain, with reduced/lower sodium varieties (Market Indicator Report 2011). The required ingredients are mixed in a ribbon blender, and then fed to an extruder through a bucket elevator. The ribbon comes out of the extruder (wavy surface/uneven), run through smooth rolls (make smooth thin sheet), and then the chip cutter cuts into the desired shape. The chips are sent to a rotary fryer where moisture is removed and oil is added followed by addition of flavoring and seasoning to the chips in a coating tumbler before packaging. Frying is not required for other directly expanded snacks such as Cheetos and corn curls. Extrusion of corn grits, with 15% moisture, screw compression ratio of 2:1 or less at 160°C–180°C, come out of the die as a puffed rope and a revolving cutter cuts into desirable size for producing corn curls. At this stage, the curls not only contain high moisture (about 6%) but also a bland taste. Thus, the curls must be dried and flavored by spraying with oil and coating with various flavoring mixtures (cheese, hydrolyzed yeast, spices, and peanut butter) and cooled to room temperature, at which the product becomes crisp and crunchy before packaging (Berk 2009).

4.7.1.2 Baby Foods

Direct expanded or flaked baby foods should be highly soluble (once it contacts a child's saliva) and fully gelatinized (for easy digestion) and are produced using high shear extruders. In general, the formulation consists of fruit juices, cane sugar, fructose as sweetener, fortified with stable vitamins and minerals, organic and natural preservatives such as vitamin E, rosemary, and the like. Cereals are formed into specific shape and design with very short bowl life and very short crunch.

Coextruded snacks, introduced in 1984 with two different flavors, colors, or textures, are now available in different shapes (pillow, trapezoid, and bars) for babies also. The contrasting texture of outer cereal casing and inner sweet/sour fruit filling (complete, partial, and coating) adds appeal to a child due to smooth transition in the taste. Buhler says that coextruded snacks in the market are of three kinds: (1) cereal-based tubes with cereal-based fillings, (2) cereal-based tubes with fat-based fillings, and (3) cereal-based tubes with water-based fillings, and the most common coextruded snack is a cereal-based outer tube with a cheese filling inside.

4.8 HEAT TRANSFER IN EXTRUSION PROCESSING

Heat and mass transfer mechanisms play a critical role in any food manufacturing process including extrusion since thermal and shear history has a great impact on product quality (Jaluria 1996). According to Mohamed and Ofoli (1990), an understanding of heat transfer and its mechanisms are prerequisites for proper control and optimization of food extrusion. Since viscous energy dissipation within the barrel is a greater heat source than that of heat transfer into barrel, it should be an important component of heat transfer analyses in SSE and TSE. Conduction heat transfer occurs in and between barrel, melt/dough, and screw; heat transfer in barrel has a significant effect on the flow and thermal transport (Lin and Jaluria 1998). Most of the food materials behave like non-Newtonian fluids in the extruder; therefore, the application of non-Newtonian analysis requires the knowledge of the rheological behaviors of the dough to be extruded (Bouvier et al. 1987). Modeling of heat transfer in a food extruder is difficult due to numerous variables, and it is impractical to include all of them for complete analysis. That is the reason why researchers make assumptions for given conditions through neglecting unimportant variables; the model cannot be used for an unjustified extrapolation to other circumstances (Lin and Jaluria 1998). Most of the researchers made assumptions such as steady state, negligible inertia and gravity forces and fully developed incompressible fluid flow; however, most of the TSE operate under starved feeding that reduces the degree of filling.

In order to develop the heat transfer model for food extrusion, one has to describe the flow pattern inside the extruder, that is, feed, compression, and sections. The metering section usually contains only melt/dough, whereas as the feed section contains only the blend and the compression section where the blend is slowly converted into melt/dough, hence, it is a combination of both. van Zuilichem et al. (1990) reported that predicting heat transfer in feed and compression sections is far more complex and well-fitting models have not been found. In order to obtain exact flow pattern, equations of motion and equations of energy should be solved; because of the convective term involved in an energy equation it is fairly difficult to solve. Most of heat transfer models consider the metering section within the flow which is non-isothermal and non-Newtonian. Earlier, different approaches have been used for developing heat transfer models such as numerical analysis, differential equations (Griffith 1962), equations of motion and energy (Pearson 1966), and finite difference techniques-equations of flow of non-Newtonian fluids (Yankov 1978). Griffith and Pearson considered power law fluid in their model, meaning that the formulation should have high moisture; as Mohamed and Ofoli (1990) reported soy polysaccharide with moisture content 70% would significantly deviate from power law behavior. Pearson presented the results in terms of dimensionless parameters of output, pressure, and Brinkman numbers. van Zuilichem et al. (1990) suggested that the introduction of dimensionless numbers (Peclet, Nusselt, Reynolds, Graetz, and Brinkman) not only describe the heat transfer in an extruder properly but also allows to check whether particular assumptions are correct or not. Yacu (1983) proposed a very thorough description of non-Newtonian behavior of mass; he considered different zones in a TSE such as solid conveying, melt pumping, and melt shearing and predicted the temperature and pressure profile for each zone separately; of course with a number of assumptions (Moscicki 2011). One of his assumptions is that the moment the feed enter the melt pumping section, the screw becomes completely filled, whereas most of the TSE operate under starved fed, lowering the degree of filling; thus, the shear the melt receives is low. Considering the problems in Yacu's model, van Zuilichem developed a model consisting of two parts, the first part is the calculation of heat transfer from barrel to extrudate, and the second part is the calculation of heat generation through viscous dissipation; thus, the total heat transfer in the extruder can be calculated for every position (Moscicki 2011).

In general, the literature survey reveals that the research on heat transfer modeling of food extrusion is very scanty due to the complex nature of computational effort when compared to the volume of food extrusion literature published (mostly product development, physical properties, nutritional properties, and extruder parameters such as SME, torque, and pressure). Moscicki (2011) suggested that gathering rheological data on various biopolymers would allow development of new models, and they have to be verified through experiments to make it more reliable and closer to reality. Still there are research scopes on heat transfer that should consider the heat loss to the environment. Levine (2009) reported that the heat loss from an industrial extruder is smaller compared to the lab extruder due to difference in area/unit throughput. For producing third-generation snacks, the vapor should be released to avoid expansion at the die exit and melt has to be cooled for forming.

According to Levine (2010), product development targeted for commercialization should not be done at lab-scale if significant heating or cooling is involved, as in pellets production. He also pointed out that the limiting factor for scale-up is heat transfer, notpuming, or shearing capacity. Knowledge of the heat transfer in food extrusion is essential for the scale-up and design of food extruders and the associated temperature control systems (van Zuilichem et al. 1990).

4.9 PROCESS CONTROL

Food extrusion exhibits strong interactions among mass, energy, and momentum coupled with complex physical and chemical transformations (starch gelatinization, protein denaturation/texturization, hydrogen bond rupture, forming, and puffing) that govern the final product qualities. Therefore, food extrusion is the most difficult process to control due to complex dynamics and the inconsistent nature of food materials (Wang and Tan 2000). The physio-chemical state changes during extrusion, which affects rheological properties that are poorly understood due to lack of appropriate on-line sensors that are able to check the final product quality. According to Cayat et al. (1995), extrusion process control is still often manual and empirical, and it is directly linked to economic, qualitative, and scientific interests. Once upon a time, process control in food industries was considered as an expensive investment; now it is accepted as a necessity to produce products with consistent qualities.

In order to control extrusion, the process has to be modeled. Different approaches can be grouped under white-box modeling and black-box modeling. White-box modeling requires as much knowledge as possible for incorporating all the internal physical, chemical, and biological laws that governs the system; the resulting model coefficients have physical significance. Simplification is inevitable during model development due to the complex nature of the extrusion process, products, and their interactions; further measuring intermediate parameters are also difficult (Cayat et al. 1995). Black-box modeling is basically an empirical model that involves simple mathematical relationship between process inputs and outputs; however, it holds good only for range of the experimental conditions that is sufficient to develop process control strategies. Techniques used to optimize formulation of operating conditions such as residence time distribution, response surface analysis, and dynamical identification comes under black-box modeling.

From a control point of view, it is necessary to classify extrusion variables into controlled and manipulated variables. In general, continuous on-line monitoring of density, texture, color, and degree of cooking is not possible; therefore, they are correlated to measurable variables such as die pressure, die temperature, and motor torque or SME, hence these measurable variables become controlled variables (Singh and Mulvaney 1994). Several researchers have identified SME as a powerful, convenient, and critical variable for process control in twin-screw extrusion since it correlates well with melt temperature at die and extrudate density correlates well with SME and moisture content (Onwulata et al. 1994). Screw speed has a great impact on SME especially when the moisture content is less than 30% (Haley and Mulvaney 2000).

Earlier it was hypothesized that pressure has a linear relation with screw speed, moisture content, and feed rate when Mulvaney et al. (1988) conducted step tests, whereas Moreira et al. (1990) demonstrated through step tests and back steps that the relation between pressure and screw speed and moisture was nonlinear. Later, Onwulata et al. (1992) and Lu et al. (1992) conducted step tests by adding temperature on the effect of product variable/expansion and process variable/melt pressure and concluded the relation was nonlinear. Die pressure is thought to be a sensitive indicator of melt shear history that affects extrudate quality; accordingly, Levine et al. (1986), Moreira et al. (1990), and Kulshreshta et al. (1995) presented controller design to regulate die pressure. However, Onwulata et al. (1994) showed that extrudate density can be changed by varying melt moisture content under certain operating conditions without affecting die pressure. According to Singh and Mulvaney (1994), the transfer function modeling approach well characterizes the nonlinear and interactive behaviors of the extrusion process.

It is well known that manipulation of extruder operating variables results in physiochemical characteristics of extrudates, and the most commonly used manipulated variables are screw speed, feed rate, liquid injection rate, moisture content, and barrel temperature profile. For commercial production, process efficiency/throughput is a critical factor that depends on feed rate, thus it becomes unavailable for manipulation. Wang et al. (2001) found that motor torque, SME, die pressure, and melt temperature at the ninth zone had a strong impact on the product quality attributes such as bulk density and moisture content. Later, Wang et al. (2008) used screw speed and liquid injection rate as the manipulated variables, and SME and motor torque as the controlled variables for developing predictive models.

Heating, cooling, shaft work input, and feed rate affect extrudate temperature and consequently product quality. It is better to have stable and specifiable extrudate temperature profile along the barrel length; in reality monitoring and controlling of extrudate temperature is not directly done. Extruder barrel consists of a number of zones that are regulated independently by setting the barrel temperature; there is no simple relation between barrel temperature and extrudate temperature, thus obtaining the latter is not easy. Even small variations in the feed would change the extrudate temperature. According to Hofer and Tan (1993), a significant time delay exists between each manipulative variable and extrudate temperature that demands accurate models with sophisticated control strategies to regulate extrudate temperature. Accordingly, Hofer and Tan (1993) developed a predictive regulator for extrudate temperature using auto-regressive moving-average models with auxiliary inputs. Model coefficients varied significantly between experiments with the same operating conditions indicating the necessity of frequent tuning of the regulator. Later, they developed a self-tuning regulator for temperature control based on feedforward and internal-model-based predictive control strategies (Tan and Hofer 1995).

Transfer functions for extrusion cooking processes have been derived using continuous-time transfer functions. Transfer function modeling is an excellent technique to characterize the nonlinear and interactive behavior of extrusion (Singh and Mulvaney 1994), and it has been used successfully in the past in food extrusions for developing dynamic models to characterize observed dynamic responses of TSE. Development of transfer function models are based on the input–output observations of a process. Typically, the transfer function is identified by conducting an experiment wherein some manipulated variables are varied for measuring the responses of one or more output variables (Haley and Mulvaney 2000). A continuous-time or discrete-time transfer function that describes the observed input–output responses can be found using a parameter estimation algorithm as several researchers have done (Moreira et al. 1990, Hofer and Tan 1993, Lu et al. 1993, Cayot et al. 1995). Hofer and Tan (1993) and Tan and Wang (1998) reported, discrete-time linear transfer function models for developing model predictive control strategies when time delay is part of the dynamic response. Moreira et al. (1990) designed and simulated a feedforward control scheme for controlling die pressure while varying feed moisture and feed rate. Weidmann and Strecker (1988) used multiple control loops for automatic start-up and controlling extruder through controlling SME, product temperature, and mass pressure. These authors have not reported either the degree of interactions or coupling in the control loop. The relative gain array can be used to quantify the level of interactions. Singh and Mulvaney (1994) used relative gain array and found the MIMO system, with screw speed and barrel heating as manipulated variables and product temperature and motor torque as controlled variables, best for minimal interaction. When they used PID controller without decoupling it showed adequate performance results for various controlled system requirements. Later, Popescu et al. (2001) used ANN for modeling the start-up process of food extrusion.

In most of the earlier studies, first off-line experiments are used for measuring all input–output responses followed by off-line development of transfer function models; subsequently, the functions were used to derive automatic controllers. The limitation of those controllers is that they are only suitable for the limited process and operating conditions studied. It is well known that food extrusion has a wide range of dynamic behavior depending on the product and processing conditions; therefore, many researchers have concluded that some means of on-line system identification for automatic tuning and adaptive control is desirable. Keeping the above in view, Haley and Mulvaney

(2000a,b) developed an on-line system identification procedure for use in automatic tuning and/or adaptive control of high-shear twin-screw extrusion processes and successfully demonstrated the use of such a system for automatic control of puffed corn snack density. They used inferential model to correlate SME with product density and melt moisture content. In the absence of on-line measurement of product properties, inferential models are required that relate product quality variables to process operating conditions in addition to the models that describe the dynamic operating behavior of the extruder. If melt moisture set point is established, it can be easily controlled by sensing feed moisture, feed rate, and regulating water feed rate using ratio control scheme; thereby, extrudates with desirable density, crispiness, and texture can be obtained (Onwulata et al. 1994, Strahm 1998). Haley and Mulvaney (2000b) demonstrated that an inferential model can be used to regulate extrudate density without on-line sensors; however, it has to be verified under actual food extrusion conditions. Earlier, Elsey et al. (1997) developed a dynamic/adaptive inferential model for predicting variables such as gelatinization, SME, specific volume, and residence time.

Owing to the complex nature of food extrusion, it is difficult to operate in a stable, reproducible manner especially during start-up and shutdown where the product deviates from specifications (Mulvaney et al. 1988, Nejman et al. 1990). Because a number of parameters changes during start up and shutdown more than steady state operation, it becomes very complex; therefore, it is not possible to program. In the absence of sufficient process information and nonlinearities nature, most of the start-up procedures depend on the operator's experience. Fuzzy logic, an advanced knowledge-based control system seems to be an alternative, and Lee et al. (2002) demonstrated the possibility of applying it to start-up of TSE. They used an algorithm based on inverse relation between screw speed and torque to avoid overtorque that causes blockage during start-up. Lee et al. (2002) made an improvement on automatic control of startup using fuzz amount to control screw speed over. Ferdinand et al. (1988, 1989) used deterministic magnitude for changing the screw speed to prevent the torque percentage of maximum exceeding the screw speed percentage of maximum. Farid et al. (2007) expanded fuzzy control for developing an intelligent controller structure for start-up of cornstarch extrusion where in water and bulk feed rate were also considered as control variables in addition to screw speed. The authors mentioned that obtaining a set of optimal control parameters such as linguistic control rules and scaling factors of the fuzzy controller are difficult. They also suggested that future research should explore genetic algorithms or neurofuzzy network to develop nonlinear mapping for scaling factors virtually covering all situations that can occur during food extrusion.

In food extrusion, starch undergoes various changes that are currently controlled through SME, die pressure, and temperature; the interaction of these variables often makes it difficult to relate to the critical changes in starch due to shear of screw elements and ingredients themselves. Continuous monitoring of the state of the components in the melt is highly desirable and forms the basis of a feedback control part of such system; however, an extruder provides great challenge for placing any on-line measurement devices. NIR is usually used for compositional analysis of foods; NIR with fiber optic would allow remote noninvasive measurements on the hot melt in the extruder barrel (Guy et al. 1996). The changes in starch strongly influence the rheology, thereby product quality, size, shape, appearance, and texture. The authors successfully used NIR spectral data over a wide range of 1300–1800 nm for predicting SME while extruding wholemeal wheat flour in a TSE at 15°C/25°C/55°C/70°C/90°C/120°C, 150–450 rpm, 15%–25% wb adjusting the solid and water feed rate.

Examining NIR absorbance spectrum between 1300–1800 nm indicates not only the degree of cooking when the melt passes through the die exit in the extruder but also can be related to product qualities (Evans et al. 1999, Huang 1999). Lab/off-line product quality measurements take up to 3 days, resulting in slow quality control loop and high wastage, whereas on-line NIR probe can estimate the product quality and reduce the quality control loop time under a minute. However, the cereal industry is reluctant to its use due to the use of a full spectrum probe that is cost prohibitive. If the desired wavelengths are known, the full-spectrum probe could be replaced with a small selection of monochromators, and on-line probe for measurement of product quality would become feasible. Accordingly, Dodds and Heath (2005) have shown (Table 4.4) only

TABLE 4.4
Effect of Extrusion Parameters on Expansion Ratio of Extrudates as Reported in the Literature

Extruder Conditions	Blend	Variables vs. ER	References
	SSE—Feed		
Lab-scale SSE with l/d of 20:1, screw compression ratio of 3:1, rod-shaped die of 0.3 mm, temperature profile of 100°C/135°C/140°C, and screw speed of 160 rpm,	Sweet potato flour (0%–100%), whole wheat bran (0%–100%), brown sugar (10%), baking soda (1%), maple syrup (3%), cinnamon (1%), and 15% water	Extrudates made from sweet potato flour have better ER than when formulated with whole wheat bran due to incomplete breakdown of bran during extrusion, further bran's sharp edges would pop the air cells as they form due to low moisture content of sweet potato flour	Dansby and Bovell-Benjamin (2003)
Lab-scale SSE with l/d of 15:1, a 2:1 screw compression ratio, and die diameter of 5 mm, temperature 120°C–125°C, screw speed of 520 rpm, feed moisture of 12% db, and feed rate of 18 kg/h	Pigeonpea flour to cassava flour at 0%–15%:100%–85% (db)	ER decreases with the addition of pigeon flour to cassava starch due to changes in blend composition.	Rampersad et al. (2003)
Lab-scale SSE with screw compression of 4:1 and round die of 3 mm, temperature profile of 100°C/140°C/160°C, screw speed of 250 rpm, and feed moisture of 16%	Mixtures of rice flour: cactus pear pulp concentrate of 95%:5%, 90%:10%, 85%:15%, and 80%:20%	ER decreases with the addition of cactus pear concentrates due to change in starch, sugar, and DF content.	El-Samahy et al. (2007)
Lab-scale SSE with l/d of 20:1, screw compression ratio of 1:1, die of 3 mm diameter, temperature profile of 50°C/100°C/140°C–180°C, screw speed of 60 rpm, feed moisture content of 18% wb, and feed rate of 1.68 kg/h	Taro flour, maize flour, nixtamalized maize flour, average particle size less than 0.5 mm, blend consists of taro flour (0%–100%) with maize flour (0%–100%) or nixtamalized maize flour (0%–100%)	An increase in taro flour in the blend increases ER due to increase in starch content of taro flour leading to gelatinization, decrease in amylose/increase in amylopectin, and decrease in fiber.	Rodríguez-Miranda et al. (2011)
Lab-scale SSE with l/d of 26:1, screw compression ratio of 3:1 and a die with six 2 mm diameter holes, temperature profile of 80°C/100°C/120°C/120°C, screw speed of 70 rpm, and moisture content of 25% wb	Cornstarch (20%–30%), oat bran and defatted soy flour (20%–50%), inulin (0.05%)	ER decreases with an increase in defatted soy flour and oat bran due to fiber ruptures air cell walls and external surface in extrudates thereby preventing the full expansion of air bubbles; protein affects water distribution in the matrix and through its macromolecular structure and conformation, which affect the extensional properties of the extruded melts. Extrusion temperature has a positive influence on ER due to change in rheological properties/decrease in viscosity of the melt.	Lobato et al. (2011)

(Continued)

TABLE 4.4 (*Continued*)
Effect of Extrusion Parameters on Expansion Ratio of Extrudates as Reported in the Literature

Extruder Conditions	Blend	Variables vs. ER	References
SSE with l/d of 1.38:1 and 8.27 cm die opening, temperature of 120°C–160°C, and feed moisture content of 18%–25% db	Feed composition of 60:40 and 70:30 acha and cowpea; Acha flour with 300–500 μm size while cowpea flour with 400–750 μm	ER increases with temperature and moisture content. The amount of cowpea has a negative influence on ER due to the dilution effect of protein on starch gelatinization with the increased firmness of plasticized extrudates, whereas a positive effect for longitudinal expansion.	Olapade and Aworh (2012)
Lab-scale SSE with l/d of 20:1, screw compression ratio of 2:1 and die opening 2.4 mm, temperature of 140°C–80°C, screw speed of 150–250 rpm, and feed moisture content of 18%–22%, wb	Pinto bean flour with particle size less than 40 mesh	Not only temperature and feed moisture have a positive influence on ER but also their interaction. An increase in moisture reduces viscosity that results in less mechanical damage to starch, thus enabling dough to expand more and faster. An increase in temperature facilitates cooking of starch, thus expands well. Beyond certain temperature (180°C) and moisture content (22%), ER reduces probably due to starch dextrinization.	Balandrán-Quintana et al. (1998)
SSE—Extruder parameters			
Lab-scale SSE with l/d of 20:1, screw compression ratio of 2:1 and a die of 3 mm diameter, temperature profile of 150°C/180°C/180°C, screw speed of 158–242 rpm, and feed moisture content of 13%–25% wb	Fermented corn and finger millet at 1:1, 5%–20% of glucose, fructose, sucrose, maltose, or dextrose	Increasing sugar levels compete with starch granules for water reduces the degree of gelatinization and limits the ER. Increase in feed moisture content has a negative effect on ER.	Onyango et al. (2005)
Lab-scale SSE with screw compression ratio of 1:1, die of 2 mm, temperature profile of 70°C/80°C/90°C/140°C–190°C, screw speed of 80 rpm, and feed moisture of 12%–16%	Arrowroot starch	Low moisture content of starches may restrict flow inside the barrel, increase shear rate and residence time, thus increase degree of starch gelatinization and expansion; shear strengths of the extruded starch products are inversely proportional to ER.	Jyothi et al. (2008)
SSE—Feed and extruder parameter			
SSE with screw compression ratio of 4:1 and a 3.5 mm diameter × 20 mm long, temperature of 150°C–180°C, screw speed of 150 rpm, feed moisture content of 15%–19%, and feed rate of 12 kg/h	Corn flour with cowpea inactivated or non-inactivated for lipoxygenase at 85:15	Moisture content negatively affects ER due to lubricating effect and causes less mechanical energy dissipation.	Sosa-Moguel et al. (2009)

(Continued)

TABLE 4.4 (*Continued*)
Effect of Extrusion Parameters on Expansion Ratio of Extrudates as Reported in the Literature

Extruder Conditions	Blend	Variables vs. ER	References
Industrial-scale SSE with l/d of 3.8:1, 12 hole die with 2 mm diameter, temperature of 150°C–175°C, screw speed of 200–280 rpm, and moisture content of 12%–18%	Rice grit (67%), durum clear flour (8%–20%), partially defatted hazelnut flour (5%–15%), and fruit waste (3%–7%) consists of milled orange peel (80.0% db), grape seeds (10.0% db), and tomato pomace (10.0% db)	Addition of partially defatted hazelnut reduces ER due to dilution of total starch available for expansion; affects the extent of starch gelatinization and the rheological properties of the melt in the extruder because of high protein content (absorb moisture thus lower gelatinization); though fat improves texture, it reduces ER. ER increases with increase in moisture due to more moisture requirement by protein and gelatinization; ER decreases with increase in temperature due to increase in dextrinization and weakening of structure.	Yagci and Gogus (2009)
SSE with screw compression ratio of 4:1, cylindrical die of 3 mm diameter and 20 mm length, temperature of 175°C, screw speed of 150 rpm, and moisture content of 14% wb	Fortuna rice (a low-amylose rice variety), hard red corn, lathyrus seeds. Average particle size of rice and corn are between 420 and 1119 μm, whereas lathyrus seeds are 210 and 570 μm.	ER of corn-based extrudate is lower than that of rice due to high oil content of whole corn (4.72% vs. 2.38%) that acts as lubricants, reducing the degree of cooking. Addition of legumes decreases ER due to decrease in starch content.	Pastor-Cavada et al. (2011)
SSE with short barrel and 2-hole die with 4 mm orifice, temperature of 105°C–150°C, screw speed of 110–170 rpm, feed moisture content of 17%–20%, and feed rate of 20 kg/h	Hybrid maize R-2303 and R-2207	Not only extrusion parameters such as screw speed, temperature, and feed moisture affect ER but also maize varieties. High temperature and screw speed contribute a suitable cell uniformity for expansion. An increase in feed rate increases ER, whereas an increase in moisture decreases ER. Low ER of maize variety R-2303 is because of the presence of greater amounts of opaque and finer particles which act as capillaries and result in non-homogeneous dough; hence, less water is available for hydration of vitreous particles.	Razzaq et al. (2012)

(*Continued*)

TABLE 4.4 (*Continued*)
Effect of Extrusion Parameters on Expansion Ratio of Extrudates as Reported in the Literature

Extruder Conditions	Blend	Variables vs. ER	References
Lab-scale SSE with l/d of 20:1, screw compression ratio of 3:1 and die diameter of 3 mm, temperature of 100°C/110°C/105°C–165°C, screw speed of 90–210 rpm, and feed moisture content of 12%–20%	Cottonseed meal: corn flour of 5:93, 10:88, 15:83, 20:78, and 25:73 with 1.5% of chili flour, 0.5% of salt and all ingredients' particle size less than 2 mm	ER increases with increase in moisture that increases pressure in the extruder, causing the product to expand at the die, largely due to flashing water vapor caused by abrupt pressure changes. At high temperatures, starch may dextrinize, reducing ER, especially with low starch content mixtures. Addition of cotton seed meal has a negative influence on ER due to high protein unless moisture increases. The combination of high temperature and high screw speed reduces ER, whereas high screw speed combined with high cotton seed meal enhances ER.	Reyes-Jáquez et al. (2012)
SSE with l/d of 3.3:1, screw compression ratio of 1:1 and die of four round holes of 1.8 mm diameter, screw speed of 324–387 rpm, moisture content of 15%, and feed rate of 20 kg/h	Corn grits and semi-defatted sesame cake flour (coarse) at 100: 0%–20%	Addition of semi-defatted sesame cake. ER increases with increase in screw speed (>360 rpm) due to mechanical shearing reduces viscosity of the molten starch favoring bubble growth. Addition of semi-defatted sesame cake increases the protein content of the blend that influences water-binding capacities, thereby starch gelatinization; further, protein has limited or non-puffing capacity compared with starch, thus dilutes starch content, resulting in low ER. Further, semi-defatted sesame cake consists of a remarkably high amount of fat (11.2%) and DF (22.7%) reducing the viscosity of the melt and fiber competing for moisture and causing premature rupture of gas cells reduces the ER.	Nascimento et al. (2012)
Lab-scale SSE with l/d of 16.43:1, temperature of 100°C–110°C, and screw speed of 100–140 rpm	Yam and brewer's spent grain at 95:5, 90:10, and 85:15	ER increases with temperature and screw speed can be attributed to increase in gelatinization and strengthening of structures, however addition of brewer's spent grain can lessen the impact of screw speed on ER.	Sobukola et al. (2013)
SSE with screw compression ratio of 3:1, die of 1.5 mm diameter with 3 holes, temperature of 80°C–100°C, screw speed of 60 rpm, and moisture content of 28%–34%, wb	Corn flour (0.191 and 0.490 mm) and broad beans (150 to 560 μm) at 70:30	ER increases with increase in temperature and decrease in moisture due to decrease in flow viscosity of the melt.	Gimenez et al. (2013)

(*Continued*)

TABLE 4.4 (*Continued*)
Effect of Extrusion Parameters on Expansion Ratio of Extrudates as Reported in the Literature

Extruder Conditions	Blend	Variables vs. ER	References
		TSE—Feed	
Lab-scale conical TSE with screw compression ratio of 3:1 and 3 mm die, temperature profile of 80°C/100°C/120°C, screw speed of 140 rpm, and feed moisture content of 22% db	Cornstarch with varying amylose content (0%–70%), whey protein concentrate with varying protein (80%–85.7%). Cornstarch to WPC is 70%–100%:0%–30%	An increase in amylose and protein decreases ER due to physico-chemical interaction between amylose and protein, considering expansion a maximum of 10% WPC can be added to blend.	Matthey and Hanna (1997)
Corotating TSE with die diameter of 5 mm, die temperature of 130°C or 165°C, screw speed of 200 rpm, feed rate of 25.8 kg/h, feed moisture content of 20%, and SME of 530 kJ/kg	Sorghum and cowpea (0%–100%), average sorghum particle size between 0.21–1 mm and cowpea particle size of 480 μm	ER decreases with increase in cowpea due to high protein content, and further lowers ratio of amylose/amylopectin in sorghum. Higher temperature facilitates starch gelatinization, thereby better expansion.	Pelembe et al. (2002)
Lab-scale corotating TSE with l/d of 32:1 and circular die of 4 mm, temperature profile of 60°C/90°C/97°C/115°C/135°C/157°C, screw speed of 150 and 250 rpm, feed moisture content of 21%–23% wb, and feed rate of 13.5 kg/h	Cornmeal to potato flakes of 2:1, crab-processing by-product (wet 10%–25% or dry 10%–40%), spices, particle sizes less than 2 mm	ER decreases with crab-processing by-products either wet or dry due to reduction in starch content, protein encloses available starch, thereby lesser gelatinization, higher ash levels.	Murphy et al. (2003)
Corotating TSE with l/d of 32:1, die diameter of 4 mm, temperature 150°C–175°C, screw speed of 225 rpm, blueberry concentrate pumping rate of 3.6 kg/h, and water feed rate of 4 kg/h	Cornmeal, ascorbic acid (0.1%–1%), blueberry concentrate (17%), or sucrose (15%)	Addition of blueberry concentrate and ascorbic acid modifies the viscoelastic characteristics of the dough during extrusion, thereby ER decreases.	Chaovanalikit et al. (2003)
Corotating, intermeshing, self-wiping TSE with l/d of 32:1, circular die of 5 mm, temperature profile of 0°C/30°C/30°C/30°C/70°C/100°C/150°C/150°C, screw speed of 400 rpm, moisture content of 15% (wb), and feed rate of 12 kg/h	Rice flour (85%–95%), hydrolyzed/unhydrolyzed fish flour < 24 mesh (5%–15%)	ER increases with hydrolyzed fish added to rice flour compared to rice flour with or without unhydrolyzed fish. In general, addition of protein would reduce ER, whereas hydrolyzed fish flour increases ER.	Choudhury and Guatam (2003)

(*Continued*)

TABLE 4.4 (*Continued*)
Effect of Extrusion Parameters on Expansion Ratio of Extrudates as Reported in the Literature

Extruder Conditions	Blend	Variables vs. ER	References
Corotating TSE with l/d of 9:1 and a die diameter of 5 mm, barrel temperature of 160°C, screw speed of 120–190 rpm and total feed rate of 25.4–43 kg/h, and SME of 500 kJ/kg	Maize flour (average particle size of 300 μm; 11%–14% protein, 50% amylose), wheat flour (average particle size of 120 μm; 7%–9% protein, 20% amylose)	Extrudate of maize flour has better ER compared to wheat flour due to difference in composition in terms of viscous characteristics time, starch, and protein.	Arhaliass et al. (2009)
Lab-scale TSE with l/d of 25:1 and die diameter of 4.5 mm, temperature profile of 30°C/80°C/120°C/160°C/160°C, screw speed of 150 rpm, feed rate of 1.8 kg/h, and feed moisture of 22%	Cornstarch replaced by navy or small red bean flour (15%–45%)	Increase in bean flour level significantly reduces ER due to interactions between protein, starch, and protein, that is, fiber would rupture cell walls and prevent air bubbles.	Anton et al. (2009)
Corotating TSE with l/d of 27:1 and circular die of 4 mm, temperature profile of 80°C/120°C, screw speed of 200 rpm, feed moisture 12%–17%, and feed rate of 25 kg/h	Wheat flour + brewer's spent grain; cornstarch + brewer's spent grain; wheat flour + red cabbage, cornstarch + red cabbage	An increase in feed moisture decreases ER as change in amylopectin molecular structure of the starch reduces the melt elasticity, resulting in low degree of starch gelatinization; further, ER varies with formulation. Moisture influences both thermal energy and shear stress, thus results in starch gelatinization.	Stojceska et al. (2009)
Corotating, self-wiping TSE with l/d of 25, circular die of 2.5 mm diameter, temperature profile of 50°C/50°C/70°C/110°C/150°C/150°C/160°C, screw speed of 300 rpm, feed rate of 25 kg/h, and water feed rate of 0.42 kg/h	Fenugreek flour (2%–10%) or debittered fenugreek polysaccharides (5%–20%), cowpea, and basmati rice (particle size less than 0.25 mm)	Increase in the proportion of chickpea in the blend upto 70% increases ER, further chickpea increases decrease ER due to high protein and DF contents in chickpea compared with rice. Proteins have the ability to affect water distribution in the matrix through their macromolecular structure and confirmation, which affects the extensional properties of the extruded melts. Protein would modify the viscoelastic properties of the melt as a result of competition for the available water between the starch and protein fractions, leading to a delay in starch gelatinization and consequently, lower moisture and expansion in the products. Inclusion of fenugreek has a negative impact on ER as fiber competes for the free water found in the matrix; further, the effect of fibers on extrudate expansion seems to be concentration dependent.	Shirani and Ganesharanee (2009)

(*Continued*)

TABLE 4.4 (*Continued*)
Effect of Extrusion Parameters on Expansion Ratio of Extrudates as Reported in the Literature

Extruder Conditions	Blend	Variables vs. ER	References
Lab-scale TSE with l/d of 28.2:1, temperature profile of 80°C/80°C/80°C/80°C/140°C/160°C/180°C/110°C, 125°C, or 140°C, screw speed of 350 rpm, feed rate of 11.5 kg/h, and water feed rate of 0.5 L/h	Rice flour, wheat semolina, corn grit, tomato paste, or skin powder at 20% with starch sources	Addition of tomato derivatives reduces ER because they lubricate the melt, and therefore, drop the SME and torque. Among the starch sources, rice flour has the highest ER due to the higher starch and lower fiber and fat content. The effect of temperature depends on the ingredients.	Dehghan-Shoar et al. (2010)
Corotating, self-wiping TSE with l/d of 25:1, die diameter of 2.5 mm, temperature of 90°C/110°C/80°C, screw speed of 130 rpm, and feed moisture content of 21%–22% wb	Foxtail millet flour (50%–70%), rice (5%–10%), amaranth (5%), Bengal gram (10%–30%), cowpea (5%–10%), particle size less than 2.5 mm	Addition of foxtail millet flour has negative impact on ER due to increase in DF. Protein affects water distribution in the matrix and its macro molecular structure and confirmation, thus influences expansion.	Deshpande and Poshadri (2011)
Corotating TSE with l/d of 25:1, single round die of 2.5 mm diameter, temperature profile of 50°C/50°C/70°C/110°C/150°C/150°C/160°C/118°C, screw speed of 306 rpm, motor, feed rate 11.5 kg/h, and water feed rate of 0.39 L/h	Yellow pea flour and rice flour (<1 mm) at 70:30 with 0%–20% guar gum (78% DF), locust bean gum (80% DF), or debittered fenugreek gum (85% DF)	Addition of gums decreases ER and can be attributed to premature breaking up of the expanding matrix film and also by the fiber effect of competing for free water found in the matrix but it is not different from control (without gums).	Ravindran et al. (2011)
Lab-scale TSE with 2.5 mm die, barrel temperature of 184°C, screw speed of 285 rpm, and feed moisture of 18% wb	Ragi (40%–50%), sorghum (10%–20%), and soy (5%–15%)	Ragi and sorghum levels negatively affect ER. Not only the barrel temperature facilitates interaction of high starch protein and formation of intermolecular disulfide bonds in the protein but also fiber enriched ingredients, results in extrudate, thereby ER.	Seth and Rajamanickam (2012)
		TSE—Extruder parameters	
Corotating, self-wiping, intermeshing TSE with l/d of 24:1, temperature profile of 0°C/70°C/100°C/150°C/150°C/150°C, screw speed of 400 rpm, moisture content of 15% wb, feed rate of 12 kg/h, type (kneading/reverse screw or their combination), position, and spacing (0–250 mm from die) of mixing elements	Rice flour	The kneading element facilitates the extent of starch breakdown in favor of product expansion. Severe screw profiles with reverse screw element, in combination with the kneading element, cause excessive molecular degradation, resulting in poor expansion. The positioning of mixing element also has influence on product expansion, which needs further investigation to establish the positioning. Among the mixing elements, the kneading element facilitates the expansion irrespective of positioning and spacing, indicating that there is no need for more than one mixing element for maximizing the expansion.	Choudhury and Gautam (1998b)

(*Continued*)

TABLE 4.4 (*Continued*)
Effect of Extrusion Parameters on Expansion Ratio of Extrudates as Reported in the Literature

Extruder Conditions	Blend	Variables vs. ER	References
Lab-scale corotating TSE with l/d of 35:1 and one slit (1 × 20 mm), screw speed of 250–350 rpm, and feed moisture content of 13%–17%	Brown rice (Patumthani 1/Supanburi 1) with a particle size of 40–60 mesh 70%, defatted soy flour 17.5%, full-fat soy flour 6%, sugar 6%, and calcium carbonate 0.5%	An increase in screw speed and decrease in moisture increases ER for both brown rice varieties.	Charunuch et al. (2003)
Conical, counter-rotating TSE with l/d of 62.5:1, and die diameter of 2 mm, product temperature of 100°C–260°C, screw speed of 150–250 rpm, residence time of 1–2.5 s, and feed moisture content of 13.2%–25% wb	Corn powder and sugar (10%)	Temperature and residence time have a positive influence on ER, whereas feed moisture influence is negative. An increase of residence time results in a degradation of amylopectin networks in the material that changes the ER and again it depends on temperature. ER decreases with moisture content increase which changes the amylopectin molecular structure in the starch reducing the melt elasticity.	Thymi et al. (2005)
Corotating TSE with l/d of 27:1 and a circular die of 4 mm, temperature profile of 80°C–120°C, screw speed of 250–350 rpm, solid feed rate of 20–25 kg/h, and water feed of 9%–11%	Wheat (16.4%–36.4%), cauliflower (0%–20%), corn (20%), oat, egg white, and milk powder (each 10%), tomato powder (5%), carrot powder (5%), onion powder (3%), salt (0.4%), dill (0.1%), and mint (0.1%)	The level of cauliflower has a negative correlation with ER due to fiber disrupting continuous structure of the melt that impedes elastic deformation during expansion.	Stojceska et al. (2008)
Lab-scale corotating TSE with l/d of 13:1 and slit die of 1.47 × 20 × 150 mm, temperature profile of 30°C/60°C/100°C/130°C/150°C, screw speed of 175 rpm, feed rate of 1.11 kg/h, feed moisture of 20.5% wb, screw configurations—medium (twin lead feed-kneading element-discharge), severe (twin lead feed-kneading element-single lead feed-kneading element-discharge)	Particle size distributions of barley grits are 81.6% (on 0.833 mm); 12.80% (on 0.420 mm); 4.2% (on 0.250 mm); 0.8% (on 0.177 mm); 0.3% (on 0.149 mm); 0.2% (on 0.125 mm); and 0.1% (<0.125 mm), whereas barley flour distributions are 12.1% (on 0.420 mm); 42.9% (on 0.250 mm); 38.9% (on 0.177 mm); 5.5% (on 0.149 mm); 0.4% (on 0.125 mm); and 0.2% (<0.125 mm).	Severe screw configuration improves mixing, and increase in chain splitting of starch and protein by severity of screw configuration increases the number of nucleation sites in the die and facilitates bubble growth and expansion. Severe screw configuration consists of kneading elements providing optimum mechanical and thermal energy input, which facilitates expansion. Authors found that SME has a positive effect on ER, whereas the particle sizes have no effect on ER.	Altan et al. (2009)

(*Continued*)

TABLE 4.4 (*Continued*)
Effect of Extrusion Parameters on Expansion Ratio of Extrudates as Reported in the Literature

Extruder Conditions	Blend	Variables vs. ER	References
Corotating TSE with l/d of 27:1 and a circular die of 4 mm, feed rate of 15–25 kg/h, temperature profile of 80°C/80°C–150°C, screw speed of 200–350 rpm, and water feed rate of 12%	Authors formulated the blend with 50% rice flour, 12.5% milk powder, 12.5% potato starch, 12.5% cornstarch, and 12.5% soy flour. They replaced rice flour with dry cranberry, beetroots, apple, carrot, and teff flour at a level of 30%.	ER decreases with addition of apple, cranberry, carrot, and beetroot, teff was similar to control. Increase in barrel temperature decreases ER due to excessive softening and structural degradation of the starch melt. In general, extrusion changes the aspect ratio, thereby forming a smaller particle that might interfere with bubble expansion, reducing the extensibility of the cell walls and causing premature rupture of steam cells in the extrudate microstructure.	Stojceska et al. (2010)
Corotating intermeshing TSE with l/d of 40:1 and two cylindrical dies of 2 mm, temperature profile of *appendix*, screw speed of 150–300 rpm, feed rate of 2 kg/h, and moisture content of 25%–40%	Sorghum passed through 2 mm with particle size distribution of (v/v) as: 10th percentile, 30 mm; 50th, 390 mm; 90th, 950 mm; volume weighted mean, 440 mm	ER depends on the moisture flash-off from the die and generation and retention of the resulting bubbles. Increase in moisture would reduce viscosity and increase melt elasticity, resulting in a thin cell wall that would not retain the generated bubbles/increase the collapsibility and thereby ER decrease; further it can be explained with glass transition concept. Glass transition temperature decreases with an increase in moisture that facilitates starch retrogradation/re-crystallization, resulting in less ER.	Mahasukhonthachat et al. (2010)
TSE with l/d of 25:1 and die diameter of 3 mm, zone 5 barrel temperature of 110°C–140°C, screw speed of 200 and 300 rpm, and feed moisture of 25%–35%	Wheat flour with 10% w/w ginseng powder blend	In general, increase in feed moisture decreases ER due to reduction in viscosity of the melt, resulting in small pressure differences between the die exit and atmosphere, facilitates the flow of the melt, reduces the mean residence time and degree of gelatinization, thus lowers ER. An increase in screw speed reduces ER due to decrease in mean residence time, melt viscosity, and pressure difference. An increase in barrel temperature increases the ER due to better degree of gelatinization, resulting in greater flashing-off of moisture upon die exit.	Chang and Ng (2011)
Corotating TSE with a circular die of 4 mm, barrel temperature profile of 80°C and 120°C, screw speed of 250 rpm, feed rate of 20 kg/h, and moisture content of 13%	Control: 30% wheat flour, 25% cornstarch, 25% potato starch, and 20% milk powder Fruit powder: 11% of the control formulation	Sugar and soluble fiber present in the food powder absorb moisture that affect the degree of gelatinization, thereby ER decreases. The order of ER is banana, apple, strawberry, and tangerine.	Potter et al. (2013)

(*Continued*)

TABLE 4.4 (*Continued*)
Effect of Extrusion Parameters on Expansion Ratio of Extrudates as Reported in the Literature

Extruder Conditions	Blend	Variables vs. ER	References
	TSE—Feed and extruder parameters		
Corotating TSE with l/d of 27:1 and two circular die of 4 mm, temperature profile of 80°C/110°C/110°C, screw speed of 100–300 rpm, feed moisture content of 11.8% wb, and feed rate of 42 kg/h	Chickpea flour (30%), maize flour (30%), oat flour (20%), cornstarch (15%), onion powder (5%), and brewer's spent grain (0%–30% of the maize flour)	ER increases with screw speed because of decrease in starch content due to addition of brewer's spent grain in the formulation.	Ainsworth et al. (2007)
Corotating TSE with l/d of 27:1 and circular die of 4 mm, temperature profile of 80°C/120°C, screw speeds of 250–350 rpm, feed rate of 20–25 kg/h, and water feed rate 9%–11%	Wheat flour (16.4%–36.4%), cauliflower (0%–20%), cornstarch (20%), egg white (10%), oat flour (10%), milk powder (10%), tomato powder (5%), carrot powder (5%), onion powder (3%), and paprika and salt (0.1%) with average cauliflower particle size less than 0.5 mm	The level of cauliflower has negative correlation with ER due to fiber molecules disrupting the continuous structure of the melt, impending its elastic deformation during expansion. At lower levels of cauliflower, the long and stiffer fiber molecules might align themselves in the extruder in the flow direction that reinforce the expanding matrix and increasing the mechanical resistance in the longitudinal direction.	Stojceska et al. (2008)
Corotating TSE with l/d of 8:1 and circular die of 4 mm diameter, temperature of 110°C–130°C, screw speed of 270–310 rpm, and the, feed moisture of 17%–21% wb, and feed rate of 4 kg/h	Rice flour (70%–90%), carrot pomace powder and pulse powder (10%–30%) with particles less than 2 mm	Rice flour proportion has a positive effect on ER due to starch content, thereby resulting in gelatinization; ER increases with moisture because of gelatinization, and beyond certain moisture levels ER decreases due to reduction of elasticity of dough through plasticization of melt; because of high mechanical shear ER increases with screw speed.	Kumar et al. (2010)
Corotating TSE with l/d of 27:1 and 4 mm circular die, temperature profile of 80°C/80°C–150°C, screw speed of 200 and 350 rpm, feed rate of 15–25 kg/h, and water feed of 12%	Rice flour (50%), milk powder (12.5%), soy flour (12.5%), cornstarch (12.5%), and potato starch (12.5%). Replace rice flour with dry cranberry, beetroots, apple, carrot, and teff flour at a level of 30%.	ER decreases with addition of DF from gluten-free sources, except teff flour (similar to control), due to difference in starch and DF components. Increasing temperature increases total DF levels, while decreasing ER, due to excessive softening and potential structural degradation of the starch melt, further breaking down components into smaller particles, which may interfere with bubble expansion, reducing the extensibility of the cell walls and causing premature rupture of steam cells in the extrudate microstructure.	Stojceska et al. (2010)

(Continued)

TABLE 4.4 (*Continued*)
Effect of Extrusion Parameters on Expansion Ratio of Extrudates as Reported in the Literature

Extruder Conditions	Blend	Variables vs. ER	References
Corotating TSE, temperature of 80°C/120°C–140°C, screw speed of 200–300 rpm, feed moisture of 17%–25% wb, and feed rate of 2.7 kg/h	White potato flour/split yellow pea flours 35/65, 50/50, 65/35 (w/w) with mean particle size of 220 μm	Starch (potato flour) has a positive effect on increasing expansion, while fiber and/or protein (peas) have a negative effect in lowering ER.	Nayak et al. (2011)
Corotating TSE with l/d of 40:1 and die with two opening of 2 mm diameter each, low-temperature profile of 25°C/50°C/50°C/75°C/90°C/100°C/100°C/100°C/100°C/100°C/100°C, high-temperature profile of 25°C/50°C/50°C/75°C/90°C/140°C/140°C/140°C/105°C/100°C/95°C, screw speed of 200 rpm, feed rate of 1.2 kg/h (barley) and 1.5 kg/h (sorghum), and dough moisture of 55% (barley) and 50% (sorghum)	Barley and sorghum—fine, <0.5 mm; medium, 0.5–1.0 mm; coarse, >1.0 mm	In general, interaction between temperature and moisture is critical for gelatinization. Direct addition of water during extrusion might not provide enough time for penetrating into large particles, leading to incomplete gelatinization and low melt viscosity, thus a decrease in ER. Because of water penetration issues, temperature effect on ER is masked.	Al-Rabadi et al. (2011)
Corotating TSE with 4 mm die, temperature of 108°C–141°C, screw speed of 242–377 rpm, feed moisture content of 18%, and feed rate of 4 kg/h	Rice flour (60%), wheat flour (40%), honey (1.6%–18.4%), particle size less than 2 mm	An increase in the amount of honey would reduce ER due to reduction in water availability; an increase in die temperature leads to high ER because of increase in degree of superheating of water in the extruder, encouraging bubble formation with a decrease in melt viscosity, similarly an increase in screw speed increases ER due to high mechanical shear.	Juvvi et al. (2012)
Lab-scale corotating TSE with l/d of 27:1 and circular die of 3 mm, barrel temperature of 80°C ± 5°C/75°C–105°C/100°C ± 10°C, screw speed of 300–350 rpm, feed rate of 15 ± 2 kg/h, and two blade cutter speed of 100 rpm	Corn flour, rice flour, and egg albumin/cheese powder at 40:40:20	Both egg albumin and cheese powder increase ER, egg albumin (35%) has a stronger influence on ER than that of cheese powder (30%) considering the blend of corn and rice flour at 50:50.	Kocherla et al. (2012)

(*Continued*)

TABLE 4.4 (*Continued*)
Effect of Extrusion Parameters on Expansion Ratio of Extrudates as Reported in the Literature

Extruder Conditions	Blend	Variables vs. ER	References
Lab-scale TSE with l/d of 29.3:1 and circular die of 3.1 mm diameter, temperature profile of 50°C/65°C/80°C/90°C/110°C/120°C, screw speed of 300 rpm, feed moisture content of 17.5%–25% wb, and feed rate of 2.1 kg/h	Corn flour and apple pomace at 100%:0%, 83%:17%, 78%:22%, and 72%:28%	ER decreases with increase in moisture due to low melt temperature, SME, and viscosity. Addition of apple pomace replaces starch, resulting in less extensibility. Insoluble fiber components exist with starch in a dispersed phase due to longitudinal alignment of fiber during extrusion causing rupture of air cells; further, fiber reduces the elastic properties of starch, lowers die swell, and reduces momentum for ER. The presence of pectin (apple pomace 11%–22%) enhances the longitudinal expansion, even as starch reduces due to apple pomace addition.	Elisa et al. (2012)
Corotating TSE with l/d of 8:1 and 4 mm die, barrel temperature of 110°C, screw speed of 280 rpm, feed moisture content of 13%–17% wb, and feed rate of 4 kg/h	Maize flour and spirulina flour at 85–95:5–15, average particle size < 2 mm	An increase of spirulina would decrease ER because of dilution in starch content and increase in protein content. Increase in screw speed decreases residence time and energy imparted to the melt thereby decreases ER.	Joshi et al. (2014)

TABLE 4.5
Wavelengths Used for Final Regression Model to Measure Extrudate Moisture Content and Bulk Density

Wavelength	Band Vibration	Moisture Content	Bulk Density
1400	O–H stretch	Oil	
1870	O–H stretch, C–O stretch cellulose	Cellulose	Cellulose
1908	O–H stretch,	—	P–OH
1942	O–H stretch, O–H bend	—	Water
1992	O–H stretch, C–O bend	Starch	Starch
2072	O–H stretch, O–H bend	—	Sucrose, starch, oil
2148	C–H bend	Oil	—
2462	C–H stretch, C–C stretch	—	Starch

Source: Dodds, S.A. and Heath, W.P., *Chemomet. Intell. Lab. Syst.*, 76, 37, 2005. With permission.

4 wavelengths, enough for estimating moisture content, and 6 for bulk density, with 2 wavelengths for both from the full spectrum of 698 wavelengths while producing corn-based expanded snack food in a pilot-scale twin-screw extrusion. However, initially a full spectrum probe is required (can be leased for short-term to reduce financial burden) for collecting full spectrum data; once reduced spectrum is constructed from full spectrum calibration, it is no longer necessary. The authors mentioned that calibration set was incomplete due to time constraint on collecting full spectrum data (Table 4.5).

It is a well-established fact that color is an important visual quality that influences consumer preference and purchase decision. Color correlates well with physical, chemical, and sensory quality indicators; further, it also plays a role in assessing the internal quality of foods. Existing commercial colorimeters can analyze external color of the foods and measure quite small areas; if the food surface is curved or uneven then the accuracy is limited. Computer vision system/image analysis technique can be a simple and powerful tool for extracting color information from curved or uneven food surfaces with nonhomogeneous coloration for food processing and quality control. Accordingly, Fan et al. (2013) extruded a blend of rice flour, cheese, edible oil, rice bran, and noli (color) powder at 80°C/100°C/110°C, 480 rpm, 60% wb, and 60 kg/h using a TSE. They analyzed the sample for surface images and textural characteristics using computer vision system (six color values h, s, I, L*, a*, and b*) and texture analyzer and correlated using linear fitting model. Based on the linear model, they reported that the hardness and gumminess score reflects directly by a* and *intensity* based on correlation coefficient of 0.9558, 0.9741, 0.9429, and 0.9619, respectively. The springiness can be calculated from hardness and gumminess scores, thus it reflects color values, indirectly. Further they used the collected data for the simulation processing in ANN showed a higher correlation coefficient of 0.9671 and 0.9856 than linear fitting model. The authors suggested that mounting computer vision system on a conveyor to monitor product quality combined with ANN would result in on-line quality control. However, the relationship between other product quality parameters and color should be considered before proceeding further.

Valadez-Blanco et al. (2007) introduced virtual white reference to overcome the sensitivity and reproducibility problems related to dark color measurements using fiber optic equipped spectrophotometer. In order to predict the final product color based on in-line measurements, multilayer feedforward artificial neural network with three hidden neurons provided an acceptable correlation between the in-line and off-line color measurements. In another study, an image analysis of extrudate produced through twin-screw extrusion of rice flour–glucose–lysine (70°C–130°C, 350–580 rpm, 22.5%–28% wb, 46.5–50.5 kg/h, and four different screw configurations) revealed that SME and product temperature had a significant influence on the color of the extrudates; and the relation between them is a fourth-degree polynomial (Lei et al. 2007). Extrudate images provide information about expansion, surface characteristics, and texture that can be related to product formulations, extrusion process (moisture content, barrel temperature, screw speed, and screw configurations), and system parameters (product temperature, and SME); thus, image analysis can be a useful tool for establishing an on-line control system and it is nondestructive and accurate.

Literature survey indicates that few researchers have worked on process control of food extrusion; therefore, there are several research opportunities for process control of food extrusion process as mentioned in this section in addition to the following: When the melt exits at tdie outlet, the vapor flash-off makes good sound. It can be explored as an inferential model for product expansion. Similarly, the amount of moisture escaping at the die exit should be quantified for inferential modeling of product density, expansion, and texture/hardness. A few on-line and in-line measurements of viscosity, rheology, residence time of the melt (Ainsworth et al. 1997, Robin et al. 2011, Horvat et al. 2012) have been found in the literature and that should be connected to final product quality; thereby process control can be established.

4.10 CONCLUSION

Literature on extrusion processing is growing especially on formulation or product development, nutrients of various foods. The chapter not only emphasized the importance of raw materials, effect of extrusion on expansion, and nutrients but also future potential research areas. Incorporation of natural colors through utilization of food industries' by-product residue after juice extraction would enhance the economics of the industry. Research on improving the retention of antioxidant, vitamins, and natural pigments would allow new healthy products. In addition, development of high protein and high fat, healthy and functional extruded products would meet the consumer demand. The need of the hour is that more research should be undertaken for understanding heat transfer during extrusion. Addressing those research areas would expand the extruded snack foods.

ACKNOWLEDGMENTS

The authors acknowledge the extruder manufacturers who have provided photos and drawings and permitted to be used in this chapter. Also, funding provided by Agricultural Experiment Station, South Dakota State University, Brookings, SD 57007 and Food and Nutrition Department, University of Wisconsin-Stout, Menomonie WI 54751 is greatly appreciated.

REFERENCES

Aboagye, Y. and Stanley, D.W. 1987. Thermoplastic extrusion of peanut flour by twin-screw extruder. *Canadian Institute of Food Science and Technology* 20: 148–153.

Adamidou, S., Nengas, I., Grigorakis, K., Nikolopoulou, D., and Jauncey, K. 2011. Chemical Composition and antinutritional factors of field peas (*Pisum sativum*), chickpeas (*Cicer arietinum*), and faba beans (*vicia faba*) as affected by extrusion preconditioning and drying temperatures. *Cereal Chemistry* 88(1): 80–86.

Agama, A.E., Ottenhof, M.A., Farhat, I.A., Paredes, O., Ortíz, D.J., and Bello, L.A. 2004. Efecto de la nixtamalización sobre las características moleculares del almidón de variedades pigmentadas de maíz. *Interciencia* 29(11): 643–649.

Ahn, H., Hsieh, F., Clarke, A.D., and Huff, H.E. 1999. Extrusion for producing low-fat pork and its use in sausage as affected by soy protein isolate. *Journal of Food Science* 64(2): 267–271.

Ainsworth, P., Ibanoglu, S., and Hayes, G D. 1997. Influence of process variables on residence time distribution and flow patterns of tarhana in a twin-screw extruder. *Journal of Food Engineering* 32: 101–108.

Ainsworth, P., Ibanoglu, S., Andrew Plunkett, A., Ibanoglu, E., and Stojceska, V. 2007. Effect of brewers spent grain addition and screw speed on the selected physical and nutritional properties of an extruded snack. *Journal of Food Engineering* 81: 702–709.

Alavi, S. and Rizvi, S.S.H. 2010. Supercritical fluid extrusion: A novel method for producing microcellular structures in starch based matrices. In: *Novel Food Processing: Effects on Rheological and Functional Properties*, Ahmed, J., Ramaswamy, H.S., Kasapis, S., and Boye, J.I., eds. CRC Press, Boca Raton, FL, pp. 403–420.

Al-Awwadi, N.A., Araiz, C., Bornet, A., Delbosc, C., Cristol, J.P., Linck, N., Ajay, J., Teissedre, P.L., and Cros, G. 2005. Extracts enriched in different polyphenolic families normalize increased cardiac NADPH oxidase expression while having differential effects on insulin resistance, hypertension, and cardiac hypertrophy in high-fructose-fed rats. *Journal of Agricultural and Food Chemistry* 53: 151–157.

ALINORM. 2009. Report of the 30th session of the Codex committee on nutrition and foods for special dietary uses 3–7. Appendix II, Cape Town, South Africa, 46p.

Alonso, R., Aguirre, A., and Marzo, F. 2000. Effect of extrusion and traditional processing methods on antinutrients and in vitro digestibility of protein and starch in faba and kidney beans. *Food Chemistry* 68: 159–165.

Altan, A., Mccarthy, K.L., and Maskan, M. 2009. Effect of extrusion cooking on functional properties and in vitro starch digestibility of barley-based extrudates from fruit and vegetable by-products. *Journal of Food Science* 74(2): E77–E86.

Altomare, R.E. and Ghossi, P. 1986. An analysis of residence time distribution patterns in a twin-screw extruder. *Biotechnology Progress* 2: 157–163.

Al-Rabadi, G.J., Torley, P.J., Williams, B.A., Bryden, W.L., and Gidley, M.J. 2011. Particle size of milled barley and sorghum and physico-chemical properties of grain following extrusion. *Journal of Food Engineering* 103: 464–472.

American Association of Cereal Chemists. 2001. The definition of dietary fibre—A report. *Cereal Foods World* 46: 1–112.

Anderson, A.K. and Ng, P.K.W. 2003. Physical and microstructural properties of wheat flour extrudates as affected by vital gluten addition and process conditions. *Food Science and Biotechnology* 12: 23–28.

Anton, A.A., Fulcher, R.G., and Arntfield, S.D. 2009. Physical and nutritional impact of fortification of corn starch-based extruded snacks with common bean (*Phaseolus vulgaris* L.) flour: Effects of bean addition and extrusion. *Food Chemistry* 113: 989–996.

Areas, J.A.G. 1992. Extrusion of food proteins. *Critical Reviews in Food Science and Nutrition* 32(4): 365–392.

Arhaliass, A., Legrand, J., Vauchel, P., Fodil-Pacha, F., Lamer, T., and Bouvier, J. 2009. The effect of wheat and maize flours properties on the expansion mechanism during extrusion cooking. *Food and Bioprocess Technology* 2: 186–193.

Artz, W.E., Warren, C., and Villota, R. 1990. Twin screw extrusion modification of a corn fiber and corn starch extruded blend. *Journal of Food Science* 55(3): 746–750.

Artz, W.E., Rao, S.K., and Sauer, R.M. 1992. Lipid oxidation in extruded products during storage as affected by extrusion temperature and selected antioxidants. In: *Food Extrusion Science and Technology*, Kokini, J.L., Ho, C.-T., and Karwe, M.V., eds. Marcel Dekker, Inc., New York, pp. 449–461.

Ascheri, J.L.R., Couri, S., and Madeira, E. 2006. *Características físicas de extrusados de arroz e inulina.* Comunicado Técnico e Embrapa Agroindústria de Alimentos, Rio de Janeiro e RJ.

Asp, N.G. and Bjorck, I. 1989. Nutritional properties of extruded foods. In: *Extrusion Cooking*, Mercier, C., Linko, P., and Harper, J.M., eds. American Association of Cereal Chemists, Inc., St. Paul, MN, pp. 399–434.

Asp, N.G. and Bjorck, I. 1981. Influence of extrusion cooking on the nutritional value. In: *Nordforsk-SIK-SNF Seminar: Extrusion Cooking of Food*, Gothenburg, Sweden.

Balandrán-Quintana, R.R., Barbosa-Cánovas, G.V., Zazueta-Morales, J.J., Anzaldúa-Morales, A., and Quintero-Ramos, A. 1998. Functional and nutritional properties of extruded whole pinto bean meal (*Phaseolus vulgaris* L.). *Journal of Food Science* 63(1): 113–116.

Batista, K.A., Prudencio, S.H., and Fernandes, K.F. 2010. Changes in the functional properties and antinutritional factors of extruded hard-to-cook common beans (*Phaseolus vulgaris, L.*). *Journal of Food Science* 75(3): C286–C290.

Beck, E.J., Tosh, S.M., Batterham, M.J., Tapsell, L.C., and Huang, X.-F. 2009. Oat β-glucan increases postprandial cholecystokinin levels, decreases insulin response and extends subjective satiety in overweight subjects. *Molecular Nutrition and Food Research* 53: 1343–1351.

Berk, Z. 2009. *Extrusion in Food Process Engineering and Technology*. Elsevier, Inc. San Diego, CA. ISBN: 978-0-12-373660-4.

Berrios, J.J. 2006. Extrusion cooking of legumes: Dry bean flours. *Encyclopedia of Agricultural, Food and Biological Engineering* 1: 1–8.

Bjorck, I. and Asp, N.G. 1983. The effects of extrusion cooking on nutritional value. *Journal of Food Engineering* 2: 281–308.

Bjorck, I., Asp, N.G., and Dahlqvist, A. 1984b. Protein nutritional value of extrusion-cooked wheat flours. *Food Chemistry* 15: 203–214.

Björck, I., Noguchi, A., Asp, N.G., Cheftel, J.C., and Dahlqvist, A. 1983. Protein nutritional value of a biscuit processed by extrusion cooking: Effects on available lysine. *Journal of Agricultural and Food Chemistry* 31: 488–492.

Blanche, S. and Sun, X. 2004. Physical characterization of starch extrudates as a function of melting transitions and extrusion conditions. *Advances in Polymer Technology* 23: 277–290.

Boonyasirikool, P. and Charunuch, C. 2000. Development of nutritious soy fortified snack by extrusion cooking. *Kasetsart Journal (Natural Science)* 34: 355–365.

Bouvier, J.M., Fayardt, G., and Clayton, J.T. 1987. Flow rate and heat transfer modeling in extrusion cooking of soy protein. *Journal of Food Engineering* 6: 123–141.

Brennan, M.A., Monro, J.A., and Brennan, C.S. 2008. Effect of inclusion of soluble and insoluble fibres into extruded breakfast cereal products made with reverse screw configuration. *International Journal of Food Science and Technology* 43: 2278–2288.

Brown, L., Rosner, B., Willett, W.W., and Sacks, F.M. 1999. Cholesterol lowering effects of dietary fiber: A meta-analysis. *American Journal of Clinical Nutrition* 69: 30–42.

Cabrera-Chávez, F., Calderón de la Barca, A.M., Islas-Rubio, A.R., Marti, A., Marengo, M., Pagani, M.A., Bonomi, F., and Iametti, S. 2012. Molecular rearrangements in extrusion processes for the production of amaranth-enriched, gluten-free rice pasta. *LWT—Food Science and Technology* 47: 421–426.

Camire, M.E. and King, C.C. 1991. Protein and fiber supplementation: Effects on extrudate cornmeal snack quality. *Journal of Food Science* 56(3): 760–763.

Camire, M.E., Camire, A., and Krumhar, K. 1990. Chemical and nutritional changes in foods during extrusion. *Critical Reviews in Food Science and Nutrition* 29(1): 35–57.

Camire, M.E., Dougherty, M.P., and Briggs, J.L. 2007. Functionality of fruit powders in extruded corn breakfast cereal. *Food Chemistry* 101: 765–770.

Camire, M.E. and Dougherty, M.P. 1998. Added phenolic compounds enhance lipid stability in extruded corn. *Journal of Food Science* 63(3): 516–518.

Camire, M.E. and Flint, S.I. 1991. Thermal processing effects on dietary fiber composition and hydration capacity in corn meal, oat meal, and potato peels. *Cereal Chemistry* 68(6): 645–647.

Camire, M.E., Chaovanalikit, A., Dougherty, M.P., and Briggs, J. 2002. Blueberry and grape anthocyanins as breakfast cereal colorants. *Journal of Food Science* 67(1): 438–441.

Camire, M.E., Violette, D., Dougherty, M.P., and McLaughlin, M.A. 1997. Potato peel dietary fiber composition: Effects of peeling and extrusion cooking processes. *Journal of Agricultural and Food Chemistry* 45: 1404–1408.

Camire, M.E., Zhao, J., and Violette, D. 1993. in vitro binding of bile acids by extruded potato peels. *Journal of Agricultural and Food Chemistry* 41: 2391–2394.

Carine, S., Xiangzhen, K., and Yufei, H. 2010. Optimization of extrusion on blend flour composed of corn, millet and soybean. *Pakistan Journal of Nutrition* 9(3): 291–297.

Carvalho, C.W.P. and Mitchell, J.R. 2000. Effect of sugar on the extrusion of maize grits and wheat flour. *International Journal of Food Science and Technology* 35: 569–576.

Cayot, N., Bounie, D., and Baussartz, H. 1995. Dynamic modelling for a twin screw food extruder: Analysis of the dynamic behaviour through process variables. *Journal of Food Engineering* 25: 245–260.

Cha, J.Y., Suparno, M., Dolan, K.D., and Ng, P.K.W. 2003. Modeling thermal and mechanical effects on retention of thiamin in extruded foods. *Journal of Food Science* 68(8): 2488–2496.

Chaiyakul, S., Jangchud, K., Jangchud, A., Wuttijumnong, P., and Winger, R. 2009. Effect of extrusion conditions on physical and chemical properties of high protein glutinous rice-based snack. *LWT—Food Science and Technology* 42: 781–787.

Chakraborty, S.K., Singh, D.S., Kumbhar, B.K., and Singh, D. 2009. Process parameter optimization for textural properties of ready-to-eat extruded snack food from millet and legume pieces blends. *Journal of Texture Studies* 40: 710–726.

Chang, Y.H. and Ng, P.K.W. 2011. Effects of extrusion process variables on quality properties of wheat-ginseng extrudates. *International Journal of Food Properties* 14: 914–925.

Chaovanalikit, A., Dougherty, M.P., Camire, M.E., and Briggs, J. 2003. Ascorbic acid fortification reduces anthocyanins in extruded blueberry-corn cereals. *Journal of Food Science* 68(6): 2136–2140.

Charunuch, C., Boonyasirikool, P., and Tiengpook, C. 2003. Physical properties of direct expansion extruded snack in utilization from Thai brown rice. *Kasetsart Journal (Natural Science)* 37: 368–378.

Chávez-Jáuregui, R.N., Silva, M.E.M.P., and Arêas, J.A.G. 2000. Extrusion cooking process for amaranth (*Amaranthus caudatus* L.). *Journal of Food Science* 65(6): 1009–1015.

Cheftel, J.C. 1986. Nutritional effects of extrusion cooking. *Food Chemistry* 20: 263–283.

Cheftel, J.C., Kitagawa, M., and Queguiner, C. 1992. New protein texturization processes by extrusion cooking at high moisture levels. *Food Reviews International* 8(2): 235–275.

Chinnaswamy, R. and Hanna, M.A. 1990. Relationship between viscosity and expansion properties of various extrusion-cooked grain components. *Food Hydrocolloids* 3: 423–434.

Chinnaswamy, R. and Hanna, M.A. 1988. Optimum extrusion-cooking conditions for maximum expansion of corn starch. *Journal of Food Science* 53: 834–840.

Choudhury, G.S. and Gautam, A. 1998a. Comparative study of mixing elements during twin-screw extrusion of rice flour. *Food Research International* 31(1): 7–17.

Choudhury, G.S. and Gautam, A. 1998b. On-line measurement of residence time distribution in a food extruder. *Journal of Food Science* 63(3): 529–534.

Choudhury, G.S. and Gautam, A. 2003. Hydrolyzed fish muscle as a modifier of rice flour extrudate characteristics. *Journal of Food Science* 68(5): 1713–1721.

Colonna, P. and Mercier, C. 1983. Macromolecular modifications of manioc starch components by extrusion cooking with and without lipids. *Carbohydrate Polymer* 3: 87–108.

Colonna, P., Tayeb, J., and Mercier, C. 1998. Extrusion cooking of starch and starchy products. In: *Extrusion Cooking*, Mercier, C., Linko, P., and Harper, J.M., eds. AACC International, St. Paul, MN, pp. 247–319.

Dansby, M.Y. and Bovell-Benjamin, A.C. 2003. Physical properties and sixth graders' acceptance of an extruded ready-to-eat sweet potato breakfast cereal. *Journal of Food Science* 68(8): 2607–2612.

Dehghan-Shoar, Z., Hardacre, A.K., and Brennan, C.S. 2010. The physico-chemical characteristics of extruded snacks enriched with tomato lycopene. *Food Chemistry* 123: 1117–1122.

Delgado, E., Vences-Montaño, M.I., Hernández Rodríguez, J.V., Rocha-Guzman, N., Rodriguez-Vidal, A., Herrera-Gonzalez, S.M., Medrano-Roldan, H., Solis-Soto, A., and Ibarra-Perez, F. 2012. Inhibition of the growth of rats by extruded snacks from bean (*Phaseolus vulgaris*) and corn (*Zea mays*). *Journal of Food Agriculture* 24(3): 255–263

Derby, R.I., Miller, B.S., Miller, B.F., and Trimbo, H.B. 1975. Visual observation of wheat–starch gelatinization in limited water systems. *Cereal Chemistry* 52: 702–713.

Deshpande, H.W. and Poshadri, A. 2011. Physical and sensory characteristics of extruded snacks prepared from Foxtail millet based composite flours. *International Food Research Journal* 18: 751–756.

Ding, Q., Ainsworth, P., Plunkett, A., Tucker, G., and Marson, H. 2006. The effect of extrusion conditions on the functional and physical properties of wheat-based expanded snacks. *Journal of Food Engineering* 73: 142–148.

Dodds S.A. and Heath, W.P. 2005. Construction of an online reduced-spectrum NIR calibration model from full-spectrum data. *Chemometrics and Intelligent Laboratory Systems* 76: 37–43.

EFSA Panel on Dietetic Products, Nutrition and Allergies (NDA). 2009. Scientific opinion on the substantiation of health claims related to β-glucans and maintenance of normal blood cholesterol concentrations. *EFSA Journal* 7: 1254–1272.

Elleuch, M., Bedigian, D., Roiseux, O., Besbes, S., Blecker, C., and Attia, H. 2011. Dietary fibre and fibre-rich by-products of food processing: Characterisation, technological functionality and commercial applications: A review. *Food Chemistry* 124(2): 411–421.

El-Samahy, S.K., Abd El-Hady, E.A., Habiba, R.A., and Moussa-Ayoub, T.E. 2007. Some functional, chemical, and sensory characteristics of cactus pear rice-based extrudates. *Journal of the Professional Association for Cactus Development* 9: 136–147.

Elsey, J., Riepenhausen, J., McKay, B., Barton, G.W., and Willis, M. 1997. Modeling and control of a food extrusion process. *Computers in Chemical Engineering* 21: S361–S366.

Emin, M.A., Mayer-Miebach, E., and Schuchmann, H.P. 2012. Retention of b-carotene as a model substance for lipophilic phytochemicals during extrusion cooking. *LWT—Food Science and Technology* 48: 302–307

Esposito, F., Arlotti, G., Bonifati, A.M., Napolitano, A., Vitale, D., and Fofliano, V. 2005. Antioxidant activity and dietary fibre in durum wheat bran by-products. *Food Research International* 38: 1167–1173.

Euromonitor International. 2012. *Breakfast Cereals*. Agri-Food Trade Service, Agriculture and Agri-Food Canada. http://www.ats-sea.agr.gc.ca/amr/6238-eng.htm. Accessed on May 15, 2013.

Evans, A.J, Huang, S., Osborne, B.G., Kotwal, Z., and Wesley I.J. 1999. Near infrared on-line measurement of degree of cook in extrusion processing of wheat flour. *Journal of Near Infrared Spectroscopy* 7: 77–84.

Fan, F.H., Ma, Q., Ge, J., Peng, Q.Y., Riley, W.W., and Tang, S.Z. 2013. Prediction of texture characteristics from extrusion food surface images using a computer vision system and artificial neural networks. *Journal of Food Engineering* 118(4): 426–433.

Fan, J., Mitchell, J.R., and Blanshard, J.M.V. 1996. The effect of sugars on the extrusion of maize grits: I. The role of the glass transition in determining product density and shape. *International Journal of Food Science and Technology* 31: 55–65.

Farid, F-P, Abdellah, A., Nadia, A-A., Lionel, B., and Jack, L. 2007. Fuzzy control of the start-up phase of the food extrusion process. *Food Control* 18: 1143–1148.

Faubion, J.M. and Hoseney, R.C. 1982. High-temperature short-time extrusion cooking of wheat starch and flour. II. Effect of protein and lipid on extrudate properties. *Cereal Chemistry* 59(6): 533–537.

FDA. 21 CFR Part 101. 1997. Food labeling, health claims; soluble dietary fiber from certain foods and coronary heart disease. *Federal Register* 62(15): 3584–3601.

Fellows, P. 2000. *Food Processing Technology: Principles and Practice*, 2nd edn. CRC Press, Boca Raton, FL. ISBN: 978-084-9308-87-1.

Ferdinand, J.M., Holly, M.L., Prescott, E.H.A., and Smith, A.C. 1989. Monitoring and control of the extrusion cooking process. In: *Process Engineering in the Food Industry: Development and Opportunities*, Field, R.W. and Howell, J.A., eds. Elsevier Applied Science, London, U.K., pp. 77–93.

Ferdinand, J.M., Holly, M.L., Prescott, E.H.A., Richmond, P., and Smith, A.C. 1988. Monitoring and optimization of the extrusion cooking process. In: *Automatic Control and Optimization of Food Process*, Renard, M. and Bimbenet, J.J., eds. Elsevier Applied Science, London, U.K., pp. 519–530.

Fernandes, M.S.F., Wang, S.H., Ascheri, J.L.R., Oliveira, M.F., and Costa, S.A.J. 2002. Produtos expandidos de misturas de canjiquinha e soja para uso como petiscos. *Pesquisa Agropecuária Brasileira* 37(10): 1495–1501.

Filli, K.B., Nkama, I., Jideani, V.A., and Abubakar, U.M. 2011. Application of response surface methodology for the study of composition of extruded millet-cowpea mixtures for the manufacture of fura: A Nigerian food. *African Journal of Food Science* 5(17): 884–896.

Fornal, L., Soral-Smietana, M., Smietana, Z., and Szpendowski, J. 1987. Chemical characteristics and physicochemical properties of the extruded mixtures of cereal starches. *Starch/Starke* 39(3): 75–78.

Gajula, H., Alavi, S., Adhikari, K., and Herald, T. 2008. Precooked bran-enriched wheat flour using extrusion: Dietary fiber profile and sensory characteristics. *Journal of Food Science* 73(4): S173–S179.

Giles, H.F., Wagner, J.R., and Mount, E.M. 2005. *Extrusion: The Definitive Processing Guide and Handbook*, Vol. 1. William Andrew, Inc., New York.

Gimenez, M.A., Gonzalez, R.J., Wagner, J., Torres, R., Lobo, M.O., and Samman, N.C. 2013. Effect of extrusion conditions on physicochemical and sensorial properties of corn-broad beans (*Vicia faba*) spaghetti type pasta. *Food Chemistry* 136: 538–545.

Gogoi, B.K. 1994. Effects of reverse screw elements on energy inputs, residence times and starch conversion during twin-screw extrusion of corn meal, Ph.D. dissertation. Rutgers University, New Brunswick, NJ.

Gogoi, B.K., Oswalt, A.J., and Choudhury, G.G. 1996. Reverse screw element(s) and feed composition effects during twin-screw extrusion of rice flour and fish muscle blends. *Journal of Food Science* 61: 590.

Gordon, T.D. 1989. Functional properties vs. physiological action of total dietary fiber. *Cereal Foods World* 34: 517–525.

Griffith, R.M. 1962. Fully developed flow in screw extruders, theoretical and experimental. *Industrial Engineering and Chemistry Fundamental* 2: 180–187.

Guth, H. and Grosch, W. 1994. Flavor changes of oatmeal extrusion products during storage. In: *Trends in Flavor Research*, Maarse, H. and van der Heij, D.G., eds. Elsevier Science Publishers, Amsterdam, the Netherlands, pp. 395–399.

Guy, R. 2001. *Extrusion Cooking: Technologies and Applications*. Woodhead Publishing, Cambridge, U.K. ISBN: 978-185-5735-59-0.

Guy, R.C.E. 1985. The extrusion revolution. *Food Manufacture* 60: 26–27.

Guy, R.C.E. and Horne, A.W. 1988. Extrusion and co-extrusion of cereals. In: *Food Structure—Its Creation and Evaluation*, Blanshard, J.M.V. and Mitchell, J.R., eds. Butterworths, London, U.K., pp. 331–349.

Guy, R.C E., Osborne, B.G., and Robert, P. 1996. The application of near infrared reflectance spectroscopy to measure the degree of processing in extrusion cooking processes. *Journal of Food Engineering* 21: 241–258

Guzman-Tello, R. and Cheftel, J.C. 1987. Thiamine loss during extrusion cooking as an indicator of the intensity of thermal processing. International *Journal of Food Science and Technology* 22: 549–562.

Haley, T.A. and Mulvaney, S.J. 1995. Advanced process control technique for the food industry. *Trends in Food Science & Technology* 6(4): 103–110.

Haley, T.A. and Mulvaney, S.J. 2000a. Online system identification and control design of an extrusion cooking process. Part I. System identification. *Food Control* 11: 103–120.

Haley, T.A. and Mulvaney, S.J. 2000b. Online system identification and control design of an extrusion cooking process. Part II. Model predictive and inferential control design. *Food Control* 11: 121–129.

Hampl, J.S., Betts, N.M., and Benes, B.A. 1998. The 'Age +5' rule; comparisons of dietary fiber intake among 4–10 year old children. *Journal of the American Dietetic Association* 98: 1418–1423.

Harper, J.M. 1988. Effect of extrusion processing on nutrients. In: *Nutritional Evaluation of Food Processing*, Karmas, E. and Harris, R.S., eds. Van Nostrand Reinhold, New York, pp. 365–391.

Harper, J.M. 1981. *Extrusion of Foods*, Vol. I. CRC Press, Boca Raton, FL.

Harper, J.M. 1994. Extrusion processing of starch. In: *Developments in Carbohydrate Chemistry*, 2nd edn., Alexander, R.J. and Zobel, H.F., eds. American Association of Cereal Chemists, St. Paul, MN, pp. 37–64.

He, J., Streiffer, R.H., Muntner, P., Krousel-Wood, M.A., and Whelton, P.K. 2004. Effect of dietary fiber intake on blood pressure: A randomized, double-blind, placebo controlled trial. *Journal of Hypertension* 22: 73–80.

Hofer, J. and Tan, J. 1993. Extrudate temperature control with disturbance prediction. *Food Control* 4: 17–24.

Horvat, M., Azad Emin, M., Hochstein, B., Willenbacher, N., and Schuchmann, H.P. 2012. A multiple-step slit die rheometer for rheological characterization of extruded starch melts. *Journal of Food Engineering* 116(2): 398–403.

Hsieh, F., Huff, H.E., and Peng, I.C. 1990. Studies of whole cottonseed processing with a twin-screw extruder. *Journal of Food Engineering* 12(4): 293–306.

Hsieh, F., Mulvaney, S.J., Huff, H.E., Lue, S., and Brent, J., Jr. 1989. Effect of dietary fiber and screw speed on some extrusion processing and product variables. *LWT—Food Science and Technology* 22: 204–207.

Huang, S. 1999. Using on-line NIR spectroscopy to measure product quality of extruded food products. In: *FSA NIR Workshop*, Sydney, New South Wales, Australia.

Huber, G.R. 2000. Twin screw extruders. In: *Extruders in Food Applications*, Riaz, M.N., ed. Technomic Publishing Co., Lancaster, PA, pp. 81–114.

Huber, G.R. 2001. Snack foods from cooking extruders. In: *Snack Foods Processing*, Lusas, E.W. and Rooney, R.W., eds. CRC Press, Boca Raton, FL, pp. 315–368.

Ilo, S. and Berghofer, E. 1998. Kinetics of thermomechanical loss of thiamin during extrusion cooking. *Journal of Food Science* 63(2): 312–316.

Ilo, S. and Berghofer, E. 2003. Kinetics of lysine and other amino acids loss during extrusion cooking of maize grits. *Journal of Food Science* 68(2): 496–502.

Ilo, S., Schoenlechner, R., and Berghofe, E. 2000. Role of lipids in the extrusion cooking processes. *Grasas y Aceites* 51(1–2): 97–110.

Iwe, M.O. 2000. Effects of extrusion cooking on some functional properties of soy-sweet potato mixtures—A response surface analysis. *Plant Food for Human Nutrition* 55(2): 169–184.

Iwe, M.O., Van zuilichem, D.J., Ngoddy, P.O., and Lammers, W. 2001. Amino acid and protein digestibility index of mixtures of extruded soy and sweet potato flours. *LWT—Food Science and Technology* 34: 71–75.

Iwe, M.O., Van zuilichem, D.J., Ngoddy, P.O., Lammers, W., and Stolp, W. 2004. Effect of extrusion cooking of soy–sweet potato mixtures on available lysine content and browning index of extrudates. *Journal of Food Engineering* 62: 143–150.

Jain, N.P. 2007. Formulation and extrusion of snack products for school children, M.S. thesis. University of Georgia, Athens, GA.

Jaluria, Y. 1996. Heat and mass transfer in the extrusion of non-Newtonian materials. *Advances in Heat Transfer* 28: 145–230.

Janes, D.A. and Guy, R.C.E. 1995. Metastable states in a food extrusion cooker. II: The effects of die resistance and minor ingredients with rice flour. *Journal of Food Engineering* 26: 61–175.

Jin, Z., Hsieh, F., and Huff, H.E. 1995. Effects of soy fiber, salt, sugar and screw speed on physical properties and microstructure of corn meal extrudate. *Journal of Cereal Science* 22: 185–194.

Joshi, S.S., Kuszynski, C.A., and Bagchi, D. 2001. The cellular and molecular basis of human health benefits of grape seed proanthocyanidin extract. *Current Pharmaceutical Biotechnology* 2: 187–200.

Joshi, S.M.R., Bera, M.B.B., and Panesar, P.S. 2014. Extrusion cooking of maize/spirulina mixture: Factors affecting expanded product characteristics and sensory quality. *Journal of Food Processing and Preservation* 38(2): 655–664.

Juvvi, P., Sharma, S., and Nanda, V. 2014. Optimization of process variables to develop honey based extruded product. *African Journal of Food Science* 6(10): 253–268.

Jyothi, A.N., Sheriff, J.T., and Sajeev, M.S. 2008. Physical and functional properties of arrowroot starch extrudates. *Journal of Food Science* 74(2): E97–E104.

Elisa, L. Karkle, E.L., Alavi, S., and Dogan, H. 2012. Cellular architecture and its relationship with mechanical properties in expanded extrudates containing apple pomace. *Food Research International* 46: 10–21.

Kervinen, R., Linko, P., Sourtti, T., and Olkku, J. 1984. Wheat starch extrusion cooking with acid or alkali. In: *Thermal Processing and Quality of Foods*, Zauthen, P., ed. Elsevier, London, U.K., pp. 175–179.

Khanal, R.C., Howard, L.R., and Prior, R.L. 2009a. Procyanidin content of grape seed and pomace, and total anthocyanin content of grape pomace as affected by extrusion processing. *Journal of Food Science* 74(6): H174–H182.

Khanal, R.C., Howard, L.R., Brownmiller, C.R., and Prior, R.L. 2009b. Influence of extrusion processing on procyanidin composition and total anthocyanin contents of blueberry. *Pomace Journal of Food Science* 74(2): H52–H58.

Killeit, U. 1994. Vitamin retention in extrusion cooking. *Food Chemistry* 49: 149–155.

Kokini, J.L., Chang, C.N., and Lai, L.S. 1992. The role of rheological properties on extrudate expansion. In: *Food Extrusion Science and Technology*, Kokini, J.L., Ho, C.T., and Marwe, M.V., eds. Mercel Dekker, New York, pp. 631–652.

Kong, J., Dougherty, M.P., Perkins, L.B., and Camire, M.E. 2008. Composition and consumer acceptability of a novel extrusion-cooked salmon snack. *Journal of Food Science* 73(3): S118–S123.

Konstance, R.P., Onwulata, C.I., Smith, P.W., Lu, D., Tunick, M.H., Strange, E.D., and Holsinger, V.H. 1998. Nutrient-based corn and soy products by twin-screw extrusion. *Journal of Food Science* 63(5): 1–5.

Kulshreshta, M.K., Zaror, C.A., and Jukes, D.J. 1995. Simulating the performance of a control system for food extruders using model-based set-point adjustment. *Food Control* 6(3): 135–141.

Kumar, N., Sarkar, B.C., and Sharma, H.K. 2010. Development and characterization of extruded product of carrot pomace, rice flour and pulse powder. *African Journal of Food Science* 4(11): 703–717.

Lai, C.S., Guetzlaff, J., and Hoseney, R.C. 1989. Role of sodium bicarbonate and trapped air in extrusion. *Cereal Chemistry* 66: 69–71.

Larrea, M.A., Chang, Y.K., and Bustos, F.M. 2005b. Effect of some operational extrusion parameters on the constituents of orange pulp. *Food Chemistry* 89: 301–308.

Larrea, M.A., Chang, Y.K., and Martınez Bustos, F. 2005. Effect of some operational extrusion parameters on the constituents of orange pulp. *Food Chemistry* 89: 301–308.

Lazou, A. and Krokida, M. 2010. Functional properties of corn and corn–lentil extrudates. *Food Research International* 43: 609–616.

Lee, S.J., Hong, C.G., Han, T.S., Kang, J.Y., and Kwon, Y.A. 2002. Application of fuzzy control to start-up of twin screw extruder. *Food Control* 13: 301–306.

Lei H., Fulcher R.G., Ruan, R., and van Lengerich, B. 2007. Assessment of color development due to twin-screw extrusion of rice–glucose–lysine blend using image analysis. *LWT—Food Science and Technology* 40(2007): 1224–1231.

Levine, L. 2010. Engineering: Heat transfer in extruders II—Simultaneous heating and cooling**.** *Cereal Foods World* 55(1): 41–42.

Levine, L., Symes, S., and Weimer, J. 1986. Automatic control of moisture in food extruders. *Journal of Food Engineering* 8: 97–115.

Li, H., Long, D., Peng, J., Ming, J., and Zhao, G. 2012. A novel in-situ enhanced blasting extrusion technique—Extrudate analysis and optimization of processing conditions with okara. *Innovative Food Science and Emerging Technologies* 16: 80–88

Li, S.-Q., Zhang, H.Q., Jin, Z.T., and Hsieh, F.-H. 2005. Textural modification of soya bean/corn extrudates as affected by moisture content, screw speed and soya bean concentration. *International Journal of Food Science and Technology* 40: 731–741.

Lima, I., Guraya, H., and Champagne, E. 2002. The functional effectiveness of reprocessed rice bran as an ingredient in bakery products. *Nahrung/Food* 46(2): 112–117.

Lin, P. and Jaluria, Y. 1998. Conjugate thermal transport in the channel of an extruder for non-Newtonian fluids. *International Journal of Heat and Mass Transfer* 41: 3239–3253.

Lin, S., Hsieh, F., and Huff, H.E. 1997. Effects of lipids and processing conditions on degree of starch gelatinization of extruded dry pet food. *LWT—Food Science and Technology* 30: 754–761.

Liu, Y., Hsieh, F., Heymann, H., and Huff, H.E. 2000. Effect of process conditions on the physical and sensory properties of extruded oat–corn puff. *Journal of Food Science* 65(7): 1253–1259.

Lobato, L.P., Anibal, D., Lazaretti, M.M., and Grossmann, M.V.E. 2011. Extruded puffed functional ingredient with oat bran and soy flour. *LWT—Food Science and Technology* 44: 933–939.

Lu, Q., Hsieh, F., Mulvaney, S.J., Tan, J., and Huff, M.E. 1992. Dynamic analysis of process variables for a twin-screw food extruder. *Lebensmittel Wissenschafi und Technologie* 25: 261–270.

Lu, Q., Mulvaney, S.J., Hsieh, F., and Huff, H.E. 1993. Model and strategies for computer control of a twin-extruder. *Food Control* 4: 25–33.

Lue, S., Hsieh, F., and Huff, H.E. 1991. Extrusion cooking of corn meal and sugar beet fiber: Effects on expansion properties, starch gelatinization, and dietary fiber content. *Cereal Chemistry* 68: 227–234.

Lue, S., Hsieh, F., Peng, I.C., and Huff, H.E. 1990. Expansion of corn extrudates containing dietary fibre: A microstructure study. *LWT—Food Science and Technology* 23: 165–173.

Mahasukhonthachat, K., Sopade, P.A., and Gidley, M.J. 2010. Kinetics of starch digestion and functional properties of twin-screw extruded sorghum. *Journal of Cereal Science* 51: 392–401.

Market Indicator Report. 2011. Consumer trends salty snack food in the united states 2011. Agriculture and Agri-Food Canada, Ottawa, ON, Canada .

Martelli, F.G. 1983. *Twin-Screw Extruders: A Basic Understanding*. Van Nostrand Reinhold, New York.

Martinez-Flores, H.E., Chang, Y.K., Martínez-Bustos, F., and Sanchez-Sinencio, F. 2008. Extrusion-cooking of cassava starch with different fiber sources: Effect of fibers on expansion and physicochemical properties. In: *Advances in Extrusion Technology*, Chang, Y.K. and Wang, S.S., eds. Technomic Publishing Co., Inc., Lancaster, PA, pp. 271–278.

Matthey, F.P. and Hanna, M.A. 1997. Physical and functional properties of twin-screw extruded whey protein concentrate–corn starch blends. *LWT—Food Science and Technology* 30: 359–366.

Mehta, R.S. 2005. Dietary fiber benefits. *Cereal Foods World* 50(1): 66–71.

Méndez-García, S., Martínez-Flores, H.E., and Morales-Sánchez, E. 2011. Effect of extrusion parameters on some properties of dietary fiber from lemon (*Citrus aurantifolia* Swingle) residues. *African Journal of Biotechnology* 10(73): 16589–16593.

Miller, R.C. and Mulvaney, S.J. 2000. Unit operations and equipment IV. Extrusion and extruders. In: *Breakfast Cereals and How They Are Made*, 2nd edn. Fast, R.B. and Caldwell, E.F., eds. American Association of Cereal Chemists, Inc., St. Paul, MN, pp. 215–277.

Mitchell, J.R. and Areas, J.A. 1992. Structural changes in biopolymers during extrusion. In: *Food Extrusion Science and Technology*, Kokini, J.L., Ho, C.T., and Karwe, M.V., eds. Mercel Dekker, New York, pp. 345–360.

Mohamed, I.O. and Ofoli, R.Y. 1990. Prediction of temperature profiles in twin screw extruders. *Journal of Food Engineering* 12: 145–164.

Moley, W. 2013. The challenges for ingredients in foods. Retrieved from http://www.foodstuffsa.co.za/news-stuff/food-science-and-technology-stuff/2398-the-challenges-for-ingredients-in-foods, accessed on February 11, 2013.

Moreira, R.G., Srivastava, A.K., and Gerrish, J.B. 1990. Feedforward control model for a twin-screw food extruder. *Food Control* 1(7): 179–184.

Moscicki, L. 2011. *Extrusion-Cooking Techniques: Applications, Theory and Sustainability*. Wiley-VCH Verlag and Co. KGaA, Weinheim, Germany.

Mukhopadhyay, N. and Bandyopadhyay, S. 2003. Extrusion cooking technology employed to reduce the anti-nutritional factor tannin in sesame (*Sesamum indicum*) meal. *Journal of Food Engineering* 56: 201–202.

Mulvaney, S.J., Hsieh, F., Onwulata, C., Brent, J., Jr., and Huff, H.E. 1988. Computer control and modeling of an extruder: Dynamics. In: *International Winter Meeting of the American Society of Agricultural Engineers*, Paper No. 88–6517, Chicago, IL, December 13–16, 1988.

Murphy, M.G., Skonberg, D.I., Camire, M.E., Dougherty, M.P., Bayer, R.C., and Briggs, J.L. 2003. Chemical composition and physical properties of extruded snacks containing crab-processing by-product. *Journal of the Science of Food and Agriculture* 83: 1163–1167.

Muthukumarappan, K. and Karunanithy, C. 2012. Extrusion process design. In: *Handbook of Food Process Design*, Ahmed, J. and Rahman, S., eds. Wiley-Blackwell, Hoboken, New Jersy. pp. 710–742. ISBN: 978-1-4443-3011-3.

Namiki, M. 1990. Antioxidants/antimutagens in food. *Critical Reviews in Food Science and Nutrition* 29: 273–300.

Nascimento, E.M.G.C., Carvalho, C.W.P., Takeiti, C.Y., Freitas, D.D.G.C., and Ascheri, J.L.R. 2012. Use of sesame oil cake (*Sesamum indicum* L.) on corn expanded extrudates. *Food Research International* 45: 434–443.

Nayak, B., Berrios, J.D.J, Powers, J.R., and Tang, J. 2011. Effect of extrusion on the antioxidant capacity and color attributes of expanded extrudates prepared from purple potato and yellow pea flour mixes. *Journal of Food Science* 76(6): C874–C883.

Nejman, N., Shakourzadeh, K., Bouvier, J.M., and Martin, T. 1990. State reproducibility and relative gain analysis in twin-screw extrusion cooking. In: *Proceedings of the International Symposium ACOFOP II*, Paris, France, November 14–15, 1990.

Niness, K. 1999. Breakfast foods and the health benefits of inulin and oligofructose. *Cereal Foods World* 44(2): 79–81.

Noguchi, A., Mosso, K., Aymanrd, C., Jevnink, J. and Cheftel, J.C. 1982. Millard reactions during extrusion cooking of protein enriched biscuits. *Lebensm Wissen and Technology* 15: 105–110.

O'Brien, J. and Morrissey, P.A. 1989. Nutritional and toxicological aspects of the Maillard browning reactions in foods. *Critical Reviews in Food Science and Nutrition* 28: 211–248.

Oduro-Yeboah, C., Onwulata, C., Tortoe, C., and Thomas-gahring, A. 2014. Functional properties of plantain, cowpea flours and oat fiber in extruded products. *Journal of Food Processing and Preservation* 38(1): 347–355.

Olapade, A.A. and Aworh, O.C. 2012. Evaluation of extruded snacks from blends of acha (*Digitaria exilis*) and cowpea (*Vigna unguiculata*) flours. *Agricultural Engineering International: CIGR Journal* 14(3): 210–217.

Onwulata, C.I. 2010. Texturization of dairy proteins for food applications. *Food Engineering Ingredients* 35: 8–11.

Onwulata, C.I., Konstance, R.P., Smith, P.W., and Holsinger, V.H. 1998. Physical properties of extruded products as affected by cheese whey. *Journal of Food Science* 63(5): 1–5.

Onwulata, C.I., Konstance, R.P., Smith, P.W., and Holsinger, V.H. 2001a. Co-extrusion of dietary fiber and milk proteins in expanded corn products. *LWT—Food Science and Technology* 34: 424–429.

Onwulata, C.I., Mulvaney, S.J., and Hsieh, F. 1994. System analysis as the basis for control of density of extruded cornmeal. *Food Control* 5: 39–48.

Onwulata, C.I., Smith, P.W., Konstance, R.P., and Holsinger, V.H. 2001b. Incorporation of whey products in extruded corn, potato or rice snacks. *Food Research International* 34: 679–687.

Onyango, C., Noetzold, H., Ziems, A., Hofmann, T., Bley, T., and Henle, T. 2005. Digestibility and antinutrient properties of acidified and extruded maize–finger millet blend in the production **of** uji. *LWT—Food Science and Technology* 38: 697–707.

Özer, E.A., Herken, E.N., Guzel, S., Ainsworth, P., and Ibanoglu, S. 2006. Effect of extrusion process on the antioxidant activity and total phenolics in a nutritious snack food. *International Journal of Food Science and Technology* 41: 289–293.

Ozer, E.A., Tbanoglu, S., Ainsworth, P., and Yagmur, C. 2004. Expansion characteristics of a nutritious extruded snack food using response surface methodology. *European Food Research and Technology* 218: 474–479.

Paes, M.C.D. and Maga, J. 2004. Effect of extrusion on essential amino acids profile and color of whole-grain flours of quality protein maize and normal maize cultivars. *Revista Brasileira de Milho e Sorgo* 3(1): 10–20.

Pan, Z., Zhang, S., and Jane, J. 1998. Effects of extrusion variables and chemicals on properties of starch-based binders and processing conditions. *Cereal Chemistry* 75(4): 541–546.

Pastor-Cavada, E., Drago, S.R., González, R.J., Juan, R., Pastor, J.E., Alaiz, M., and Vioque, J. 2011. Effects of the addition of wild legumes (*Lathyrus annuus* and *Lathyrus clymenum*) on the physical and nutritional properties of extruded products based on whole corn and brown rice. *Food Chemistry* 128: 961–967.

Pelembe, L.A.M., Erasmusw, C., and Taylor, J.R.N. 2002. Development of a protein-rich composite sorghum–cowpea instant porridge by extrusion cooking process. *LWT—Food Science and Technology* 35: 120–127.

Pearson, J.R.A. 1966. Mechanical principles of polymer melt processing. *Transactions of the Society of Rheology* 13: 357–365.

Pilli, T.D., Derossi, A., Talja, R.A., Jouppila, K., and Severini, C. 2011. Study of starch-lipid complexes in model system and real food produced using extrusion-cooking technology. *Innovative Food Science and Emerging Technologies* 12: 610–616.

Pokorny, J. and Schmidt, S. 2006. Natural antioxidant functionality during food processing. In: *Antioxidants in Food*, Pokorny, J., Yanishlieva, N., and Gordon, M., eds. CRC Press, Washington, DC, pp. 331–351.

Popescu, O., Popescu, D.C., Wilder, J., and Karwe, M.V. 2001. A new approach to modeling and control of a food extrusion process using neural networks and expert system. *Journal of Food Process Engineering* 24(1): 17–36.

Poskitt, M.E. and Morgan, J.B. 2005. Infancy, childhood and adolesence. In: *Human Nutrition*, Geissler, C. and Powers, H., eds. Elsevier, London, U.K., pp. 275–298.

Potter, R., Stojceska, V., and Plunkett, A. 2012. The use of fruit powders in extruded snacks suitable for Children's diets. *LWT—Food Science and Technology* 51(2): 537–544.

Priyanka, K., Aparna, K., and Lakshmi, D.N. 2012. Development and evaluation of ready to eat extruded snack using egg albumin powder and cheese powder. *Agricultural Engineering International: CIGR Journal* 14(4): 179–187.

Prudencio-Ferreira, S.H. and Areas, J.A.G. 1993. Protein-protein interactions in the extrusion of soya at various temperatures and moisture contents. *Journal of Food Science* 58: 378–381.

Pszczola, D.E. 2010. Making snack statements. *Food Technology* 64: 59–75.

Rampersad, R., Badrie, N., and Comissiong, E. 2003. Physico-chemical and sensory characteristics of flavored snacks from extruded cassava/pigeonpea flour. *Journal of Food Science* 68(1): 363–367.

Ravindran, G. and Hardacre, A. 2010. Galactomannans enhance the functionality of cereal-legume based extruded nutritional snacks. In: *International Conference on Food Innovation*, Spain, October 25–29.

Ravindran, G., Carr, A., and Hardacre, A. 2011. A comparative study of the effects of three galactomannans on the functionality of extruded pea–rice blends. *Food Chemistry* 124: 1620–1626.

Razzaq, M.R., Anjum, F.M., Khan, M.I., Khan, M.R., Nadeem, M., Javed, M.S., and Sajid, M.W. 2012. Effect of temperature, screw speed and moisture variations on extrusion cooking behavior of maize (*Zea mays*. L). *Pakistan Journal of Food Science* 22(1): 12–22.

Reyes-Jáquez, D., Casillas, F., Flores, N., Andrade-González, I., Solís-Soto, A., Medrano-Roldán, H., Carrete, F., and Delgado, E. 2012. The effect of glandless cottonseed meal content and process parameters on the functional properties of snacks during extrusion cooking. *Food and Nutrition Sciences* 3: 1716–1725.

Riaz, M.N., ed. 2000. *Extruders in Food Applications*. Technomic Publishing, Lancaster, PA.

Riaz, M.N. 2005. Extrusion processing of oilseed meals for food and feed production. In: *Bailey's Industrial Oil and Fat Products*, 6th edn., Shahidi, F., ed. John Wiley and Sons, Inc., Hoboken, New Jersy.

Robin, F., Bovet, N., Pineau, N., Schuchmann, H.P., and Palzer, S. 2011. Online shear viscosity measurement of starchy melts enriched in wheat bran. *Journal of Food Science* 76(5): E405–E412.

Robin, F., Schuchmann, H.P., and Palzer, S. 2012. Dietary fiber in extruded cereals: Limitations and opportunities. *Trends in Food Science & Technology* 28: 23–32.

Rokey, G.J. 2000. Single screw extruder. In: *Extruders in Food Applications*, Riaz, M.N, ed. Technomic Publishing, Lancaster, PA, pp. 30–31.

Rodríguez-Miranda, J., Ruiz-López, I.I., Herman-Lara, E., Martínez-Sánchez, C.E., Delgado-Licon, E., and Vivar-Vera, M.A. 2011. Development of extruded snacks using taro (*Colocasia esculenta*) and nixtamalized maize (*Zea mays*) flour blends. *LWT—Food Science and Technology* 44: 673–680.

Semasaka, C., Kong, X., and Hua, Y. 2010. Optimization of extrusion on blend flour composed of corn, millet and soybean. *Pakistan Journal of Nutrition* 9(3): 291–297.

Schweizer, T.F., Reimann, S., Solms, J., Eliasson, A.C., Asp, N.-G. 1986. Influence of drum drying and twin screw extrusion cooking on wheat carbohydrates. II. Effects of lipids on physical properties, degradation and complex formation of starch in wheat flour. *Journal of Cereal Science* 4: 249–260.

Seiz, K. 2006. Easily boost fiber content. *Baking Manage* 10(5): 42–44.

Senanayake, S.A.M.A.N.S. and Clarke, B. 1999. A simplified twin screw co-rotating food extruder: Design, fabrication, and testing. *Journal of Food Engineering* 40: 129–137.

Seth, D. and Rajamanickam, G. 2012. Development of extruded snacks using soy, sorghum, millet and rice blend—A response surface methodology approach. *International Journal of Food Science and Technology* 47: 1526–1531.

Sharma, P., Gujral, H.S., and Singh, B. 2012. Antioxidant activity of barley as affected by extrusion cooking. *Food Chemistry* 131: 1406–1413.

Sharma, P. and Gujral, H.S. 2011. Effect of sand roasting and microwave cooking on antioxidant activity of barley. *Food Research International* 44: 235–240.

Shirani, G. and Ganesharanee, R. 2009. Extruded products with Fenugreek (*Trigonella foenum-graecium*) chickpea and rice: Physical properties, sensory acceptability and glycaemic index. *Journal of Food Engineering* 90: 44–52.

Singh, B. and Mulvaney, S.J. 1994. Modeling and process control of twin screw cooking food extruders. *Journal of Food Engineering* 23(4): 403–428.

Singh, B., Sekhon, K.S., and Singh, N. 2007. Effects of moisture, temperature and level of pea grits on extrusion behaviour and product characteristics of rice. *Food Chemistry* 100: 198–202.

Singh, S., Gamlath, S., and Wakeling, L. 2007. Nutritional aspects of food extrusion: A review. *International Journal Food Science Technology* 42(8): 916–929.

Skerget, M., Kotnik, P., Hadolin, M., Hras, A.R., Simonic, M., and Knez, Z. 2005. Phenols, proanthocyanidins, flavones and flavonols in some plat materials and their antioxidant activities. *Food Chemistry* 89: 191–198.

Slavin, J. 2003. Impact of the proposed definition of dietary fibre on nutrient databases. *Journal of Consumption and Analysis* 16: 287–291.

Smith, A.C. 1992. Studies on the physical structure of starch-based materials in the extrusion cooking process. In: *Food Extrusion Science and Technology*, Kokini, J.L., Ho, C., and Karwe, M.V., eds. Marcel Dekker, New York, pp. 573–618.

Sobota, A., Sykut-Domanska, E., and Rzedzicki, Z. 2012. Effect of extrusion-cooking process on the chemical composition of corn–wheat extrudates, with particular emphasis on dietary fibre fractions. *Polish Journal of Food and Nutrition Sciences* 60(3): 251–259.

Sobukola, O.P., Babajide, J.M., and Ogunsade, O. 2013. Effect of brewers spent grain addition and Extrusion parameters on some properties of extruded yam starch-based pasta. *Journal of Food Processing and Preservation* 37(5): 734–743.

Sokhey, A.S., Kollengode, A.N., and Hanna, M.A. 1994. Screw configuration effects on corn starch expansion during extrusion. *Journal of Food Science* 59(4): 1365–2621.

Sopade, P.A. and Le Grys, G.A. 1991. Effect of added sucrose on extrusion cooking of maize starch. *Food Control* 2: 103–109.

Sosa-Moguel, O., Ruiz-Ruiz, J., Nez-Ayala, A.M., González, R., Drago, S., Betancur-Ancona, D., and Chel-Guerrero, L. 2009. Effect of extrusion conditions and lipoxygenase inactivation treatment on the physical and nutritional properties of corn/cowpea (*Vigna unguiculata*) blends. *International Journal of Food Sciences and Nutrition* 60(S7): 341–354.

Steel, C.J., Leoro, M.G.V., Schmiele, M., Ferreira, R.E., and Chang, Y.K. 2012. *Thermoplastic Extrusion in Food Processing, Thermoplastic Elastomers*. InTech. Rijeka, Croatia. ISBN: 978-953-51-0346-2, doi: 10.5772/36874.

Stojceska, V., Ainsworth, P., Plunkett, A. and Ibanoglu, S. 2010. The advantage of using extrusion processing for increasing dietary fibre level in gluten-free products. *Food Chemistry* 121: 156–164.

Stojceska, V., Ainsworth, P., Plunkett, A., and Ibanoglu, S. 2009. The effect of extrusion cooking using different water feed rates on the quality of ready-to-eat snacks made from food by-products. *Food Chemistry* 114: 226–232.

Stojceska, V., Ainsworth, P., Plunkett, A., Ibanoglu, E., and Ibanoglu, S.I. 2008. Cauliflower by-products as a new source of dietary fibre, antioxidants and proteins in cereal based ready-to-eat expanded snacks. *Journal of Food Engineering* 87: 554–563.

Strahm, B. 1998. Fundamentals of polymer science as an applied extrusion tool. *Cereal Foods World* 43(8): 621–625.

Suknark, K., Lee, J., Eitenmiller, R.R., and Phillips, R.D. 2001. Stability of tocopherols and retinyl palmitate in snack extrudates. *Journal of Food Science* 66(6): 897–902.

Suknark, K., McWatters, K.H., and Phillips, R.D. 1998. Acceptance by American and Asian consumers of extruded fish and peanut snack products. *Journal of Food Science* 63(4): 721–725.

Suknark, K., Phillips, R.D., and Huang, Y.W. 1999. Tapioca-fish and tapioca-peanut snacks by twin-screw extrusion and deep-fat frying. *Journal of Food Science* 64(2): 303–308.

Tan, J. and Hofer, J.M. 1995. Self-tuning predictive control of processing temperature for food extrusion. *Journal of Process Control* 5(3): 183–189.

Tan, J. and Wang, Y. 1998. Dual-target predictive control and application in food extrusion. In: *Automatic Control of Food and Biological Processes*, SIK, Goteborg, Sweden.

Tapola, N., Karvonen, H., Niskanen, L., Mikola, M., and Sarkkinen, E. 2005. Glycemic responses of oat bran products in type 2 diabetic patients. *Nutrition, Metabolism & Cardiovascular Diseases* 15: 255–261.

Tappy, L., Gugolz, E., and Wursch, P. 1996. Effects of breakfast cereals containing various amounts of β-glucan fibers on plasma glucose and insulin responses in NIDDM subjects. *Diabetes Care* 19: 831–834.

Theander, O. and Westerlund, E. 1987. Studies on chemical modification in heat processed starch and wheat flour. *Starch/Starke* 39: 88–93.

Thomas, R G., Pehrsson, P R., Ahuja, J KC, Smieja, E., and Miller, K.B. 2013. Recent trends in ready-to-eat breakfast cereals in the U.S. *Procedia Food Science* 2: 20–26.

Tosh, S.M. and Yada, S. 2010. Dietary fibres in pulse seeds and fractions: Characterization, functional attributes, and applications. *Food Research International* 43(2): 450–460.

Thymi, S., Krokida, M.K., Pappa A., and Maroulis, Z.B. 2005. Structural properties of extruded corn starch. *Journal of Food Engineering* 68: 519–526.

Valadez-Blanco, R., Virdi, A.I.S., Balke, S.T., and Diosady, L.L. 2007. In-line color monitoring during food extrusion: Sensitivity and correlation with product color. *Food Research International* 40: 1129–1139.

van Zuilichem, D.J., Jager, T. and Stolp, W. 1988. Residence time distributions in extrusion cooking. Part II: Single-screw extruders processing maize and soya. *Journal of Food Engineering* 7: 197–210.

van Zuilichem, D.J., Jager, T., Spaans, E.J., and De Ruigh, P. 1990. The influence of a barrel valve on the degree of fill in a co-rotating twin screw extruder. *Journal of Food Engineering* 10(4): 241–254.

Vasanthan, T., Gaosong, J., Yeung, J., and Li, J. 2002. Dietary fibre profile of barley flour as affected by extrusion cooking. *Food Chemistry* 77(1): 35–40.

Velioglu, Y.S., Mazza, G., Gao, L., and Oomah, B.D. 1998. Antioxidant activity and total phenolics in selected fruits, vegetables, and grain products. *Journal of Agricultural and Food Chemistry* 46: 4113–4117.

Veronica, A.O., Olusola, O.O., and Adebowale, E.A. 2006. Qualities of extruded puffed snacks from maize/soybean mixture. *Journal of Food Process Engineering* 29: 149–161.

Volman, J.J., Ramakers, J.D., and Plat, J. 2008. Dietary modulation of immune function by β-glucans. *Physiological Behaviors* 94: 276–284.

Wang, L., Chessari, C., and Karpiel, E. 2001. Inferential control of product quality attributes: Application to food cooking extrusion process. *Journal of Process Control* 11: 621–636.

Wang, L., Smith, S., and Chessari, C. 2008. Continuous-time model predictive control of food extruder. *Control Engineering Practice* 16: 1173–1183.

Wang, N., Bhirud, P.R., and Tyler, R.T. 1999. Extrusion texturization of air-classified pea protein. *Journal of Food Science* 64(3): 509–513.

Wang, Y. and Tan, J. 2000. Dual-target predictive control and application in food extrusion. *Control Engineering Practice* 8: 1055–1062.

Weidmann, W. and Strecker, J. 1988. Process control of cooker-extruder. In: *Automatic Control and Optimization of Food Processes*, Renard, M. and Bimbert, J.J., eds. Elsevier Science Publishers Ltd., New York, pp. 201–214.

Wood, P.J., Braaten, J.T., Scott, F.W., Riedel, K.D., Wolynetz, M.S., and Collins, M.W. 1994. Effect of dose and modification of viscous properties of oat gum on plasma glucose and insulin following an oral glucose load. *British Journal of Nutrition* 72: 731–743.

Yacu, W.A. 1983. Modelling of a two-screw co-rotating extruder. In: *Thermal Processing and Quality of Foods*. Elsevier Applied Science Publishers, London, U.K.

Yagci, S. and Gögüs, F. 2009. Development of extruded snack from food by-products: A response surface analysis. *Journal of Food Process Engineering* 32: 565–586.

Yam, K.L., Gogoi, B.K., Karwe, M.V. and Wang, S.S. 1994. Shear conversion of corn meal by reverse screw elements during twin-screw extrusion at low temperature. *Journal of Food Science* 59(1): 113–114.

Yankov, V.I. 1978. Hydrodynamics and heat transfer of apparatus for continuous high speed production of polymer solutions. *Heat Transfer Soviet Research* 10: 67–71.

Zasypkin, D.V. and Lee, T.C. 1998. Extrusion of soybean and wheat flour as affected by moisture content. *Journal of Food Science* 63(6): 1058–1061.

Zhang, M., Bai, X., and Zhang, Z. 2011. Extrusion process improves the functionality of soluble dietary fiber in oat bran. *Journal of Cereal Science* 54(1): 98–103.

Zielinski, H., Kozlowska, H., and Lewczuk, B. 2001. Bioactive compounds in the cereal grains before and after hydrothermal processing. *Innovative Food Science and Emerging Technologies* 2: 159–169.

5 Dehydration
Spray Drying—Freeze Drying

Athanasia M. Goula

CONTENTS

5.1 INTRODUCTION

Drying or dehydration of foods is an extremely important food-processing operation used to preserve foods for extended periods of time. Although both these terms are applied to the removal of water from food, drying usually refers to natural desiccation, such as spreading fruit on racks in the sun, and dehydration designates drying by artificial means, such as a blast of hot air. The distinguishing features between drying and concentration are the final level of water and nature of the product. Concentration leaves a liquid food, whereas drying typically produces product with water content sufficiently low to give solid characteristics.

In general, food dehydration implies the removal of water from the foodstuff. In most cases, dehydration is accomplished by vaporizing the water that is contained in the food, and to do this latent heat of vaporization must be supplied. There are two important process-controlling factors (Figure 5.1):

- The transfer of heat to provide the necessary latent heat of vaporization
- The movement of water or water vapor through the food material and then away from it to effect the separation of water from foodstuff

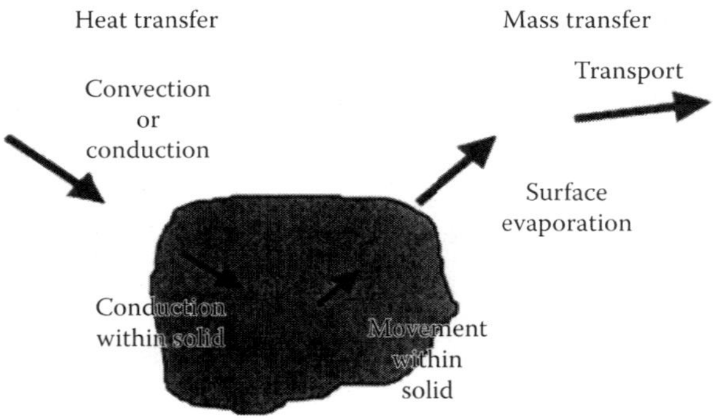

FIGURE 5.1 Processes involved in evaporative drying. (From Brammer, J.G. and Bridgwater, A.V., *Renew. Sust. Energy Rev.*, 3, 243, 1999. With permission.)

Drying processes fall into three categories:

1. Air and contact drying under atmospheric pressure. Heat is transferred through the foodstuff either from heated air or from heated surfaces. The water vapor is removed with the air.
2. Vacuum drying. Advantage is taken of the fact that evaporation of water occurs more readily at lower pressures than at higher ones. Heat transfer is generally by conduction, sometimes by radiation.
3. Freeze drying. The water vapor is sublimed off from frozen food. The food structure is better maintained under these conditions.

Food dehydration is not limited to the selection of a drying method. The physicochemical concepts associated with food dehydration need to be understood for an appropriate assessment of the drying phenomena in any food product. Water activity, glass transition temperature, dehydration mechanisms and theories, and chemical and physical changes should be recognized as key elements for any food dehydration operation. Thus, prior to examining selected examples of drying systems—spray drying and freeze drying—the basic principles of dehydration are presented.

5.2 BASIC PRINCIPLES OF DEHYDRATION

5.2.1 State of Water in Foods

The fundamental purpose of food dehydration is to lower the availability of water in the food to a level at which there is no danger of growth of undesirable microorganisms. A secondary purpose is the lowering of the water content in order to minimize rates of chemical reactions, and to facilitate distribution and storage. The availability of water for microbial growth and chemical activity is determined not only by the total water content but also by the nature of its binding to foods (Karel, 1979).

Solids can be classified as follows (Mujumdar and Menon, 1995):

1. Nonhygroscopic capillary-porous media, where
   a. There is a clearly recognizable pore space.
   b. The amount of physically bound moisture is negligible.
   c. The medium does not shrink during drying.
2. Hygroscopic-porous media, where
   a. There is a clearly recognizable pore space.
   b. There is a large amount of physically bound liquid.
   c. Shrinkage often occurs in the initial stages of drying.
3. Colloidal (nonporous) media, where
   a. There is no pore space (evaporation can take place only at the surface).
   b. All liquid is physically bound.

The moisture content of a solid is usually expressed as the moisture content by weight of bone-dry material in the solid. Sometimes a wet-basis moisture content, which is the moisture–solid ratio based on the total mass of wet material, is used. Water may become bound in a solid by retention in capillaries, solution in cellular structures, and solution with the solid, or chemical or physical adsorption on the surface of the solid. Unbound moisture in a hygroscopic material is the moisture in excess of the equilibrium moisture content corresponding to saturation humidity. All the moisture content of a nonhygroscopic material is unbound moisture. Free moisture content is the moisture content removable at a given temperature and may include both bound and unbound moisture.

For half a century, scientists have realized that relative vapor pressure, water activity (a_w), is much more important to the quality and stability of foods than the total amount of water present.

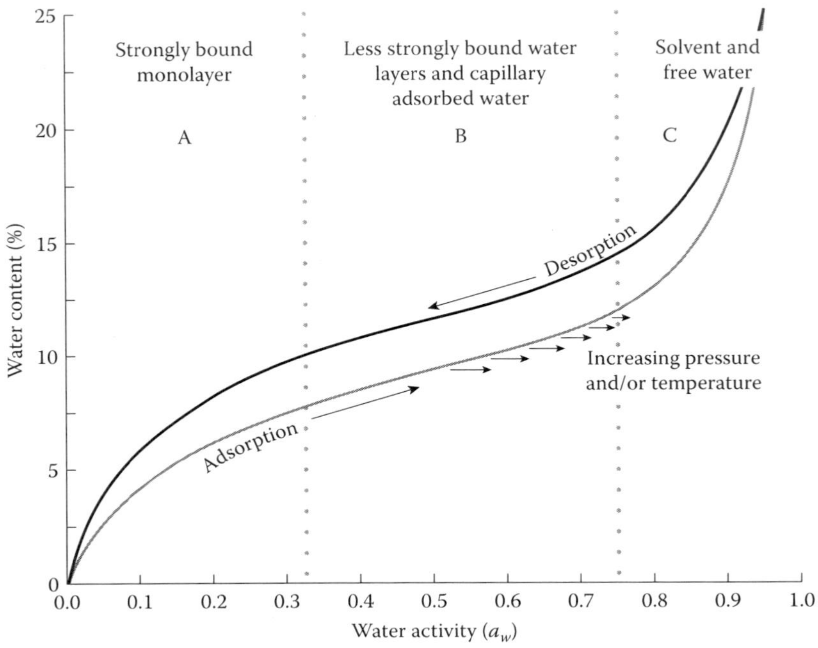

FIGURE 5.2 Typical isotherm curve.

The water sorption isotherm, that is the curve relating to the partial pressure of water in the food to the water content, is an important tool, especially in low-moisture foods. It can be applied in order to optimize the drying or rehydration conditions and determine the stability of the product during storage. Sorption isotherms can be generated from an adsorption process or a desorption process; the difference between these curves is defined as hysteresis, as it is shown in Figure 5.2. Water adsorption by food products is a process in which water molecules progressively and reversibly mix together with food solids via chemisorption, physical adsorption, and multilayer condensation. An isotherm can be typically divided into three regions; the water in region A represents strongly bound water, and the enthalpy of vaporization is considerably higher than the one of pure water. The bound water includes structural water and monolayer water, which is sorbed by the hydrophilic and polar groups of food components (polysaccharides, proteins, etc.). Bound water is unfreezable, and it is not available for chemical reactions or as a plasticizer. In region B, water molecules bind less firmly than in the first zone; they usually fill in small capillaries. The vaporization enthalpy is slightly higher than that of pure water. This class of constituent water can be looked upon as the continuous transition from bound to free water. The properties of water in region C are similar to those of the free water that is held in voids, large capillaries, crevices; and the water in this region loosely binds to food materials (Kinsella and Fox, 1985; Mohsenin, 1985). Moreover, hysteresis is related to the nature and state of the components of food, reflecting their potential for structural and conformational rearrangements, which alters the accessibility of energetically favorable polar sites. The presence of capillaries in food results in considerable decrease in water activity. The explanation for the occurrence of moisture sorption hysteresis comprises the ink bottle theory, the molecular shrinkage theory, the capillary condensation, and the swelling fatigue theory (Raji and Ojediran, 2011).

Brunauer et al. (1938) classified sorption isotherms according to their shape establishing five different types as it is shown in Figure 5.3. Type 1: Langmuir and/or similar isotherms that present a characteristic increase in water activity related to the increasing moisture content; the first derivative of this plot increases with moisture content, and the curves are convex upward. This type of sorption isotherm is typically applicable in the process of filling the water monomolecular layer at the

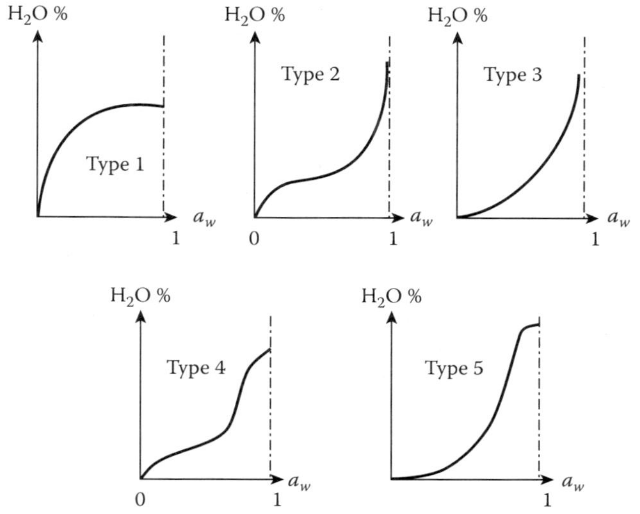

FIGURE 5.3 Types of isotherms curves. (From Brunauer, S. et al., *J. Am. Chem. Soc.*, 50, 309, 1938. With permission.)

internal surface of a material. Type 2: Sigmoidal sorption isotherms in which the curves are concave upward; it takes into account the existence of multilayers at the internal surface of a material. Type 3: Known as the Flory–Huggins isotherm, it accounts for a solvent or plasticizer such as glycerol above the glass transition temperature. Type 4: It describes the adsorption of a swellable hydrophilic solid until a maximum of site hydration is reached. Type 5: The Brunauer–Emmett–Teller (BET) multilayer adsorption isotherm, it is the one observed in the adsorption of water vapor on charcoal, and it is related to the isotherm types 2 and 3. Moisture sorption data have been reported for several foods (Maroulis et al., 1988; Kim et al., 1991; Chen and Jayas, 1998; McLaughlin and Magee, 1998; Akanbi et al., 2005), and the two isotherms most frequently found are types 2 and 4 (Mathlouthi and Rogé, 2003; Basu et al., 2005; Blahovec and Yanniotis, 2009).

It is common to present sorption isotherms by mathematical models based on empirical and theoretical criteria. In literature, there is a long list of available isotherm models, which can be divided into several categories; kinetic models based on an absorbed water mono layer (BET model; Brunauer et al., 1938), kinetic models based on a multilayer and condensed film (GAB model; van den Berg and Bruin, 1981), semi empirical (Halsey model; Halsey, 1948), and purely empirical models (Oswin and Smith models; Oswin, 1945; Smith, 1947). The most common equations describing water sorption in food products are presented in Table 5.1.

The thermodynamic properties of foods, such as differential enthalpy and differential entropy, provide an understanding of water properties and energy requirements associated with the sorption behavior. Differential heat of sorption, often referred as isosteric heat of sorption, is used as an indicator of the state of water adsorbed by the solid particles, and its knowledge is of great importance when designing equipment for dehydration processes. The differential entropy of a material is proportional to the number of available sorption sites at a specific energy level, whereas Gibbs free energy is indicative of the affinity of sorbents for water and provides a criterion whether water sorption occurs as a spontaneous process (Telis et al., 2000).

Over the past decades, the concepts related to water activity have been enriched by those of glass transition temperature (T_g), thus providing an integrated approach to the role of water. T_g defines a second-order phase change temperature at which a solid "glass" is transformed to a liquid-like "rubber". As the temperature increases above T_g various changes, such as increase of free volume, decrease of viscosity, increase of specific heat, and increase of thermal expansion, are noticed.

TABLE 5.1
Isotherm Models

Model	Mathematical Expression	
GAB (Van den Berg and Bruin, 1981)	$X = X_m C K a_w / \left[(1 - K a_w)(1 - K a_w + C K a_w)\right]$	(5.1)
BET (Brunauer et al., 1938)	$X = X_m C a_w / \left[(1 - a_w)(1 + (C - 1) a_w)\right]$	(5.2)
Halsey (Halsey, 1948)	$X = \left[-A/(T \ln a_w)\right]^{1/B}$	(5.3)
Smith (Smith, 1947)	$X = A + B\log(1-a_w)$	(5.4)
Oswin (Oswin, 1945)	$X = A\left[a_w/(1 - a_w)\right]^B$	(5.5)
Peleg (Peleg, 1993)	$X = K_1 a_w^{n_1} + K_2 a_w^{n_2}$	(5.6)

Note: X, moisture content (% dry basis); X_m, monolayer value (% dry basis), T, temperature (K), C, K, A, B, K_1, K_2, n_1, n_2, constants (—).

The most important changes affecting food behavior are related to the exponential increase of molecular mobility and decrease of viscosity. These factors govern various time-dependent and often viscosity-related structural transformations, such as stickiness, collapse, and crystallization during food processing and storage. The importance of T_g of amorphous food materials for processing and storage stability has been recognized and emphasized by many researchers, and a wide range of potential food applications of the glass transition phenomenon have been identified (Roos et al., 1995; Matveev et al., 2000; Moraga et al., 2005). Generally, a disagreement exists involving the role of the glass transition temperature on the rates of chemical reactions. The question arises as to whether molecular mobility of reactants as dictated by the state of the system (glassy versus rubbery) or the chemical potential of water (i.e., water activity) controls reaction rates. It is the opinion of several authors that both aspects are compatible and perfectly complementary, and knowledge of both is necessary to understand the food–water relationships (Chirife and Buera, 1994). According to Frias and Oliveira (2001), it can be dangerous to neglect one approach in detriment of the other without perfectly understanding the implications, but in some specific situations, it can be interesting to separate the a_w approach or the T_g one to foresee, which dominates the decay in a specific matrix. Nicoleti et al. (2007) reported that using the glass transition temperature approach to describe chemical reaction rates during drying showed to be advantageous from a mathematical point of view in the high moisture content domain, whereas in the low moisture content domain, this advantage was lost. A different behavior was observed by other authors, who concluded that the physical state of the product would be important to degradation rates at the low moisture content range (Tsimidou and Biliaderis, 1997; Frias and Oliveira, 2001; Serris and Biliaderis, 2001; Goula and Adamopoulos, 2010).

5.2.2 Types of Water Movement

Moisture in a drying food can be transferred both in liquid and in gaseous phases. The water movement within the solid may be explained by different mechanisms like diffusion of liquid due to concentration gradient, vapor diffusion due to partial vapor pressure, and liquid movement due to capillary forces. The main modes of moisture transport are presented in Table 5.2. Apart from these types of movement, mass transport due to osmotic pressure, pressure gradient, gravity, shrinkage, or external pressure can also take place (Van Arsdel et al., 1973; Strumillo and Kudra, 1985; Barbosa-Canovas and Vega-Mercado, 1995).

TABLE 5.2
Main Modes of Water Movement during Air Drying

Liquid diffusion	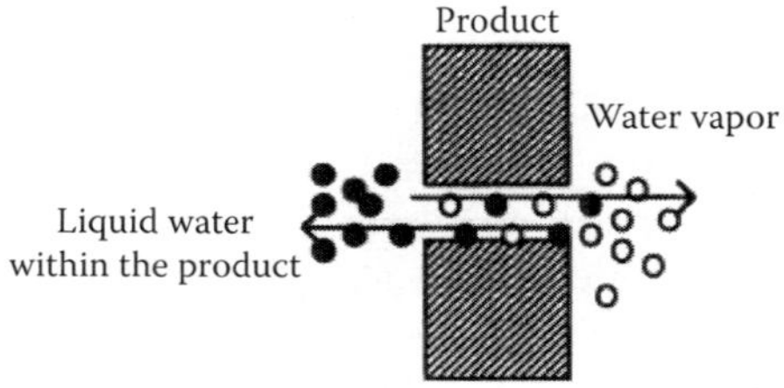	The water vapor flux is a function of the water concentration gradient within the product. The liquid water reaches the surface and is evaporated.
Vapor diffusion	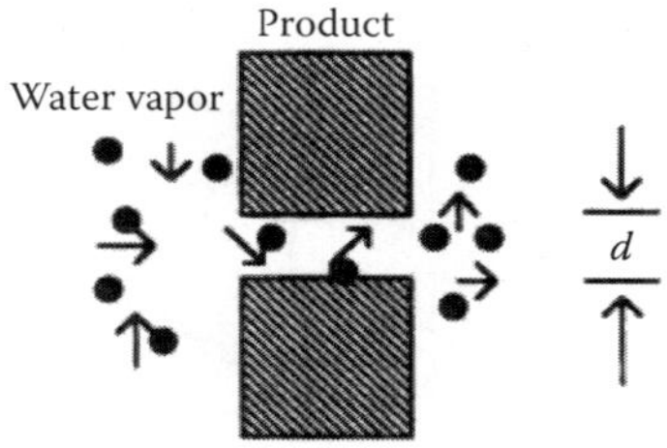	The water flux is a function of the vapor density within the product. The size and amount of pores, tortuosity, and the geometry of the solid affect the vapor flux.
Capillary forces	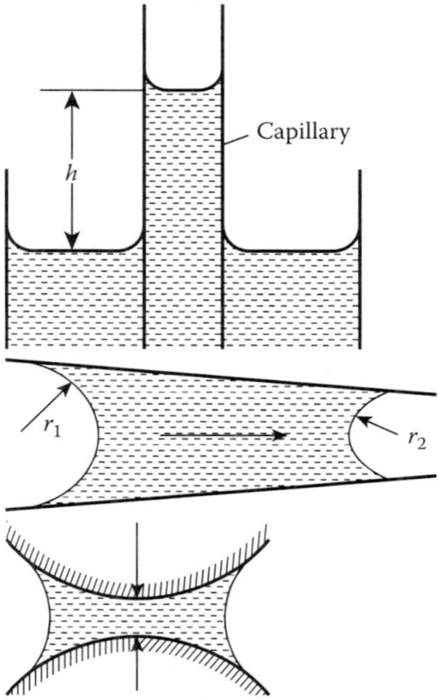	The water flux through the interstices and over the surface of a solid is due to molecular attraction between the liquid and the solid.
Evaporation–condensation	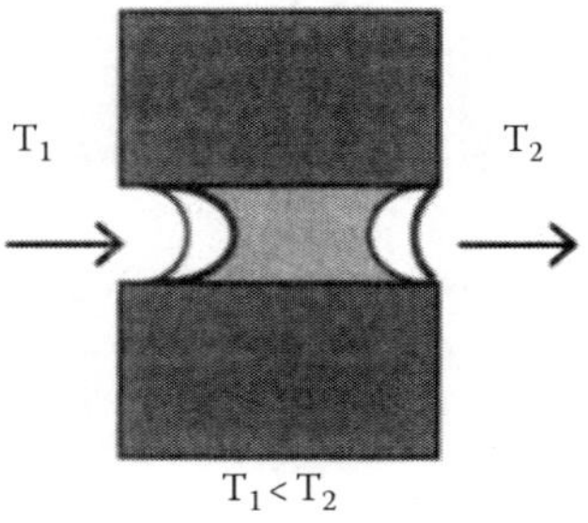	The rate of condensation is equal to the rate of evaporation at the surface of the product and allows no accumulation of water in the pores near the surface.

Sources: Barbosa-Canovas, G.V. and Vega-Mercado, H., *Dehydration of Foods*, Chapman & Hall, New York, 1995; Bruin, S. and Luyben, K.C., Drying of food materials: A review of recent developments, in *Advances in Drying*, A.S. Mujumdar, (ed.), Hemisphere, New York, 1980.

5.2.3 Psychrometry

The capacity of air for moisture removal depends on its humidity and its temperature. The study of relationships between air and its associated water is called psychrometry (Figure 5.4). Humidity, *Y*, is the measure of the water content of the air. The absolute humidity is the mass of water vapor per unit mass of dry air. Air is said to be saturated with water vapor at a given temperature and pressure if its humidity is at maximum under these conditions. If further water is added to saturated air, it must appear as liquid water in the form of a mist or droplets. Under conditions of saturation, the partial pressure of the water vapor in the air is equal to the saturation vapor pressure of water at that temperature. The relative humidity, *RH*, is defined as the ratio of the partial pressure of the water vapor to the partial pressure of saturated water vapor at the same temperature and is often expressed as percentage.

The wet-bulb temperature is the temperature reached by a water surface, such as that registered by a thermometer bulb surrounded by a wet wick, when exposed to air passing over it. The wick, and therefore the thermometer bulb, decreases in temperature below the dry-bulb temperature until the rate of heat transfer from the warmer air to the wick is just equal to the rate of heat transfer needed to provide for the evaporation of water from the wick into the air stream. As the relative humidity of the air decreases, the difference between the wet-bulb and dry-bulb temperatures, called the wet-bulb depression, increases and a line connecting the wet-bulb temperature and relative humidity can be plotted on a suitable chart. When the air is saturated, the wet-bulb temperature and the dry-bulb temperature are identical.

A further important concept is that of the adiabatic saturation condition. This is the situation reached by a stream of water, in contact with the humid air, both of which ultimately reach a temperature at which the heat lost by the humid air on cooling is equal to the heat of evaporation of the water leaving the stream of water by evaporation (Earle, 1983). In the drying process, where the heat for evaporation is supplied by the hot air passing over a wet solid surface, the system behaves like the adiabatic saturation system, since no heat is obtained from any source external to the air and the wet product and the latent heat must be obtained by cooling the air. In dryers, the air is usually reheated so as to reduce the relative humidity and, thus, to give an additional capacity to evaporate more water from the product.

5.2.4 Drying Rate

5.2.4.1 Drying Curves

The rate-of-drying curve is obtained from data generated by weighing a sample of food undergoing drying and relating weight loss to moisture content. Moisture content is usually expressed as kg or lb of water per kg or lb of dry product, rather than percentage of moisture in the food product. As moisture content changes, both values change. However, the kg of dry matter is always constant during drying, so a constant reference point is used when referring to drying in kg water/kg dry matter. Figure 5.5 shows a typical curve of loss of moisture during drying of a food product. As it can be shown, after a short equilibration period, the moisture content decreases rapidly with time. This initial drying period is followed by a much slower rate of drying as the moisture content of the product decreases. The drying rate is the slope of the moisture content change with time.

Figure 5.6 shows a typical drying rate curve for a constant drying condition. Point B represents an equilibrium temperature condition of the product surface. Section BC, known as constant rate period, represents the removal of unbound water from the product. During this period, the surface of the product is very wet, and the water acts as if the solid is not present (Barbosa-Canovas and Vega-Mercado, 1995). The surface temperature is approximately that of the wet-bulb temperature, whereas the drying rate is determined by external conditions of temperature, humidity, and air velocity. In this stage of drying, the rate-controlling step is the diffusion of the water

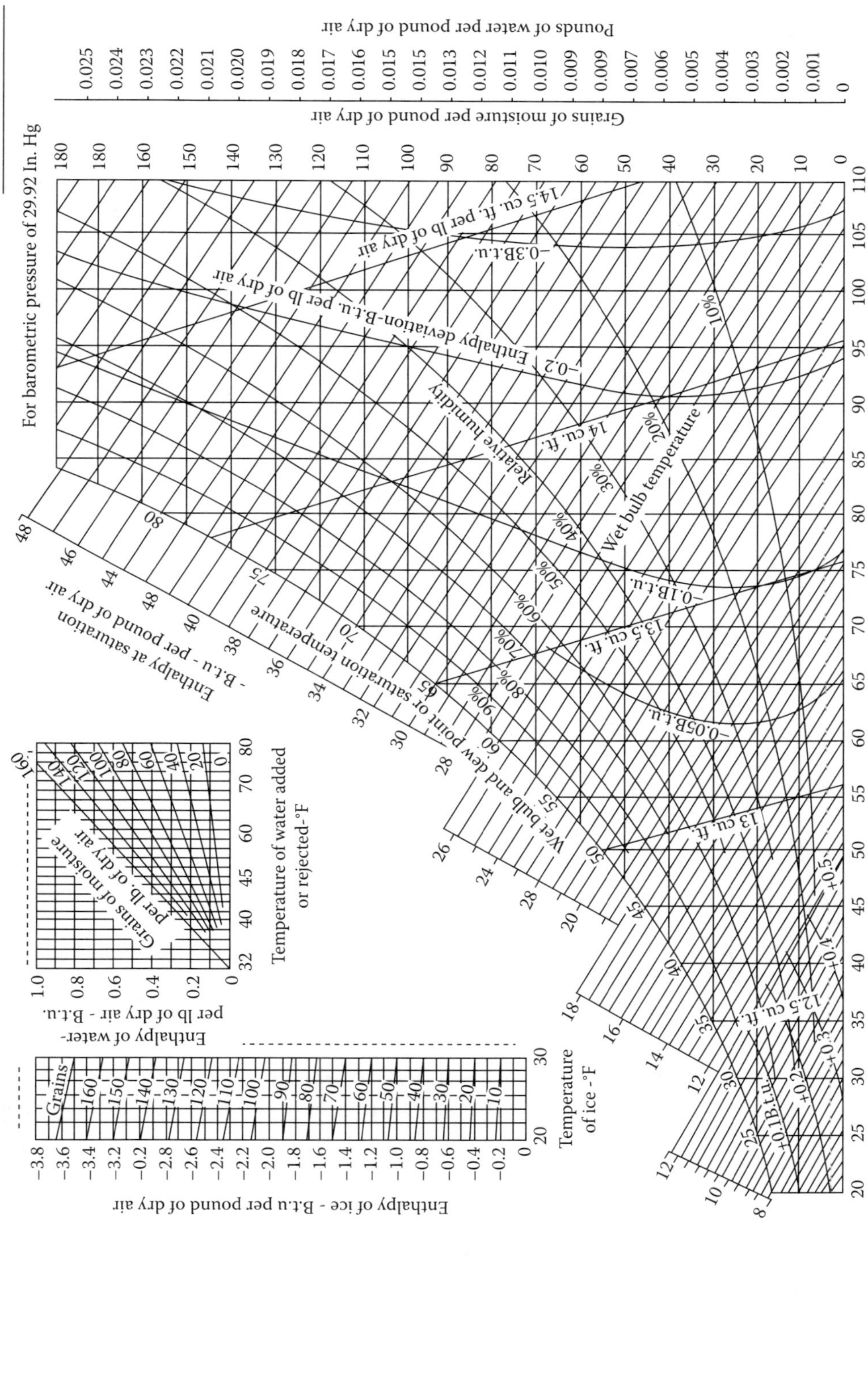

FIGURE 5.4 Psychrometric chart. (From Brammer, J.G. and Bridgwater, A.V., *Renew. Sust. Energy Rev.*, 3, 243,1999. With permission.)

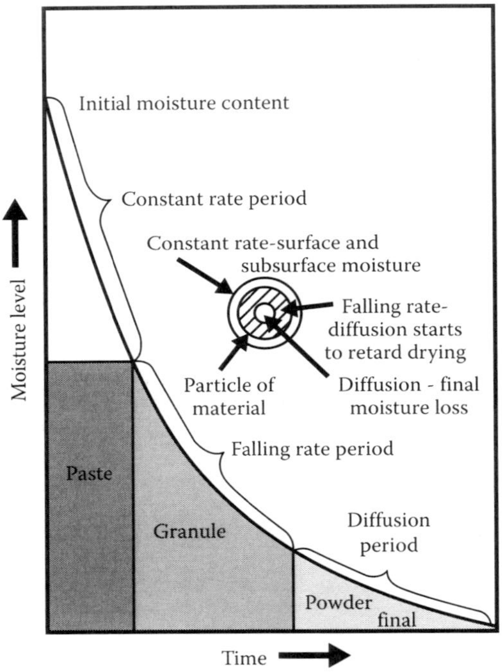

FIGURE 5.5 Typical curve of loss of moisture during drying of a food product.

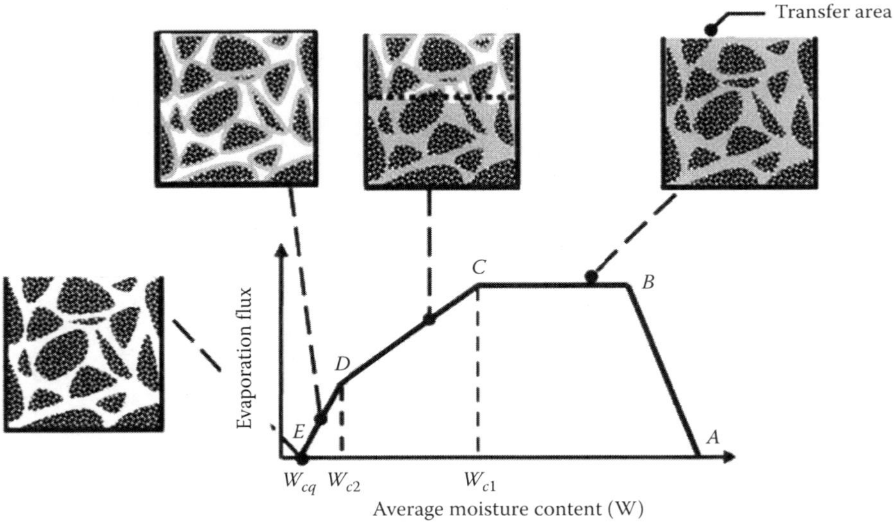

FIGURE 5.6 Typical drying rate curve. (From Vaxelaire, J. and Cézac, P., *Water Res.*, 38, 2215, 2004. With permission.)

vapor across the air–moisture interface and the rate at which the surface for diffusion retrieves. Toward the end of the constant rate period, moisture has to be transported from the inside of the solid to the surface by capillary forces, and the drying rate may still be constant. When the average moisture content has reached the critical moisture content X_c, the surface film of moisture has been so reduced by evaporation that further drying causes dry spots to appear upon the surface (point D). The drying rate falls even though the rate per unit wet solid surface area

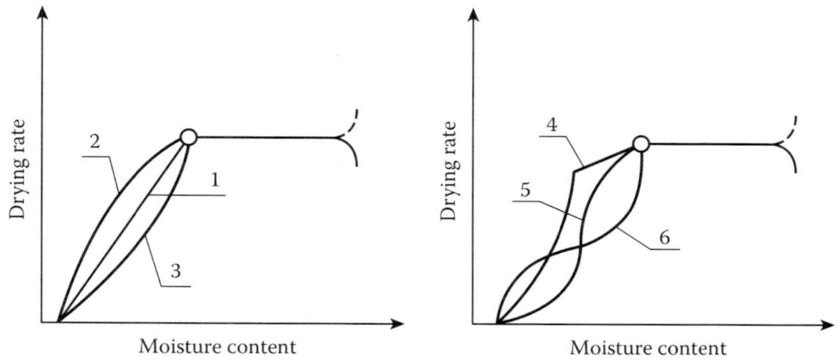

FIGURE 5.7 Types of drying curves. (From Luikov, A.V., *Analytical Heat Diffusion Theory*, Academic Press, New York, 1958. With permission.)

remains constant. This second drying stage, called the first falling rate period, proceeds until the surface film of liquid is entirely evaporated. On further drying, second falling rate period (section DE), the rate at which moisture may move through the solid as a result of concentration gradients between the deeper parts and the surface is the controlling step. During this stage, some of the moisture bound by sorption is being removed, and as the moisture concentration is lowered by the drying, the rate of drying falls even more rapidly than before and continues until the moisture content falls down to the equilibrium value X_e for the prevailing air humidity and then drying stops (Mujumdar and Menon, 1995).

It has been stated that the shape of the falling drying rate period curve depends, among others, on the type of material being dried. Six types of drying curves are reported for the second drying period (Figure 5.7) (Luikov, 1958). The first two curves are characteristic of those for capillary-porous bodies with large specific evaporation surfaces. The other curves are characteristic of those for capillary-porous bodies with small specific evaporation surfaces and for colloidal bodies (e.g., starch). The curves 4, 5, and 6 represent the colloidal-capillary-porous bodies like corn, bread, and peat. According to Strumillo and Kudra (1985), the shape of the drying curves in the first and second drying rate periods as well as the relation between the two periods depend on the mass transfer conditions. Figure 5.8 presents sketches of the drying rate curves for cases of external conditions, internal conditions, and external–internal conditions. It is characteristic that no constant drying rate period will be observed for the drying of hygroscopic solids and also when the drying is internally controlled for all types of solids.

5.2.4.2 Constant Rate Period

During the constant rate drying period, the rate of moisture removal from the product is limited only by the rate of evaporation from water surface on or within the product. The water content at the surface is considered to be constant, as water migration from the interior of the product is sufficiently rapid to maintain constant surface moisture. The drying rate can be determined by the driving force for convective mass transfer of water molecules from the product surface into the drying air. The driving force for drying is the difference between the vapor pressure of water at the surface of the food and that of the drying air. The rate of heat input to the product just balances the amount of water being vaporized. In the simplest case, all of the heat for drying comes from convective heat transfer between the drying air blowing across the food and the product surface. However, there may be radiation heat transfer into the product. If the food product sits on a solid tray, only the top surface is exposed to the drying airflow, and heat transfer into the bottom of the product occurs by a combination of convection and conduction heat transfer. If drying occurs only by convection, the surface temperature stabilizes at the wet-bulb temperature for the used drying air. If, however, other

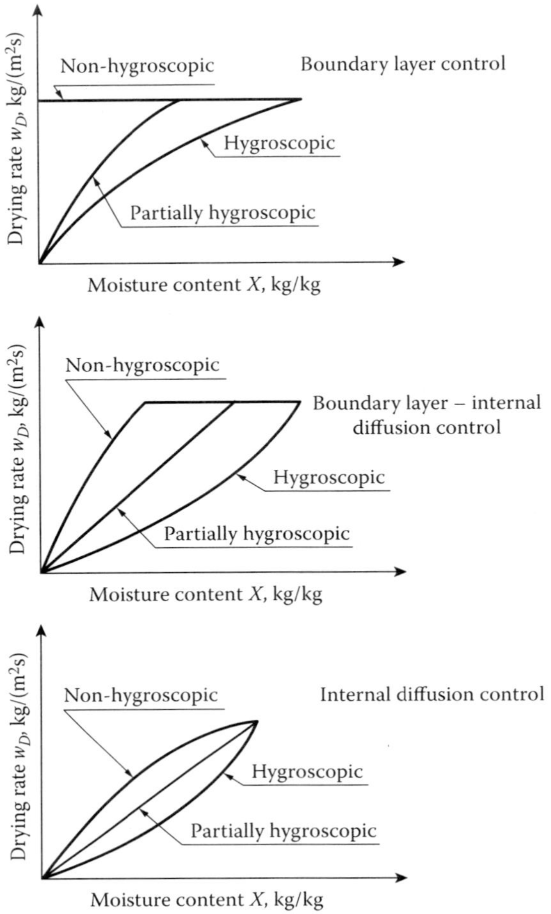

FIGURE 5.8 Drying rate curves for the cases of external conditions, internal conditions, and external-internal conditions. (From Strumillo, C. and Kudra, T., *Drying: Principles, Applications and Design*, Gordon & Breach Science Publishers, New York, 1986. With permission.)

heat transfer mechanisms supply a portion of the heat into the product, a slightly higher, but still constant, temperature is attained, called pseudo-wet-bulb temperature (Heldman and Hartel, 1997).

The drying rate, R_c, is expressed as (Barbosa-Canovas and Vega-Mercado, 1995):

$$R_c = -\frac{m_s}{A}\frac{\mathrm{d}X}{\mathrm{d}t} = k_y M_b \left(Y_w - Y\right) = \frac{h\left(T - T_w\right)}{\lambda_w} \tag{5.7}$$

where

m_s is the total mass of dry matter in the food being dried
A is the product surface area exposed to drying air
X is the product moisture content
k_y is the mass transfer coefficient
M_b is the molecular weight of air
Y_w is the humidity at wet-bulb temperature
Y is the bulk air humidity
h is the heat transfer coefficient

T_w is the wet-bulb temperature
T is the drying temperature
λ_w is the latent heat of water evaporation at T_w

The mass transfer coefficient can be obtained from the following correlation for laminar flow parallel to a flat plate (Okos et al., 1992):

$$\frac{k_y l}{D_{AB}} = 0.664 Re^{0.5} Sc^{0.33} \tag{5.8}$$

$$Re = \frac{du\rho}{\mu} \quad Sc = \frac{\mu}{\rho D_{AB}} \tag{5.9}$$

where
l is the plate length in the direction of the flow
D_{AB} is the molecular diffusivity of an air–water mixture
Re is the Reynolds number
Sc is the Schmidt number
d is the characteristic dimension
u is the fluid velocity
ρ is the density
μ is the viscosity

The heat transfer coefficient can be expressed by Equation 5.4 (Geankoplis, 1983) or 5.8 (Chirife, 1983):

$$h = 0.0204 G^{0.8} \tag{5.10}$$

$$Nu = \frac{hd}{k} = 2 + aRe^{0.5} Pr^{0.33} \tag{5.11}$$

$$Pr = \frac{C_p \mu}{k} \tag{5.12}$$

where
G is the mass velocity of air
k is the thermal conductivity
a is a constant
Pr is the Prandlt number
C_p is the heat capacity

The drying time in the constant rate period can be found by integrating Equation 5.7 from the initial moisture content, X_i, to the critical moisture content, X_c (Heldman and Hartel, 1997):

$$t_c = \frac{X_i - X_c}{\left(\frac{h}{\lambda_w}\right)\left(\frac{A}{m_s}\right)(T - T_w)} \tag{5.13}$$

5.2.4.3 Falling Rate Period

After reaching the critical moisture content, the drying rate may decrease linearly with decreasing moisture content for the remaining portion of the process. In some products, there may be more than one falling rate period. In the first falling rate period, the saturated surface area decreases as the moisture movement within the solid can no longer supply enough moisture. The drying rate decreases as the unsaturated surface area increases. The factors that influence the drying rate include those that effect moisture movement away from the solid in addition to the rate of internal moisture movement. When the entire surface area reaches the unsaturated state, the internal moisture movement becomes the controlling factor (Heldman and Singh, 1981). During the falling rate period, a moisture profile exists within the food, depending on the drying conditions. Moisture content is highest at the center of the piece and lowest at the surface (Heldman and Hartel, 1997). Surface temperature increases slowly and once the second falling rate period has begun, the temperature approaches that of the drying air.

The movement of the water within the solid may be explained by different mechanisms: diffusion of liquid due to concentration gradient, vapor diffusion due to partial vapor pressure, liquid movement due to gravity, and surface diffusion (Barbosa-Canovas and Vega-Mercado, 1995). The relative contribution of each mechanism may change during the drying process. In the early stages of the falling rate period, liquid diffusion may be the controlling mechanism of internal mass transfer, whereas during the latter stages of drying, a combination of vapor diffusion and thermal flow may control drying. In addition, as many food products lose moisture, they shrink or a surface layer or crust (case hardening) may form, and these phenomena may affect mass transfer mechanisms.

A number of studies have tried to model mass transfer during falling rate period and different types of models have been proposed. Two main approaches can be identified: mechanistic models based on Fick's law and empirical or semiempirical models, such as those presented in Table 5.3. In Table 5.3, MR is the dimensionless moisture ratio = $(X - X_e)/(X_i - X_e)$, where X is the moisture content of the product at each moment, X_i is the initial moisture content, and X_e is the equilibrium moisture content. The values of X_e, are relatively small compared to X or X_i. Thus, MR can be reduced to $MR = X/X_i$ (Thakor et al., 1999).

One common approach is to empirically describe drying in the falling rate period using an effective diffusivity, D_{eff}, which is a combination for all internal mass transfer mechanisms. D_{eff} is often determined by measuring actual drying rate data and fitting these data to the unsteady state diffusion equation to calculate an effective rate of diffusion. For drying of a thin film in one dimension, this equation may be written as (Heldman and Hartel, 1997):

$$\frac{\partial X}{\partial t} = D_{\text{eff}} \frac{\partial^2 X}{\partial x^2} \tag{5.14}$$

where x is the dimension of thickness of the drying film.

TABLE 5.3
Drying Curve Models

Model	Equation	
Lewis	$MR = \exp(-kt)$	Bruce (1985)
Page	$MR = \exp(-kt^n)$	Page (1949)
Modified page	$MR = \exp(-kt)^n$	Overhults et al. (1973)
Henderson and Pabis	$MR = a_1\exp(-kt)$	Henderson and Pabis (1951)
Logarithmic	$MR = a_1\exp(-kt) + a_2$	Togrul and Pehlivan (2002)
Approximation of diffusion	$MR = a_1\exp(-kt) + (1 - a_1)\exp(-ka_2t)$	Yaldiz et al. (2001)
Wang and Singh	$MR = 1 + a_1t + a_2t^2$	Wang and Singh (1978)

Note: k, n, a_1, a_2, constants.

Analytical solutions of the diffusion equation are available for standard solid shapes (sphere, slab, and infinite cylinder). However, the diffusivity varies with the moisture content and, at the same time, significant shrinkage takes place. Under these conditions, D_{eff} can be estimated at various moisture contents using simplified solutions of the diffusion equation and numerical methods (Karathanos et al., 1990). The moisture diffusivity can also be estimated as a function of the moisture content and the temperature using empirical models, such as the exponential equation (Marinos-Kouris and Maroulis, 1995). The regular regime theory of drying (Gekas, 1993; Barbosa-Canovas and Vega-Mercado, 1995) has been applied to limited cases of food drying, yielding diffusivity values similar to those obtained by the simplified or numerical solutions of the diffusion equation. According to Marinos-Kouris and Maroulis (1995), diffusivities in foods have values in the range of 10^{-13}–10^{-5}, and most of them are accumulated in the region of 10^{-11}–10^{-8}.

In the falling rate period, the experimentally obtained drying rate curve can be again used to estimate the time of drying:

$$t_f = \frac{m_s}{A}\int_{X_e}^{X_c} \frac{dX}{R_f} \tag{5.15}$$

where R_f is the drying rate during the falling rate period. From the experimental data, a plot of $1/R_f$ versus X is drawn. The area under the curve is determined to calculate the drying rate.

5.2.5 Quality Changes during Drying

Over the past three decades, there has been an increased concern for food quality with a significant amount of work accomplished in the area of kinetics of nutrient destruction or general quality degradation during drying processes (Koca et al., 2007). Because a good understanding of various quality changes can provide a means to optimize the drying process and hence minimize the degradation of these important quality attributes, a number of predictive models have been proposed and tested.

5.2.5.1 Nutritional and Color Changes

Reactions occurring during drying can result in quality losses, particularly nutrient losses, and in other deleterious changes caused by nonenzymatic browning. Attempts have been made to analyze these changes in a quantitative manner and to relate them to process conditions. A summary of published data on nutrient degradation of foods during drying is given in Table 5.4.

Empirical models are among the most common types of models that may be used to predict changes of vitamins and other bioactive compounds during food drying (Devahastin and Niamnuy, 2010). These models simply express relationships between the content of a particular nutrient and variables or parameters involved in the drying process. This type of correlations can be classified further into linear, nonlinear, or polynomial correlations. In most cases, however, changes of nutrients are fitted better to nonlinear correlations. Among the nonlinear empirical models, one may cite, for example, Suvarnakuta et al. (2005) for the prediction of β-carotene degradation in carrots during hot air drying, vacuum drying, and low-pressure superheated steam drying (LPSSD); Goula and Adamopoulos (2005a) for the prediction of lycopene stability in tomato pulps during spray drying; and Khazaei et al. (2008) for the prediction of losses of ascorbic acid in tomato slices during hot air drying.

A more fundamental modeling approach, which involves the use of various kinetic models, is also widely used to predict changes of nutrients and color of foods during drying. The number of possible reactions in foodstuffs is very large and several reaction mechanisms may be involved. Simply, first-order degradation reactions have been studied many times. An example is the destruction of heat

TABLE 5.4
Published Data on Kinetics of Nutrient Degradation of Some Foods

Nutrient	Product	Drying Method	Reference
Ascorbic acid	Amaranth	Sun drying	Mosha et al. (1995)
	Cowpea	Sun drying	Mosha et al. (1995)
	Pumpkin	Sun drying	Mosha et al. (1995)
	Potato	Air drying (40°C–50°C)	Rovedo and Viollaz (1998)
	Carrot slice	Vacuum microwave, air, freeze drying	Lin et al. (1998)
	Okra (whole)	Sun (33°C), air (40°C–80°C), vacuum (500 mmHg) drying	Inyang and Ike (1998)
	Apple slice	Microwave vacuum drying	Erle and Schubert (2001)
	Strawberry halve	Microwave vacuum drying	Erle and Schubert (2001)
	Strawberry (puree)	Freeze, drum (138°C), spray (inlet 195°C)	Abonyi et al. (2002)
	Carrot (puree)	Freeze, drum (138°C), spray (inlet 195°C)	Abonyi et al. (2002)
	Potato chip	Impingement drying (115°C–145°C)	Caixeta et al. (2002)
	Asparagus slice	Tray, spouted bed, freeze drying	Nindo et al. (2003)
	Indian gooseberry	Vacuum drying, low pressure superheated steam drying	Methakhup et al. (2005)
	Red pepper	Air drying (50°C–70°C)	Di Scala and Crapiste (2008)
	Pineapple slice	Air drying	Karim and Adebowale (2009)
	Sweet potato	Air drying (30°C–50°C)	Orikasa et al. (2010)
Carotene	Carrot slice	Vacuum microwave, air, freeze drying	Lin et al. (1998)
	Strawberry (puree)	Freeze, drum (138°C), spray (inlet 195°C)	Abonyi et al. (2001)
	Carrot (puree)	Freeze, drum (138°C), spray (inlet 195°C)	Abonyi et al. (2001)
	Carrot, pumpkin	Osmotic dehydration	Pan et al. (2003)
	Carrot slice	Air, microwave vacuum, freeze drying	Regier et al. (2005)
	Carrot cube	Superheated steam, vacuum, air drying (50°C–80°C)	Suvarnakuta et al. (2005)
	Carrot slice	Air drying (50°C–80°C)	Goula and Adamopoulos (2010)
Lycopene	Tomato	Air (95°C), vacuum (55°C) drying	Shi et al. (1999)
	Tomato pulp	Spray drying (inlet 110°C–140°C)	Goula and Adamopoulos (2005a,b)
	Carrot slice	Air, microwave vacuum, freeze drying	Regier et al. (2005)

labile components, such as vitamins and chlorophyll. The hydrolysis of sugars and esters can also be considered as first-order reactions. Limited research, however, has been achieved with regard to the kinetics of food quality degradation during dehydration. The problem of chemical conversions during drying is extremely complicated. During the drying process, concentrations also change and it is not clear to what extent, thereby, the chemical changes are influenced. The role of water in quality decay kinetics in foods has been an important subject of discussion and several hypotheses on how water content and molecular mobility affect the chemical reactions involved have been proposed. These hypotheses include: (1) increasing mobility of reaction partners and catalysts with increasing water activity, (2) increasing energy of activation with decreasing water activity, (3) changing concentration of water soluble reaction partners (Karel, 1979; Sokhansanj and Jayas, 1995). In general, chemical reactions are slower as the water activity decreases. However, with nonenzymatic browning reactions one observes a maximum rate of reaction at intermediate water activity levels, whereas various oxidation reactions show a minimum rate of reaction at a certain water activity (Bluestein and Labuza, 1988). As during drying, the product temperature increases and the water activity decreases, reaction rate constant may first increase when the temperature effect is dominating, whereas it may decrease later when the influence of the lower water activity becomes the dominating factor (Leniger and Bruin, 1977).

In the case of a first-order rate equation, the local current concentration of a component somewhere in a drying particle is given by:

$$\frac{C}{C_0} = \exp\left(-\int_0^t k\mathrm{d}t\right) \tag{5.16}$$

where t is the drying time. Since k is a function of moisture content, which in turn depends on the place in the drying material, the current concentration will also be a function of the place. For an exact calculation of the average current concentration, one would have to know the water concentration distribution and the temperature at each moment, calculate the local conversion as a function of time, and finally calculate the average concentration by volume integration. According to Escher and Blanc (1977), only the average water concentration and temperature in the drying material as functions of time need to be known. The simplest approximation is to assume a uniform value of k over the material, as corresponding with the average moisture content. This leads to

$$\frac{\bar{C}}{C_0} = \exp\left(-\int_0^t \bar{k}\mathrm{d}t\right) \tag{5.17}$$

with $\bar{k}$ the value of k at $\bar{X}$.

In addition to the ability to model changes of various bioactive compounds, kinetic modeling can also be used to follow and predict changes of color of foods during drying. Among many types of reactions that lead to the changes of color of fruits and vegetables, pigment destruction, oxidation, enzymatic browning, nonenzymatic browning, and phenol polymerization are most widely studied and modeled (Dandamrongrak et al., 2003; Bahloul et al., 2009).

According to Nicoleti et al. (2007), the degradation kinetics of food constituents may be related to molecular mobility of food matrix, which is affected by the free volume and by the molecular relaxation time of the food structure, and the glass transition temperature (T_g) has been used as the main indicator of this mobility. There is an important increase in molecular mobility across T_g that can affect reactions kinetics in situations where diffusion of reactants or products is the controlling factor. During drying, the removal of water increases the glass transition temperature of the matrix substantially. Thus, as drying progresses, the matrix approaches the onset of glass transition and may even vitrify during drying (Frias and Oliveira, 2001). Several studies have been published that analyze reaction kinetics in foods according to the food matrix state. Karmas et al. (1992) found that the reaction rates of nonenzymatic browning were influenced by glass transition temperature in diffusion-limited systems. Buera et al. (1995) reported that the stabilization at the glassy state was effective for the hydrolysis of sucrose in polyvinylpyrrolidone (PVP) systems. According to Kerr et al. (1993), the DNPP-hydrolysis reaction in a homologous series of frozen maltodextrin solutions was related to ($T - T_g$). The glass transition temperature approach was also found to adequately describe the thermal degradation of β-carotene during air drying of carrots (Goula and Adamopoulos, 2010) and of ascorbic acid during hot air drying of maltodextrin solutions (Frias and Oliveira, 2001) and convective drying of whole persimmons (Nicoleti et al., 2007). On the contrary, Bell and Hageman (1994), who studied the degradation of aspartame in a solid-state system, concluded that water activity could be more influential than glass transition temperature on this reaction.

Other types of models that may be used to predict changes of nutrients and color of foods during drying are also available. Kaminski and Tomczak (2000), for example, discussed the capability of three models, namely, first-order reaction model, multilayer perceptron (MLP) model, and hybrid model, which is a combination of the two former models, to predict changes of quality index, which is the relative content of ascorbic acid in various vegetables during drying. The inputs to the MLP network are the quality index, material moisture content, and material temperature at any time t, while the output from the network is the quality index at time $t + 1$. Zenoozian et al. (2008) used an artificial neural network (ANN) model to predict the color intensity of osmotically dehydrated and air-dried pumpkins; the ANN-based model was selected as it has proved to have much better performance than conventional empirical or semi empirical models when being used to predict quality changes of fruits and vegetables undergoing drying (Chen et al., 2001). To improve the predictability of the ANN-based model while retaining its nonlinear mapping capability, Zenoozian and Devahastin (2009) later used combined wavelet transform–artificial neural network (WT-ANN) to predict the color intensity of pumpkins based on inputs of wavelet coefficients and drying time.

5.2.5.2 Physical Changes

Drying causes notorious physical and structural modifications of foods. The most pronounced macroscopic modification is the shrinkage and deformation. Transient thermal and moisture gradients develop tensional and compressional stresses. Tensional stresses are greater than compressional ones, especially at the boundary of the dried material. Stresses cause tissue breakage and fracturing during drying (Lewicki and Pawlak, 2005). Further drying induces formation of cracks, the inner tissue is pulled apart, and numerous holes are produced (Ramos et al., 2003; Lewicki and Pawlak, 2005). Loss of water and segregation of components occurring during drying could cause rigidity, damage and disruption of cell walls, and even collapse of cellular tissue. These changes are associated with volume reduction of the product (Mattea et al., 1989). Fast drying leads to cracking, resulting in final rigid products with more volume and a surface crust; slow drying rates result on uniform and denser products (Brennan, 1994). Shrinkage and T_g are interrelated in that significant change in volume can be noticed only if the temperature of the process is higher than the T_g of the material at that particular moisture content. In effect, above T_g the viscosity drops considerably to a level that facilitates deformation (Roos and Himberg, 1994).

Two substantial different approaches have been taken in order to model shrinkage during drying of food materials. The first one consists on an empirical fitting of experimental shrinkage data as a function of moisture content. The second approach is more fundamental and based on a physical interpretation of the food system and tries to predict geometrical changes based on conservation laws of mass and volume. In both cases, linear and nonlinear models describe shrinkage behavior versus moisture content (Mayor and Sereno, 2004). Among the linear models, one may cite, for instance, Simal et al. (1995) for green peas, Tulasidas et al. (1997) for grapes, McMinn and Magee (1997) for potatoes, and Mavroudis et al. (1998) and McLaughlin and Magee (1998) for apples. Nonlinear models exist for a variety of fruits and vegetables (e.g., Lozano et al., 1983). It is noted that if development of pores during drying is not negligible, a linear model may not be adequate to model the volumetric shrinkage behavior. This is the case for drying at higher temperatures or lower moisture contents (Katekawa and Silva, 2005) or during vacuum drying and superheated steam drying (Panyawong and Devahastin, 2007).

The aforementioned correlations can be used in combination with any drying model to predict the evolution of shrinkage with time. However, in many cases, a drying model and a shrinkage model are not coupled. A simple but adequate approach that may be used to model shrinkage of foods during drying involves alteration of a computational domain of the transport (heat and mass transfer) equations. The domain is considered to be fixed at each computational time step and is updated continuously through the whole computation. A uniform deformation of the computational domain is assumed in this case (Devahastin and Niamnuy, 2010).

The kinetic modeling approach can be used to model all textural changes of a material during drying (Devahastin and Niamnuy, 2010). While most published studies have indicated that texture degradation followed a first-order kinetic model (Nisha et al., 2005), other researchers (Rahardjo and Sastry, 1993) have fitted their data to the dual mechanism first-order kinetic model. Other researchers (Rizvi and Tong, 1997; Sila et al., 2004) have applied the fractional conversion technique, which takes into account the nonzero equilibrium texture property, for kinetic data reduction and found that the kinetics of softening of foods followed a simple first-order kinetic model rather than the dual mechanism model. Moyano et al. (2007) studied textural changes during frying of potato showing the existence of an initial tissue softening stage followed by a progressively hardening stage.

5.3 DRYING EQUIPMENT

In order to select between dryers effectively, it is important to classify them accurately and to identify clearly the essential differences between them. Kemp and Bahu (1995) have developed a detailed classification system based on three major and five minor criteria:

1. Mode of operation
   a. Batch (including semi-batch operation of continuous dryers)
   b. Continuous (including semi-continuous operation)
2. Form of feed and product
   a. Particles (including granules, agglomerates, pellets)
   b. Film or sheet
   c. Block, slab, or artefact
   d. Paste, slurry, or solution
3. Mode of heating
   a. Conduction or contact drying
   b. Cross-circulated, through circulated or dispersion convective drying
   c. Radiation; infrared, solar, or flame radiation
   d. Dielectric; radio-frequency or microwave radiation
   e. Combinations, for example, conductive/convective, radio-frequency enhancement
4. Operating pressure—vacuum or atmospheric
5. Gas flow pattern—none, cross-flow, cocurrent, countercurrent, complex
6. Solids flow pattern—stationary, well mixed, plug flow, complex
7. Solids transport method—stationary, mechanical, airborne, combined
8. Solids mixing—undisturbed layer, mechanical agitation, rotary, airborne

Kemp and Bahu (1995) give classifications of the principal types of batch and continuous dryers using mode of heating as the primary criterion, and these are reproduced as Figure 5.9. The selection of a complete drying installation includes many considerations other than the drying characteristics of the wet feedstock. These factors include the storage and delivery of the feed material, any equipment for performing it or blending back dried fine particles, the means of conveying the material as it dries, the equipment for collecting the dried product, and ancillary plants for the supply of heat, vacuum or refrigeration. Past experience in operating equipment will be a guide in the case of an existing product or drying process, and careful consideration of past choices normally reveals some deficiencies which can be rectified. Simple bench tests can reveal considerable semiquantitative information about a material's drying behavior under proposed drying conditions, leading to the elimination of some types of dryer. There have been many attempts to provide first guides for the selection of a dryer for a particular job. Simple decision trees are set out by van't Land (1984) for batch and continuous dryers, and these are reproduced in Figures 5.10 and 5.11.

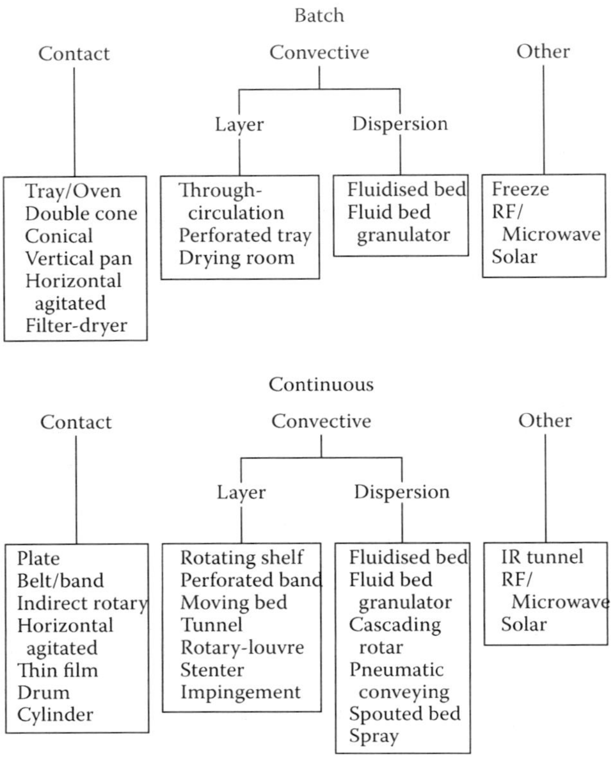

FIGURE 5.9 Principal types of batch and continuous dryers (From Kemp, I.C. and Bahu, R.E., *Drying Technol.*, 13, 1553, 1995. With permission.)

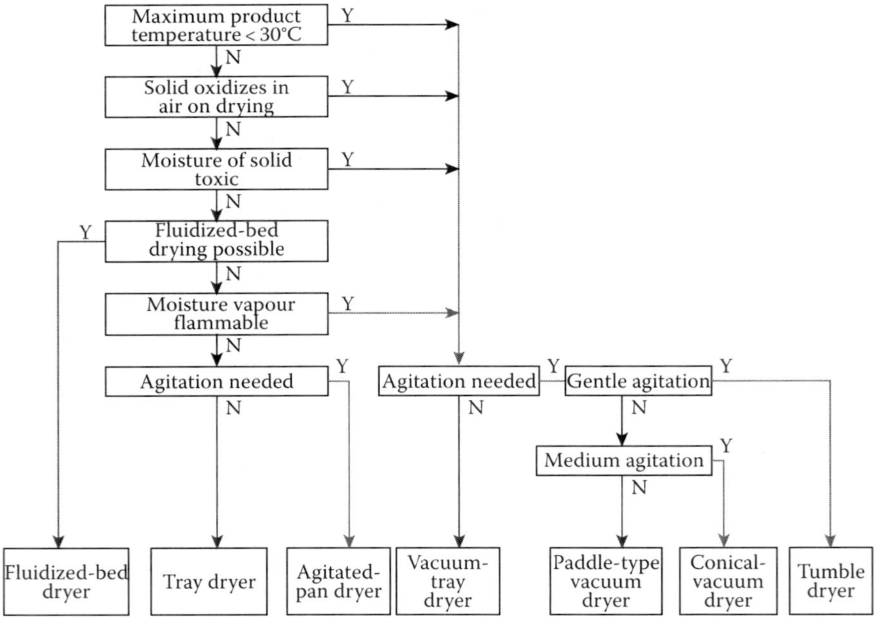

FIGURE 5.10 Decision tree for the selection of a batch dryer. (From van't Land, C.M., *Chem. Eng.*, 91, 53, 1984. With permission.)

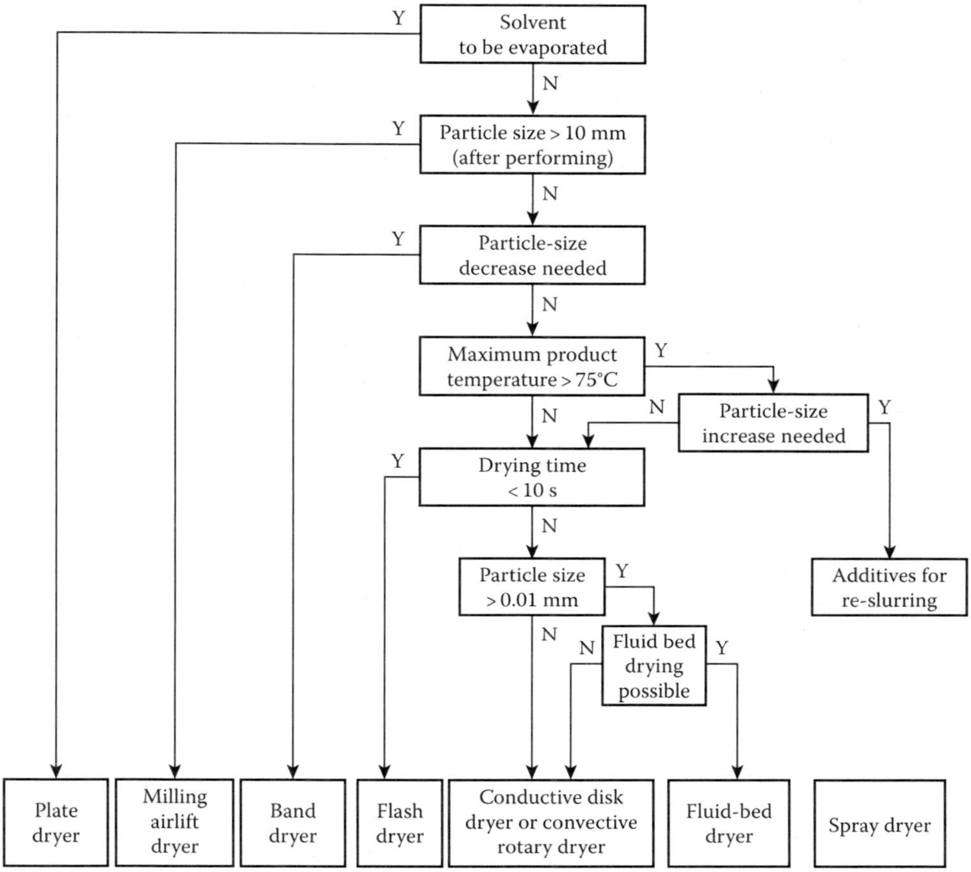

FIGURE 5.11 Decision tree for the selection of a continuous dryer. (From van't Land, C.M., *Chem. Eng.*, 91, 53, 1984. With permission.)

5.3.1 Sun Dryers

Large quantities of fruits, particularly grapes, apricots, figs, prunes, dates, coffee beans, cereal grains, and fish are sun dried prior to storage and preservation. Sun-dried fruits contain about 15%–20% moisture (wet basis), and they can be stored in bulk without the danger of microbial spoilage (Saravacos and Kostaropoulos, 2002). However, dried fruits, especially figs and apricots, may require fumigation treatment with sulfur dioxide or other permitted insecticide during storage and before packaging (Saravacos and Maroulis, 2011).

Sun drying is only possible in areas where, in an average year, the weather allows foods to be dried immediately after harvest. The main advantages of sun drying are low capital and operating costs and the fact that little expertise is required. The main disadvantages of this method are as follows: contamination, theft or damage by birds, rats, or insects; slow or intermittent drying and no protection from rain or dew that wets the product encourages mould growth and may result in a relatively high final moisture content; low and variable quality of products due to over- or under-drying; large areas of land needed for the shallow layers of food; laborious since the crop must be turned, moved if it rains; direct exposure to sunlight reduces the quality (color and vitamin content) of some fruits and vegetables. Moreover, since sun drying depends on uncontrolled factors, production of uniform and standard products is not expected.

5.3.2 Solar Dryers

Solar drying is a form of convective drying in which the air is heated by solar energy in a solar collector. The air movement is by natural convection, but addition of an electrical fan will increase considerably the collector efficiency and the drying rate of the product. In broad terms, solar drying systems can be classified into two major groups, namely (Ekechukwua and Norton, 1999):

1. Active solar-energy drying systems (most types of which are often termed hybrid solar dryers) and
2. Passive solar-energy drying systems (conventionally termed natural-circulation solar drying systems)

Three distinct subclasses of either the active or passive solar drying systems can be identified, namely:

1. Integral-type solar dryers
2. Distributed-type solar dryers, and
3. Mixed-mode solar dryers

The main features of typical designs of the various classes of solar-energy dryers are illustrated in Figure 5.12. Apart from the earlier basic types, several dryers have been developed with improved technology.

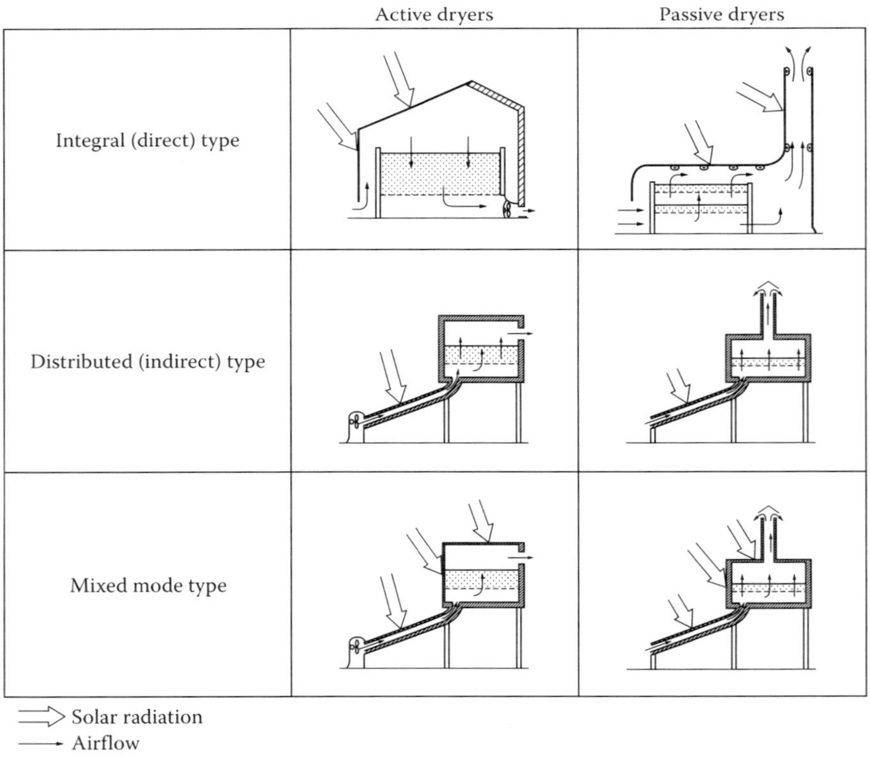

FIGURE 5.12 Typical solar energy dryer designs. (From Ekechukwua, O.V. and Norton, B., *Energy Convers. Manage.*, 40, 515, 1999. With permission.)

5.3.2.1 Staircase Type Dryer (Hallak et al., 1995)

This type is in the shape of a metal staircase with its base and sides covered with double-walled galvanized metal sheets with a cavity filled with nondegradable thermal insulation. The upper surface is covered with transparent polycarbon sheet to allow the sun's rays to pass through and be trapped. Air moves by natural convection as it enters through the bottom and leaves from the top.

5.3.2.2 Reverse Flat Plate Absorber Cabinet Dryer (RACD) (Goyal and Tiwari, 1997)

The absorber plate is horizontal and downward facing. A cylindrical reflector is placed under it to introduce solar radiation from below. The area of the aperture is the same as that of the absorber plate. The cabinet dryer is mounted on top of the absorber to maintain a gap for air to flow above the absorber, which becomes heated, and the hot air enters the dryer from the bottom. First, hot air heats the product spread over wire mesh, and then moisture starts moving from the interior of the kernel to the surface and then to the chamber. Second, the moisture-laden air exits the chamber through the vent because of the vapor pressure difference between the chamber and the outside, in the natural mode of operation.

5.3.2.3 Rotary Column Cylindrical Dryer (Sarsilmaz et al., 2000)

It contains essentially three parts—air blow region (fan), air heater region (solar collector), and drying region (rotary chamber). A fan with a variable speed of airflow rate is connected to the solar collector using a tent fabric. The dryer is manufactured from wooden plates at the top and bottom and thin plywood plates at the sides to make a cylindrical shape. A rectangular slot is opened on the side wall where it faces the solar air heater for the passage of hot air via tent fabric.

5.3.2.4 Multi-shelf Portable Solar Dryer (Singh et al., 2004)

It has four main parts, that is, multi-tray rack, trays, movable glazing, and shading plate. The ambient air enters from the bottom and moves up through the material loaded in different trays. After passing through the trays, the air leaves from the top. The multi-rack is inclined depending upon the latitude of the location. There are seven perforated trays, which are arranged at seven different levels one above the other. The product to be dried is loaded in these trays.

5.3.3 Vacuum Dryers

Vacuum drying is used to dry heat-sensitive products because the evaporation of water proceeds at temperatures as low as 30°C. Heat is supplied by conduction, and the temperature of the product can easily be controlled. Due to molecular transport of evaporated water, the process is long and can take as long as 24 h. Dry products are of very good quality, but the shelf life is dependent on the post-drying processes applied. The drying time can be reduced by application of pressure-regulatory system. The system operates by changing pressure in intermittent or prescribed cyclic pattern in the drying chamber (Maache-Rezzoug et al., 2001). An excellent quality of dry material was obtained in a fluctuating pressure system in which heat was supplied by microwaves (Szarycz et al., 2002).

5.3.4 Bin, Silo, and Tower Dryers

Bin and silo dryers are used widely in the drying of agricultural products, notably grain from an average harvest moisture of 25% to a storage moisture of about 15%. Bin dryers are also used in finishing drying of some vegetable materials, when that product is difficult to dry in the primary dryer without raising the temperature. Dehumidified air at near ambient temperature may be needed to finish drying of hygroscopic materials (Saravacos and Kostaropoulos, 2002). Airflow rates in conventional bolted steel silos are limited by vertical depth and food characteristics to low aeration airflow.

New technology horizontal, cross-flow air movement where air paths radiate from a center vertical aerator to perforated sidewall plenums allows airflow rates capable of in-silo drying without high temperature heat. Jayas et al. (1991) found that horizontal airflow resistance through elongated kernels (maize, wheat, sunflower, etc.) was 40%–50% lower than the same vertical airflow through the product. Combining a center vertical perforated aerator pipe to provide cross-flow aeration with controlled sidewall exhaust outlets in sealed silos can allow drying of partially filled silos and continuation of drying while the silo is being filled. Controlling airflow rates through sidewall and roof exhaust vent air-valve openings allows wetter layers to be dried, while a partially or completely dried product receives reduced airflow or no airflow. The addition of low burner heat to drying air with modulating controls allows for seed drying day and night without germination damage. In-silo dryers with hopper bottoms allow rapid dry-grain transfer to storage bins for efficient, economical drying (Noyes, 2005).

5.3.5 Tray/Cabinet Dryers

Tray or cabinet dryers are the simplest convective dryers, and they are used for drying relatively small batches of food materials in the form of pieces, which must be supported on trays. A typical device consists of a cabinet containing removable trays on which the solid to be dried is spread. After loading, the cabinet is closed, and team-heated air is blown across and between the trays to evaporate the moisture. When the solid has reached the desired degree of dryness, the cabinet is opened and the trays replaced with a new batch.

One of the most important difficulties in the use of these dryers is the nonuniformity of moisture content found in the finished product taken from various parts of the dryer. This is largely the result of inadequate and nonuniform air movement inside the drier. It is important to eliminate stagnant air pockets and to maintain reasonably uniform air humidity and temperature throughout the dryer. In order to do this, large volume of air must be blown over the trays, if possible at velocity ranging up to 3 or 4 m/s if the solid does not blow from the trays at these rates. But this is done at the expense of the loss of large quantities of heated fresh air. The loss of heat in the discharge air will then usually be prohibitive in cost. Instead, it is the practice to admit only relatively small quantities of fresh air and to recirculate the bulk of it. Generally, the greater the gas velocity over, through, or impinging upon a material, the greater is the convective heat transfer coefficient (Treybal, 1958). Furthermore, the better the solids are dispersed in a gas for a surface exposure, the greater is the heat transfer rate. In direct heat dryers, more gas is needed to transport heat than to purge vapor.

5.3.6 Tunnel Dryers

Tunnel dryers can be constructed from low-cost materials and are simple to operate. They are suitable for economical dehydration of fruits and vegetables near the production farms. Since the production of most fruits and vegetables is seasonal, the dryers should be used for various products to increase their operating time. In addition, the thermal efficiency of the dryer is improved by recirculation (Saravacos and Kostaropoulos, 2002).

These dryers consist of relatively long tunnels through which trucks, loaded with trays filled with the drying solid, are moved in contact with a current of gas to evaporate the moisture. The trucks may be pulled continuously through the dryer by a moving chain to which they are attached. In a simpler arrangement, the loaded trucks are introduced periodically at one end of the dryer, each displacing a truck at the other end. Parallel or countercurrent flow of gas and solid may be used. For a relatively low-temperature operation, the gas is usually steam-heated air, while for higher temperatures, flue gas from the combustion of a fuel may be used. Operation may be adiabatic, or the gas may be heated by steam coils along its path through the dryer, and operation may then be substantially at constant temperature (Treybal, 1958).

5.3.7 Conveyor Belt Dryers

Conveyor belt dryers are used extensively in food processing for continuous drying of food pieces. A conveyor belt dryer is made up of drying chambers placed in series. A drying chamber is actually the elementary module whose repetition forms the whole plant. For best performance, drying chambers are grouped together into drying sections. All chambers participating in a drying section are provided with a common conveyor belt, on which the product to be dried is uniformly distributed at the entrance. Obviously, redistribution of the product takes place when it leaves a drying section and enters the one that follows. Each drying chamber is equipped with an individual heating utility and fans for air circulation through the product. Air is heated by means of heat exchange units that operate with steam. On entering the chamber, fresh air is mixed with the recirculated air at a point below the heat exchangers. It is common practice that within each drying chamber, the temperature and humidity of the drying air stream entering the product, as well as its temperature difference while leaving it, are controlled. In this case, the final control elements are the steam valve and the chamber dampers that regulate the exchanged heat rate and the flow rate of fresh air entering the chamber, respectively.

5.3.8 Rotary Dryers

Rotary dryers are suitable for handling free-flowing granular materials, which may be tumbled about without concern over breakage, such as granulated sugar and some grains. The solid must clearly be one which is neither sticky nor gummy, which might stick to the sides of the dryer or tend to "ball" up. In such cases, recycling of a portion of the dried product may, nevertheless, permit use of a rotary dryer (Treybal, 1958). Rotary dryers are used mostly in drying food by-products and wastes (e.g., citrus peel and pulp), where high temperatures are permissible and economics is important (Saravacos and Kostaropoulos, 2002).

Rotary dryers consist of an inclined long cylinder rotating slowly (Figure 5.13). The solid to be dried is continuously introduced into one end of the cylinder, while heated air flows into the same or the other. Inside the dryer, lifting flights extending from the cylinder wall for the full length of the dryer lift the solid and shower it down in a moving curtain through the air. At the feed end of the dryer, a few short spiral flights assist in imparting the initial forward motion to the solid before the principal flights are reached. The air is heated either in heat exchangers or by mixing with combustion gases of suitable fuel, for example, natural gas. An exhaust fan is used to pull the gas through the dryer, since this provides more complete control of the gas flow. An exhaust fan is usually used to pull the gas through the dryer, whereas in some cases a blower is provided at the gas entrance to

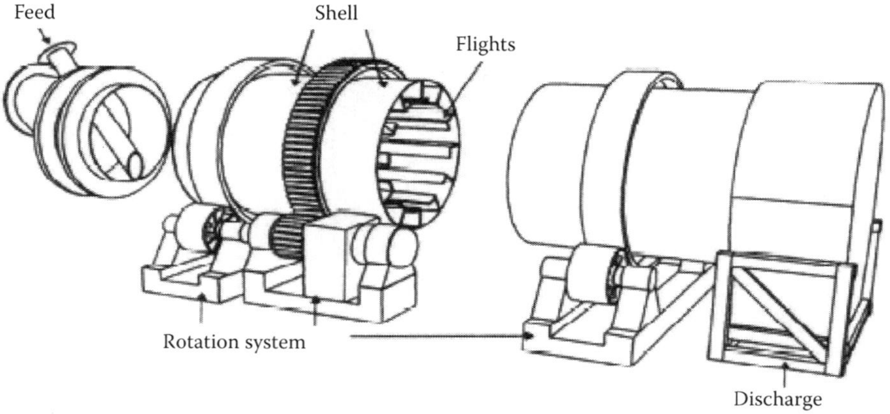

FIGURE 5.13 Typical rotary dryer. (From Silva, M.G. et al., *Braz. J. Chem. Eng.*, 29, 359, 2012. With permission.)

maintain a pressure close to the atmospheric pressure. A dust collector of the cyclone, filter or washing type may be interposed between the fan and the gas exit.

5.3.9 Fluid Bed Dryers

Fluidized bed drying has found many practical applications in the drying of granular solids in the food, ceramic, pharmaceutical, and agriculture industries, is easy to implement, and has the following advantages (Mujumdar and Devahastin, 2000):

- High drying rates due to good gas–particle contact, leading to improved heat and mass transfer rates
- Small flow area
- High thermal efficiency
- Low capital and maintenance costs

In a typical fluidized bed drying system, hot air is forced through a bed at a sufficiently high velocity to overcome gravitational effects on the particles, while ensuring that the particles are suspended in a fluidized manner. A typical layout of a simple fluidized bed system is shown in Figure 5.14. Hot air passes from the bottom through the perforated plate and interacts with the wet feed in a cross-flow manner. This interaction causes the particles to fluidize, enabling efficient gas–particle contact to take place and resulting in the particles being dried effectively. The dried particles are then discharged through the exit port of the fluidized bed. Such a system is cheap and easy to design, requiring low capital cost. The effectiveness of the heat and mass between fluidized particles and the drying air can usually be improved by increasing the drying gas velocity. However, if the gas flow is too high, there is a possibility that the gas passes around the particles' surfaces without allowing sufficient time for heat and mass exchange to take place between the particles' surface moisture and gas flow. One possible way to minimize this contact problem is via the employment of intermittent fluidization (Chua and Chou, 2003).

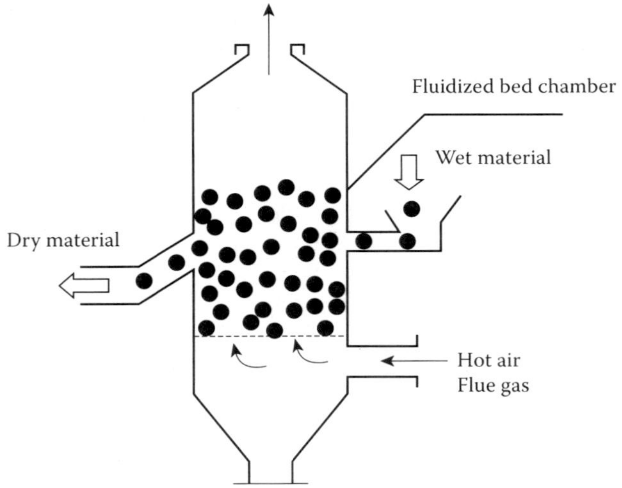

FIGURE 5.14 Longitudinal schematic of the fluidized bed system. (From Hovmand, S., Fluidized bed drying, in *Handbook of Industrial Drying*, A.S. Mujumdar, (ed.), Marcel Dekker, New York, 1995. With permission.)

5.3.10 Pneumatic or Flash Dryers

Pneumatic dryers are used for fast and efficient drying of food particles that can be suspended and transported in the stream of heating air. The mixture of hot air/gases moves the wet material from the bottom through the drying tube at a high velocity and temperatures up to 130°C. The material is dried fast due to the high heat and mass transfer rates. However, since the once-through residence time in the drying tube is very short, recirculation of the product is often required or a special cyclone, which acts as an additional dryer, is used.

5.3.11 Drum Dryers

Drum drying is extensively used in commercial drying of different types of starchy food products, sludge, suspensions, and pastes of a wide range of viscosities. Since exposure to high temperature is limited to a few seconds, drum drying is suitable for many heat-sensitive products. Daud and Armstrong (1987) presented a comprehensive study on the types, principles of operation, capacity, feeding and spreading of drying products, and the control of drum dryers. To optimize the design and operating variables associated with the drum dryers, extensive experimental and theoretical studies have been conducted by various researchers. Gavrielidou et al. (2002) listed the main variables as steam pressure, rotation speed of drum, drum clearance, pool level between the drums, and condition of the feed materials; that is, the concentration, physical characteristics, and temperature at which the material reaches the drum surface.

A slowly revolving internally steam-heated metal drum continuously dips into a trough containing the substance to be dried. A thin film of the substance is retained on the drum surface. The thickness of the film is regulated by a spreader knife and as the drum revolves, moisture is evaporated into the surrounding air by heat transferred through the metal of the drum. The dried material is then continuously scraped from the drum surface by a knife. Double drum dryers consisting of two drums placed close together and revolving in opposite directions, may be fed from above by admitting the feed into the depression between the drums. The drum dryers are usually operated at atmospheric pressure, but vacuum operation is possible for heat sensitive products by enclosing the drying system in a vacuum chamber.

5.3.12 Osmotic Dryers

Osmotic dehydration is widely used to remove water from fruits and vegetables by immersion in aqueous solution of sugars and/or salts at high concentration. This process is usually used to partially remove water from vegetable tissues obtaining stabilization without acidification or pasteurization treatments. Moreover, osmotic dehydration is used as a pretreatment prior to freezing, freeze drying, vacuum drying, and air drying (Dixon and Jen, 1977).

A driving force for water removal is set up because of a difference in osmotic pressure between the food and its surrounding solution. The complex cellular structure of food acts as a semipermeable membrane. Because the membrane responsible for osmotic transport is not selective, other solutes in the cells can also be leached into the osmotic solution (Dixon and Jen, 1977; Lerici et al., 1985; Giangiacomo et al., 1987). The rate of diffusion of water from any material made up of such tissues depends upon factors, such as temperature and concentration of the osmotic solution, the size and geometry of the material, the ratio of fruit to solution, and the level of agitation of the solution. Various osmotic agents, such as sucrose, glucose, fructose, corn syrup, sodium chloride, and so on plus their combination have been used for osmotic dehydration. Generally, sucrose solutions are used for fruits, and sodium chloride solution is used for vegetables. The addition of small quantities of sodium chloride to osmotic solutions increased the driving force of the drying process, and synergistic effects of sucrose and sodium chloride have been reported (Lerici et al., 1985).

Among the products usually treated by osmotic dehydration are some meat products; and especially vegetables, such as potatoes (Lenart and Flink, 1984), peas (Kayamak-Ertekin and Çakaloz, 1995), carrots (Rastogi and Raghavarao, 1997), red paprikas (Ade-Omowaye et al., 2002), and cherry tomatoes (Azoubel and Murr, 2004); and fruits, such as bananas (Rastogi et al., 1997), apples (Sereno et al., 2001), pears (Park et al., 2002), pineapples (Rastogi and Raghavarao, 2004), and apricots (Riva et al., 2005).

5.3.13 Heat Pump Dryers

Recently, there has been a great interest in utilizing Heat Pump Drying (HPD) for drying fruits, vegetables, and biological materials (Hawlader et al., 2005). Heat pumps (HPs) are devices for raising the temperature of low grade heat energy to a more useful level using a relatively small amount of high grade energy (Eisa, 1995). HPDs consume 50%–80% less energy than conventional dryers operating at the same temperature. This makes such dryers a feasible option for users who are not satisfied with the comparatively high energy consumption of directly heated dryers (Schmidt et al., 1998).

A HP drying system consists mainly of two subsystems; a HP system and a drying chamber. HPs can transfer heat from natural heat sources in the surroundings, such as the air, ground or water, or from industrial or domestic waste, chemical reaction or dryer exhaust air. The drying chamber can be formed as a tray, fluid bed, rotary or band conveyor. The main components of the general HP unit are an evaporator, a condenser, a compressor, and an expansion valve. In a HP drying system, the working fluid (refrigerant) at low pressure is vaporized in the evaporator by heat drawn from the dryer exhaust air. The compressor raises the enthalpy of the refrigerant of the HP and discharges it as superheated vapor at high pressure. Heat is removed from the working fluid and returned to the process air at the condenser. In the drying system, the hot air at the exit of the condenser is allowed to pass through the drying chamber where it gains latent heat from the product to be dried. The working fluid from the condenser is then throttled to the low pressure line and enters the evaporator to complete the cycle. The humid air at the dryer exit then passes through the evaporator where condensation of moisture occurs as the air goes below dew point temperature (Colak and Hepbasli, 2009).

5.3.14 Infrared Dryers

Another low-cost drying method suitable for employment in rural farming areas is Infrared (IR) drying. IR drying can be considered to be an artificial sun-drying method, which can be sustained throughout the entire day. Sandu (1985) described the advantages of applying IR to foodstuffs, including: versatility; simplicity of the required equipment; fast response of heating and drying; easy installation to any drying chamber; and low capital cost. Ginzburg (1959) has shown the suitability of applying IR drying to foodstuffs, such as grains, flour, vegetables, pasta, meat, and fish.

During IR drying, radiative energy is transferred from the heating element to the product surface without heating the surrounding air. Figure 5.15 shows a simple IR dryer design that can be easily constructed for use in rural farming areas. The design includes a manual-conveyor system whereby the food product enters from the inlet hoop and is dried as it moves parallel to the IR lamps. The level of irradiation can be adjusted via the voltage regulator, while intermittent IR drying can be implemented by turning the timer relay knob.

5.3.15 Spray Dryers

Spray drying is used to dehydrate liquid foods or food suspensions into dry powders or agglomerates and involves both particle formation and drying. The feed is transformed from the fluid state into droplets and then into dried particles by spraying it continuously into a hot drying medium. Section 5.4 describes the spray drying process.

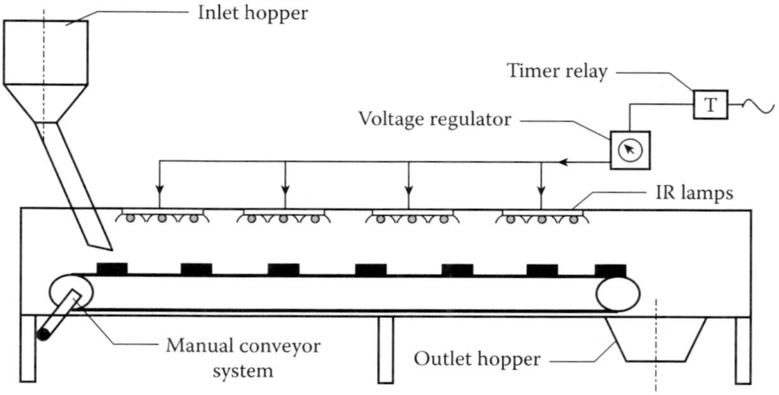

FIGURE 5.15 Schematic diagram of a conveyor IR drying system. (From Ratti, C. and Mujumdar, A.S., Infrared drying, in *Handbook of Industrial Drying*, A.S. Mujumdar, (ed.), Marcel Dekker, New York, 1995. With permission.)

5.3.16 Freeze Dryers

Freeze drying, or lyophilization, is the sublimation/removal of water content from frozen food. The dehydration occurs under a vacuum, with the product solidly frozen during the process. Shrinkage is eliminated or minimized, and a near-perfect preservation results. Freeze-dried food lasts longer than other preserved food and is very light, which makes it perfect for space travel. The purpose of freeze drying is to remove a solvent (usually water) from dissolved or dispersed solids. Freeze drying is the method for preserving materials which are unstable in solution. In addition, freeze drying can be used to separate and recover volatile substances and to purify materials. Section 5.5 concerns the freeze drying process.

5.4 SPRAY DRYING

5.4.1 Basic Principles

Spray drying produces particles by atomizing a solution or slurry and evaporating moisture from the resulting droplets by suspending them in a hot gas. As the rapid evaporation keeps the droplets' temperature relatively low, the product quality is not significantly or negatively affected (Roustapour et al., 2005). Figure 5.16 is a schematic of a typical spray dryer. The liquid feed is pumped to an atomizer, which breaks up the feed into a fine spray and ejects it into the spray chamber, where it is mixed with hot drying gas. Moisture is evaporated from the spray, and the droplets are transformed into dry particles. The separation of the particles from the drying gas and their subsequent collection will either take place in external gas cleaning equipment (i.e., cyclones or filters) or, in the case of coarser particles, within the chamber itself.

Although the vast majority of cases employ hot atmosphere to drive moisture from each spray droplet, there are cases in which the delicacy of the operation demands that the drying medium is first dehumidified and then just warmed over atmospheric temperatures. This is a variation of the basic spray drying concept and is termed *low-temperature spray drying*. Another variation is *foam spray drying*, which involves the introduction of a gas into the feed stock before atomization to produce spherical foamed particles containing vacuoles so as to achieve a low bulk density. A further variation, which is termed *spray freeze drying*, is to spray the product into freezing air, whereupon the individual droplets are frozen for subsequent moisture removal through sublimation under vacuum. Should the temperature of the air permit only solidification of the spray droplets, the process is termed *spray cooling* (Masters, 1985).

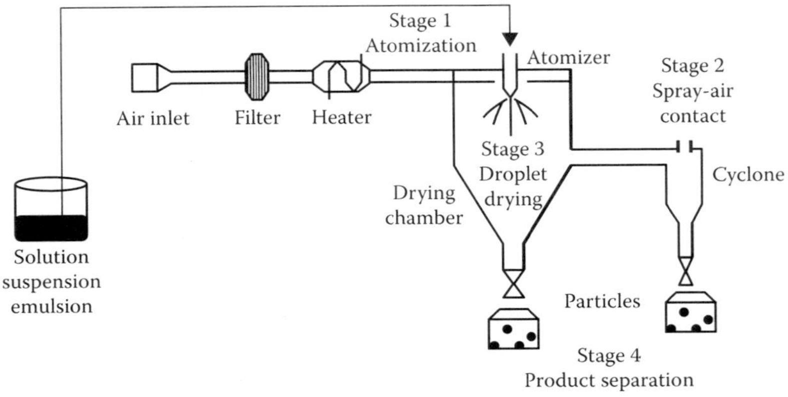

FIGURE 5.16 Typical spray drying system.

TABLE 5.5
Spray-Dried Food Products

Product	References	Product	References
Milk	Birchal et al. (2005); Nijdam and Langrish (2005a,b); Fonseca et al. (2011); Rogers et al. (2012)	Tomato	Al-Asheh et al. (2003); Goula and Adamopoulos (2004, 2005a,b, 2008)
Soymilk	Telang and Thorat (2010); Giri and Mangaraj (2012)	Pomegranate	Youssefi et al. (2009); Vardin and Yasar (2012)
Skim milk	Shrestha et al. (2007); Amiri-Rigi et al. (2012)	Watermelon	Quek et al. (2007)
Whey protein	Anandharamakrishnan et al. (2007)	Apricot	Bhandari et al. (1993)
Egg	Franke and Kieβling (2002); Ayadi et al. (2008); Wenzel et al. (2010); Rannou et al. (2013)	Guava	Chopda and Barret (2001)
Honey	Bhandari et al. (1997); Boonyai et al. (2005)	Cactus pear	Rodríguez-Hernández et al. (2005)
Apple	Boonyai et al. (2005)	Blackberry	Ferrari et al. (2012)
Orange	Chegini and Ghobadian (2005); Shrestha et al. (2007); Goula and Adamopoulos (2010)	Raisin	Papadakis et al. (2005)
Pineapple	Bhandari et al (1997); Abadio et al. (2004)	Mango	Cano-Chauca et al. (2005)
Lime	Dolinsky et al. (2000); Zareifard et al. (2012)	Soybean	Georgetti et al. (2008)
Ginger extract	Jangam and Thorat (2010)	Sourdough	Tafti et al. (2013)

Spray drying is widely applied in several industrial sectors, including food, pharmaceutical, and chemical. Several biological and thermal-sensitive materials, liquid materials, such as milk, fruit juices and pulps, herbal extracts, enzymes, essential oils, aromas, and various pharmaceuticals have been dried by this process. Some of the spray-dried foods are listed in Table 5.5.

The principal advantages of spray drying are as follows (Filkova and Mujumdar, 1995):

- Product properties and quality are more effectively controlled.
- Heat-sensitive foods can be dried at atmospheric pressure and low temperature.
- Spray drying permits high-tonnage production in continuous operation and relatively simple equipment.

- The product comes in contact with the equipment surfaces in an anhydrous condition, thus simplifying corrosion problems and selection of materials of construction.
- Spray drying produces relatively uniform, spherical particles with nearly the same proportion of nonvolatile compounds as in the liquid feed.
- Since the operating gas temperature may range from 150°C to 500°C, the efficiency is comparable to that of other types of direct dryers.

5.4.2 Stages of Spray Drying

Spray drying consists of four process stages: (1) atomization, (2) contact of atomized spray with drying air, (3) evaporation of moisture from the spray, and (4) separation of dried product from the exhausted air.

5.4.2.1 Atomization

The atomization process generally refers to the formation of powder or liquid suspension in a gas as well as subsequent reduction in particle size (Cal and Sollohub, 2010). An effective atomization process converts the fluid feed to tiny droplets with equal size, which leads to uniform heat and mass transfer during the drying process. Due to the subsequent reduction in particle size and dispersion of the particles in the drying gas, the surface area of the particles increases exponentially. This increment in surface area of the particles helps to dry the feed in seconds. With the small size of droplets and the even distribution of the fluid feed, the moisture removal occurs without disturbing the integrity of the material. The atomization is achieved by atomizers, which are generally classified as rotary atomizers, pressure nozzles, pneumatic nozzles, and sonic nozzles (Cal and Sollohub, 2010). Atomizers are classified based upon the type of energy, which acts upon the bulk fluid. For example, rotary atomizers use centrifugal energy to atomize the feed, while pressure nozzles use a pressure buildup (Masters, 1985; Filkova et al., 2007).

5.4.2.1.1 Rotary Atomizer

Figure 5.17 illustrates the principles of operation of a rotary atomizer. The liquid to be atomized is fed onto a rapidly rotating vaned wheel or disk. The liquid is accelerated outward by centrifugal forces and ejected as a thin sheet of liquid, which is subsequently broken up into droplets.

Rotary atomizers are flexible with a good turndown ability and are ideal for many spray drying operations. All orifices and clearances along the fluid flow path are large. This makes for an atomizer that requires very low feed pressures and is virtually nonclogging, even for the difficult solutions and slurries encountered in spray drying (Oakley, 1997). A significant drawback of rotary

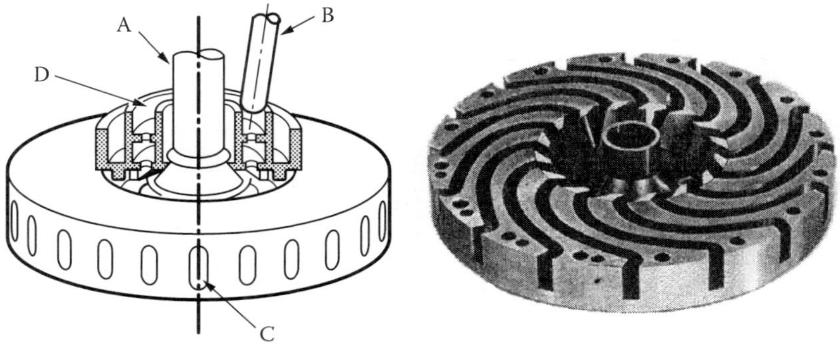

FIGURE 5.17 Rotary atomizer. (A: shaft, B: feed pipe, C: radial channel, D: dispenser). (From Brennan, J.G. et al., *Food Engineering Operations*, 2nd edn., Applied Science Publishers, London, U.K., 1976; Masters, K., *Drying Technol.*, 12, 235, 1985. With permission.)

atomization is that the droplets are thrown horizontally outward at high velocity. Hence, a wide chamber is required to ensure that there is sufficient distance for droplets to be deflected before striking the chamber wall.

5.4.2.1.2 Pressure Nozzle

The operation of pressure nozzles, also called one-fluid nozzles, is based on the principle of the fluid flowing, under pressure, through a conduit with a decreasing diameter (Figure 5.18). The fluid leaving such a pipe loses some part of its pressure in favor of particle velocity and undergoes atomization. In such a case, the orifice through which the fluid leaves the pipe is of 0.4–4 mm in diameter. The angle at which the atomization occurs varies from device to device, but is usually in the range of 40°–150°, which allows for the use of narrow chambers (Cal and Sollohub, 2010).

With this type of nozzle, it is generally possible to produce the droplets within a narrow range of diameters, and the dried particles are usually hollow spheres. Pressure nozzles are not suitable for highly concentrated suspensions and abrasive materials because of their tendency to clog and erode the nozzle orifice (Filkova and Mujumdar, 1995). Hydraulic nozzles provide the operator with only a limited ability to control the properties of the obtained particles, beside the feed composition. The only parameter of the nozzle that can be changed or adjusted is the rate of feeding, upon which the pressure of the supplied fluid subjected to atomization is dependent. Even the smallest changes from the normative values may cause atomization and drying disorders.

5.4.2.1.3 Pneumatic Nozzle

Pneumatic nozzles utilize high-velocity gas streams to impinge on and break up low-velocity liquid streams to form the divided liquid droplets (Figure 5.19). Two types of two-fluid-phase-spraying

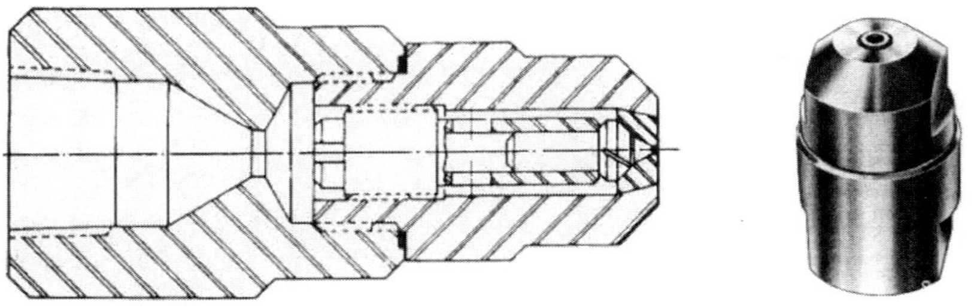

FIGURE 5.18 Pressure nozzle. (From Brennan, J.G. et al., *Food Engineering Operations*, 2nd edn., Applied Science Publishers, London, U.K., 1976; Masters, K., *Drying Technol.*, 12, 235, 1985. With permission.)

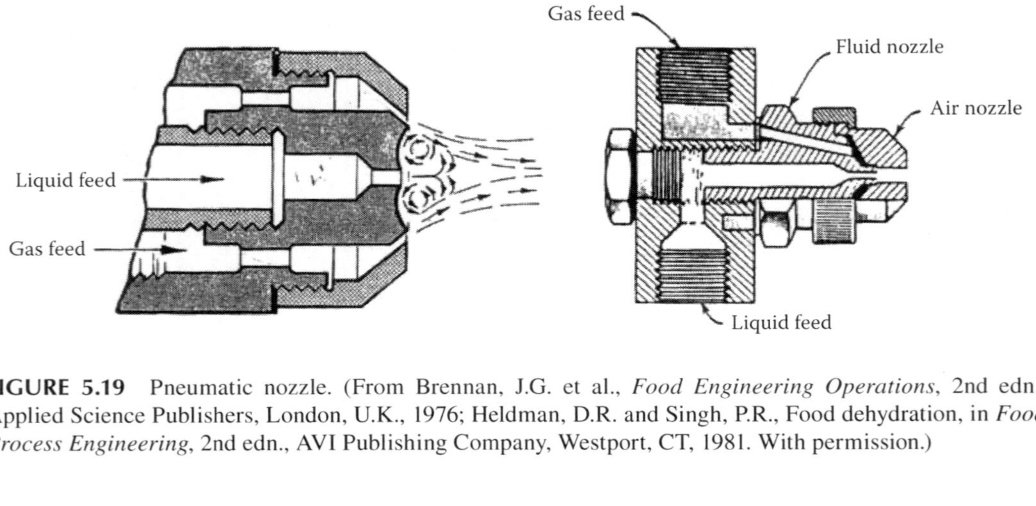

FIGURE 5.19 Pneumatic nozzle. (From Brennan, J.G. et al., *Food Engineering Operations*, 2nd edn., Applied Science Publishers, London, U.K., 1976; Heldman, D.R. and Singh, P.R., Food dehydration, in *Food Process Engineering*, 2nd edn., AVI Publishing Company, Westport, CT, 1981. With permission.)

nozzles have been developed so far for blowing off fine droplets. One is the external mixing type, which is composed of concentric tubes. This is the most popular type. Liquid blown off from the inner tube is sheared by the airflow from the outer one. The shear force makes a fine mist. However, there is some difference in the shear rate depending on the location in the liquid stream. This fact results in a broad distribution of the droplet size. Therefore, the diameter of the inner orifice should be as small as possible in order to produce homogeneous fine droplets. When small droplets of 10 μm mean diameter are required, the diameter of the inner orifice should be less than 0.2 mm and the weight-based flow ratio of air to liquid about 2.5. If the inner diameter is increased in order to supply large amount of liquid for industrial scale, the flow ratio should be extremely large up to 100.

The other type is the internal mixing type. The liquid and air are intermixed before blowing off. The primary droplets blown from the nozzles collide with each other and are broken down to a small size. In industrial production, it is empirically known that the combination of 1 L/min of the liquid flow rate and 3 kg/min of the airflow will produce droplets of 10 μm in a mean diameter (Hino et al., 2000). This nozzle is not available for suspensions, concentrated solutions, and solutions containing hard water. When it is used for such systems, some droplets containing solid components might be dried in the mixing chamber. The feeding of the solution, therefore, will be interrupted and stopped by the sludge, which gradually accumulates on the chamber wall.

5.4.2.1.4 Other Types

There are, however, feeds (especially non-Newtonian fluids with high viscosity) that are not suitable for atomization performed by such conventional devices. One of the possible solutions for working with such feeds is to use an ultrasonic nozzle. In such nozzles, a high-frequency electric signal is applied to two electrodes placed between two piezoelectric transducers, causing vibrations that are further transferred and amplified by a titanium nozzle tip. The nozzle outlet in the atomization spot vibrates at the ultrasonic frequency and causes the feed to be atomized. An unquestionable advantage of such nozzles is their ability to self-clean.

Atomization by conventional electro-hydrodynamic nozzles is another common solution. The main principle of such a construction lies in the effect of an electrostatic field on the fluid particles supplied by the capillary under low pressure. The low pressure pushes the fluid out of the nozzle, forming a meniscus. The application of an external electric field causes the acceleration of the charged particles within the fluid, which results in the fluid particles gradually acquiring momentum. When the fluid gains the appropriate acceleration, the meniscus turns into a long filament, the end of which subsequently breaks up into tiny charged droplets (Wu and Clark, 2008; Li and Lai, 2011).

Van Deventer et al. (2013) introduced a new atomization nozzle based on inkjet technology. The print head is able to process material feed containing higher concentrations of solids, resulting in an even higher overall system performance. The produced monodisperse droplets undergo less shear force in the nozzle, and all droplets experience the same heat load during drying. The monodisperse droplets lead to monodisperse powder particles, a completely different product compared to conventional atomization nozzles that deliver particles with a wide spread in size and form. The absence of fines results in a more economic process due to a reduction in the intensity of exhaust air filtration. The resulting powders are spherical without air enclosures, resulting in high-density powders. The monodisperse nature of the droplets not only results in increased powder properties, it also enables more exact dosing of drying air, resulting in an increase in energy efficiency.

5.4.2.2 Droplet–Air Contact

The fluid spray and air contact time are important components of spray drying as they determine the drying rate and the intensity of drying. The duration of stay of each droplet inside the drying chamber is determined by the spray's flow rate and size of drying chamber, which governs the air contact time. Hence, it is important to design the drying chamber and air dispenser in such a way to create easy flow of the product to prevent deposition of partially dried product on the chamber wall and on the atomizer. The main reason for the wall deposit is the overly rapid travel of the droplets, which leads to a reduction of the contact time with the drying air, which prevents adequate escape of moisture into the drying air. Generally, the spray and air contact patterns are classified as cocurrent, countercurrent, and mixed flow (Masters, 1985; Cal and Sollohub, 2010).

Cocurrent spray dryers are the most common and widely used dryers when compared to other systems (Zbicinski et al., 2002). Drying kinetics and particle behavior in cocurrent systems are well unknown compared to countercurrent or mixed flow systems. In cocurrent dryers, the atomizer and the drying gas stream inlet are placed in the upper part of the drying chamber. The feed droplets travel in the same direction of the drying gas flow while losing moisture content. In countercurrent dryers, the air inlet is placed opposite the descending dispersion droplets. In other words, the drying gas is provided from the bottom of the drying chamber, and the atomization occurs in the top of the drying chamber.

Countercurrent systems are a less commonly used method, and relatively very few products are dried by using this method. These systems account for only 5% of all spray dryers used. The hydrodynamics of recirculation of continuous phase and particle agglomeration is complicated in the countercurrent drying system. Due to this complicated flow, the drying phenomena in these systems are not well understood (Piatkowski and Zbicinski, 2007).

In *mixed-flow* dryers, the feed is atomized upwards, in the direction of the top of the chamber, while the drying air inlet is placed in the upper part of the chamber; thus, the drying air passes countercurrent relative to the atomized feed. This method produces fair-sized particles. This is the most economical method that is used to dry thermostable products (Cal and Sollohub, 2010).

5.4.2.3 Evaporation of Moisture

The drying of feed droplets in a spray drying process is a result of simultaneous heat and mass transfer. The heat from the drying medium is transferred to droplets by convection and then converted to latent heat during the evaporation of the droplets' moisture content. The rate of heat and mass transfer depends upon the droplet diameter and the relative velocity of the air and droplets (Masters, 1985; Filkova et al., 2007).

The drying kinetics of droplets containing solid components is usually divided in two drying stages. In the first drying stage, the droplet with excess of liquid is subjected to the flow of drying gas, gains sensible heat, and begins to evaporate from the surface. This liquid evaporation results in droplet diameter shrinking and simultaneous increase of solids concentration near the droplet surface. The latter leads to the precipitation of solid fraction at the droplet surface and, eventually, the surface becomes covered with a layer of solid component called crust. At this moment, the droplet turns into a wet particle, and the second drying stage commences. In this stage, the drying process is hindered by additional resistance to mass transfer because of the crust layer. The second drying stage continues until the particle moisture content reduces to equilibrium value with the drying medium. After that the drying process actually stops, the particle is heated up to the equilibrium temperature (Mezhericher et al., 2010). Typical droplet temperature and moisture content evolutions during the spray drying process are shown in Figure 5.20. In this figure, the interval between points 0 and 1 is corresponding to the droplet initial heating, the period between points 1 and 2 is related to the droplet evaporation period, the interval between points 2 and 3 to the second drying stage, and, finally, the period between points 3 and 4 designates the particle heating up to equilibrium with the surrounding drying agent.

The published theoretical models of droplet drying kinetics can be related to the following categories: models based on a semi empirical approach that utilizes the concept of characteristic drying curve (CDC) (Langrish and Kockel, 2001); deterministic drying models, which describe the processes within the droplet using the continuity, momentum, energy, and species conservation equations (Mezhericher et al., 2008); recently appeared deterministic models with distribution of solid

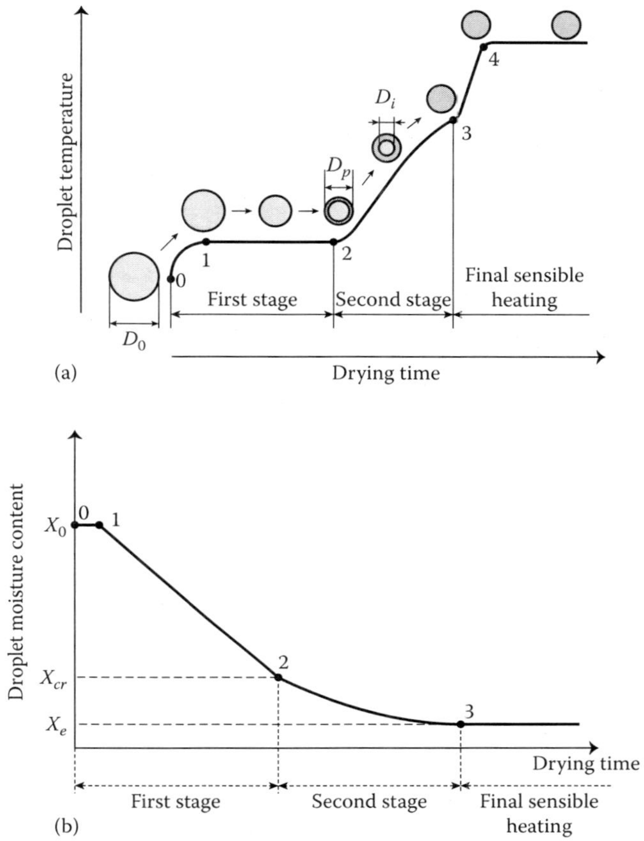

FIGURE 5.20 Typical drying evolutions of droplet temperature (a) and moisture content (b). (From Mezhericher, M. et al., *Chem. Eng. Process. Process Intensif.*, 49, 1205, 2010. With permission.)

component described by population balance (Seydel et al., 2005); and novel drying models based on reaction engineering approach (REA) (Patel and Chen, 2008). The semi empirical CDC models of droplet drying kinetics are usually represented by a small set of simplified equations allowing fast computations and are known to work well for small particle sizes (Mezhericher et al., 2010). The deterministic analytical models of single droplet drying describe the process by a set of differential equations with corresponding initial and boundary conditions. The solution of this set of equations is complicated by the presence of a moving domain boundary because of the shrinking droplet radius in the first drying stage and the receding interface between the dry crust and wet core regions of the wet particle in the second drying stage (Mezhericher et al., 2007). As a result, the numerical solution of such models is a complex problem, and the simulation of the drying process can demand significant computer resources and time. The REA models, which have appeared in recent years, demonstrate a fine agreement with experiments as well as fast calculations and reduced demands of the computer resources. However, at the present time, the application of REA for droplet drying kinetics is still limited by the range of materials whose drying behavior was already experimentally studied (Mezhericher et al., 2010).

However, the evaporation characteristics of droplets within a spray differ from the evaporation characteristics of single droplets. Although the basic theory applies in both cases, it is difficult to apply this theory to the case of a large number of droplets evaporating close to the atomizer. Any analysis of spray evaporation depends upon defining the spray in terms of a representative mean diameter and size distribution, the relative velocity between the droplet and its surrounding air, droplet trajectory, and the number of droplets present at any given time per given volume of drying air. Furthermore, there are grave difficulties in determining these factors in the vicinity of the atomizer (Masters, 1985).

In recent years, computational fluid dynamics (CFD) has been increasingly applied to food processing operations. In spray drying operations, CFD simulation tools are now often used because measurements of airflow, temperature, particle size, and humidity within the drying chamber are very difficult and expensive to obtain in a large-scale dryer. CFD is a simulation tool, which uses powerful computers in combination with applied mathematics to model fluid flow situations and aid in the optimal design of industrial processes. The method comprises solving equations for the conservation of mass, momentum, and energy, using numerical methods to give predictions of velocity, temperature, and pressure profiles inside the system. Powerful graphics can be used to show the flow behavior of fluid with 3D images (Figure 5.21) (Kuriakose and Anandharamakrishnan, 2010).

Crowe (1980) was probably the first who presented the so-called particle source in cell (PSI-cell) model for gas–droplet flows in a spray dryer. Goldberg (1987) developed this numerical technique making assumptions such as isotropic turbulence at a point and k–ε model for estimating the turbulent eddy viscosity. Reay (1988) discussed the work of Goldberg, emphasizing the detail that can be extracted from CFD programs. According to his conclusions, the most likely areas for wall deposition are an annular area of the dryer roof, corresponding to the small recirculation eddy and a region below the atomizer where large particles are likely to deposit. Oakley and Bahu (1991) predicted the trajectories of water droplets from a hollow-cone pressure nozzle in the same chamber as modeled by Goldberg. They noted a strong cooling effect of the evaporating droplets on the central gas jet and a significant evaporation in a recirculation zone in the outer regions of the chamber. Langrish and Zbicinski (1994) demonstrated the use of a CFD program to explore methods for decreasing the wall deposition rate, including simple modification to the air inlet geometry and a reduction in the spray cone angle. Masters (1994) reported the use of a CFD program enabling the scale up of a dryer modification, which involved a secondary airflow around the walls to prevent powder settling on them. Southwell et al. (1999) have used CFD to improve the flow distribution in a pilot spray dryer; Kieviet and Kerkhof (1997) simulated the airflow pattern (no spray) and the temperature and humidity pattern (water spray)

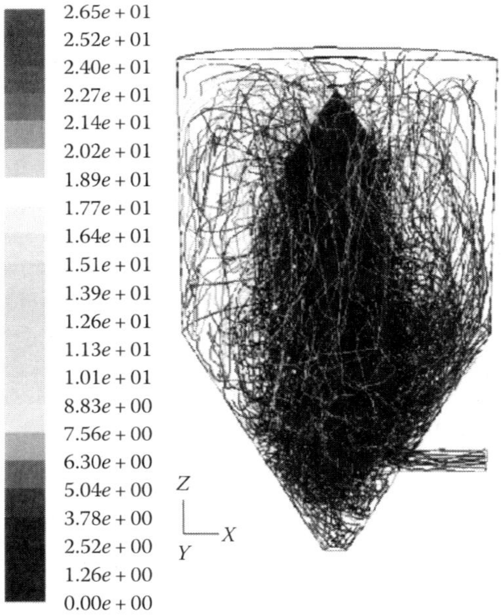

FIGURE 5.21 Particle trajectories colored by residence time(s). (From Kuriakose, R. and Anandharamakrishnan, C., *Trend Food Sci. Technol.*, 21, 383, 2010. With permission.)

in a cocurrent pilot plant spray dryer. Frydman et al. (1999) presented a numerical model for the prediction of gas velocity and temperature and particle trajectories in a superheated steam spray dryer. Kievet et al. (1997) noted the interaction of substantial wall deposition with the residence time distribution. The history of ideas and designs proposed for the reduction of wall deposition has been reviewed by Langrish and Fletcher (2003), Langrish (2009), and Kuriakose and Anandharamakrishnan (2010), who concluded that CFD techniques offer assistance in finding practical solutions.

A CFD code predicts fluid flow by numerically solving the partial differential equations, which describe the conservation of mass and momentum and are known as the Navier–Stokes equations. A grid is placed over the flow region of interest, and discrete equations are derived by applying the conservation of mass and momentum over each cell of the grid in turn. In the case of turbulent flows, the conservation equations are solved to obtain time-averaged information. Since the time-averaged equations contain additional terms, which represent the transport of momentum, heat, and mass by turbulence, turbulence models that are based on a combination of empiricism and theoretical considerations are introduced to calculate these quantities from details of the mean flow. The standard k–ε turbulence model is generally considered the simplest model, which can adequately predict a wide range of flows. This model uses a specific relation to determine the local value of the eddy viscosity from the turbulent kinetic energy (k), and its dissipation (ε). It assumes that the turbulence is isotropic (the same in all directions). This assumption sharply reduces the intensity of the calculation, while the accuracy is hardly reduced (Straatsma et al., 1999). That is why the k–ε model is currently one of the most popular turbulence models for technical calculations.

The behavior of droplets and particles is usually calculated by the discrete phase model, in which a number of individual particulates are chosen to represent the spray leaving the atomizer. Each simulated particulate will represent a specified mass flow rate and hence particulate number per second. Each of these particulates is tracked through the flow using the local time-averaged gas velocities predicted by the gas phase model and empirical drag coefficients. At the same time, equations are solved to determine the temperature and evaporation rate of the particulates, using the local time-averaged temperature and humidity predicted and the empirical heat and mass transfer coefficients.

The dispersion of the particles due to turbulence is modeled using a stochastic discrete-particle approach. The trajectory equations for individual particles are integrated, using the instantaneous fluid velocity along the particle path during the integration. By computing the trajectory in this manner for a sufficient number of representative particles, the random effects of turbulence on the particle dispersion may be accounted for. Also, the discrete phase model calculates the effect of the particulates on the gas phase by the particle-source-in-cell method. Within each control volume of the gas phase calculation through which a particulate passes, the amount of momentum, mass, and heat that is transferred from the particulate to the gas phase is calculated. This is multiplied by the number of actual particulates per second, representing the simulated particulates, to give the total source terms transferred from the particulates to the gas phase. These source terms are then included in the gas phase simulation, which is then repeated. The entire process of calculating the gas phase, particulate behavior and source terms, and recalculating the gas phase is repeated until convergence is reached (Goula and Adamopoulos, 2004).

The partial differential equation, which describes the conservation of mass for the continuous phase can be expressed as follows:

$$\frac{\partial \rho_a}{\partial t} + \vec{\nabla} \cdot \left(\rho_a \vec{u}_a \right) = S_M \tag{5.18}$$

where

ρ_a is the air density

u_a is the air velocity

S_M is the source term of mass, coming from the dispersed phase due to droplets evaporation and is computed by examining the change in mass of a particle as it passes through each control volume.

The mass change is computed simply as

$$M = \frac{\Delta m_p}{m_{p0}} \dot{m}_{p0} \tag{5.19}$$

where

m_p is the droplet mass

m_{p0} is the initial droplet mass

Conservation of momentum is expressed as

$$\frac{\partial \rho_a \vec{u}_a}{\partial t} + \vec{\nabla} \cdot \left(\rho_a \vec{u}_a \vec{u}_a \right) = -\vec{\nabla} P + \rho_a \vec{g} + \vec{F} - \vec{\nabla} \cdot \tau \tag{5.20}$$

where

P is the pressure

τ is stress tensor

F is the overall external force coming from interactions with the dispersed phase and is computed as the sum of the changes in momentum of the particles as they pass through each control volume

$$\vec{F} = \sum \frac{3C_{\text{drag}}\rho_a}{4\rho_p D_p} \left(\vec{u}_a - \vec{u}_p \right)^2 \dot{m}_p \Delta t \tag{5.21}$$

where

C_{drag} is the drag coefficient

ρ_p is the particle density

D_p is the particle diameter

u_p is the particle velocity

The energy equation is solved in the following form:

$$\frac{\partial \rho_a H_a}{\partial t} + \vec{\nabla} \cdot \left(\rho_a \vec{u}_a H_a \right) = \lambda_a \nabla^2 T_a + \frac{\partial P}{\partial t} + \vec{u}_a \vec{\nabla} P + \tau : \vec{\nabla} \vec{u}_a + S_H \tag{5.22}$$

where

H_a is the enthalpy of the air

λ_a is the thermal conductivity of the air

T_a is the temperature of the air

S_H is the heat exchanged between the continuous and the dispersed phases

The change in thermal energy of a particle as it passes through each control volume is computed as

$$Q = \left[\frac{\overline{m}_p}{m_{p0}} c_p \Delta T_p + \frac{\Delta m_p}{m_{p0}} \left(-h_{fg} + \int_{T_{\text{ref}}}^{T_p} c_{p,i} \mathrm{d}T \right) \right] \dot{m}_{p0} \tag{5.23}$$

where

c_p is the heat capacity

h_{fg} is the latent heat of vaporization

T_p is the temperature of the particles

the subscript i refers to the volatiles

Particle trajectories are calculated by integrating the force balance on the particles:

$$\frac{d\vec{u}_p}{dt} = \frac{3C_{\text{drag}}\rho_a\left(\vec{u}_a - \vec{u}_p\right)\cdot\left|\vec{u}_a - \vec{u}_p\right|}{4D_p\rho_p} + \bar{g}\left(\frac{\rho_p - \rho_a}{\rho_p}\right) \tag{5.24}$$

When the particle temperature is less than the vaporization temperature, the inert heating law is applied, and a simple heat balance is used to relate the particle temperature to the convective heat transfer at the particle surface:

$$m_p c_p \frac{dT_p}{dt} = hA_p\left(T_\infty - T_p\right) \tag{5.25}$$

The heat transfer coefficient, h, is usually evaluated using the correlation of Ranz and Marshall:

$$Nu = 2.0 + 0.6\,\text{Re}^{1/2}\,\text{Pr}^{1/3} \tag{5.26}$$

When the temperature of the droplet reaches the vaporization temperature and until the droplet reaches the boiling point, the vaporization law is applied to predict the vaporization from a droplet. The rate of vaporization is governed by gradient diffusion, with the flux of droplet vapor into the gas phase related to the gradient of the vapor concentration between the droplet surface ($C_{i',s}$) and the bulk gas ($C_{i',\infty}$):

$$N_{i'} = k_c\left(C_{i',s} - C_{i',\infty}\right) \tag{5.27}$$

The concentration of vapor at the droplet surface is evaluated by assuming that the partial pressure of vapor at the interface is equal to the saturated vapor pressure at the droplet temperature (Equation 5.28), whereas the vapor concentration in the bulk gas is known from the solution of the transport equation for species i' (Equation 5.29):

$$C_{i',s} = \frac{P_{\text{sat}}(T_p)}{RT_p} \tag{5.28}$$

$$C_{i',\infty} = X_{i'}\frac{P_{\text{op}}}{RT_\infty} \tag{5.29}$$

where X is the mole fraction of species i'. The mass transfer coefficient, k_c, is usually calculated from the correlation:

$$Sh = 2.0 + 0.6\,\text{Re}^{1/2}\,Sc^{1/3} \tag{5.30}$$

The droplet temperature is updated according to a heat balance that relates the sensible heat change in the droplet to the convective and latent heat transfer between the droplet and the continuous phase:

$$m_p c_p \frac{dT_p}{dt} = hA_p\left(T_\infty - T_p\right) + \frac{dm_p}{dt}h_{fg} \tag{5.31}$$

When the temperature of the droplet reaches the boiling temperature, a boiling rate equation is applied, and the droplet remains at fixed temperature (T_{bp}):

$$\frac{\mathrm{d}D_p}{\mathrm{d}t} = \frac{4\lambda_a}{\rho_p c_{p,\infty} D_p}\left(1+0.23\sqrt{\mathrm{Re}}\right)\ln\left[1+\frac{c_{p,\infty}\left(T_\infty - T_p\right)}{h_{fg}}\right] \quad (5.32)$$

The coupled two-phase simulation is accomplished as follows:

- The extra source terms in the Navier–Stokes equations are set to zero.
- An approximate solution for the continuous phase is calculated.
- The discrete phase is introduced by calculating the particle trajectories for each discrete phase injection. At every location on the trajectory, the exchange of water vapor, momentum, and enthalpy to the gas phase is calculated.
- The continuous phase is recalculated, using the interphase exchange of water vapor, momentum, and enthalpy determined during the previous particle calculation.
- The discrete phase trajectories are recalculated in the modified continuous phase flow field.
- The previous two steps are repeated until a converged solution is achieved, in which both the continuous phase flow field and the discrete phase particle trajectories are unchanged with each additional calculation.

5.4.2.4 Separation of Dried Product

This separation is often done through a cyclone, placed outside the dryer, which reduces product loss in the atmosphere. The dense particles are recovered at the base of the drying chamber, while the finest ones pass through the cyclone to separate from the humid air. In addition to cyclones, spray dryers are commonly equipped with filters, called "bag houses" that are used to remove the finest powder, and the chemical scrubbers remove the remaining powder or any volatile pollutants (e.g., flavorings). The obtained powder is made up of particles, which originate from spherical drops after shrinking. The drop of water and gas content depends on the composition, and these particles can be compact or hollow (Gharsallaoui et al., 2007).

5.4.3 Process Parameters

The spray drying parameters, such as inlet temperature, airflow rate, feed flow rate, atomizer speed, types of carrier agent and their concentration are influencing properties, such as particle size, bulk density, moisture content, yield, and hygroscopicity in spray-dried foods.

5.4.3.1 Inlet Temperature

Generally, moisture content decreases with an increase in drying temperature due to the faster heat transfer between the product and the drying air. At higher inlet air temperatures, there is a greater temperature gradient between the atomized feed and the drying air, and this results in a greater driving force for water evaporation. This observation was obtained for different powders, such as orange juice (Chegini and Ghobadian, 2005), watermelon juice (Quek et al., 2007), tomato juice (Goula and Adamopoulos, 2008), and acai juice (Tonon et al., 2008, 2011).

Additionally, an increase in inlet air temperature often results in a decrease in bulk density due to the rapid formation of a dry layer on the droplet surface, and it causes the skinning over or casehardening on the droplets. This leads to the formation of vapor-impermeable films on the droplet surface, followed by the formation of vapor bubbles and, consequently, the droplet expansion (Chegini and Ghobadian, 2005; Tonon et al., 2008, 2011). Walton (2000) reported that the increase of drying air temperature generally causes the decrease in particle density and provides a greater tendency to the particles to hollow.

In addition, the particle size is affected by the inlet air temperature as reported by Tonon et al. (2011). The use of a higher inlet air temperature leads to the production of larger particles and causes higher swelling. A similar finding was also obtained by other authors (Chegini and Ghobadian, 2005; Nijdam and Langrish, 2005a). According to them, drying at higher temperatures results in faster drying rates, which lead to the early formation of a structure that does not allow the particles to shrink during drying. When the inlet air temperature is low, the particles remain more shrunk and smaller.

Furthermore, the inlet air temperature also influenced the morphology in acai juice powder as reported by Tonon et al. (2008). Figure 5.22 shows the micrographs of particles at different

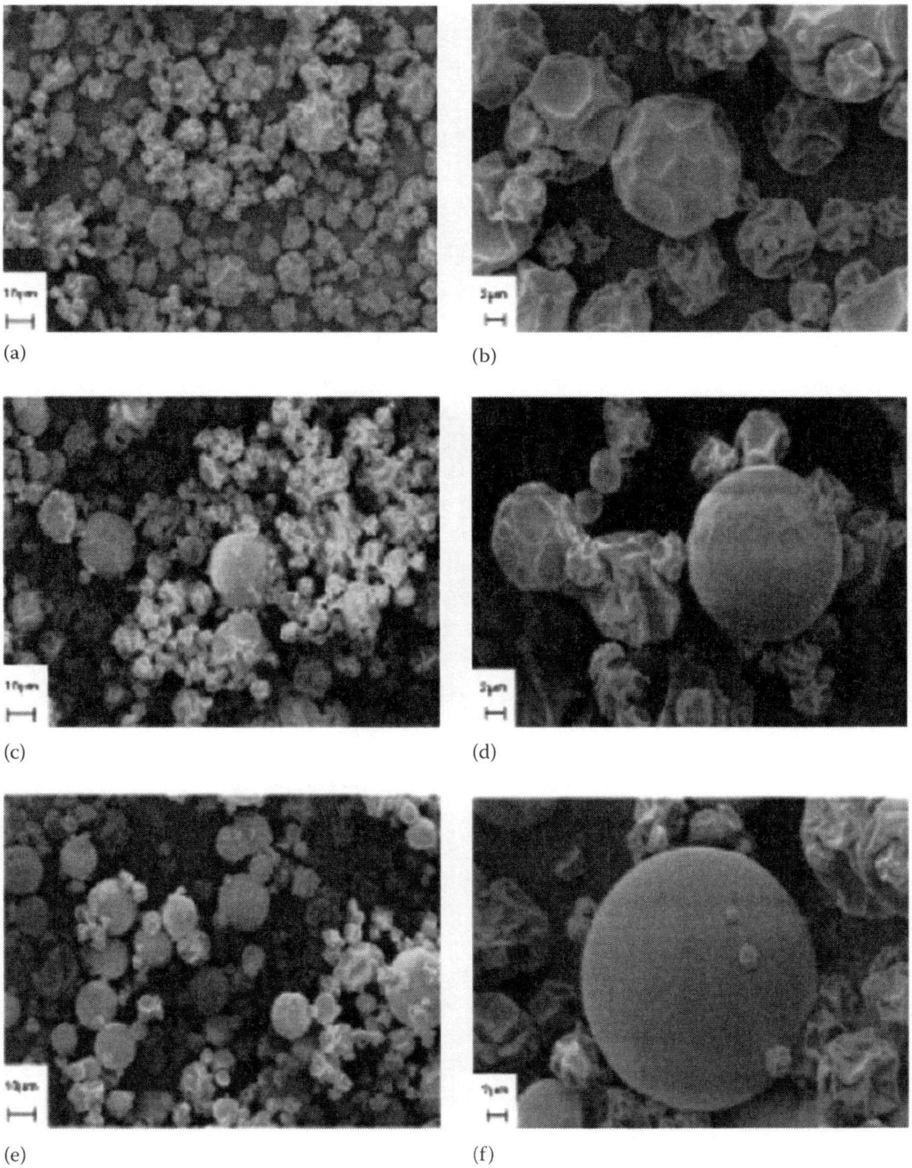

FIGURE 5.22 Micrographs of particles of acai juice powder at different temperatures and in different magnifications (a) 138°C, 2000×; (b) 138°C, 7000×; (c) 170°C, 2000×; (d) 170°C, 7000×; (e) 202°C, 2000×; (f) 202°C, 7000×. (From Tonon, R.V. et al., *J. Food Eng.*, 88, 411, 2008. With permission.)

temperatures and in different magnifications. When the inlet air temperature is low, the particles show a shriveled surface, while increasing the drying temperature results in a larger number of particles with a smooth surface. This is associated with the differences in the drying rate, which had higher values at higher temperatures. Alamilla-Beltrán et al. (2005) reported that when low inlet air temperatures are used, the crust is more pliable and collapsed, while the use of higher drying temperatures results in a more rigid and porous crust. Nijdam and Langrish (2005a) also confirmed the formation of more rigid particles with the use of higher temperatures in the spray drying of milk. The vacuole forms within the particle rapidly after a skin develops on the surface, and it inflates once the particle temperature exceeds the native ambient boiling point and the vapor pressure within the vacuole rises above the local ambient pressure. When the drying temperature is sufficiently high and the moisture is evaporated very quickly, the skin becomes dry and hard. As a result, the hollow particle cannot deflate when vapor condenses within the vacuole as the particle moves into cooler regions of the dryer. However, when the drying temperature is lower, the skin remains moist and supple for longer, so that the hollow particle can deflate and shrivel as it cools.

5.4.3.2 Drying Airflow Rate

Generally, the energy available for evaporation varies according to the amount of drying air. This could give the impression that the drying airflow rate must be at a maximum in all cases. However, the movement of air predetermines the rate and degree of droplet evaporation by influencing (1) the passage of spray through the drying zone, (2) the concentration of the product in the region of the dryer walls, and (3) the extent to which semi dried droplets reenter the hot areas around the air disperser (Goula and Adamopoulos, 2005b). A lower drying airflow rate causes an increase in product sojourn time in the drying chamber (Masters, 1985) and enforces circulation effects (Oakley and Bahu, 1991; Goula and Adamopoulos, 2004). Increased residence times lead to a greater degree of moisture removal. As a result, an increase in drying airflow rate, decreasing the residence time of the product in the drying chamber, leads to higher moisture contents.

According to Goula and Adamopoulos (2005b), the effect of drying airflow rate on powder bulk density depends on its effect on moisture content due to the sticky nature of the product. The higher the powder moisture content, the more particles tend to stick together, leaving more interspaces between them and consequently resulting in a larger bulk volume. As a result, the airflow rate increases, leading to an increase in powder moisture content and a decrease in powder bulk density. Masters (1985) reported that increasing residual moisture content increases bulk density of a dry product. However, this trend was reported for non-thermoplastic products.

5.4.3.3 Atomizer Speed or Compressed Airflow Rate

Increasing the atomizer speed, residual moisture decreases (Chegini and Ghobadian, 2005). At a higher atomizer speed, smaller droplets are produced and more moisture is evaporated resulting from the increased contact surface. In addition, higher atomizer speed results in smaller particle size and quicker drying due to the larger surface area and, consequently, prevents the "skinning" over of the droplets. Increasing the atomizer speed, the insoluble solid is reduced and as a result, the solubility of the powder is improved.

Moisture content shows a decrease with an increase in compressed airflow rate due to the effect of this flow rate on mean particle size (Goula and Adamopoulos, 2005b). Increase in air–liquid flow ratio in a two-fluid nozzle atomizer decreases the mean size of the spray droplets (Nath and Satpathy, 1998). With smaller particles, narrow spray cones are formed, and air may not penetrate the center of the spray pattern until droplets have travelled quite some distance from the nozzle (Liang and King, 1991). This reduced mixing of hot air should make the drying rates decrease. However, drying is facilitated by smaller particle sizes for two reasons. First, a larger surface area provides more surface in contact with the heating medium and more surface from which the moisture can escape.

Second, smaller particles reduce the distance heat must travel to the center of the particles and reduce the distance through which moisture in the center of the particles must travel to reach the surface and escape.

5.4.3.4 Feed Solids Concentration

The higher the feed solids concentration, the slower is the formation of the particulate solid surface. According to Downton et al. (1982), if dehydration conditions do not permit surface solidification before particles collide with each other, stickiness results in agglomeration. Hence, a smaller surface area, which provides less surface in contact with the heating medium, and less surface from which the moisture can escape. Second, the percentage of moisture decrease is highly correlated to residue formation (Goula et al., 2004). The more intense the problem of residue accumulation, the lower is the moisture decrease resulting from the increase in feed solids content.

An increase in feed concentration leads to an increase in powder particle size (Goula et al., 2004). Generally, in a spray drying system the size of the dried particles depends on the size of the atomized droplets. The droplet size on atomization depends upon the mode of atomization, physical properties of the feed, and feed solids concentration. Droplet size usually increases as the feed concentration or viscosity increases and the energy available for atomization (i.e., rotary atomizer speed, nozzle pressure, air–liquid flow ratio in a pneumatic atomizer) decreases. Thus, the effect of feed solids concentration on dried particle size is due to its effect on spray droplets size. The powder particle size depends upon the degree of moisture removal during drying, in addition to the atomized droplet size. A high degree of moisture removal is associated with a high size difference in the product before and after drying. In addition, the percentage of size increases inversely with that of moisture decrease. These observations prove the dependence of particle size on the degree of moisture removal during drying.

Bulk density shows a decrease with an increase in feed solids concentration. This is due to the effect of feed solids content on particle size. According to Nath and Satpathy (1998), as a general rule, larger particles will usually be less dense, so the bulk density of a powder with a large particle size will be lower.

5.4.4 Stickiness

The products to be spray dried can be categorized into two major groups: nonsticky and sticky products. Sticky products are generally difficult to be spray dried. During the drying process, they may remain as syrup or stick on the dryer wall or form unwanted agglomerates in the dryer chamber and conveying system, resulting in lower product yields and operating problems. Some of the examples of such sticky products are fruit and vegetable juice powders, honey powders, and amorphous lactose powder. Nonsticky products can be dried using a simpler dryer design, and the powder obtained is relatively less hygroscopic and more free flowing.

The problem of powder stickiness is mainly due to the low glass transition temperature (T_g) of the low molecular weight sugars present in such products, essentially sucrose, glucose, and fructose (Roos et al., 1995). Spray drying is a fast process, which produces a dry product in an amorphous (glassy) form. Solids in an amorphous state have a very high viscosity (>10^{12} Pa s) and as the temperature rises during drying, the viscosity decreases to a critical value of around 10^7 Pa s, where they first become sticky (Bhandari et al., 1997). This critical viscosity is reached at temperatures 10°C–20°C above T_g and these temperatures decrease with an increase in water content (Roos and Karel, 1991). It can be, therefore, assumed that the temperature of the particle surface during drying should not reach 10°C–20°C above T_g (Bhandari and Howes, 1999). As a consequence, high molecular weight additives, which have a very high T_g and raise the T_g of the feed, are usually added to the spray dryer feed to achieve successful drying at feasible drying temperature conditions. Maltodextrins are the most common drying aids used at present (Roos and Karel, 1991; Bhandari et al., 1997).

Various methods capable of producing a free-flowing fruit juice powder have been proposed: addition of drying aids (maltodextrins, glucose, soybean protein, sodium chloride, skim milk powder) (Bhandari et al., 1993, 1997; Adhikari et al., 2004; Jaya and Das, 2004; Papadakis et al., 2005; Roustapour et al., 2005; Quek et al., 2007; Chegini et al., 2008), scrapping of dryer surfaces (Karatas and Esin, 1994), cooling of the drying chamber walls (Jayaraman and Das Gupta, 1995; Chegini and Ghobadian, 2005; Chegini et al., 2008), and admission of atmospheric air near the chamber bottom, allowing transport of the powder to a collector having a low humidity atmosphere (Ponting et al., 1973).

According to Bhandari et al. (1997), drying aids have a very high T_g and raise the T_g of the feeds. Maltodextrins are the most common drying aids used at present. The T_g of maltodextrins vary from 100°C to 243°C according to their dextrose equivalent (DE) property. Several researchers have added maltodextrins to sugar-rich foods to reduce wall depositions problems. Bhandari et al. (1993) produced blackcurrant, raspberry, and apricot juice powders by spray drying a mixture of the juice and maltodextrin with a DE of 35 at 150°C. They fixed the proportion of juice solid and maltodextrin in the ratio 55:35, 55:45, and 50:40, respectively. Jaya and Das (2004) carried out drying of mango pulp at various levels of maltodextrin and obtained an optimum feed mix composition of 0.43–0.57 kg/kg of mango solids. Papadakis et al. (2005) overcame the problem of stickiness during spray drying of raisin juice concentrate using 21 DE, 12 DE, and 5 DE maltodextrins and the maximum ratio of (raisin solids)/(maltodextrin solids) they achieved was 57/33. Roustapour et al. (2005) reported that an addition of 20% maltodextrin (5 DE) to lime juice is the optimum amount for complete and successful drying of the juice. Quek et al. (2007) produced watermelon powder using maltodextrin concentrations between 3% and 5%. Grabowski et al. (2008) reported that maltodextrin with a DE of 11 is an effective drying aid for spray drying of sweet potato puree when it is added in a concentration of 10 kg/kg of potato puree. Bhandari et al. (1993) found that a certain amount of maltodextrin has to be added to prevent excessive amount of products from sticking to the dryer walls. They also found that more wall deposition occurred when the ratio of fruit to maltodextrin increased and reported that the acceptable range of this ratio could be increased by decreasing the inlet air temperature or by using a higher molecular weight maltodextrin. Tsouroufiis et al. (1975) also mentioned that when used as drying aids for orange juice, low dextrose equivalent maltodextrins were found to give higher collapse temperatures than high DE maltrins at the same concentrations. In general, the amount of maltodextrins necessary for successful drying depends upon three major factors, the composition of the product, the drying temperature, and the maltodextrin type and is largely based on trial and error and operator experience, rather than on any a priori methods based on the product components. Bhandari et al. (1997) developed a semitheoretical drying aid index based on product recovery, which was successfully used to determine the optimum fruit juice/maltodextrin ratio in a pilot scale spry dryer. However, according to Adhikari et al. (2004), questions arise such as, what is the effect of the addition of a drying aid on the drying kinetics of products containing low molecular weight sugars, and how is the surface stickiness of these materials affected when the drying agent is added.

The major limitations of the use of drying aids are the subsequent change in the product properties and the cost. As far as the cooling of the drying chamber walls is concerned, the cool wall will be favorable to minimize the thermoplastic particles from sticking, as the wall will be cold enough to cool and solidify the outer surface of the thermoplastic particles coming in contact. This method, however, was found to improve the process but not to resolve the problem. The reason is that the cold chamber wall will also cool the surrounding environment and cause an increase in the relative humidity of the air close to the wall surface.

Goula and Adamopoulos (2005b, 2008, 2010) modified an experimental spray dryer for drying tomato concentrate and orange juice. The modification made to the original dryer design consisted of connecting the spray dryer inlet air intake to an absorption air dryer. The modified spray drying system was proved advantageous over the standard laboratory spray dryer. The much lower outlet temperatures and humidities of the dehumidified drying air resulted in the formation of a solid particle surface, which decreased residue accumulation or dryer fouling and minimized the number of thermoplastic particles sticking to the dryer wall. In addition, preliminary air dehumidification, promoting rapid particulate skin formation, decreased powder moisture content and increased powder bulk density and solubility.

5.5 FREEZE DRYING

5.5.1 Basic Principles

Freeze drying is the most prominent example of separation by sublimation. In freeze drying, water is removed as vapor from a frozen substance. The water passes from the solid phase directly into the vapor phase without becoming a liquid *en route*; consequently, it is necessary that the temperature of the sublimation zone in a material being freeze dried be held below the triple point temperature of the water or aqueous solution in the material being dried (Figure 5.22).

The freeze drying process is used as a benchmark regarding product quality. Because of the absence of liquid water and the low temperatures required for the process, most of the deterioration reactions and microbiological activities are prevented. Quality losses due to enzymatic and nonenzymatic browning reactions are also reduced; however, the oxidation of lipids, caused by low moisture levels achieved during drying is higher in freeze-dried products. Packing the products in packages impermeable to oxygen can control this lipid oxidation (Ibarz and Barbosa-Canovas, 2003). The solid state of water during freeze drying protects the primary structure and the shape of the product, with minimal reduction in volume (Ratti, 2001). This was confirmed by Marques et al. (2005), who investigated the physical properties of vacuum freeze drying of tropical fruits. As expected, vacuum freeze-dried foods have high values of porosity and low values of apparent density and conserve color, flavor, and taste (Claussen et al., 2007). Thus, freeze-dried products have a long shelf life without refrigeration, 2 years for a product with a 2% residual moisture content being usual.

The efficiency of the freeze-drying process in terms of productivity and product quality is defined by various process variables, the product temperature being the most determinant one. Although sublimation phenomenon is faster when the temperature is increased, an overheating of the product could result in the collapse of the pore structure (Pikal and Shah, 1990) and the loss of quality properties. This collapse phenomenon is usually associated with the glass transition temperature of the maximally freeze-concentrated phase. According to Ratti (2001), product freeze drying temperatures are below T_g only at the end of the process. Nevertheless, to compare temperatures at the average moisture content of the product during freeze drying would lead to a misunderstanding of the problem since, in this particular process, the solid is divided into dry and frozen regions separated by a receding front. The portion of the solid that is in contact with the highest temperatures for long periods of time is thus the amorphous dry matrix, which has a low moisture content and a T_g corresponding closely to that of the dry solids. Thus, the glass transition temperature of dry solids would be a crucial optimization parameter for the freeze-drying process.

The main disadvantage of freeze drying is the high fixed and operational costs. Vacuum operations are mainly batchwise, which represents another additional cost, together with the demanding requirements of apparatus operated under vacuum (Di Matteo et al., 2003). An important part of operational costs is the energy cost, which is high in vacuum freeze drying. The specific moisture extraction rate of industrial vacuum freeze drying is in the range of 0.4 or below. An important part of previous investigations aimed to reduce freeze drying times and consequently to lower the energy consumption, analyzing the possibilities of controlling the heat intensity, and the vacuum pressure and therefore investigating possible ways of optimizing the freeze-drying process. Several studies were carried out in laboratories and in pilot scale plants (Kuu et al., 1995; Liapis et al., 1995).

TABLE 5.6
Freeze-dried Food Products

Product	References	Product	References
Egg	Jaekel et al. (2008); Pignoli et al. (2009); Liu et al. (2011)	Apple	Krokida and Philippopoulos (2005); Lewicki and Wiczkowska (2005); Li et al. (2008)
Rice	Yu et al. (2011)	Banana	Kar et al. (2003); Bera et al. (2012)
Mushroom	Argyropoulos et al. (2011)	Strawberry	Shishehgarha et al. (2002); Shih et al. (2008)
Carrot	Litvin et al. (1998); Kerdpiboon et al. (2005)	Citrus	Lee et al. (2012)
Sweet potato	Ahmed et al. (2010)	Berry	Aiyer et al. (2011)
Garlic	Rahman et al. (2005); Fante and Noreña (2015)	Raspberry	Mejia-Meza et al. (2010)
Onion	Abbasi and Azari (2009)	Blueberry	Reyes et al. (2011)
Broccoli	Mahn et al. (2012)	Apricot	Fahloul et al. (2009)
Cabbage	Duan et al. (2007)	Pear	Komes et al. (2007)
Tomato	George et al. (2011)	Orange	Koroishi et al. (2005)
Pepper	Ade-Omowaye et al. (2003)	Flour	Cepeda et al. (1998)
Coffee	Sagara et al. (2005)	Beef	Wang and Shi (1999)
Fish	Sablani et al. (2001)		

Freeze drying has been applied with success to diverse biological material, such as meat, coffee, juices, dairy products, cells, and bacteria and is now standard practice in the production of protein hydrolysates, hormones, blood plasma, and vitamin preparations. The application of freeze drying to food products has traditionally been confined to the production of heat- or oxygen-sensitive foodstuffs or those foods having a special end use, such as space foods, military or extreme-sport foodstuffs, and instant coffee (Ratti, 2001). Recently, however, the market for "natural" and "organic" products has been increasing strongly, along with consumer demand for foods with minimal processing and high quality, but without the presence of preservatives. Some of the freeze-dried foods are listed in Table 5.6.

5.5.2 Stages of Freeze Drying

Three stages can be identified in the complete freeze drying process (Lopez-Quiroga et al., 2012), during which different physical phenomena take place (Figures 5.23 and 5.24):

1. The first stage (freezing) involves a quick decrease of the sample temperature (reaching values below water triple point) in order to control the ice crystals size growth and to avoid possible damage to the material.
2. The second step, the so-called primary drying, consists of heating of the sample under partial vacuum conditions (always below the triple point) to force ice sublimation. This leads to an interconnected porous structure, which can be later rehydrated very effectively while preserving the organoleptic and nutritional properties of the product. During this stage (the longest one of the cycle), which conditions most of the quality properties of the product, almost all frozen water is sublimated.
3. Finally, the last step, the secondary drying, is an ordinary drying process where the water still bound to the porous matrix is desorbed by increasing the temperature. Typical figures for final moisture levels in the product are around 0.5% w/w.

Freezing is an efficient desiccation step where most of the solvent, typically water, is separated from the solutes to form ice. As freezing progresses, the solute phase becomes highly

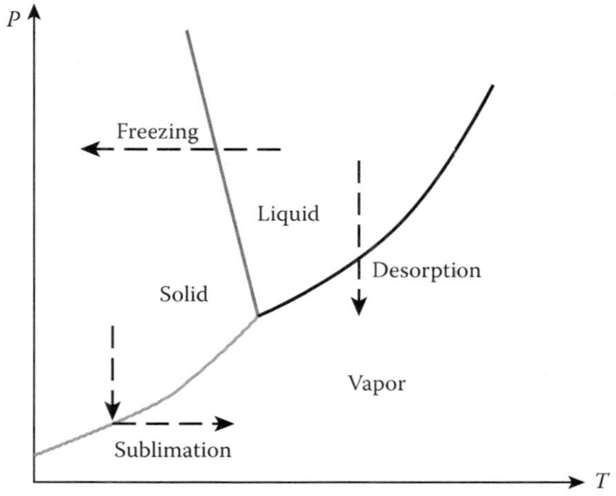

FIGURE 5.23 Freeze–drying physical phenomena represented on the water phase diagram. (From Lopez-Quiroga, E. et al., *J. Food Eng.*, 111, 555, 2012. With permission.)

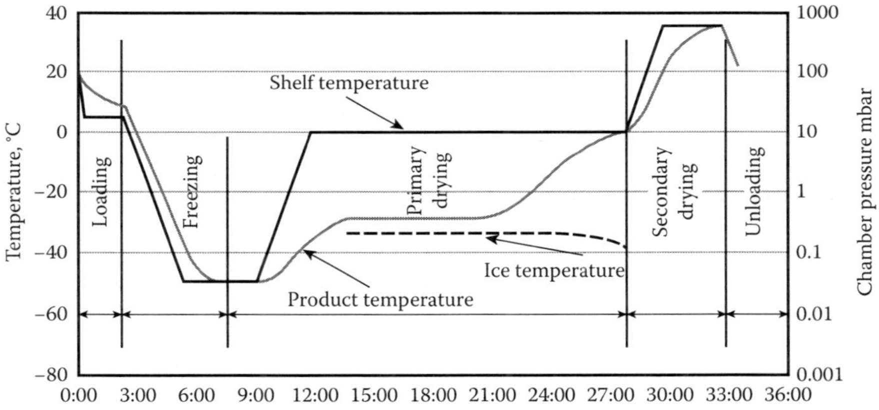

FIGURE 5.24 Freeze drying stages.

concentrated and is termed the "freeze concentrate." By the end of freezing, the freeze concentrate usually contains only about 20% of water (w/w) or less than 1% of total water in the solution before ice formation. The freezing stage typically takes several hours to finish (Tang and Pikal, 2004). The performance of the overall freeze-drying process depends significantly on this stage because the shape of the pores, the pore size distribution, and pore connectivity of the porous network of the dried layer formed by the sublimation of frozen water during the primary drying stage depend on the ice crystals that formed during the freezing stage. This dependence is of extreme importance because the parameters that characterize the mass and heat transfer rates in the dried product during primary and secondary drying are influenced significantly by the porous structure of the dried product. If the ice crystals are small and discontinuous, then the mass transfer rate of water vapor in the dried layer could be limited. On the other hand, if appropriate size dendritic ice crystals are formed and homogeneous dispersion of the pre-eutectic and post-eutectic frozen solution can be realized, the mass transfer rate of water vapor in the dried layer could be, high and the product could be dried more quickly (Liapis et al., 1995).

Freezing often induces many destabilizing stresses. One practical process approach to stabilization is to minimize the surface area of ice by growing large ice crystals, which can be achieved by reduced supercooling. The degree of supercooling is the temperature difference between the thermodynamic or equilibrium ice formation temperature and the actual temperature at which ice begins to form, which is usually around 10°C–25°C lower but changes with cooling rate and other factors. Higher supercooling results in more/smaller ice crystals and larger ice specific surface area. Different freezing methods, like liquid nitrogen freezing, loading vials onto precooled shelves, or ramped cooling on the shelves, give different supercooling effects with normally the highest supercooling with liquid nitrogen freezing of small volumes and the lowest supercooling for the precooled shelf method. It was reported that slow cooling (0.5°C/min) causes larger supercooling effects than the precooled shelf method. However, the precooled shelf method gives large heterogeneity in supercooling between vials, which is undesirable (Jiang and Nail, 1998).

The *first drying stage* involves ice sublimation under vacuum. The vapor generated in the sublimation interface is eliminated through the pores of the product due to low pressure in the drying chamber, and the condenser prevents the vapor from returning to the product. The driving force of sublimation is the pressure difference between the water vapor pressure in the ice interface and the partial water vapor pressure in the drying chamber (Ibarz and Barbosa-Canovas, 2003). The philosophy of primary drying is to choose the optimum target product temperature (T_p), bring the product to the target product temperature quickly, and hold the product temperature roughly constant at the target temperature throughout all of primary drying. The product temperature should always be several degrees below macroscopic collapse temperature (T_c) in order to obtain a dry product with an acceptable appearance. The temperature difference between T_p and T_c is called the temperature safety margin. An optimized freeze-drying process runs with the product temperature as high as possible (Pikal and Shah, 1990). In other words, the target product temperature should be as close as possible to T_c. Consequently, the optimum target product is a compromise between safety and freeze-drying time. A small safety margin (2°C) should be used if freeze-drying time is long (e.g., more than 2 days), whereas a large safety margin (5°C) is proposed if freeze-drying time is short (<10 h), and a safety margin of 3°C should be used if primary drying time is somewhere between 2 days and 10 h (Tang and Pikal, 2004). In general, T_p should not be higher than −15°C, or the heat and mass transfer capabilities of the freeze dryer may be overloaded. Overloading of the freeze dryer typically causes loss of chamber pressure control and product temperatures in excess of the target.

Primary drying is carried out at low pressure to improve the rate of ice sublimation. The chamber pressure impacts both heat and mass transfer and is an important parameter for freeze-drying process design. At a given product temperature (i.e., given ice vapor pressure), the smallest chamber pressure gives the highest ice sublimation rate. However, very low chamber pressure may cause problems, such as contamination of the product with volatile stopper components or pump oil (Pikal and Lang, 1978) and also produce larger heterogeneity in heat transfer, thereby giving larger product temperature heterogeneity (Pikal et al., 1984). The optimum chamber pressure is a compromise between high sublimation rate and homogenous heat transfer and, in most applications, varies from 50 to 200 mTorr (Tang and Pikal, 2004).

The *second drying stage* begins when the ice in the product has been removed, and moisture comes from water partially bound to the material being dried. Secondary drying is not easily quantifiable. The rate is governed by diffusion of water from the product filaments and its subsequent desorption and condensation (Franks, 1998). The diffusion process is not subject to simple kinetics but, as might be expected, drying is accelerated by an increase in the temperature (Aldous et al., 1997). The drying rate is diffusion-limited and tends to plateau at each temperature. In contrast to primary drying, the chamber pressure does not appear to affect the secondary drying rate to any marked extent (Pikal et al., 1984). Ideally, therefore, secondary drying conditions should be such as to track the glass transition profile, starting from T_g up to the desired storage temperature and water content (Franks, 1998).

5.5.3 Heat and Mass Transfer—Modeling

Several theoretical models concerning the heat and mass transfer phenomena during freeze drying can be found in the literature (Liapis and Bruttini, 1995; Lombraña and Izkara, 1995; Lombraña, 1997). More recently, numerical models with highly detailed equations have been developed (Brülls and Rasmuson, 2002; George and Datta, 2002). However, in most cases, adjustable parameters are needed to match the model predictions with experimental data (Sadikoglu and Liapis, 1997; Sheehan and Liapis, 1998). In other cases, no comparison with experimental data is presented (Liapis and Bruttini, 1995). In addition, most of the models were developed for liquids and not for solid products (Sheehan and Liapis, 1998; Brülls and Rasmuson, 2002). Khalloufi et al. (2005) have built a model using mass and energy balances in the dried and frozen regions and taking into account both sublimation and desorption in the set of coupled nonlinear partial differential equations. These equations were solved numerically by using a finite element scheme, and all the parameters involved in the model were obtained independently from experimental data cited in the literature.

During the freeze-drying process, the product is placed between heating plates. Therefore, a radiant source as well as heat by conduction from the bottom is imposed to the material slab. There are two mechanisms of water elimination during freeze drying: sublimation (which eliminates the frozen water) and desorption (which eliminates the bounded unfrozen water). Sublimation occurs at the interface of the ice front as a result of the transferred heat (Simatos et al., 1975). As dehydration proceeds, the ice front retreats, and the sublimated water vapor (at the ice front) is removed by diffusion through the porous layer (Liapis and Bruttini, 1995). The water vapor flows through the dried layer countercurrent to the heat flow. Then, it passes through the chamber to be finally collected on the condenser plate.

The flow of vapor through the dry matrix is represented through a permeability-type equation (Khalloufi et al., 1999):

$$\frac{\partial T_d}{\partial t} = \frac{\partial}{\partial x}\left(\varphi_1 \frac{\partial T_d}{\partial x}\right) + \frac{\partial}{\partial x}\left(\varphi_2 \frac{\partial P_v}{\partial x}\right) \tag{5.33}$$

$$\varphi_1 = \frac{k_d}{Cp_d \rho_d} \quad \text{and} \quad \varphi_2 = \frac{PerCp_v}{Cp_d \rho_d} T_d \tag{5.34}$$

where
- P_v is the vapor pressure
- k is the thermal conductivity
- Cp is the specific heat
- ρ is the density
- the subscript d refers to the dry matrix

The permeability coefficient, *Per*, and the Knudsen constant, *Kn*, are calculated as follows (Liapis and Litchfield, 1979):

$$Per = \frac{M}{RT} \frac{C_2 D_{AB} Kn}{C_2 D_{AB} + Kn(P_T - P_v)} \tag{5.35}$$

$$Kn = C_1 \left(\frac{M}{RT_d}\right)^{0.5} \tag{5.36}$$

$$C_1 = \frac{2}{3}\left(\frac{8}{\pi}\right)^{0.5} r \tag{5.37}$$

where
- M is the molecular weight
- C_1 is the Knudsen constant value
- C_2 is the dimensionless constant
- P_T is the total pressure
- r is the pore radius of dried layer

The molecular diffusivity, D_{AB}, of the binary mixture is estimated by the Fuller equation (Skelland, 1974):

$$D_{AB} = 1.16 \times 10^{-4} T^{1.75} \tag{5.38}$$

The pressure profile is represented through the following equation (Khalloufi et al., 2005):

$$\frac{\partial P_v}{\partial t} = \frac{\partial}{\partial x}\left(\beta_1 \frac{\partial P_v}{\partial x}\right) + \beta_2 \frac{\partial T_d}{\partial t} \tag{5.39}$$

$$\beta_1 = \frac{PerR}{M\varepsilon} \quad \text{and} \quad \beta_2 = \frac{P_v}{T_d} \tag{5.40}$$

where ε is the porosity.

In the frozen region (f), there is only heat transfer by conduction. Thus, the temperature profile is represented by the following differential expression (Khalloufi et al., 2005):

$$\frac{\partial T_f}{\partial t} = \frac{\partial T_f}{\partial x}\left(\omega \frac{\partial T_f}{\partial x}\right) \tag{5.41}$$

$$\omega = \frac{k_f}{Cp_f \rho_f} \tag{5.42}$$

The boundary condition at the material surface can be described by:

$$k_d \frac{\partial T_d}{\partial x} = -h_T \left(T_\infty - T_d[0,t]\right) \tag{5.43}$$

where h_T is the heat transfer coefficient calculated as:

$$h_T = \sigma F \left(T_\infty^2 + T_d^2[0,t]\right) \tag{5.44}$$

where
- σ is the Stefan–Boltzmann constant
- F is the radiation view factor

A convection resistance coefficient, h_P, which depends on the total pressure (Lombraña and Izkara, 1995), can be considered for mass transfer at the surface:

$$Per\frac{\partial P_v}{\partial x} = h_P\left(P_\infty - P_v\left[0,t\right]\right) \quad (5.45)$$

The amount of sublimated vapor, $m_{sub}(t)$, can be expressed as follows:

$$m_{sub}\left(t\right) = \rho_f WCFFlLS\left(t\right) \quad (5.46)$$

where

WC is the initial water content
FF is the fraction of frozen water
l is the width
L is the length
$S(t)$ is the interface position

The amount of remaining water, X_r, is the sum of the bound water and the frozen water that has not been sublimated yet:

$$X_r\left(t\right) = WC\rho_f lL\left(E - FFS\left(t\right)\right) \quad (5.47)$$

where E is the half of thickness.

Lopez-Quiroga et al. (2012) developed a novel low-dimensional model based on a time-scale simplification approach (the matrix-scale model). This model constitutes the core of the proposed optimal control approach, which defines the operation conditions for minimizing freeze-drying cycle time while preserving product quality (final water content) through the solution of a dynamic nonlinear programming. The main difference with respect to the other models is related to the treatment of the heat transfer phenomena in the dried layer. It must be remarked that separated energy transfer mechanisms for vapor and porous matrix in the dried region are considered. In order to both achieve a better comprehension of the process as well as to identify its leading roles, the coupled mass and energy balances are described by taking into account the inherent thermophysical properties of the system. Thermal diffusivities, desorption rate, and the mass flux velocity define a set of characteristic times in which different physical phenomena take place. Based on this time-scale analysis, a simplification of the governing equations is performed. For the freeze-drying case, the relevant time scale is the one related to the temperature distribution within the porous matrix.

As far as the atmospheric freeze-drying (AFD) process is concerned, literature review revealed that a few investigators have reported their modeling effort on AFD. So far, two simulation approaches for AFD have been developed: uniformly retreating ice front (URIF) model and a diffusion model. Kutsakova et al. (1984) modeled AFD of granulated products of cylindrical shape using cold air as the carrier gas. The model was used in a simple case to predict granule residence time for a given final granule radius. A further kinetics study of ice sublimation in a fluidized-bed dryer operating under atmospheric conditions was demonstrated by Boeh-Ocansey (1985). Wolff and Gibert (1990) also proposed a model for atmospheric freeze drying using a fluidized bed of particulate adsorbents (starch) of different masses. Menshutina et al. (2004) modeled an AFD process in a spouted bed using the heterogeneous media concept for modeling of the hydrodynamics and drying kinetics. Haida et al. (2005) studied a batch fluid-bed AFD process and predicted the heat and mass transfer phenomena between the bed and product using a 1D heterogeneous two-phase fluid-bed model. Claussen et al. (2007) developed a simplified mathematical model based on uniformly retreating ice front (URIF) considerations to calculate theoretical drying curves of atmospheric freeze-dried foods in a tunnel dryer.

5.5.4 Freeze-Drying Systems

Conventional freeze-drying plants in use today generally support the food material to be dried in particulate form on a series of trays within the drying chamber. Such a dryer is shown schematically in Figure 5.25. Heat is supplied by a circulating heat transfer agent within platens, which may be in direct contact with the trays or may transmit heat by radiation. Steam may also be used as a heating agent in radiant-heat freeze dryers. Although steam jet ejectors are used to some extent in Europe for removing water vapor as well as inerts, water vapor removal is generally accomplished by ice formation onto chilled, metal surface condensers, which are defrosted after the conclusion of freeze drying. The condenser unit is backed up by a vacuum pump or steam jet for inerts removal and may be located within the drying chamber or in one or more separate chambers, which may be closed off intermittently for defrosting (King and Labuza, 1970). Tunnel freeze dryers utilize large vacuum cabinets where trolleys carrying the trays are loaded at intervals through a large vacuum lock located at the entrance to the freeze dryer and discharges in a similar way at the exit (Liapis and Bruttini, 1995).

The first approach toward a continuous freeze drying process in large-scale practice is a process in which particulate foods tumble along a slowly rotating, inclined tube, with the tube being polygonal in cross section so as to provide good tumbling and mixing action. Heat is supplied continuously from a steam chamber through which a number of such tubes pass, and water vapor is removed by a conventional vacuum system. In another variant of this process, the heat is supplied through finned tubes, which act as baffles to promote mixing of the particles of the substance being dried within a rotating chamber. Another approach to continuous freeze drying uses gravity loading and unloading through vapor locks into many vertical cells in parallel, which provide means of heating distributed vertically along each cell and which provide small dimensions across which vapor can escape from the material being dried. This approach is designed to overcome the problems of continuous loading and of distribution of material for even drying within a continuous freeze dryer (King and Labuza, 1970).

Figure 5.26 shows a diagram of a continuous tray dryer, which is suitable for lumps, slices, or granules. It consists of a tunnel with a vacuum lock at each end, one for loading prefrozen lumps of

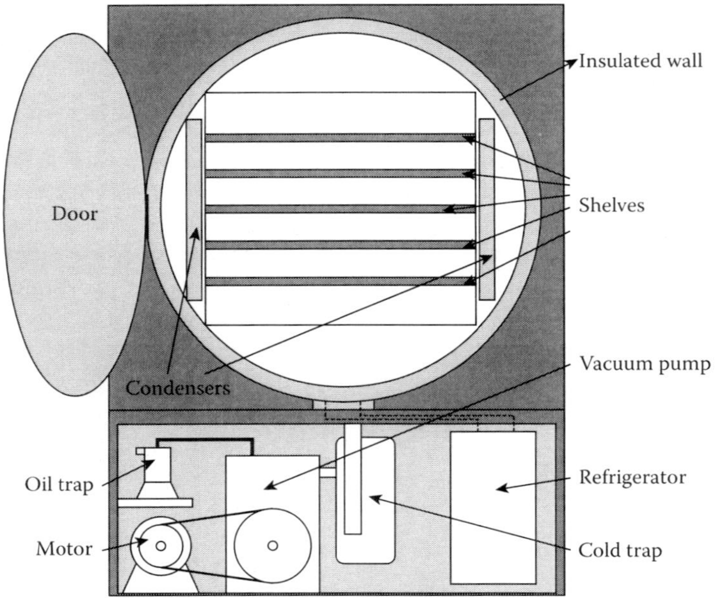

FIGURE 5.25 Batch freeze dryer. (From Ahmed, J. and Rahman, M.S., *Handbook of Food Process Design*, Wiley, New York, 2012, 1600pp. With permission.)

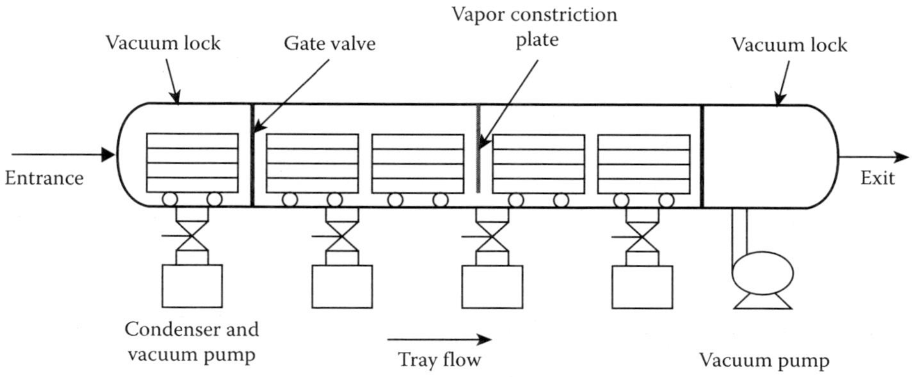

FIGURE 5.26 Continuous tray freeze dryer. (From Mellor, J.D., *Fundamentals of Freeze-Drying*, Academic Press, London, 1978. With permission.)

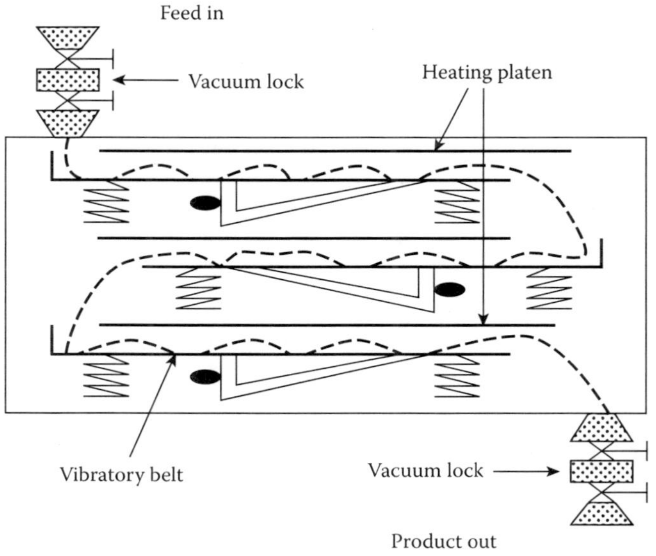

FIGURE 5.27 Trayless continuous freeze dryer. (From Barbosa-Canovas, G.V. and Vega-Mercado, H., *Dehydration of Foods*, Chapman & Hall, New York, 1995. With permission.)

food and the other for discharging the dried product. The chamber pressure is in the range of 0.1–1 mbar, and the dryer capacity is about 50–80 kg of fresh product/m^2 day (Barbosa-Canovas et al., 2005). Another type of continuous freeze dryer, suitable for free flowing and granular foods, is presented in Figure 5.27. In this type, the prefrozen material enters at one end of a horizontal, cylindrical vacuum chamber via an entrance lock, onto a vibrating deck. This carries them to the other end of the chamber. They then fall onto a second deck, which transports them back to the front end of the chamber, where they fall onto another vibrating deck. In this way, the granules move back and forth in the chamber until they are dry. They are then discharged from the chamber through a vacuum lock. Heat is supplied by radiation from heated platens above the decks (Brennan and Grandison, 2012).

Methods which have been used or tested for vacuum generation and water vapor removal include multistage steam jet ejectors, various sorts of vacuum pumps, refrigerated metal surface condensers, liquid and solid desiccants, and cold, ice-immiscible liquids. Refrigerated condensers are currently used in most large-scale applications. Condensers are critical "pumps" maintaining the freeze drying conditions, while the vacuum pump just removes the noncondensable gases of the environment.

In the design of a refrigerated metallic-surface condenser system, it is important to provide for a minimum pressure drop of water vapor between the food surface and the condensing surface. This calls for large vapor transfer lines from the drying chamber to the condenser chamber, or for putting the condenser surfaces inside the drying chamber. If the condenser is inside the drying chamber, the design should guard against excessive heat loss by radiation from the heating surfaces direct to the condenser surface. It is also important to design the condenser system so that the condenser surface will frost up more or less evenly. A system of baffles, forcing all vapors to flow over the condensing surfaces on their way to the pump, has been found to improve operation and provide more efficient utilization of the entire condensing surface (King and Labuza, 1970).

5.5.5 Technical Improvements

In order to reduce manufacturing costs, *atmospheric freeze drying* (AFD) was developed. Dr Harold Meryman is considered the inventor of the science of atmospheric freeze drying. He showed experimentally that the diffusion of water vapor from the drying boundary through the dried shell is facilitated primarily by the vapor pressure gradient rather than by the absolute pressure in the system. The advantages of the atmospheric freeze-drying process, in comparison with vacuum freeze drying, is given in the work of Li (2005) as follows: (1) low initial investment cost since expensive vacuum auxiliary equipment could be eliminated, (2) the process could be designed as a continuous system with higher productivity and lower operating cost, (3) the application of a heat pump system and different process temperature elevating modes in AFD would decrease energy consumption and drying time, and (4) inert gas drying environment, such as nitrogen or helium, can be applied to minimize the product degradation caused by oxidation. The main conclusion from his work is that vacuum freeze drying still offers the best product quality, although AFD products show similar characteristics of rehydration kinetics and hygroscopicity. Only 15% deterioration of the antioxidant properties was measured compared to vacuum freeze drying.

Conventional fluid beds can be used for atmospheric freeze drying. The heat and mass transfer is very good in a fluid-bed dryer where the drying agent passes each unit of the product. Liquid foods can be frozen and granulated before drying using a traditional granulator. Another way of making a new high-quality instant powder is to first put the product under high pressure together with liquid CO_2 at –30°C to –40°C and then expand it to the surrounding temperature and pressure. This leads to considerable foaming of the product. The foamed matrix is frozen, granulated, and dried at freezing temperatures (Claussen et al., 2007). Di Matteo et al. (2003) proposed a freeze-drying operation at atmospheric pressure that utilizes a fluidized bed of absorbent particles. They reported that approximately 34% energy reduction can be obtained, whereas drying times are increased by up to threefold since the use of atmospheric pressure changes the process from one involving heat transfer to one involving mass transfer. In addition, the quality of the obtained products is inferior, since the risk of product collapse is increased (Lombraña and Villarán, 1995). To avoid size reduction caused by mechanical cracking, tunnel drying is a suitable alternative to fluid-bed freeze drying. The heat and mass transfer, however, is not as good as in fluid bed drying (Claussen et al., 2007).

Another method that has recently been developed is the *adsorption freeze drying*. In this method, a desiccant is used to create a high vapor drive at low temperatures. The adsorbent replaces the condenser and lets a reduction of 50% in total costs as compared to traditional freeze drying. Despite the many advantages, the quality of adsorption freeze-dried foods is slightly reduced and sometimes poor as compared to that obtained by traditional freeze drying (Ratti, 2001).

Within the pharmaceutical industry, *spray freeze drying* seems to be a good alternative for producing free-flowing powder, with high surface area, porous end product, and good instant characteristics. The methods reported may be classified as (1) spray freeze drying into liquids, (2) spray freeze drying into gases, and (3) spray freeze drying into gases over a fluidized bed. Spray freeze drying into liquid is a new technology designed to produce nanostructure particles with high surface area (Claussen et al., 2007).

REFERENCES

Abadio, F.D.B., A.M. Domingues, S.V. Borges, and V.M. Oliveira, Physical properties of powdered pineapple (*Ananas comosus*) juice—Effect of malt dextrin concentration and atomization speed. *J. Food Eng.*, 2004; 54: 285–287.

Abbasi, S. and S. Azari, Novel microwave-freeze drying of onion slices. *Int. J. Food Sci. Technol.*, 2009; 44: 974–979.

Abonyi, B.I., H. Feng, J. Tang, C.G. Edwards, B.P. Chew, D.S. Mattinson, and J.K. Fellman, Quality retention in strawberry and carrot purees dried with Refractance Window™ system. *J. Food Sci.*, 2002; 57: 1051–1055.

Ade-Omowaye, B.I.O., N.K. Rastogi, A. Angerbach, and D. Knorr, Osmotic dehydration behaviour of red paprika (*Capsicum anuum* L.). *J. Food Sci.*, 2002; 57: 1790–1795.

Ade-Omowaye, B.I.O., K.A. Taiwo, N.M. Eshtiaghi, A. Angersbach, and D. Knorr, Comparative evaluation of the effects of pulsed electric field and freezing on cell membrane permeabilisation and mass transfer during dehydration of red bell peppers. *Innov. Food Sci. Emerg. Technol.*, 2003; 4: 177–188.

Adhikari, B., T. Howes, B.R. Bhandari, and V. Troung, Effect of addition of maltodextrin on drying kinetics and stickiness of sugar and acid-rich foods during convective drying: Experiments and modelling. *J. Food Eng.*, 2004; 52: 53–58.

Ahmed, J. and M.S. Rahman, *Handbook of Food Process Design*, New York: Wiley, 2012, 1600pp.

Ahmed, M., A.M. Sorifa, and J.B. Eun, Effect of pretreatments and drying temperatures on sweet potato flour. *Int. J. Food Sci. Technol.*, 2011; 45: 725–732.

Aiyer, H.S., Y. Li, Q.H. Liu, N. Reuter, and R.C.G. Martin, Dietary freeze-dried black raspberry's effect on cellular antioxidant status during reflux-induced esophagitis in rats. *Nutrition*, 2011; 27: 182–187.

Akanbi, C.T., R.S. Adeyemi, and A. Ojo, Drying characteristics and sorption isotherm of tomato slices. *J. Food Eng.*, 2005; 73: 157–153.

Alamilla-Beltrán, L., J.J. Chanona-Pérez, A.R. Jiménez-Aparicio, and G.F. Gutiérrez-Lopez,Description of morphological changes of particles along spray drying. *J. Food Eng.*, 2005; 57: 179–184.

Al-Asheh, S., R. Jumah, F. Banat, and S. Hammad, The use of experimental factorial design for analysing the effect of spray dryer operating variables on the production of tomato powder. *Food Bioprod. Process.*, 2003; 81: 81–88.

Aldous, B.J., F. Franks, and A.L. Greer, Diffusion of water within an amorphous carbohydrate. *J. Mater. Sci.*, 1997; 32: 301–308.

Amiri-Rigi, A., Z. Emam-Djomeh, M.A. Mohammadifar, and M. Mohammadi, Spray drying of low-phenylalanine skim milk: Optimisation of process conditions for improving solubility and particle size. *Int. J. Food Sci. Technol.*, 2012; 47: 495–503.

Anandharamakrishnan, C., C.D. Rielly, and A.G.F. Stapley, Effects of process variables on the denaturation of whey proteins during spray drying. *Drying Technol.*, 2007; 25(5): 799–807.

Argyropoulos, D., M.T. Khan, and J. Müller, Effect of air temperature and pre-treatment on color changes and texture of dried *Boletus edulis* Mushroom. *Drying Technol.*, 2011; 29: 1890–1900.

Ayadi, M.A., M. Khemakhem, H. Belgith, and H. Attia, Effect of moderate spray drying conditions on functionality of dried egg white and whole egg. *J. Food Sci.*, 2008; 73: E281–E287.

Azoubel, P.M. and F.E.X. Murr, Mass transfer kinetics of osmotic dehydration of cherry tomato. *J. Food Eng.*, 2004; 51: 291–295.

Bahloul, N., N. Boudhrioua, M. Kouhila, and N. Kechaou, Effect of convective solar drying on colour, total phenols and radical scavenging activity of olive leaves (*Olea europaea* L.). *Int. J. Food Sci. Technol.*, 2009; 44: 2551–2557.

Barbosa-Canovas, G.V. and H. Vega-Mercado, *Dehydration of Foods*, New York: Chapman & Hall, 1995.

Barbosa-Cánovas, G.V., E. Ortega-Rivas, P. Juliano, and H. Yan, *Food Powders. Physical Properties, Processing, and Functionality*, New York: Kluwer Academic/Plenum Publishers, 2005.

Basu, S., U.S. Shivhare, and A.S. Mujumdar, Models for sorption isotherms for foods: A review. *Drying Technol.*, 2005; 24: 917–930.

Bell, L.N. and M.J. Hageman, Differentiating between the effects of water activity and glass transition-dependent mobility on a solid state chemical reaction: Aspartame degradation. *J. Agric. Food Chem.*, 1994; 42: 2398–2401.

Bera, M., R. Chakraborty, and P. Bhattacharya, Optimization of intensification of freeze-drying rate of banana: Combined applications of IR radiation and cryogenic freezing. *Sep. Sci. Technol.*, 2012; 48: 345–358.

Bhandari, B.R. and T. Howes, Implication of glass transition for the drying and stability of dried foods. *J. Food Eng.*, 1999; 40: 71–79.

Bhandari, B.R., N. Datta, R. Crooks, T. Howes, and S. Rigby, A semi-empirical approach to optimise the quantity of drying aids required to spray dry sugar-rich foods. *Drying Technol.*, 1997; 15: 2509–2525.

Bhandari, B.R., A. Senoussi, E.D. Dumoulin, and A. Lebert, Spray drying of concentrated fruit juices. *Drying Technol.*, 1993; 11: 33–41.

Birchal, V.S., M.L. Passos, G.R.S. Wildhagen, and A.S. Mujumdar, Effect of spray-dryer operating variables on the whole milk powder quality. *Drying Technol.*, 2005; 23: 511–535.

Blahovec, J. and S. Yanniotis, Modified classification of sorption isotherms. *J. Food Eng.*, 2009; 91: 72–77.

Bluestein, P.M. and T.P. Labuza, Effects of moisture removal on nutrients. In: *Nutritional Evaluation of Food Processing*, E. Karmas and R.S. Harris, (eds.). New York: Van Nostrand Reinhold, 1988.

Boeh-Ocansey, O. Some factors influencing the freeze drying of carrot discs in vacuo and at atmospheric pressure. *J. Food Eng.*, 1985; 4: 229–243.

Boonyai, P., T. Howes, and B. Bhandari, Applications of the cyclone stickiness test for characterization of stickiness in food powders. *Drying Technol.*, 2005; 24: 703–709.

Brammer, J.G. and A.V. Bridgwater, Drying technologies for an integrated gasification bio-energy plant. *Ren. Sust. Energ. Rev.*, 1999; 3: 243–289.

Brennan, J.G. *Food Dehydration: A Dictionary and Guide*, Oxford, U.K.: Butterworth-Heinemann Ltd., 1994.

Brennan, J.G., J.R. Butters, N.D. Cowell, and A.E.V. Lilly, *Food Engineering Operations*, 2nd edn, London, U.K.: Applied Science Publishers, 1976.

Brennan, J.G. and A.S. Grandison, *Food Processing Handbook*, New York: Wiley, 2012.

Bruce, D.M. Exposed-layer barley drying, three models fitted to new data up to 150°C. *J. Agric. Eng. Res.*, 1985; 32: 337–347.

Bruin, S. and K.C. Luyben, Drying of food materials: A review of recent developments. In: *Advances in Drying*, A.S. Mujumdar, (ed.). New York: Hemisphere, 1980.

Brülls, M. and A. Rasmuson, Heat transfer in vial lyophilization. *Int. J. Pharm.*, 2002; 245: 1–15.

Brunauer, S., P.H. Emmett, and E. Teller, Adsorption of gases in multimolecular layers. *J. Am. Chem. Soc.*, 1938; 50: 309–320.

Buera, M.P., J. Chirife, and M. Karel, A study of acid catalyzed sucrose hydrolysis in an amorphous polymeric matrix at reduced moisture contents. *Food Res. Int.*, 1995; 28: 359–355.

Caixeta, A.T., R. Moreira, and M.E. Castell-Perez, Impingement drying of potato chips. *J. Food Process Eng.*, 2002; 25: 53–90.

Cal, K. and K. Sollohub, Spray drying technique. I: hardware and process parameters. *J. Pharm. Sci.*, 2010; 99: 575–585.

Cano-Chauca, M., P.C. Stringheta, A.M. Ramos, and J. Cal-Vidal, Effect of the carriers on the microstructure of mango powder obtained by spray drying and its functional characterization. *Innov. Food Sci. Emerg. Technol.*, 2005; 5: 420–428.

Cepeda, E., M.C. Villarán, and N. Aranguiz, Functional properties of faba bean (*Vicia faba*) protein flour dried by spray drying and freeze drying. *J. Food Eng.*, 1998; 35: 303–310.

Chegini, G.R. and B. Ghobadian, Effect of spray-drying conditions on physical properties of orange juice powder. *Drying Technol.*, 2005; 23: 557–558.

Chegini, G.R., J. Khazaei, B. Ghobadian, and A.M. Goudarzi, Prediction of process and product parameters in an orange juice spray dryer using artificial neural networks. *J. Food Eng.*, 2008; 84: 534–543.

Chen, C. and D.S. Jayas, Dynamic equilibrium moisture content for grain drying. *Can. Agric. Eng.*, 1998; 40: 299–303.

Chen, C.R., H.S. Ramaswamy, and I. Alli, Prediction of quality changes during osmo-convective drying of blueberries using neural network models for process optimization. *Drying Technol.*, 2001; 19: 507–523.

Chirife, J. Fundamentals of the drying mechanism during air dehydration of foods. In: *Advances in Food Drying*, A.S. Mujumdar, (ed.). New York: Hemisphere Publishing Corporation, 1983.

Chirife, J. and M.P. Buera, Water activity glass transition and microbial stability in concentrates/semimoist food systems. *J. Food Sci.*, 1994; 59: 921–927.

Chopda, C.A. and D.M. Barrett, Optimization of guava juice and powder production. *J. Food Process. Preserv.*, 2001; 25: 411–430.

Chua, K.J. and S.K. Chou, Low-cost drying methods for developing countries. *Trends Food Sci. Technol.*, 2003; 14: 519–528.

Claussen, I.C., I. Strømmen, A.K.T. Hemmingsen, and T. Rustad, Relationship of product structure, sorption characteristics, and freezing point of atmospheric freeze-dried foods. *Drying Technol.*, 2007; 25: 853–855.

Colak, N. and A. Hepbasli, A review of heat pump drying: Part 1—Systems, models and studies. *Energy Convers. Manage.*, 2009; 50: 2180–2185.

Crowe, C.T. Modelling spray-air contact in spray drying systems. In: *Advances in Drying*, A.S. Mujumdar, (ed.). New York: Hemisphere Publishing Company, 1980.
Dandamrongrak, R., R. Mason, and G. Young, The effect of pretreatments on the drying rate and quality of dried bananas. *Int. J. Food Sci. Technol.*, 2003; 38: 877–882.
Daud, W.R.B.W. and W.D. Armstrong, Pilot plant study of the drum dryer. In: *Drying '87*, A.S. Mujumdar, (ed.). New York: Hemisphere, 1987.
Devahastin, S. and C. Niamnuy, Modelling quality changes of fruits and vegetables during drying: A review. *Int. J. Food Sci. Technol.*, 2010; 45: 1755–1757.
Di Matteo, P., G. Donsì, and G. Ferrari, The role of heat and mass transfer phenomena in atmospheric freeze-drying of foods in a fluidised bed. *J. Food Eng.*, 2003; 59: 257–275.
Di Scala, K. and G. Crapiste, Drying kinetics and quality changes during drying of red pepper. *LWT—Food Sci. Technol.*, 2008; 41: 789–795.
Dixon, G.M. and J.J. Jen, Changes of sugar and acids of osmotic-dried apple slices. *J. Food Sci.*, 1977; 42: 1125–1127.
Dolinsky, A., K. Maletskaya, and Y. Snezhkin, Fruit and vegetable powders production technology on the bases of spray and convective drying methods. *Drying Technol.*, 2000; 18: 747–758.
Downton, D.P., J.L. Flores-Luna, and C.J. King, Mechanism of stickiness in hygroscopic, amorphous powders. *Ind. Eng. Chem. Res.*, 1982; 21: 447–451.
Duan, X., M. Zhang, and A.S. Mujumdar, Studies on the microwave freeze drying technique and sterilization characteristics of cabbage. *Drying Technol.*, 2007; 25: 1725–1731.
Earle, R.L. *Unit Operations in Food Processing*, Oxford, UK: Pergamon Press, 1983.
Eisa, M.A.R. Applications of heat pumps in chemical processing. *Energy Convers. Manage.*, 1995; 37: 359–377.
Ekechukwua, O.V. and B. Norton, Review of solar-energy drying systems II: An overview of solar drying technology. *Energy Convers. Manage.*, 1999; 40: 515–555.
Erle, U. and H. Schubert, Combined osmotic and microwave-vacuum dehydration of apples and strawberries. *J. Food Eng.*, 2001; 49: 193–199.
Escher, F. and B. Blanc, Quality and nutritional aspects of food dehydration. In: *Food Quality and Nutrition*, W.K. Downey, (ed.). London, U.K.: Applied Science Publishers, 1978.
Fahloul, D., M. Lahbari, H. Benmoussa, and S. Mezdour, Effect of osmotic dehydration on the freeze drying kinetics of apricots. *J. Food Agric. Environ.*, 2009; 7: 117–121.
Fante, L. and C.P.Z. Noreña, Quality of hot air dried and freeze-dried of garlic (*Allium sativum* L.). *J. Food Sci. Technol.*, 2015; 52: 211–220.
Ferrari, C.C., S.P.M. Germer, I.D. Alvim, F.Z. Vissotto, and de Aguirre, Influence of carrier agents on the physicochemical properties of blackberry powder produced by spray drying. *Int. J. Food Sci. Technol.*, 2012; 47: 1237–1245.
Filkova I. and A.S. Mujumdar, Industrial spray drying systems. In: *Handbook of Industrial Drying*, 2nd edn. A.S. Mujumdar, (ed.). New York: Marcel Dekker, 1995.
Filkova, I., L.X. Huang, and A.S. Mujumdar, Industrial spray drying systems. In: *Handbook of Industrial Drying*, 3rd edn. A.S. Mujumdar, (ed.). Boca Raton, FL: CRC/Taylor & Francis Group, 2007.
Fonseca, C.R., M.S.G. Bento, E.S.M. Quintero, A.L. Gabas, and C.A.F. Oliveira, Physical properties of goat milk powder with soy lecithin added before spray drying. *Int. J. Food Sci. Technol.*, 2011; 45: 508–511.
Franke, K. and M. Kießling, Influence of spray drying conditions on functionality of dried whole egg. *J. Sci. Food Agric.*, 2002; 82: 1837–1841.
Franks, F. Freeze-drying of bioproducts: Putting principles into practice. *Eur. J. Pharm. Biopharm.*, 1998; 45: 221–229.
Frias, J.M. and J.C. Oliveira, Kinetic models of ascorbic acid thermal degradation during hot air drying of maltodextrin solutions. *J. Food Eng.*, 2001; 47: 255–252.
Frydman, A., J. Vasseur, F. Ducept, M. Sionneau, and J. Moureh, Simulation of spray drying in superheated steam using computational fluid dynamics. *Drying Technol.*, 1999; 17: 1313–1325.
Gavrielidou, M.A., N.A. Vallous, T.D. Karapantsios, and S.N. Raphaelides, Heat transport to a starch slurry gelatinising between the drums of a double drum dryer. *J. Food Eng.*, 2002; 54: 45–58.
Geankoplis, C.J. *Transport Processes and Unit Operations*, 3rd ed. Engelwood Cliffs, NJ: Prentice-Hall, Inc., 1983.
Gekas, V., *Transport Phenomena of Foods and Biological Materials*, Boca Raton, FL: CRC Press, 1993.
George, J.P. and A.K. Datta, Development and validation of heat and mass transfer models for freeze-drying of vegetable slices. *J. Food Eng.*, 2002; 52: 89–93.
Georgé, S., F. Tourniaire, H. Gautier, P. Goupy, E. Rock, and C. Caris-Veyrat, *Food Chem.*, 2011; 124: 1503–1511.

Georgetti, S.R., R. Casagrande, C.R.F. Souza, W.P. Oliveira, and M.J.V. Fonseca, Spray drying of the soybean extract: Effects on chemical properties and antioxidant activity. *LWT—Food Sci. Technol.*, 2008; 41: 1521–1527.

Gharsallaoui, A., G. Roudaut, O. Chambin, A. Voilley, and R. Saurel, Applications of spray-drying in microencapsulation of food ingredients: An overview. *Food Res. Int.*, 2007; 40: 1107–1121.

Giangiacomo, R., D. Torreggianni, and E. Abbo, Osmotic dehydration of fruit. Part 1. Sugar exchange between fruit and extracting syrup. *J. Food Process. Preserv.*, 1987; 11: 183–195.

Ginzburg, A.S. Application of infrared radiation in food processing. In: *Chemical and Process Engineering Series*, A. Grochowski, (ed.). London, U.K.: Leonard Hill, 1959.

Giri, S.K. and S. Mangaraj, Processing influences on composition and quality attributes of soymilk and its powder. *Food Eng. Rev.*, 2012; 4: 149–154.

Goldberg, J.E. Prediction of spray dryer performance, PhD thesis, Oxford, U.K.: University of Oxford, 1987.

Goula, A.M. and K.G. Adamopoulos, Influence of spray drying conditions on residue accumulation—Simulation using CFD. *Drying Technol.*, 2004; 22: 1107–1128.

Goula, A.M. and K.G. Adamopoulos, Stability of lycopene during spray drying of tomato pulp. *LWT—Food Sci. Technol.*, 2005a; 38: 479–487.

Goula, A.M. and K.G. Adamopoulos, Spray drying of tomato pulp in dehumidified air. II. The effect on powder properties. *J. Food Eng.*, 2005b; 55: 35–42.

Goula, A.M. and K.G. Adamopoulos, Effect of maltodextrin addition during spray drying of tomato pulp in dehumidified air: II. Powder properties. *Drying Technol.*, 2008; 25: 725–737.

Goula, A.M. and K.G. Adamopoulos, Kinetic models of β-carotene degradation during air drying of carrots. *Drying Technol.*, 2010; 28: 752–751.

Goula, A.M., K.G. Adamopoulos, and N.A. Kazakis, Influence of spray conditions on tomato powder properties. *Drying Technol.*, 2004; 22: 1129–1151.

Goyal, R.K. and G.N. Tiwari, Performance of a reverse flat plate absorber cabinet dryer: A new concept. *Energy Convers. Manage.*, 1999; 40: 385–392.

Grabowski, J.A., Truong, V.D., and C.R. Daubert, Nutritional and rheological characterization of spray dried sweetpotato powder. *J. Food Sci. Techn.*, 2008; 41: 206–216.

Haida, H., P. Tomova, W. Behns, M. Ihlow, and L. Mörl, Investigations on adsorption drying at low temperatures. *Chem. Eng. Technol.*, 2005; 28: 28–31.

Hallak, H., J. Hilal, F. Hilal, and R. Rahhal, The staircase solar dryer: Design and characteristics. *Renew. Energy*, 1995; 7: 177–183.

Halsey, G. Physical adsorption on non-uniform surfaces. *J. Chem. Phys.*, 1948; 15: 931–937.

Hawlader, M.N.A., C.O. Perera, and M. Tian, Comparison of the retention of 5-gingerol in drying under modified atmosphere heat pump drying and other drying methods. *Dry Technol.*, 2005; 24: 51–55.

Heldman, D.R. and R.W. Hartel, *Principles of Food Processing*, New York: Chapman & Hall-International Thomson Publishing, 1997.

Heldman, D.R. and P.R. Singh, Food dehydration. In: *Food Process Engineering*, 2nd ed. Westport, CT: AVI Publishing Company, 1981.

Henderson, S.M. and S. Pabis, Grain drying theory. II. Temperature effects on drying coefficients. *J. Agric. Eng. Res.*, 1951; 5: 159–174.

Hino T., S. Shimabayashi, N. Ohnishi, M. Fujisaki, H. Mori, O. Watanabe, K. Kawashima, and K. Nagao Development of a new type nozzle and spray-drier for industrial production of fine powders. *Eur. J. Pharm. Biopharm.*, 2000; 49: 79–85.

Hovmand, S. Fluidized bed drying. In: *Handbook of Industrial Drying*, A.S. Mujumdar, (ed.). New York: Marcel Dekker, 1995.

Ibarz, A and G. Barbosa-Canovas, *Unit Operations in Food Engineering*, Boca Raton, FL: CRC/Taylor & Francis Group, 2003.

Inyang, U.E. and C.I. Ike, Effect of blanching, dehydration method and temperature on the ascorbic acid, colour, sliminess and other constituents of okra fruit. *Int. J. Food Sci. Nutr.*, 1998; 49: 125–130.

Jaekel, T., K. Dautel, and W. Ternes, Preserving functional properties of hen's egg yolk during freeze-drying. *J. Food Eng.*, 2008; 87: 522–525.

Jangam, S.V. and B.N. Thorat, Optimization of spray drying of ginger extract. *Drying Technol.*, 2010; 28: 1425–1434.

Jaya, S. and H. Das, Effect of maltodextrin, glycerol monostearate and tricalcium phosphate on vacuum dried mango powder properties. *J. Food Eng.*, 2004; 53: 125–134.

Jayaraman, K.S. and D.K. Das Gupta, Drying of fruits and vegetables. In: *Handbook of Industrial Drying*, 2nd edn. New York: Marcel Dekker, 1995.

Jayas, D.S., S. Cenkowski, S. Pabis, and W.E. Muir, Review of thin layer drying and wetting equations. *Drying Technol.*, 1991; 9: 551–588.

Jiang, S. and S.L. Nail, Effect of process conditions on recovery of protein activity after freezing and freeze-drying. *Eur. J. Pharm. Biopharm.*, 1998; 45: 249–257.

Kaminski, W. and E. Tomczak, Degradation of ascorbic acid in drying process—A comparison of description methods. *Drying Technol.*, 2000; 18: 777–790.

Kar, A., P. Chandra, R. Prasad, D.V.K. Samuel, and D.S. Khurdiya, Comparison of different methods of drying for banana (Dwarf Cavendish) slices. *J. Food Sci. Technol.*, 2003; 40: 378–381.

Karatas, S. and A. Esin, Determination of moisture diffusivity and behavior of tomato concentrate droplets during drying in air. *Drying Technol.*, 1994; 12: 799–822.

Karathanos, V.T., G. Villalobos, and G.D. Saravacos, Comparison of two methods of estimation of the effective moisture diffusivity from drying data. *J. Food Sci.*, 1990; 55: 218–223.

Karel, M. Prediction of nutrient losses and optimization of processing conditions. In: *Nutritional and Safety Aspects of Food Processing*, S.R. Tannenbaum, (ed.). New York: Marcel Dekker, 1979.

Karim, O.R. and A.A. Adebowale, A dynamic method for kinetic model of ascorbic acid degradation during air dehydration of pretreated pineapple slices. *Int. Food Res. J.*, 2009; 15: 555–550.

Karmas, R., Buera, M.P., and M. Karel, Effect of glass transition on rates of nonenzymatic browning in food systems, *J. Agric. Food Chem.*, 1992; 40: 873–879.

Katekawa, M.E. and M.A. Silva, A review of drying models including shrinkage effects. *Drying Technol.*, 2005; 24: 5–20.

Kayamak-Ertekin, F. and T. Çakaloz, Osmotic dehydration of peas. II. Influence of osmosis on drying behavior and product quality. *J. Food Process. Preserv.*, 1995; 20: 105–119.

Kemp, I.C. and R.E. Bahu, A new algorithm for dryer selection. *Drying Technol.*, 1995; 13: 1553–1578.

Kerdpiboon, S., W.L. Kerr, and S. Devahastin, Neural network prediction of physical property changes of dried carrot as a function of fractal dimension and moisture content. *Food Res. Int.*, 2005; 39: 1110–1118.

Kerr, W.L., M.H. Lim, D.S. Reid, and H. Chen, Chemical reaction kinetics in relation to glass transition temperatures in frozen food polymer solutions. *J. Sci. Food Agric.*, 1993; 51: 51–55.

Khalloufi, S., J.-L. Robert, and C. Ratti, Solid foods freeze-drying simulation and experimental data. *J. Food Process Eng.*, 2005; 28: 107–132.

Khazaei, J., G.-R. Chegini, and M. Bakhshiani, A novel alternative method for modeling the effects of air temperature and slice thickness on quality and drying kinetics of tomato slices: Superposition technique. *Drying Technol.*, 2008; 25: 759–775.

Kieviet, F.G. and P.J.A. Kerkhof, Air flow, temperature and humidity patterns in a co-current spray dryer: Modelling and measurements. *Drying Technol.*, 1997; 15: 1753–1773.

Kieviet, F.G., J. Van Raaij, P.P.E.A. De Moor, and P.J.A.M. Kerkhof, Measurement and modelling of the air flow pattern in a pilot-plant spray dryer. *Trans. Inst. Chem. Eng.*, 1997; 75: 321–328.

Kim, K.H., Y. Song, and L.K. Yam, Water sorption characteristics of dried red peppers (*Capsicum annum* L.). *Int. J. Food Sci. Technol.*, 1991; 29: 339–345.

King, C.J. and T.P. Labuza, Freeze-drying of foodstuffs. *CRC Crit. Rev. Food Technol.*, 1970; 1: 379–451.

Kinsella, J.E. and P.F. Fox, Water sorption by proteins: Milk and whey proteins. *CRC Crit. Rev. Food Sci. Nutr.*, 1985; 24: 91–103.

Koca, N., S.H. Burdurlu, and F. Karadeniz, Kinetics of colour changes in dehydrated carrots. *J. Food Eng.*, 2007; 78: 449–455.

Komes, D., T. Lovrić,, and K. Kovačević Ganić, Aroma of dehydrated pear products. *LWT—Food Sci. Technol.*, 2007; 40: 1578–1585.

Koroishi, E.T., E.A. Boss, and R. Maciel Filho, Development of orange juice freeze drying process. *CHISA 2005—17th International Congress of Chemical and Process Engineering*, Praha, Czech Republic, 2011.

Krokida, M.K. and C. Philippopoulos, Volatility of apples during air and freeze drying. *J. Food Eng.*, 2005; 73: 135–141.

Kuriakose, R. and C. Anandharamakrishnan, Computational fluid dynamics (CFD) applications in spray drying of food products. *Trend Food Sci. Technol.*, 2010; 21: 383–398.

Kutsakova, V.E., Y.V. Utkin, and I.A. Makeeva, Characteristics of sublimation (freeze) drying of granulated products in cylinder dryers at atmospheric pressure. *J. Appl. Chem. USSR*, 1984; 57: 791–793.

Kuu, W.-Y., J. McShane, and J. Wong, Determination of mass transfer coefficients during freeze drying using modeling and parameter estimation techniques. *Int. J. Pharm.*, 1995; 124: 241–252.

Langrish, T.A.G., Degradation of vitamin C in spray dryers and temperature and moisture content profiles in these dryers. *Food Bioprocess Technol.*, 2009; 2: 400–408.

Langrish, T.A.G. and D.F. Fletcher, Prospects for the modelling and design of spray dryers in the 21st century. *Drying Technol.*, 2003; 21: 197–215.

Langrish, T.A.G. and T.K. Kockel, The assessment of a characteristic drying curve for milk powder for use in computational fluid dynamics modeling, *Chem. Eng.*, 2001; *J.* 84: 59–74.

Langrish, T.A.G. and I. Zbicinski, The effects of air inlet geometry and spray cone angle on the wall deposition rate in spray dryers. *Trans. Inst. Chem. Eng.*, 1994; 72: 420–430.

Lee, C.-W., H.-J. Oh, S.-H. Han, and S.-B. Lim, Effects of hot air and freeze drying methods on physicochemical properties of citrus "hallabong" powders. *Food Sci. Biotech.*, 2012; 21: 1533–1539.

Lenart, A. and J.M. Flink, Osmotic dehydration of potato. II. Spatial distribution of the osmotic agent. *J. Food Technol.*, 1984; 19: 55–89.

Leniger, H.A. and S. Bruin, The state of the art of food dehydration. In: *Food Quality and Nutrition*, W.K. Downey, (ed.). London, U.K.: Applied Science Publishers, 1977.

Lerici, C.L., G. Pinnavaia, M. Dalla Rosa, and L. Bartolucci, Osmotic dehydration of fruits: Influence of osmotic agents on drying behavior and product quality. *J. Food Sci.*, 1985; 50: 1217–1219.

Lewicki, P.P. and G. Pawlak, Effect of mode of drying on microstructure of potato. *Drying Technol.*, 2005; 23: 847–859.

Lewicki, P.P. and J. Wiczkowska, Rehydration of apple dried by different methods. *Int. J. Food Prop.* 2005; 9: 217–225.

Li, L.H. and W.H. Lai, Energy form of cone-jets in electrohydrodynamic atomization. *Int. J. Turbo Jet-Engines*, 2011; 28: 199–207.

Li, S. Atmospheric freeze drying of food products in a closed system, PhD thesis, Lodz, Poland: In Faculty of Process and Environmental Engineering, Technical University of Lodz, 2005.

Li, S., I. Zbicinski, H. Wang, J. Stawczyk, and Z. Zhang, Diffusion model for apple cubes atmospheric freeze-drying with the effect of shrinkage. *Int. J. Food Eng.*, 2008; 4: 10.

Liang, B. and C.J. King, Factors influencing flow patterns, temperature fields and consequent drying rates in spray drying. *Drying Technol.*, 1991; 9: 1–25.

Liapis, A.I. and R. Bruttini, A theory for the primary and secondary drying stages of the freeze-drying of pharmaceutical crystalline and amorphous solutes: Comparison between experimental data and theory. *Sep. Technol.*, 1995; 4: 144–155.

Liapis, A.I. and R.J. Litchfield, Optimal control of a freeze dryer-I Theoretical development and quasi steady state analysis. *Chem. Eng. Sci.*, 1979; 34: 975–981.

Liapis, A.I., M.J. Pikal, and R. Bruttini, Research and development needs and opportunities in freeze drying. *Drying Technol.*, 1995; 14: 1255–1300.

Lin, T.M.D. T. Durance, and C.H. Scaman, Characterization of vacuum microwave, air and freeze dried carrot slices. *Food Res. Int.*, 1998; 31: 111–117.

Litvin, S., C.H. Mannheim, and J. Miltz, Dehydration of carrots by a combination of freeze drying, microwave heating and air or vacuum drying. *J. Food Eng.*, 2008; 35: 103–111.

Liu, J., S. Ma, B. Liu, X. Yang, Y. Zhang, and E. Wang, Effects of different drying methods on solubility of whole egg powder. *Trans. Chin. Soc. Agric. Eng.*, 2011; 27: 383–388.

Lombrana, J.I., The influence of pressure and temperature on freeze-drying in an adsorbent medium and establishment of drying strategies. *Food Res. Int.*, 1997; 30(3–4): 213–222.

Lombraña, J.I. and J. Izkara, Experimental estimation of effective transport coefficients in freeze drying for simulation and optimization purposes. *Drying Technol.*, 1995; 14: 743–753.

Lombraña, J.I. and M.C. Villarán, Interaction of kinetic and quality aspects during freeze drying in an adsorbent medium. *Ind. Eng. Chem. Res.*, 1996; 35: 1957–1975.

Lopez-Quiroga, E., L.T. Antelo, and A. Alonso, Time-scale modeling and optimal control of freeze–drying. *J. Food Eng.*, 2012; 111: 555–555.

Lozano, J.E., E. Rotstein, and M.J. Urbicain, Shrinkage, porosity and bulk density of foodstuffs at changing moisture contents. *J. Food Sci.*, 1983; 48: 1497–1502.

Luikov, A.V. *Analytical Heat Diffusion Theory*, New York: Academic Press, 1958.

Maache-Rezzoug, Z., S.A. Rezzoug, and K. Allaf, Development of a new drying process particularly adapted to thermo-sensitive products dehydration by successive pressure drops—Application to collagen gel. *Drying Technol.*, 2001; 19: 1951–1974.

Mahn, A.V., P. Antoine, and A. Reyes, Optimization of drying kinetics and quality parameters of broccoli florets. *Int. J. Food Eng.*, 2011; 7: 14.

Marinos-Kouris, D. and Z.B. Maroulis, Thermophysical properties of the drying of Solids. In: *Handbook of Industrial Drying*, A.S. Mujumdar, (ed.). New York: Marcel Dekker, 1995.

Maroulis, Z.B., E. Tsami, D. Marinos-Kouris, and G.D. Saravacos, Application of the GAB model to the moisture sorption isotherms for dried fruits. *J. Food Eng.*, 1988; 7: 53–78.

Marques, L.G., A.M. Silveira, and J.T. Freire, Freeze-drying characteristics of tropical fruits. *Drying Technol.*, 2005; 24: 457–453.

Masters, K. *Spray Drying Handbook*, 4th edn. New York: Wiley, 1985.

Masters, K. Scale-up of spray dryers. *Drying Technol.*, 1994; 12: 235–257.

Mathlouthi, M, and B. Rogé, Water vapor sorption isotherms and the caking of food powders. *Food Chem.*, 2003; 82: 51–71.

Mattea M., M.J. Urbicain, and E. Rotstein, Computer model of shrinkage and deformation of cellular tissue during dehydration. *Chem. Eng. Sci.*, 1989; 44: 2853–2859.

Matveev, Y.I., V.Y. Grinberg, and V.B. Tolstoguzov, The plasticizing effect of water on proteins, polysaccharides and their mixtures. Glassy state of biopolymers, food and seeds. *Food Hydrocolloid*, 2000; 14: 425–437.

Mavroudis, N.E., V. Gekas, and I. Sjo holm, Osmotic dehydration of apples. Shrinkage phenomena and the significance of initial structure on mass transfer. *J. Food Eng.*, 1998; 38: 101–123.

Mayor, L. and A.M. Sereno, Modelling shrinkage during convective drying of food materials: A review. *J. Food Eng.*, 2004; 51: 373–385.

McLaughlin, C.P. and T.R.A. Magee, The determination of sorption isotherms and the isosteric heats of sorption for potatoes. *J. Food Eng.*, 1998; 35: 257–280.

McLaughlin, C.P. and T.R.A. Magee, The effects of shrinkage during drying of potato spheres and the effect of drying temperature on vitamin C retention. *Food Bioprod. Process.*, 1998; 75: 138–142.

McMinn, W.A.M. and T.R.A. Magee, Kinetics of ascorbic acid degradation and non-enzymic browning in potatoes. *Food Bioprod. Process.*,1997; 75: 223–231.

Mejia-Meza, E.I., J.A. Yáñez, C.M. Remsberg, J.K. Takemoto, N.M. Davies, B. Rasco, and C. Clary, Effect of dehydration on raspberries: Polyphenol and anthocyanin retention, antioxidant capacity, and antiadipogenic activity. *J. Food Sci.*, 2010; 75: H5–H12.

Mellor, J.D. *Fundamentals of Freeze-Drying*. London: Academic Press, 1978.

Menshutina, N.V., M.G. Gordienko, A.A. Voynovskiy, and T. Kudra, Dynamic analysis of drying energy consumption. *Drying Technol.*, 2004; 22: 2281–2290.

Methakhup, S., N. Chiewchan, and S. Devahastin, Effects of drying methods and conditions on drying kinetics and quality of Indian gooseberry flake. *LWT—Food Sci. Technol.*, 2005; 38: 579–587.

Mezhericher, M., A. Levy, and I. Borde, Theoretical drying model of single droplets containing insoluble or dissolved solids. *Drying Technol.*, 2007; 25: 1025–1032.

Mezhericher, M., A. Levy, and I. Borde, Droplet–droplet interactions in spray drying using 2D computational fluid dynamics. *Drying Technol.*, 2008; 25: 255–282.

Mezhericher, M., A. Levy, and I. Borde, Spray drying modelling based on advanced droplet drying kinetics. *Chem. Eng. Process. Process Intensif.*, 2010; 49: 1205–1213.

Mohsenin, N.N. *Physical Properties of Plant and Animal Materials*, 2nd edn. New York: Gordon and Breach Science Publishers, 1985.

Moraga, G., N. Martinez-Navarrete, and A. Chiralt, Water sorption isotherms and phase transitions in kiwifruit. *J. Food Eng.*, 2005; 72: 147–155.

Mosha, T.C., R.D. Pace, S. Adeyeye, K. Mtebe, and H. Laswai, Proximate composition and mineral content of selected Tanzanian vegetables and the effect of traditional processing on the retention of ascorbic acid, riboflavin and thiamine. *Plant Foods Hum. Nutr.*, 1995; 48: 235–245.

Moyano, P.C., E. Troncoso, and F. Pedreschi, Modeling texture kinetics during thermal processing of potato products. *J. Food Sci.*, 2007; 72: E102–E107.

Mujumdar, A.S. and S. Devahastin, *Fundamental Principles of Drying*, Brossard, Quebec, Canada: Exergex Corporation, 2000.

Mujumdar, A.S. and A.S. Menon, Drying of solids. In: *Handbook of Industrial Drying*, A.S. Mujumdar (ed.), 2nd edn. New York: Marcel Dekker, 1995.

Nath, S. and G.R. Satpathy, A systematic approach for investigation of spray drying processes. *Drying Technol.*, 1998; 15: 1173–1193.

Nicoleti, J.F., V. Silveira, J. Telis-Romero, and V.R.N. Telis, Influence of drying conditions on ascorbic acid during convective drying of whole persimmons. *Drying Technol.*, 2007; 25: 891–899.

Nijdam, J.J. and T.A.G. Langrish, An investigation of milk powders produced by a laboratory-scale spray dryer. *Drying Technol.*, 2005a; 23: 1043–1055.

Nijdam, J.J. and T.A.G. Langrish, The effect of surface composition on the functional properties of milk powders. *J. Food Eng.*, 2005b; 77: 919–925.

Nindo, C.I., T. Sun, S.W. Wang, J. Tang, and J.R. Powers, Evaluation of drying technologies for retention of physical quality and antioxidants in asparagus (*Asparagus officinalis* L.) *LWT—Food Sci. Technol.*, 2003; 35: 507–515.

Nisha, P., R.S. Singhal, and A.B. Pandit, Kinetic modeling of texture development in potato cubes (*Solanum tuberosum*), green gram whole (*Vigna radiate* L.) and red gram splits (*Cajanus cajan* L.). *J. Food Eng.*, 2005; 75: 524–530.

Noyes, A.V. et al., Domestic fabric article refreshment in integrated cleaning and treatment, US Patent 7,033,985, 2005.

Oakley, D.E. Produce uniform particles by spray drying. *Chem. Eng. Progr.*, 1997; 93: 48–54.

Oakley, D.E. and R.E. Bahu, Spray/gas mixing behaviour within spray dryers. In: *Drying*, A.S. Mujumdar and I. Filkova, (eds.). Amsterdam, the Nethrelands: Elsevier Applied Science, 1991.

Okos, M.R., G. Narasimhan, R.K. Singh, and A.C. Witnauer, Food dehydration. In: *Handbook of Food Engineering*, D.R. Heldman and D.B. Lund, (eds.). New York: Marcel Dekker, 1992.

Orikasa, T., L. Wu, Y. Ando, Y. Muramatsu, P. Roy, T. Yano, T. Shiina, and A. Tagawa, Hot air drying characteristics of sweet potato using moisture sorption isotherm analysis and its quality changes during drying. *Int. J. Food Eng.*, 2010; 5: 12.

Oswin, C.R. The kinetics of package life III. The isotherm. *J. Chem. Ind.*, 1945; 55: 419–421.

Overhults, D.G., G.M. White, H.E. Hamilton, and I.J. Ross, Drying soybeans with heated air. *Trans. Am. Soc. Agric. Eng.*, 1973; 15: 112–113.

Page, G.E. Factors influencing the maximum rates of air drying shelled corn in thin layers, MS thesis, West Lafayette, IN: Department of Mechanical Engineering, Purdue University, 1949.

Pan, Y.K., L.J. Zhao, Y. Zhang, G. Chen, and A.S. Mujumdar, Osmotic dehydration pretreatment in drying of fruits and vegetables. *Drying Technol.*, 2003; 21: 1101–1114.

Panyawong, S. and S. Devahastin, Determination of deformation of a food product undergoing different drying methods and conditions via evolution of a shape factor. *J. Food Eng.*, 2007; 78: 151–151.

Papadakis, S., C. Gardeli, and C. Tzia, Spray drying of raisin juice concentrate. *Drying Technol.*, 2005; 24: 173–180.

Park, K.J., A. Bin, F.P.R. Brod, and T.H.K.B. Park, Osmotic dehydration kinetics of pear D'anjou (*Pyrus communis* L.). *J. Food Eng.*, 2002; 52: 293–298.

Patel, K.C. and X.D. Chen, Sensitivity analysis of the reaction engineering approach to modeling spray drying of whey proteins concentrate. *Drying Technol.*, 2008; 25: 1334–1343.

Peleg, M.Assessment of a semi-empirical four parameter general model for sigmoid moisture sorption isotherms. *J. Food Process Eng.*, 1993; 15: 21–37.

Piatkowski, M. and I. Zbicinski, Analysis of the mechanism of counter-current spray drying. *Transport Porous Med.*, 2007; 55: 89–101.

Pignoli, G., M.T. Rodriguez-Estrada, M. Mandrioli, L. Barbanti, L. Rizzi, and G. Lercker, Effects of different rearing and feeding systems on lipid oxidation and antioxidant capacity of freeze-dried egg yolks. *J. Agric. Food Chem.*, 2009; 57: 11517–11527.

Pikal, M.J. and J.E. Lang, Rubber closures as a source of haze in freeze dried parenterals: Test methodology for closure evaluation. *J. Parenteral Drug Assoc.*, 1978; 32: 152–173.

Pikal, M.J. and S. Shah, The collapse temperature in freeze drying: Dependence on measurement methodology and rate of water removal from the glassy phase. *Int. J. Pharm.*, 1990; 52: 155–185.

Pikal, M.J., M.L. Roy, and S. Shah, Mass and heat transfer in vial freeze-drying of pharmaceuticals: Role of the vial. *J. Pharm. Sci.*, 1984; 73: 1224–1237.

Ponting, J.D., W.L. Stanley, and M. J. Copley, Fruit and vegetables juices. In: *Food Dehydration*, B.S. Van Arsdel, M.J. Copley, and A.I. Morgan, (eds.), 2nd edn. Westport, CT: AVI Publishing Company, 1973.

Quek, S.Y., N.K. Chok, and P. Swedlund, The physicochemical properties of spray-dried watermelon powders. *Chem. Eng. Process. Process Intensif.*, 2007; 45: 385–392.

Rahardjo, B. and S.K. Sastry, Kinetics of softening of potato tissue during thermal treatment. *Trans. IChemE*, 1993; 71(Part C): 235–241.

Rahman, M.S., H.I. Al-Sheibani, M.H. Al-Riziqi, A. Mothershaw, N. Guizani, and G. Bengtsson, Assessment of the anti-microbial activity of dried garlic powders produced by different methods of drying. *Int. J. Food Prop.*, 2005; 9: 503–513.

Raji, A.O. and J.O. Ojediran, Moisture sorption isotherms of two varieties of millet. *Food Bioprod. Process.*, 2011; 89: 178–184.

Ramos, I.N., T.R.S. Brandao, and C.L.M. Silva, Structural changes during air drying of fruits and vegetables. *Food Sci. Technol. Int.*, 2003; 9: 201–205.

Rannou, C., F. Texier, M. Moreau, P. Courcoux, A. Meynier, and C. Prost, Odour quality of spray-dried hens' egg powders: The influence of composition, processing and storage conditions. *Food Chem.*, 2013; 138: 905–914.

Rastogi, N.K. and K.S.M.S. Raghavarao, Water and solute diffusion coefficients of carrot as a function of temperature and concentration. *J. Food Eng.*, 1997; 34: 429–440.

Rastogi, N.K. and K.S.M.S. Raghavarao, Mass transfer during osmotic dehydration of pineapple: Considering Fickian diffusion in cubical configuration. *Lebensm.-Wiss. Technol.*, 2004; 37: 43–47.

Rastogi, N.K., K.S.M. Raghavarao, and K. Niranjan, Mass transfer during osmotic dehydration of banana: Fickian diffusion in cylindrical configuration. *J. Food Eng.*, 1997; 31: 423–432.

Ratti, C. Hot air and freeze-drying of high-value foods: A review. *J. Food Eng.*, 2001; 49: 311–319.

Ratti, C. and A.S. Mujumdar, Infrared drying. In: *Handbook of Industrial Drying*, A.S. Mujumdar, (ed.). New York: Marcel Dekker, 1995.

Reay, D. Fluid flow, residence time simulation and energy efficiency in industrial dryers. In: *Proceedings of the Fifth International Drying Symposium* (*IDS'88*), Versailles, France, 1988.

Regier, M., E. Mayer-Miebach, D. Behsnilian, E. Neff, and H.P. Schuchmann, Influences of drying and storage of lycopene-rich carrots on the carotenoid content. *Drying Technol.*, 2005; 23: 989–998.

Reyes, A., A. Evseev, A. Mahn, V. Bubnovich, R. Bustos, and E. Scheuermann, Effect of operating conditions in freeze-drying on the nutritional properties of blueberries. *Int. J. Food Sci. Nutr.*, 2011; 52: 303–305.

Riva, M., S. Campolongo, A.A. Leva, A. Maestrelli, and D. Torreggiani, Structure–property relationships in osmo-airdehydrated apricot cubes. *Food Res. Int.*, 2005; 38: 533–542.

Rizvi, A.F. and C.H. Tong, Fractional conversion for determining texture degradation kinetics of vegetables. *J. Food Sci.*, 1997; 52: 1–7.

Rodríguez-Hernández, G.R., González-García, R., Grajales-Lagunes, A., Ruiz-Cabrera, M.A., and M. Abud-Archila, Spray-drying of cactus pear juice (Opuntia streptacantha): Effect on the physicochemical properties of powder and reconstituted product. *Drying Technol.*, 2005; 23: 955–973.

Rogers, S., Y. Fang, S.X. Qi Lin, C. Selomulya, and X. Dong Chen, A monodisperse spray dryer for milk powder: Modelling the formation of insoluble material. *Chem. Eng. Sci.*, 2012; 71: 75–84.

Roos, Y.H. and M.J. Himberg, Non-enzymatic browning behavior, as related to glass transition of a food model at chilling temperatures. *J. Agric. Food Chem.*, 1994; 42: 893–898.

Roos, Y.H. and M. Karel, Plasticizing effect of water on thermal behavior and crystallization of amorphous food materials. *J. Food Sci.*, 1991; 55: 38–43.

Roos, Y.H., M. Karel, and J.L. Kokini, Glass transitions in low moisture and frozen foods: Effects on shelf life and quality. Scientific status summary. *Food Technol.*, 1995; 50: 95–108.

Roustapour, O.R., M. Hosseinalipour, and B. Ghobadian, An experimental investigation of lime juice drying in a pilot plant spray dryer. *Drying Technol.*, 2005; 24: 181–188.

Rovedo, C.O. and P.E. Viollaz, Prediction of degrading reactions during drying of solid foodstuffs. *Drying Technol.*, 1998; 15: 551–578.

Sablani, S.S., R.M. Myhara, Z.H. Al-Attabi, and M.M. Al-Mugheiry, Water sorption isotherms of freeze dried fish sardines. *Drying Technol.*, 2001; 19: 573–580.

Sadikoglu, H. and A.I. Liapis, Mathematical modelling of the primary and secondary drying stages of bulk solution freeze-drying in trays: Parameter estimation and model discrimination by comparison of theoretical results with experimental data. *Drying Technol.*, 1997; 15: 791–810.

Sagara, Y., K. Kaminishi, E. Goto, T. Watanabe, Y. Imayoshi, and H. Iwabuchi, Characteristic evaluation for volatile components of soluble coffee depending on freeze-drying conditions. *Drying Technol.*, 2005; 23: 2185–2195.

Sandu, C. Infrared radiative drying in food engineering: A process analysis. *Biotechnol. Progress*, 1985; 2: 109–119.

Saravacos, G.D. and A.E. Kostaropoulos, *Handbook of Food Processing Equipment*, New York: Springer, 2002.

Saravacos, G.D. and Z.B. Maroulis, *Food Process Engineering Operations*, Boca Raton, FL: CRC/Taylor & Francis Group, 2011.

Sarsilmaz, C., C. Yildiz, and D. Pehlivan, Drying of apricots in a rotary column cylindrical dryer (RCCD) supported with solar energy. *Renew. Energy*, 2000; 21: 117–127.

Schmidt, E.L., K. Klocker, N. Flacke, and F. Steimle, Applying the transcritical CO_2 process to a drying heat pump. *Int. J. Refrig.*, 1998; 21: 202–211.

Sereno, A.M., R. Moreira, and E. Martinez, Mass transfer coefficients during osmotic dehydration of apple in single and combined aqueous solutions of sugar and salt. *J. Food Eng.*, 2001; 47: 43–49.

Serris, G.S. and C.G. Biliaderis, Degradation kinetics of beet-root pigment encapsulated in polymeric matrices. *J. Sci. Food Agric.*, 2001; 81: 1–10.

Seydel, P., J. Blömer, and J. Bertling, Modeling particle formation at spray drying using population balances. *Drying Technol.*, 2005; 24: 137–145.

Sheehan, P. and A.I. Liapis, Modeling of the primary and secondary drying stages of the freeze drying of pharmaceutical products in vials: Numerical results obtained from the solution of a dynamic and spatially multi-dimensional lyophilization model for different operational policies. *Biotechnol. Bioeng.*, 1998; 50: 712–728.

Shi, J., M.L. Maguer, Y. Kakuda, A. Liptay, and F. Niekamp, Lycopene degradation and isomerization in tomato dehydration. *Food Res. Int.*, 1999; 32: 15–21.

Shih, C., Z. Pan, T.H. McHugh, D. Wood, and E. Hirschberg, Sequential infrared radiation and freeze-drying method for producing crispy strawberries. *Trans. ASABE*, 2008; 51: 205–215.

Shishehgarha, F., J. Makhlouf, and C. Ratti, Freeze-drying characteristics of strawberries. *Drying Technol.*, 2002; 20: 131–145.

Shrestha, A.K., T. Howes,, B.P. Adhikari, and B.R. Bhandari, Water sorption and glass transition properties of spray dried lactose hydrolysed skim milk powder. *LWT—Food Sci. Technol.*, 2007; 40: 1593–1500.

Sila, D.N., C. Smout, T.S. Vu, and M.E. Hendrickx, Effects of high-pressure pretreatment and calcium soaking on the texture degradation kinetics of carrots during thermal processing. *J. Food Sci.*, 2004; 59: E205–E211.

Silva, M.G., T.S. Lira, E.B. Arruda, V.V. Murata, and M.A.S. Barrozo, Modelling of fertilizer drying in a rotary dryer: parametric sensitivity analysis. *Braz. J. Chem. Eng.*, 2012; 29: 359–359.

Simal, S., A. Mulet, J. Tarrazo, and C. Rossello, Drying models for green peas. *Food Chem.*, 1995; 55: 121–128.

Simatos, D., M. Faure, E. Bonjour, and M. Couach, The physical state of water at low temperature in plasma with different water contents as studied by differential thermal analysis and differential scanning calorimetry. *Cryobiology*, 1975; 12: 202–208.

Singh, S., P.P. Singh, and S.S. Dhaliwal, Multi-shelf portable solar dryer. *Renew. Energy*, 2004; 29: 753–755.

Skelland, A.H.P. Molecular diffusivities. In: *Diffusional Mass Transfer*. New York: Wiley, 1974.

Smith, S.E. The sorption of water vapour by high polymers. *J. Am. Chem. Soc.*, 1947; 59: 545.

Sokhansanj, S. and D.S. Jayas, Drying of foodstuffs. In: *Handbook of Industrial Drying*, A.S. Mujumdar, (ed.). New York: Marcel Dekker, 1995.

Southwell, D.B., T.A.G. Langrish, and D.F. Fletcher, Use of computational fluid dynamics techniques to assess design alternatives for the plenum chamber of a small spray dryer. In: *Proceedings of the First Asian-Australian Drying Conference*, Bali, Indonesia, 1999.

Straatsma, T.P., G. van Houwelingen, A.E. Steenbergen, and P. De Jong, Spray drying of food products: 1. Simulation model. *J. Food Eng.*, 1999; 42: 57–72.

Strumillo, C. and T. Kudra, *Drying: Principles, Applications and Design*, New York: Gordon & Breach Science Publishers, 1985.

Suvarnakuta, P., S. Devahastin, and A.S. Mujumdar, Drying kinetics and b-carotene degradation in carrot undergoing different drying processes. *J. Food Sci.*, 2005; 70: S520–S525.

Szarycz, M., R. Kramkowski, E. Kaminski, and K. Jaloszynski, Kinetics of carrot drying in the conditions of reduced pressure with microwave heating. *Acta Agrophys.*, 2002; 77: 147–154.

Tafti, A.G., S.H. Peighambardoust, J. Hesari, A. Bahrami, and E.S. Bonab, Physico-chemical and functional properties of spray-dried sourdough in breadmaking. *Food Sci. Technol. Int.*, 2013; 19: 271–278.

Tang, X. and M.J. Pikal, Design of freeze-drying processes for pharmaceuticals: Practical advice. *Pharm. Res.*, 2004; 21: 191–200.

Telang, A.M. and B.N. Thorat, Optimization of process parameters for spray drying of fermented soy milk. *Drying Technol.*, 2010; 28: 1445–1455.

Telis, V.R.N., A.L. Gabas, F.C. Menegalli, and J. Telis-Romero, Water sorption thermodynamic properties applied to persimmon skin and pulp. *Thermochim. Acta*, 2000; 343: 49–55.

Thakor, N.J., S. Sokhansanj, F.W. Sosulski, and S. Yannacopoulos, Mass and dimensional changes of single canola kernels during drying. *J. Food Eng.*, 1999; 40: 153–150.

Togrul, I.T. and D. Pehlivan, Mathematical modeling of solar drying of apricots in thin layers. *J. Food Eng.*, 2002; 55: 209–215.

Tonon, R.V., C. Brabet, and M.D. Hubinger, Influence of process conditions on the physicochemical properties of açai (*Euterpe oleraceae* Mart.) powder produced by spray drying. *J. Food Eng.*, 2008; 88: 411–418.

Tonon, R.V., S.S. Freitas, and M.D. Hubinger, Spray drying of açai (*Euterpe oleraceae* Mart.) juice: Effect of inlet air temperature and type of carrier agent. *J. Food Process. Preserv.*, 2011; 35: 591–700.

Treybal, R.E. *Mass Transfer Operations*, Tokyo, Japan: McGraw-Hill, Kogakusha, 1958.

Tsimidou, M. and C.G. Biliaderis, Kinetic studies of Saffron (*Crocus sativus* L.) quality deterioration. *J. Agric. Food Chem.*, 1997; 45: 2890–2898.

Tsourouflis, S., J.M. Flink, and M. Karel, Loss of structure in freeze-dried carbohydrates solutions: Effect of temperature, moisture content and composition. *J. Sci. Food Agric.*, 1975; 27: 509–519.
Tulasidas, T.N., C. Ratti, and G.S.V. Raghavan, Modelling of microwave drying of grapes. *Can. Agric. Eng.*, 1997; 39: 57–57.
Van Arsdel, W.B., M.J. Copley, and A.I. Morgan, *Food Dehydration*, Westport, CT: AVI Publishing Company, 1973.
Van den Berg, C. and S. Bruin, Water activity and its estimation in food systems. In: *Water Activity: Influences on Food Quality*, L.B. Rockland and G.F. Stewart, (eds.). New York: Academic Press, 1981.
Van Deventer, H. and R. Houben, and R. Koldeweij, New atomization nozzle for spray drying. *Drying Technol.*, 2013; 31: 89–897.
Van't Land, C.M. Selection of industrial dryers. *Chem. Eng.*, 1984; 91: 53–51.
Vardin, H. and M. Yasar, Optimisation of pomegranate (*Punica granatum* L.) juice spray-drying as affected by temperature and maltodextrin content. *Int. J. Food Sci. Technol.*, 2012; 47: 157–175.
Vaxelaire, J. and P. Cezac Moisture distribution in activated sludges: A review. *Water Res.*, 2004; 38: 2215–2230.
Walton, D.E. The morphology of spray-dried particles. A qualitative view. *Drying Technol.*, 2000; 18: 1943–1985.
Wang, C.Y. and R.P. Singh, Use of variable equilibrium moisture content in modeling rice drying. *Trans. Am. Soc. Agric. Eng.*, 1978; 11: 558–572.
Wang, Z.H. and M.H. Shi, Microwave freeze drying characteristics of beef. *Drying Technol.*, 1999; 17: 433–447.
Wenzel, M., I. Seuss-Baum, and E. Schlich, Influence of pasteurization, spray- and freeze-drying and storage on the carotenoid content in egg yolk. *J. Agric. Food Chem.*, 2010; 58: 1725–1731.
Wolff, E. and H. Gibert, Atmospheric freeze-drying part 1: Design, experimental investigation and energy-saving advantages. *Drying Technol.*, 1990; 8: 385–404.
Wu, Y.Q. and R.L. Clark, Electrohydrodynamic atomization: A versatile process for preparing materials for biomedical applications., *J. Biomater. Sci. Polym. Edn.*, 2008; 19: 573–501.
Yaldiz, O., C. Ertekin, and H I. Uzun, Mathematical modeling of thin layer solar drying of sultana grapes. *Energy*, 2001; 25: 457–455.
Youssefi, S., Z. Emam-Djomeh, and S.M. Mousavi, Comparison of artificial neural network (ANN) and response surface methodology (RSM) in the prediction of quality parameters of spray-dried pomegranate juice. *Drying Technol.*, 2009; 27: 910–917.
Yu, K.-C., C.-C. Chen, and P.-C. Wu, Research on application and rehydration rate of vacuum freeze drying of rice. *J. Appl. Sci.*, 2011; 11: 535–541.
Zareifard, M.R., M. Niakousari, Z. Shokrollahi, and S. Javadian, A feasibility study on the drying of lime juice: The relationship between the key operating parameters of a small laboratory spray dryer and product quality. *Food Bioprocess Technol.*, 2012; 5: 1895–1905.
Zbicinski, I., A. Delag, C. Strumillo, and J. Adamiec, Advanced experimental analysis of drying kinetics in spray drying. *Chem. Eng. J.*, 2002; 85: 207–215.
Zenoozian, M.S., S. Devahastin, M.A. Razavi, F. Shahidi, and H.R. Poreza, Use of artificial neural network and image analysis to predict physical properties of osmotically dehydrated pumpkin. *Drying Technol.*, 2008; 25: 132–144.

6 Chilling

E. Dermesonlouoglou, Virginia Giannou, and Constantina Tzia

CONTENTS

6.1 INTRODUCTION

Food quality is determined by a combination of intrinsic, such as food's composition and physicochemical properties (i.e., moisture and nutrient content, pH), and extrinsic (storage/environmental conditions) parameters. Consequently, the retention of foods under inappropriate conditions can irreversibly alter their characteristics (Armstrong, 2004).

Chilling refers to storage of perishable goods at temperatures above freezing, at the range of 15°C to −2°C, depending on the products' nature (i.e., animal tissues, fresh milk and fruits, and vegetables not subject to chilling injury require temperatures just above freezing for maximum storage life, while several fruits, such as bananas and avocados, are best preserved at temperature above 7°C) (Karel and Lund, 2003). Most common commercial and household refrigerators usually operate at temperatures between 4°C and 7°C. The considerable advantage of chilling over other techniques is that it is classified among the mildest preservation methods with minor effects on the quality, textural, and sensory characteristics of food products. Its principal feature is that it slows down the growth and reproduction rate of food spoilage microorganisms; it retards lipids oxidation, impedes temperature-dependent activities, such as respiration, transpiration, and ethylene production, and inhibits enzymatic reactions, thus extending shelf life. For the reasons stated above, it is considered the most popular method for the preservation of fresh foods, especially meat, fish, dairy products, fruit, vegetables, and ready-made meals (Hui et al., 2004; Mishra and Gamage, 2007).

6.2 PRECOOLING

Most fresh produce start to degrade immediately after harvest. Respiration due to enzymatic oxidation occurs in plant tissues resulting in the consumption of sugars, starch, and moisture with the parallel production of gases (mainly carbon dioxide) and heat. Unless the heat is removed, these phenomena are accelerated along with the growth of spoilage microorganisms and loss of nutrients. Precooling usually precedes shipping, cold storage, or processing of foods, and its main purpose is to rapidly remove the undesired heat (known as field heat) from freshly harvested/slaughtered goods in order to retard the rate of the products' degradation. In this way it minimizes quality and product losses, extends the shelf life of highly perishable goods, expands their market opportunities, and also reduces the refrigeration capacity required for their transportation or storage. Precooling differs from cold storage where the temperature is simply maintained at a predetermined level, raising, however, the energy cost of the cooling process.

In order to minimize the refrigeration cost, there is a common practice not to precool products to the ideal temperature but to reduce it to half or 7/8 the desired value (usually the operating temperature of the coolant) and then transfer products to a cold room for further cooling. As a general rule, the 7/8 of the desired cooling time requires 3 times longer cooling when compared with half-cooling time.

Several precooling methods can be used to reduce the heat load of fresh produce. Current practices include room cooling, forced-air cooling, hydrocooling, ice, and vacuum cooling. These are described in the following paragraphs (Karel and Lund, 2003; Kienholz and Edeogu, 2002; Mishra and Gamage, 2007; Ramaswamy and Marcotte, 2006; Thompson et al., 2008).

6.2.1 Room Cooling

This is the simplest precooling method commonly applied in products sensitive to surface moisture with a relatively long storage life. It is a slow-cooling method in which products are exposed to cold air in a refrigerated room or transportation truck. Cold air is usually discharged into the room horizontally just below the ceiling, sweeps the ceiling, and returns to the cooling coils after circulating through the produce on the floor by the refrigeration fans (Figure 6.1). For satisfactory results, the air velocity should be kept between 60 and 120 m/min around and between cooling containers, while cooling can also be accelerated through the use of ceiling jets or cooling bays.

The major benefit of room cooling is that the produce is cooled and stored in the same place, thus requiring less handling. It can also be used for the curing of produce. In addition, it is characterized by simple design and operation and requires moderate refrigeration load. However, room cooling presents several disadvantages that may limit its use. It necessitates a relatively large empty floor space between stacked containers, enough open spaces between products, and good air speed to achieve the optimal cooling effect, otherwise cooling is unreasonably long. Nevertheless, excessively high air velocities are not recommended as they may lead to serious water loss for fresh produce.

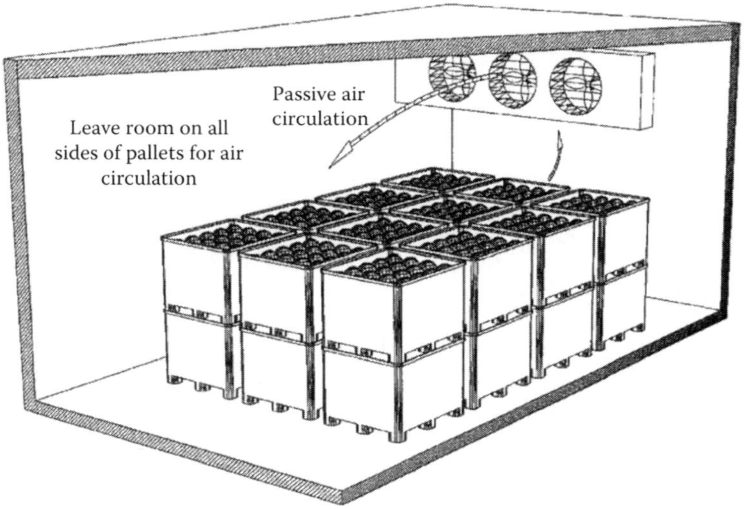

FIGURE 6.1 Room cooling diagram. (Adapted from Kienholz, J. and Edeogu, I., *Fresh Fruit and Vegetable Pre-Cooling for Market Gardeners in Alberta,* Alberta Agriculture, Food and Rural Development, Edmonton, Alberta, Canada, 2002. With permission.)

6.2.2 Forced-Air Cooling

It is the most versatile and widely used precooling method mainly used for bulk produce and palletized products. In forced-air cooling, high velocity cold air is distributed by air ducts and forced to flow through the containers in direct contact with each piece of produce. In this way, an air pressure difference is developed between the opposite faces of stacks of vented containers carrying heat away and resulting in rapid and even cooling. Cold air can be distributed either horizontally (Figure 6.2) or vertically (Figure 6.3). In horizontal flow systems, air is forced to flow horizontally from one side to the other through holes in the sides of the stacks. On the contrary, in vertical flow systems, airflow is vertically directed from the bottom to the top of the stacks through holes in the bottom. In both cases, the holes between containers should be aligned to facilitate airflow, while the opposite sides (top and bottom in horizontal and sides in vertical systems) should preferably be kept sealed to prevent air from bypassing the products. Different techniques are available in forced-air cooling. These include tunnel type, cold wall, serpentine cooling, and evaporative forced-air cooling.

The significant advantage of forced-air cooling over room cooling is that it can be 10 times faster, while it also enables better control of the cooling process. However, there is a high potential for water loss from the fresh produce due to air movement, unless high airflow rates are applied and humidity is kept near 100%. Condensation problems on the produce can be minimized by placing a cover on top of the stack of containers.

6.2.3 Hydrocooling

It is the most economic method of precooling, being simple and efficient at the same time. It is frequently employed in fruits and vegetables such as celery, asparagus, peas, radishes, carrots, and peaches. In this case, fresh products are immersed in, flooded with, sprinkled, or sprayed with cold water at a temperature of around 0°C, either in batch or continuous modes. Conveyor hydrocoolers are the most commonly used. In those produce, either in bulk or in containers, hydrocoolers are placed on a conveyor through a shower of water.

Hydrocooling prevails over the aforementioned techniques as far as the cooling speed is concerned because water has a higher heat removal capacity than air. However, water must be kept cold and

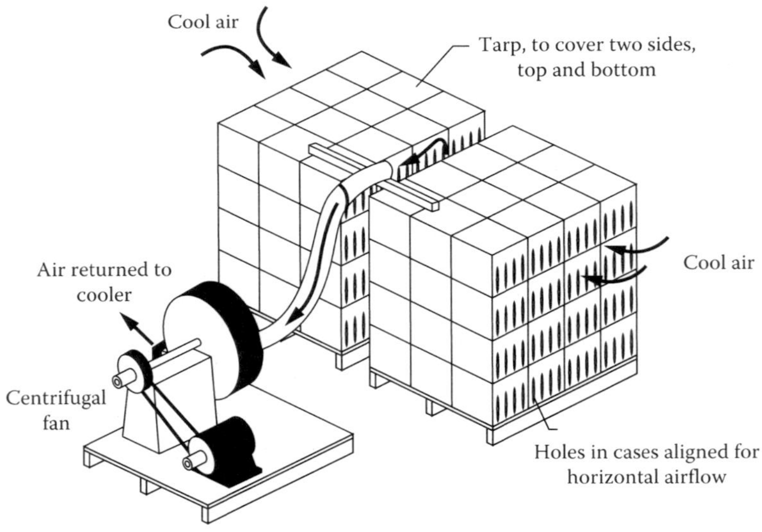

FIGURE 6.2 Forced-air cooling diagram (horizontal airflow). (Adapted from Kienholz, J. and Edeogu, I., *Fresh Fruit and Vegetable Pre-Cooling for Market Gardeners in Alberta,* Alberta Agriculture, Food and Rural Development, Edmonton, Alberta, Canada, 2002. With permission.)

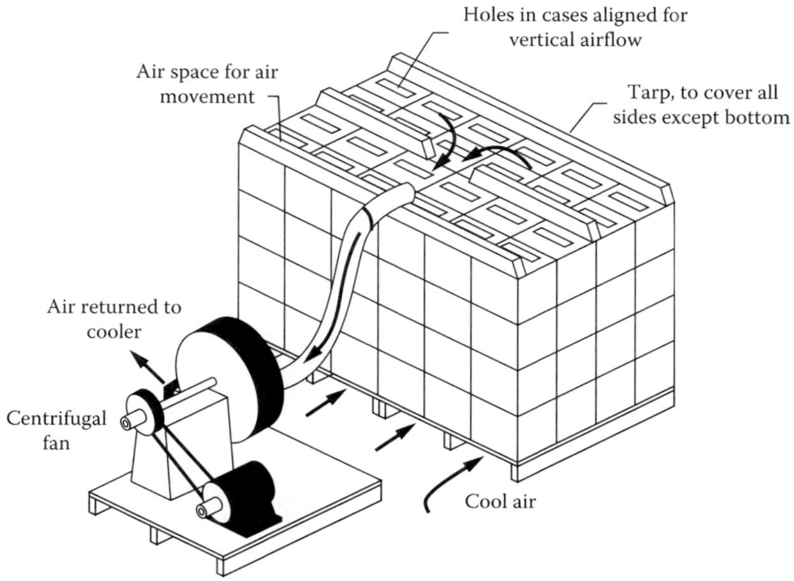

FIGURE 6.3 Forced-air cooling diagram (vertical airflow). (Adapted from Kienholz, J. and Edeogu, I., *Fresh Fruit and Vegetable Pre-Cooling for Market Gardeners in Alberta,* Alberta Agriculture, Food and Rural Development, Edmonton, Alberta, Canada, 2002. With permission.)

should come in contact with as much of the surface of the produce as possible. This is why it is not appropriate for commodities that cannot tolerate wetting such as strawberries, mushrooms, berries, potatoes, bulb onions, or garlic. For optimal cooling and energy saving, hydrocoolers should also be adequately insulated. Hydrocooling does not induce water loss. On the contrary, it may even revive slightly wilted produce. The major limitation of hydrocooling is that it requires particular attention to water quality and sanitation. Mild disinfectants, such as chlorine (in the form of sodium hypochlorite solution or as dry-powdered calcium hypochlorite), ozone, or hydrogen peroxide are used for this purpose. It also requires containers tolerant to water and chemicals, increasing therefore the packaging cost. Finally, in order to maintain the cooling effect, immediate rehandling of the products is usually necessary by shipping or transferring them to a cold storage room.

6.2.4 Ice Cooling

It is one of the oldest and fastest precooling methods commonly used for chilling field-packed vegetables such as broccoli, cabbage, cantaloupes, peaches, root crops (i.e., radishes and carrots), as well as poultry and fish. For this reason, ice in crushed or fine granular form, or ice-slush (an agitated mixture of ice and water) is packed around or placed in direct contact with the produce. Thereby, the cooling effect is maintained during and after transportation for a sufficient time since it is often a marketing requirement for several products that ice remains until they are received by the retailer.

Ice cooling also provides a high relative humidity environment for fresh produce. Thus, unlike precooling in air, it may result in an increase in the weight of several products such as poultry. It is even faster than hydrocooling since ice has a higher heat removal capacity than water (latent heat of ice melting is 334 kJ/kg). They both share, however, similar limitations. More specifically, water sanitation is of major importance as melted ice could be a potential source of contamination for fresh produce; chlorine is usually added to address this issue. The containers employed must be resistant to water for a prolonged time as well and should have drainage vents for the melted water. Finally, ice cooling is associated with considerably higher labor cost and freight load as the typical weight of the ice for initial cooling is equivalent to 30% of the product weight.

6.2.5 Vacuum Cooling

Vacuum cooling is the fastest and most uniform method of precooling. It is highly recommended for leafy and floral vegetables, such as lettuce and celery, with high surface-to-volume ratio and ability to readily release internal water. In this case, fresh produce are loaded in a vacuum chamber, equipped with a vacuum device (mechanical vacuum pump or steam-jet pump) and a condenser (necessary when vacuum is achieved through a mechanical vacuum pump), and air is drawn out creating a high vacuum (4–4.6 mmHg). This reduction in the atmospheric pressure causes a substantial decrease in the boiling temperature of water generating moisture evaporation which withdraws heat from the products. The extent of cooling is proportional to the water evaporated (evaporation of 1 kg of water removes approximately 2200 kJ), while the temperature is lowered to about 5°C for each 1% reduction in water content.

Precooling times in vacuum cooling usually range between 20 min and 2 h if appropriate loading and packaging is applied. However, it is considered as an expensive method, requires high capital investment and skilled operators, and is designed for batch operation only. However, this can be partly offset by their ability to be portable and readily available for use. Therefore, in order to be economically advantageous, vacuum coolers should be preferably built in productions plants with a large annual workload and as closely as possible to the harvest area (if not portable). Finally, with regard to moisture loss (2%–3%) induced during cooling, this can be overcome either by pre-wetting fresh produce before vacuum is applied or by using a water spray system during cooling (hydrovacuum cooling). In both cases, particular attention should be given to water hygiene.

6.2.6 Precooling Method Selection Criteria

Apart from the particular characteristics of each product, the selection of the appropriate precooling method depends on several factors, such as the temperature reduction requirements, the refrigeration load, the desirable cooling rate, and equipment and operating costs. Therefore, in cases where multiple products with diverse optimal cooling conditions are received, the use of different precooling means may be required. As far as the capital cost is concerned, liquid ice coolers can be the most expensive, followed by vacuum coolers, forced-air coolers, hydrocoolers, and room coolers. Moreover, vacuum coolers exhibit the highest energy cost, followed by hydrocoolers, ice coolers, and forced-air coolers. Labor, maintenance, and other equipment costs should also be contemplated when comparing different cooling systems, especially if special packaging (e.g., waxed boxes or reusable plastic containers) is required. Operating conditions (environmental temperature and humidity) and product parameters (type, size, shape, and composition) should also be considered (Pham, 2001).

6.3 COOLING

6.3.1 Cooling Methods

Heat can be withdrawn from an object by convection, conduction, radiation, or evaporation. In convection, heat is transferred due to the fluid motion of a gas or a liquid. As the hotter part of the fluid is lighter, it is forced to rise upward while the colder part is moving downward, thus creating convection currents that relocate heat. In case this flow is not spontaneous but induced by other means, it is called forced convection. In conduction, heat is transferred within an object through the contact of particles (i.e., vibration of atoms and molecules). Heat can also be carried through electrons, which explains why metals are in general good conductors of heat. It does not involve any motion of a substance but rather is a transfer of energy within an object or between objects in contact. This energy travels from the more energetic molecules, which are found at the hotter areas, to those in contact that exhibit lower molecular energy. In radiation, heat is removed through electromagnetic waves, without the need of a transfer medium, based on the fact that warmer objects emit larger amounts of energy than they receive. Finally, in evaporation, heat is extracted by water evaporation from the surface of the product to the environment. Conduction and convention, however, are predominant in food applications, while combinations of two or more methods may occur (James, 2006; Jensen et al., 2004).

6.3.2 The Refrigeration Cycle

Most common chilling equipment used to remove heat is classified into mechanical or cryogenic refrigerators. The latter use solid carbon dioxide (dry-ice pellets), liquid carbon dioxide, or most commonly in chilling operations, liquid nitrogen. These can either be used to chill another refrigerant or directly chill the products. However, due to their significantly higher cost, cryogenic refrigerators have a limited range of applications (i.e., in mechanically formed meat products, cryogenic grinding, and dough products).

The mechanical refrigeration system is characterized by the circulation of a fluid (called refrigerant) in a closed cycle and consists of four basic components namely a compressor, a condenser, an expansion valve, and an evaporator (Figure 6.4). Any other component, except these, is considered an accessory. The refrigerators' parts are manufactured from materials presenting high thermal conductivity, such as copper, in order to achieve high rates of heat transfer and high thermal efficiencies.

The refrigeration system can be divided into the high pressure area (compressor and condenser) and the low pressure area (evaporator and expansion valve). The refrigerant, in the vapor state, is passed with suction through the compressor, which is typically a vapor compression pump that increases pressure. It is compressed at constant entropy and exits the compressor superheated. It is then sent to the condenser (a heat exchanger) where it is cooled releasing heat and condensed into liquid at constant pressure and temperature. The liquid refrigerant is then routed to the expansion valve where it is forced to pass through a small hole. As a result, its pressure quickly drops; the boiling point is lowered and evaporation is therefore facilitated resulting in a mixture of liquid and vapor at a lower temperature and pressure. The cold liquid–vapor mixture is directed through the evaporator coil or tubes, under reduced pressure, where it is completely vaporized by cooling the warm air from the space being refrigerated. The resulting saturated vapor is finally routed back to the compressor inlet to complete the

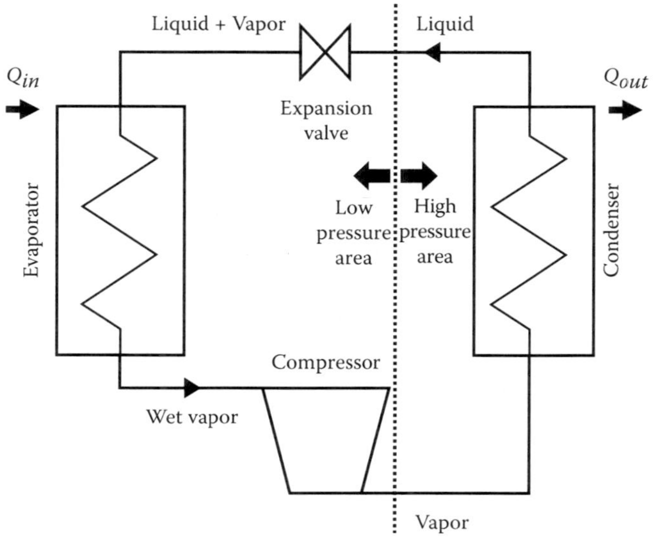

FIGURE 6.4 Schematic diagram of the mechanical refrigeration system.

thermodynamic cycle. The refrigerant repeats this cycle again and again, absorbing heat from one area (cold room) and relocating it to another (condenser). This of course remains the most simplified form of a refrigeration system, and more sophisticated approaches have been developed. Different types of refrigerators' parts are also available, according to load requirements and characteristics. These are explained in the following paragraphs (Fellows, 2000; Heap, 2000; Ibarz and Barbosa-Cánovas, 2003).

6.3.2.1 Compressors

Compressors can be distinguished as positive displacement and dynamic compressors. In positive displacement compressors, the vapors of the refrigerant are drawn and trapped inside, while the rise in pressure is achieved by decreasing the volume. They can be further categorized according to power requirements as

- Rotary (<30 kW)
- Reciprocating (10–300 kW) and
- Screw compressors (100–500 kW)

Rotary compressors are designed for very small to small capacities, such as domestic refrigerators, and are suitable for refrigerants with moderate or low condensing pressures (i.e., R-21 and R-114). These can either be of the rolling piston or the rotating vane type. Reciprocating compressors are best suited in small to medium capacity systems and are very popular due to their lower cost. Along with screw compressors, they best operate with refrigerants that exhibit comparatively low volume per kg and condense at relatively high pressure (i.e., R-12, R-22, and ammonia).

The most common representatives of dynamic compressors are centrifugal compressors, in which kinetic energy is produced by impellers turning at a high speed and converted into pressure energy through a diffuser. These are used for large capacities (>200 kW) and are suitable for refrigerants with large displacement and low condensing pressure (i.e., R-11 and R-113). Although their capital cost is higher, they are advantageous in that they have lower maintenance costs and occupy less space (Arora, 2010; Maroulis and Saravacos 2003; Pardo and Niranjan, 2006).

6.3.2.2 Condensers

The most common types of condensers are

1. Air-cooled condensers, in which air is used as the cooling medium. These can rely either on natural air circulation or forced convection (i.e., with propellers or fans) with the airflow being either vertically upward or horizontal.
2. Water-cooled condensers, which use water for cooling and can be further grouped into
   a. Tube-in-tube or double pipe, where concentric pipes with different diameters are used and shaped to the desired form (i.e., straight or coiled). Refrigerant flows in one pipe and water on the other pipe (usually the inner one) in counterflow. These systems are mainly used for smaller scale applications.
   b. Shell-and-coil, which consist of a continuous coil inside a welded or flanged outer shell. Water flows in the coil, while the refrigerant vapors are condensed inside the shell.
   c. Shell-and-tube, where a group of straight tubes is mounted inside a cylindrical shell. Water runs on the inner side of the tubes, while the refrigerant condenses inside the shell and around those tubes. This is the most popular type of condenser.
3. Evaporative condensers, which use both air and water as the cooling medium. Water is sprayed over the condenser tubes where the refrigerant vapors flow, while air is thrusted through fans from the bottom of the condenser to the top. As water comes in contact with the hot tube surface, it removes heat from the refrigerant and evaporates into steam (Hundy et al., 2008; Trott and Welch, 2000).

6.3.2.3 Evaporators

Most evaporators cool air or a liquid which is subsequently used to remove heat from the required load. They are classified into

- *Dry expansion evaporators (also called direct expansion or DX evaporators)*: The liquid refrigerant enters the evaporator (through an expansion valve) and as it runs through it, within tubes or coils, it is completely vaporized and superheated to a certain degree before reaching the exit. The air or liquid to be cooled passes over the outside of those tubes or coils. These are the most widely used.
- *Flooded evaporators*: They consist of a vessel (shell) containing the refrigerant, while the liquid intended for cooling flows within tubes or channels either immersed or in direct contact with the refrigerant. As indicated by their names, these operate at a constant refrigerant liquid level regulated by a float valve which compensates the amount of liquid lost due to evaporation. These evaporators are considered more efficient since they provide a larger heat transfer surface in contact with the liquid refrigerant. However, due to higher operational cost, they are mostly applicable in larger units (Arora, 2006; Stoecker, 1998; Wang, 2000; Whitman et al., 2009).

6.3.2.4 Expansion Devices

Their primary function is to remove pressure from the liquid refrigerant in order to allow/control expansion or change of state (from liquid to vapor) in the evaporator. They can be grouped into two types as follows:

1. *Variable restriction*: In these valves, the extent of opening and therefore the flow of the refrigerant can be modulated in response to operational parameters, such as liquid level, pressure, or temperature. They can be further distinguished into
   a. *Thermostatic expansion valves (TXV)*: They mainly comprise of a power head which contains a diaphragm, a capillary tube, and a sensor bulb connected on top of the power head, and the main body which accommodates a superheat adjustment spring as well as a valve seat and needle (Figure 6.5). These are dynamic valves, designed to regulate the flow of the liquid refrigerant according to the heat load requirements of

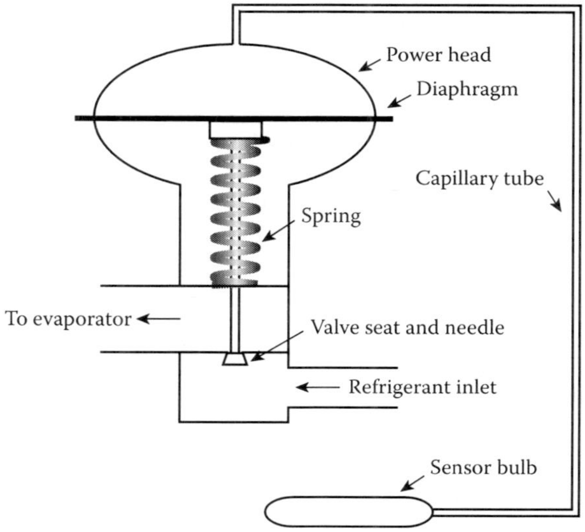

FIGURE 6.5 Simplified diagram of the thermostatic expansion valve.

the evaporator (by maintaining a constant superheat of the refrigerant at the outlet of the evaporator), while they simultaneously prevent its flood back to the compressor. When the liquid portion of the refrigerant in the evaporator rises, the temperature in the suction line drops. This triggers the sensor to release pressure on the diaphragm and therefore close the valve. Due to their automatic operation and high efficiency, they are the most commonly used. Electronic expansion valves (EEV) are a much more sophisticated type of TXV with their operation being controlled through signals sent by an electronic controller.

b. *Automatic expansion valves* (*AXV*): They consist of a diaphragm that controls the operation of a needle valve through an adjustable spring placed on top of it. They are designed to maintain a constant pressure in the evaporator. Therefore they permit the flow of the refrigerant through them when the pressure inside the evaporator drops and restrict the flow when the pressure increases. These types of valves should preferably be used in systems with constant loads, while both TXV and AXV are recommended for dry expansion evaporators.

c. *Low-side float expansion valves*: These along with high-side float expansion valves are used with flooded evaporators and differ in the placement of the valve. They consist of a hollow ball with an arm which floats on the liquid refrigerant. This is connected to a needle valve located at the inlet of the float chamber on the low-pressure side of the system. As the refrigerant evaporates, its level descends dragging the float downward; the motion being transmitted to the needle valve, which opens allowing more liquid to enter the chamber.

d. *High-side float expansion valves*: In this case, the liquid chamber with the float ball is located at the high-pressure side of the system, between the condenser and evaporator, while the valve seat and needle are found at the outlet of the chamber. As the level of the liquid refrigerant arriving from the condenser rises, the valve opens to allow it to pass into the evaporator. This is, of course, based on the principle that the refrigerant vapor is condensed in the condenser at the same rate at which the liquid vaporizes in the evaporator.

2. *Constant restriction*: These typically consist of a capillary tube, which can be described as a long "hair-like" tube, coiled to several turns in order to occupy less space, with a bore diameter of 0.5–2.28 mm (0.02–00.090 in.). The pressure drop achieved in the refrigerant flow is proportional to the length of the capillary tube and inversely proportional to its bore diameter. They are the simplest and less expensive types of expansion devices; however, they are not adjustable to considerable load changes and can be susceptible to clogging by foreign particles. They are most commonly used in small units, such as domestic refrigerators/freezers or water coolers and air conditioners (Arora, 2006; Stoecker, 1998; Whitman et al., 2005).

6.3.3 Refrigerants

They are the primary working fluids used in a cooling mechanism. They absorb heat at a low temperature and low pressure and release it at a higher temperature and pressure. During this cycle most refrigerants undergo phase changes from liquid to gas and vice versa. They can either be single chemical compounds or a mixture of multiple compounds, usually in a fluid form (ASHRAE, 2009; Wang, 2000).

The ideal refrigerant would exhibit the following properties:

- Convenient thermodynamic properties such as low boiling point, high latent heat of vaporization, moderate density in liquid form, high density in vapor form (to achieve a reduced compressor's size), and high critical temperature.

- Reasonable working pressure, preferably a little higher or below atmospheric pressure.
- It should not be flammable or poisonous/toxic and should be easily detected in case of leakage.
- Be noncorrosive and compatible with common design materials and mechanical components used in a refrigeration system.
- It should not induce the decomposition of the lubricating oil used and their miscibility to be low.
- Be chemically stable at working conditions and easy to handle.
- Cause the least possible environmental impact (low ozone depletion (ODP) and global warming (GWP) potential).
- Be of low cost.

However, none of the existing refrigerants meets all of the criteria mentioned above. The most important refrigerants used are described in the following paragraphs. Apart from their chemical formula, they are also known by their three-digit number designation (proposed by the DuPont Company) in which the first digit (from left to right) represents the number of carbon atoms in the molecule minus one (when zero it is omitted), the second represents the number of hydrogen atoms in the molecule plus one and the third one, the number of fluorine atoms. This applies in the case of hydrocarbons and halocarbons, while for inorganic compounds, the last two digits correspond to their molecular weight. When two refrigerants have the same code, based on the number of their elements, a letter is added at the end which represents how symmetrical the molecular arrangement is. Other coding has also been proposed for blends of refrigerants (Fellows, 2000; Trott and Welch, 2000).

6.3.3.1 Ammonia (R-717)

Ammonia as well as hydrocarbons, CO_2, water, and air are considered a "natural refrigerants" as they are naturally occurring, non-synthetic substances, that do not deplete the ozone layer and make no or negligible contribution to global warming. It has long been used as a refrigerant for industrial applications and remains the most predominant as it has excellent heat transfer properties and is not miscible with oil. Leakage of ammonia can also be easily detected due to its characteristic, objectionable odor, and as it is lighter than air it moves upward. However, it is toxic, flammable, and corrosive on copper or copper alloys.

6.3.3.2 Carbon Dioxide (R-744)

Carbon dioxide is an odorless, colorless, nonflammable, and nontoxic refrigerant which makes it safer for use, for example, on refrigerated ships. It is also compatible with almost all metallic materials and polyester oil. Furthermore, it is inexpensive and widely available worldwide. However, it requires considerably higher operating pressures compared to ammonia, while in high concentration may cause asphyxiation (safety limit is set at 0.5% in air, while it is immediately dangerous to life at >4%) (Padalkar and Kadam, 2010).

6.3.3.3 Hydrocarbons (HCs)

They are low-toxic substances containing only hydrogen and carbon. The most characteristic of them are ethane (R-170), propane (R-290), propylene (R-1270), butane (R-600), and isobutene (R-600a). They have been introduced, mainly in Europe, as replacers for chlorofluorocarbon (CFC) refrigerants. They can be applied in small, sealed refrigerant systems, such as domestic refrigerators. However, due to their highly flammable characteristics, they are not easily recommended for large-scale industrial applications.

6.3.3.4 Halogen Refrigerants (CFCs)

Halogen refrigerants (chlorofluorocarbons) are nontoxic, nonflammable, and chemically very stable compounds containing chlorine, fluorine, and carbon. They have good heat transfer properties and lower costs compared to other refrigerants. However, their long atmospheric lifetimes, ranging between 60 and 400 years, allow them, when they reach the stratosphere, to interact with ozone molecules. CFC molecules also absorb and emit radiation in the infrared part of the spectrum (greenhouse gases), thus contributing to global warming. These resulted in the international ban on their use as refrigerants (definitely in 1995) under the Montreal Protocol. Trichlorofluoromethane (R-11) and dichlorodifluoromethane (R-12) were the ones more extensively used.

6.3.3.5 Partially Halogenated CFCs (or HCFCs)

They were introduced and used for a limited period as substitutes for CFCs. They contain a small amount of chlorine, fluorine, carbon, and hydrogen which makes them less stable in the atmosphere and enables them to partly break down in the lower levels before reaching the stratosphere. Chlorodifluoromethane or difluoromonochloromethane (R-22) has been the most popular among them but is expected to be phased out completely by the year 2020.

6.3.3.6 Hydrofluorocarbons (HFCs)

These refrigerants have been developed as alternatives to CFCs and HCFCs. They are partially fluorinated methanes and ethanes and do not contain chlorine atoms but only hydrogen, fluorine, and carbon. Although their atmospheric lifetimes in the lower atmosphere are even shorter than those of HCFCs, they are still in question as they are considered as greenhouse gases. Most popular among these is CF_3CH_2F (R-134a) which was introduced as an alternative for R-12 systems with some modifications (mainly oil change). It is noncorrosive, nonflammable, and exhibits extremely low toxicity, but because it requires polyolester-based synthetic lubricants, which are hygroscopic, it necessitates careful control of moisture. It is best suitable for medium temperature applications.

6.3.3.7 Azeotropic Mixtures

These are usually binary mixtures of substances which individually have different properties but behave as a pure fluid when mixed together and cannot be clearly separated through distillation. Their composition does not substantially vary between the liquid and vapor state or change in case of leak. Common azeotropic refrigerants include R-502 (48.8% R-22 and 51.2% R-115), which can be replaced by R-507 (50% R-125 and 50% R-143a).

6.3.3.8 Near-Azeotropic Mixtures

These mixtures consist of two or more refrigerants with different boiling points. They present a very low temperature glide during evaporation or condensation and very small differences between their liquid and vapor composition. R404A and R407C are two of the most popular near-azeotropic refrigerant mixtures. R404A (44% R-125, 52% R-143a, and 4% R-134a) is used for medium and low-temperature refrigeration applications, while R407C (23% R-32, 25% R-125, and 52% R-134a) can be applied in supermarket refrigeration, refrigerated transports, and food processing.

6.3.3.9 Non-Azeotropic (or Zeotropic) Mixtures

They are multicomponent mixtures with varying composition during evaporation or condensation and a temperature glide greater than 10°F. They offer the advantage of modifying the composition to suit various temperature requirements; however, they require more strict control to prevent leakage.

6.3.3.10 Secondary Refrigerants

They are used as an intermediate medium in order to facilitate heat transfer between the object to be cooled and the primary refrigerant and reduce refrigeration cost. They should preferably have a low freezing point and high thermal conductivity, and not be flammable, toxic, or corrosive. Common materials used for this purpose are water, air, brines (i.e., with sodium chloride, magnesium chloride, or calcium chloride), glycols (propylene glycol or ethylene glycol), and oils (Fellows, 2000; Heap, 2000; Ibarz and Barbosa-Cánovas, 2003; Wang, 2000; Whitman et al., 2009).

6.3.4 Chill Storage

After precooling, food products are transferred to cold storage for further processing or maintenance prior to retail display. In several cases, *in situ* cooling is not feasible or not absolutely necessary and can be performed during transport or directly in the processing or storage plants.

6.3.4.1 Transport Prior to Chill Storage

Perishable products can be transported by trucks, trains, airplanes, or ships. Water ice or dry ice is used in suitable containers usually for short distances. Liquid nitrogen or liquid carbon dioxide can also be applied. These are stored in properly insulated tanks under low pressure and released to cool the air inside the truck/chamber when the pressure rises above a certain limit. Another alternative refrigeration method is the use of cold plates around the truck walls filled with an eutectic solution (usually a brine). This is frozen before loading and as it absorbs heat from the products changes its state from solid to liquid. Finally, mechanical compression systems are also used. These are either generated by the vehicle's engine, and work only when the truck is in operation, or are powered by an independent engine mounted on the vehicle. Important considerations that should be taken into account during transport are adequate loading and packaging to allow airflow between products as well as insulation and seal to prevent heat uptake from the environment (Mishra and Gamage, 2007).

As far as cold storage is concerned, in order to achieve optimal operation both constructional and operational parameters should be carefully designed and inspected.

6.3.4.2 Constructional Parameters of Cold Rooms

Constructional parameters include building design and layout. Cold stores typically consist of a refrigerated chamber (possibly with different compartments) designed to chill and store perishable products. Separation of products is necessary when different storage temperature and/or relative humidity is required or in the case of products with odor migration problems (Hundy et al., 2008). Prefabricated industrial panels of the sandwich type can be used for their construction, which contain an insulation material injected between metal claddings made of steel (stainless or galvanized) or aluminum. The panels are tightly joined through a slip joint system. Polyurethane foam is most frequently used for insulation; however, cork or polystyrene can also be applied. In general, materials with high impermeability are preferred as they can also serve as vapor barriers and control migration of water vapor. Paints when applied should contain antimicrobial compounds, while the floor should preferably be coated with anti-slippery films (i.e., polyurethane resins) of high resistance in chemicals or aggressive compounds, microbial contamination, and mechanical damage in order to achieve high sanitation/hygiene and cleanability standards. Bricks or stones and concrete can be used as an outer cell. Cold stores are most commonly cooled through cold air circulation generated by a mechanical refrigeration unit.

The dimensions and therefore the capacity of a refrigerated storage should be carefully calculated according to the expected storage time and load, taking into account the space required for equipment, stacking material, easy handling of products, and comfortable movement of the personnel. Based on the capacity, cold rooms can be classified into walk-in cold rooms and warehouse cold stores. The first are mainly used in the retail industry and are designed with limited construction requirements for smaller scale applications such as supermarkets. Warehouse cold stores, on the contrary, are built up for heavy duty industrial applications and are usually equipped with multiple refrigeration units. In any case, walls, ceiling, and floor material should provide maximum insulation and ease of cleaning and maintenance. It is preferable to be equipped with the same insulation material. Doors should also provide airtight closing. Several types of doors' design are available, such as sliding, hinged, hydraulic, or controlled atmosphere doors (Pardo and Niranjan, 2006).

6.3.4.3 Operational Parameters of Cold Rooms

Operational parameters mainly include the control of temperature, moisture levels, air distribution, lightning, and personnel traffic (Ramaswamy and Marcotte, 2006).

6.3.4.3.1 Temperature Control

Temperature control is considered a key factor during cold storage as most activities that influence the shelf life and quality of fresh produce, such as respiration, transpiration, and microbiological, enzymatic, and chemical activity, are temperature dependent. Therefore in order to prevent deterioration and prolong storage time, food products should preferably be stored at the lowest possible temperature without causing chilling injury. The optimum storage temperatures for fresh fruits and vegetables can be found in the literature (Aked, 2002; Gross et al., 2004; Salunkhe et al., 1991; Tabil and Sokhansanj, 2001). For most perishable goods and chilled products, storage temperatures of less than 5°C are usually recommended, while the growth of most food-poisoning microorganisms is halted below 3°C.

However, most often, it is not feasible or economical to individually store each product. In these cases, it is recommended to keep the temperature limits closer to the optimum for most sensitive or perishable products. Dairy and meat products in general require lower storage temperatures compared to fresh fruits and vegetables. Care should also be taken to avoid co-storage of incompatible products. For example, fresh fish or citrus fruits should not be closely stored with dairy products or meat as odor migration may seriously diminish the sensory and quality characteristics of the latter. Temperature fluctuations must also be controlled to less than 1°C–2°C as they can be detrimental for the quality of food products and are associated with moisture migration. These are not only caused by problems in the operation of the refrigeration equipment but can also be due to excessive doors opening and staff activity inside the cold room or co-storage of products with large temperature differential. Fluctuations in temperature can be diminished by providing adequate cooling capacity, maintaining uniform air distribution, and minimizing the temperature differential between the evaporator coil and air temperature. Temperature controllers and recorders should be used, located away from hot or cold spots such as doors or fans (Trott and Welch, 2000).

6.3.4.3.2 Moisture Control

Along with temperature, moisture control is also important during cold storage as it is directly associated with critical safety and quality aspects, such as the transpiration rate (moisture loss) of fresh fruits and vegetables or microbial growth. When the moisture levels inside the cooling chamber are excessive, the risk of microbial growth or rotting increases. Condensation of water vapors on the cold room walls can also lead to mold buildup. On the contrary, moisture loss is associated with economic damage and quality degradation (i.e., dehydration or skin cracking in fruits and vegetables). Again optimum moisture levels differ between products and co-storage of goods with different requirements should be avoided. For example, most fruits and vegetables require high humidity levels during cold storage (>85%). This does not apply though for dried fruits, onions, and garlic which require low moisture levels (<70%) otherwise their rotting process accelerates. Low moisture is also desired in meat products in order to avoid slime formation.

Best practices to avoid excessive moisture loss or gain are through the control of the temperature difference between the evaporator coil and the air inside the room, and temperature variations. High humidity levels are obtained by using an evaporator with a high surface area and maintaining the refrigerant at its highest possible temperature. Water loss can additionally be manipulated in several cases through adequate packaging, waxing, or use of edible coatings (Grandison, 2006).

6.3.4.3.3 Atmospheric Composition

Especially with regard to the microbial growth during cold storage, apart from adjusting the temperature and humidity limits, the atmospheric composition is also considered important. The germination of certain decay microorganisms can be hindered by controlling the oxygen levels in their environment. Therefore, cold rooms working under controlled atmosphere are often proposed. Carbon dioxide or carbon monoxide has been used to replace oxygen, with encouraging results regarding the preservation of the qualitative characteristics of certain fresh produce. More specifically, low oxygen and high carbon dioxide levels may preserve color and texture and slow ripening and respiration rate as well as the development of specific storage disorders, such as scald in apples or woolliness in nectarines (Zhou et al., 2000). Typical controlled atmospheric concentrations include 2%–3% oxygen and 5% carbon dioxide. It should be mentioned however that, in some cases, it may promote internal discoloration, off-flavors formation, or irregular ripening in fruits and vegetables, while protective equipment (oxygen masks) is required for personnel working inside the cold room to prevent suffocation.

6.3.4.3.4 *Airflow Distribution*

Airflow distribution should be uniform across the room. This is accomplished both by appropriate design and loading. Lanes of 10–15 cm should be allowed between containers or pallets and walls, while stack alignment should be perpendicular to the airflow direction. Airflow speed should be adjusted according to load requirements, the characteristics of the products being stored, and the packaging used. Too rapid air circulation can result in the dehydration of certain products that are susceptible to moisture loss, while too low airspeeds result to temperature stratification and ineffective cooling. However, fans' operation is proposed to be continuous and, when the desired temperature is reached, to reduce their speed. Adequate ventilation or air filters should also be applied to clean air especially when the products to be stored bring out strong odors or are sensitive to odor uptake due to migration. Such filters may absorb several air contaminants or volatile organic compounds (VOC's) and trap excessive moisture as well as gases (i.e., ethylene) produced by fruits and vegetables, thus retarding spoilage. These are usually made of activated carbon.

6.3.4.3.5 *Lighting*

The illumination of cold rooms should follow all the standards that are generally applicable to the food industry. Lighting fixtures should be adequately designed and manufactured so as to prevent products' contamination with foreign material, leaks, corrosion, accumulation of bacteria, fire, or electrical problems. Lighting should also be uniform and facilitate working and movement of the personnel, allowing them to perform tasks in a safe, efficient, and productive manner. Lighting systems should preferably offer high energy efficiency and require minimum maintenance. Apart from these, cold rooms require waterproof lighting, able to withstand increased moisture conditions, and operate satisfactorily under low temperatures. The lights should only be switched on when work is carried out inside the cold chamber and remain switched off during the rest of the time in order to impede photochemical reactions and sprouting in some vegetables. As mentioned before, electric motors, lights, and personnel may generate heat inside the cold room and therefore increase its internal load. For lighting, heat dissipation is proportional to their wattage and working hours.

6.3.4.3.6 *Personnel*

Personnel moving in and out the cold room should conform to good hygienic practices and be properly trained. In several cases, special clothing is also required. Input frequency in the cold chamber and the residence time in it should be kept to a minimum to avoid cold stress in the employees and excessive heat generation in the room. Heat dissipation by personnel ranges, in average, between 200 and 300 W per person and depends on several factors such as body type and weight, level of activity, clothing, and room temperature (Ramaswamy and Marcotte, 2006).

6.3.5 Retail Display

Refrigerated display cabinets are extensively used in retail stores (i.e., supermarkets or convenience stores) as they may serve multiple purposes, such as preserving and merchandizing products, extending their shelf life, and attracting consumers. Different types of display cabinets are commercially available according to temperature and loading requirements. They can be of chest type or most commonly of upright type with shelves and be equipped with either an integral or a nonintegral (when parts of the refrigeration system are located at a different location from the cabinet) refrigeration system. Integral systems are more versatile and easy to relocate. Their main drawback, however, is that heat is directly rejected from the condenser into the surrounding area, which can be undesirable especially in warm climates during summer. They are also more susceptible to grease and dust buildup or to attract insects (Whitman et al., 2009).

Furthermore, display cabinets can either be of open or closed display type. Open type cabinets are more appealing and handy for consumers as they can easily reach products. However, they are less efficient, more susceptible to excessive moisture condensation, and usually more energy consuming than the closed type. The latter can either be designed with hinged or sliding doors. Hinged doors tend to close and seal more efficiently, but they exhibit comparatively higher infiltration heat load as warmer air from the external environment is more easily dragged and pushed inside the cabinet (Evans, 2014).

Display cabinets are most susceptible to temperature or humidity fluctuations, especially during rush hours and loading, and suitable sensors/controllers should be used for their monitoring (Fellows, 2000).

6.4 HEAT AND MASS TRANSFER CONSIDERATIONS DURING COOLING PROCESS AND STORAGE

In order to provide food products of superior sensorial, quality, and safety characteristics, attention must be paid to every aspect during the cooling process and storage. However, uniform storage conditions in cold stores are difficult to attain in practice. Several studies have shown temperature or moisture heterogeneity, with nonuniform airflow in cold rooms (Chourasia and Goswami, 2007; Mirade and Daudin, 2006), in refrigerated trucks (Moureh et al., 2009), in display cabinets (Laguerre et al., 2002, 2011, 2012), or in domestic refrigerators (Laguerre et al., 2002, 2010). This uneven distribution of airflow is related to the presence of the product and the cooling equipment (Ho et al., 2010). Variation of heat transfer coefficient between the air and the product at different positions in the cold room was observed leading to different product cooling rates (Mirade, 2007). Although some novel cooling technologies, such as vacuum cooling, have been introduced to the refrigeration industry, cooling in the present industrial practice is still usually accomplished by using air blast coolers, water immersion coolers, or slow air coolers (cold rooms) which can offer different cooling environments (Heap, 2000).

Heat and mass transfer in cold rooms is a complex phenomenon because of the presence of the product (i.e., airflow modification, heat of respiration) and the coupling between heat transfer and airflow. The heat transfer phenomena involved during product cooling and storage are conduction (within the product), convection (between cold air and product surface), and radiation (between product surface and cold room walls) (Hu and Sun, 2000). Another source of temperature heterogeneity is the heat of respiration of the product. Simultaneously, moisture evaporation from the product surface can be significant, which causes product weight loss. Both the heat and moisture transfers are influenced by flow characteristics (such as cooling air temperature and velocity), air properties (viscosity, density, conductivity, and specific heat), product properties, shape, dimension, and arrangement of the load. The comprehension of the heat/mass transfer and airflow in cold stores is a complex task (Smale et al., 2006). Failure to understand the phenomena taking place in the equipment can result in excessive weight loss, reduced shelf life, or deterioration in product quality. This deterioration rate is more significant at the warm zone due to high product respiration rate or at the cold zone due to chilling injury (James, 1996).

6.5 COOLING LOAD CALCULATION

6.5.1 Introduction

In order to calculate the refrigeration load, several parameters must be taken into account. These include product load and initial temperature, the desired temperature shift, the required cooling rate, the refrigeration capacity and insulation of the coolant, and the heat generated from other sources such as motors/equipment, lighting, and personnel. An important factor is also products' properties (thermal conductivity, density, and specific heat), size, shape, weight and surface-to-volume ratio, the surface heat transfer coefficient, and the thermal resistance of the packaging (if any) (Mishra and Gamage, 2007; Pham, 2002). These are explained in the following paragraphs.

The cooling load Q (kJ), that is, the heat that is removed from the product upon cooling by ΔT (°C or K), is calculated from Equation 6.1:

$$m_p C_p \Delta T = Q \tag{6.1}$$

where

m is the mass of the product (kg)

C_p is its specific heat (kJ/kg K)

Average values for specific heat of different foods at different temperatures are given in the literature. Empirical relations for the determination of the specific heat by the food composition have also been given (e.g., the empirical "mixing rule," Equation 6.2):

$$C_p = X_w C_{pw} + X_c C_{pc} + X_f C_{pf} + X_s C_{ps} \tag{6.2}$$

where

X_w, X_c, X_f, X_s are the mass fractions

C_{pw}, C_{pc}, C_{pf}, C_{fs} are the specific heats of water, carbohydrate, fat, and salt components of the food product

Typical values of specific heat for foods used in refrigeration and freezing calculations are the following: High-moisture foods above freezing ($T > 0$°C): 3.5–3.9 kJ/kg K; High-moisture foods below freezing ($T < 0$°C): 1.8–1.9 kJ/kg K; Dehydrated foods: 1.3–2.1 kJ/kg K; Fat ($T > 40$°C): 1.7–2.2 kJ/kg K; Fat ($T < 40$°C): 1.5 kJ/kg K. The enthalpy of foods, such as fruits and vegetables, meats, and fats, are given in the Riedel diagrams as a function of moisture content and temperature (Maroulis and Saravacos, 2003; Saravacos and Kostaropoulos, 2002).

6.5.2 Modeling Product Heat Load during Cooling

In designing a refrigeration system, several other sources of heat, in addition to the demand by the food itself, should be considered. Table 6.1 summarizes some of the most relevant factors.

TABLE 6.1
Basic Sources of Heat to be Considered in a Cold Room Design

Source of Heat	Considerations
Heat transferred by the food material	Sensible heat; respiration heat (applied to refrigerated fruits and vegetables); latent heat (when freezing occurs)
Heat introduced through ceiling, floor, and walls	Conduction of heat through walls and insulation materials
Heat transferred by engines working inside the cold room	Working time; number of engines; total power
Heat transferred by illumination bulbs	Working time; number of bulbs; total power
Heat transferred by people	Working time; number of people
Heat introduced by air renewal	Number of times the door is opened; other programming activities

Considering that the air is completely miscible with uniform temperature and the change in air density does not change with temperature change, the general form of Equation 6.1 is given for the cooling load calculation as (Equation 6.3):

$$m_pC_p\Delta T = \sum(Q_{introduced\ by}) - \sum(Q_{transferred\ by}) \tag{6.3}$$

6.5.3 Heat Transferred by the Food Material

The heat transferred by the food consists of the sensible heat, the heat of the product moisture evaporation, and the respiration heat (for fresh fruits and vegetables). In cold stores the temperature of the food material is reduced, and the heat is transferred by the food material to the air. The temperature of the food varies at different positions in the precooling cabinet. The size and shape of the pack or container affects the rate of heat transfer to the cooling air (or, in some cases, water). The temperature and speed of the air also affects this. Within the pack, the weight, density, water content, specific heat capacity, thermal conductivity, latent heat content, and initial food temperature each plays a part. In the case of unpackaged foods, the factors leading to rapid cooling also lead to rapid loss of moisture, so it may seem that slow cooling is better. Generally, this is not the case as the extended cooling time is also an extended drying-out time. More rapid chilling is possible with thinner packs, higher airspeeds, and lower air temperatures.

During the precooling stage, the total heat which must be removed to reduce the temperature from T_{p1} to T_{p2} (°C) is given by Equation 6.4:

$$m_pC_p(T_{p1} - T_{p2}) = Q \tag{6.4}$$

where

m_p the mass of product (kg)
C_p the average specific heat in the temperature range from T_{p1} to T_{p2} (J/kg°C)

The heat transfer rate by the food to the air, without considering water evaporation, is also given by Equations 6.5 and 6.6:

$$m_pC_p\frac{dT_p}{dt} = hA_p\left(T_a - T_p\right) \Rightarrow T_p = T_a + \left(T_{pIT} - T_a\right)exp\left(-\frac{hA_pt}{m_pC_p}\right) \tag{6.5}$$

$$q_p = hA_p(T_p - T_a) \tag{6.6}$$

This approach is only accurate when the Biot number (N_{Bi}) is below 0.1. To improve estimates for $N_{Bi} > 0.1$, the use of the half-life or temperature reduced in half was proposed. This time is calculated for a given h, shape and size of the product, and the value is used to calculate a new value (hA_p) that can be used in Equations 6.5 and 6.6.

Water evaporation during the precooling stage is considered to have a constant rate since the changes of both humidity and water activity of the food is small (Equations 6.7 and 6.8):

$$N_p = kA_p(p_a - p_s) \tag{6.7}$$

$$q_{evap} = N_pL \tag{6.8}$$

TABLE 6.2
Constants (*a* and *b*) of Fruits and Vegetables for Respiration Heat Calculation (Equation 6.10, where q_r in mW/kg and *T* in °C)

Food Material	*a*	*b*
Apple	19.4	0.108
Beet	38.1	0.056
Broccoli	97.7	0.121
Cabbage	16.8	0.074
Carrot	29.1	0.083
Celery	20.3	0.104
Grapefruit	11.7	0.092
Green beans	861	0.115
Lettuce	59.1	0.07
Onion	6.9	0.099
Orange	13.4	0.106
Peach	14.8	0.133
Pear	12.1	0.173
Peas	111.0	0.106
Pepper	33.4	0.072
Spinach	65.6	0.131
Strawberry	50.1	0.106
Tomato	13.2	0.103

Source: Toledo, R.T., *Fundamentals of Food Process Engineering*, 2nd edn., The AVI Publishing Co. Inc., Westport, CT, 1991. With permission.

If the water evaporation is significant, Equations 6.7 and 6.8 become (Equation 6.9):

$$m_p C_p \frac{dT_p}{dt} = q_p + q_{evap} = hA_p(T_a - T_p) + q_{evap} \tag{6.9}$$

During chilled storage, a significant part of the cooling load can be the heat produced during respiration of fruits and vegetables. The heat per unit time produced by the respiration is a function of temperature (T) and can be calculated by Equation 6.10:

$$q_r = ae^{bT} \tag{6.10}$$

where a and b are constants (Table 6.2).

The heat produced during respiration can be calculated as following if the function of temperature by time is known (Equation 6.11):

$$Q_r = \int_0^t ae^{bT(t)}dt \tag{6.11}$$

The temperature at any point of the food varies exponentially with the cooling time, and there is a temperature difference between the surface and the center of the food. In Equation 6.11, the average temperature of the food is used and linear change with time is assumed. During chilled storage, changes in temperature and food moisture are small and, the sensible heat as well as the evaporation heat can be vented or offered to the food (Equation 6.12) (Brown and Gould, 1992; Heldman and Singh, 1981).

$$m_pC_p\frac{dT_p}{dt}=hA_p\left(T_a-T_p\right)+q_{evap}+q_r \tag{6.12}$$

6.5.4 Heat Introduced through Ceiling, Floor, and Walls

The losses of cooling through cabinets (i.e., ceiling, floor, walls) are calculated for each surface separately, when differences in heat transfer exist, and then accumulated. The differences are due to different construction and insulating materials, different thicknesses, contact with the outside air or with other material, and orientation of the surface. For each side the heat transfer coefficients are calculated for the inner wall, through the insulating material of the wall, and from the outer wall to the air. Such average coefficients for various conditions and materials are given in the literature. The simplest case is to calculate the total heat transfer coefficient for each surface and the heat loss without taking into account the temperature variation in the wall:

$$q_w=UA(T_{ext}-T_a) \tag{6.13}$$

where

q_w is the heat entering through the wall (W)
U is the overall heat transfer coefficient (W/m^2 °C)
A is the wall surface (m^2)
T_{ext} is the outdoor temperature (°C) (Equation 6.13)

The outside temperature varies during the day and due to weather conditions. For small changes, and when the difference (T_{ext} − T_a) is large, a mean value of T_{ext} can be used. Moreover, for the most refrigeration chambers the q_w is a small term in Equation 6.13, and its total effect to T_a is not significant. A significant amount of heat is introduced by the cracks due to the change in pressure due to temperature change. To calculate the required cooling load for the removal of the heat based on the difference of temperature and pressure, the volume of air entering the chamber must be calculated (Brown and Gould, 1992; Heldman and Singh, 1981).

6.5.5 Heat Transferred by Engines Working Inside the Cold Room and People

The motors inside the rooms produce energy at a rate of 1025 J/s (*hp*). The bulbs emit energy proportional to their power. People entering the cold room emit energy depending on their activity and the air temperature. However, this can be neglected; it is a common practice to take an average heat load, independent of temperature, equal to 300 J/s per person. Incoming water vapor by humans has usually a small contribution in Equation 6.3 and is not taken into account. Water from other sources is also negligible, unless hot water was used (i.e., in an air-conditioned meat processing area). If there is a need to take into account the water from other sources, an estimation of the total quantity (N_t) is taken and added to Equation 6.3 (Brown and Gould, 1992; Heldman and Singh, 1981).

6.5.6 Heat Introduced by Air Renewal (Door Openings)

The door opening introduces hot air, and the quantity depends on the size of the door, the density difference between the outdoor area and the room, the existence of barriers (plastic curtains), and the circulation of the air inside the cold room. Data on the rate of heat transfer through the door of refrigerated chambers are determined empirically and for temperatures between 22°C and 66°C Equation 6.14 is used

$$q_d = 2126 W_d e^{0.048\Delta T} H_d^{1.71} \quad (6.14)$$

where

q_d is the rate of heat transfer (W)
W_d is door width (m)
H_d is the door height (m)
ΔT is temperature difference between the outside area and the chamber (°C)

More precisely, the heat introduced by air renewal can be calculated if the speed of the air was measured or estimated (Equation 6.15).

$$q_d = \mathbf{0.5} A v_a p_a C_a (T_{ext} - T_a) \quad (6.15)$$

where

A is the door surface (m²)
v_a is the average velocity of air through the opening (m/s)
p_a is the density of air in the chamber (kg/m³)
c_a is the specific heat of air (J/kg°C)
T_{ext} is the outdoor temperature (°C)

The vapor flow by air renewal is given by Equation 6.16:

$$N_d = \mathbf{0.5} A v_a p_a (W_{ext} - W_a) \quad (6.16)$$

where

W_{ext} is the absolute humidity of the outside air (kg water vapor/kg dry air)
W_a is the absolute humidity in the chamber (kg steam/kg dry air) (Brown and Gould, 1992; Heldman and Singh, 1981; Singh and Mannapperuma, 1990)

6.5.7 Heat Introduced through Ceiling, Floor, and Walls

The floor has a regulatory role for the changes of the temperature of the cold room, especially when it is solid with substantial thickness. The heat balance and heat transfer from the floor or any other material that has a balancing effect on temperature can be expressed as following, by considering the temperature of all materials as the same (Equations 6.17 and 6.18):

$$m_b C_b \frac{dT_b}{dt} = hA_b(T_a - T_b) \quad (6.17)$$

$$q_b = hA_b(T_b - T_a) \quad (6.18)$$

where

m_b is the mass of material (kg)
c_b is the specific heat of the material (J/kg°C)
T_b is the material temperature (°C)
A_b is the exposed surface of the material (m²)

The surface heat transfer coefficient (h) depends on the average velocity of the air above the material surface and the surface itself. The specific heat when using a variety of materials must be calculated as an average based on the percentage by weight of each material (Brown and Gould, 1992; Heldman and Singh, 1981).

6.5.8 Numerical Models

With the advent of computers, numerical methods were introduced into food research, and the whole temperature field inside the product could be modeled. The use of computers in plant design stimulated research into the prediction of dynamic heat load from food, and integrated models are being developed to take into account the interactions of the food with the processing equipment. Calculation of the heat removal process may be complicated by phase change (freezing of foods), during which the product's thermal properties undergo large changes over a small temperature range. The product undergoing cooling often has very complex shape and composition. Heat transfer may be coupled with moisture transfer, and the equations governing the two processes should be solved simultaneously. The heat transfer coefficient is often difficult to determine for the infinite variety of real-life situations, such as packaged products, cryogenic cooling, highly turbulent flow, swirling, and nonparallel flow, and within the same chiller the heat transfer coefficient will vary greatly from place to place. The heat transfer coefficient may also vary along the surface of a product due to the complex turbulent flow pattern and the development of the flow boundary layer, a problem still largely unsolved even by the most advanced computational fluid dynamics (CFD) packages. Food products often have inconsistent compositions, shapes, and sizes, which lead to variable thermal behavior and quality after processing, so that it is difficult to quantify accurately the effects of different processing practices.

At the same time, it was quickly realized that numerical methods would enable the prediction of many other quantities of interest related to the quality and shelf life of the food. Microbial growth, physical change (such as weight loss), biochemical changes (such as those determining the tenderness of meat or ripeness and flavor of fruit), and subjective factors (such as surface appearance, flavor, and texture) are often highly sensitive to temperature and moisture changes. An ability to predict accurately the whole temperature and moisture fields in a product would allow the food technologist to optimize the product quality factors as well as the economics. Steady progress is being made with recent advances in computer software and hardware (Defraeye et al., 2014; Lijun and Sun, 2002; Lovatt et al., 1993).

6.5.8.1 Computational Fluid Dynamics (CFD) Models

Numerical methods such as finite differences (FD) and finite elements (FE) have been developed and CFD have been applied to calculate the heat transfer (Davey and Pham, 1997; Lijun and Sun, 2002). CFD which can be used to study different cooling conditions and package designs can be very useful to understand and improve cooling processes (Ambaw et al., 2013; Defraeye et al., 2013) and reduce energy consumption (Defraeye et al., 2014). CFD models (Mirade and Daudin, 2006; Nahor et al., 2005) or simplified models (Wang and Touber, 1990) were developed to predict temperature, humidity, and air velocity in refrigerated cold rooms. However, in most of the cases, only experimental temperatures were used to compare with the numerical values. CFD models can confirm experimental results and help correct a dysfunction revealed in the experimental diagnosis as shown in the study of industrial chillers of beef carcass carried out by Mirade and Picgirard (2001).

Some disadvantages can make FE and FD inappropriate for many heat load estimation tasks. FD may be conveniently used for only a small range of regular shapes. FE can deal with a wider range of shapes but is still limited to shapes that can be described using only a few parameters. It can be time-consuming to prepare an FD calculation and even more so for an FE calculation. The FD model is not completely satisfactory for engineering analysis of biological materials which have large variations in properties. FE analysis has proved to be useful on nonlinear heat transfer problems during cooking, cooling, and thawing processes of foods (Lovatt et al., 1993).

6.5.8.2 Ordinary Differential Equations (ODE) Models

Simplified methods using ordinary differential equations (ODEs) have been developed. ODE models, lumped parameter models, or stirred tank models are those where an object undergoing thermal changes is represented by a small number of components, each of which is at uniform temperature. Because of the simplification of the physical situation that this entails, ODE models usually have to incorporate empirical parameters to improve prediction accuracy. They are often used to calculate the variation of product heat loads with time because they are much faster than more rigorous partial differential equation (PDE) models, such as FD. ODE models are not suitable for product temperature calculations, except in the case of very low Biot numbers (uniform product temperature) (Lovatt et al., 1993; Pham, 2000).

6.6 CHILLING TIME PREDICTION

6.6.1 Introduction

The food refrigeration literature contains many reports of methods for predicting temperature changes in foods undergoing chilling processes (Cleland, 1990; Wang and Sun, 2003). These methods range in complexity and applicability from exact analytical solutions, for a small number of regular geometries under restricted conditions, to sophisticated numerical methods applicable to any geometry under any set of constant or varying conditions. Most of the early work methods for predicting chilling time are based on the assumption of constant, uniform thermal properties and on the mean or center temperature of the food. More recent efforts have concentrated on the properties and geometry of the product. Other researchers developed models to predict moisture loss from food undergoing refrigeration. Selection of the most appropriate chilling time prediction method for a given product depends on the nature of the product and the conditions under which it is typically processed (Merts et al., 2007). When the food product is placed in a container or package prior to the process, the cooling characteristics are described by unsteady-state or transient heat transfer relationships. During preservation processes, the unsteady-state heat transfer relationships provide temperature distribution histories within the product.

6.6.1.1 Unsteady-State Cooling

Often, it is vital to know the change in temperature with time during the unsteady-state period of cooling of foods. The governing equation describing unsteady-state heat transfer is (Equation 6.19)

$$\frac{\partial T}{\partial t} = a\left(\frac{\partial^2 T}{\partial x^2} + \frac{\partial^2 T}{\partial y^2} + \frac{\partial^2 T}{\partial z^2}\right) \tag{6.19}$$

Before considering the solution of the governing equation, it is desirable to determine the relative importance of internal versus external resistance to heat transfer. For this purpose, the dimensionless Biot number, N_{Bi}, is useful (Equation 6.20):

$$N_{Bi}\frac{hD}{k} = \frac{D/k}{1/h} = \frac{internal\ resistance}{external\ resistance} \tag{6.20}$$

A low Biot number (smaller than 0.2) indicates that the internal resistance to heat transfer is negligible, and thus, the temperature within the object is uniform at any given instant in time. A large Biot number (greater than 40) indicates that the internal resistance to heat transfer is not negligible, and thus, a temperature gradient may exist within the object. Between a Biot number of 0.2 and 40 there is a finite resistance to heat transfer both internally and at the surface of the object undergoing cooling (Singh, 2007).

In typical blast cooling or freezing operations for foods, the Biot number is large, ranging from 0.2 to 20. Thus, the internal resistance to heat transfer is generally not negligible during food cooling and freezing, and a temperature gradient will exist within the food item (Cleland, 1990).

6.6.1.2 Negligible Internal Resistance to Heat Transfer

The dimensionless temperature ratio, also called the unaccomplished temperature ratio, can be expressed with the following exponential equation (Equation 6.21):

$$\frac{T - T_m}{T_0 - T_m} = \exp\left(\frac{-hA_S}{pC_pV}\right)t \tag{6.21}$$

In a completely dimensionless form (Equation 6.22),

$$\frac{T - T_m}{T_0 - T_m} = \exp\left(-N_{Bi}N_{Fo}\right) \tag{6.22}$$

6.6.1.3 Negligible Surface Resistance to Heat Transfer

The governing Equation 6.19 may be solved analytically for some regular shaped objects, such as an infinite plate, infinite cylinder, and a sphere. The following set of equations allows calculation of the variable temperature, T, with time at any location within the object, when the initial temperature is uniform, T_0, and the surface temperature, T_s is constant.

For a plane wall of thickness much smaller than length (L) and height,

$$\frac{T - T_S}{T_0 - T_S} = 1 - \frac{4}{\pi}\sum_{n=0}^{\infty}\frac{(-1)^n}{2n+1}\exp\left(-\frac{(2n+1)^2}{4}\pi^2 N_{Fo}\right)\cos\frac{(2n+1)\pi x}{2L} \tag{6.23}$$

where the wall's thickness is $2L$, and x is the variable distance from the center axis (Equation 6.23).

For a cylinder

$$\frac{T - T_S}{T_0 - T_S} = 1 - \frac{2}{a}\sum_{n=1}^{\infty}\frac{\exp(-a\mu_n^2 t)J_0(r\mu_n)}{\mu_n J_1(\alpha\mu_n)} \tag{6.24}$$

where

a is the radius of the cylinder
r is the variable distance from the axis
α is the thermal diffusivity
t is the time
J_0 and J_1 are Bessel functions of the first kind of zero and first order, respectively

The discrete values of μ_n are the roots of the transcendental equation (Equation 6.25).

$$J_0(\alpha\mu_n) = 0 \tag{6.25}$$

For a sphere

$$\frac{T - T_S}{T_0 - T_S} = 1 - \frac{2a}{\pi r}\sum_{n=1}^{\infty}\frac{(-1)^n}{2n+n}\sin\frac{n\pi r}{a}\exp(-n^2\pi^2 N_{Fo}) \tag{6.26}$$

where

a is the radius of the sphere

r is the variable distance from the center

The temperature at the center axis is given by the limit at $r \to 0$.

$$\frac{T - T_S}{T_0 - T_S} = 1 - 2\sum_{n=1}^{\infty}(-1)^n \exp(-n^2\pi^2 N_{Fo}) \tag{6.27}$$

6.6.1.4 Finite Surface and Internal Resistance to Heat Transfer

The preceding equations describe temperature distribution in an object when the surface temperature is constant. However, many cases involve convection at the boundary between the solid object and the surrounding fluid. The following expressions are useful to calculate the temperature, T, anywhere in the object with a uniform initial temperature, T_0, when immersed in a fluid of temperature, T_m, and a convection boundary condition.

For a plane wall with convection heat transfer at the surface (where thickness is much smaller than length and height) (Equation 6.28):

$$\frac{T - T_m}{T_0 - T_m} = \sum_{n=1}^{\infty} \frac{4\sin\mu_n}{2\mu_n + \sin(2\mu_n)} \exp(-\mu_n^2 N_{Fo})\cos(\mu_n x^*) \tag{6.28}$$

where $N_{Fo} = \alpha t/L^2$, $2L$ the thickness of plane wall, $x^* = x/L$, at the center line, $x^* = 0$, and the eigenvalues are positive roots of the transcendental equation (Equation 6.29)

$$\mu_n \tan\mu_n = N_{Bi} \tag{6.29}$$

For an infinite cylinder with convection at the surface (Equation 6.30)

$$\frac{T - T_m}{T_0 - T_m} = \sum_{n=1}^{\infty} \frac{2J_1(\mu_n)}{\mu_n\left(J_0^2(\mu_n) + J_1^2(\mu_n)\right)} \exp(-\mu_n^2 N_{Fo}) J_0(\mu_n r^*) \tag{6.30}$$

where $r^* = r/R$ and the discrete values of μ_n are the positive roots of the transcendental equation (Equation 6.31)

$$\mu_n \frac{J_1(\mu_n)}{J_0(\mu_n)} = N_{Bi} \tag{6.31}$$

where J_0 and J_1 are Bessel functions of first kind and order zero and one, respectively.

For a sphere with convection at the surface (Equation 6.32):

$$\frac{T - T_m}{T_0 - T_m} = \sum_{n=1}^{\infty} \frac{4[\sin(\mu_n) - \mu_n\cos(\mu_n)]}{2\mu_n + \sin(2\mu_n)} \exp(-\mu_n^2 N_{Fo}) \frac{1}{\mu_n r^*}(\mu_n r^*) \tag{6.32}$$

where $r^* = r/R$ and the discrete values of μ_n are the positive roots of the transcendental equation (Equation 6.33):

$$1 - \mu_n \cot \mu_n = N_{Bi} \quad (6.33)$$

Roots of the transcendental equation are tabulated in mathematical handbooks.

6.6.1.5 Use of Charts to Estimate Temperature History during Unsteady-State Cooling

The use of analytical solutions to determine temperature history is cumbersome because of the need to evaluate numerous terms in the series. These solutions have been reduced to charts that are much easier to use. These charts are presented in Figures 6.6 through 6.8 for infinite slab, infinite cylinder, and sphere, respectively (Heldman, 2011). The temperature–time chart in Figure 6.6 presents the relationship between the temperature ratio and the Fourier number, N_{Fo}, at various magnitudes of the inverse Biot number. The Fourier number is defined as given in Equation 6.34:

$$\frac{at}{d_c^2} = \frac{kt}{pC_p d_c^2 N_{Fo}} \quad (6.34)$$

where a is the thermal diffusivity and m²/s = k/pC_p. The d_c, for an infinite sphere, is the radius of the sphere, for an infinite slab, the half-thickness of the slab, and for an infinite cylinder, the radius of the cylinder. When evaluating unsteady-state heat transfer in a finite geometry, several relationships have been developed. Singh and Heldman (2009) describe and illustrate the applications of these expressions of typical food containers (cans, pouches, brick shapes, and boxes).

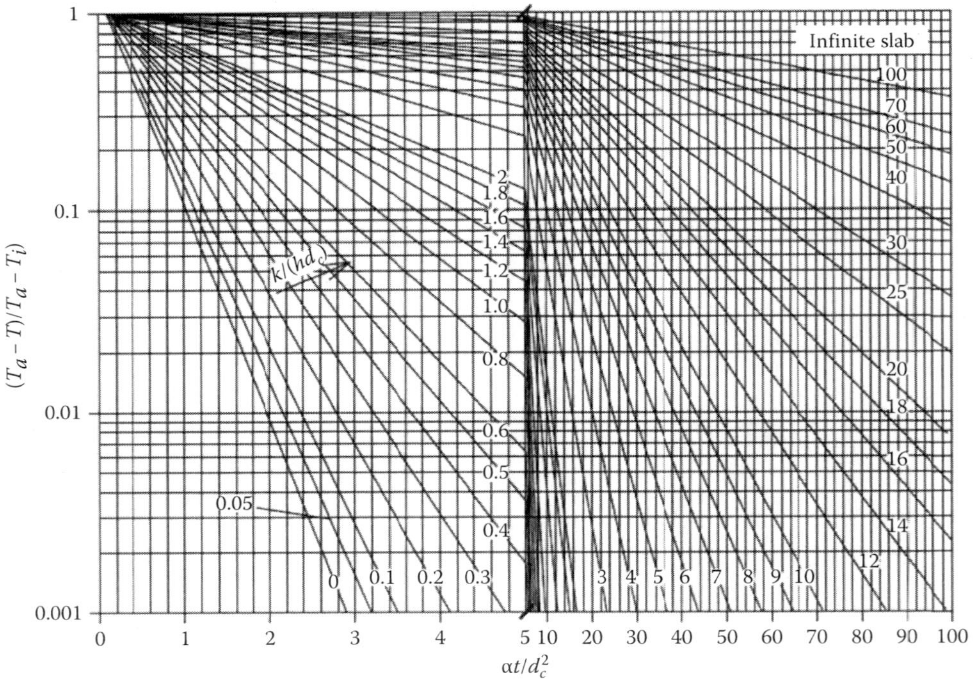

FIGURE 6.6 Unsteady temperature distributions in an infinite slab. (From Singh, R.P. and Heldman, D.R., *Introduction to Food Engineering*, 4th edn., Elsevier Academic Press, San Diego, CA, 2009.)

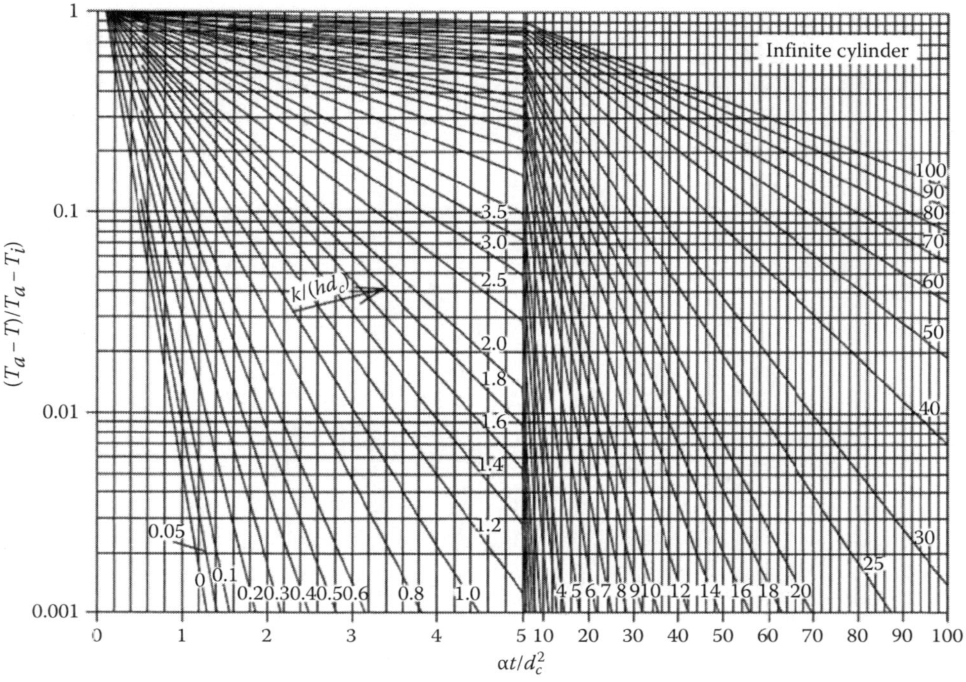

FIGURE 6.7 Unsteady temperature distributions in an infinite cylinder. (From Singh, R.P. and Heldman, D.R., *Introduction to Food Engineering*, 4th edn., Elsevier Academic Press, San Diego, CA, 2009.)

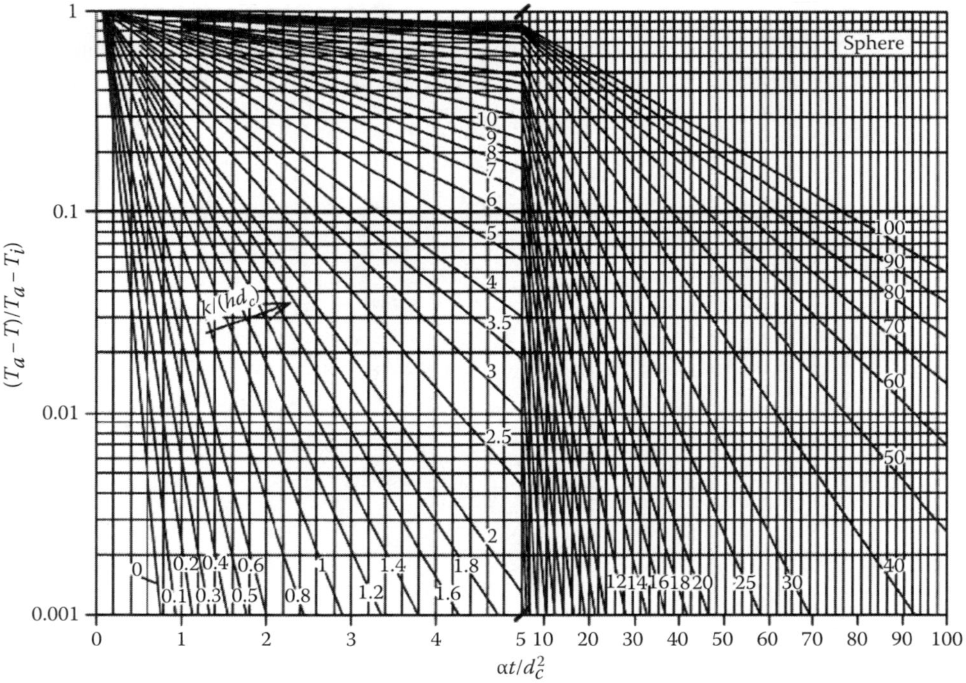

FIGURE 6.8 Unsteady temperature distributions in a sphere. (From Singh, R.P. and Heldman, D.R., *Introduction to Food Engineering*, 4th edn., Elsevier Academic Press, San Diego, CA, 2009.)

6.6.1.6 Predicting Temperature during Transient Heat Transfer

In many food preservation situations, the temperature can be predicted when the temperature ratio is less than 0.7. For these situations, Ball (1923) and Ball and Olson (1957) recognized a simplified approach that has been incorporated into the prediction of temperatures during thermal process design. For this approach, two parameters are defined: a time factor (f_h) and a lag factor (j_c). When these parameters are introduced into the general solution, the following expression is obtained (Equation 6.35):

$$\text{Ln}\frac{(T_a - T)}{J_c(T_a - T_i)} = -\frac{2.303}{f_h}t \qquad (6.35)$$

After conversion of the logarithms, the expression becomes (Equation 6.36)

$$\log(T_a - T) = -\frac{t}{f_h} + \log[J_c(T_a - T_i)] \qquad (6.36)$$

This expression illustrates that the magnitudes of the two parameters (f_h, j_c) can be evaluated experimentally by the measurement of temperature as a function of time. The heating rate constant (f_h) is the time required for a one log cycle change in the temperature difference on the linear portion of the temperature–time relationship. The lag constant (j_c) describes the portion of the temperature–time curve at the beginning of cooling and prior to the linear portion of the log–linear relationship between temperature and time. Pflug et al. (1965) have analyzed the relationships between the parameters in Equation 6.33 and the key factors influencing heating or cooling rates for conduction-heating objects. The results of the analysis were presented in a series of charts to allow for prediction of heating or cooling times. One of the charts presented a relationship between a dimensionless number, incorporating the rate constant (f_h), and the Biot number, as presented in Figure 6.9. The relationships presented in Figure 6.4 are for the three standard geometries. The range of Biot numbers considered is to conditions of finite internal and surface resistance to heat transfer and illustrates that there is negligible change to the magnitude of the time factor (f_h) at Biot numbers above 40.

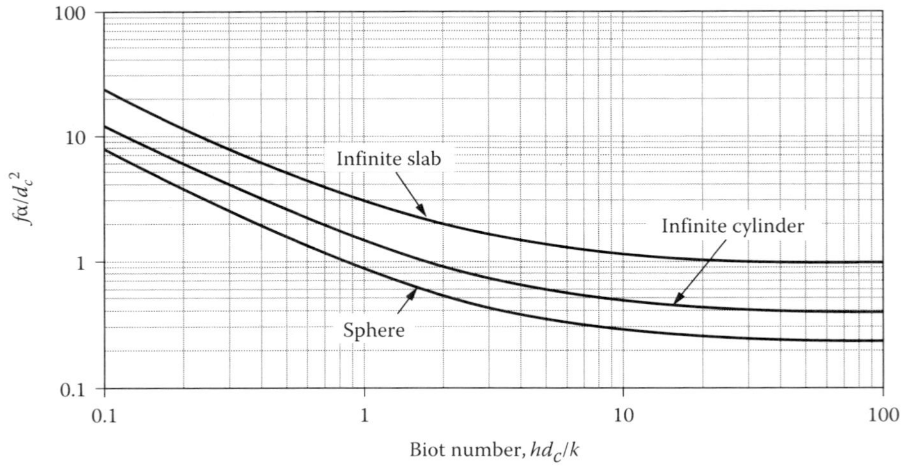

FIGURE 6.9 The relationship between the heating rate constant (f_h) and Biot number. (From Pflug, I.J. et al., *ASHRAE Trans.*, 71(1), 238, 1965.)

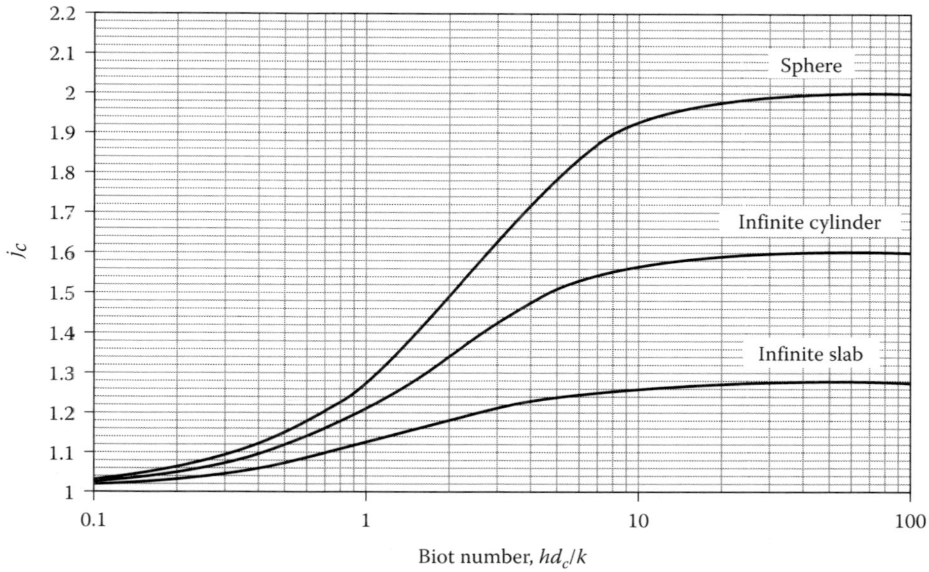

FIGURE 6.10 The relationship between the lag constant (j_c) at the geometric center and the Biot number. (From Pflug, I.J. et al., *ASHRAE Trans.*, 71(1), 238, 1965.)

The cooling lag constant varies with location within the object, and j_c represents the influence of the Biot number on the magnitude at the geometric center of the object. This relationship is presented in Figure 6.5. The relationships in Figure 6.10 are for the three standard geometries and indicate that the magnitude of the lag constants are near 1.0 at low Biot numbers before increasing to higher constant values at Biot numbers above 40. Equation 6.34 predicts the mass average temperature within the object by using the appropriate value of the lag constant. The relationships between the mass average lag constant (j_m) and the Biot number are presented in Figure 6.11. As indicated by Figure 6.6, the mass average lag constant has magnitudes of 1.0 at low Biot numbers and decreases as the Biot number increases.

Equation 6.34 can be used to predict temperatures within finite geometries that are more similar to food product containers and packages. When the shape of the object is a finite cylinder, the values of f_h and j_c are estimated from Equations 6.37 and 6.38:

$$\frac{1}{f_{\text{finite cylinder}}} = \frac{1}{f_{\text{infinite cylinder}}} + \frac{1}{f_{\text{infinite slab}}} \tag{6.37}$$

$$j_c(\text{finite cylinder}) = j_c(\text{infinite cylinder}) * j_c(\text{infinite slab}) \tag{6.38}$$

A brick-shaped object gets the relationships (Equations 6.39 and 6.40):

$$\frac{1}{f_{\text{brick}}} = \frac{1}{f_{\text{infinite slab }\mathbf{1}}} + \frac{1}{f_{\text{infinite slab }\mathbf{2}}} + \frac{1}{f_{\text{infinite slab }\mathbf{3}}} \tag{6.39}$$

$$j_c = (\text{brick}) = j_c(\text{infinite slab }\mathbf{1}) * j_c(\text{infinite slab }\mathbf{2}) * j_c(\text{infinite slab }\mathbf{3}) \tag{6.40}$$

Although there are limitations to the use of these relationships, estimations of cooling times or temperatures after short periods of cooling are acceptable for most situations encountered during thermal processes for food (Heldman, 2011).

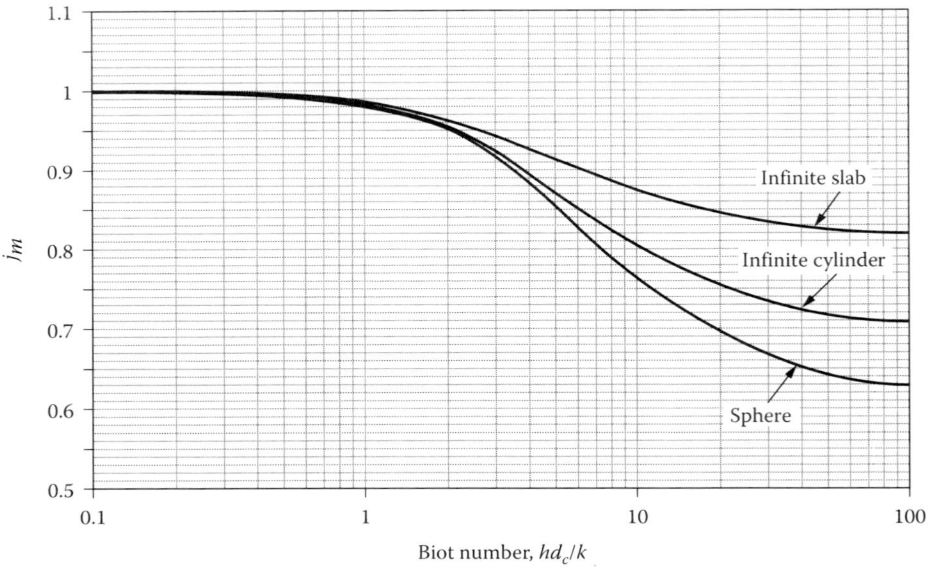

FIGURE 6.11 The relationship between the mass average lag constant and Biot number. (From Pflug, I.J. et al., *ASHRAE Trans.*, 71(1), 238, 1965.)

6.6.1.7 Use of Cooling Curves to Estimate the Chilling Time

One method for obtaining the surface heat transfer coefficient of a food product with an internal temperature gradient involves the use of cooling curves. For simple, one-dimensional food geometries, such as infinite slabs, infinite circular cylinders, or spheres, empirical and analytical solutions to the one-dimensional transient heat equation can be found. The slope of the cooling curve may be used in conjunction with these solutions to obtain the Biot number for the cooling process. The heat transfer coefficient may then be determined from the Biot number. All cooling processes exhibit similar behavior. After an initial "lag," the temperature at the thermal center of the food item decreases exponentially (Cleland, 1990). As shown in Figure 6.12, a cooling curve depicting this behavior can be obtained by plotting, on semi-logarithmic axes, the fractional unaccomplished temperature difference versus time. The fractional unaccomplished temperature difference, Y, is defined as follows (Equation 6.41):

$$Y = \frac{T_m - T}{T_m - T_i} = \frac{T - T_m}{T_i - T_m} \tag{6.41}$$

The "lag" between the onset of cooling and the exponential decrease in the temperature of the food item is measured by the j factor, as shown in Figure 6.12. From Figure 6.12, it can be seen that the linear portion of the cooling curve can be described as follows (Equation 6.42):

$$Y = j\exp(-C\theta) \tag{6.42}$$

where C is the cooling coefficient, which is minus the slope of the linear portion of the cooling curve. For simple geometrical shapes, such as infinite slabs, infinite circular cylinders, and spheres, analytical expressions for cooling or freezing time may be derived. To derive these expressions, the following assumptions are made: (1) the thermophysical properties of the food item and the cooling medium are constant, (2) the internal heat generation and moisture loss from the food item are neglected,

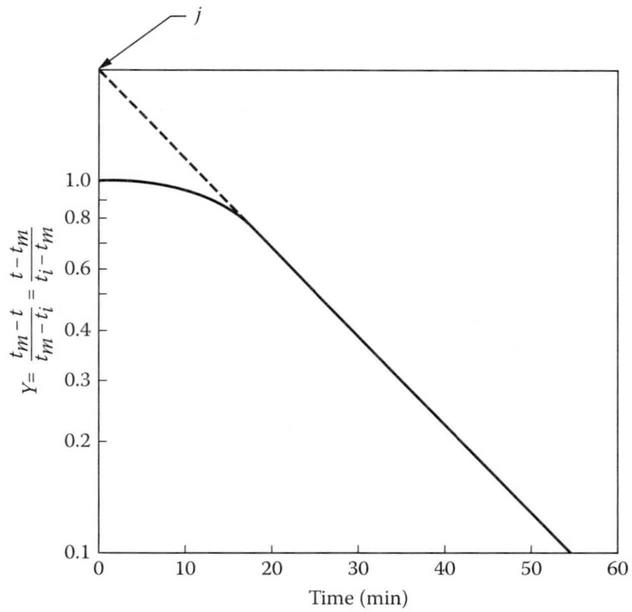

FIGURE 6.12 Typical cooling curve. (From Heldman, D.R., Physical transport models, in: *Food Preservation Process Design*, 2nd edn., The AVI Publishing Co. Inc., Westport, CT, 2011.)

(3) the food item is homogeneous and isotropic, (4) the initial temperature distribution within the food item is uniform, (5) heat conduction occurs only in one dimension, and (6) convective heat transfer occurs between the surface of the food item and the cooling medium (Bryan and Fricke, 2004).

6.7 QUALITY DETERIORATION AND SHELF-LIFE DETERMINATION DURING CHILLING STORAGE

6.7.1 Quality Deterioration and Shelf Life

The quality of food is not a simple property; it depends not only upon the initial integrity of the raw materials but also on the changes occurring during processing and subsequent storage that may result in potential losses and decreased bioavailability. Foods are physicochemically and biologically active systems and their quality is a dynamic state moving continuously to reduced levels. Any food product can retain a required level of sensory, quality, and safety characteristics for a finite time after production under defined storage conditions. The shelf life of a food can be defined as the time period within which the food is safe to consume and/or has an acceptable quality to consumers. The definition of shelf life and the criteria for the determination of the end of shelf life are dependent on specific commodities and on the definition's intended use (i.e., for regulatory vs. marketing purposes). Practical storage life (PSL) is the period of proper storage after processing of an initially high-quality product during which the sensorial quality remains suitable for consumption (Taoukis et al., 1997a,b). According to EU Regulation 1169/2011 (Article 2, §2), time of minimum durability of a food means the date until which the food retains its specific properties when properly stored. Microbiological spoilage, lipid oxidation, nonenzymatic browning, enzymatic activity, and vitamin loss are among the most significant quality deteriorating factors that signal the end of food shelf life both for chilled and frozen food products. The values of these parameters can be correlated to sensory results for the same food, and a limit that corresponds to the lowest acceptable sensory quality can be set. The relative contribution of each factor to the overall quality may vary at different levels of quality or at different storage conditions.

Chilling storage generally results in effective short-term preservation by retarding the growth of microorganisms, pathogenic (i.e., *Aeromonas hydrophila, Bacillus cereus, Clostridium botulinum, Escherichia coli, Listeria monocytogenes, Salmonella, Vibrio parahaemolyticus, Yersinia enterocolitica*) and spoilage (e.g., pseudomonas, lactic acid bacteria, yeasts, molds), postharvest metabolic activities of intact plant tissues and post-slaughter metabolic activities of animal tissues, deteriorative chemical reactions, including enzyme-catalyzed oxidative browning or oxidation of lipids, and chemical changes associated with color degradation (i.e., fruits and vegetables), lipid hydrolysis associated with off-odors and rancidity development (i.e., meat, fish, and dairy products), glycolysis associated with pH decline (i.e., fruits, vegetables, meat, and milk), proteolysis associated with the loss/deterioration of texture (i.e., bitter flavor development (i.e., meat, fish, milk, and dairy products) and loss of nutritive value and moisture loss (Brown and Hall, 2000). Chilling is also used for such non-preservative purposes as crystallization; aging of beef, wine, and cheese; and to facilitate such operations as pitting of cherries and peaches, cutting of meat, and slicing of bread (Drummond and Sun, 2006; Hui et al., 2003).

6.7.2 Modeling of Quality Deterioration and Shelf Life

A food product starts to degrade once it is produced. The rate and the degree of degradation depend on both the composition and the environmental conditions during storage and distribution. Food kinetics is based on the thorough study of the rates at which physicochemical reactions proceed. Food quality loss and shelf life can be evaluated by monitoring a characteristic quality index, A. The change of quality index A with time (dA/dt) can be represented by the following kinetic equation (Equation 6.43):

$$-\frac{dA}{dt} = kA^n \tag{6.43}$$

where

k is the rate constant depending on temperature, product, and packaging characteristics

n is a power factor called reaction order which defines whether the rate of change is dependent on the amount of A present

The Arrhenius equation is often used to describe the temperature dependence of deterioration rate (Equation 6.44):

$$k(T)_t = k_0 \exp\left[\frac{-E_a}{RT}\right] \tag{6.44}$$

where

k_0 is a pre-exponential factor

E_a is the activation energy (Taoukis et al., 1997b)

One may conveniently use this information to get a reasonable estimate of how a temperature change may affect the rate of reaction. To simplify the process further, one may get over the need to evaluate k_0 by using a ratio between the rates of reaction when the temperature is changed by any arbitrary value. The most commonly used value is 10°C, and therefore the ratio between the rate of reactions is known as Q_{10}. The value of Q_{10} may be calculated using Equation 6.43 to express the rate of reaction first for a temperature of (T + 10) and then for T and dividing the two, namely (Equation 6.45):

$$Q_{10} = \frac{dD_2/dt}{dD_1/dt} = \frac{\exp\left(\dfrac{E_a}{R(T+10)}\right)}{\exp\left(\dfrac{E_a}{RT}\right)} = \exp\left(\frac{10E_a}{RT(T+10)}\right) \tag{6.45}$$

Thus, by studying a deterioration process and measuring the rate of loss at two or three temperatures (higher than storage temperature), one could extrapolate on an Arrhenius plot with a straight line to predict the deterioration rate at the desired storage temperature. This is the basis for accelerated shelf-life testing (ASLT) (Mizrahi, 2000; Taoukis et al., 1997a,b). It should be noted that in some cases a straight line will not ensue, especially if a phase change occurs. The most common way of accelerating the rate of reaction is by placing the product at elevated constant temperatures. However, non-isothermal procedures, using programmed changes in conditions, are necessary to be tested. For chilled products the Arrhenius equation cannot be suitable. Instead the use of the square root model (Equation 6.46) was proposed.

$$\sqrt{r} = b(T - T_0) \tag{6.46}$$

The criterion to define the end of shelf life of a particular food may be variable depending on the definition of product quality grade, so the shelf life may also be variable. The shelf life of most perishable foods is based mainly on sensorial quality. For example, quality degradation of fresh meat can be attributed to microbial activity and chemical oxidations that cause off-flavor development and loss of color. These changes are directly identified by consumers. On the other hand, most chilled/frozen food and many other longer shelf-life food products degrade mainly by slow chemical reactions correlated to nutritional value loss. For instance, vitamin C concentration in chilled/frozen fruits and vegetables can fall below the required standards as listed on the label before sensory characteristics become inadequate. The criteria for shelf-life determination may also vary depending on the consumer sensitivity. The most obvious sensory characteristics for the consumer are appearance, taste, and odor. Sensory evaluation is often correlated with instrumental measurements of a specific quality index (i.e., vitamin C). In general, the limits for shelf-life determination of a particular food product depend on legal requirements, (e.g., zero tolerance for *botulinum toxin*), consumer preferences, marketing requirements, and cost (Fu and Labuza, 1997).

Traditional methods for the determination of shelf life include storage of the product at different temperatures and determining spoilage by sensory evaluation or microbial count. This will involve the natural flora of the product, which may vary between batches. Depending on the product, process, and storage conditions the microbiological shelf life may be determined by the growth of either spoilage or pathogenic microorganisms. For products where the shelf life may be set by the growth of pathogenic microorganisms (i.e., *Listeria monocytogenes, Clostridium botulinum, Staphylococcus aureus,* and *Bacillus cereus*), this may involve challenge testing the product with the organism prior to storage and microbial analysis at regular intervals. In recent years, with the increasing capabilities and widespread availability of personal computers, predictive microbiology has become an abundant area for research and software development and application. In generic terms there are two categories of predictive models. Mechanistic models describe the theoretical basis of the microbial response, but owing to the complexity of microbial physiology and our current level of understanding, these types of models are rare. In contrast, there is a plethora of empirical models that mathematically describe the data but do not give insight into the underlying process. Empirical models can be further subdivided into probabilistic and kinetic models. Probabilistic models describe the probability of a microbiological event occurring that is independent of time (i.e., the probability of growth or toxin formation ever occurring) or that is time dependent (probability at a given time of an event occurring). Probabilistic models are most relevant for determining whether certain microorganisms will grow when they are close to their growth boundaries. This type of model is commonly used to predict the growth or toxin formation by *C. botulinum*. Kinetic models describe the rate and extent of growth or inactivation as described before. In practice, the different types of kinetic models have included growth, survival (conditions at non-lethal temperatures that will not support growth), and thermal inactivation. The use of mathematical models can help to reduce the need for storage trials, challenge tests, product reformulations, and process modifications, which are labor intensive, time consuming, and expensive (Betts and Everis, 2000).

REFERENCES

Aked J., Maintaining the post-harvest quality of fruits and vegetables (Ch. 7), in *Fruit and Vegetable Processing Improving Quality*, W. Jongen (ed.). Boca Raton, FL: Woodhead Publishing Ltd., 2002, pp. 119–149.

Ambaw A., M.A. Delele, T. Defraeye, Q.T. Ho, L.U. Opara, B.M. Nicolaı, P. Verboven, The use of CFD to characterize and design post-harvest storage facilities: Past, present and future. *Comput. Electron. Agric.*, 2013; 93: 184–194.

Armstrong G.A., Minimal thermal processing: Cook-chill and sous vide technology (Chapter 24), in *Handbook of Preservation and Processing*, Y.H. Hui, S.D.M. Chazala, K.D. Graham, W.-K. Nip, K.D. Murrell (eds.). New York: Marcel Dekker, Inc., 2004.

Arora C.P., *Refrigeration and Air-Conditioning,* 3rd edn. New Delhi, India: Tata McGraw Hill, 2006.

Arora R.C., *Refrigeration and Air Conditioning*, New Delhi, India: PHI Learning Private, 2010.

ASHRAE, Refrigerants (Ch. 29), in *ASHRAE Handbook—Fundamentals,* I-P edn. Atlanta, GA: ASHRAE Inc., 2009.

Ball C.O., F.C.W. Olson, *Sterilization in Food Technology.* New York: McGraw-Hill, 1957.

Betts G., L. Everis, Shelf life determination and challenge testing, in *Chilled Foods: A Comprehensive Guide,* 2nd edn. M. Stringer, C. Dennis (eds.). Cambridge, U.K.: CRC Woodhead Publishing Ltd, 2000.

Brown M.H., G.W. Gould, Processing, in *Chilled Foods: A Comprehensive Guide*, C. Dennis, M. Stringer (eds.). New York: Ellis Horwood, 1992.

Brown M.H., M.N. Hall, Non-microbiological factors affecting quality and safety, in *Chilled Foods: A Comprehensive Guide,* 2nd edn. M. Stringer, C. Dennis (eds.). Cambridge, U.K.: CRC Woodhead Publishing Ltd, 2000.

Bryan R.B., B.A. Fricke, Heat transfer coefficients of forced-air cooling and freezing of selected foods. *Int. J. Refrig.*, 2004; 27: 540–551.

Chourasia M.K., T.K. Goswami, Steady state CFD modeling of airflow, heat transfer and moisture loss in a commercial potato cold store. *Int. J. Refrig.*, 2007; 30: 672–689.

Cleland A.C., *Food Refrigeration Processes Analysis, Design and Simulation.* London, U.K.: Elsevier Science, 1990.

Davey L.M., Q.T. Pham, Predicting the dynamic product heat load and weight loss during beef chilling using a multi-region finite difference approach. *Int. J. Refrig.*, 1997; 20: 470–482.

Defraeye T., R. Lambrecht, M.A. Delele, A.A. Tsige, U.L. Opara, P. Cronje, P. Verboven, B. Nicolai, Forced-convective cooling of citrus fruit: Cooling conditions and energy consumption in relation to package design. *J. Food Eng.*, 2014; 121: 118–127.

Defraeye T., R. Lambrecht, A.A. Tsige, M.A. Delele, U.L. Opara, P. Cronje, P. Verboven, B. Nicolai, Forced-convective cooling of citrus fruit: Package design. *J. Food Eng.*, 2013; 118: 8–18.

Drummond L., D.-W. Sun, Effects of chilling and freezing on safety and quality of food products, in *Handbook of Food Science, Technology and Engineering,* Y.H. Hui (ed.). Boca Raton, FL: Taylor & Francis Group, LLC, Vol. 4, 2006.

Evans J., *Are Doors on Fridges the Best Environmental Solution for the Retail Sector?* Surrey, U.K.: The Institute of Refrigeration, 2014.

Fellows P., Chilling (Ch. 19), in *Food Processing Technology, Principles and Practice,* 2nd edn. Cambridge, U.K.: Woodhead Publishing Limited and CRC Press LLC, 2000, pp. 387–405.

Fu B., T.P. Labuza, Shelf-life testing: Procedures and prediction methods, in *Quality in Frozen Food*, M.C. Erickson, Y.-C. Hung (eds.). New York: Chapman & Hall, 1997.

Grandison A.S., Postharvest handling and preparation of foods for processing (Ch. 1), in *Food Processing Handbook*, J.G. Brennan (ed.). Weinheim, Germany: Wiley-VCH Verlag GmbH & Co. KGaA, 2006, pp. 9–13.

Gross K.C., C.Y. Wang, M. Saltveit, The commercial storage of fruits, vegetables, and florist and nursery stocks, in *Agricultural Handbook Number 66.* K.C. Gross, C.Y. Wang, M. Saltveit (eds.). Beltsville, MD: USDA, 2004 (www.ba.ars.usda.gov/hb66/index.html).

Heap R.D., The refrigeration of chilled foods (Ch. 4), in *Chilled Foods—A Comprehensive Guide,* 2nd edn. M. Stringer, C. Dennis (eds.). Cambridge, U.K.: Woodhead Publishing Limited, 2000.

Heldman D.R., Physical transport models, in *Food Preservation Process Design,* 2nd edn. D.R. Heldman (ed). Westport, CT: The AVI Publishing Co. Inc., 2011.

Heldman D.R., R.P. Singh, *Food Process Engineering,* 2nd edn. Westport, CT: The AVI Publishing Co. Inc., 1981.

Ho S.H., L. Rosario, M.M. Rahman, Numerical simulation of temperature and velocity in a refrigerated warehouse, *Int. J. Refrig.*, 2010; 33: 1015–1025.

Hu Z., D.-W. Sun, CFD simulation of heat and moisture transfer for predicting cooling rate and weight loss of cooked ham during air-blast chilling process. *J. Food Eng.*, 2000; 46: 189–197.

Hui Y.H., S. Ghazala, D.M. Graham, K.D. Murrell, W.-K. Nip, Storage at chilling temperatures, in *Handbook of Vegetable Preservation and Processing.* Y.H. Hui, S. Ghazala, D.M. Graham, K.D. Murrell, W.-K. Nip (eds.). New York: Marcel Dekker, Inc., 2003.

Hui Y.H., M.-H. Lim, W.-K. Nip, J.S. Smith, P.H.F. Yu, Principles of food processing (Ch. 1), in *Food Processing—Principles and Applications*, J. Scott Smith, Y.H. Hui (eds.). Oxford, U.K.: Blackwell Publishing, 2004.

Hundy G.F., A.R. Trott, T.C. Welch, *Refrigeration and Air-Conditioning,* 4th edn. Oxford, U.K.: Elsevier Ltd., 2008.

Ibarz A., G.V. Barbosa-Cánovas, Food preservation by cooling (Ch. 16), in *Unit Operations in Food Engineering.* A. Ibarz, G.V. Barbosa-Cánovas (eds.). Boca Raton, FL: CRC Press LLC, 2003.

James S., The chill chain "from carcass to consumer". *Meat Sci.*, 1996; 43(1): 203–216.

James S.J., Principles of food refrigeration and freezing (Ch. 112), in *Handbook of Food Science, Technology, and Engineering,* Y.H. Hui (ed.). Boca Raton, FL: Taylor & Francis Group, LLC., Vol. 3, 2006.

Jensen W.K., C. Devine, M. Dikeman, *Encyclopedia of Meat Sciences.* Oxford, U.K.: Elsevier Ltd., 2004, pp. 1131–1161.

Karel M., D.B. Lund, Storage at chilling temperatures (Ch. 7), in *Physical Principles Preservation of Food,* 2nd edn. M. Karel, D.B. Lund (eds.). New York: Marcel Dekker, Inc., 2003, pp. 237–273.

Kienholz J., I. Edeogu, *Fresh Fruit & Vegetable Pre-Cooling for Market Gardeners in Alberta.* Edmonton, Alberta, Canada: Alberta Agriculture, Food and Rural Development, 2002.

Laguerre O., S. Benamara, D. Flick, Study of water evaporation and condensation in a domestic refrigerator loaded by wet product. *J. Food Eng.*, 2010; 97: 118–126.

Laguerre O., E. Derens, D. Flick, Temperature prediction in a refrigerated display cabinet: Deterministic and stochastic approaches. *Electron. J. Appl. Stat. Anal.*, 2011; 4: 191–202.

Laguerre O., E. Derens, B. Palagos, Study of domestic refrigerator temperature and analysis of factors affecting temperature: A French survey. *Int. J. Refrig.*, 2002; 25: 653–659.

Laguerre O., M.H. Hoang, V. Osswald, D. Flick, Experimental study of heat transfer and air flow in a refrigerated display cabinet. *J. Food Eng.*, 2012; 113: 310–321.

Lijun W., D.-W. Sun, Modelling three conventional cooling processes of cooked meat by finite element method, *Int. J. Refrig.*, 2002; 25: 100–110.

Lovatt S.J., Q.T. Pham, A.C. Cleland, M.P.E. Loeffen, A new model of predicting the time-variability of product heat load during food cooling—Part 1: Theoretical considerations. *J. Food Eng.*, 1993; 18: 13–36.

Maroulis Z.B., G.D. Saravacos, Refrigeration and freezing (Ch. 5), in *Food Process Design.* Z.B. Maroulis, G.D. Saravacos (eds.). New York: Marcel Dekker, Inc., 2003.

Merts I., E.D. Bickers, T. Chadderton, Application and testing of a simple method for predicting chilling times for hoki (*Macruronus novaezelandiae*). *J. Food Eng.*, 2007; 78: 162–173.

Mirade P.S., CFD modeling of indoor atmosphere and water exchanges during the cheese ripening process (Ch. 28), in *Computational Fluid Dynamics in Food Processing* (*Contemporary Food Engineering series*). D.-W. Sun (ed.). Cambridge, U.K.: CRC Woodhead Publishing Ltd, 2007.

Mirade P.-S., J.-D. Daudin, Computational fluid dynamics prediction and validation of gas circulation in a cheese ripening room. *Int. Dairy J.*, 2006; 16: 920–930.

Mirade P.-S., L. Picgirard, Assessment of airflow patterns inside six industrial beef carcass chillers. *Int. J. Food Sci. Technol.*, 2001; 36: 463–475.

Mishra V.K., T.V. Gamage, Postharvest handling and treatments of fruits and vegetables (Ch. 3), in *Handbook of Food Preservation,* 2nd edn., M.S. Rahman (eds.). Boca Raton, FL: Taylor & Francis Group, LLC, 2007.

Mizrahi S., Accelerated shelf-life tests, in *The Stability and Shelf Life of Food*, D. Kilcast, P. Subramaniam (eds.). Cambridge, U.K.: CRC Woodhead Publishing Ltd., 2000.

Moureh J., S. Tapsoba, E. Derens, D. Flick, Air velocity characteristics within vented pallets loaded in a refrigerated vehicle with and without air ducts. *Int. J. Refrig.*, 2009; 32: 220–234.

Nahor H.B., M.L. Hoang, P. Verboven, M. Baelmans, B.M. Nicolai, CFD model of the airflow, heat and mass transfer in cool stores. *Int. J. Refrig.*, 2005; 28: 368–380.

Padalkar A.S., A.D. Kadam, Carbon dioxide as natural refrigerant. *Int. J. Appl. Eng. Res., Dindigul*, 2010; 1(2): 261–272.

Pardo J.M., K. Niranjan, Freezing (Ch. 4), in *Food Processing Handbook*, J.G. Brennan (ed.). Weinheim, Germany: Wiley-VCH Verlag GmbH & Co. KGaA, 2006, pp. 125–145.

Pflug I.J., J.L. Blaisdell, I. Kopelman, Developing temperature-time curves for objects that can be approximated by a sphere, infinite plate or infinite cylinder. *ASHRAE Trans.*, 1965; 71(1): 238–248.

Pham Q.T., Modelling thermal processes: Cooling and freezing, in *Food Process Modelling*, L.M.M. Tijskens, M.L.A.T.M. Hertog, B.M. Nicolai (eds.). Cambridge, U.K.: CRC Woodhead Publishing Ltd., 2000.

Pham Q.T., Modelling thermal processes: Cooling and freezing (Ch. 15), in *Food Process Modeling*, P. Tijskens, M. Hertog, B. Nicolai (eds.). Cambridge, U.K.: Woodhead Publishing Limited, 2001.

Pham Q.T., Calculation of processing time and heat load during food refrigeration. AIRAH Conference "Food for Thought-Cool", Sydney, Australia, May 24, 2002.

Ramaswamy H., M. Marcotte, Low-temperature preservation (Ch. 4), in *Food Processing, Principles and Applications.* Boca Raton, FL: Taylor & Francis Group, LLC, 2006.

Salunkhe D.K., H.R. Bolin, N.R. Reddy, *Storage, Processing, and Nutritional Quality of Fruits and Vegetables—Fresh Fruits and Vegetables.* Boca Raton, FL: CRC Press, LLC., Vol. 1, 1991.

Saravacos G.D., A.E. Kostaropoulos, *Handbook of Food Processing Equipment*, New York: Kluwer Academic/Plenum Publishers, 2002.

Singh R.P., Heating and cooling processes for foods (Ch. 5), in *Handbook of Food Engineering,* 2nd edn., D.R. Heldman, D.B. Lund (eds.). Boca Raton, FL: Taylor & Francis Group, LLC, 2007.

Singh, R.P., D.R. Heldman, *Introduction to Food Engineering,* 4th edn. San Diego, CA: Elsevier Academic Press, 2009.

Singh R.P., J.D. Mannapperuma, Developments in food freezing, in *Biotechnology and Food Process Engineering*, H.G. Schwartzberg, M.A. Rao (eds.). New York: Marcel Dekker, 1990.

Smale N.J., J. Moureh, G. Cortella, A review of numerical models of airflow in refrigerated food applications. *Int. J. Refrig.*, 2006; 29: 911–930.

Stoecker W., *Industrial Refrigeration Handbook.* New York: McGraw Hill, 1998.

Tabil Jr. L.G., S. Sokhansanj, Mechanical and temperature effects on shelf life stability of fruits and vegetables (Ch. 2), in *Food Shelf Life Stability Chemical, Biochemical, and Microbiological Changes*, N.A.M. Eskin, D.S. Robinson (eds.). Boca Raton, FL: CRC Press, LLC, 2001.

Taoukis P.S., T.P. Labuza, I.S. Saguy, Kinetics of Food Deterioration and Shelf-Life Prediction (Ch. 9), in *Handbook of Food Engineering Practice*, K.J. Valentas, E. Rolstein, R.P. Singh (eds.). Boca Raton, FL: CRC Press, LLC, 1997a.

Taoukis P., Labuza T.P., Saguy I., Kinetics of food deterioration and shelf-life prediction, in *The Handbook of Food Engineering Practice*, K.J. Valentas, E. Rolstein, R.P. Singh (eds.). Boca Raton, FL: CRC Press, LLC, 1997b.

Thompson J.F., F.G. Mitchell, T.R. Rumsey, R.F. Kasmire, C.H. Crisosto, *Commercial Cooling of Fruits, Vegetables and Flowers.* Davis, CA: Agricultural and Natural Resources, University of California, 2008.

Toledo R.T., *Fundamentals of Food Process Engineering,* 2nd edn. Westport, CT: The AVI Publishing Co. Inc, 1991.

Trott A.R., T.C Welch, *Refrigeration and Air Conditioning,* 3rd edn. Oxford, U.K.: Reed Educational and Professional Publishing, Ltd., 2000, pp. 28–35.

Wang H., S. Touber, Distributed dynamic modelling of a refrigerated room. *Int. J. Refrig.*, 1990; 13: 214–222.

Wang L., D.-W. Sun, Recent developments in numerical modelling of heating and cooling processes in the food industry—A review. *Trends Food Sci. Technol.*, 2003; 14(10): 408–423.

Wang S.K., *Handbook of Air Conditioning and Refrigeration,* 3rd edn. New York: McGraw Hill, 2000.

Whitman W.C., W.M. Johnson, J.A. Tomczyk, *Refrigeration and Air-Conditioning Technology,* 5th edn. New York: Thomson Delmar Learning, 2005.

Whitman W.C., W.M. Johnson, J.A. Tomczyk, E. Silberstein, *Refrigeration and Air Conditioning Technology,* 6th edn. New York: Delmar Cengage Learning, 2009.

Zhou H.-W., S. Lurie, A. Lers, A. Khatchitski, L. Sonego, R. Ben Arie, Delayed storage and controlled atmosphere storage of nectarines: Two strategies to prevent woolliness. *Postharvest Biol. Technol.*, 2000; 18(2): 133–141.

7 Freezing

Maria Giannakourou

CONTENTS

7.1 INTRODUCTION

Freezing is one of the ancient methods of preservation, but its commercialization took place later than canning due to the lack of commercial refrigeration equipment (Maroulis and Saravacos 2003). A rapid increase in sales of frozen foods in recent years is closely associated with increased ownership and better conditions of domestic freezers and frequent use of microwave ovens, due to the lack of time for preparing a "fresh" meal. Frozen foods, as well as chilled foods and ready-to-eat or ready-to-cook foods, give the impression of high quality and "freshness" to the consumer and, particularly in meat, fruit, and vegetable sectors, outsell canned or dried products (Fellows 2000).

The first important factor in food freezing is the rate of change of temperature. It is dependent in part on external conditions, such as the equipment used, and on the size of the object to be frozen (Reid 1998). In recent literature, there is a debate concerning the effect of the freezing rate on different tissues; some researchers suggest that the freezing rate for vegetables is very important for maintaining an acceptable quality, while others point out that the freezing rate of animal tissue is of little importance. Considering the nature of food, food matrices consist of cell walls and cell membranes that act as barriers that separate the interior of the cell from the external environment. In this context, during freezing, due to water gradient, an osmotic phenomenon occurs, leading to water transfer from the interior of the cell into the extracellular medium, which results in the formation of ice crystals. Depending on the speed of heat removal, either slow freezing, leading only to extracellular ice crystals, or fast freezing, leading to both intracellular and extracellular ice, is chosen.

When addressing the advantages of the freezing process, the effects of both the freezing rate (that are related to the freezing process and equipment used) and the conditions of the subsequent frozen storage must be studied in a combined way. A very well-designed, effective freezing process that gives way to a high-quality frozen product can be easily counterbalanced and obscured by an inadequate frozen storage.

In this chapter, the main areas of interest are the freezing equipment used, novel methods proposed for freezing, new approaches for the control and optimization of the current cold chain in frozen food distribution, as well as the latest trends in this industry.

7.2 FREEZING METHODS

Various methods and equipment can be used for industrial freezing. Food products can be either frozen in a separate processing step during the freezing procedure (as in a cold storage area or in a commercial blast freezer) or as part of a continuous process so that the exit line of the manufacturing assembly produces the frozen product (Ramaswamy and Marcotte 2006).

The main criteria used for choosing the most appropriate method/equipment involve the properties and the packaging requirements of the specific food matrix; the availability of simple, easy to clean equipment; the cost of operation; and the effective freezing of the food, combined with the least possible quality loss of the final frozen product. Additionally, one should consider the rate of freezing required, the type of operation (batch or continuous), the scale of production, and the range of products to be processed.

Generally speaking, during freezing, the product to be frozen is brought into contact with a medium that can remove heat. This medium, for example, might be cold air in an air-blast freezer, or a cryogenic fluid such as CO_2 or N_2. These cryogenic refrigerants are at temperatures lower than the temperatures typically attained by the circulation of mechanically refrigerated air.

The type of equipment chosen will have a significant effect on the profitability of the business. Systems that allow for continuous temperature control will minimize product spoilage and can increase the shelf life and the value of the product, but if they use more energy to run, or if they require frequent maintenance and adjustment, they will add unnecessary costs to the production process. Often, there is a trade-off between energy and maintenance in the overall operating cost of the plant (Pearson 2008).

In the literature, there are several ways proposed in order to categorize freezing methods. The most popular one groups the different freezing methods as follows (Ramaswamy and Marcotte 2006):

1. Air freezing
2. Plate freezing
3. Liquid immersion freezing
4. Cryogenic freezing

An alternative classification based on the rate of movement of the ice front is (Fellows 2000)

- Slow freezers and sharp freezers (0.2 cm/h) including still-air freezers and cold stores
- Quick freezers (0.5–3 cm/h) including air-blast and plate freezers
- Rapid freezers (5–10 cm/h) including fluidized-bed freezers
- Ultrarapid freezers (10–100 cm/h), that is, cryogenic freezers

Air freezing, with the use of cold air either by natural or by forced convection, is the most common method employed. The cold air is continuously passed over the food product, and in doing so, it removes heat. On the other hand, in cryogenic freezing, liquid nitrogen (LN_2) or liquid carbon dioxide (LCO_2) is used by the freezing industry; the cryogen is piped as a liquid into the freezer unit (either by "spraying" or by immersing food) (North and Lovatt 2012) and applied directly to the product in a variety of modes, depending on the cryogen, freezer type, or food product. As it will be discussed in this section, each method has pros and cons that should be taken into account when designing the freezing plant.

7.2.1 Air Freezing

This method, which is the most popular one, is mainly used for freezing food products of any shape, prepackaged or not, and of products containing small independent items of similar size, such as green peas.

In this method, the temperature of food is reduced with cold air flowing at a relatively high speed. Air velocities between 2.5 and 5 m/s give the most economical freezing. Lower air velocities result in slow product freezing, and higher velocities increase unit freezing costs considerably (Rahman and Velez-Ruiz 2007). Additionally, in this type of freezing, since convection is the main mechanism of heat transfer, the value of heat transfer coefficient, h, is of crucial importance. In still-air (natural convection), h is typically 5–10 W/(m^2 °C); in blast freezers with very rapid air movement, h may be 20–30 W/(m^2 °C) (Karel and Lund 2003), and subsequently in this latter case, the rate of freezing is much higher.

Air freezers may be either batch or continuous ones, both using cold air of about −23°C to −30°C, that circulates either naturally or with the aid of fans (forced convection). In the latter case, where air velocities obtained are much higher, the process is more controllable since the temperature is more uniform within the freezer, and the air velocity can be modified so as to obtain the desired heat transfer coefficient at the surface of the food product (North and Lovatt 2012).

Batch blast air freezers (Figure 7.1) are mainly used in industrial units of low performance and are the simplest common form of forced convection freezers. In this case, food is stacked on trays in rooms or cabinets where cold air is circulated with the aid of fans. Products, often packaged, are introduced inside the insulated room on trolleys and are usually hung or stacked in racks, so as to allow air to pass over the surface of each item. Blast freezing is relatively economical and highly flexible in that foods of different shapes and sizes can be frozen. The equipment is compact and has a relatively low capital cost and a high throughput (200–1500 kg/h). However, moisture from the food is transferred to the air and builds up as ice on the refrigeration coils, and this necessitates frequent defrosting. Another important disadvantage of this method is that high dehydration losses of up to 5% are observed due to the large volumes of recycled air. Other issues to be kept in mind are freezer burn and oxidative changes to unpackaged or individually quick frozen (IQF) foods (Fellows 2000).

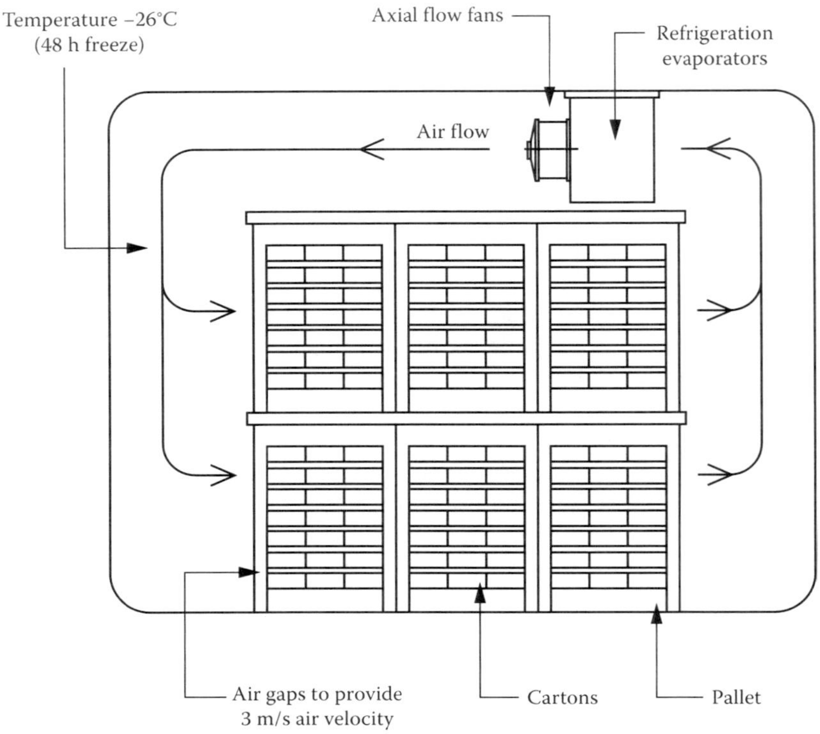

FIGURE 7.1 Batch air blast freezers. (Adapted from Bowater, F.J., Design of carton air blast freezing systems, in *Rapid Cooling of Food, Proceeding of Bristol Meeting*, The International Institute of Refrigeration, Bristol, U.K., 2001.

Continuous air blast freezers frequently include a system that transports the food products through an environment containing air moving at high velocity, meaning that the freezing process may constitute a step into the online process of food production.

Food items may be transferred either in tunnels using continuously moving belts or in fluidized-bed freezers.

7.2.1.1 Tunnel Freezers

In this case, the product is usually driven using a belt or a moving chain through an insulated tunnel through which cold air is forced to flow at high velocity (Figure 7.2a). The moving speed of the belt is specifically calculated in order to obtain complete freezing of the product until the time it emerges out of the tunnel. The direction of the air flow may be the same or the opposite of that of the food or even vertical to that; usually, a countercurrent flow is employed. In many cases, one can distinguish two separate compartments, one used for rapid prefreezing (air temperature −4°C to −10°C) and one for the final freezing (air temperature −32°C to −40°C).

An optimized version of this category is continuous spiral conveyor freezer (Figure 7.2b), where food products that require long freezing times (generally 10 min to 3 h) can be frozen without needing to use extremely long tunnels. A spiral belt freezer consists of a long belt wrapped cylindrically in two tiers, thus requiring minimal floor space. It is also suitable for products that require gentle handling during freezing.

7.2.1.2 Fluidized-Bed Freezers

A fluidized-bed freezer forces cold air up under the product at a high enough velocity to "fluidize" the product, as shown in Figure 7.3. Therefore, air acts as cooling medium and the transport

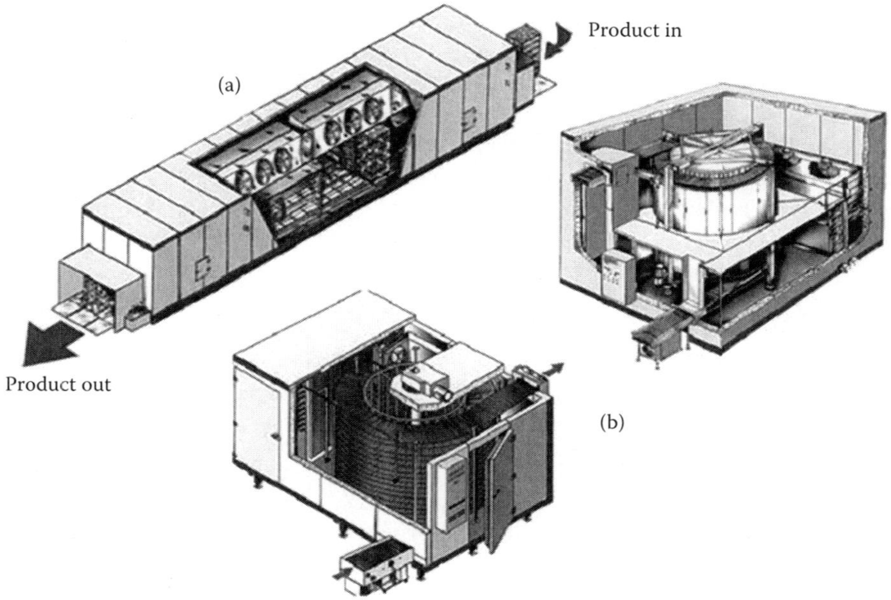

FIGURE 7.2 Continuous air blast freezers. (a) With a moving belt in a tunnel and (b) spiral conveyor freezer. (Adapted from Fikiin, K., Emerging and novel freezing processes, in: Evans, J.A., ed., *Frozen Food Science and Technology*, Blackwell Publishing, Oxford, U.K., 2008, pp. 101–123.)

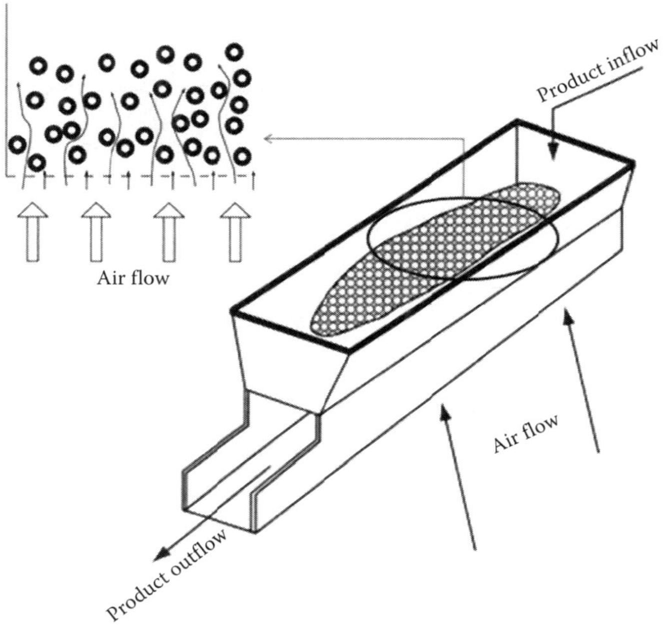

FIGURE 7.3 Fluidization principle and fluidized bed system. (Michael Boast Associates). (Reproduced from Blast and plate freezing, in: Macrae, R., Robinson, R.K., and Sadler, M.J., eds., *Encyclopaedia of Food Science, Food Technology and Nutrition*, Academic Press, London, U.K., 1993.)

medium at the same time, thus products suited to fluidized bed freezing are small and uniform in size (to get easily fluidized) and are not susceptible to damage due to the high velocity mixing that occurs in a fluidized bed (North and Lovatt 2012). Air also prevents individual particles from undesired aggregation into a uniform mass. Common examples are vegetables, such as peas green beans, strawberry slices, corn kernels, and diced carrots.

Parameters that must be controlled in order to obtain complete individual quick freezing (IQF) of products are the circulation speed of the cold air that influences the thickness of the bed formed as well as the rate of input of the product within the freezer.

The main advantages of this type of freezer are: (1) the low requirements for room space, (2) the easy and rapid freezing due to the small sizes and thermal resistance of the IQF products, great overall heat transfer surface of the fluidized foods, and high surface heat transfer coefficients of small particles of uniform size and shape (IQF freezing) (Fikiin 2008), (3) good quality of the frozen products that have an attractive appearance and do not stick together, and (4) minimum possibility of suffering from freezer burn due to the formation of a layer of ice on the entire surface of the product, hindering sublimation.

7.2.2 Plate Freezing

In these freezers, the refrigerant or the cooling medium is separated from the materials to be frozen by a conducting material, often a steel plate. Contact freezing offers several advantages over air cooling, that is, much better heat transfer and significant energy savings. However, the need for regularly shaped products with large flat surfaces is a major hindrance with plate systems and the need to wrap and wash off the immersion liquid in immersion systems (James 2008). The major problem arises from the difficulty in maintaining good contact between the materials being frozen and the heat exchange surfaces. Therefore, this method cannot be applied for all food products; it can be used for flat packages or flat food portions (e.g., blocks of fish flesh) by maintaining pressure upon the plates during the freezing process, thus, enforcing contact with the packages (Karel and Lund 2003). Overall heat transfer coefficients for plate freezers are in the range of 50–100 W/(m^2 °C) in systems maintaining good contact.

There are several subcategories of plate freezers of batch or continuous operation, with horizontal or vertical plates (Figure 7.4a and b, respectively). Horizontal plate contact freezers are usually

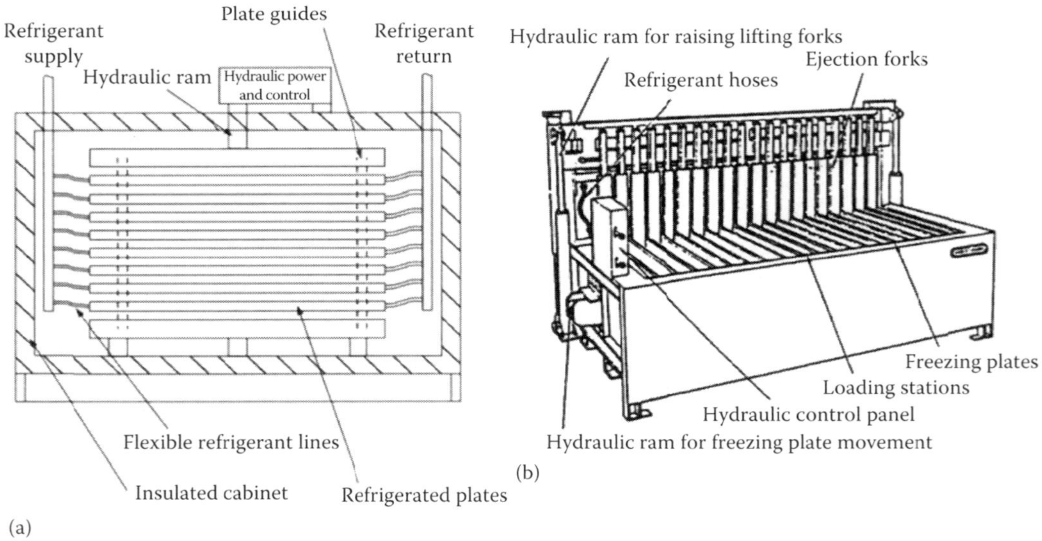

FIGURE 7.4 Plate freezers with (a) horizontal (Michael Boast Associates) and (b) vertical plates. (From www.fao.org.)

used for products already packaged in appropriate regular shapes in order to assure the best contact with the metallic surface, and, thus, the most effective heat removal.

Compared to the horizontal plate freezer mainly used for processed fish products, the vertical version developed later is appropriate for unpackaged material with a flexible shape, such as fish. Due to the vertical openings, the products can be filled in between the plates by weight or volume. For most products, bulk product is required, and the products are frozen to the plates (Magnussen et al. 2008). Products are fed from the top, and the finished block of frozen products are discharged to the side, top, or bottom. Usually, this operation is mechanized. The main advantages of this method are the relatively low initial cost and the good quality retention, whereas the main drawback stated is the slow freezing rate obtained.

7.2.3 Liquid Immersion Freezing

In this method, food is immersed in a low-temperature solution to achieve fast temperature reduction through direct heat exchange (Rahman and Velez-Ruiz 2007). Alternatively, food product, usually prepackaged, can be sprayed with the cold solution, that remains in a liquid form throughout the process. The refrigerating media commonly used are propylenoglycol, glycerol, salt solutions ($CaCl_2$ or NaCl), and sugar or alcohol solutions. These solutions are cooled down to low temperatures without being frozen. The solutes used must be safe to the product in terms of health, and compatible, as far as sensory attributes are concerned.

The immersion freezing in non-boiling liquid refrigerating media is a well-known method having several important advantages: high heat transfer rate, fine ice crystal system in foods, ideal for products of nonuniform shape, easy to introduce in an online continuous procedure, low investments, and operational costs (Lucas and Raoult-Wack 1998; Fikiin 2008). Despite the aforementioned benefits, its applicability is rather limited due to technological reasons, such as the uncontrolled solid uptake from the refrigerating medium, properties of the liquid per se (high viscosity, etc.), the need to protect the food due to direct contact with the refrigerant and high danger of freezer burn. Moreover, when immersion in liquid cooling media is used to freeze packaged foods, some of the advantages of improved heat transfer are lost because of resistance of the package and additional technical difficulties, and costs arise from the need to wash or otherwise remove the media from package surfaces (Karel and Lund 2003).

The main areas of application are for freezing orange juice condensates, poultry (especially in the initial stages of freezing), and fish tissues (Blucas 2004).

7.2.4 Cryogenic Freezing

This type of freezing involves the immediate contact of the food with the refrigerant (or cryogen), while the latter changes phase during freezing (e.g., liquid nitrogen changes from a liquid to a gas, while solid CO_2 changes from the solid to the gas phase). The heat from the food, therefore, provides the latent heat of vaporization or sublimation of the cryogen. Three advantages of cryogenic freezers, compared to mechanical systems, are the very rapid freezing obtained (heat transfer coefficients reported in excess of 200 W/(m^2 °C)), lower capital cost, and flexibility to process a number of different products without important changes to the system (James and James 2003), but they are easily counterbalanced by the high cost of cryogens and their environmental impact and safety. The setup cost of a cryogenic freezing system is approximately one-fourth of the cost of its mechanical counterpart; however, the operating costs are almost eight times. Other benefits include increased production capacity, better quality of food in terms of texture, taste, and appearance, reduced losses due to dehydration and drip, and longer shelf life of the processed food (Shaikh and Prabhu 2007).

Besides liquid nitrogen and solid carbon dioxide, Freon 12 freezers were used in a number of applications in the 1960s and 1970s, including in freezing IQF vegetables and shrimp. However, the

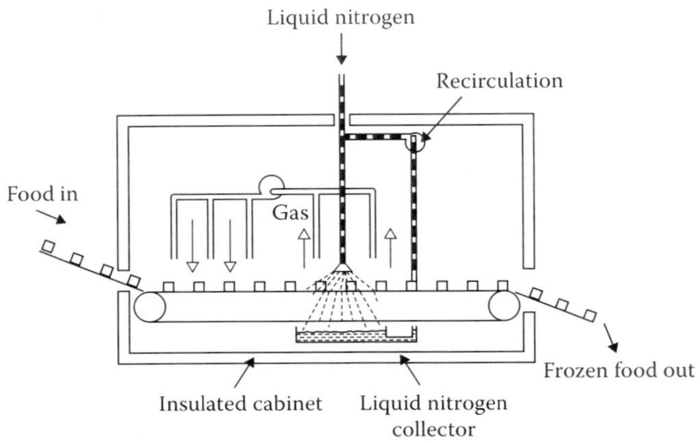

FIGURE 7.5 Liquid nitrogen freezer. (From Fellows, P., Freezing, in: Fellows, P., *Food Processing Technology*, Woodhead Publishing Limited and CRC Press LLC, Boca Raton, FL, 2000, Chapter 21.)

recognition that fluorinated hydrocarbons can severely damage the atmospheric ozone layer led to the complete abandonment of this method of direct freezing (Karel and Lund 2003).

Cryogenic gases can also be used to produce a hard, frozen crust on a soft product to allow for easier handling, packaging, or further processing (Rahman and Velez-Ruiz 2007). Therefore, initially, cryogenic freezing is applied in order to create a frozen crust on a fluid product, after which the product may then be conveyed to a conventional mechanical freezer.

Liquid hydrogen, commonly used, is odorless, colorless, chemically inert, and boils at −195.8°C. When a food tissue is brought into immediate contact with this kind of cooling media, there is a high risk of food cracking due to the instantaneous heat transfer. Therefore, to avoid this phenomenon, in liquid-nitrogen freezers, packaged or unpackaged food travels on a perforated belt through a tunnel (Figure 7.5) where there are two separate compartments, one for prefreezing (using gaseous nitrogen) and the other for the main freezing process. The use of gaseous nitrogen reduces the thermal shock of the food, and recirculation fans increase the rates of heat transfer. For this method, equipment is simple and dehydration losses are significantly limited. Oxygen is removed from food tissue, leading to less oxidative phenomena. Moreover, the sensory attributes are better retained, and there are not many problems of freezer burn. However, due to the high operational cost, its use is limited to high-value products, such as shrimps.

On the other hand, carbon dioxide freezers operate in two ways: either the product is mixed with solid CO_2 (dry ice) that sublimates and freezes the food, or the food is sprayed with liquid CO_2 that evaporates and freezes the food. Taking into account the higher boiling point of CO_2 (−78.5°C) compared to the corresponding value of nitrogen, it can be assumed that food tissues do not undergo such a temperature shock. Nevertheless, gaseous CO_2 is toxic and needs to be immediately removed from the freezing plant. Other problematic issues are the high cost and the texture loss of sensitive food tissues.

7.2.5 Emerging Freezing Techniques

Some new freezing techniques or combinations are being developed for their potential benefits, technical and economical advantages, and quality enhancements (Rahman and Velez-Ruiz 2007, Kennedy 2003).

7.2.5.1 High-Pressure Freezing

High-pressure application during the freezing or thawing of diverse food products has been studied. The application of high hydrostatic pressure to control and enhance the freezing process has been an interesting subject of research in recent decades (Zhu et al. 2005, Otero and Sans 2006, Norton et al.

2009, Tironi et al. 2010, Kiani and Sun 2011, Kiani et al. 2013). The use of high pressure makes high degrees of supercooling possible resulting in an even and fast ice nucleation and growth all over the sample on pressure release. As a result, in contrast with the conventional methods in which an ice front moving through the sample is produced, fine ice crystals are formed, improving the quality of the final frozen product (Kiani and Sun 2011).

According to the phase diagram of water, three different types of high-pressure freezing processes can be distinguished in terms of the way in which the phase transition occurs (Fernandez et al. 2006): high-pressure-assisted freezing (HPAF), high-pressure shift freezing (HPSF), and high-pressure-induced freezing (HPIF). Pressure-assisted means phase transition under constant pressure, higher than the atmospheric, pressure shift means phase transition due to a pressure release, and pressure-induced means phase transition initiated by a pressure increase and continued at constant pressure (Knorr et al. 1998). In HPSF, releasing the pressure once the temperature of the food reduces to the modified freezing point results in a high supercooling effect, and the ice nucleation rate is greatly increased. The main advantage is that the initial formation of ice is instantaneous and homogeneous throughout the whole volume of the product. Therefore, high-pressure shift freezing can be especially useful to freeze foods with large dimensions where the effects of freeze cracking caused by thermal gradients can become harmful (Norton et al. 2009).

Additionally, the use of high pressure in freezing may involve the maintenance of nonfrozen materials at low temperatures and high pressures (temperature in the range of 0°C to −20°C without freezing by maintaining pressure at appropriately high levels of 20–200 MPa).

As shown in Figure 7.6, the freezing point of water and the types of ice formed depend on pressure. Therefore, instead of controlling the temperature, it is possible to affect the ice–water transition by controlling the pressure (Karel and Lund 2003). The water phase diagram shows that at atmospheric pressure, ice crystals will initialize at around 0°C and this water crystallization usually leads to significant damage of the biological food tissues. In order to avoid this major drawback of the freezing process, very rapid freezing rates are necessary. When freezing is realized at atmospheric pressure, these extremely fast freezing rates can only be achieved for very thin layers of 5–25 μm; to overcome this problem, the initial freezing (cryoscopic) point of water can be depressed by adding chemical cryoprotectants or by increasing the ambient pressure. At a pressure of 200 MPa, the freezing point drops to about −22°C (see Figure 7.6), which enables a depth of vitrification of about 200 μm so that objects with a thickness of up to 0.4–0.6 mm could be well frozen (Fikiin 2008). The main advantages of this kind of alternative freezing, short freezing times are accomplished, better ice structure (microcrystalline or vitreous), less mechanical stress during formation of ice crystals, along with much better quality retention.

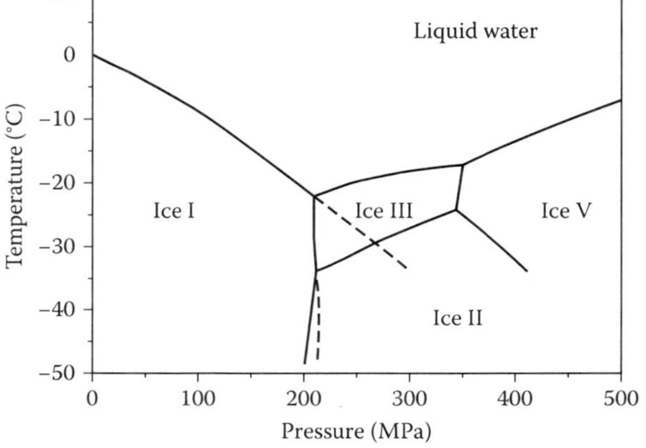

FIGURE 7.6 Phase diagram for water and ice.

The effects of HPSF on the quality of different food tissues has been extensively studied over the past years, focusing mainly on comparing this novel technique with conventional ones (Otero and Sanz 2012). Most studies investigate the effect of HPSF on ice crystals in matrices, such as food gels or vegetables that suffer from a significant drip and shape loss, due to the ice crystals formed. HPSF treatment has been also shown to be effective for animal tissue, such as pork (Zhu et al. 2004) or lobster (Chevalier et al. 2000), since it produces small, rounded ice crystals, evenly distributed in the matrix, protecting in a more adequate way its initial shape and size.

Even if this novel technique sounds quite promising, a lot of extra work is necessary to fully comprehend the effect of low temperature/high pressure on food tissue, to improve the corresponding equipment, and to design a process that is economically viable for the food industry.

7.2.5.2 Ultrasound-Accelerated Freezing

It is generally accepted that high freezing rates are beneficial for frozen food quality as it promotes intensive nucleation and leads to the formation of small ice crystals (Delgado and Sun 2012). Therefore, optimization of the freezing process in terms of ice crystals formation is of major concern for the current food industry. Among other novel freezing techniques proposed, the use of ultrasound, and especially power ultrasound, a kind of ultrasound wave with low frequency (18–20 to 100 kHz) and high intensity (generally higher than 1 W/cm^2) in food processing and preservation seems promising as it is found to improve the freezing and crystallization process (Zheng and Sun 2006, Kiani et al. 2011).

The transmitting of sound waves across the aqueous phase can cause the occurrence of cavitation if its amplitude exceeds a certain level. Briefly, the principles of ultrasound are based on the cavitation phenomenon that strongly depends on a number of parameters of the acoustic waves and the produced bubbles (frequency, pressure amplitude of the sound field, etc.) as well as on the temperature and dissolved gas content. Cavitation has a double role: it can lead to the production of gas bubbles but also the occurrence of microstreaming. The former can promote ice nucleation while the latter is able to accelerate the heat and mass transfer process related to the freezing process. Ice crystals will fracture when subject to alternating acoustic stress, consequently leading to products of smaller crystal size distribution, which is one of the most important targets of the freezing industry regarding the retention of high quality of frozen food stuffs (Zheng and Sun 2006).

This technique is already used in accelerating ice nucleation process, but recently it has been shown that its applicability can extend to freeze concentration and freeze drying processes in order to control crystal size distribution of frozen food products. When applied to fresh food tissues, ultrasound can not only increase the freezing rate but also improve the quality of the frozen products. An example of effective use of this method is in the ice cream industry, where ice crystal size is adequately reduced; high heat transfer rates are accomplished, thanks to the elimination of encrustation phenomena, etc. (Zheng and Sun 2006, Awad et al. 2012).

7.2.5.3 Magnetic Resonance Freezing

One of the main issues in the freezing process is undesirable cell dehydration and water transfer; therefore, retaining water within the cell is one of the most important aspects when designing a freezing process, as it ensures good preservation of the initial attributes of the raw material. A system for magnetic resonance freezing (MRF) preventing such cellular dehydration could be composed of a common freezer and a special magnetic resonance device (Fikiin 2008).

Initially, the food undergoes continuous magnetic wave vibrations, which impedes crystallization and obtains supercooling below the initial freezing point. In the next step, after a suitable product-specific period of time, the magnetic fields are abruptly removed, and a flash freezing of the entire food volume is accomplished with many resulting quality benefits for the frozen end product. By applying this technique, the critical zone of water crystallization is passed rapidly, cellular dehydration is significantly reduced, fine ice crystals are formed, tissue damage is minimized, and the integrity of food tissue is maintained.

At present, MRF data are still kept as a confidential know-how of a number of companies, while MRF equipment still needs to prove its claimed advantages and capabilities through extensive tests within a sufficiently representative industrial environment (Fikiin 2008).

7.2.5.4 Hydrofluidization and Ice Slurries

The hydrofluidization method (HFM) for fast freezing of foods was suggested in order to overcome the drawbacks and to bring together the advantages of both air fluidization and immersion food freezing techniques (Fikiin 2008). This novel technique uses a circulating system that pumps the refrigerating liquid upwards through orifices and/or nozzles into a refrigerating vessel, thereby creating agitating jets and increasing heat transfer to foods during freezing (Verboven et al. 2003, Peralta et al. 2009). These form a fluidized bed of highly turbulent liquid and moving products, and thus evoke extremely high surface heat transfer coefficients. The main advantages are: the use of equipment of small size and the improved IQF freezing (individual quick freezing of small particles). HF enhances the transfer coefficients involved. The combination of these high coefficients with the use of small food samples leads to processes in which the transport phenomena within the food is affected by the fluid flow over the food samples (Fikiin, 2008, Peralta et al. 2010).

Ice slurries are mixtures of microcrystals of ice in a carrier liquid that do not freeze under the operating conditions used (Torres de Maria et al. 2005, Fumoto et al. 2013). The liquid can be either pure freshwater or a binary solution consisting of water and a freezing point depressant. Sodium chloride, ethanol, ethylene glycol, and propylene glycol are the four most commonly used freezing point depressants in industry (Kauffeld et al. 2010). The latent heat due to melting of the ice makes it possible to improve the superficial coefficients of exchange with respect to those obtained using conventional single-phase fluids.

Ice slurries have been only recently used in the food industry in the chilling and freezing of foods, using direct contact in fluidized bed systems. Preliminary studies show a dramatic increase in freezing rates compared with alternative technologies. Davies (2005) reports, for example, freezing small fish and various vegetables and fruits, freezing times can be as short as 1–10 min compared with 60 or more minutes for simple immersion cooling. The reason for such a high heat transfer coefficient is the high rate of agitation created in the fluidized bed, which constantly supplies fresh slurry to be in contact with the surface of the object being cooled.

The ice slurries reveal a great energy potential as hydrofluidization method refrigerating media whose small ice particles absorb latent heat when thawing on the product surface (Fikiin 2008). As mentioned earlier, the goal of the ice slurry introduction is to provide an enormously high surface heat transfer coefficient (of the order of 1000–2000 $W/m^2/K$ or more), excessively short freezing time, and uniform temperature distribution in the whole volume of the freezing apparatus. Besides the aforementioned advantages, according to Fikiin (2008), there are more important issues obtained through HFM:

- The critical zone of water crystallization (from −1°C to −8°C) is quickly passed through, protecting cellular tissues from damage.
- As the product freezes immediately in a solid crust, appearance is well retained, as osmotic transfer is significantly limited.
- By appropriately selecting composition of the HFM media, new attractive products can be formulated with extended shelf life and improved sensory characteristics.
- The operation is continuous, easy to maintain, convenient for automation, and further processing or packaging of the HFM-frozen products is considerably easier.
- When ice slurry is used as HFM agents, they may easily be integrated into systems for thermal energy storage, accumulating ice slurry during the night at cheap electricity charges.
- The HFM freezers use environmentally friendly secondary coolants (for instance, syrup type aqueous solutions and ice slurries), and the refrigerant is closed in a small, isolated system, in contrast to the conventional and harmful air fluidization technology of HCFCs and expensive HFCs where there is a much greater risk of emission to the environment.

7.2.5.5 Application of Antifreeze Proteins and Ice Nucleation Proteins

As a part of their protective system or as a source that provides the basic nutrients, antifreeze proteins (AFPs) are produced by many cold climate organisms, ranging from plant, to bacteria, to animal (Kiani and Sun 2011, Wang and Sun 2012). Antifreeze proteins (AFPs), antifreeze glycoproteins (AFGPs), and thermal hysteresis proteins (THPs) have the ability to bind ice and modify the normal growth of ice crystals. These proteins are found to inhibit ice crystallization at lower concentrations or even to inhibit complete ice growth at higher concentrations over a temperature range; the latter is of major importance for the quality of frozen foods, since high supercooling during the freezing process leads to fine ice crystals. However, this supercooling does not follow Raoult's law, meaning that supercooling in this case does not depend on concentration, instead proteins interact in an active way with ice. Actually, proteins, bind themselves—with their hydrophilic portion—to specific planes of ice crystals when they encounter a growing ice front via van der Waals interactions and/or hydrogen bonds (Cruz et al. 2009). The basis for adsorption specificity lies in a hydrogen-bonding match between groups on the ice-binding site of the AFP and oxygen atoms on the ice lattice. This leads to the suppression of growth of ice nuclei and decreases the freezing point, hence inhibiting ice formation and changing the growth rate.

Natural antifreeze proteins and glycoproteins have been found to protect from freezing damage and, thus, extend shelf life for many species, such as fish, plants, insects, and bacteria living in subzero temperatures (Hassas-Roudsari and Goff 2012). AFPs may inhibit recrystallization during freezing, storage, transport, and thawing, thus preserving food texture by reducing cellular damage and also minimizing the loss of nutrients by reducing drip. Besides naturally occurring antifreeze proteins, AFPs may be introduced into other food products either by physical processes, such as mixing and soaking, or by gene transfer (Griffith and Ewart 1995, Zhang et al. 2008).

During freezing, ice formation is initiated by ice nucleation, which can be promoted by the presence of foreign particles that act as ice nucleation activators (INAs) (Wang and Sun 2012). Ice nucleation proteins have been found to be produced by certain gram-negative bacteria that promote the nucleation of ice at temperatures higher than the initial freezing point. In other words, they can catalyze the formation of ice in undercooled water at high subzero temperatures (Zhang et al. 2008). Moreover, INAs have been shown to create large and long ice crystals in ordered directions (Kiani and Sun 2011).

As far as their application in the food sector, many INA bacterial cells have been studied for their effect on overall quality of frozen foods during cold storage (Li and Lee 1995, Hassas-Roudsari and Goff 2012). The common species of ice nucleation active (INA) bacteria found to produce ice nucleation activator belong to genera *Pseudomonas*, *Erwinia*, and *Xanthomonas*. The main advantages of INA are that they elevate the temperature of ice nucleation, shorten the freezing time, and change the texture of frozen foods, thus decreasing energy cost and improving the quality (Li and Sun 2002). For example, samples of egg white when frozen at −10°C underwent supercooling lower than −6°C, but when INA bacterial cells (*Erwinia ananas*) were added, the samples showed only a slight degree of supercooling (Arai and Watanabe 1986).

Despite the very encouraging results concerning the effectiveness of INA bacteria on the quality of frozen foods, one major concern to their applications in the food industry is that bacterial ice nucleators must be environmentally safe, nontoxic and nonpathogenic, and palatable (Li and Lee 1995). If whole bacterial cells are added, it is necessary to ensure that inedible microorganism is completely removed from the food prior to consumption. As an alternative, an enzyme-modification procedure has been proposed (Wand and Sun 2012) in order to produce small linear peptides with antifreeze activity (enzyme-modified antifreeze proteins, EMAFPs).

7.2.5.6 Dehydrofreezing

Dehydrofreezing is a variant of freezing in which a food is dehydrated to a desirable moisture and then frozen (Li and Sun 2002). Dehydrofreezing provides a promising way to preserve fruits and vegetables by removing a part of water from food materials prior to freezing (Biswal et al. 1991). A reduction in moisture content would reduce the amount of water to be frozen, thus lowering refrigeration load during freezing. In addition, dehydrofrozen products could lower the cost of packaging, distribution, and storage and maintain product quality comparable to conventional products (Biswal et al. 1991).

The first part of the process involves partial dehydration of the product by immersion in a hypertonic solution in order to decrease the amount of crystals formed during the freezing process (Dermesonlouoglou et al. 2007). Minor damage of the cellular membranes occurs, and therefore a better conservation of the fruit and vegetable properties is assumed (Marani et al. 2007, Ramallo and Mascheroni 2010). In this sense, osmotic dehydration was reported as a pretreatment in freezing (Giannakourou and Taoukis 2003, Torreggiani and Bertolo 2004), hence it will be further detailed in the following section, concerning treatments prior to freezing. Briefly, the osmotic step before freezing aims at the lowering of the water activity and the partial dehydration of the food matrix; at the same time, by carefully designing the composition of the osmotic solution in which the food is immersed, a significant enrichment can be obtained with physiologically active components, such as prebiotics, vitamins and minerals, dietary fiber, fish oils, and plant sterols, and novel food products with functional properties may be produced (Fito et al. 2001).

7.2.5.7 Other Novel Technologies for Food Freezing

Electrostatic field-assisted freezing is a very recent proposal that investigates the impact of electrical disturbance in terms of phase change. Supercooling of several foods with the application of an external electric field is studied, and the preliminary results seem really promising regarding the area of improvement of the size of ice crystals (Le Bail et al. 2012).

The application of microwave irradiation has also been proposed in order to suppress ice nucleation; in fact, Jackson et al. (1997) found that microwave irradiation results in the formation of an ice-free (vitrified) region adjacent to the cooling block (Kiani and Sun 2011). In the same paper, it was shown, that the combined use of both microwave and cryoprotectant (ethylene glycol) was successfully used to influence ice formation and enhance vitrification even at relatively low cooling rates and for relatively low cryoprotectant concentrations.

7.3 TREATMENTS PRIOR TO FREEZING

As discussed in detail in a previous chapter, freezing almost always causes physical and chemical changes in food due to ice formation and a subsequent loss of quality. In that sense, there is an increased interest in using pretreatments prior to the detrimental process of freezing in order to enhance the quality of frozen/thawed fruit and stabilize the frozen tissue (Chassagne-Berces et al. 2010). Conventional pretreatments concern washing, blanching, and soaking and treatments such as comminuting, coating, grinding, and packaging. Apart from the traditional ones, some of the pretreatments suggested are quite novel and include the addition of different sugars (Chiralt et al. 2001, Marani et al. 2007) and the addition of calcium or low-methoxyl pectin (Buggenhout et al. 2006).

7.3.1 Washing

In order to remove soil and other contaminants from vegetable or fruit tissues, washing is really important for final product quality. The most crucial role of this step of the process is to reduce substantially the microbial load of the fresh produce. On the other hand, if the washing treatment has not been applied properly, this step can cause cross-contamination during the following steps of the process (Olaimat and Holley 2012). Vegetables are washed typically with water that generally contains free chlorine from approximately 0 to 30 ppm. Chlorine and chlorinated compounds are still the most widely used sanitizers in the food industry (Beuchat et al. 2004), despite scientific data published showing that excessive use of chlorine can be harmful due to the formation of carcinogenic disinfection by-products (trihalomethanes, chloramines, haloketones, chloropicrins, and haloacetic acids) caused by the reaction of residual chlorine with organic matter (Cao et al. 2010, Hernandez et al. 2010). In view of this risk, these compounds have been forbidden in many European countries, and there is a trend in eliminating chlorine-based compounds from the decontamination and disinfection process; instead, innovative and emerging technologies are proposed to be applied in the food industry. For example, ultrasound technology can be adapted in the washing tank for decontamination of fruit and vegetables where the ultrasonic waves can be generated from the surface of the tank (Bilek and Turantas 2013). For more effective application, it is suggested that ultrasound be combined with other methods. This multiple hurdle concept constitutes an attractive approach to enhance microbial inactivation as previous works have demonstrated the hurdle effect in different fruits and vegetables, such as plum fruit (Chen and Zu 2011), strawberries (Cao et al. 2010, Alexandre et al. 2012), apples and lettuce (Huang et al. 2006), and red bell pepper (Alexandre et al. 2013).

7.3.2 Heat Treatments: Blanching

Blanching is a very common surface heat treatment of vegetables and fruits prior to freezing. It is achieved by immersion in hot water or steaming at temperatures close to 100°C. Steam blanching has the advantage of avoiding the leaching out of solids, which is important regarding wastewater and environmental regulations. At present, blanching continues to pose serious environmental issues to the food industry (Torreggiani et al. 2000).

The aim of blanching is the inactivation of enzymes through the denaturation of proteins. Since enzymes are responsible for the majority of degradation reactions in food matrices, the effect of blanching is crucial for retaining food quality during subsequent freezing and frozen storage. This operation is a thermal process designed to inactivate the enzymes responsible for generating off-flavors and odors and to achieve the stabilization of texture and nutritional quality and the destruction of microorganisms (Olivera et al. 2008). Blanching has also other beneficial side effects, such as the enhancement of color of green vegetables and carrots (Patras et al. 2011, Martinez et al. 2013), an initial, low reduction of microorganisms, a decrease in pesticide and nitrate levels in spinach, carrots, etc. It may also have specific applications in specific tissues, such as in potato, in which it adjusts water content, reduces frying time, and improves its texture (Abu-Ghannam and Crowley 2006). Especially in vegetables, the effect of blanching consists in inhibiting enzyme activity and, thus, reducing the enzymatic browning effect caused by enzymes, such as peroxidase, lipoxygenase, and chlorophyllase, blocking the development of the foul smells for which lipoxygenase and protease are responsible, stabilizing the nutritional value of the product, and preventing the oxidating activity of ascorbic acid (Barrett and Theerakulkait 1995, Bevilaqua et al. 2004).

However, being a mild heat treatment, there are important disadvantages concerning the application of blanching, especially on fruit tissue that is more sensitive to texture loss. The most prevalent effect is the loss of turgor in cells due to thermal destruction of membrane integrity and partial degradation of cell wall polymers (Bahçeci et al. 2005, Petzold et al. 2013). Apart from cellular tissue damage, there is also an increased risk of microbial contamination due to the removal of natural microflora and absorption of water by the food, which is usually undesirable since it alters the yield. Moreover, some initial quality deterioration is sometimes observed, which may however be counterbalanced by protecting the food at the following steps of freezing and frozen storage.

In order to assess the effectiveness of blanching, the enzyme of peroxidase is conventionally used to monitor and evaluate the blanching extent since it is one of the most heat stable enzymes, occurring in a considerable number of vegetables (Concalves et al. 2007).

Current food industry is interested in designing the blanching step so as to minimize its negative effect and maximize its beneficial results during the subsequent freezing process and storage (Bevilaqua et al. 2004). This can be accomplished by using high-temperature short-time exposures rather than longer times at milder conditions. Additionally, there are numerous recent publications proposing the immersion of foods in aqueous solutions containing specific molecules that can improve the effectiveness of blanching (Martinez et al. 2013). Citric acid may be added in order to lower the pH value, ascorbic acid or other antibrowning agents to minimize color degradation (for instance, mushroom browning or cabbage yellowing), $CaCl_2$ to enhance and protect tissue integrity, etc.

In a number of published articles, a short microwave treatment is proposed prior to the thermal procedure of blanching in order to minimize its negative side effects (Dorantes-Alvareza et al. 2011, Zheng and Lu 2011).

Due to the negative effect of heat treatment on food quality, a cooling step is strongly advised immediately after blanching and before the freezing procedure. Rapid cooling is preferred, and higher yields have been observed with immersion blanching-air blast chilling (Torreggiani et al. 2000).

7.3.3 Partial Dehydration and Formulation Pretreatments

Treatments prior to freezing can minimize the damaging effect of texture degradation of fruits at thawing (Huxsoll 1982, Maestrelli et al. 2001). Water removal through partial air drying, dewatering–impregnation soaking (DIS) in concentrated solution (Torreggiani et al. 2000) or their combination has been shown to improve significantly frozen food cell integrity and protect its structure even in the case of delicate tissues.

Another technique is immersion chilling and freezing in concentrated aqueous solutions, which makes it possible to combine formulation (dewatering and impregnation) with precooling.

7.3.3.1 Partial Air Drying

Partial dehydration is generally achieved by air drying. When followed by freezing, the resulting process is termed dehydrofreezing (Kennedy 2000, Torreggiani et al. 2000). The main advantages include energy savings (reduced water load for freezing and subsequent transport) and better quality and stability (Huxsoll 1982). After partial air drying, water activity of food products remains relatively high, $a_w > 0.96$ since water removal does not exceed 50%–60% of the initial water content. Therefore, appropriate pretreatments such as blanching or dipping in antioxidant solutions are necessary to avoid browning (Giangiacomo et al. 1994). Dehydrofrozen fruits may be used as fresh-like substitutes in frozen fruit salads, surface garniture or as fillings in pastry. This partial dehydration step has been proven effective as a pretreatment to the following freezing process for apple, pear, and clingstone peach (Torreggiani et al. 2000).

7.3.3.2 Osmotic Dehydration or Dewatering–Impregnation Soaking

Instead of conventional air drying (or in combination with this traditional treatment), recently there is a lot of interest in applying osmotic dehydration as a prefreeze treatment. This process involves immersing the solid food material (as a whole or in pieces) into hypertonic solutions, usually of high carbohydrate or salt concentration and the type of solute used depends on the desired purpose

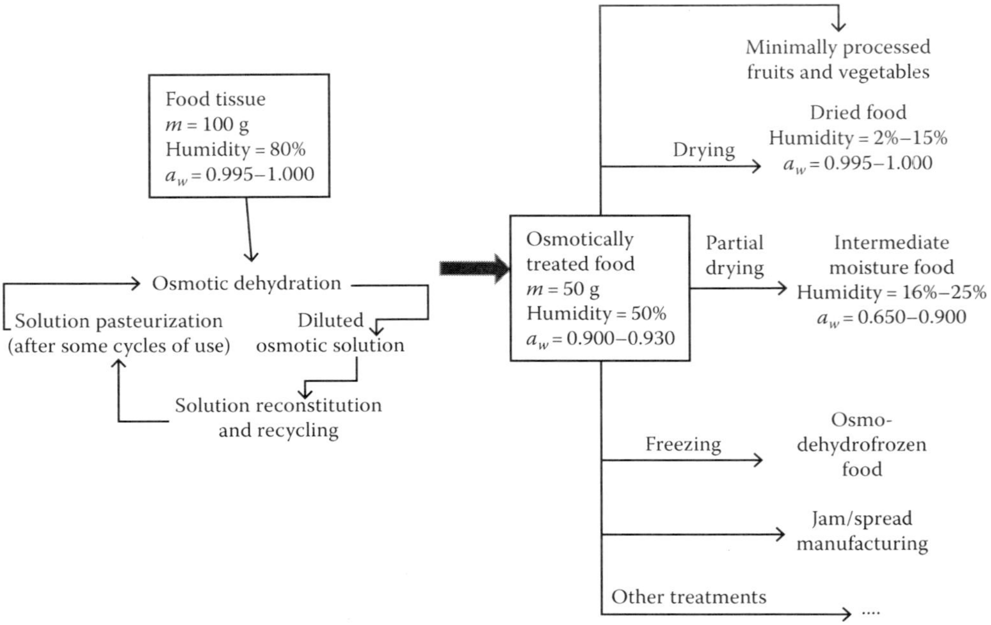

FIGURE 7.7 Representative flow chart for the application of osmotic dehydration as a pretreatment.

(Rahman and Velez-Ruiz 2007). This kind of treatment is recently proposed as a preliminary procedure prior to numerous subsequent preservation methods, following the principles of hurdle technology (Figure 7.7).

During the osmotic procedure, there are two main countercurrent flows that take place through the food membrane due to mass gradients: water flows out of the food into the concentrated solution and an opposite transfer of solute towards the interior of the food matrix. Apart from these main mass transfer flows, other secondary exchanges take place, such as loss of water soluble food components, which may significantly alter food sensory attributes.

This method, when applied prior to freezing, leads to concentration of the intracellular material, reduces significantly the "unbound" water, reduces the freezing point, and has a lot of advantages that make it a very effective pretreatment. The main advantages are (Pinavaia 1988, Torreggiani 1995):

- Reduction of the total latent heat of freezing
- Less energy required to freeze the osmosed food tissue
- Higher freezing rates and improved crystallization, favoring the formation of small ice crystals
- Significant reduction of volume and weight of frozen food that, in the case of high initial water content (i.e., in fruit tissues), may reach 50%
- Better sensory attributes, such as texture, taste, and flavor after thawing
- Less drip loss during subsequent thawing
- Control of food consistency

In addition to obtaining reduction of the energy load, costs of packaging and transport are also significantly reduced at the same time (Huxsoll 1982). Final products that occur by this combined procedure (osmotic dehydration-freezing) are called "osmo-dehydrofrozen."

The most interesting feature of this treatment is the uptake of solutes that modifies food tissue depending on the composition of the osmotic solution decided (Kennedy 2000). Therefore, after DIS, a new, improved product, with optimized functional, sensory, and nutritional attributes may occur by:

- Adjusting the composition of food by reducing water activity or by adding water activity lowering agents
- Introducing ingredients into the food tissue with antioxidant, antimicrobial, or other preservative properties
- Incorporating in the osmotic solution additives with well-established nutritional, health, or sensory benefits (Fito et al. 2001)
- Providing increased consistency to the food tissue due to dewatering (Torreggiani et al. 2000)

Recently, there is a lot of interest in a new approach that relates frozen food stability with the glass transition theory (Giannakourou and Giannou 2015), which is based mainly on kinetic instead of energetic criteria to study a number of changes. In this context, during osmotic dehydration, due to the solute uptake, the kinetic hindering of diffusion-controlled reactions and molecular mobility should have also been considered. According to the glass transition theory (Slade and Levine 1991, Levine and Slade 1992), chemical and physical stability is related to the molecular mobility of the unfrozen phase, which in turn is strongly related to the glass transition temperature. Numerous studies have been published that focus on the effect of the osmotic step to the change of the glass transition temperature of the modified tissue (T_g') that significantly affects its overall quality and stability.

Due to the increased interest of the application of an osmotic pretreatment prior to conventional freezing, in Table 7.1, some representative literature data is summarized, mainly for vegetative tissues (fruits and vegetables). Additionally, there are many important publications concerning the modeling of the procedure based on mass and thermal balances (Agnelli et al. 2005, Goula and Lazarides 2012) and/or diffusion kinetics (Dermesonlouoglou et al. 2008).

7.4 POST-FREEZING PROCESSES

7.4.1 Frozen Food Packaging

As in all preservation processes, post-procedure treatments are of increased importance in order to maintain the quality and stability of the final product. Therefore, the main aim of packaging is to protect processed food from after-freezing contamination and to ensure its integrity throughout the frozen chain from the producer to the final consumer. Additionally, since the product will be exposed to very low temperatures, the packaging material needs to withstand these conditions and protect food tissue from quality problems, such as freezer burn or color fading (Cooksey and Krochta 2012). An important issue is that the packaging material chosen should meet certain minimum technical, legislative, and environmental requirements. Other required attributes are convenience and health safety to the consumer, simplicity and attractiveness to the producer, all combined to minimum environmental impact and sensory modification of the product.

The basic requirements a material should meet in order to be used for frozen food packaging are (George 2000):

- Chemical and physical stability in a wide temperature range (−40°C up to almost room temperature) or even at higher temperatures if it is designed to be cooked inside its packaging, as in ready-to-cook meat products.
- Adequate barrier attributes, considering mainly permeability to air, water, light, and vapors. The packaging material must be chosen in order to minimize moisture loss, that usually leads to weight and quality loss. If packaging is designed so as to practically eliminate free space around the food, frost formation, desiccation, and formation of off-flavors can be significantly minimized.

TABLE 7.1
Representative Publications on the Application of Osmotic Dehydration Prior to Freezing

Food Tissue-Parameters Measured	Parameters of the Osmotic Treatment	Temperature Conditions of Freezing (and Thawing)	Combination with Other Technique	Reference
Mango, kiwi, and strawberry—mechanical properties	30°C, in sugar syrup 35°Brix, 45°Brix, 55°Brix, 60°Brix, and 65°Brix	−18°C for 1 and 30 days (also after thawing)	Vacuum	Chiralt et al. (2001)
Green peas—color, texture, hardness, and sensory attributes	8°C, 20°C and 40°C, in a solution of 5%, 10%, and 17% NaCl	−17°C for 10 days	—	Biswal et al. (1991)
Kiwi—color and mechanical properties	30°C, in sugar syrup 35°Brix, 45°Brix, 55°Brix, and 65°Brix	−18°C for 1 and 30 days (also after thawing)	Vacuum	Talens et al. (2002)
Melon—texture, color, flavor, and sensory attributes	25°C for 1 h, in sugar syrup 60% w/w	−20°C for 48 months	Combination with air drying	Maestrelli et al. (2001)
Apricots—color, vitamin C, and T_g'	25°C for 45 and 120 min (atm. pressure), in a solution 65% w/w saccharose, maltose, and sorbitol + 0.1% ascorbic acid + 0.1% NaCl	−20°C for 8 months	Vacuum	Forni et al. (1997)
Strawberry—texture, visualization with microscope of thawed samples	25°C, 37.5°C, and 50°C for 0.25, 7.625 και 15 h in solutions of sugar and $CaCl_2$	−20°C for 2 months and subsequent thawing	—	Suutarinen et al. (2000)
Strawberry—drip loss after thawing	5°C for 21 h, in solutions of glycerol, glucose and saccharose (10%–60% w/v)	−20°C for 1 day	—	Garrote and Bertone (1989)
Strawberry, apricots, and cherries—texture and sensory for thawed samples	25°C for 8 and 16 h, in corn syrup + 1% NaCl	−20°C and thawing		Pinnavaia et al. (1988)
Apples—color, microscopy, and texture after thawing	50°C for 40 min, in 75% corn syrup, 52% saccharose, and 50% sorbitol	−35°C and thawing	Vacuum	Tregunno and Goff (1996)
Apples—drip loss of thawed samples	50°C for 5 h, in 55% corn syrup (38DE)	−40°C for 1 h and thawing	—	Lazarides et al. (1995)
Kiwi—texture and color of thawed samples	30°C, in sugar syrup 60°Brix and 72°Brix	−40°C for 1 h and thawing	Combination with air drying	Robbers et al. (1997)

(*Continued*)

TABLE 7.1 (*Continued*)
Representative Publications on the Application of Osmotic Dehydration Prior to Freezing

Food Tissue-Parameters Measured	Parameters of the Osmotic Treatment	Temperature Conditions of Freezing (and Thawing)	Combination with Other Technique	Reference
Pineapple—mechanical properties, ascorbic acid, and drip loss of thawed samples	40°C, in sucrose syrup 60°Brix for 30–240 min (kinetic study)	−31.5°C for 2 h and thawing at 20°C for 2 h	—	Ramallo and Mascheroni (2010)
Pears, kiwi, strawberry, and apples—color, texture, and drip loss of thawed samples	30°C, in solutions of sucrose, fructose, HMW (high molecular weight sugars), 47°Brix–69°Brix for 1–24 h (kinetic study)	Freezing at −40°C (conventional air-blast tunnel)		Marani et al. (2007)
Carrots—texture, microscopy (and after thawing)	10°C for 5 h, in a 50% w/v sucrose solution	−18°C for 7 days and thawing at 25°C for 1 h		Ando et al. (2012)
Green peas—color, texture, DCS (glass transition temperature measurement)	35°C for 4 h or 5°C for 12 h, in a 56.5% w/w oligofructose or maltitol or oligofructose/trehalose solution, adding NaCl and $CaCl_2$	Freezing at −40°C for 2 h. Kinetic study at −3°C to −24°C	—	Giannakourou and Taoukis (2003b)
Tomatoes—color, texture, lycopene, ascorbic acid, and sensory evaluation	35°C for 1 h, in a 56.5% w/w high DE maltodextrin solution, adding NaCl and $CaCl_2$	Freezing at −40°C for 2 h. Kinetic study at −5°C to −20°C		Dermesonlouoglou et al. (2007)
Cucumber—color, texture, and sensory evaluation	15°C for 360 min or 35°C for 300 min or 55°C for 180 min, in a 56.5% w/w high DE maltodextrin or oligofructose solution, adding NaCl and $CaCl_2$	Freezing at −40°C. Kinetic study at −5°C to −15°C		Dermesonlouoglou et al. (2008)
Strawberry—phenol content, volatiles, and consumer acceptance	30°C for 4 h, in sucrose solution (50% w/w)	Stored at −18°C for 1 month	Vacuum	Blanda et al. (2009)

- Another important prerequisite is to maintain as long as possible low food temperature, when being distributed or transported in the cold chain. Thus, the packaging material needs to own good insulation properties.
- Attractiveness to the consumer is very important. The key issue is the ability to print onto the surface of the product, which means that the appropriate outer layer should be chosen, based on its smoothness, wettability, absorbency, and compatibility to the ink used, Additionally, newer forms of presenting products to the consumer make frozen food items quite appealing, such as shrink-wrap films that combine good gas barrier properties with the desired transparency.

Compatibility to packaging equipment, since in most cases, packaging is an additional step in a continuous automated production line. There is a compromise between food packaging material attributes and existing machinery, including filling lines, conveyors, sealing systems, etc.

The main material categories used for frozen foods include plastics, paperboard, metals (aluminum), and corrugated paperboard. Usually, depending on special requirements of frozen food tissue, food packaging is a combination of different materials, each of them meeting a different requirement. Briefly, the most common materials used are:

- *Plastics*: Polyethylene is available in different densities (LDPE, low density and HDPE, high density), usually used in vegetable bags. Other polymers used are polypropylene (PP) and polyester (PET) that have good moisture resistance but lack toughness and heat resistance (Cooksey and Krochta 2012). Trays of polyester are suitable for reheating in conventional and microwave ovens, with stability at high temperatures, exceeding 250°C. Polystyrene (PS), polyamide (PA), and polyvinyl chloride (PVC) are also used in different applications. However, the most common practice is to combine films of different polymers with different properties in order to tailor the packaging materials and gain the specific attributes required through laminates and co-extrusions.
- *Metals*: Aluminum properties make it suitable for packaging frozen food. Aluminum trays or pans are common for frozen foods that are designed to be immediately heated in an oven by the consumer at home, such as pizzas and pies (Hasselman and Scheer 2012). Aluminum foil can also be used laminated to plastic films and paperboard due to its superior light and moisture barrier properties.
- *Paperboard*: It is often used either in the form of dual ovenable trays for susceptor materials or as a secondary packaging in the form of boxes. Depending on their thickness, the materials used can be paper (thickness up to 3 mm), board (thickness between 3–11 mm), or fiberboard, all made from wood pulp. A basic advantage of this packaging material is that it is made from a renewable source. When paperboard is used in the primary packaging and food is designed to be cooked within its package, paperboard is commonly coated with polyester (PET) to provide heat resistance for microwave and conventional oven heating (referred to as dual ovenable).

Paper and board can also be laminated with PE or waxed to reinforce water barrier properties or used in a laminate as a surface coating to provide a smooth surface for ink and printing of superior quality.

- Corrugated paperboard is often used as a tertiary packaging of frozen foods, and the basic designing parameters are the weight of linerboard, corrugating medium used, and the height of the flutes, as well as the number of flutes used per unit length of the board (Robertson 2012).

Finally, the choice of the packaging material is limited and often detected by legislative and environmental regulations. There are detailed directives that involve all sectors of the food chain in order to protect the environment from packaging waste and ensure minimum energy and raw materials' consumption for packaging.

As far as developments and future trends are concerned, active packaging, designed to perform a specific function while acting as a physical barrier, seems a really promising area for the frozen food sector. Recently, the application of modified atmosphere packaging (MAP) has obtained significant extensions to the quality shelf life of fresh and chilled agricultural and horticultural produce (George 2000). Other alternative forms of packaging are edible films and coatings, usually made of natural constituents. Such films and coatings are applied directly on to the food product surface and become an integrated part of the product enclosed.

Intelligent or smart packaging is another development in food packaging; intelligent systems are capable of providing information about the time–temperature history and the properties of a packaged food and can provide evidence of pack integrity, tamper evidence, product safety and quality. These systems include time–temperature indicators (TTIs), gas sensing dyes, microbial growth indicators, and physical shock indicators. A lot of studies have been recently published focusing on the effectiveness of TTI use as a part of food packaging in order to assess product quality status and therefore optimize the existing inventory systems. The use of TTIs in the frozen food cold chain will be further discussed in the following section that presents the current frozen food life cycle from production to consumption level.

The recent application of vacuum packaging in frozen food has the aim of minimizing deteriorative reactions, including microbiological and chemical changes and is especially used in ready meals (Rachtanapun and Ractanapun 2012). This technique is actually a form of modified atmosphere packaging since food is placed in a gas-permeable, heat-stable pouch or film, most of the oxygen in the vicinity of the food item is removed, and the package is hermetically sealed. These sous vide packed foods are usually cooked under controlled conditions of temperature and time (low temperature–long time). After cooking, the products are rapidly chilled and kept frozen and reheated according to specific instructions before final consumption (ready-to-cook). Vacuum packaging inhibits the growth of aerobic microorganisms, however, favoring at the same time the growth of anaerobic ones, which means that the thermal process of sous vide procedure must be carefully designed so as to inactivate vegetative pathogens and botulinum spores.

7.4.2 Thawing

Thawing is not the adverse process to freezing due to the different physicochemical properties of water and ice. Thawing is a substantially longer process than freezing when temperature differences and other conditions are similar (Fellows 2000). Water has a significantly lower thermal conductivity and a lower thermal diffusivity than ice; therefore, as heat is induced and the food starts thawing, surface ice melts and the surface layer of water reduces the rate at which heat is conducted to the frozen interior.

During thawing, there are significant drip losses, which can offer an ideal substrate for enzyme activity and microorganism growth, a risk that makes the thawing process of crucial importance for final food safety and quality. Commercially, foods are often thawed just below the freezing point, to retain a firm texture for subsequent processing.

Some foods are cooked immediately and are therefore heated rapidly to a temperature which is sufficient to destroy microorganisms. Others (e.g., ice cream, frozen cakes) are not cooked and should therefore be consumed within a short time after thawing (Fellows 2000).

Food thawing is typically undertaken in the air, in cold water, in tap water, and in hot water (Eastridge and Bowker 2011); techniques that are time and energy consuming, leading to substantial food quality degradation. Novel methods include high-pressure thawing, microwave thawing, ohmic thawing, acoustic thawing (Li and Sun 2002), and high voltage electrostatic thawing (He et al. 2013).

7.5 MONITORING AND CONTROL OF THE CURRENT COLD CHAIN

Handling of frozen food after its packaging in the industry until it reaches its final destination, that is, consumer's freezer is of major importance for its overall quality. The most common stages in a frozen product lifecycle are depicted in Figure 7.8 and include storage at special areas in the industry, the distribution center, the retail and the consumer domestic freezer, as well as various transport steps.

Knowledge of the real time–temperature conditions at each stage would be an important tool to assess the shelf life of frozen products, while there is also an obligation to meet all legislative

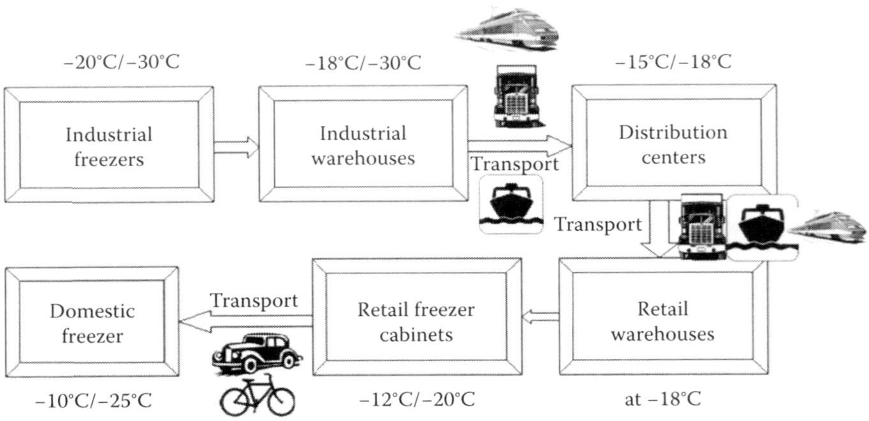

FIGURE 7.8 Representative stages of frozen food distribution chain and representative temperatures in the current cold chain.

requirements described in detail in both national and European laws (in EU Directives 89/108, 92/1, FDA, Codex Alimentarius, etc.).

The success and applicability of effective low-temperature preservation depends strongly on the capacity to integrate distinct operational steps and facilities in order to maintain storage conditions from food production to consumption. In particular, the storage temperature should be kept as invariable as possible, since fluctuations can have potentially serious consequences for product quality and safety. Therefore, "cold chain" relates to the maintenance, monitoring, and control of low-temperature storage conditions from initial freezing to product consumption, including product transport and display (Woolfe 1992, Wright and Taub 1998, Drummond and Sun 2010). According to a definition provided by the Concerted Action FAIR-CT 96-1180 sponsored by the European Commission "the cold chain is the part of the Food Industry, which deals with the transport, storage, distribution, and selling of frozen foods. It includes equipment and the operation of that equipment to maintain frozen food in a fully frozen condition at the correct temperature." Alternatively, according to the International Institute of Refrigeration, the "cold chain" refers to the continuity of frozen distribution, that is, the means successively employed to ensure the frozen preservation of perishable foodstuffs from the production to the consumption stage (Taoukis et al. 2012).

Several studies have been recently carried out to assess the importance of low temperature handling of frozen food, focusing on the effect of temperature fluctuations or temperature abuses during handling on product quality (Gormley et al. 2002, Giannakourou and Taoukis 2003, Hansen et al. 2004, Tsironi et al. 2009, Phimolsiripol et al. 2011). When temperature fluctuations occur during frozen storage, while the amount of ice in a system will generally remain constant, the number of ice crystals will reduce and the average size will increase (Bogh-Sorensen 2002). Especially when the temperature is not constant, recrystallization takes place, increasing the size of ice crystals. The growth in size of ice crystals can significantly influence the quality loss during frozen storage and handling of perishable foods. Additionally, temperature variation within a product can cause moisture migration, relocating the water within the product so as to move toward surfaces and to leave the denser regions of the product. Consequently, when there is void space around a product in a package, moisture will transfer into this space and tend to accumulate on the surface of the product and the internal surface of the package. When looking into the literature, most studies are performed under constant temperature conditions, and the effect of temperature fluctuations is only theoretically addressed (Skrede 1996).

In this context, the required temperature conditions need to be maintained all the way from the producer to the consumer, assuring a maximum low temperature of −18°C, a limit set by the majority of international and national regulations. Any increase in the temperature of the environment in which the product is held above that marginally accepted temperature is proven to have a significant adverse effect on the quality, and sometimes even on the safety of the product. Especially when the food is inadvertently thawed, microbiological issues become serious and may lead to food rejection (Devine et al. 1996, Mandigo and Osburn 1996). To put that in practical perspective, when a frozen product is held, even for a few minutes, in warmer than –18°C air, it will start to thaw, despite its "frozen-like" appearance. Restoring the temperature at the appropriate levels will lead the product to slower freezing because the equipment in the cold chain is designed to maintain product at –18°C and not freeze product down to that temperature (Taoukis et al. 2012).

Considering the multiple parameters that affect the efficiency of the current cold chain and the importance of a steady and adequately cold logistic path for the product acceptance, in terms of both safety and quality, it becomes evident that monitoring and control of the cold chain is a prerequisite for reliable quality management and optimization (Browne and Allen 1998, Dubelaar et al. 2001, Tijkens et al. 2001). Good temperature control is essential in all sections of the frozen food chain and can be obtained through improved equipment design, quality assurance systems application, and by an increased operator awareness. The current philosophy, however, for food quality optimization is to introduce temperature monitoring in an integrated, structured quality assurance system, based mostly on prevention, through the entire lifecycle of the product (Panozzo et al. 1999, Taoukis 2001).

7.5.1 Requirements, Conditions, and Control of the Stages of the Cold Chain

7.5.1.1 Cold Store

When referring to "cold store" this describes an enclosed chamber or box, completely insulated (meaning that walls, ceiling, and floor are fully insulated), fitted with an insulated door. In order to maintain a prefixed temperature inside the chamber, refrigerating machinery is required. Some important parameters regarding the function of this cold store are its size (that depends on what facility it belongs to, i.e., industry, hypermarket, small retail facility, etc.), the products that will be stored inside (amount-nature), the temperature required, the temperature/humidity of the surroundings, and whether its purpose is only to maintain a certain temperature or it will be also used to lower the food temperature.

As far as frozen foods are concerned, the frozen product is stored at different points of the chain inside chambers of different characteristics and performance during its marketing route to the home freezer and its final consumption. As mentioned earlier, the size of the cabinet, initial temperature of the incoming food, temperature required, temperatures of the surroundings, mechanical characteristics (location of refrigeration machinery, compressors, ventilation, and insulation), and energy/cost matters are issues of first priority when considering cold store requirements. In this context, an effective stock rotation and a safe inventory management within any storage area are of significant importance for an optimized frozen food chain control. Until now, most systems typically rely on time-based criterion such that items within a cold storage area are scheduled for distribution according to the length of time that an item has been in storage (Wells and Singh 1998). The two most common stock rotation policies used to assess the priority in which an item will be promoted to the following stage of the cold chain are the first-in, first-out (FIFO) policy, and the last-in first-out (LIFO) policy. Both policies are based on the age of an item (e.g., the time that a product is retained in storage), not taking into account the real conditions that the product was subjected to, that is, FIFO policy requires the oldest item within a stock to be issued first, and the LIFO policy allows the youngest item on hand to take highest priority. That means that the only criterion is the production date or the lot number and not the real quality state of the product that depends on the real time–temperature conditions of its storage, which could even render a product unsuitable for distribution.

As the most traditional one, it is always stated that the "first-in, first-out" management approach must be strictly adhered to in all stages of the freezer chain through fully automatic handling procedures in the freezer storage rooms (Taoukis et al. 2012).

As a more effective alternative to these conventional stock management systems, Wells and Singh (1989), Taoukis et al. (1991), Giannakourou et al. (2001, 2002) proposed a different issue policy based on the real quality status and the maximum expected remaining shelf life of a product. These innovative systems, usually called the least shelf-life, first-out (LSFO) or shortest remaining shelf-life (SRSL) inventory issue policy, are based on giving priority to the items with the shortest (but still acceptable) remaining shelf life. The development of LSFO is based on validated shelf life models of the monitored food, specification of the initial value of the quality index, the value of the selected quality index at the limit of acceptability, and constant time–temperature monitoring in the distribution chain with TTIs. TTI application to improve the existing cold chain, reduce the out-of-date stock, and lead to a cost-effective product management will be discussed in a following section.

Regarding temperature requirements during frozen storage, according to EU Directive 89/108 (Quick Frozen Food Directive, QFF), after quick freezing, the product temperature should be maintained at –18°C or colder after thermal stabilization. Some frozen foods, for example, beef, broilers, and butter, have a fairly long storage life even at –12°C, while foods such as lean fish require storage temperature around –28°C in order to reduce quality loss and prolong their storage life (Bogh-Sorensen 2002).

In the United States, a temperature of –18°C or colder is recommended, adding that some products, for example, ice cream and frozen snacks require –23°C or colder.

The EU Directive 92/1 requires that a temperature recording device must be installed in each storage facility in order to register and store for at least a year the temperature data of air surrounding perishable food.

7.5.1.2 Transport

The different points of transport, from the cold store to the retail outlet and then to the consumer frozen storage, are critical points for a product's overall quality and safety. A significant factor is the temperature inside the transport vehicles and the fluctuations occurring during transit. Inside the transport vehicles, proper temperature conditions are more difficult than in large cold stores due to several factors:

1. Supplementary heat is introduced during loading and unloading the vehicle.
2. Defrosting has a more severe effect on foods compared to refrigerated stores (more restricted space for the coils, more humid air inlet).
3. Possible close contact between the foods and the lateral walls, through loading and displacements of the cargo, connected to the forces that the road circulation generates in the cargo (Panozzo 2008).

The vehicle must be provided with a good refrigerated system, operating constantly during transportation to maintain the product frozen. The most widely used system to refrigerate the inlet of the vehicle is a vapor compression mechanical system. Another important issue is to avoid undesirable heat infiltration, which may occur due to hot weather, sunny conditions, inadequate insulation, or air leakage. Additionally, caution should be given during the loading/unloading procedure (as short as possible), as well as to the protection needed to the cargo during these processes. When taking precautions to avoid the above, it should be possible to achieve good quality, healthy, and safe frozen food products.

Legislation on control of transport equipment and temperatures during transport has been increasingly stricter, especially for intra-European transports of frozen foods (Taoukis et al. 2012, Agreement on the International Carriage of Perishable Foodstuffs). The QFF Directive requires that the temperature of quick frozen food must be maintained at –18°C or colder at all points in the product, with possible brief upward fluctuations of not more than 3°C during transport (Article 5). Directive 92/1 requires that transport equipment must have an appropriate temperature recording device installed, which should be approved by the authorities in the EU member state where the vehicle is registered. The temperature data should be dated and stored for at least a year by a responsible person.

The Agreement on Transport of Perishables, the so-called ATP agreement, has been ratified by about 30 countries, mainly in Europe, but also by Russia, the United States, and other countries. In cold transport between countries participating in the ATP agreement, special equipment must be used which should be inspected and tested for compliance with the standards in Annex 1, Appendices 1, 2, 3, and 4. In ATP, Annex 2, it is stated that "for the carriage of frozen and quick frozen foodstuffs, the transport equipment has to be selected and used in such a way that during carriage the highest temperature of the foodstuff in any point of the load does not exceed −20°C (for ice cream), −18°C (for quick frozen food, frozen fish, etc.), −12°C (for all frozen foodstuffs, except butter), and −10°C (for butter)."

The ATP agreement includes precise and strict requirements on the technical properties of transport equipment (quality of insulation, construction, etc.). In most EU countries, these rules are not enforced, allowing for the transport and distribution of frozen foods to occasionally take place in unsuitable equipment, that is, inadequate insulation, insufficient cooling capacity of the refrigeration machinery, etc. In France, however, ATP certified equipment must be used for the transport of frozen foods, prescribing the exact ATP category for different groups of foods.

The United States Code of Recommended Practices (Frozen Food Roundtable 1999) suggests that temperature should be measured in an appropriate place and recorded in vehicles used for frozen food transport.

According to a definition assigned by the UK authority, local distribution is the part of the distribution chain in which the product is delivered to the point of retail sale, including sale to a catering establishment (Bogh-Sorensen 2002). In France, local distribution is limited to 8 h, and the United States Code of Practice recommends that a frozen food measured with a temperature above −12°C should be rejected, or, at least, examined for acceptable quality prior to being offered for sale.

As far as modes of transport in the cold chain are concerned, this assumes

- Road transport, where the road tracks are divided according to DIN 8959/2000 as short-, medium-, and long-distance.
- Rail transport: for rail transport the wagons are refrigerated by either mechanical or stored energy. Mechanical refrigeration for each single wagon is the predominant technology (Panozzo 2008).
- Water transport: the most usual form of water transport is sea transport. In this case, ships can be insulated and refrigerated or can be container ships.
- Intermodal transport: intermodal transport can be carried out using standard vehicles that can be carried on rail wagons (e.g., the piggyback system) or ships. Most intermodal transport uses insulated boxes that can be transferred easily from ships to rail wagons to road vehicles, without interrupting the continuity of the cold chain. However, these systems are heavy and not easy to use (Panozzo 2008).

Finally, one should not overlook the fact that one of the weakest links in the distribution chain is the transport period from the product purchase to the consumer domestic freezer. When this time period is not part of the thawing process, meaning that the product will not be immediately consumed but it will be stored in a home freezer, the effect of this time might be significant for product quality and wholesomeness. According to the results of a consumer survey conducted in Greece, 26% of people need more than 20 min to carry food from the point of purchase to the home freezer, with 2% exceeding 45 min. Considering the usual temperatures during summer months (>32°C), this temperature abuse might lead frozen food to significant thawing and consequently to major deterioration.

7.5.1.3 Retail Display

Display cabinets are intended to be used for displaying frozen foods to consumers but not for lowering their temperature. For display at the stage of the retail sale of frozen foods, specially designed refrigerated cabinets are employed, aimed mainly at an effective display—from the consumer point of view—and, at the same time, correct and safe storage of food. Unfortunately, the two functions required are contradictory. For effective display, a cabinet with high product visibility and wide front opening, possibly in the absence of any kind of door or lid, and equipped with a bright lighting system is required (Rigot 1990). However, this design can cause a significant heat load on the cabinet, often leading to an unacceptably high temperature for the product, usually in excess of the recommended values for the storage of the goods (Cortella 2008).

In current practice, several types of cabinet are used. Numerous criteria can be adopted for the classification of this equipment (Cortella 2008), for example, the load temperature, the cabinet geometry, the presence of doors or lids, the type of air distribution, the type of refrigerating equipment, etc. Based on geometrical criteria, they can be classified as (Taoukis et al. 2012): (1) vertical multi-deck with or without glass doors, using refrigerated air circulated by fans throughout the cabinet, (2) open top cabinets, which lower food temperature by forced air circulation and/or natural convection, and (3) combined, for example, consisting of a horizontal and a vertical cabinet. A common display cabinet consists of a thermally insulated body that will bear the food load and the cooling equipment. The refrigeration unit may be totally within the cabinet (integral cabinet) or partially situated in a remote location, with only the heat-exchanging coils and the fan being inside the cabinet.

As far as regulation is concerned, according to QFF EU Directive (Article 5.2(b)) "tolerances in the product temperature in accordance with good practice are permitted. These tolerances may reach 3°C (to a product temperature of −15°C), if and to the extent that the Member States so decide. The Member State shall select the temperature in the light of stock or product rotation in the retail trade. The Commission shall be informed of the measures taken." According to Directive 92/1, temperature recording is not mandatory, and the temperature is measured by at least one clearly visible thermometer, which in open (gondola type) cabinets must indicate the temperature of the return air at the load line level.

Cooling equipment is certified by the manufacturer to comply with European Standard EN441 for a specific "climatic class." Direct exposure to sunlight and draughts must be avoided. The required cabinet performance will only be achieved if the ambient conditions are cooler and less humid than limits specified for the climatic class shown in the nameplate. Air conditioning is advisable if proper conditions cannot be guaranteed.

It is important that the cabinet is only loaded with products at −18°C or below, following the maker's instructions and within the load limit line clearly marked on the equipment. If this line is violated, products outside this limit will be kept in a higher, inappropriate temperature and will disturb air circulation, warming all food.

Most countries maintain that the temperature for frozen products in retail cabinets should be −20°C with a possible temporary tolerance up to −18°C or −15°C (Jul 1982). In a recent survey carried out in the context of the temperature database of the FRISBEE project (Food Refrigeration Innovations for Safety, Consumers' Benefit, Environmental Impact and Energy Optimization Along the Cold Chain in Europe, FP7, Food, Agriculture and Fisheries, and Biotechnology) at almost 320 retail display points for frozen foods, it was shown that the aforementioned temperature requirement is not met. As shown in Figure 7.9, in almost 36% of retail storage the temperature was higher than −18°C, with 8% exceeding the −14°C.

As discussed by Jul (1984), a systematic stock rotation and maintenance of the first-in, first-out principle could contribute more to product quality than expensive cabinet modifications. Going further, as it was mentioned earlier, an innovative stock management system, based on product actual time–temperature history (LSFO) and the response of TTIs, could further optimize the distribution chain, minimizing the unacceptable products (Giannakourou and Taoukis 2002). Similarly, loading procedures and handling of products before stocking in freezer cabinets are points of potential improvement.

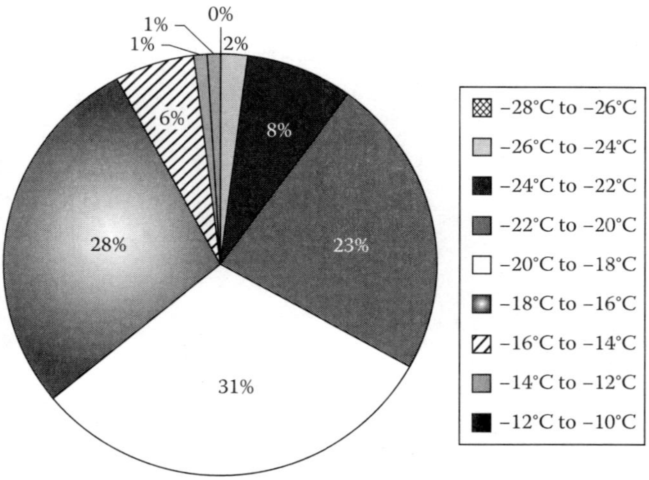

FIGURE 7.9 Temperature distribution in retail display of frozen foods from the database of the FRISBEE project (2013) based on 320 cases.

7.5.1.4 Home Storage

In spite of technological progress, energy efficiency, and environmentally friendly alternatives assumed at equipment level, domestic refrigeration, which is not covered by regulations, is still a source of concern; some indications show that food is stored at temperatures that are too high (Laguerre et al. 2002); for instance, the findings of James et al. (2008) and many other studies show that domestic refrigerators throughout the world operate above the recommended temperatures and are not properly maintained (Johnson et al. 1998, Ovka and Jevsnik 2009, Taoukis et al. 2012).

A consumer survey, mentioned by Laguerre (2008), published in the Grand Froid magazine showed that the freezer has several uses within the family: to preserve purchased frozen products (74.5% of surveyed people), to produce frozen food from purchased fresh food (51.8%), to produce frozen food from personal production (18%), and to preserve left-overs (14.7%). The products which the consumer freezes at home are primarily meats, chicken (72% of surveyed people), and vegetables (60.5%).

Unfortunately, the last stage of the freezer chain is the least studied, probably due to difficulties in data collection, concerning temperature conditions in domestic freezers, consumer habits, and approximate storage periods before consumption. However, when addressing the quality issue of frozen foods from production to final consumption in an integrated and structured way, such a period should be included in the overall assessment of quality degradation in the cold chain. In a survey conducted in 100 Greek households, not only almost 25% of freezers were found to operate at temperatures >−14°C, but also significant fluctuations were observed, possibly due to door opening, product replenishment, or inefficient refrigeration system (Taoukis et al. 2012).

Although there has been a significant improvement in technological aspects regarding household freezing equipment, as well as better information/education concerning appropriate treatment of frozen foods, recent findings were not optimistic; from data collected by the temperature database of the FRISBEE project (Food Refrigeration Innovations for Safety, Consumers' Benefit, Environmental Impact and Energy Optimization Along the Cold Chain in Europe, FP7, Food, Agriculture and Fisheries, and Biotechnology), at almost 200 household freezers, more than 30% was found to operate at temperatures inappropriate to preserve frozen foods. In Figure 7.10a and b, results from both surveys are shown. Furthermore, when assessing the performance of a freezing module, averages are not sufficient; in many cases, fluctuations may be detrimental for food

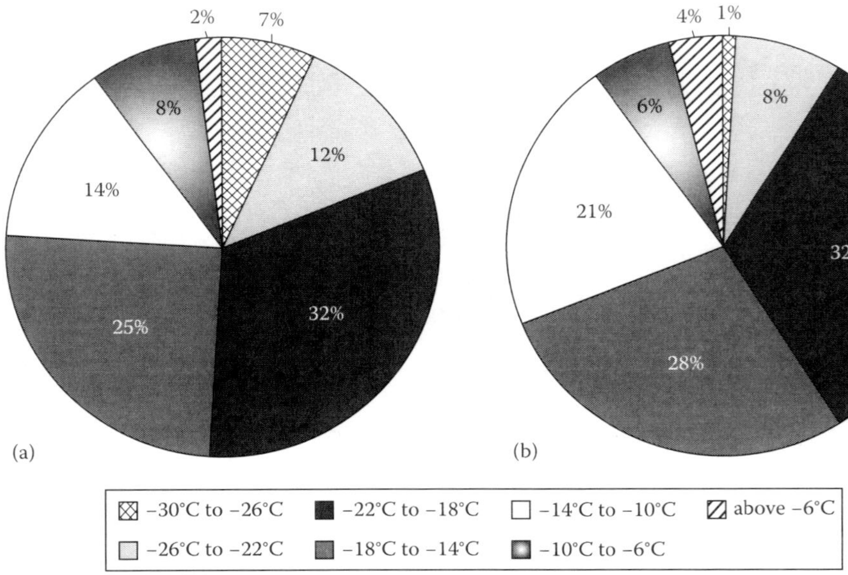

FIGURE 7.10 Temperature distribution in domestic freezers from (a) a previous survey in 100 households (Giannakourou and Taoukis 2002) and (b) the database of FRISBEE project (2013).

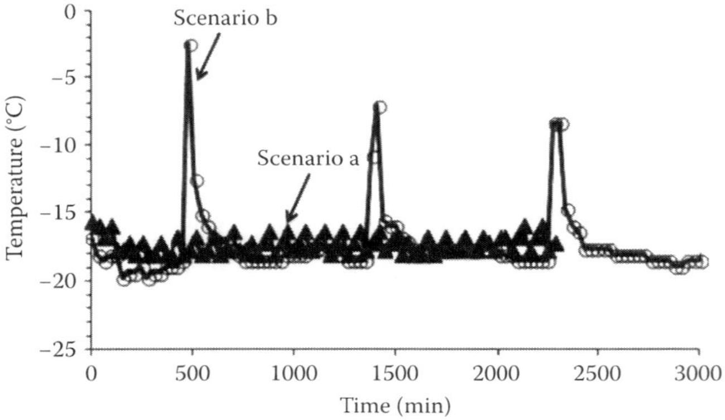

FIGURE 7.11 Indicative temperature distribution in domestic freezers with the same average value (from the database of FRISBEE project).

quality and safety. In Figure 7.11, two temperature profiles are shown, where average temperature was almost the same, very close to the ideal condition for preserving frozen food quality (at −18°C). However, it is obvious that in the case of scenario b (open circles with solid line), food is not equally well preserved due to significant fluctuations when compared to scenario a (closed triangles with solid line).

Another issue of major concern is the energy consumption in the household environment, especially in refrigerators and freezers (Bansal and Kruger 1995). Taking into account the global concern for energy saving, there is a lot of study done aiming at the establishment of minimum energy performance standards (MEPS) and various utility objectives to encourage consumers to use more efficient units. MEPS are the mandatory levels of energy efficiency for household appliances required to be met by all the models on sale to ensure overall product quality.

As an example given by Bansal and Kruger (1995), the classification of freezer compartments according to their storage temperatures is based on the "star system"

1. "One-star" compartment (*): compartment in which the storage temperature is not higher than −6°C.
2. "Two-star" compartment (**): compartment in which the storage temperature is not higher than −12°C.
3. "Three-star" compartment (***): compartment in which the storage temperature is not higher than −18°C.

In this context, studies on more efficient appliances have led to the widespread proliferation of intelligent control of domestic appliances. Most modern appliances now are fully controlled by microprocessor-based sensors (Bansal et al. 2011). Concerning domestic freezers, the latest advances refer to adaptive defrost sensor, automatic control of anti-condensation heaters, door open alarm sensor, sensor to control temperatures under different operating regimes for energy savings, and smart grid interoperability. Other improvements concern door sealing materials, gasket design, alternative, environmentally-friendly refrigerants, improved insulation, improved fan motors, etc.

7.5.1.5 Transfer Points

Transfer points, that is, points where frozen products are moved from a cold area to another, are known to be responsible for temperature abuse, mishandling, and significant fluctuations. At these points, since personnel responsibility is not clearly defined, temperature recording does not take place, leading to a possible "break" of the cold chain (Taoukis et al. 2012). As Jul (1984) describes, a frequent occurrence is that a truck has to be emptied completely in order to gain access to a particular shipment, due to ineffective loading. It is then almost certain that there will be an undue delay in placing the rest of the shipment back and restoring the appropriate temperature conditions. There are also a lot of cases where, frozen products transported by sea are left on the pier due to delays, subjected possibly to abusive temperature conditions.

A first necessary step for minimizing the negative effect of these points is the identification, the control, and the assessment of the potential hazard that transfer points may represent. The personnel involved should be trained to record temperatures, minimize the delay time, and ensure the continuity of the freezer chain. Finally, these points should be part of an integral, continuous recording system of the cold chain (perhaps with the use of TTIs attached on the food item) so that any temperature abuse is reflected on the final quality status of the product at its final destination (retailer or consumer).

7.5.1.5.1 Monitoring Temperature

In order to obtain a full record of product history, the temperature of both the food and its surroundings should be monitored. Additionally, multiple measurements at different locations should be taken in case of a large batch or varying conditions in the chamber. Measurements can be realized either by a mechanical or an electronic equipment, with or without the potential of recording and maintaining an electronic file of data (Taoukis et al. 2012).

A common way to continuously measure temperatures, either inside food or inside freezer compartments, is by using sensors. The three principal types of sensors commercially available are thermocouples, platinum resistance, and semiconductors (thermistor). The choice depends on requirements for accuracy, speed of response, range of temperatures to be monitored, robustness, and cost. The predominant types of thermocouples are of type K (with nickel–chromium and nickel–aluminum alloy wires) and type T (with copper–nickel alloy). The main advantages of the thermocouples are their low cost, facility to be hand-prepared, and a very wide range of temperatures measured (from −184°C to 1600°C).

Another category includes read out and recording systems; the most common device is the electronic digital readout instrument, which is powered by batteries and allows for storing and printing out, or even an alarm notification when the temperature goes outside a preset limit. The recent miniaturization of circuit systems has produced some compact and powerful data logging systems, which can potentially follow the food within the food case or pallet throughout all stages of the cold chain (e.g., COX Tracer™, Cox recorders, Belmont, NC; "Diligence"™, Comark, Hertfordshire, England; KoolWatch™, Cold Chain Technologies, MA, United States; DL200-T, Telatemp, California, United States; i-Button, Dallas semiconductor, Maxim, TX; Dickson TK-500, and Dickson Addison, IL). The development in this area is oriented in further decrease of data loggers' size in order to have the opportunity to monitor the actual temperature of foods by placing the logger between food packs (Taoukis et al. 2012).

A recent advance in this area is the use of TTIs; they are an alternative, cost-effective way to individually monitor the temperature conditions of food products throughout the chain. Actually, TTIs are simple devices with an easily measurable response that reflect the accumulated time–temperature history of the product on which they are attached. The principle of TTI application lies in the use of a physicochemical mechanism and a measurable change to display (1) the current temperature, (2) the crossover of a preset temperature, or (3) the integrated time–temperature history of the frozen food. Their operation is based on irreversible reactions that are initiated at the time of their activation and proceed with an increasing rate as the temperature is elevated in a manner that resembles the temperature dependence of most quality loss reactions of foods (Taoukis 2001, Taoukis and Labuza 2003). These devices are attached on the food itself or outside the packaging and actually follow the food during its circuit from manufacturer to final consumption. The ultimate purpose of their application is the "translation" of their reading to the quality status of the food through the appropriate algorithm as discussed previously. Such TTI devices are MonitorMark™, FreezeWatch™ (3M, St. Paul, Minnesota), Fresh-Check® and Transtracker (Temptime, previously Lifelines, Morristown, NJ), ColdMark™ and WarmMark (Cold Ice Technologies, MA, United States), VITSAB (Malmo, Sweden), OnVu (Bizerba GmbH & Co, KG, Balingen, Germany), Timestrip® (Timestrip UK Ltd), and Freshpoint (Tel Aviv, Israel) (Figure 7.12).

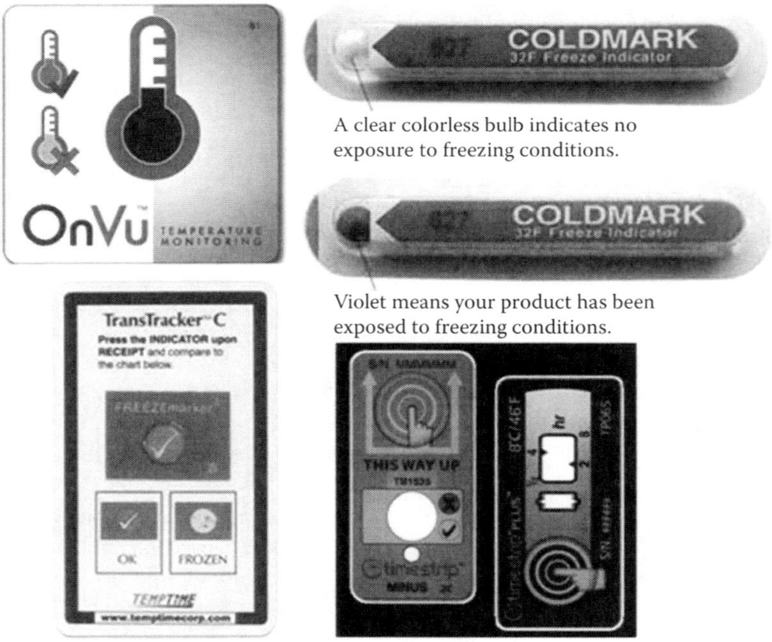

FIGURE 7.12 Examples of time temperature indicators used in the current cold chain to monitor the real time temperature history of frozen and chilled products.

Considering the difficulties and the prolonged time periods needed for a thorough kinetic study of frozen foods, there are few studies published correlating frozen food quality with TTI (Schubert 1977, Dolan et al. 1985, Singh and Wells 1985, Yoon et al. 1994, Giannakourou et al. 2001, 2002, 2003). In these studies, TTIs of different type were found to be effective monitoring, controlling, and quality predictive tools for the real distribution of frozen food products.

Taking into account the real conditions of the cold chain (Figures 7.9 and 7.10) and the occurrence of inappropriate temperatures during the distribution of frozen foods, as described in the previous section, it is obvious that the final quality of frozen foods is not identical for all items when reaching the final consumer; therefore, the existence of a single, uniform expiration date is not adequate when aiming at precisely assessing the remaining shelf life of a product. Thus, TTI considered as temperature history recorders could be reliably used to indicate quality and remaining shelf life, and potentially to introduce an optimized cold chain management system. This improved system, coded as LSFO, which has been previously assessed by Taoukis et al. (1998) for chilled products has been also evaluated for frozen vegetables (Giannakourou and Taoukis 2002, 2003). This novel approach proposed as an alternative FIFO policy is based on the assortment of products according to their quality status, as it is predicted by the attached TTI at designated points of their marketing route. The key point in LSFO implementation is the estimation of each product time–temperature history through the reading of the attached TTI at the predefined points of the cold chain, in order to classify products and forward the ones closer to their actual expiration. Overall, this system would optimize the current inventory management system leading to products of more consistent quality and nutritional value at the time of consumption.

REFERENCES

Abu-Ghannam, N. and Crowley, H. 2006. Modelling the kinetics of peroxidase inactivation, colour and texture changes of pumpkin (*Cucurbita maxima* L.) during blanching. *J. Food Eng.*, 74: 335–344.

Agnelli, M.E., Marani, C.M., and Mascheroni, R.H. 2005. Modelling of heat and mass transfer during (osmo) dehydrofreezing of fruits. *J. Food Eng.*, 69: 415–424.

Agreement on the International Carriage of Perishable Foodstuffs and on the Special Equipment to be Used for Such Carriage (ATP) (Accord relatif aux transports internationaux de denrees perissables). 1970. E/ECE 810 rev. 1, E/ECE/TRANS/563 Rev. 1: New York: United Nations.

Alexandre, E.M.C., Brandao, T.R.S., and Silva, C.L.M. 2012. Efficacy of non-thermal technologies and sanitizer solutions on microbial load reduction and quality retention of strawberries. *J. Food Eng.*, 108: 417–426.

Alexandre, E.M.C., Brandao, T.R.S., and Silva, C.L.M. 2013. Impact of non-thermal technologies and sanitizer solutions on microbial load reduction and quality factor retention of frozen red bell peppers. *Innov. Food Sci. Emerg. Technol.*, 17: 199–205.

Ando, H., Kajiwara, K., Oshita, S., and Suzuki, T. 2012. The effect of osmotic dehydrofreezing on the role of the cell membrane in carrot texture softening after freeze-thawing. *J. Food Eng.*, 108: 473–479.

Arai, S. and Watanabe, M. 1986. Freeze texturing of food materials by ice-nucleation with the bacterium *Erwinia ananas*. *J. Biol. Chem.*, 50(1): 169–175.

Awad, T.S., Moharram, H.A., Shaltout, O.E., Asker, D., and Youssef, M.M. 2012. Applications of ultrasound in analysis, processing and quality control of food: A review. *Food Res. Int.*, 48: 410–427.

Bahçeci, K., Serpen, A., Gokmen, V., and Acar, J. 2005. Study of lipoxygenase and peroxidase as indicator enzymes in green beans: Change of enzyme activity, ascorbic acid and chlorophylls during frozen storage. *J. Food Eng.*, 66: 187–192.

Bansal, P.K. and Kruger, R. 1995. Test standards for household refrigerators and freezers I: Preliminary comparisons. *Int. J. Refr.* 18(1): 4–20.

Bansal, P.K., Vineyard, E., and Abdelaziz, O. 2011. Advances in household appliances—A review. *Appl. Therm. Eng.*, 31: 3748–3760.

Barrett, D.M. and Theerakulkait, C. 1995. Quality indicators in blanched, frozen, stored vegetables. *Food Technol.*, 49, 62–65.

Beuchat, L.R., Pettigrew, C.A., Tremblay, M.E., Roselle, B.J., and Scouten, A.J. 2004. Lethality of chlorine, chlorine dioxide, and a commercial fruit and vegetable sanitizer to vegetative cells and spores of *Bacillus cereus* and spores of Bacillus thuringiensis. *J. Food Protect.*, 67: 1702–1708.

Bevilacqua, M., D'Amore, A., and Polonara, F. 2004. A multi-criteria decision approach to choosing the optimal blanching–freezing system. *J. Food Eng.*, 63: 253–263.

Bilek, S.E. and Turantas, F. 2013. Decontamination efficiency of high power ultrasound in the fruit and vegetable industry: A review. *Int. J. Food Microbiol.*, 166: 155–162.

Biswal, R.N., Bozorgmehr, K., Tompkins, F.D., and Liu, X. 1991. Osmotic concentration of green beans prior to freezing. *J. Food Sci.*, 56(4): 1008–1011.

Blanda, G., Cerretani, L., Cardinali, A., Barbieri, S., Bendini, A., and Lercker, G. 2009. Osmotic dehydrofreezing of strawberries: Polyphenolic content, volatile profile and consumer acceptance. *LWT—Food Sci. Technol.*, 42: 30–36.

Blucas, J.G. 2004. Freezing. In: *Food Processing and Preservation*. Ed.: J.G. Blucas, pp. 247–286. Athens, Greece: Stamoulis Editions.

Bøgh-Sørensen, L. 2002. Frozen food legislation. *Bull. Int. Inst. Refrig.*, 4: 4–18.

Boast, M.F.G. 2003. Blast and plate freezing. In: *Encyclopedia of Food Sciences and Nutrition*, 2nd edn. Eds.: L.C. Trugo and P.M. Finglas, pp. 2718–2725, Orlando, FL: Academic Press.

Bowater, F.J. 2001. Design of carton air blast freezing systems. In: *Rapid Cooling of Food, Proceeding of Bristol Meeting*, The International Institute of Refrigeration, Bristol, U.K.

Browne, M. and Allen, J. 1998. Logistics of food transport. In: *Food Transportation*, Eds.: R. Heap, M. Kierstan, and G. Ford, pp. 22–50. London, U.K.: Blackie Academic & Professional.

Buggenhout, S.V., Messagie, I., Maes, V., Duvetter, T., Van Loey, A., and Hendrickx, M. 2006. Minimizing texture loss of frozen strawberries: Effect of infusion with pectin methyl esterase and calcium combined with different freezing conditions and effect of subsequent storage/thawing conditions. *Eur. Food Res. Technol.*, 223(3): 395–404.

Cao, S., Hu, Z., Pang, B., Wang, H., Xie, H., and Wu, F. 2010. Effect of ultrasound treatment on fruit decay and quality maintenance in strawberry after harvest. *Food Control*, 21(4): 529–532.

Chassagne-Berces, S., Fonseca, F., Citeau, M., and Marin, M. 2010. Freezing protocol effect on quality properties of fruit tissue according to the fruit, the variety and the stage of maturity. *LWT—Food Sci. Technol.*, 43: 1441–1449.

Chevalier, D., Sentissi, M., Havet, M., and Le Bail, A. 2000. Comparison of air-blast and pressure shift freezing on Norway lobster quality. *J. Food Sci.*, 65: 329–333.

Chiralt, A., Martinez-Navarrete, N., Martinez-Monzo, J., Talens, P., Moraga, G., Ayala, A., and Fito, P. 2001. Changes in mechanical properties throughout osmotic processes: Cryoprotectant effect. *J. Food Eng.*, 49(2–3): 129–135.

Cooksey, K. and Krochta, J. 2012. Introduction to frozen food packaging. In: *Handbook of Frozen Food Processing and Packaging*. Ed.: D.-W. Sun, pp. 711–730. Boca Raton, FL: CRC Press, Taylor & Francis Group.

Cortella, G. 2008. Frozen retail display. In: *Frozen Food Science and Technology*. Ed.: J.A. Evans, pp. 303–324. Oxford, U.K.: Blackwell Publishing.

Cruz, R.M.S., Vieira, M.C., and Silva C.L.M. 2009. The response of watercress (*Nasturtium officinale*) to vacuum impregnation: Effect of an antifreeze protein type I. HJ. *Food Eng.*, 95: 339–345.

Davies, T.W. 2005. Slurry ice as a heat transfer fluid with a large number of application domains. *Int. J. Refrig.*, 28: 108–114.

Delgado, A. and Sun, D.W. 2012. Ultrasound-accelerated freezing. In: *Handbook of Frozen Food Processing and Packaging*. Ed.: D.-W. Sun, pp. 645–666. Boca Raton, FL: CRC Press, Taylor & Francis Group.

Dermesonlouoglou, E., Giannakourou, M., and Taoukis, P.S. 2007. Stability of dehydrofrozen tomatoes pretreated with alternative osmotic solutes. *J. Food Eng.*, 78: 272–280.

Dermesonlouoglou, E.K., Pourgouri, S., and Taoukis, P.S. 2008. Kinetic study of the effect of the osmotic dehydration pre-treatment to the shelf life of frozen cucumber. *Innov. Food Sci. Emerg. Technol.*, 9: 542–549.

Devine, C.E., Bell, R.G., and Lovatt, S. 1996. Red meats. In: *Freezing Effects on Food Quality*. Ed.: L.E. Jeremiah, pp. 73–76. New York: Marcel Dekker.

Dolan, K.D., Singh, R.P., and Wells, J.H. 1985. Evaluation of Time-temperature related quality changes in ice cream during storage. *J. Food Process Preserv.*, 9: 253–271.

Dorantes-Alvareza, L., Jaramillo-Floresa, E., Gonzáleza, K., Martinez, R., and Parada, L. 2011. Blanching peppers using microwaves. *Proc. Food Sci.*, 1: 178–183.

Drummond, L. and Sun, D.-W. 2010. Effects of chilling and freezing on safety and quality of food products. In: *Processing Effects on Safety and Quality of Foods*. Ed.: E. Ortega-Rivas, pp. 295–321. New York: CRC Press, Taylor & Francis Group.

Dubelaar, C., Chow, G., and Larson, P. 2001. Relationships between inventory, sales and service in a retail chain store operation. *Int. J. Phys. Distrib. Logist. Manag.*, 31(2): 96–108.

Eastridge, J.S. and Bowker, B.C. 2011. Effect of rapid thawing on the meat quality attributes of USDA select beef strip loin steaks. *J. Food Sci.*, 76(2): 156–162.

EU Directive 89/108 of the Council of 21 December 1988. Approximation of the laws of the member states relating to quick frozen foodstuffs for human consumption (89/108/CEE). OJ L40, p. 34.

EU Directive 92/1. Monitoring of temperatures in the means of transport, warehousing and storage of quick frozen foods intended for human consumption. OJ L34, pp. 28–29.

Fellows, P. 2000. Freezing, Chapter 21. In: *Food Processing Technology*. Ed.: P. Fellows. Boca Raton, FL: Woodhead Publishing Limited and CRC Press LLC.

Fernandez, P.P., Otero, L., Guignon, B., and Sanz, P.D. 2006. High-pressure shift freezing versus high-pressure assisted freezing: Effects on the microstructure of a food model. *Food Hydrocolloids*, 20: 510–522.

Fikiin, K. 2008. Emerging and novel freezing processes. In: *Frozen Food Science and Technology*. Ed.: J.A. Evans, pp. 101–123. Oxford, U.K.: Blackwell Publishing.

Fito, P., Chiralt, A., Betoret, N., Gras, M., Chafer, M., Martinez-Monzo, J., Andres, A., and Vidal, D. 2001. Vacuum impregnation and osmotic dehydration in matrix engineering: Application in functional fresh food development. *J. Food Eng.*, 49: 175–183.

Forni, E., Sormani, A., Scalise, S., and Torreggiani, D. 1997. The influence of sugar composition on the colour stability of osmodehydrofrozen intermediate moisture apricots. *Food Res. Int.*, 30(2): 87–94.

Frozen Food Roundtable. 1999. Frozen food handling and merchandizing. American Frozen Food Institute and 14 Other Associations/Organizations in the USA, Virginia, US.

Fumoto, K., Sato, T., Kawanami, T., Inamura, T., and Shirota, M. 2013. Ice slurry generator using freezing-point depression by pressurization—Case of low-concentration NaCl aqueous solution. *Int. J. Refrig.*, 36: 795–800.

Garrote, R.L. and Bertone, R.A. 1989. Osmotic concentration at low temperature of frozen strawberry halves. Effect of glycerol, glucose and sucrose solution on exudates loss during thawing. *Food Sci. Technol.*, 22: 264–267.

George, M. 2000. Selecting packaging for frozen food products, Chapter 10. In: *Food Preservation Techniques*, Eds.: P. Zeuthen and L. Bøgh-Sørensen. Cambridge, U.K.: Woodhead Publishing Ltd.

Giangiacomo, R., Torreggiani, D., Erba, M.L., and Messina G. 1994. Use of osmodehydrofrozen fruit cubes in yogurt. *Ital. J. Food Sci.*, 3: 345–350.

Giannakourou, M.C., Koutsoumanis, K., Dermesonlouoglou, E., Taoukis, P.S. 2001. Applicability of the intelligent shelf life decision system for control of nutritional quality of frozen vegetables. *Acta Hort.*, 566: 275–280.

Giannakourou, M.C. and Taoukis, P.S. 2002. Systematic application of time temperature integrators as tools for control of frozen vegetable quality. *J. Food Sci.*, 67(6): 2221–2228.

Giannakourou, M.C. and Taoukis, P.S. 2003a. Application of a TTI-based distribution management system for quality optimisation of frozen vegetables at the consumer end. *J. Food Sci.*, 68(1): 201–209.

Giannakourou, M.C. and Taoukis, P.S. 2003b. Stability of dehydrofrozen green peas pretreated with non conventional osmotic agents. *J. Food Sci.*, 68(6): 2002–2010.

Giannakourou, M.C. and Giannou, V. 2015. Chilling and freezing. In: *Handbook of Food Processing and Engineering, Vol II: Food Process Engineering*. Eds.: K. Tzia and T. Varzakas, pp. 319–374. Boca Raton, FL: CRC Press.

Gonçalves, E.M., Pinheiro, J., Abreu, M., Brandão, T.R.S., and Silva, C.L.M. 2007. Modelling the kinetics of peroxidase inactivation, colour and texture changes of pumpkin (*Cucurbita maxima* L.) during blanching. *J. Food Eng.*, 81(4): 693–701.

Gormley, R., Walshe, T., Hussey, K., and Buttler, F. 2002. The effect of fluctuating vs. constant frozen storage temperature regimes on some quality parameters of selected food products. *Lebensm. Wiss. Technol.*, 35: 190–200.

Goula, A.M. and Lazarides, H.N. 2012. Modeling of mass and heat transfer during combined processes of osmotic dehydration and freezing (Osmo-Dehydro-Freezing). *Chem. Eng.*, 82: 52–61.

Griffith, M. and Ewart, K.V. 1995. Antifreeze proteins and their potential use in frozen foods. *Biotechnol. Adv.*, 13(3): 375–405.

Hansen, E., Lauridsen, L., Skibsted, L.H., Moawad, R.K., and Andersen, M.L. 2004. Oxidative stability of frozen pork patties: Effect of fluctuating temperature on lipid oxidation. *Meat Sci.*, 68(2): 185–191.

Hassas-Roudsari, M. and Goff, H.D. 2012. Ice structuring proteins from plants: Mechanism of action and food application-review. *Food Res. Int.*, 46: 425–436.

Hasselman, G. and Sceer, A.K. 2012. Packaging of frozen foods with other materials. In: *Handbook of Frozen Food Processing and Packaging*. Ed.: D.-W. Sun, pp. 759–778. Boca Raton, FL: CRC Press, Taylor & Francis Group.

He, X., Liu, R., Nirasawa, S., Zheng, D., and Liu, H. 2013. Effect of high voltage electrostatic field treatment on thawing characteristics and post-thawing quality of frozen pork tenderloin meat. *J. Food Eng.*, 115: 245–250.

Hernandez, A.F., Robles, P.A., Gomez, P.A., Callejas, T.A., and Artes, F. 2010. Low UV-C illumination for keeping overall quality of fresh-cut watermelon. *Postharvest Biol. Technol.*, 55(2): 114–120.

Huxsoll, C.C. 1982. Reducing the refrigeration load by partial concentration of foods prior to freezing. *Food Technol.*, 36(5): 98–102.

Jackson, T.H., Ungan, A., Critser, J.K., and Gao, D.Y. 1997. Novel microwave technology for cryopreservation of biomaterials by suppression of apparent ice formation. *Cryobiology*, 34: 363–372.

James, C. and James, S. 2003. Freezing: Cryogenic freezing. In: *Encyclopedia of Food Science and Nutrition.* Eds: L.C. Trugo and P.M. Finglas, pp. 2725–2732. Orlando, FL: Academic Press.

James, S. 2008. Freezing of meat. In: *Frozen Food Science and Technology*. Ed.: J.A. Evans, pp. 124–150. Oxford, U.K.: Blackwell Publishing.

James, S.J., Evans, J., and James, C. 2008. A review of the performance of domestic refrigerators. *J. Food Eng.*, 87: 2–10.

Johnson, A.E., Donkin, A.J., Morgan, K., Lilley, J.M., Neale, R.J., and Page, R.M. et al. 1998. Food safety knowledge and practice among elderly people living at home. *J. Epidemiol. Commun. Health*, 52(11): 745–748.

Jul, M. 1982. The intracacies of the freezer chain. *Revue Internationale du froid*, 5(4): 226–230.

Jul, M. 1984. *The Quality of Frozen Foods*. Orlando, FL: Academic Press.

Karel, M. and Lund, D.B. 2003. Freezing, Chapter 8. In: *Physical Principles of Food Preservation*, 2nd edn. Eds.: M. Karel and D.B. Lund. New York: Marcel Dekker Inc.

Kauffeld, M., Wang, M.J., Goldstein, V., and Kasza, K.E. 2010. Ice slurry applications—A review. *Int. J. Refrig.*, 33: 1491–1505.

Kennedy, C. 2000. Developments in freezing, Chapter 12. In: *Food Preservation Techniques*. Eds.: P. Zeuthen and L.B. Sorensen. Cambridge, U.K.: Woodhead Publishing Limited.

Kennedy, C. 2003. Developments in freezing, Chapter 12. In: *Food Preservation Techniques*, Eds.: P. Zeuthen and L. Bøgh-Sørensen. Cambridge, U.K.: Woodhead Publishing Ltd.

Kiani, H. and Sun, D.W. 2011. Water crystallization and its importance to freezing of foods: A review. *Trends Food Sci. Technol.*, 22: 407–426.

Kiani, H., Sun, D.W., Zhang, Z., Al-Rubeai, M., and Naciri, M. 2013. Ultrasound-assisted freezing of *Lactobacillus plantarum* subsp. plantarum: The freezing process and cell viability. *Innov. Food Sci. Emerg. Technol.*, 18: 138–144.

Kiani, H., Zhang, Z., Delgado, A., and Sun, D.W. 2011. Ultrasound assisted nucleation of some liquid and solid model foods during freezing. *Food Res. Int.*, 44: 2915–2921.

Knorr, D., Schluter, O., and Heinz, V. 1998. Impact of high hydrostatic pressure on phase transition of foods. *Food Technol.*, 52(9): 42–45.

Laguerre, O. 2008. Consumer handling of frozen foods. In: *Frozen Food Science and Technology*. Ed.: J.A. Evans, pp. 325–346. Oxford, U.K.: Blackwell Publishing.

Laguerre, O., Derens, E., and Palagos, B. 2002. Study of domestic refrigerator temperature and analysis of factors affecting temperature: A French survey. *Int. J. Refrig.*, 25: 653–659.

Lazarides, H.N. and Mavroudis, N.E. 1996. Kinetics of osmotic dehydration of a highly shrinking vegetable tissue in a salt-free medium. *J. Food Eng.*, 30: 61–74.

Le Bail, A., Orlowska, M., and Havet, M. 2012. Electrostatic field-assisted food freezing. In: *Handbook of Frozen Food Processing and Packaging*. Ed.: D.-W. Sun, pp. 685–691. Boca Raton, FL: CRC Press, Taylor & Francis Group.

Lee, K.H. and Sun, D.-W. 2012. Plastic packaging of frozen foods. In: *Handbook of Frozen Food Processing and Packaging*. Ed.: D.-W. Sun, pp. 731–742. Boca Raton, FL: CRC Press, Taylor & Francis Group.

Levine, H. and Slade, L. 1992. Glass transitions in foods. In: *Physical Chemistry of Foods*. Eds.: H.G. Schwartzberg and R.W. Hartel, pp. 83–221. New York: Marcel Dekker, Inc., IFT.

Li, B. and Sun, D.-W. 2002. Novel methods for rapid freezing and thawing of foods—A review. *J. Food Eng.*, 54: 175–182.

Li, J. and Lee, T.C. 1995. Bacterial ice nucleation and its potential application in the food industry. *Trends Food Sci. Technol.*, 6: 259–265.

Lucas, T. and Raoult-Wack, A.L. 1998. Immersion chilling and freezing in aqueous refrigerating media: Review and future trends. *Int. J. Refrig.*, 21(6): 419–429.

Maestrelli, A., Lo Scalzo, R., Lupi, D., Bertolo, G., and Torreggiani, D. 2001. Partial removal of water before freezing: Cultivar and pre-treatments as quality factors of frozen muskmelon (*Cucumis melo*, cv reticulates Naud.). *J. Food Eng.*, 49: 255–260.

Magnussen, O.M., Hemmingsen, A.K.T., Hardarrson, V., Sordtvedt, T.S., and Eikevik, T.M. 2008. Freezing of fish. In: *Frozen Food Science and Technology*. Ed.: J.A. Evans, pp. 151–164. Oxford, U.K.: Blackwell Publishing.

Mandigo, R.W. and Osburn, W.N. 1996. Cured and processed meats. In: *Freezing Effects on Food Quality*. Ed.: L.E. Jeremiah, pp. 171–177. New York: Marcel Dekker.

Marani, C.M., Agnelli, M.E., and Mascheroni, R.H. 2007. Osmo-frozen fruits: Mass transfer and quality evaluation. *J. Food Eng.*, 79: 1122–1130.

Maroulis, Z.B. and Saravacos, G.D. 2003. Chapter 5: Refrigeration and freezing. In: *Food Process Design*. Ed.: Z.B Maroulis and G.D. Saravacos. New York: Marcel Dekker Inc.

Martínez, S., Pérez, N., Carballo, J., and Franco, I. 2013. Effect of blanching methods and frozen storage on some quality parameters of turnip greens ("grelos"). *LWT—Food Sci. Technol.*, 51: 383–392.

North, M.F. and Lovatt, S.J. 2012. Freezing methods and equipment. In: *Handbook of Frozen Food Processing and Packaging*. Ed.: D.-W. Sun, pp. 187–200. Boca Raton, FL: CRC Press, Taylor & Francis Group.

Norton, T., Delgado, A., Hogan, E., Grace, P., and Sun, D.W. 2009. Simulation of high pressure freezing processes by enthalpy method. *J. Food Eng.*, 91: 260–281.

Olaimat, A.N. and Holley, R.A. 2012. Factors influencing the microbial safety of fresh produce: A review. *Food Microbiol.*, 32(1): 1–19.

Olivera, D.F., Viña, S.Z., Marani, C.M., Ferreyra, R.M., Mugridge, A., Chaves, A.R., and Mascheroni, R.H. 2008. Effect of blanching on the quality of Brussels sprouts (*Brassica oleracea* L. *gemmifera DC*) after frozen storage. *J. Food Eng.*, 84: 148–155.

Otero, L. and Sanz, P.D. 2006. High pressure shift freezing: Main factors implied in the phase transition time. *J. Food Eng.*, 72: 354–363.

Otero, L. and Sanz, P.D. 2012. High-pressure shift freezing. In: *Handbook of Frozen Food Processing and Packaging*. Ed.: D.-W. Sun, pp. 667–683. Boca Raton, FL: CRC Press, Taylor & Francis Group.

Ovka, A. and Jevšnik, M. 2009. Maintaining a cold chain from purchase to the home and at home: Consumer opinion. *Food Control*, 20: 167–172.

Panozzo, G. 2008. Frozen food transport. In: *Frozen Food Science and Technology*. Ed.: J.A. Evans, pp. 276–302. Oxford, U.K.: Blackwell Publishing.

Panozzo, G., Minotto, G., and Barizza, A. 1999. Transport and distribution of foods: Today's situation and future trends. *Int. J. Refrig.*, 22: 625–639.

Patras, A., Tiwari, B.K., and Brunton, N.P. 2011. Influence of blanching and low temperature preservation strategies on antioxidant activity and phytochemical content of carrots, green beans and broccoli. *LWT—Food Sci. Technol.*, 44: 299–306.

Pearson, A. 2008. Specifying and selecting refrigeration and freezer plant. In: *Frozen Food Science and Technology*. Ed.: J.A. Evans, pp. 81–100. Oxford, U.K.: Blackwell Publishing.

Peralta, J.M., Rubiolo, A.C., and Zorrilla, S.E. 2009. Design and construction of a hydrofluidization system. Study of the heat transfer on a stationary sphere. *J. Food Eng.*, 90: 358–364.

Peralta, J.M., Rubiolo, A.C., and Zorrilla, S.E. 2010. Mathematical modeling of the heat transfer and flow field of liquid refrigerants in a hydrofluidization system with a stationary sphere. *J. Food Eng.*, 99: 303–313.

Petzold, G., Caro, M., and Moreno, J. 2013. Influence of blanching, freezing and frozen storage on physicochemical properties of broad beans (*Vicia faba* L.). *Int. J. Refrig.*, 40(2): 429–434. doi: 10.1016/j.ijrefrig.2013.05.007.

Phimolsiripol, Y., Siripatrawan, U., and Cleland, D.J. 2011. Weight loss of frozen bread dough under isothermal and fluctuating temperature storage conditions. *J. Food Eng.*, 106(2): 134–143.

Pinnavaia, G., Dalla Rosa, M., and Lerici, C.R. 1988. Dehydrofreezing of fruit using direct osmosis as concentration process. *Acta Alimen. Pol.*, 14(38): 51–57.

Rachtanapun, P. and Rachtanapun, C. 2012. Vacuum packaging. In: *Handbook of Frozen* Food *Processing and Packaging*. Ed.: D.-W. Sun, pp. 861–873. Boca Raton, FL: CRC Press, Taylor & Francis Group.

Rahman, M.S. and Velez-Ruiz, J.F. 2007. Food preservation by freezing. In: *Handbook of Food Preservation*. Ed.: M.S. Rahman, pp. 636–665. Boca Raton, FL: CRC Press, Taylor & Francis Group, LLC.

Ramallo, L.A. and Mascheroni, R.H. 2010. Dehydrofreezing of pineaaple. *J. Food Eng.*, 99: 269–275.

Ramaswamy, H. and Marcotte, M. 2006. Low-temperature preservation. In: *Food Processing: Principles and Applications*. Eds.: H. Ramaswamy and M. Marcotte, pp. 169–232. Boca Raton, FL: CRC Press, Taylor & Francis Group, LLC.

Reid, D. 1998. Freezer preservation of fresh foods: Quality aspects, Chapter 14. In: *Food Storage Stability*. Eds.: I.A. Taub and P.S. Reid. Boca Raton, FL: CRC Press.

Robbers, M., Singh, R.P., and Cunha, L.M. 1997. Osmotic-Convective dehydrofreezing process for drying kiwifruit. *J Food Sci.*, 62(5): 1039–1047.

Robertson, G.L. 2012. Paper-based packaging of frozen foods. In: *Handbook of Frozen Food Processing and Packaging*. Ed.: D.-W. Sun, pp. 743–758. Boca Raton, FL: CRC Press, Taylor & Francis Group.

Schubert, H. 1977. Criteria for the application of T-TI indicators to quality control of deep frozen food products. *I.I.F.-I.I.R.—Commissions C1/C2*, Ettlingen, Germany, pp. 407–423.

Shaikh, N.I. and Prabhu, V. 2007. Modeling and simulation of cryogenic tunnel freezer. *J. Food Eng.*, 80: 701–710.

Singh, R.P. and Wells, J.H. 1985. Use of time-temperature indicators to monitor quality of frozen hamburger. *Food Technol.*, 39(12): 42–50.

Skrede, G. 1996. Fruits. In: *Freezing Effects on Food Quality*. Ed.: L.E. Jeremiah, pp. 183–245. New York: Marcel Dekker.

Slade, L. and Levine, H. 1991. Beyond water activity: Recent advances based on an alternative approach to the assessment of food quality and safety. *Food Sci. Nutr.*, 30(2,3): 115–357.

Suutarinen, J., Heiska, K., Moss, P., and Autio, K. 2000. The effects of calcium chloride and sucrose prefreezing treatments on the structure of strawberry tissues. *Lebensm. Wiss. Technol.*, 33: 89–102.

Talens, P., Martinez-Navarette, N., Fito, P., and Chiralt, A. 2002. Changes in optical and mechanical properties during osmodehydrofreezing of kiwi fruit. *Innov. Food Sci. Emerg. Technol.*, 3(2): 191–199.

Taoukis, P.S. 2001. Modelling the use of time-temperature indicators. In: *Food Process* Modelling. Eds.: L.M.M. Tijskens, M.L.A.T.M. Hertog, and B.M. Nicolaï, pp. 402–431. New York: CRC Press.

Taoukis, P.S., Fu, B., and Labuza, T.P. 1991. Time-temperature indicators. *Food Technol.*, 45(10): 70–82.

Taoukis, P.S. and Labuza, T.P. 2003. Time-temperature indicators (TTIs). In: *Novel Food Packaging Techniques*. Ed.: R. Ahvenainen, pp. 103–126. Oxford, U.K.: Woodhead Publishing Limited.

Taoukis, P.S., Bili, M., and Giannakourou, M. 1998. Application of shelf life modelling of chilled salad products to a TTI based distribution and stock rotation system. *Acta Hort.*, 476: 131–140.

Taoukis, P.T., Giannakourou, M.C., and Tsironi, T.N. 2012. Monitoring and control of the cold chain. In: *Handbook of Frozen Food Processing and Packaging*. Ed.: D.-W. Sun, pp. 273–299. Boca Raton, FL: CRC Press, Taylor & Francis Group.

Tijkens, L.M.M., Koster, A.C., and Jonker, J.M.E. 2001. Concepts of chain management and chain optimisation. In: *Food Process Modelling*. Eds.: L.M.M. Tijkskens, M.L.A.T.M. Hertog, and B.M. Nicolaï, pp. 448–469. New York: CRC Press.

Tironi, V., de Lamballerie, M., and Le Bail, A. 2010. Quality changes during the frozen storage of sea bass (*Dicentrarchus labrax*) muscle after pressure shift freezing and pressure assisted thawing. *Innov. Food Sci. Emerg. Technol.*, 11: 565–573.

Torreggiani, D. 1995. Technological aspects of osmotic dehydration in foods. In: *Food Preservation by Moisture Control*. Eds.: G.V. Barbosa-Canovas and J. Welti-Chanes. New York: CRC Press.

Torreggiani, D. and Bertolo, G. 2004. Present and future in process control and optimization of osmotic dehydration from unit operation to innovative combined process: An overview. *Adv. Food Nutr. Res.*, 48: 173–238.

Torreggiani, D., Lucas, T., and Raoult-Wack, A. 2000. The pretreatments of fruits and vegetables, Chapter 4. In: *Managing Frozen Foods*. Ed.: C. Kennedy. Cambridge, U.K.: Woodhead Publishing Limited.

Torres de Maria, G., Abril, J., and Casp, A. 2005. Coefficients d'échanges superficiels pour la réfrigération et lacongélation d'aliments immergés dans un coulis de glace. *Int. J. Food Refrig.*, 28: 1040–1047.

Tregunno, N.B. and Goff, H.D. 1996. Osmodehydrofreezing of apples: Structural and textural effects. *Food Res. Int.*, 29(5–6): 471–479.

Verboven, P., Scheerlinck, N., and Nicolaï, B.M., 2003. Surface heat transfer coefficients to stationary spherical particles in an experimental unit for hydrofluidisation freezing of individual foods. *Int. J. Refrig.*, 26: 328–336.

Wang, S. and Sun, D.-W. 2012. Antifreeze proteins. In: *Handbook of Frozen Food Processing and Packaging*. Ed.: D.-W. Sun, pp. 693–708. Boca Raton, FL: CRC Press, Taylor & Francis Group.

Wells, J.H. and Singh, R.P. 1989. A quality-based inventory issue policy for perishable foods. *J. Food Process. Preserv.*, 12: 271–292.

Wells, J.H. and Singh, R.P. 1998. Quality management during storage and distribution, Chapter 13. In: *Food Storage Stability*. Eds.: I.A. Taub and R.P. Singh. New York: CRC Press.

Woolfe, M.L. 1992. Temperature monitoring and measurement. In: *Chilled Foods: A Comprehensive Guide*. Eds.: C. Dennis and M. Stringer, pp. 77–109. Great Britain, U.K.: Ellis Horwood Ltd.

Wright, B.B. and Taub, I.A. 1998. Stored product quality: Open dating and temperature monitor, Chapter 12. In: *Food Storage Stability*. Eds.: I.A. Taub and R.P. Singh. New York: CRC Press.

Yoon, S.H., Lee, C.H., Kim, D.Y., Kim, J.W., and Park, K.H. 1994. Time-Temperature indicator using phospholipid-phospholipase system and application to storage of frozen pork. *J. Food Sci.*, 59(3): 490–493.

Zhang, C., Zhang, H., Wang, L., and Guo, X. 2008. Effect of carrot (*Daucus carota*) antifreeze proteins on texture properties of frozen dough and volatile compounds of crumb. *LWT*, 41: 2019–1036.

Zheng, H. and Lu, H. 2011. Effect of microwave pretreatment on the kinetics of ascorbic acid degradation and peroxidase inactivation in different parts of green asparagus (*Asparagus officinalis* L.) during water blanching. *Food Chem.*, 128: 2087–2093.

Zheng, L. and Sun, D.W. 2006. Innovative applications of power ultrasound during food freezing processes—A review. *Trends Food Sci. Technol.*, 17: 16–23.

Zhu, S., Ramaswamy, H.S., and Le Bail, A. 2005. High-pressure calorimetric evaluation of ice crystal ratio formed by rapid depressurization during pressure-shift freezing of water and pork muscle. *Food Res. Int.*, 38: 193–201.

8 Microwave Heating Technology

M. Benlloch-Tinoco, A. Salvador, D. Rodrigo, and N. Martínez-Navarrete

CONTENTS

8.1 INTRODUCTION

Thermal technologies have been at the core of food preservation and production for many years. However, despite the fact that heat treatments provide the required safety profile and extension of shelf life (Osorio et al., 2008), some more recent thermal technologies, for example, microwave energy, are being explored in an attempt to find alternatives to conventional heating methods that essentially rely on conductive, convective, and radiative heat transfer and lead to dramatic losses of both desired sensory properties and nutrients and bioactive compounds (Picouet et al., 2009). Currently, given the recent increased demand for health-promoting foods with fresh-like characteristics (Elez-Martínez et al., 2006), the industrial sector is showing a greater interest in the development and optimization of novel food preservation processes, intending to meet consumer expectations by marketing a variety of high-quality, minimally processed food products in which the required safety and shelf-life demands are achieved but the negative impact on quality attributes is minimized (Señorans et al., 2003).

Microwave energy might replace traditional heating methods, at least partially, providing food products of superior quality with extended shelf life (Elez-Martínez et al., 2006; Picouet et al., 2009; O'Donnell et al., 2010). This technology can be considered as a key factor in food innovation to successfully differentiate products (Deliza et al., 2005) or to find new uses for foods by helping to develop novel ways to process them.

In this chapter, the fundamental mechanisms of microwave heating are presented, followed by a review of the microwave systems and equipment used at an industrial level and the applications of this technology in unit operations in the food industry. This chapter also deals with the kinetic data analysis of microwave processes and describes the impact of microwaves on microorganisms, enzymes, nutrients, and bioactive compounds, and also on the shelf life of a fruit-based product. Finally, conclusions and future improvements for the application of microwave energy to industrial food processing are discussed.

8.2 PRINCIPLES OF MICROWAVE HEATING

Microwave energy is transported as an electromagnetic wave (0.3–300 GHz). When intercepted by dielectrical materials, microwaves produce an increase in product temperature associated with dipole rotation and ionic polarization (Schubert and Regier, 2010). Molecular friction of permanent dipoles within the material takes place as they try to reorient themselves with the electrical field of the incident wave, generating heat that is dissipated throughout the food material (Salazar-González et al., 2012). Additionally, microwave interaction with polar molecules results in the rotation of molecules in the direction of the oscillating field and, in turn, collisions with other polar molecules occur, a fact that also contributes to heat generation (Salazar-González et al., 2012). All these molecular movements occur to a greater extent in a liquid than in a solid medium. Microwaves are nonionizing, and their quantum energy is several orders of magnitude lower than other types of electromagnetic radiation, meaning that microwave energy is sufficient to move the atoms of a molecule but insufficient to cause chemical changes by direct interaction with molecules and chemical bonds (Vadivambal and Jayas 2007; Schubert and Regier 2010).

Typically, microwave food processing uses a frequency of 2450 MHz for home ovens, and 915 and 896 MHz for industrial heating in the United States and Europe, respectively (Wang and Sun, 2012). This type of technology involves volumetric heating, which means that the materials can absorb microwave energy directly and internally. For this reason, in comparison with conventional heating methods, microwaves lead to a faster heating rate, thus reducing process time (Huang et al., 2007; Queiroz et al., 2008; Igual et al., 2010).

8.3 MICROWAVE HEATING SYSTEMS AND EQUIPMENT

Microwave technology has been steadily gaining importance in the food processing area. Evidence of this is the enormous sales rates of household ovens and the increase in the spread of microwave ovens throughout the industrialized world. In the last few years, approximately 10 million microwave ovens have been sold annually in the United States and Europe (Schubert and Regier, 2006).

Basically, a microwave system consists of three parts: (1) the microwave source, (2) the waveguide, and (3) the applicator. The magnetron tube is by far the most commonly used microwave source for industrial and domestic applications (Schubert and Regier, 2005). A magnetron consists of a vacuum tube with a central electron-emitting cathode with a highly negative potential. This cathode is surrounded by a structured anode that forms cavities, which are coupled by the fringing fields and have the intended microwave resonant frequency (Schubert and Regier, 2006). The waveguides are elements that are used to guide the electromagnetic wave, consisting principally of hollow conductors, normally with a constant cross section, rectangular and circular forms being of most practical use. Within the waveguide, the wave may spread out in so-called modes which define the electromagnetic field distribution within the waveguide (Schubert and Regier, 2005). The applicator is basically the element that contains and distributes the microwave energy that surrounds the food product to be heated. Common applicators can be classified by type of field configuration into three types: (1) near-field, (2) single-mode, and (3) multi-mode applicators. Multi-mode applicators play by far the most important role in industrial and domestic uses because of the typical dimensions of microwave ovens (Schubert and Regier, 2005).

To date, microwave heating has not been used as successfully in the food industry as in households. The development of a nonhomogeneous field distribution has been one of the main factors that is limiting the exploitation of this technology to its fullest potential in the food industry. Despite the fact that the number of working installations increases every year, it is still considered to be quite low (Hebbar and Rastogi, 2012). Inhomogeneous field distribution may lead to an undesired inhomogeneous heating pattern, producing hot spots that damage the item being heated and cold spots where the item may be under-heated or under-processed, thereby compromising product quality, stability, and repeatability (IMS, 2014). Bearing in mind that the homogeneity of the electromagnetic field distribution depends strongly both on microwave equipment features and on food properties, improvement of industrial microwave systems design could be a key factor to promote a greater spread of this technology in the industrialized world.

In fact, it could be claimed that microwave systems design has shown a spectacular evolution over the years. Early operational systems included batch processing of, for example, yogurt in cups (Anonymous, 1980), their primary drawback being their inability to heat materials in a predictable and uniform manner. Then continuous microwave applicators were developed in an attempt to solve these problems, which allowed continuous processing to improve heating uniformity and at the same time accomplish the high throughputs desired by the food industry (Hebbar and Rastogi, 2012). Since then, microwave equipment has improved remarkably. Figure 8.1 shows a model of the industrial microwave equipment currently used by American and European companies to heat, cook, and pasteurize different kind of food products. Nowadays, there is a variety of continuous microwave systems with features that address the major obstacles to the commercialization of microwave heaters for many industrial applications. For example, the fact that microwave energy can optionally be irradiated in modern industrial ovens by one high-power magnetron or by several low-power magnetrons, or be used under vacuum conditions, as in microwave-assisted air-drying and microwave-assisted freeze-drying operations, in order to improve the efficiency of the process, can be taken as proof of this substantial evolution (Schubert and Regier, 2005; Vadivambal and Jayas, 2007).

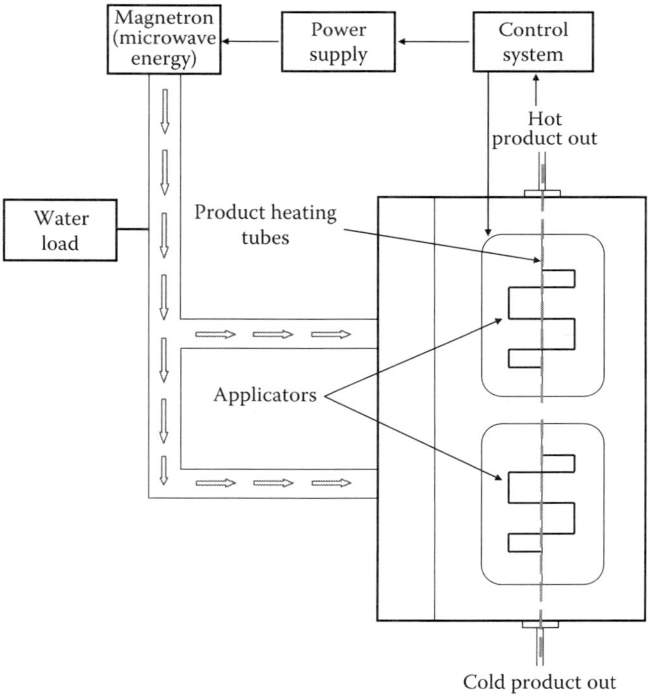

FIGURE 8.1 Major components of a model microwave heating system currently used in the food industry.

However, the applicator must be considered the essence of these novel systems, given that the unique structures and geometries of the applicators employed in the currently used operational systems may be considered as the key factor that allows the target material to pass through a uniform microwave field and efficiently absorb the available microwave energy, thereby providing evident competitive advantages (IMS, 2014).

8.4 INDUSTRIAL APPLICATIONS IN FOOD PROCESSING

Microwave energy has been extensively used in the area of food processing for various commercial purposes (Vadivambal and Jayas, 2007). On one hand, microwave heating has become very popular at the level of household applications, there being a very large number of microwave ovens in households. In fact, the success of household microwave ovens is such that some products have been developed especially for them, for example, microwave popcorn (Schubert and Regier, 2006). On the other hand, this technology is also widespread in the industrialized world. Although microwave heating has not been as effectively used in the food industry as in households (Hebbar and Rastogi, 2012), industrial applications for both fluids and solid foods do exist (Salazar-González et al., 2012). Nevertheless, in the case of solid or frozen foods, it is normally combined with convective heating. Drying, cooking, blanching, pasteurization, thawing and tempering, microwave vacuum-drying, and microwave freeze-drying are some of the commercially proven applications of this technology (Vadivambal and Jayas, 2010; Hebbar and Rastogi, 2012; Salazar-González et al., 2012).

Among the aforementioned food processing operations in which microwave energy may be used competitively, some specific applications are worth highlighting. First, the finish drying of potato chips was one of the first large-scale applications of microwave heating in the food processing industry. In this case, microwaves helped significantly to overcome the difficulties found in conventional potato chip processing by accelerating the dehydration step. Potato chips can be dried conventionally to a moisture content of 6–8 g water/100 g product and then microwaved to the desired final moisture level in the product (Salazar-González et al., 2012). Nowadays, however, tempering of meat for further processing, precooking of bacon, pasta drying, and microwave vacuum drying of fruit juice concentrates are considered to be the most important industrial applications of microwave heating (Vadivambal and Jayas, 2010; Salazar-González et al., 2012). Microwave tempering of meat results are particularly profitable because microwaves can easily penetrate the frozen product, reaching the inner regions in a short time, whereas conventional tempering leads to a large temperature gradient and takes several days (Vadivambal and Jayas, 2010). Recently, microwave technology has also been proven to be suitable for industrial sterilization of food products. In this respect, FDA (Food and Drugs Administration) acceptance has been newly granted for a sweet potato puree product sterilized using continuous flow microwave processing and aseptic packaging. This first industrial implementation of continuous flow microwave sterilization for low-acid products was carried out by Yamco in Snow Hill, NC (Salazar-González et al., 2012) and several microwave-sterilized products, such as pasta dishes, pasta sauces, and rice dishes are currently being marketed in Europe by companies such as Top's Foods.

8.5 ESTABLISHING A NOVEL THERMAL PRESERVATION PROCESS: KINETIC DATA ANALYSIS

Designing a sound thermal treatment, either novel or conventional, requires extensive understanding of the process, the heating behavior of the product, and its impact on a target microorganism, the establishment of thermal processes being based on two premises: (1) the heat resistance of microorganisms for each specific product formulation and composition and (2) the heating rate of the specific product (Awuah et al., 2007).

Since safety must always be the primary concern, thermal treatments are constrained by the requirement to achieve the target lethality, but lethality is not the only aspect to be considered; quality loss must also be taken into account. Thermal processes tend to be optimized to maximize microbial and enzyme inactivation and minimize degradation of sensory attributes and loss of nutritional value (Awuah et al., 2007; Wang and Sun, 2012). To perform this optimization, knowledge of the processing parameters and of the inactivation kinetics of target microorganisms, enzymes, and quality attributes is of utmost importance (Valdramidis et al., 2012).

The calculations involved in the kinetic studies used to design and optimize conventional heat processes are well established. However, when it comes to microwave processes, the issue becomes more complicated, and the main concern lies in the particular form of heating that takes place during microwave exposure (Banik et al., 2003). In conventional heating, a holding period is expected, but in case of microwaves, the heating that takes place is exclusively non-isothermal (Matsui et al., 2008). Furthermore, it is usually not possible to fix the parameters that affect the heating process, such as (1) the heating rate, (2) the range of temperatures at which the samples are exposed, or (3) provision of appropriate sample homogenization. At present, little is known kinetically about the general basic relationship between microbial and enzyme inactivation and quality retention in foods and microwave exposure.

More specifically, focusing on microbial inactivation, Fujikawa et al. (1992), Tajchakavit et al. (1998), Cañumir et al. (2002), Yaghmaee and Durance (2005), and Pina-Pérez et al. (2014) have conducted some of the few studies regarding the kinetics of destruction of foodborne pathogens and spoilage microorganisms by microwave irradiation. According to their results, microbial inactivation due to microwave processing can be fitted using first-order kinetics, which has been successfully employed to describe destruction of *Cronobacter sakazakii*, *Saccharomyces cerevisiae*, and *Lactobacillus plantarum* under microwave processing (Fujikawa et al., 1992; Tajchakavit et al., 1998; Pina-Pérez et al., 2014).

When first-order kinetics models are used to describe the inactivation process, the existence of a linear relationship between the logarithm of the microbial population and time is assumed. Two key parameters (D and z values) are then determined from the survival and resistance curves, respectively (Awuah et al., 2007; Tajchakavit and Ramaswamy, 1997). The D-value represents the heating time required to reduce 90% of the existing microbial population under isothermal conditions (Equation 8.1). The z-value represents the temperature change that results in a 90% reduction of the D-value (Equation 8.2).

$$\log \frac{N}{N_0} = -\frac{t}{D} \tag{8.1}$$

where

N is the survivor counts after treatment (CFU/g)
N_0 is the initial microorganism population (CFU/g)
t is the processing time (s)
D is the D-value at the temperature studied (s)

$$\log \frac{D}{D_{ref}} = \frac{T_{ref} - T}{z} \tag{8.2}$$

where

D is the D-value at each temperature studied (s)
D_{ref} is the D-value at reference temperature (s)
T is the processing temperature
T_{ref} is the reference temperature (s)
z is the z-value or temperature sensitivity (°C)

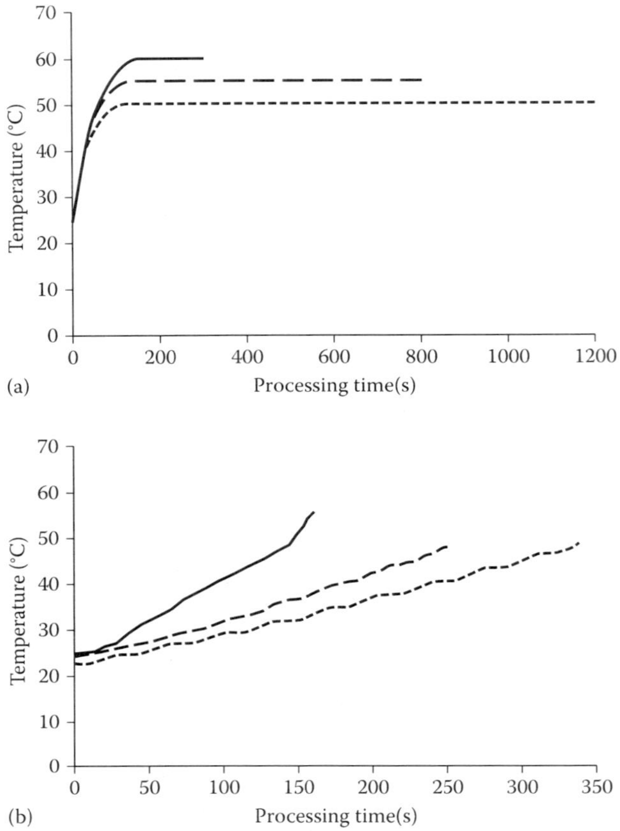

FIGURE 8.2 Mean kiwifruit puree temperature profile for conventional thermal processing (a) at 60°C (–), 55°C (– –·), and 50°C (-----) and microwave processing (b) at 1000 W (–), 900 W (– –·), and 600 W (-----).

As previously mentioned, one of the main aspects to take into consideration when performing kinetic data analyses in microwave processes is the fact that, as opposed to conventional heat treatments, the heating that takes place is exclusively non-isothermal. This can be seen in Figure 8.2, which shows temperature profiles of a kiwifruit puree sample subjected to different conventional and microwave treatments. Accordingly, correction of processing time values for come-up periods is essential prior to kinetic data analyses. Time–temperature profiles have to be used to calculate the effective time (t_e) (Equation 8.3), which represents the isothermal holding time at the selected reference temperature that causes the same level of microbial destruction as the heating actually applied, as if the microwave treatments had been performed under isothermal conditions (Tajchakavit and Ramaswamy, 1997; Awuah et al., 2007; Matsui et al., 2008; Latorre et al., 2012). Since no holding period at a preset temperature is expected in microwave processes, the maximum temperature reached during the treatment is considered as T_{ref} (reference temperature) (Matsui et al., 2008; Latorre et al., 2012).

$$t_e = \int_0^\infty 10^{\left(T(t)-T_{ref}/z\right)} dt \tag{8.3}$$

where

t_e is the effective time (s)
$T(t)$ is the processing temperature at each processing time
T_{ref} is the reference temperature (s)
z is the z-value or temperature sensitivity (°C)

Matsui et al. (2008) proposed a method for calculating *D* and *z*-values under microwave heating by nonlinear regression. According to their reports, the predicted surviving microbial population for each microwave experimental run can be calculated from Equation 8.4. Then a nonlinear estimation procedure can be used to minimize the sum of squared errors (SSE) between experimental and predicted surviving microorganisms, defined in Equation 8.5.

$$\log\left(\frac{N}{N_0}\right)_{predicted} = -\frac{t_e}{D_{T_{ref}}} = \frac{\int_0^{\infty} 10^{((T_{ref}-T(t))/z)} \cdot dt}{D_{T_{ref}}} \quad (8.4)$$

where
- N is the survivor counts after treatment (CFU/g)
- N_0 is the initial microorganism population (CFU/g)
- t_e is the effective time (s)
- $D_{T_{ref}}$ is the D-value at reference temperature (s)
- T_{ref} is the reference temperature (s)
- $T(t)$ is the processing temperature at each processing time
- z is the z-value or temperature sensitivity (°C)

$$SSE = \sum_{i=1}^{n}\left[\log\left(\frac{N}{N_0}\right)_{experimental} - \log\left(\frac{N}{N_0}\right)_{predicted}\right] \quad (8.5)$$

where
- N is the survivor counts after treatment (CFU/g)
- N_0 is the initial microorganism population (CFU/g)
- N is the number of experimental runs

A further aspect to be taken into consideration is that the ability to properly understand and carry out kinetic data analysis in microwave heating is important not only for the accurate design of preservation processes but also for the establishment of appropriate comparisons between microwave and conventional heat treatments (Latorre et al., 2012). A comparison of microwave and conventional heating has been the basis of many studies dealing with microwave process applications, such as those performed by Gentry and Roberts (2005) or Igual et al. (2010). Nevertheless, poor correction of processing time values for come-up periods prior to kinetic data analysis owing to the nonisothermal nature of microwave processes may lead to mistaken interpretations, hinder comparison of different research works, and cause conflicting opinions regarding the superiority, of microwave technology over conventional heat treatments.

8.6 MICROWAVE PRESERVATION OF FRUIT-BASED PRODUCTS: APPLICATION TO KIWIFRUIT PUREE

As mentioned previously, microwave energy could potentially replace conventional heating for some specific purposes, and commercially proven applications of this technology in several food processing operations are a matter of fact (Awuah et al., 2007; Vadivambal and Jayas, 2010). Nevertheless, microwaves have not yet been exploited to their fullest potential in the food industry (Picouet et al., 2009).

Nowadays, there is still a gap in knowledge concerning fundamental understanding of the interactions of microwaves when applied to food, and published information on the impact of this technology on food safety, stability, and quality aspects is currently both scarce and inconsistent.

Bearing in mind that the application of microwave heating would be justified only from the standpoint of obtaining high-quality products (Vadivambal and Jayas, 2007), in-depth research work on the impact of microwaves on microorganisms, enzymes, bioactive compounds, and sensory properties of a variety of foods might make an important contribution to expanding the use of this technology on an industrial level.

In the present chapter, the particular case of a kiwifruit puree has been selected as a model fruit-based product to evaluate the impact of a microwave preservation process on some safety and quality issues.

8.6.1 Microbial Decontamination

Thermal preservation treatments are particularly designed to minimize public health hazards and to extend the useful shelf life of food products, information regarding thermal resistance of microorganisms, pathogenic or otherwise, being crucial to a correct understanding of their lethal effect.

In the present study, the safety of a ready-to-eat kiwifruit puree subjected to microwave heating was investigated by checking how effective microwaves are at inactivating *Listeria monocytogenes*, taken as the pathogen of greatest concern in the product (Figure 8.3). Although fruit products of an acidic nature, such as kiwifruit (pH = 3.4), have not been recognized as being potentially the main vehicles for foodborne illnesses, there has been increasing concern because some outbreaks have been caused by consumption of unpasteurized juices contaminated with *Escherichia coli* or *Salmonella* spp. (Buffler, 1993; Picouet et al., 2009) or of salad vegetables or mixed salads with *L. monocytogenes* (EFSA, 2013). *L. monocytogenes* is currently recommended by the National Advisory Committee on Microbiological Criteria for Foods as an appropriate target organism to be used for fruit juices. Despite the fact that the minimum pH allowing growth of this pathogen in food products has been reported to be pH 4.6 (Carpentier and Cerf, 2011), ready-to-eat fruit-based acidic products may still represent a potential hazard to health, given the well-known ability of *L. monocytogenes* to proliferate in products stored under refrigeration for long periods.

Microwave inactivation of *L. monocytogenes* in the kiwifruit product (*Actinidia deliciosa* var. Hayward) was determined by using the following experimental procedure. The puree was inoculated by adding 1 mL of a concentrated suspension of the microorganism so as to give an initial *L. monocytogenes* concentration of 10^7 CFU/g. The product was then processed in a microwave oven (model: 3038GC, Norm, China) provided with a turntable plate and a fiber-optic probe

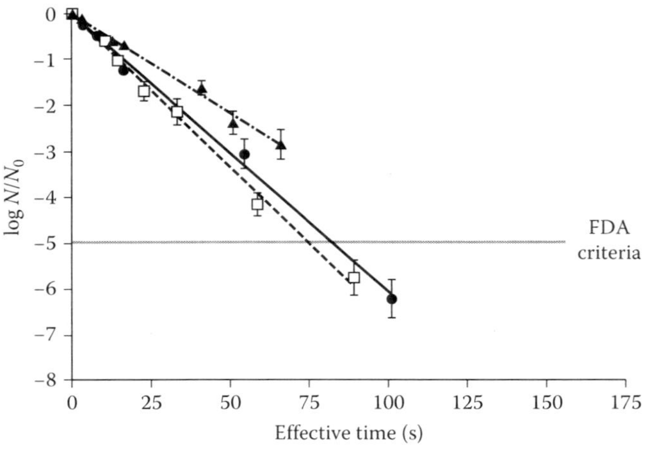

FIGURE 8.3 Survival of *L. monocytogenes* under microwave processing at 1000 W (experimental (•), model (–)), 900 W (experimental (□), model (----)) and 600 W (experimental (▲), model (–·–)). The plotted values and error bars represent the average of three replicates and the corresponding standard deviation.

(CR/JP/11/11671, OPTOCOM, Germany) which was connected to a temperature datalogger (model: FOTEMP1-OEM, OPTOCOM, Germany) to continuously record the time–temperature history at different points in the sample. The safety of the process was evaluated on a sample placed in the coldest spot, previously identified (data not shown), since contaminating pathogenic microorganisms may survive in cold spots (Nicolaï, 1998). For each treatment, a 500 g sample was tempered to an initial temperature of 25°C and then heated in the microwave oven in a standard glass beaker (9 cm inner diameter and 12 cm height) (BKL3-1K0-006O, Labbox, Spain). Treated samples taken from the coldest spot were immediately cooled in ice-water until the puree reached 35°C (for 10–15 s). Then, immediately after they had been inoculated or subjected to different MW power–time processes, respectively, serial decimal dilutions of the treated and untreated samples were performed in 0.1% (w/v) sterile peptone water (Scharlab Chemie S.A., Barcelona, Spain). The medium used for enumerating viable cells was tryptic soy agar (Scharlab Chemie S.A., Barcelona, Spain). The selected dilutions were incubated at 37°C for 48 h, after which the counting step was carried out. The reduction of viable cells was expressed as the decimal logarithm of the quotient of the treated and untreated cells.

Survival curves were obtained at three power levels (600, 900, and 1000 W) with processing times varying between 50 and 340 s (Figure 8.3). Since *L. monocytogenes* inactivation under microwave heating was close to linearity, as previously reported by other authors for *S. cerevisiae* and *L. plantarum* (Fujikawa et al., 1992; Tajchakavit et al., 1998), the data obtained were fitted to first-order kinetics (see Section 8.5). As mentioned previously, in order to make it possible to compare the kinetic parameters (*D*-values) obtained under microwave processing at the preset power levels, kinetic data transformation was performed. Treatment times were corrected, and effective time values were obtained. Calculated effective times represented the equivalent holding time at each processing temperature as if the treatments had been performed under isothermal conditions (Awuah et al., 2007; Matsui et al., 2008; Latorre et al., 2012).

From the inactivation data presented in Figure 8.3, it can be claimed that in the kiwifruit puree samples subjected to effective times higher than 75 and 82 s at 900 and 1000 W, respectively, the pasteurization objective of 5D established by the FDA (2004) was accomplished. To the best of our knowledge, the only published study on microwave *Listeria* spp. inactivation in fruit-based products is the one conducted by Picouet et al. (2009). They found a 7-$\log_{10}$ cycle reduction of *Listeria innocua* in an apple puree subjected to 900 W for 35 s. However, conventional heat inactivation of *L. monocytogenes* in different fruit substrates has been evaluated by several authors. For example, Hassani et al. (2005) reported that 5-$\log_{10}$ cycles of *L. monocytogenes* were inactivated in a reference medium (pH = 4) when it was subjected to 58°C for 84 s, and Fernández et al. (2007) found a 4-$\log_{10}$ cycle reduction when a sucrose solution (pH = 7, $a_w = 0.99$) was maintained at 60°C for 60 s.

The effect of the processing parameters, power (W) and time (s), on inactivation of *L. monocytogenes* was evaluated statistically. Both factors were shown to affect the *L. monocytogenes* reduction level achieved significantly ($p < 0.05$), although no significant differences were found between 1000 and 900 W. Both higher power level and higher effective time led to significantly higher *L. monocytogenes* inactivation ($p < 0.05$) (Figure 8.3). In this respect, the higher the microwave power, the lower the effective time necessary to reach the same level of inactivation. For example, in order to achieve the FDA recommendations for pasteurized products (5-$\log_{10}$ cycle inactivation), a considerably longer effective time was required at 600 W ($t_e = 116$ s) than at 1000 W ($t_e = 82$ s).

Kinetic parameters describing *L. monocytogenes* inactivation under microwave processing were calculated (see Section 8.5), providing the following *D*-values: $D_{60°C} = 42.85 \pm 0.13$ (R^2-adjusted = 0.992) at 600 W, $D_{60°C} = 17.35 \pm 0.34$ (R^2-adjusted = 0.993) at 900 W, and $D_{60°C} = 17.04 \pm 0.34$ at 1000 W (R^2-adjusted = 0.996). Although the kinetics of *L. monocytogenes* inactivation by thermal treatment has been studied extensively in various foodstuffs such as beef, milk, chicken, carrot, cantaloupe, and watermelon juice (Chhabra et al., 1989; Bolton et al., 2000; Sharma et al., 2005), in reference medium (Hassani et al., 2005, 2007) and in sucrose solutions (Fernández et al., 2007), there is no information available about the survival behavior of this pathogen in fruit-based products under microwave heating. Cañumir et al. (2002) reported higher D-values for microwave apple juice pasteurization when the inactivation kinetics of *E. coli* was evaluated, ranging between $D_{70.3°C}$ = 25.2 s to $D_{38.3°C}$ = 238.8 s for 900 and 270 W, respectively. Yaghmaee and Durance (2005) found similar D-values for microwave inactivation of *E. coli* in peptone water at 510 W, with $D_{55.6°C}$ = 30 s and $D_{60.5°C}$ = 18 s. Once more, the power level effect can be evaluated by comparing the $D_{60°C}$-values. Microwave processing performed at 900 and 1000 W led to considerably faster bacterium reduction than processing at 600 W.

Like the results of other authors (Fujikava et al., 1992), the results obtained in this study proved the effectiveness of microwave heating against foodborne pathogens of concern, such as *L. monocytogenes*, showing that safety can be properly ensured in fruit-based products by means of this technology.

8.6.2 Enzyme Inactivation

Enzymes are naturally present in fruit and vegetables and can cause product deterioration in many ways (Whitaker et al., 2003). Enzymes such as peroxidase (POD) and polyphenol oxidase (PPO) are principally responsible for the degradation of color and nutritive value of most food products of vegetable origin (Queiroz et al., 2008), while pectin methylesterase (PME) causes changes in the rheological properties of foods by means of pectin de-esterification (Jolie et al., 2010). In view of the very negative impact that enzymes of this kind could have on kiwifruit-based products, POD, PPO, and PME were selected to check how effective microwave heating is at inactivating enzymes in the product. To study the effect of microwave power and process time on the inactivation of POD, PPO, and PME in the product using the minimum number of experimental trials (Beirão-da-Costa et al., 2006), an experimental design based on a central composite design was applied (Cochran and Cox, 1957). Power and time were designed to vary between 300 and 900 W and between 100 and 300 s, respectively. Each microwave treatment was carried out as described in Section 8.6.1. The temperature of the sample was recorded continuously, in this case in the hottest spot, previously identified (data not shown). Enzyme activity was measured in all the treated samples and also in the untreated sample, which was used as the control, following the methods described by De Ancos et al. (1999) for POD and PPO and by Rodrigo et al. (2006) for PME. The percentage of enzyme inactivation (I) was then calculated by using Equation 8.6.

$$I = \frac{A_F - A_T}{A_F} \times 100 \tag{8.6}$$

where

A_F is the enzyme activity of fresh kiwifruit puree

A_T is the enzyme activity of treated kiwifruit puree

The results obtained showed that the inactivation of POD, PPO, and PME in the kiwifruit puree produced by processing in the desired range of microwave power (300–900 W) and time (100–300 s) varied from 43% ± 6% to 88.0% ± 0.7%, from 11.4% ± 0.5% to 81% ± 2%, and from −19.0% ± 1.3% to 57% ± 6%, respectively. These results indicate that, in kiwifruit, PME and POD were the enzymes that were most resistant and most sensitive to microwaves, respectively, while PPO showed an intermediate behavior. Similar results have been reported by other authors for this fruit as well as for strawberry when subjected to conventional heat processes (McFeeters et al., 1985; De Ancos et al., 1999; Beirão-da-Costa et al., 2008; Terefe et al., 2010). Despite the fact that POD was the most sensitive enzyme in this case, it could still be considered as a suitable indicator of treatment efficiency since it has been reported to be very important in kiwifruit because of its high activity and extensive contribution to the quality of this fruit (Fang et al., 2008).

One of the microwave treatments applied (300 W–100 s) led to a promotion of PME activity. This might be related with the low temperature reached by the sample in this case, around 43°C, and the short exposure time. A similar phenomenon was observed by Beirão-da-Costa et al. (2008), who found a significant ($p < 0.05$) increase in PME activity in kiwifruit slices subjected to mild heat treatment prior to inactivation. Another sample subjected to 300 W reached 45°C, but the treatment time was 300 s. Under these conditions, inactivation of PME was only 4.3% (standard deviation 0.7). The temperature reached by the other samples was in the range of 60°C–100°C.

The results obtained from the enzyme inactivation study were also analyzed by means of the Response Surface Methodology, yielding 3D plots for POD, PPO, and PME inactivation (Figures 8.4 through 8.6). As can be observed in Figure 8.4, there was a significant ($p < 0.05$) increase in POD inactivation up to a power of 800 W, decreasing slightly when a higher microwave power was applied. De Ancos et al. (1999) observed that inactivation of papaya POD behaved similarly under microwave heating. They reported an increase in peroxidase inactivation when the microwave power increased from 285 to 570 W for 30 s of processing time. Thereafter, a higher power level (800 W) did not increase POD inactivation. In accordance with other authors, as the process time increased, there was a linear increase in POD inactivation (Matsui et al., 2008).

Figure 8.5 shows the PPO inactivation behavior as related to microwave power and process time. As can be observed, the level of PPO inactivation rose significantly ($p < 0.05$) as the microwave power increased. However, the increase in the PPO inactivation observed was smaller at greater powers. Process time also had a significant ($p < 0.05$) effect, leading to a greater inactivation of this enzyme. Latorre et al. (2012) and Matsui et al. (2008) found that there was a greater

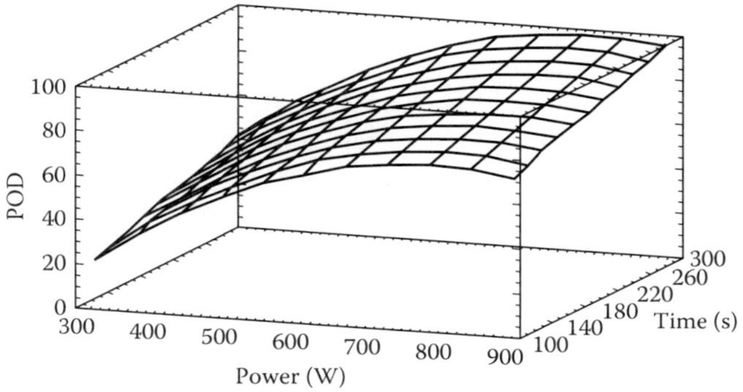

FIGURE 8.4 Response surface plot for the percentage of peroxidase (POD) inactivation in kiwifruit puree as a function of microwave power and process time.

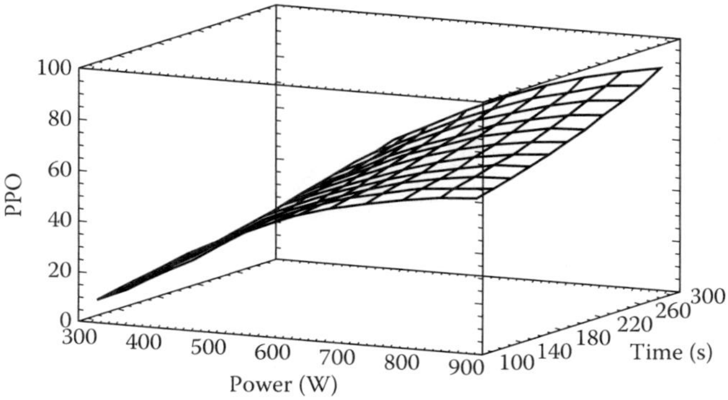

FIGURE 8.5 Response surface plot for the percentage of polyphenoloxidase (PPO) inactivation in kiwifruit puree as a function of microwave power and process time.

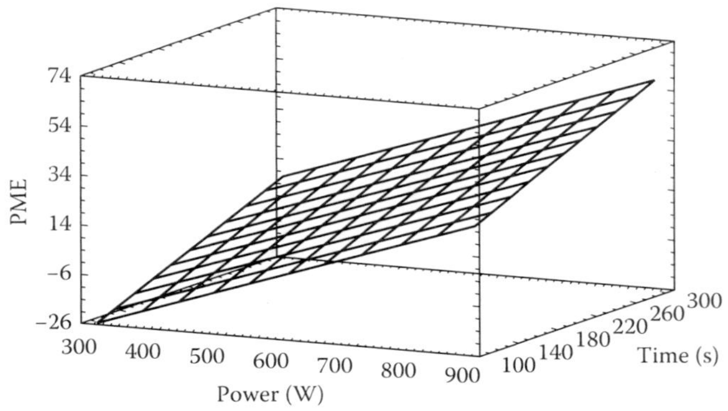

FIGURE 8.6 Response surface plot for the percentage of pectin methylesterase (PME) inactivation in kiwifruit puree as a function of microwave power and process time.

level of PPO inactivation in red beet and green coconut water, respectively, after longer microwave exposure. De Ancos et al. (1999) observed that PPO inactivation in kiwifruit and strawberry was controlled better by prefixing the power rather than the exposure time. In addition, an interactive effect on PPO inactivation was observed between microwave power and process time. As expected, as greater microwave power was applied, the level of PPO inactivation rose faster in samples subjected to longer treatment times than in kiwifruit puree subjected to shorter treatment times.

From Figure 8.6, it can be said that PME inactivation increased significantly ($p < 0.05$) as the microwave power level rose and the processing time lengthened. Similarly, Tajchakavit and Ramaswamy (1997) reported a linear relationship between time and PME inactivation during microwave heating of orange juice, and Kratchanova et al. (2004) found, when microwaving orange peel, that as microwave power increased, PME inactivation also increased.

Summarizing, in accordance with what has been reported by other authors, the results of the present study highlight the suitability of microwave heating for enzyme inactivation (De Ancos et al., 1999; Matsui et al., 2008; Latorre et al., 2012), which means that stability and quality can be properly ensured during the shelf life of fruit-based products by means of this technology (Igual et al., 2010; Zheng and Lu, 2011).

8.6.3 Impact on Sensory Properties

Despite the fact that sensory assessment must be considered as an essential tool to guide any modification of the food processing step (Di Monaco et al., 2005), there still seems to be a need for sensory analyses that focus on the impact that alternative technologies, such as microwaves, have on food product characteristics (Da Costa et al., 2000).

In the present work, the following experimental procedure was performed to evaluate the impact of microwave processing on the most important sensory characteristics of kiwifruit puree. The effect of the two processing variables (microwave power and process time) was investigated simultaneously by means of a rotatable central composite design (Section 8.6.2). Each microwave treatment was carried out as described in Section 8.6.1. Cooked purees were then cold stored (4°C) for 24 h before sensory assessment.

A sensory panel with 11 assessors (four men and seven women), recruited from students and employees of the Food Technology Department (Universitat Politècnica de València) aged between 25 and 50, was trained over a period of 2 months (12 training sessions). Samples were tempered at 25°C and served in disposable standard-size plastic containers identified with three-digit codes. In all cases, training and formal assessment were performed in a normalized

TABLE 8.1
Attributes, Scale extremes, and Evaluation Technique Used in Descriptive Sensory Assessment of Kiwifruit Puree Treated with Microwaves

Attribute and Scale Extremes	Technique
Kiwi odor intensity (low/high)	Observe
Atypical odor (low/high)	Observe
Typical kiwi color (low/high)	Observe
Tone (green/brown)	Observe
Lightness (light/dark)	Observe
Granularity (low/high)	Evenness of the sample's surface. Take a spoonful of the sample and observe its surface.
Visual consistency (low/high)	Take enough quantity of kiwi puree with a spoon and drop it to evaluate its visual consistency.
Sweetness (low/high)	Taste the necessary quantity of kiwi puree to notice the intensity of sweetness.
Acidity (low/high)	Taste the necessary quantity of kiwi puree to notice the intensity of acidity.
Astringency (low/high)	Taste the necessary quantity of kiwi puree to notice the intensity of astringency.
Kiwi taste intensity (low/high)	Taste the necessary quantity of kiwi puree to notice the intensity of typical kiwi taste.
Atypical taste (low/high)	Taste the necessary quantity of kiwi puree to notice the intensity of typical kiwi taste.
Aftertaste (low/high)	Assess the persistence of taste after ingesting kiwi puree.
Mouth consistency (low/high)	Taste the sample and evaluate its consistency during ingestion.

tasting room. The selection of descriptors was made over two 1 h sessions using the checklist method (Table 8.1) (Lawless and Heymann, 1998). During the training period, all the treated and untreated samples were tasted. Tests of three different samples in each session were used by the panelists for each descriptor until the panel was homogeneous in the ranking of the samples. Panel members were then trained in the use of scales by using reference samples (10 cm unstructured scales for all the attributes). Panel performance was checked by an analysis of variance (ANOVA) for the discrimination ability of the panelists and the reproducibility of their assessments. Once the training period was over, the formal assessment was performed. To this end, a balanced complete block experimental design was carried out in duplicate (two different sessions), using the Compusense® program release five 4.6 software (Compusense Inc., Guelph, Ontario, Canada) to evaluate the samples. The intensity of the sensory attributes was scored on a 10 cm unstructured line scale. Samples were selected randomly and served with a random three-digit code. All the treated samples were subjected to formal analysis, as well as the untreated sample.

The results obtained from the sensory assessment indicated that significant differences ($p < 0.05$) among samples were only found in the sensory descriptors "typical kiwifruit color," "tone," "visual consistency," "lightness," and "atypical taste." As a general rule, for these five descriptors (Figure 8.7), noticeable differences increased in treated samples compared with untreated samples when heating intensity increased. In fact, significant differences were not found ($p > 0.05$) between fresh kiwifruit puree and samples processed at 200 W–200 s, 300 W–100 s, and 600 W–60 s as the lines in the spider plot nearly overlapped (Figure 8.7a). Figure 8.7b shows greater differences in each significant attribute between treated samples and fresh kiwifruit puree, except in "visual consistency." Panelists considered that samples 600 W–200 s and 900 W–100 s had less lightness and a lower "typical kiwifruit color intensity" than fresh puree ($p < 0.05$). These samples and also the 300 W–300 s one seemed to be significantly ($p < 0.05$) browner, or rather less green, than the fresh kiwifruit puree. Figure 8.7c shows greater differences in the assessments given to samples 600 W–340 s, 900 W–300 s, and 1000 W–200 s as compared with fresh kiwifruit puree. In general, the panelists considered that the three processed samples had significantly ($p < 0.05$) less lightness and greenness, with a lower typical kiwifruit color intensity and higher atypical taste intensity; however, they had the same visual consistency as fresh kiwifruit puree.

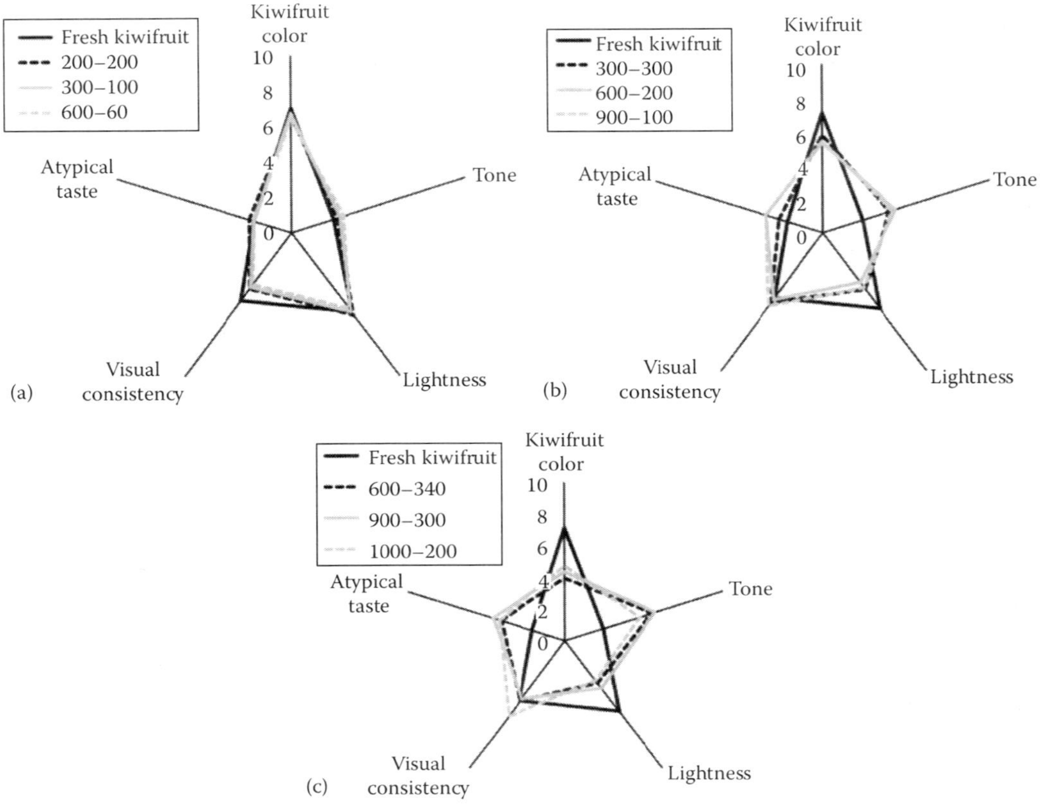

FIGURE 8.7 Average values (on a 0–10 scale) of panel member assessments of kiwifruit color, tone, lightness, visual consistency, and atypical taste of treated samples: 200 W–200 s, 300 W–100 s, and 600 W–60 s (a); 300 W–300 s, 600 W–200 s, and 900 W–100 s (b); and 600 W–340 s, 900 W–300 s, and 1000 W–200 s (c), compared with fresh sample.

Additionally, the XLSTAT 2009 program was employed to make a principal component analysis (using a correlation matrix), with the aim of studying the correlation between the various microwave treatments applied in the present study and the sensory attributes of kiwifruit puree. Figure 8.8 shows the first two component maps of the principal component analysis constructed using the sensory data. Two components were extracted that explain 80.59% of the data variability. The first component explained most of this variance (63.83%); for this reason, it has been used to describe all the kiwifruit puree characteristics. This component showed a positive correlation with the sensory attributes "typical kiwifruit color intensity," "kiwifruit odor intensity," "lightness," "acidity," "astringency," and "kiwi taste intensity" and a negative correlation with the sensory attributes "atypical odor," "tone," "atypical taste," "visual consistency," and "mouth consistency." Samples 200 W–200 s, 300 W–100 s, and 600 W–60 s were characterized by a similar acidity, astringency, color, odor, and taste to the fresh kiwifruit, owing to the less intensive treatments that were applied to these samples. On the other hand, when the most severe treatments were applied (600 W–340 s, 900 W–300 s, and 1000 W–200 s), the samples were characterized by a higher atypical odor and taste, higher visual and mouth consistency, and more browning. Finally, the granularity and consistency of samples 300 W–300 s, 600 W–200 s, and 900 W–100 s were higher than those of the other samples.

On the whole, it can be said that the application of intense treatments of high microwave power mainly affected the color and taste of the kiwifruit puree. Significant perceivable differences

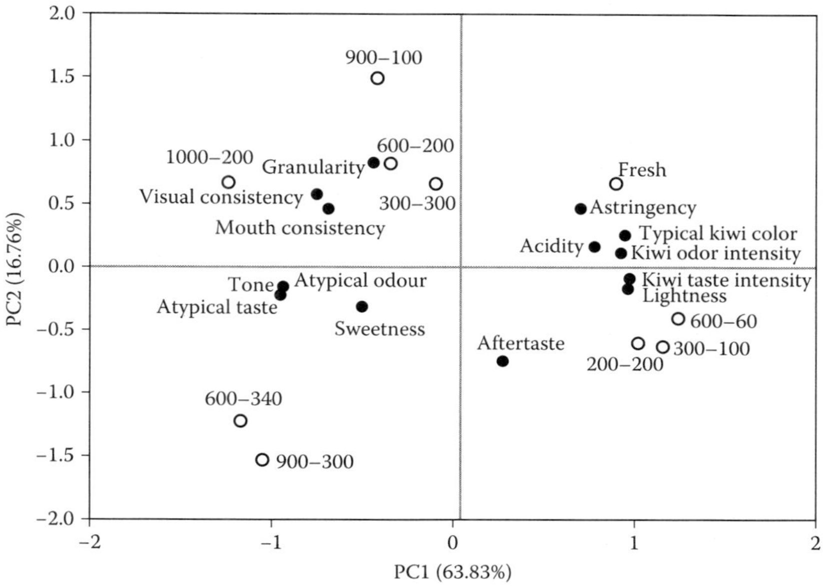

FIGURE 8.8 Plot of the first two components of the Principal Component Analysis carried out on fresh and treated samples and sensory attributes.

between kiwifruit puree samples were only found in the five descriptors mentioned earlier, and they increased when microwave power increased. As expected, the most severely treated samples showed the highest variation in these parameters.

8.6.4 Impact on Nutrients and Functional Compounds

Epidemiological studies suggest that the consumption of fruit and vegetables may play an important role in the protection against many chronic diseases. In addition to the well-established benefits of the essential vitamins and minerals found in these products, they also provide the diet with a good source of fiber and a diverse array of phytochemicals (Barret and Lloyd, 2012). More specifically, kiwifruit has high vitamin C and E contents and marked antioxidant activity; its vitamin C content being even higher than that found in grapefruit and orange (Igual et al., 2010), citric fruits that are widely recognized as good sources of this bioactive compound. In fact, given its excellent nutritional and functional characteristics, Fiorentino et al. (2009) defined kiwifruit as a unique and precious cocktail of protective phytochemicals.

The main goal of fruit processing is to create microbiologically safe products and to extend their shelf life so that they can be consumed all year round and transported safely to consumers all over the world. However, processors also strive to produce the highest-quality food, attempting to minimize losses of nutritional and functional value (Barret and Lloyd, 2012).

In the present work, the impact of microwaves on the nutritive and functional value of kiwifruit was investigated by evaluating changes produced in the main bioactive compounds of this fruit by microwave pasteurization treatment. Processing conditions for puree pasteurization were chosen (1000 W–340 s) on the basis of preliminary experiments, considered in terms of the enzyme (90% of POD) and microbial (5D of *L. monocytogenes*) inactivation to be achieved. Then the microwave treatment was carried out as described in Section 8.6.1. Vitamin C, A, and E contents, total phenols, total tannins, total flavonoids, and antioxidant activity were measured in the treated sample and also in the untreated sample, which was used as the control, following the methodology described by Taira (1995), Djeridane et al. (2006), Igual et al. (2010), and García-Martínez et al. (2012).

TABLE 8.2
Mean Values and Standard Deviation of the Vitamin C, Vitamin A, and Vitamin E Contents, Total Phenols, Total Flavonoids, and Total Tannins of Fresh and Microwaved Kiwifruit Puree

	Fresh	Microwaved
Vitamin C (mg/100 g)	75.9 ± 1.3[a]	75.5 ± 1.1[a]
Vitamin A (mg/100 g)	0.057 ± 0.007[b]	ND[a]
Vitamin E (mg/100 g)	2.45 ± 0.06[b]	2.22 ± 0.07[a]
Total phenols (mg GAE/100 g)	22 ± 2[b]	17.2 ± 0.5[a]
Total flavonoids (mg RE/100 g)	1.16 ± 0.05[b]	0.74 ± 0.06[a]
Total tannins (mg GAE/100 g)	14.40 ± 0.10[a]	10.6 ± 0.8[a]
Antioxidant activity (mM Trolox/g)	5.81 ± 0.05[b]	1.99 ± 0.06[a]

[a] ND, not detected.
[b] In rows, different letters denote significant differences ($p < 0.05$) according to the Tukey test.

The bioactive compound contents of the microwaved and control samples are summarized in Table 8.2. Unexpectedly, vitamin C was not shown to be the compound that was most sensitive to microwaves in kiwifruit, given that all the components analyzed in the puree decreased significantly ($p < 0.05$) as a result of the microwave pasteurization treatment except vitamin C and total tannins, which remained significantly ($p > 0.05$) unchanged after processing. Variations in the other component contents due to processing were calculated as the difference in each compound in the treated puree in comparison with the fresh puree, with reference to 100 g of fresh puree. The calculated values are −100% of vitamin A, which was shown to be the compound that was most sensitive to microwaves, −36.2% of total flavonoids, −21.8% of total phenols, −9.4% of vitamin E, and −65.7% of antioxidant activity. The losses observed in the pasteurized kiwifruit are in the range typically expected for pasteurization processes, which are considered as treatments severe enough to reduce the levels of most bioactive compounds present in fruit, with vitamins found to be among the most heat-sensitive food components (Awuah et al., 2007; Rawson et al., 2011). In both microwave and conventional heat processes, simple thermal decomposition would appear to be the most likely cause for these losses, but this degradation may be a complex phenomenon that is also dependent on oxygen, light, pH, water solubility, and the presence of chemical, metal, or other compounds that could catalyze deteriorative reactions (Awuah et al., 2007). On the other hand, it is worth highlighting that the concentration of vitamin C, one of the most important bioactive compounds in kiwifruit because of its particularly high amount and its attributable antioxidant activity, was significantly ($p < 0.05$) unaffected by microwaves. Barrett and Lloyd (2012) reviewed the effect of microwave processing on bioactive compounds in products of vegetable origin and reported that the use of microwaves leads to a higher retention of vitamin C in most fruits and vegetables than the application of conventional heating.

8.6.5 Shelf Life of Fruit-Based Products

To date, many comparative studies have been conducted on the effect of microwaves and conventional heating on various quality aspects of fruits (Barrett and Lloyd, 2012), pointing out advantages of microwave heating (Huang et al., 2007). However, it should be taken into consideration that, although published data on the effect of microwaves on safety and quality are available for various food systems, little still seems to be known about the impact of microwaves on the shelf life and post-processing quality loss of fruit products. Marketing of these products frequently involves a storage step, which might also contribute considerably to their final quality, the evolution of their properties and the growth of microorganisms (pathogens or otherwise) during shelf life being an important issue to study (Rodrigo et al., 2003).

Having shown the suitability of microwave energy for effective inactivation of microorganisms and enzymes without strongly affecting the nutritive and functional value of the kiwifruit puree, it was proposed to investigate whether microwaves could potentially replace conventional heating for pasteurization purposes. To this end, the shelf life of a microwave-pasteurized kiwifruit puree was compared with that of a conventionally heat-pasteurized kiwifruit puree, on the basis of their microbial stability and the impact on the bioactive compounds and antioxidant activity of the product when stored at 4°C.

Accordingly, a microwave pasteurization treatment designed on the basis of the enzyme (90% of POD) and microbial (5D of *L. monocytogenes*) inactivation to be achieved was applied (1000 W–340 s) in the same way as described in Section 8.6.1. Additionally, a conventional thermal pasteurization treatment, which was equivalent to the microwave process in terms of POD and *L. monocytogenes* inactivation, was carried out to establish a comparison between the two technologies. The conventional thermal treatment consisted in heating the sample at 97°C for 30 s in a thermostatic circulating water bath (Precisterm, Selecta, Barcelona, Spain). After the kiwifruit had been triturated, 20 g of puree was placed in TDT stainless steel tubes (1.3 cm inner diameter and 15 cm length) and closed with a screw stopper. A thermocouple that was connected to a datalogger was inserted through the sealed screw top in order to record the time–temperature history of the sample during treatment. Prior to this heating step, the samples were preheated to 25°C to shorten and standardize the come-up time (150 s). Treated samples were immediately cooled in ice-water until the puree reached 35°C. The treated and untreated kiwifruit purees were then packaged in clean, sterile plastic tubes (1.7 cm inner diameter and 11.8 cm length) and then stored in darkness at 4°C for 188 days. The microbial population and the concentrations of the main bioactive compounds and antioxidant activity were measured in the samples during storage. Survival of *L. monocytogenes* was evaluated as previously explained in Section 8.6.1. The total mesophilic bacteria (TMB) and yeast and mold (Y&M) counts were examined by diluting the uninoculated samples in 0.1% (w/v) sterile peptone water (Scharlab Chemie S.A., Barcelona, Spain) and enumerating the viable cells in plate count agar (PCA, Scharlab Chemie S.A., Barcelona, Spain) and potato dextrose agar (PDA, Scharlab Chemie S.A., Barcelona, Spain) acidified with tartaric acid (10%) (Sigma-Aldrich, Germany) by adding 1 mL of tartaric acid per 10 mL of PDA, respectively. The selected dilutions were incubated at 30°C for 48 h for TMB and at 25°C for 5 days for Y&M. Additionally, the vitamin C content, total phenols and flavonoids, and antioxidant activity were determined as previously described (see Section 8.6.3).

The shelf life of the treated products was determined, taking into account the acceptable limit established by EU legislation (*L. monocytogenes* $\leq 2.0 \log_{10}$ CFU/g, and TMB and Y&M $\leq 3.0 \log_{10}$ CFU/g) (EU, 2005). On this basis, the shelf life of the microwaved and conventionally pasteurized purees was found to be 123 and 81 days, respectively (Figure 8.9). These results are in the range of those published by other authors for various fruits subjected to conventional thermal processes. The shelf life of heat-pasteurized orange and carrot juice (98°C for 21 s) stored at 2°C, thermally pasteurized pomegranate (90°C for 5 s) stored at 5°C, and conventionally heat-pasteurized orange juice (90°C for 50 s) stored at 4°C was found to be 70, 120, and 105 days, respectively (Leizerson and Shimoni, 2005; Rivas et al., 2006; Vegara et al., 2013). Picouet et al. (2009) reported that an apple puree preserved by gentle microwave heating (652 W–35 s) had a shelf life of at least 14 days under refrigeration conditions.

The nutritional and functional value of the microwaved and conventionally heat-treated kiwifruit purees at the beginning and end of their shelf life is presented in Table 8.3. Variations in the components due to the combined effects of processing and storage were calculated as the difference in each compound in the treated puree at the end of its shelf life in relation to the fresh puree, with reference to 100 g of fresh puree. Losses of 43%, 23%, and 62% in vitamin C, total phenols, and total flavonoids were found for the microwave-pasteurized sample (123 days at 4°C), while losses of 61%, 58%, and 56% in vitamin C, total phenols, and total flavonoids were observed for the conventionally thermally pasteurized puree (81 days at 4°C), respectively. However, antioxidant activity

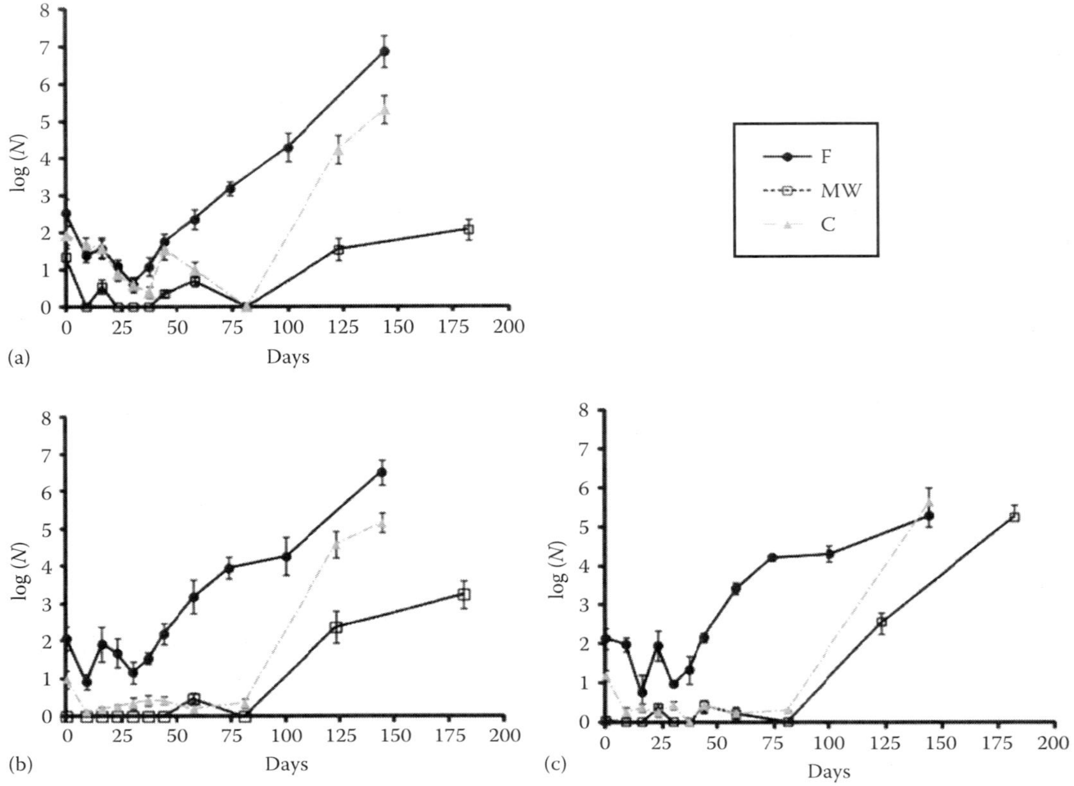

FIGURE 8.9 Survival of (a) *L. monocytogenes*, (b) total mesophilic bacteria, (c) and Y&M in the kiwifruit puree (F, fresh; MW, microwaved; C, conventionally heated) during storage at 4°C.

TABLE 8.3
Mean Values and Standard Deviation of the Vitamin C Content (mg/100 g), Total Phenols (mg GAE/100 g), Total Flavonoids (mg RE/100 g), and Antioxidant Activity (mM Trolox/100 g) of Microwaved and Conventionally Heated Kiwifruit Puree, at the Beginning and at the End of Their Shelf Life at 4°C

	Beginning of Shelf-Life		End of Shelf-Life	
	0 days		123 days	81 days
	Microwaved	Conventionally heated	Microwaved	Conventionally heated
Vitamin C	64.2 ± 0.7^a	62.3 ± 0.7^a	37.2 ± 0.6^b	25.4 ± 1.5^c
Total phenols	25.50 ± 0.07^a	22.2 ± 0.3^b	13.92 ± 0.08^c	9.3 ± 0.3^d
Total flavonoids	0.825 ± 0.004^a	0.67 ± 0.02^b	0.437 ± 0.013^c	0.505 ± 0.010^d
Antioxidant activity	1211 ± 37^a	1117 ± 27^b	478 ± 35^c	463 ± 41^c

Note: In rows, different letters (a, b, c or d) denote significant differences ($p < 0.05$) according to the Tukey test.

was reduced by 62% in both cases. These results clearly indicate the superiority of microwaves to preserve the nutritional and functional value of the product. Igual et al. (2010, 2011) found losses of similar magnitude and reported that microwave-pasteurized grapefruit juices stored at −18°C preserved total phenols, individual flavonoids, and antioxidant capacity better when compared with fresh or conventionally pasteurized ones.

In view of the results obtained in the present study, microwave heating not only seems to provide greater microbial stability than conventional heat processing, allowing longer preservation of kiwifruit puree, but also maintains the bioactive compound contents and antioxidant activity of the product to an equal or greater extent.

8.7 CONCLUSIONS AND FUTURE TRENDS

Microwave heating is an interesting technique for processing fruit-based products and preserving their safety and quality during storage. The use of this technology represents a good alternative to conventional heating methods, allowing efficient inactivation of both microorganisms and enzymes to be achieved, while it does not greatly affect product quality. Considering the increased demand for high-quality foods as well as cost competitiveness, microwave technology might be taken as an innovative tool to help consumer expectations to be addressed by the marketing of safe, high-quality, minimally processed fruit-based products. Currently, with the rapidly changing scenario, the future prospects of microwaves in food processing are bright. However, the use of microwave technology is still limited at present to selected categories of high-value food products. Some of the key issues that should be adequately addressed to make microwave processing more attractive are improvement in equipment design, reduction in equipment cost, and improvement in process control. Moreover, in general terms, there is a need for further studies to bridge the gap between laboratory research and industrial applications.

ACKNOWLEDGMENTS

The authors thank the Ministerio de Educación y Ciencia for the financial support given through Projects AGL 2010-22176 and AGL2013-48993-C2-2-R and the Generalitat Valenciana for the financial support given through Project ACOMP/2012/161 and the grant awarded to the author María Benlloch.

REFERENCES

Anonymous. (1980). The potential of Bach. *Food Manufacture*, *55*(10), 53–53.

Awuah, G. B., Ramaswamy, H. S., and Economides, A. (2007). Thermal processing and quality: Principles and overview. *Chemical Engineering and Processing*, *46*, 584–602.

Banik, S., Bandyopadhyay, S., and Ganguly, S. (2003). Bioeffects of microwave—A brief review. *Bioresource Technology*, *87*, 155–159.

Barrett, D. M. and Lloyd, B. (2012). Advanced preservation methods and nutrient retention in fruits and vegetables. *Journal of Food Science and Agriculture*, *92*, 7–22.

Beirão-da-Costa, S., Cardoso, A., Martins, L. L., Empis, J., and Moldão-Martins, M. (2008). The effect of calcium dips combined with mild heating of whole kiwifruit for fruit slices quality maintenance. *Food Chemistry*, *108*, 191–197.

Beirão-da-Costa, S., Steiner, A., Correia, L., Empis, J., and Moldão-Martins, M. (2006). Effects of maturity stage and mild heat treatments on quality of minimally processed kiwifruit. *Journal of Food Engineering*, *76*, 616–625.

Bolton, D. J., McMahon, C. M., Doherty, A. M., Sheridan, J. J., McDowell, D. A., Blair, I. S. et al. (2000). Thermal inactivation of *Listeria monocytogenes* and *Yersinia enterocolitica* in minced beef under laboratory conditions and in sous-vide prepared minced and solid beef cooked in a commercial retort. *Journal of Applied Microbiology*, *88*, 626–632.

Buffler, C. R. (1993). *Microwave Cooking and Processing: Engineering Fundamentals for the Food Scientist.* Van Nostrand Reinhold, New York.

Cañumir, J. A., Celis, J. E., de Brujin, J., and Vidal, L. V. (2002). Pasteurization of Apple juice by using microwaves. *LWT—Food Science and Technology*, *35*, 389–392.

Carpentier, B. and Cerf, O. (2011). Review—Persistence of *Listeria monocytogenes* in food industry equipment and premises. *International Journal of Food Microbiology*, *145*, 1–8.

Chhabra, A. T., Carter, W. H., Linton, R. H., and Cousin, M. A. (1999). A predictive model to determine the effects of pH, milk fat, and temperature on thermal inactivation of *Listeria monocytogenes*. *Journal of Food Protection*, *62*, 1143–1149.

Cochran, W. G. and Cox, G. M. (1957). Experimental designs. John Wiley & Sons Inc., New York.

Da Costa, M. C., Deliza, R., Rosenthal, A., Hedderley, D., and Frewer, L. (2000). Non-conventional technologies and impact on consumer behaviour. *Trends in Food Science and Technology*, *11*, 188–193.

De Ancos, B., Cano, M. P., Hernández, A., and Monreal, M. (1999). Effects of microwave heating on pigment composition and color of fruit purees. *Journal of the Science of Food and Agriculture*, *79*, 663–670.

Deliza, R., Rosenthal, A., Abadios, F. B. D., Silva, C. H. O., and Castillo, C. (2005). Application of high pressure technology in the fruit juice processing: benefits perceived by consumers. *Journal of Food Engineering*, *67*(1), 241–246.

Di Monaco, R., Cavella, S., Torrieri, E., and Masi, P. (2005). Consumer acceptability of vegetable soups. *Journal of Sensory Studies*, *22*, 81–98.

Djeridane, A., Yousfi, M., Nadjemi, B., Boutassouna, D., Stocker, P., and Vidal, N. (2006). Antioxidant activity of some Algerian medicinal plants extracts containing phenolic compounds. *Food Chemistry*, *97*, 654–660.

Elez-Martínez, P., Aguiló-Aguayo, I., and Martín-Belloso, O. (2006). Inactivation of orange juice peroxidase by high-intensity pulsed electric fields as influenced by process parameters. *Journal of the Science of Food and Agriculture*, *87*, 71–81.

EU. (2005). Commission regulation (EC) No. 2073/2005 of 15 November 2005 on the microbiological criteria of foodstuffs. *Official Journal of the European Union, L338*, 1–26.

EFSA. European Food Safety Authority. (2013). Scientific opinion on the risk posed by pathogens in food of non-animal origin. Part 1 (outbreak data analysis and risk ranking of food/pathogen combinations). *EFSA Journal*, *11*(1), 3025–3163.

Fang, L., Jiang, B., and Zhang, T. (2008). Effect of combined high pressure and thermal treatment in kiwifruit peroxidase. *Food Chemistry*, *109*, 802–807.

Fernández, A., López, M., Bernardo, A., Condón, S., and Raso, J. (2007). Modeling thermal inactivation of *Listeria monocytogenes* in sucrose solutions of various water activities. *Food Microbiology*, *24*, 372–379.

Fiorentino, A., D'Abrosca, B., Pacifico, S., Mastellones, C., Scognamiglio, M., and Monaco, P. (2009). Identification and assessment of antioxidant capacity of phytochemicals from kiwi fruits. *Journal of Agricultural and Food Chemistry*, *57*, 4148–4155.

Food and Drug Administration, U. S. Department of Health and Human Services (2004). *Juice HACCP Hazards and Controls Guidance*. www.fda.gov., United States.

Fujikawa, H., Ushioda, H., and Kudo, Y. (1992). Kinetics of Escherichia coli destruction by microwave irradiation. *Applied and Environmental Microbiology*, *58*(3), 920–924.

García-Martínez, E., Igual, M., Martín-Esparza, M. E., and Martínez-Navarrete, N. (2012). Assessment of the bioactive compounds, color, and mechanical properties of apricots as affected by drying treatment. *Food and Bioprocess Technology*, *6*, 3247–3255.

Gentry, T. S. and Roberts, J. S. (2005). Design and evaluation of a continuous flow microwave pasteurization system for apple cider. *LWT—Food Science and Technology*, *38*, 227–238.

Hassani, M., Álvarez, I., Raso, J., Condón, S., and Pagán, R. (2005). Comparing predicting models for heat inactivation of *Listeria monocytogenes* and *Pseudomonas aeruginosa* at different pH. *International Journal of Food Microbiology*, *100*, 213–222.

Hassani, M., Mañas, P., Pagán, R., and Condón, S. (2007). Effect of a previous heat shock on the thermal resistance of *Listeria monocytogenes* and *Pseudomonas aeruginosa* at different pHs. *International Journal of Food Microbiology*, *116*, 228–238.

Hebbar, H. U. and Rastogi, N. K. (2012). Microwave heating of fluid foods. In *Novel Thermal and Non-Thermal Technologies for Fluid Foods*, Cullen, P. J., Tiwari, B. K., and Valdramidis, V. P. (eds.), Academic Press, San Diego, CA.

Huang, Y., Sheng, J., Yang, F., and Hu, Q. (2007). Effect of enzyme inactivation by microwave and oven heating on preservation quality of green tea. *Journal of Food Engineering*, *78*, 687–692.

Igual, M., García-Martínez, E., Camacho, M. M., and Martínez-Navarrete, N. (2010). Effect of thermal treatment and storage on the stability of organic acids and the functional value of grapefruit juice. *Food Chemistry*, *118*, 291–299.

Igual, M., García-Martínez, E., Camacho, M. M., and Martínez-Navarrete, N. (2011). Changes in flavonoid content of grapefruit juice caused by thermal treatment and storage. *Innovative Food Science and Emerging Technologies*, *12*, 153–162.

IMS. (2014). Industrial microwave systems. www.industrialmicrowave.com. Accessed June 20, 2014.

Jolie, R. P., Duvetter, T., Houben, K., Vandevenne, E., Van Loey, A. M., Declerck, P. J. et al. (2010). Plant pectin methylesterase and its inhibitor from kiwi fruit: Interaction analysis by surface plasmon resonance. *Food Chemistry*, *121*, 207–214.

Kratchanova, M., Pavlova, E., and Panchev, I. (2004). The effect of microwave heating of fresh orange peels on the fruit tissue and quality of extracted pectin. *Carbohydrate Polymers*, *56*, 181–185.

Latorre, M. E., Bonelli, P. R., Rojas, A. M., and Gerschenson, L. N. (2012). Microwave inactivation of red beet (*Beta vulgaris* L. var. conditiva) peroxidase and polyphenoloxidase and the effect of radiation on vegetable tissue quality. *Journal of Food Engineering*, *109*, 676–684.

Lawless, H. and Heymann, H. (1998). *Sensory Evaluation of Food: Principles and Practices*. Chapman & Hall, New York.

Leizerson, S. and Shimoni, E. (2005). Stability and sensory shelf-life of orange juice pasteurized by continuous ohmic heating. *Agricultural and Food Chemistry*, *53*, 4012–4018.

Matsui, K. N., Gut, J. A. W., de Oliveira, P. V., and Tadini, C. C. (2008). Inactivation kinetics of polyphenol oxidase and peroxidase in green coconut water by microwave processing. *Journal of Food Engineering*, *88*, 169–176.

McFeeters, R. F., Fleming, H. P., and Thompson, R. L. (1985). Pectinesterase activity, pectin methylation, and texture changes during storage of blanched cucumber slices. *Journal of Food Science*, *50*, 201–219.

Nicolaï, B. (1998). Optimal control of microwave combination ovens for food heating. In *Third Karlsruhe Nutrition Symposium. European Research towards Safer and Better Food. Review and Transfer Congress. Proceedings Part 2*, Gaukeland, V. and Spieß, W. E. L. (eds.), Bundesforschungsanstalt für Ernährung. Karlsruhe , Germany, pp. 328–332.

O'Donnell, C. P., Tiwari, B. K., Bourke, P., and Cullen, P. J. (2010). Effect of ultrasonic processing on food enzymes of industrial importance. *Trends in Food Science and Technology*, *21*, 358–367.

Osorio, O., Martínez-Navarrete, N., Moraga, G., and Carbonell, J. V. (2008). Effect of thermal treatment on enzymatic activity and rheological and sensory properties of strawberry purees. *Food Science and Technology International*, *14*(5), 103–108.

Picouet, P. A., Landl, A., Abadias, M., Castellari, M., and Viñas, I. (2009). Minimal processing of a Granny Smith apple purée by microwave heating. *Innovative Food Science and Emerging Technologies*, *10*(4), 545–550.

Pina-Pérez, M. C., Benlloch-Tinoco, M., Rodrigo, D., and Martínez, A. (2014). *Cronobacter sakazakii* inactivation by microwave processing. *Food and Bioprocess Technology*, *7*(3), 821–828.

Queiroz, C., Mendes, M. L., Fialho, E., and Valente-Mesquita, V. L. (2008). Polyphenol oxidase: Characteristics and mechanisms of browning control. *Food Reviews International*, *24*, 361–375.

Rawson, A., Patras, A., Tiwari, B. K., Noci, F., Koutchma, T., and Brunton, N. (2011). Effect of thermal and non-thermal processing technologies on the bioactive content of exotic fruits and their products: Review of recent advances. *Food Research International*, *44*, 1875–1887.

Rivas A., Rodrigo, D., Martínez, A., Barbosa-Cánovas, G.V., and Rodrigo, M. (2006). Effect of PEF and heat pasteurization on the physical–chemical characteristics of blended orange and carrot juice. *LWT—Food Science and Technology*, *3*, 1163–1170.

Rodrigo, D., Arranz, J. I., Koch, S., Frígola, A., Rodrigo, M. C., Esteve, M. J., Calvo, C., and Rodrigo, M. (2003). Physicochemical characteristics and quality of refrigerated Spanish orange-carrot juices and influence of storage conditions. *Journal of Food Science*, *68*(6), 2111–2116.

Rodrigo, D., Cortés, C., Clynen, E., Schoofs, L., Van Loey, A., and Hendrickx, M. (2006). Thermal and high-pressure stability of purified polygalacturonase and pectinmethylesterase from four different tomato processing varieties. *Food Research International*, *39*, 440–448.

Salazar-González, C., San Martín-González, M. F., López-Malo, A., and Sosa-Morales, M. E. (2012). Recent studies related to microwave processing of fluid foods. *Food Bioprocess and Technology*, *5*, 31–46.

Schubert, H. and Regier, M. (2005). *The Microwave Processing of Foods*. Woodhead Cambridge, U.K.

Schubert, H. and Regier, M. (2006). Novel and traditional microwave applications in the food industry. In *Advances in Microwave and Radio Frequency Processing: Report from the Eighth conference on Microwave and High-Frequency Heating*. Willert-Porada, M. (eds.), Springer, Berlin, Germany.

Schubert, H. and Regier, M. (2010). *The Microwave Processing of Foods*. Woodhead, Cambridge, U.K.

Señorans, F. J., Ibáñez, E., and Cifuentes, A. (2003). New trends in food processing. *Critical Reviews in Food Science and Nutrition*, *43*(5), 507–526.

Sharma, M., Adler, B. B., Harrison, M. D., and Beuchat, L. R. (2005). Thermal tolerance of acid-adapted and unadapted *Salmonella*, *Escherichia coli* O157:H7, and *Listeria monocytogenes* in cantaloupe juice and watermelon juice. *Letters in Applied Microbiology*, *41*, 448–453.

Taira, S. (1995). Astringency in persimmon. In *Fruit Analysis*, Linskens, H.-F. and Jackson, J. F. (eds.), Springer, Berlin, Germany.

Tajchakavit, S. and Ramaswamy, H. S. (1997). Thermal vs. microwave inactivation kinetics of pectin methylesterase in orange juice under batch mode heating conditions. *LWT—Food Science and Technology*, *30*, 85–93.

Tajchakavit, S., Ramaswamy, H. S., and Fustier, P. (1998). Enhanced destruction of spoilage microorganisms in apple juice during continuous flow microwave heating. *Food Research International*, *31*(10), 713–722.

Terefe, N. S., Yang, Y. H., Knoerzer, K., Buckow, R., and Versteeg, C. (2010). High pressure and thermal inactivation kinetics of polyphenol oxidase and peroxidase in strawberry puree. *Innovative Food Science and Emerging Technologies*, *11*, 52–60.

Vadivambal, R. and Jayas, D. S. (2007). Changes in quality of microwave treated agricultural products. *Biosystems Engineering*, *98*, 1–16.

Vadivambal, R. and Jayas, D. S. (2010). Non-uniform temperature distribution during microwave heating of food materials—A review. *Food Bioprocess Technology*, *3*, 161–171.

Valdramidis, V. P., Taoukis, P. S., Stoforos, N. G., and Van Impe, J. F. M. (2012). Modeling the kinetics of microbial and quality attributes of fluid food during novel thermal and non-thermal processes. In *Novel Thermal and Non-Thermal Technologies for Fluid Foods*, Cullen, P. J., Tiwari, B. K., and Valdramidis, V. P. (eds.), Academic Press, San Diego, CA.

Vergara, S., Martí, N., Mena, P., Saura, D., and Valero, M. (2013). Effect of pasteurization process and storage on color and shelf-life of pomegranate juices. *LWT—Food Science and Technology*, *54*(2), 592–596.

Yaghmaee, P. and Durance, T. D. (2005). Destruction and injury of *Escherichia coli* during microwave heating under vacuum. *Journal of Applied Microbiology*, *98*, 498–506.

Wang, L. and Sun, D-W. (2012). Heat and mass transfer in thermal food processing. In *Thermal Food Processing: New Technologies and Quality Issues*, Sun, D-W. (ed.), CRC Press.

Whitaker, J. R., Voragen, A. G. J., and Wong, D. W. S. (2003). *Handbook of Food Enzymology*. Marcel Dekker, Inc., New York.

Zheng, H. and Lu, H. (2011). Effect of microwave pretreatment on the kinetics of ascorbic acid degradation and peroxidase inactivation in different parts of green asparagus (*Asparagus officinalis* L.) during water blanching. *Food Chemistry*, *128*, 1087–1093.

9 Advances in Food Additives and Contaminants

Theodoros Varzakas

CONTENTS

9.1 INTRODUCTION

Food additives have been used for centuries to improve food quality. Smoke, alcohols, and spices have been extensively used for the last 10,000 years as additives for food preservation. The above-mentioned additives as well as a restricted number of additives comprised the main food additives until the Industrial Revolution. The latter brought so many changes in foods and asked for improved quality as well as quantity of manufactured foods. For this reason, many chemical substances have been developed either for preservation and for color and/or odor enhancement. In the 1960s, over 2500 different chemical substances were used toward food manufacturing. In the United States, over 2,500 different additives were used to manufacture over 15,000 different foods. The desire for nutritional, functional, and tasty foods is an on-going process. An additive is used to improve the shape, color, aroma, and extend the shelf life of a food. The following categories of additives have been described:

- Acids
- Coloring agents
- Aromatic substances
- Toxic metals
- Biocides
- Preservatives

There has been an intense skepticism regarding the safe use of additives in foods. In the 1960s and 1970s, the increase of toxicological information caused an increase in the knowledge of possible risks derived from the consumption of foods containing additives. It was observed that the use of food additives has toxicological effects in humans. It is for this reason that in this chapter the limits of food additives use as well as pesticide residues control have been mentioned. It is well known that there is a plethora of risks derived from additives but also benefits from their use in food production. Additives will continue to play a significant role in food production since consumers continue to desire healthier, tastier, and occasionally functional foods and the population of the earth continues to increase.

9.2 SWEETENERS-GENERAL PHILOSOPHY OF DIRECTIVE 94/35/EC ON SWEETENERS FOR USE IN FOODSTUFFS

For sweeteners to be included in this directive, they first had to comply with the general criteria set out in Annex II of the Food Additives Framework Directive 89/107/EEC (OJL 40, 11,2,89, pp. 27–33). Under these criteria, food additives may only be approved if it has been demonstrated that they perform a useful purpose, are safe, and do not mislead the consumer. The recitals of Directive 94/35/EC on sweeteners for use in foodstuffs further explain that the use of sweeteners to replace sugar is justified for the production of

1. Energy-reduced foods
2. Noncarcinogenic foods (i.e., foods that are unlikely to cause tooth decay)
3. Foods without added sugars, for the extension of shelf life through the replacement of sugar and for the production of dietetic products

9.2.1 Structure Definition of Sweetener (FSA, Regulation 2(1))

For the purposes of these regulations, a sweetener is defined as a food additive that is used or intended to be used either to impart a sweet taste to food or as a table-top sweetener. Table-top sweeteners are products that consist of, or include, any permitted sweeteners and are intended for sale to the ultimate consumer, normally for use as an alternative to sugar. Foods with sweetening properties, such as sugar and honey, are not additives and are excluded from the scope of this legislation. Food Regulations in 1995 report that when a substance listed as a permitted sweetener, cannot be used for purposes other than sweetening, for example, where sorbitol is used as a humectant in accordance with the Miscellaneous Food Additives Regulations 1995 and parallel Northern Ireland legislation.

9.2.2 Permitted Sweeteners

(*Regulations* 2(1), 3(1), 4(*a*), *and Schedule* 1)

The only sweeteners permitted for sale to the ultimate consumer or for use in or on food are those listed in Schedule 1 to the Regulations whose specific purity criteria are in compliance with that stated in the Annex to Directive 95/31/EC.

9.2.3 Foods Allowed to Contain Permitted Sweeteners

(*Regulations* 3(2), 3(3), *and Schedule* 1)

Permitted sweeteners are only allowed to be used in or on foods that fall within one of the categories listed in Schedule 1 to the Regulations. A maximum usable dose for each permitted sweetener, varying according to the food category, is also specified within Schedule 1, and this must be respected. The use of two or more sweeteners in a single food is permitted, provided suitable categories exist, and the maximum level for each individual sweetener is observed. The sale of foods that do not comply with these provisions is illegal.

9.2.4 Sweeteners in Compound Foods—Carry-Over

(*Regulation* 2(1) (*Amended* 1997) *and Regulation* 5A)

The regulations have been amended to include provisions on carry-over (Regulation 5A) to bring them into line with the GB Regulations on Colors and Miscellaneous Food Additives. These provisions allow the presence of a permitted sweetener in a compound food, to the extent that the sweetener is allowed by the Regulations in one of the ingredients of the compound food. However, the definition of "compound foods" in Regulation 2(1) means that permitted sweeteners are only allowed in the following compound foods:

1. Those with no added sugar or that are energy-reduced
2. Dietary foods intended for a low-calorie diet (excluding those specifically prepared for infants and young children)
3. Those with a long shelf life

The regulations also provide for what is commonly known as "reverse carry-over." This means permitted sweeteners can be present in foods (such as intermediary products) in which they would not otherwise be permitted, provided that these foods are to be used solely in the preparation of a compound food that will conform to the regulations.

9.2.5 Foods Not Allowed to Contain Sweeteners

(*Regulation* 3(4) (*Amended* 1997) *and Regulation* 5)

The use of sweeteners in any foods for infants and young children is prohibited. This is specified in Council Directive 89/398/EEC on the approximation of the laws of the member states relating to foodstuffs intended for particular nutritional uses (OJL 186, 30.6.89, pp. 27–32), and this prohibition now includes foods for infants and young children not in good health. The sale of such products containing sweeteners is also prohibited. Foods for infants and young children, generally known as "baby foods," include foods specially prepared for infants and young children who are in good health; or whose digestive processes or metabolism are disturbed; or who have a special physiological condition where they would be able to obtain benefit from controlled consumption of certain substances in foods. For the purposes of this prohibition, Regulation 2(1) defines "infants" as children under the age of 12 months and "young children" as children aged between 1 and 3 years. These definitions reflect those given in Article 1(2) of Directive 91/321/EEC on infant formulae and follow-on formulae (OJL 175, 4.7.91, pp. 35–49) that are made under Directive 89/398/EEC.

The terms "maximum usable dose" and "*quantum satis*"

(*Regulations* 2(3)(*c*) *and* 2(3)(*d*))

These expressions are explained in general guidance notes, section 1, paragraphs 19 and 15, respectively.

The terms "with no added sugar" and "energy-reduced"

(*Regulations* 2(3)(*a*) *and* 2(3)(*b*))

Many of the categories listed in Schedule 1 to the Regulations are described as "with no added sugar" or "energy-reduced." The final product must comply with the definitions of these terms, and the effect is to further restrict the type of foods in which sweeteners may be used. However, the actual terms "with no added sugar" or "energy-reduced" are not required by these regulations to be used in the labeling of such products. Whatever description is used for those products must be in accordance with the Food Labeling Regulations 1996.

"Energy-reduced" foods are foods with an energy value reduced by at least 30% compared with the original or a similar food. The legislation does not define the precise basis for this comparison, but wherever possible it should be by reference to one or more products that are currently on the market. If it is not possible to identify a comparable product that is currently on the market, the comparison could be made on the basis of previously marketed products. In an extreme case where it is not possible to identify an actual product, the comparison might be made with a hypothetically equivalent product, the composition of which is based on the use of sucrose rather than permitted sweeteners.

9.2.6 Additional Labeling Requirements for Table-Top Sweeteners

(*Regulation* 4(*b*))

The Regulations include labeling requirements that apply to *table-top sweeteners* only, in addition to the requirements contained within existing UK labeling legislation; table-top sweeteners must include on their labels the phrase:

"[*Name of sweetener(s)*]-based table-top sweetener"

Furthermore, where table-top sweeteners contain polyols and/or aspartame, the following phrases must also be included on their labels:

For polyols—"excessive consumption may induce laxative effects"

For aspartame—"contains a source of phenylalanine"

For the purposes of these regulations, polyols are considered to be sorbitol and sorbitol syrup (E420 (i) and (ii)), mannitol (E421), isomalt (E953), maltitol and maltitol syrup (E965(i) and (ii)), lactitol (E966), and xylitol (E967).

The Additive and Food Contaminants (AFC) Panel of the European Food Safety Authority[1] (EFSA) has evaluated the new long-term study on the carcinogenicity of aspartame conducted by the European Ramazzini Foundation (ERF) (European Foundation of Oncology and Environmental Sciences) in Bologna, Italy. In its opinion, the panel concluded, on the basis of all the evidence currently available, that there is neither need to further review the safety of aspartame nor to revise the previously established Acceptable Daily Intake (ADI) for aspartame (40 mg/kg body weight).

The panel also noted that intakes of aspartame in Europe, with levels up to 10 mg/kg body weight/day, are well below the ADI.

Aspartame, an intense sweetener, has been authorized for use in foods and as a table-top sweetener for more than 20 years in many countries throughout the world. Extensive investigations have been carried out on aspartame and its breakdown products through experimental animal and human studies, intake studies, and post-marketing surveillance.

In addition to a number of safety evaluations conducted in the past, the Scientific Committee on Food (SCF) carried out a review of all original and more recent studies on aspartame in 2002 and reconfirmed that aspartame is safe for human consumption.

9.3 SWEETENERS-CLASSIFICATIONS, USES, SAFETY, MANUFACTURING, QUALITY CONTROL, AND REGULATORY ISSUES

9.3.1 Overview

According to World Health Organization (WHO), approximately 1 billion people in the planet are overweight and 300 million obese. Obesity is closely related to diabetes type II, heart failure, and cancer. It is required to take measures to improve population health and reduce health costs. Hence, people follow a balanced diet, low in calories.

Health professionals try to reduce the weight of their patients by reducing their caloric intake and advising them to use sweeteners in their diet especially when they have to control their weight or they are diabetic. However, media frequently argue on their safety, the effect of sweeteners on appetite and the appearance of neurological problems as well as cancer and mood disorders.

We currently consume foods that are in stake. Consumers ask for certified organic foods, nongenetically modified with no herbicides, pesticides, etc. They do not want to face more scandals such as BSE or adulterated olive oils or hear more food recalls.

Consumers do not want to hear published studies such as those by the ERF, according to which there is a close relation between aspartame consumption and cancer in rats. On the other hand, EFSA reviewed the studies of the institute and concluded that data showing that aspartame causes cancer was not sufficient.

According to Professor Ragnar Lofstedt of King's College, London, the ERF data refer to a pseudo-terror and was based on no documented evidence; however, media coverage was high at the world-level raising concerns to consumers and health professionals regarding the safety of aspartame.

The same happened in the early 1970s with saccharin when it was found that it causes cancer in male rats. Recent research has not proved that correlation even when consuming large quantities.

EFSA assures that all approved sweeteners in food and drinks are safe to use.

Health professionals and consumers should read peer-reviewed articles in reputable journals and consult the scientific committees of their national food and drug authorities which consult with EFSA (www.efsa.europa.eu).

Sweeteners are defined as food additives that are used or intended to be used either to impart a sweet taste to food or as a table-top sweetener. Table-top sweeteners are products that consist of, or include, any permitted sweeteners and are intended for sale to the ultimate consumer, normally for use as an alternative to sugar. Foods with sweetening properties, such as sugar and honey, are not additives and are excluded from the scope of official regulations. Sweeteners are classified as either high intensity or bulk. High-intensity sweeteners possess a sweet taste but are noncaloric, provide essentially no bulk to food, have greater sweetness than sugar, and are therefore used at very low levels.

9.3.2 Sugars

Sugar is a carbohydrate found in every fruit and vegetable. All green plants manufacture sugar though photosynthesis, but sugar cane and sugar beets have the highest natural concentrations. Beet sugar and cane sugar—identical products that may be used interchangeably—are the most common sources for the sugar used in the United States. Understanding the variety of sugars available and their functions in food will help consumers determine when sugar can be replaced or combined with nonnutritive sweeteners.

Sugar comes in many forms. The following includes many different sugars, mostly made from sugar cane or sugar beets such as table sugar, fruit sugar, crystalline fructose, superfine or ultrafine sugar, confectioner's or powdered sugar, coarse sugar, sanding sugar, turbinado sugar, brown sugar, and liquid sugars.

9.3.3 Classification of Sweeteners

Sweeteners can be categorized as low calorie, reduced calorie, intense calorie, bulk calorie, caloric alternative calorie, natural sugar-based calorie, sugar polyols calorie, low calorie nonnutritive calorie, nonnutritive calorie, nutritive calorie, natural calorie, syrups, intense sweeteners, and others.

Caloric alternative calorie sweeteners characterize crystalline fructose, high fructose corn syrup, isomaltulose, trehalose.

Intense sweeteners include acesulfame K, alitame, aspartame, brazzeine, cyclamate, glycyrrhizin, neohesperidine, neotame, saccharin, stevioside, sucralose, thaumati.

Bulk sweeteners include crystalline fructose, erythritol, isomalt, isomaltulose, lactitol, maltitol, maltitol syrup, mannitol, sorbitol, sorbitol syrup, trehalose, xylitol, crystalline fructose.

Nonnutritive, high-intensity sweeteners include acesulfame K, aspartame and neotame, saccharin and cyclamate, sucralose.

Reduced-calorie bulk sweeteners include erythritol, isomalt, lactitol, maltitol and maltitol syrups, sorbitol and mannitol, tagatose, xylitol.

Low-calorie sweeteners include acesulfame K, alitame, aspartame, cyclamate, neohesperidin dihydrochalcone, tagatose, teotame, saccharin, stevioside, sucralose, less common high-potency sweeteners.

Reduced-calorie sweeteners include erythritol, hydrogenated starch hydrolysates and maltitol syrups, isomalt, maltitol, lactitol, sorbitol and mannitol, xylitol.

Caloric alternatives characterize crystalline fructose, high fructose corn syrup, isomaltulose, trehalose.

Other sweeteners characterize brazzeine, glycyrrhizin, thaumatin, polydextrose, sucrose, polyols, dextrose, fructose, galactose, lactose, maltose, stafidin, glucose, saccharoze, D-tagatose, thaumatin, glycerol, glycerizim.

High-intensity sweeteners (also called nonnutritive sweeteners) can offer consumers a way to enjoy the taste of sweetness with little or no energy intake or glycemic response, and they do not support growth of oral cavity microorganisms. Therefore, they are targeted toward treatment of obesity, maintenance of body weight, management of diabetes, and prevention and reduction of dental caries. There are several different high-intensity sweeteners. Some of the sweeteners are naturally occurring, while others are synthetic (artificial) or semisynthetic. For example, acesulfame K and saccharin are not metabolized and are excreted unchanged by the kidney. Sucralose, stevioside, and cyclamate undergo degrees of metabolism, and their metabolites are readily excreted. Acesulfame K, aspartame, and saccharin are permitted as intense sweeteners for use in food. High-intensity sweeteners are often used as mixtures of different, synergistically compatible sweeteners.

Bulk sweeteners are disaccharides and monosaccharides of plant origin delivered in solid or liquid form and used in sweeteners per se or in foods in quantities greater than 22.5 kg. Sucrose from sugarcane and sugar beet and starch-derived glucose and fructose from maize (corn), potato, wheat, and cassava are the major sweeteners sold in bulk to the food and beverage manufacturing industry or packers of small containers for retail sale.

Unrefined sweeteners include all natural, unrefined, or low-processed sweeteners. Sweeteners are usually made with the fruit or sap of plants but can also be made from the whole plant or any part of it; some sweeteners are also made from starch with the use of enzymes. Sweeteners made by animals, especially insects, are put in their own section as they can come from more than one part of plants.

From sap: The sap of some species is concentrated to make sweeteners, usually through drying or boiling.

Cane juice, syrup, molasses, and raw sugar are all made from sugarcane, sweet sorghum syrup made from *Sorghum* spp., maize, agave syrup made from the sap of *Agave* spp., birch syrup made from the sap of birch trees, maple syrup made from tapped maple trees, palm sugar and sweet resin from sugar pine.

From roots: The juice extracted from the tuberous roots of certain plants is concentrated to make sweeteners, usually through drying or boiling. For example, sugar beet syrup and yacon syrup.

9.3.4 Sucrose and Fructose

Sucrose, or table sugar, has been the most common food sweetener. In the late 1960s, a new method was introduced that converts glucose in corn syrup to fructose. High fructose corn syrup is as sweet as sucrose but less expensive, so soft-drink manufacturers switched over to using it in the mid-1980s. Now it has surpassed sucrose as the main added sweetener in the American diet.

However, fructose has many disadvantages such as its metabolism is carried out almost exclusively in the liver. It is more likely to result in the formation of fats which increase the risk for heart disease.

9.3.5 Sugar Alcohols

Sugar alcohols are used in candies, baked goods, ice creams, and fruit spreads. They do not affect blood-sugar levels as much as sucrose, a real advantage for people with diabetes, and they do not contribute to tooth decay.

9.3.6 Sweetened Beverages

Sweeteners added to sports and juice drinks are problematic with adverse health outcomes. Not surprisingly, they are found to lead to excess weight.

When children regularly consume sweetened beverages, they are getting used to a level of sweetness that could affect their habits for a lifetime.

9.3.7 Artificial Sweeteners

The Food and Drug Administration (FDA)-approved artificial, calorie-free sweeteners include acesulfame K (Sunett), aspartame (NutraSweet, Equal), neotame, saccharin (Sweet'N Low, others), and sucralose (Splenda). All are intensely sweet.

Some fears regarding aspartame are based on animal experiments using doses many times greater than any person would consume. Lack of long-term studies in humans supports the fear regarding their safety.

9.3.8 New Sweeteners Study Shows No Link with Cancer

A study of more than 16,000 patients has found no link between sweetener intake and the risk of cancer. More recent research in rats found that sweetener intakes similar to those consumed by humans could increase the risk of certain types of cancer. These findings were not replicated in studies of humans. After evaluating these and other studies in 2006, EFSA concluded that no further safety reviews of aspartame were needed and that the ADI of 40 mg/kg body weight should remain.

Professor Carlo La Vecchia from the Institute of Pharmaceutical Research Mario Negri in Milan, according to the research conducted by Silvano Gallus et al. analyzing more than 11,000 cases over a period of 13 years, concluded that consumers of low-nutritive sweeteners such as saccharin and aspartame did not show any signs of cancer.

Finally in the States, a study conducted with the support of National Research Institute for Cancer, sampling approximately 450 volunteers, did not show any statistically significant change between aspartame consumption and leukemia or lymphoma or brain tumors.

9.3.9 Sweeteners and Diabetes

Low-calorie sweeteners can help in the nutrition of diabetics. Sweeteners exert no effect on insulin, and sugar levels could help in the control of weight of people suffering from diabetes type II. These people continue to enjoy food with pleasure when consuming foods supplemented with sweeteners.

9.3.10 Sweeteners and Appetite

Recent research has shown that consumption of sweeteners does not cause changes in appetite.

Appetite is affected by calorie intake and volume of the consumed food/drink.

A zero-calorie drink will suppress appetite for an hour but will not affect the quantity of food consumed next.

In a study published by the American Journal of Clinical Nutrition analyzing 224 studies on the effect of low-caloric sweeteners on appetite, it was reported that these ingredients can reduce energy value. They also concluded that sweeteners replacing ingredients of high calories can help in weight control.

9.3.11 Sweeteners, Pregnancy, and Children

Worldwide organizations have carried out toxicological studies and have concluded that low-calorie sweeteners are safe for human consumption during pregnancy; however, the recommended daily intake dose should not be exceeded.

Regarding children, studies have shown that aspartame use has no neurological or behavioral effects on adults and children.

9.3.12 Sweeteners and Neurological Problems

There is no scientific evidence to conclude that consumption of aspartame or other sweeteners is related to Parkinson's or Alzheimer's disease.

9.3.13 Sweeteners and Weight Control

Replacement of sugar by sweeteners in food and drinks reduces calorie intake, hence it acts synergistically. More specifically, soft drinks produced with the addition of sweeteners have a negligible nutritional value. Sweets still maintain their high calories since a large number of them come from lipids or proteins and replacement by sweeteners reduces total calories by only a small percentage.

According to Professor Drewnowski of Washington State University and recent clinical and epidemiological studies, nonnutritive sweeteners can be a powerful tool for management of weight, obesity, and diabetes. Of course all these should be used in the context of a balanced diet.

9.3.14 Safety of Low-Calorie Sweeteners

All low-calorie sweeteners are subjected to comprehensive safety evaluation by regulatory authorities; any unresolved issues at the time of application have to be investigated before they are approved for use in the human diet. Definitive independent information on the safety of sweeteners can be obtained from the websites of the European Food Safety Authority (EFSA) (http://www.efsa.europa.eu/EFSA/efsa_locale-1178620753812_home.htm), the European Scientific Committee on Foods (SCF) (http://ec.europa.eu/food/fs/sc/scf/reports_en.html), and Joint WHO/FAO Expert Committee on Food Additives (JECFA, http://www.who.int/ipcs/food/jecfa/en/).

The safety testing of food additives involves *in vitro* investigations to detect possible actions on DNA and *in vivo* studies in animals, to determine what effects the compound is capable of producing when administered at high doses, or high-dietary concentrations, every day.

The daily dose levels are increased until either some adverse effect is produced or until 5% of the animal's diet has been replaced by the compound. The dose levels are usually very high because a primary purpose of animal studies is to find out what effects the compound can produce on the body irrespective of dose level (hazard identification). The dose–response data are analyzed to determine the most sensitive effect (the so-called critical effect). The highest level of intake that does not produce the critical effect, the no-observed-adverse-effect level (NOAEL), is used to establish a human intake with negligible risk, which is called the ADI. The NOAEL is normally derived from chronic (long-term) studies in rodents. The ADI is usually calculated as the NOAEL (in mg/kg body weight per day) divided by a 100-fold uncertainty factor, which is to allow for possible species differences and human variability.

Low-calorie sweeteners are often added to foods as mixtures or blends because mixtures can provide an improved taste profile, and in some cases the combination is sweeter than predicted from the amounts present. The only property that is common to all low-calorie sweeteners is their activity at the sweet-taste receptor. They do not share similar metabolic fates or high-dose effects.

Therefore, no interactions would arise if different low-calorie sweeteners are consumed together in a blend, and each sweetener would be as safe as if it were consumed alone.

9.3.15 Manufacturing of Sweeteners

Sweeteners can be manufactured by hydrolysis; for example, digestion of maltitol requires hydrolysis before absorption.

Hydrolysis phenomena in organic chemistry are ester, polysaccharide, glucosides, proteins, or amides hydrolysis.

Another method is by evaporation concentration. Concentration is carried out by cooling or by pressure increase. This is used in the production of isomaltose, maltitol, mannitol, sorbitol, and xylitol.

Discoloration is another manufacturing method of sweeteners. It is a type of adsorption during which undesired substances in some liquids that give them color are removed. It is carried out by active carbon and natural hydrosilicic derivatives of aluminum, activated by acid.

Active carbon can be manufactured by steam and carbonic anhydrate at 900°C–1000°C. Discoloration is used to neutralize free radicals of glucose, sorbitol, and citric acid.

Crystallization is also used in the manufacture of sweeteners. It is the most difficult step to control during all stages of maltitol preparation, and to be able to control the size distribution and the purity of the produced crystals is the important task. Crystallization is an important unit of operation for the manufacture of many sugars, polyols, salts, etc.

The presence of impurities in industrial sugar syrups leads to important modifications in the crystal shape. Indeed, the presence of maltotriitol in the crystallization medium has been shown to control the formation of bipyramidal or prismatic maltitol crystals.

Sorbitol is a natural sugar–alcohol (polyol) and is abundant in nature, and its commercial production depends on catalytic hydrogenation of the appropriate reducing sugar where the reaction of aldehyde and ketone groups is replaced by stable alcohol groups. Maltitol, maltitol syrups, and polyglycitols are manufactured in the same way.

Manitol production from algae is a cost effective and large entrepreneurial activity in China and is not comparable to the traditional chemical way of extraction.

The raw materials for sorbitol and mannitol production are mainly manufactured from either starch or sugar. Starches could arise from maize, wheat flour, or tapioca starch.

Sorbitol and maltitol are sold in liquid and solid forms. Mannitol is available in crystallic form due to its low solubility.

Lactitol is a synthetic alcohol belonging to polyols. It is manufactured by the catalytic hydrogenation of galactose found in whey milk. Lactitol is crystallized in different anhydrous and hydrated forms. It is also well known as lactate, lactositol, and lactobiosite.

Xylitol is also a polyol found in many plants. It is used as a sweetener since 1960. It is commercially manufactured from xylan, a fraction of polysaccharides found in wood pulp.

Catalytic hydrogenation was discovered in 1920 by Senderens, a French food chemist. He and his professor Paul Sabatier founded the modern process of hydrogenation used in lactitol production. Karrer and colleagues made the first preparation, and lactitol is commercialized since 1980. Lactitol is a disaccharide made from sorbitol and galactose, produced from lactose, a milk sugar, with catalytic hydrogenation using Raney nickel as an enzyme. A solution of 30%–40% lactose is heated at 100°C. Reaction is carried out in a sterilized equipment under hydrogen pressure at 40 bars or higher. Sedimentation of the catalyst follows and then hydrogenated solution is filtered and purified from ion resins and active carbon. The clean solution of lactitol is then crystallized. Hydrogenation under strict conditions (130°C, 90 bar) ends in sectional separation to lactulose, partial hydrolysis in galactose and glucose, and hydrogenation to reaction alcohol sugars of lactitol, sorbitol, and galactitol.

Acesulfame K (6-methyl-1,2,3-oxathiazin-4(3*H*)-one-2,2-dioxide) was an accidental discovery in 1967 from studies at Hoechst Corporation in West Germany on novel ring compounds. Its full name is potassium acesulfame and consists of a 1,2,3-oxathiazine ring, a six-heterocyclic system in which oxygen, sulfur, and nitrogen atoms are adjacent to each another.

The ADI has been defined by the World Health Organization (WHO) as "an estimate by JECFA of the amount of a food additive, expressed on a bodyweight basis, that can be ingested daily over a lifetime without appreciable health risk" and is based on an evaluation of available toxicological data. For example, in Europe the ADI is set at 9 mg/kg of bodyweight/day for acesulfame K. For aspartame there is a safety margin, even in high consuming diabetics. The FDA has set the ADI for aspartame at 50 mg/kg of body weight/day. An ADI of 40 mg/kg body weight/day set by the committee of experts of the Food and Agriculture Organization (FAO) and the WHO is not likely to be exceeded, even by children and diabetics.

Aspartame (*N*-L-a-aspartyl-L-phenylalanine-1-methylester) is a dipeptide. This sweetener was discovered accidentally in 1965 by James M. Schlatter at G.D. Searle Co. This nutritive sweetener provides 4 cal/g, but the amount required to give the same sweetness as sugar is only 0.5% of the calories.

Ajinomoto company developed many methods and processes for the commercial production of aspartame. Then new processes were developed to replace the initial process. In the late 1970s, Toyo Soda Company finalized new techniques with the aid of enzymes to connect the N-protected aspartate with β-phenylalanine methylester and followed crystallization and purification steps. This biocatalytic method shows better results.

The raw materials for aspartame production are L-phenylalanine (produced by fermentation) and L-aspartate. Toyo soda process can use DL-phenylalanine, manufactured synthetically and can offer better costs compared to the cleaner optical form of L-phenylalanine. Aspartate is manufactured chemically. Crystallization methods could be either static or sterilized, and the technique depends on type, size, and shape of the crystallized form.

Neotame is a derivative of the dipeptide composed of the amino acids aspartic acid and phenylalanine Neotame (NTM) is the generic name for *N*-[*N*-(3,3-dimethylbutyl)-l±a-aspartyl]-l-phenylalanine-1-methyl ester. NTM is conveniently prepared from aspartame and 3,3-dimethylbutyraldehyde, in a one-step high-yield reduction process through purification, drying, and emulsification. NTM is an odorless white crystalline compound which may be obtained anhydrous or, more usually, as a hydrate.

Saccharin (1,2-benzisothiazol-3(2*H*)-on-1,1-dioxide) is the oldest high-intensity, nonnutritive sweetener discovered in 1879 by Fahlberg and cyclamate in 1937. It is commercially available in three forms: acid saccharin, sodium saccharin, and calcium saccharin.

Until the end of 1950s, these two sweeteners were used as sugar substitutes in patients suffering from diabetes. In 1957 Helgren from Abbott laboratories discovered that the mixing of cyclamate with saccharin at a ratio of 10:1, since saccharin is 10 times more intense than cyclamate, improves quality in taste of the final product. Soft drinks were thus produced without any deterioration in taste.

However, at low pH, saccharin and its salts can slowly hydrolyze to 2-sulfobenzoic acid and 2-sulfoamylobenzoic acid. Saccharin was initially produced by the synthesis-oxidation of toluene derivatives.

Cyclamate (cyclohexyl sulfamic acid monosodium salt) is an artificial sweetener that is 35 times sweeter than sugar. The ADI value for cyclamate has been set at 11 mg/kg body weight by the JECFA and 7 mg/kg body weight by the SCF. Cyclamate is used in foods either as sodium or calcium salts.

Sucralose, 1,6-dichloro-1,6-dideoxy-β-D-fructofuranosyl 4-chloro-4-deoxy-α-D-galacto-pyranoside or 4,1′,6′-trichloro-4,1′,6′-trideoxy-galacto-sucrose, is a chlorinated derivative of sucrose, derived from a patented multistep process, involving selective chlorination of sugar at the 4, 1′, and 6′ positions substituting three hydroxyl groups on the sucrose molecule. It was discovered in 1976 by carbohydrate research chemists at Queen Elizabeth College and Tate and Lyle, United Kingdom. The basic process of its production focuses on the selective protection of main hydroxyl groups followed by chlorination and purification. Sucralose can be crystallized from an aqueous solution and can be produced at a high purity and stability level.

Fermentation of naturally occurring yeast *Moniliella pollilnis* is used to manufacture erythritol ((2*R*,3*S*)-butane-1,2,3,4-tetraol). This yeast was first isolated from fresh pollen found in honeycombs. Then, it was concluded that under oxygen conditions this yeast produced erythritol at relatively high concentrations.

Except for erythritol, which is a four-carbon symmetrical polyol and exists only in the meso form, the majority of sugar alcohols are produced industrially by hydrogenation in the presence of Raney nickel as a catalyst from their parent reducing sugar. Both disaccharides and monosaccharides can form sugar alcohols; however, sugar alcohols derived from disaccharides (e.g., maltitol and lactitol) are not entirely hydrogenated because only one aldehyde group is available for reduction.

Erythritol is an acyclic carbohydrate consisting of four carbon atoms, each carrying a hydroxyl group. Since it exhibits a meso structure, the molecule is achiral, although exhibiting two asymmetric carbon atoms (2*R*,3*S*).

Osmotolerant microorganisms ferment the D-glucose resulting from hydrolyzed starch. As a result, a mixture of erythritol and minor amounts of glycerol and ribitol is formed. Cell material, polyols, and other dirt are removed by purification and separation steps. Erythritol then crystallizes from the concentrated solution and gets dried in crystals with a purity higher than 99%.

During fermentation process, the microorganisms have the characteristics of tolerating high-sugar concentrations resulting in high-erythritol yields.

Isomalt is a mixture of the diastereomers 6-*O*-α-D-glucopyranosyl-D-sorbitol (isomaltitol, 1,6-GPS) and 1-*O*-α-D-glucopyranosyl-D-mannitol (GPM; the 1-*O* and 6-*O*-bound mannitol derivatives are identical because of the symmetry of the mannitol part of the molecule) obtained by hydrogenating isomaltulose, which is enzymatically derived from sucrose.

Isomalt is manufactured in a two-stage process in which sugar is first transformed into isomaltulose, a reducing disaccharide (6-*O*-α-D-glucopyranosido-D-fructose). During this step, α-(1–2) glycosidic chain between glucose and fructose is repositioned by enzymes to the α-(1–6) glycosidic chain. Isomaltulose is then hydrogenated, using a Raney nickel catalytic converter. The final product—isomalt—is an equimolar composition of 6-*O*-α-D-glucopyranosido-D-sorbitol (1,6-GPS) and 1-*O*-α-D-glucopyranosido-D-mannitol-dihydrate (1,1-GPM-dihydrate).

D-Tagatose ((3*S*,4*S*,5*R*)-1,3,4,5,6-Pentahydroxy-hexan-2-one) is a ketohexose C-4 fructose epimer potentially obtainable by the oxidation of the corresponding hexitol D-galactitol. It is obtained from D-galactose by isomerization under alkaline conditions in the presence of calcium. Tagatose is a sugar naturally found in small amounts in milk. It can be produced commercially from lactose, which is first hydrolyzed to glucose and galactose. Glucose gives glucose syrup as a final product. Galactose is isomerized under alkaline conditions to D-tagatose by calcium hydroxide and a catalyst. The resulting mixture can then be purified and solid tagatose produced by crystallization in white, crystallic form.

Alitame [L-α-aspartyl-*N*-(2,2,4,4-tetramethyl-3-thioethanyl)-D-alaninamide] is an amino acid-based sweetener developed by Pfizer Central Research from L-aspartic acid, D-alanine, and 2,2,4,4-tetraethylthioethanyl amine.

The final product is purified by crystallization with a methylbenzenesulphonic acid followed by an extra purification step in order to get recrystallized from water.

9.3.16 Quality Control of Sweeteners

Quality control of sweeteners involves

(a) The control of relative sweetness of the sweetener (RS).

In the calibration of sweetness of different sugars, it is acceptable to rate saccharose as 100. Sugars that are sweeter than saccharose are ranked in numbers higher than 100 and those with a less sweet taste are ranked lower than 100.

(a) FPA—Flavor Profile Analysis is a technique that identifies the taste and odor of a sample. This is estimated through consumer research techniques. However, techniques such as FPA are used to predict quality in taste. According to this technique, calibration of the intensity of food ingredients is carried out. More specifically, calibration starts from sweet followed by bitter, sour, salty, metallic, cold, and licorice taste. Every taste or flavor has a unique characteristic and has its own rating intensity.
(b) FPA differs from Threshold Odor Test (TOT). This test is used to evaluate the quantity of odor in the water. During this test, water is dissolved in water that has no odors and is then smelled. Dilutions continue until no odor is detected. The last dilution in which odor is detected from water determines the threshold odor number (TON).
(c) Another parameter of control of sweetening taste is the sweetening power (P). This differs from RS. P is defined by the ratio of the concentration of sweetener in the food and the reaction of the sweetener in equivalent solutions of sucrose.
(d) Temporal Profile (TP) of the sweetener is the main factor contributing to the difference in taste between products containing a nonnutritive sweetener and those containing sucrose. Hence, a method was created to compare the temporal profiles of nonnutritive sweeteners. In this method, the time needed for a sweetener to reach the power of an equivalent solution of 10% sucrose is defined as appearance time (AT) of sweet taste and is equal to 4 s. Another parameter defined by this method is the time required to reduce the perceived intensity of sweetness from the intensity equivalent to a solution of 10% sucrose to that equivalent to 2% sucrose. This is defined as extinction time (ET). A 10% sucrose solution has an ET equal to 14 s. AT and ET values for saccharin are similar to those of sucrose; hence saccharin is like sucrose. AT values for aspartame are slightly higher than those of sucrose, and ET aspartame values are significantly higher showing that aspartame is sweeter with a slight delay in the start of sweetness and with a noteworthy relative sweetness. Temporal profile of monoamine glycorrhizins is interesting since it delays sweetness; hence it is not used in foods and soft drinks.

9.3.17 EU Regulatory Issues for Use of Sweeteners in Food Products

The use of sweeteners in the EU is regulated by (Council Directive 89/107/EEC of December 21, 1988 on the approximation of the laws of the Member States concerning food additives authorized for use in foodstuffs intended for human consumption, as amended by Directive 94/34/EC) and a specific directive, European Parliament and Council Directive 94/35/EC of June 1994 on sweeteners for use in foodstuffs, amended by Directives 96/83/EC and 2003/115/EC. The annexes to the specific directives provide the information on which sweeteners are permitted in different foodstuffs or groups of foodstuffs together with the maximum permitted levels (MPL).

The specific directives have provisions for periodic monitoring of the use of food additives. The EU monitoring system is based on recommendations given in the report of the working group on "development of methods for monitoring intake of food additives in the EU, task 4.2".

The review of published data on intake of intense sweeteners in the EU up to 1997 indicated that their average intakes were below the relevant ADI values. The highest estimated intakes of cyclamate by diabetics and children were close to or slightly above the ADI.

Studies on the intake of intense sweeteners in different countries of the EU published since 1999 indicate that the average and 95th percentile intakes of acesulfame K, aspartame, cyclamate, and saccharin by adults are below the relevant ADIs.

Finally, bulk sweeteners (polyols) permitted in the EU are not included in the EU monitoring system. No upper limits of their use for sweetening purposes have been specified since monitoring of the exposure to the bulk sweeteners is not feasible.

9.4 ADDITIVE CATEGORIES

9.4.1 Acids

9.4.1.1 Lactic Acid

Dubos (1950) concluded that lactic acid has a bacteriostatic effect on *Mycobacterium tuberculosis*, which increased as pH decreased. Experiments carried out with *Bacillus coagulans* in tomato paste showed that lactic acid was four times more effective regarding the inhibition of bacterial growth compared to malic, citric, propionic, and acetic acid (Rice and Pederson, 1954).

Lactic acid is used in the jams, sweets, and drinks industries. This acid is the best acid to control the acidity and assure the transparency of brine in pickles (Gardner, 1972). Calcium lactate can be used as a taste enhancer, as a baking processing aid to proof the dough, as an inhibitor of decolorization of fruits and vegetables, as a gelatinization factor during pectins dehydration, or as an improver of the properties of milk powder or condensed milk (CFR, 1988). The ethyl esters of lactic acid can be used to enhance taste, and calcium lactate can be used in dietetic foods as well as nutrition supplements.

Lactic acid is an intermediate product of human metabolism. In cases of pneumonia, tuberculosis, and heart failure, a nonphysiological quantity of acid was detected in human blood. A growth problem was detected when acid was injected into water at a quantity of 40 mg/100 mL or in food at 45.6 mg/100 mL in hamsters. Lactic acid proved to be lethal in newborn–fed milk with an unknown quantity of acid (Young et al., 1982). The poisonous effect of acid has to do with its isomeric form. Babies fed with milk that has been acidified with D(−) or DL form suffered acidosis, lost weight, and became dehydrated. Hence, only the L(+) form can be used in premature newborn babies. Acid is a food ingredient and an intermediate metabolite of human beings, therefore, there are no established limits of daily consumption for humans (FAO/WHO, 1973).

9.4.1.2 Succinic Acid

FDA allowed the use of succinic acid as a tasty enhancer and as a pH regulator (CFR, 1988). Succinic acid reacts with proteins and is used to modify the dough plasticity (Gardner, 1972). Succinic acid derivatives can be used as taste enhancing agents or in combination with parafins as a protective layer for fruits and vegetables. They can be used in pills production at a percentage not greater than 4.5%–5.5% of gelatin percentage and 15% of the total weight of capsule. Many derivatives of succinic acid are used as ingredients of paper and paperboard in food packaging.

Succinic acid anhydrite is ideal for baking powder production of Allied Chemical Co. (Gardner, 1966, 1972). The low level of acid hydrolysis is important during mixing of dough, since it is required that the additional acid not to react with soda during mixing until the product swells.

In an acute toxicity study in rats with a daily subcutaneous injection of 0.5 mg for a period of 60 days and with a dose increase of 2 mg/day during the fourth week, no detrimental effect was observed. Acid is produced in some fruits and constitutes an intermediate product of the cycle of Krebs. Hence, no established limits can be determined for humans.

9.4.1.3 Fumaric Acid

Fumaric acid is responsible for a sour taste in foodstuffs. It is widely used in fruit juices, in desserts, in the frozen dough of biscuits, wild cherry liqueur, and in wines. It can also be used as a coating agent in caramel and bread. Fumaric acid contributes to the extension of the shelf life of baking powder due to its restricted solubility and the low humidity absorption. It attains very good antioxidant properties and is used to avoid rancidity of pork fat, butter, milk powder, sausages, bacon, walnuts, and chips (Lewis, 1989). It could also be used as a preservative in green foodstuffs and fishes in the same way as sodium benzoate does. CFR (1988) allowed the use of fumaric acid and its salts as dietetic products, nutritional additives. It can also be used as a source of available iron in the human organism if fumaric acid is combined with iron. Many fumaric acid derivatives have been approved for use in foodstuffs.

9.4.1.4 Malic Acid

CFR (1988) allowed the use of malic acid in foodstuffs as an acidifier, aroma enhancer, and pH regulator. Malic acid also contributes to the nonbrowning of fruits and acts synergistically with antioxidant substances (Gardner, 1972).

Malic acid is used in the production of iced fruits, marmalades, nonalcoholic carbonated drinks, and drinks originating from fruits. The limits of use have been defined accordingly: 3.4% in nonalcoholic drinks, 35% in chewing gums, 0.8% in gelatins, puddings, 6.9% in hard candies, 3.5% in processed fruits and juices, 3% in soft candies, and 0.7% in all other foodstuffs.

9.4.1.5 Tartaric Acid-(Carboxylmethylcellulose: CMC)

CFR (1988) allowed the use of tartaric acid in foodstuffs as an acidifier, solidifying agent, taste enhancing agent, as a material for maintenance of humidity, and pH regulator. Tartaric acid can be used in drinks manufacturing as an enhancing agent of the red color of wine (Gardner, 1972). It can also be used in tarts, marmalades, and candies. A mixture of tartaric acid and citric acid can be used in the production of hard candies with special flavors such as apple, wild cherry. This acid reacts with other antioxidant substances to avoid rancidity and discoloration in cheeses. Tartaric acid and its salts with potassium constitute ingredients of baking powders; L(+) isomer and its salts can be consumed up to 30 mg/kg body weight (FAO/WHO, 1973).

Tartrate crystals (potassium hydrogen tartrate and calcium tartrate), developed naturally in wine, are the major cause of sediment in bottled wines. In order to prevent its occurrence, some treatments have been used such as metatartaric acid, cold stabilization, and electrodialysis (OIV, 2012). Addition of mannoproteins obtained from hydrolysis of yeast cell wall was authorized by the European Community since 2005 (OIV, 2012). Mannoproteins inhibit the crystallization of tartrate salts by lowering the crystallization temperature.

CMC can also be used to prevent tartaric stability (OIV, 2012). However, before application and approval in the European Union in 2009 as an oenological product, CMC was used as emulsifiers (E466) in the food industry since the 1940s (Ribereau-Gayon et al., 2006) at levels to 10 g/L or 10 g/kg.

Several studies have displayed the positive effect of wine tartaric stabilization with CMC, reporting that this addition had identical effect on tartaric stability as metatartaric acid (Greeff et al., 2012). Furthermore, it was observed that CMC maintains its inhibitory efficiency 2 months after being exposed to 30°C, unlike metatartaric acid which is hydrolyzed (Gerbaud et al., 2010).

CMC is a cellulose derivative obtained by etherification of the free primary alcohol groups of the glucopyranose units linked by β-(1–4)-glycosidic linkages. CMC used in oenology, can be obtained with different degree of substitution (DS—number of glucose units substituted with carboxyl groups in relation to total glucose units), and different degree of polymerization (DP—chain length), that is, the average number of glucose units in the polymer molecule (Crachereau et al., 2001; Stojanovic et al., 2005).

Carboxylmethylcellulose (CMC) is authorized to prevent wine tartaric instability. The effect of CMC's structural characteristics on their effectiveness is not well understood. Guise et al. (2014) compared the impact of CMCs with different degrees of substitution and molecular weight, on tartaric stability, tartaric acid, mineral concentration, phenolic compounds, chromatic and sensory characteristics in white wines, and compared its effectiveness with other oenological additives. Mini-contact test showed that all CMCs and metatartaric acid stabilized the wines; however, some Arabic gums and mannoproteins do not stabilize the wines. CMCs had no significant effect on tartaric acid, potassium, calcium, and sensory attributes. Tartaric stabilization effectiveness depends on CMC's degree of substitution, and also on wine matrix, probably its initial potassium content. Results suggest that CMC is a good alternative to white wine tartaric stabilization; nevertheless deeper structure knowledge is necessary in order to choose the appropriate CMC for a given tartaric instability.

9.4.1.6 Citric Acid

Citric acid is a common metabolite of plants and animals. It is widely used in foodstuffs and the pharmaceutical industry as well as the chemical industry for isolation of ions and bases neutralization. Citric acid esters can be used as plasticizers in the manufacture of polymers and as adhesives. Citric acid can be used as acidifier and taste enhancer. It acts synergistically with other antioxidant substances. Citric acid and its salts are used in ice creams, drinks, fruits, jams, and as an acidifier in the manufacturing of canned vegetables. Calcium citrate is used as a stabilizer in pepper, potatoes, tomato, and beans during their processing. It has also been widely applied in dairy products manufacturing (Gardner, 1972). Moreover, it can be used in creamy sauces and soft cheeses; Sodium citrate can be used as an emulsifier. This proves to be the main acidifying factor in carbonated drinks entailing a rancid taste (Gardner, 1966). It can be used in line with other substances as an antioxidant and as a decelerator of fruits' browning. It is more hygroscopic than fumaric acid and could cause problems during storage of powdered products. Citric acid can be found in animal tissues and constitutes an intermediate product of Krebs cycle; hence, the daily consumption levels for the acid and its salts have not been determined yet (FAO/WHO, 1963; Varzakas et al., 2010).

9.4.1.7 Ascorbic Acid

Ascorbic acid is being used as an antimicrobial and antioxidative factor and enhances the uniformity and color stability. Ascorbic acid and its salts with Na and Ca are used as food additives (CFR, 1988). The D isomer of ascorbic acid compared to the L isomer has no biological value and gets oxidated faster than ascorbic acid with the result of protection of vitamin C from oxidation (Gardner, 1966). It is used as a pH regulator to avoid the enzymatic browning of fruits and vegetables. Plants and all mammals except man, apes, and guinea pigs synthesize ascorbic acid, hence, these three need to consume alternative sources of acid such as vegetables. High doses of ascorbic acid are recommended for cancer therapy and common cold. Ascorbic acid does not cause problems if it is consumed in high doses compared to vitamins A and D that might cause problems.

9.4.1.7.1 Cholinium or Choline

Cholinium-based ionic liquids are quaternary ammonium salts with a wide range of potential industrial applications. Based on the fact that cholinium is a complex B vitamin and widely used as food additive, cholinium-based ionic liquids are generically regarded as environmentally "harmless" and, thus, accepted as "nontoxic," although their ecotoxicological profile is poorly known. This work by Ventura et al. (2014) provides new ecotoxicological data for 10 cholinium-based salts and ionic liquids, aiming to extend the surprisingly restricted body of knowledge about the ecotoxicity of this particular family and to gain insight on the toxicity mechanism of these compounds. The results reported show that not all the cholinium tested can be considered harmless toward the test organism adopted. Moreover, the results suggest that the cholinium family exhibits a different mechanism of toxicity as compared to imidazolium ionic liquids previously described in the literature.

Cholinium, also known as choline, *N*,*N*,*N*-trimethylethanolammonium cation, an essential nutrient (Zeisel and da Costa, 2009), is receiving a considerable attention (Domínguez de María and Maugeri, 2011; Gorke et al., 2010) due to its claimed "benign" (Petkovic et al., 2011), biocompatible (Petkovic et al., 2010; Sekar et al., 2012), "environmentally friendly" or "nontoxic" nature (Gorke et al., 2010; Li et al., 2012; Nockemann et al., 2007a; Petkovic et al., 2010). This family is derived from quaternary ammonium salts described as important structures in living processes and used as precursor for the synthesis of vitamins (e.g., vitamin B complexes and thiamine) and enzymes that participate in the carbohydrate metabolism (Meck and Williams, 1999; Zeisel, 1999).

9.4.1.8 Acetic Acid

Acetic acid is used as a pH regulator, solvent, and as a pharmaceutical ingredient (CFR, 1988). It is safe when consumed in combination with the proper processing conditions. It can also be used as an additive in mustard, ketchup, mayonnaise, and sauces.

Acetic acid is well known for the various unpleasant effects on humans such as allergic symptoms, ulcer (Tuft and Ettelson, 1956), and anesthesia (Wiseman and Adler, 1956) and epidermic reactions (Weil and Rogers, 1951) until death (Palmer, 1932). Acid is made up of plant and animal tissues. There is no restriction on the daily consumption of acid for humans.

9.4.1.9 Sorbic Acid

Sorbic acid and its salts are widely used as fungicides and preservatives in pickles, mayonnaise, salads, spices, fruit and vegetables pulping, jams, frozen salads, syrups, beer, wines, sweets, cheeses, yoghurt, fishes, meat, poultry, and in various bakery products (Sofos and Busta, 1993). Sorbic acid is the least harmful preservative. According to a subchronic study of 2 months, carried out in 25 female and 25 male mice, which consumed 40 mg acid/kg body weight, sorbic acid did not lead to severe effects on the guinea pigs. According to studies carried out in rats fed food containing 1% or 2% sorbic acid for 80 days, no histologic abnormalities or growth problems were observed (FAO/WHO, 1974) (Table 9.1).

9.4.1.10 Propionic Acid

Propionic acid is extensively used in cheeses, sweets, gelatins, puddings, jams, drinks, soft candies, and Swedish cheese at a percentage of 1%. Calcium propionate can be used as an antimicrobial agent and acts as an inhibitor of the mould formation in bread dough. No daily consumption limits were determined due to the fact that propionic acid is an intermediate metabolite of human metabolism (FAO/WHO, 1973).

9.4.1.11 Benzoic Acid

CFR (1988)-approved benzoic acid for use at a concentration of 0.1%. The acid and its sodium salt can be used in processed foods, as food and drink preservatives with pH less than 4.5. It has an inhibitory action on mould growth and bacteria belonging to the following species: Bacillaceae, Enterobacteriaceae, Micrococcaceae. Benzoic acid and its sodium salt can be used in the preservation of carbonated drinks and nondrinks, fruit juices, jams, mayonnaise, mustard, pickles, bakery products, and ketchup (Chichester and Tanner, 1972; Luck, 1980).

Sodium benzoate, the sodium salt of benzoic acid, is generally used as a chemical preservative to prevent alteration or degradation caused by microorganisms during storage (Luck, 1980). Both sodium benzoate and benzoic acid exhibited inhibitory activity against a wide range of fungi, yeasts, molds, and bacteria (Mota et al., 2003). Sodium benzoate is more widely used in a great variety of foods and beverages because of its good stability and excellent solubility in water. However, excessive intake of these preservatives might be potentially harmful to the consumers because they have

TABLE 9.1
Uses and Indicated Quantities of Sorbic Acid Use

Dairy products (cheese, sour cream, yoghurt)	0.05–0.30
Bakery (cakes, dough, sugar crust, garnish)	0.03–0.30
Vegetables (fresh salads, boiled vegetables, pickles, olives, starters)	0.02–0.20
Fruits (dry fruits, fruit juices, fruit salad, jam, syrup)	0.02–0.25
Drinks (wine, carbonated drinks, fruit drinks)	0.02–0.10
Miscellaneous (smoked and salted fish, mayonnaise, margarine, sweets)	0.05–0.20

the tendency to induce allergic contact dermatitis, convulsion, hives, and hepato-cellular damage among others (Nair, 2001).

There are various techniques studied for benzoates determination. Traditionally, benzoates are analyzed mainly by thin layer chromatography (TLC), gas chromatography (GC), capillary electrophoresis (CE), and micellar electrokinetic chromatography.

A rapid and sensitive fluorescence polarization immunoassay (FPIA), based on a polyclonal antibody, has been developed by Ren et al. (2014) for the detection of sodium benzoate in spiked samples. The immunogen and fluorescein-labeled analyte conjugate were successfully synthesized, and the tracer was purified by TLC. Under the optimal assay conditions, the FPIA shows a detection range of 0.3–20.0 μg/mL for sodium benzoate with a detection limit of 0.26 μg/mL in the borate buffer. In addition, the IC_{50} value was 2.48 μg/mL, and the cross-reactivity of the antibodies with 10 structurally and functionally related analogs were detected, respectively. Four kinds of food samples (energy drink, candy, ice sucker, RIO™ cocktail) were selected to evaluate the application of FPIA in real systems. The recoveries were 96.68%–106.55% in energy drink; 95.78%–100.80% in candy, 86.97%–102.70% in ice sucker, and 103.58%–109.87% in benzoate-contained sample RIO™ cocktail, and coefficients of variation of this method were all lower than 11.25%. Comparing with the detection results of HPLC, the developed FPIA has comparative performance in the real sample determination. The results suggest that the FPIA developed in this study is a rapid, convenient, and simple method, which is suitable to be used as a screening tool for homogeneous detection of sodium benzoate in food products.

9.4.2 Coloring Agents

The synthetic dyes used nowadays are divided into the following three categories (Table 9.2):

1. *FD&C dyes*: Certified for food use, medicines, and cosmetics.
2. *D & C dyes*: Dyes considered as safe for medicinal and cosmetic use when they come into contact with muciferous glands or when they get absorbed.
3. *External D & C dyes*: Dyes that due to their toxicity do not get certified for product use. However, they are considered as safe for use in products of exterior applications.

Moreover, coloring agents used in foodstuffs can be divided into the following two classes:

1. Dyes requiring certification
2. Dyes not requiring certification

TABLE 9.2
Quantity of Coloring Agents Used in Certain Foods

Food Categories	Dye Concentration (mg/kg)	Average Quantity (mg/kg)
Caramel-sweets	10–400	100
Drinks	5–200	75
Cereals	200–500	350
Maraschino cherry	100–400	200
Pet foods	100–400	200
Ice creams	10–200	30
Sausage	40–250	125
Snack	25–500	200

Source: From Varzakas, T.H. et al., Food additives and contaminants, Chapter 13, in: Yildiz, F., ed., *Advances in Food Biochemistry*, CRC Press/Taylor & Francis Group, 2010, pp. 409–457.

TABLE 9.3
Representative Examples of Tint Arising from Mixing of Basic Dyes

Tint	FD&C Blue No. 1	FD&C Blue No. 2	FD&C Red No. 3	FD&C Red No. 40	FD&C Yellow No. 5	FD&C Yellow No. 6
Strawberry			5	95		
Black	36			22		42
Egg white					85	15
Cinnamon	5			35	60	
Green	3				97	
Green mint	25				75	
Orange						100
Grape	20			80		
Black cherry	5			95		
Chocolate	10			45	45	
	8			52	40	
Caramel	6		21		64	9
Peach				60		40
Blackberry	5		75			20

Source: From Varzakas, T.H. et al., Food additives and contaminants, Chapter 13, in: Yildiz, F., ed., *Advances in Food Biochemistry*, CRC Press/Taylor & Francis Group, 2010, pp. 409–457.

9.4.2.1 Dyes Requiring Certification for Their Use

9.4.2.1.1 FD&C Red No. 2 (Amaranth)

This dye was one of the first seven dyes allowed for use in 1906. Amaranth is a red–brown powder easily dissolved in water and produces a deep purple or a sea red solution (Table 9.3).

In experiments where amaranth was used at a percentage over 5% of the total dye, no pathological findings were reported with the exception of mutations and tumor development in rats. Long-term experiments for 7 years in dogs injected with the above-mentioned dyes at 2% did not cause any pathological problems. Similar experiments in mice, rats, rabbits, hamsters, cats, and dogs showed neither reproduction problems nor teratogenic problems or other unpleasant effects (WHO, 1973). Hence, it was decided that it can be included in the list of approved dyes. However, two experiments carried out in the ex-Soviet Union reported that the use of this dye caused carcinogenesis as well as embryotoxic reactions in mice that consumed 0.8%–1.6% amaranth.

9.4.2.1.2 FD&C Red No. 4

This dye is allowed in foods since 1929. It is a red-colored powder easily dissolved in water, producing an orange–yellow solution. It was originally used as an additive in butter and margarine. It did not cause carcinogenesis in mice when contained in food at 5% for a period of 2 years. It was concluded that this dye was toxic when consumed at 1% of the food for a period of 7 years. In another experiment carried out in dogs that consumed the dye at 2% for a period of 6 months, three out of four died. Due to the above-mentioned experiments the use of this dye was forbidden in 1976 (US FDA, 1983b) (Tables 9.4 and 9.5).

9.4.2.1.3 Citrus Red No. 2

The use of this dye is restricted; it is used mainly in leather dying. Nutritional studies showed that its use causes carcinogenesis in cats and dogs. Further experiments showed that this dye causes carcinogenesis.

TABLE 9.4
Dyes Used for Food; Name, Origin, Functionality, Adverse Effects, ADI, and Applications

E	Name	Origin	Functionality	Effects	ADI	Product Uses
E102	Tartazine	Synthetic dye	Yellow color	Headaches in children, allergic reactions in adults. Forbidden in Norway and Austria	0–7.5	Fruit juices, cake mixtures, soups, ice creams, sauces, jams, yoghurt, sweets, gums, lollipops
E104	Cinolin yellow	Synthetic dye of coal tar	Cloudy yellow color	Low absorption from gastroenteric system. Forbidden in Norway, United States, Australia, and Japan	0–0.5	Smoked cod, ice creams
E107	Yellow 2G	Synthetic dye of coal tar	Yellow color	Allergic reactions. Forbidden in Norway, United States, Switzerland, Japan, and Sweden	—	
E120	Carmine	Natural color from extraction of dried insects	Red	Cancer	0–2.5	Ice creams, alcoholic drinks
E122	Azorubin	Synthetic color	Red	Cause allergic reactions to asthmatic people and those who are sensitive in aspirin, edema, retention of gastric juices	0–4.0	
E128	Red 2G	Synthetic dye	Red	In the intestine it gets transformed to aniline, which causes methemoglobinemia	0–0.1	Sausages, cooked meat products, jams, drinks
E150 δ	Ammonium sulfite, caramel color	Synthetic dye derived from carbohydrates	Brown	Gastroenteric problems, reduction in white blood cells of patients with B6 vitamin in low levels	—	Glucose tablets, ice cream, baking flour, total milled bread
E151	Bright Black BN ή PN	Synthetic color of coal tar and nitrogen dye	Black	Intestine cysts in mice	0–1	Black sauces, chocolate mouse

Source: From Varzakas, T.H. et al., Food additives and contaminants, Chapter 13, in: Yildiz, F., ed., *Advances in Food Biochemistry*, CRC Press/Taylor & Francis Group, 2010, pp. 409–457.

9.4.2.1.4 FD&C Red No. 40

This dye is allowed in the United States and Canada following thorough experiments. However, it is not allowed in the United Kingdom, Sweden, and Europe.

9.4.2.1.5 FD&C Yellow No. 3, FD&C Yellow No. 4

These dyes are orange colored and fat soluble. They were used in 1918 in the United States as additives in margarine production. They were reported to be liver toxic in dogs and cats (Allamark

TABLE 9.5
Maximum Amount of Organic Acids in Various Foods

Foods	Acetic	Acerate Calcium	Acerate Sodium	Diacatate Sodium	Adipic	Caprilic	Malic	Succinic
Baked foods	0.25	0.2		0.4	0.05	0.013		
Drinks. nonalcoholic drinks					0.005		3.4	
Breakfast cereals			0.007					
cheese	0.8	0.02				0.04		
Chewing gum	0.5						3.0	
spices	9.0				5.0			0.084
Dairy products	0.8				0.45			
Oils–fats	0.5		0.5	0.1	0.3	0.005		
Frozen dairy desserts					0.004	0.005		
Gelatin		0.2			0.55	0.005	0.8	
Sauces	3.0			0.25		0.1		
Hard candies			0.15				6.9	
Jam			0.12					2.6
Meat products	0.6		0.12	0.1	0.3	0.005		0.0061
Fruit juices							3.5	
Snacks		0.6	0.05		1.3	0.016		
Soft candies			0.2	0.1		0.005	3.0	

Source: From Varzakas, T.H. et al., Food additives and contaminants, Chapter 13, in: Yildiz, F., ed., *Advances in Food Biochemistry*, CRC Press/Taylor & Francis Group, 2010, pp. 409–457.

et al., 1955). These dyes caused weight loss in guinea pigs even at 0.05% (Hansen et al., 1963). Mice consuming a quantity of dye with their food higher than 0.05% for a period greater than 2 years showed heart atrophy or hypertrophy. Dyes are metabolized under the effect of gastric acid conditions (Harrow and Jones, 1954; Radomaski and Harrow, 1966) and their metabolite caused liver cancer in rats as well as urinary bladder cancer in dogs (Clayson and Gardner, 1976). Due to these reported experimental results these dyes were rejected from the list of dyes allowed to be used.

9.4.2.2 Dyes Not Requiring Certification for Their Use

9.4.2.2.1 Annatto Extracts

Annatto extracts is one of the older dyes used in foods, textiles, and cosmetics. The extraction of these dyes is carried out from the pericarp of the seeds of the Annatto tree (*Bixa orellana* L.); annatto oil is produced from distillation of the coloring agents from the seeds, using an edible vegetable oil. The main coloring agent is norbixin (E160b). It is supplied as an oil-soluble extract which contains mainly bixin (a mono methyl ester) or as a water-soluble product, which is produced by extraction with aqueous alkali and therefore contains mainly norbixin, the free acid.

Processing is primarily carried out by abrading away the pigment in a suspending agent. Abrasion may be followed by aqueous alkaline hydrolysis with simultaneous production of norbixin. Traditionally, water or vegetable oil is used as a suspending agent, although solvent processing is now also employed to produce more purified annatto extracts. Annatto is usually marketed as an extract of the annatto seed, containing amounts of the active pigments bixin or norbixin that can vary from less than 1% to over 85%.

In 2002, the Secretariat of the FAO/WHO Joint Expert Committee on Food Additives (JECFA) requested information relating to the toxicity, intake, and specifications of annatto extracts. Previous intake estimates for annatto had provided ambiguous results because the bixin/norbixin content of the annatto extract was unclear. As a consequence, many of the stated use levels, such as some of those appearing in the draft CCFAC General Standard for Food Additives (GSFA) (CAC, 1984), over-estimated true use levels by more than an order of magnitude. This resulted in estimates of intake of annatto extracts that appeared to exceed the JECFA ADI of 0.065 mg/kg bw/day (based on pure bixin). Previous intake estimates for annatto provided ambiguous results because the bixin/norbixin content of annatto extract was unclear. European annatto producers consulted with the food industry to determine usage levels of specific annatto extracts. These data were combined with the levels of bixin/norbixin in particular extracts to estimate the concentration of bixin/norbixin in foods. Concentrations in food were combined with data about food consumption using various methods to estimate consumer intakes, which ranged from less than 1%–163% of the ADI (0.065 mg/kg bw/day). Higher intake estimates are conservative because they assume that a consumer always chooses a food that is colored with annatto extracts. In practice this is extremely unlikely, since annatto is associated only with certain product/flavor combinations (Tennant and O' Callaghan, 2005).

Food colors differ from many other food additives in that the level required to meet a technological need can vary considerably from food to food. This is because colors are usually associated with particular flavors and so the need for a coloring agent, for example, an orange color such as annatto, will vary greatly depending on whether the flavor is vanilla, lemon, orange, or mango. Another source of variability is the opacity of foods. The more opaque a food is, the higher the amount of color required to achieve a given shade. As a consequence, typical use levels (i.e., use in the majority of colored food products) do not necessarily reflect the maximum use level required to achieve the color density required for certain color/flavor/food combinations. It is therefore necessary to take this potential source of variability carefully into account when estimating consumer intakes.

9.4.2.2.2 Betalains (Dehydrated Sugar Beets)

These dyes belong to the plants of the family *Centrospermae* and more specifically plants such as beetroot and bougainvillea. Beetroot contains both dyes, the red (betacyanins) and the yellow (betaxanthins). Betalains are sensitive to light, pH, and heat (Marmion, 1984; Newsome, 1990; Sankaranaranan, 1981). They are dissolved easily in water giving solutions of blackberry or cherry coloration. Beetroot dyes are used in foods with low storage time and do not require high and extended heat during their processing. In heat processing cases dye addition could be carried out just before or after the end of processing (Marmion, 1984). Betalains are used at 0.1%–1% of the total dye in hard candies, yoghurt, sauces, cakes, meat substitutes, mild drinks, and gelatinous desserts.

9.4.2.2.3 Anthocyanins

Anthocyanins are used as pH indicators. The dye is red at an acidic environment, whereas it becomes blue as pH increases. Anthocyanins have more intense color at pH = 3.5; hence, these dyes are used in foods with low pH. They get discolored in the presence of amino acids (Sankaranaranan, 1981), while they get oxidized in the presence of ascorbic acid.

9.4.2.2.4 Saffron

Saffron is an expensive dye since it is required to get 165,000 flower stains for the production of 50 g of pure dye. The most important Saffron dyes are crocin and crocetin. Crocin has a yellow–orange color and dissolves in hot water, whereas it dissolves less in alcohol. Crocetin dissolves less in water and more in organic solvents. Saffron is used for its aroma and its color, is resistant to sunlight, moulds, pH, and has a high dying capacity. It is used at a concentration of 1–260 ppm in cooked food, soups, and confectionery products (Table 9.6).

The "coloring strength" of aqueous saffron extracts, expressed as $E_{1\text{cm}}^{1\%}$ 440 nm, is considered to be of utmost importance for the retail price as well as consumer acceptance (Ordoudi and Tsimidou, 2004).

TABLE 9.6
Saturated Organic Acids and Characteristics with Regard to Aroma

Acid	Characteristics
Acetic	Strong aroma, fine, carmine
Propionic	Rancid, fatty, milky
Butyric	Fatty, milky, rancid
Valeric	Rancid, sweet, chocolate
Caprylic	Fatty, fruit smell, soapy
Myristic	Soapy
Palmitic	Soapy, waxy
Stearic	Soapy, waxy

In addition to color, the taste and aroma of saffron, associated with the presence of picrocrocin and its dehydration product safranal, are expressed as $E_{1cm}^{1\%}$ 257 nm and $E_{1cm}^{1\%}$ 330 nm values, respectively. For example, saffron samples with $E_{1cm}^{1\%}$ 257 nm values less than 40 units and $E_{1cm}^{1\%}$ 330 nm values less than 20 units are considered substandard (ISO, 2011).

Fourier transform infrared (FT-IR) spectroscopy has a strong potential in the analysis and quality control of foods because of its sensitivity, versatility, and speed (Sun, 2009). Recent advances in FT-IR instrumentation along with the development of multivariate data analysis have resulted in an increase in applications of mid-infrared (MIR) spectroscopy to food safety, end-product quality/process control, and authenticity studies (Karoui et al., 2010).

One application of the FT-MIR technique, along with chemometrics, was recently reported by Anastasaki et al. (2010) describing differentiation of saffron samples from four countries (Iran, Spain, Sardinia, and Greece); an objective also investigated using Fourier transform near-infrared (FT-NIR) some years ago (Zalacain et al., 2005).

Ordoudi et al. (2014) extended the application of the FT-MIR technique to the quality control of traded saffron that suffers various types of fraud or mislabeling. Spectroscopic data were obtained for samples stored for different periods in the dark. Samples with the highest quality according to ISO 3632 specifications produced a typical spectrum profile (reference set). Principal component analysis (PCA) of spectroscopic data for this set, along with HPLC-DAD analysis of major apocarotenoids, assisted identification of FT-IR bands that carry information about desirable sensory properties that weaken during storage. The band at 1028 cm^{-1}, associated with the presence of glucose moieties, along with intensities in the region 1175–1157 cm^{-1}, linked with breakage of glycosidic bonds, were the most useful for diagnostic monitoring of storage effects on the evaluation and test set samples. FT-IR was found to be a promising, sensitive, and rapid tool in the fight against saffron fraud (Table 9.7).

9.4.2.2.5 Cochineal Extracts

They are derived from the body of female insects of species *Coccus cacti*. This insect can be found in the Canary islands, and the dye is expensive. It can be used in the production of candies, sweets, alcoholic and nonalcoholic drinks, jams, eye shadows, rouge, and as a coating for pills at a percentage of 0.04%–0.2%.

9.4.2.2.6 Chlorophyll

Chlorophyll is the most abundant natural dye of plants. However, it is not widely used in foods (Sankaranaranan, 1981). Its green color is decomposed easily even at mild processing conditions. Due to its structure during processing, magnesium gets replaced by hydrogen destroying the purple ring giving a dark brown color (Francis, 1985). Chlorophyll stabilizes with replacement of magnesium ions with copper ions. Chlorophyll extracts are forbidden for use in foods in the United States. The use of complex chlorophyll-copper is allowed in the EU and Canada.

TABLE 9.7
Aldehydes Found in Bovine, Chicken, and Pork Fat

Aldehydes	Bovine	Chicken	Pork
C5	+	−	−
C6	+	+	+
C7	+	+	−
C7 2t	+	+	+
C7 2t 4C	−	+	+
C8	+	−	+
C8 2t	+	+	+
C9	+	+	+
C9 2t	+	+	+
C9 2t 4C	−	+	+
C10	−	+	+

Source: From Varzakas, T.H. et al., Food additives and contaminants, Chapter 13, in: Yildiz, F., ed., *Advances in Food Biochemistry*, CRC Press/Taylor & Francis Group, 2010, pp. 409–457.

Chlorophyll is a natural pigment produced by green plants and algae. It absorbs light energy from the sun, which is then used to synthesize carbohydrates from carbon dioxide in a process called photosynthesis (Raven et al., 2005). Aside of its role in photosynthesis, chlorophyll has many other applications. It is used as a natural coloring agent both in foods and cosmetics (Humphery, 2004).

Its rich green color enhances many products from candy to soap and has also been used as an internal deodorant and mouthwash.

The demand for chlorophyll has dramatically increased due to the recent popularity of using natural instead of artificial additives.

Additionally, chlorophyll and its derivatives are generally considered to have health benefits due to their antioxidant and antimutagenic activities, and thus play a role in the prevention of chronic diseases such as cancer (Lanfer-Marquez et al., 2005; Mario et al., 2007). They have been included in dietary recommendations and incorporated into many health foods. Chlorophyll has also long been used for therapeutic purposes, such as wound healing and anti-inflammatory applications (Edwards, 1954; Larato and Pfau, 1970). Their ability to act as photosensitizers has also enabled their application in photodynamic therapy of cancer (Sternberg et al., 1998).

The basic structure of a chlorophyll molecule is a porphyrin ring with a magnesium atom in the center and a long hydrocarbon tail as a side chain. Two main types of chlorophyll are found in higher plants, chlorophylls a and b. They differ only slightly in the composition of a single side chain (Fleming, 1967). In nature, chlorophyll is localized in a complex structure of plant leaves.

Chlorophyll has long been considered to have health benefits and is widely used as a food additive. The conventional extraction method using organic solvents involves penetration and diffusion through many layers of a leaf structure, making it hard to predict and control. A diffusion equation with boundary conditions based on the wetting theory was proposed by Hung et al. (2014). Three kinds of solvent—acetone, ethanol, and dimethyl sulfoxide—were used to extract chlorophyll from creeping oxalis. Chlorophyll of three kinds of plants (creeping oxalis, nodal flower synedrella, and broomjute sida) were extracted with 80% acetone. The amounts of extracted chlorophyll were in good agreement with the theoretical prediction. The effective diffusion coefficient of the chlorophyll transport increases with increasing energy barrier of chlorophyll wetting at the epidermis for all solvent types, acetone concentrations, and plant species. The barrier height of chlorophyll wetting increases with the surface energy of plants.

9.4.2.2.7 Carotenoids

Carotenoids are sensitive to alkali and very sensitive to air, light, and high temperature. They are undissolved in water, ethyl alcohol, and glycerol. β-carotene has nutritive value because it gets transformed into provitamin A in the human organism and can be used without quantitative limitations. It takes a yellow–orange color in foods when used at concentrations of 2–50 ppm. β-carotene constitutes one of the basic ingredients of butter, margarine, cheeses, ice cream, yoghurt, and pasta products.

Carotenoids are one group of natural pigments responsible for yellow, orange, and red colors in many fruits and vegetables. Among them, three carotenoids, lutein, β-carotene, and lycopene, are prevalent in our daily diets. In addition to the color attribute, they have been confirmed to possess some health benefit functions, such as supplying vitamin A to the body, reducing the damage of the eye retina from exposure to near-ultraviolet light, preventing the risk of prostate cancer, etc. (Rodriguez-Amaya, 2010).

β-carotene (E160a (i)) is obtained by solvent extraction of edible plants (e.g., lucerne) or from algae (*Dunaliella salina*), and can be naturally accompanied by minor carotenoids (e.g., α-carotene). Nearly pure—β-carotene (E160a (ii)) is available from the fungus *Blaskeslea trispora*. The coloring principle of red pepper extracts (E160c) is capsanthin, accompanied with capsorubin and various minor carotenoids. Lycopene (E160d) is found in several fruits (e.g., papaya) and represents the main carotenoid of red tomatoes. Lutein (E161b) is found in several plants (e.g., alfalfa) or flowers (e.g., marigold; *Tagetes erecta* L.). In marigold, lutein is acylated with various fatty acids.

A sensitive HPLC multimethod was developed by Breithaupt (2004) for the determination of the carotenoid food additives (CFA) norbixin, bixin, capsanthin, lutein, canthaxanthin, β-apo-80-carotenal, β-apo-80-carotenoic acid ethyl ester, β-carotene, and lycopene in processed food using an RP C30 column. For unequivocal identification, the mass spectra of all analytes were recorded using LC–(APcI) MS. For extraction, a manual process as well as accelerated solvent extraction (ASE) was applied. Important ASE parameters were optimized. ASE was used for the first time to extract CFA from various food matrices. Average recoveries for all analytes ranged from 88.7%–103.3% (manual extraction) and 91.0%–99.6% (ASE), with the exception of norbixin using ASE (67.4%). Limits of quantitation (LOQ) ranged from 0.53 to 0.79 mg/L. The presented ASE method can be used to monitor both the forbidden application of CFA and the compliance of food with legal limits.

As yellow and red artificial colorants have a similar tint to carotenoid pigments, they have been used totally or partially to replace the natural pigments in food products (Kiseleva et al., 2003). Their high stability and solubility made them widely used in various drinks and processed food products. However, the main chemical structure of artificial food colorants consists of azo groups and aromatic rings which may have potential toxicity to human health, especially when they are excessively consumed daily (Minioti et al., 2007).

A method for simultaneously determining four artificial food colorants [Red Nos. 2 (R2) and 40 (R40), Yellow Nos. 5 (Y5) and 6 (Y6)] and three carotenoids (lycopene, lutein, and β-carotene) was developed by Shen et al. (2014).

These colorants were successfully separated by the developed high-pressure liquid chromatography (HPLC) method combined with a photo diode array detector. The detection limit (at signal to noise >4) was from the lowest of 0.2 ng/mL for lutein to the highest of 50.0 ng/mL for R40. With a two-phase solvent and ultrasound-assisted extraction, the recoveries of the artificial and natural pigments in 15 different types of food products were between 80.5%–97.2% and 80.1%–98.4%, respectively. This HPLC method with the ultrasound-assisted extraction protocol could be used as a sensitive and reliable analysis technique in simultaneously identifying and quantifying the reddish and yellowish pigments in different foods, regardless whether they are artificial food colorants or/ and natural carotenoids.

9.4.2.2.8 Titanium Dioxide (No. 77891)

This dye exists in nature in three crystallic forms; however, only one is used as an additive (anatase, brookite, and rutile). It possesses an intense white color and is resistant to sunlight, oxidation, pH, and the presence of microbes. Only the synthetically produced titanium dioxide can be used as a food additive (Marmion, 1984). This dye cannot be dissolved in all solvents. The allowed quantity of dye in foods is up to 1% and can be used in confectionery products in the formation of white parts as well as a background. Finally, it can be use in the production of pills and cosmetics.

9.4.2.2.9 Iron Oxides and Hydroxides

These synthetic dyes impart a large range of colors varying from red, yellow, and black with exceptional stability in light and temperature. Natural dyes are not acceptable as food additives due to the difficulties encountered to isolate the pure substance. They are insoluble in most solvents; however, they can be dissolved in hydrogen chloride. The maximum allowed quantity of these dyes is lower than 0.25% w/w in fish paste and in pet food.

9.4.2.2.10 Other Dyes

Chrysoidine is an azo dye which is usually used for dyeing leather, fibers, and paper (Lei et al., 2011). Because chrysoidine may cause several adverse effects on human health, it is banned from being used as a food additive in China (Gui et al., 2010). However, some illegal merchants still dye bean products and yellow croaker with chrysoidine (Lu et al., 2012). In addition, since it is widely used in the dye industry, the presence of chrysoidine in water has become an environmental issue due to the adverse human health and ecological impacts (Bayramoglu and Arica, 2012; Tan et al., 2011).

Chrysoidine is an industrial azo dye, and the presence of chrysoidine in water and food has become an environmental concern due to its negative effects on human beings. Binding of dyes to serum albumins significantly influences their absorption, distribution, metabolism, and excretion properties. Yang et al. (2014) explored the interactions of chrysoidine with bovine serum albumin (BSA). Isothermal titration calorimetry results reveal the binding stoichiometry of chrysoidine to BSA is 1:15.5, and van der Waals and hydrogen bonding interactions are the major driving force in the binding of chrysoidine to BSA.

Molecular docking simulations show that chrysoidine binds to BSA at a cavity close to Sudlow site I in domain IIA. However, no detectable conformational change of BSA occurs in the presence of chrysoidine, as revealed by UV–vis absorption, circular dichroism (CD), and fluorescence spectroscopy studies.

Curcumin: Curcumin, (diferuloylmethane; 1,7-bis[4-hydroxy-3-methoxyphenyl]-1,6-heptadiene-3,5-dione) along with its mono and di-demethoxy derivatives, collectively called curcuminoids, constitute the major coloring agent and the biologically active constituents of *Curcuma longa* L. or turmeric. Extensive *in vitro* and *in vivo* studies have suggested that curcumin has anticancer, antiviral, antiarthritic, antiamyloid, antioxidant, and anti-inflammatory properties.

Curcumin is also practically insoluble in water at acidic or neutral pH and is not stable in alkali/high pH conditions, contributing to its poor absorption and low bioavailability.

Solid dispersion is one of the methods used to improve the dissolution (Vo et al., 2013) and bioavailability of poorly water soluble compounds (Kohri et al., 1999; Pan et al., 2000). Solid dispersion involves the incorporation of water insoluble compound (s) into a hydrophilic carrier matrix, resulting in a release profile of the active compound that is governed by the polymers properties (Chiou and Riegelman, 1971).

The poor oral bioavailability of curcumin poses a significant pharmacological barrier to its use therapeutically and/or as a functional food. Chuah et al. (2014) reported the evaluation of the bioavailability and bio-efficacy of curcumin as an amorphous solid dispersion (ASD) in a matrix consisting of hydroxypropyl methylcellulose (HPMC), lecithin, and isomalt using hot melt extrusion for application in food products. Oral pharmacokinetic studies in rats showed that ASD curcumin was ~13-fold more bioavailable compared to unformulated curcumin. Evaluation of the anti-inflammatory activity of ASD curcumin *in vivo* demonstrated enhanced bio-efficacy compared to unformulated curcumin at a 10-fold lower dose. Thus, ASD curcumin provides a more potent and efficacious formulation of curcumin which may also help in making the color, taste, and smell which currently limit its application as a functional food ingredient.

Sudan dyes: Sudan I–IV are synthetic azoic compounds that appear orange or red, and they are industrial dyes used as toners for petroleum, engine oil, polishing agent, and some other industrial solvents. However, these dyes can be accumulated in the fatty tissues of the liver (lipotrophic). The International Agency for Research on Cancer (IARC) considers them to be third category carcinogens, which are animal carcinogens.

Sudan dyes are widely used in industry and sometimes illegally used as food additives despite their potential toxicity. Sun et al. (2014) investigated the interactions of Sudan II and Sudan IV with bovine hemoglobin (BHb) by fluorescence, synchronous fluorescence, resonance light scattering (RLS), UV–vis absorption, CD, and molecular modeling techniques. Binding of Sudan dyes to BHb could cause static quenching of the fluorescence, indicating changes in the microenvironment of tryptophan and tyrosine residues. The binding constants estimated for Sudan II and IV were 1.84 × 104 and 2.54 × 104 L/mol, respectively, at 293 K (20°C). Each protein molecule bound one Sudan molecule approximately. Sudan II and IV were held at the hydrophobic cavity of BHb mainly by hydrophobic interaction. The decrease of α-helix and the increase of β-sheet seen in the CD spectra revealed a conformational alteration of the protein. From all the results, they concluded that Sudan IV has a stronger impact on the structure and function of BHb than Sudan II (Sun et al., 2014).

9.4.2.3 Aromatic Substances—Flavorings

Aroma is a food and drink property causing stimulation of the olfactory centers. The role of the aromatic substances in foods is summarized below:

1. Impart a characteristic aroma, for example, vanillin gives a vanillin aroma to ice creams.
2. Enhance, complement, or modify the already existing aroma, for example, vanillin addition to moderate chocolate or cocoa's smell.
3. Cover a nondesirable odor.

The odor of a chemical substance has a direct relationship with its chemical formula. Compounds such as nitrogen, oxygen, and sulfur play a significant role. More specifically, the addition of a hydroxyl group into the molecule of a compound reduces or restricts its odor, whereas replacement of a hydroxyl group with a ketone group enhances the odor.

Citral, one of the most important natural flavoring compound having an intense lemon aroma and flavor is widely used as an additive in foods, beverages, and cosmetics with high consumer acceptance (Maswal and Dar, 2014). Citral is chemically unstable and degrades over time in aqueous solutions due to acid catalyzed and oxidative reactions, leading to loss of desirable flavor and formation of off-flavors. Therefore, incorporation of citral into foods and beverages is a major challenge for the food industry because their chemical deterioration needs to be inhibited to minimize the loss of product quality. The task to find the appropriate delivery system is most challenging for the food industry. The encapsulation and delivery techniques of citral, mostly based on colloidal systems, have been reviewed by the authors. Moreover, the remaining technical challenges of such delivery systems like insignificant stabilization of citral, use of non-biocompatible constituents, instability to the environmental stress and difficulty of their preparation are discussed for prospective development of such formulations.

Citral or 3,7-Dimethyl-2,6-octadienal is an acyclic monoterpenoid consisting of two geometrical isomers. The *E*-isomer is specifically referred to as geranial or citral A and the *Z*-isomer as neral or citral B. Citral is one of the most important flavoring compound used widely in beverages (Piorkowski and McClements, 2013), foods, and fragrances for its characteristic flavor profile (Choi et al., 2009). It is also used commercially in the production of vitamin A, ionones, and methyl ionones (Pihlasalo et al., 2007).

For the stabilization and delivery of citral and other lemon oil derivatives in food, cosmetics, and pharmaceutical industries, the following encapsulation and delivery techniques have been employed: Spray drying, oil-in-water emulsions, multilayer emulsions, nanoemulsions, molecular complexes, and self-assembly delivery systems.

Simple oil-in-water emulsions are most widely used for encapsulation of citral, although being susceptible to breakdown over time or exposed to various environmental stresses.

Multilayer emulsions have also been used having better stabilization capacity of citral but requiring at the same time more complicated fabrication techniques. The efficiency and availability of citral has been greatly enhanced when delivered in the form of a nanoemulsion instead of a normal emulsion.

Molecular hosteguest delivery systems involving cyclodextrins have emerged as an excellent delivery vehicle for citral having capability of reducing the off-flavor formation to a greater extent but have a disadvantage of requiring longer time for the equilibration (Maswal and Dar, 2014).

9.4.2.3.1 Polycyclic Aromatic Hydrocarbons

Polycyclic aromatic hydrocarbons (PAHs) comprise the largest class of chemical compounds known to be cancer-causing agents. Some, while not carcinogenic, may also act as synergically. PAHs are being found in water, air, soil, and therefore also in food. They originate from diverse sources such as tobacco smoke, engine exhausts, petroleum distillates, and coal-derived products, with combustion sources predominating. However, PAHs may also form directly in food as a result of several heat processes such as grilling (Panek et al., 1995), roasting, smoke drying, and smoking.

PAHs are compounds consisting of two or more condensed aromatic rings, lineared together, either *cata*-annellated (linearly or angularly) or *peri*-condensed.

Formation, factors affecting concentrations, legal limits, and occurrence of PAHs in smoked meat products and smoke flavor additives are briefly reviewed by Simko (2002). The most widely employed techniques such as TLC, GC, and HPLC are evaluated. Moreover, sample preparation, pre-separation procedures, separation and detection systems being used for determination have been evaluated with emphasis on the latest developments in applied food analysis, and the chosen data regarding the concentration of PAHs in smoked meat products and smoke flavor additives are summarized.

9.4.2.3.2 Flavoring Substances

Munro and Danielewska-Nikiel (2005) conducted a study to determine the margins of safety between non-observed-effect levels (NOELs) and estimates of daily intake for 809 flavoring substances evaluated by the JECFA between 2000 and 2004. Estimates of daily intake were calculated by means of two methods, the maximized survey-derived daily intake (MSDI) and the possible average daily intake (PADI). The MSDI estimates were based on the production volume of flavoring agents as reported by industry, whereas the higher more conservative PADI estimates were derived by multiplying the anticipated average use level of a flavoring substance in each of 33 food categories by the average amount of food consumed daily from that food category and summing the intake over all 33 food categories. These intake estimates were used to calculate the margins of safety for the flavoring agents to determine whether adequate margins of safety would still exist in the event that the MSDIs used by JECFA to evaluate the safety of flavoring substances underestimated daily intakes. Based on the calculation of the margins of safety using the MSDI values, 99.9% of the 809 flavoring substances evaluated by JECFA have margins of safety of greater than 100. In comparison, 98% of flavoring substances have margins of safety of greater than 100 when the margins of safety were calculated according to PADI values. The results indicate that if the MSDI estimates used by JECFA for the evaluation of the safety of flavoring substances were underestimated, a wide margin of safety exists for all but a few of the flavoring substances even when intakes were estimated from PADI values.

Although the procedure for the safety evaluation of flavoring agents (PSEFI) does not require toxicological data in every instance but rather relies primarily on structure–activity relationships to assess safety, this study confirmed that on the basis of a more traditional toxicological process of evaluating chemical safety, the procedure showed that the evaluated substances did not present a concern for safety at current estimates of intake (Munro and Danielewska-Nikiel, 2005).

9.4.2.4 Toxic Substances

It is difficult to distinguish clearly between the essential metals (micronutrients) and the toxic substances in human diet. Nearly all metals are toxic for humans if they get absorbed by the organism at high concentrations. However, it is possible to differentiate between those elements that are essential and cause toxicosis at extremely low concentrations and do not possess known therapeutic normal functions and those that do cause toxicosis. The most dangerous heavy metals, which occur nearly everywhere and particularly in agricultural products are mercury, lead, and cadmium. Other toxic metals are arsenic, boron, beryllium, selenium, and others. The most probable sources of food contamination with toxic substances are soil, agricultural chemicals and fertilizers, contaminated water from industrial wastes, the various food processing stages, plants that can absorb toxic substances from the contaminated land and store them in their tissues, as well as the muddy depositions of pit wastes.

The most common metals contaminating foods are mentioned below:

9.4.2.4.1 Lead

The presence of lead in the human food chain constitutes a major hygiene problem for the world. It exists in every organ or tissue of the human body and varies between 100 and 400 mg or approximately 1.7 μg/g tissue (Barry, 1975). Over 90% lead is present in the bones. It constitutes a physiological ingredient of human diet (Table 9.8). The daily food uptake of lead is estimated to be approximately equal to 100–300 μg, with particularly high percentages during intense environmental pollution (WHO, 1976). Lead absorption contained in foods is estimated to be approximately equal to 10% for adults and 40% for children (Reilly, 1991). Many dietetic factors seem

TABLE 9.8
Lead Concentration in Food and Drinks

Food	Range (mg/kg)	Average (mg/kg)
Cereals	<0.01–0.81	0.17
Meat and fish	<0.01–0.70	0.17
Fresh fruits	<0.01–0.76	0.12
Canned fruits	0.04–10.0	0.40
Fresh vegetables	<0.01–1.5	0.22
Canned vegetables	0.01–1.5	0.24
Milk	<0.01–0.08	0.03
Water	1–50[a]	5[a]
Alcoholic drinks	50–100[a]	

Source: From Varzakas, T.H. et al., Food additives and contaminants, Chapter 13, in: Yildiz, F., ed., *Advances in Food Biochemistry*, CRC Press/Taylor & Francis Group, 2010, pp. 409–457.

[a] μg/L.

to affect the uptake levels of lead. Hence, low calcium levels, iron deficiency, and diets rich in carbohydrates but low in proteins, as well as diets containing high levels of vitamin D, lead to an increased absorption of lead. In the healthy adult, approximately 90% of the absorbed lead gets excreted by urine and feces.

Lead infection can be distinguished in acute and chronic. Symptoms of acute infection include blood, nervous, gastroenteric, and hepatic failures (Reilly, 1991). In general, phenomena such as anorexia, indigestion, and constipation occur followed by colic with intense paroxysmal abdominal pain. Although lead encephalopathy has been observed in little children (NAS, 1972; Reilly and Reilly, 1972), very little is known regarding the symptoms of chronic encephalopathy. Some of the known typical clinical symptoms are mild anemia, overactivity, aggressive behavior, mental disorder, peripheral neuropathy, paralysis, and renal failure (WHO, 1976).

9.4.2.4.2 Mercury (Hg)

Mercury is considered as the most hazardous heavy metal in the food chain due to its continuous presence in the environment, its bioaccumulation and transportation into the water chain, and its high levels in a large variety of foods. It exists in three different forms; elementary mercury, mercuric mercury, and acetyl mercury. The chemical form affects at a great percentage the absorption, its distribution in the body tissues, and its biological half-life. The standard human diet contains less than 50 μg mercury/kg food (Bouquiaux, 1974), whereas seafood represents the main source of mercury.

Mercury is a cumulative poison stored mainly in the liver and the kidneys. The concentration levels depend on the type of organism and the chemical form. Mercury, in its metallic form, is less absorbed, easily excreted from the organism, and hence it is improbable to cause poisoning. On the contrary, both organic and inorganic substances of mercury are particularly toxic for humans. Hence, methyl mercury is considered as one of the six most dangerous chemicals existing in the environment (Bennet, 1984). It is absorbed by the intestine, enters the bloodstream rapidly, and binds to the plasma proteins. Methyl mercury could prove to be neurotoxic for adults and embryos since it accumulates in the brain (Berlin et al., 1963). It can cause irreversible damage to the central nervous system leading to disorder, terror, restriction of the optical field, blindness, loss of hearing, and finally death (Reilly, 1991). Selenium seems to react to mercury poisoning in many animal species (Stoewsand et al., 1974).

The most important source of mercury or methyl mercury uptake by man is fishery products (fish and seafood) and mainly hunted fish, found on the top of the food chain, such as swordfish, tuna, bass, mackerel, and shark. Big hunted fish with the highest percentage of methyl mercury are often migratory and can be found in waters of high or low infection in mercury.

The European Union has implemented Regulation 466/2001 of the Commission, according to which the maximum permitted tolerance values in total mercury in foodstuffs amounts 0.5 mg/kg fresh weight product in fishery products, except for certain fishes where the maximum level is 1 mg/kg fresh weight product.

At an international level in June 2003, the temporary permitted weekly uptake percentage was determined at 1.6 μg/kg body weight, decreased from 3.3 μg/kg body weight.

Mercury uptake greatly depends on the quantity of fish consumed and varies significantly between member states. According to data published by SCOOP (Scientific Co-operation on questions relating to food) the average rate of exposure to total mercury through nutrition is estimated to be 109 μg/kg food. In different member states the average day consumption of fishery products per person varies between 10 and 80 g (weekly consumption varies between 70 and 560 g). Hence, the weekly exposure to total mercury via nutrition varies from 7 to 61 μg/person, and in a person weighing 60 kg the weekly uptake varies from 0.1 to 1 μg. People and children consuming large quantities of fishery products may take methyl mercury near or exceeding the determined temporary percentage of weekly uptake.

Inclusion of determination of heavy metals (mercury, cadmium, and lead) in fresh and frozen fish in the official food control programs by member states is of outmost importance under the current circumstances.

EFSA published an opinion in 2004 regarding the danger to human health from the consumption of food that may contain mercury. It is estimated that the European consumers' uptake of methyl mercury is close to the international uptake level. EFSA suggested the realization of additional studies, mainly in particular groups such as pregnant women, or women due to be pregnant, women suckling, and children. Until completion of these studies, EFSA recommended that consumers and sensitive population groups should get informed of mercury. According to EFSA's opinion these groups should not consume more than a small portion (<100 g) of a big hunting fish per week, such as shark or swordfish. Along the same line, tuna fish should not be consumed more than twice a week.

9.4.2.4.3 Cadmium (Cd)

Cadmium poisoning is very frequent due to the increased solubility in organic acids. It is considered as one of the most hazardous micronutrients in foods and the environment due to its high toxicity (Vos et al., 1987). Levels of cadmium in food are normally very low if contamination has not occurred (Table 9.9). The total range varies between 0.095 and 0.987 mg/kg. Increased cadmium levels are mainly observed in meat and seafood. The daily consumption of cadmium has been estimated in many countries to be 10–80 µg/day (Dabeka et al., 1987). According to WHO, cadmium level in drinkable water is defined to be 10 µg/L. (WHO, 1963). The most reported food poisoning cases due to cadmium come from Japan and Australia (Rayment et al., 1989).

It has been estimated that under normal conditions approximately 6% of cadmium existing in consumed foods is absorbed by the human organism (Reilly, 1991). It has also been reported that high calcium and protein levels in the diet tend to considerably increase cadmium absorption. The highest percentage of cadmium that gets absorbed is withheld in the kidneys. Hence, long-term cadmium absorption results in serious kidney failure, as well as bone attack, leading to brittleness and possible breakdown of the skeleton (Frieberg et al., 1974). Cadmium toxicity constitutes the main cause of the disease itai-itai observed in some population groups in Japan (Rayment et al., 1989). High levels of cadmium in the diet are considered responsible for the increase of the rate of different cancers in humans (Browning, 1969). Cadmium toxicity is significantly inhibited by the presence of selenium, zinc, and cobalt.

TABLE 9.9
Cadmium Concentration in Selected Foods

Food	Cadmium Concentration (µg/kg)
Bread	<2–43
Potatoes	<2–51
Cabbage	<2–26
Apples	<2–19
Poultry	<2–69
Minced beef	<2–28
Kidney (sheep)	13–2000
Prawns	17–913
Seafood	50–3660
Drinkable water	<1–21 µg/L

Source: From Varzakas, T.H. et al., Food additives and contaminants, Chapter 13, in: Yildiz, F., ed., *Advances in Food Biochemistry,* CRC Press/Taylor & Francis Group, 2010, pp. 409–457.

9.4.2.4.4 Arsenic (As)

Arsenic is traditionally closely linked to homicides and suicides. Its toxicity has to do with the chemical form of this element. Hence, inorganic compounds of arsenic are more toxic, followed by organic compounds, and finally arsenic in gas form (Buck, 1978). In the past, products based on arsenic such as parasiticidals, insecticides, fungicides, wood preservatives, and other similar products have been widely used. Nowadays, all these products have been forbidden in many countries due to the proved toxicity of arsenic.

Arsenic currently exists in most foods because of its wide dispersion into the environment and its past use as a chemical in agriculture. With the exemption of seafood, it is usually present in foodstuffs at a concentration lower than 0.5 mg/kg. According to FAO/WHO the maximum permissible day consumption uptake level is prescribed at a concentration of 2 μg/kg body weight (CAC, 1984). Arsenic also exists in all drinkable waters at a concentration varying between 0 and 0.2 mg/L. The maximum concentration in drinkable water is 0.01 mg arsenic/L (Drinking water standards, 1962).

Arsenic with three or five valences contained in foods can be easily absorbed by the gastroenteric tube. Then it can be easily transported into all tissues and organs. It is mainly accumulated in the skin, hair, nails, and to a certain degree in the bones and the muscles. The total arsenic level in human organism has been estimated to be 14–20 mg (Schroeder and Balassa, 1966).

Arsenic in general is a poison with the five valences form less toxic than the three valences form. It can bind to organic sulfhydryl groups, so it inhibits the action of many enzymes, especially those involved in cellular metabolism and respiration (Reilly, 1991). Clinical symptoms are associated with dilation and increased permeability of capillary tubes, especially in the intestine. Chronic arsenic poisoning means appetite loss leading to weight loss, gastrointestinal disorders, peripheral neuritis, conjunctivitis, hyperkeratosis, and skin melanosis. Moreover, it is a cancer suspect agent (IARC, 1973).

9.4.2.4.5 Selenium (Se)

Despite being one of the essential micronutrients in human and animal diet, its uptake in particularly high concentrations often leads to the appearance of the toxic syndrome. Selenosis of farm animals has been widely reported in many parts of the world (Reilly, 1991).

Selenium presence in the human food chain is mainly affected by its levels in agricultural soils. Hence, its daily uptake varies depending heavily on the geographical area. The dose of 50–200 μg/day were proposed as normal uptake selenium levels, with the exception of infants and children under 6 years of age where it is suggested the 10–40 and 20–120 μg/day, respectively.

Selenium is present in foodstuffs in the form of selenocysteine and selenomethionine. Approximately 80% of the organic selenium present in foodstuffs seems to be absorbed by the organism. Absorption is greater in foods of plant origin compared to foods of animal origin (Young et al., 1982).

The safety limit between selenium, as an important micronutrient in human diet, and the appearance of toxic symptoms is quite small. The main symptoms from the uptake of an increased selenium quantity are dermatitis, vertigo, fragile nails, gastric disorders, hair loss, and garlic odor during breathing.

Selenium as the essential trace element has attracted attention for the last decades due to many evidences indicating that Se deficiency states may be related to a variety of degenerative diseases (Arthur et al., 2003). Dietary daily selenium intakes in many European countries are currently much lower than recommended ones (Rayman, 2004; Stabnikova et al., 2008).

Sprouted seeds possess the potential to stimulate the growth and acidifying activity of lactic acid bacteria. Simultaneous enrichment of the sprouted seeds deficient in selenium may create a multifunctional additive to sourdough fermentation with consequent supplementation for human diet. Bread being a staple diet seems to be a perfect carrier of this micronutrient.

Diowksz et al. (2014) reported that germinating seeds revealed high ability to accumulate selenium from watering solution, up to 188 mg Se/kg of dried sprouts in the case of lentil. The process of selenium accumulation strongly depended on the plant species resulting in total Se content in soy and lentil biomass twice as high as in rye or wheat sprouts. A high correlation ($R_2 = 0.97$) between selenium concentration in water and Se accumulation in biomass was observed. Raising of pH from 7.0 to 8.0 increased 2.5 times selenium uptake.

The addition of Se-enriched plant biomass has a stimulating effect on the sourdough fermentation process allowing to shorten it by 8–16 h, depending on the dose and temperature. Se-fortified rye sprouts used in amounts 2.5 g/100 g of flour did not deteriorate sensory quality of bread and raised five times the selenium content in it.

9.4.2.4.6 Antimonium (Sb)

Antimonium is toxic and constitutes a major ingredient of many foodstuffs. High levels of antimonium in foods arise from contamination due to containers glazed with enamel containing antimonium. These containers might be used for food storage or as cooking utensils.

Very little is known about the nutritional uptake of antimonium. The daily uptake is about 0.25–1.25 mg for American children (Murthy et al., 1971). The allowed permissible concentration of antimonium in drinkable water is 0.1 mg/L. Antimonium is stored mainly in the liver, kidneys, and skin. Poisoning symptoms include colic, nausea, weakness, collapse with slow or irregular respiration, and reduced body temperature.

9.4.2.4.7 Aluminum (Al)

Aluminum is widely used in many industrial applications. Aluminum compounds are used in the food industries both as additives and as different utensils (cooking utensils or storage containers). Moreover, they can be used in pharmaceutical and cosmetic industries. Its wide use is due to the low cost; it does not undergo oxidation; it is recyclable, flexible, and easy to handle; and it is manufactured from bauxite, which is abundant in nature.

The concentration of aluminum in different foodstuffs varied from 0.05 to 129 mg/kg with an average concentration of approximately 12.6 mg/kg (Pennington and Jones, 1988). With the exception of certain spices and tea leaves, the normal aluminum levels in foodstuffs are very low. Contamination could also occur from the use of aluminum containers and aluminum cans in the food industries. Water does not constitute a significant source of aluminum uptake.

The chemical form of aluminum affects aluminum absorption. Furthermore, parathyroid hormones, vitamin D, and iron seem to affect aluminum absorption (Reilly, 1991). In the human bloodstream aluminum is stored mainly in the liver, kidneys, spleen, bones, heart, and brain tissues.

9.4.2.4.8 Tin (Sn)

Tin is widely distributed in low quantities in most lands. It exists in lower than 1.0 mg/kg concentrations in all main food groups, except canned vegetables (9–80 mg/kg) and fruits (12–129 mg/kg) (Sherlock and Smart, 1984). However, these values today are considered too high since the industry uses polyacrylic resin as a coating agent. The use of tinned cans in the canning industry could be a source of contamination with tin.

Tin found in foodstuffs seems to be less absorbed by the human organism and is excreted mainly by the stool (WHO, 1973). Low quantities of the absorbed tin could be withheld by the kidneys, liver, and bones. High tin levels in foodstuffs could cause acute poisoning. The lethal toxic dose for humans is 5–7 mg/kg body weight. Chronic poisoning leads to delay of growth, anemia, and histopathologic lesions in liver. Moreover, tin affects iron absorption and the formation of hemoglobin (Reilly, 1991).

9.4.2.5 Medicinal Residues

The use of medicines as additives in animal nutrition has been approved since the 1950s. These nonnutritional additives contain hormones, antibiotics, sulfonamides, nitrofurans, and arsenic compounds. These additives have a very significant effect on the increase of yield of dairy products.

Antibiotics and other medicines are given to productive animals in therapeutic doses (200–1000 g medicine/tone food, 220–1100 mg/kg) for disease treatment, in preventative doses (100–400 g medicine/tone food, 110–440 mg/kg) for the prevention of diseases caused by bacteria or protozoa, and finally at doses higher than or equal to 200 g (2.2–220 mg/kg) medicine/tone food for 2 weeks or more with the aim to increase the food yield and the acceleration of animal growth (Moorman and Koenig, 1992; NAS, 1980).

Despite the fact that the mechanisms with which medicines enrich the animal food and accelerate their growth are hardly tangible, the following have been reported concerning their use (Franco et al., 1990):

- They suppress the responsible microorganisms for mild but nonrecognizable infections.
- They reduce the microbial production of toxins, which decelerate growth.
- Antimicrobial factors reduce the destruction of the essential nutrients from microbes in the gastroenteric or increase the biosynthesis of vitamins or other growth factors.
- They improve the absorption capacity and use of nutrients.

However, the use of these antimicrobial medicines could result in the presence of residues in foods of animal origin. Some representative examples of possible sources of contamination of dairy and meat products from medicinal residues are the extended use or the high dose of approved medicines, the frequent mammary gland therapy with antibiotics, the misuse of these medicines (e.g., the wrong mixing of contaminated milk with noncontaminated), the twofold increase of the dose, the use of dry period therapy in milking cows, the nongood application of the directions of use of these medicines (Booth and Harding, 1986; Jones and Semour, 1988; McEwen et al., 1991).

Medicinal residues in foods should be avoided for the following reasons (Brady and Katz, 1988):

- Some residues could cause temperamental reactions, in particular in sensitive groups of the population, which could be very serious.
- In general, the presence of medicinal residues exceeding the determined allowed levels is illegal.
- Some medicinal residues in liquid dairy products have the ability to intervene in starter cultures which are used in processed dairy products.
- The presence of residues shows that the food could be coming from an animal that had suffered a severe infection.
- The concern of the public regarding food safety, hygiene, and food quality has greatly increased nowadays.
- The most important argument is that the presence of residues leads to the formation of microorganisms resistant to medicines, human pathogens.

9.4.2.6 Germicides

The term germicides refers to a group of chemicals used at a world level in agricultural products with the aim to control, destruct, or inhibit weeds, insects, fungi, and other harmful plants or animals. Approximately 320 active ingredients of parasiticides are available in a thousand different combinations (Hotchkiss, 1992).

Out of the 500,000 estimated plant or animal species and microorganisms, the well-known harmful plants and animals are less than 1%. However, this small percentage is enough to cause great-scale economic destructions. Hence, it is reported that insects, pathogenic plants, and weeds destroy approximately 37% of the American agricultural production, with losses reaching 50%–60% in the developing world. The total cost for challenging harmful plants and animals is too high since it requires the use of huge amounts of germicides and becomes even higher if one estimates the indirect cost from the use of germicides associated with the destruction of beneficial organisms, the perturbation of ecological systems, and the appearance of infections and diseases in humans (Pimentel, 1991).

Germicides consist of herbicides, insecticides, and fungicides. There exists three main chemical groups of insecticides: organochlrorine compounds, organophosphorous, and carbamides. A fourth group constitutes the synthetic pyrethroids (synthetic chemicals associated with natural pyrethroids abundant in chrysanthemum). These compounds have a generally low toxicity in mammals, including humans, and most of them are biodegradable. The most common herbicides are organochlrorine compounds, whereas the most common fungicides are pentachlorophenol, cadmium chloride, and commercially Benomyl, Captan, Maneb, Thiram, Zineb, etc. Most of these germicides are based on genetic mutations and in heavy metals as their active ingredients.

Most germicides have a carcinogenic, teratogenic, and embryotoxic action at a high degree, while some of them affect even the central nervous system. Organochlorine germicides bioaccumulate on human tissues from liver enzymes instead of detoxifying and get excreted. The result is the production of epoxides and peroxides, which cause membrane damage and lead to free radicals formation. These free radicals interact with DNA acting as mutation material (Pryor, 1980). Moreover, organochlorine compounds restrict the transfer of inorganic material through the cell membranes and inhibit cell respiration. Organophosphoric compounds inhibit acetylcholinesterase, a key enzyme involved in the process of transfer of nervous impulses. These germicides are often particularly toxic and because of their volatile character, they have an immediate effect upon inhaling. They are more biodegradable compared to organochlorine compounds; however, they are still suspected of chronic toxic effects. Moreover, carbamides are poisons of the nervous system with varying toxicity.

When germicides are not used in the right way, they result in the presence of residues in food and may cause severe problems in consumers' health. Their use in agricultural products presents three relevant but distinct hazards that determine the possibility a failure or injury might occur as described below (Hotchkiss, 1992):

1. Environmental hazards, associated with unpleasant effects on nontarget organisms and contamination of ground water.
2. Occupational hazards associated with the farmers working in the fields who breathe the germicides. These are the most significant ones because human exposure to germicides is direct; hence, health damage is more probable.
3. The presence of residues in the foodstuffs.

Occupational hazards could be considerably reduced with strict controls and application of the required preventive technology. Environmental effects are associated with certain germicides and, more specifically, heavy metal compounds and organochlorine compounds. These are particularly toxic and resistant to biodegradation.

Over the last few years, public have begun to worry about the presence of germicide residues in foodstuffs. The discovery that certain germicides, considered safe some years ago, cause hazardous effects in human health has increased anxiety even more. This pushed the US National Research Council (NRC) to start in 1985 the study of methods to determine the residue limits of these germicides in foods. Hence, according to NRC (1987), a very important parameter was used to express the oncogenic strength of a germicide. This is expressed as Q* (tumors/mg germicide/kg body weight/day) and shows the possibility of a germicide to cause tumors in guinea pigs. Hence, a high value of Q* shows a strong oncogenic reaction (i.e., formation of additional tumors) at a given dose. These estimations for different germicides are average values derived from many positive oncogenic studies in animals. The nutritional oncogenic hazard is estimated by multiplying Q* with the exposure, for example, food consumption × germicide residue.

Finally, it should be mentioned that there are tables comprising the limits FAO/WHO has determined regarding the ADI and the maximum residue limits (MRLs).

9.4.2.7 Preservatives

9.4.2.7.1 Antimicrobial Substances

It is well known that the growth of microorganisms in foods is undesirable with the exception of the useful microbial fermentations in certain products, which are essential for the product to acquire its organoleptic characteristics and its final structure. The development of undesirable microorganisms in foods leads to the change of appearance, structure, consistency, taste, color, and nutritional value. Moreover, certain microorganisms are toxic for humans, causing infections or endotoxin formation following their growth in foods. Thus, it is required to inhibit the microorganisms' growth in foods with the aim of food preservation and enhancement and assurance of their quality.

Food preservation has been accomplished through the application of physical or chemical methods (Sofos and Busta, 1993). The natural methods of inhibition or destruction of microorganisms consist of the change of temperature through the application of high (e.g., pasteurization, sterilization) or low temperatures (refrigeration, freezing), the exposure to ionizing radiation and water removal (e.g., drying). The chemical methods of food preservation include the use of desirable microorganisms for fermentations or the direct use of chemical additives, which act as antimicrobial factors. Moreover, some chemical additives such as salt, phosphoric salts, and antioxidants sometimes exert a direct or indirect antimicrobial action. In this chapter, we will only refer to the chemicals used exclusively as antimicrobial substances in foods, pharmaceuticals, or other materials (Sofos and Busta, 1992).

Nowadays, the needs of the modern market as well as the requirements of the modern consumer have led to the wide use of chemical antimicrobial substances in foods. In general, the modern marketing system is largely based upon the use of antimicrobial substances.

There are limits for all used antimicrobial substances. Antimicrobial action depends on the following factors:

- The physical and chemical properties (e.g., solubility, toxicity)
- The types of microorganisms involved
- The type and the properties of the preserved product

The use of combinations of inhibitory factors is well known (Leistner, 1985) as the principle of microbial hurdles. Finally, it seems that pH affects the range of the antimicrobial action of a chemical antimicrobial substance.

The use of antimicrobial substances is regulated and approved by the competent authorities of each countrys while at an international level FAO and WHO deal with such matters. Some of these antimicrobial substances are characterized as generally recognized as safe (GRAS) by FDA, and are acquitted from the food additives regulations. These substances should be approved for a specific use. Moreover, they must obey good manufacturing practices and be properly labeled.

Burdock and Carabin (2004) reported on the history and description of GRAS, a system for review and approval of ingredients for addition to food. The GRAS approval process for a food ingredient relies on the judgment of "... experts qualified by scientific training and experience to evaluate its safety..." the end product of which is no better or worse than that by FDA but often more expeditious. The process and requirements for a successful GRAS determination are discussed and compared with that of the food additive petition (FAP) process. The future of the GRAS process is assured by its history of successful performance, bringing safe food ingredients to the consumer in a timely manner.

GRAS is probably more useful now than before in examining, for example, (1) macronutrients, which cannot be tested by conventional means (at more than 5% of the diet of test animals), (2) biotechnologically produced or novel foods and responding to the concept "as safe as" conventional food, (3) high doses of nutrients (e.g., vitamins or specific foods such as meat in "Atkins-type" diets) for which no animal models exist.

The most important approved chemicals acting as antimicrobial substances and food preservatives are described below (Table 9.10).

TABLE 9.10
Preservatives in Food; Name, Origin, Functionality, Adverse Effects, ADI, and Applications

E	Name	Origin	Functionality	Effects	ADI	Product Uses
E200	Sorbic acid	Natural from fruits or synthetic from ketene	Preservative	Skin irritant on direct contact	0–25	Yoghurt, soft drinks, sweets, wine
E210	Benzoic acid	Natural substance, synthetically manufactured	Preservative	Allergic reactions. Not recommended for hyperactive children	0–5	Jams, beer, fruit juices, yoghurts, soft drinks
E211	Sodium benzoate	From benzoic acid	Preservative	Allergic reactions	0–5	Caviar, sweets, sauces, (soya, barbeque) soft drinks
E212	Potassium benzoate	From benzoic acid	Preservative	Allergic reactions	0–5	Margarine, concentrated fruit juices
E213	Calcium benzoate	From benzoic acid	Preservative	Allergic reactions	0–5	Fruit condensed juices
E214	ρ-Hydroxybenzoic ethyl	From benzoic acid	Preservative	Allergic reactions. Similar to E215, E216, E217, E218, E219	0–10	Beer, fruit juices, aromatic substances, syrups, yams
E216	ρ-Hydroxybenzoic propyl	From benzoic acid	Preservative	Similar to E214		Similar to E214
E218	ρ-Hydroxybenzoic methyl	Synthetic	Preservative	Causes allergic and skin reactions	0–10	Cooked beets, sauces, chocolate fillings
E220	Sulfur dioxide	Chemically produced	Stabilizer, preservative, aromatic compound	Destroys vitamin B, possible gastric problems	0–0.7	Fruit salads, dry fruits, beer, gelatin, fruit juices, sausages
E221	Sodium thiosulfate	From sulfurous acid	Equipment sterilization, antioxidant	Allergic reactions. Not recommended for those having problems with liver	0–0.7	Frozen prawns, potatoes, beer, wine
E222	Sodium hydrogen sulfite	From sulfurous acid	Preservative	Allergic reactions in patients suffering from asthma	0–0.7	Beer, wine, milk, dairy products, fruit juices
E223	Sodium bisulfite	Sulfurous acid salt	Antioxidant, preservative	Allergic and skin reactions	0–0.7	Frozen prawns, potatoes, alcoholic drinks, pickles, orange juice
E224	Potassium bisulfite	Synthetic	Preservative	Allergic reactions	0–0.7	Frozen potatoes, seafood, wines
E226	Calcium sulfite	Synthetic	Preservative	Allergic reactions	—	cider, fruit juices
E227	Calcium hydrogen sulfite	Synthetic	Preservative	Allergic reactions	—	Beer, jams, gel
E230	Diphenyl	Synthetic, produced from benzene	Preservative	Exposure to diphenyl causes nausea, vomiting, eye and nose irritations	0–0.5	Processing of fruit skin
E236	Formic acid	Synthetically produced	Preservative	No consequences Forbidden in the United Kingdom	0–3	No technological applications

(Continued)

TABLE 9.10 (*Continued*)
Preservatives in Food; Name, Origin, Functionality, Adverse Effects, ADI, and Applications

E	Name	Origin	Functionality	Effects	ADI	Product Uses
E239	Hexamethylene-tetramine	Derived from formaldehyde and ammonia	Preservative	Gastroenteric perturbation. Possible carcinogenesis	0–0.15	Provolone cheese, marinated foods, red herrings
E249	Potassium nitrite	Synthetic salt	Meat preservative	It enters the bloodstream and affects hemoglobin. Headaches, methemoglobinemia. If it reacts with amines it gives carcinogenic nitrosamines.	0–0.2	Cooked meat, sausages, smoked fish
E250	Sodium nitrite	Synthetic salt	Preservative	Similar to E249	0–0.2	Meat products, sausages, bacon, frozen pizza
E251	Sodium nitrate	Natural salt	Preservative, color stabilizer	Nitric acid could be converted into nitrates. Similar to E249	0–5	Bacon, ham, cheeses, frozen pizza
E252	Potassium nitrate	Natural salt or synthetic	Preservative, color stabilizer	Anemia, kidney infection. Similar to E249	—	Sausages, cooked pork meats, dutch cheeses, canned meat
E261	Potassium acetate	Synthetic	Natural colors preservative	Low toxicity, gets excreted in urine	—	Provolone cheese, marinated foods, red herrings
E270	Lactic acid	Natural substance	Preservative	Gastroenteric disorders in infants	—	Margarine, carbonated drinks, infant milk
E281	Sodium propionate	Natural propionic acid salt	Preservative	Possible correlation with migraines	—	Processed cheese, bakery and dairy products
E282	Calcium propionate	Propionic acid salt	Preservative	Skin irritant in pure form in bakery workers	—	Processed cheese, dairy and baking products, frozen pizza
E290	Carbon dioxide	Natural gas	Preservative, packaging gas	In the stomach, it increases gastric juice secretions	—	Juices, soft drinks

Source: From Varzakas, T.H. et al., Food additives and contaminants, Chapter 13, in: Yildiz, F., ed., *Advances in Food Biochemistry*, CRC Press/Taylor & Francis Group, 2010, pp. 409–457.

9.4.2.7.2 Sorbic Acid and its Salts

Sorbic acid is a compound existing both in nature and produced synthetically. Nowadays, it is widely used due to its low cost. Sorbic acid and its salts constitute very effective antimicrobial substances against many yeasts, molds, and bacteria. Certain microorganisms such as lactic acid bacteria as well as specific yeasts and molds are resistant to the inhibition by sorbic acid. The activity of sorbic acid against microorganisms is an operation derived from synergic or antagonistic effects with product composition, pH, water activity, chemical additives, storage temperature, microbial flora, atmospheric gases, and packaging (Sofos and Busta, 1981). A significant factor affecting the antimicrobial action of sorbic acid is the pH of the substrate. The maximum pH for activity is 6.5 whereas measureable inhibition of microbial growth has been reported even at pH 7.0. Sorbic acid concentrations (<0.3%) can only inhibit and not inactivate microorganisms in foods. Higher concentrations can inactivate microorganisms; however, they cannot be used or allowed in foods due to their unpleasant effect on product taste. Finally, it is well known that sorbic acid is one of the less harmful preservatives used in foods.

9.4.2.7.3 Benzoic Acid and its Salts

Benzoic acid can be widely used as preservatives in the food, pharmaceutical, and cosmetics industries. They possess very powerful antimicrobial and antioxidant properties. In the food industries, they are widely used as preservatives in foods with pH less than 4.5 due to the low cost and the easy incorporation into products. The narrow pH range in which they act, their undesirable taste and their toxicological profile, which is less desirable than other antimicrobial substances, constitute limitations in their use (Chipley, 1993). Preservatives concentrations usually vary from 0.05 up to 0.1%. However, either lower concentrations are employed, or they can be used in combination with other antimicrobial substances when taste problems are evident (Jermini and Schmidt-Lorenz, 1987). As antimicrobial substances, benzoic acid, and its salts are more effective against yeasts and bacteria compared to molds (Sofos and Busta, 1992). Their antimicrobial action varies depending on the food, its pH and water activity, as well as the type and microorganism species. Some yeasts, which destroy food of medium humidity, seem to be resistant in the inhibition caused by benzoic acid and its salts (Warth, 1977, 1985, 1988, 1989). Moreover, there are certain microorganisms capable of metabolizing benzoic acid. Reports on its use as an antimicrobial substance showed that these compounds do not cause unpleasant effects on human health (Chipley, 1993). It is generally considered as a safe food preservative.

9.4.2.7.4 Popionic Acid and its Salts

They exist naturally in Swedish cheeses at a percentage of 1%, derived from Propionibacterium bacteria, which are involved in ripening of these cheeses. Besides the mould inhibition in cheeses, they can be used as preservatives in baked products, where they cause inhibition of fungi as well as some bacterial species. Moreover, they could be used as antimicrobial factors at levels between 0.1% and 0.38%. Their antimicrobial action depends on the type of microbe and the product pH (highest activity at pH = 6.0). Propionic acid and its salts can be categorized as GRAS when used in concentrations that do not surpass the normal amount required for the accomplishment of the desired effect.

9.4.2.7.5 Esters of Parahydroxybenzoic Acid

They are used as antimicrobial substances in cosmetics, medicines, and in foods. Compared to benzoic acid their solubility in water is higher and decreases with the increase of the number of carbon atoms. These esters are stable on the air and resistant to cold and heat and steam sterilization. They are often used in concentrations varying between 0.05% and 0.1%. They exert antimicrobial action against yeasts, molds, and bacteria. The microbial inhibition increases when the carbon chain of the ester is not branched. The use of these compounds is favored in high pH products where other antimicrobial factors are ineffective. They could be also effectively used in conjunction with benzoic acid depending on the cost, pH, and taste, and more particularly in products with slightly acidic pH (Sofos and Busta, 1992).

Methyl paraben (CAS No. 99-76-3) is a methyl ester of *p*-hydroxybenzoic acid. It is a stable, nonvolatile compound used as an antimicrobial preservative in foods, drugs, and cosmetics for over 50 years. Methyl paraben is readily and completely absorbed through the skin and from the gastrointestinal tract. It is hydrolyzed to *p*-hydroxybenzoic acid, conjugated, and the conjugates are rapidly excreted in the urine. There is no solid evidence of accumulation. Acute toxicity studies in animals indicate that methyl paraben is practically nontoxic by both oral and parenteral routes. In a population with normal skin, methyl paraben is practically nonirritating and nonsensitizing. In chronic administration studies, NOEL as high as 1050 mg/kg have been reported and a NOAEL in the rat of 5700 mg/kg is posited. Methyl paraben is not carcinogenic or mutagenic. It is not teratogenic or embryotoxic and is negative in the uterotrophic assay. The mechanism of cytotoxic action of parabens may be linked to mitochondrial failure dependent on induction of membrane permeability transition accompanied by the mitochondrial depolarization and depletion of cellular ATP through uncoupling of oxidative phosphorylation.

Parabens are reported to cause contact dermatitis reactions in some individuals on cutaneous exposure. Parabens were implicated in numerous cases of contact sensitivity associated with cutaneous exposure, but the mechanism of this sensitivity is still unknown (Soni et al., 2002). Sensitization occurred when medications containing parabens were applied to damaged or broken skin. Allergic reactions to ingested parabens were reported, although rigorous evidence of the allergenicity of ingested paraben is lacking.

9.4.2.7.6 Sulfur Dioxide and Salts of Sulfuric Acid

Sulfuric acid is one of the most traditional antimicrobial substances, widely employed in products such as wines (Banks et al., 1987). It is soluble in water and gives sulfurous acid and its ions when it comes into contact with water in foods. The ion's percentage increases with the reduction of pH value. Sulfuric acid salts can be used as preservatives and are more useful since they are available in dry forms. The antimicrobial action against yeasts and molds and bacteria is selective, with certain species to be more sensitive in inhibition compared to other species (Sofos and Busta, 1992). Bacteria are generally more sensitive in inhibition (Chichester and Tanner, 1972). Sulfur dioxide and sulfuric acid salts are GRAS substances; however, their level of use in wines is restricted to 0.035%. Its presence at higher levels lead to undesired taste and is generally forbidden in foodstuffs, which are considered sources of thiamine due to the fact that they inactivate it (Daniel, 1985; Walker, 1988; Ough, 1993).

9.4.2.7.7 Carbon Dioxide

It is a gas being solidified at −78.5°C; thus, forming dry ice. It is used in the maintenance of carbonated drinks, vegetables, fruits, meat, fish, and wines. The dry ice form can be used to store or transport products at low temperatures. As an ingredient of modified temperatures, that is used to store fruits and vegetables, it delays the respiration and ripening and inhibits the growth of yeasts and molds. It can also be used in carbonated drinks as a fermentation factor and as an inhibitor of microbial growth. It inhibits oxidation changes in beer. The antimicrobial activity depends on its concentration, the type of microorganisms, the water activity, and storage temperature. Since these factors vary, carbon dioxide could not exert any action, stimulate growth, or inhibit growth, or be lethal for microorganisms (Clark and Takacs, 1980; Davidson et al., 1983; Enfors and Molin, 1978; Foegeding and Busta, 1983). From the toxicological point of view, it is well known that respiration of 30%–60% carbon dioxide in the presence of 20% oxygen could cause death in animals, whereas man's exposure in atmospheres containing more than 10% could lead to loss of senses. Moreover, respiration of smaller quantities for an extended period of time could be deleterious.

9.4.2.7.8 Epoxides

Ethylene and propylene oxides are cyclic ethers with only one oxygen atom connected with two neighboring carbon atoms of the same chain (Davidson et al., 1983). Ethylene oxide gas could go through most of the organic materials without causing any damage, which makes it very useful for sterilization of materials sensitive to temperature. Epoxides are drastic against yeasts, mold, and insects, but less drastic against bacteria. The antimicrobial action of the two epoxides depends on the humidity and their ease in the penetrability of organic materials.

9.4.2.7.9 Hydrogen Peroxide

It is an oxidative factor as well as discoloration factor with antimicrobial properties. It is an unstable compound and dissociates easily to form water and oxygen. The dissociation and the antimicrobial activity of this compound increase at higher temperatures. The antimicrobial action arises from its oxidative properties and depends on the concentration, pH, temperature, and exposure time (El-Gendy et al., 1980; Smith and Brown, 1980). Besides its use as an additive, hydrogen peroxide is produced by lactobacilli in foods. It is permitted to be used in cheese manufacturing, whey powder processing, and in other applications (Cords and Dychdala, 1993; Stevenson and Shafer, 1983).

9.4.2.7.10 Antioxidants

Antioxidants are substances used in food preservation together with antimicrobial substances (Table 9.11). It is important to realize that besides microbial destruction oxidative degradation of polyunsaturated fatty acids contributes significantly at the time of food preservation. Fatty acid oxidation is a complex process of chemical and biological reactions leading to the formation of a large number of products (Figure 9.1). Fat oxidation results in changes in taste and aroma of foods with a high concentration in oils and fats, changes in their structure due to the reaction of products derived from oxidation of fats with proteins, and loss of their nutritional value due to the destruction of vitamins, amino acids, and essential fatty acids (Dziezak, 1986). Furthermore, fat oxidation products are directly related to the development of a number of diseases such as arteriosclerosis, coronary heart disease, and cancer, as well as the process of ageing of cells.

TABLE 9.11
Antioxidants in Food; Name, Origin, Functionality, Adverse Effects, ADI, and Applications

E	Name	Origin	Functionality	Effects	ADI.	Product Uses
E302	L-Ascorbic calcium	Synthetic salt	Preservative, antioxidant	Low possibility for stone formation in the kidneys	—	Madrilene
E310	Gaelic propyl ester	Synthetic substance	Antioxidant	Gastric and skin irritations. Not allowed in baby foods	0–0.5	Vegetable oils, margarine, snacks, gums
E311	Gaelic oktylester	Similar to E310	Similar to E310	Similar to E310	0–0.5	Similar to E310
E312	Gaelic dodecyl ester	Similar to E310	Similar to E310	Similar to E310	0–0.5	Similar to E310
E320	Butylichydroxyanisol (BHA)	Synthetic, derived from ρ-methoxyphenol and isobutane	Antioxidant	Not allowed in baby foods. Forbidden in Japan	0–5	Biscuits, sweets, sauces, chips, soft drinks, margarine
E321	Butylichydroxytoluene (BHT)	Synthetic, derived from cresol and isobutylene	Antioxidant	In pigs, 1 g causes death in 2 weeks. Not allowed in baby foods	0–0.5	Gums, chips, margarine, peanuts, sauce cubes, mash
E325	Lactic sodium	Lactic acid salt	Humectant. It maintains pH levels.	None	—	Cheese, gel, jam, ice creams, margarine
E326	Lactic potassium	Similar to E311	Similar to E311	Similar to E311	0–0.5	Similar to E325

Source: From Varzakas, T.H. et al., Food additives and contaminants, Chapter 13, in: Yildiz, F., ed., *Advances in Food Biochemistry*, CRC Press/Taylor & Francis Group, 2010, pp. 409–457.

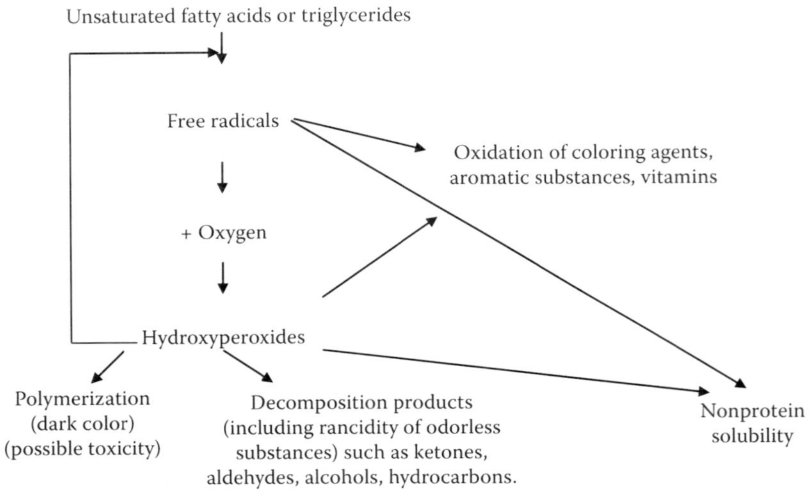

FIGURE 9.1 Mechanisms of fatty acid oxidation. (From Varzakas, T.H. et al., Food additives and contaminants, Chapter 13, in: Yildiz, F., ed., *Advances in Food Biochemistry*, CRC Press/Taylor & Francis Group, 2010, pp. 409–457.)

Antioxidants play an important role in the deceleration of lipid oxidation reactions in foodstuffs. According to FDA they are defined as "substances used as preservatives, with the aim to reduce spoilage, rancidity, or food discoloration, which are derived from oxidations." Antioxidants addition in foodstuffs is either intentional (direct addition into product) or symptomatic (migration of antioxidants from packaging material into product). The right and effective use of antioxidants depends on the understanding of (1) the chemistry of oils and fats and (2) the mechanism of oxidation and their operation as substances, which react in food oxidation (Stuckey, 1972).

The approval of an antioxidant for food use requires extended toxicological studies, which consist of studies for possible transformation, carcinogenic, and teratogenic effects. Finally, there are some antioxidant substances that have been recognized as safe (GRAS).

Twelve food additives and six secondary plant products were analyzed with regard to their antioxidant activity by using three different test systems (Trolox equivalent antioxidant capacity, photochemiluminescence, ferric reducing antioxidant power) as reported by Kranl et al. (2005). The results differed depending on the assay. All the food additives showed antioxidant activities comparable to the calibration substance Trolox. In contrast, the secondary plant products had an up to 16 times higher antioxidant potential. This might be a good reason for the food industry to opt for natural antioxidants instead of synthetic ones in order to get storage stability for processed food items—which, according to recent surveys, is in the interest of consumers.

Lipid peroxidation is a reaction occurring in food products which leads to off-flavor and other quality losses (e.g., changes in color and texture). To stabilize products, the food industry uses food additives endowed with antioxidant activity. Typical antioxidants are butylated hydroxytoluene (BHT) and butylated hydroxyanisole (BHA) as well as ascorbic acid and their derivatives. During the last decade, natural antioxidants have been gaining ground at the expense of synthetic ones. Secondary plant products, widely known for their health promoting effects, have then been evaluated for their use as food ingredients with stabilizing effects (Britt et al., 1998).

An overview of most of the analytical methods existing to determine the antioxidant activity was compiled by Bohm and Schlesier (2004).

The three recent EU directives which fixed MPL for food additives for all member states (EC, 1995) also include the general obligation to establish national systems for monitoring the intake of these substances in order to evaluate their safe use. Leclercq et al. (2000) considered additives with primary antioxidant technological function or for which an ADI was established by the SCF: gallates, BHA, BHT, and erythorbic acid. The potential intake of these additives in Italy was estimated by means of a hierarchical approach using either step-by-step or more refined methods. The likelihood of the current ADI to be exceeded was very low for erythorbic acid, BHA, and gallates. On the other hand, the theoretical maximum daily intake (TMDI) of BHT was above the current ADI. The three food categories found to be main potential sources of BHT were "pastry, cake and biscuits," "chewing gums," and "vegetables oils and margarine." They overall contributed 74% of the TMDI. Actual use of BHT in these food categories is discussed together with other aspects such as losses of this substance in the technological process and percentage of ingestion in the case of chewing gums.

Tert-butylhydroquinone (TBHQ) is an approved food-grade antioxidant, which has been used as an effective preservative for unsaturated vegetable oils, numerous edible animal fats, and meat products at concentrations less than 0.02% (Kashanian and Dolatabadi, 2009). It does not change the flavor or odor of the material and does not lead in discoloration, even in the presence of iron (Okubo et al., 2003).

TBHQ was tested for potential cytotoxicity and genotoxicity upon A549 lung cancer cells and human umbilical vein endothelial cells (HUVEC) by Eskandani et al. (2014). Cytotoxicity was evaluated by MTT assay and flow cytometry analysis, while genotoxicity was assessed *in vitro* by alkaline comet, DNA fragmentation, and DAPI staining assays. Dose- and time-dependent TBHQ decreased the growth of A549 and HUVEC cells. Flow cytometry analyses determined early and late apoptosis in the treated cells. Also, single strand DNA breaking has been observed through comet assay technique. In addition, morphology of DAPI stained cells and DNA fragmentation assay using gel electrophoresis showed clear fragmentation in the chromatin and DNA rings within the nucleus of cell's treated TBHQ.

Recently, the protective effect of the TBHQ on nephrotoxicity in rats and traumatic brain injury-induced brain oedema and cortical apoptosis in mice had been reported (Jin et al., 2011; Perez-Rojas et al., 2011).

Controversially, the effects of TBHQ on mutagenesis and carcinogenesis can be either enhancing or suppressing, and the antioxidative and cytoprotective properties of TBHQ can be changed into prooxidative, cytotoxic, and genotoxic properties (Nagai et al., 1996; Okubo et al., 1997; Perez-Rojas et al., 2011).

9.4.2.7.11 Antibiotics in Animal Feed

Animals and humans constitute overlapping reservoirs of resistance, and consequently use of antimicrobials in animals can have an impact on public health. For example, the occurrence of vancomycin-resistant enterococci in food-animals is associated with the use of avoparcin, a glycopeptide antibiotic used as a feed additive for the growth promotion of animals. Vancomycin-resistant enterococci and vancomycin resistance determinants can therefore spread from animals to humans. The bans on avoparcin and other antibiotics as growth promoters in the EU have provided scientists with a unique opportunity to investigate the effects of the withdrawal of a major antimicrobial selective pressure on the occurrence and spread of antimicrobial resistance. Data shows that although the levels of resistance in animals and food, and consequently in humans, were markedly reduced after the termination of use, the effects on animal health and productivity were very minor (Wegener, 2003).

Food animal production in intensive systems and, within those, various types of chemical additives included in the compound diets fed has received much adverse attention from the media, consumers, politicians, legislators, and advisory groups in recent years according to Pugh (2002). No additive class received more adverse comments than the antibiotics used for the purpose of enhancing the efficiency of animal production. Pugh (2002) considered the safety of the antibiotic feed additives (AFAs) against the background of the regulatory measures in place, defined their role and described the relevant concerns. Regarding the health risk which underpinned the AFA bans, he made a warning about the precedent created by the use of the precautionary principle in the recent banning of six of their number.

McEvoy (2002) reviewed the legislative framework controlling the use of veterinary medicines and zootechnical food additives in the EU. From a contamination perspective, "problem" compounds include sulfonamides, tetracyclines, nitroimidazoles, nitrofurans, ionophore coccidiostats, and nicarbazin. The literature on each of these was reviewed, and examples of interventions to minimize contamination were given. Examples of contaminants include naturally occurring and synthetic toxic environmental compounds (e.g., mycotoxins and dioxins) which may contaminate feed raw materials. Zootechnical feed additives and veterinary medicines may also contaminate unmedicated feeding stuffs due to carry over during feed production. Contaminated feed can cause deleterious health effects in the animals and, through "secondary exposure" of consumers to products deriving from these animals, may be harmful to people.

9.4.2.8 Packaging Materials

Materials widely used in food packaging consist of paper, cellulose products, cellophane, and metals (aluminum, tin, stainless steel, etc.), glass, plastic, and various materials such as wood, textiles. Most of these materials have been used for years and have caused minor problems. On the contrary, plastics bring into contact with food a large number of chemical ingredients, which have not been used before by the food industry; however, they offer new advantages as new packaging materials. Migration of these substances, being used in plastics processing, into food causes the biggest worry and concern regarding food safety problems and dangers.

All plastics, besides the basic polymer derived from the oil industry, consist of a number of other substances either added intentionally during their preparation or processing or found inside them unavoidably as residues from polymerization reactions. Polymers are high molecular weight compounds, inert and restricted solubility in aqueous and oily systems; hence, it is improbable to migrate into food at a high concentration (Crosby, 1981). Even if packaging fragments get consumed by mistake, they will not react with body fluids present in the gastroenteric system. The concern on the safe use of plastics as packaging materials mainly arises from the possible toxicity of other lower molecular weight compounds which may appear in the final product and get filtered into the food during its storage.

The two main reasons of food adulteration from plastic packaging materials are the following:

- Polymerization residues, which consist of monomers, oligomers (with molecular weight higher than 200), catalysts (mainly metallic salts and organic peroxides), solvents, emulsifiers and water proof materials, impurities, infectious plant materials, inhibitors, and degradation products
- The various auxiliary processing materials, such as antioxidants, plasticizers, light and heat stabilizers, coloring agents, lubricants, fungicides, etc.

The more volatile the monomers, the lower their concentration with time, however, low levels of these monomers could continue to be found in the final product.[105,106] Styrene and acrylonitrile residues are generally removed with higher difficulties.

9.4.2.8.1 Antimicrobial Edible Coatings

The development of antimicrobial edible coatings as a cost effective means to increase the shelf life of fresh-cut fruit has been investigated and its potential demonstrated (Brasil et al., 2012; Franssen and Krochta, 2003; Gutierrez, Barry-Ryan and Bourke, 2008; Mantilla et al., 2013; Sipahi et al., 2013; Vargas et al., 2009). Any material used for wrapping foods to extend the shelf life of the product that may be eaten together with food, with or without further removal, is considered an edible film or coating (Pavlath and Orts, 2009). Alginate is among the polysaccharides that have been successfully used to coat fresh-cut fruits (Sadili-Bico et al., 2010).

Coatings based on sodium alginate, pectins, and gellan gum have been shown to be effective, not only in retarding water loss but also in incorporating different active agents such as probiotic microorganisms or natural antioxidants and antimicrobials (Soliva-Fortuny, 2010).

Martinon et al. (2014) developed a multilayered edible coating with antimicrobial agent to extend the shelf life of fresh-cut cantaloupe stored at 4°C. Three different sets of experiments were designed to test the effect of different concentrations of chitosan (0.5, 1, 2 g/100 g), pectin (0.5, 1, 2 g/100 g), and encapsulated trans-cinnamaldehyde (1, 2, 3 g/100 g) on the quality of fresh-cut cantaloupe.

The first set (chitosan concentrations) provided the optimum concentration of chitosan based on preferences by the panelists. The second set (pectin concentrations) produced an acceptable coating that maintained shelf life. The third set (antimicrobial concentrations) helped establish the optimum concentration of trans-cinnamaldehyde in the coating. Changes in fruits texture, color, moisture, acidity, and pH were monitored. Sensory testing was carried out to support the objective quality data, and microbiological analysis helped verify the antimicrobial effectiveness of trans-cinnamaldehyde. Uncoated fresh-cut cantaloupes stored at 4°C served as controls. To test for the antimicrobial activity of chitosan alone, a second set of controls consisted of samples coated only with chitosan.

Application of a multilayered edible coating composed of 2 g/100 g trans-cinnamaldehyde, 2 g/100 g chitosan, and 1 g/100 g pectin helps extend the shelf life of fresh-cut cantaloupe (up to 9 days).

The performances of edible coatings depend on not only the coating methods employed but also the properties of the coating materials (type, amount, density, viscosity, and surface tension). Many natural materials have the potential to make well performing edible coatings, including proteins, polysaccharides, and lipids (Al-Hassan and Norziah, 2012). Among these natural biopolymers, chitosan, sodium alginate, and soy protein isolate are the three most promising coating materials (Janjarasskul and Krochta, 2010).

The applications of these three materials as edible coating on various food products have been widely discussed (Elsabee and Abdou, 2013; Pizato et al., 2013; Ramos et al., 2012).

Spraying is another widely used method for applying coatings. This technique offers uniform coating, thickness control, and the possibility of successive applications which does not contaminate the coating solution (Andrade et al., 2012). It has been reported that bovine gelatin is successfully spray-coated onto fresh meat including beef tenderloins, pork loins, salmon fillets, and chicken breasts to improve their storage qualities (Antoniewski et al., 2007).

Nam et al. (2011) indicated that electrostatic spraying of ascorbic acid at 500 mg/kg could efficiently prevent both lipid oxidation and color change in ground beef. Ganesh et al. (2012) pointed out that electrostatic spray of food-grade acids and plant extracts is more effective compared with conventional spray in decontaminating *Escherichia coli* O157:H7 on spinach and iceberg lettuce.

The performance of edible coating is influenced by the properties of coating materials and execution methods. Zhong et al. (2014) studied three different coating materials (chitosan, sodium alginate, and soy protein isolate) and four different coating application methods (dipping, enrobing, spraying, and electrostatic spraying) and tested their performance for coating mozzarella cheese. The properties of coating solutions, morphology, and basic quality changes of the cheese during storage at 4°C were evaluated. Results showed that sodium alginate solution was the most viscous ($\eta = 0.155$ Pa s) and had small contact angle on hydrophobic substrate surface indicating its better spreadability on cheese. Film thickness displayed obvious differences based on the coating methods (ranging from 30.6 to 83.3 mm), with two spraying methods leading to thinner coatings. Sodium alginate-coated cheese possessed the best overall physicochemical properties during storage, whereas the preservation effects were not significantly different among four coating methods. This study provided valuable new information about the effective coating application methods for different coating materials.

9.4.2.8.2 Bisphenol and Food Packaging

Bisphenol A (BPA) is an indirect food additive which is used as a material for the production of phenol resins, polyacrylates, and polyesters but mainly for the production of epoxy resins and polycarbonate plastics. Epoxy resins are used in engineering applications, in paints and adhesives, and in a variety of protective coatings in metal cans for foods, bottle tops, and water supply pipes. The polycarbonate plastics are used to make a variety of common products including baby bottles and water bottles, medical and dental devices, eyeglass lenses, and household electronics because they have high impact strength, hardness, toughness, transparency, and resistance to many acids and oils

(Pulgar et al., 2000). Unfortunately, BPA has been known to be leached from these plastics, especially those that are cleaned with harsh detergents or those that contain acidic or high-temperature liquids. The excessive level of BPA leads to coronary heart disease and diabetes (Vandenberg et al., 2010). The United States Environmental Protection Agency has set a maximum acceptable level of BPA as 0.05 mg/kg body weight/day (Alonso-Magdalena et al., 2010).

A rapid, environmental friendly, and sensitive sensor for the detection of BPA was developed at glassy carbon electrode (GCE) modified with Tannic acid functionalized N-doped graphene (TA/N-G) immobilized by Nafion. Compared with other sensors, the proposed sensor greatly enhanced the response signal of BPA due to the active surface area of N-G and high absorption efficiency of TA. Under optimal conditions, the oxidation current increased linearly with increasing the concentration of BPA in the range of 0.05–13 mM with the detection limit of 4.0 nM. The fabricated electrode showed good reproducibility, stability, and anti-interference. The developed electrochemical sensor was successfully applied to determine BPA in food package (Jiao et al., 2014).

9.4.2.8.3 Additives and Low-Fat Meat Sausages

Manufacture of low-fat sausages is usually achieved by two basic principles: the use of lean meats (which increases cost) and/or reduction of fat and caloric content by adding water ingredients that introduce less or no calories.

In emulsified meat products, diminishing fat without increasing water content results in a harder product (Hand et al., 1987), while fat replacement with water increases exudative and cooking losses (Su et al., 2008) and also affects texture and juiciness of the product (Cierach et al., 2009). However, Ordoñez et al. (2001) indicated that low-fat (10%) frankfurters with a texture profile similar to standard ones could be manufactured.

In order to reduce these problems, different strategies had been applied such as: add proteins (soy proteins, whey protein concentrate, gluten, sodium caseinate), gums (xanthan, locust bean gum, carrageenans, microcellulose, pectins, konjac), and/or sodium polyphosphate in different combinations (Ayadi et al., 2009; Brewer, 2012; Candogan and Kolsarici, 2003; Hsu and Sun, 2006; Jiménez-Colmenero et al., 2012; Lurueña-Martínez et al., 2004; Muchenje et al., 2009; Pietrasik, 2003; Youssef and Barbut, 2010, 2011).

In a previous work, Marchetti et al., (2013) studied low-fat meat emulsions with pre-emulsified deodorized fish oil as a source of unsaturated fatty acids.

The addition of 1 g/100 g of several binders (milk proteins, whey protein concentrate (WPC), thermally treated WPC, ovoalbumin, HPMC, methylcellulose, mixtures of κ:ι carrageenans or xanthan-locust bean gums (XLBG)) to meat emulsions containing 5 g of fish oil/100 g, 25 g of water/100 g and 1.4 g NaCl/100 g, were compared to control formulations with 5 g/100 g of beef tallow or fish oil without additives.

Response surface methodology was used by Marchetti et al. (2014) to analyze the effect of milk proteins and 2:1 κ:ι-carrageenans on cooking loss (CL), weight lost by centrifugation (WLC) and texture attributes of low-fat meat sausages with pre-emulsified fish oil. A central-composite design was used to develop models for the objective responses. Changes in carrageenans affected the responses more than the milk proteins levels. Convenience functions were calculated for CL, WLC, hardness, and springiness of the product. Responses were optimized simultaneously minimizing CL and WLC; ranges for hardness and springiness corresponded to commercial products (20 g of pork fat/100 g). The optimum corresponded to 0.593 g of carrageenans/100 g and 0.320 g of milk proteins and its total lipid content was 6.3 g/100 g. This formulation was prepared and evaluated showing a good agreement between predicted and experimental responses. These additives could produce low-fat meat sausages with pre-emulsified fish oil with good nutritional quality and similar characteristics than traditional ones.

9.4.2.8.4 Hydrocolloids

Nowadays, there is a great interest for naturally derived hydrocolloids with low cost and proper functionality. *Lepidium sativum* seed gum is one which could meet these demands. The *L. sativum*, also known as cress seed, is an annual herb from Cruciferae family growing in Middle East countries, Europe, and United States. It is a culinary herb and has some health promoting properties. Cress seeds have been used in traditional medicine for a long time to treat asthma, hypertension, hepatotoxicity, hyperglycemia, enuresis, and fractures (Ghante et al., 2011; Gokavi et al., 2004).

A number of studies have been carried out to quantify the intrinsic viscosity, and some other studies have been devoted to elucidate the effects of additives such as salts and sugars on the intrinsic viscosity of a broad range of hydrocolloids (Gómez-díaz and Navaza, 2003; Higiro et al., 2006, 2007; Ma and Pawlik, 2007; Mohammad Amini and Razavi, 2012; Nickerson et al., 2004; Razavi et al., 2012; Samavati et al., 2007).

Intrinsic viscosity of the hydrocolloids in different solvent/cosolute systems provides deep insight on the fundamental molecular properties of biopolymers in food systems. In the research by Behrouzian et al. (2014), the influence of some salts and sugars were investigated on intrinsic viscosity ($[\eta]$) of cress seed gum (CSG) as a new potential source of hydrocolloid. The concentration range of the additives were 0%–40% w/w, 0%–15% w/w, 0–100 mM and 0–15 mM for sucrose, lactose, NaCl, and $CaCl_2$, respectively. Various models, that is, Huggins, Kraemer, Tanglertpaibul-Rao, and Higiro were used to estimate the intrinsic viscosity. The polyelectrolyte behavior of the CSG is confirmed by the decrease of $[\eta]$ with addition of salts. The addition of sucrose and lactose was shown to initially decrease the intrinsic viscosity, possibly due to either a reduction in solvent quality or a reduction in polymer/polymer association, followed by an increase in $[\eta]$ at higher concentrations. Berry number and the slope of master curve demonstrated that CSG samples in all salt and sugar concentrations were in dilute domain without entanglement occurrence. The exponent b (slope of power-law model) also revealed that CGS samples had the conformation between random coil and rigid rod; however, selected salts and sugars slightly changed the CSG conformation to random coil.

9.4.2.8.5 Stabilizers

Stabilizers are natural or chemically modified substances with botanical origin. They form gel structures in water and hence are called hydrocolloids. They are divided into alginates, carob, locust beans, guar gum, carrageenans, gelatin, and CMC.

Hydrocolloids can be added to mixes for bakery applications to get a better structure of the batter. Starch, carrageenan, and calcium acetate make the protein network in the pudding stronger.

Calcium disodium ethylenediaminetetraacetate is an additive recognized as suitable for use in some foods in conformance with the legislation of the European Union (EU) and US FDA, and the recommendations of the Codex Alimentarius Commission established by the WHO and the Food and Agriculture Organization of the United Nations. The food categories in which it is authorized depend on the country. For instance, the EU allows the addition of calcium disodium ethylenediaminetetraacetate (identified as E385) in products such as fat and oils emulsions, sauces, some canned or bottled fruits and vegetables, and fish, and fishery products including molluscs and crustaceans.

This compound is the most common chelating agent used as antioxidant synergist in food to improve the stability of foods. It is useful to control the reaction of trace transition metal ions such as cooper and iron ions, which prevents the initiation of the oxidation of unsaturated lipids and the subsequent deterioration of color and texture. Ethylenediaminetetraacetic acid (EDTA) is more prone to form complexes with cooper, iron, and other metals than with calcium at pH close to 7 (Jimenez, 2014).

As regards the determination of E385 only in the liquid portion of a canned product, it is suitable to remark that article 11 of the regulation of the EU on food additives (Regulation (EC) No. 1333/2008) indicates that the maximum levels of food additives shall apply to the food as marketed.

The sample preparation in fatty samples such as margarine, butter, mayonnaise, and salad dressing is basically similar to that of vegetables and mushrooms. Aliquots of sample are diluted or extracted with water (Campana et al., 1996; Krokidis et al., 2005) and, sometimes, subjected to a liquid–liquid partitioning with a nonmiscible organic solvent such as benzene (Krokidis et al., 2005) or dichloromethane (Ruiz et al., 2007) to remove fatty compounds from the diluted sample.

An original method to determine the food additive calcium disodium ethylenediaminetetraacetate in bottled food is proposed by Jimenez (2014). The method involves the solid–liquid extraction of a portion of the whole content of legume or artichoke bottles or the dilution of sauce samples with water followed by an evaporation step by heating. Finally, EDTA is methylated and determined by GC. Recoveries obtained on spiked samples were acceptable (96%–108%) with RSDs comprised from 4.3% to 10%. Results suggest that the determination of an additive only in the liquid phase of legume or artichoke bottles is not suitable to know its total amount because the additive is distributed between the liquid and solid phases.

The contribution of each step of the analytical method to the uncertainty of the measured concentration has been assessed by a "bottom-up" approach, including the heterogeneity of the sample which appeared to be very variable after studying 20 samples.

In consumers' diet, a range of processed products contain pectin ranging from fruit and vegetable juices (Sila et al., 2009), purees and pastes (Van Buggenhout et al., 2009) where pectin originates from the raw material up to products such as yogurt, jams, and many others where pectin is added as a thickener, stabilizer or gelling agent (Corredig et al., 2000; Corredig and Wicker, 2001).

Pectin, a heterogeneous group of polysaccharides, is structurally and functionally the most complex polysaccharide in plant cell walls with various functions during plant growth and development.

It also has a diverse range of food and biomedical uses (Mohnen, 2008). Pectin is composed of approximately 70% galacturonic acid (GalA) and is suggested to be a triad component encompassing homogalacturonan (HG), rhamnogalacturonan II (RG-II), and rhamnogalacturonan I (RG-I) domains (Mohnen, 2008).

Pectin is a common, extremely complex, and process-sensitive polysaccharide in plant cell walls with many uses as an additive in the food and biomedical industry. Process-induced chemical changes in pectin result in various effects on its functionality. An in-depth study is presented of the effects of thermal, compared to high hydrostatic, pressure combined with high temperature (HP/HT) processing on pectin nanostructure and characteristics. The results obtained emphasized the necessity of taking into account pectin association and conformation in solution when analyzing molecular weight changes. At a pH of 6.3, a decrease in molecular weight was observed for both thermal and HP/HT treated samples but with partially different reasons. While for the thermally treated samples the reduction in molecular weight was mostly due to pectin depolymerization for the HP/HT treated samples, a significant effect was observed for conformational changes induced by electrostatic repulsion caused by the complete demethoxylation of the polymer. On the contrary due to conformational changes, an increase in the observed molecular weight was noticed for HP/HT treated samples at a pH of 4.4. The study also clearly shows the necessity of combining an absolute molar mass determination method like multiangle laser light scattering (MALLS) in studies on the effect of processing on pectin Shpigelman et al. (2014).

9.4.2.8.6 Emulsifiers

Emulsifiers contain a molecule with a hydrophobic and a hydrophilic end. These concentrate in the interface between the fat and the continuous aqueous phase of the system, reducing the interfacial tension. This results in finely dispersed fat globules surrounding the air bubbles. The emulsifier is affected by the concentration, other ingredients, the mix, and the processing time. The main emulsifiers are mono and diglycerides of fatty acids and lecithin, which constitute of egg yolk, milk solids, and soya beans (Table 9.12).

Polyglycerol polyricinoleate (PGPR) is used as an emulsifier in the food industry, especially in chocolate coatings and chocolate bars. PGPR improves the characteristics of molten chocolate by reducing yield stress, facilitating the coating of confectionery pieces, while limiting the amount of cocoa butter involved. The enzymatic synthesis of PGPR catalyzed by lipases presents several advantages over chemical synthesis, including enzyme specificity and the mild conditions needed, thereby avoiding undesirable side-reactions and by-products. A novel process to synthesize PGPR using a biocatalyst, Novozym® 435, is presented by Ortega-Requena et al. (2014). Novozym 435 is

TABLE 9.12
Emulsifiers in Food; Name, Origin, Functionality, Adverse Effects, ADI, and Applications

E	Name	Origin	Functionality	Effects	ADI	Product Uses
E400	Alginic acid	Natural extract	Insoluble in cold water	None	0–5	Flans, yoghurt, jams
E407	Carrageenan	Extract from algae	Stabilizer, leavening agent	Maybe not be safe in large quantities	—	Condensed milk, chocolate milk, ice cream, yoghurt, biscuits, and toothpastes
E412	Guar gum	Natural extract	Stabilizer, leavening agent	Large quantities cause nausea	—	Ice creams, sauces, yoghurt, cream cheese
E414	Acacia gum	Natural substance	Stabilizer, leavening agent	None	—	Wine, beer, canned vegetables
E420	Sorbitol	Natural substance	Stabilizer, sweetener	Large quantities cause flatulence, abdominal pains, diarrhea	—	Chocolate, sweets, ice creams, jams for diabetics, ready cakes, gums
E422	Glycerin (Glycerol)	Natural substance. Also synthetically manufactured	Aromatic	None	—	Jams, gel, ice creams, desserts, drinks
E461	Methylcellulose	From cellulose	Emulsifier, Stabilizer, leavening agent	Large quantities might cause nausea	—	Drinks, dietetic products, sauces, soft drinks

Source: From Varzakas, T.H. et al., Food additives and contaminants, Chapter 13, in: Yildiz, F., ed., *Advances in Food Biochemistry*, CRC Press/Taylor & Francis Group, 2010, pp. 409–457.

appropriate for catalyzing both the reactions involved in this process. A PGPR fulfilling European specifications for this food additive as well as recommendations set out in the Food Chemical Codex was obtained using a discontinuous vacuum reactor with a dry nitrogen flow. In addition, the biocatalyst reuse would decrease costs. Moreover, it was confirmed that the ability to obtain PGPR in a one-step reaction significantly shortens the time required.

Although fatty acid esters are generally produced by lipases from various sources using organic solvents, solvent toxicity, and cost are problems for many applications. The major advantages of a solvent-free system (SFS) are that the absence of solvents facilitates downstream processing, offers significant cost saving, and minimizes environmental impact (Torrelo et al., 2009; Ye and Hayes, 2012).

PGPR is a food additive used to keep emulsions of oil and water systems with a high water content stable and used as a viscosity modifier (Mun et al., 2010; Su et al., 2006). In the chocolate industry, PGPR is used because it causes a noticeable reduction in the yield stress of molten chocolate. This allows chocolate to be molded without any air bubbles and easier coating of particulate ingredients, while the thickness of the chocolate coating can be optimally adjusted (Schantz and Rohm, 2005; Wilson et al., 1998). An additional property of PGPR in chocolate is its ability to limit fat bloom (Lonchampt and Hartel, 2004; Schenk and Peschar, 2004).

PGPR is traditionally prepared using chemical methods which involve two alkali-catalyzed stages: firstly, the condensation of ricinoleic acid; secondly, the esterification of PGPR acid. The main disadvantage of this process is the energy costs due to long operating times and high temperatures.

Furthermore, this method increases side reactions that produce by-products, providing the final product with organoleptic features that make it unsuitable for use as a food additive (Denecke et al., 1981).

9.4.2.8.7 *Milk Proteins*

Milk proteins are an important part of the manufacture of stable milk foams and emulsions, where long-term dispersion stability is often essential. Whey proteins are known to alter their adsorption at fluid interfaces, responding to both different aqueous environmental conditions and the presence of several food additives such as lipids, sugars, electrolytes, polypeptides, and polysaccharides. However, some studies reported mutual incompatibility with the use of both high- and low-molecular mass surfactants in milk (Kamath et al., 2008; Pérez et al., 2010).

It is also known that caseins adsorb rapidly into the air–water interface due to high surface activity and flexible structure, while globular proteins such as β-lactoglobulin and α-lactalbumin adsorb more slowly (Marinova et al., 2009). Moreover, caseinate produces adsorbed layers similar to those found for β-casein, which is the individual casein with the highest surface activity (Carrera-Sánchez and Rodríguez-Patino, 2005).

The effect of different concentrations of added sodium caseinate (Na–Cas) and WPC on the foaming properties of reconstituted skim milk powder (SMP) was studied by Martinez-Padilla et al. (2014) to quantify practical applications as additives in aerated foods. The density and viscosity of the samples increased with increasing concentration of SMP or protein. The pH of fortified products was close to those of the reconstituted SMP at 10%. The surface tension of both the reconstituted SMP alone and that fortified with WPC was higher than that of reconstituted SMP fortified with Na–Cas; these samples also took more time to reach equilibrium. Foaming capacity increased with increased SMP or protein. Despite the low surface tension attained at equilibrium of reconstituted SMP with increasing Na–Cas concentration, foam stability decreased. The kinetics of drainage, Ostwald ripening, and the collapse of bubbles were calculated. The most stable foams were those from reconstituted SMP fortified with WPC.

9.4.2.8.8 *Iron Phosphates*

Water-soluble iron salts (e.g., ferrous sulfate) are easily formulated and absorbed (good bioavailability) but often produce substantial changes in the color and taste of foods.

Decreasing the size of insoluble iron-containing compounds, such as iron(III) pyrophosphate (ferric pyrophosphate, or FePP) (Fidler et al., 2004) or iron phosphate (Hilty et al., 2010; Rohner et al., 2007), to nanoscale colloidal particles can significantly increase the iron absorption without giving rise to organoleptic changes in the product (Velikov and Pelan, 2008).

Despite the wide use of ferric pyrophosphate, many of the basic properties of its colloidal form are not well established, and only very recently insights into the colloidal stability and morphology of ferric pyrophosphate mixed salts were obtained (Van Leeuwen et al., 2012a,b,c).

Ferric pyrophosphate is a widely used material in the area of mineral fortification, but its synthesis and properties in colloidal form are largely unknown. Rossi et al. (2014) reported on the synthesis and characterization of colloidal iron(III) pyrophosphate particles with potential for application as a food additive in iron-fortified products. They presented a convenient and food grade synthetic method yielding stable colloids of nanometer size with a distinctive white color, a unique characteristic for iron-containing colloids. Physical properties of the colloids were investigated using different techniques to assess particle crystallinity, surface charge, mass density, refractive index, internal structure, elemental composition, and magnetic properties. The findings of this research are especially relevant for food and beverage science and technology and will help develop a more effective use of these fortifiers in colloidal form.

9.4.2.8.9 *Micellar Systems*

Micellar systems allow the coexistence of clear, isotropic, and thermodynamic stable mixtures of water, oil, and a surfactant, the presence of a cosurfactant being frequently necessary to ensure the stability of the system. These systems have notable industrial and food applications due to their capability to solubilize a large range of hydrophilic and hydrophobic substances (Cid et al., 2013; Esmaili et al., 2011; Fratter and Semenzato, 2011; Gaysinsky et al., 2008; Kumar et al., 2002; Sugiura et al., 2001).

FDA has reviewed the safety of salicylic acid (SA) and methyl salicylate and permits their use as indirect food additive in food contact materials (FDA, 2012a). For instance, SA is authorized for use as preservative in adhesives and also as catalyst and cross-linking agent for epoxy resins. In addition, SA attracted much interest because several experimental studies have confirmed that SA could be used for postharvest handling of fruits and vegetables as a food additive. SA treatment was demonstrated to delay fruit ripening and/or reduce decay of several fruits and vegetables (Li and Han, 2000; Srivastava and Dwivedi, 2000; Wei et al., 2011; Zhang et al., 2003), to alleviate chilling injury of tomato and cucumber storage at low temperature (Han et al., 2002), or to be used as antibrowning agent in fresh-cut Chinese water chestnut stored at low temperature (Peng and Jiang, 2006).

Micellar systems have excellent food applications due to their capability to solubilize a large range of hydrophilic and hydrophobic substances. Cid et al. (2014) studied the mixed micelle formation between the ionic surfactant sodium dodecyl sulfate (SDS) and the phenolic acid SA at several temperatures in aqueous solution. The critical micelle concentration and the micellization degree were determined by conductometric techniques and the experimental data used to calculate several useful thermodynamic parameters, like standard free energy, enthalpy, and entropy of micelle formation.

SA helps the micellization of SDS, both by increasing the additive concentration at a constant temperature and by increasing temperature at a constant concentration of additive. The formation of micelles of SDS in the presence of SA was a thermodynamically spontaneous process and is also entropically controlled. SA plays the role of a stabilizer and gives a pathway to control the three-dimensional water matrix structure. The driving force of the micellization process is provided by the hydrophobic interactions. The isostructural temperature was found to be 307.5 K for the mixed micellar system. This article explores the use of SDS–SA based micellar systems for their potential use in fruits postharvest.

9.4.2.8.10 Leavening Agents

Sodium bicarbonate and sodium hydropyrophosphate are usually used to improve structure. They often combine with citric acid to form gases during preparation of batter and during baking. During baking they decompose to form carbon dioxide, which is entrapped in the dough by the formed protein network (Table 9.13).

9.4.2.8.11 Antilumping Agents

Anticaking agents like silicon dioxide or calcium silicate are added in vending machines to minimize bridging and lumping.

9.4.2.8.12 Toxicological Concerns

The threshold of toxicological concern (TTC) refers to the possibility of establishing a human exposure threshold value for all chemicals, below which there is no significant risk to human health. The concept that exposure thresholds can be identified for individual chemicals in the diet, below which no appreciable harm to health is likely to occur, is already widely embodied in the practice of many regulatory bodies in setting ADIs for chemicals for which the toxicological profile is known. However, the TTC concept goes further in proposing that a *de minimis* value can be identified for any chemical, including those of unknown toxicity, taking the chemical structure into consideration.

This concept forms the scientific basis of the US Food and Drug Administration (FDA) 1995 threshold of regulation for indirect food additives. The TTC principle has also been adopted by the Joint FAO/WHO Expert Committee on Food Additives (Commission of the European Communities, 1997; JECFA, 1997) in its evaluations of flavoring substances (Barlow et al., 2001).

Hattan and Kahl (2002) reported on current developments in food additive toxicology in the United States. They mentioned that a recently published proposal (Fed. Reg. 66, 2001) for mandatory submission of information on all plant-derived bioengineered foods fed to humans or animals

TABLE 9.13
Leavening Agents in Food; Name, Origin, Functionality, Adverse Effects, ADI, and Applications

E	Name	Origin	Functionality	Effects	ADI	Product Uses
E503	Ammonium carbonate	Synthetic	Leavening agent	Stimulation of muciferous glands	—	Cocoa, baby foods
E510	Ammonium chloride	Synthetic	Flour improver	Should be avoided by atoms suffering hepatic failure and liver cirrosis due to diuretic properties	—	Bakery products
E514	Sodium sulfate	Natural (from minerals)	Flour improver	Due to sodium increase in the body, it should be avoided by patients suffering cardiac and kidney failure	—	Brewing
E518	Magnesium sulfate	Natural (sea water)	Hardening and noncaking properties	It could cause toxic phenomena in patients suffering kidney failure if it does not get excreted	—	Table salt and pharmaceuticals
E540	Calcium pyrophos-phate	Natural (minerals) or artificial	Nutrition supplement	Danger in perturbation of the balance between phosphorous–calcium	0–70	Processed cheeses and healthy diet products
E541	Aluminum sodium phosphate (acidic)	Artificial	Flour improver	Accumulates in brain nervous tissues of patients with Alzheimer	0–6	Cakes and fries
E541	Aluminum sodium Phosphate (basic)	Synthetic	Emulsifier and leavening agent	Similar to E541 (acidic)		Similar to E541 (acidic)
E544	Polyphos-phoric calcium	Artificial	Emulsifier and leavening agent	Danger in perturbation of the balance between phosphorous–calcium	0–70	Dairy products
E545	Polyphos-phoric ammonium	Artificial ammonium	Complex properties	Similar to E544		
E553	B-talk (French chalk)	Natural	Anticaking and leavening properties	Wheeziness and cough due to chemical inflammation of small bronchi	—	Imported salt, garlic and onion powder, sugar fine, and chewing gums
E554	Aluminum sodium silicate	Natural or/and artificial	Nonagglomerating agent	Neurotoxic and responsible for Alzheimer's disease	—	Dried food in powder form, salt, hard cheeses in pieces, rice
E556	Calcium–aluminum silicate	Natural (mineral)	Nonagglomerating agent	Similar to E554		Rice, chewing gums, hard cheeses in pieces, salt

Source: From Varzakas, T.H. et al., Food additives and contaminants, Chapter 13, in: Yildiz, F., ed., *Advances in Food Biochemistry*, CRC Press/Taylor & Francis Group, 2010, pp. 409–457.

will be reviewed. Under this proposal, information such as data on identity, level, and function of the introduced substance(s); an estimate of dietary exposure; allergenic potential of the protein; data relevant to other safety issues that may be associated with the substance; selection of a comparable food; historic uses of comparable food; composition, and characteristics of bioengineered food versus those of the comparable food should be provided. In addition, characterization of the parent plant; construction of the transformation vector and introduced genetic material along with number of insertion sites and genes; data on the genetic material and any newly inserted genes for antibiotic resistance should be submitted with the notification. The Interagency Coordinating Committee for Validation of Alternative Methods (ICCVAM) was identified by the US Congress as the organization to review and validate new alternative toxicological test methods for 14 US government agencies. Validated and accepted alternative toxicity tests will be incorporated into toxicity testing recommendations for regulatory agencies.

9.5 DO ADDITIVES CAUSE HEALTH PROBLEMS?

Crohn's disease is a chronic granulomatous inflammation of the gastrointestinal tract which was first described in the beginning of the twentieth century. The histological similarity with intestinal tuberculosis has led to the assumption of an involvement of mycobacteria and mycobacterial antigens, respectively, in the etiology. A major defense mechanism against mycobacterial lipid antigens is the CD1 system which includes CD1 molecules for antigen presentation and natural killer T cells for recognition and subsequent production of cytokines like interferone-gamma and tumor necrosis factor-alpha. These cytokines promote granulomatous transformation. Various food additives, especially emulsifiants, thickeners, surface-finishing agents and contaminants like plasticizers share structural domains with mycobacterial lipids. It was therefore hypothesized, by Traunmuller (2005), that these compounds are able to stimulate by molecular mimicry the CD1 system in the gastrointestinal mucosa and to trigger the proinflammatory cytokine cascade.

These are diglycerides of long-chained fatty acids esterified with acetic, lactic, citric, or tartaric acid (E472a–d) and fatty acid esters of propylene glycol (E477), polyglycerol (E476), or polyoxyethylene sorbitan (E432–E436). They are frequently used in the industrialized production of sweets, cream desserts, sauces, spreads, and margarine. Increased consumption of the latter already has been suspected to play a role in the etiopathogenesis of CD (Guthy, 1982). Furthermore, waxes of animal or vegetable origin like beeswax, carnauba wax, and candelilla wax (E901–E914) used as surface-finishing agents for sweets and citrus fruits contain long-chained fatty acid esters (Hamilton, 1995).

Other candidates are dialkyl phthalate esters (di[2-ethylhexyl]phthalate, DEHP; diisononyl phthalate, DINP), used as plasticizers for polyvinyl chloride (PVC) since the 1930s (Bouma and Schakel, 2002; Latini et al., 2004). They were found in concentrations up to 45% in children's toys and teething rings made of soft PVC and migrate into the saliva during chewing and mouthing activities. Ready-to-serve meals are heavily contaminated with plasticizers when PVC gloves were worn during cooking and packaging (Tsumura et al., 2001).

The understanding of Crohn's disease as a CD1-mediated delayed-type hypersensitivity to certain food additives would lead to strong emphasis on a dietary treatment.

Van den Brandt et al. (1995) concluded that epidemiology can contribute significantly to hazard identification, hazard characterization, and exposure assessment. Epidemiologic studies directly contribute data on risk (or benefit) in humans as the investigated species, and in the full food intake range normally encountered by humans. Areas of contribution of epidemiology to the risk assessment process are identified, and ideas for tailoring epidemiologic studies to the risk assessment procedures are suggested, dealing with data collection, analyses, and reporting of both existing and new epidemiologic studies. The paper by Van den Brandt et al. (1995) described a scheme to classify epidemiologic studies for use in risk assessment and deals with combining evidence from multiple studies. Using a matrix approach, the potential contribution to each of the steps in the risk assessment process is evaluated for categories of food substances. The contribution to risk assessment of specific food substances depends on the quality of the exposure information.

Sarıkaya and Cakır (2005) evaluated four food preservatives (sodium nitrate, sodium nitrite, potassium nitrate, and potassium nitrite) and then five combinations at a concentration of 25 mM for genotoxicity in the somatic mutation and recombination test (SMART) of *Drosophila melanogaster.* Three-day-old larvae transheterozygous, including two linked recessive wing hair mutations (multiple wing hairs and flare), were fed at different concentrations of the test compounds (25, 50, 75, and 100 mM) in standard Drosophila instant medium. Wings of the emerging adult flies were scored for the presence of spots of mutant cells, which can result from either somatic mutation or mitotic recombination. Moreover, lethal doses of food preservatives used were determined in the experiments. A positive correlation was observed between total mutations and the number of wings having undergone mutation. For the evaluation of genotoxic effects, the frequencies of spots per wing in the treated series were compared to the control group, which is distilled water. Chemicals used were ranked as sodium nitrite, potassium nitrite, sodium nitrate, and potassium nitrate according to their genotoxic and toxic effects. Moreover, the genotoxic and toxic effects produced by the combined treatments considerably increased, especially when the four chemicals were mixed. That study revealed that correct administration of food preservatives/additives may have a significant effect on human health.

Although potassium sorbate (PS), ascorbic acid, and ferric or ferrous salts (Fe-salts) are used widely in combination as food additives, the strong reactivity of PS and oxidative potency of ascorbic acid in the presence of Fe-salts might form toxic compounds in food during its deposit and distribution. Kitano et al. (2002) evaluated the reaction mixture of PS, ascorbic acid, and Fe-salts for mutagenicity and DNA-damaging activity by means of the Ames test and rec-assay. Effective lethality was observed in the rec-assay. No mutagenicity was induced in either *Salmonella typhimurium* strains TA98 (with or without S-9 mix) or TA100 (with S-9 mix). In contrast, a dose-dependent mutagenic effect was obtained when applied to strain TA100 without S-9 mix. The mutagenic activity became stronger increasing with the reaction period. Furthermore, the reaction products obtained in a nitrogen blanket did not display any mutagenic and DNA-damaging activity. PS, ascorbic acid, and Fe-salts were inactive when they were used separately. Omission of one component from the mixture of PS, ascorbic acid, and Fe-salt turned the reaction system inactive. These results demonstrate that ascorbic acid and Fe-salt oxidized PS and the oxidative products caused mutagenicity and DNA-damaging activity.

Nowadays there has been a strong shift to natural compounds. The interest in the possible use of natural compounds to prevent microbial growth has notably increased in response to the consumer pressure to reduce or eliminate chemically synthesized additives in foods. Minimally processed fruits are an important area of potential growth in rapidly expanding fresh cut produce. However, the degree of safety obtained with the currently applied preservation methods seems to be sinufficient. Lanciotti et al. (2004) gave an overview on the application of natural compounds, such as hexanal, 2-(*E*)-hexenal, hexyl acetate and citrus essential oils, to improve the shelf life, and the safety of minimally processed fruits as well as their mechanisms of action.

Plants and plant products can represent a source of natural alternatives to prolong the shelf-life and the safety of food. Some of these volatile compounds were found to play a key role in the defense systems of fresh produce against decay microorganisms. Some of these compounds are produced throughout the lipoxygenase pathway that catalyzes the oxygenation of unsaturated fatty acids, forming fatty acid hydroperoxides.

Aldehydes and the related alcohols are produced by the action of hydroperoxide lyases, isomerases, and dehydrogenases. Many of the natural aromas of fruits and vegetables responsible for their "green notes," such as hexanal, hexanol, 2-(*E*)-hexenal, and 3-(*Z*)-hexenol, are carbon compounds formed through this pathway (Table 9.14).

These compounds are also important constituents of the aroma of tomatoes, tea, strawberry, olive oil, grape, apples, and pear. Moreover, plant essential oils, composed mainly by terpenoides,

TABLE 9.14
Aroma and Taste Enhancers in Food; Name, Origin, Functionality, Adverse Effects, ADI, and Applications

E	Name	Origin	Functionality	Effects	ADI	Product Uses
E256	Malic acid	Natural (potatoes, apples, etc.) or artificial (commercial use)	Aromatic agent/ acidifying agent	Avoid the use of artificially made malic acid in baby foods.		Canned tomatoes, peas, orange puree with low calories, concentrated tomato juice, fruit juices
E620	L-glutamate	Natural (amino acid of plants and animals) and artificial (commercial use)	Aroma and taste enhancer	Its accumulation could cause damage to brain cells of guinea pigs.	0–120	Sauces, soups, aromatic lasagna and processed cheeses
E621	Acidic monosodium glutamate	Artificial	Taste enhancer	Its accumulation could cause brain tissue disorders.	0–120	Instant soups and sauces
E622	Acidic potassium glutamate	Artificial	Taste enhancer	High doses could cause phenomena related to high potassium.	—	Vegetable soups
E627	Disodium Guanylate	Artificial	Taste enhancer	It should be avoided by patients with high uric acid due to the fact of being raw material of the synthesis of uric acid.	—	Canned meat products, crackers, vegetable soups, and sauces
E631	Sodium inosinate	Natural or artificial	Taste enhancer	Similar to E627		Processed meat products, sauces, and soups
E635	Ribonucleic sodium	Artificial	Taste enhancer	Similar to E627		Similar to E627
E421	Mannitol	Natural or artificial	Sweetener and noncaking	Some people may develop nausea, vomiting, diarrhea.		Sweets, ice creams, gums
E905	Microcrystalline wax	Artificial	Antifoaming agent and glazing agent	High doses cause diarrhea and irritation.		Dried fruits, citrus fruits, gums
E951	Aspartame	Artificial	Sweetener	Phenylketonurea patients should avoid it.		Soft drinks, milk drinks, fruit juices, dry nuts, nutrition supplements, mustard
E952	Cyclamic acid and its sodium and calcium salts	Artificial	Sweetener	Cyclamic get excreted unalterable by the human body except some others that could disrupt it and cause undesirable side effects.		Low calorie products

Source: From Varzakas, T.H. et al., Food additives and contaminants, Chapter 13, in: Yildiz, F., ed., *Advances in Food Biochemistry*, CRC Press/Taylor & Francis Group, 2010, pp. 409–457.

TABLE 9.15
Suggested Additives to be Banned from School Meals

Colorings
E102—Tartrazine
E104—Quinoline Yellow
E107—Yellow 2G
E110—Sunset Yellow
E120—Cochineal
E122—Carmoisine
E123—Amaranth
E124—Ponceau 4R
E128—Red 2G
E131—Patent Blue V
E132—Indigo Carmine
E133—Brilliant Blue FCF
E151—Black PN
E154—Brown FK
E155—Brown -HT
Preservatives
E211—Sodium benzoate
E212—Potassium benzoate
E213—Calcium benzoate
E214—Ethyl-4-hydroxybenzoate
E215—Ethyl-4-hydroxybenzoate sodium salt
E216—Propyl 4-hydroxybenzoate
E217—Propyl 4-hydroxybenzoate sodium salt
E218—Methyl 4-hydroxybenzoate
E219—Methyl 4-hydroxybenzoate sodium salt
E220—Sulfur dioxide
E221—Sodium sulfite
E222—Sodium hydrogen sulfite
E223—Sodium metabisulfite
E224—Potassium metabisulfite
E226—Calcium sulfite
E227—Calcium hydrogen sulfite
E250—Sodium nitrite
E251—Sodium nitrate
Flavorings/enhancers
E621—Monosodium glutamate
E635—Sodium 5-ribonucleotide
Sweeteners
Aspartame
Acesulfame K
Sodium Saccharine

Source: From Varzakas, T.H. et al., Food additives and contaminants, Chapter 13, in: Yildiz, F., ed., *Advances in Food Biochemistry*, CRC Press/Taylor & Francis Group, 2010, pp. 409–457.

were extensively studied for their antimicrobial activity against many microorganisms including several pathogens (Delaquis et al., 2002). In particular, the activity of oils from Labiatae and citrus fruits were investigated.

According to the Hyperactive Children's Support Group and Organix Research the additives shown in Table 9.15 should be banned from school meals.

9.6 RISK ASSESSMENT OF FOOD ADDITIVES

Edler et al. (2002) presented a review on the mathematical methods and statistical techniques presently available for hazard assessment. Existing practices of JECFA, FDA, EPA, etc., were examined for their similarities and differences. A framework is established for the development of new and improved quantitative methodologies. They concluded that mathematical modeling of the dose–response relationship would substantially improve the risk assessment process. An adequate characterization of the dose–response relationship by mathematical modeling clearly requires the use of a sufficient number of dose groups to achieve a range of different response levels. This need not necessarily lead to an increase in the total number of animals in the study if an appropriate design is used. Chemical-specific data relating to the mode or mechanism of action and/or the toxicokinetics of the chemical should be used for dose–response characterization whenever possible. It is concluded that a single method of hazard characterization would not be suitable for all kinds of risk assessments, and that a range of different approaches is necessary so that the method used is the most appropriate both for the data available and for the risk characterization. Future refinements to dose–response characterization should incorporate more clearly the extent of uncertainty and variability in the resulting output.

FAO/WHO encourages member countries to develop national food control measures based on risk assessment in order to assure proper protection level to consumers and facilitate fair trade. This is particularly important for developing countries as WTO members because it is clearly stated in the sanitary and rhytosanitary measures (SPS) agreement that: (1) SPS measures should be based on risk assessment techniques developed by relevant international organizations; and (2) Codex standards which is based on risk assessment are regarded as the international norm in trade dispute settlement. When conducting risk assessment on food chemicals (including additives and contaminants) in developing countries, in most cases it is not necessary to conduct their own hazard characterization because the ADIs or PTWIs of food chemicals developed by international expert groups (e.g., JECFA) are universally applicable and also developing countries do not have the resources to repeat those expensive toxicological studies. On the other hand, it is necessary to conduct exposure assessment in developing countries because exposure to food chemicals varies from country to country (Chen, 2004). This is not only crucial in setting national standards but also very important for developing countries to participate in the process of developing Codex standards. In addition to food standard development, risk assessment is equally useful in setting up priorities in imported food inspection and evaluating the success of various food safety control measures.

Renwick et al. (2003) presented a review of risk characterization, the final step in risk assessment of exposures to food chemicals. The report is the second publication of the project "Food Safety in Europe": Risk Assessment of Chemicals in the Food and Diet (FOSIE, 2002). The science underpinning the hazard identification, hazard characterization and exposure assessment steps has been published in a previous report (Food Safety in Europe, 2002). Risk characterization is the stage of risk assessment that integrates information from exposure assessment and hazard characterization into advice suitable for use in decision making. The focus of this review was primarily on risk characterization of low molecular weight chemicals but consideration was also given to micronutrients and nutritional supplements, macronutrients and whole foods. Problem formulation, as discussed here, is a preliminary step in risk assessment that considers whether an assessment is needed, who should be involved in the process and the further risk management, and how the information will provide the necessary support for risk management. The report described good evaluation practice as an organizational process and the necessary condition under which risk assessment of chemicals should be planned, performed, scrutinized, and reported. The outcome of risk characterization may be quantitative estimates of risks, if any, associated with different levels of exposure, or advice on particular levels of exposure that would be without appreciable risk to health, for example, a guidance value such as an ADI. It should be recognized that risk characterization is often an iterative and evolving process.

ADI is derived from the NOAEL or other starting point, such as the benchmark dose (BMD), by the use of an uncertainty or adjustment factor. In contrast, for non-threshold effects, a quantitative hazard estimate can be calculated by extrapolation, usually in a linear fashion, from an observed incidence within the experimental dose–response range to a given low incidence at a low dose. This traditional approach is based on the assumption that there may not be a threshold dose for effects involving genotoxicity. Alternatively, for compounds that are genotoxic, advice may be given that the exposure should be reduced to the lowest possible level.

A case-by-case consideration and evaluation is needed for hazard and risk characterization of whole foods. The initial approach to novel foods requires consideration of the extent to which the novel food differs from any traditional counterparts, or other related products; hence, whether it can be considered as safe as traditional counterparts/related products (the principle of substantial equivalence). As for macronutrients, epidemiological data identifying adverse effects, including allergic reactions, may also exist. Human trials on whole foods, including novel foods, will only be carried out when no serious adverse effects are expected.

REFERENCES

Al-Hassan, A. A. and Norziah, M. H. (2012). Starch-gelatin edible films: Water vapour permeability and mechanical properties as affected by plasticizers. *Food Hydrocolloid,* 26(1), 108–117.

Allamark, M. G., Grice, H. C., and Lu, F. C. (1955). Chronic toxicity studies on food colors. Part I. Observations on the toxicity of FD&C Yellow No. 3 (Oil Yellow AB) and FD& C Yellow No. 4 (Oil Yellow OB) in rats. *J. Pharm. Pharmacol.*, 7, 591.

Alonso-Magdalena, P., Ropero, A. B., Soriano, S., Quesada, I., and Nadal, A. (2010). *Horm. Int. J. Endocrinol. Metab.*, 9, 118–126.

Anastasaki, E., Kanakis, C., Pappas, C., Maggi, L., del Campo, C. P., Carmona, M. et al. (2010). Differentiation of saffron from four countries by mid-infrared spectroscopy and multivariate analysis. *Eur. Food Res. Technol.*, 230, 571–577.

Andrade, R. D., Skurtys, O., and Osorio, F. A. (2012). Atomizing spray systems for application of edible coatings. *Compr. Rev. Food Sci. Food Safety*, 11(3), 323–337.

Antoniewski, M. N., Barringer, S., Knipe, C., and Zerby, H. (2007). Effect of a gelatin coating on the shelf-life of fresh meat. *J. Food Sci.*, 72(6), E382–E387.

Armenta, S., Garrigues, S., and De la Guardia M. (2004). FTIR determination of aspartame and acesulfame-K in tabletop sweeteners, *J. Agric. Food Chem.*, 52(26), 7798–7803.

Arthur, J. R., McKenzie, R. C., and Beckett, G. J. (2003). Selenium in the immune system. *J. Nutr.*, 133, 1457S–1459S.

Arvanitoyannis, I. S., Tserkezou, P., and Varzakas, T. (2006). An update of US food safety, food technology, GM food and water protection and management legislation. *Int. J. Food Sci. Technol.*, 41(Supplement 1), 130–159.

Ayadi, M. A., Kechaou, A., Makni, I., and Attia, H. (2009). Influence of carrageenan addition on turkey meat sausages properties. *J. Food Eng.*, 93, 278–283.

Banks, J. G., Nychas, G. J., and Board R. G. (1987) Sulphite preservation of meat products. In *Preservatives in the Food, Pharmaceutical and Environmental Industries,* Board, R. B., Allwood, M. C., and Banks J. G. (eds.), Society for Applied Bacteriology Technical Series No. 22, Blackwell Scientific Publications, Oxford, U.K., pp. 17–33.

Barlow, S., Dybing, E., Edler, L., Eisenbrand, G., Kroes, R., and van den Brandt, P. (eds.). (2002). Food Safety in Europe (FOSIE): Risk assessment of chemicals in food and diet. *Food Chem. Toxicol.*, 40(2/3), 137.

Barlow, S. M. et al. Threshold of toxicological concern for chemical substances present in the diet, Report of a Workshop held in Paris, France, 1999, Organised by the ILSI Europe Threshold of Toxicological Concern Task Force. *Food Chem. Toxicol.*, 39, 893, 2001.

Barry, P. S. (1975). Lead concentration in human tissues. *Br. J. Ind. Med.*, 32, 119.
Bayramoglu, G. and Arica, M. Y. (2012). Preparation of comb-type magnetic beads by surface-initiated ATRP: Modification with nitrilotriacetate groups for removal of basic dyes. *Ind. Eng. Chem. Res.*, 51, 10629–10640.
Beadle, J. R. et al. (1992). Process for Manufacturing Tagatose. Bean/xanthan gum addition and replacement of pork fat with olive oil on the quality, U.S. Patent 5,078,796.
Behrouzian, F., Razavi, S. M. A., and Karazhiyan, H. (2014). Intrinsic viscosity of cress (*Lepidium sativum*) seed gum: Effect of salts and sugars. *Food Hydrocolloid*, 35, 100–105.
Bennet, B. G. (1984). Six most dangerous chemicals named, Monitoring and Assessment Research Center, London, U.K., on behalf of UNEP/ILO/WHO International Program on Chemical Safety.
Berlin, M. H., Clarkson, T. W., and Frieberg, L. T. (1963). Maximum allowable concentrations of mercury compounds. *Arch. Environ. Health*, 6, 27.
Bohm, V. and Schlesier, K. (2004). Methods to evaluate the antioxidant activity. In *Production Practices and Quality Assessment of Food Crops*, Dris, R. and Jain, S. M. (eds.), Quality Handling and Evaluation, Kluwer Academic Publishers, Dordrecht, the Netherlands, Vol. 3. 1-4020-1700-6, pp. 55–71.
Booth, J. M. and Harding, F. (1986). Testing for antibiotic residues in milk. *Vet. Rec.*, 119, 565.
Bouma, K. and Schakel, D. J. (2002). Migration of phthalates from PVC into saliva simulant by dynamic extraction. *Food Addit. Contam.*, 19, 602.
Bouquiaux, J. (1974). *CEC European Symposium on the Problems of Contamination of Man and His Environment by Mercury and Cadmium,* CID, Luxemborg.
Brady, M. S. and Katz, S. E. (1988). Antibiotic/antimicrobial residues in milk. *J. Food Protect.,* 51, 8.
Brasil, I. M., Gomes, C., Puerta-Gomez, A., Castell-Perez, M. E., and Moreira, R. G. (2012). Polysaccharide-based multilayered antimicrobial edible coating enhances quality of fresh-cut papaya. *LWT Food Sci. Technol.,* 47(1), 39–45.
Breithaupt, D. E. (2004). Simultaneous HPLC determination of carotenoids used as food coloring additives: Applicability of accelerated solvent extraction. *Food Chem.,* 86, 449.
Brewer, M. S. (2012). Reducing the fat content in ground beef without sacrificing quality: A review. *Meat Sci.,* 91, 385–395.
Britt, C. et al. (1998). Influence of cherry tissue on lipid oxidation and heterocyclic aromatic amine formation in ground beef patties. *J. Agric. Food Chem.*, 46, 4891.
Browning, E. (1969). *Toxicity of Industrial Metals,* Butterworths, London, U.K.
Buck, W. B. (1978). Toxicity of inorganic and aliphatic organic arsenicals. In *Toxicity of Heavy Metals in the Environment,* Ohm, F. W. (ed.), Marcel Dekker, New York, pp. 357–369.
Burdock, G. A. and Carabin, I. G. (2004). Generally recognized as safe (GRAS): History and description. *Toxicol. Lett.*, 150, 3.
CAC, Contaminants, Joint FAO/WHO Food Standards Program, Codex Alimentarius, Vol. XVII, World Health Organization, Geneva, Switzerland, 1984.
Campana, A. M. G., Barrero, F. A., and Ceba, M. R. (1996). Sensitive spectrofluorimetric method for the determination of ethylenediaminetetraacetic acid and its salts in foods with zirconium ions and Alizarin Red S in a micellar medium. *Anal. Chim. Acta,* 329, 319–325.
Candogan, K. and Kolsarici, N. (2003). The effects of carrageenan and pectin on some quality characteristics of low-fat beef frankfurters. *Meat Sci.,* 64, 199–206.
Carr, B. T., Pecore, S. D., Gibes, K. M., and DuBois, G. E. (1993). Sensory methods for sweetener evaluation. In *Flavor Measurement*, Ho, C.-T. and Manley, C. H. (eds.), Marcel Dekker, Inc., New York, pp. 219–237.
Carrera-Sánchez, C. and Rodríguez-Patino, J. M. (2005). Interfacial, foaming and emulsifying characteristics of sodium caseinate as influenced by protein concentration in solution. *Food Hydrocolloid,* 19, 407–416.
CCIC. (1968). Guidelines for good manufacturing practices: Use of certified FD&C colors in food, Certified Color Industry Committee. *Food Technol.,* 22(8), 14.
CFR. (1998). Code of Federal Regulations Title 21, Office of the Federal Register, U.S. Government Printing Office, Washington, DC.
Chen, J. (2004). Challenges to developing countries after joining WTO: Risk assessment of chemicals in food. *Toxicology,* 198, 3.
Chichester, D. F. and Tanner, F. W. Jr. (1972). Antimicrobial food additives. In *Handbook of Food Additives*, 2nd edn., Furia, T. E. (ed.), CRC Press, Cleveland, OH, Vol. 1, pp. 115–184.
Chiou, W. L. and Riegelman, S. (1971). Pharmaceutical applications of solid dispersion systems. *J. Pharm. Sci.,* 60(9), 1281–1302.
Chipley, J. R. (1993). Sodium benzoate and benzoic acid. In *Antimicrobials in Foods*, 2nd edn., Davinsonand, P. M. and Branen, A. L. (eds.), Marcel Dekker, New York, pp. 11–48.

Choi, S. J., Decker, E. A., Henson, L., Popplewell, L. M., and McClements, D. J. (2009). Stability of citral in oil-in-water emulsions prepared with medium-chain triacylglycerols and triacetin. *J. Agric. Food Chem.*, 57, 11349–11353.

Chuah, A. M. et al. (2014). Enhanced bioavailability and bioefficacy of an amorphous solid dispersion of curcumin. *Food Chem.*, 156, 227–233.

Cid, A., Mejuto, J. C., Orellana, P. G., Lopez-Fernandez, O., Rial-Otero, R., and Simal-Gandara, J. (2013). Effects of ascorbic acid on the microstructure and properties of SDS micellar aggregates for potential food applications. *Food Res. Int.,* 50, 143–148.

Cid, A., Morales, J., Mejuto, J. C., Briz-Cid, N., Rial-Otero, R., and Simal-Gandara, J. (2014). Thermodynamics of sodium dodecyl sulphate-salicylic acid based micellar systems and their potential use in fruits postharvest. *Food Chem.,* 151, 358–363.

Cierach, M., Modzelewska-Kapitula, M., and Szaciło, K. (2009). The influence of carrageenan on the properties of low-fat frankfurters. *Meat Sci.*, 82, 295–299.

Clark, D. S. and Takacs, J. (1980). Gases as preservatives. In *Microbial Ecology of Foods, Factors Affecting life and Death of Microorganisms, International Commission on Microbiological Specifications of Foods*, Academic Press, New York, Vol. I, pp. 170–192.

Clayson, D. B. and Gardner, R. K. (1976). Carcinogenic aromatic amine and related compounds. In *Chemical Carcinogens,* Searle, C. E. (eds.), ACS Monograph 173, American Chemical Society, Washington, DC.

Cords, B. R. and Dychdala, G. R. (1993). Sanitizers: Halogens, surface-active agents, and peroxides. In *Antimicrobials in Foods,* 2nd edn., Davinson, P. M. and Branen, A. L. (eds.), Marcel Dekker, New York, pp. 469–538.

Corredig, M., Kerr, W., and Wicker, L. (2000). Molecular characterization of commercial pectins by separation with linear mix gel permeation columns in-line with multi-angle light scattering detection. *Food Hydrocolloid,* 14(1), 41–47.

Corredig, M. and Wicker, L. (2001).Changes in themolecularweight distribution of three commercial pectins after valve homogenization. *Food Hydrocolloid,* 15(1), 17–23.

Crachereau, J. C., Gabas, N., Blouin, J., Hıbrard, B., and Maujean, A. (2001). Stabilisation tartrique des vins par la carboxymıthylcellulose. *Bulletin de l'OIV*, pp. 841–842.

Crompton, T. R. (1979). *Additive Migration from Plastics into Food*, Pergamon Press, Oxford, U.K.

Crosby, N. T. (1981). *Food Packaging Materials*, Applied Science Publishers, London, U.K.

Dabeka, R. W., McKenzie, A. D., and Lacroix, G. M. A. (1987). Dietary intakes of lead, cadmium, arsenic and fluoride by Canadian adults, a 24-hours duplicate diet study, *Food Addit. Contam.*, pp. 89–101.

Daniel, J. W. (1985). Preservatives. In *Food Toxicology-Real or Imaginary Problem*? Gimson, G. G. and Walker, R. (eds.), Taylor & Francis, London, U.K., pp. 229–237.

Davidson, P. M. et al. (1983). Naturally occurring and miscellaneous food antimicrobials. In *Antimicrobials in Foods*, Branen, A. L. and Davinson, P. M. (eds.), Marcel Dekker, New York, pp. 371–419.

Delaquis, P. J., Stanich, K., Girard, B., and Mazza, G. (2002). Antimicrobial activity of individual and mixed fractions of dill, cilantro, coriander and eucalyptus essential oils. *Int. J. Food Microbiol.*, 74, 101.

Denecke, P., Börner, G., and Von Allmen, V. (1981). Method of preparing polyglycerol polyri-cinoleic fatty acids esters, UK Patent 2,073,232.

Diowksz, A., Kordialik-Bogacka, E., and Ambroziak, W. (2014). Se-enriched sprouted seeds as functional additives in sourdough fermentation. *LWT—Food Sci. Technol.*, 56, 524–528.

Domínguez de María, P. and Maugeri, Z. (2011). Ionic liquids in biotransformations: From proof-of-concept to emerging deep-eutectic-solvents. *Curr. Opin. Chem. Biol.*, 15, 220–225.

Drinking Water Standards (1962). Public Health Service Publication No. 956, U.S. Government Printers, Washington, DC.

DuBois, G. E., Walters, D. E., Schiffman, S. S., Warwick, Z. S., Booth, B. J., Pecore, S. D., Gibes, K. M., Carr, B. T., and Brands, L. M. (1991). In *Sweeteners: Discovery, Molecular Design and Chemoreception*, Walters, D. E., Orhoefer, F. T., and DuBois, G. E. (eds.), American Chemical Society, Washington, DC, pp. 261–276.

Dubos, R. J. (1950). The effect of organic acids on mammalian tubercle bacilli. *J. Exp. Med.*, 92, 319.

Dziezak, J. D. (1986). Antioxidants, the ultimate answer to oxidation. *Food Technol.*, 40(9), 94.

EC (1996). *Technical Guidance Documents in Support of Directive 96-67-EEC on Risk Assessment of New Notified Substances and Regulation (EC) No. 1488/94 on Risk Assessment of Existing Substances (Parts I, II, III and IV).* EC catalogue numbers CR-48-96-001, 002, 003, 004-EN-C. Office for Official Publications of the European Community, Luxemburg.

Edler, L. et al. (2002). Mathematical modelling and quantitative methods. *Food Chem. Toxicol.*, 40, 283.

Edwards, B. J. (1954). Treatment of chronic leg ulcers with ointment containing soluble chlorophyll. *Physiotherapy*, 40, 177–179.

EFSA. (May 5, 2006). EFSA assesses new aspartame study and reconfirms its safety, Press Release, Parma, Italy.

El-Gendy, S. M. et al. (1980). Survival and growth of *Clostridium* species in the presence of hydrogen peroxide. *J. Food Protect.*, 43, 431.

Elsabee, M. Z. and Abdou, E. S. (2013). Chitosan based edible films and coatings: A review. *Mater. Sci. Eng. C: Mater. Biol. Appl.*, 33(4), 1819–1841.

Enfors, S. O. and Molin, G. (1978). The influence of high concentrations of carbon dioxide on the germination of bacteria spores. *J. Appl. Bacteriol.*, 45, 279.

Eskandani, M., Hamishehkar, H., and Dolatabadi, J. E. N. (2014). Cytotoxicity and DNA damage properties of tert-butylhydroquinone (TBHQ) food additive. *Food Chem.,* 153: 315–320.

Esmaili, M. et al. (2011). Beta casein-micelle as a nano vehicle for solubility enhancement of curcumin, food industry application. *LWT—Food Sci. Technol.,* 44, 2166–2172.

EU, 2003. *European Parliament and Council Directive 2003/115/EC of 22 December 2003 Amending Directive 94/35/EC on Sweeteners for Use in Foodstuffs.*

European Commission, European Parliament and Council Directive No. 95/2/EC of 20 February 1995 on additives other than colours and sweeteners for use in foodstu€s. Official Journal No. L61, 18.3, 95.

FAO/WHO. (1963). Specifications for the identity and purity of food additives and their toxicoligical evaluation: Emulsifiers, stabilizers, bleaching and maturing agents, 7th Report of the Joint Food and Agriculture Organization of the United Nations/World Health Organization Expert Committee on Food Additives, WHO Technical Report Series No. 281, FAO Nutrition Meetings Report Series No. 35.

FAO/WHO. (1973). Toxological evaluation of certain food additives with a review of general principles and of specifications, 17th Report of the Joint Food and Agriculture Organization of the United Nations/World Health Organization Expert Committee on Food Additives, WHO Technical Report Series No. 539, FAO Nutrition Meetings Report Series No. 53.

FAO/WHO. (1974). Evaluation of certain good additives, 18th Report of the Joint Food and Agriculture Organizatio of the United Nations/World Health Organization Expert Committee on Food Additives, WHO Technical Report Series No. 557, FAO Nutrition Meetings report Series No. 54.

FDA. (2012a). Code of Federal Regulations Title 21—Food and Drugs. Part 175—Indirect food additives: Adhesives and components of coatings. http://cfr.regstoday.com/21cfr.aspx. Accessed 1.03.2013.

Fidler, M. C. et al. (2004). A micronised, dispersible ferric pyrophosphate with high relative bioavailability in man. *Br. J. Nutr.,* 91, 107–112. http://dx.doi.org/10.1079/BJN20041018.

Fleming, I. (1967). Absolute configuration and the structure of chlorophyll. *Nature,* 216, 151–152.

Foegeding, P. M. and Busta, F. F. (1983). Effect of carbon dioxide, nitrogen, and hydrogen gases on germination of Clostridium botulinum spores. *J. Food Protect.*, 46, 987.

Francis, F. J. (1985). Pigments and other contaminants. In *Food Chemistry,* Fennema, O. R. (ed.), Marcel Dekker, Inc., New York, 1985.

Franco, D. A., Webb, J., and Taylor, C. E. (1990). Antibiotic and sulfonamide residues in meat: Implications for human health. *J. Food Protect.*, 53, 178.

Franssen, L. R. and Krochta, J. M. (2003). Edible coatings containing natural antimicrobials for processed foods. In *Natural Antimicrobials for the Minimal Processing of Foods*, Roller, S. (ed.), Woodhead Publishing Limited, Abington, MA, pp. 250–262.

Fratter, A. and Semenzato, A. (2011). New association of surfactants for the production of food and cosmetic nanoemulsions: Preliminary development and characterization. *Int. J. Cosmet. Sci.,* 33, 443–449.

Frieberg, L. et al. (1974). *Cadmium in the Environment*, 2nd edn., CRC Press, Boca Raton, FL.

Ganesh, V., Hettiarachchy, N. S., Griffis, C. L., Martin, E. M., and Ricke, S. C. (2012). Electrostatic spraying of food-grade organic and inorganic acids and plant extracts to decontaminate Escherichia coli O157: H7 on spinach and iceberg lettuce. *J. Food Sci.,* 77(7), M391–M396.

Gardner, W. H. (1966). *Food Acidulants*, Allied Chemical Corporation, New York.

Gardner, W. H. (1972). Acidulants in food processing. In *Handbook of Food Additives,* 2nd edn., Furia, T. E. (ed.), CFC Press, Cleveland, OH, Vol. 1, pp. 225–270.

Gaysinsky, S., Davidson, P. M., McClements, D. J., and Weiss, J. (2008). Formulation and characterization of phytophenol-carrying antimicrobial microemulsions. *Food Biophys.*, 3, 54–65.

Gerbaud, V., Gabas, N., Blouin, J., and Crachereau, J. C. (2010). Study of wine tartaric acid salt stabilization by addition of carboxymethylcellulose (CMC): Comparison with the "Protective colloids" effect. *J. Int. des Sci. de la Vigne et du Vin,* 44, 231–242.

Ghante, M. H., Badole, S. L., and Bodhankar, S. L. (2011). Health benefits of garden cress (*Lepidium sativum* Linn.). In *Nuts and Seeds in Health and Disease Prevention*, Preedy, V. R., Watson, R. R., and Patel, V. B. (eds.), Elsevier Press, London, U.K., pp. 521–527.

Gokavi, S. S., Malleshi, N. G., and Guo, M. (2004). Chemical composition of garden cress (Lepidium sativum) seeds and its fractions and use of bran as a functional ingredient. *Plant Foods Hum. Nutr.*, 59, 105–111.

Gómez-Díaz, D. and Navaza, J. M. (2003). Rheology of aqueous solutions of food additives effect of concentration, temperature and blending. *J. Food Eng.,* 56, 387–392.

Gorke, J., Srienc, F., and Kazlauskas, R. (2010). Toward advanced ionic liquids. Polar, enzyme- friendly solvents for biocatalysis. *Biotechnol. Bioprocess Eng.*, 15, 40–53.

Gostner, A. et al. (2006). EFFECT of isomalt consumption on fecal microflora and colonic metabolism in healthy volunteers. *Br. J. Nutr.,* 95, 40–50.

Greeff, A. E., Robillard, B., and du Toit, W. J. (2012). Short- and long-term efficiency of carboxymethylcellulose (CMC) to prevent crystal formation in South African wine. *Food Addit. Contam.: Part A*, 29, 1374–1385.

Gui, W., Xu, Y., Shou, L., Zhu, G., and Ren, Y. (2010). Liquid chromatography–tandem mass spectrometry for the determination of chrysoidine in yellow-fin tuna. *Food Chem.,* 122, 1230–1234.

Guise, R., Filipe-Ribeiro, L., Nascimento, D., Bessa, O., Nunes, F. M., and Cosme, F. (2014). Comparison between different types of carboxylmethylcellulose and other oenological additives used for white wine tartaric stabilization. *Food Chem.,* 156, 250–257.

Guthy, E. (1982). Morbus Crohn und Nahrungsfette. *Dtsch Med. Wochenschr.*, 107, 71.

Gutierrez, J., Barry-Ryan, C., and Bourke, P. (2008). The antimicrobial efficacy of plant essential oil combinations and interactions with food ingredients. *Int. J. Food Microbiol.,* 124(1), 91–97.

Hamilton, J. K. (ed.). (1995). *Waxes: Chemistry, Molecular Biology and Functions*, The Oily Press, Dundee, U.K., pp. 1–90, 257–310.

Han, T., Li, L. P., and Feng, S. Q. (2002). Effect of exogenous salicylic acid on physiological parameters of cucumber and tomato fruits stored at chilling injury temperature. *Sci. Agric. Sin.,* 35, 571–575.

Hand, L. W., Hollingsworth, C. A., Calkins, C. R., and Mandigo, R. W. (1987). Effects of preblending, reduced fat and salt levels on frankfurter characteristics. *J. Food Sci.,* 52, 1149–1151.

Hansen, W. H. et al. (1963). Chronic oral toxicity of ponceau 3R. *Toxicol. Appl. Pharmacol.*, 5, 105.

Harrow, L. S. and Jones, J. H. (1954). The decomposition of azo colors in acid solution. *J. Assoc. Off. Agric. Chem.*, 37, 1012.

Hattan, D. G. and Kahl, L. S. (2002). Current developments in food additive toxicology in the USA. *Toxicology*, 181, 182, 417–420.

Higginbotham, J. D. (1983). Recent developments in non-nutritive sweeteners. In *Developments in Sweeteners,* Grenby, T. H. et al. (eds.), Applied Science Publishers, London, U.K., Vol. 2, pp. 119–156.

Higiro, J., Herald, T. J., and Alavi, S. (2006). Rheological study of xanthan and locust bean gum interaction in dilute solution. *Food Res. Int.,* 39, 165–175.

Higiro, J., Herald, T. J., Alavi, S., and Bean, S. (2007). Rheological study of xanthan and locust bean gum interaction in dilute solution: Effect of salt. *Food Res. Int.,* 40, 435–447.

Hilty, F. M. et al. (2010). Iron from nanocompounds containing iron and zinc is highly bioavailable in rats without tissue accumulation. *Nature Nanotechnol.,* 5(5), 374–380. http://dx.doi.org/10.1038/NNANO.2010.79.

Hotchkiss, J. H. (1992). Pesticide residue controls to ensure food safety. *Crit. Rev. Food Sci. Nutr.*, 31, 191.

Hsu, S. Y. and Sun, L.-Y. (2006). Comparisons on 10 non-meat protein fat substitutes for low-fat Kung-wans. *J. Food Eng.,* 74, 47–53.

Humphery, A. M. (2004). Chlorophyll as a color and functional ingredient. *J. Food Sci.,* 69, C422–C425.

Hung, S-M., Hsu, B-D., and Lee, S. (2014). Modelling of isothermal chlorophyll extraction from herbaceous plants. *J. Food Eng.,* 128, 17–23.

IARC. (1973). *Evaluation of Carcinogenic Risk of Chemicals to Man, Some Inorganic and Organometallic Compounds*, International Agency for Research on Cancer, Lyon, France, Vol. 2, 1973.

Ilbäck, N.-G. et al. (2003). ESTIMATED intake of the artificial sweeteners acesulfame-K, aspartame, cyclamate and saccharin in a group of Swedish diabetics. *Food Addit. Contam.,* 20(2), 99–114.

ISO. (2011). International Standard ISO 3632-1:2011. Saffron (*Crocus sativus* L.) Specifications, ISO, Geneva, Switzerland.

Janjarasskul, T. and Krochta, J. M. (2010). Edible packaging materials. In *Annual Review of Food Science and Technology*, Doyle, M. P. and Klaenhammer, T. R. (eds.), Palo Alto: Annual Reviews, Vol. 1, pp. 415–448.

Jermini, M. F. G. and Schmidt-Lorenz, W. (1987). Activity of Na-benzoate and ethyl-paraben against osmotolerant yeasts at different water activity values. *J. Food Protect.*, 50, 920.

Jiao, S., Jin, J., and Wang, L. (2014). Tannic acid functionalized N-doped grapheme modified glassy carbon electrode for the determination of bisphenol A in food package. *Talanta,* 122, 140–144.

Jimenez, J. J. (2014). Determination of calcium disodium ethylenediaminetetraacetate (E385) in marketed bottled legumes, artichokes and emulsified sauces by gas chromatography with mass spectrometric detection. *Food Chem.,* 152, 81–87.

Jiménez-Colmenero, F., Cofrades, S., Herrero, A. M., Fernández-Martín, F., Rodríguez-Salas, L., and Ruiz-Capillas, C. (2012). Konjac gel fat analogue for use in meat products: Comparison with pork fats. *Food Hydrocolloid,* 26, 63–72.

Jin, W. et al. (2011). Protective effect of tertbutylhydroquinone on cerebral inflammatory response following traumatic brain injury in mice. *Injury,* 42(7), 714–718.

Joint FAO/WHO Expert Committee on Food Additives. (2002). List of substances scheduled for evaluation and request for data, Rome, Italy.

Jones, G. M. and Seymour, E. H. (1988). Cowside antibiotic residue testing. *J. Dairy Sci.,* 71, 1691–1699.

Kamath, S., Huppertz, T., Houlihan, A. V., and Deeth, H. C. (2008). The influence of temperature on the foaming of milk. *Int. Dairy J.,* 18, 994–1002.

Karoui, R., Downey, G., and Blecker, C. (2010). Mid-infrared spectroscopy coupled with chemometrics: A tool for the analysis of intact food systems and the exploration of their molecular structure-quality relationships: A review. *Chem. Rev.,* 110, 6144–6168.

Kashanian, S. and Dolatabadi, J. E. N. (2009). DNA binding studies of 2-tertbutylhydroquinone (TBHQ) food additive. *Food Chem.,* 116(3), 743–747.

Kinghorn, D. A. (2002). Overview. In Stevia the Genus *Stevia (Medicinal and Aromatic Plants—Industrial Profiles),* Kinghorn, A. D. (Ed.), Taylor & Francis/CRC Press, New York, pp. 1–17.

Kiseleva, M. G., Pimenova, V. V., and Eller, K. I. (2003). Optimization of conditions for the HPLC determination of synthetic dyes in food. *J. Anal. Chem. URSS,* 5(7), 685–690.

Kitano, K. et al. (2002). Mutagenicity and DNA-damaging activity caused by decomposed products of potassium sorbate reacting with ascorbic acid in the presence of Fe salt. *Food Chem. Toxicol.,* 40, 1589.

Kohri, N. et al. (1999). Improving the oral bioavailability of albendazole in rabbits by the solid dispersion technique. *J. Pharm. Pharmacol.,* 51(2), 159–164.

Kranl, K. et al. (2005). Comparing antioxidative food additives and secondary plant products—Use of different assays, *Food Chem.,* 93, 171.

Krokidis, A. A., Megoulas, N. C., and Koupparis, M. A. (2005). EDTA determination in pharmaceutical formulations and canned foods based on ion chromatography with suppressed conductimetric detection. *Anal. Chim. Acta,* 535, 57–63.

Kumar, R., Thakur, L., Mozumdar, S., and Patanjali, P. K. (2002). Solubilization of some hydrophobic food flavouring agents in single and mixed micellar systems. *J. Surface Sci. Technol.,* 18, 153–161.

Lanciotti, R. et al. (2004). Use of natural aroma compounds to improve shelflife and safety of minimally processed fruits, *Trends Food Sci. Technol.,* 15, 201.

Lanfer-Marquez, U. M., Barros, R. M. C., and Sinnecker, P. (2005). Antioxidant activity of chlorophylls and their derivatives. *Food Res. Int.,* 38, 885–891.

Larato, D. C. and Pfau, F. R. (1970). Effects of water soluble chlorophyllin ointment on gingival inflammation. *NY Dental J.,* 36, 291–293.

Latini, G., De Felice, C., and Verrotti, A. (2004). Plasticizers, infant nutrition and reproductive health. *Reprod. Toxicol.,* 19, 27.

Leclercq, C., Arcella, D., and Turrini, A. (2000). Estimates of the theoretical maximum daily intake of erythorbic acid, gallates, butylated hydroxyanisole (BHA) and butylated hydroxytoluene (BHT) in Italy: A stepwise approach. *Food Chem. Toxicol.,* 38, 1075.

Lei, H. T., Liu, J., Song, L. J., Shen, Y. D., Haughey, S. A., Guo, H. X., Yang, J. Y., Xu, Z. L., Jiang, Y. M., and Sun, Y. M. (2011). Development of a highly sensitive and specific immunoassay for determining chrysoidine, a banned dye, in soybean milk film. *Molecules,* 16, 7043–7057.

Leistner, L. (1985). Hurdle technology applied to meat products of the shelf stable product and intermediate moisture food types. In *Properties of Water in Foods,* Sinators, D. and Multon, J. C. (eds.), Martinus Nijhoff Publishers, Dordredt, the Netherlands, pp. 309–329.

Lewis, R. J. Sr. (1989). *Food Additives Handbook,* Van Nostrand Reinhold, New York.

Li, L. P. and Han, T. (2000). Effects of salicylic acid on the quality of peaches stressinduced oxidative damage in Arabidopsis involves calcium, abscisic acid, ethylene, and salicylic acid. *Fruit Sci.,* 17, 97–100.

Li, Z., Liu, X., Pei, Y., Wang, J., and He, M. (2012). Design of environmentally friendly ionic liquid aqueous two-phase systems for the efficient and high activity extraction of proteins. *Green Chem.,* 14, 2941–2950.

Lindley, M. G., Beyts, P. K., Canales, I., and Borrego, F. (1993). Flavour modifying characteristics of the intense sweetener neohesperidin dihydrochalcone. *J. Food Sci.,* 58, 592–594, 666.

Lonchampt, P. and Hartel, R. W. (2004). Fat bloom in chocolate and compound coatings. *Eur. J. Lipid Sci. Technol.*, 106, 241–274.

Lu, F. G., Sun, M., Fan, L. L., Qiu, H. M., Li, X. J., and Luo, C. N. (2012). Flow injection chemiluminescence sensor based on core–shell magnetic molecularly imprinted nanoparticles for determination of chrysoidine in food samples. *Sens. Actuat. B*, 173, 591–598.

Lück, E. (1980). *Antimicrobial Food Additives, Characteristics, Uses, Effects*, Springer-Verlag, Berlin, Germany.

Lück, E. (1985). Chemical preservation of food. *Zentralbl. Bakteriol. Mikrobiol. HygB*, 180 (2–3), 311–318.

Lurueña-Martínez, M. A., Vivar-Quintana, A. M., and Revilla, I. (2004). Effect of locust bean/xanthan gum addition and replacement of pork fat with olive oil on the quality characteristics of low-fat frankfurters. *Meat Science*, 68, 383–389.

Ma, X. and Pawlik, M. (2007). Intrinsic viscosities and Huggins constants of guar gum in alkali metal chloride solutions. *Carbohydr. Polym.*, 70, 15–24

Mäkinen, K. K., Saag, M., Isotupa, K. P., Olak, J., Nömmela, R., Söderling, E., and Mäkinen, P. L. (2004). SIMILARITY of the effects of erythritol and xylitol on some risk factors of dental caries. *Caries Res.*, 39 (3), 207–215.

Mantilla, N., Castell-Perez, M. E., Gomes, C., and Moreira, R. G. (2013). Multilayered antimicrobial edible coating and its effect on quality and shelf-life of fresh-cut pineapple (*Ananas comosus*). *LWT Food Sci. Technol.*, 51(1), 37–43.

Marchetti, L., Andrés, S. C., and Califano, A. N. (2013). Textural and thermal properties of low-lipid meat emulsions formulated with fish oil and different binders. *Lebensm. Wiss. Technol.*, 51, 514–523.

Marchetti, L., Andrés, S. C., and Califano, A. N. (2014). Low-fat meat sausages with fish oil: Optimization of milk proteins and carrageenan contents using response surface methodology. *Meat Sci.*, 96, 1297–1303.

Marinova, K. G. et al. (2009). Physico-chemical factors controlling the foam ability and foam stability of milk proteins: Sodium caseinate and whey protein concentrates. *Food Hydrocolloid,* 23, 1864–1876.

Mario, G., Ferruzzi, M. G., and Blakeslee, J. (2007). Digestion, absorption, and cancer preventative activity of dietary chlorophyll derivatives. *Nutr. Res.*, 27, 1–12.

Markarian, J. (2002). Additives in food packaging. *Plast. Addit. Compound.*, 9: 22–27.

Marmion, D. M. (1984). *Handbook of U.S. Colorants for Foods, Drugs and Cosmetics*, Wiley, New York.

Martínez-Padilla, L. P., García-Mena, V., Casas-Alencáster, N. B., and Sosa-Herrera, M. G. (2014). Foaming properties of skim milk powder fortified with milk proteins. *Int. Dairy J.*, 36, 21–28.

Martiñon, M. E., Moreira, R. G., Castell-Perez, M. E., and Gomes, C. (2014). Development of a multilayered antimicrobial edible coating for shelf-life extension of fresh-cut cantaloupe (*Cucumis melo* L.) stored at 4°C. *LWT—Food Sci. Technol.*, 56, 341–350.

Maswal, M. and Dar, A. A. (2014). Formulation challenges in encapsulation and delivery of citral for improved food quality. *Food Hydrocolloid,* 37, 182–195.

Mattes, R. D. and Popkin, B. M. (2009). Nonnutritive sweetener consumption in humans: Effects on appetite and food intake and their putative mechanisms. *Am. J. Clin. Nutr.*, 89(1), 1–14.

McEvoy, J. D. G. (2002). Contamination of animal feeding stuffs as a cause of residues in food: A review of regulatory aspects, incidence and control. *Anal. Chim. Acta,* 473, 3.

McEwen, S. A., Black, W. D., and Meek, A. H. (1991). Antibiotic residue prevention methods, farm management, and occurrence of antibiotic residues in milk. *J. Dairy Sci.*, 74, 2128–2137.

Meck, W. H. and Williams, C. L. (1999). Choline supplementation during prenatal development reduces proactive interference in spatial memory. *Dev. Brain Res.*, 118, 51–59.

Meilgaard, M., Vance Civille, G., and Carr, B. T. (1987). *Sensory Evaluation Techniques,* CRC Press, Boca Raton, FL, Vol. II, pp. 5–6.

Minioti, K. S., Sakellariou, C. F., and Thomaidis, N. S. (2007). Determination of 13 synthetic food colorants in water-soluble foods by reversed-phase high performance liquid chromatography coupled with diode-array detector. *Anal. Chim. Acta,* 583(1), 103–110.

Mohammad Amini, A. and Razavi, S. M. A. (2012). Dilute solution properties of Balangu (*Lallemantia royleana*) seed gum: Effect of temperature, salt, and sugar. *Int. J. Biol. Macromol.*, 51, 235–243.

Mohnen, D. (2008). Pectin structure and biosynthesis. *Curr. Opin. Plant Biol.*, 11(3), 266–277.

Moorman, M. A. and Koenig, E. (1992). Antibiotic residues and their implications in foods. *Scope,* 7, 4.

Mota, F. J. et al. (2003). Optimisation of extraction procedures for analysis of benzoic and sorbic acids in foodstuffs. *Food Chem.*, 82(3), 469–473.

Muchenje, V., Dzama, K., Chimonyo, M., Strydom, P. E., Hugo, A., and Raats, J. G. (2009). Some biochemical aspects pertaining to beef eating quality and consumer health: A review. *Food Chem.*, 112, 279–289.

Mun, S., Choi, Y., Rho, S. J., Kang, C. G., Park, C. H., and Kim, Y. R. (2010). Preparation and characterization of water/oil/water emulsions stabilized by polyg-lycerol polyricinoleate and whey protein isolate. *J. Food Sci.,* 75, E116–E125.

Munro, I. C. and Danielewska-Nikiel, B. (2005). Comparison of estimated daily intakes of flavouring substances with no-observed-effect levels. *Food Chem. Toxicol,* 44(6), 758–809.

Murthy, G. K., Rhea, U., and Peeler, J. R. (1971). Antimony in the diet of children. *Environ. Sci. Technol.*, 5, 436.

Nagai, F., Okubo, T., Ushiyama, K., Satoh, K., and Kano, I. (1996). Formation of 8-hydroxydeoxyguanosine in calf thymus DNA treated with tertbutylhydroquinone, a major metabolite of butylated hydroxyanisole. *Toxicol. Lett.,* 89(2), 163–167.

Nair, B. (2001). Final report on the safety assessment of benzyl alcohol, benzoic acid, and sodium benzoate. *Int. J. Toxicol.,* 20(3), 23–50.

Nam, K., Seo, K., Jo, C., and Ahn, D. (2011). Electrostatic spraying of antioxidants on the oxidative quality of ground beef. *J. Animal Sci.,* 89(3), 826–832.

NAS. (1972). *Airborne Lead in Perspective*, National Academy of Sciences, Washington, DC.

NAS. (1980). *The Effects on Human Health of Subtherapeutic Use of Antimicrobials in Animal Feeds*, National Academy of Sciences, Washington, DC.

Newsome, R. L. (1990). Natural and synthetic coloring agents. In *Food Additives*, Branen, A. L., Davidson, P. M., and Salminen, S. (eds.), Marcel Dekker, New York.

Nickerson, M. T., Paulson, A. T., and Hallett, F. R. (2004). Dilute solution properties of kcarrageenan polysaccharides: Effect of potassium and calcium ions on chain conformation. *Carbohydr. Polym.,* 58, 25–33.

Nockemann, P., Thijs, B., Driesen, K., Janssen, C. R., Van Hecke, K., Van Meervelt, L., Kossmann, S., Kirchner, B., and Binnemans, K. (2007a). Choline saccharinate and choline acesulfamate: Ionic liquids with low toxicities. *J. Phys. Chem. B*, 111, 5254–5263.

NRC. (1987). *Regulating Pesticides in Food*, The Delaney Paradox, National Academy Press, Washington, DC.

O'Brien-Nabors, L. (1991). *Alternative Sweeteners*, Marcel Dekker, New York.

O'Donnell, K. (2005). Carbohydrate and intense sweeteners. In *Chemistry and Technology of Soft Drinks and Fruit Juices,* 2nd edn., Ashurst, P. R. (ed.). Blackwell Publishing Ltd., Oxford, U.K., pp. 68–89.

OIV (Organisation International de la Vigne et du Vin). (2012). *International Oenological Codex*, Edition Officielle, Paris, France.

Okubo, T., Nagai, F., Ushiyama, K., and Kano, I. (1997). Contribution of oxygen radicals to DNA cleavage by quinone compounds derived from phenolic antioxidants, tert-butylhydroquinone and 2,5-di-tert-butylhydroquinone. *Toxicol. Lett.,* 90(1), 11–18.

Okubo, T., Yokoyama, Y., Kano, K., and Kano, I. (2003). Cell death induced by the phenolic antioxidant tertbutylhydroquinone and its metabolite tertbutylquinone in human monocytic leukemia U937 cells. *Food Chem. Toxicol.,* 41(5), 679–688.

Ordóñez, M., Rovira, J., and Jaime, I. (2001). The relationship between the composition and texture of conventional and low-fat frankfurters. *Int. J. Food Sci. Technol.*, 36, 749–758.

Ordoudi, S. A., de los Mozos Pascual, M., and Tsimidou, M. Z. (2014). On the quality control of traded saffron by means of transmission Fourier-transform mid-infrared (FT-MIR) spectroscopy and chemometrics. *Food Chem.*, 150, 414–421.

Ordoudi, S. A. and Tsimidou, M. Z. (2004). Saffron quality: Effect of agricultural practices, processing and storage. In *Production Practices and Quality Assessment of Food Crops*, Dris, R. and Jain, S. M. (eds.), Kluwer Academic Publication, Dordrecht, the Netherlands, pp. 209–260.

Ortega-Requena, S., Bódalo-Santoyo, A., Bastida-Rodríguez, J., Máximo-Martín, M. F., Montiel-Morte, M. C., and Gómez-Gómez, M. (2014). Optimized enzymatic synthesis of the food additive polyglycerol polyricinoleate (PGPR) using Novozym® 435 in a solvent free system. *Biochem. Eng. J.,* 84, 91–97.

Ough, C. S. (1993a). Sulfur dioxide and sulfites. In *Antimicrobials in Foods*, 2nd edn., Davidson, P. M. and Branen, A. L. (eds.), Marcel Dekker, New York, pp. 137–190.

Palmer, A. A. (1932).Two fatal cases of poisoning by acetic acid. *Med. J. Aust.*, 1, 687.

Pan, R. N., Chen, J. H., and Chen, R. R. (2000). Enhancement of dissolution and bioavailability of piroxicam in solid dispersion systems. *Drug Dev. Ind. Pharm.*, 26(9), 989–994.

Panek, J., Davıdek, J., and Jehlickova, Z. (1995). In *Natural Toxic Compounds of Foods*: *Formation and Change during Food Processing and Storage*, Davıdek, J. (eds.), CRC Press, Boca Raton, FL, p. 195.

Pavlath, A. E. and Orts, W. J. (2009). Edible films: Why, what and how! Chapter 1. In *Edible Films and Coatings for Food and Other Applications*, Embuscado, M. E. and Huner, K. (eds.), Springer, New York, pp. 1–24.

Peng, L. and Jiang, Y. (2006). Exogenous salicylic acid inhibits browning of fresh-cut Chinese water chestnut. *Food Chem.*, 94, 535–540.

Pennington, J. A. T. and Jones, J. W. (1988). Aluminum in American diets. In *Aluminum in Health: A Critical Review,* Gitelman, H.J. (ed.), Marcel Dekker, New York, pp. 67–100.

Pérez, A. A., Carrera-Sánchez, C., Rodríguez-Patino, J. M., Rubiolo, A. C., and Santiago, L. G. (2010). Milk whey proteins and xanthan gum interactions in solution at the air-water interface: A rheokinetic study. *Colloids Surf. B Biointerf.,* 81, 50–57.

Perez-Rojas, J. M., Guerrero-Beltrαn, C. E., Cruz, C., Sαnchez-Gonzαlez, D. J., Martνnez-Martνnez, C. M., and Pedraza-Chaverri, J. (2011). Preventive effect of tert-butylhydroquinone on cisplatin-induced nephrotoxicity in rats. *Food Chem. Toxicol.*, 49(10), 2631–2637.

Petkovic, M., Ferguson, J.L., Gunaratne, H.Q.N., Ferreira, R., Leitao, M.C., Seddon, K.R., Rebelo, L.P.N., and Pereira, C.S. (2010). Novel biocompatible cholinium-based ionic liquids-toxicity and biodegradability. *Green Chem.,* 12, 643–649.

Petkovic, M., Seddon, K.R., Rebelo, L.P.N., and Silva Pereira, C. (2011). Ionic liquids: A pathway to environmental acceptability. *Chem. Soc. Rev.*, 40, 1383–1403.

Pietrasik, Z. (2003). Binding and textural properties of beef gels processed with κ-carrageenan, egg albumin and microbial transglutaminase. *Meat Sci.*, 63, 317–324.

Pihlasalo, J., Klika, D. K., Murzin, Y. D., and Nieminen, V. (2007). Conformational equilibria of citral. *J. Mol. Struct. Theochem.*, 814, 33–41.

Pimentel, D. (1991). The dimensions of the pesticide question. In *Ecology, Economics, Ethics: The Broken Circle*, Birmann, F. H. and Kellert, S. R. (eds.), Yale University Press, New Haven, CT, pp. 59–69.

Piorkowski, D. T. and McClements, D. J. (2013). Beverage emulsions: Recent developments in formulation, production, and applications. *Food Hydrocolloid,* 42(1), 5–41.

Pizato, S., Cortez-Vega, W. R., de Souza, J. T. A., Prentice-Hernandez, C., and Borges, C. D. (2013). Effects of different edible coatings in physical, chemical and microbiological characteristics of minimally processed peaches (*Prunus persica* L. Batsch). *J. Food Saf.,* 33(1), 30–39.

Pryor, W. A. (1980). *Free Radicals in Biology*, Academic Press, New York, Vol. IV.

Pugh, D. M. (2002). The EU precautionary bans of animal feed additive antibiotics. *Toxicol. Lett.*, 128, 35.

Pulgar, R., Olea-Serrano, M. F., Novillo-Fertrell, A., Rivas, A., Pazos, P., Pedraza, V., Navajas, J. M., and Olea, N. (2000). Determination of bisphenol A and related aromatic compounds released from bis-GMA-based composites and sealants by high performance liquid chromatography. *Environ. Health Perspect.,* 108, 21–27.

Pιrez-Rojas, J. M., Guerrero-Beltrαn, C. E., Cruz, C., Sαnchez-Gonzαlez, D. J., Martνnez-Martνnez, C. M., and Pedraza-Chaverri, J. (2011). Preventive effect of tert-butylhydroquinone on cisplatin-induced nephrotoxicity in rats. *Food Chem. Toxicol.,* 49(10), 2631–2637.

Radomaski, J. L. and Harrow, L. S. (1966). The metabolism of 1-(o-talylazo)-2-naphthy-lamine (Yellow OB) in rats. *Ind. Med. Surg*., 35, 882, 1966.

Ramos, O. L., Fernandes, J. C., Silva, S. I., Pintado, M. E., and Malcata, F. X. (2012). Edible films and coatings from whey proteins: A review on formulation, and on mechanical and bioactive properties. *Crit. Rev. Food Sci. Nutr.,* 52(6), 533–552.

Raven, P. H., Evert, R. F., and Eichhorn, S. E. (2005). *Photosynthesis, Light, and Life: Biology of Plants*, 7th edn., W.H. Freeman, San Francisco, CA, pp. 119–127.

Rayman, M. P. (2004). The use of high-Se yeast to raise Se status: How does it measure up? *Br. J. Nutr.,* 92, 557–573.

Rayment, G. E., Best, E. K., and Hamilton, D. J. (August 28–September 2, 1989). Cadmium in fertilizers and soil amendments. In *Chemistry International Conference*, Royal Australian Chemical Institute, Brisbane, Queensland, Australia.

Rayner, P. (1991). Colors. In *Food Additive User's Handbook*, Smith, J. (ed.), Blackie, London, U.K.

Razavi, S. M. A., Mohammadi Moghaddam, T., Emadzadeh, B., and Salehi, F. (2012). Dilute solution properties of wild sage (Salvia macrosiphon) seed gum. *Food Hydrocolloid,* 29, 205–210.

Regulation (EC) No. 1333/2008 of the European Parliament and of the Council, of 16 December 2008, on food additives. (December 31, 2008). Official Journal of European Union, L354, pp. 16–33.

Reilly, A. and Reilly, C. (1972). Patterns of lead pollution in the Zambian environment. *Med. J. Zambia*, 6, 125.

Reilly, C. (1991). *Metal Contamination of Food*, 2nd edn., Elsevier Applied Science, London, U.K.

Ren, L., Meng, M., Wang, P., Xu, Z., Eremin, S. A., Zhao, J., Yin, Y., and Xi, R. (2014). Determination of sodium benzoate in food products by fluorescence polarization immunoassay. *Talanta*, 121, 136–143.

Renwick, A. et al. (2003). Risk characterisation of chemicals in food and diet. *Food Chem. Toxicol.*, 41, 1211.

Renwick, A. G. et al. (2003). RISK characterization of chemicals in food and diet. *Food Chem. Toxicol.,* 41, 1211–1271.

Ribιreau-Gayon, P., Glories, Y., Maujean, A., and Dubourdieu, D. (2006). *Handbook of Enology: The Chemistry of Wine Stabilization and Treatments*, 2nd edn., Wiley & Sons Ltd., Bordeaux, France, Vol. 2.

Rice, A. C. and Pederson, C. S. (1954). Factors influencing growth of *Bacillus coagulans* in canned tomato juice and specific organic acids. *Food Res.*, 19, 124.

Rodriguez-Amaya, D. B. (2010). Quantitative analysis, in vitro assessment of bioavailability and antioxidant activity of food carotenoids—A review. *J. Food Compos. Anal.*, 23(7), 726–740.

Rohner, F. et al. (2007). Synthesis, characterization, and bioavailability in rats of ferric phosphate nanoparticles. *J. Nutr.*, 137(3), 614–619.

Rossi, L., Velikov, K. P., and Philipse, A. P. (2014). Colloidal iron(III) pyrophosphate particles. *Food Chem.*, 151, 243–247.

Ruiz, T. P., Lozano, C. M., and Garcia, M. D. (2007). High-performance liquid chromatography-post-column chemiluminescence determination of aminopolycarboxylic acids at low concentration levels using tris (2,20-bipyridyl) ruthenium(III). *J. Chromatogr. A*, 1169, 151–157.

Sadili-Bico, S. L., Raposo, M. F. de J., Santos Costa de Morais, R. M., and Miranda, A. M. (2010). Chemical dips and edible coatings to retard softening and browning of fresh-cut banana. *Int. J. Postharvest Technol. Innovat.*, 2(1), 3–24.

Salih, F. M. Risk assessment of combined photogenotoxic effects of sunlight and food additives. *Sci. Total Environ*, 362(1–3), 68–73.

Samavati, V., Razavi, S. H., Rezaei, K. A., and Aminifal, M. (2007). Intrinsic viscosity of locust bean gum and sweeteners mixture in dilute solutions. *EJEAFChe*, 6(3), 1879–1889.

Sankaranaranan, R. (1981). Food colors. *Ind. Food Packer*, 35, 25.

Sardesai, V. M. and Waldshan, T. H. (1991). Natural and synthetic intense sweeteners. *J. Nutr. Biochem.*, 2, 236–244.

Sarıkaya, R. and Cakır, S. (2005). Genotoxicity testing of four food preservatives and their combinations in the *Drosophila* wing spot test. *Environ. Toxicol. Pharmacol.*, 20, 424.

Schantz, B. and Rohm, H. H. (2005). Influence of lecithin-PGPR blends on the rheological properties of chocolate. *LWT—Food Sci. Technol.*, 38, 41–45.

Schenk, H. and Peschar, R. (2004). Understanding the structure of chocolate. *Radiat. Phys. Chem.*, 71, 829–835.

Schiffman, S. S., Booth, B. J., Carr, B. T., Losee, M. L., Sattely-Miller, E. A., and Grahma, B. G. (1995). Investigation of synergism in binary mixtures of sweeteners. *Brain Res. Bull.*, 38(2), 105–120.

Schiweck, H. (1994). *Isomalt, Ullmann's Encyclopedia of Industrial Chemistry*, Chemie, Weinheim, Germany, Vol. A25, pp. 426–429.

Schroeder, H. A. and Balassa, J. J. (1966). Abnormal trace metals in man: Arsenic. *J. Chron. Dis.*, 19, 85.

Sekar, S., Surianarayanan, M., Ranganathan, V., MacFarlane, D. R., and Mandal, A. B. (2012). Choline-based ionic liquids-enhanced biodegradation of azodyes. *Environ. Sci. Technol.*, 46, 4902–4908.

Shen, Y., Zhang, X., Prinyawiwatkul, W., and Xu, Z. (2014). Simultaneous determination of red and yellow artificial food colourants and carotenoid pigments in food products. *Food Chem.*, 157, 553–558.

Sherlock, J. C. and Smart, G. A. (1984). Tin in foods and the diet. *Food Addit. Contam.*, 1, 277.

Shpigelman, A., Kyomugasho, C., Christiaens, S., Van Loey, A. M., and Hendrickx, M. E. (2014). Thermal and high pressure high temperature processes result in distinctly different pectin non-enzymatic conversions. *Food Hydrocolloid*, 39, 251–263.

Sila, D. N. et al. (2009). Pectins in processed fruits and vegetables: Part II-structure-function relationships. *Compr. Rev. Food Sci. Food Saf.*, 8(2), 86–104.

Simko, P. (2002). Determination of polycyclic aromatic hydrocarbons in smoked meat products and smoke flavouring food additives. *J. Chromatogr. B*, 770, 3.

Sipahi, R. E., Castell-Perez, M. E., Moreira, R. G., Gomes, C., and Castillo, A. (2013). Improved multilayered antimicrobial alginate-based edible coating extends the shelf-life of fresh-cut watermelon (*Citrullus lanatus*). *LWT Food Sci. Technol.*, 51(1), 9–15.

Smith, Q. J. and Brown, K. L. (1980). The resistance of dry spore of *Baccilus subtilis* var, globibii (NCIB 80958) to solutions of hydrogen peroxide in relation to aseptic packaging. *J. Food Technol.*, 15, 169.

Sofos, J. N. and Busta, F. F. (1981). Antimicrobial activity of sorbate. *J. Food Protect*, 44, 614.

Sofos, J. N. and Busta, F. F. (1992). Chemical food preservatives. In *Principles and Practice of Disinfection: Preservation and Sterilization*, 2nd edn., Russell, A. D., Hugo, W. B., and Ayliffe, G. A. J. (eds.), Blackwell Scientific Publications, London, U.K., pp. 351–397.

Sofos, J. N. and Busta, F. F. (1993). Sorbic acid and sorbates. In *Antimicrobials in Foods*, 2nd edn., Davidson, P. M. and Branen, A. L. (eds.), Marcel Dekker, New York, pp. 49–94.

Soliva-Fortuny, R. C. (2010). Polysaccharide coatings extend fresh-cut fruit shelf-life. *Emerg. Food Res. Develop. Rep.*, 21, 1–2.

Soni, M. G. et al. (2002). Evaluation of the health aspects of methyl paraben: A review of the published literature. *Food Chem. Toxicol.*, 40, 1335.

Srivastava, M. K. and Dwivedi, U. N. (2000). Delayed ripening of banana fruit by salicylic acid. *Plant Sci.*, 158, 87–96.

Stabnikova, O., Ivanov, V., Larionova, I., Stabnikov, V., Bryszewska, M. A., and Lewis, J. (2008). Ukrainian dietary bakery product with selenium-enriched yeast. *LWT Food Sci. Technol.*, 41, 890–895.

Sternberg, E. D., Dolphin, D., and Brqckner, C. (1998). Porphyrin-based photosensitizers for use in photodynamic therapy. *Tetrahedron,* 54, 4151–4202.

Stevenson, K. E. and Shafer, B. D. (1983). Bacterial spore resistance to hydrogen peroxide. *Food Technol.*, 37(11), 111.

Stoewsand, G. S., Bache, C. A., and Lisk, D. J. (1974). Dietary selenium protection of methylmercury intoxication of Japanese quail. *Bull. Environ. Contam. Toxicol.*, 11, 152.

Stojanovic, Z., Jeremic, K., Jovanovic, S., and Lechnerb, M. D. (2005). A comparison of some methods for the determination of the degree of substitution of carboxymethylcellulose starch. *Starch/Storke*, 57, 79–83.

Stuckey, B. N. (1972). Antioxidants as food stabilizers. In *CRC Handbook of Food Additives,* Furia, T. E. (ed.), CRC Press, Boca Raton, FL, Vol. I, pp. 185–223.

Su, J., Flanagan, J., Hemar, Y., and Singh, H. (2006). Singh, Synergistic effects of polyglycerolester of polyricinoleic acid and sodium caseinate on the stabilisation of water–oil–water emulsions. *Food Hydrocolloid*, 20, 261–268.

Su, Y. K., Bowers, J. A., and Zayas, J. F. (2008). Physical characteristics and microstructure of reduced-fat frankfurters as affected by salt and emulsified fats stabilized with nonmeat proteins. *J. Food Sci.*, 65, 123–128.

Sugiura, S. et al. (2001). Formation and characterization of reversed micelles composed of phospholipids and fatty acids. *J. Colloid Interface Sci.*, 240, 566–572.

Sun, D.-W. (2009). *Infrared Spectroscopy for Food Quality Analysis and Control*, 1st edn., Elsevier, New York (Chapter 4).

Sun, H., Xia, Q., and Liu, R. (2014). Comparison of the binding of the dyes Sudan II and Sudan IV tobovine haemoglobin. *J. Lumin.*, 148, 143–150.

Tan, C. Y., Li, M., Lin, Y. M., Lu, X. Q., and Chen, Z. L. (2011). Biosorption of basic orange from aqueous solution onto dried A. filiculoides biomass: Equilibrium, kinetic and FTIR studies. *Desalination,* 266, 56–62.

Tennant, D. R. and O'Callaghan, M. (2005). Survey of usage and estimated intakes of annatto extracts. *Food Res. Int.*, 38, 911.

Torrelo, G., Torres, C. F., Senorans, F. J., Blanco, R. M., and Reglero, G. (2009). Solvent-free prepa-ration of phytosteryl esters with fatty acids from butterfat in equimolecular conditions in the presence of a lipase from Candida rugosa. *J. Chem. Technol. Biotechnol.*, 84, 745–750.

Traunmuller, F. (2005). Etiology of Crohn's disease: Do certain food additives cause intestinal inflammation by molecular mimicry of mycobacterial lipids? *Med. Hypotheses,* 65, 859.

Tsumura, Y. et al. (2001). Eleven phthalate esters and di(2-ethylhexyl) adipate in one-week duplicate diet samples obtained from hospitals and their estimated daily intake. *Food Addit. Contam.*, 18, 449.

Tuft, L. and Ettelson, L. N. (1956). Canker sores from allergy to weak organic acid (citric and acetic). *J. Allergy*, 27, 536.

U.S. FDA. (1983b). Termination of provisional listing of FD&C Red No. 1. Title 21, Code of Federal Regulation, Part 81, Sect. 81.10. Office of the Federal Register, General Services Administration, Washington, DC.

Van Buggenhout, S., Sila, D. N., Duvetter, T., Van Loey, A., and Hendrickx, M. E. (2009). Pectins in processed fruits and vegetables: Part III e texture engineering. *Compr. Rev. Food Sci. Food Saf.*, 8(2), 105–117.

Van den Brandt, P. et al. (1995a). Evaluation of national assessments of intake of annatto extract (bixin). Evaluation of certain food additives and contaminants (44th report of the Joint FAO/WHO Expert Committee on Food Additives). WHO Technical Report Series No. 859, pp. 485–492.

van den Ouweland, G. A. M. and Swaine, J. R. L. (1980). Investigation of the species specific flavor of meat. *Perfumer Flavorist,* 5, 15.

Van Leeuwen, Y. M., Velikov, K. P., and Kegel, W. K. (2012a). Morphology of colloidal metal pyrophosphate salts. *RSC Adv.*, 2(6), 2534. http://dx.doi.org/10.1039/c2ra00449f.

Van Leeuwen, Y. M., Velikov, K. P., and Kegel, W. K. (2012b). Repeptization by dissolution in a colloidal system of iron(III) pyrophosphate. *Langmuir*, 28(48), 16531–16535. http://dx.doi.org/10.1021/la303668a.

Van Leeuwen, Y. M., Velikov, K. P., and Kegel, W. K. (2012c). Stabilization through precipitation in a systemof colloidal iron(III) pyrophosphate salts. *J. Colloid Interface Sci.*, 381(1), 43–47. http://dx.doi.org/10.1016/j.jcis.2012.05.018.

Vandenberg, L. N., Chahoud, I., Heindel, J. J., Padmanabhan, V., Paumgartten, F. J., and Schoenfelder, G. (2010). Urinary, circulating, and tissue biomonitoring studies indicate widespread exposure to bisphenol A. *Environ. Health Perspect.*, 118, 1055–1070.

Vargas, M., Chiralt, A., Albors, A., and Gonzalez-Martinez, C. (2009). Effect of chitosan-based edible coatings applied by vacuum impregnation on quality preservation of fresh-cut carrot. *Postharvest Biol. Technol.*, 51(2), 263–271.

Varzakas, T., Labropoulos, A., and Anestis, S. (eds.) (2012). *Sweeteners: Nutritional Aspects, Applications and Production Technology*, CRC Press, Taylor & Francis Group.

Varzakas, T. H., Arvanitoyannis, I. S., and Labropoulos, A. E. (2010). Food additives and contaminants, Chapter 13. In *Advances in Food Biochemistry*, Yildiz, F. (ed.), CRC Press, Taylor & Francis Group, Boca Raton, FL, pp. 409–457.

Velikov, K. P. and Pelan, E. (2008). Colloidal delivery systems for micronutrients and nutraceuticals. *Soft Matter*, 4(10), 1964. http://dx.doi.org/10.1039/b804863k.

Ventura, S. P. M., Silva, F. A. E., Gonçalves, A. M. M., Pereira, J. L., Gonçalves, F., and Coutinho, J. A. P. (2014). Ecotoxicity analysis of cholinium-based ionic liquids to *Vibrio fischeri* marine bacteria. *Ecotoxicol. Environ. Saf.*, 102, 48–54.

Vo, C. L., Park, C., and Lee, B. J. (2013). Current trends and future perspectives of solid dispersions containing poorly water-soluble drugs. *Eur. J. Pharma. Biopharm.*, 85(3B), 799–813.

Vos, G., Hovens, J. P. C., and Delft, W. V. (1987). Arsenic, cadmium, lead and mercury in meat, livers and kidneys of cattle slaughtered in The Netherlands during 1980–1985. *Food Addit. Contam.*, 4, 73.

Walker, R. (1988). Toxicological aspects of food preservatives. In *Nutritional and Toxicological Aspects of Food Processing*, Walker, R. and Quarttrucci, E. (eds.), Taylor & Francis, London, U.K., pp. 25–49.

Warth, A. D. (1977). Mechanism of resistance of *Saccharomyces bailli* to benzoic, sorbic and other weak acids used as food preservatives. *J. Appl. Bacteriol.*, 43, 215–230.

Warth, A. D. (1985). Resistance of yeast species to benzoic and sorbic acids and to sulfur dioxide. *J. Food Protect.*, 48, 564.

Warth, A. D. (1988). Effect of benzoic acid on growth yield of yeasts differing in their resistance to preservatives. *Appl. Environ. Microbiol.*, 54, 2091.

Warth, A. D. (1989). Relationships among cell size, membrane permeability, and preservative resistance in yeast species. *Appl. Environ. Microbiol.*, 55, 2995.

Washuett, J., Riederer, P., and Bancher, E. (1973). A qualitative and quantitative study of sugar-alcohols in several foods. *J. Food Sci.*, 38, 1262–1263.

Wegener, H. C. (2003). Antibiotics in animal feed and their role in resistance development. *Curr. Opin. Microbiol.*, 6, 439.

Wei, Y., Liu, Z., Liu, D., and Ye, X. (2011). Effect of salicylic acid treatment on postharvest quality, antioxidant activities, and free polyamines of Asparagus. *J. Food Sci.*, 76, S126–S132.

Weil, A. J. and Rogers, H. E. (1951), Allergic reactivity to simple aliphatic acids in man. *J. Invest. Dermatol.*, 17, 227.

WHO. (1963). *International Standards for Drinking Waters*, 2nd edn., World Health Organization, Geneva, Switzerland.

WHO. (1973). Trace Elements in Human Nutrition, Technical Report Series No. 532, World Health Organization, Geneva, Switzerland.

WHO. (1976). *Environmental Health Criteria 3: Lead*, World Health Organization, Geneva, Switzerland.

WHO Food Additives Series 16. Acesulfame potassium. Monographs and Evaluations. Joint FAO/WHO Expert Committee on Food Additives.

Wilson, L. A. et al. (1999). Urinary monitoring of saccharin and acesulfame-K as biomarkers of exposure to these additives. *Food Addit. Contam.*, 16(6), 227–238.

Wilson, R., Van Schie, B. J., and Howes, D. (1998). Overview of the preparation, use and bio-logical studies on polyglycerol polyricinoleic acid (PGPR). *Food Chem. Toxicol.*, 36, 9–10.

Wiseman, R. D. and Adler, D. K. (1956). Acetic acid sensitivity as a cause of cold urticatia. *J. Allergy*, 27, 50.

Yang, B., Hao, F., Li, J., Wei, K., Wang, W., and Liu, R. (2014). Characterization of the binding of chrysoidine, an illegal food additive to bovine serum albumin. *Food Chem. Toxicol.*, 65, 227–232.

Ye, R. and Hayes, D. G. (2012). Solvent-free lipase-catalysed synthesis of saccharide-fattyacid esters: Closed-loop bioreactor system with in situ formation of metastable suspensions. *Biocatal. Biotransform.*, 30, 209–216.

Young, V. R., Nahapetian, A., and Janghorbani, M. (1982). Selenium bioavailability with reference to human nutrition. *Am. J. Clin. Nutr.*, 35, 1076.

Youssef, M. K. and Barbut, S. (2010). Effects of caseinate, whey and milk proteins on emulsified beef meat batters prepared with different protein levels. *J. Muscle Foods*, 21, 785–800.

Youssef, M. K. and Barbut, S. (2011). Effects of two types of soy protein isolates, native and preheated whey protein isolates on emulsified meat batters prepared at different protein levels. *Meat Sci.*, 87, 54–60.

Zalacain, A. et al. (2005). Near-infrared spectroscopy in saffron quality control: Determination of chemical composition and geographical origin. *J. Agric. Food Chem.*, 53, 9337–9341.

Zeisel, S. H. (1999). Choline and phosphatidylcholine. In *Nutrition in Health and Disease*, 9th edn., Shils, M., Olson, J. A., Shike, M., Ross, A. C. (Eds.), Williams & Wilkins, Baltimore, MD.

Zeisel, S. H. and da Costa, K.-A. (2009). Choline: An essential nutrient for public health. *Nutr. Rev.*, 67, 615–623.

Zhang, Y., Chen, K. S., Zhang, S. L., and Ferguson, I. (2003). The role of salicylic acid in postharvest ripening of kiwifruit. *Postharvest Biol. Technol.*, 28, 67–74.

Zhong, Y., Cavender, G., and Zhao, Y. (2014). Investigation of different coating application methods on the performance of edible coatings on Mozzarella cheese. *LWT—Food Sci. Technol.*, 56, 1–8.

LEGISLATION

Council Directive 1996 (96/51/EC of July 23, 1996) amending Directive 70/524/EEC concerning additives in feedingstuffs. Official Journal of European Commission. L 235, 39–58.

Directive 89/107/EEC (OJL 40, 11,2,89, pp. 27–33).

Directive 89/398/EEC (OJL 186, 30.6.89, pp. 27–32).

Directive 91/321/EEC on infant formulae and follow-on formulae (OJL 175, 4.7.91, pp. 35–49).

INTERNET SOURCES

http://europa.eu.int/comm./food/food/animalnutrition/feedadditives/authoadditivesen.htm. Accessed on October 2013.

http://europa.eu.int/comm./food/food/animalnutrition/feedadditives/update.pdf. Accessed on October 2013.

http://europa.eu.int/comm./consumers/cons_safe/prodsafe/gpsd/rapexweekly/2004week30.htm. Accessed on October 2013.

http://www.cspinet.org/reports/chemcuisine.htm. Accessed on October 2013.

http://www.cfsan.fda.gov/1rd. Accessed on October 2013.

http://www.x-sitez.com/allergy/additives/colors100–181.htm. Accessed on October 2013.

http://www.x-sitez.com/allergy/additives/preservatives200–290.htm. Accessed on October 2013.

http://www.x-sitez.com/allergy/additives/296–385.htm. Accessed on October 2013.

http://www.x-sitez.com/allergy/additives/vege400–495.htm. Accessed on October 2013.

http://www.x-sitez.com/allergy/additives/misa500–579.htm. Accessed on October 2013.

http://www.cfsan.fda.gov/_lrd/foodaddi.html. Accessed on October 2013.

http://www.nationalaglawcenter.org/readindrooms/foodsafety/. Accessed on October 2013.

http://www.epa.gov/history/topics/fqpa/02.htm. Accessed on October 2013.

http://www.fcs.uga.edu/pubs/current/fdns-e-3.html. Accessed on October 2013.

http://www.cspinet.org/reports/chemcuisine.htm. Accessed on October 2013.

10 Ohmic Heating
Principles and Application in Thermal Food Processing

M. Reza Zareifard, M. Mondor, S. Villeneuve, and S. Grabowski

CONTENTS

10.1 INTRODUCTION

The application of high-temperature short-time (HTST) technology by means of aseptic-processing technique for filling is limited by the time required to conduct sufficient heat to the center of large particles to ensure sterilization. These techniques rely on indirect heating with some deficiencies that led to the development of different technologies based upon direct resistance heating, such as Ohmic heating (Eliot-Godéreaux et al., 2001). Ohmic heating, which cannot be considered as an emerging technology any longer, has obvious advantages in saving energy, shortening the heating time, and improving processed food quality, especially when processing foods containing large particulates (Chen et al., 2010).

Ohmic heating is comparable to microwave heating without an intermediary step of converting electricity into microwaves through the magnetron before applying to the product. Varghese et al. (2014) reviewed Ohmic heating technology and its application in food processing and concluded that this technique is especially advantageous in processing semi solid, particulate foods. It has proven advantages over conventional thermal processing and novel thermal alternative technologies. However, the success of Ohmic heating depends on the rate of heat generation in the system, the electrical conductivity (EC) of the food, electrical field strength, residence time, and the method by which the food flows through the system.

Ohmic heating can be used for heating liquid foods containing large particulates, such as soups, stews, and fruit slices in syrups and sauces, and heat-sensitive liquids. The technology is useful for the treatment of proteinaceous foods, which tend to denature and coagulate when thermally processed. For example, liquid eggs can be ohmically heated in a fraction of a second without coagulating them (Bozkurt and Icier, 2012). Juices can be treated to inactivate enzymes without affecting the flavor (Leizerson and Shimoni, 2005).

A large number of potential future applications exist for Ohmic heating including its use in blanching, thawing, evaporation, dehydration, fermentation, extraction, sterilization, pasteurization, and heating of foods to serving temperature, including in the military field or long-duration space missions. One of the main advantages claimed for Ohmic heating is rapid and relatively uniform heating. Ohmic heating is currently being used for processing of whole fruits in Japan and the United Kingdom. One commercial facility in the United States uses Ohmic heating for the processing of liquid eggs (FDA, 2011). It is believed that Ohmic heating is one of the most demanding advanced green technologies by the food industries.

The heating occurs in the form of internal energy transformation (from electric to thermal) within the materials (Sastry and Barach, 2000) due to their electrical resistance properties. Therefore, Ohmic heating can be seen as an internal thermal energy generation technology, and not only as thermal energy transfer, meaning that it does not depend on heat transfer either through a solid–liquid interface or inside a solid in a two-phase system (Knirsch et al., 2010).

In general, Ohmic processing enables to heat materials at extremely rapid rates, from a few seconds to a few minutes (Sastry, 2005). It also enables, under certain circumstances, large particulates and carrier fluids to heat at comparable rates, thus making it possible to use HTST and ultrahigh temperature (UHT) techniques on solids or suspended materials (Imai et al., 1995).

The heating rate of particles in a fluid depends on the relative conductivities of the system's phases and the relative volume of those phases (Sarang et al., 2007). Low-conductivity solid particles, compared to fluid conductivity particles, tend to lag behind the fluid at low concentrations related to the volume of the fluid. However, in conditions where the concentration of the particles is high, those same low-conductivity particles may heat faster than the surrounding fluid. So, the phenomenon of particle lagging or particle leading depends on the significance of particle resistance to the overall circuit resistance (Sastry and Palaniappan, 1992a). This phenomenon occurs because, with the increase of the particles' concentration, the electric current path through the fluid becomes more tortuous, forcing a greater percentage of the current to flow through the particles. This can result in higher energy generation rates within the particles and consequently in a greater relative particle heating rate (Sastry and Palaniappan, 1992b; Sarang et al., 2007). This fact indicates that it may be possible to adjust the heating pattern of solid–fluid systems by adjusting the overall influence of particles' resistance in the system through setting the particles' concentration in the fluid.

In a nonhomogeneous material, such as soups containing slices of solid foods, the electrical conductivity (EC or σ) of the particles and its relation to the fluid conductivity is pointed as a critical parameter. This parameter can be adjusted to the desired level by soaking the materials in proper solution prior to processing. There are critical values of σ (below 0.01 S/m and above 10 S/m) where Ohmic heating is not applicable (Piette et al., 2001, 2004).

Due to the lack of sufficient basic data, such as electrical conductivities of solid food that are necessary for a high quality, energy efficient, and safe process design, the technique has not been so commercialized for particulate foods. However, during the last two decades, attempts have been made by several researchers to understand the heating behavior and evaluate the EC of different food components and matrices. This helps food processors to apply the Ohmic heating technique for commercial processing of particulate foods and for the researchers to cross-check the values and fill up the gaps by developing more accurate equations to predict the electrical conductivities of foods for a given condition in a specific Ohmic heating system.

To predict the EC of food materials, mathematical and empirical models have been developed as a function of different influencing parameters over the last two decades by many researchers (Palaniappan and Sastry, 1991a,b; Yongsawatdigul et al., 1995; Wang and Sastry, 1997; Wu et al., 1998; Moura et al., 1999; Marcotte et al., 2000; Castro et al., 2003; Ayadi et al., 2004; Icier and Ilicali, 2004, 2005a,b; Assiry et al., 2006; Pongviratchai and Park, 2007; Singh et al., 2008). In case of nonhomogenous materials, effective models were proposed to estimate the electrical conductivities of two-phase mixtures (Maxwell, 1881; Meredith and Tobias, 1960; Brailsford and Major, 1964; Kopelman, 1966; Assiry et al., 2006; Zhu et al., 2010); however, applications of these models are limited only to dilute spherical particle dispersions (Sastry and Palaaniappan, 1992a). Reported values for the EC of food materials are valid for the specified conditions (for example: a given temperature, a known salt or solid content, a set voltage and frequency), whereas this property is strongly dependent on several factors, and that is why multi-regression models are required for its prediction and application. Investigators have been exploring some of these influencing factors on the EC of foodstuffs over years, and therefore linear or nonlinear and multiple regression models are available.

In a recent study, Chen et al. (2010) analyzed the process sensitivity of continuous Ohmic heating process for soup products containing liquid and large particulates by the use of a validated computer modeling package to determine the critical control factors. Based on this study, it can be concluded that for the Ohmic heating process used for producing liquid-particle foods, EC of both carrier fluid and particles is the most sensitive variable to the process temperature and target lethality values. Zareifard et al. (2013a) reviewed and reported up-to-date information concerning the existing models for the prediction of the EC of food materials.

When heterogeneous foods (liquids with suspended solid particles) are processed, the liquid acts as an intermediate heat transfer medium, and thermal lags exist within particles. Modeling is a suitable method to provide useful information and examine different scenarios, as well as to predict the particle's temperature and monitor the coldest point during the process. Particle temperatures were conservatively estimated by mathematical models. However, while the particle orients in different directions, the problem is becoming more difficult (Sastry and Palaaniappan, 1992b). There have been some computational modeling works for the last two decades (de Alwis and Fryer, 1990; Sastry, 1992; Zhang and Fryer, 1993; Khalaf and Sastry, 1996; Davies et al., 1999; Ye et al., 2004; Salengke and Sastry, 2007a; Shynkaryk and Sastry, 2012) concerning thermal behavior of food materials undergoing Ohmic heating process. These models mainly simulate the temperature profiles of the liquid and solid components in the food system during Ohmic heating which is necessary for the process design. However, one of the main inputs to these computational models is the EC of the materials. Therefore, knowledge of EC is essential for a safe process design. The worst-case scenario or the extreme condition should be taken into account while designing the process. For this purpose, it is necessary to have a true understanding of the component's resistance that represents extreme conditions or slowest heating conditions.

Concerning food safety and rate of microbial destruction, research is still ongoing with some doubt if there is any additional effect of electrical field on the kinetic parameters of microorganisms of public health concern. Like thermal processing, Ohmic heating inactivates microorganisms by heat. Additional nonthermal electroporation-type effects have been reported at low frequency (50–60 Hz), when electrical charges can build up and form pores across microbial cells (Kuang and Nelson, 1998), however, it is not yet clear; the contribution of the additional effect, and therefore, heating is the main mechanism to be claimed during Ohmic heating.

High energy cost, consumer awareness of the food quality, keen competition in the local and global market, worldwide free business, and intention to use advanced technology have forced the food industry investors to think seriously and apply the new technologies in order to be on top of the market. In recent years, application of Ohmic heating technology in food industries is spreading from European countries and Japan to the North American producers and slowly to the other parts of the world. Food machinery manufacturing companies such as APV Baker (Crawley, U.K.) have designed and manufactured appropriate unit operation systems like Ohmic heating plate heat exchanger for continuous processing of liquids with an affordable price for the industries. Other major equipment suppliers providing commercial-size Ohmic heaters for the food industry include Emmepiemme SRL (Piacenza, Italy); C-Tech Innovation (Capenhurst, U.K.); Agro-Process (Boucherville, Canada), and Raztek Corp. (Patterson, CA). The first Ohmic heating system in commercial operation was installed in the United Kingdom in the early 1990s followed by Japan, France, and United States to process a variety of foods (Parrott, 1992). This technology has been used more in liquid foods such as milk and juices than multi-phase food systems containing liquid and solid; however, recent applications of this technology for the solid foods have also been considered notably. Therefore, since early 2000 major food companies have shown their interest in Ohmic heating processing as a green technology, following the intensive research activities performed in the last two decades discovering critical processing points. In this chapter, the principle of Ohmic heating for the design of a process will be presented, followed by the factors considered as key parameters. Exemplary information about the industries that are invested and applied in this technology will be provided, and current opinions concerning microbial inactivation of pathogens in food undergoing Ohmic heating will be debated.

10.2 PRINCIPLES

The principles of Ohmic heating are simple, and its concept is not new. It is based on the passage of electrical current through a body as a result of applying potential voltage to the electrodes at both ends of the body. The material that is placed between the electrodes acts as an electrical resistance in which heat is generated. Other synonyms used in the literature to describe this principle of heating are direct resistance heating, Joule effect heating, electroconductive heating, and electroresistive heating. Ohmic heating is comparable to microwave heating without the intermediary step of converting electricity into microwaves through the magnetron before heating the product (Ruan et al., 2001). However, as a heating technology, Ohmic heating has a very high coefficient of performance, close to one, meaning that every 1 W of electrical power is converted to almost 1 W of heat. Ohmic heating technology offers an alternate way to rapidly heat the food considered as the resistance. In this method, when an alternating current is passed through a food that has appropriate EC, heat generation takes place through the foodstuff, and the electrical energy is directly converted into heat, causing a temperature rise. Classical heat transfer mechanisms such as convection or conduction are minimal.

In food application, the system is similar to an electrical circuit as shown in Figure 10.1, which is comprised of a resistance and a source of current with appropriate voltage gradient. The food product which is placed between the two electrodes acts as the resistance when an alternating current passes through it. However, the most important factor is the EC of the product which is a temperature-dependent parameter.

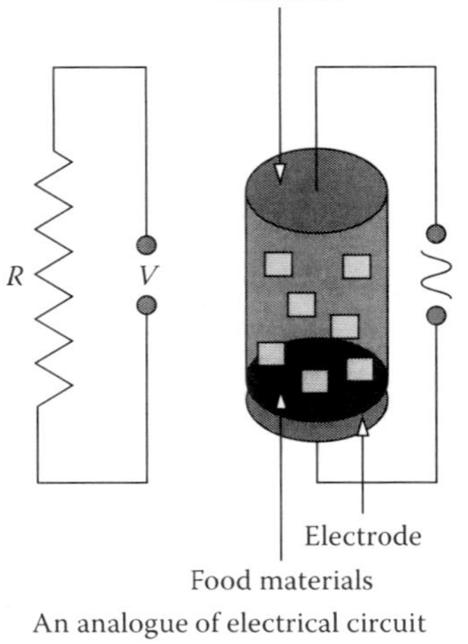

FIGURE 10.1 Principle of Ohmic heating process for food materials.

The rate of heating is directly proportional to the square of the electric field strength and the EC or the resistance of material based on Joule's first law: $P = V^2/R$; or $P = I^2R$ considering Ohm's law: $V = IR$; where P is the energy per unit time in W or J/s; V is the potential difference in V; R is the electrical resistance measured in Ω; and I is the electrical current in A. The electric field strength can be varied by adjusting the electrode gap or the applied voltage.

The EC (σ, in Siemens per meter, S/m) of food materials, which is one of the most important parameters to know in Ohmic heating process, can be measured from the resistance of the sample and the geometry of a simple Ohmic heating cell using the following equation (Palaniappan and Sastry, 1991a; Zareifard et al., 2003):

$$\sigma = \left(\frac{1}{R}\right) \times \left(\frac{L}{A}\right) \times 100 = \left(\frac{I}{V}\right) \times \left(\frac{L}{A}\right) \times 100$$

where

L is the gap between the electrodes (cm)

A is the cross-section surface area of the electrodes (cm^2)

The resistance of the sample (R) in Ω was determined from voltage (V) and current (I) data as $R = V/I$.

EC refers to the ability of a material such as food to transport an electric charge. By introducing an electrical voltage difference across a conducting material, the transferable charges will flow through the conductor that produces an electrical current (primarily alternating). The existence of the electrolytic component, such as salt and acids in food materials, allows the electric current to pass through them. The passage of an electric current through food systems is the basis of Ohmic heating technique, which generates heat internally within both liquid and solid phases and raises the temperature that can be used to pasteurize or sterilize food. The amount of generated heat is directly proportional to the current that is produced by the voltage gradient and the EC of the foodstuff (Skudder and Biss, 1987; Sastry and Li, 1996; Icier and Ilicali, 2004). Furthermore, the amount of current that passes through the system depends on the EC or electrical resistance of the materials presented in a food system. Therefore, in an Ohmic heating process, EC can be considered as the most important and key parameter as indicated by many researchers (de Alwis et al., 1989; Palaniappan and Sastry, 1991a; de Alwis and Fryer, 1992; Tulsiyan et al., 2008). The suitability of the materials for Ohmic heating can be determined by their electrical conductivities. EC can be considered as a fundamental characteristic of all materials that may depend on other materials' properties such as composition, soluble salt percentage, electrolyte mobility, and temperature (de Alwis and Fryer, 1992; Ayadi et al., 2004). Moreover, EC measurements can be used to detect compositional discrepancy and structural changes in food products (Zhuang et al., 1997; Guerin et al., 2004). A liquid-particle food system is mainly being controlled by the EC of both solid and liquid phases.

The importance of EC of liquid and solid food phases during Ohmic heating have been demonstrated by de Alwis et al. (1989, 1990) through experiments and simulation works. Proper knowledge of electrical conductivities of foods under Ohmic heating conditions is essential for designing a process for an Ohmic heating system. The design of a continuous Ohmic heating processing line is principally determined by the selected configuration of the system, the heating rate, the flow rate, and the desired temperature rise of the food product (Reznik, 1996). The Ohmic heating rate is calculated from the electric field, EC, density, and specific heat of the food product. Therefore, the heating rate depends largely on the physical properties of the food and the EC. Ohmic heating works effectively for foods because most pumpable foodstuffs contain dissolved ionic salts and acids and water in excess of 30%, which render the material electrically conductive (de Alwis and Fryer, 1992). Pure fats, oils, alcohols, and sugars are not suitable materials for Ohmic heating. These substances are not sufficiently electrically

TABLE 10.1
Values of Electrical Conductivity (EC) for Some Materials and Level of Heating Rate

Product Type	EC at 25°C (S/m)	Heating Rate (°C/s)	Category
Copper[a]	5.96×10^7	Not applicable	
Stainless steel[a]	1.45×10^6		
Titanium[a]	2.38×10^6		
Sea water (3.5% salt)[a]	4.8	1–5	Rapid heating
Starch 4.3% + 1% salt[b]	2.1		
Starch 4.3% + 0.5% salt[b]	1.07		
Pickles and chutneys[c]	2.0–3.0		
Savory sauces[c]	1.6–1.8		
Various soups[c]	1.4–1.8		
Minced beef, lamb[c]	0.8–1.2		
Pet foods[c]	0.7–1.2		
Full-cream milk[c]	0.52	7–50	Very rapid heating
Dairy desserts[c]	0.38–0.50		
Beaten egg[c]	0.4		
Vegetable pieces[c]	0.06–0.1		
Fruit pieces[c]	0.05–0.15		
Margarine[c]	0.027	Unsuitable products	Too low
Sugar syrup[c]	0.001		
Drinking water[a]	5×10^{-4} to 5×10^{-2}	Not applicable	
Glass[a]	10^{-15} to 10^{-11}		
Teflon[a]	10^{-25} to 10^{-23}		

[a] Helmenstine (2013).
[b] Marcotte (1999).
[c] Stirling (1987).

conductive. Table 10.1 shows the values of EC for some food and nonfood materials. Generally, biological materials and foodstuff are not suitable for Ohmic heating treatment if the value of their electrical conductivities is very low (less than 0.1 S/m) or too high (more than 3 S/m). Zareifard et al. (2013c) collected and classified existing published data on different food materials in detail.

10.3 FACTORS AFFECTING OHMIC HEATING

The Ohmic heating process is more complicated than conventional heating methods as it involves more factors, such as EC, influencing the heating behaviors of carrier fluid and particles. The achievements from research works in the past two or three decades are more valued in academic purpose, but there are still some limitations for the industrial application and commercialization of this technology. Most research works were based on the batch Ohmic heating unit instead of continuous Ohmic heating system which contains heating, holding, and cooling sections. The resistance of a material to pass the current indicates the amount of generated heat in the material, for instance, a food matrix undergoing Ohmic heating process (Skudder and Biss, 1987; Sastry and Li, 1996; Icier and Ilicali, 2004). This resistance or level of conductivity depends on many factors which can be mainly categorized as product and system parameters. The major processing variables investigated over past decades included electrical conductivities of carrier fluid and particles, particle thermal diffusivity, surface heat transfer coefficient, particle size, particle concentration, flow rate, and initial product temperature. Some parameters such as residence time and heat transfer have been well studied for aseptic processing condition prior to Ohmic heating application in food processing. The results of those studies are still useful and applicable in Ohmic heating process due to the similarities in continuous tube-flow process conditions. Other factors related to the product include shape and size, percentage, and orientation of the solid content in the electrical field which basically affect the overall resistance of the system. EC of food materials has been considered as the most important influencing parameter on process temperature and target lethality values as indicated by many researchers (de Alwis et al., 1989; Palaniappan and Sastry, 1991a; de Alwis and Fryer, 1992; Tulsiyan et al., 2008; Chen et al., 2010). This property has to be measured experimentally like other thermo-physical properties of materials such as thermal conductivity, thermal diffusivity, and heat capacity. These properties are not fixed values and could be varied under different conditions especially for biological materials and foodstuffs. There are several parameters that can affect and change the values of EC for a given commodity. Measuring EC of biological materials also depends on the specification of measurement system, the so called system parameters. Other parameters, like under pressure or vacuum, which can change the rate of heating during Ohmic heating process, directly affect the EC of the materials as well. Food processing investigators have investigated some of these influencing factors on the EC of foodstuffs. Simple linear or nonlinear and multiple regression models have been developed and proposed by a group of researchers in the literature.

Some of the product parameters can be named as temperature, ionic strength, free water (Lima et al., 1999), material shape, size, and orientation (Sastry, 1992; Sastry and Palaniappan, 1992a), as well as the ingredient of the food matrix, and the ratio of solid to liquid in complex food matrices (Palaniappan and Sastry, 1991b; Marcotte et al., 1998; Castro et al., 2003; Zareifard et al., 2003; Keshavarz et al., 2011), or solid content in liquid foods such as juices and dairy products (Moura et al., 1999; Icier and Ilicali, 2004, 2005a; Kong et al., 2008; Sun et al., 2008). de Alwis and Fryer (1990), Sastry and Palaniappan (1992a), and Zareifard et al. (2003) reported that particle size, concentration, and orientation have a significant effect on the heating rate and the overall conductivities of the foodstuffs, containing solids. Furthermore, they found that food systems with lower EC required a longer heating time to achieve the target temperature and showed that the EC decreased as particle size and concentration increased while it increased linearly with temperature.

Therdthai and Zhou (2001) briefly reviewed some specific points concerning the influencing factors on the EC of liquid food such as milk. In liquids, EC is considered as a property involving the movement of anions to anode and cations to cathode and the electron transportation to complete the current path (Loveland, 1986). Factors influencing the movement of ions include concentration, electric potential, temperature, and mechanical stirring (Crow, 1994). The chemical structure of a material also affects its EC. For instance, protein has a positive influence on EC (St-Gelais et al., 1995), whereas fat and lactose cannot conduct current. It has been reported that the EC of milk decreases as the concentration of fat and lactose increase (Prentice, 1962), while it always increases with temperature.

In addition to the above mentioned factors concerning the material or product specifications, liquid and solid, there are other parameters related to Ohmic heating system which can directly or indirectly affect the EC of materials. These parameters include power supply specifications such as voltage gradient and frequency, electrodes design, cell size in static Ohmic unit, or fluid flow characteristics in continuous tube-flow conditions (Halden et al., 1990; Palaniappan and Sastry, 1991a; Park et al., 1995; Yongsawatdigul et al., 1995; Wu et al., 1998; Lima et al., 1999; Icier and Ilicali, 2004, 2005a,b; Pongviratchai and Park, 2007). It is well known that the EC of biological materials is a strong function of temperature, and therefore the attempts have been made to evaluate the EC of a variety of food commodities at different temperatures and to develop empirical and regression equations to model EC as a function of temperature.

Each of the above mentioned parameters can affect EC. But this property could be more sensitive to some than the others. There have been some reports about sensitivity analysis related to

Ohmic heating process (Lee et al., 2007). Investigation results showed that particle size, particle thermal properties such as density and specific heat were the most sensitive parameters among product properties which influenced holding tube length required, while fluid properties such as fluid thermal conductivity and viscosity were less susceptible to affect the model prediction regardless of simulation types (Chen et al., 2010). Among the process parameters, product flow rate and product initial temperature seemed to be the most critical parameters within the range covered in a study by Lee et al. (2007).

A good amount of reliable data on EC has been published; however, discrepancies do exist among the available data or the estimations from the proposed empirical models. In some cases, the EC of food materials has been measured at room temperature, which is not the temperature experienced by the food during Ohmic heating. For proper design of Ohmic heating system, it is essential to measure and collect the EC data of foods as well as how they are influenced by parameters such as frequency, voltage applied, voltage gradient, and the composition of food matrix. Some data have been published for high-frequency processes (e.g., microwave), but they are not applicable to low-frequency Ohmic heating processes. Only few publications (Fryer et al., 1993; Zareifard et al., 2003; Salengke and Sastry, 2007b; Tulsiyan et al., 2008) have reported electrical conductivities of real food matrices as a mixture of both liquid and solid components. Usually, as the percentage of solids increases (except for salt), the EC decreases. It is generally observed that the EC of liquids is higher than that of solids, and in liquid foods, the effect of the proportion of solids is of less importance but still significant, according to Sastry and Palaniappan (1992a).

In a recent published effort, Zareifard et al. (2013d) classified the influencing factors on the EC of the food mainly as product and system parameters and highlighted that EC is an indication of the flow of electrical current through the material. It is the opposite of electrical resistance which causes heat generation within nonconducted material while exposed to an electrical field. Any parameters that vary the resistance of a material can change the electrical property of the material. EC is a temperature-dependent thermo-physical property of materials. Other parameters such as: composition of the material, texture, viscosity, shape, and size also significantly affect the EC of biological materials. Furthermore, applied voltage, distance between the electrodes, and the frequency of the power supply unit have been reported as the influencing parameters arising from the systems employed in its measurement. They have mentioned that a real food matrix is usually composed of solid particles and liquid, and concluded that the components which receive the minimum heat are the critical ones to be considered for a safe process. In contrast to the traditional heating process where liquids heat up faster than solids, in Ohmic heating, solid particles may heat up faster than liquids as heating depends on the EC of each component. It is worth mentioning that the EC of solid foods can be modified and adjusted by soaking in salt solution if necessary.

10.4 DESIGN OF OHMIC HEATING SYSTEM

Ohmic heating system design needs several critical factors to be considered for each individual application. These factors are depended on the complexity of the food matrix to be processed and also the unit operation system. Early commercial scale applications of Ohmic heating failed due to the absence of inert electrode materials and control equipment accurate enough to keep the temperature within the necessary range and sufficiently strong to withstand the conditions of commercial production (Ruan et al., 2001). There are endless possibilities for the design of an Ohmic heating system, but some key components are present in each system: a power supply, heating units, electrodes configuration, a flow control system, and a data acquisition system. Currently, Ohmic heating systems are mainly engineered to work in continuous operation mode, as opposed to batch or static mode. Consequently, a flow control system is required and usually consists of pumps, connecting tubes, T-joints, and valves.

10.4.1 Power Supply and Heating Units

Electrical power supply is needed to produce the required energy. The heating system consists of electrodes positioned at various points in the heating section, which may take the form of a cylinder or plate. The electrodes are connected to the power supply and must be in physical contact with the substance to be heated to pass the electric current through it. The distance between the electrodes will determine the strength of the electric field. The surface area of the electrodes is calculated from the required power, the maximum voltage at the power supply, and the current density of the electrodes. Most commercial Ohmic heating systems are operated under constant voltage conditions. Electrodes must be carefully designed to avoid the electrolytic effect that causes the dissolution of the metallic electrodes (Remik, 1988; Reznik, 1996). Early Ohmic heating systems used a range of electrode materials from graphite to aluminum or stainless steel. In recent systems, the use of food-compatible electrode material such as titanium and the correct current density has eliminated contamination problems. Other ways to overcome this problem include using high-power frequency at alternating frequencies above 100 kHz since there is no apparent metal dissolution (Ruan et al., 2001). However, recently Lee et al. (2013) reported that electrode corrosion did not occur when the frequency exceeded 1 kHz while studing the effect of frequency and waveform on inactivation of *Escherichia coli* and *Salmonella* in salsa by Ohmic heating process.

In a continuous system, two possible geometries for operation in a continuous mode can be considered: transverse field and collinear field configurations (Stirling, 1987). In the transverse field configuration, the applied electric field and current flux are perpendicular to the food material flow, while in the collinear configuration, the applied electric field and current flux are parallel to the mass flow. Although the transverse configuration is simple from a mechanical construction point of view, it poses two electrical problems. The first problem is related to the presence of electrodes close to both the inlet and outlet of the heating chamber, which can result in large leakage currents to earth through the product material (Stirling, 1987). The second problem is that the current density in the direction of mass flow can be nonuniform at the electrode edges. Consequently, it is not uncommon to observe localized overheating or boiling in the product, as well as electrode erosion (Varghese et al., 2014). For these reasons, the transverse configuration Ohmic heating system is mainly used for electrical treatment of liquid foods, while the collinear configuration is preferred for particulate foods because of its internal clearance since electrodes are inserted into an electrically insulated tube (Skudder, 1988; Marcotte, 1999). Normally, the Ohmic heating system is completed by holding and cooling parts and data acquisition system used to measure temperature distribution, applied voltage, and current.

10.4.2 Mass and Heat Transfer during Ohmic Heating

As pointed by Varghese et al. (2014), only few studies have been reported in the literature concerning mass transfer during Ohmic heating. These reports indicate that Ohmic heating pretreatment significantly increases extraction yields of soymilk from soybeans (Kim and Pyun, 1995) and apple juice from apples (Lima and Sastry, 1999). More specifically, Lima and Sastry (1999) reported that Ohmic pretreatments of apple samples at 4 and 60 Hz resulted in a significantly higher juice yield than that of untreated samples (609.4 ± 23.1, 586.9 ± 9.0, and 486.4 ± 30.9 mL/kg for the 4, 60 Hz and untreated juices, respectively). The 4 Hz pretreated apple samples also resulted in higher juice yield than that of 60 Hz pretreated samples. No visual difference in color or clarity was observed among the juices. Wang and Sastry (2000) also reported that Ohmic pretreatment (40 V/cm, 60 Hz) of cylindrical samples of carrot, potato, and yam heated up to average temperatures of 50°C or 80°C significantly enhanced the drying rate (in most situations, more than conventional and microwave heating pretreatment) when the samples were dried in a hot-air dehydrator. They observed that ohmically pretreated samples clearly showed moisture diffusion from intra to intercellular regions, while this was not obvious for samples pretreated by the other two heating techniques. So, it was concluded that the enhancement of drying rate observed for Ohmically pretreated samples was due to this movement of moisture through cell walls. Schreier et al. (1993) also reported enhanced diffusion of beet dye from beetroot tissue into a fluid when the tissues were submitted to Ohmic heating (50 Hz), as compared to conventional heating, for samples, heated to a final temperature of 80°C. Electroosmosis has been proposed to explain the increase in dye diffusion. From the aforementioned studies, we can conclude that the study of mass transfer during Ohmic heating is of interest since it could have important commercial applications.

Heat and mass transfer studies are essential in any food processing design and manufacturing unit operation. Heating rates of solid and liquid phases during Ohmic heating depend on the electrical conductivities of both phases. If the heating rate of the phases can be predicted, an Ohmic heating process can be easily designed. When the electrical conductivities of both phases are equal, both phases will heat at the same rate which represents an ideal case (Varghese et al., 2014). Ignoring convection and conduction effects and neglecting heat loss to the surroundings, the rate of heating in an Ohmic system is derived from energy balance as follows (Stirling, 1987):

$$\text{Heating Rate} = \frac{\text{Power Density}}{(\text{Specific Heat} \times \text{Density})}$$

Assuming that the Ohmic heating system is operated under constant voltage conditions, the heating rate can be expressed as (Marcotte, 1999):

$$\frac{dT}{dt} = \frac{\sigma(\nabla V)^2}{\rho C_p}$$

where
- T is the temperature
- t is the time
- σ is the local EC
- ∇V is the voltage gradient
- ρ is the density
- C_p is the specific heat capacity

Once the target processing temperature and product specification are known, the required electrical power which will determine the size of the transformer for the power supply can be determined from the following equation:

$$P = \dot{m} C_p (T_{\text{out}} - T_{\text{in}})$$

where
- T_{out} is the outlet temperature
- T_{in} is the inlet temperature
- $\dot{m}$ is the product mass flow rate
- P is the power

Marcotte (1999) recommended that the maximum power at the power supply must be 30% greater than the required power estimated in the aforementioned equation.

10.4.3 Example of Continuous System Design

Problem statement: To pasteurize 0.5 kg/s (1800 kg/h) of cocktail vegetable juice from refrigerated temperature 10°C to 70°C ($\Delta T = 60$°C).

Assuming specific heat of juice C_p = 4.2 kJ/kg °C, required power is 126 kW ($P = \dot{m}C_p (T_{out} - T_{in}) = 0.5 \times 4.2 \times 60$). Considering the efficiency of the system at about 90%, the total required power will be 140 kW. (In practice, 20%–30% extra power is recommended for the system performance and operation.)

If the system runs for an hour, to process 1800 kg/h of the juice, the required energy is 140 kW h, and if the cost of energy is \$0.15/kW h, the total cost will be 140 × 0.15 = \$21, which is about 1.2¢/kg of juice.

For a practical situation, the flow can be divided in 10 tubes with the same diameter.

Assuming 10 processing tubes for the total flow:

Mass flow rate (Q) in each line = 1800/10 = 180 kg/h = 0.05 kg/s = 5×10^{-5} m³/s
Liquid density ρ = 1000 kg/m³
Inside diameter tube = 2 in. (cross-section surface area = 20.25 cm² = 20.25×10^{-4} m²)
Average liquid velocity in the main tube = $V = Q/A = 5 \times 10^{-5}$ m³/s/20.25×10^{-4} m² = 0.025 m/s = 2.5 cm/s

Electrode specification:

Cylinder shape
Diameter = 2 cm
Length = 4.8 cm
Surface area = 2 × 3.14 × 1 × 4.8 = 30.1 cm²
Distance between the electrodes in each tube = 1.0 m
Time to heat up (t) = L/V = 1.0 m/0.025 m/s = 40 s

Power supply specification:

Required power for each line = 140 kW/10 = 14 kW
EC of the food = 1.2 S/m = 0.012 S/cm
Finding R:
$\sigma = L/AR$ or $R = L/A\sigma$
$R = 100/30.1 \times 0.012 = 276\ \Omega$
Voltage and amperes required for each line;
$P = I^2R$ or $I = (P/R)^{0.5}$
$I = (14{,}000/276)^{0.5} = 7.1$ A
$V = P/I = 14{,}000/7.1 = 1980$ V
Voltage gradient is $V/L = 2000/1.0 = 2000$ V/m = 20 V/cm

For the given condition, the total required power (10 parallel lines to process 1800 kg/h in total) is 140 kW which can be obtained through a strong power supply of 2000 V and 70 A. However, if it is desired to change the power supply specifications, it is possible to change some parameters. For example, if the length of the tube or the distance between the electrodes reduces from 1.0 to 0.5 m the resistance (R) will be decreased; therefore, more current can pass through the food system and the required amperage increases from 7 to 10 A, and hence the required voltage decreases from 2 to 1.4 kV according to the following calculations:

$\sigma = L/AR$ or $R = L/A\sigma$
$R = 50/30 \times 0.012 = 138\ \Omega$
$I = (P/R)^{0.5} = (14{,}000/138)^{0.5} \cong 10$ A
$V = P/I = 14{,}000$ W/10 A = 1400 V

In this case, the heating time will be shorter since the flow rate and the flow velocity remain the same; therefore,

Distance between the electrodes in each tube = 0.5 m
Time to heat up (t) = L/V = 0.5 m/0.025 m/s = 20 s

In summary, to heat-up 0.5 kg/s (30 kg/min or 1800 kg/h) of vegetable juice from 10°C to 70°C, 140 kWh energy is required which can be provided by a power supply with 2 kV and 70 A using 10 parallel resistance lines, each needing 7 A (columns of the liquid with 1 in. in diameter and 1.0 m long, $R = 276\ \Omega$). Another alternative is to provide the 140 kW power through 1.4 kV and 100 A (columns of the liquid 1 in. in diameter but 0.5 m long, with less resistance $R = 138\ \Omega$). If low-frequency power supply is used (60 Hz), to avoid of contamination with metal corrosion, the electrodes could be made from titanium in a cylindrical shape with 2 cm outside diameter and 4.8 cm length which can be installed perpendicular to the flow direction within the heating tube lines. If high-frequency (10 kHz and more) power supply is used, regular stainless steel materials used in food industry can be also used for the electrodes. The typical design of such a system is shown in Figure 10.2, in which only one segment of a continuous Ohmic heating system equipped with 10 kW (1 kV and 10 A, for example) variable power supply is detailed.

Table 10.2 shows the key parameters in an exemplary Ohmic heating system. As it can be seen from the numbers in the table, any change in these parameters results in different power supply specifications in terms of the voltage and amperage requirements, while the required power is constant as long as the flow rate remains the same. However, if the flow specifications change (changing the fluid velocity for the given tube size or vice versa) the required power, voltage, and amperage change. Of course, EC of the food undergoing processing plays a remarkable role. Please note that the aforementioned example has been presented for illustration purposes only as it was proven in practice.

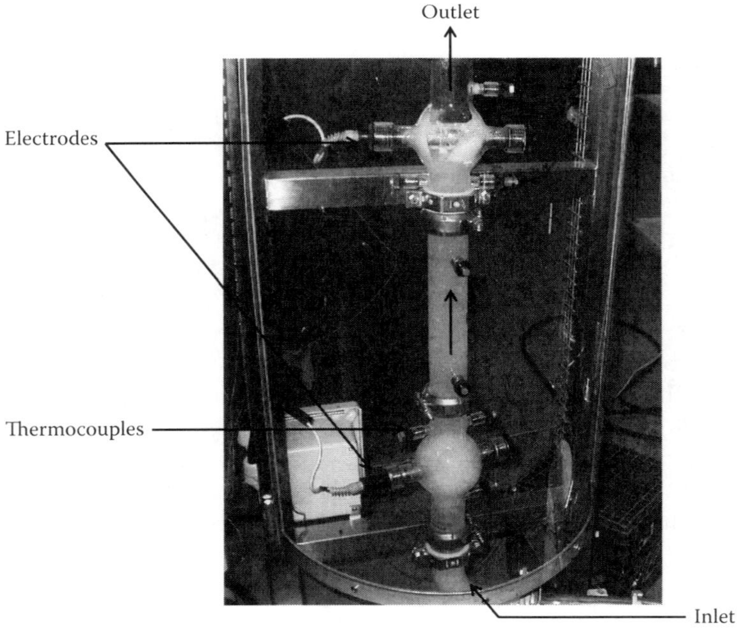

FIGURE 10.2 One heating segment of a continuous Ohmic heating system in Food Research and Development Center, Agriculture and Agri-Food Canada.

TABLE 10.2
Sensitivity Analysis: Effect of Ohmic Heating System and Product Parameters on Power, Voltage, and Amperage Requirements

			Changing the Parameters						
			Once at a Time				Two or More at a Time		
Parameters	**Unit**	**Selected**	L_t	**ID**	Q	σ	L_t, **ID**	L_t, **ID**, Q	L_t, **ID**, Q, σ
Tube length, L_t	cm	100	**50**	100	100	100	**50**	**50**	**50**
Tube size, ID	in.	2	2	**3**	2	2	**3**	**3**	**3**
Tube diameter, ID	cm	5.08	5.08	**7.62**	5.08	5.08	**7.62**	**7.62**	**7.62**
Tube cross section, A_t	cm^2	20.3	20.3	**45.6**	20.3	20.3	**45.6**	**45.6**	**45.6**
Flow rate-volumetric, Q	l/s	0.05	0.05	0.05	**0.03**	0.05	0.05	**0.03**	**0.03**
Flow rate-volumetric, Q	l/h	180	180	180	**108**	180	180	**108**	**108**
Flow velocity, V	cm/s	2.5	2.5	1.1	**1.5**	2.5	1.1	**0.7**	**0.7**
Fluid density, ρ	kg/m^3	1000	1000	1000	1000	1000	1000	1000	1000
Flow rate-mass	kg/s	0.05	0.05	0.05	0.03	0.05	0.05	0.03	0.03
Heating time, t_H	s	41	**20**	91	**68**	41	**46**	**76**	**76**
Fluid viscosity at 20°C, μ	kg/m s	0.002	0.002	0.002	0.002	0.002	0.002	0.002	0.002
Reynolds no. at 20°C		627	627	418	376	627	418	251	251
Fluid viscosity at 80°C, μ	kg/m s	0.001	0.001	0.001	0.001	0.001	0.001	0.001	0.001
Reynolds no. at 80°C		1254	1254	836	752	1254	836	502	502
Fluid electrical conductivity, σ	S/cm	0.01	0.01	0.01	0.01	**0.02**	0.01	0.01	**0.02**
Electrode-cylindrical Shape									
Electrode diameter, d	cm	2	2	**3**	2	2	**3**	**3**	**3**
Electrode length, L_e	cm	4.9	4.9	**7.4**	4.9	4.9	**7.4**	**7.4**	**7.4**
Electrode surface area, A_e	cm^2	30.6	30.6	**69.9**	30.6	30.6	**69.9**	**69.9**	**69.9**
Electrodes distance, L	cm	100	**50**	100	100	100	**50**	**50**	**50**
Resistance, R	Ω	326	**163**	**143**	326	**163**	**72**	**72**	**36**
Specific heat, C_p	J/kg °C	4200	4200	4200	4200	4200	4200	4200	4200
Temperature difference, ΔT	°C	60	60	60	60	60	60	60	60
Power-calculated, P	kW	12.6	12.6	12.6	**7.6**	12.6	12.6	**7.6**	**7.6**
Efficiency	%	95	95	95	95	95	95	95	95
Power-required, P	kW	13.3	13.3	13.3	**8.0**	13.3	13.3	**8.0**	**8.0**
Amperage, A	A	6.4	**9.0**	**9.6**	**4.9**	**9.0**	**13.6**	**10.5**	**14.9**
Voltage, V	V	2080	**1471**	**1378**	**1611**	**1471**	**974**	**754**	**534**
Voltage gradient, dV/dL	V/cm	21	**29**	**14**	**16**	**15**	**19**	**15**	**11**
Assumed running time, t_R	h	1	1	1	1	1	1	1	1
Energy-required, E	kW h	13.3	13.3	13.3	8.0	13.3	13.3	**8.0**	**8.0**
Electrical energy cost, $	¢/kW h	15	15	15	15	15	15	15	15
Cost/kg	¢/kg	1.11	1.11	1.11	1.11	1.11	1.11	1.11	1.11
Cost/h	$/h	2.0	2.0	2.0	1.2	2.0	2.0	1.2	1.2

Numbers in bold are the values changed as compared with the initial selected column.

10.5 INDUSTRIAL APPLICATIONS

The number of industrial applications of Ohmic heating processes is growing year-by-year; however, sometimes it is still considered as an emerging technology. Ohmic heating equipment can be a part of whole food processing installation in both new and already existing systems. As it is a gentle and continuous process to volumetrically heat different food products, it is especially recommended for the following general applications:

- Products in lump form such as meat, fruit, vegetables, etc.
- Pasteurization/sterilization of vegetable soups and puree
- Pasteurization of dairy products
- Heating or preheating of liquid food products such as juices, syrups, etc.

Continuous as well as batch systems (rather more rarely) are available for commercial food processing. There are several manufacturers of Ohmic heating equipment and units in the world, easy to find through the Internet searching machines. By contact with the manufacturers, it can be easy to find the most suitable industrial, pilot, or laboratory-scale units. It can vary by type, available heating power, efficiency, etc. In most cases of continuous processing, holding tubes of aseptic processing system are connected to the outlet of Ohmic heating unit to assure sufficient residence time and lethality. Significant reduction of necessary cooking time in Ohmic heating systems as compared to conventional cooking is expected (C-Tech Innovation, 2013). Potatoes, pasta, rice, onion, apple, beef, pork, or chicken have reduced cooking time (as compared to standard process) by about 50%, while carrots, peas, or mushrooms by about 20%–30%.

Due to the advantages of rapid heating rate and light weight of Ohmic heating system, NASA was interested in the application of this technique for heating and sterilization of food in space to serve the astronauts during space mission. Pandit et al. (2007) explored the feasibility of designing a light weight compact food warming unit. They have developed an Ohmic heating system suitable for space shuttles to warm the food for a crew of four–six persons on a crew exploration vehicle within half an hour. Based on their finding, one 250 W Ohmic heating system was enough to heat four pouches of food in parallel, containing eight ounces of food sample in each pouch. The temperature at the geometric center of each pouch was monitored by T-type thermocouples for a target temperature of 65°C ± 2°C. Pouches were designed with two electrodes within them and optimized to the shape of a rectangular prism for a greater uniformity of heating and stack ability. Also, Jun and Sastry (2005, 2007) designed and developed efficient light weight Ohmic food warming and sterilization units for the Mars exploration vehicle. They have reported the design of a pouch fitted with electrodes to permit Ohmic heating and applicable to warm food to serving temperature.

Table 10.3 presents some examples of industrial applications of Ohmic heating in food processing industries in the world. As these are examples only, we know that much more units are already working in the food processing industry with strong tendency to apply much more. It is based on high convenience and efficiency, especially for continuous processes. Figures 10.3 and 10.4 present two classical examples of industrial-scale Ohmic heating units.

10.6 MATHEMATICAL MODELING

Modeling is a useful tool for the design of Ohmic heating system and also a beneficial approach to estimate the effect of influencing process parameters on target lethality achieved during processing of food material. The fundamental problem in continuous flow sterilization of food matrix containing solid and liquid components is the lack of knowledge of the coldest point within the entire system. Temperature measurement of moving objects in Ohmic heating system appears as an even

TABLE 10.3
Examples of Commercial Ohmic Heating Units for Food Products

Who	Where	Power	Product
Sous Chef Ltd. (H.J. Heinz Division)	UK	75 kW	Prepared meals of meat and vegetables
Confidential	Europe	75 kW	High-acid products such as fruits in syrup and low-acid products such as vegetables and meats
Confidential	Europe	300 kW	Low-acid particulate foods
Advanced Food Science (AFS) Land O' Lakes	US, IL	5 kW	R&D assistance for product development (continuous system)
Ohio State Un.	US, OH	60 V, 60 Hz	Pilot scale for R&D (liquids, solids and mixtures)
NCFST/FDA/APV Bedford Park	US, IL	5 kW	Dynamic batch Ohmic heating unit for research tests
Nissei Co. Wildfruit Division	Japan	75 kW	High-acid particulate foods such as whole strawberries in syrup (10 L units or 10 kg bags in box)
Confidential	Japan	75 kW	Prepared meals
Confidential	Japan	300 kW	High-acid particulate foods
Nestlé	US, MO	300 kW	Shelf stable low-acid beef stew and ravioli
	Switzerland	60 kW	Milk foods
CTCPA Dury-lès-Amiens	France	10 kW	Various products, with or without particles, pilot-scale installation.
UTC Compiègne	France	5 kW	Capacity of 1 kg/batch at a laboratory scale Testing facilities for a full range of products
Odin Packaging Systems, Parma	Italy	NA	NA
EPRI, Palo Alto	US, CA	NA	Liquid, batch & continuous system
Agriculture Canada Food R&D Center	Canada	5 kW 0–30 kHz	Continuous liquid and solid food
Centro di Tramariglio	Italy	50 kW	Continuous aseptic Ohmic heating system
Sala Baganza, Parme	Mexico	250 kW	NA
Confidential	France	50 kW	Products containing meat
Confidential	France	3 × 150 kW	Cheese
		2 × 100 kW	Cheese
Confidential	Italy	NA	Stewed fruits, Ohmic sterilizer of 2000 kg/h
La Fruitierie du Val Evel	France	60 kW	Fruit sauce, Ohmic sterilizer of 1500 kg/h
Diana Naturals	France	NA	Concentrates, Ohmic sterilizer of 2000 kg/h
Daregal	France	NA	Liquid aromatic herbal preparations, automated Ohmic pasteurizer of 600 kg/h
SFAN (previously Fructalys)	France	NA	Fruits preparation for ice-cream with pieces, semi-automated Ohmic pasteurizer of 500 kg/h
Concept fruits (previously oxades)	France	NA	Fruits preparation for yoghurt and ice-cream with pieces, automated Ohmic sterilizer of 1800 kg/h
CTCPA Avignon	France	NA	Various products with and without particles, Ohmic sterilizer of 300-500 kg/h
I.U.P. Genie des Systemes Agro-industriels	France	NA	Various products with and without particles, Ohmic sterilizer of 500 kg/h

Sources: Updated from Zareifard et al 2013b. Zareifard, M.R. et al., Electrical conductivity: Importance and methods of measurement, in: Ramaswamy, H.S., Marcotte, M., Sastry, S., and Abdelrahim, K., eds., *Ohmic Heating in Food Processing*, CRC Press, New York, NY, 2013b.

NA—data not available.

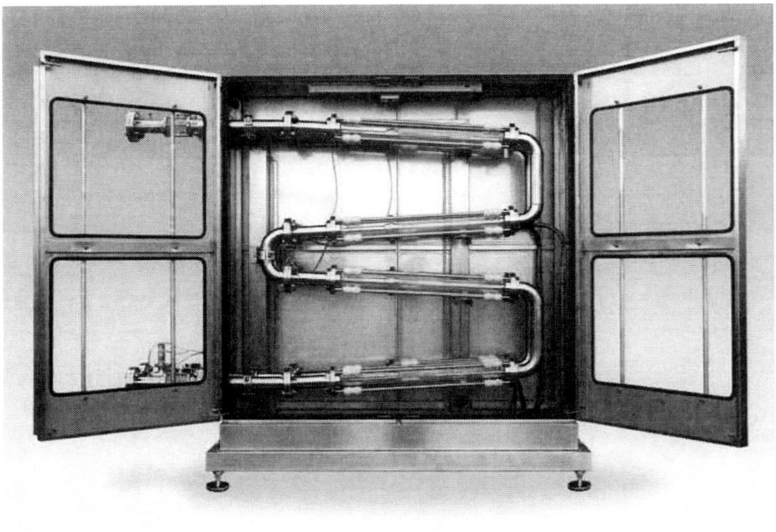

FIGURE 10.3 Ohmic heater of 240 kW for soup processing. (Courtesy of Emmepiemme, Piacenza, Italy; From Municino, F., Industrial applications of Ohmic heating of foods, in: *Workshop on Ohmic Heating*, Abrantes, Portugal, December 19, 2012.)

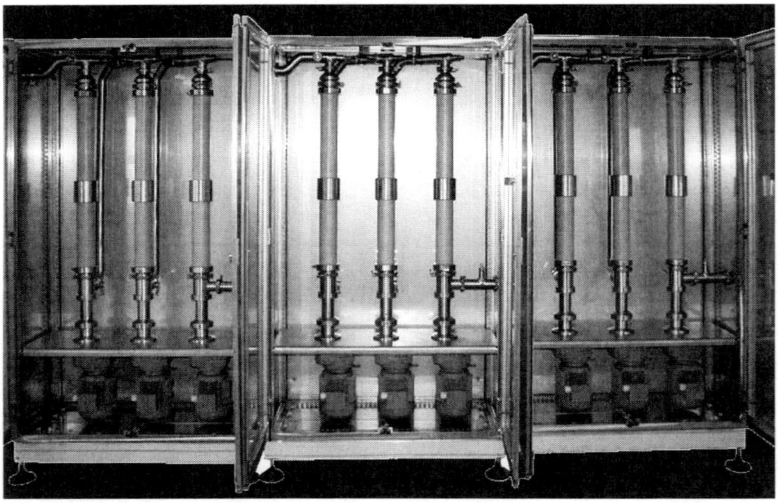

FIGURE 10.4 Example of Ohmic heating unit with heating tubes in series. (From Agro-Process, Technologie chauffage direct, *ZIP Technologique Cintech Agroalimentaire*, 27th edn., April 8, 2010, Saint-Hyacinthe, Quebec, Canada, 2010, also available at Internet: www.agro-process.com. Accessed on October 10, 2013.)

greater challenge than conventional heating process due to the presence of an electric field in the system. Monitoring the temperature of a moving particle due to the motion and moving conditions for the purpose of process validation has always been problematic. However, the cold spot of the food matrix which is a critical controlling parameter can be predicted using the mathematical models followed by validation procedure. Basic information has been built up over the last two decades for the required inputs to the computer simulation packages in order to predict the processed temperature of both solid and liquid during process.

Ohmic heating rates are complex functions of liquid and solid electrical conductivities, particle geometry and size, particle orientation, thermal variation of physical properties, and flow regime surrounding the particle (de Alwis et al., 1989). In practice, for particulate foods, most solid particles exhibit lower EC than liquids. It is well known that the particle temperature of a moving object can be influenced by other processing parameters such as fluid velocity, density, and viscosity. As a result, it is very difficult to apply simple design rules; therefore, different mathematical modeling approaches have been proposed as useful tools for designing an Ohmic heating system. Fryer and de Alwis (1989) and Sastry (1986, 1989) are the first, in the last two or three decades, to attempt to model and to identify the worst-case heating scenario in Ohmic heating process, followed by others (de Alwis et al., 1989, 1992; de Alwis and Fryer, 1990; Sastry, 1992; Sastry and Palaniappan, 1992a,b; Fryer et al., 1993; Khalaf and Sastry, 1996; Sastry and Li, 1996; Murakami and Ramanauskas, 1997; Orangi et al., 1998; Sastry and Salengke, 1998; Ye et al., 2004; Salengke and Sastry, 2007a; Chen et al., 2010; Zhu et al., 2010; Shynkaryk and Sastry, 2012; Kamonpatana et al., 2013). However, there is still lack of systematic information on process sensitivity analysis which covers major processing variables (Chen et al., 2010), and the proposed models are always limited to some conditions; therefore, biological validation is must prior to any further industrial application of such models.

Modeling of Ohmic heating process has been an interesting and challenging subject; therefore, attempts have been made to understand the heating behavior of solid–liquid mixtures under worst-case heating scenarios. Worst-case scenarios are being associated with a single solid piece (within the food matrix) of substantially different EC than its surroundings (Kamonpatana et al., 2013). In this approach, it is considered that the Ohmic heating system must be designed to adequately heat the slowest-heating location within the entire system. Safe sterilization processing of foods containing solid and liquid requires the assurance of all parts of the food treated adequately for the purpose of pathogenic microorganism destruction. The mathematical details of this approach are beyond the scope of this chapter, but it is worth mentioning that the worst-case scenario in Ohmic heating may involve many possibilities, while only few cases will be reviewed here.

In a set of experiments and modeling practice, Salengke and Sastry (2007a,b) indicated that the two possible situations that may constitute worst-case scenarios in specific situations are: (1) an inclusion particle with a static medium surrounding the solid; and (2) an inclusion particle with a mixed fluid. They applied two modeling approaches to analyze potentially hazardous situations, both concerning a single particle in liquid, but one involving a static medium surrounding the solid and the other involving a mixed fluid. The two above mentioned models were compared for extreme conditions where the EC of a single particle could be either significantly lower or higher than that of the medium. Laplace's equation was applied to determine the electrical field distribution within the Ohmic heating system in static fluid (first model), while the temperature distribution was governed by the classical unsteady-state heat equation by conduction plus a generation term (second model). The EC was assumed to be linearly dependent on the temperature according to Palaniappan and Sastry (1991b). For the second model (mixed fluid), a circuit theory analysis was applied to compute the voltage distributions along the Ohmic heating. Solutions of the heat transfer equation involved a separate solution for the liquid and solid phases that was described in detail elsewhere (Sastry and Palaniappan, 1992b; Salengke and Sastry, 2007a).

The Salengke and Sastry (2007a) indicated that the mixed fluid model provided a more conservative prediction of cold-spot temperatures than the static model when the cold spot occurs within the particle; typically occurring when the medium is more conductive than the solid. However, the static fluid model provided more conservative prediction of the cold-spot temperatures when the cold spot is within the fluid; typically when the solid is more conductive than the medium. They also emphasized that the cold spot is within the solid when the medium is more conductive, except when the solid size becomes sufficiently large to intercept a large fraction of the current. Under this condition, the cold zone is within the medium at shadow zones immediately in front and back of the particle. When the particle is more conductive, the coldest zone is within the medium (where the particle size is small compared to the surface area of the electrodes); however, at large particle sizes, the particle cold spot approaches the medium cold spot, except in the case of the mixed fluid model. In an experimental trial, the same authors (Salengke and Sastry, 2007b) measured and monitored the cold-spot zones in a mixture of solid–liquid matrix confirming the prediction of under-processed regions in an Ohmic heating system. In spite of the general belief and common assumptions that worst-case heating scenario is represented by a static situation, they showed that the most conservative situation may frequently be associated with mixed fluid. Static situations may result in worst-case heating (typically within the fluid) when particles are more conductive than the fluid. Finally, they concluded that in Ohmic heating process of solid–liquid mixtures under continuous flowing condition, the mixed fluid model is more conservative when the cold spot is within the solid particle, but the static model is generally conservative when the cold spot is within the medium. For a safe process design, it is therefore necessary to investigate all the possibilities and potential under-process conditions due to the complexity in food processing.

In another approach, Marra et al. (2009) developed a three-dimensional mathematical model using finite element FEMLAB software for a cylindrical batch Ohmic heating unit to process reconstituted potato. Modeling involved the simultaneous solution of the heat transfer equation with a term considering the Ohmic power source and Laplace's equation. Reconstituted potato was chosen to represent a uniform solid food material, and physical and electrical properties were determined prior to the experiment as a function of temperature. The potato-based food stuff was heated using the Ohmic cell at 100 V. Temperature profiles and temperature distribution of the Ohmic heating process were simulated and validated via experimental data involving the measurement of temperature profiles at nine different locations arranged symmetrically inside the unit. The authors reported a good correlation between the model and their experimental data. No cold spots within the product were detected, but both experimental and model data analysis showed that heat loss occurs in the cell wall and electrode surfaces, resulting in slightly cooler regions which would be the critical areas to be monitored when the Ohmic heating unit is used for the sterilization of solid foods.

Despite the fact that considerable effort has been put into simulation of foods and predicting the heating pattern during Ohmic heating process, the majority of models for multiphase food mixtures have dealt with a single solid particle system surrounded by liquid fluid, which is an over-simplified case. Using the computational fluid dynamics (CFD) software Fluent in a 2D environment, Shim et al. (2010) tried to model a more complex solid–liquid food system containing three different solid particles with substantially different electrical conductivities and 3% NaCl solution. The solid food samples used in their experiment were potato, meat, and carrot, which are less conductive than the carried medium. The samples were 1 cm^3 cubes. Uniform ionic concentration was assumed at any location within all solid cubes, and heat loss to the environment was neglected. Electrical conductivities of solid and liquid samples were considered temperature dependent. Modeling involved the solution of Laplace's equation for the heat transfer phenomena considering the internal energy source due to the electrical resistance. The transient heating patterns of each solid food sample and of the carrier medium were obtained. With the maximum prediction error of 6°C, the authors reported good agreement between the predicted temperature values and the experimental data. Under-processed and over-heated regions were also reported as cold spots between the particles where the current density lacks, and the hot spots on the continuous phase in zones perpendicular to the solid cubes.

Recently, Shynkaryk and Sastry (2012) have applied a pilot-scale Ohmic heating chamber with sidewise parallel electrodes for continuous flow treatment of highly viscous chicken chow mein sauce for the purpose of developed model verification. Simultaneous solution of the Navier–Stokes, continuity, electrostatic, and energy transfer partial differential equations was achieved using commercial finite element modeling software. For an accurate simulation, the electrical, thermal, and rheological properties of chicken chow mein sauce (EC; viscosity; thermal conductivity; specific heat capacitance; density) were measured experimentally. They also studied the residence time of sauce in the Ohmic heating chambers by solving the convection and diffusion equation for a tracer species injected in the system. For the model validation, the predicted current delivered by each power generator and the temperatures in selected locations were compared with those obtained from trials performed under the same conditions. A good agreement was reported between experimental and simulated data with the coefficients of determination $R^2 = 0.987$.

More recently, Kamonpatana et al. (2013) developed a mathematical model for solid–liquid mixture in a commercial Ohmic heater with electrical field perpendicular to the flowing of food material. From the safety point of view, they have emphasized that the slowest heating solid particle needs to receive sufficient heat treatment at the outlet of the holding section to ensure sterility of a solid–liquid mixture processed in continuous flow Ohmic systems. Therefore, the fastest moving particle velocity was identified using over 299 particles through a radio-frequency identification technique and the results were used as an input to the model for the worst-case heating scenario. Thermal verification was conducted by comparing predicted and measured fluid temperatures at heater and hold tube outlets. In a case study using chicken alginate particles, their model predicted the length of the holding tube to be 15.85 m for a process temperature of 134°C to achieve a target lethal effect at the cold spot of the slowest-heating particle. The following microbiological verification tests were conducted using at least 299 chicken/alginate particles inoculated with *Clostridium sporogenes* spores per run and proved the absence of viable microorganisms at the target treatment. It was concluded that the procedures could be used to validate the sterilization capability of a continuous Ohmic heating system for producing low-acid foods containing particulates.

Chen et al. (2010) studied the process sensitivity analysis of continuous Ohmic heating process for soup products containing large particulates by using a computer modeling tool to determine the critical controlling points during Ohmic heating process. In order to evaluate the process temperatures and accumulated lethality for both carrier fluid and particles at the end of the holding tube, they investigated the main influencing processing parameters (electrical conductivities of carrier fluid and particles, particle thermal diffusivity, surface heat transfer coefficient, particle size, particle concentration, flow rate, and initial product temperature). They showed the results of their study showed that under given basic processing conditions with a control target process temperature of 133°C (the carrier fluid temperature at the entrance of the holding tube or the end of the heating tube), increasing carrier fluid EC from 1.42 to 2.13 S/m at a reference temperature of 70°C, particle size from 16 to 24 mm, and flow rate from 1600 to 2400 L/h could result in lower particle center temperature than surrounding liquid at the end of the holding tube. However, increasing particle EC from 0.96 to 1.44 S/m (at a reference temperature of 70°C), thermal diffusivity from 1.16×10^{-7} to 1.74×10^{-7} m^2/s, surface heat transfer coefficient from 112 to 168 W/m^2 °C, particle concentration from 52% to 78%, and initial product temperature from 56°C to 84°C could generate higher particle center temperature than liquid. They have concluded that EC of both carrier liquid and particles are the most sensitive variables to process temperature and target lethality. Based on their analysis, under a given control temperature, increasing carrier fluid EC would result in lower process temperatures for carrier fluid and particles at the end of the holding tube, while increasing particle conductivity would generate higher process temperatures for them. An interesting finding is the possibility of particles to reach a higher temperature during Ohmic heating process than the carrier fluid if the particle conductivity is increased close to the carrier fluid conductivity. Other processing variables such as particle size, particle concentration, flow rate, and initial product temperature are also needed to be controlled and monitored during the process in order to achieve the expected process temperature and lethality value at the end of the holding tube. Chen et al. (2010) reported that thermal diffusivity of the processed materials and fluid-to-particle heat transfer coefficient have less effect on the process temperature and accumulated lethality value for both carrier fluid and particles as compared with other influencing factors in their study. It is worth mentioning that the results from their study were based on the effects from the individual processing variable deviations, while multi-variable changes simultaneously may reflect differently, and biological validation is a must to confirm the modeling prediction.

10.7 MICROBIAL DESTRUCTION

A serious question of a possible additional effect of electric field on food-borne microorganisms in Ohmic heating was present since an early step of its development. The thermal effect is very clear and typical as for conventional heating in food pasteurization, cooking, and sterilization (Koutchma, 2009). There are many discrepancies and controversies in the results already published, and some of them were summarized by Sastry (2005), Sun et al. (2008), Knirsch et al. (2010), FDA (2011), Deak and Farkas (2013), and Kamonpatana et al. (2013).

According to these reviews, nonthermal effects have been inconclusive. Most studies either did not specify food sample temperatures or failed to eliminate temperature as a variable. It is critically important that any studies comparing conventional and Ohmic heating were conducted under similar temperature histories. Table 10.4 presents some experimental data of Ohmic heating for selected food microorganisms. For example, Palaniappan et al. (1992) found no difference between the effects of Ohmic and conventional heat treatments on the death kinetics of yeast cells (*Zygosaccharomyces bailii*) under identical thermal histories. In some cases, however, a mild electrical pretreatment of *E. coli* decreased the subsequent inactivation requirement.

Some studies suggest that a mild electroporation mechanism may contribute to microbial cells inactivation during Ohmic heating. For example, studies on fermentation with *Lactobacillus acidophilus* under the presence of a mild electric field (Cho et al., 1996) have indicated that although the fermentation lag phase can be significantly reduced, the productivity of the fermentation is also lowered by the presence of the electrical field. According to the authors, this may be due to the presence of mild electroporation, which improves the transport of substrates at the early stages of fermentation, thereby accelerating it. At the later stages, the electroporation effect would improve the transport of metabolites into the cell and thereby inhibit fermentation. The presence of pore-forming mechanisms on cellular tissue of microorganisms has been

TABLE 10.4
Examples of Existing Information Concerning the Additional Effect of Ohmic Heating on Selected Food Microorganisms

No.	Food Microorganism	*T* (°C)	Frequency (Hz)	Additional Effect	*D* (Ohmic) (min)	*D* (Thermal) (min)	*z* (Ohmic) (°C)	Source
1.	*S. thermophilus* 2646	70	20 k	Yes	6.59 + 0.35	7.54 + 0.37	—	Sun et al. (2008)
		75			3.09 + 0.55	3.30 + 0.42		
		80			0.16 + 0.03	0.20 + 0.03		
2.	*B. subtilis*	88	60	Yes	30.2	32.8	9.16	Cho et al. (1999)
		92.3			9.87	9.87		
3.	*Z. bailii*	49.8	60	No	274	294.6	7.68	Palaniappan and Sastry (1992)
		52.3			113	149.7		
		55.8			43.11	47.21		
		58.8			17.84	16.88		
4.	*E. coli* (vegetative cells)	60	60	No	180.4	165	9.36	Sastry and Barach (2000)
		64.5			51.17	48.2		
		68.5			21.19	22.97		
		71			11.79	11.83		
5.	*C. sporogenes*	110	60–20 k	No	1.4	1.38	12.7	Marcotte et al. (2009)
6.	*E. coli* O157:H7; *Salmonella enterica* serovar *Typhimurium*	60–90.9	60–20 k	Yes	0.4 at 60 Hz 0.25 at 1 kHz			Lee et al. (2012, 2013)

confirmed by work, for example, of Imai et al. (1995). Another similar study (Cho et al., 1999), conducted under near-identical temperature conditions, indicates that the kinetics of inactivation of *Bacillus subtilis* spores can be accelerated by an Ohmic treatment. A two-stage Ohmic treatment (Ohmic heating, followed by a holding time prior to a second heat treatment) further accelerated death rates. Yoon et al. (2002) have indicated that leakage of intracellular constituents of *Saccharomyces cerevisiae* was enhanced under Ohmic heating as compared to conventional heating. The authors suggested, like some other researchers (Uemura and Isobe, 2003; Praporscic et al., 2006), that mild electroporation might occur during Ohmic heating with low-voltage electrical treatment (less than 500 V/cm) as well as at high-voltage treatment (14–16.3 kV/cm). The principal reason for the additional effect of these Ohmic treatments may be the low frequency (50–60 Hz) used, which allows cell walls to build up charges and form pores (electroporation mechanism). This is in contrast to high-frequency methods such as radio or microwave frequency heating, where the electric field is essentially reversed before sufficient wall charge buildup occurs. Some contrary evidence has also been noted. In particular, the work of Lee and Yoon (1999) has indicated that greater leakage of *S. cerevisiae* intracellular constituents is detected under high frequencies.

Sun et al. (2008) published some experimental data of microbial counts of *Streptococcus thermophilus* in raw milk processed for comparison in Ohmic heating and conventional thermal units. The *D*-value of these bacteria in Ohmic heating at 70°C–80°C was lower than that in conventional heating. These results clearly show that Ohmic heating causes a higher microbial death rate than conventional heating does. It is interesting that such an effect was achieved with high-frequency Ohmic heating of 20 kHz, i.e., much higher than that described in previous references (for 50–60 Hz). Figure 10.5 presents a graphical illustration of some of the results achieved in this study. It is interesting to observe that an additional effect of electric field becomes relatively smaller with increase of the processing temperature.

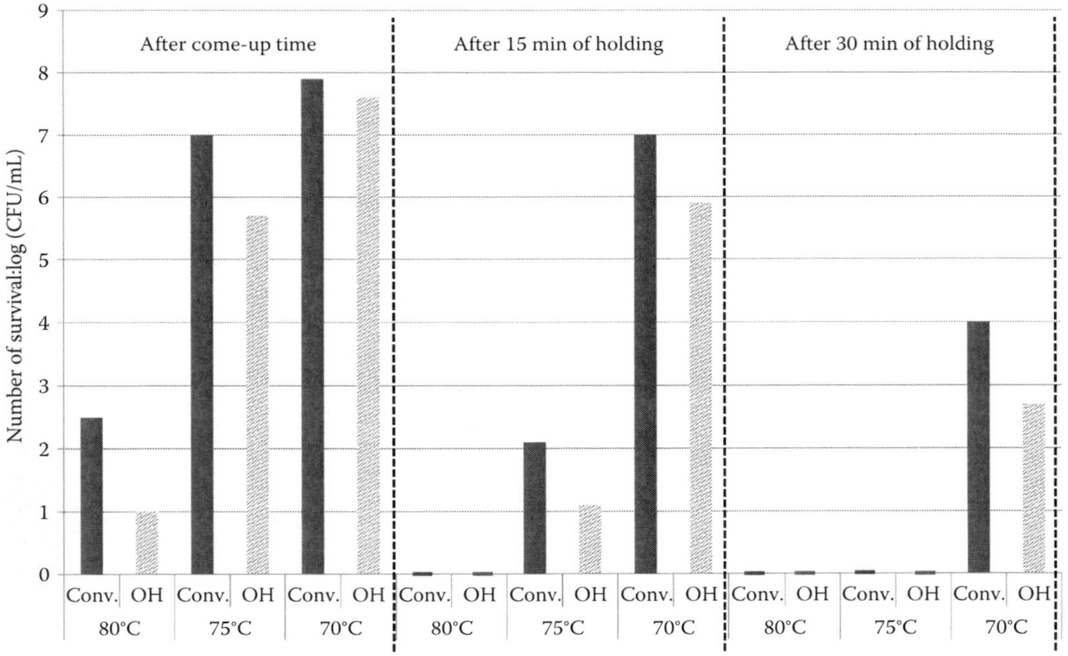

FIGURE 10.5 Comparison of thermal reduction of *Streptococcus thermophiles* in raw milk processed by Ohmic heating and conventional thermal treatment. (Data from Sun, H.-X. et al., *Food Sci. Technol. Res.*, 14(2), 117, 2008.)

Little or no information is available regarding the effects on survivor counts from Ohmic heating processes on post-processing storage. In some experimental study, survivor counts in a pasteurized liquid egg product decreased over storage time, apparently due to injury effects (Reznik, 1999). In this study, initial post-pasteurization plate counts from conventional and Ohmic heating were similar; however, after 12 weeks of storage, many conventionally heated samples were reported to reach counts of 10,000 cfu/mL, while the Ohmic-heated samples exhibited less than 10 cfu/mL in most samples. As expected, electrical field-related injuries can be the explanation for this observation. Contrary to this, in another pilot-plant study, Marcotte et al. (2009) reported that there was no additional Ohmic heating effect for *C. sporogenes* spores suspended in a buffer as compared with conventional thermal treatment. However, it is possible that such an additional effect was covered by a strong thermal effect at 110°C.

In conclusion, the principal mechanism of microbial inactivation in Ohmic heating is the thermal one. Of course, it is possible that for some food microorganisms different additional nonthermal effects exist as presented in Table 10.4. However, it is worth mentioning that although some evidence exists of such the nonthermal effects, it may be unnecessary for processors to claim this additional effect in their process filings (Marcotte et al., 2009; Deak and Farkas, 2013; Kamonpatana et al., 2013). More data is still needed before this additional microorganism destruction effect could be added to the typical thermal effect.

10.8 CLOSING REMARKS

The advantages of Ohmic heating technology have been listed, such as uniform heating; internal heat generation without the limitation of conventional heat transfer and the nonuniformity commonly associated with microwave heating due to limited dielectric penetration; volumetrically heating process, and not a large temperature gradient; possibility of faster heating of particulates than liquid, which is impossible in conventional heating; reducing risks of fouling on heat transfer surface and burning of the food product, resulting in better nutrients and vitamin retention and minimal mechanical damage; high energy efficiency because almost all of the electrical energy is converted into heat; ease of process control; and more quiet working environment.

Fryer and de Alwis (1989) already concluded that Ohmic heating technique is suitable for thermal treatment of heterogeneous foods like fruits, meat mixtures, and tomato particles while there was some technical limitation. Later on Legrand et al. (2007) reported that the developments of new Ohmic sterilizers overcome the technical problems associated with the electrodes and have reduced the investment costs by a factor of 10 between 1993 and 2003. Recently, Fryer and Versteege (2008) illustrated that Ohmic heating process, along with the high-pressure technique, are the leading advanced technologies in the food processing field. These two techniques already pass the feasibility and capability stages, scale up to commercial applications was achieved, and we are in the technology transfer phase to the industrial needs. They also underlined that this process is more energy efficient than microwaves as most of the energy goes to the food and mentioned that the uniformity of heating depends on the uniformity in the EC of the foodstuff. More recently, Varghese et al. (2014) has reviewed the application of Ohmic heating technology and highlighted the advantage of the technology as it uniformly heats in contrast with the nonuniform distribution of microwave heating (Decareau, 1985; Datta and Hu, 1992). In the absence of a hot wall, Ohmic heating provides a considerable advantage for foodstuff applications by not only avoiding the degradation of thermosensitive compounds through overheating but also by reducing the fouling of treated food surfaces during processing (Ayadi et al., 2005). Special advantages are that Ohmic heating can improve food quality and saves cost and energy to processors.

Although the technique appears both simple and advantageous, several difficulties are encountered in its application. Ruan et al. (2001) specified that the disadvantages of Ohmic heating are associated with its unique electrical heating mechanisms. For example, the heat generation rate may be easily affected by the electrical heterogeneity of the particle, heat channeling, complex coupling between temperature and electrical field distributions, and particle shape and orientation. All these make the process complex and contribute to some, rather small, nonuniformity in temperature, which may be difficult to monitor and control. It is already known that Ohmic heaters powered by low-frequency (50–60 Hz) alternating currents induced a corrosion of stainless steel electrodes and apparent electrolysis of the heating medium. However, these electrochemical phenomena can be effectively suppressed by using high-frequency alternating currents (Armatore et al., 1998; Wu et al., 1998; Samaranayake and Sastry, 2005) and many industrial unit operations have been already developed and applied by food processing companies.

10.9 CONCLUSIONS

Ohmic heating, a very promising food processing technique, is one of the fastest-growing advanced thermal processing technologies among the food processing industries due to very high energy efficiency, environmentally friendly conditions, low operating cost, and possible better processed food quality due to high-temperature short-time conditions. It is applicable to a wide range of foods with electrical conductivities between 0.1 and 3 S/m. In some cases, by application of specific additives or pretreatment, an adjustment of EC of food to the above practical range and then Ohmic heating of such a food is also possible. Thermal effects on microbial survival are well documented. However, the additional effects of electrical field are not yet confirmed, and therefore, cannot be included in the calculation of process lethality. In other words, while processing under Ohmic heating conditions, D and z values for food pathogen of public health concern have to be assumed to be the same as for the standard thermal treatment. High-frequency power supply (1 kHz or higher) is recommended to prevent erosion of electrodes and food contamination.

REFERENCES

Agro-Process, 2010. Technologie chauffage direct. *ZIP Technologique Cintech Agroalimentaire*, 27th edn., April 8, 2010, Saint-Hyacinthe, Québec, Canada (also available at Internet: www.agro-process.com). Accessed on October 10, 2013.

Armatore, C., Berthou, M., and Hebert, S., 1998. Fundamental principles of electrochemical ohmic heating of solutions. *Journal of Eelctroanalytical Chemistry*, 457:191–203.

Assiry, A. M., Sastry, S. K., and Samaranayake, C. P., 2006. Influence of temperature, electrical conductivity, power and pH on ascorbic acid degradation kinetics during ohmic heating using stainless steel electrodes. *Bioelectrochemistry*, 68:7–13.

Ayadi, M. A., Leuliet, J. C., Chopard, F., Berthou, M., and Lebouche, M., 2004. Electrical conductivity of whey protein deposit xanthan gum effect on temperature dependency. *Food and Bioproducts Processing*, 82(C4):320–325.

Ayadi, M. A., Leuliet, J. C., Chopard, F., Berthou, M., Lebouché, M., 2005. Experimental study of hydrodynamics in a flat ohmic cell: Impact on fouling by dairy products. *Journal of Food Engineering*, 70:489–498.

Bozkurt, H. and Icier, F., 2012. The change of apparent viscosity of liquid whole egg during ohmic and conventional heating. *Journal of Food Process Engineering*, 35:120–133.

Brailsford, A. D. and Major, K. G., 1964. The thermal conductivity of aggregates of several phases including porous materials. *British Journal of Applied Physics*, 15:313.

Castro, I., Teixeira, J. A., Salengke, S., Sastry, S. K., and Vicente, A. A., 2003. The influence of field strength, sugar and solid content on electrical conductivity of strawberry products. *Journal of Food Process Engineering*, 26:17–29.

Chen, C., Abdelrahim, K., and Beckerich, I., 2010. Sensitivity analysis of continuous ohmic heating process for multiphase foods. *Journal of Food Engineering*, 98(2):257–265.

Cho, H.-Y., Yousef, A. E., and Sastry, S. K., 1996. Growth kinetics of *Lactobacillus acidophilus* under ohmic heating. *Biotechnology and Bioengineering*, 49:334–340.

Cho, H.-Y., Yousef, A. E., and Sastry, S. K., 1999. Kinetics of inactivation of *Bacillus subtilis* spores by continuous or intermittent ohmic heating and conventional heating. *Biotechnology and Bioengineering*, 62:368–372.

Crow, D. R., 1994. *Principles and Application of Electro Chemistry*, 4th edn. Blackie Academic & Professional, Glasgow, U.K.

C-Tech Innovation, 2013. Brochure: Ohmic heating range. A revolutionary, low cost way of heating food. Internet page: www.ctechinnovation.com, recent accessed: April 30, 2013.

Datta, A. K. and Hu, W., 1992. Optimization of quality in microwave heating. *Food Technology*, 46(12):53–56.

Davies, L. J., Kemp, M. R., and Fryer, P. J., 1999. The geometry of shadows: Effects of inhomogeneities in electrical field processing. *Journal of Food Engineering*, 40:245–258.

de Alwis, A. A. P. and Fryer, P. J., 1990. A finite element analysis of heat generation and transfer during ohmic heating of food. *Chemical Engineering and Science*, 45(6):1547–1559.

de Alwis, A. A. P. and Fryer, P. J., 1992. Operability of the ohmic heating process: Electrical conductivity effects. *Journal of Food Engineering*, 15:21–48.

de Alwis, A. A. P., Halden, K., and Fryer, P. J., 1989. Shape and conductivity effects in the ohmic heating of foods. *Chemical Engineering Research and Design*, 67:159–168.

de Alwis, A. A. P., Zhang, L., and Fryer, P. J., 1992. Modeling sterilization and quality in the ohmic heating process. In: *Advances in Aseptic Processing Technologies*, R. K. Singh and P. E. Nelson, eds. Elsevier, New York, pp. 103–142.

Deak, T. and Farkas, J., 2013. *Microbiology of Thermally Preserved Foods: Canning and Novel Physical Methods*. DEStech Publ. Inc., Lancaster, PA, pp. 135–137.

Decareau, R. V., 1985. *Microwave in the Food Processing Industry*. Academic, Orlando, FL.

Eliot-Godéreaux, S. C., Zuber, F., and Goullieux, A., 2001. Processing and stabilisation of cauliflower by ohmic heating technology. *Innovative Food Science and Emerging Technologies*, 2(4):279–287.

FDA, 2011. Kinetics of microbial inactivation for alternative food processing technologies: Ohmic and inductive heating. FDA Internet page: http://www.fda.gov/Food/FoodScienceResearch/SafePracticesforFoodProcesses/ucm101246.htm (last updated: September 11, 2011), recent accessed: April 30, 2013.

Fryer, P. J. and de Alwis, A. A. P., 1989. Validation of the APV ohmic heating process. *Chemistry and Industry*, 16:630–634.

Fryer, P. J., de Alwis, A. A. P., Koury, E., Stapley, A. G. F., and Zhang, L., 1993. Ohmic processing of solid–liquid mixtures: Heat generation and convection effects. *Journal of Food Engineering*, 18:102–125.

Fryer, P. J. and Versteege, C., 2008. Processing technology innovation in the food industry. *Innovation: Management, Policy and Practice*, 10(1):74–90.

Guerin, R., Delplace, G., Dieulot, J.-Y., Leuliet, J. C., and Lebouche, M., 2004. A method for detecting in real time structure changes of food products during a heat transfer process. *Journal of Food Engineering*, 64:289–296.

Halden, K., de Alwis, A. A. P., and Fryer, P. J., 1990. Changes in the electrical conductivity of foods during ohmic heating. *International Journal of Food Science Technology*, 25:9–25.

Helmenstine, A. M., 2013. Table of electrical resistivity and conductivity. http://chemistry.about.com/od/moleculescompounds/a/Table-Of-Electrical-Resistivity-And-Conductivity.htm, recent accessed: November 30, 2013.

Icier, F. and Ilicali, C., 2004. Electrical conductivity of apple and sour cherry juice concentrates during ohmic heating. *Journal of Food Process Engineering*, 27:159–180.

Icier, F. and Ilicali, C., 2005a. The effects of concentration on electrical conductivity of orange juice concentrates during ohmic heating. *European Food Research and Technology*, 220:406–414.

Icier, F. and Ilicali, C., 2005b. The use of tylose as a food analog in ohmic heating studies. *Journal of Food Engineering*, 69:67–77.

Imai, T., Uemura, K., Ishida, N., Yoshizaki, S., and Noguchi, A., 1995. Ohmic heating of Japanese white radish *Rhaphanus sativus*. *International Journal of Food Science and Technology*, 30:461–472.

Jun, S. and Sastry, S. K., 2005. Modeling and optimization of pulsed ohmic heating of foods inside the flexible package. *Journal of Food Process Engineering*, 28(4):417–436.

Jun, S. and Sastry, S. K., 2007. Reusable pouch development for long term space mission: A 3D ohmic model for verification of sterilization efficacy. *Journal of Food Engineering*, 80:1199–1205.

Kamonpatana, P., Hussein, M., Shynkaryk, M., Heskitt, B., Yousef, A., and Sastry, S., 2013. Mathematical modeling and microbiological verification of ohmic heating of a multicomponent mixture of particles in a continuous flow ohmic heater system with electric field perpendicular to flow. *Journal of Food Engineering*, 118:312–325.

Keshavarz, M., Davarnejad, R., and Ghaderi, E., 2011. Effective parameters consideration in ohmic heating process in two phase static system of bio-particle-liquid. *International Journal of Food Engineering*, 7(1):1–17.

Khalaf, W. G. and Sastry, S. K., 1996. Effect of fluid viscosity on the ohmic heating rate of solid–liquid mixtures. *Journal of Food Engineering*, 27:145–158.

Kim, J. and Pyun, Y., 1995. Extraction of soy milk using ohmic heating. In: *Abstract, Ninth Congress on Food Science and Technology*, Budapest, Hungary.

Knirsch, M. C., Alves dos Santosa, C., Augusto Martins de Oliveira Soares Vicente, A., and Vessoni Penna, T. C., 2010. Ohmic heating—A review. *Trends in Food Science and Technology*, 21:436–441.
Kong, Y., Li, D., Wang, L., Bhandari, B., Chen, X. D., and Mao, Z., 2008. Ohmic heating behavior of certain selected liquid food materials. *International Journal of Food Engineering*, 4(3):Article 2.
Kopelman, I. J., 1966. Transient heat transfer and thermal properties in food systems. Ph.D. dissertation, Michigan State University, East Lansing, MI.
Koutchma, T., 2009. Traditional and high-technology approaches to microbial safety in foods, Chapter 23. In: *Microbiologically Safe Foods*, Heredia, N., Wesley, I., and Garcia, S., eds. John Wiley & Sons Inc., Hoboken, NJ.
Kuang, W. and Nelson, S. O., 1998. Low-frequency dielectric properties of biological tissue: A review with some new insights. *Transactions of the American Society of Agricultural Engineers*, 41(1):173–178.
Lee, C. H. and Yoon, S. W., 1999. Effect of ohmic heating on the structure and permeability of the membrane of *Saccharomyces cerevisiae*. In: *IFT Annual Meeting*, Chicago, IL, July 24–28.
Lee, J. H., Singh, R. K., and Chandrana, D. I., 2007. Sensitivity analysis of aseptic process simulations for foods containing particulates. *Journal of Food Process Engineering*, 12(4):295–321.
Lee, S.-Y., Ryu, S., and Kang, D.-H., 2013 Effect of frequency and waveform on inactivation of *Escherichia coli* O157:H7, *Salmonella enterica* serovar *Typhimurium* in salsa by ohmic heating. *Applied and Environmental Microbiology*, 79:10–17.
Lee, S.-Y., Sagong, H.-G., Ryu, S., and Kang, D.-H., 2012. Effect of continuous ohmic heating to inactivate *Escherichia coli* O157:H7, *Salmonella typhimurium* and *Listeria monocytogenes* in orange juice and tomato juice. *Journal of Applied Microbiology*, 112:723–731.
Legrand, A., Leuliet, J.-C., Duquesne, K. P., Winterton, P., Fillaudeau, L., 2007. Physical, mechanical, thermal and electrical properties of cooked red ban for continuous Ohmic heating process. *Journal of Food Engineering*, 81:447–458.
Leizerson, S. and Shimoni, E., 2005. Effect of ultrahigh-temperature continuous Ohmic heating treatment on fresh orange juice. *Journal of Agricultural and Food Chemistry*, 53(9):3519–3524.
Lima, M., Heskitt, B. F., and Sastry, S. K., 1999. The effect of frequency and wave form on the electrical conductivity–temperature profiles of turnip tissue. *Journal of Food Process Engineering*, 22:41–54.
Lima, M. and Sastry, S. K., 1999. The effects of ohmic heating frequency on hot-air drying rate and juice yield. *Journal of Food Engineering*, 41:115–119.
Loveland, J. W., 1986. Conductance and oscillometry. In: *Instrumental Analysis*, 2nd edn., Christian, G. D. and O'Reilly, J. E., eds. Allyn & Bacon Inc., Boston, MA, pp. 122–143.
Marcotte, M., 1999. Ohmic heating of viscous liquid foods. Ph.D. Thesis, Department of Food Science and Agricultural Chemistry, McGill University, Montreal, Quebec, Canada.
Marcotte, M., Piette, J. P. G., and Ramaswamy, H. S., 1998. Electrical conductivities of hydrocolloid solutions. *Journal of Food Process Engineering*, 21:503–520.
Marcotte, M., Trigui, M., and Ramaswamy, H. S., 2000. Effect of salt and citric acid on electrical conductivities and ohmic heating of viscous liquids. *Journal of Food Processing and Preservation*, 24:389–406.
Marcotte, M., Zareifard, R., Pecheux, E., Zhu, S., and Grabowski, S., 2009. Measurement and simulation of electric conductivity of particle-fluid mixtures during ohmic heating. In: *CoFE Annual Meeting*, Columbus, OH, April 6–9, 2009.
Marra, F., Zell, M., Lyng, J. G., Morgan, D. J., and Cronin, D. A., 2009. Analysis of heat transfer during ohmic processing of a solid food. *Journal of Food Engineering*, 91:56–63.
Maxwell, J. C., 1881. *A Treatise on Electricity and Magnetism*, Vol. 1, 27th edn. Clarendon Press, Oxford, U.K.
Meredith, R. E. and Tobias, C. W., 1960. Resistance to potential flow through a cubical array of spheres. *Journal of Applied Physics*, 31:1270–1273.
Moura, S. C. S. R., Vitali, A. A., and Hubinger, M. D., 1999. A study of water activity and electrical conductivity in fruit juices: Influence of temperature and concentration. *Brazilian Journal of Food Technology*, 2(1–2):31–38.
Municino, F., 2012. Industrial applications of Ohmic heating of foods. In: *Workshop on Ohmic Heating*, Abrantes, Portugal, December 19.
Murakami, E. G. and Ramanauskas, P., 1997. Development of cold spot during electrical resistance heating of food materials. In: *Conference of Food Engineering* (*CoFE'97*), Los Angeles, CA, November 19–21.
Orangi, S., Sastry, S. K., and Li, Q., 1998. A numerical investigation of electroconductive heating in solid–liquid mixtures. *International Journal of Heat and Mass Transfer*, 41(14):2211–2220.
Palaniappan, S. and Sastry, S. K., 1991a. Electrical conductivities of selected solid foods during ohmic heating. *Journal of Food Process Engineering*, 14:221–236.
Palaniappan, S. and Sastry, S. K., 1991b. Electrical conductivity of selected juices: Influences of temperature, solids content, applied voltage, and particle size. *Journal of Food Process Engineering*, 14:247–260.

Palaniappan, S., Sastry, S. K., and Richter, E. R., 1992. Effect of electroconductive heat treatment and electrical pretreatment on thermal death kinetics of selected microorganisms. *Biotechnology and Bioengineering*, 39:225–232.

Pandit, R. B., Somavat, R., Jun, S., Heskitt, B., and Sastry, S., 2007. Development of a light weight ohmic food warming unit for a Mars exploration vehicle. *World of Food Science*, 2:1–12.

Park, S. J., Kim, D., Uemura, K., and Noguchi, A., 1995. Influence of frequency on ohmic heating of fish protein gel. *Nippon Shokuhin Kagaku Kogaku Kaishi*, 42(8):569–574.

Parrott, D. L., 1992. Use of ohmic heating for aseptic processing of food particulates. *Food Technology*, 46:68–72.

Piette, G., Buteau, M. L., De Halleux, D., Chiu, L., Raymond, Y., Ramaswamy, H. S. et al., 2004. Ohmic heating of processed meat and its effects on product quality. *Journal of Food Science*, 69(2):FEP71–FEP77.

Piette, G., Dostie, M., and Ramaswamy, H. S., 2001. Ohmic cooking of processed meats-state of art and prospects. *Procedures of the International Congress of Meat Science and Technology*, 47:62–67.

Pongviratchai, P. and Park, J. W., 2007. Electrical conductivity and physical properties of surimi–potato starch under ohmic heating. *Journal of Food Science*, 72(9):503–507.

Praporscic, I., Lebovka, N. I., Ghnimi, S., and Vorobiev, E., 2006. Ohmically heated, enhanced expression of juice from apple and potato tissues. *Biosystems Engineering*, 93:199–204.

Prentice, J. H., 1962. The conductivity of milk: The effect of the volume and degree of dispersion of fat. *Journal of Dairy Research*, 29:131–134.

Remik, D., 1988. Apparatus and method for electrical heating of food products. U.S. Patent No. 4,739,140.

Reznik, D. L., 1999. Personal communication. *In*: Sastry and Barach, 2000, Ohmic and Inductive Heating, *Journal of Food Science*, Reztek Corp., Sunnyvale, CA, 65(4):42–46.

Reznik, R., 1996. Electroheating apparatus and methods. U.S. Patent No. 5,583,960.

Ruan, R., Ye, X., Chen, P., Dooona, C., and Taub, I., 2001. Ohmic heating. *In*: *Thermal Technologies in Food Processing*, Richardson, P., ed. Woodhead Publishing Limited, Cambridge, U.K.

Salengke, S. and Sastry, S. K., 2007a. Models for ohmic heating of solid–liquid mixtures under worst-case heating scenarios. *Journal of Food Engineering*, 83(3):337–355.

Salengke, S. and Sastry, S. K., 2007b. Experimental investigation of ohmic heating of solid–liquid mixtures under worst-case heating scenarios. *Journal of Food Engineering*, 83(3):324–336.

Samaranayake, C. P. and Sastry, S. K., 2005. Electrode and pH effects on electrochemical reactions during ohmic heating. *Journal of Electroanalytical Chemistry*, 577:125–135.

Sarang, S., Sastry, S. K., Gaines, J., Yang, T. C. S., and Dunne, P., 2007. Product formulation for ohmic heating: Blanching as a pretreatment method to improve uniformity in heating of solid–liquid food mixtures. *Journal of Food Science E: Food Engineering and Physical Properties*, 72(5):E227–E234.

Sastry, S. K., 1986. Mathematical evaluation of process schedules for aseptic processing of low-acid foods containing discrete particulates. *Journal of Food Science*, 51:1323–1328, 1332.

Sastry, S. K., 1989. A model for continuous sterilization of particulate foods by ohmic heating, Cologne, Germany, May 28–June 3.

Sastry, S. K., 1992. A model for heating of liquid–particle mixtures in a continuous flow ohmic heater. *Journal of Food Process Engineering*, 15:263–278.

Sastry, S. K., 2005. Advances in Ohmic heating and moderate electric field (MEF) processing. In: *Novel Food Processing Technologies*, Barbosa-Canovas, G.V., Tapia, M.S., Cano, M.P., Martin-Belloso, O., and Martinez, A., eds. CRC Press, London, U.K.

Sastry, S. K. and Barach, J. T., 2000. Ohmic and inductive heating. *Journal of Food Science*, 65(Supplement 4):42–46.

Sastry, S. K. and Li, Q., 1996. Modeling the ohmic heating of foods. *Food Technology*, 50(5):246–248.

Sastry, S. K. and Palaniappan, S., 1992a. Ohmic heating of liquid–particle mixtures. *Food Technology*, 46(12):64–67.

Sastry, S. K. and Palaniappan, S., 1992b. Mathematical modeling and experimental studies on ohmic heating of liquid-particle mixtures in a static heater. *Journal of Food Processing Engineering*, 15:241–261.

Sastry, S. K. and Salengke, S., 1998. Ohmic heating of solid–liquid mixtures: A comparison of mathematical models under worst-case heating conditions. *Journal of Food Process Engineering*, 21:441–458.

Schreier, P., Reid, D., and Fryer, P., 1993. Enhanced diffusion during the electrical heating of foods. *International Journal of Food Science and Technology*, 28:249–260.

Shim, J. Y., Lee, S. H., and Jun, S., 2010. Modeling of ohmic heating patterns of multiphase food products using computational fluid dynamics codes. *Journal of Food Engineering*, 99:136–141.

Singh, S. P., Tarsikka, P. S., and Singh, H., 2008. Study on viscosity and electrical conductivity of fruit juices. *Journal of Food Science and Technology*, 45(4):371–372.

Skudder, P. J., 1988. Ohmic heating: New alternative for aseptic processing of viscous foods. *Food Engineering*, 60:99–101.

Skudder, P. J. and Biss, C., 1987. Aseptic processing of food products using ohmic heating. *Chemical Engineering*, 433:26–28.

Stirling, R., 1987. Ohmic heating: A new process for the food industry. *Journal of Power Engineering*, 6:365–371.

St-Gelais, B., Champagne, C. P., Erpmoc, F., and Audet, P., 1995. The use of electrical conductivity to follow acidification of dairy blends. *International Dairy Journal*, 5:427–438.

Sun, H.-X., Kawamura, S., Himoto, J.-I., Itoh, K., Wada, T., and Kimura, T., 2008. Effects of ohmic heating on microbial counts and denaturation of proteins in milk. *Food Science and Technological Research*, 14(2):117–123.

Therdthai, N. and Zhou, W., 2001. Artificial neural network modeling of the electrical conductivity property of recombined milk. *Journal of Food Engineering*, 50:107–111.

Tulsiyan, P., Sarang, S., and Sastry, S. K., 2008. Electrical conductivity of multicomponent systems during ohmic heating. *International Journal of Food Properties*, 11:233–241.

Uemura, K. and Isobe, S., 2003. Developing a new apparatus for inactivating *Bacillus subtilis* spore in organge juice with a high electric field AC under pressurized conditions. *Journal of Food Engineering*, 56:325–329.

Varghese, K. S., Pandey, M. C., Radhakrishna, K., and Bawa, A. S., 2014. Technology, applications and modelling of ohmic heating: A review. *Journal of Food Science Technology* (published online April 2012). doi: 10.1007/s13197-012-0710-3.

Wang, W. and Sastry, S. K., 1997. Changes in electrical conductivity of selected vegetables during multiple thermal treatments. *Journal of Food Process Engineering*, 20:499–516.

Wang, W. and Sastry, S. K., 2000. Effects of thermal and electro-thermal pre-treatments on hot air drying rate of vegetable tissue. *Journal of Food Process Engineering*, 23(4):229–319.

Wu, H., Kolbe, E., Flugstad, B., Park, J. W., Yongsawatdigul, J., 1998. Electrical properties of fish mince during multi-frequency ohmic heating. *Journal of Food Science*, 63:1028–1032.

Ye, X., Ruan, R., Chen, P., and Doona, C., 2004. Simulation and verification of ohmic heating in a static heater using MRI temperature mapping. *Lebensmittel-Wissenschaft und -Technologie* (*Food Science and Technology*), 37:49–58.

Yongsawatdigul, J., Park, J. W., and Kolbe, E., 1995. Electrical conductivity of Pacific whiting surimi paste during ohmic heating. *Journal of Food Science*, 60(5):922–925, 935.

Yoon, S. W., Lee, C. Y. J., Kim, K. M., and Lee, C. H., 2002. Leakage of cellular material from *Saccharomyces cerevisiae* by ohmic heating. *Journal of Microbiology and Biotechnology*, 12:183–188.

Zareifard, M. R., Marcotte, M., Ramaswamy, H. S., and Karimi, Y., 2013a. Modeling of electrical conductivity in the context of Ohmic heating. In: *Ohmic Heating in Food Processing*, Ramaswamy, H.S., Marcotte, M., Sastry, S., and Abdelrahim, K., eds. CRC Press, New York, NY, pp. 67–89.

Zareifard, M. R., Marcotte, M., Ramaswamy, H. S., and Karimi, Y., 2013b. Electrical conductivity: Importance and methods of measurement. In: *Ohmic Heating in Food Processing*, Ramaswamy, H.S., Marcotte, M., Sastry, S., and Abdelrahim, K., eds. CRC Press, New York, NY, pp. 17–36.

Zareifard, M. R., Ramaswamy, H. S., Marcotte, M., and Karimi, Y., 2013c. The electrical conductivity of foods. In: *Ohmic Heating in Food Processing*, Ramaswamy, H.S., Marcotte, M., Sastry, S., and Abdelrahim, K., eds. CRC Press, New York, NY, pp. 37–51.

Zareifard, M. R., Ramaswamy, H. S., Marcotte, M., and Karimi, Y., 2013d. Factors influencing electrical conductivity. In: *Ohmic Heating in Food Processing*, Ramaswamy, H.S., Marcotte, M., Sastry, S., and Abdelrahim, K., eds. CRC Press, New York, NY, pp. 53–65.

Zareifard, M. R., Ramaswamy, H. S., Trigui, M., and Marcotte, M., 2003. Ohmic heating behavior and electrical conductivity of two-phase food systems. *Innovative Food Science and Emerging Technologies*, 4:45–55.

Zhang, L. and Fryer, P. J., 1993. Models for the electrical heating of solid-liquid food mixtures. *Chemical Engineering Science*, 48(4):633–642.

Zhu, S. M., Zareifard, M. R., Chen, C. R., Marcotte, M., and Grabowski, S., 2010. Electrical conductivity of particle–fluid mixtures in ohmic heating: Measurement and simulation. *Food Research International*, 43:1666–1672.

Zhuang, Y., Zhou, W., Nguyen, M. H., and Hourigan, J. A., 1997. Determination of protein content of whey powder using electrical conductivity measurement. *International Dairy Journal*, 7:647–653.

11 High-Pressure Process Design and Evaluation

Eleni Gogou and Petros Taoukis

CONTENTS

11.1 INTRODUCTION

High-pressure (HP) processing, especially HP pasteurization, is one of the most interesting nonthermal processes of foods due to its growing commercial application over the past few years (Tonello-Samson, 2014); the number of industrially installed HP processing lines has reached 261 worldwide. More than 50% of the industrial HP units are located in North America, and 25% is installed in Europe, mainly in the Mediterranean countries: Spain, France, Italy, Greece, and Portugal (Tonello-Samson, 2014).

The commercial success of HP processing can be mainly attributed to the ability to produce fresh-like food products with superior nutritional value and extended shelf life when compared with the respective thermally treated food products. These two basic aspects of HP-processed food products successfully fulfill both consumers' demand for fresh-like foods and the food industry's requirement for shelf-life extension. Apart from retaining food quality and extending shelf life, HP processing applications have been proven to assure food safety in terms of inactivating vegetative pathogens. Vegetative pathogens like *Escherichia coli*, *Listeria monocytogenes*, and *Salmonella* can be inactivated under HP conditions. On the other hand, bacterial spores are more pressure and temperature resistant; thus, more intense HP treatment conditions in combination with high temperatures need to be applied. HP processing is thus distinguished in HP pasteurization and pressure-assisted thermal sterilization (PATS) according to the magnitude of pressure applied and intensity of combined temperature. HP pasteurization processes are usually performed in the pressure domain of 300–650 MPa either at room temperature or at moderate temperature of 30°C–50°C. PATS processes involve treatment of foods at the pressure domain of 600–1400 MPa combined with elevated temperature in the range of 70°C–130°C (Barbosa-Canovas and Juliano, 2008).

Process deviations can be expected in any routine manufacturing process including HP processing; thus, pressure and temperature monitoring during processing is crucial to guarantee the efficiency of the process. Process evaluation is the determination of the process impact on the process target. In HP processes, the target is either an enzyme or a microorganism or both. In general, in order to perform a HP process evaluation, the target should be selected taking into account the most pressure and temperature resistant among the food attributes under consideration. In HP processing, the most important processing parameters are pressure, time, and temperature. In general, the high hydrostatic pressure is in principle assumed to be uniform and not affected by geometry or uniformity of the processed food product. This has been theoretically questioned (Minerich and Labuza, 2003; Maldonado et al., 2014) but in practice is accepted to be held. Process time is fixed, taking into account come-up time and time for release of pressure. On the other hand, temperature nonuniformity can be expected in HP processes due to adiabatic heating during pressure buildup and to heat transfer phenomena during the treatment holding. Consequently, HP process nonuniformity mainly lies in temperature gradients inside the HP vessel.

11.2 TEMPERATURE UNIFORMITY OF HP PROCESSES

The uniformity of the HP process is under consideration. Studies have demonstrated that during a HP process, the adiabatic heat (during buildup) and the heat loss through the HP vessel wall (during holding time) can cause temperature gradients in the processing unit and the processed products (Denys et al., 2000b; Otero et al., 2000; Delgado et al., 2008).

It has been recognized that pressurization and depressurization during pressure buildup and at the end of the pressure holding time induce a temperature change due to compression and expansion in both the food and the pressure transmitting fluid. The magnitude of these temperature changes depends on

- HP process conditions; pressure and temperature applied
- Rate of compression/decompression
- Composition of the food and more specifically the water content
- Type of pressure transmitting fluid
- Type of HP vessel (vertical or horizontal)

One important aspect when designing and evaluating a HP process is to identify the thermal effect implied by the pressure treatment. During compression, temperature changes in the sample and the pressurizing fluid is expected to be different due to the different thermophysical properties of the food sample and the pressure transmitting fluid.

Grauwet et al. (2012) published an extensive review on the two promising methods for assessing and evaluating temperature uniformity in HP processing:

1. Computational thermal fluid dynamics (CTFD) modeling and simulation of temperature fields
2. Pressure–temperature–time indicators (PTTIs) for experimental temperature uniformity assessment and process evaluation

11.2.1 Development of Heat Transfer Phenomena Models

Temperature changes during HP processes are mainly attributed to heat transfer phenomena that take place between food samples, pressure transmitting fluid, pressure vessel, and pressure surrounding equipment. As expected, heat transfer phenomena impose temperature gradients in the HP-treated food products. This means that HP treatments of food products take place at temperatures different from the initial HP vessel temperature, and this temperature difference is not homogeneous within the HP vessel, resulting in HP-treated food products treated at a pressure in combination with quite different treatment temperatures (Otero et al., 2002). Thus, the development of mathematical models to describe the heat transfer phenomena during HP processes is a very interesting field for HP process evaluation that has been addressed by several researchers (Denys et al., 2000a; Otero et al., 2006; Knoerzer et al., 2007; Smith et al., 2014a,b). The use of numerical heat transfer models has been introduced, and in most cases, the predicted temperature profiles during a HP process based on the developed models have been compared to the experimentally recorded temperature profiles. Discrete numerical modeling has been used to predict temperature and flow distribution inside the HP vessel. Numerical heat transfer models have been also used in conjunction with computational fluid dynamics (CFD) by several authors (Denys et al., 2000a,b; Hartmann and Delgado, 2002, 2003; Hartmann, et al., 2003, 2004; Ghani and Farid, 2007; Knoerzer et al., 2007; Otero et al., 2007; Juliano et al., 2008; Infante et al., 2009) to predict temperature along with flow distributions within the HP vessel. In most of these works, the distribution of enzyme and/or microbial inactivation in food products or food system models were investigated in parallel throughout the HP vessel with the aim to evaluate process uniformity.

Denys et al. (2000b) developed a modeling approach to evaluate process uniformity during batch HP processing. Heat transfer models were developed for apple sauce and tomato paste. The results confirmed a satisfactory agreement between experimental and predicted temperature profiles (Figure 11.1).

The developed heat transfer models were incorporated in available enzyme kinetics of *Bacillus subtilis* α-amylase inactivation (Ludikhuyze et al., 1997) and soybean lipoxygenase inactivation (Ludikhuyze et al., 1998). The uniformity of HP-induced inactivation of the selected enzymes during batch HP processing was evaluated. The residual enzyme activity distribution appeared to be dependent on the inactivation kinetics of the enzyme under consideration and the selected pressure–temperature combinations used.

The combined use of the heat transfer model with inactivation kinetics was recognized to be a useful tool to evaluate enzyme inactivation uniformity during batch HP processing of foods. It was demonstrated that the uniformity of *B. subtilis* α-amylase and soybean lipoxygenase inactivation (Figure 11.2) during batch HP processing is dependent on the inactivation kinetics of the enzyme under consideration and on the temperature uniformity during the treatment time.

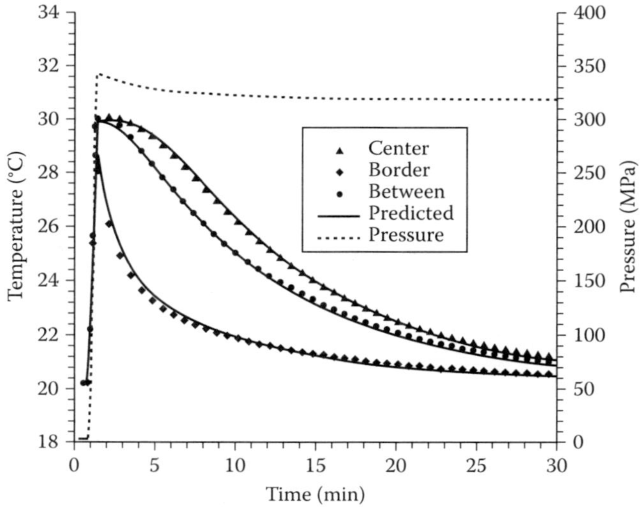

FIGURE 11.1 Experimental and predicted temperature profiles for a "conventional" batch HP process of tomato paste initial temperature 20.2°C ± 0.1°C; maximum pressure 342 MPa. (From Denys, S. et al., *Innov. Food Sci. Emerg. Technol.*, 1, 5, 2000b.)

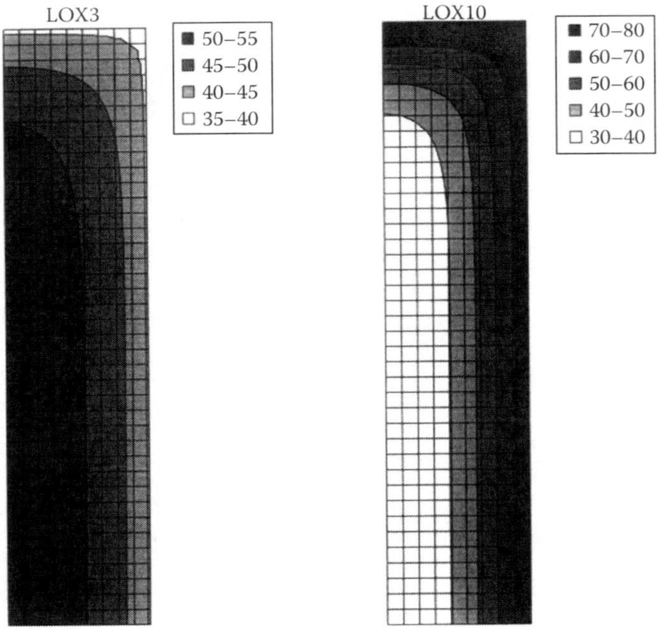

FIGURE 11.2 Spacial distribution of the lipoxygenase activity retention (%) for two HP treatments LOX3 (initial temperature: 15°C, pressure: 539 MPa, time 25 min) and LOX10 (initial temperature: 50°C, pressure: 354.9 MPa, time 25 min) (upper quarter of the pressure vessel). (From Denys, S. et al., *Innov. Food Sci. Emerg. Technol.*, 1, 5, 2000b.)

Another mathematical model was developed by Otero et al. (2006) to describe the heat and mass transfer phenomena occurring during HP treatments of foods, using agar as a food model system. The authors developed the heat and mass transfer model including factors like the filling ratio as the models were developed for one high and one low filling ratio, 71% and 12%, respectively. As depicted in Figure 11.3, temperature gradients within the same HP vessel are quite different when the filling ratios are different. According to this work, when the filling ratio is low, thermal re-equilibrium is reached sooner. However, this could not be proposed for HP industrial practices as low filling ratios are not feasible from an economical point of view.

Smith et al. (2014a) recently published a work where they evaluated the temperature differences taking place during HP processes in vertical and horizontal HP equipment based on developed heat transfer models. As it was demonstrated, horizontal and vertical flows inside a solvent food undergoing a HP process are different. This flow difference leads to different temperature distributions, which is potentially challenging for the food manufacturing industry as long as uniformity of HP effects on enzymatic and/or microbial inactivation is required. At this point, it has to be noted that most available industrial-scale HP units have a horizontal orientation. Smith et al. (2014a) developed a horizontal heat transfer model by adapting an existing vertical one developed by Infante et al. (2009).

The results, as depicted in Figures 11.4 and 11.5, showed that temperature uniformity is different for the two orientations. For a liquid-type food (in this work water) and for a long and thin machine, it was shown that the temperature in general was more uniform for the horizontal case, for the processes discussed. For the vertical model temperatures changed along height, while for the horizontal model they changed more radially.

The numerical simulations available through the developed heat transfer models can be coupled with kinetic models of a food quality and/or safety attribute. This approach can be used to evaluate the nonuniform inactivation of a selected quality and/or safety food attribute when subjected to nonuniform temperature distribution within the HP treatment vessel. This approach

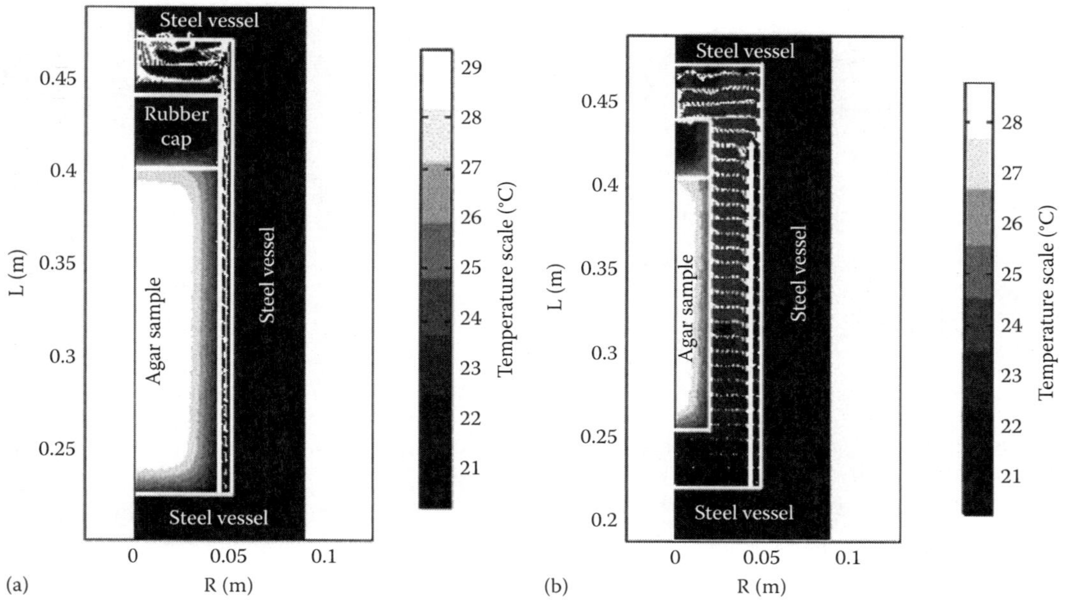

FIGURE 11.3 Temperature and velocity fields inside the high-pressure vessel 5 min after compression. (a) HP vessel high filling ratio of 71% and (b) HP vessel low filling ratio of 12%. (From Otero, L. et al., *J. Food Eng.*, 78, 1463, 2006.)

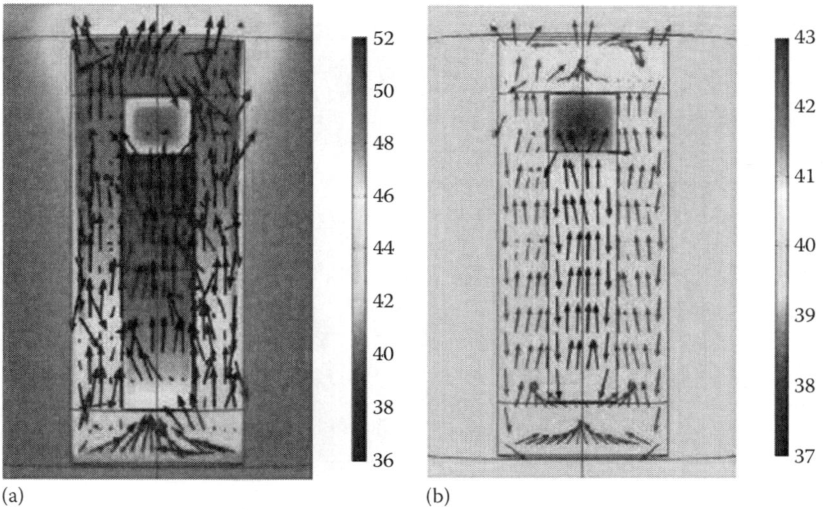

FIGURE 11.4 Slice plots of 3D vertical model; HP process conditions of initial temperature: 40°C, pressure: 360 MPa, and time: 15 min. (a) End of pressure buildup (P_{up}) and (b) end of pressure holding time (P_{hold}). (From Smith, N.A.S. et al., *Innov. Food Sci. Emerg. Technol.*, 22, 51, 2014a.)

was very well exploited by Rauh et al. (2009) focusing on the effect of temperature gradients within the HP vessel on food quality-related enzymes; α-amylase, lipoxygenase, and polyphenol oxidase. Enzymes with different pressure and temperature kinetics were chosen to examine the extent of the impact of temperature inhomogeneities during HP processes to the final target, that is, α-amylase, lipoxygenase, and polyphenol oxidase. The kinetics of inactivation of *B. subtilis* α-amylase and soybean lipoxygenase were taken from Denys et al. (2000b), who examined the

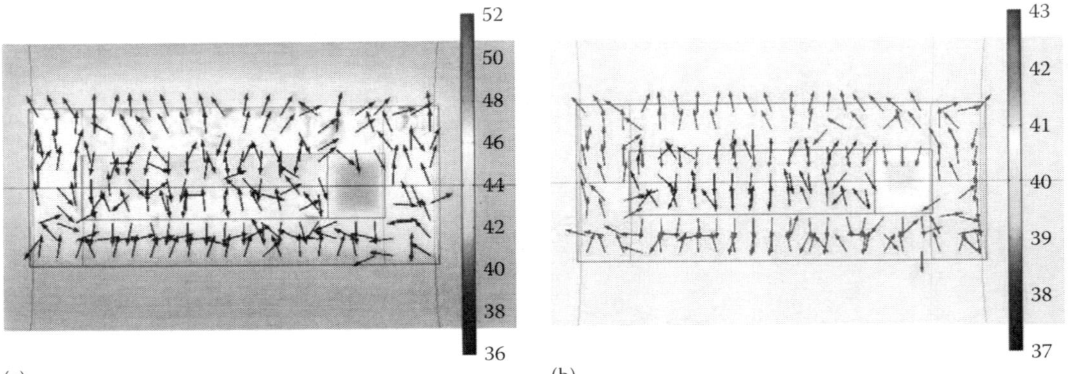

FIGURE 11.5 Slice plots of 3D horizontal model; HP process conditions of initial temperature: 40°C, pressure: 360 MPa, and time: 15 min. (a) End of pressure buildup and (b) end of pressure holding time. (From Smith, N.A.S. et al., *Appl. Math. Comput.*, 226, 20, 2014b.)

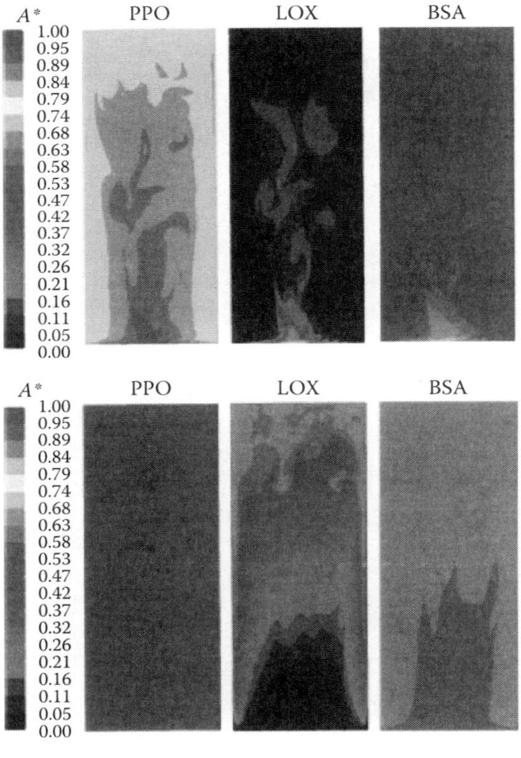

FIGURE 11.6 Dimensionless enzyme activity A^* at the end of the pressure holding time for processes: (a) upper figure and (b) lower figure. (From Rauh, C. et al., *J. Food Eng.*, 91, 154, 2009.)

kinetics in 0.01 M Tris buffer at pH 8.6 and 9, respectively. Pressure-induced inactivation of polyphenol oxidase from avocado follows a description of Weemaes et al. (1998), who determined the kinetics in phosphate buffer (pH 7; 0.1 M).

The results depicted in Figure 11.6 confirm that temperature gradients within the HP vessel (Figure 11.7) can result in nonuniform inactivation of the studied enzymes. Gradients in enzyme activities are expected to be broader for enzymes with higher temperature sensitivity, that is,

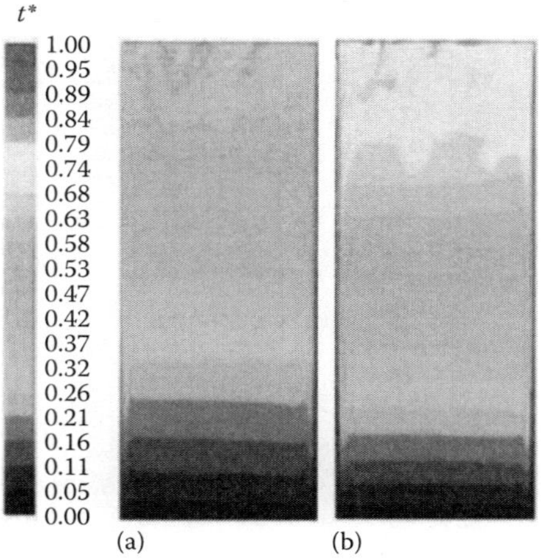

FIGURE 11.7 Dimensionless temperature field at the end of the pressure holding time ($t^* = t/t_{process} = 0.99$) for a high-pressure process with initial temperature of (a) 50°C and (b) 20°C. (From Rauh, C. et al., *J. Food Eng.*, 91, 154, 2009.)

enzymes characterized by inactivation kinetics with strong temperature dependence. The extent of nonuniform process impact on inactivation of enzymes or microorganisms may be also dependent on thermophysical properties of the treated food products, for example, viscosity, water content.

11.2.2 Artificial Neural Networks Development

The use of artificial neural networks (ANN) has been explored earlier for thermal processes of foods by Afaghi et al. (2001) and Sablani et al. (1995) and for drying processes by Farkas et al. (2000). Torrecilla et al. (2005) explored the use of ANN with the objective to predict temperature after pressure buildup as well as the time needed for the sample under HP to reach isothermal conditions.

An ANN is a mathematical algorithm, which has the capability of relating the input and output parameters, learning from examples through iteration without requiring prior knowledge of the relationships between the process variables. The ANN developed by Torrecilla et al. (2005) was designed to have an input layer of five nodes that corresponded to five input processing variables:

1. Applied pressure
2. Pressure increase rate
3. Initial set temperature
4. High-pressure vessel temperature
5. Temperature of the air surrounding the HP system

The output layer consisted of two neurons; one representing the maximum or minimum temperature reached in the sample after compression and one representing the time needed to re-equilibrate the temperature in the sample after decompression. The ANN structure is depicted in Figure 11.8.

The available numerical simulations based either on heat transfer models or ANN can serve as valuable tools to calculate temperature at any position inside the HP vessel and the food product, not only at individual positions. The aim of securing temperature uniformity in HP-processed products is crucial to be able to extract valid conclusions about the effects that a given pressure and temperature treatment produces in a specific food. These different temperature–time profiles detected during the process at different sample locations may result in a pronounced nonuniform distribution of enzyme and/or microbial inactivation and nutritional and/or sensorial quality

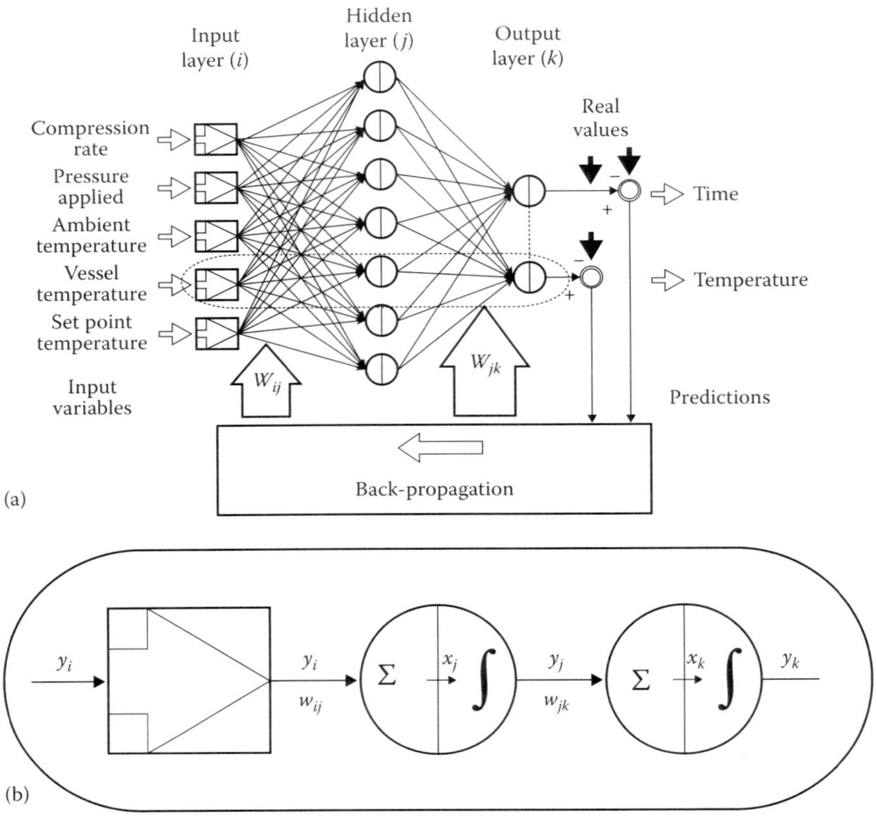

FIGURE 11.8 (a) Structure of the neural network model schematically shown. (b) Detail of the function of a neuron. (From Torrecilla, J.S. et al., *J. Food Eng.*, 69(3), 299, 2005.)

degradation within the processed product. Depending on the pressure–temperature degradation and the kinetics of the component that is focused on, the effect will be more or less pronounced (Denys et al., 2000a).

11.3 KINETICS UNDER DYNAMIC HIGH-PRESSURE PROCESS CONDITIONS

Industrial applications of HP processing are usually performed at the pressure range of 300–600 MPa. When pressure magnitude of 300–600 MPa is applied, adiabatic heating can lead to a significant temperature increase, which plays a significant role and should not be ignored when developing kinetic models of food quality attributes. Most published results in the field of kinetic models development under HP conditions generate isobaric curves of food quality attributes, deterioration, survival curves of spoilage microorganisms, and isorate contour plots of remaining enzymatic activity (Figure 11.9).

Most of these curves are considered to be developed both at isobaric and isothermal conditions. Although researchers state that these curves were generated from isobaric pressure conditions, the same thing cannot be stated for temperature conditions. However, adiabatic heating effect is actually involved; thus, most kinetic models have been generated using data obtained under varying temperature conditions (Peleg, 2006). The effect of varying temperature magnitude depends on the magnitude of pressure and temperature synergism on the selected food attribute. This attribute can be an endogenous enzyme or a spoilage and/or pathogenic microorganism.

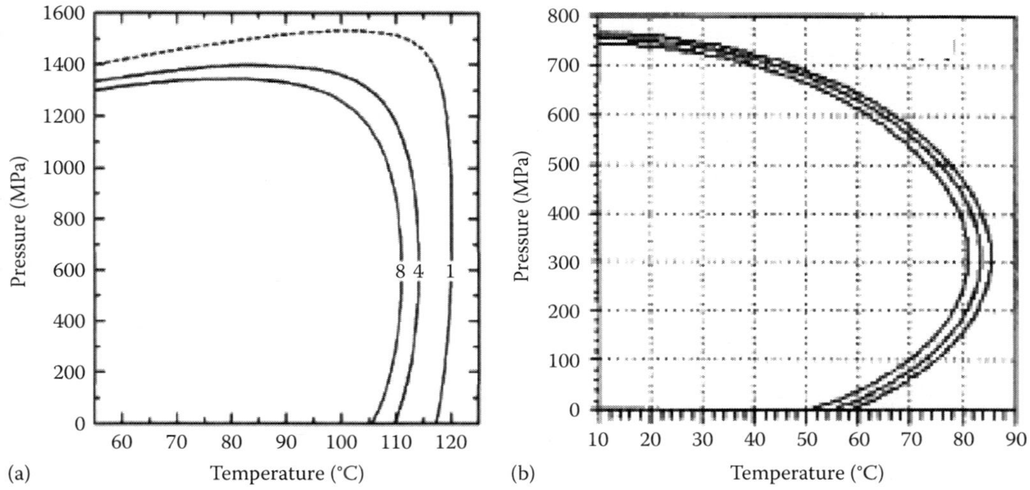

FIGURE 11.9 Isorate contour plots of (a) *Clostridium botulinum* death. (From Margosch, D. et al., *Appl. Environ. Microbiol.*, 72, 3476, 2006.) (b) Pectinmethyleastarase inactivation in carrot puree. (From Ly-Nguyen, B. et al., *J. Food Sci.*, 68, 1377, 2003.)

11.3.1 Kinetic Modeling Taking into Consideration Nonisothermal Conditions during Isobaric HP Processes

The impact of varying temperature in a HP treatment efficacy can be accessed through the development of proper mathematical modeling strategy where temperature variations and gradients are taken into consideration.

In Figure 11.8, a typical time–temperature–pressure profile of a HP treatment is depicted. The recording of pressure and temperature corresponds to a HP treatment performed at 450 MPa and initial HP vessel temperature of 50°C, using the high-pressure unit (Food Pressure Unit FPU 1.01, Resato International BV, Roden, Holland, pressure transmitting fluid: polyglycol ISO viscosity class VG 15, pressure buildup: 15 MPa/s). As depicted in Figure 11.10, the maximum temperature is recorded when the desired pressure is achieved (pressure buildup).

A typical experimental protocol of collecting kinetic data to develop a mathematical kinetic model involves samples treated for different times leading to different levels of a selected food quality attribute, that is, remaining enzyme activity. According to the recorded temperature during an isobaric HP treatment, samples of different treatment times are actually subjected to different nonisothermal time–temperature profiles, and this should be taken into account in the development of a mathematical kinetic model.

In a recent study, Gogou et al. (2010b) developed a kinetic model for xylanase inactivation using the effective temperature concept instead of constant nominal temperature for given HP treatments. The effective temperature concept was used in the Arrhenius equation and first-order inactivation kinetics (Equation 11.1).

$$\left(\frac{A}{A_0}\right)_P = \exp\left(-k_{T_{ref}} \cdot \exp\left(\frac{-E_a}{R} \cdot \left(\frac{1}{T_m(t)} - \frac{1}{T_{ref}}\right)\right) \cdot t\right) \quad (11.1)$$

where

$(A/A_0)_P$ is the remaining enzyme activity after processing for a time period t at isobaric conditions
$k_{T_{ref}}$ is the inactivation rate constant at a constant reference temperature (min^{-1})
E_a is the activation energy (J/mol)

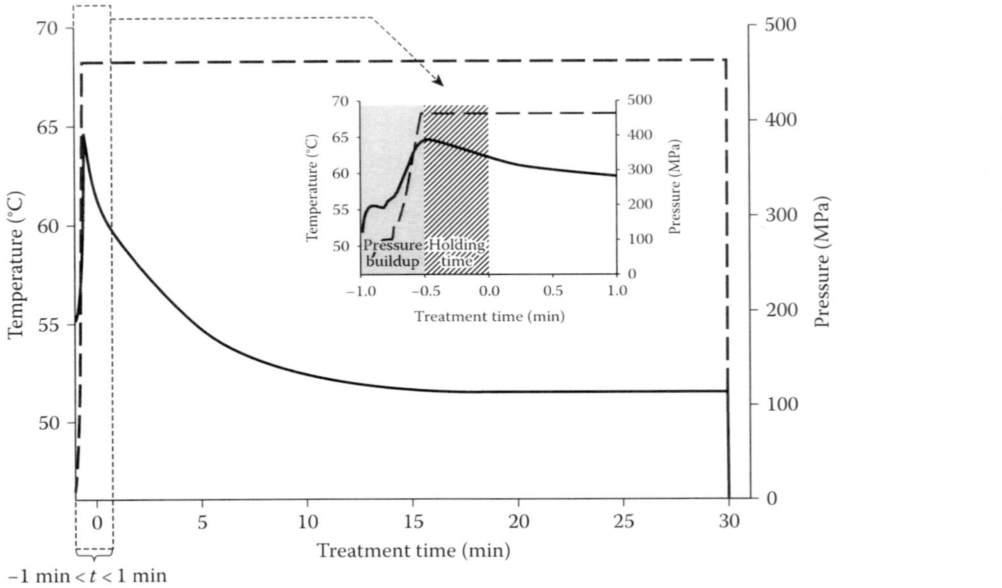

FIGURE 11.10 Pressure (– –) and temperature (—) profile during an isobaric HP treatment at 450 MPa combined with initial vessel temperature of 50°C.

R is the universal gas constant (8.314 J/mol·K)
$T_m(t)$ is the mean value of the recorded nonisothermal temperature profile for the time period t (K)
T_{ref} is a reference temperature (333.16 K = 60°C) and t is the processing time (min)

Using a rough estimation of the activation energy value (E_a) for HP enzyme inactivation of 100 kJ/mol the effective temperature, T_{eff}, can be used for each processing time instead of the temperature recorder at the different sampling times. The effective temperature is the constant temperature that results in the same remaining enzyme activity as the variable temperature profile and depends on the E_a value of the inactivation (Equation 11.2)

$$T_{eff} = \int_0^t - \frac{E_a \cdot T_{ref}}{R \cdot T_{ref} \cdot \ln \dfrac{k(t)}{k_{T_{ref}}} - E_a} \times dt \tag{11.2}$$

The comparison between the recorded temperature and the effective temperature at the different sampling times depicted in Table 11.1 is actually describing the necessity of taking into account the dynamic nonisothermal nature of HP processes for the estimation of HP kinetic parameters to be precise. The difference between the effective temperature and initial vessel temperature values is generally expected to be more pronounced at the high pressures domain due to increased adiabatic heating. As it can be assumed, samples remaining for shorter times in the pressure vessel are actually subjected to higher effective temperatures than those remaining for longer times.

In order to use the commonly applied kinetic approach of depicting and calculating inactivation rate constant at constant temperature and pressure conditions, the concept of equivalent time for each experimental point can be used:

$$t_{eq} = \exp\left(-\frac{E_a}{R}\left(\frac{1}{T_{eff}} - \frac{1}{T_{in}}\right)\right) \cdot t \tag{11.3}$$

TABLE 11.1
Comparison Between The Recorded Temperature and The Effective Temperature at Different Treatment Times During a High Pressure Process Performed at 450 MPa with Initial Vessel Temperature of 50°C (Figure 11.10)

Time (min)	Initial HP Vessel Temperature (T_{in})	Recorded HP Vessel Temperature (T_t)	Effective Temperature ($T_{eff,t}$)
3	50.0	56.8	55.4
6	50.0	54.0	54.3
12	50.0	52.0	53.2
20	50.0	51.4	52.6
30	50.0	51.4	52.2

where

t_{eq} is the equivalent time, which is defined as the time of treatment at a constant temperature (T_{in}) that results in the same remaining enzyme activity as the variable temperature distribution over the experimental time of treatment (t)

T_{in} is the initial temperature of the pressure vessel before the adiabatic heating and t is the time of treatment (min)

For first-order kinetics, enzyme inactivation under HP processing can be described by:

$$\frac{A}{A_0} = \exp(-k_{T,P} \cdot t_{eq}) \tag{11.4}$$

Pressure effect on the inactivation rate constant can be expressed through the activation volume value, V_a:

$$k = k_{P_{ref}} \cdot \exp\left[-\frac{V_{a,T}}{RT} \cdot (P - P_{ref})\right] \tag{11.5}$$

where

P_{ref} is the reference pressure

$k_{P_{ref}}$ is the inactivation rate constant at P_{ref}

T is the temperature (K)

Negative activation volumes indicate that the observed inactivation is favored by pressure; the increase of temperature usually leads to in reduced absolute values of activation volume, indicating that inactivation rate constant becomes less pressure dependent. Temperature dependence of the inactivation rate constant can be expressed by the E_a values calculated by Equation 11.1 by the described iterative process. A lot of published data agree that activation energy decreases with increasing pressure indicating less temperature dependence of the inactivation rate at the high pressures domain.

11.3.2 *F*-Value Determination

The traditional approach of modeling the inactivation kinetics of microorganisms and enzymes under HP has been based on the assumption that inactivation kinetics follows first-order kinetics. Based on this assumption, an analogy between the lethal effect of HP and thermal processes can be alleged. In the literature, there are a lot of kinetic parameters available (Table 11.2) describing

TABLE 11.2
Decimal Reduction Times Reported in High Pressure Inactivation of Microorganisms in High Pressure Treated Food Products

		High Pressure Conditions Tested			
Microorganism	**Food Product**	**Pressure (MPa)**	**Temperature (°C)**	***D*-Value (Min)**	**References**
Clostridium sporogenes PA3679	Meat broth	800	108	0.695	Rovere et al. (1996)
Total microbial flora	Milk	300	Ambient	22.9	Mussa and Ramaswamy (1997)
L. monocytogenes	Pork	414	25	2.17	Ananth et al. (1998)
Salmonella typhimurium				1.48	
Saccharomyces cerevisiae	Orange juice	350	Ambient	0.63	Parish (1998)
L. monocytogenes	Pork	400	Ambient	3.5	Mussa et al. (1999)
C. sporogenes	Meat broth	800	90	5.3	Rovere et al. (1999)
S. cerevisia ascospores	Orange juice	400	37.2	0.97	Zook et al. (1999)
	Apple juice			0.88	
E. coli	Egg	300	5	3.8	Lee (2002)
L. monocytogenes	Milk	600	Ambient	2.43	Dogan and Erkmen (2004)
	Peach juice			1.52	
	Orange juice			0.87	
E. coli	Milk	600	25	1.66	Erkmen and Dogan (2004)
	Peach juice			1.22	
	Orange juice			0.68	
Total aerobic count	Milk	600	25	3.19	Erkmen and Dogan (2004)
	Peach juice			1.50	
	Orange juice			0.83	
E. coli	Carrot juice	300	40	7.9	Van Opstal et al. (2005)
E. coli	Cashew apple juice	400		1.21	Lavinas et al. (2008)
S. typhimurium	Milk	300	25	9.21	Erkmen (2009)
	Orange juice			1.50	

the effect of pressure on the inactivation parameters either in terms of D_P and z_P values (Equation 11.6) or in terms of rate constants (k) at specific isobaric pressure conditions and corresponding activation volume values (V_a); a term used to describe pressure dependence of inactivation kinetics (Equation 11.7) in analogy with activation energy (E_a) used to describe the corresponding temperature dependence (Equation 11.8).

$$D_P = D_{P_{ref}} \cdot 10^{(P_{ref} - P/z_P)} \tag{11.6}$$

$$k_P = k_{P_{ref}} \cdot \exp\left[-\frac{V_a}{R} \cdot \frac{P - P_{ref}}{T}\right] \tag{11.7}$$

$$k_T = k_{T_{ref}} \cdot \exp\left[-\frac{E_a}{R} \cdot \left(\frac{1}{T} - \frac{1}{T_{ref}}\right)\right] \tag{11.8}$$

where

k_P is the inactivation rate constant at pressure P
k_T is the inactivation rate constant at temperature T
P_{ref} is the reference pressure
$k_{P_{ref}}$ is the inactivation rate at P_{ref}
T_{ref} is the reference temperature
$k_{T_{ref}}$ is the inactivation rate at T_{ref}
V_a is the activation volume
E_a is the activation energy and R the universal gas constant

The above equation kinetics should be applied to experimental findings only when isothermal and isobaric conditions can be achieved. Several published studies (Rovere et al., 1998; Stoforos and Taoukis, 1998; de Heij et al., 2003; Koutcma et al., 2005) have demonstrated that a linear model is suitable to predict the microbial inactivation of *G. stearothermophilus* and *C. sporogenes* PA3679 considered as classical surrogates in a thermally assisted HP process (when processed at isobaric and isothermal conditions during holding time). Consequently, the approaches used in thermal processing can be applied for HP process evaluation. Pflug's concept (Pflug and Zeghman, 1985) can be adapted for determining the F-value for the HP pasteurization and sterilization of foods:

$$F_{P_{ref},T_{ref}} = \frac{1}{k_{ref_{P,T}}} \int_0^t k(T,P) \cdot dt = \frac{-\ln(A/A_0)}{k_{ref_{P,T}}} \tag{11.9}$$

11.4 DEVELOPMENT OF PRESSURE–TEMPERATURE–TIME INDICATORS

11.4.1 PTTIs Definition and Requirements

The control of a HP process used for the preservation of food requires effective and reliable tools that can monitor the cumulative impact of pressure and temperature on food safety and quality attributes. In analogy to time–temperature–integrators (TTIs) for thermal processes (Maesmans et al., 1994; Hendrickx et al., 1995), the use of PTTIs has been proposed for impact evaluation of HP processes (Van der Plancken et al., 2008). A potential PTTI can be a component that has an irreversible change, dependent on both pressure and temperature, which can be correlated to the changes of a target quality or safety attribute undergoing the same HP process, without the knowledge of the actual pressure–temperature–time profile. Van der Plancken et al. (2008) defined the PTTI as "*a small, wireless device that shows a pressure, temperature, and time dependent, easily and accurately measurable, irreversible read-out to the HP treatment.*"

In order to investigate a system as a PTTI, a full kinetic characterization of the system is required. A candidate PTTI ideally should display a sensitivity of its kinetic parameter to both small differences in temperature and in pressure. Enzymes and microorganisms are considered to be good candidates for the development of PPTI systems.

For a candidate PTTI to be used in evaluating HP processes, its temperature and pressure sensitivity should be similar to the respective temperature and pressure sensitivity of the HP process target. The HP process target can be either an enzyme or microorganisms (spoilage and/or pathogens) in the case of HP pasteurization. In the case of HP sterilization processes, the target is the more thermotolerant spores. As pressure is assumed to be constant and uniform while temperature is on the one hand not constant (due to adiabatic heating and heat transfer phenomena) and on the other hand has been reported to be nonuniform.

11.4.2 Proposed PTTI Systems

Several researchers have proposed different systems to be used as PTTIs; the proposed PTTIs are summarized in Table 11.3. Although most of the proposed PTTI systems are enzymes, other systems like powdered copper tablet (Minerich and Labuza, 2003), microorganisms like *C. sporogenes* (Koutcma et al., 2005), and starch (Bauer and Knorr, 2005) have also been investigated to be used as PTTIs with the aim to be used as HP process evaluation tools.

Minerich and Labuza (2003) published one of the first studies in developing a pressure indicator for HP processing of foods. They developed a tablet consisting of compressed powdered copper. The specific system cannot be considered as a PTTI as the tablet density, serving as the indicator response, was affected only by the applied pressure and treatment time and not temperature. However, reference to the specific study should not be overlooked when studying the PTTIs concept as it is considered to be one of the first approaches to develop an indicator to evaluate the pressure applied in HP processing. As described by the authors, the tablet comprised of uniaxially compressed powdered copper; the density of the tablet was found to increase with increasing pressure in the range of 400–600 MPa. The tablet density was also found to increase as treatment time increased. However, the performance of HP treatments at different temperatures in the range of 7°C–28°C had no significant effect in tablet density. The indicator response was the observed density of the tablet that can be further translated to the Heckel value (Heckel, 1961a,b) leading to the indicator response function as given in Equation 11.6.

$$H = \ln\frac{1}{1-\Phi} = KP + A \tag{11.6}$$

where

H is the Heckel value
Φ is the relative density ($\rho_o/\rho_{absolute}$)
ρ_o, $\rho_{absolute}$ are the observed and absolute density
$\rho_{absolute}$ is 8.96 g/cm^3 for copper
P is the applied pressure
K, A are the constants

TABLE 11.3
Proposed Systems Serving as PTTIs for HP Process Evaluation

PTTI System	PTTI Response	HP Conditions Tested		References
		P (MPa)	*T* (°C)	
B. subtilis α-amylase		250–550	25–55	Ludikhuyze et al. (1997)
Milk alkaline phosphatase	Remaining enzyme activity	100–800	10–60	Claeys et al. (2003)
Compressed powdered copper tablet	Density	400–600	7–24	Minerich and Labuza (2003)
C. sporogenes	Color	600–800	91–108	Koutchma et al. (2005)
Starch	Starch gelatinization	100–700	29–57	Bauer and Knorr (2005)
Coenzyme Q(0)	Degradation of ubiquinone Q(0)	400–800	40, 80	Fernández García et al. (2009)
B. subtilis α-amylase	Remaining enzyme activity	400–600	10–40	Grauwet et al. (2009a)
Bacillus amyloliquefaciens α-amylase	Remaining enzyme activity	150–680	10–45	Grauwet et al. (2009b)
Thermomyces lanuginosus β-xylanase	Remaining enzyme activity	100–600	50–70	Gogou et al. (2010b)
Ovomucoid system	Residual trypsin inhibitor activity	400–700	99–111	Grauwet et al. (2010, 2011)
Thermotoga maritima β-xylanase	Remaining enzyme activity	500–700	99.4–113.8	Vervoort et al. (2011)

The developed indicator was further used in experiments where it was placed in the geometric center of ham samples and further treated with HP. According to the results, ham samples were found to receive approximately 9 MPa less pressure than the applied pressure (400 MPa) with the HP equipment. This pressure receiving difference is rather small and can be attributed to the fact that ham has a considerable structure and cannot be considered to perform as a liquid.

Bauer and Knorr (2005) investigated the potential use of pressure-induced starch gelatinization as a PTTI. The impact of HP processing on different types (A-type, B-type, and C-type) of starches crystallinity was examined. As anticipated, starch gelatinization was sensitive to the changes of HP processing conditions of temperature, pressure, and time. The velocity rate of gelatinization increased, the higher the constant pressure, indicating that pressure-induced gelatinization is a time-dependent process. Although the authors recognized that starches could be considered as potential PTTIs, it was noted that starches have to be carefully selected taking into account maximum gelatinization at selected HP processing conditions. Moreover, more work needs to be done on starch gelatinization in order to use it as PTTIs as the indicator response should be validated, and a PTTI response function should be developed and kinetically studied. This work can only be considered as the basis for starch-based PTTIs proposition that has to be further explored and challenged.

So far, the proposed PTTI systems of copper tablet and starch can be considered as extrinsic PTTI systems. Claeys et al. (2003) on the other hand investigated the effect of HP on some milk proteins and the potential of using milk alkaline phosphatase as an intrinsic PTTI for milk HP pasteurization. Claeys et al. (2003) used the available kinetic model of alkaline phosphatase by Ludikhuyze et al. (2000). Although the specific indicator has been successfully used as a thermal pasteurization indicator for milk, its application as PTTI has some restrictions according to the results; pressure and temperature sensitivity of alkaline phosphatase is not adequate for HP pasteurization processes at ambient temperature.

Fernández García et al. (2009) studied the Diels–Alder reactions between coenzyme Q(0) as dienophile and sorbic acid as diene compounds under HP treatments combined with temperature of 40°C and 80°C. The proposed PTTI response was the remaining concentration of coenzyme Q(0) that was kinetically modeled as a function of pressure and temperature:

$$\ln\frac{C}{C_0} = -k(t+t_0) = -\exp\left[\alpha_1 - \frac{V_\alpha(P-P_0)}{RT}\right]\cdot(t+t_0) \tag{11.7}$$

where

R is the universal gas constant (8.314472 J/K/mol)

t_0, which is fixed to be $t_0 = 2.5$ min, is the time to achieve the nominal pressure P

T is the temperature

C is the concentration of coenzyme Q(0) at time t, with C_0 its initial value at time $t = -t_0$

The proposed PTTI by Fernández García et al. (2009) was reported to show a pressure dependence (expressed by the z_P-value of 222 MPa at 80°C) comparable to the z_P-values described for some microorganisms that are suggested as surrogates of *C. botulinum* in HP treatments. This is the case of *B. amyloliquefaciens* spores at 95°C (Rajan et al., 2006), with a z_P-value of 170 MPa. However, it must be pointed out that the proposed PTTI was calibrated for HP treatments combined with temperature up to a maximum temperature of 80°C. Its use at HP treatments with elevated temperature of 95°C designed to inactivate more pressure-resistant compounds such as spores seems to be restricted as noted by the authors.

Another very promising system to serve as a PTTI is amylase. Ludikhuyze et al. (1997) were the first to investigate the potential use of *B. subtilis* α-amylase as a tool to assess the impact of HP treatments. A kinetic model was earlier developed by Ludikhuyze et al. (1996) describing the combining effect of pressure and temperature on the remaining enzyme activity. The derived kinetic model (Equation 11.8) was further validated by Ludikhuyze et al. (1997) under dynamic HP conditions including multiple successive HP cycles.

$$\ln\left(\frac{A}{A_0}\right) = -\int_0^t \left[k_{refP,T}\cdot\exp\left(-B\left(\frac{1}{T}-\frac{1}{T_{ref}}\right)\right)\cdot\exp(-C(P-P_{ref}))\right]\cdot dt \qquad (11.8)$$

where

A is the enzyme activity at time t
A_0 is the enzyme activity at time 0
B is the parameter expressing the temperature dependence of the k value
C is the parameter expressing the pressure dependence of the k value
k is the inactivation rate constant at P and T (min^{-1})
$k_{refP,T}$ is the inactivation rate constant at P_{ref} and T_{ref} (min^{-1})
P_{ref} is the reference pressure (500 MPa)
P is the pressure (MPa)
T_{ref} is the absolute reference temperature (313 K)
T is the absolute temperature (K) and dt is the time interval (min)

The authors reported that validation tests at dynamic HP conditions led to the conclusion that amylase remaining enzyme activity can be used as a PTTI response to predict the impact of HP processes. However, the repeatability of the PTTI response is an issue as the enzyme used was a commercially available enzyme that was found to show different kinetic parameters between different production lots. Another issue raised by the authors was the observation that in order to have accurate predictions, the kinetic model should be developed and calibrated within the pressure and temperature range of interest.

Grauwet et al. (2009b) continued studying the kinetics of *B. subtilis* α-amylase inactivation under HP treatments in the pressure range of 400–600 MPa combined with temperatures of 10°C–40°C. Solvent engineering was used to optimize the required enzyme stability of α-amylase (1 g/L—MES 0.05 M pH 5.0) in the pressure–temperature range of interest. The authors developed a model derived from the thermodynamic behavior of the difference in free energy in the pressure–temperature domain:

$$\ln\frac{A}{A_{unt}} = -\int_0^\tau \left[\ln k_{ref} + \frac{i}{T}(T-T_{ref}) + \frac{j}{T}(P-P_{ref}) + \frac{l}{T}(P-P_{ref})(T-T_{ref}) + \frac{n}{T}(P-P_{ref})^2\right]\cdot dt \qquad (11.9)$$

where

A is the enzyme activity at processing time t
A_{unt} is the initial enzyme activity of the untreated enzyme sample
h–j and l–n are unknown, empirical model parameters
k_{ref} is the rate constant at reference temperature of 308 K (T_{ref}) and reference pressure of 500 MPa (P_{ref})

Another α-amylase (*B. amyloliquefaciens*) was also kinetically modeled by Grauwet et al. (2009b). The remaining enzyme activity was modeled in order to serve as a reliable PTTI response:

$$\ln\frac{A}{A_{unt}} = -\int_0^{\tau}\left[\begin{aligned}&\ln k_{ref} + \frac{h}{T}(T-T_{ref}) + \frac{i}{T}(P-P_{ref}) + \frac{j}{T}(P-P_{ref})(T-T_{ref})\\ &+\frac{l}{T}\left(T\left(\ln\frac{T}{T_{ref}}-1\right)+T_{ref}\right)+\frac{m}{T}(P-P_{ref})^2\end{aligned}\right]\cdot dt \tag{11.10}$$

where

- A is the enzyme activity at processing time t
- A_{unt} is the initial enzyme activity of the untreated enzyme sample
- h–j and l–m are unknown, empirical model parameters
- k_{ref} is the rate constant at reference temperature of 308 K (T_{ref}) and reference pressure of 500 MPa (P_{ref}).

The determined kinetic parameters of Equations 11.9 and 11.10 were reestimated under dynamic conditions and subsequently successfully validated under dynamic HP conditions relevant for the food industry. The α-amylase-based PTTIs proposed by Grauwet et al. (2009a,b) were used to map temperature uniformity in industrial-scale HP equipment. On the one hand, based on numerical integration of different pressure–temperature profiles and kinetic data, the potential amylase–based PTTIs were assessed in simulated HP pasteurization processes. The potential of both PTTIs to detect temperature differences was experimentally assessed in horizontally oriented industrial-scale HP equipment. It was experimentally demonstrated that the proposed PTTIs response readings were adequate to detect different temperature zones in a HP vessel intended for pasteurization processes. However, it must be pointed out that for successive and reliable use of any proposed PTTI system, another very important prerequisite should not be neglected. Only when pressure and temperature dependence of the PTTI is similar to the process target (i.e., endogenous enzyme, spoilage or pathogen microorganism) pressure and temperature dependence, the PPTIs response can be reliably used for HP process impact evaluation.

Enzymes apart from being candidate PTTIs for HP pasteurization, can be also used as PTTIs for high-pressure high-temperature (HPHT) processes. In a HPHT process, food products are subjected to high pressure combined with elevated temperatures including product preheating, adiabatic heating during pressure buildup, pressure holding possibly accompanied by heat exchange between components with a different compression heat, and finally temperature decrease during pressure release accompanied with subsequent cooling (Barbosa-Canovas and Juliano, 2008). For an enzyme-based PTTI to be applicable in HPHT processes, it must be assured that the candidate enzyme is thermotolerant enough to undergo intense HP conditions combined with temperatures exceeding 80°C. The use of thermo and piezo-tolerant enzymes as PTTIs can ensure that the PTTI response; remaining enzyme activity, after the HPHT process can be assayed within the respective detection limits in order to be used to determine the impact of the process. Thermotolerance of enzymes can be further enhanced by means of pH adjustment or water activity decrease. Ovomucoid, a trypsin inhibitor present in chicken egg-white, is a protein that was investigated by Grauwet et al. (2010) as a candidate for PTTI development. In this study, different pH values were used to assess the enhancement of ovomucoid thermotolerance and the effect of pH shift in temperature sensitivity of the protein when subjected to HPHT processes. According to the results, the most suitable pH was 8.0 (adjusted with the temperature-stable phosphate buffer). The residual trypsin inhibitor activity served as the PTTI response, which was kinetically modeled (Equation 11.11) as a function of applied temperature in the pressure range of 400–700 MPa and temperature range of 95°C–117°C.

$$resTIA = (a + bT) \cdot \exp\left[-k_{ref} \cdot \exp\left[\frac{E_a}{R}\left(\frac{1}{T_{ref}} - \frac{1}{T}\right)\right] \cdot t\right] \quad (11.11)$$

where
$resTIA$ is the residual trypsin inhibitor activity at treatment time t
a and b are empirical model parameters
E_a is the activation energy (J/mol) expressing the temperature dependency at constant pressure
R is the universal gas
k_{ref} is the reaction rate constant (min^{-1}) at temperature T (K) and reference temperature T_{ref} (373 K)

The ovomucoid system showed clear temperature sensitivity under increased pressure enabling its use to assess temperature uniformity or to detect the lowest or highest temperature zones inside a HP vessel. With the assumption that the kinetics of target attributes (safety and/or quality) and more specifically the temperature dependency of the target attributes show a similar dependency with that of the proposed PTTI, process impact evaluation can be performed. However, the response of the proposed ovomucoid-based PTTI is restricted to HPHT processes where the temperature does not exceed 111°C in combination with holding times up to 3 min. This is limiting the use of the proposed PTTI as it could not be used for process evaluation where the target is spores inactivation. However, the Grauwet et al. (2010) approach was the first one to address the challenge of HPHT process evaluation. The specific protein-based PTTI was further optimized and challenged by Grauwet et al. (2011). The kinetic model was optimized to include both temperature and pressure effect on the reaction rate constant of residual trypsin inhibitor activity:

$$k_{T,P} = k_{refT,P} \cdot \exp\left[-\frac{E_a}{R}\left(\frac{1}{T} - \frac{1}{T_{ref}}\right)\right] \exp\left[-\frac{V_a}{RT}(P - P_{ref})\right] \quad (11.12)$$

where
$k_{refT,P}$ is the reaction rate constant of the trypsin-inhibitor inactivation at reference temperature (T_{ref}) and reference pressure (P_{ref})
E_a is the activation energy
V_a is the activation volume
R the universal gas constant

The optimized ovomucoid-based PTTI was treated at six different coordinates in a pilot-scale HPHT vessel (Figure 11.11), and it was verified that it can be used to assess temperature uniformity issues of HPHT processes of 600 MPa, initial temperature of 85°C, and treatment time of 5 min by detecting the lower and higher temperature zones within the HP vessel.

Xylanase is another enzyme that has been considered to be a good candidate for PTTI development (Gogou et al., 2010b; Vervoort et al. 2011). The specific enzyme originating from the hyperthermophile *Thermotoga maritima* was studied by Vervoort et al. (2011) with the aim to monitor temperature gradients in HPHT processes. The thermal inactivation at isothermal–isobaric conditions was described by a first-order model at atmospheric pressure as well as at elevated pressure. At HPHT conditions, the pressure dependence of the D-values was negligible, while the temperature dependence was distinct. This pressure independence suggests that a more accurate specific proposed system can serve more as a time–temperature indicator (TTI) rather than a PTTI. The authors suggested that the specific xylanase-based indicator can serve as a TTI for HP sterilization processes as the observed z-value of 17.16°C ± 0.04°C matches the temperature sensitivity of *C. sporogenes* spores' inactivation. Xylanases originating from thermophilic

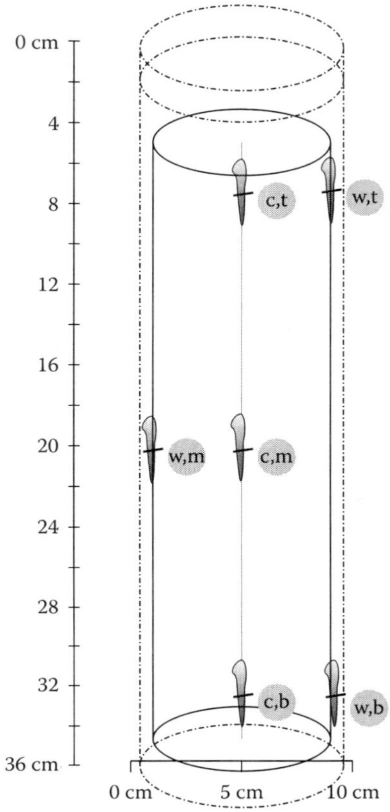

FIGURE 11.11 Schematic overview of PTTI positioning throughout the HPHT vessel (dashed line: POM container; bold line: sample holder). Different distances from the vessel bottom (3.5 cm bottom [b]; 16 cm middle [m]; 29 cm top [t]) and different distances from the vessel wall (0.8 cm wall [w]; 5 cm center [c]) were studied. (From Grauwet, T. et al., *J. Food Eng.*, 105, 36, 2011.)

microorganisms have been earlier proposed (Gogou et al., 2010a) as TTIs for thermal processes due to their enhanced thermal resistance.

Another xylanase originating from *Thermomyces lanuginosus* was studied by Gogou et al. (2010b) in pursuit of PTTI development applicable for HP pasteurization processes. The synergistic effect of pressure and temperature on xylanase inactivation rate constant was found to be adequately described as a function of pressure and temperature by Equation 11.13, usually applied to model enzyme inactivation (Polydera et al., 2004; Katsaros et al., 2009).

$$k_{P,T} = k_{ref_{P,T}} \cdot \exp\left[-\frac{(E_{a,P_{ref}} - B\cdot(P-P_{ref}))}{R}\cdot\left(\frac{1}{T}-\frac{1}{T_{ref}}\right) - \frac{V_{a,T_{ref}} + A\cdot(T-T_{ref})}{RT}\cdot(P-P_{ref})\right] \quad (11.13)$$

The developed kinetic model, having as variables pressure and temperature, allows the calculation of remaining enzyme activity (serving as the PTTI response) at any combination of HP processing conditions within the studied domain. The xylanase remaining enzyme activity after a HP process and the developed kinetic model can be used to estimate the F-value for the specific process and be correlated to the impact on the target attribute (e.g., an endogenous enzyme). Based on the mathematical model and the parameters estimation, an isorate contour plot can be constructed. The isokinetic diagram of xylanase is illustrated in Figure 11.12. From

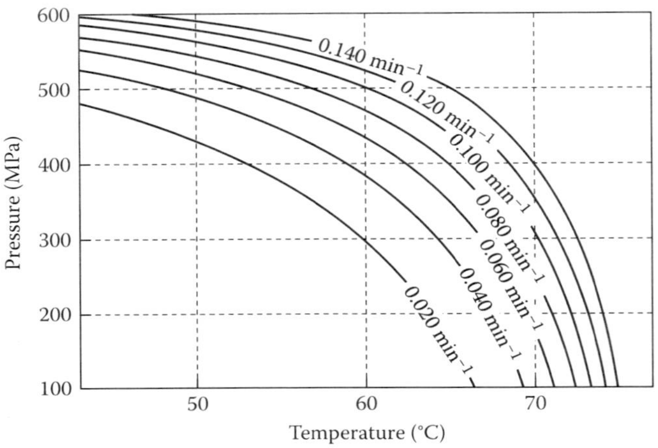

FIGURE 11.12 Simulated isorate contour plot of xylanase inactivation under HP processing. (From Gogou, E. et al., *J. Food Sci.*, 75(6), 379, 2010b.)

the isorate contour graph, it is clear that xylanase inactivation is strongly dependent on both pressure and temperature. High-pressure inactivation at low levels of pressure (<300 MPa) is dominantly affected by temperature with the inactivation rate constant to be sensitive to small temperature gradients.

The developed kinetic model of the proposed xylanase-based PTTI was challenged under various HP conditions (within the studied pressure–temperature domain). The remaining enzyme activity was used to predict the temperature history (expressed by the effective temperature value) for the given pressure and time combinations. The actual effective temperature, T_{eff} (°C), was calculated integrating the recorded time–temperature profile of each HP treatment and compared to the predicted effective temperature values, T_{eff} (°C). The error in predicting the effective temperature based on the PTTI response was calculated for all HP conditions. The predictive performance of the PTTI was acceptable with the average error value equal to 3.4%. In 67% of the cases, the absolute error value was less than 10%. The maximum deviations observed were −15% and 27% based on single measurements. Such error values can be avoided by multiple measurements.

The application of the proposed xylanase-based PTTI in HP processing of orange juice was evaluated. The juice cloud is considered to be a desirable characteristic of orange juice, since it favorably affects turbidity, flavor, and the characteristic color of orange juice. Cloud loss accompanied by gelation of juice concentrates is a major problem associated with orange juice quality deterioration, which has been attributed primarily to the activity of PME, a cell wall-bound pectic enzyme released into the juice during juice extraction (Versteeg et al., 1980; Oakenfull and Scott, 1984; Cameron et al., 1999). Several studies have dealt with the impact of HP on PME inactivation in orange juice (Irwe and Olsson, 1994; Basak and Ramaswamy, 1996; Cano et al., 1997; Nienaber and Shellhammer, 2001; Bull et al., 2004). Polydera et al. (2004) studied the inactivation kinetics of PME in Greek navel orange juice during combined high pressure (100–800 MPa) and moderate temperature (30°C–60°C) treatments. The activation energy values ranged from 177 kJ/mol at 100 MPa to 95 kJ/mol at 750 MPa. The activation volume values ranged from −36.7 mL/mol at 30°C to −14.2 mL/mol at 60°C. A mathematical model that describes the PME inactivation rate as a function of pressure and temperature conditions was developed. Process conditions of 600 MPa at 40°C for 4 min for sufficient inactivation of the pressure labile PME isoenzyme were proposed, that is, the minimum target *F*-value of an adequate HP process must be 4 min.

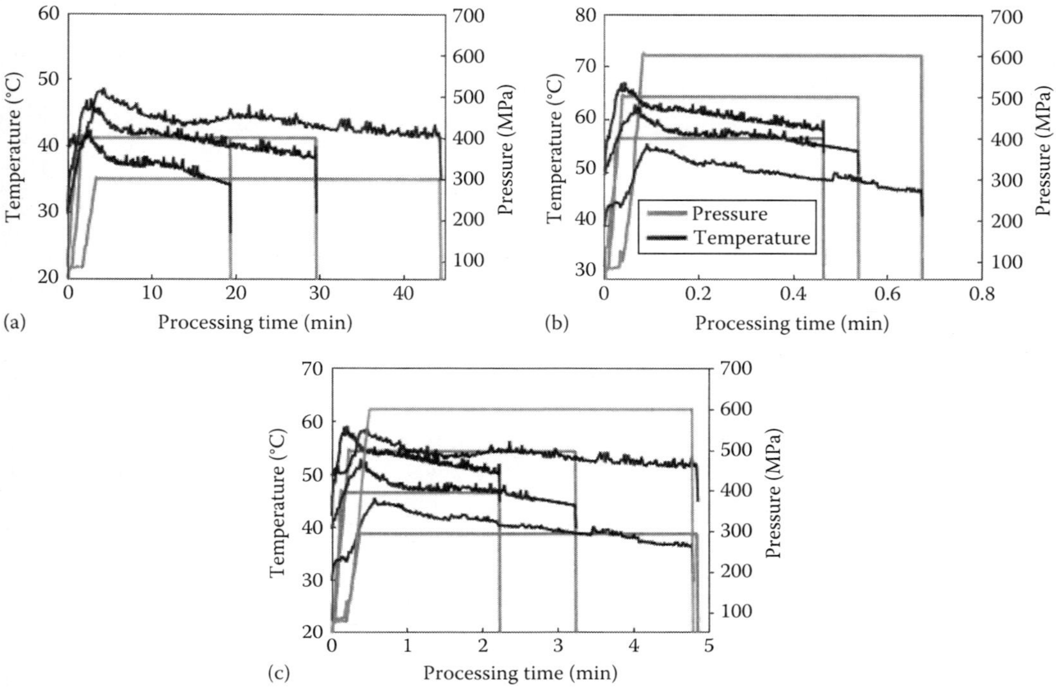

FIGURE 11.13 Pressure–time–temperature profiles used for the evaluation of the xylanase-based PTTI application in the case study of PME inactivation in Navel orange juice (a) HP1, HP3, HP6; (b) HP5, HP8, HP10; (c) HP2, HP4, HP7, HP9. (From Gogou, E. et al., *J. Food Sci.*, 75(6), 379, 2010b.)

The application of the xylanase-based PTTI was evaluated in simulated HP processes of navel orange juice at different pressure–temperature conditions designed to achieve the targeted *F*-value (Figure 11.13).

The effectiveness of each of these processes could be evaluated by the PTTI response, that is, the value of xylanase remaining activity, $(A/A_0)_{xylanase}$. The PTTI responses that would be obtained for the HP processes were calculated from the integration of the developed kinetic model of xylanase inactivation. Taking into account that in most HP processes pressure is constant and well defined, from the PTTI response the effective temperature, T_{eff}, of the process is determined by solving Equation 11.14.

$$\left(\frac{A}{A_0}\right)_P = \exp\left(-k_{T_{ref}} \cdot \exp\left(\frac{-E_a}{R} \cdot \left(\frac{1}{T_{eff}} - \frac{1}{T_{ref}}\right)\right) \cdot t\right) \tag{11.14}$$

where

$(A/A_0)_P$ is the remaining enzyme activity after processing for a time period t at isobaric conditions

$k_{T_{ref}}$ is the inactivation rate constant at a constant reference temperature (min^{-1})

E_a is the activation energy (J/mol)

R is the universal gas constant (8.314 J/mol·K)

T_{eff} is the effective temperature that is the equivalent constant temperature corresponding to the nonisothermal temperature profile for the time period t (K)

T_{ref} is a reference temperature (333.16 K = 60°C) and t is the processing time (min)

The PTTI-derived effective temperature, $T_{eff,PTTI}$, is used to calculate the F-value of the process from the inactivation kinetic model of the target attribute (PME in navel orange juice) from Equation 11.15:

$$F_{T_{ref},P_{ref}} = \frac{k(T_{eff}, P)}{k_{ref}} \cdot t_{process} \tag{11.15}$$

The estimated F-values based on the PTTI response for different HP processes are listed in Table 11.2 as F_{PTTI}. They are compared to F_{PME}, the F-values directly calculated from the PME kinetics and the actual P–T profile of each process (usually not available in a real process). The F_{PTTI} slightly underpredicted the F-value for PME as the kinetic parameters of xylanase inactivation used to calculate the T_{eff} did not match the ones of PME. Based on the PTTI response, the adequacy of the processes (F-values of 4 min) can be verified. Similarly, in high HP conditions, an error in the effective temperature estimation in the range of 10% leads to deviations in the F-value calculation of 14% for the inactivation of *L. monocytogenes* Scott A in bovine milk based on a large number of simulated P–T scenarios and unpublished kinetic data (Katsaros et al., 2007) (Table 11.4).

Gogou et al. (2010b) demonstrated that the use of the proposed xylanase-based PTTI allows the prediction of the temperature history of isobaric-nonisothermal HP treatments. The prediction of temperature history during HP processes is essential for processing optimization and quality control of the final food product. According to the results of this study, the effective temperature concept was satisfactorily applied in the laboratory-scale HP equipment.

The application of the proposed concept in commercial-size equipment is of practical interest. Separate enzyme-based PTTIs can be used to estimate the effective temperature of a given HP process for product units in representative locations in the vessel. Usually in commercial batch processes, HP treatment is applied to packaged food products. A PTTI can be part of the packaging, allowing the effective temperature estimation for specific processed products. An approach based on the estimation of the process' effective temperature from the PTTIs' response can be used in order to evaluate process impact on the target attribute.

TABLE 11.4
HP Process F-values for PME Enzyme Inactivation in Orange Juice Calculated from the Xylanase PTTI Response (F_{PTTI}) and the Integration of the Process Conditions (F_{PME})

Process[a]	P (MPa)	T_{eff} (°C)	F_{PTTI}[b] (min)	F_{PME}[b] (min)
HP1	300	43.8	4.10	4.37
HP2	300	53.7	4.01	4.24
HP3	400	41.2	4.49	5.20
HP4	400	53.2	3.91	4.50
HP5	400	61.2	3.81	4.36
HP6	500	37.1	3.78	5.08
HP7	500	47.1	4.40	5.73
HP8	500	57.1	4.56	5.66
HP9	600	39.9	4.86	6.50
HP10	600	49.9	3.96	4.92

[a] HP process codes are referring to Figure 11.13.
[b] F-values are referred to 600 MPa and 40°C.

11.5 CONCLUSION

HP processing of food products is a process that has been implemented industrially over the last few years. Although applications are successful and promising for delivering high quality, safe food with extended shelf-life design, optimization and evaluation of HP processes is still under consideration. The development of reliable tools to evaluate the impact of HP processing is a very interesting research field that a lot of research groups worldwide have addressed. A lot of published data have proven that temperature nonuniformity is expected during HP processes (pasteurization and sterilization). Two promising tools in evaluating temperature nonuniformity and its impact in food safety and quality attributes were thoroughly described in this chapter: development of heat transfer models and PTTIs. However, further work needs to be done for these tools to be validated and their accuracy to be determined and finally applied for HP processing evaluation.

REFERENCES

Afaghi M., Ramaswamy H.S., Prasher S.O. 2001. Thermal process calculations using artificial neural network models. *Food Research International*, 34: 55–65.

Ananth V., Dickson J.S., Olson D.G., Murano E.A. 1998. Shelf-life extension, safety and quality of fresh pork loin treated with high hydrostatic pressure. *Journal of Food Protection*, 61(12): 1649–1656.

Antonio M.J., Nairb A., Cuitiño A., Schaffner D., Karwe M.V. 2014. Process non-uniformity during high hydrostatic pressure processing of heterogeneous foods. In: *2014 International Nonthermal Processing Workshop Book of Abstracts*, 10, October 21–24, 2014, Columbus, OH.

Barbosa-Canovas G.V., Juliano P. 2008. Food sterilization by combining high pressure and thermal energy. In: *Food Engineering: Integrated Approaches*. eds., Gutierrez-Lopez G., Barbosa-Canovas G.V., Welti-Chanes J., Parada-Arias E. Springer, New York, pp. 9–46.

Basak S., Ramaswamy H.S. 1996. Ultra high pressure treatment of orange juice: A kinetic study on inactivation of pectin methyl esterase. *Food Research International*, 29(7): 601–607.

Bauer B.A., Knorr D. 2005. The impact of pressure, temperature and treatment time on starches: Pressure-induced starch gelatinization as pressure time temperature indicator for high hydrostatic pressure processing. *Journal of Food Engineering*, 68: 329–334.

Bull M., Zerdin K., Howe E., Goicoechea D., Paramanandhan P., Stockman R., Sellahewa J., Szabo E., Johnson R., Stewart C. 2004. The effect of high pressure processing on the microbial, physical and chemical properties of Valencia and Navel orange juice. *Innovative Food Science and Emerging Technologies*, 5(2): 135–149.

Cameron R.G., Baker R.A., Buslig B.S., Grohmann K. 1999. Effect of juice extractor settings on juice cloud stability. *Journal of Agricultural Food Chemistry*, 47(7): 2865–2868.

Cano M.P., Hernández A., De Ancos B. 1997. High pressure and temperature effects on enzyme inactivation in strawberry and orange products. *Journal of Food Science*, 62: 85–88.

Claeys W.L., Van Loey I.A.M., Hendrickx M.E. 2003. Review: Are intrinsic TTIs for thermally processed milk applicable for high-pressure processing assessment? *Innovative Food Science and Emerging Technologies*, 4: 1–14.

de Heij W.B.C., Van Schepdael L.J.M.M., Moezelaar R., Hoogland H., Matser A.M., Van den Berg R.W. 2003. High-pressure sterilization: Maximizing the benefits of adiabatic heating. *Food Technology*, 57(3): 37–42.

Delgado A., Rauh C., Kowalczyk W., Baars A. 2008. Review of modelling and simulation of high pressure treatment of materials of biological origin. *Trends in Food Science and Technology*, 19(6): 329–336.

Denys S., Ludikhuyze L.R., Van Loey A.M., Hendrickx M.E. 2000a. Modeling conductive heat transfer and process uniformity during batch high-pressure processing of foods. *Biotechnology Progress*, 16: 92–101.

Denys S., Van Loey A.M., Hendrickx M.E. 2000b. A modelling approach for evaluating process uniformity during batch high hydrostatic pressure processing: Combination of a numerical heat transfer model and enzyme inactivation kinetics. *Innovative Food Science and Emerging Technologies*, 1: 5–19.

Dogan C., Erkmen O. 2004. High pressure inactivation kinetics of *Listeria monocytogenes* inactivation in broth, milk, and peach and orange juices. *Journal of Food Engineering*, 62(1): 47–52.

Erkmen O. 2009. High hydrostatic pressure inactivation kinetics of *Salmonella typhimurium*. *High Pressure Research*, 29(1): 129–140.

Erkmen O., Dogan C. 2004. Kinetic analysis of *Escherichia coli* inactivation by high hydrostatic pressure in broth and foods. *Food Microbiology*, 21: 181–185.

Farkas I., Reményi P., Biró A. 2000. A neural network topology for modelling grain drying. *Computers and Electronics in Agriculture*, 26: 147–158.

Fernandez Garcia A., Butz P., Corrales M., Lindauer R., Picouet P., Rodrigo G. 2009. A simple coloured indicator for monitoring ultra high pressure processing conditions. *Journal of Food Engineering*, 92: 410–415.

Ghani A.G.A., Farid M.M. 2007. Numerical simulation of solid–liquid food mixture in a high pressure processing unit using computational fluid dynamics. *Journal of Food Engineering*, 80(4): 13031–1042.

Gogou E., Katapodis P., Christakopoulos P., Taoukis P.S. 2010a. Effect of water activity on the thermal stability of *Thermomyces lanuginosus* xylanases for process time–temperature integration. *Journal of Food Engineering*, 100(4): 649–655.

Gogou E., Katapodis P., Taoukis P.S. 2010b. High pressure inactivation kinetics of a *Thermomyces lanuginosus* xylanase evaluated as a process indicator. *Journal of Food Science*, 75(6): 379–386.

Grauwet T., Rauh C., Van der Plancken I., Vervoort L., Hendrickx M., Delgado A., Van Loey A. 2012. Potential and limitations of methods for temperature uniformity mapping in high pressure thermal processing. *Trends in Food Science and Technology*, 23: 97–110.

Grauwet T., Van der Plancken I., Vervoort L., Hendrickx M., Van Loey A. 2009a. Investigating the potential of *Bacillus subtilis* α-amylase as a pressure–temperature–time indicator for high hydrostatic pressure pasteurization processes. *Biotechnology Progress*, 4: 1184–1193.

Grauwet T., Van der Plancken I., Vervoort L., Hendrickx M., Van Loey A. 2009b. Solvent engineering as a tool in enzymatic indicator development for mild high pressure pasteurization processes. *Journal of Food Engineering*, 97: 301–310.

Grauwet T., Van der Plancken I., Vervoort L., Hendrickx M., Van Loey A. 2010. Protein-based indicator system for detection of temperature differences in high pressure high temperature processing. *Food Research International*, 43: 862–871.

Grauwet T., Van der Plancken I., Vervoort L., Hendrickx M., Van Loey A. 2011. Temperature uniformity mapping in a high pressure high temperature reactor using a temperature sensitive indicator. *Journal of Food Engineering*, 105: 36–47.

Hartmann C., Delgado A. 2002. Numerical simulation of convective and diffusive transport effects on a high-pressure-induced inactivation process. *Biotechnology and Bioengineering*, 79: 94–104.

Hartmann C., Delgado A. 2003. The influence of transport phenomena during high-pressure processing of packed food on the uniformity of enzyme inactivation. *Biotechnology and Bioengineering*, 82: 725–735.

Hartmann C., Delgado A., Szymczyk J. 2003. Convective and diffusive transport effects in a high pressure induced inactivation process of packed food. *Journal of Food Engineering*, 59(1): 33–44.

Hartmann C., Schuhholz J.-P., Kitsubun P., Chapleau N., Le Bail A., Delgado A., 2004. Experimental and numerical analysis of the thermofluiddynamics in a high-pressure autoclave. *Innovative Food Science and Emerging Technologies*, 5(4): 399–411.

Heckel R.W. 1961a. An analysis of powder compaction phenomena. *Transactions of the Metallurgical Society of AIME*, 221: 1001–1008.

Heckel R.W. 1961b. Density-pressure relationships in powder compaction. *Transactions of the Metallurgical Society of AIME*, 221: 671–675.

Hendrickx M., Maesmans G., De Cordt S., Noronha J., Van Loey A., Tobback P. 1995. Evaluation of the integrated time–temperature effect in thermal processing of foods. *Critical Reviews in Food Science and Nutrition*, 35: 231–262.

Infante J.A., Ivorra B., Ramos A.M., Rey J.M. 2009. On the modelling and simulation of high pressure processes and inactivation of enzymes in food engineering. *Mathematical Models Methods Applied Science*, 19(12): 2203–2229.

Irwe S., Olsson I. 1994. Reduction of pectinesterase activity in orange juice by high pressure treatment. In: *Minimal Processing of Foods and Process Optimisation—An Interface*. eds., Singh R.P., Oliveira F A.R. CRC Press, Boca Raton, FL, pp. 35–42.

Juliano P., Knoerzer K., Fryer P., Versteeg C. 2008. *C. botulinum* inactivation kinetics implemented in a computational model of a high pressure sterilization process. *Biotechnology Process*, 25(1): 163–175.

Katsaros J.G., Alexandrakis Z., Tsevdou M., Taoukis P.S. 2007. High hydrostatic pressure inactivation of *Listeria monocytogenes* in bovine and sheep milk. In: *IFT-2007 Annual Meeting, Institute of Food Technologists*, Chicago, IL.

Katsaros G.I., Tsevdou M., Panagiotou T., Taoukis P.S. 2009. Kinetic study of high pressure microbial and enzyme inactivation and selection of pasteurization conditions for Valencia orange juice. *International Journal of Food Science and Technology*, 45(6): 1119–1129.

Knoerzer K., Juliano P., Gladman S., Versteeg C., Fryer P.J. 2007. A computational model for temperature and sterility distributions in a pilot-scale high-pressure high-temperature process. *American Institute of Chemical Engineers*, 53: 2996–3010.

Koutchma T., Guo B., Patazca E., Parisi B. 2005. High pressure–high temperature sterilization: From kinetic analysis to process verification. *Journal of Food Process Engineering*, 28(6): 610–629.

Lavinas F.C., Miguel M.A.L., Lopes M.L.M., Valentemesquita V.L. 2008. Effect of high hydrostatic pressure on cashew apple (*Anacardium occidentale* L.) juice preservation. *Journal of Food Science*, 73(6): M273–M277.

Lee D.-U. 2002. Application of combined non-thermal treatments for the processing of liquid whole egg. Ph.D. Thesis, Technical University of Berlin, Berlin, Germany.

Ludikhuyze L., Claeys W., Hendrickx M. 2000. Combined pressure–temperature inactivation of alkaline phosphatase in bovine milk: A kinetic study. *Journal of Food Science*, 65(1): 155–160.

Ludikhuyze L., De Cordt S., Weemaes C., Hendrickx M., Tobback P. 1996. Kinetics for heat and pressure-temperature inactivation of *Bacillus subtilis* α-amylase. *Food Biotechnology*, 10(2): 93–103.

Ludikhuyze L., Oey I., Van den Broeck I., Weemaes C., Hendrickx M. 1998. Effect of combined pressure and temperature on soybean lipoxygenase. Influence of extrinsic and intrinsic factors on isobaric-isothermal inactivation kinetics. *Journal of Agricultural and Food Chemistry*, 46: 4074–4080.

Ludikhuyze L.R., Van den Broeck I., Weemaes C.A., Hendrickx M.E. 1997. Kinetic parameters for pressure–temperature inactivation of *Bacillus subtilis* α-amylase under dynamic conditions. *Biotechnology Progress*, 13: 617–623.

Ly-Nguyen B., Van Loey A.M., Smout C., Özcan S.E., Fachin D., Verlent I. 2003. Mild-heat and high-pressure inactivation of carrot pectin methylesterase: A kinetic study. *Journal of Food Science*, 68: 1377–1383.

Maesmans G., Hendrickx M., De Cordt S., Van Loey A., Noronha J., Tobback P. 1994. Evaluation of process value distribution with time temperature integrators. *Food Research International*, 27: 413–423.

Margosch D., Ehrmann M.A., Buckow R., Heinz V., Vogel R.F., Ganzle M.G. 2006. High-pressure-mediated survival of *Clostridium botulinum* and *Bacillus amyloliquefaciens* endospores at high temperature. *Applied and Environmental Microbiology*, 72: 3476–3481.

Minerich P.L., Labuza T.P. 2003. Development of a pressure indicator for high hydrostatic pressure processing of foods. *Innovative Food Science and Emerging Technologies*, 4: 235–243.

Mussa D.M., Ramaswamy H.S. 1997. Ultra high pressure pasteurization of milk: Kinetics of microbial destruction and changes in physico-chemical characteristics. *Lebensmittel-Wissenschaft und-Technologie*, 30: 551–557.

Mussa D.M., Ramaswamy H.S., Smith J.P. 1999. High-pressure destruction kinetics of *Listeria monocytogenes* on pork. *Journal of Food Protection*, 62(1): 40–45.

Nienaber U., Shellhammer T.H. 2001. High-pressure processing of orange juice: Kinetics of pectinmethylesterase inactivation. *Journal of Food Science*, 66(2): 328–331.

Oakenfull D., Scott A. 1984. Hydrophobic interaction in the gelation of high methoxyl pectins. *Journal of Food Science*, 49: 1093–1098.

Otero L., Molina-Garcia A.D., Sanz P.D. 2000. Thermal effect in foods during quasi-adiabatic pressure treatments. *Innovative Food Science and Emerging Technologies*, 1: 119–126.

Otero L., Molina-Garcia A.D., Sanz P.D. 2002. A model for real thermal control in high-pressure treatment of foods. *Biotechnology Progress*, 18(4): 904–908.

Otero L., Ramos A.M., de Elvira C., Sanz P.D. 2007. A model to design high pressure processes towards a uniform temperature distribution. *Journal of Food Engineering*, 78: 1463–1470.

Otero L., Ramos A.M., Elvira C., Sanz P.D. 2006. A model to design high-pressure processes towards an uniform temperature distribution. *Journal of Food Engineering*, 78: 1463–1470.

Parish M.E. 1998. High pressure inactivation of *Saccharomyces cerevisiae*, endogenous microflora and pectinmethylesterase in orange juice. *Journal of Food Protection*, 18(1): 57–65.

Peleg M. 2006. High CO_2 and ultrahigh hydrostatic pressure preservation (Chapter 6). In: *Advanced Quantitative Microbiology for Foods and Biosystems*. ed., Peleg M. CRC Press, Taylor & Francis Group, Boca Raton, FL, pp. 177–184.

Pflug I.J., Zeghman L.G. 1985. Microbial death kinetics in the heat processing of food: Determining an F-value. In: *Proceedings of Aseptic Processing and Packaging of Foods*, September 9–12, 1985, Tylosand, Sweden, pp. 211–220.

Polydera A.C., Galanou E., Stoforos N.G., Taoukis P.S. 2004. Inactivation kinetics of pectin methylesterase of Greek Navel orange juice as a function of high hydrostatic pressure and temperature process conditions. *Journal of Food Engineering*, 62(3): 291–298.

Rajan S., Ahn J., Balasubramaniam V.M., Yousef A.E. 2006. Combined pressure-thermal inactivation kinetics of *Bacillus amylolequefaciends* spores in egg patty mince. *Journal of Food Protection*, 69(4): 853–860.

Rauh C., Baars A., Delgado A. 2009. Uniformity of enzyme inactivation in a short-time high-pressure process. *Journal of Food Engineering*, 91: 154–163.

Rovere P., Carpi G., Dall'Aglio G., Gola S., Maggi A., Miglioli L., Scaramuzza N. 1996. High-pressure heat treatments: Evaluation of the sterilizing effect and of thermal damage. *Industria Conserve*, 71: 473–483.

Rovere P., Gola S., Maggi A., Scaramuzza N., Miglioli L. 1998. Studies on bacterial spores by combined pressure-heat treatments: Possibility to sterilize low-acid foods. In: *High Pressure Food Science, Bioscience and Chemistry*. ed., Isaacs N.S. The Royal Society of Chemistry, Campbridge, U.K., pp. 354–363.

Rovere P., Lonnerborg N.G., Gola S., Miglioli L., Scaramuzza N., Squarcina N. 1999. Advances in bacterial spores inactivation in thermal treatments under pressure. In: *Advances in High Pressure Bioscience and Biotechnology*. ed., Ludwig H. Springer-Verlag, Berlin, Germany, pp. 114–120.

Sablani S.S., Ramaswamy H.S., Prasher S.O. 1995. A neural network approach for thermal processing applications. *Journal of Food Processing and Preservation*, 19: 283–301.

Smith N.A.S., Knoerzer K., Ramos A.M. 2014a. Evaluation of the differences of process variables in vertical and horizontal configurations of High Pressure Thermal (HPT) processing systems through numerical modelling. *Innovative Food Science and Emerging Technologies*, 22: 51–62.

Smith N.A.S., Mitchell S.L., Ramos A.M. 2014b. Analysis and simplification of a mathematical model for high-pressure food processes. *Applied Mathematics and Computation*, 226: 20–37.

Stoforos N.G., Taoukis P.S. 1998. A theoretical procedure for using multiple response time-temperature integrators for the design and evaluation of thermal processes. *Food Control*, 9: 279–287.

Tonello-Samson C. 2014. Industrial applications of high pressure processing in food industry. In: *8th International Conference on High Pressure Bioscience and Biotechnology*, July 15–18, 2014, Nantes, France, Book of Abstracts, p. 48.

Torrecilla J.S., Otero L., Sanz P.D. 2005. Artificial neural networks: A promising tool to design and optimize high-pressure food processes. *Journal of Food Engineering*, 69(3): 299–306.

Van der Plancken I., Grauwet T., Indrawati O., Van Loey A., Hendrickx M. 2008. Impact evaluation of high pressure treatment on foods: Considerations on the development of pressure-temperature-time integrators (pTTIs). *Trends in Food Science & Technology*, 19: 337–348.

Van Opstal I., Vanmuysen S.C.M., Wuytack E.Y., Masschalck B., Michiels C.W. 2005. Inactivation of *Escherichia coli* by high hydrostatic pressure at different temperatures in buffer and carrot juice. *International Journal of Food Microbiology*, 98: 179–191.

Versteeg C., Romouts F.M., Spitansen C.H., Pilnik W. 1980. Thermostability and orange juice cloud destabilizing properties of multiple pectinesterases from orange. *Journal of Food Science*, 45(4): 969–998.

Vervoort L., Van der Plancken I., Grauwet T., Verjans P., Courtin C.M., Hendrickx M.E. 2011. Xylanase B from the hyperthermophile *Thermotoga maritima* as an indicator for temperature gradients in high pressure high temperature processing. *Innovative Food Science and Emerging Technologies*, 12: 187–196

Weemaes, C.A., Ludikhuyze, L.R., Van den Broeck, I., Hendrickx, M.E. 1998. Kinetics of combined pressure-temperature inactivation of avocado polyphenoloxidase. *Biotechnology and Bioengineering*, 60(3): 292–300.

Zook C.D., Parish M.E., Braddock R.J., Balaban M.O. 1999. High pressure inactivation kinetics of *Saccharomyces cerevisiae* ascospores in orange and apple juices. *Journal of Food Science*, 64(3): 533–535.

12 High-Pressure Processing of Foods
Technology and Applications

George Katsaros, Z. Alexandrakis, and Petros Taoukis

CONTENTS

12.1 DEMAND FOR NONTHERMAL PROCESSING OF FOODS—THE TECHNOLOGY OF HP

The potential of applying high pressure (HP) as the main physical process for food treatment and preservation has been reported at the beginning of the last century (Hite, 1899). In the 1990s, several researchers showed that HP treatment can lead to food products free from undesirable microorganisms and enzymes with minor changes in their quality and sensory attributes (Grant et al., 2000). Additionally, HP-treated products such as jams, fruit products, fruit jellies, salad dressings, sauces, etc., were introduced and launched onto the Japanese, American, and European market (Tewari et al., 1999). In 2000, more innovative products (guacamole, peeled oysters, fruit juices, and poultry) appeared in the United States (Meyer, 2000). Food preservation using high-pressure pasteurization as an alternative to thermal pasteurization leads to products with extended shelf life and improved sensory and nutritional characteristics (preservation of thermally sensitive flavor compounds and nutrients, color, and texture) (Fondberg-Broczek, 1998). HP treatments, in general, are effective in inactivating most vegetative pathogenic and spoilage microorganisms at pressures above 200 MPa at chilled or process temperatures less than 45°C, but the rate of inactivation is strongly influenced by the peak pressure and the process time (Raoult-Wack, 1994). Sterilization, that is, inactivation of spores such as *Clostridium botulinum* and Bacillus could be achieved through synergies of elevated heat and pressure (Ahn et al., 2007). High pressure processing (HPP) technology, as commercially defined today, is unable to produce low-acid shelf stable products since bacterial spore inactivation requires high pressures of at least 800–1700 MPa at room temperature, far in excess of what is commercially available today. High-pressure high-temperature (HPHT) processing (HP sterilization), or pressure-assisted thermal sterilization (PATS), involves the use of moderate initial chamber temperatures between 60°C and 90°C in which, through internal compression heating at pressures of 600 MPa or greater, in-process temperatures can reach 90°C–130°C. The process has been proposed as a high-temperature short-time process, where both pressure and compression heat contribute to the process's lethality (Leadley, 2005). In this case, compression heat developed through pressurization allows instantaneous and volumetric temperature increase, which, in combination with high pressure, accelerates spore inactivation in low-acid media. The main advantage of HPHT treatment is the shorter processing time compared to conventional thermal processing in eliminating spore-forming microorganisms (Matser et al., 2004). This shorter process time and ultimate pressurization temperatures lower than 121°C have resulted in higher quality and nutrient retention in selected products. For example, better retention of flavor components in fresh basil, firmness in green beans, and color in carrots, spinach, and tomato puree have been found after HPHT processing (Krebbers et al., 2002, 2003). Nutrients such as vitamins C and A have also shown higher retention after HPHT processing in comparison to retort methods (Matser et al., 2004).

Apart from HP cold pasteurization and sterilization, use of HP for homogenization of liquid foods is another approach. High-pressure homogenization (HPH) is a technique in which a fluidic product is forced through a narrow gap at high pressures (150–400 MPa). Thus, the product is subjected to very high shear stress causing the formation of fine emulsion droplets. The resistance to droplet deformation and breakup caused by the Laplace pressure has to be reduced. The high pressures applied cause a temperature increase due to adiabatic heating, and it has to be controlled in order to avoid food texture deterioration. HPH also enhances product texture and mouthfeel. Using nanoemulsions in food products can facilitate the use of less fat without a compromise in creaminess, thus offering the consumer a healthier option. Some applications of HPH include the formation of the acid gelation properties of recombined whole milk, the viscosity of fermented dairy beverages, the microstructure of low-fat yoghurt and low-fat emulsions, as well as the rheological properties of ice cream mixtures.

Despite the substantial research activity of the last years and the continuously growing application of HP technology, scientific, technological, and technical issues should be answered for each product before HP industrial application. Food industry is in need of fully substantiated and validated answers regarding the applicability and the benefits of HP as a physical process for better preservation of the various food products. The optimization of the conditions (safety, quality retention, nutritional value, and consumer acceptability being the main parameters) combined with the accurate design and control of the process is essential. The systematic study of the technical parameters (that would allow for the use of the scientific achievements in a controlled and effective industrial process) is based on the kinetic approach of the destructive reactions of several factors that lead to spoilage or degradation of quality and functional properties of the food during HP treatments (Hendrickx and Knorr, 2002). This approach is detailed by Stoforos and Taoukis (1998) in analogy to the conventional thermal processing and constitutes the basis for the design, the treatment, and the modeling of the procedure that would offer the possibility of a safe and effective industrial use. The degree of deterioration of an index sensitive to the HP process is a function of its resistance at various pressure–temperature conditions, depending also on the treatment duration. Assuming that the deterioration is described by a first-order reaction:

$$-\frac{dC}{dt} = kC, \quad k = f(T, P, \ldots) \tag{12.1}$$

where

C is the concentration (or activity or population, etc.) of the HP sensitive index (e.g., number of microorganisms/mL, g/L, etc.)

k is the reaction rate constant at constant conditions of the procedure (min^{-1})

t the time of the treatment (min)

T the temperature during the HP treatment (K)

P the pressure (MPa)

Integrating Equation 12.1 leads to:

$$\ln(C) - \ln(C_o) = \int_{t_a}^{t_b} -k\,dt \tag{12.2}$$

where

C_o is the initial concentration

subscripts a and b refer to the initial and the final condition, respectively (in this case, the beginning and the end of the procedure)

If the effect of pressure and temperature on the reaction rate is known, then the integral of Equation 12.2 can be calculated using the appropriate Arrhenius (Equation 12.3) and Eyring (Equation 12.4) expressions:

$$k = k_{T_{ref}} \exp\left[-\frac{E_a}{R}\left(\frac{1}{T} - \frac{1}{T_{ref}}\right)\right] \tag{12.3}$$

$$k = k_{P_{ref}} \exp\left[-\frac{V_a}{R}\frac{(P - P_{ref})}{T}\right] \tag{12.4}$$

where

T_{ref} and P_{ref} represent a reference temperature and a reference pressure, respectively

E_a (J/mol) and V_a (mL/mol) the activation energy and volume, respectively

R the universal gas constant (8.314 J/(mol K)

Using Equations 12.3 and 12.4 in Equation 12.2, assuming that the activation energy and volume depend on the pressure and the temperature respectively, Equation 12.2 is finally expressed as follows:

$$\ln\left[\frac{C}{C_o}\right] = \int_0^t \left[-k_{ref} \exp\left[-\frac{E_a(P)}{R}\left(\frac{1}{T} - \frac{1}{T_{ref}}\right) - \frac{V_a(T)}{R}\frac{(P - P_{ref})}{T}\right]\right] dt \tag{12.5}$$

This equation forms the basis for the modeling of the HP treatment (e.g., for the determination of an equivalent time of treatment, for example, the calculation of the *F*-value) in a similar way as this is done for conventional thermal treatments.

The quantification of the effect of the process conditions and the subsequent storage on the shelf life of products allows for the optimization of the design of the overall production. Products that have been manufactured under "optimized" conditions must be tested for their acceptability by representative consumer groups and appropriate sensory methodology.

Effective implementation of HP technology in the food industry requires knowledge of the optimal conditions of processing. Hence, the following steps for each product need to be taken before the product is marketed:

1. Determination of the main factors that cause safety issues (growth of pathogens) and quality deterioration of the product (specific spoilage microorganisms and undesirable enzymes)
2. Determination of the kinetics inactivation of the above mentioned factors in food models and real food matrices in order to find an adequate HP process for the preservation of the food in question

3. Evaluation of HP-treated food products' quality and nutritional characteristics
4. Determination and modeling of the shelf life of HP-treated products and comparison with products that have been subjected to conventional preservation methods (heating, addition of preservatives)
5. Consumer acceptability and preference test

Several case studies will be presented in this chapter to illustrate the potential of HP for the development of food products of high quality with extended shelf life.

In HP treatments, the process conditions are a function of the following extrinsic parameters:

- Applied pressure (MPa)
- Temperature (C)
- Holding time (min)
- Pressure buildup time (min)
- Pressure release time (min)
- Number of pulses and time interval between pulses when pulsed pressures are used

Adiabatic heating should be considered when applying high pressures.

12.1.1 High-Pressure Equipment

HP processing in an industrial scale is conducted at various pressures (usually from 100 to 600 MPa) and room temperature 25°C or chill temperatures (4°C–10°C), for various processing times (ranging from 1 to 15 min). The high-pressure units available, comprise one or a set of pressure intensifiers and usually one vessel (volume depending on the models available ranging from 40 up to 525 L), with a maximum operating pressure and temperature of usually 600 MPa. The pressure transmitting fluid used is water. The desired value of pressure is set and after pressure buildup (rate depending on the number of intensifiers and the volume of the vessel), the pressure vessel is isolated. This point defines the time zero of this process. Pressure of the vessel is released ($t_{release}$ < 1 min) after a preset time interval (according to the production design). The temperature in the chamber is monitored during the process. The initial adiabatic temperature increase during pressure buildup should be taken into consideration in order to achieve the desired operating temperature during pressurization. Products are usually precooled at appropriate temperatures in order to reach the target temperature after pressure buildup, taking into consideration the monitored adiabatic heating (approximately 3°C/100 MPa).

The cost of HP processing is dependent on the volume of the units and is presented in Figure 12.1.

Fruits and vegetables and meat and meat products are the main food sectors that HP technology has found application, as evidenced by the number of products produced worldwide (Figures 12.2 and 12.3).

The United States is the leader in food production using HP technology for cold pasteurization, followed by Europe (Figure 12.4).

The growing interest in HP-treated food products is also depicted in the number of HP units installed worldwide, and the rate is continuously growing. Figure 12.5 shows the continuous installed units all over the world, while the trend shows that in a few years (2018) more than 600 units will be installed in industries all over the world.

Nowadays, there are HP units available for vessel volumes ranging from 50 to more than 500 L, depending on the industry needs. In Figure 12.6, a unit developed and produced by Hiperbaric (www.hiperbaric.com) is shown. It is a unit comprising a 525 L capacity vessel, large 380 mm diameter, and throughput of over 3000 kg of product per hour.

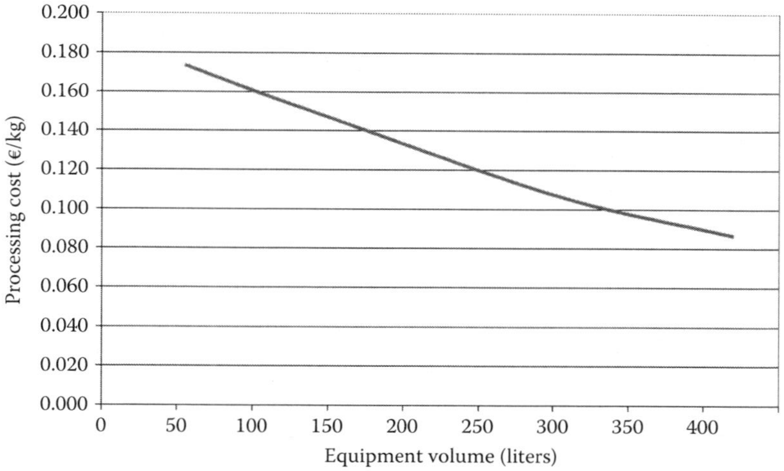

FIGURE 12.1 Effect of equipment volume on the processing cost of food products.

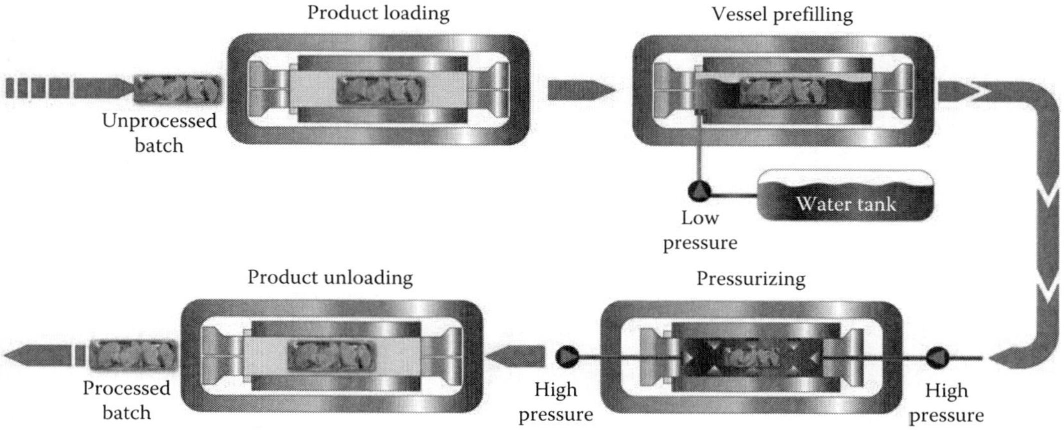

FIGURE 12.2 Diagram of the operation of a HP unit. (From NC Hiperbaric web page http://www.hiperbaric.com/en/high-pressure, HIPERBARIC, Miami.)

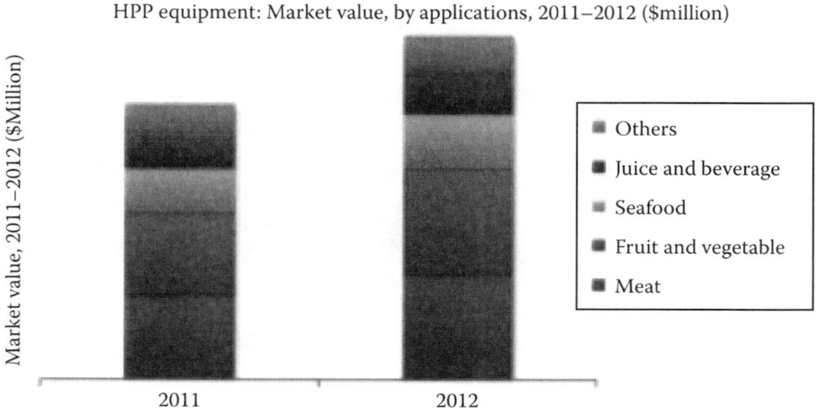

FIGURE 12.3 Market value of food sectors in which HP technology is being applied over the last years. (From MarketsandMarkets Analysis, November 2013.)

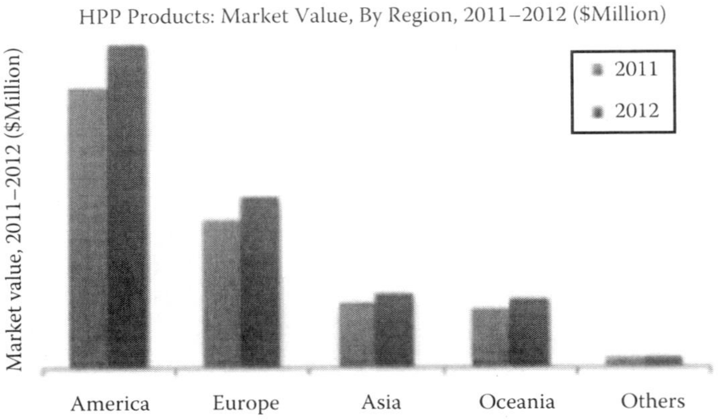

FIGURE 12.4 Market value by region of food products produced applying HP technology over the last years. (From MarketsandMarkets Analysis, November 2013.)

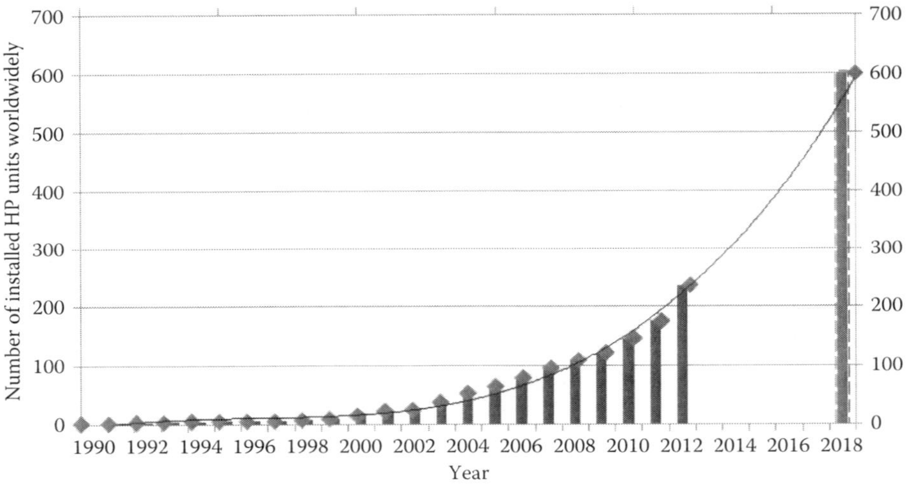

FIGURE 12.5 World growth of the food industry use of HP processing technology.

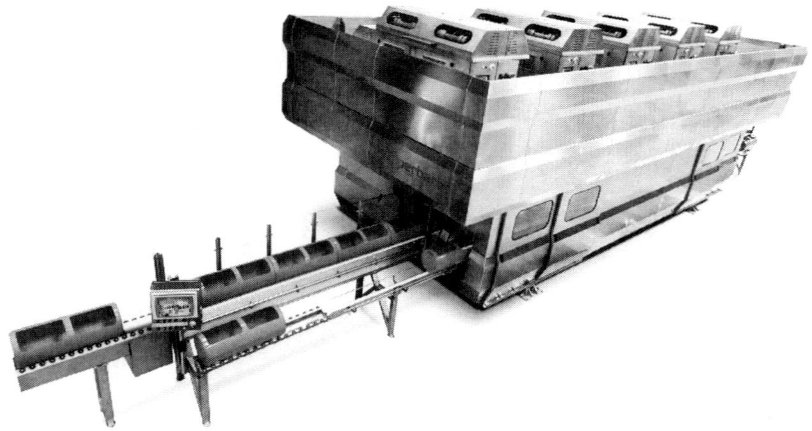

FIGURE 12.6 Model 525 HP unit developed by Hiperbaric with a vessel of 525 L capacity and 380 mm diameter.

12.2 EFFECT OF HP ON MICROORGANISMS OF FOODS

HP inactivates microorganisms by altering their morphology and several vulnerable components such as cell membranes, ribosomes, and enzymes, including those involved in the reproduction and survival procedures (the replication and transcription of DNA) (Yuste et al., 2001; Norton and Sun, 2008). The primary site of pressure damage is usually the cytoplasmic membrane. Cell permeability and ion exchange are altered (Hoover et al., 1989). These modifications in the microbial cell membrane are probably the main cause of sublethal injury generated by the pressure treatment (Patterson et al., 1995). A variety of factors can affect the sensitivity of microorganisms to HP, including the species and strain of the microorganism, the growth temperature and growth phase prior to HP, the composition of the suspending medium, and the magnitude and duration of the applied pressure (Patterson et al., 1995).

Bacterial vegetative cells, yeasts, and molds are relatively sensitive to pressurization since they are inactivated in the range of pressure 400–600 MPa at ambient temperature, while the bacterial spores (several Clostridium species) appeared to be relatively resistant to it. Spores are highly pressure resistant since pressures exceeding 1200 MPa may be needed to obtain a significant level of inactivation (Sale et al., 1970; Hoover et al., 1989; Knorr, 1995; Smelt, 1998). In general, gram-positive bacteria (*Listeria monocytogenes*, *Staphylococcus aureus*) are more resistant to pressure than gram-negative (Pseudomonas, Salmonella spp, *Yersinia enterocolitica*, *Vibrio parahaemolyticus*) (Hayakawa et al., 1994; Cheftel, 1995; Mackey et al., 1995) due to the rigidity of teichoic acids in the peptidoglycan layer of the gram-positive cell wall (Lado and Yousef, 2002).

Furthermore, cultures in the exponential growth phase have been shown to be far more sensitive than cultures in the logarithmic growth or stationary phase (Hoover et al., 1989). Microbial cells surviving pressurization become sublethally injured and develop sensitivity to physical and chemical environments to which the normal cells are resistant (Kalchayanand et al., 1995, 1998; Hauben et al., 1996). In addition, pressurization can inactivate some parasites such as *Trichinella spiralis*, but its efficiency on inactivation of viruses is limited (Cheftel, 1995). Extensive studies have been concentrated on several species including gram-negative pathogens Salmonella and *Escherchia coli O157:H7*; gram-positive microorganisms *S. aureus* and *L. monocytogenes*; spore-forming pathogens *Bacillus cereus*; and nonpathogenic flora like *Pediococcus* spp., *Lactobacillus brevis*, and *Lactobacillus plantarum* strains.

The approach in the thermal processing literature for the quantification of microbial inactivation uses the decimal reduction time (D) and the thermal resistant constant z. Equivalently for HP processing, the D-value is defined as the treatment time at a constant pressure and temperature that is required to reduce the microbial count by 90%, z_T: the thermal resistance constant, the temperature increase for 90% decrease of D-value (at a constant reference pressure), z_P: the pressure resistance constant, the pressure increase needed for 90% decrease of D-value (at a constant reference temperature).

The effect of HP (300–600 MPa) combined with temperature (25°C–40°C) on the inactivation of *L. monocytogenes* in bovine and sheep milk and the baroprotective effect of their constituents on the viability of this microorganism is presented to demonstrate the kinetic approach to be employed (unpublished own data). Inactivation followed first-order reaction kinetics at all pressures and temperatures tested. The decimal reduction times (D-values) were estimated for the microorganism at all conditions tested (Table 12.1). An increase of HP and temperature led to a reduction of the D-values of the microorganism, indicating a synergistic effect of these process parameters on the inactivation. The inoculated *L. monocytogenes* (strain Scott A) in the sheep milk appeared to be more pressure-resistant compared to the same microorganism in the bovine milk due to the baroprotective effect of its higher lipid content. D-values of 0.9 and 2.6 min for inactivation of *L. monocytogenes* in bovine and sheep milk, respectively, were calculated after processing at 600 MPa and 25°C. At the highest applied pressure–temperature condition (600 MPa, 40°C), the D-value of *L. monocytogenes* was lesser than 1 min for both tested media.

TABLE 12.1
Estimated Decimal Reduction Times, *D* (min), for HP Inactivation of *L. monocytogenes* in Bovine and Sheep Milk in the Range of Temperature from 25°C to 40°C

T (°C)	25	32	40	
***P* (MPa)**	***D* (Min)**	***D* (Min)**	***D* (Min)**	**Z_T (°C)**
Bovine milk				
300	10.80 ± 1.32[1,a]	9.75 ± 0.86[1,a]	3.33 ± 0.74[2,a]	28.6 ± 8.9[a]
400	8.34 ± 0.52[1,b]	2.75 ± 0.48[2,b]	2.54 ± 0.24[2,a]	29.6 ± 10.3[a]
500	1.96 ± 0.46[1,c]	1.32 ± 0.26[1,c]	1.01 ± 0.36[1,b]	50.3 ± 21.4[a]
600	0.92 ± 0.14[1,d]	0.43 ± 0.08[2,d]	0.40 ± 0.12[2,c]	41.1 ± 16.3[a]
Z_P (MPa)	286.4 ± 65.1[1]	262.5 ± 63.7	369.0 ± 128.0[1]	
Sheep Milk				
300	12.40 ± 1.27[1,a]	11.20 ± 1.26[1,a]	9.12 ± 0.44[2,a]	112.7 ± 40.3[a]
400	9.09 ± 0.55[1,b]	4.10 ± 1.41[2,b]	3.09 ± 0.28[2,b]	32.4 ± 10.6[b]
500	3.31 ± 0.33[1,c]	2.34 ± 0.39[2,b]	1.27 ± 0.25[3,c]	35.5 ± 8.5[b]
600	2.60 ± 0.26[1,d]	1.04 ± 0.07[2,c]	0.61 ± 0.13[2,d]	23.8 ± 4.6[b]
Z_P (MPa)	432.0 ± 90.7[1]	339.0 ± 77.2[1]	284.3 ± 66.5[1]	

Average value ± standard error between three measurements.
Different superscript letters indicate significantly different means ($P < 0.05$) within a column (differences between pressures).
Different superscript numbers indicate significantly different means ($P < 0.05$) within a row (differences between temperatures).

Similar *D*-values have been reported in the literature (Styles et al., 1991; Mussa and Ramaswamy, 1999). *D*-values are in the range of 9.3 min (Styles et al., 1991) and 8.1 min (Mussa and Ramaswamy, 1999) for inactivation of *L. monocytogenes* in raw milk by applying pressure of 340 and 350 MPa, respectively, in the range of the the *D*-values presented in Table 12.1 for bovine milk and ovine milk.

Koseki et al. (2008) reported that 5 min treatment of 500 MPa at 25°C resulted in an approximately 5 log10 reduction of *L. monocytogenes* in UHT whole milk compared to 600 MPa and 25°C or 500 MPa and 40°C in Table 12.1.

A model that can predict the *D*-values of a microorganism in the range of process pressure–temperature combinations studied could be used (Equation 12.6), taking into account all the data obtained from *L. monocytogenes* inactivation as well as the temperature effect on the z_P-value and the pressure effect on the z_T-value.

$$D = D_o \cdot \left[\exp\left\{ -\frac{2.303 \cdot T \cdot T_{\text{ref}}}{Z_T} \cdot \exp\left[-A(P - P_{\text{ref}})\right] \cdot \left(\frac{1}{T} - \frac{1}{T_{\text{ref}}}\right) - \left(\frac{B \cdot (T - T_{\text{ref}}) - \left(\frac{2303 \cdot R \cdot T}{Z_p}\right)}{R} \right) \cdot \frac{(P - P_{\text{ref}})}{T} \right\} \right]^{-1} \quad (12.6)$$

where

D_o is the decimal reduction time at $P_{\text{ref}} = 500$ MPa and $T_{\text{ref}} = 32$°C
z_T is the thermal resistance constant (temperature needed for 90% increase or decrease of the *D*-value)
z_P is the pressure resistance constant (pressure needed for 90% increase or decrease of the *D*-value)

A is parameter that expresses the effect of pressure on the z_T-value
B is parameter that expresses the effect of temperature on the z_P-value
R is the universal gas constant

The parameters of Equation 12.6 were estimated by nonlinear regression (Table 12.2).

HP can also inactivate spoilage microorganisms such as *Pediococcus* spp. found in sea bream fillets or lactic acid bacteria (LAB) found in orange juices. HP inactivation of *Pediococcus* spp. followed first-order kinetics (Figures 12.3 through 12.5 and 12.7) with a tailing region at all treatment conditions. The decimal reduction times (D-values) as a function of pressure and temperature conditions are presented in Table 12.3.

The pressure–temperature combinations necessary to achieve the targeted inactivation of *Pediococcus* spp., the main deteriorative factor of high-pressured gilthead sea bream fillets, can be estimated by Equation 12.6, enabling the proper design of high pressure combined with temperatures treatment of the fillets (Table 12.4).

Quality degradation of untreated orange juice is due to the presence of spoilage microorganisms and the activity of pectinolytic enzymes. The inactivation of orange juice spoilage microflora by HP has been reported by many researchers. Yeasts have been shown to be considerably more sensitive to HP than LAB, while moulds are the most labile of the spoilage microflora of orange juice

TABLE 12.2
Estimated Parameters of the Multiparameter Equation (Equation 12.6) that Describe the Decimal Reduction Time *D* (Min) of *L. monocytogenes* at Any Combination of Pressures and Temperatures (T_{ref} = 305 K and P_{ref} = 500 MPa)

Parameter	Estimated Values (*L. monocytogenes* in Bovine Milk)	95% Confidence Interval		Estimated Values (*L. monocytogenes* in Sheep Milk)	95% Confidence Interval	
		Lower[a]	Upper[a]		Lower[a]	Upper[a]
D_o (min)	1.50	0.11	2.92	2.17	1.40	2.93
Z_T (°C)	21.5	6.7	50.6	22.5	2.3	63.0
Z_P (MPa)	278.3	108.9	447.7	281.3	63.5	485.2
A (MPa^{-1})	0.001	−0.001	0.002	0.0012	−0.032	0.019
B (mol/KJ)	−0.889	−3.256	1.482	−1.359	−2.09	−0.423
R^2	0.95			0.98		

[a] Upper and lower values were estimated for 95% confidence intervals by the nonlinear regression routine (SYSTAT 8.0).

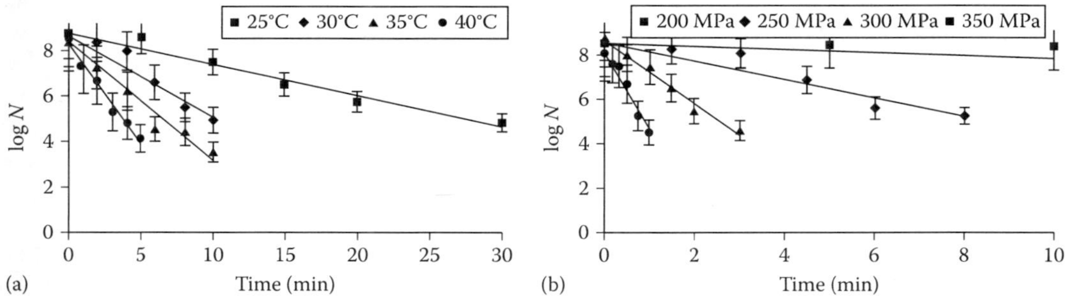

FIGURE 12.7 (a) *Lactobacillus brevis* inactivation during HP processing at 250 MPa combined with temperatures in the range of 25°C–40°C and (b) *L. plantarum* inactivation during HP processing at 200–350 MPa at 35°C and 200–350 MPa. (Bars denote the standard deviation of triplicate measurements).

TABLE 12.3
Estimated Decimal Reduction Times, *D* (Min), for High-Pressure Inactivation of *Pediococcus* spp. of Sea Bream Fillets in the Range of Temperature from 20°C to 40°C Using the Baranyi Model

	Sea Bream Fillets			
T (°C)	20	30	40	
P (MPa)	*D* (Min)	*D* (Min)	*D* (Min)	Z_T (°C)
150	205	138	50	32.8
300	8.7	5.0	1.5	26.5
450	3.6	0.9	0.5	23.4
600	0.8	0.3	0.2	34.5
Z_P (MPa)	196	172	196	

TABLE 12.4
Parameters of the Multiparameter Equation (Equation 12.6) that Describe the Decimal Reduction Time *D* (Min) of *Pediococcus* spp. at Any Combination of Pressures and Temperatures (T_{ref} = 303 K and P_{ref} = 300 MPa)

Parameter	D_o (Min)	Z_T (°C)	Z_P (MPa)	*A* (MPa^{-1})	R^2
Estimated values (*Pediococcus* spp. of Sea Bream fillets)	4.38	31	178	−0.002	0.99

(Hoover et al., 1989). Several studies on the inactivation of the microflora of orange juice have been published, and they mainly deal with the issue of inactivation of certain foodborne pathogens and spoilage bacteria (Alpas and Bozoglu, 2003; Erkmen and Dogan, 2004; Bayindirli et al., 2006). Basak et al. (2002) investigated *Leuconostoc mesenteroides* and *Saccharomyces cerevisiae* inactivation in single-strength and concentrated orange juice in the range of 100–400 MPa. Kinetic analysis revealed the presence of a dual effect of pressure inactivation, an instantaneous pressure kill (dependent on the pressure level), and a first-order inactivation (dependent on the holding time). Both microorganisms revealed similar z_P values when tested in the same type of juice.

Inactivation kinetics of *L. plantarum* and *L. brevis* in neutral and acid media (pH value of 7.0 and 4.2, respectively) was studied at a pressure range of 400–534 MPa and ambient temperature. Both bacteria exhibited similar z_P values when tested in neutral medium. In acid medium, *L. plantarum* seemed to be more resistant than *L. brevis* (Mallidis et al., 2003). Katsaros et al. (2010) investigated *L. brevis* and *L. plantarum* strains inactivation in valencia orange juice in the range of 200–350 MPa.

In Figure 12.7a, the effect of process temperature at 250 MPa on the reduction of *L. brevis* counts and in Figure 12.7b, the effect of pressure processing at 35°C on the reduction of *L. plantarum* counts are shown. HP inactivation of both LAB species was described by first-order kinetics. The *D*-values were estimated for each microorganism within the studied domain (200–350 MPa and 25°C–40°C). The *D*-values were estimated for each microorganism at all studied pressure and temperature combinations (Table 12.5). The *D*-values decreased with increasing processing pressure and temperature at all temperature and pressure levels tested, respectively. A synergistic effect of temperature and pressure was observed.

The multiparameter Equation 12.6 was applied to describe the effect of pressure and temperature process conditions on the *D*-value of each microorganism, combining the aforementioned effect of pressure on the z_T values and considering that the effect of process temperature on the z_P values is small and follows no clear trend (z_P values ranged from 80 to 105 MPa at 25°C–40°C) (predicted parameters of the model presented in Table 12.6). LAB appeared to be very sensitive to pressures above 300 MPa. Thus, a moderate high pressure–temperature process condition, sufficient for

TABLE 12.5
Decimal Reduction Times, D (Min), of HP-Treated Valencia Orange Dominant LAB Species *L. brevis* (*L. b*) and *L. plantarum* (*L. p*)

	25°C		30°C		35°C		40°C		Z_T	
Pressure	$D_{L.b}$	$D_{L.p}$	$D_{L.b}$	$D_{L.p}$	$D_{L.b}$	$D_{L.p}$	$D_{L.b}$	$D_{L.p}$	*L. b*	*L. p*
200 MPa	15.53	17.82	14.93	14.90	9.27	14.73	3.77	10.31	31.5	77.5
250 MPa	7.21	8.65	2.82	3.00	2.02	2.30	0.84	2.21	18.9	29.0
300 MPa	1.07	1.97	0.68	1.11	0.49	0.71	0.41	0.32	40.0	19.5
350 MPa	0.67	0.50	0.31	0.32	0.28	0.29	0.19	0.17	32.5	36.6
Z_P	105	81	82	95	97	92	79	80		

Note: Z_P and Z_T values are also presented. Bold values indicate the treatment time (D, min) adequate for the inactivation of both microorganisms.

TABLE 12.6
Parameters of the Model of Valencia Orange LAB Inactivation (Equation 12.6) as a Function of Pressure and Temperature for T_{ref} = 30°C and P_{ref} = 300 MPa

Parameter	*L. brevis*	*L. plantarum*
D_o (min)	342 ± 0.51	1.32 ± 0.11
z_T (°C)	23.8 ± 1.4	18.8 ± 1.3
z_P (MPa)	94.7 ± 7.8	95.0 ± 11
A (MPa^{-1})	−0.009 ± 0.001	−0.013 ± 0.002

Notes: ± Values represent the standard error of regression.

inactivation of both microorganisms, offers the ability of a minimal process design with a minimum nutritional and sensorial characteristic deterioration.

12.3 EFFECT OF HP PROCESSING ON ENDOGENOUS AND EXOGENOUS ENZYMES OF FOOD PRODUCTS

Enzymes participate in cellular metabolic processes with the ability to enhance the rate of reaction between biomolecules. The changes in their active site or enzyme denaturation can lead to a reversible or an irreversible loss of their activity. HP mechanism for enzyme denaturation is governed by the *Le Chatelier* principle, which predicts that application of pressure shifts an equilibrium to the state that occupies the smallest volume, so any reaction accompanied by volume decrease is accelerated by elevated pressures (Cano et al., 1997). The decrease of the enzyme activity can be attributed to the fact that high pressure affects the substrate–enzyme interaction. In the case that the substrate is a macromolecule, the effects may be on the structure of the macromolecule making the enzymic action easier or more difficult. The effect of HP on the activity of enzymes that are important to food quality, such as polyphenoloxidases (Rapeanu et al., 2005; Buckow et al., 2009), pectin methylesterases (Polydera et al., 2004; Guiavarch et al., 2005; Plaza et al., 2007; Boulekou et al., 2009; Katsaros et al., 2010; Alexandrakis et al., 2014a,b), peroxidases (Garcia-Palazon et al., 2004; Terefe et al., 2010), cysteine proteases (Katsaros et al., 2009b,c), and aminopeptidases (Katsaros et al., 2009a), has been studied and reported in several publications.

The enzymes found in different systems (buffer, juice, tissue) often follow first-order inactivation kinetics (Equation 12.7). However, the existence of several isoenzymes that show different heat or/and pressure resistance has been observed. The biphasic model signifies the coexistence of at least two isoenzymes, a pressure-resistant and a pressure-labile one. In the case of fractional conversion model (Equation 12.8), first-order inactivation is applied taking into account a nonzero residual activity upon prolonged processing.

Specifically, the inactivation of enzymes can often be described by a first-order kinetic model (Equation 12.7) (Eagerman and Rouse, 1976):

$$Ln\left(\frac{A_t}{A_o}\right) = -k \cdot t \tag{12.7}$$

where

A_o and A_t are the initial activity and the remaining activity at time t, respectively
k the inactivation rate constant (min^{-1})

In the case of the fractional-conversion model for all pressure–temperature conditions, Equation 12.8 could be used:

$$Ln\left(\frac{A - A_f}{A_o - A_f}\right) = -k \cdot t \tag{12.8}$$

where

A is the enzyme activity after processing for a treatment duration t
A_f the residual activity after processing
A_o the initial activity
t the processing time (min)
k the inactivation rate constant (min^{-1})

The kinetic model adequately described the loss of enzyme activity during processing, showing a first-order inactivation of the sensitive portion of the enzyme (labile isoenzyme) and the presence of a resistant enzyme fraction that is hardly inactivated by the pressure or temperature applied.

One of the key enzymes in fruit and vegetable processing is pectin methylesterase (PME), which mainly affects the texture of fruits and vegetables, lowering their viscosity as well as deteriorating the quality of citrus juices by the destabilization of clouds. In the literature, there is a significant number of papers describing the effect of HP and temperature on PME activity from different fruits and vegetables, such as citrus-based foods (Versteeg et al., 1980; Basak and Ramaswamy, 1996; Cano et al., 1997; Goodner et al., 1998, 1999; Van den Broeck et al., 2000b; Polydera et al., 2004; Guiavarch et al., 2005; Sampedro et al., 2008; Katsaros et al., 2010; Alexandrakis et al., 2014a), tomato-based foods (Van den Broeck et al., 2000a; Stoforos et al., 2002; Rodrigo et al., 2006; Plaza et al., 2007; Verlent et al., 2007), peach (Boulekou et al., 2009), strawberry (Cano et al., 1997; Ly-Nguyen et al, 2002c), fruits and vegetables and other materials (Castro et al., 2006, 2008), carrot (Ly-Nguyen et al., 2002b, 2003a; Balogh et al., 2004; Sila et al., 2007), banana (Ly-Nguyen et al., 2002a, 2003b), apple (Riahi and Ramaswamy, 2003), persimmon (Katsaros et al., 2006), and sea buckthorn (Alexandrakis et al., 2014b). The results obtained from these studies indicate that the specific origin of the enzyme, both fruit and variety of fruits, results in different inactivation kinetics. For example, it was found that the pressure stability of PMEs can vary by several orders of magnitude ranging from pressure-sensitive types like orange juice (*Valencia* cv.) PME (Katsaros et al., 2010) to significantly barotolerant ones like persimmon juice (*Hachiya* cv.) PME. Overall, persimmon PME showed a high thermal and pressure stability requiring intense process conditions for adequate inactivation (500–800 MPa at 40°C–70°C). The controlled inactivation of this enzyme is necessary for the optimal production design of persimmon juice rich in antioxidants that could be consumed as is or mixed with other fruit juices.

In general, more intense pressure and temperature process conditions enhance enzyme inactivation. In some cases, there is a synergistic effect of pressure and temperature (process combining pressure at a certain temperature results in faster inactivation when compared to the additive enzyme inactivation by thermal treatment at the same temperature at 1 MPa and by pressure treatment at low temperatures). However, at high temperatures (close to temperatures resulting in thermal inactivation of enzymes at atmospheric pressure) an antagonistic effect of pressure and temperature could be observed. In these cases, the enzyme inactivation is slower when the enzyme is treated at a certain temperature combined with pressure compared to the inactivation by only thermal processing. Polydera et al. (2004) studied the HP inactivation of orange juice (*Navel* cv.) PME and found a synergistic effect of pressure and temperature on this enzyme under HP processing conditions, except in the high temperature–low pressure region where an antagonistic effect was noted. Boulekou et al. (2009) investigated the inactivation of endogenous PME in Greek commercial peach pulp under high hydrostatic pressure (100–800 MPa) combined with moderate temperature (30–70°C). High pressure and temperature acted synergistically on PME inactivation, except at the high temperature of 70°C at the middle pressure range (100–600 MPa), where an antagonistic effect of pressure and temperature was observed. At this specific middle pressure range, an increase of pressure processing led to increased inactivation rate constants of peach PME. The effect of pressure and temperature on the inactivation rate constant of peach PME is presented in Figure 12.8.

The multiparameter model (Equation 12.5) could be used to express the enzyme inactivation rate constant as a function of temperature and pressure process conditions, taking into account the dependence of both activation energy and activation volume on pressure and temperature, respectively. This modeling approach enables the quantitative estimation of the HP–temperature conditions needed to achieve targeted enzyme inactivation in different plant sources.

Pressure stability of enzymes is largely varied depending on the type of enzyme, presence of other enzymes, the type of substrates, ionic strength, pH, nature of the medium in which the enzyme is dispersed, pressure, temperature, and treatment time (Cheftel, 1991; Irwe and Olsson, 1994).

The selection of the optimal process conditions for the inactivation of the two more deteriorative factors for food products, that is, microorganisms and enzymes is necessary. Katsaros et al. (2010)

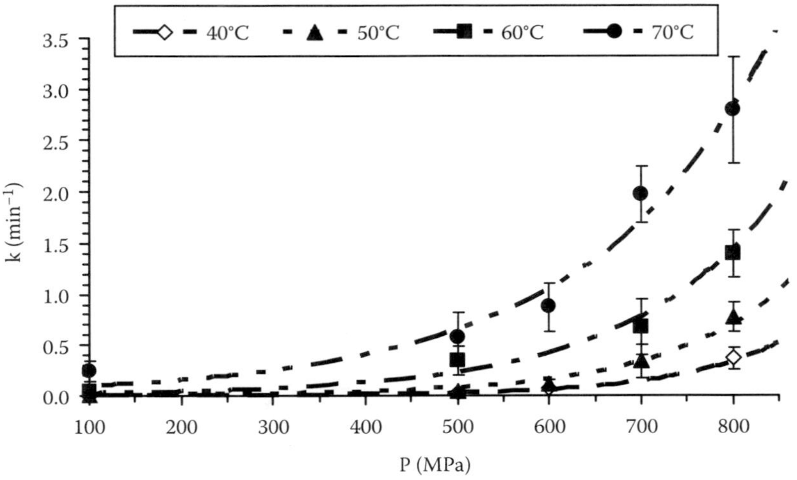

FIGURE 12.8 PME inactivation rate constant as a function of pressure and temperature. Lines represent values predicted from Equation 12.5.

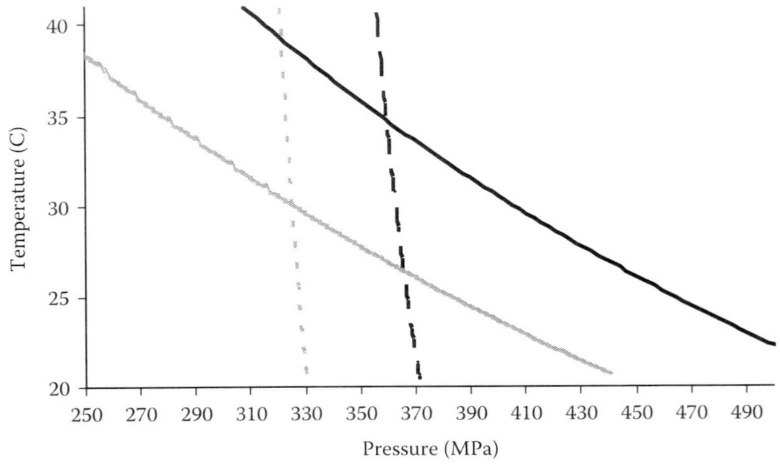

FIGURE 12.9 Required processing time for the inactivation of PME and LAB as a function of pressure at 25°C and 30°C. Dashed lines represent a 7D LAB reduction and solid lines represent 90% PME inactivation. Black lines show processing at 25°C, gray lines show processing at 30°C.

studied the effect of pressure on these two factors, and they estimated the necessary HP process conditions for the cold pasteurization of valencia orange juice. They concluded that both LAB and PME appeared to be relatively sensitive to pressures above 300 MPa. Valencia orange PME sensitivity is of high interest because at certain processing conditions, it does not comply with the general statement that PMEs are generally more resistant than microorganisms and that treatment for PME inactivation is sufficient for juices' pasteurization. So, in the case of valencia orange juice, a selection of an adequate, moderate high pressure–temperature process condition, sufficient for inactivation of both factors is necessary for the process design of the pasteurization. For valencia orange juice pasteurization, inactivation of 90% of the pressure/temperature-labile PME fraction was considered as the process target. For LAB, a 7D reduction of the most resistant strain was considered.

The required processing times for pasteurization of valencia orange juice PME at different process pressures, at 25°C and 30°C are shown in Figure 12.9. Processing at 25°C requires longer times for the inactivation of LAB species up to 250 MPa, while for higher pressures the process time for inactivation of PME is longer. At 30°C, the respective pressure at which the PME process time exceeds the time required for microbial inactivation is 300 MPa. The cross of PME and LAB curves for processing at 25°C (250 MPa pressure for 34 min) and 30°C (310 MPa pressure for 4 min) could be considered the optimum process conditions for these process temperatures, that is, the minimum times achieving the targeted PME and LAB inactivation.

12.4 EFFECT OF HP ON NUTRITIONAL CHARACTERISTICS OF FOODS

L-Ascorbic acid, carotenoids, flavonoids, and other polyphenolic compounds are the most important substances of orange juice contributing to its total antioxidant activity (Miller and Rice-Evans, 1997; Rapisarda et al., 1999; Gardner et al., 2000). The effect of processing or storage on the antioxidant activity of orange juice is an important issue for investigation. A number of different phenomena (loss of naturally occurring antioxidants, improvement of their antioxidant activity, formation of various compounds having antioxidant or prooxidant properties) can occur during processing or postprocessing storage of orange juice, affecting its overall antioxidant activity (Lee and Alfred, 1992; Nicoli et al., 1997; Goyle and Ojha, 1998; Lee and Chen, 1998; Lindley, 1998; Arena et al., 2001).

Few works have been published on the effect of HP on postprocessing total antioxidant activity of orange juice or other food systems. Fernández-García et al. (2001a) found that HP treatment and storage up to 21 days at 4°C caused no significant differences in antioxidant capacity, vitamin C, or carotene content of orange juice and an orange–lemon–carrot juice product. Similar or better retention of antioxidant capacity of high-pressure treated products after 1 month storage at 4°C, compared to untreated samples, was reported in apple juice and tomato puree (Fernández-García et al., 2000, 2001b). Butz et al. (2002) found that high pressure did not induce loss of health-promoting substances (e.g., vitamins, antioxidants, antimutagens) of vegetables. De Ancos et al. (2000) reported stability or even improvement of radical scavenging activity of persimmon fruit purees after HP treatment due to the stability of carotenoids. Polydera et al. (2004a), working with reconstituted from frozen juice concentrate, reported higher antioxidant activities for high-pressurized orange juice compared to thermally pasteurized orange juice during their storage.

The comparative kinetics of the effect of HP and thermal pasteurization on postprocessing antioxidant activity of fresh navel orange juice was also studied. Ascorbic acid loss was found to follow apparent first-order kinetics during storage of both high-pressure and thermally pasteurized orange juice as depicted in Figure 12.10a and b, respectively.

Lower ascorbic acid loss rates were observed for HP orange juice compared to conventionally pasteurized juice especially at refrigeration temperatures. The retention of ascorbic acid after storage of high-pressurized orange juice for 1 month at 5°C was 84%, in contrast to thermal treatment that led to retention of ascorbic acid equal to 72%. A possible explanation for the lower ascorbic acid degradation rates during storage of high-pressure-treated orange juice could be a loss of availability of metal ions (e.g., iron, copper), catalyzing the ascorbic acid degradation, due to their hydration or the formation of complexes with chelating agents, such as citric acid, reported to be favored by high pressure (Cheah and Ledward, 1995, 1997). The destruction of peroxides by HP application (Cheah and Ledward, 1995) may also be a possible reason for the retardation of ascorbic acid degradation after HP treatment of orange juice.

The effect of storage temperature on ascorbic acid degradation rate was described by Arrhenius kinetics (Equation 12.3), as illustrated in Figure 12.11 for both high-pressure and thermally treated orange juice.

The activation energy was determined to be 68.5 kJ/mol ($R^2 = 0.97$) and 53.1 kJ/mol ($R^2 = 0.97$) for high-pressurized and thermally pasteurized orange juice, respectively, suggesting greater temperature dependence of the ascorbic acid degradation rate for HP orange juice. The same trend was observed for valencia orange juice reconstituted from frozen concentrate (Polydera et al., 2003).

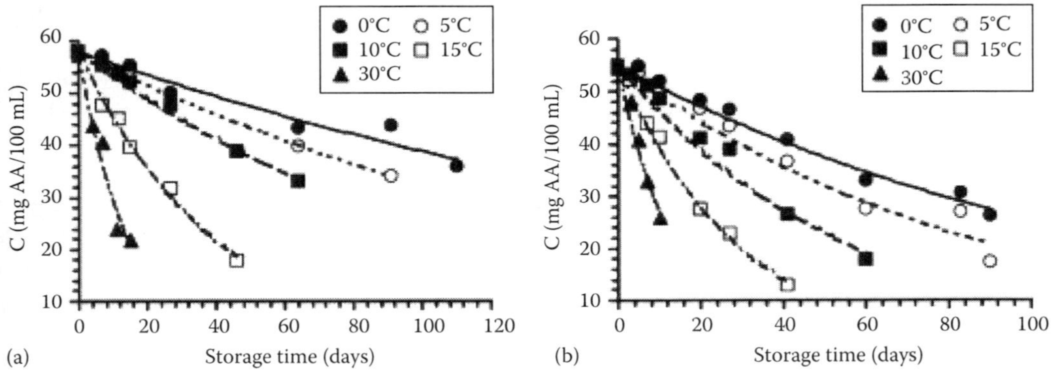

FIGURE 12.10 Ascorbic acid loss during storage of (a) high-pressurized and (b) thermally pasteurized orange juice at 0°C–30°C. (From Polydera, A.C. et al., *J. Food Eng.*, 62, 291, 2004, doi: 10.1016/S0260-8774(03)00242-5.)

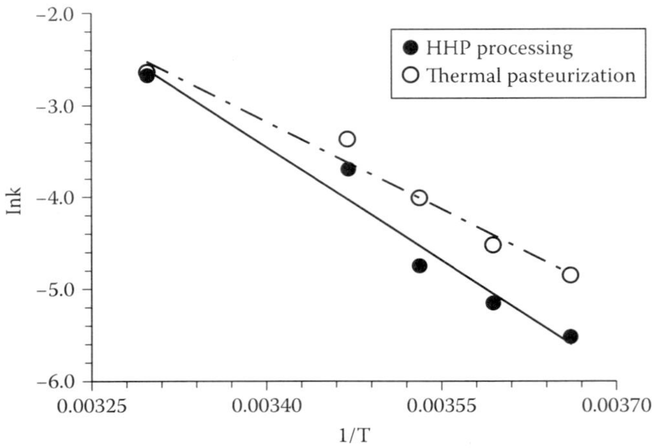

FIGURE 12.11 Effect of storage temperature (0°C–30°C) on ascorbic acid loss rate of HP and thermally pasteurized orange juice. (From Polydera, A.C. et al., *J. Food Eng.*, 62, 291, 2004, doi: 10.1016/S0260-8774(03)00242-5.)

It is worth noting that the respective rates of ascorbic acid degradation in the reconstituted juice at chill chain temperatures (0°C–10°C) were significantly higher than the ones in the fresh juice for both high pressure and thermal treatments.

12.5 EFFECT OF HP ON THE SHELF LIFE OF FOOD PRODUCTS

12.5.1 Shelf-Life Extension and Quality Improvement of Processed Meats

In-pack HP pasteurization of meat products can lead to a significant shelf life extension and organoleptic quality improvement. Prepacked bratwurst sausages have a limited durability of few days in chilled storage. Nitrates are used as preservatives, although not needed for color fixation. Bratwurst sausages can significantly benefit from cold-pasteurization in-pack by HP. HP processing can be used effectively for the production of extended shelf life, preservative-free, packed bratwurst sausages. Bratwurst sausages vacuum packed in laminate polymer film (EVOH, LDPE) were HP processed (600 MPa, 25°C, 5 min) and were stored at different temperatures in the range of 0°C–15°C. Change of selected quality indices was kinetically studied. Color, texture, sensory, and microbiological measurements were conducted for preservative-free HP-treated samples and for nontreated samples with and without nitrates. The HP process did not alter the color and the texture of the treated samples when compared to untreated ones without nitrates. The use of nitrates leads to uncharacteristic-for-this-product pink coloration. LAB growth, the main spoilage mechanism, was correlated to sensory hedonic scores that exhibited an apparent zero-order behavior. The rates of microbiological and organoleptic deterioration were estimated, and their temperature dependence was modeled by Arrhenius. Activation energies, E_a, were calculated at 90 kJ/mol and 102 kJ/mol for the untreated samples without and with nitrates, respectively, and 101 kJ/mol for the HP-treated samples. For the HP cold pasteurized samples, shelf life was practically tripled compared to untreated preservative-free ones and doubled compared to ones with nitrate days at 5°C, respectively (Table 12.7 (unpublished data)). The results confirm the potential for significant shelf-life extension and quality improvement of HP-processed bratwurst sausages.

TABLE 12.7
Shelf Life (Days) of Preservative-Free HP-Treated Sausages and Nontreated Sausages with and without Nitrite at 0°C–15°C

Storage Temperature (°C)	Control Samples	Nitrite-Made Samples	HP-Treated Samples
0	27	42	120*
5	20	32	70*
10	14	16	26
15	7	9	15

TABLE 12.8
Shelf Life (Days) of HP-(500 and 600 MPa) Treated and Nontreated Packed Sliced Ham at 0°C–15°C

Storage Temperature (°C)	Nontreated Packed Sliced Ham	HP-Treated Packed Sliced Ham at 500 MPa	HP-Treated Packed Sliced Ham at 600 MPa
0	27	82	224
5	15	75	164
10	12	42	63
15	7	22	50

Similarly, sliced, cooked ham was treated at 500 and 600 MPa (25°C, 5 min) and its shelf life was estimated and compared to conventional samples for storage temperatures 0°C–15°C. The growth of LAB was determined and modeled.

The shelf life of samples processed at 500 and 600 MPa for 5 min and the corresponding conventional ones was determined and presented in Table 12.8, where it is apparent that the treated ones' shelf life was extended by about three to eight times when compared to untreated ones.

12.5.2 Production of a High-Quality NFC Orange Juice with a Long Shelf Life at Chilled Conditions

Thermal pasteurization conditions (80°C, 60 s) were chosen among different conditions used in industrial practice also taking into consideration PME thermal inactivation kinetics (Polydera et al., 2004). About 95% inactivation of initial PME activity of untreated juice was achieved. The remaining 5% of the enzyme activity corresponds to the more heat-resistant isoenzyme, which can cause cloud loss after long periods of storage (Cameron et al., 1997; Versteeg et al., 1980). More intense thermal treatment in order to inactivate this portion would greatly affect sensory characteristics of orange juice without contributing substantially to further PME inactivation. Selection of HP processing conditions was mainly based on PME inactivation kinetics (Polydera et al., 2004). From the conditions providing adequate PME inactivation, the ones resulting in optimum sensory quality were selected. These conditions also exceeded process requirements for microbial stability of orange juice (Polydera et al., 2003). A treatment of fresh Greek navel orange juice at 600 MPa and 40°C for 4 min can cause inactivation of the sensitive isoenzyme, leading to a remaining PME activity equal to approximately 7% of the initial activity of untreated juice (similar to that after thermal pasteurization). Application of more intense HP conditions (pressure, temperature, or time) resulted in faster or further inactivation of PME, affecting, however, negatively the sensory quality of orange juice. The effect on the organoleptic characteristics was comparatively evaluated through sensory testing of orange juice after treatment at various combinations of pressure (400–900 MPa), temperature (30°C–60°C), and time (2–10 min), leading to inactivation of the labile portion of the enzyme. The selected processing conditions of 600 MPa and 40°C for 4 min also did not affect nutritional parameters of orange juice like ascorbic acid concentration or total antioxidant activity.

The decrease of ascorbic acid concentration to levels unacceptable by legislation or industrial practice often defines orange juice shelf life. Shelf life of orange juice was estimated as the time period in which there is a 50% ascorbic acid loss. As discussed previously, the slower ascorbic acid loss rates during storage of high-pressurized orange juice led to a significant extension of its shelf life compared to that of the conventionally pasteurized juice (Table 12.9). The shelf life increase of high-pressure-processed juice compared to pasteurized juice ranged from 13 days (49% increases) for storage at 15°C to 99 days (112% increase) for storage at 0°C. When stored at 30°C, similar shelf-life values were found for juices of both treatments. The above determined shelf-life values for fresh orange juice (Table 12.9) were up to double the respective values reported by Polydera et al. (2003) for reconstituted valencia orange juice.

High-pressure treatment of 600 MPa at 408°C for 4 min led to a better retention of ascorbic acid during postprocessing storage of fresh orange juice at 0°C–30°C compared to conventional thermal pasteurization (80°C, 60 s). An extension of shelf life was therefore achieved for HP-treated orange juice. Immediately after processing, HP orange juice retained the flavor of untreated fresh juice better, while its sensory characteristics were also judged superior during storage compared to thermally pasteurized juice. Due to the above-described benefits of extension of shelf life, superior organoleptic quality, and better nutrient retention, HP technology emerges as an advantageous alternative process for high-valued products like orange juice.

Apart from HP application as a nonthermal pasteurization technique, there are other potential applications; HP processing ranging from 200 to 350 MPa may denature proteins from the adduct or muscle of mollusks such as oysters and clams. The treated muscle that is responsible for closing the shell will not be able to contract, and the oyster will open. This exposes the meat for easy extraction, resulting in a significant yield increase (He et al., 2002). Another potential application is cheese maturation enhancement (Malone et al., 2003) by increasing the aminopeptidases activity responsible for the maturation process (Katsaros et al., 2009c). The gelatinization of starch under pressure is significantly different from that induced by heat, and hence they offer unique functional properties, like for example, a formation of weak gels, which could be used as fat replacer in dietary foods (Zhang et al., 2008). Pressure-induced protein gels open up to possible generation of new textures as they additionally retain their original flavor and color accompanied by a glossy appearance. Such gels can be applied for the manufacturing of milk products, for example, to improve yoghurt texture (Johnston et al., 1993) or increase cheese yield (López-Fandino et al., 1996).

TABLE 12.9
Shelf Life (Days) of High-Pressure or Thermally Pasteurized Orange Juice Stored at 0°C, 5°C, 10°C, 15°C, and 30°C

	High-Pressurized Orange Juice		Thermally Pasteurized Orange Juice	
Storage Temperature (°C)	Based on 50% Ascorbic Acid Loss	Based on Sensory Evaluation	Based on 50% Ascorbic Acid Loss	Based on Sensory Evaluation
0	187	147	88	111
5	109	89	58	66
10	64	55	39	40
15	39	35	26	24
30	9	10	9	6

12.6 ENVIRONMENTAL AND ECONOMIC ASPECTS OF THE USE OF HP ON THE FOOD INDUSTRY

HP pasteurization, apart from retaining the fresh-like characteristics of foods better than commonly used thermal treatments, is environmental friendly. A high-pressure process requires power to increase pressure, and part of the power consumed is converted to heat for the temperature increase due to pressurization. Theoretically, the compression work and energy required for temperature increase due to pressurization are about 52 kJ/kg and 70 kJ/kg upon compression of pure water up to 600 MPa, respectively. In addition, it should be noted that high-pressure sterilization/pasteurization does not require a cooling process, and decompression will decrease the temperature of the product. The theoretical total energy input into a high-pressure process at 600 MPa for processing pure water is thus about 122 kJ/kg. During HP processing, water is used as a medium for pressure buildup. This water is normally recycled since it is not in contact with the food (HP processing of packaged products), allowing for minimum water consumption. Although high-pressure equipment is considered to be generally more expensive than conventional processing/packaging systems, significant energy cost savings may be accumulated over time through the use of high pressure rather than high temperature.

12.7 CONCLUSIONS—FUTURE OUTLOOK

High-pressure technology is, among the nonthermal technologies, the one with the higher potential for industrial applications. Several food products nowadays are produced using this technology for shelf-life extension. It is of high importance for the process that it is an in-pack pasteurization technique not allowing for cross-contamination. It is independent of the shape and size of treated products since pressure is considered to be applied uniformly in the whole food mass. Apart from pasteurization purposes, HP could be used for opening oysters resulting in yield increase of meat extraction, reduction of cheese maturation time, as well as help in the formation of weak gels to be used as fat replacer in dietary foods, etc.

In Table 12.10, the advantages, limitations (disadvantages), potential products to be treated, and products already available in the worldwide market are presented for this technology.

Strengths, weaknesses, opportunities and threats (SWOT) analysis could be used to evaluate the strengths, weaknesses, opportunities, and threats involved in HP technology (Katsaros and Taoukis, 2015). SWOT analysis involves specifying the objective of the new technology implications and identifying the internal and external factors that are favorable and unfavorable to achieving that objective. Identification of SWOTs is important because they can inform later steps in planning to achieve the objective. This would allow for informing potential technology users of the importance of the technology application.

- *Strengths*: Characteristics of the business or project that give it an advantage over others
- *Weaknesses*: Characteristics that place the team at a disadvantage relative to others
- *Opportunities*: Elements that the project could exploit to its advantage
- *Threats*: Elements in the environment that could cause trouble for the business or project

In all new product development in minimal processing, it is not only the improved eating quality of products that is important. The process must be capable of operating in a factory environment and not just in a laboratory with highly qualified staff, it should ensure a financial benefit to the manufacturer, and it should be sufficiently flexible to accommodate a wide range of products, often having short production runs and brief product life cycles (Manvell, 1996). In Table 12.11, the SWOT analysis for HP technology compared to conventional thermal treatment is presented (Katsaros and Taoukis, 2015).

Food industries' interest is growing over the last years for the application of HP for the production of novel products or superior products compared to conventional ones (longer shelf life and

TABLE 12.10
Advantages, Limitations (Disadvantages), Potential Products to Be Treated, and Products Already Available in the Worldwide Market for HP Technology

Technology	Advantages of Technology Application	Disadvantages of Technology Application	Potential Products for Processing Using Specific Technology	Products Already Available in the World Market Processed Using This Specific Technology
High hydrostatic pressure	Inactivation of microorganisms and enzymes, while simultaneously retaining organoleptic, nutritional, and textural characteristics. Instant and uniform pressure transfer to the food. The pressure transfer is independent of the size and shape of the processed product. Food products high in water concentration do not deform due to pressure applied. In-pack cold pasteurization of food products increasing significantly the shelf life of foods. May be applied to liquid, semiliquid, and not liquid foods. This technology may be applied to various food industries since significant number of studies have been conducted and cited in the literature demonstrating the potentials for its application. Brined cheeses ripening time reduction. Selective increase of specific enzymes activity. Production of food products with novel rheological characteristics. Energy efficient and environmental friendly technology.	Different optimal process conditions for different food products. High capital cost. Batch-type processing (semicontinuous processing could be achieved if more than one HP unit is used in parallel).	Pasteurization of fruits, vegetables, and their juices. Pasteurization of milk and milk products. Pasteurization of meat and meat products (the most promising application of this technology). Ripening time reduction and pasteurization of cheeses. Pasteurization of fish and fish products. Crustacean meat extraction yield increase. Pasteurization of fruit preparations and jams, retention of fresh-like characteristics. Wines and beers clarification and pasteurization. Pasteurization of deli salads, dips, and ready meals.	In the worldwide market, lots of HP-treated products may be found such as fruit juices, dairy and meat products, fruit preparations, wet salads, crustaceans, avocado-based products, etc.

better quality and organoleptic characteristics). The HP pasteurization technology has a high technology readiness level (TRL). The TRL is dependent on the research published in the cited literature of the maturation level of the equipment producing industries for industrial scale equipments and the advantages of the technology application when compared to conventional technologies. The primary purpose of using TRL is to help management in making decisions concerning the development and transitioning of technology. The TRL scoring for HP technology is depicted in Table 12.12.

TABLE 12.11
SWOT Analysis for HP Technology Compared to Conventional Thermal Treatment

Strengths	Weaknesses
Significant increase of the shelf life of nonthermally pasteurized food products. Nutritional and organoleptic characteristics similar to untreated products. Potential of in-pack cold pasteurization avoiding cross-contaminations. Availability of food products at longer distance markets. Production of safe products minimizing the risk for foodborne illnesses. Energy-efficient and environmental-friendly technologies.	Different optimal process conditions for treated products. Significantly high capital cost. Training of personnel to handle the equipment of the novel technology.
Opportunities	**Threats**
The increased shelf life of the HP-treated products resulting in the reduction of food waste can counterbalance the increased capital cost. Opportunities for spin-off industries development as well as for novel technologies equipment production industries.	The effectiveness of the application and adaptation of the novel technologies is a function of: The investment cost for a food industry. The added cost for processing compared to a conventional product. The consumer acceptance for the new products.

TABLE 12.12
TRL Level for HP Technology

Technology Readiness Level (TRL)	Score	Technology
TRL 1—basic principles observed TRL 2—technology concept formulated TRL 3—experimental proof of concept TRL 4—technology validated in LAB TRL 5—technology validated in relevant environment (industrially relevant environment in the case of key enabling technologies) TRL 6—technology demonstrated in relevant environment (industrially relevant environment in the case of key enabling technologies) TRL 7—system prototype demonstration in operational environment TRL 8—system complete and qualified TRL 9—actual system proven in operational environment (competitive manufacturing in the case of key enabling technologies; or in space)	9	*High hydrostatic pressure technology* has already been applied industrially for the production of various food products with prolonged shelf life worldwide. Many research teams study the potentials of its application and present their works by a significant number of published papers cited in the literature. Some of the results may be directly applied in the food industry. Combining the above with the high maturation level of the appropriate HP equipment production industries, the TRL may be scored as *TRL* = 9.

Source: http://ec.europa.eu/research

REFERENCES

Ahn J, Balasubramaniam VM and Yousef AE (2007). Inactivation kinetics of selected aerobic and anaerobic bacterial spores by pressure-assisted thermal processing. *International Journal of Food Microbiology* 113:321–329.

Alexandrakis Z, Katsaros G, Stavros P, Katapodis P, Nounesis G, Taoukis P (2014a). Comparative structural changes and inactivation kinetics of pectin methylesterases from different orange cultivars processed by high pressure. *Food Bioprocess Technology* 7:853–867. doi: 10.1007/s11947-013-1087-7

Alexandrakis Z, Kyriakopoulou K, Katsaros G, Krokida M, Taoukis P (2014b) Process condition optimization of high pressure pasteurized sea buckthorn juice with long shelf-life and antioxidant activity. *Food Bioprocess Technology* 7:3226–3234.

Alpas H, Bozoglu F (2003) Efficiency of high pressure treatment for destruction of *Listeria monocytogenes* in fruit juices. *FEMS Immunology and Medical Microbiology* 35:269–273.

Arena E, Fallico B, Maccarone E, Sofia VS (2001) Thermal damage in blood orange juice: Kinetics of 5-hydroxymethyl-2-furancarboxaldehyde formation. *International Journal of Food Science and Technology* 36:145–151.

Balogh T, Smout C, Ly-Nguyen B, Van Loey A, Hendrickx M (2004) Thermal and high pressure inactivation kinetics of carrot pectinmethylesterase (PME): From model systems to real foods. *Innovative Food Science and Emerging Technologies* 5:429–436.

Basak S, Ramaswamy HS (1996) Ultra high pressure treatment of orange juice: A kinetic study on inactivation of pectin methyl esterase. *Food Research International* 29(7):601–607.

Basak S, Ramaswamy HS, Piette JPG (2002) High pressure destruction kinetics of *Leuconostoc mesenteroides* and *Saccharomyces cerevisiae* in single strength and concentrated orange juice. *Innovative Food Science and Emerging Technologies* 3:223–231.

Bayindirli A, Alpas H, Bozoglu F, Hizal M (2006) Efficiency of high pressure treatment on inactivation of pathogenic microorganisms and enzymes in apple, orange, apricot and sour cherry juices. *Food Control* 17:52–58.

Boulekou SS, Katsaros GJ, Taoukis PS (2009) Inactivation kinetics of peach pulp pectin methylesterase as a function of high hydrostatic pressure and temperature process conditions. *Food Bioprocess Technology* 3:699–706. doi: 10.1007/s11947-008-0132-4

Buckow R, Weiss U, Knorr D (2009) Inactivation kinetics of apple polyphenol oxidase in different pressure-temperature domains. *Innovative Food Science and Emerging Technologies* 10:441–448.

Butz P, Tauscher B (2002) Emerging technologies: Chemical aspects. *Food Research International* 35:279–284.

Cameron RG, Baker RA, Grohmann K. (1997) Citrus tissue extracts affect juice cloud stability. *Journal of Food Science* 62(2):242–245.

Cano MP, Hernandez A, De Ancos B (1997) High pressure and temperature effects on enzyme inactivation in strawberry and orange products. *Journal of Food Science* 62(1):85–88.

Castro SM, Saraiva JA, Lopes-da-Silva JA et al. (2008) Effect of thermal blanching and of high pressure treatments on sweet green and red bell pepper fruits (*Capsicum annuum* L.). *Food Chemistry* 107:1436–1449. doi: 10.1016/j.foodchem.2007.09.074

Castro SM, Van Loey A, Saraiva JA, Smout C, Hendrickx M (2006) Inactivation of pepper (*Capsicum annuum*) pectin methylesterase by combined high-pressure and temperature treatments. *Journal of Food Engineering* 75:50–58.

Cheah PB, Ledward DA (1995) High-pressure effects on lipid oxidation. *Journal of the American Oil Chemists' Society* 72(9):1059–1063.

Cheah PB, Ledward DA (1997) Catalytic mechanism of lipid oxidation following high pressure treatment in pork fat and meat. *Journal of Food Science* 62:1135.

Cheftel J-C (1991) Applications des hautes pressions en technologie alimentaire. *Industrial Alimentaire Agriculture* 108(3):141–153.

Cheftel J-C (1995) High pressure, microbial inactivation and food preservation. *Food Science Technology* 1:75–90.

De Ancos B, Gonzalez E, Cano MP (2000) Effect of high-pressure treatment on the carotenoid composition and the radical scavenging activity of persimmon fruit purees. *Journal of Agricultural and Food Chemistry* 48(8):3542–3548.

Eagerman BA, Rouse AH (1976) Heat inactivation temperature-time relationships for pectinesterase inactivation in citrus juices. *Journal of Food Science* 41(6):1396–1397.

Erkmen O, Dogan C (2004) Kinetic analysis of *E. coli* inactivation by high hydrostatic pressure in broth and food. *Food Microbiology* 21:181–185.

Fernández-García A, Butz P, Bognar A, Tauscher B (2001a) Antioxidative capacity, nutrient content and sensory quality of orange juice and an orange-lemon-carrot juice product after high pressure treatment and storage in different packaging. *European Food Research and Technology* 213(4–5):290–296.

Fernández-García A, Butz P, Tauscher B (2001b) Effects of high pressure processing on carotenoid extractability, antioxidant activity, glucose diffusion, and water binding of tomato puree (*Lycopersicon esculentum* Mill.). *Journal of Food Science* 66(7):1033–1038.

Fernández García A, Butz P, Tauscher B (2000) Does the antioxidant capacity of high pressure treated apple juice change during storage? *High Pressure Research* 19:153–160.

Fonberg-Broczek M, Arabas J, Kostrzewa E, Reps A, Szczawiński J, Szczepek J, Windyga B, Porowski S (1999) High pressure treatment of fruit, meat, and cheese products: Equipment, methods and results. In *Processing Foods. Quality Optimisation and Process Assessment* (pp. 281–300), Oliveira FAR, Oliveira JC (Eds.). CRC Press LLC, Boca Raton, FL.

Garcia-Palazon A, Suthanthangjai W, Kajda P, Zabetakis I (2004) The effects of high hydrostatic pressure on beta-glucosidase, peroxidase and polyphenol oxidase in red raspberry (*Rubus idaeus*) and strawberry (Fragariaxananassa). *Food Chemistry* 88:7–10.

Gardner PT, White TAC, Mcphail DB, Duthie GG (2000) The relative contributions of vitamin C, carotenoids and phenolics to the antioxidant potential of fruit juices. *Food Chemistry* 68:471–474.

Goodner JK, Braddock RJ, Parish ME (1998) Inactivation of pectinesterase in orange and grapefruit juices by high pressure. *Journal of Agricultural and Food Chemistry* 46(5):1997–2000.

Goodner JK, Braddock RJ, Parish ME, Sims CA (1999) Cloud stabilization of orange juice by high pressure processing. *Journal of Food Science* 64(4):699–700.

Goyle A, Ojha P (1998) Effect of storage on vitamin C, microbial load and sensory attributes of orange juice. *Journal of Food Science and Technology* 35(4):346–348.

Grant S, Patterson M, Ledward D (2000) Food processing gets freshly squeezed. *Chemistry and Industry* (24 January 2000):55–58.

Guiavarch Y, Segovia O, Hendrickx M, Van Loey A (2005) Purification, characterization, thermal and high-pressure inactivation of a pectin methylesterase from white grapefruit (*Citrus paradisi*). *Innovative Food Science of Emerging Technologies* 6:363–371. doi: 10.1016/j.ifset.2005.06.003

Hauben KJA, Wuytac EY, Soontjens CF, Michiels CW (1996) High pressure transient sensitization of *Escherichia coli* to lysozyme and nisin by disruption of outer membrane permeability. *Journal of Food Protection* 59:350–359.

Hayakawa I, Kanno T, Tomita M, Fujio Y (1994) Application of high pressure for spore inactivation and protein denaturation. *Journal of Food Science* 59:159–163.

He H, Adams RM, Farkas DF, Morrissey MT (2002) Use of high-pressure processing for oyster shucking and shelf-life extension. *Journal of Food Science* 67(2):640–645.

Hendrickx MEG, Knorr D (2002) *Ultra High Pressure Treatment of Foods*. New York: Kluwer Academic. Aspen food engineering series.

Hite B (1899) The effect of pressure in the preservation of milk. Bulletin, West Virginia University Agricultural Experiment Station Vol. 58, pp. 15–35.

Hoover DG, Metrick C, Papineau AM, Farkas DF, Knorr D (1989) Biological effects of high hydrostatic pressure on food microorganisms. *Food Technology* 43:99–107.

Irwe S, Olsson I (1994) Reduction of pectinesterase activity in orange juice by high pressure treatment. In RP Singh, FAR Oliveira (Eds.) *Minimal Processing of Foods and Process Optimisation—An Interface* (pp. 35–42). Boca Raton, FL: CRC Press.

Johnston DE, Austin BA, Murphy RJ (1993) Properties of acid-set gels prepared from high-pressure treated skim milk. *Milchwissenschaft* 48:206–209.

Kalchayanand N, Sikes T, Dunne CP, Ray B (1995) Bacteriocin based biopreservatives add an extra dimension in food preservation by hydrostatic pressure. In: *Activities Report of Rand D Associates* (Vol. 48(1), pp. 280–286). San Antonio, TX: Research and Development Associates for Military Food and Packaging Systems, Inc.

Kalchayanand N, Sikes T, Dunne CP, Ray B (1998) Factors influencing death and injury of foodborne pathogens by hydrostatic pressure-pasteurization. *Food Microbiology* 15:207–214.

Katsaros GI, Apseridis I, Taoukis PS (2006) Modelling of high hydrostatic pressure inactivation of pectinmethylesterase from persimmon (*Diospyros virginiana*). IUFoST Edpsciences, doi: 10.1051/IUFoST:20060753

Katsaros GI, Giannoglou MN, Taoukis PS (2009a) Kinetic study of the combined effect of high hydrostatic pressure and temperature on the activity of *Lactobacillus delbrueckii* ssp. bulgaricus aminopeptidases. *Journal of Food Science* 74(5):E219–E225. doi: 10.1111/j.1750-3841.2009.01148.x

Katsaros GI, Katapodis P, Taoukis PS (2009b) High hydrostatic pressure inactivation kinetics of the plant proteases ficin and papain. *Journal of Food Engineering* 91:42–48.

Katsaros GI, Katapodis P, Taoukis PS (2009c) Modeling the effect of temperature and high hydrostatic pressure on the proteolytic activity of kiwi fruit juice. *Journal of Food Engineering* 94:40–45. doi: 10.1016/j.jfoodeng.2009.02.026

Katsaros GI, Tsevdou M, Panagiotou T, Taoukis PS (2010) Kinetic study of high pressure microbial and enzyme inactivation and selection of pasteurisation conditions for Valencia Orange Juice. *International Journal of Food Science and Technology* 45:1119–1129. doi: 10.1111/j.1365-2621.2010.02238.x

Katsaros G, Taoukis P (2015) New innovative technologies. In *Food Engineering Handbook. Food Process Engineering* (pp. 595–628), T Varzakas, C Tzia, (Eds.). CRC Press, Boca Raton, FL.

Knorr D (1995) High-pressure effects on plant derived foods. In *High-Pressure Processing of Foods*, (pp. 123–125) DA Ledward, DE Johnston, RG Earnshaw, APM Hasting, (Eds.). Nottingham, U.K.: Nottingham University Press.

Knorr E Hendrickx M (2003) *High Pressure Treatment of Foods*, Kluwer Academic/Plenum Publishers, New York.

Koseki S, Mizuno Y, Yamamoto K (2008) Use of mild-heat treatment following high-pressure processing to prevent recovery of pressure-injured *Listeria monocytogenes* in milk. *Food Microbiology* 25:288–293.

Krebbers, B, Matser AM, Hoogerwerf SW, Moezelaar R, Tomassen MMM, van den Berg RW (2003) Combined high-pressure and thermal treatments for processing of tomato puree: Evaluation of microbial inactivation and quality parameters. *Innovative Food Science & Emerging Technology* 4(4):377–385.

Krebbers B, Matser AM, Koets M, van den Berg RW (2002). Quality and storage-stability of high-pressure preserved green beans. *Journal of Food Engineering* 54(1):27–33.

Lado BH, Yousef AE (2002) Alternative food-preservation technologies: Efficacy and mechanisms. *Microbes and Infection* 4:433–440.

Leadley C (2005) High pressure sterilisation: A review. *Campden & Chorleywood Food Research Association* 47:1–42.

Lee HS, Alfred L (1992) Antioxidative activity of browning reaction products isolated from storage-aged orange juice. *Journal of Agricultural and Food Chemistry* 40:550–552.

Lee HS, Chen CS (1998) Rates of vitamin C loss and discoloration in clear orange juice concentrate during storage at temperatures of 4–24°C. *Journal of Agricultural and Food Chemistry* 46:4723–4727.

Lindley MG (1998) The impact of food processing on antioxidants in vegetable oils, fruits and vegetables. *Trends in Food Science and Technology* 9:336–340.

Lopez Fandino R, Carrascossa AV, Olano A (1996) The effects of high pressure on whey protein denaturation and cheese making properties of raw milk. *Journal of Dairy Science* 79:929–936.

Ly-Nguyen B, Van Loey A, Fachin D, Verlent I, Duvetter T, Vu ST, Smout C, Hendrickx ME (2002a) Strawberry pectin methylesterase (PME): Purification, characterization, thermal and high-pressure inactivation. *Biotechnology Progress* 18:1447–1450.

Ly-Nguyen B, Van Loey A, Fachin D, Verlent I, Indrawati, Hendrickx M (2002b) Purification, characterization, thermal and high-pressure inactivation of pectin methylesterase from bananas (cv. Cavendish). *Biotechnology and Bioengineering* 78:683–691.

Ly-Nguyen B, Van Loey A, Fachin D, Verlent I, Indrawati, Hendrickx M (2002c) Partial purification, characterization and thermal and high-pressure inactivation of pectin methylesterase from carrots (*Daucus carrota* L.). *Journal of Agricultural and Food Chemistry* 50:5437–5444.

Ly-Nguyen B, Van Loey AM, Smout C, Ozcan SE, Fachin D, Verlent I, Truong SV, Duvetter T, Hendrickx ME (2003a) Mild-heat and high-pressure inactivation of carrot pectin methylesterase: A kinetic study. *Journal of Food Science* 68:1377–1383.

Ly-Nguyen B, Van Loey AM, Smout C, Verlent I, Duvetter T, Hendrickx M (2003b) Effects of mild heat and high pressure processing on banana pectinmethylesterase: A kinetic study. *Journal of Agricultural and Food Chemistry* 51:7974–7979.

Mackey BM, Forestiera K, Isaacs N (1995) Factors affecting the resistance of Listeria monocytogenes to high hydrostatic products. *Food Biotechnology* 9(1/2):1–11.

Mallidis C, Galiatsatou P, Taoukis PS, Tassou C (2003) The kinetic evaluation of the use of high hydrostatic pressure to destroy *Lactobacillus plantarum* and *Lactobacillus brevis*. *International Journal of Food Science and Technology* 38:579–585.

Malone AS, Wick C, Shellhammer TH, Courtney PD (2003) High pressure effects on proteolytic and glycolytic enzymes involved in cheese manufacturing. *Journal of Dairy Science* 86:1139–1146.

Manvell C (1996) Opportunities and problems of minimal processing and minimally processed foods. EFFOST/GDI. Congress on Minimal Processing of Foods. November 6–8, 1996.

MarketsandMarkets report (2013) HPP (high pressure processing) market by equipment type (orientation, vessel size), application (meat, seafood, beverage, fruit & vegetable), product type (meat & poultry, seafood, juice, ready meal, fruit & vegetable) & geography—Forecast to 2018, Report code: FB 2151, November 2013.

Matser AM, Krebbers B, van den Berg RW, Bartels PV (2004) Advantages of high pressure sterilization on quality of food products. *Trends in Food Science and Technology* 15(2):79–85.

Meyer P (2000) Ultra high pressure, high temperature food preservation process. US Patent 6,017,572.

Miller NJ, Rice-Evans CA (1997) The relative contributions of ascorbic acid and phenolic antioxidants to the total antioxidant activity of orange and apple fruit juices and blackcurrant drink. *Food Chemistry* 60:331–337.

Mussa DM, Ramaswamy HS, Smith JP (1999) High pressure (HP) destruction kinetics of *Listeria monocytogenes* Scott A in raw milk. *Food Research International* 31:343–350.

Nicoli M, Anese M, Parpinel MT, Franceschi S, Lerici CR (1997) Loss and/or formation of antioxidants during food processing and storage. *Cancer Letters* 114:71–74.

Norton T, Sun D-W (2008) Recent advances in the use of high pressure as an effective processing technique in the food industry. *Food Bioprocess Technology* 1(1):2–34.

Patterson MF, Quinn M, Simpson R, Gilmour A (1995) Sensitivity of vegetative pathogens to high hydrostatic pressure treatment in phosphate-buffered saline and foods. *Journal of Food Protection* 58:524–529.

Plaza L, Duvetter T, Monfort S, Clynen E, Schoofs L, Van Loey AM et al. (2007) Purification and thermal and high-pressure inactivation of pectinmethylesterase isoenzymes from tomatoes (*Lycopersicon esculentum*): A novel pressure labile isoenzyme. *Journal of Agricultural and Food Chemistry* 55(22):9259–9265.

Polydera AC, Galanou E, Stoforos NG, Taoukis PS (2004) Inactivation kinetics of pectin methylesterase of greek Navel orange juice as a function of high hydrostatic pressure and temperature process conditions. *Journal of Food Engineering* 62:291–298. doi: 10.1016/S0260-8774(03)00242-5

Polydera AC, Stoforos NG, Taoukis PS (2003) Comparative shelf life study and vitamin C loss kinetics in pasteurised and high pressure processed reconstituted orange juice. *Journal of Food Engineering* 60:21–29. doi: 10.1016/S0260-8774(03)00006-2

Raoult-Wack AL (1994) Recent advances in the osmotic dehydration of foods. *Trends in Food Science and Technology* 5:255–260.

Rapeanu G, Van Loey AM, Smout C, Hendrickx M (2005) Effect of pH on thermal and/or pressure inactivation of victoria grape (*Vitis vinifera* sativa) Polyphenol Oxidase: A kinetic study. *Journal of Food Science* 70(5):E301–E307.

Rapisarda P, Tomaino A, Lo Cascio R et al. (1999) Antioxidant effectiveness as influenced by phenolic content of fresh orange juices. *Journal of Agricultural and Food Chemistry* 47(11): 4718–4723.

Riahi E, Ramaswamy HS (2003) High-pressure processing of apple juice: Kinetics of pectin methyl esterase inactivation. *Biotechnology Progress* 19:908–914. doi: 10.1021/bp025667z

Rodrigo D, Cortes C, Clynen E, Schoofs L, Van Loey A, Hendrickx M (2006) Thermal and high-pressure stability of purified polygalacturonase and pectinmethylesterase from four different tomato processing varieties. *Food Research International* 39:440–448.

Sale AJH, Gould GW, Hamilton WA (1970) Inactivation of bacterial spores by hudrostatic pressure. *Journal of General Microbiology* 60:323–334.

Sampedro F, Rodrigo D, Hendrickx M (2008) Inactivation kinetics of pectin methyl esterase under combined thermal high-pressure treatment in an orange juice–milk beverage. *Journal of Food Engineering* 86(1):133–139.

Sila DN, Smout C, Satara Y, Truong V, Van Loey A, Hendrickx M (2007) Combined thermal and high pressure effect on carrot pectinmethylesterase stability and catalytic activity. *Journal of Food Engineering* 78:755–764.

Smelt JPPM (1998) Recent advances in the microbiology of high pressure processing. *Trends in Food Science and Technology* 9:152–158.

Stoforos NG, Crelier S, Robert MC, Taoukis PS (2002) Kinetics of tomato pectin methylesterase inactivation by temperature and high pressure. *Journal of Food Science* 67(3):1026–1031.

Stoforos NG, Taoukis PS (1998) A theoretical procedure for using multiple response time-temperature integrators for the design and evaluation of thermal processes. *Food Control* 9(5):279–287. doi.org/10.1016/S0956-7135(98)00017-6

Styles MF, Hoover DG, Farkas DF (1991) Response of *Listeria monocytogenes* and *Vibrio parahaemolyticus* to high hydrostatic pressure. *Journal of Food Science* 56(5):1404–1407.

Terefe NS, Yang YH, Knoerzer K, Buckow R, Versteeg C (2010) High pressure and thermal inactivation kinetics of polyphenol oxidase and peroxidase in strawberry puree. *Innovative Food Science and Emerging Technologies* 11:52–60.

Tewari G, Jayas DS, Holley RA (1999) High pressure processing of foods: An overview. *Science des Aliments* 19:619–661.

Van Den Broeck I, Ludikhuyze LR, Van Loey AM, Hendrickx ME (2000a) Effect of temperature and/or pressure on tomato pectinesterase activity. *Journal of Agricultural and Food Chemistry* 48(2):551–558.

Van Den Broeck I, Ludikhuyze LR, Van Loey AM, Hendrickx ME (2000b) Inactivation of orange pectinesterase by combined high-pressure and -temperature treatments: A kinetic study. *Journal of Agricultural and Food Chemistry* 48(5):1960–1970.

Verlent I, Hendrickx M, Verbeyst L, Van Loey F (2007) Effect of temperature and pressure on the combined action of purified tomato pectinmethylesterase and polygalacturonase in presence of pectin. *Enzyme and Microbial Technology* 40:1141–1146.

Versteeg C, Rombouts FM, Spaansen CH, Pilnik W (1980) Thermostability and orange juice cloud destabilizing properties of multiple pectinesterases from orange. *Journal of Food Science* 45(4):969–998.

Yuste J, Capellas M, Pla R, Fung D, Mor-mur M (2001) High pressure processing for food safety and preservation: A review. *Journal of Rapid Methods & Automation in Microbiology* 9(1):1–10.

Zhang G, Sofyan M, Hamaker BR (2008) Slowly digestible state of starch: Mechanism of slow digestion property of gelatinized maize starch. *Journal of Agricultural and Food Chemistry* 56:4695–4702.

13 Pulsed Electric Fields

Gulsun Akdemir Evrendilek and Theodoros Varzakas

CONTENTS

13.1 PRINCIPLES OF PEF TECHNOLOGY

Application of pulsed electric field (PEF) technology is based on the fact that many foods can conduct electricity when they are placed between the electrodes of an electrical circuit because of the presence of ions (Zhang et al., 1994a,c). Foods are subjected to different electrochemical reactions or changes under the applied electric current, and these reactions can cause microbial inactivation. Because of the electrical resistance of the foods, several reactions such as ohmic heating, electrolysis, cell membrane disruption, and shock waves caused by arc discharge can occur (Hulsheger and Niemann, 1980; Sitzmann, 1995; Sastry and Barach, 2001; Zuckermann et al., 2002). Ohmic heating is formed by the conversation of applied electrical energy to the heat instantly inside the food, and the amount of the heat is directly related to the current induced by the voltage gradient in the field and the electrical conductivity (Sastry and Li, 1996). It should be noted that these reactions are not independent from each other, and the application of electrical energy determines the individual effect on microorganisms. In order to minimize the undesirable effect of each reaction such as temperature increase, electrolytic oxidative effects, and disintegration of food particles, which have adverse effect on foods, duration of the high-voltage pulses were applied with relatively long intervals (Hulsheger and Niemann, 1980; Palaniappan and Sastry, 1990; Zhang et al., 1994a; Sitzmann 1995), pulses applied during process is practiced with extremely short duration (1–100 μs), and pulse intervals between discharges is adjusted from 1 ms to several seconds (Qin et al., 1995b). On the other hand, applied electric field is kept between 10 and 80 kV/cm in order to obtain maxiumum amount of microbial and enzyme inactivation (Barbosa-Canovas et al., 1999).

13.2 PEF PROCESSING SYSTEM

Basic components of the PEF processing systems include a high-voltage repetitive pulser, a treatment chamber(s), a cooling system(s), voltage- and current-measuring devices such as an oscilloscope, a control unit (trigger generator where pulse width, pulse delay time, and frequency are set up), and a data acquisition system (Figure 13.1). In order to obtain PEFs, a fast discharge of electrical energy within a short period of time is required. This is done by the pulse-forming network (PFN), an electrical circuit consisting of one or more power supplies with the ability to charge voltages

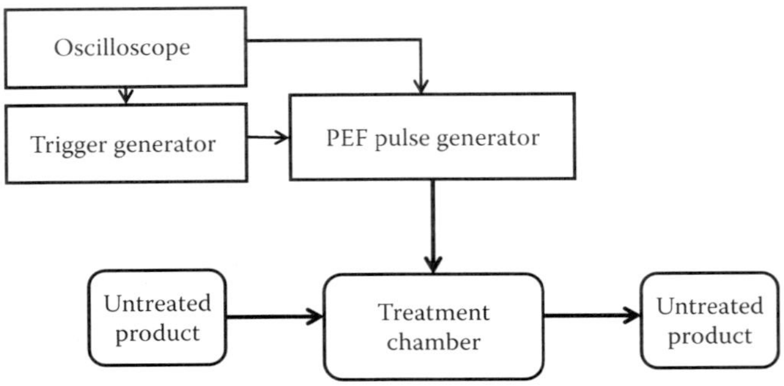

FIGURE 13.1 Basic components of PEF systems.

(up to 60 kV), switches (ignitron, thyratron, tetrode, spark gap, and semiconductors), capacitors (0.1–10 μF), resistors (2–10 MΩ), and treatment chambers (Gongora-Nieto et al., 2002; Mohamed and Eissa, 2012).

The main issue in PEF systems is to obtain high voltage from the low utility level voltage, and this is done by the pulsed power supply. Low utility level voltage is used to charge a capacitor bank, and a switch is used to discharge energy from the capacitor to the treatment chamber, and thus, across the food in. Food is placed in treatment chambers during the processing and they house the discharging electrodes. PEF-processed products need to be stored at refrigeration temperature; thus, after processed product is cooled, if necessary it is packed aseptically and then stored at refrigerated or ambient temperatures (Qin et al., 1995b; Zhang et al., 1997; Mohamed and Eissa, 2012).

Treatment chambers are evolved from static to continuous with recent developments, and researchers have designed various types of chambers with different materials (United States Patent, 2014). Although static chambers were used in early studies where mostly static PEF processing system were constructed, continuous flow treatment chambers are being used with continuous PEF processing systems that resemble pilot scale or industrial scale systems. Processed liquid food is pumped through pulsing electrodes in the same or opposite direction of electric current in continuous flow PEF treatment chambers; therefore, continuous flow PEF treatment chambers are more suitable for large-scale operations. Several types of continuous flow treatment chambers such as coaxial chamber (composed of an inner cylinder surrounded by an outer annular cylindrical electrode that allows food to flow between them), cofield flow (electric field and food flow in the same direction), and parallel-plate treatment chambers were designed and developed (Martin-Belloso et al., 1997; Yin et al., 1997; Barbosa-Canovas et al., 1998; Fiala et al., 2001; Fox et al., 2007; Shamsi and Sherkat, 2009) (Figure 13.2).

Based on the high-voltage pulse generator circuit design, electric field can be applied in different shapes such as exponential decay, square wave pulses (in bipolar or monopolar form), and instant charge reversal (Figure 13.3). Exponential decay pulse has a rapid increase to a maximum value and then decays slowly to the minimum value. Due to the slow decay, it has a long tail, which causes heat generation that does not have bactericidal effect. Square wave pulses are more energy efficient and have more lethal effects than that of exponential decay pulses. Bipolar pulses because of the reversal in the orientation or polarity of the electric field results in a corresponding change in the direction of charged molecules on microbial cell membrane to further damage. In addition, bipolar pulses have minimum energy utilization and reduced deposition of solids on the electrode surface (Qin et al., 1994; Ho et al., 1995; Barbosa-Canovas et al., 1998).

PEF treatment is mostly indicated by electric field strength (kV/cm), treatment time (from ns to μs), or energy (kJ/kg). PEF processing has lots of variables; however, the key parameters include 15–50 kV/cm PEF intensity, 1–5 μs pulse width, 200–500 Hz (pulses/s) pulse frequency, treatment time (μs), and total applied energy (W) (Evrendilek et al., 2004c; Wan et al., 2009).

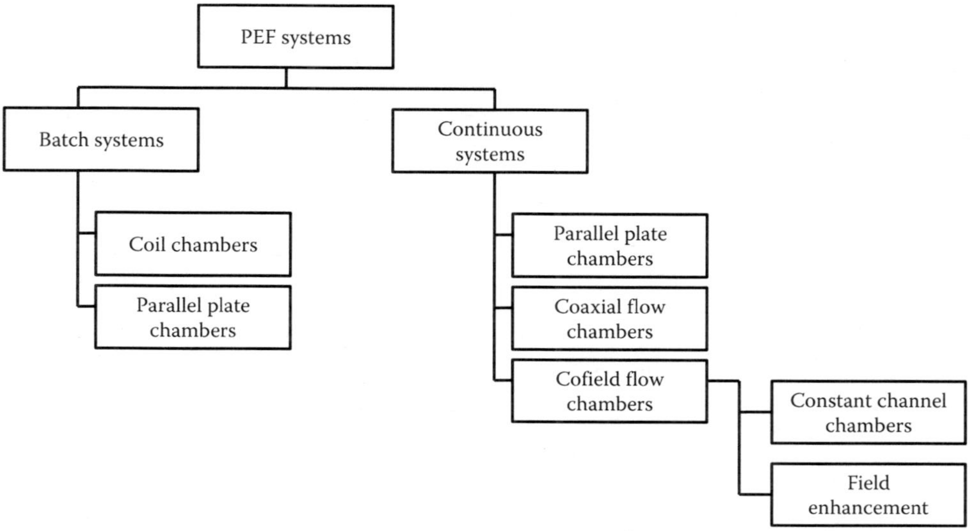

FIGURE 13.2 PEF treatment chamber classification. (Adapted from Shamsi, K. and Sherkat, F., *Asian J. Food Agro Ind.*, 2(3), 216, 2009.)

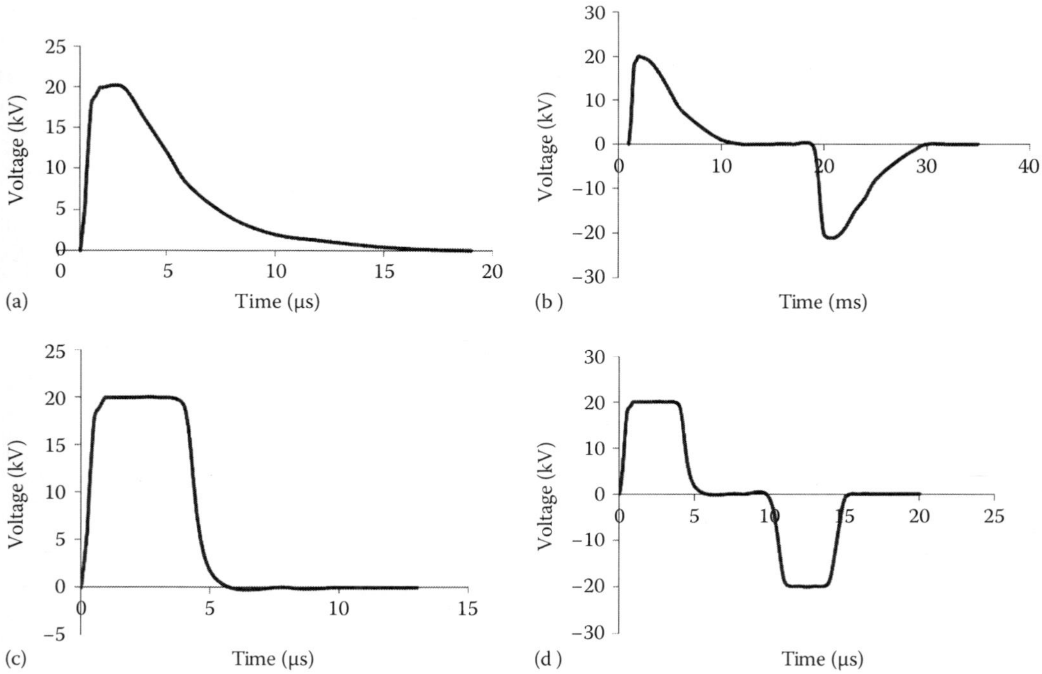

FIGURE 13.3 (a) Exponential decay, (b) instant charge reversal, (c) square wave, and (d) bipolar pulses.

Applied electric field (E_p), treatment time (T_t), and total energy input (W) are calculated by the following expressions:

$$E_p = \frac{V}{D} \tag{13.1}$$

$$T_t = n_p \, \tau \, n \, n_t \tag{13.2}$$

$$W = V \, I \, \tau \, f \, 10 \tag{13.3}$$

where
- V is the applied voltage in kV
- D is the gap distance of the treatment chamber in cm
- n_p is thenumber of pulses per chamber
- τ is the pulse width in μs
- n is the number of chambers
- n_t is the number of times that the sample has been processed
- I is the current in A
- f is the frequency in Hz (Evrendilek et al., 2004c)

Different electrical parameters that influence the quality of PEF-processed juice have been extensively studied over the past years, including pulse profile, pulse polarity, pulse duration, pulse frequency, electric field strength, etc. (Aguilar-Rosas et al., 2007; Riener et al., 2008; Aguiló-Aguayo et al., 2010; Evrendilek et al., 2000; Morales de la Peña, et al., 2010).

13.3 INACTIVATION OF MICROORGANISMS AND ENZYMES

Application of electric power for food processing was reported in the early 1900s; however, first studies regarding PEF processing of foods did not eliminate ohmic heating, and thus, inactivation energy generated by heat. Later studies have shown inactivation of microorganisms with the application of 3000–4000 V (Beattie, 1915; Beattie and Lewis, 1925; Bendicho et al., 2002a). Processing of milk by the so-called electropure process included heating of milk up to 70°C and then passing it through carbon electrode to inactivate *Mycobacterium tuberculosis* and *Escherichia coli*. Even though previously attempts have been made to pasteurize food samples by electric power or electric field (Getchell, 1935), most of the studies give credit to a study involved processing of milk by heating at 71°C for 15 s with an alternating current of 220 V (Hall and Trout, 1968; Paleniappan and Sastry, 1990; Martin et al., 1994; Bendicho et al., 2002a). Even though earlier studies regarding PEF treatment of milk samples focused on microbial inactivation, it is still the main achievement in present studies.

Although numerous studies were performed to inspect and explore the microbial inactivation by PEF, the exact mechanisms are yet to be fully understood (Pagan et al., 2005). Inactivation by PEF—as a general acceptance—includes the electroporation theory (Zimmermann, 1986; Tsong, 1991), explaining the destruction of the semipermeable barrier of the cell membrane and irreversible pore formation when PEF treatment at an electric field intensity greater than a critical threshold of transmembrane potential of 1 V across the target cells is applied. This theory has a big ground and acceptance as a general mechanism of inactivation since it is also visually proven by transmission (TEM) and scanning electron microscopy (SEM) (Harrison et al., 1997; Calderon-Miranda et al., 1999; Wan et al., 2009). Membrane damage was also shown by the increased uptake of the fluorescent dye propidium iodide (PI) and leakage of intracellular compounds of *E. coli*, *Listeria innocua*, and *Saccharomyces cerevisiae* cells (Aronsson et al., 2005; Garcia et al., 2005; Wan et al., 2009).

It was proposed that electroporation of cell membrane causes cell destruction (Pothakamury et al., 1997). Although pore formation that causes cell death is known, it is still not clear whether the pores are formed in the lipid or the protein matrices (Barbosa-Canovas et al., 1999). As it is mentioned, the magnitude of applied electric field strength is an important factor for cell damage, and damage is increased with increased field strength. Images from TEM show that *Staphylococcus aureus* cells exhibited rough surfaces when they were suspended in simulated milk ultrafiltrate (SMUF) with 64-pulse treatments of 20, 30, and 40 kV/cm. When *S. aureus* cells were exposed to more severe conditions, the treated cells showed small holes in the membrane and leakage of cellular contents. Therefore, it was concluded that the increase in microbial inactivation with the applied electric field strength is related to the increase in cell deterioration (Pothakamury et al., 1997).

On the other hand, recent studies revealed that cytomembrane and RNA were primary targets of PEF-induced damage on *S. cerevisiae*. According to proposed mechanism, cell death occurs due to the accumulation of injury. Decrease in cytomembrane fluidity (lower cytomembrane fluidity and higher cytomembrane viscosity), increase in microviscosity, change in membrane lipid composition (decrease in ratio of contents of unsaturated fatty acids versus saturated fatty acids), and disruption of RNA are associated with the sublethally injured *S. cerevisiae* in PEF treatment defined as the damage of cytomembrane. Although PEF treatment caused no significant change in DNA, disruption occurred in RNA. It is also found that the injured *S. cerevisiae* can repair the damage and survive during PEF treatment if the culture conditions are favorable to cell growth, such as in nonselective medium or when supplemented with RNA stabilizer (Zhao et al., 2014). It is also proven that pore formation occurs due to the changes in the lipid matrix of the cell membrane; PEF can cause damage in the genetic material and intracellular nucleic acids such as RNA, which is more adversly affected than DNA.

Even though studies revealed the inactivation mechanisms of vegetative cells by PEF, it is known that microbial inactivation is affected by different factors. Today, the food industry is more interested in getting a reasonable amount of inactivation on the microorganism of significance on certain foods; thus, another challenge is to obtain at least 5 log reduction without adversely affecting the physical, nutritional, and sensory properties of foods. Although pasteurization has been based on thermal processing traditionally, it has been shown that there is potential for several nonthermal technologies to obtain the same goal (Barbosa-Cánovas and Bermúdez-Aguirre, 2011; Saldana et al., 2014). As a result, redefinition of pasteurization is made by the National Advisory Committee on Microbiological Criteria for Foods as “any process, treatment, or combination thereof that is applied to food to reduce the most resistant microorganism(s) of public health significance to a level that is not likely to present a public health risk under normal conditions of distribution and storage” (NACMCF, 2006; Saldana et al., 2014). Therefore, optimization PEF processing has to be realized in order to achieve maximum inactivation. It is a known fact that the effectiveness of PEF on microbial cells changes depending on the differences in the cell wall structure and cell size, process parameters, and treatment medium. Now, it is also known that compared to gram positive cells, gram negatives are easier to inactivate; yeasts and molds due to the larger size of the cells are easier to inhibit by high-intensity electric pulses. As the stationary state cells are the most resistant to PEF, cells in the log phase are the most sensitive to PEF.

On the other hand, the reason that this technology is not successful in inactivating bacterial spores is because spore form is structured from the envelope involving the coat and the cortex, and this rigid structure prevents the permeabilization effects of PEF on the spore cytoplasmatic membrane (Pagan et al., 1998; Saldana et al., 2014). Due to the very limited amount of water, the conductivity of the spore coat is very low and, therefore, high-intensity electric pulses cannot be transmitted through the spore structure. The best way to inactivate the spore by PEF is either to break the spore coat with other agents and/or heat higher than >50°C and then apply the PEF or to allow the germination of spores and to apply PEF after the germination (Su et al., 1996; Marquez et al., 1997; Barbosa-Canovas et al., 1999; Barsotti and Cheftel, 1999).

An increase in the magnitude of PEF process parameters usually causes an increase in cell inactivation. Food composition such as protein, lipid, carbohydrate, mineral, and water content as well as pH, conductivity, and presence of inhibitors are important factors to increase or decrease the level of inactivation. In general, higher water activity, lower pH, and certain level of conductivity are desired for maximum inactivation (Su et al., 1996; Marquez et al., 1997; Barbosa-Canovas et al., 1999; Barsotti and Cheftel, 1999).

In addition to microorganisms, a significant amount of enzyme inactivation should be achieved by PEF in order to be an alternative or replacement to heat treatment. In contrast to vegetative cells, enzymes are harder to inactivate, and usually longer treatment times and higher electric field strength are required for significant amount of inactivation for the enzymes naturally found or synthesized by microorganisms. Localized joule heating of cell membrane components may cause thermal denaturation of membrane-bound enzymes present in microorganism during PEF treatment (Simpson et al., 1999). As the structure and size of the enzymes are different, effectiveness of the PEF changes among the enzymes, and magnitude of high voltage applied need to be much higher than that of the inactivation of microorganisms because some enzymes used in food processing may remain viable under PEF processing depending on processing conditions while microbial control is obtained (Ho et al., 1995; Giner et al., 2000; Deeth et al., 2007; Tewari and Juneja, 2007).

13.4 PEF PROCESSING OF FOODS

Fruit juices due to their low viscosity, high acidity, and suitable conductivity are the most appropriate food products to be processed by PEF. Therefore, most of the studies were performed with fruit juices to inactivate various microorganisms including foodborne pathogens, food spoilage bacteria, and plant pathogenic bacteria (Table 13.1); inactivation of enzymes including pectin methyl esterase (PME), lipoxygenase, polyphenol oxidase (PPO), peroxidase (PO), and poly galacturonase (PG) (Table 13.2); changes in the nutritive value such as vitamins, antioxidant properties, phenolic compounds, anthocyanin compounds, and minerals as well as sensory properties and shelf-life extension.

Even though most of the inactivation studies are summarized in Table 13.1, there are some interesting approaches on microbial inactivation by PEF that need to be mentioned. Propidium monoazide (PMA) is a DNA-intercalating dye. PMA-qPCR (quantitative PCR) has been reported as a novel method to detect alive bacteria in complex samples. Xing-long et al. (2013) employed this method to monitor the sterilization effects of UHP, ultrasound, and high PEF on *E. coli* O157:H7. The results showed that all three sterilization techniques are successful in killing viable *E. coli* O157:H7 cells under appropriate conditions. PMA-qPCR can effectively monitor the amount of DNA released from viable *E. coli* O157:H7 cells, and the results from PMA-qPCR were highly consistent with those from plate counting after treatment with UHP, ultrasound, and high PEF. The maximal ΔCt between PMA-qPCR and qPCR obtained in this study was 10.39 for UHP, 5.76 for ultrasound, and 2.30 for high PEF. The maximal sterilization rates monitored by PMA-qPCR were 99.92% for UHP, 99.99% for ultrasound, and 100% for high PEF.

PEF is also used to treat different water samples, for example, hospital wastewaters (Rieder et al., 2008; Gusbeth et al., 2009), that are usually contaminated with pathogenic bacteria (Kummerer, 2001). The main advantages of PEF technology in water treatment are less unwanted by-products compared to other inactivation techniques such as chlorination, ozonation, and UV irradiation (Rook, 1977; Paraskeva and Graham, 2002; Schwartz et al., 2003) and no developed adaptation to electric field by the descendants of treated bacteria (Gusbeth et al., 2009).

Inactivation of microorganisms with PEFs is one of the nonthermal methods most commonly used in biotechnological applications such as liquid food pasteurization and water treatment. Zgalin et al. (2012) studied the effects of microsecond and nanosecond pulses on inactivation of *E. coli* in distilled water. According to these results, when using microsecond pulses, the level of inactivation increases with the application of more intense electric field strengths and with the number of pulses delivered. Almost 2-log reductions in bacterial counts were achieved at the field strength of

TABLE 13.1
Inactivation of Microorganisms by PEF in Different Food Samples

Media	Treatment Conditions	Microorganism	Microbial Reduction (log cfu/mL)	References
Orange juice	29.5 kV/cm, 60 μs of treatment time, square wave, cofield flow tubular	Aerobic microorganisms	4.2	Qiu et al. (1998)
	30 kV/cm, 240 μs of treatment time, 2 μs of pulse width, 1000 Hz, 2 mL/s, cofield flow tubular	Aerobic microorganisms; yeasts and molds	2.5; 2.5	Jia et al. (1999)
	30 or 50 kV/cm, 100 L/h	*Leuconostoc mesenteroides*; *Escherichia coli*; *L. innocua*; *S. cerevisiae* ascospore	2–6; 5.5–6.6; 5–5.5; 2	McDonald et al. (2000)
	90 kV/cm, 55°C, 50 pulses	*S. typhimurium*	5.9	Liang et al. (2002)
	35 kV/cm, 59 μs of treatment time, 1.4 μs of pulse width, 600 pps, 98 L/h, pilot plant scale system, cofield flow tubular	Aerobic microorganisms; yeasts and molds	7; 7	Yeom et al. (2000b)
	40 kV/cm, 97 μs of treatment time, 2.6 μs of pulse width, 1000 pps, 500 L/h, commercial scale system, cofield flow tubular	Aerobic microorganisms; yeasts and molds	6; 6	Min et al. (2003a)
	P1: 120 pulses/mL, 46 kV/cm P2: 120 pulses/mL, 46 kV/cm 45°C, thyratron-based pulse power supply	Initial microorganisms	P1: 2 P2: 3	El-Hag et al. (2006)
	34.3 kV/cm, 20°C	*Zygosaccharomyces bailii* ascospores; *Z. bailii* vegetative cells	3.8; 4.7	Raso et al. (1998b)
	8–11 kV/cm, 3 pulses, 50 μF capacitance or 12.5 kV/cm, 40 pulses, 1 μF capacitance, exponential decay pulses	*S. cerevisiae*	2–5.8	Molinari et al. (2004)
	1000 μs (4 μs pulse width), 35 kV/cm, 200 Hz, bipolar mode	*S. cerevisiae*	5.1	Elez-Martinez et al. (2004)
	8 chambers, 1–10 μs pulse width, 1000 Hz, 1–12 kV electric field strength, mono-bipolar pulse	*Lactobacillus brevis*	5.8	Elez-Martinez et al. (2005)
	P1: 120 pulses/mL, 46 kV/cm P2: 120 pulses/mL, 46 kV/cm 45°C, thyratron-based pulse power supply	Initial microorganisms	P1: 2 P2: 3	El-Hag et al. (2006)

(Continued)

TABLE 13.1 (*Continued*)
Inactivation of Microorganisms by PEF in Different Food Samples

Media	Treatment Conditions	Microorganism	Microbial Reduction (log cfu/mL)	References
	30, 35, 40 kV, 19°C–34°C, 2–3 μs pulse width, pilot size, continuous coaxial treatment chamber, exponential decay pulses	*Byssochlamys fulva* conidiospores; *Neosartoria fischeri* ascospores	6; negligible inactivation	Raso et al. (1998b)
Orange–carrot juice	28, 25, 22 × 10^5 V/m, 11.5, 7 and 9 × 10^{-5} s total treatment time	*L. plantarum*; *E. coli*; *R. rubra*	1.3; 2.6; 6.5	Selma et al. (2004)
	22 kV/cm, 1000 μs	*L. plantarum*	5.8	Gomez et al. (2005)
	28 kV/cm, <35°C	*Yersinia enterocolitica*	6	
Whey protein fortified orange juice	32 kV/cm, 92 μs of treatment time, 3.3 μs of pulse width, 800 Hz, 79 L/h, cofield flow tubular	Aerobic microorganisms; yeasts and molds	0.5; 3.5	Sharma et al. (1998)
Tomato juice	40 kV/cm, 57 μs of treatment time, 2 μs of pulse width, 1000 pps, 500 L/h, commercial scale system, cofield flow tubular	Aerobic microorganisms; yeasts and molds	6; 6	Min et al. (2003b)
	80 kV/cm, 20 pulses, 50°C + nisin (100 U/mL), circular treatment chamber, square, exponential decay or bipolar shape	Naturally occurring microorganisms	4.4	Nguyen and Mittal (2007)
	30, 35, 40 kV, 19°C–34°C, 2–3.3 μs pulse width, pilot size, continuous coaxial treatment chamber, exponential decay pulses	*B. fulva* conidiospores; *N. fischeri* ascospores	6; <1	Raso et al. (1998b)
Pineapple juice	33.0 kV/cm, 20°C	*Z. bailii* ascospores; *Z. bailii* vegetative cells	3.4; 4.3	
Cranberry juice	6.5 kV/cm, 22°C	*Z. bailii* ascospores; *Z. bailii* vegetative cells	4.2; 4.6	
	0–40 kV/cm, 69–80 μs treatment time	*E. coli*	6.4	Sen Gupta et al. (2005)
	36.5 kV/cm, 22°C, coaxial treatment chamber; 51.0 kV/cm, 34°C, coaxial treatment chamber	*B. fulva* conidiospores; *N. fischeri*	5.9; Not inactivated	Raso et al. (1998a)
	40 kV/cm, 150 μs of treatment time, square wave, cofield flow tubular	Aerobic microorganisms; yeasts and molds	4.8; 4.9	Jin and Zhang (1999)

(*Continued*)

TABLE 13.1 (*Continued*)

Inactivation of Microorganisms by PEF in Different Food Samples

Media	Treatment Conditions	Microorganism	Microbial Reduction (log cfu/mL)	References
Longan juice	32 kV/cm, 90 s tr. time, 3 μs pulse duration, 10 Hz, laboratory unit, square wave bipolar pulse	*E. coli*; yeasts	2–3; 6–7	Zhang et al. (2010)
Sour cherry juice	30 kV/cm, 3 μs pulse duration, 20 μs pulse delay time, 500 pps frequency, bipolar square wave pulses, batch scale cofield flow tubular	*E. coli* O157:H7; *S. aureus*; *Listeria monocytogenes*; *Pseudomonas syringae* subs. *syringae*; *Botyritis cinerea*; *Erwinia carotovara*; *Penicillum expansum*	4.53; 3.82; 3.17; 4.69; 6.62; 4.90; 5.68	Altuntas et al. (2010)
	30 kV/cm, 3 μs pulse duration, 20 μs pulse delay time, 500 pps frequency, batch scale cofield flow tubular, bipolar square wave pulses	*P. expansum*; *B. cinerea*	100% inactivation in spore germination rate and germination tube elongation	Evrendilek et al. (2008, 2009)
Apricot nectar	30 kV/cm, 3 μs pulse duration, 20 μs pulse delay time, 500 pps frequency, batch scale cofield flow tubular, bipolar square wave pulses	*E. coli* O157:H7; *S. aureus*; *L. monocytogenes*; *Pseudomonas syringae* subs. *syringae*; *Erwinia carotovara*; *P. expansum*; *B. cinerea*	3.10; 2.69; 2.91; 4.07; 4.73; 5.04; 5.74	Evrendilek et al. (2013)
	30 kV/cm, 3 μs pulse duration, 20 μs pulse delay time, 500 pps frequency, batch scale cofield flow tubular, bipolar square wave pulses	*P. expansum*; *B. cinerea*	100% inactivation in spore germination rate and germination tube elongation	Evrendilek et al. (2008, 2009)
Peach nectar	30 kV/cm, 3 μs pulse duration, 20 μs pulse delay time, 500 pps frequency, batch scale cofield flow tubular, bipolar square wave pulses	*E. coli* O157:H7; *S. aureus*; *L. monocytogenes*; *Pseudomonas syringae* subs. *syringae*; *Erwinia carotovara*; *P. expansum*; *B. cinerea*	4.11; 4.02; 4.07; 3.59; 4.27; 4.73; 4.28	Altuntas et al. (2011)
	30 kV/cm, 3 μs pulse duration, 20 μs pulse delay time, 500 pps frequency, batch scale cofield flow tubular, bipolar square wave pulses	*P. expansum*; *B. cinerea*	100% inactivation in spore germination rate and germination tube elongation	Evrendilek et al. (2008, 2009)
Melon juice	35 kV/cm for 1709 μs at 193 Hz and 4 μs pulse duration, batch scale bipolar and square wave	*E. coli* O157:H7; *S. enteritidis*; *L. monocytogenes*	3.71; 3.70; 3.56	Mosqueda-Melgar et al. (2008)
Watermelon juice	35 kV/cm for 1682 μs at 193 Hz and 4 μs pulse duration, batch scale bipolar and square wave	*E. coli* O157:H7; *S. enteritidis*; *L. monocytogenes*	3.56; 3.60; 3.41	Mosqueda-Melgar et al. (2008)

(*Continued*)

TABLE 13.1 (*Continued*)
Inactivation of Microorganisms by PEF in Different Food Samples

Media	Treatment Conditions	Microorganism	Microbial Reduction (log cfu/mL)	References
Pomegranate juice	17, 23, 30 kV/cm at 5°C, 15°C, 25°C, 35°C, bench scale, bipolar square wave pulses	*E. coli* O157:H7 *S. aureus*	1.1, 1.7, 2.3 (5°C) 1.5, 2.0, 2.8 (15°C) 2.0, 2.6, 3.6 (25°C) 2.3, 4.3, 5.3 (35°C) 1.0, 1.4, 2.1 (5°C) 1.4, 1.9, 2.5 (15°C) 1.9, 2.2, 2.8 (25°C) 2.0, 2.6, 3.0 (35°C)	Evrendilek (2009)
Carrot juice	1.5-μs pulse duration, 10 Hz pulse frequency, 0.029 μF capacitor, 6 mL tr. chamber, 52.5 mL/min flow rate, parallel-plate electrodes, exponentially decaying wave	*E. coli*	3.8	Zhong et al. (2005)
Formulated carrot juice	27 kV/cm, 3 μs pulse duration, 20 μs pulse delay time, batch scale cofield flow tubular, bipolar square wave pulses	Total aerobic mesophilic bacteria; total mold and yeast; total enterobactericeae; *E. coli* O157:H7	4.30; 3.42; 4.46; 3.57	Akin and Evrendilek (2009)
Grape juice	4 μs pulse width 35 kV/cm, 1000 Hz, 1 ms tr. time, bench scale, bipolar square wave pulses	*S. cerevisiae*	4	Garde-Cerdán et al. (2007)
	35 kV/cm, 303 Hz, pulse width for 1 ms, bench scale, bipolar square wave pulses	*Kloeckera apiculata*; *S. cerevisiae*; *L. plantarum* + *Lactobacillus hilgardii* mixture; *Gluconobacter oxydans*	3.88; 3.94; 3.54; 2.24	Marselles-Fontanet et al. (2009)
	35 kV/cm, 20°C	*Z. bailii* ascospores; *Z. bailii* vegetative cells	3.5; 5.0	Raso et al. (1998b)
	30, 35, and 40 kV, 19–34°C, 2–3.3 μs pulse width, pilot size, continuous coaxial treatment chamber, exponential decay pulses	*B. fulva* conidiospores; *N. fischeri* ascospores	6; negligible inactivation	
	34 kV/cm, initial temperature 55°C, 40 kJ/kg	*Aspergillus niger*; *E. coli*; *Lactobacillus rhamnosus*; *R. rubra*	4.6; 6.4; 4.6; 5.4	Heinz et al. (2003)
Red and white grape juices	P1: 20 pulses 65 kV/cm (peak-to-peak), 50°C with 2 h incubation of the juice with 1:3 lyso:chrisin (1:3 ratio of lysozyme and nisin; 0.4 g/100 mL).	*S. cerevisiae*	P1: 5.9	Wu et al. (2005)
	P2: 51°C, 20 pulses of 80 kV/cm (peak-to-peak) with nisin (400 U/mL)		P2: 6.2	
	P3: 20, 65 kV/cm (peak-to-peak), 50°C to white grape juice with (0·4 g/1000 mL)		P3: 4.4	

(*Continued*)

TABLE 13.1 (*Continued*)
Inactivation of Microorganisms by PEF in Different Food Samples

Media	Treatment Conditions	Microorganism	Microbial Reduction (log cfu/mL)	References
Cranberry juice	30, 35, 40 kV, 19°C–34°C, 2–3.3 μs pulse width, pilot size, continuous coaxial treatment chamber, exponential decay pulses	*B. fulva* conidiospores; *N. fischeri* ascospores	6; negligible inactivation	Raso et al. (1998b)
Pineapple juice	30, 35, 40 kV, 19°C–34°C, 2–3.3 μs pulse width, pilot size, continuous coaxial treatment chamber, exponential decay pulses	*B. fulva* conidiospores; *N. fischeri* ascospores	6; negligible inactivation	Raso et al. (1998b)
Milk	33 kV/cm, 35 pulses, 43°C; 36.7 kV/cm, 36 μs of treatment time, 40 pulses, 63°C	*E. coli*; *S. dublin*	3; 4	Dunn and Pearlman (1987)
	20–80 kV/cm, 0–100 pulses, 27–169 μs treatment time, exponential decay, square wave, oscillatory, bipolar (instant-charge-reversal) pulses	*E. coli* K12	6.25	Sen Gupta et al. (2003)
	17–46 kV/cm electric field strength, 545 pulses, 1.1 or 100 Hz frequency, exponentially decay pulses	*L. innocua*	~5	Picart et al. (2002)
	25°C, 30 kV/cm, 400 pulses, 600 μs of treatment time, square wave pulses	*L. monocytogenes*	2.5	Reina et al. (1998)
UHT milk	22.4 kV/cm, 300 μs of treatment time	*E. coli*	4.8	Grahl and Markl (1996)
	60 kV/cm, 26–210 μs	*B. stearothermophilus*	3	Shin et al. (2007)
	35 kV/cm, 64 pulses of bipolar square wave for 188 μs	*P. fluorescens*; *Bacillus cereus*; *L. lactis*	0.3–3	Michalac et al. (2003)
Homogenized milk	36.7 kV/cm, 40 pulses, 25 min	*S. dublin*; *E. coli*	Full reduction; 3	Dunn and Pearlman (1987)
SMUF	25 kV/cm, 20 pulses, exponential decay, <25°C	*E. coli*	4	Zhang et al. (1994b)
	7°C, 36 kV/cm, 16 pulses; 20°C, 36 kV/cm, 8 pulses; 33°C, 36 kV/cm, 8 pulses	*E. coli*	2–3; 2.5; 2.5	Zhang et al. (1995)
	40 kV/cm, oscillatory decay, <30°C		3	Qin et al. (1994)
	16 kV/cm, 180 μs of treatment time, bipolar; 50 kV/cm, 48 pulses, square wave, <30°C	*Bacillus subtilis*; *E. coli*	5.5; 3.6	
	16 kV/cm, 300 μs of treatment time, 40 pulses, exponential decay, <30°C	*Lactobacillus delbrueckii* ATCC 11842; *L. subtilits* ATCC 9372	4–5; 4–5	Pothakamury et al. (1995)

(*Continued*)

TABLE 13.1 (*Continued*)
Inactivation of Microorganisms by PEF in Different Food Samples

Media	Treatment Conditions	Microorganism	Microbial Reduction (log cfu/mL)	References
	16 kV/cm, 300 μs of treatment time, 40 pulses or 50 pulses (*L. subtilits* ATCC 9372), exponential decay, <30°C	*Lactobacillus delbrueckii* ATCC 11842; *L. subtilits* ATCC 9372	4–5	
	16 kV/cm, 300 μs of treatment time, 40 pulses or 50 pulses (*L. subtilits* ATCC 9372), exponential decay, <30°C	*E. coli*	4	Pothakamury et al. (1995)
	7°C, 36 kV/cm, 64 pulses; 20°C, 36 kV/cm, 64 pulses	*E. coli*	4, 5	Pothakamury et al. (1996)
Skim milk	40 kV/cm, exponential decay, 15°C	*E. coli*	6	Martin et al. (1997)
	30°C, 25 kV/cm, 100 pulses; 50°C, 25 kV/cm, 100 pulses	*S. dublin*	1, 2	Sensoy et al. (1997)
	50 kV/cm, 62 pulses, square wave, <30°C	*E. coli*	2.5	Qin et al. (1995b)
	17–46 kV/cm electric field strength, 545 pulses, 1.1 or 100 Hz frequency, exponentially decay pulses	*L. innocua*	~5	Picart et al. (2002)
	3.7 μs pulse duration time, 250 Hz frequency, 35 kV/cm electric field strength, 450 μs of treatment time (stepwise and circulation mode fluid handling system)	*S. aureus*	3.3 and 3.5	Evrendilek et al. (2004a)
Raw SM	30 and 40 kV/cm, 1–30 pulses, 20°C–72°C, <10 s. Best balance of inactivation at 55°C with 40 kV/cm.	*L. innocua*	4.3	Guerrero-Beltran et al. (2010)
	50 kV/cm, 64 μs of treatment time, 36°C	*L. innocua*	2.5	Calderon-Miranda et al. (1999)
SM gel	3.25 μs pulse width, 15, 20 or 30 kV/cm electric field strength, 5, 10, or 50 pulses	*L. monocytogenes*	4.5	Fleischman et al. (2004)
Phosphate buffer	2.5 μs pulse width, 3 Hz frequency, 0–60 pulses 41 kV/cm electric field strength	*E. coli*; *L. innocua*	2.3–6.5; 0.7–2.8	Dutreux et al. (2000)

(*Continued*)

TABLE 13.1 (*Continued*)
Inactivation of Microorganisms by PEF in Different Food Samples

Media	Treatment Conditions	Microorganism	Microbial Reduction (log cfu/mL)	References
	17–46 kV/cm electric field strength, 545 pulses, 1.1 or 100 Hz frequency, exponentially decay pulses; 30, 40, or 50 kV/cm plus 10 IU nisin application	*L. innocua*	~5; 2.0, 2.7, and 4.3	Picart et al. (2002)
Fat-free milk	2.5 µs pulse width, 3 Hz frequency, 0–60 pulses 41 kV/cm electric field strength,	*E. coli*; *L. innocua*	2.3–6.5, 0.7–2.8	Dutreux et al. (2000)
Dairy cream	17–46 kV/cm electric field strength, 545 pulses, 1.1 or 100 Hz frequency, exponentially decay pulses	*L. innocua*	~5	
Mcllvaine buffer	1 Hz, 2 µs pulse width, 15–28 kV/cm electric field strength, square waveform	*L. monocytogenes*	~5	Alvarez et al. (2003a)
	1–5 Hz, 1–15 µs pulse width, 5.5–28 kV/cm electric field strength, square waveform	*Yersinia enterocolitica*	6	Alvarez et al. (2003b)
	20 pulses, 40 kV/cm at 65°C, 25 pulses, 36 kV/cm at 61°C, or 31 pulses, 31 kV/cm at 56°C	*L. innocua*	6	Sepulveda et al. (2005)
Yogurt drink	30 kV/cm electric field strength, 1.4 µs pulse width, 500 pps frequency, 32 µs total treatment time, 100 L/h flow rate	Initial microflora		Evrendilek et al. (2004b)
Yogurt	18 kV/cm, 55°C	*S. cerevisiae*	3	Dunn and Pearlman (1987)
Cheese whey	40 kV/cm, 4937 µs PEF plus UV (with 7.7 s, 229 mJ/mL dosage)	*L. innocua*, *Z. bailii*	3.0–5.0; 7.9–8.8	Dave et al. (2012)
Liquid whole egg	26 kV/cm, 2 and 4 µs pulse duration, 1.25 and 2.50 Hz pulsing rates, up to 100 pulses/unit volume, temperature below <30°C	*E. coli*	6	Martin-Belloso et al. (1997)
	2 µs pulse duration, 3.5 Hz, 10.6, 21.3, and 32 pulses, 30, 35, and 40 kV, exponentially decay pulse	*L. innocua*	3.5	Calderon-Miranda et al. (1999)
	20–35 kV/cm, 100–900 Hz pulse frequency, 2–8 pulse number, 4°C–30°C temperature, 7–9 pH	*S. enteritidis*	3.5	Jeantet et al. (1999)

(*Continued*)

TABLE 13.1 (*Continued*)
Inactivation of Microorganisms by PEF in Different Food Samples

Media	Treatment Conditions	Microorganism	Microbial Reduction (log cfu/mL)	References
	9–15 kV/cm, 138 pulses, 1 Hz frequency, 2 μs pulse duration, 138 s treatment time	*E. coli* O157:H7	4	Bazhal et al. (2006)
	20–45 kV/cm, 106–472 kJ/kg	*S. enteritidis* and the heat resistant *Senftenberg* 775 W	3	Monfort et al. (2010)
	25 kV/cm, 75–100 kJ/kg + heat (52°C/3.5′, 55°C/2′, 60°C/1′) with 2% triethyl citrate	*Salmonella serovars* (*dublin*, *enteritidis* 4300, *enteritidis* 4396, *typhimurium*, *typhi*, *Senftenberg*, *Virchow*)	5	Monfort et al. (2012)
	25 kV/cm, 200 kJ/kg followed by heat (60°C/3.5 min) to LWE added with 10 mM EDTA or 1% triethyl citrate	*S. senftenberg* 775W; *L. monocytogenes*	5; 5	Monfort et al. (2013)
Liquid-dialyzed egg	L5 kV/cm at 0°C, 1 Hz, 500 pulses, square voltage fields	*E. coli* O157:H7	3.5	Amiali et al. (2004)
Egg white	200 pps frequency, 2.12 μs pulse duration, 25 kV/cm electric field strength, 250 μs total treatment time, PEF + 55°C for 3.5 min	*S. enteritidis*	1 and 4.3	Hermawan et al. (2004)
Spices (dry)	65 kV/cm, 750 μs of treatment time	Yeasts	4.2	Keith et al. (1997)
Pea soup	33 kV/cm, 0.5 L/min, 4.3 Hz, and 30 pulses	*E. coli*; *Bacillus subtilis*	6.5; 5.3	Vega-Mercado et al. (1996a)
Model beer	28.8–34.8 kV/cm and 50 kJ/kg	*L. plantarum*	2–4	Ulmer et al. (2002)
Beer	1 mL/s flow rate, 41 kV/cm, 600 pps pulse repetition rate, 4 μs pulse duration, 175 μs total treatment time; 10.5 mL/s flow rate, 22 kV/cm, 14 μs pulse duration time, 800 pps pulse repetition rate, 216 μs total treatment time	*L. plantarum*; *P. damnosus*; *Bacillus subtilis*; *S. uvarum*; *R. rubra*	4.7; 5.8; 4.8; 4.1; 4.3	Evrendilek et al. (2004c)
	45 kV/cm, 804 μs treatment time	*Bacillus subtilis*; *L. plantarum*; *S. cerevisiae*	3.7; 4.8; 4.8	

(*Continued*)

TABLE 13.1 (*Continued*)
Inactivation of Microorganisms by PEF in Different Food Samples

Media	Treatment Conditions	Microorganism	Microbial Reduction (log cfu/mL)	References
Must and wine	186 kJ/kg, 29 kV/cm	*Dekkera anomala*; *Brettanomyces bruxellensis*; *Lactobacillus hilgardii*; *L. plantarum*	3; 3; 3, 3	Puertolas et al. (2009)
Red wine	17–31 kV/cm, 10°C–30°C, 40 mL/min flow rate, 3 μs pulse duration, 500 pps frequency	*Escherichia coli* O157:H7; *C. lipolytica*; *S. cerevisiae*; *H. anomala*; *L. bulgaricus*	3.5; 5.3; 5.95; 5.29; 4.0	Uysal (2010)
Liquid media and	65 kV/cm, 5 μs, 500 Hz	*Campylobacter jejuni* isolates and three *C. coli* isolates	2.41–5.19	Haughton et al. (2012)
	65 kV/cm, 5 μs, 500 Hz	*Escherichia coli* (ATCC 25922); *S. enteritidis* (ATCC 13076)	4.33–7.22	
Raw chicken meat	65 kV/cm, 5 μs, 500 Hz	Total viable counts, *Enterobacteriaceae*, *C. jejuni*, *Escherichia coli*, *S. enteritidis*	No significant reduction	
Smothie	25 kV, 1 kHz, 4–32 μs pulse width combined with pasteurization (72 for 26 s) and MTS (20 kHz, 31 mm amplitude, 40 W/cm^2 intensity	*L. innocua*	PEF: 2.7 MTS:3 PEF + MST: 4.2 MST + PEF:5.6 PAST:6.5	Palgan et al. (2012)
Peptone water	10,000; 20,000; and 30,000 μs pulse period, of 5, 15, and 25 kV/cm electric field strength, 3, 6, and 9 μs pulse widths	*Lactobacillus acidophilus* LA-K	Slower growth rate	Cueva and Aryana (2012)
MRS broth	Monopolar square pulses at varying nominal electric field strengths and number of pulses, energies of 34.6, 65.8, and 658.1 J/cm^3	*L. plantarum* 564	Slower growth rate	Seratlić et al. (2013)

TABLE 13.2
Inactivation of Enzymes in Different Food Samples

Medium	Enzyme	PEF Treatment Condition	Reduction (%)	Source
Apple juice	POD	P1: 60°C + PEF (30 kV/cm, 100 kJ/kg) bench scale	100	Schilling et al. (2008a)
	PPO	P1: 60°C + PEF (30 kV/cm, 100 kJ/kg) bench scale P2: 40°C + PEF (pilot plant scale)	P1: 100 P2: 48	Schilling et al. (2008a)
	PPO	38.5 kV/cm and 300 pps combined with 50°C, bipolar pulses	70	Sanchez-Vega et al. (2009)
Orange juice	PME	35 kV/cm, 59 μs of treatment time, 1.4 μs of pulse width, 600 pps, 98 L/h, pilot plant scale	88	Yeom et al. (2000a)
	PME	20–35 kV/cm, 2.0 or 2.2 of pulse width, 700 pps, 0.42, 0.31 mL/s	90	Yeom et al. (2002)
	PME	24 kV/cm, 1048 μs treatment time, 29.9 kJ, <50°C	95	Hitit (2011)
	PME	25.26 kV/cm, 1033.9 and 1206.2 ms treatment time, 51.32 J	93.8	Agcam (2012)
Tomato juice	POD	35 kV/cm, 1500 μs 4 μs pulses at 100 Hz	97	Aguilo-Aguayo et al. (2008)
	PPO	24 kV/cm for 320 and 962 μs	69	Luo et al. (2010)
	PME	35 kV/cm, 1500 μs treatment time, 4 μs pulse duration at 100 Hz	82	Aguilo-Aguayo et al. (2008)
	PME	24 kV/cm, 800 μs of treatment time, 400 pulses	93.8	Giner et al. (2001)
	PME	40 pulses of 87 kV/cm at 50°C	55	Nguyen and Mittal (2007)
	Lypoxygenase	40 kV/cm, 57 μs of treatment time, 2 μs of pulse width, 1000 pps, 500 L/h	54	Min and Zhang (2007)
	Lypoxygenase	30 kV/cm, 60 μs of treatment time, 3 μs of pulse width, 1 mL/s, 50°C	88.1	Min et al. (2003c)
	Lypoxygenase	40 kV/cm for 57 μs	53	Min et al. (2003b)
	Lypoxygenase	24 kV/cm for 320 and 962 μs	88	Luo et al. (2010)
	Lypoxygenase	24 kV/cm, 1048 μs treatment time, 29.9 kJ, <50°C	89.1	Hitit (2011)
	PG	40 pulses of 87 kV/cm at 50°C	No inactivation	Nguyen and Mittal (2007)
	PG	35 kV/cm, 1500 μs 4 μs pulses at 100 Hz	12% PEG	Aguilo-Aguayo et al. (2008)
Carrot juice	POD	35 kV/cm for 1000 μs applying 6 μs pulse width at 200 Hz	73	Quitao-Teixeira et al. (2008)
		35 kV/cm for 1500 μs	93	Quintao-Teixeira et al. (2013)

(Continued)

TABLE 13.2 (*Continued*)
Inactivation of Enzymes in Different Food Samples

Medium	Enzyme	PEF Treatment Condition	Reduction (%)	Source
Grape juice	PPO	24 kV/cm, 1048 μs treatment time, 29.9 kJ, <50°C	93–98	Hitit (2011)
White grape juice	POD	25–35 kV/cm, 200–1000 Hz, 1–5 ms treatment time	50	Marselles-Fontanet et al. (2007)
	PPO	25–35 kV/cm, 200–1000 Hz, 1–5 ms treatment time	100	Marselles-Fontanet et al. (2007)
	Lipoxygenase	21.5 kV/cm, <50°C, 22 Hz, exponential decay	Trace	Grahl and Markl (1996)
Whole milk	Lipoxygenase	19 kV/cm, 70°C, 1 Hz, monopolar square wave shape	0	Van Loey et al. (2002)
	Lipase	21.5 kV/cm, <50°C, 22 Hz, exponential decay	60	Grahl and Markl (1996)
	Lipase	80 pulses, 27.4 kV/cm	62	Bendicho et al. (2003)
	Alkaine phosphatase	21.5 kV/cm, <50°C, 22 Hz, exponential decay	60	Grahl and Markl (1996)
		19 kV/cm, 70°C, 1 Hz, monopolar square shape pulses	74	Van Loey et al. (2002)
	Protease	35.5 kV/cm, 866 μs, 111 Hz	57.1	Bendicho et al. (2003)
	Plasmin	50 pulses, 45 kV/cm, 15°C processing temperature	90	Vega-Mercado et al. (2006)
Raw milk	Lipase	21.5 kV/cm, 400 kJ/L	65	Castro et al. (2001)
	Peroxidase	21.5 kV/cm, 400 kJ/L	25	Castro et al. (2001)
	Alkaline phosphatase	21.5 kV/cm, 400 kJ/L	<5	
	Alkaline phosphatase	2.2 kV/mm	60	
SM	Alkaline phosphatase	21.8 kV/cm, 43.9°C, 700 pulses of 400 μs	65	Castro (1994)
	Protease	35.5 kV/cm, 866 μs, 111 Hz	81.1	Bendicho et al. (2003)
	Protease (*Bacillus subtilis*)	19.7–37.3 kV/cm, 34°C and 40°C, 22 pulses of 67 Hz and 80 pulses of 4 μs at 0.1 Hz, monopolar square	Depending on the frequency	Bendicho et al. (2003)
SMUF/SM	Protease (*Bacillus subtilis*)	19.7–35.5 kV/cm, <46°C, 67.89 and 111 Hz, monopolar square	62.7–81	Bendicho et al. (2003)
SMUF	Protease (*Bacillus subtilis*)	16.4–27.4 kV/cm, 34°C and 40°C, 22 pulses of 67 Hz and 80 pulses of 4 μs at 0.1 Hz, monopolar square	Depending on the frequency	Bendicho et al. (2003)
	Plasmin	15–45 kV/cm, 60°C–80°C, 0.1 Hz	90	Vega-Mercado et al. (1995)
	Lipase (*P. fluorosens*)	16.1–27.1 kV/cm, <31°C, 2–3.5 Hz, exponential decay	62.1	Bendicho et al. (2002b)

30 kV/cm with 8 pulses, and a 4.5-log reduction was observed at the same field strength using 48 pulses. Extending the duration of microsecond pulses from 100 to 250 μs showed no improvement in inactivation. Nanosecond pulses alone did not have any detectable effect on inactivation of *E. coli* regardless of the treatment time, but a significant 3-log reduction was achieved in combination with microsecond pulses (Zgalin et al., 2012).

Inactivation of *E. coli* and *L. innocua* by combinations of high-intensity light pulses (HILP), ultrasound (US), PEF, and sublethal concentrations of nisin (2.5 mg/L) or lactic acid (500 mg/L) was investigated by Muñoz et al. (2012) in two different buffer systems (pH 4 for *E. coli* and pH 7 for *L. innocua*). Individually, HILP (3.3 J/cm^2), US (126 s residence time, 500 W, 40°C), and PEF (24 kV/cm, 18 Hz, and 1 μs of pulse width) did not induce a microbial reduction of greater than 2.7 or 3.6 log units for *L. innocua* and *E. coli*, respectively. Combined treatment using HILP + PEF sufficiently inactivated *E. coli* without antimicrobial addition. The addition of either antimicrobial enhanced the effect of US + PEF for both *E. coli* and *L. innocua* whereas the addition of lactic acid enhanced the effect of HILP + US. For *L. innocua*, the addition of nisin enhanced the effect of HILP + PEF. This confirms the potential of selected nonthermal technologies for microbial inactivation when combined with antimicrobials.

Orange juice due to the popularity, and thus, the economic importance, is one of the most studied juice among others. One of the earliest study conducted by PEF treatment of orange juice at 36 kV/cm electric field strength and 250 μs treatment time provided microbial stability during more than 3 weeks (Qin et al., 1995a). PEF processing of freshly squeezed orange juice at 30 kV/cm electric field strength with 480 μs treatment time was found very effective to inactivate the native microflora without significantly altering the measured sensory characteristics (Jia et al., 1999). Similar results were also reported from the processing of orange juice at 35 kV/cm electric field strength with 59 μs treatment time, and it was found that PEF and heat treatment at 94.6°C for 30 s had similar effects on orange juice quality. PEF processing provided greater amount of vitamin C retention and better preservation of flavor compounds than heat treatment during shelf-life studies at 4°C. Measured quality attributes such as browning index, higher whiteness, and higher hue angle values were obtained by PEF treatment, in addition to no significant change on pH and °Brix values. Processing of orange juice by PEF and heat provided a significant degree of microbial inactivation, and microbial count was reported below 1 log cfu/mL during shelf-life studies conducted at both refrigeration and ambient temperature (Yeom et al., 2000a,b). Application of 40 kV/cm electric field strength with 97 μs treatment time provided microbiologically stable juices during 112 days at 4°C (Min et al., 2003a). Processing of fruit juice at 35 kV/cm electric field strength and 1000 μs treatment time provided both significant amount of microbial and enzyme inactivation, and thus, the PEF-processed orange juice was stable at least 56 days of refrigerated storage at 4°C. A slight decrease in the antioxidant capacity of PEF-treated orange juice throughout storage was reported, and these changes were not significantly different to those observed in the fresh juice (Elez-Martinez et al., 2006; Martin-Belloso and Soliva-Fortuny, 2010).

Studies with PEF treatment of orange juices provide a significant increase on the shelf life. Although different shelf-life extensions were reported due to the differences on process parameters, reported shelf-life extensions for orange juice were up to 28 days at 4°C (80 kV/cm for 20 pulses, without aseptic packaging) (Hodgins et al., 2002). Moreover, better preservation of ascorbic acid content and the retention of flavor compounds in juices are reported as benefits of PEF over heat treatment (Hodgins et al., 2002). Little or no effect on the vitamin C content of orange juice, which is sensitive to heat—less than 80% retention after thermal pasteurization, is reported by the processing that includes the combination of PEF (80 kV/cm) and thermal treatment up to 50°C (Wu et al., 2005; Torregrosa et al., 2006).

When compared with control samples, PEF processing did not cause significant changes on pH, °Brix, electric conductivity, viscosity, nonenzymatic browning index (NEBI), hydroxymethylfurfural (HMF), color, organic acid content, and volatile flavor compounds in squeezed citrus juices of grapefruit, lemon, orange, and tangerine (Cserhalmi et al., 2006). Commercial-scale PEF treatment at 40 kV/cm electric field strength and 97 μs treatment time provided higher amount of microbial inactivation and better preservation of quality parameters than that of heat treatment at 90°C for 90 s (Min et al., 2006). No significant difference was detected in the HMF content between PEF and heat-treated orange juice samples. Moreover, PEF-treated orange juice samples showed lesser amount of changes in the color values than heat-pasteurized samples (Cortes et al., 2008).

PEF processing was also successful for enzyme inactivation in orange juice. Inactivation studies of PME showed 88% reduction in orange juice as compared to 98% reduction for thermally processed orange juice. After PEF processing and during subsequent storage, no increase in PME activity was reported (Yeom et al., 2000b). PEF processing successfully inactivated both PME and POD enzymes, and no changes were reported in the residual activities of both enzymes during 56 days of storage (Elez-Martinez et al., 2006). Combination of PEF with moderate heat (<50°C) caused 90%–92.7% reduction of PME activity in orange juice. It was reported that this amount of inactivation is sufficient to prevent development of cloudiness in orange juice (Hodgins et al., 2002; Yeom et al., 2002). Inactivation kinetics of PME by PEF and heat pasteurization (90°C for 10 and 20 s), right after both treatments and during shelf-life studies at 4°C for 180 days in freshly squeezed-in orange juice samples, showed 93.82% inactivation after PEF applications and 6.82% and 4.15% residual PME activities after heat pasteurizations. PME activity of PEF-processed samples decreased or did not change during storage, while that of heat-pasteurized samples increased during storage (Agcam et al., 2014).

While heat processing caused 10%–41.7% loss of volatile flavor compounds, the maximum loss was 9.7% for PEF-processed orange juice samples (Jia et al., 1999). PEF treatment also caused less browning and a brighter color during storage at 4°C compared to a heat-treated sample. It is possible that the better properties obtained might be due to a higher retention of vitamin C (Min et al., 2003a).

Compared to heat treatment, PEF processing provided more stable vitamin C with the half-life of up to 60 days, 2–5 times longer, depending on the PEF treatment and storage temperature (Min et al., 2003a; Torregrosa et al., 2006). PEF-processed orange juice was found more bioavailable for human absorption than that of the fresh orange juice (Sanchez-Moreno et al., 2005). In addition to vitamin C, PEF-treated orange juice also had better texture, flavor, and overall acceptability than thermally processed juice. PEF processing provided significantly higher content of the flavor compounds D-limonene, alpha-pinene, myrcene, and valencene as compared to thermally treated and untreated samples (Min et al., 2003a).

Freshly squeezed orange juice was PEF-processed using four electric field strengths (13.82, 17.06, 21.50, and 25.26 kV/cm), two different treatment times (1033.9 and 1206.2 μs), and eight different electric energy intensities (10.89, 12.70, 17.37, 20.26, 29.57, 34.50, 43.99, and 51.32 J) compared with heat pasteurization at 90°C for 10 s (HP1) and for 20 s (HP2) during storage at 4°C for 180 days and total phenolic concentration was found higher after mild PEF and heat pasteurization treatments than both the other treatments and fresh orange juice. At the end of the storage, a higher amount of total phenolic concentration than that of the initial concentration was measured when processed after the low-intensity PEF treatments. Total phenolic compounds of orange juice started to degrade after being processed with high-intensity PEF treatments and heat pasteurization applications (Agcam, 2012).

PEF processing of a blended orange–carrot juice mixture by 25 kV/cm electric field strength, 280 and 330 μs treatment time, and conventional HTST at 98°C for 21 s presented that even though heat process provided more microbial inactivation, PEF provided better preservation of important quality properties as well as less ascorbic acid degradation with shelf-life extension of 50 days at 2°C (Torregrosa et al., 2006).

Apple juice is also one of the most studied juices by PEF. Processing of fresh apple juice by PEF treatment at 30, 26, 22, and 18 kV/cm electric field strengths and 172, 144, 115, and 86 μs total treatment times provided 5 log cfu/mL reduction on *E. coli* O157:H7 and *E. coli* 8739 (Evrendilek et al., 1999). The effect of PEF treatment on the inactivation of *E. coli* O157:H7 with 34, 31, 28, 25, and 22 kV/cm electric field strengths and 166 μs mean total treatment time by bench scale system and shelf-life evaluation of aseptically packaged apple juice and cider with 35 kV/cm electric field strength and 94 μs mean total treatment time by pilot plant scale were investigated with PEF and heat treatment (60°C for 30 s) combination to process fresh apple cider. Bench scale PEF system provided 4.5 log cfu/mL reduction of *E. coli* O157:H7 cells, and the pilot plant scale PEF system improved the microbial shelf life of apple cider without any alteration in the natural food color and vitamin C for both samples (Evrendilek et al., 2000). Processing of apple juice by PEF (4 μs wide bipolar pulse, 35 kV/cm electric field strength and 1200 pps frequency) or HTST (90°C for 30 s) revealed that measured attributes were less affected by PEF treatment than by thermal pasteurization (Aguilar-Rosas et al., 2007). Again, processing of apple juice by PEF or HTST showed that PEF was more effective than HTST process to preserve pH and color (Charles-Rodríguez et al., 2007). Apple juice processed by UHT (115°C, 125°C, and 135°C for 3 or 5 s) or PEF (33–42 kV/cm with frequencies of 150, 200, 250, or 300 pps) showed that PEF processing provided better preservation of color, pH, acidity, and soluble solids (Sanchez-Vega et al., 2009).

Effects of electric field strength (0–35 kV/cm) and pulse rise time (PRT) of 2 and 0.2 μs during PEFs on enzymatic activity, vitamin C, total phenols, antioxidant capacities, color, and rheological characteristics of fresh apple juice showed that the residual activity (RA) of PPO and peroxidase (POD) decreased with increased electric field strength and PRT, and almost complete inactivation of both enzymes was achieved at 35 kV/cm with 2 μs-PRT. Vitamin C content in apple juice decreased significantly during PEF treatment, and the largest loss was 36.6% at 30 kV/cm and 2 μs-PRT. The content of total phenols was not affected by PEF with 2 μs-PRT but decreased significantly by PEF with 0.2 μs-PRT. The antioxidant capacity of apple juice was evaluated by DPPH radical scavenging activity, ferric-reducing antioxidant power (FRAP), and oxygen radical absorbance capacity (ORAC). The DPPH value was not affected by PEF, whereas FRAP and ORAC values increased by increasing the electric field strength and decreasing the PRT. PEF-treated apple juice had a significantly higher lightness and yellowness than the controlled sample. The apparent viscosity and consistency index (K) of apple juice decreased, while the flow behavior index (n) increased by increasing the electric field strength, and apple juice treated at 2 μs-PRT had a significantly higher apparent viscosity than that treated at 0.2 μs-PRT (Bi et al., 2013).

Current bench-scale PEF treatment systems in different institutions are different; therefore, conflicting results are obtained sometimes because of the fact that even though electric field strength is similar, total energy and total treatment time might be different. As a result of this situation, reports on the color of PEF-treated apple juice are conflicting. While Ortega-Rivas et al. (1998) reported color fading after 2–16 pulses at 50–66 kV/cm, Evrendilek et al. (2000) observed no color loss after 94 μs at 35 kV/cm in PEF-treated apple juice and cider.

Processing of freshly squeezed apple juice from golden delicious fruit by PEF (4 μs pulse width, 35 kV/cm electric field strength, 1200 pps frequency) and HTST at 90°C for 30 s resulted in minimal variability in pH and no significant changes in acidity, phenolics content, and volatile compounds (Aguilar-Rosas et al., 2013). Apple juice processing with the same processing parameters ended up with a significant difference in the residual activities of PME and PPO as well as the chroma index for color compared to control samples. No statistically significant difference was observed for the hue colour index (Aguilar-Rosas et al., 2007).

Complete deactivation was achieved when PEF treatment and preheating of the juices to 60°C were combined to inactivate POD and PPO enzymes. Maximum PPO deactivation of 48% was achieved when the apple juice was preheated to 40°C and PEF-treated at 30 kV/cm (100 kJ/kg) by the pilot plant PEF system (Schilling et al., 2008a).

UHT (115°C, 125°C, and 135°C for 3 and 5 s) and PEF processing (between 33 and 42 kV/cm with frequencies of 150, 200, 250, and 300 pps) were compared to each other for the processing of apple juice, and UHT was found more efficient for the inactivation of PPO reducing 95% of the RA at the maximum temperature and time. However, 70% reduction of residual PFO activity was achieved by the PEF treatment at 38.5 kV/cm and 300 pps combined with 50°C. Color, pH, acidity, and soluble solids were all less affected by PEF than by UHT when compared with the untreated juice (Sanchez-Vega et al., 2009).

In order to determine the effect of PEF processing on the shelf life and quality characteristics of tomato juice, both PEF and thermal treatments were compared to each other, and it was found that both thermally and PEF-processed juices showed microbial shelf life at 4°C for 112 days. The residual lipoxygenase activities of thermally and PEF-processed juices after the treatments were 0% and 47%, respectively; however, PEF-processed juice retained more ascorbic acid than thermally processed juice at 4°C for 42 days. Thermally and PEF-processed juices had no significant difference in the concentration of lycopene, °Brix, pH, or viscosity during the storage. The flavor and overall acceptability of PEF-processed juice were preferred to those of thermally processed juice according to sensory evaluations (Min et al., 2007). PEF also caused less browning in tomato juice, with a much slower browning rate during storage, decreasing little over 60 days. PEF-treated tomato samples were redder than heat-treated samples and similar to the untreated sample (Min and Zhang, 2003).

PEF, moderate temperature (<50°C), and antimicrobial compounds were tested to reduce naturally growing microorganisms in tomato juice. The microbial count decreased with the increase in pulse number and treatment temperature at constant field strength, and a significant difference in the reduction of microbial count with temperature increase from 45°C to 50°C was found. No reduction was detected in vitamin C content due to the treatment. Polygalacturonase (PG) activity present in tomato juice was not affected by PEF, but the activity of PME was reduced by 55%. Antimicrobials such as clove oil and mint extract provided large microbial decay at low concentration and mild heat without PEF (Nguyen and Mittal, 2007).

PEF provided significant degree of PPO and LOX inactivation in tomato juice that activities of both enzymes decreased with the increase of the applied electric field and the number of pulses (Luo et al., 2010). Effects of PEF (35 kV/cm for 1500 μs using bipolar 4 μs pulses at 100 Hz) processing of tomato juice color parameters and viscosity, as well as POD, PME, and PG activity during 77 days of storage at 4°C was measured and compared to the thermal treatments at 90°C for 1 min or 30 s for unprocessed tomato juice. PEF-treated tomato juice showed higher values of lightness than both the thermally processed and the untreated juice throughout storage time. POD of PEF-treated tomato juice was inactivated by 97%, whereas in the case of the thermally treated juice, 90% and 79% inactivation was achieved after 1 min and 30 s, respectively. The highest PME inactivation in tomato juice was obtained by PEF (82%) and heat treatment at 90°C for 1 min (96%). PG of PEF-treated tomato juice was inactivated by 12%, whereas thermal treatments at 90°C for 1 min or 30 s achieved 44% and 22% inactivations, respectively. Despite the low rates of PG inactivation obtained, the pattern followed in the RA along the storage time was similar in the tomato juice treated by PEF than the thermally processed juice (Aguilo-Aguayo et al., 2008).

Sour cherry juice was successfully processed with increased electric field strengths (0, 17, 20, 23, 27, and 30 kV/cm) and treatment times (0, 66, 105, 131, 157, and 210 μs), and measured properties of sour cherry juice such as pH, TA, °Brix, conductivity, color (L^*, a^*, and b^*), NEBI, metal ion concentration, ascorbic acid as well as in the metal ion concentrations did not get adversly affected by PEF treatment (Altuntas et al., 2010). PEF processing of pomegranate juice with 0, 17, 23, and 30 kV/cm electric field strengths at 5°C, 15°C, 25°C, and 35°C ended up with a significant decrease in the amount of total phenolic substances (TPS), antioxidant capacity, and total anthocyanin content (TAC) with increased processing temperature. Especially, the PEF processing at lower temperatures such as 5°C and 15°C caused less degradation than that of the treatment at 25°C and 35°C (Evrendilek, 2009).

Grape juice physicochemical properties did not change with PEF processing; however, the concentration of lauric acid diminished after PEF processing, and the concentration of some amino acids varied after both PEF and thermal treatments (Garde-Cerdan et al., 2007). PEF processing of grape juice by 524, 655, 786, 917, and 1048 μs treatment time and 24 kV/cm electric field strength did not cause any significant difference on pH, °Brix, conductivity, titratable acidity, color (L^*, a^*, b^*, and chroma), TPS, TAC, and FT-IR spectra of the samples. Increased treatment time caused a significant decrease in the total antioxidant activity and PPO activity (Hitit, 2011). PEF and thermal processing of different varieties of freshly squeezed grape juice evidenced that grape variety is a factor to be taken into account when comparing the processing effects of PEF and heat treatments. Results of general and specific microbial populations were not affected by each processing treatment. Soluble solids, pH, acidity, and the electrical conductivity of grape juice were not affected by PEF processing. On average, PEF treatment reduced radical scavenging activity by 9% in front of the 13% of the heat treatment, whereas both treatments halved the protein content (Marsellés-Fontanet et al., 2013).

Cranberry juice processing either by PEF (20 and 40 kV/cm for 50 and 150 μs) or by thermal treatment (at 90°C for 90 s) showed that higher field strength and longer treatment time caused more reduction in viable microbial cells. Moreover, the overall volatile profile of the juice was not affected by PEF treatment, but it was affected by thermal treatment (Jin and Zhang, 2007).

A better knowledge of the effect of refrigerated storage on the nutritional and physicochemical characteristics of foods processed by emerging technologies with regard to unprocessed juices is necessary. Thus, blueberry juice was processed by HHP (600 MPa/42°C/5 min) and PEF (36 kV/cm, 100 μs) and the stability of physicochemical parameters, antioxidant compounds (ascorbic acid, total phenolics, and total anthocyanins), and antioxidant capacity was studied just after treatment and during 56 days at refrigerated storage at 4°C. Just after treatment, all treated blueberry juices showed a decrease lower than 5% in ascorbic acid content as compared with the untreated one. At the end of refrigerated storage, unprocessed and PEF-treated juices showed similar ascorbic acid losses (50%) in relation to untreated juice, although HHP juices maintained better ascorbic acid content during storage time (31% losses). All juices exhibited fluctuations in total phenolic values with a marked decrease after 7 days in refrigerated storage; however, prolonged storage of the juices at 4°C, up to 56 days resulted in a change in the total phenolic content for all juices in comparison with day 7. HHP preserved antioxidant activity (21% losses) more than unprocessed (30%) and PEF-(48%) processed juices after 56 days at 4°C. Color changes (a^*, b^*, L, Chroma, $h°$, and ΔE) were slightly noticeable after refrigerated storage for all juices (Barba et al., 2012).

The application of sublethal, nonthermal processing and GRAS antimicrobial hurdle combinations has the potential to allow for the production of safe, stable products while also maintaining the desired organoleptic characteristics of a minimally processed product. An initial step to assessing the suitability of nonthermal treatments is to evaluate their efficacy in model solutions prior to their study in food systems. The effect of high-intensity PEFs (HIPEF) in combination with antimicrobial substances against spoilage microorganisms in fruit juices has been the focus of a few studies (Hodgins et al., 2002; Wu et al., 2005; Liang et al., 2006; Nguyen and Mittal, 2007; Mosqueda-Melgar et al., 2008). The effect of combining HIPEF with citric acid or cinnamon bark oil, as antimicrobial substances, on the microbiological shelf life of strawberry, orange, apple, pear, and tomato juices was evaluated by Mosqueda-Melgar et al. (2012) in addition to the sensory properties of these products. An extension of the microbiological shelf life of fruit juices treated by HIPEF with or without antimicrobial substances was observed in comparison with those juices without processing. Naturally occurring microorganisms in the juices were successfully inactivated by HIPEF treatment. Among the HIPEF-treated juices, those from strawberry and orange did not show microbial growth during the 91 days of storage at 5°C. However, resident microbial populations in apple, pear, and tomato juices only were controlled during that time when HIPEF was combined with antimicrobials. Therefore, combinations of those treatments may be a feasible alternative to thermal pasteurization to ensure the microbiological quality and safety in juices and to avoid the risk of foodborne illness caused by the consumption of these commodities. No significant changes on the sensory attributes in all studied fruit juices processed by HIPEF were found, but when citric acid or cinnamon bark oil were added, noticeable changes on some sensory attributes such as aroma, taste, and sourness of these fruit juices were perceived (Mosqueda-Melgar et al., 2012).

Nectars in contrast to juices have higher viscosity and fruit particles, thus it is difficult to process them by PEF. Even though processing of nectars by PEF is harder, it was feasible to process both apricot and peach nectars by different electric field strengths (0, 17, 20, 23, 27, and 30 kV/cm) and different treatment times (0, 66, 105, 131, 157, and 210 μs). It was revealed that processing of apricot and peach nectad by PEF did not cause a significant difference in pH, TA, °Brix, conductivity, color (*L**, *a**, and *b**), NEBI, metal ion concentration, ascorbic acid, and beta carotene retention as well as in the metal ion concentrations (Altuntas et al., 2011; Evrendilek et al., 2013).

Tea is one of the most consumed beverages, and heat treatment has a detrimental effect on bioactive compounds. Thus, PEF processing from 20 to 40 kV/cm electric field strength for 200 μs treatment time was conducted for green tea infusions, and it was seen that the total free amino acids were efficiently retained by the application of electric field strength. There was a significant increase in the amount of polyphenols, catechins, and the total free amino acids of the samples by 7.5% after PEF treatment at 40 kV/cm. PEF treatment caused an increase in the total amino acids especially theanine, which is beneficial for the quality of commercial ready-to-drink green tea infusion products. While PEF treatment at 20 or 30 kV/cm had no significant effect on flavor compounds of green tea infusions, the total concentration of volatiles lost was approximately 10% after PEF treatment at 40 kV/cm for 200 μs (Zhao et al., 2009).

Beside fruit juices, different liquid foods having low viscosity and high acidity, drinks such as wine and beer, are successfully processed by PEF. Wine is a very delicate product as any technology applied to process wine adversely affects the flavor, color, total polyphenol index (TPI), and TAC. Evaluation of PEF processing at the end of alcoholic fermentation resulted on higher color intensity (CI), TPI, and TAC than unprocessed wine. PEF provided shorter maceration time for wine production, and CI, TPI and TAC of PEF-treated wine samples were 38%, 22%, and 11% higher than the unprocessed samples even after 4 months of aging in bottle. PEF-treated and unprocessed wines had no significant difference for the measured attributes (Puertolas et al., 2010a).

PEF processing with a maximum of 31 kV/cm at 40°C ± 2°C can be successfully used to process red wine samples with no significant change on pH, °Brix, TA, color (*L**, *a**, *b**, hue, and chroma), TMAC, TPSC, and TAC. Changes in these properties will also affect the sensory characteristics of the red wine. Since no significant difference was detected for these quality characteristics, sensory properties of the samples did not reveal any significant difference. Moreover, increased electric field strength and treatment temperature caused a significant inactivation on *E. coli* O157:H7, *Lactobacillus bulgaricus*, *Candida lipolytica*, *S. cerevisiae*, and *Hansenula anomala* cells (Abca, 2010).

Processing of beer, although it is a big challenge to be processed by PEF due to the amount of carbon dioxide present, was successfully performed. Inactivation induced by the PEF treatment of beer for the inactivation of natural flora and inoculated cultures of *Saccharomyces uvarum*, *Rhodotorula rubra*, *Lactobacillus plantarum*, *Pediococcus damnosus*, and *Bacillus subtilis* resulted in a significant decrease. However, there was a significant increase in the amount of Cr, Zn, Fe, and Mn ions in the beer samples resulting from the migration of electrode materials after PEF treatment leading to a statistically significant degradation in flavor and mouth feel (Evrendilek et al., 2004a).

Higher protein, carbohydrate, and fat content are not desirable for PEF processing due to poor electrical conductivity. Thus, it is sometimes a big challenge for PEF to be applied to high protein content foods such as milk and liquid whole egg (LWE). The application of 12.5 kV/cm high-intensity electric fields and 10 pulses in the order of milliseconds to process egg products caused the structure modification of egg white proteins and β-lactoglobulin concentrate, and both proteins were partially denatured (26%–40%). Denaturation temperatures of the β-lactoglobulin concentrate were reduced by approximately 4°C–5°C, and the thermo stability of egg white proteins was partially increased in addition to the increase in the gelation rate of β-lactoglobulin concentrate and the gelation rate of egg white at 63°C lowered by PEF treatment (Perez and Pilosof, 2004).

While LWE samples treated at 19 and 32 kV/cm electric field strength showed a similar foaming capacity as fresh untreated egg, thermal processing and PEF treatments of 37 kV/cm caused a substantial decrease in the foaming capacity of untreated liquid egg. It was further reported that

PEF processing caused less changes in the microstructure, and the lipoprotein matrix appeared to be less affect by the PEF than by heat treatment when compared to the control. PEF processing did not significantly affect both the water-soluble protein content of the LWE samples (19.5%–23.6% decrease) and the mechanical properties of the egg gels (up to 21.3% and 14.5% increase in hardness and cohesiveness, respectively); however, these properties were significantly affected by heat pasteurization. PEF-treated samples—compared to heat pasteurization—did not exhibit significant difference in the water holding capacity (WHC) (Marco-Moles et al., 2011).

PEF processing of LWE with 1.2 mL/s flow rate, 200 pps frequency, 2.12 μs pulse duration, 25 kV/cm electric field strength, and 250 μs total treatment time caused no significant change in viscosity, electrical conductivity, color, pH, and °Brix relative to control samples. The PEF processing resulted in 1 log cfu/mL reduction in *Salmonella* E*nteritidis* population in LWE. The PEF-treated samples were subjected to heat at 55°C for 3.5 min to inactivate the remaining bacteria without denaturing the LWE. The combination of PEF and heat treatments led to a 4.3 log cfu/mL reduction in *S.* E*nteritidis* population, and the PEF + 55°C-treated LWE samples presented significantly longer shelf life at 4°C as compared with the control and heat-treated samples (Hermawan et al., 2004).

PEF processing of ovalbumin solutions (2%, pH 7.0, 200 Ω cm) and dialyzed fresh egg white (pH 9.2, 200–250 Ω·cm) was performed with 50–400 exponential decay pulses, 27–33 kV/cm electric field strength, and 0.3 μs pulse width (at a capacitance of 20 nF) or 0.9 μs (at 80 nF) with processing temperature of 29°C. It was reported that while the four sulfhydryl groups of native ovalbumin did not react with 5,5′-dithiobis(2-nitrobenzoic acid) (DTNB), they became reactive immediately after pulse processing, indicating either partial protein unfolding or enhanced SH ionization. Dissipated energy provided the extent of SH reactivity, and 3.7 SH groups becoming reactive after 100 or 200 pulses at 31.5 kV/cm and 80 nF. Since only 0.79 or 0.2 SH group was found to remain reactive for 30 min or 8 h after pulse processing, SH reactivity was reversible. Before and after 15–30 min pulse processing, the fourth derivatives of UV spectra of ovalbumin were determined. As a result, it was concluded that electric pulses known to induce significant microbial inactivation did not cause notable changes in the proteins investigated (Fernandez-Diaz et al., 2000).

Thermal processing of milk—even though it is the most common technology— is not desired due to the changes in the physical properties and aroma compounds, and thus, the potential of PEF processing for milk processing is being searched. It was revealed that PEF processing of milk up to 40 kV/cm provided no apparent changes in physical and chemical properties in addition to no significant difference in sensory attributes between heat pasteurized and PEF-treated milk (Qin et al., 1995c). PEF processing (40 kV/cm, 30 pulses, and 2 μs pulse width using exponential decaying pulses) of raw skim milk (SM) (0.2% milk fat) provided shelf-life extension of 2 weeks at 4°C; however, combination of heat treatment (80°C for 6 s) followed by PEF (30 kV/cm, 30 pulses, and 2 μs pulse width) increased the shelf life up to 22 days, with 3.6 log cfu/mL total aerobic plate, and no viable coliform count (Fernandez-Molina et al., 2000, 2001). Similarly, PEF processing (40 kV/cm electric field strength) of milk (2% milk fat) extended the shelf life of the samples to 2 weeks at refrigeration temperature (Qin et al., 1995a).

PEF processing (30.76–53.84 kV/cm electric field strength and 12, 24, and 30 pulse numbers) of SM and whole milk (WM) in combination with heat treatment (20°C, 30°C, and 40°C) caused a decrease in solids nonfat (SNF) of SM samples. As treatment became stronger, similar behavior was observed for fat and protein content resulting in 0.18% and 0.17% decrease, respectively, under the strongest conditions. A 0.11% decrease in protein and an even higher decrease in fat content were obtained after 40°C heat treatment. Higher stability with minor variations was observed in PEF-treated milk samples at 4°C after 33 days of storage (Bermudez-Aguirre et al., 2011). PEF processing with 30–60 kV/cm electric field strength and 26–210 μs treatment time provided no significant difference in pH and titration acidity, although 8 log cfu/mL reductions in *E. coli* and *Pseudomonas fluorescens* and 3 log cfu/mL reductions in *Bacillus stearothermophilus* were obtained with 210 μs treatment time and 60 kV/cm electric field strength (Shin et al., 2007).

B group vitamins, cholecalciferol, tocopherol, ascorbic acid, vitamin A, and other vitamins were not adversely affected by 1.8 and 2.7 kV/mm electric field strength, and application of electric field strength up to 80 kV/cm did not significantly alter the fat content and protein integrity of milk samples (Deeth et al., 2007). Similarly, PEF processing of milk with 400 μs treatment time at 22.6 kV/cm electric field strength provided more ascorbic acid retention (93.4%), as well as water-soluble vitamins, such as thiamine, riboflavin, ascorbic acid, and fat-soluble vitamins, such as cholecalciferol and tocopherol, as compared to heat treatment (63°C for 30 min or 75°C for 15 s) (Bendicho et al., 2002a). However, Floury et al. (2006) reported that PEF processing at the range of 45–55 kV/cm electric field strength with 2.1–3.5 μs treatment time caused enhancement of coagulation properties and decrease in viscosity. Moreover, no adverse effect on the bovine milk IgG in a protein-enriched soymilk was observed by the PEF treatment with 35 kV/cm for 73 μs (Li et al., 2003). PEF processing with 36.7 kV/cm electric field strength and 40 pulses over a 25 min time period resulted in less flavor degradation and no chemical or physical changes in milk quality attributes for cheese making (Dunn, 1996).

Nonthermal approaches such as HHP, PEF and ultrasound (US) were tested and used to increase the retention of omega-3 in Queso fresco (QF), cheddar (C) and mozzarella (M) fortified with omega-3. Three stages of cheese making were evaluated for fortification, after milk pasteurization, during curdling, and salting. Better retention was observed with microencapsulated oil, after milk pasteurization (8.49 mg/g) in QF, during salting (8.69 mg/g) in C, and during curdling (2.69 mg/g) in M. PEF, and US achieved the highest retention (5.20–5.12 mg/g) in QF samples, whereas HHP was the best method (5.49 and 6.64 mg/g) in C and M samples, respectively. Higher weight (up to 19% more), increased moisture (5%), and increased pH (6.35) were observed in QF after sonication processing. Faster spoilage was observed in both QF and M samples during storage at 4°C, even though PEF was able to delay microbial growth in QF and HHP in M, consequently (Bermudez-Aguirre and Barbosa-Canovas, 2012).

Although most of the studies reported no significant changes in measured properties of milk samples, it is shown that depending on the severity of PEF processing conditions, some changes in physical properties of milk may be observed. It is shown that PEF processing (square wave pulses, electric field strength, and pulse width applications of 45 kV/cm/500 ns and 55 kV/cm/250 ns, with increasing pulse frequencies from 40 to 120 Hz and energy input varying from range 0 to 100 kJ/kg) caused a decrease in viscosity and an enhancement of coagulation properties in addition to effects on casein micelles (Floury et al., 2006).

Besides processing of milk samples by PEF, changes in the technological properties of milk samples processed by PEF were also reported. For example, PEF processing of milk with 36.7 kV/cm electric field strength and 40 pulses for more than 25 min treatment time used in cheese-making presented no chemical or physical changes in quality attributes and lesser amount of flavor degradation (Dunn, 1996). Cheese made from PEF-processed milk had better flavor profile, better textural properties such as hardness and springiness of the cheese, while the other textural attributes (i.e., adhesiveness and cohesiveness) remained unchanged as compared to cheddar cheese produced from heat pasteurization (63°C for 30 min) (Sepúlveda-Ahumada et al., 2000).

Yogurt-based products have gained popularity due to positive health effects; however, heat processing destroys beneficial bacteria while extending shelf life, thus alternatives are also being searched for yogurt-based products. Formulated yogurt-based products processed by mild heat treatment at 60°C for 30 s and 30 kV/cm electric field strength for 32 μs total treatment time significantly decreased the total viable aerobic bacteria and total mold and yeast of yogurt-based products during storage at both 4°C and 22°C, with no significant difference between the control and processed products in addition to no significant difference on color, pH, and °Brix (Yeom et al., 2004). Similarly, compared to unprocessed control samples, 60°C + PEF at processing of yogurt-based drink samples revealed no significant difference in the L^*, a^*, b^* values, °Brix, in addition to no significant difference in the selected sensory attributes (Evrendilek et al., 2004b). No changes in total solids, color, pH, proteins, moisture, particle size, and conductivity were detected in SM samples with PEF treatment (35 kV/cm field strength with 64 pulses of bipolar square wave for 188 μs), while 0.3–3.0 log cfu/mL reductions of *P. fluorescens*, *Lactococcus lactis* were obtained (Michalac et al., 2003).

Some studies include a combination of antimicrobial agents with PEF to enhance efficacy of processing. For example, PEF processing caused a maximum of 2.5 log cfu/mL reduction in endogenous bacteria by 30, 40, and 50 kV/cm electric field strength. Inactivation of the bacteria changed to 2, 2.7, or 3.4 log cfu/mL with the same PEF intensities and subsequent exposure to 10 IU nisin/mL (Calderon-Miranda, 1998).

Enzymes in milk systems are important for the determination of processing as well as the stability of processed product. Alkaline phosphatase treated by PEF at 22.3 kV/cm electric field strength with 0.78 ms pulse width showed a tendency to associate and aggregate. It is known that the polarization created by electrical charges of dipoles on the enzyme can cause the aggregate formation in enzymes; thus, proposed mechanism of the inactivation of alkaline phosphatase by PEF is the polarization of the molecule, which leads to the aggregation of the enzyme (Castro et al., 2001). Plasmin was inactivated by 90% with PEF processing at 45 kV/cm with 50 pulses in SMUF (Vega-Mercado et al., 1995). After application of PEF treatment at 21.5 kV/cm electric field strength with 400 kJ/L energy, 65%, 25%, and <5% inactivation of lipase, peroxidase, and alkaline phosphatase were obtained (Grahl and Markl, 1996; Castro et al., 2001).

Different food products were successfully processed by PEF, and it can be concluded that generally PEF treatment alone does not adversely affect the physical, chemical, and sensory properties of fruit juices and nectars as well as milk and low-viscosity dairy products. Even though it was reported in some studies that PEF provides better sensory properties and longer and/or equal shelf life as compared to heat processing, one of the main issues is the temperature increase during PEF processing; thus, it must be monitored during processing.

Inulin is widely used in different food products for fat replacement and calorie reduction (Villegas and Costell, 2007). Inulin, a kind of fructan, is present in various plants such as chicory root, dahlia tuber, Jerusalem artichoke, and burdock root (Milani et al., 2011). Among these inulin-containing plants, chicory root is considered as the most adapted plant for the industrial production of inulin. The influences of PEF parameters 600 V/cm for 10–50 ms treatment time and diffusion temperature (varied between 30°C and 80°C) on soluble matter extraction kinetics, inulin content of juice, and pulp exhaustion are investigated. The draft (liquid to solid mass ratio) was fixed at 140%, similar to the industrial conditions. PEF treatment facilitates extraction of inulin at conventional diffusion temperature (70°C–80°C), and the diffusion temperature can even be reduced by 10°C–15°C with comparable juice inulin concentration. Less energy consumption can be achieved by reducing PEF treatment duration to 10 ms, which is observed sufficient for effective extraction (Zhu et al., 2012).

Currently, greater attention has been paid to side effects occurring during PEF treatment and the influences on food qualities and food components such as investigation of the electrochemical reaction and oxidation of lecithin under PEF processing. Results showed that electrochemical reaction of NaCl solutions at different pH values occurred during PEF processing. Active chlorine, reactive oxygen, and free radicals were detected, which were related to the PEF parameters and pH values of the solution. Lecithin extracted from yolk was further selected to investigate the oxidation of food lipids under PEF processing, confirming the occurrence of oxidation of lecithin under PEF treatment. The oxidative agents induced by PEF might be responsible for the oxidation of extracted yolk lecithin. Moreover, this study found that vitamin C as a natural antioxidant could effectively quench free radicals and inhibit the oxidation of lipid in NaCl and lecithin solutions as model systems under PEF processing, representing a way to minimize the impact of PEF treatment on food qualities (Zhao et al., 2012).

13.5 PEF AND RECOVERY OF BIOACTIVE COMPOUNDS FROM BY-PRODUCTS AND EXTRACTION PROCESS

The use of PEF for the recovery of bioactive compounds from by-products is not well studied up to now. The technique has been mainly used to extract polyphenols from grape by-products. In red grape by-product, the level of anthocyanins was increased by 60% when PEF was applied as a pretreatment of 1 min at 25°C in combination with a conventional thermal extraction at 70°C for 1 h (Corrales et al., 2008). When white grape skins were treated with PEF at a temperature of 20°C, 10% more polyphenols were extracted (Boussetta et al., 2009).

It has been demonstrated that the combination of PEF-assisted extraction and pressing is an effective technique to obtain different products from solid plant matrices. Several studies have shown the positive effect of this combined treatment on the yield and the quality of the juice extracted from apples and carrots (El-Belghiti et al., 2005; Schilling et al., 2007; Grimi et al., 2009; Jaeger et al., 2012) and in the extraction yield of betanines from red beet (Lopez et al., 2009).

Fresh juice or citrus-based drinks are obtained by processing of orange fruits. This generates very large amounts of byproduct wastes such as orange peels. These peels are a rich source of polyphenols such as flavonoids. Extraction of these compounds from orange peels is a crucial step for use of these compounds in the food and pharmaceutical industries as antioxidants. PEF-assisted extraction by pressing of polyphenols from fresh orange peels stands as an economical and environmentally friendly alternative to conventional extraction methods which require the product to be dried, use large amounts of organic solvents, and need long extraction times (Luengo et al., 2013).

The influence of PEF treatment on the extraction by pressing of total polyphenols and flavonoids (naringin and hesperin) from orange peel was investigated by Luengo et al. (2013). A treatment time of 60 μs (20 pulses of 3 μs) achieved the highest cell disintegration index (Z_p) at the different electric field strengths tested. After 30 min of pressurization at 5 bars, the total polyphenol extraction yield (TPEY) increased 20%, 129%, 153%, and 159% for orange peel treated by PEF at 1, 3, 5, and 7 kV/cm, respectively. PEF treatment of orange peels at 5 kV/cm increased the quantity of naringin and hesperidin in 100 g of orange peels extract from 1 to 3.1 mg/100 g of fresh weight (fw) orange peel and from 1.3 to 4.6 mg/100 g fw orange peel, respectively. Compared to the untreated sample, PEF treatments of 1, 3, 5, and 7 kV/cm increased the antioxidant activity of the extract by 51%, 94%, 148%, and 192%, respectively.

PEF can enhance the efficiency of mass transfer of water or of valuable compounds from biological matrices during extraction, drying, or diffusion processes (Donsi et al., 2010; Hou et al., 2010) and enhance the chemical reactions, such as Maillard reaction and ethanol–acetic acid esterification. Thus, the kinetics in color formation, pH decline, and antioxidant activity as assessed by DPPH radical in model Maillard reaction induced by PEF was investigated glucose–glycine aqueous solutions (pH 9.0) subjected to PEF treatment at different electric intensities (of 10–50 kV/cm, respectively) and treatment times (of 0.8–4.0 ms, respectively). Results showed that color formation and DPPH radical scavenging ability followed zero-order and first-order kinetics; however, glucose consumption and pH decline followed adequately first-order kinetics. These results demonstrated that first-order kinetics was a much more reasonable model to fit the PEF promoting Maillard reaction (Wang et al., 2013). This study reinforces the idea that PEF might be a good candidate for controlling Maillard reaction in food processing (Wang et al., 2011; Lin et al., 2012).

The electrical treatment for solid–liquid extraction was also investigated on a batch pilot plant scale. The PEF treatment (E = 2 kV/cm, specific energy input W_{spec} = 6 kJ/kg) of mash obtained from the "Jona Gold" and "Royal Gala" apple varieties, followed by juice expression with a Hollmann baling press (50 kg of treated apples), resulted in a 14% and 5% increase in juice yield, respectively. The average size of apple mash was in the range of 2 cm (Toepfl, 2006). In contrast, the PEF treatment (E = 3 kV/cm, W_{spec} = 10 kJ/kg) of mash from a blend of six cider apple varieties followed by expression with a horizontal HPL 200 filter-press (220 kg of treated apples) did not enhance juice yield. Juice release amounted to an average of 84.9% ± 0.4% for the control and treated samples (Schilling et al., 2008b).

PEF treatment (E = 1000 V/cm, f = 200 Hz and t_p = 100 μs) was applied to French cider apple mash pumped into a collinear treatment chamber at the flow rate of 280 kg/h. Juices were recovered continuously using a single belt press. PEF treatment of the mash (32 ms and 46 kJ/kg) increased the juice yield by 4.1%. The content of total native polyphenols decreased by 17.8% in the treated juice due to oxidation by PPO. Meanwhile, the activity of this enzyme in PEF-treated juices was also decreased by 18.3%. It was suggested that PEF treatment enhanced the oxidation of native polyphenol compounds in cells because of electroporation of the inner cell membrane. The loss of PPO activity was related to the inhibition of the enzyme by the oxidized phenolic compounds. For this reason, one of the oxidation markers (molecular ion [M–H] – of chlorogenic acid dimer m/z = 705) was monitored in the juices, and it was observed that electric treatment caused 6.8% increase in the amount of this compound in the treated samples. A significant difference in the color (DE = 6) in the L^*, a^*, and b^* space was detected between the control and treated juices. The color of the treated juice was the most appreciated attribute among the sensorial panel when compared to the control. The overall chemical composition of the treated juices was not different when compared with the respective controls (Turk et al., 2012).

PEF pretreatment induced a reduction in drying time of up to 12% when 10 kV/cm and 50 pulses were applied. For instance, after 60 min of drying, the dimensionless moisture ratio for PEF-treated (10 kV/cm, 50 pulses) samples was 0.18 compared to 0.26 for the untreated apples. The effective moisture diffusivity, calculated on the basis of Fick's second law, was 1.04×10^{-9} m^2/s for intact samples and from 1.09×10^{-9} to 1.25×10^{-9} m^2/s for PEF-treated samples with 10 pulses at 5 kV/cm and 50 pulses at 10 kV/cm, respectively (Wiktor et al., 2013).

PEF technology has been studied either as a way to extract nutritional components leading to wines with higher nutritive content or as a pressing aid to increase yield (Lopez et al., 2008a,b; Grimi et al., 2009; Puertolas et al., 2010b). Others have studied the technology as a preservation tool reporting the effects of PEF treatments on specific components such as fatty and amino acids (Garde-Cerdan et al., 2007), microorganisms (Jaya et al., 2004; Wu et al., 2005; Marselles-Fontanet et al., 2009), and enzymes (Marselles-Fontanet and Martin-Belloso, 2007).

Several studies have shown very effective separation of cellular content in microfluidic devices using PEF (McClain et al., 2003; Gao et al., 2004). Both studies were focussed on single cell analysis, by immobilizing the cells in a micro device and in that way separating the soluble compounds from the cells. To analyze intracellular content this technique is very effective, however, to extract the content of cells on an industrial scale immobilization will not be efficient.

Microalgae are a promising source for proteins, lipids, and carbohydrates for the food/feed and biofuel industry. In comparison with soya and palm oil, microalgae can be grown in a more efficient and sustainable way. To make microalgae production economically feasible, it is necessary to optimally use all produced compounds. To accomplish this, focus needs to be put on biorefinery techniques which are mild and effective. Of the techniques described, PEF seems to be the most developed technique as compared to other cell disruption applications. For separation technology, ionic liquids seem most promising as they are able to separate both hydrophobic and hydrophilic compounds. But additional studies need to be evolved in the coming years to investigate their relevance as novel cell disruption and separation methods (Vanthoor-Koopmans et al., 2013). PEF has been successfully applied to recover lipids from microalgae at low-energy intensities, and it was reported the extraction solvent could more easily reach the lipids by PEF treatment. Thus, it was suggested as a suitable technique to be used in large-scale processes (Sheng et al., 2012). In another study, electroporation was caused using PEF in *Auxenochlorella protothecoides*, and the spontaneous release of intracellular products was studied after electroporation (Gottel et al., 2011). It was found that 10%–15% of the initial biomass could be found in the supernatant after centrifugation; however, high molecular compounds were not released from the cells. The increase in dissolved organic compounds in the supernatant after electroporation was mostly caused by small molecules like carbohydrates and amino acids.

Overall, PEF treatment shows to be a very promising technique useful for cell perforation at large scale. The energy consumption of PEF is much lower than for the conventional techniques (de Boer et al., 2012).

Different examples of acceleration of the mass transfer processes (Donsi et al., 2010; Porras-Parral et al., 2012), pressing (Lebovka et al., 2004), drying (Ade-Omowaye et al., 2001), freezing (Jalte et al., 2009a), extraction (Loginova et al., 2011a,b), and osmotic dehydration (Amami et al., 2008), were already reported.

PEF (100–1000 V electric field strength, 100 μs pulse duration) processing of potato slices for disintegration treated between parallel-plate electrodes by using a laboratory compression chamber equipped with a PEF-treatment system. The apparent density of slices (bed of particles), ρ, varied within 0.313 and 1 g/cm^3. The electrical conductivity σ (ρ) of the packing of slices versus volume fraction of particles φ was approximated by the percolation law $\sigma \infty (\varphi - \varphi_c)t$, where $\varphi_c \approx 0.290$, $t = t_i \approx 0.46$ and $t = t_d \approx 1.39$ for the intact and completely damaged tissues, respectively. The impact of electric field strength and apparent density of slices on PEF-induced damage kinetics was studied. The more accelerated kinetics of damage was observed for more dense packing of slices. The approximated relation between the applied, E, and effective, E_e, electric field strengths accounting for the σ (ρ) dependence was derived (Mhemdi et al., 2013). In this study, research toward the electroporation efficiency of PEF applied to the porous packing of sliced food particles is provided.

PEF treatment leads to pore formation in cell membrane, and thus, modifies diffusion of intra- and extracellular media. It is also suggested that PEF pretreatment prior to freezing could reduce shrinkage and help maintain the color and shape uniformity (Jalte et al., 2009b). PEF processing was utilized to prepare low-fat content potato chips, which also illustrated the utility of PEF in different food processing operations (Janositz et al., 2011).

The combination of PEF (0.5 kV/cm, 100 pulses, 4 Hz), texturizing, and antifreeze agents were applied to determine the effects of these technologies on quality retention of defrosted potato strips (10 mm thickness, 100 g) placed in different solutions (1%, w/v) of $CaCl_2$, glycerol, trehalose as well as NaCl and sucrose. Then, all the samples were soaked in the same solution for 10 min. After draining, samples were packed into polypropylene pouches and stored at −18°C for 12 h. Samples were thawed at room temperature (20°C) for 3 h. Untreated controls and PEF-treated control samples were also frozen and thawed in similar conditions. To assess the potato strip quality, the thawed samples were analyzed for moisture content, weight loss, firmness, and color attributes. The results indicate that PEF treatment by itself is not a suitable pretreatment method for frozen potato strips and should be assisted by $CaCl_2$ and trehalose treatment to prevent softening after defrosting. Firmness analyses determined that application of PEF alone results in 2.38 N. However, PEF in combination with $CaCl_2$ and trehalose results in 2.97 and 2.99 N, respectively, which are both significantly firmer than the samples solely treated with PEF. $CaCl_2$ and trehalose were effective in not only maintaining the structural integrity of the cells but also retaining color attributes. The L^* value was found to be higher in $CaCl_2$ and trehalose treated samples (58.95 and 57.21, respectively), as compared to PEF-treated samples (53.97) denoting a darker color. Application of $CaCl_2$ and trehalose in combination with PEF also resulted in significantly less weight loss after thawing (Shayanfar et al., 2013).

13.6 CURRENT STATUS OF THE TECHNOLOGY

Studies conducted with PEF processing are mostly accomplished with bench-scale systems. However, the pilot-plant systems constructed both in Washington State University and Ohio State University also add much information to literature regarding this technology. With the guidance of lab-scale as well as pilot-scale studies, industrial-scale PEF systems are currently developed (Puertolas et al., 2010a).

It is known that the capacity of PEF systems based on 80 kW modules are extended to the systems with an average power of up to 240 kW and treatment capacity of 10,000 L/h for microbial inactivation (Toepfl, 2011). Although previous reports indicated that as compared to heat treatment PEF processing is costly due to the initial cost of equipment, recent studies revealed that energy saving with PEF processing is also an important advantage compared with conventional thermal treatment due to the estimated total treatment costs, which is in a range of 1–2 €/L of material to be treated for cell disintegration and 0.01–0.02 €/L of liquid media for preservation (Toepfl, 2011).

Commercial scale PEF processing systems have also been implemented by Genesis Juices (Oregon, USA) and, in 2004, these systems were developed, based on the patented technology, into commercial-scale systems. Besides Genesis Juices, Pure Pulse Technologies developed a commercial-scale PEF processing system with a capacity of 600, 1200, or 1800 L/h, maximum power of 16, 30, and 50 kW with 1–4 μs pulse duration, and 20–40 kV/cm electric field with 5°C–15°C temperature increase in the treatment chamber (Anonymous, 2014). PEF processing systems developed by Pure Pulse technologies are designed basically for fruit juices (Shamsi and Sherkat, 2009). Recent attempts have been made by the DIL system to preserve heat-sensitive liquids such as fruit juices, preparations, beer, cocktail premixes, salsas, or plasmas. It was reported that the newly developed PEF systems were defined as cost-effective, and these systems provided short-time cell disintegration, improvement of mass transfer processes, and microbial inactivation (Anonymous, 2013). Even though the developments in PEF technology are very promising, commercial use of this technology requires more efforts from the food industry to replace heat treatment or to be used in combination with heat treatment.

ACKNOWLEDGMENT

Some data presented by the first author were collected in the framework of the project supported by Turkish Prime Ministry State Planning Organization (DPT 2009 K 120 140).

REFERENCES

Abca, E. E. 2010. Effects of pulsed electric fields on quality characteristics of red wine (in Turkish). Ms.C. thesis. Abant Izzet Baysal University, Bolu, Turkey.

Ade-Omowaye, B. I. O., Angersbach, A., Taiwo, K. A., and Knorr, D. 2001. Use of pulsed electric field pretreatment to improve dehydration characteristics of plant based foods. *Trends in Food Science and Technology*. 12(8):285–295.

Agcam, E., Akyıldız, A., and Evrendilek, G. A. 2014. Comparison of phenolic compounds of orange juice processed by pulsed electric fields (PEF) and conventional thermal pasteurisation. *Food Chemistry*. 15:354–361.

Agcam, E. 2012. Effects of pulsed electric fields and heat treatments on quality characteristics and shelf-life of orange juice (in Turkish). MSc thesis. Cukurova University, Adana, Turkey.

Aguilar-Rosas, S. F., Ballinas-Casarrubias, M. L., Efias-Ogaza, L. R., Martin-Belloso, O., and Ortega-Rivas, E. 2013. Enzyme activity and colour changes in apple juice pasteurized thermally and by pulsed electric fields. *Acta Alimentaria*. 42(1):45–54.

Aguilar-Rosas, S. F., Ballinas-Casarrubias, M. L., Nevarez-Moorillon, G. V., Martin-Belloso, O., and Ortega-Rivas, E. 2007. Thermal and pulsed electric fields pasteurization of apple juice: Effects on physicochemical properties and flavour compounds. *Journal of Food Engineering*. 83(1):41–46.

Aguilo-Aguayo, I., Soliva-Fortuny, R., and Martin-Belloso, O. 2008. Comparative study on color, viscosity and related enzymes of tomato juice treated by high-intensity pulsed electric fields or heat. *European Food Research and Technology*. 227(2):599–606.

Aguiló-Aguayo, I., Soliva-Fortuny, R., and Martin-Belloso, O. 2010. Optimizing critical high-intensity pulsed electric fields treatments for reducing pectolytic activity and viscosity changes in watermelon juice. *European Food Research and Technology*. 231:509–517.

Akin, E. and Evrendilek, G. A. 2009. Effect of pulsed electric fields on physical, chemical and microbiological properties of formulated carrot juice. *Food Science and Technology International*. 15:275–282.

Altuntas, J., Evrendilek, G. A., Sangun, M. K., and Zhang, H. Q. 2010. Effects of pulsed electric field processing on the quality and microbial inactivation of sour cherry juice. *International Journal of Food Science and Technology*. 45:899–905.

Altuntas, J., Evrendilek, G. A., Sangun, M. K., and Zhang, H. Q. 2011. Processing of peach nectar by pulsed electric fields with respect to physical and chemical properties and microbial inactivation. *Journal of Food Process Engineering*. 34:1506–1522.

Alvarez, I., Pagan, R., Condon, S., and Raso, J. 2003a. The influence of process parameters for the inactivation of *Listeria monocytogenes* by pulsed electric fields. *International Journal of Food Microbiology*. 87:87–95.

Alvarez, I., Raso, J., Sala, F. J., and Condon, S. 2003b. Inactivation of *Yersinia enterocolitica* by pulsed electric fields. *Food Microbiology*. 20:691–700.

Amami, E., Khezami, L., Vorobiev, E., and Kechaou, N. 2008. Effect of pulsed electric field and osmotic dehydration pretreatment on the convective drying of carrot tissue. *Drying Technology*. 26(2):231–238.

Amiali, M., Ngadi, M. O.,Raghavan, V. G. S., and Smith, J. P. 2004. Inactivation of *Escherichia coli* O157:H7 in liquid dialyzed egg using pulsed electric fields. *Food and Bioproducts Processing*. 82(2):151–156.

Anonymous. 2013. Food for life. http://www.wagralim.be (accesed on September 2013).

Anonymous. 2014. How does PurePulse work. http://www.purepulse.eu (accesed on December 2013).

Aronsson, K., Ronner, U., and Borch, E. 2005. Inactivation of *Escherichia coli*, *Listeria innocua* and *Saccharomyces cerevisiae* in relation to membrane permeabilization and subsequent leakage of intracellular compounds due to pulsed electric field processing. *International Journal of Food Microbiology*. 99(1):19–32.

Barba, F. J., Jäger, H., Meneses, N., Esteve, M. J., Frígola, A., and Knorr, D. 2012. Evaluation of quality changes of blueberry juice during refrigerated storage after high-pressure and pulsed electric fields processing. *Innovative Food Science and Emerging Technologies*. 14:18–24.

Barbosa-Cánovas, G. V. and Bermúdez-Aguirre, D. 2011. Introduction. In H. Q. Zhang, G. V. Barbosa-Cánovas, V. M. Balasubramaniam, C. P. Dunne, D. F. Farkas, and J. T. C. Yuan (eds.), *Nonthermal Processing Technologies for Food*. pp. 20–30. Wiley-Blackwell, Oxford, U.K.

Barbosa-Canovas, G. V., Gongora-Nieto, M. M., Pothakamury, U. R., and Swanson, B. G. 1999. *Preservation of Foods with Pulsed Electric Fields*. Academic Press, San Diego, CA.

Barbosa-Canovas, G. V., Pothakamury, U. R., Palou, E., and Swanson, B. G. 1998. *Non-thermal Preservation of Foods*. pp. 9–48, 53–110, 113–136, 139–159. Marcel Dekker, New York.

Barsotti, L. and Cheftel, J. C. 1999. Food processing by pulsed electric fields 2: Biological aspects. *Food Reviews International*. 15:181–213.

Bazhal, M. I., Ngadi, M. O., Raghavan, G. S. V., and Smith, J. P. 2006. Inactivation of *Escherichia coli* O157:H7 in liquid whole egg using combined pulsed electric field and thermal treatments. *LWT—Food Science and Technology*. 39:419–425.

Beattie, J. M. 1915. Report on the electrical treatment of milk to the city of Liverpool. Tinling and Co., Liverpool, U.K.

Beattie, J. M. and Lewis, F. C. 1925. The electric current (apart from the heat generated). Abacteriological agent in the sterilization of milk and other fluids. *Journal of Hygiene*. 24:123–137.

Bendicho, S., Barbosa-Canovas, G. V., and Martin, O. 2003. Reduction of protease activity in milk by continuous flow high-intensity pulsed electric field treatments. *Journal of Dairy Science*. 86:697–703.

Bendicho, S., Barbosa-Canovas, G. V., and Martin, O. 2002a. Milk processing by high intensity pulsed electric fields. *Trends in Food Science and Technology*. 13:195–204.

Bendicho, S., Estela, C., Fernadez-Molina, J. J., Barbosa-Canovas, G. V., and Martin, O. 2002b. Effect of high intensity pulsed electric field and thermal treatments on a lipase from *Pseudomonas fluorescens*. *Journal of Dairy Science*. 85:19–27.

Bermudez-Aguirre, D. and Barbosa-Canovas, G. V. 2012. Fortification of queso fresco, cheddar and mozzarella cheese using selected sources of omega-3 and some nonthermal approaches. *Food Chemistry*. 133:787–797.

Bermudez-Aguirre, D., Fernandez, S., Esquivel, H., Dunne, P. C., and Barbosa-Canovas, G. V. 2011. Milk processed by pulsed electric fields: Evaluation of microbial quality, physicochemical characteristics, and selected nutrients at different storage conditions. *Journal of Food Science*. 76:289–299.

Bi, X., Liu, F., Rao, L., Li, J., Liu, B., Liao, X., and Wu, J. 2013. Effects of electric field strength and pulse rise time on physicochemical and sensory properties of apple juice by pulsed electric field. *Innovative Food Science and Emerging Technologies*. 17:85–92.

Boussetta, N., Lebovka, N., Vorobiev, E., Adenier, H., Bedel-Cloutour, C., and Lanoisellé, J.-L. 2009. Electrically assisted extraction of soluble matter from Chardonnay grape skins for polyphenol recovery. *Journal of Agricultural and Food Chemistry*. 57:1491–1497.

Calderon-Miranda, M. L. 1998. Inactivation of *Listeria inocua* by pulsed electric fields and nisin. PhD thesis. Washington State University, Pullman, WA.

Calderon-Miranda, M. L., Barbosa-Canovas, G. V., and Swanson, B. G. 1999. Transmission electron microscopy of *Listeria innocua* treated by pulsed electric fields and nisin in skimmed milk. *International Journal of Food Microbiology*. 51(1):31–38.

Castro, A. J. 1994. Pulsed electric field modification of activity and denaturation of alkaline phosphatase. PhD thesis. Washington State University, Pullman, WA.

Castro, A. J., Swanson, B. G., Barbosa-Canovas, G. V., and Zhang, Q. H. 2001. Pulsed electric field modification of milk alkaline phosphatase activity. In G. V. Barbosa-Canovas and Q. H. Zhang (eds.), *Pulsed Electric Fields in Food Processing. Fundamental Aspects and Applications*, pp. 65–82. Technomic Publishing Company Inc., Lancaster, PA.

Charles-Rodríguez, A. V., Nevárez-Moorillón, G. V., Zhang, Q. H., and Ortega-Rivas, E. 2007. Comparison of thermal processing and pulsed electric fields treatment in pasteurization of apple juice. *Food Bioproducts Process Engineering*. 85(2):93–97.

Corrales, M., Toepfl, S., Butz, P., Knorr, D., and Tauscher, B. 2008. Extraction of anthocyanins from grape by-products assisted by ultrasonics, high hydrostatic pressure or pulsed electric fields: A comparison. *Innovative Food Science & Emerging Technologies*. 9:85–91.

Cortes, C., Esteve, M. J., and Frigola, A. 2008. Color of orange juice treated by high intensity pulsed electric fields during refrigerated storage and comparison with pasteurized juice. *Food Control*. 19 (2):151–158.

Cserhalmi, Zs, Sass-Kiss, A., Tóth-Markus, M., and Lechner, N. 2006. Study of pulsed electric field treated citrus juices. *Innovative Food Science and Emerging Technologies*. 7(1–2):49–54.

Cueva, O. and Aryana, K. J. 2012. Influence of certain pulsed electric field conditions on the growth of *Lactobacillus acidophilus* LA-K. *Journal of Microbial and Biochemical Technology*. 4(7):137–140.

Dave, A., Walkling-Ribeiro, M., Rodríguez-González, O., Griffiths, M. W., and Corredig, M. 2012. Effect of PEF and UV and their combination on selected microorganisms and physico-chemical properties in whey. *Journal of Animal Science Suppl. 90(3)/Journal of Dairy Science Supplement*. 95(2):168.

de Boer, K., Moheimani, N. R., Borowitzka, M. A., and Bahri, P. A. 2012. Extraction and conversion pathways for microalgae to biodiesel: A review focused on energy consumption. *Journal of Applied Phycology*. 24(69):1681–1698.

Deeth, H. C., Datta, N., Ross, A. I. V., and Dam, X. T. 2007. Pulsed electric field technology: Effect on milk and fruit juices. In G. Tewari and V. K. Juneja (eds.), *Advances in Thermal and Non-Thermal Food Preservation*. pp. 241–279. Blackwell Publishing, Ames, IA.

Donsi, F., Ferrari, G., and Pataro, G. 2010. Applications of pulsed electric field treatments for the enhancement of mass transfer from vegetable tissue. *Food Engineering Reviews*. 2(2):109–130.

Dunn, J. 1996. Pulsed light and pulsed electric field for foods and eggs. *Poultry Science*. 75(9):1133–1136.

Dunn, J. E. and Pearlman, J. S. 1987. Methods and apparatus for extending the shelf life of fluid food products. Maxwell Laboratories, Inc. U.S. Patent 4,695,472.

Dutreux, N., Notermans, S., Wijtzes, T., Gongora-Nieto, M. M., Barbosa-Canovas, G. V., and Swanson, B. G. 2000. Pulsed electric fields inactivation of attached and free-living *Escherichia coli* and *Listeria innocua* under several conditions. *International Journal of Food Microbiology*. 54:91–98.

El-Belghiti, K., Rabhi, Z., and Vorobiev, E. 2005. Kinetic model of sugar diffusion from sugar beet tissue treated by pulsed electric field. *Journal of the Science of Food and Agriculture*. 85(2):213–218.

Elez-Martinez, P., Aguilo-Aguayo, I., and Martin-Belloso, O. 2006. Inactivation of orange juice peroxidase by high-intensity pulsed electric fields as influenced by process parameters. *Journal of the Science of Food and Agriculture*. 86(1):71–81.

Elez-Martinez, P., Escola-Hernandez, J., Soliva-Fortuny, R. C., and Martin-Belloso, O. 2004. Inactivation of *Saccharomyces cerevisiae* suspended in orange juice using high-intensity pulsed electric fields. *Journal of Food Protection*. 67:2596–2602.

Elez-Martinez, P., Escola-Hernandez, J., Soliva-Fortuny, R. C., and Martin-Belloso, O. 2005. Inactivation of *Lactobacillus brevis* in orange juice by high-intensity pulsed electric fields. *Food Microbiology*. 22(4):311–319.

El-Hag, A. H., Jayaram, S. H., and Griffiths, M. W. 2006. Inactivation of naturally grown microorganisms in orange juice using pulsed electric fields. *IEEE Transactions on Plasma Science*. 34(4):1412–1415.

Evrendilek, G. A. 2009. Pulsed electric field processing of pomegranate juice. *International Conference on Bio and Food Electrotechnologies*, 22–23 October 2009, Compiegne, France.

Evrendilek, G. A., Altuntas, J., Sangun, M. K., and Zhang, H. Q. 2013. Apricot nectar processing by pulsed electric fields. *International Journal of Food Properties*. 16(1):216–227.

Evrendilek, G. A., Jin, Z. T., Ruhlman, K. T., Qiu, X., Zhang, Q. H., and Richter, E. R. 2000. Microbial safety and shelf-life of apple juice and cider processed by bench and pilot scale PEF systems. *Innovative Food Science and Emerging Technologies*. 1:77–86.

Evrendilek, G. A., Li, S., Dantzer, W. R., and Zhang, Q. H. 2004a. Pulsed electric field processing of beer: Microbial, sensory and quality analyses. *Journal of Food Science*. 69(8):228–232.

Evrendilek, G. A., Tok, F. M., Soylu, E. M., and Soylu, S. 2008. Inactivation of *Penicillum expansum* in sour cherry juice, peach and apricot nectars by pulsed electric fields. *Food Microbiology*. 25:662–667.

Evrendilek, G. A., Tok, F. M., Soylu, E. M., and Soylu, S. 2009. Inactivation of *Botyritis cinerea* in sour cherry juice, peach and apricot nectars by pulsed electric fields. *Italian Journal of Food Science*. 21(2):171–182.

Evrendilek, G. A., Yeom, H. W., Jin, Z. T., and Zhang, Q. H. 2004b. Safety and quality evaluation of yogurt-based drink processed by pilot plant PEF system. *Journal of Food Process Engineering*. 27(3):197–212.

Evrendilek, G. A., Zhang, Q. H., and Richter, E. R. 1999. Inactivation of *Escherichia coli* O157:H7 and *Escherichia coli* 8739 in apple juice by pulsed electric fields. *Journal of Food Protection*. 62:793–796.

Evrendilek, G. A., Zhang, Q. H., and Richter, E. R. 2004c. Application of pulsed electric fields to skim milk inoculated with *Staphylococcus aureus*. *Biosystems Engineering*. 87:137–144.

Fernandez-Diaz, M. D., Barsotti, L., Dumay, E., and Cheftel, J. C. 2000. Effects of pulsed electric fields on ovalbumin solutions and dialyzed egg white. *Journal of Agriculture and Food Chemistry*. 48(6):2332–2339.

Fernandez-Molina, J. J., Barbosa-Canovas, G. V., and Swanson, B. G. 2000. Pasteurization of skim milk with heat and pulsed electric fields. In *IFT Annual Meeting Technical Programme*. Institute of Food Technologists, Dallas, TX.

Fernandez-Molina, J. J., Barbosa-Canovas, G. V., and Swanson, B. G. 2001. The combined effect of pulsed electric fields and conventional heat on the microbial quality and shelf-life of skim milk. In *Annual Meeting, Book of Abstracts*. Session 59H-18. Institute of Food Technologists.

Fiala, A., Wouters, P. C., van den Bosch, H. F. M., and Creyghton, Y. L. M. 2001. Coupled electrical-fluid model of pulsed electric field treatment in a model food system. *Innovative Food Science and Emerging Technology*. 2:229–238.

Fleischman, G. J., Ravishankar, S., and Balasubramaniam, V. M. 2004. The inactivation of *Listeria monocytogenes* by pulsed electric field (PEF) treatment in a static chamber. *Food Microbiology*. 21:91–95.

Floury, J., Grosset, N., Leconte, N., Pasco, M., Madec, M.-N., and Jeantet, R. 2006. Continuous raw skim milk processing by pulsed electric field at non-lethal temperature: Effect on microbial inactivation and functional properties. *Lait*. 86(1):43–57.

Fox, M. B., Esveld, D. C., and Boom, R. M. 2007. Conceptual design of a mass parallelized PEF microreactor. *Trends in Food Science and Technology*. 18:484–491.

Gao, J., Yin, X.-F., and Fang, Z.-L. 2004. Integration of single cell injection, cell lysis, separation and detection of intracellular constituents on a microfluidic chip. *Lab on a Chip*. 4(1):47–52.

Garcia, D., Gomez, N., Manas, P., Condon, S., Raso, J., and Pagan, R. 2005. Occurrence of sublethal injury after pulsed electric fields depending on the micro-organism, the treatment medium pH and the intensity of the treatment investigated. *Journal of Applied Microbiology*. 99(1):94–104.

Garde-Cerdan, T., Arias-Gil, M., Marsellés-Fontanet, A. R., Ancín-Azpilicueta, C., and Martín-Belloso, O. 2007. Effects of thermal and non-thermal processing treatments on fatty acids and free amino acids of grape juice. *Food Control*. 18(5):473–479.

Getchell, B. E. 1935. Electric pasteurization of milk. *Agricultural Engineering*. 16:408–410.

Giner, J., Gimeno, V., Barbosa-Canovas, G. V., and Martin, O. 2001. Effects of pulsed electric field processing on apple and pear polyphenoloxidases. *Food Science and Technology International*. 7:339–345.

Giner, J., Gimeno, V., Espachs, A., Elez, P., Barbosa-Canovas, G. V., and Martin, O. 2000. Inhibition of tomato (*Licopersicon esculentum* Mill.) pectin methylesterase by pulsed electric fields. *Innovative Food Science and Emerging Technologies*. 1:57–67.

Gomez, N., Garcia, D., Alvarez, I., Raso, J., and Condon, S. 2005. A model describing the kinetics of inactivation of *Lactobacillus plantarum* in a buffer system of different pH and in orange and apple juice. *Journal of Food Engineering*. 70:7–14.

Gongora-Nieto, M. M., Sepulveda, D. R., Pedrow, P., Barbosa-Canovas, G. V., and Swanson, B. G. 2002. Food processing by pulsed electric fields: Treatment delivery, inactivation level, and regulatory aspects. *LWT—Food Science and Technology* . 35(5):375–388.

Gottel, M., Eing, C., Gusbeth, C., Frey, W., and Muller, G. 2011. Influence of pulsed electric field (PEF) treatment on the extraction of lipids from the microalgae *Auxenochlorella protothecoides*. In *Proceedings of the International Conference on Plasma Science*. Chicago, IL, pp. 1–11.

Grahl, T., and Markl, H. 1996. Killing of microorganisms by pulsed electric fields. *Applied Microbiology and Biotechnology*. 45:148–157.

Grimi, N., Lebovka, N. I., Vorobiev, E., and Vaxelaire, J. 2009. Effect of a pulsed electric field treatment on expression behavior and juice quality of chardonnay grape. *Food Biophysics*. 4(3):191–198.

Guerrero-Beltran, J. Á.,Sepulveda, D. R., Gongora-Nieto, M. M., Swanson, B., and Barbosa-Cánovas, G. B. 2010. Milk thermization by pulsed electric fields (PEF) and electrically induced heat. *Journal of Food Engineering*. 100(1):56–60.

Gusbeth, C., Frey, W., Volkmann, H., Schwartz, T., and Bluhm, H. 2009. Pulsed electric field treatment for bacteria reduction and its impact on hospital wastewater. *Chemosphere*. 75:228–233.

Hall, C. W. and Trout, G. M. 1968. *Milk Pasteurisation*. AVI, Van Nostrand Reinhold, NY.

Han, J. H. 2007. *Packaging for Nonthermal Processing of Food*, pp. 3–16. John Wiley & Sons, New York.

Harrison, S. L., Barbosa-Canovas, G. V., and Swanson, B. G. 1997. *Saccharomyces cerevisiae* structural changes induced by pulsed electric field treatment. *LWT—Food Science and Technology*. 30(3):236–240.

Haughton, P. N., Lyng, J. G., Cronin, D. A., Morgan, D. J., Fanning, S., and Whyte, P. 2012. Efficacy of pulsed electric fields for the inactivation of indicator microorganisms and foodborne pathogens in liquids and raw chicken. *Food Control*. 25:131–135.

Heinz, V., Toepfl, S., and Knorr, D. 2003. Impact of temperature on lethality and energy efficiency of apple juice pasteurization by pulsed electric fields treatment. *Innovative Food Science and Emerging Technologies*. 4:167–175.

Hermawan, N., Evrendilek, G. A., Dantzer, W. R., Zhang, Q. H., and Richter, E. R. 2004. Pulsed electric field treatment of liquid whole egg inoculated with *Salmonella enteritidis*. *Journal of Food Safety*. 24:71–85.

Hitit, B. 2011. Effects of pulsed electric fields on the quality of fruit juices (in Turkish). MSc thesis. Abant Izzet Baysal University, Bolu, Turkey.

Ho, S. Y., Mittal, G. S., Cross, J. D., and Griffiths, M. W. 1995. Inactivation of *Pseudomonas fluorescens* by high voltage electric pulses. *Journal of Food Science*. 60:1337–1340, 1343.

Hodgins, A. M., Mittal, G. S., and Griffiths, M. W. 2002. Pasteurization of fresh orange juice using low-energy pulsed electrical field. *Journal of Food Science*. 67:2294–2299.

Hou, J. G., He, S. Y., Ling, M. S., Li, W., Dong, R., Pan, Y. Q., and Zheng, Y. N. 2010. A method of extracting ginsenosides from Panax ginseng by pulsed electric field. *Journal of Separation Science*. 33:2707–2713.

Hulsheger, H. and Niemann, E. G. 1980. Lethal effects of high-voltage pulses on *E. coli* K12. *Radiation and Environmental Biophysics*. 18:281–288.

Jaeger, H., Schulz, M., Lu, P., and Knorr, D. 2012. Adjustment of milling, mash electroporation and pressing for the development of a PEF assisted juice production in industrial scale. *Innovative Food Science and Emerging Technologies*. 14:46–60.

Jalte, M., Lanoiselle, J.-L., Lebovka, N. I., and Vorobiev, E. 2009a. Freezing of potato tissue pre-treated by pulsed electric fields. *LWT—Food Science and Technology*. 42(2):576–580.

Jalte, M., Lanoiselle, J. L., Lebovka, N. I., and Vorobiev, E. 2009b. Freezing of potato tissue pre-treated by pulsed electric fields. *LWT—Food Science and Technology*. 42:576–580.

Janositz, A., Noack, A. K., and Knorr, D. 2011. Pulsed electric fields and their impact on the diffusion characteristics of potato slices. *LWT—Food Science and Technology.* 44:1939–1945.

Jaya, S., Varadharaju, N., and Kennedy, Z. 2004. Inactivation of microorganisms in the fruit juice using pulsed electric fields. *Journal of Food Science and Technology—Mysore*. 41(6):652–655.

Jeantet, R., Florence, B., Françoise, N., Michel, R., and Gérard, B. 1999. High intensity pulsed electric fields applied to egg white: Effect on *Salmonella* Enteritidis inactivation and protein denaturation. *Journal of Food Protection*. 62(12):1381–1386(6).

Jia, M., Zhang, Q. H., and Min, D. B. 1999. Pulsed electric field processing effects on flavor compounds and microorganisms of orange juice. *Food Chemistry*. 65(4):445–451.

Jin, Z. T. and Zhang, Q. H. 1999. Pulsed electric field inactivation of microorganisms and preservation of quality of cranberry juice. *Journal of Food Processing and Preservation*. 23(6):481–497.

Jin, Z. T. and Zhang, Q. H. 2007. Pulsed electric field inactivation of microorganisms and preservation of quality of cranberry juice. *Journal of Food Processing and Preservation*. 23(6):481–497.

Keith, W. D., Harris, L. J., Hudson, L., and Griffiths, M. W. 1997. Pulsed electric fields as a processing alternative for microbial reduction in spice. *Food Research International*. 30(3–4):185–191.

Kummerer, K. 2001. Drugs in the environment: Emission of drugs, diagnostic aids and disinfectants into wastewater by hospitals in relation to other source—A review. *Chemosphere* 45:957–969.

Lebovka, N. I., Praporscic, I., and Vorobiev, E. 2004. Effect of moderate thermal and pulsed electric field treatments on textural properties of carrots, potatoes and apples. *Innovative Food Science and Emerging Technologies*. 5(1):9–16.

Li, S. Q., Zhang, Q. H., Lee, Y. Z., and Pham, T. V. 2003. Effects of pulsed electric fields and thermal processing on the stability of bovine immunoglobulin G (IgG) in enriched soymilk. *Journal of Food Science*. 68:1201–1207.

Liang, Z., Cheng, Z., and Mittal, G. S. 2006. Inactivation of spoilage microorganisms in apple cider using a continuous flow pulsed electric field system. *LWT—Food Science and Technology*. 39:350–356.

Liang, Z., Mittal, G. S., and Griffiths, M. W. 2002. Inactivation of *Salmonella typhimurium* in orange juice containing antimicrobial agents by pulsed electric field. *Journal of Food Protection*. 65:1081–1087.

Lin, Z. R., Zeng, X. A., Yu, S. J., and Sun, D. W. 2012. Enhancement of ethanol–acetic acid esterification under room temperature and non-catalytic condition via pulsed electric field application. *Food Bioprocess Technology*. 5:2637–2645.

Loginova, K. V., Lebovka, N. I., and Vorobiev, E. 2011a. Pulsed electric field assisted aqueous extraction of colorants from red beet. *Journal of Food Engineering*. 106(2):127–133.

Loginova, K. V., Vorobiev, E., Bals, O., and Lebovka, N. I. 2011b. Pilot study of countercurrent cold and mild heat extraction of sugar from sugar beets, assisted by pulsed electric fields. *Journal of Food Engineering*. 102(4):340–347.

Lopez, N., Puertolas, E., Condon, S., Alvarez, I., and Raso, J. 2008a. Application of pulsed electric fields for improving the maceration process during vinification of red wine: Influence of grape variety. *European Food Research and Technology*. 227(4):1099–1107.

Lopez, N., Puertolas, E., Condon, S., Alvarez, I., and Raso, J. 2008b. Effects of pulsed electric fields on the extraction of phenolic compounds during the fermentation of must of tempranillo grapes. *Innovative Food Science and Emerging Technologies*. 9(4):477–482.

Lopez, N., Puertolas, E., Condon, S., Raso, J., and Alvarez, I. 2009. Enhancement of the extraction of betanine from red beetroot by pulsed electric fields. *Journal of Food Engineering*. 90(1):60–66.

Luengo, E., Álvarez, I., and Raso, J. 2013. Improving the pressing extraction of polyphenols of orange peel by pulsed electric fields. *Innovative Food Science and Emerging Technologies*. 17:79–84.

Luo, W., Zhang, R. B., Wang, L. M., Chen, J., and Guan, Z. C. 2010. Conformation changes of polyphenol oxidase and lipoxygenase induced by PEF treatment. *Journal of Applied Electrochemistry*. 40(2):295–301.

Marco-Moles, R., Rojas-Graü, M. A., Hernando, I., Pérez-Munuera, I., Soliva-Fortuny, R., and Martin-Belloso, O. 2011. Physical and structural changes in liquid whole egg treated with high-intensity pulsed electric fields. *Journal of Food Science*. 76(2):257–264.

Marquez, V. O., Mittal, G. S., and Griffiths, M. W. 1997. Destruction and inhibition of bacterial spores by high voltage pulsed electric field. *Journal of Food Science*. 62:399–401, 409.

Marselles-Fontanet, A. R. and Martin-Belloso, O. 2007. Optimization and validation of PEF processing conditions to inactivate oxidative enzymes of grape juice. *Journal of Food Engineering*. 83(3):452–462. http://www.sciencedirect.com/science/journal/02608774/83/3.

Marselles-Fontanet, A. R., Puig, A., Olmos, P., Minguez-Sanz, S., and Martin-Belloso, O. 2009. Optimising the inactivation of grape juice spoilage organisms by pulse electric fields. *International Journal of Food Microbiology*. 130(3):159–165.

Marsellés-Fontanet, Á. R., Puig-Pujol, A., Olmos, P., Mínguez-Sanz, S., and Martín-Belloso, O. 2013. A comparison of the effects of pulsed electric field and thermal treatments on grape juice. *Food and Bioprocess Technology*. 6:978–987.

Martin, O., Qin, B. L., Chang, F. J., Barbosa-Canovas, G. V., and Swanson, B. G. 1997. Inactivation of *Escherichia coli* in skim milk by high intensity pulsed electric fields. *Journal of Food Process Engineering*. 20:317–336.

Martin, O., Zhang, Q., Castro, A. J., Barbosa-Canovas, G. V., and Swanson, B. G. 1994. Revisión: Empleo de pulsos eléctricos de alto voltaje para la conservación de alimentos. Microbiología e ingeniería del proceso. *Spanish Journal of Food Science and Technology*. 34:1–34.

Martin-Belloso, O. and Soliva-Fortuny, R. 2010. Pulsed electric fields processing basics. In H. Q. Zhang, G. V. Barbosa-Canovas, V. M. Balasubramaniam, C. P. Dunne, D. F. Farkas, and J. T. C. Yuan (eds.), *Nonthermal Processing Technologies for Food*. Wiley-Blackwell, Oxford, U.K.

Martin-Belloso, O., Vega-Mercado, H., Qin, B. L., Chang, F. J., Barbossa-Canovas, G. V., and Swanson, B. G. 1997. Inactivation of *Escherichia coli* suspended in liquid egg using pulsed electric fields. *Journal of Food Processing and Preservation*. 21:193–208.

McClain, M. A., Culbertson, C. T., Jacobson, S. C., Allbritton, N. L., Sims, C. E., and Ramsey, J. M. 2003. Microfluidic devices for the high-throughput chemical analysis of cells. *Analytical Chemistry*. 75(21):5646–5655.

McDonald, C. J., Lloyd, S. W., Vitale, M. A., Petersson, K., and Innings, F. 2000. Effects of pulsed electric fields on microorganisms in orange juice using electric field strengths of 30 and 50 kV/cm. *Journal of Food Science*. 65:984–989.

Mhemdi, H., Grimi, N., Bals, O., Lebovka, N. I., and Vorobiev, E. 2013. Effect of apparent density of sliced food particles on the efficiency of pulsed electric field treatment. *Innovative Food Science and Emerging Technologies*. 18:115–119.

Michalac, S., Alvarez, V., Ji, T., and Zhang, Q. H. 2003. Inactivation of selected microorganisms and properties of pulsed electric field processing milk. *Journal of Food Processing and Preservation*. 27(2):137–151.

Milani, E., Koocheki, A., and Golimovahhed, Q. A. 2011. Extraction of inulin from Burdock root (*Arctium lappa*) using high intensity ultrasound. *International Journal of Food Science and Technology*. 46:1699–1704.

Min, S. and Zhang, Q. H. 2007. Packaging for high pressure processing, irradiation and pulsed electric field processing. In J. F. Han (ed.), *Packaging for Nonthermal Proccesing of Foods*. Blackwell Publishing, Ames, IA.

Min, S., Jin, T. Z., Min, S. K., Yeom, H., and Zhang, Q. H. 2006. Commercial-scale pulsed electric field processing of orange juice. *Journal of Food Science*. 68(4):1265–1271.

Min, S., Jin, Z. T., Min, S. K., Yeom, H., and Zhang, Q. H. 2003a. Commercial-scale pulsed electric field processing of orange juice. *Journal of Food Science*. 68:1265–1271.

Min, S., Jin, Z. T., and Zhang, Q. H. 2003b. Commercial scale pulsed electric field processing of tomato juice. *Journal of Agricultural and Food Chemistry*. 51:3338–3344.

Min, S., Jin, Z. T., and Zhang, Q. H. 2007. Commercial scale pulsed electric field processing of tomato juice. *Food and Bioprocess Technology*. 1(4):364–373.

Min, S., Min, S. K., and Zhang, Q. H. 2003c. Inactivation kinetics of tomato juice lipoxygenase by pulsed electric fields. *Journal of Food Science*. 68:1995–2001.

Min, Z. and Zhang, Q. H. 2003. Effects of commercial-scale pulsed electric field processing on flavor and color of tomato juice. *Journal of Food Science*. 68:1600–1606.

Mohamed, M. E. and Eissa, A. H. A. 2012. Pulsed electric fields for food processing technology. In A. H. A. Eissa (ed.), *Structure and Function of Food Engineering*, pp. 275–307. InTech Prepress, Rijeka, Croatia.

Molinari, P., Pilosof, A. M. R., and Jagus, R. J. 2004. Effect of growth phase and inoculum size on the inactivation of *S. cerevisiae* in fruit juices by pulsed electric fields. *Food Research International*. 37(8):793–798.

Monfort, S., Gayan, E., Raso, J., Condon, S., and Alvarez, I. 2010. Evaluation of pulsed electric fields technology for liquid whole egg pasteurization. *Food Microbiology*. 27(7):845–852.

Monfort, S., Sagarzazu, N., Condon, S., Raso, J., and Alvarez, I. 2013. Liquid whole egg ultrapasteurization by combination of PEF, heat, and additives. *Food and Bioprocess Technology*. 6(8):2070–2080.

Monfort, S., Saldana, G., Condon, S., Raso, J., and Alvarez, I. 2012. Inactivation of *Salmonella* spp. in liquid whole egg using pulsed electric fields, heat, and additives. *Food Microbiology*. 30(2):393–399.

Morales-de la Peña, M., Salvia-Trujillo, L., Rojas-Graü, M. A., and Martín-Belloso, O. 2010. Impact of high intensity pulsed electric field on antioxidant properties and quality parameters of a fruit juice–soymilk beverage in chilled storage. *LWT—Food Science and Technology*. 43:872–881.

Mosqueda-Melgar, J., Raybaudi-Massilia, R. M., and Martin-Belloso, O. 2012. Microbiological shelf life and sensory evaluation of fruit juices treated by high-intensity pulsed electric fields and antimicrobials. *Food and Bioproducts Processing*. 90:205–214.

Mosqueda-Melgar, J., Raybaudi-Massilia, R. M., and Martín-Belloso, O. 2008. Combination of high-intensity pulsed electric fields with natural antimicrobials to inactivate pathogenic microorganisms and extend the shelf-life of melon and watermelon juices. *Food Microbiology*. 25(3):479–491.

Muñoz, A., Palgan, I., Noci, F., Cronin, D. A., Morgan, D. J., Whyte, P., and Lyng, J. G. 2012. Combinations of selected non-thermal technologies and antimicrobials for microbial inactivation in a buffer system. *Food Research International*. 47:100–105.

National Advisory Committee on Microbiological Criteria for Foods (NACMCF). 2006. Requisite scientific parameters for establishing the equivalence of alternative methods of pasteurization. *Journal of Food Protection*. 69:1190–1216.

Nguyen, P. and Mittal, G. S. 2007. Inactivation of naturally occurring microorganisms in tomato juice using pulsed electric field (PEF) with and without antimicrobials. *Chemical Engineering and Processing*. 46(4):360–365.

Ortega-Rivas, E., Zarate-Rodriguez, E., and Barbosa-Canovas, G. V. 1998. Apple juice pasteurization using ultrafiltration and pulsed electric fields. *Food and Bioproducts Processing*. 76:193–198.

Pagan, R., Condon, S., and Raso, J. 2005. Microbial inactivation by pulsed electric fields. In G. V. Barbosa-Canovas, M. S. Tapia, and M. P. Cano (eds.), *Novel Food Processing Technologies*, pp. 45–68. CRC Press.

Pagan, R., Esplugas, S., Gongora-Nieto, M. M., Barbosa-Canovas, G. V., and Swanson, B. G. 1998. Inactivation of *Bacillus subtilis* spores using high intensity pulsed electric fields in combination with other food conservation technologies. *Food Science and Technology International*. 4:33–44.

Palaniappan, S. and Sastry, S. K. 1990. Effects of electricity on microorganisms: A review. *Journal of Food Processing and Preservation*. 14:393–414.

Palgan, A., Muñoz, F., Noci, P., Whyte, D. J., Morgan, D., Cronin, A., and Lyng, J. G. 2012. Effectiveness of combined pulsed electric field (PEF) and manothermosonication (MTS) for the control of *Listeria innocua* in a smoothie type beverage. *Food Control* 25:621–625.

Paraskeva, P. and Graham, N. J. 2002. Ozonation of municipal wastewater effluents. *Water and Environmental Research*. 74:569–581.

Perez, O. E. and Pilosof, A. M. R. 2004. Pulsed electric fields effects on the molecular structure and gelation of β-lactoglobulin concentrate and egg white. *Food Research International*. 37(1):102–110.

Picart, L., Dumay, E., and Cheftel, J. C. 2002. Inactivation of *Listeria innocua* in dairy fluids by pulsed electric fields: Influence of electric parameters and food composition. *Innovative Food Science and Emerging Technologies*. 3(4):357–369.

Porras-Parral, G., Miri, T., Bakalis, S., and Fryer, P. J. 2012. The effect of electrical processing on mass transfer in beetroot andmodel gels. *Journal of Food Engineering*. 112(3):208–217.

Pothakamury, U. R., Barbosa-Canovas, G. V., Swanson, B. G., and Spence, K. D. 1997. Ultrastructural changes in *Staphylococcus aureus* treated with pulsed electric fields. *Food Science and Technology International*. 3:113–121.

Pothakamury, U. R., Monsalve-Gonzalez, A., Barbosa-Canovas, G. V., and Swanson, B. G. 1995. Inactivation of *Escherichia coli* and *Staphylococcus aureus* in model foods by pulsed electric field technology. *Food Research International*. 28:167–171.

Pothakamury, U. R., Vega-Mercado, H., Zhang, Q., Barbosa-Canovas, G. V., and Swanson, B. G. 1996. Effect of growth stage and temperature on inactivation of *E. coli* by pulsed electric fields. *Journal of Food Protection*. 59:1167–1171.

Puertolas, E., Hernandez-Orte, P., Sladana, I., and Raso, J. 2010a. Improvement of winemaking process using pulsed electric fields at pilot-plant scale. Evolution of chromatic parameters and phenolic content of Cabernet Sauvignon red wines. *Food Research International*. 43(3):761–766.

Puertolas, E., López, N., Condón, S., Álvarez, I., and Raso, J. 2010b. Potential applications of PEF to improve red wine quality. *Trends in Food Science and Technology*. 21(5):247–255.

Puertolas, E., Lopez, N., Condon, S., Raso, J., and Alvarez, I. 2009. Pulsed electric fields inactivation of wine spoilage yeast and bacteria. *International Journal of Food Microbiology*. 130(1):49–55.

Qin, B. L., Chang, F. J., Barbosa-Canovas, G. V., and Swanson, B. G. 1995a. Non-thermal inactivation of *Saccharomyces cerevisiae* in apple juice using pulsed electric fields. *LWT—Food Science and Technology*. 28:564–568.

Qin, B. L., Pothakamury, U. R., Vega-Mercado, H., Martin-Belloso, O. M., Barbosa-Canovas, G. V., and Swanson, B. G. 1995b. Food pasteurisation using high-intensity pulsed electric fields. *Food Technology*. 49(12):55–60.

Qin, B. L., Zhang, Q., Barbosa-Cánovas, G. V., Swanson, B. G., and Pedrow, P. D. 1994. Inactivation of microorganisms by pulsed electric fields with different voltage waveforms. *IEEE Transactions on Dielectrics and Insulation*. 1:1047–1057.

Qiu, X., Sharma, S., Tuhela, L., Jia, M., and Zhang, Q. H. 1998. An integrated PEF pilot plant for continuous nonthermal pasteurisation of fresh orange juice. *Transactions of the ASAE* 41:1069–1074.

Quitao-Teixeira, L. J., Aguiló-Aguayo, I., Ramos, A. M., and Martín-Belloso, O. 2008. Inactivation of oxidative enzymes by high-intensity pulsed electric field for retention of color in carrot juice. *Food and Bioprocess Technology*. 1(4):364–373.

Quintao-Teixeira, L. J., Soliva-Fortuny, R., Mota Ramos, A. M., and Martin-Belloso, O. 2013. Kinetics of peroxidase inactivation in carrot juice treated with pulsed electric fields. *Journal of Food Science*. 78(2):222–228.

Raso, J., Calderon, M. L., Gongora, M., Barbosa-Canovas, G. V., and Swanson, B. G. 1998a. Inactivation of mold ascospores and conidiospores suspended in fruit juices by pulsed electric fields. *LWT—Food Science and Technology*. 31(7/8):668–672.

Raso, J., Calderon, M. L., Gongora, M., Barbosa-Canovas, G. V., and Swanson, B. G. 1998b. Inactivation of *Zygosaccharomyces bailii* in fruit juices by heat, high hydrostatic pressure and pulsed electric fields. *Journal of Food Science*. 63:1042–1044.

Reina, L. D., Jin, Z. T., Zhang, Q. H., and Yousef, A. E. 1998. Inactivation of *Listeria monocytogenes* in milk by pulsed electric field. *Journal of Food Protection*. 61:1203–1206.

Rieder, A., Schwartz, T., Schon-Holz, K., Marten, S., Süss, J., Gusbeth, C., Kohnen, W., Svoboda, W., Obst, U., and Frey, W. 2008. Molecular monitoring of inactivation efficiencies of bacteria during pulsed electric field treatment of clinical wastewater. *Journal of Applied Microbiology*. 105:2035–2045.

Riener, J., Noci, F., Cronin, D. A., Morgan, D. J., and Lyng, J. G. 2008. Combined effect of temperature and pulsed electric fields on apple juice peroxidase and polyphenoloxidase inactivation. *Food Chemistry*. 109:402–407.

Rook, J. J. 1977. Chlorination reactions of fulvic acids in natural waters. *Environmental Science and Technology*. 11(5):478–482.

Saldana, G., Alvarez, I., Condon, S., and Raso, J. 2014. Microbiological aspects related to the feasibility of PEF technology for food pasteurization. *Critical Reviews in Food Science and Nutrition*. 54(11):1415–1426.

Sanchez-Moreno, C., Plaza, L., Elez-Martínez, P., DeAncos, B., Martín-Belloso, O., and Cano, M. P. 2005. Impact of high pressure and pulsed electric fields on bioactive compounds and antioxidant activity of orange juice in comparison with traditional thermal processing. *Journal of Agricultural and Food Chemistry*. 53(11):4403–4409.

Sanchez-Vega, R., Mujica-Paz, H., Marquez-Melendez, R., Ngadi, M. O., and Ortega-Rivas, E. 2009. Enzyme inactivation on apple juice treated by ultrapasteurization and pulsed electric fields technology. *Journal of Food Processing and Preservation*. 3 (4):486–499.

Sastry, S. K. and Barach, J. T. 2001. Ohmic and inductive heating. *Journal of Food Science*. 65(Suppl.):42–46.

Sastry, S. K. and Li, Q. 1996. Modeling the ohmic heating of foods. *Food Technology*. 50(5):246–248.

Schilling, S., Alber, T., Toepfl, S., Neidhart, N., Knorr, D., Schieber, A., and Carle, R. 2007. Effects of pulsed electric field treatment of apple mash on juice yield and quality attributes of apple juices. *Innovative Food Science and Emerging Technologies*. 8:127–134.

Schilling, S., Schmid, S., Jäger, H., Ludwig, M., Dietrich, H., Toepfl, S., Knorr, D., Neidhart, S., Schieber, A., and Carle R. 2008a. Comparative study of pulsed electric field and thermal processing of apple juice with particular consideration of juice quality and enzyme deactivation. *Journal of Agricultural and Food Chemistry*. 56(12):4545–4554.

Schilling, S., Toepfl, S., Ludwig, M., Dietrich, H., Knorr, D., Neidhart, S., Schieber, A., and Carle, R. 2008b. Comparative study of juice production by pulsed electric field treatment and enzymatic maceration of apple mash. *European Food Research and Technology*. 226:1389–1398.

Schwartz, T., Hoffman, S., and Obst, U. 2003. Formation of natural biofilms during chlorine dioxide and UV disinfection in a public drinking water distribution system. *Journal of Applied Microbiology*. 95:591–601.

Selma, M. V., Salmeron, M. C., Valero, M., and Fernandez, P. S. 2004. Control of *Lactobacillus plantarum* and *Escherichia coli* by pulsed electric fields in MRS broth, Nutrient broth and orange–carrot juice. *Food Microbiology* 21:519–525.

Sen Gupta, B., Masterson, F., and Magee, T. R. A. 2003. Inactivation of *E. coli* K12 in apple juice by high voltage pulsed electric field. *European Food Research and Technology*. 217(5):434–437.

Sen Gupta, B., Masterson, F., and Magee, T. R. A. 2005. Inactivation of *E. coli* in cranberry juice by a high voltage pulsed electric field. *Engineering in Life Sciences*. 5:148–151.

Sensoy, I., Zhang, Q. H., and Sastry, S. K. 1997. Inactivation kinetics of *Salmonella* Dublin by pulsed electric field. *Journal of Food Process Engineering*. 20:367–381.

Sepúlveda-Ahumada, D. R., Ortega-Rivas, E., and Barbosa-Cánovas, G. V. 2000. Quality aspects of cheddar cheese obtained with milk pasteurized by pulsed electric fields. *Food and Bioproducts Processing*. 78(2):65–71.

Seratlić, S., Bugarski, B., Nedović, V., Radulović, Z., Wadsö, L., Dejmek, P., and Gómez Galindo, F. 2013. Behavior of the surviving population of *Lactobacillus plantarum* 564 upon the application of pulsed electric fields. *Innovative Food Science and Emerging Technologies*. 17:93–98.

Shamsi, K. and Sherkat, F. 2009. Application of pulsed electric field in non-thermal processing of milk. *Asian Journal of Food and Agro Industry*. 2(3):216–244.

Sharma, S. K., Zhang, Q. H., and Chism, G. W. 1998. Development of a protein fortified fruit beverage and its quality when processed with pulsed electric field treatment. *Journal of Food Quality*. 21:459–473.

Shayanfar, S., Chauhan, O. P., Toepfl, S., and Heinz, V. 2013. The interaction of pulsed electric fields and texturizing -antifreezing agents in quality retention of defrosted potato strips. *International Journal of Food Science and Technology*. 48:1289–1295.

Sheng, J., Vannela, R., and Rittmann, B. E. 2012. Disruption of Synechocystis PCC 6803 for lipid extraction. *Water Science and Technology*. 65(3):7.

Shin, J. K., Jung, K. J., Pyun, Y. R., and Chung, M. S. 2007. Application of pulsed electric fields with square wave pulse to milk inoculated with *E. coli, P. fluorescens* and *B. stearothermophilus*. *Food Science and Biotechnology*. 16:1082–1084.

Simpson, R. K., Whittington, R., Earnshaw, R. G., and Russel, N. J. 1999. Pulsed high electric causes "all or nothing" membrane damage in *Listeria monocytogenes* and *Salmonella typhimurium*, but membrane H+-ATPase is not a primary target. *International Journal of Food Microbiology*. 48:1–10.

Sitzmann, W. 1995. High-voltage pulse techniques for food preservation. In *New Methods of Food Preservation*, Gould, G.W., (ed.), pp. 236–251. Blackie Academic and Professional, London, U.K.

Su, Y., Zhang, Q. H., and Yin, Y. 1996. Inactivation of *Bacillus subtilis* spores using high voltage pulsed electric fields. In *Annual Meeting, Book of Abstracts*. Session 26A-14. Institute of Food Technologists.

Tewari, G. and Juneja, V. P. 2007. *Advances in Thermal and Nonthermal Food Preservation*. Wiley Blackwell Publications, Oxford, U.K.

Toepfl, S. 2006. Pulsed electric fields (PEF) for permeabilization of cell membranes in food- and bioprocessing-applications, process and equipment design and cost analysis. Dissertation. Berlin University, Berlin, Germany.

Toepfl, S. 2011. Pulsed electric field food treatment scale up from lab to industrial scale. *Procedia Food Science*. 1:776–779.

Torregrosa, F., Esteve, M. J., Frígola, A., and Cortés, C. 2006. Ascorbic acid stability during refrigerated storage of orange–carrot juice treated by high pulsed electric field and comparison with pasteurized juice. *Journal of Food Engineering*. 73(4):339–345.

Tsong, T. Y. 1991. Electroporation of cell membranes. *Biophysical Journal*. 60:297–306.

Turk, M. F., Billaud, C., Vorobiev, E., and Baron, A. 2012. Continuous pulsed electric field treatment of French cider apple and juice expression on the pilot scale belt press. *Innovative Food Science and Emerging Technologies*. 14:61–69.

Ulmer, H. M., Heinz, V., Ganzle, M. G., Knorr, D., and Vogel, R. F. 2002. Effects of pulsed electric fields on inactivation and metabolic activity of *Lactobacillus plantarum* in model beer. *Journal of Applied Microbiology*. 93(2):326–335.

United States Patent 4764473. 2014. Chamber for the treatment of cells in an electrical field. http://www.freepatensonline.com/4764473.html. Accessed October 2013.

Uysal, E. E. 2010. Effects of pulsed electric fields on quality characteristics of red wine. Ms.C. thesis. Abant Izzet Baysal University, Bolu, Turkey.

Van Loey, A., Verachtert, B., and Hendrickx, M. 2002. Effects of high electric fields pulses on enzymes. *Trends in Food Science and Technology*. 12:94–102.

Vanthoor-Koopmans, M., Wijffels, R. H., Barbosa, M. J., and Eppink, M. H. M. 2013. Biorefinery of microalgae for food and fuel. *Bioresource Technology*. 135:142–149.

Vega-Mercado, H., Barbosa-Canovas, G. V., Powers, J., and Swanson, B. G. 1995. Inactivation of plasmin by pulsed electric field. *Journal of Food Science*. 60:1143–1146.

Vega-Mercado, H., Martin-Belloso, O., Chang, F.-J., Barbosa-Canovas, G. V., and Swanson, B. G. 1996a. Inactivation of *Escherichia coli* and *Bacillus subtilis* suspended in pea soup using pulsed electric fields. *Journal of Food Processing and Preservation*. 20(6):501–510.

Vega-Mercado, H., Powers, J. R., Barbosa-Canvas, G. V., and Swanson, B. G. 2006. Plasminin activation with pulsed electric fields. *Journal of Food Science*. 60(5):1143–1146.

Villegas, B. and Costell, E. 2007. Flow behaviour of inulin–milk beverages. Influence of inulin average chain length and of milk fat content. *International Dairy Journal*. 17:776–781.

Wan, J., Coventry, J., Swiergon, P., Sanguansri, P., and Versteeg, C. 2009. Advances in innovative processing technologies for microbial inactivation and enhancement of food safety- pulsed electric field and low-temperature plasma. *Trends in Food Science and Technology*. 20:414–424.

Wang, J., Guan, Y.- G., Yu, S.-J., Zeng, X.-A., Liu, Y.-Y., Yuan, S., and Xu, R. 2011. Study on the Maillard reaction enhanced by pulsed electric field in a glycin–glucose model system. *Food and Bioprocess Technology*. 4:469–474.

Wang, Z., Wang, J., Guo, S., Ma, S., and Yu, S.-J. 2013. Kinetic modeling of Maillard reaction system subjected to pulsed electric field. *Innovative Food Science and Emerging Technologies*. 20:121–125.

Wiktor, A., Iwaniuk, M., Śledź, M., Nowacka, M., Chudoba, T., and Witrowa-Rajchert, D. 2013. Drying kinetics of apple tissue treated by pulsed electric field. *Drying Technology: An International Journal*. 31(1):112–119.

Wu, Y., Mittal, G. S., and Griffiths, M. W. 2005. Effect of pulsed electric field on the inactivation of microorganisms in grape juices with and without antimicrobials. *Biosystems Engineering*. 90:1–7.

Xing-long, X., Cong-cong, L., Yang, Q., Yi-gang, Y., and Hui, W. 2013. Molecular monitoring of *Escherichia coli* O157:H7 sterilization rate using qPCR and propidium monoazide treatment. *Letters in Applied Microbiology*. 56:333–339.

Yeom, H. W., Evrendilek, G. A., Jin, Z. T., and Zhang, Q. H. 2004. Processing of yogurt-based products with pulsed electric fields: Microbial, sensory and physical evaluations. *Journal of Food Processing and Preservation*. 28(3):161–178.

Yeom, H. W., Streaker, C. B., Zhang, Q. H., and Min, D. B. 2000a. Effects of pulsed electric fields in the activity of microorganisms and pectin methyl esterase in orange juice. *Journal of Food Science*. 65:1359–1363.

Yeom, H. W., Streaker, C. B., Zhang, Q. H., and Min, D. B. 2000b. Effects of pulsed electric fields on the quality of orange juice and comparison with heat pasteurization. *Journal of Agriculture and Food Chemistry*. 48(10):4597–4605.

Yeom, H. W., Zhang, Q. H., and Chism, G. W. 2002. Inactivation of pectin methyl esterase in orange juice by pulsed electric fields. *Journal of Food Science*. 67:2154–2159.

Yin, Y., Zhang, Q. H., and Sastry, S. H. 1997. High voltage pulsed electric field treatment chambers for the preservation of liquid food products. U.S. Patent 5,690,978. November 25, 1997.

Zgalin, M. K., Hodzic, D., Rebersek, M., and Kanduser, M. 2012. Combination of microsecond and nanosecond pulsed electric field treatments for inactivation of *Escherichia coli* in water samples. *Journal of Membrane Biolology*. 245:643–650.

Zhang, B. G., Zhang, M., Shi, J., and Xu, Y. 2010. Pulsed electric field processing effects on physicochemical properties, flavor compıunds and microorganisms of longan juice. *Journal of Food Processing and Preservation*. 34:1121–1138.

Zhang, Q., Barbosa-Canovas, G. V., and Swanson, B. G. 1995. Engineering aspects of pulsed electric field pasteurization. *Journal of Food Engineering*. 25:261–281.

Zhang, Q., Chang, F. J., Barbosa-Canovas, G. V., and Swanson, B. G. 1994a. Inactivation of microorganisms in a semisolid model food using high voltage pulsed electric fields. *LWT—Food Science and Technology*. 27:538–543.

Zhang, Q., Monsalve-Gonzalez, A., Barbosa-Canovas, G. V., and Swanson, B. G. 1994b. Inactivation of *E. coli* and *S. cerevisiae* by pulsed electric fields under controlled temperature conditions. *Transactions of the ASAE*. 37:581–587.

Zhang, Q., Monsalve-Gonzalez, A., Qin, B., Barbosa-Canovas, G. V., and Swanson, B. G. 1994c. Inactivation of *Saccharomyces cerevisiae* in apple juice by square-wave and exponential-decay pulsed electric fields. *Journal of Food Process Engineering*. 17:469–478.

Zhang, Q. H., Qiu, X., and Sharma, S. K. 1997. Recent developments in pulsed electric processing. In D. I. Chandrana (ed.), *New Technologies Yearbook*, pp. 31–42. National Food Processors Association, Washington, DC.

Zhao, W., Tang, Y., Lu, L., Chen, X., and Li, C. 2014. Review: Pulsed electric fields processing of protein-based foods. *Food and Bioprocess Technology*. 7:114–125.

Zhao, W., Yang, R., Liang, Q., Zhang, W., Hua, X., and Tang, Y. 2012. Electrochemical reaction and oxidation of lecithin under pulsed electric fields (PEF) processing. *Journal of Agriculture and Food Chemistry*. 60:12204–12209.

Zhao, W., Yang, R., Wang, M., and Lu, R. 2009. Effects of pulsed electric fields on bioactive components, colour and flavour of green tea infusions. *International Journal of Food Science and Technology*. 44(2):312–321.

Zhong, K., Hu, X., Zhao, G., Chen, F., and Liao, X. 2005. Inactivation and conformational change of horseradish peroxidase induced by pulsed electric field. *Food Chemistry*. 92:473–479.

Zhu, Z., Bals, O., Grimi, N., and Vorobiev, E. 2012. Pilot scale inulin extraction from chicory roots assisted by pulsed electric fields. *International Journal of Food Science and Technology*. 47:1361–1368.

Zimmermann, U. 1986. Electrical breakdown, electropermeabilization and electrofusion. *Reviews of Physiology Biochemistry and Pharmacology*. 105:176–256.

Zuckermann, H., Krasik, Y. E., and Felsteiner, J. 2002. Inactivation of microorganisms using pulsed high-current underwater discharges. *Innovative Food Science and Emerging Technologies*. 3:329–336.

Zulueta, A., Esteve, M. J., Frasquet, I., and Frνgola, A. 2007. Fatty acid profile changes during orange juice–milk beverage processing by high-pulsed electric field. *European Journal of Lipid Science and Technology*. 109(1):25–31.

14 Use of Magnetic Fields Technology in Food Processing and Preservation

Daniela Bermúdez-Aguirre, Oselys Rodriguez-Justo, Victor Haber-Perez, Manuel Garcia-Perez, and Gustavo V. Barbosa-Cánovas

CONTENTS

14.1 INTRODUCTION

In the last decades, food engineers have been looking for new alternatives to preserve and process food without the use of heat. Even though thermal treatment has been one of the main processes in the food industry, undesirable changes in the product are very well known. Thus, scientists and engineers have used alternative physical and chemical hurdles to inactivate microorganisms in food trying to minimize possible changes in the product and preserve the fresh-like characteristics. However, when these novel technologies have been tested in food, positive changes have been also observed in the product, for example, enhancement of color or texture and increase in the yield of specific products. These technologies have also been explored for further product development.

Some of the most well-known emerging technologies in food processing are high hydrostatic pressure, pulsed electric fields, and ultrasound, which have shown positive results in microbial inactivation and in the preservation of quality and sensory characteristics of food items. Another emerging technology in the food processing arena is the use of magnetic fields. This technology was under research for microbial inactivation about a decade ago, but results at that time were not favorable. However, research on the use of magnetic fields continued, and some applications have been found.

In this chapter, the basics of magnetic fields technology for food processing and preservation are presented along with some devices. The use of magnetic fields for microbial inactivation is briefly discussed and several cases are presented. Other uses of magnetic fields for food processing, such as the use of fermentation processes, are also included.

14.2 MAGNETIC FIELDS TECHNOLOGY

Magnetic fields are force fields surrounding electric current circuits or ferromagnetic materials, the first observations were made by the ancient Greeks, using iron. In the eighteenth century, Oersted showed that a magnetic field exists when an electric current flows in a wire. Later, Ampere affirmed that a magnetic field is produced from the movement of electrical charges (Grigelmo-Miguel et al. 2011).

The use of magnetic fields in biological activities is not new; the first uses were reported in 1938, and Hoffman (1985) first used magnetic fields for food preservation in 1985 (Barbosa-Cánovas et al. 2005). Most of the reports in the literature about magnetic fields are from the last decade of the last century or in the beginning of this century. During this time, a number of research groups around the world were devoted to the study of the possible effect of microbial inactivation of magnetic fields in a number of food items, trying to take advantage of its physicochemical effects to be used as a nonthermal technology. However, results showed different behavior of the microorganisms against magnetic fields during inactivation that, in some cases, were contradictory.

Now, after several years of research, it is possible to group the biological effects observed because of the presence of magnetic fields. Some of these effects are related to the direction of migration of certain microorganisms, growth and reproduction of microorganisms, increase of DNA synthesis, orientation of biomolecules, and reduction in the number of cells, among others (Barbosa-Cánovas et al. 1998).

Magnetic fields can be classified as static (SMF) or oscillating (OMF); the latter is applied in the form of pulses in which the intensity of the pulses decreases over time. Also, magnetic fields can be either homogeneous or heterogeneous depending on the intensity of the field in the area surrounded by the magnetic coil (Barbosa-Cánovas et al. 1998). Further, magnetic fields are classified as low-intensity (around 10 Gauss) or high-intensity (thousands of Gauss) fields (Barbosa-Cánovas et al. 2005).

14.3 EQUIPMENT

Magnetic fields are generated by applying current into electric coils. For microbial inactivation, there are three types of coils that are often used: superconductive coils, coils producing DC fields, and coils energized through capacitors (Barbosa-Cánovas et al. 1998).

The magnetic field is more intense close to the coil; if there are many loops of wire, the current is greater and the field is stronger. When a coil is placed inside a core, the magnetic field can be more intense, and iron cores can reach magnetic fields about 3 T; if the intensity needs to be higher, such as 20 T, then superconductive coils made of superconductive filaments placed in a copper matrix should be used. Higher fields (30 T) can be generated using cooled resistive and hybrid magnets, produced from a superconductive magnetic coil with a water-cooled resistive coil. To supply these really high magnetic fields, energy is released in a pulsed way, and such magnetic fields are known as oscillating magnetic fields. Each applied pulse is applied for a short duration (Grigelmo-Miguel et al. 2011). In Figure 14.1, the typical configuration for magnetic fields equipment is shown for a continuous system (a) and for a batch system (b).

In most reports about magnetic fields, the intensity of the treatment is reported in Tesla (T); this unit is used in the SI to define flux density, which is the relationship between Weber per unit of area (Wb/m^2). Weber is the international unit for flux, and it is equivalent to 10^8 lines. The field intensity is measured in Ampere/meter (A/m). In some reports, the flux density is also reported as Gauss (G), and the relationship between T and G is the following: $1\ T = 10^4\ G$ (Barbosa-Cánovas et al. 1998). G is used in the centimeter–gram–second (C.G.S.) system of units.

In order to process food items, the product is placed inside plastic bags and sealed, and then it is processed (as seen in Figure 14.2). Most of the reports for this technology show the use of oscillating magnetic fields. Butz and Tauscher (2002) mention the range of processing conditions between 1 and 100 pulses (5–50 T, 5–500 kHz, 0°C–50°C, 25–100 μs), according to research done

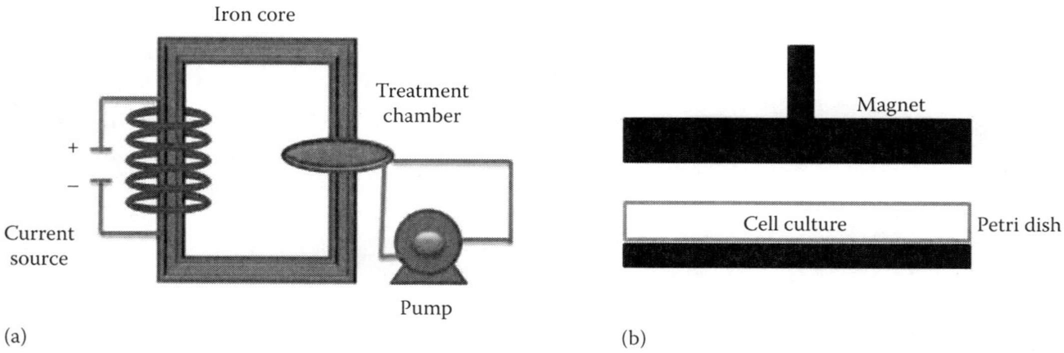

FIGURE 14.1 Basic configuration for magnetic fields treatment. (a) System to treat small volumes of cell suspension and (b) system for static treatment.

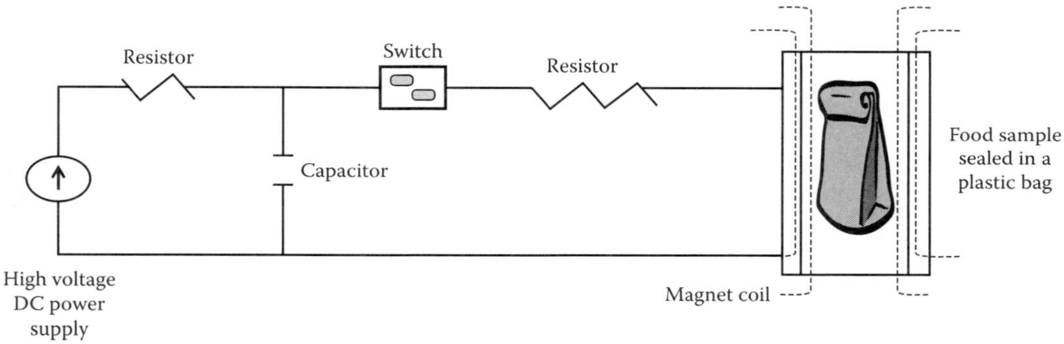

FIGURE 14.2 Circuit for Magneform 7000 series coil. (Adapted from Barbosa-Cánovas, G.V. et al., *Nonthermal Preservation of Foods*, Marcel Dekker, New York, 1998, pp. 113–138.)

at this time. The increase in the temperature of food during the processing with magnetic fields is about 2°C–5°C (Barbosa-Cánovas et al. 2000). However, it is evident that this technology is not ready to be used for industry purposes, as these authors also mentioned, since the results regarding microbial inactivation are not consistent.

14.4 INACTIVATION OF MICROORGANISMS

The effect of magnetic fields on microorganisms is noticeable when the rate of the flow of the cytoplasm in algae is accelerated or retarded depending on the magnetic field or when the microorganisms migrate in a specific direction according to the Earth's magnetic field. Magnetotatic bacteria have magnetosomes, a kind of magnetic particle composed mainly of magnetite which is the original source of iron (Barbosa-Cánovas et al. 1998). Magnetic iron is present as mineral crystals surrounded by a phospholipid membrane (Grigelmo-Miguel et al. 2011). Because of the presence of magnetosomes, bacteria can rotate and migrate according to the magnetic fields of the earth. The required flux density for microbial inactivation typically ranges from 5 to 50 T (Barbosa-Cánovas et al. 1998).

Reports in the literature about microbial inactivation using magnetic fields show contradictory results, as can be seen in Table 14.1. In some cases, inactivation has been observed under specific conditions; meanwhile, other researchers report null inactivation or even stimulation of the microbial growth. In fact, this technology has not shown progress in its development in the last few years because of the unsuccessful results for microbial inactivation.

TABLE 14.1
Examples of the Use of Magnetic Fields in the Treatment of Microorganisms

Microorganism	Processing Conditions	Media	Results	References
Yeast	0.57 T, static and oscillating magnetic fields	N/A	No inactivation with static fields, inactivation was increased using oscillating fields	Barbosa-Cánovas et al. (1998)
Saccharomyces cerevisiae	4600 Gauss (1 T = 10^4 G), 28°C and 38°C, 0% and 15% NaCl, 24, 48, and 72 h	Sabouraud dextrose broth	Salt and temperature increases the growth of the yeast	Van Nostran et al. (1967)
Bacillus subtilis	50–900 Gauss (1 T = 10^4 G), 4 h	Trypticase soya broth	No effect on spore germination	Moore (1979)
Escherichia coli ATCC 11775	18 T, 50 pulses, 30 µs, 42°C, combination with other nonthermal technologies: HHP, US, PEF, AM	Peptone solution	No inactivation	San Martín et al. (2001)
E. coli ATTC 11775 and *S. cerevisiae*	OMF: 18 T, 20–40 µs, 10–15 kHz SMF: 20 T, 1 h holding time	Phosphate buffer, peptone water or McIlvaine buffer	No inactivation	Harte et al. (2001)
E. coli B	Homogeneous magnetic field: 7 T Heterogeneous magnetic field: 5.2–6.1 T, 3.2–6.7 T 37°C	Culture broth	Growth affected depending on the phase growth of the microorganism Stronger effect of heterogeneous magnetic fields	Tsuchiya et al. (1996)

Note: N/A, not available; HHP, high hydrostatic pressure; US, ultrasound; PEF, pulsed electric fields; AM, antimicrobials; OMF, oscillating magnetic fields; SMF, static magnetic fields.

Hofmann (1985) patented a device in the United States to inactivate microorganisms using magnetic fields. The tested food items and microorganisms were milk with *S. thermophilus,* yogurt and orange juice with *Saccharomyces*, and rolls of dough containing some spores. The field intensity ranged from 7 to 40 T, and room temperature was used in most of the cases. Results showed that microbial inactivation was feasible from 2 to 4 log reduction (Barbosa-Cánovas et al. 1998).

More recently, experiments at Washington State University involved the use of three target microorganisms, two surrogates of pathogens such as *E. coli* and *Listeria innocua*, and spoilage yeast, *S. cerevisiae.* The media for inactivation were mainly buffer solutions, and the equipment used was a Magneform 7000 series™ coil, as seen in Figure 14.2 (Maxwell Laboratory, San Diego, CA). Inactivation was conducted at 18 T, using 50 magnetic-field pulses with temperature from 10°C to 50°C. Inactivation was negligible for most of the treatments, and the few observed reductions were due to the effect of thermal treatment rather than the effect of magnetic fields. Similar experiments were conducted at Los Alamos National High Magnetic Field Laboratory using two devices, one operated at 18 T and one operated at 50 T; microorganisms treated with 18 T were *E. coli* and *S. cerevisiae* for 1 h. A second set of experiments using 50 T were conducted using one, two, and three magnetic-field pulses testing *Bacillus subtilis, Escherichia coli, L. innocua,* and *Saccharomyces cerevisiae.* In all cases, no inactivation was observed because of the effect of magnetic fields (Barbosa-Cánovas et al. 2005).

In another attempt to achieve microbial inactivation, *E. coli* ATCC 11775 cells were subjected to several mild, nonthermal treatments before being treated with magnetic fields to study the effect of previous stress during inactivation. Cells were suspended in peptone water and conducted under pulsed electric fields (6.25 kV/cm, 5.6 ms), ultrasound (20 kHz, 70 W, 242 µm), high hydrostatic pressure (207 MPa, 5 min), and antimicrobials (lysozyme 1 mg/mL or nisin 77.5 mg/L); all of these treatments were applied with mild conditions in order to observe the inactivation after the use of magnetic fields. However, once the main treatment was applied (18 T, 50 magnetic-field pulses, 30 μs, 42°C), no further inactivation was observed in any of the experiments (San Martín et al. 2001). Here, it is evident that even after the cells have been stressed with other nonthermal technologies, cells are still strong enough to resist the inactivation from magnetic fields. In other words, magnetic fields are not affecting cell behavior in a significant way to conduct them to inactivation. Even though the intensity of the treatment can be considered as a high magnetic field (18 T), it is apparent that stronger intensities are required to inactivate cells. From a practical point of view, it is also worth considering whether higher magnetic fields or longer processing times should be used for cell inactivation or if other innovative nonthermal technologies should be tried instead.

In a contradictory study, *E. coli* K12 cells were subjected to magnetic fields of very low intensity and frequency (1–10 mT, 2–50 Hz) to observe the effect of this technology on the growth and protein synthesis of the cells after the treatment. Results showed that the growth of the bacteria was reduced by 3.8%. Regarding the stress proteins that bacteria can synthesize when they are subjected to some physical hurdles like heat, *E. coli* cells did not show any change in the protein pattern, even when magnetic fields treatment was applied together with heat (37° or 43°C). However, another microorganism, *Proteus vulgaris* did show a complete protein pattern because of the effect of magnetic fields (Mittenzwey et al. 1996). It is indeed an interesting study because the processing conditions were extremely light compared to the ones used for microbial inactivation mentioned previously. *E. coli*, a microorganism of interest in food microbiology, showed a slight response to these conditions regarding its growth, but the same treatment was not able to act as a stress factor together with temperature for inactivation. Although the authors mention that the reason could be the specific strain of *E. coli* used for this research, this same strain has been used in many food experiments and has been inactivated by several nonthermal technologies. Also, it is interesting to see how *P. vulgaris*, an enterobacteria that is not directly related with food, was able to show some stress after being treated with magnetic fields of very low intensity. Indeed, many parameters should be considered when the use of magnetic fields is explored, not only related with the process by itself but also regarding microbial strains and conditions and the medium of treatment.

Another strain of *E. coli* ATCC 25922 and a strain of *Pseudomonas aeruginosa* ATCC 27853 were subjected to very low frequency and intensity of magnetic fields (2 mT, 50 Hz), and the growth rate was reduced significantly when an antibiotic (kanamycin or amikacin) was also used during incubation for 4, 6, and 8 h. However, after 24 h, the number of treated cells of *E. coli* and *P. aeruginosa* was increased by 5% and 42%, respectively, suggesting a possible progressive adaptive response (Segatore et al. 2012). Indeed, these results are not favorable for the food industry since nonthermal technologies should be used to inactivate microorganisms without sublethal injuries or adaptive response.

14.4.1 Mechanisms of Microbial Inactivation

There are two main models for microbial inactivation using magnetic fields: the ion cyclotron resonance (ICR) model and the ion parametric resonance (IPR) model. The first model, ICR, is related to the effect of magnetic fields on ion behavior; once the energy is transferred from the magnetic fields to the ions, it is also transferred to the metabolic activities, for example, the transport of Ca^{2+} across the membrane. This point of the cell can be considered as the interaction site that diffuses the effect of the magnetic fields to other sites, such as the organelles or protein molecules. The second model, IPR, is based on the interaction of ions with specific molecules, such as proteins enzymes, nucleic acids, or lipids. It is believed that ions can initiate or control some enzymatic reactions as cofactors; this theory considers any kind of ion attached with a molecular structure (Barbosa-Cánovas et al. 1998).

Another theory proposed by Hofmann (1985) explains the possible inactivation of microorganisms as a result of the energy emitted by the magnetic fields and acting together with the DNA molecules. DNA possesses magnetically active parts (magnetic dipoles) in which the energy released during the treatment can act to break down certain covalent bonds that affect microbial activities such as growth (Corry et al. 1995; Hofmann, 1985).

Other possible mechanisms of inactivation include the radical recombination model, formation of metastable pores in cell membranes due to the presence of magnetite, or modified protein synthesis (San Martín et al. 2001).

14.5 OTHER USES OF MAGNETIC FIELDS IN FOOD PROCESSING

Even though the use of magnetic fields to inactivate microorganisms has not been successful and relatively no progress has been made in this area, some researchers have found the use of magnetic fields suitable for other practices in food processing.

Based on previous research, in which microbial inactivation did not occur under the treatment with magnetic fields but where cell growth was stimulated with this technology, this advantage has been used to produce some biomass of interest to the food industry, for example, in processes such as fermentation. Electrostimulation of yeast growth has been reported, specifically in the case of *S. cerevisiae* (Fiedler et al. 1995), a yeast of interest in food science, in which growth was enhanced using very low magnetic fields (about 0.7 mT, 50 Hz). Further studies have shown that the stimulation of yeast growth also depends on the frequency, intensity, time, and temperature of the experiment in addition to the biological state of the cell. This study showed the feasibility of growing yeast in the range of frequencies from 10 to 100 Hz and 0.5 mT (Mehedintu and Berg 1997). In another study, the growth of *S. cerevisiae* to ferment sugar cane molasses and produce ethanol was increased using magnetic fields (5–20 mT), showing a final efficiency 17% higher than the control yeast (Pérez et al. 2007).

Similar findings were reported for *Corynebacterium glutamicum* that is used in the food industry for the fermentation and production of soy sauce and yogurt. Electrostimulation of the bacteria was possible because of the application of very low magnetic fields (15 Hz, 3.4 mT, 8 h, and 50 Hz, 4.9 mT, 6 h), resulting in an important increase in cell growth and production of biomass (Lei and Berg 1998).

Chacón Álvarez et al. (2006) studied the use of very low-intensity magnetic fields in the production of nisin. Nisin is used as a natural antimicrobial in the food industry; it is produced by *Lactococcus lactis* subspecies *lactis* during fermentation of a modified milk medium. Nisin is widely used in several countries as an antimicrobial and can be added to cheese, milk, and other dairy products, canned vegetables, and baby foods, among others. However, there is a growing interest in increasing the production of this antimicrobial. For this reason, the use of magnetic fields has been tested to speed up the fermentation process and to study the possible increase of yield of nisin. A cheese whey permeate was used to generate nisin, and the best processing conditions were found using 4 h, 1.5 m/s, and 5 mT, which allowed an increase in nisin yield up to five times greater compared with the control (Chacón Álvarez et al. 2006). This study provides a very clear example in which the use of magnetic fields of much reduced intensity was used to stimulate the growth and production of biomass.

Spirulina is a food known as a great source of nutrients such as protein, vitamins, minerals, polyunsaturated fatty acids, zeaxanthin, and myxoxanthophyl (Li et al. 2007). This food is a cyanobacterium that contains all of the amino acids, and, because of that, it is considered an excellent nutritional supplement. The use of magnetic fields was also tested on the growth of this microorganism to examine the possible enhancement of growth and possible changes in nutrition composition. The treatment was applied in a photobioreactor with magnetic fields of about 0.25 mT using *S. platensis*, resulting in the enhancement of growth of the algae and improvement in amino acid composition (histidine) and trace elements (Ni, Sr, Cu, Mg, Fe, Mn, Ca, Co, and V) attributable to the nutritional assimilation of basic elements such as C, N, and P (Li et al. 2007).

14.6 CONCLUSIONS

The use of magnetic fields, both static and oscillating, has been tested in food science for microbial inactivation as the main target. However, several groups have previously shown that this technology, under tested conditions using high-intensity fields (5–30 T), is not able to inactivate microorganisms at the required levels. Only a few researchers found positive results, which are sometimes contradictory, but these outcomes can be explained in terms of equipment design, microorganism species, and processing conditions, among others.

Nevertheless, through the search for the best inactivation conditions, researchers have also observed the stimulation of cell growth due to the presence of magnetic fields. Most enhancement on cell growth has been observed in very low-intensity magnetic fields (in the range of mT), where possible use of this technology for other food activities has been explored. Enhancement in cell growth, with increased yields in biomass production in several fermentation processes, has been explored with successful results. Although this focus of study has shown slow development in the last few years, it is becoming more important, and future studies can be conducted using this technology for stimulation of cell growth.

REFERENCES

Barbosa-Cánovas, G.V., Pothakamury, U.R., Palou, E., and Swanson, B.G. 1998. *Nonthermal Preservation of Foods*. New York: Marcel Dekker, pp. 113–138.

Barbosa-Cánovas, G.V., Schaffner, D.W., Pierson, M.D., and Zhang, Q.H. 2000. Oscillating magnetic fields. *Journal of Food Science.* 65(Suppl.): 86–89.

Barbosa-Cánovas, G.V., Swanson, B.G., San Martín, G.M.F., and Harte, F. 2005. Use of magnetic fields as a nonthermal technology. In: *Novel Food Processing Technologies*, Eds. G.V. Barbosa Cánovas, M.S. Tapia, and P. Cano. Boca Ratón, FL: CRC Press, pp. 443–451.

Butz, P. and Tauscher, B. 2002. Emerging technologies: Chemical aspects. *Food Research International.* 35: 279–284.

Chacón Álvarez, D., Haber Pérez, V., Rodríguez Justo, O., and Monte Alegre, R. 2006. Effect of the extremely low frequency magnetic field on nisin production by *Lactococcus lactis* subsp. *lactis* using cheese whey permeate. *Process Biochemistry.* 41: 1967–1973.

Corry, J.E.L., James, C., James, S.J., and Hinton, M. 1995. *Salmonella, Campylobacter* and *Escherichia coli* O157:H7 decontamination techniques for the future. *International Journal of Food Microbiology.* 28: 187–196.

Fiedler, U., Gröbner, U., and Berg, H. 1995. Electrostimulation of yeast proliferation. *Bioelectrochemistry and Bioenergetics.* 38: 423–425.

Grigelmo-Miguel, N., Soliva-Fortuny, R., Barbosa-Cánovas, G.V., and Martín-Belloso, O. 2011. Use of oscillating magnetic fields in food preservation. In: *Nonthermal Processing Technologies for Food*, Eds. H. Zhang, G.V. Barbosa-Cánovas, B. Balasubramaniam, P. Dunne, D. Farkas, and J. Yuan. Ames, IA: Blackwell Publishing Ltd., pp. 222–235.

Harte, F., San Martín, M.F., Lacerda, A.H., Lelieveld, H.L.M., Swanson, B.G., and Barbosa-Cánovas, G.V. 2001. Potential use of 18 tesla static and pulsed magnetic fields on *Escherichia coli* and *Saccharomyces cerevisiae. Journal of Food Processing and Preservation.* 25: 223–235.

Hofmann, G.A. 1985. Deactivation of microorganisms by an oscillating magnetic field. *US Patent* 4,524,079.

Lei, C. and Berg, H. 1998. Electromagnetic window effects on proliferation rate of *Corynebacterium glutamicum. Bioelectrochemistry and Bioenergetics.* 45: 261–265.

Li, Z.Y., Guo, S.Y., Li, L., and Cai, M.Y. 2007. Effects of electromagnetic field on the batch cultivation and nutritional composition of *Spirulina platensis* in an air-lift photobioreactor. *Biosource Technology.* 98: 700–705.

Mehedintu, M. and Berg, H. 1997. Proliferation response of yeast *Saccharomyces cerevisiae* on electromagnetic fields parameters. *Bioelectrochemistry and Bioenergetics.* 43: 67–70.

Mittenzwey, R., Süβmuth, R., and Mei, W. 1996. Effects of extremely low-frequency electromagnetic fields on bacteria—The question of a co-stressing factor. *Biochemistry and Bioenergetics.* 40: 21–27.

Moore, R.L. 1979. Biological effects of magnetic fields: Studies with microorganisms. *Canadian Journal of Microbiology.* 25: 1145–1151.

Pérez, V.H., Reyes, A.F., Justo, O. R., Alvarez, D.C., and Alegre, R.M. 2007. Bioreactor coupled with electromagnetic field generator: Effects of extremely low frequency electromagnetic fields on ethanol production by *Saccharomyces cerevisiae. Biotechnology Progress.* 23: 1091–1094.

San Martín, M.F., Harte, F.M., Lelieveld, H., Barbosa-Cánovas, G.V., and Swanson, B. 2001. Inactivation effect of an 18-T pulsed magnetic field combined with other technologies on *Escherichia coli. Innovative Food Science and Emerging Technologies.* 2: 273–277.

Segatore, B., Setacci, D., Bennato, F., Cardigno, R., Amicosante, G., and Iorio, R. 2012. Evaluations of the effects of extremely low-frequency electromagnetic fields on growth and antibiotic susceptibility of *Escherichia coli* and *Pseudomonas aureginosa. International Journal of Microbiology.* 2012: 1–7, Article ID 587293.

Tsuchiya, K., Nakamura, K., Okuno, K., Ano, T., and Shoda, M. 1996. Effect of homogeneous and inhomogeneous high magnetic fields on the growth of *Escherichia coli. Journal of Fermentation and Bioengineering.* 81(4): 343–346.

Van Nostran, F.E., Reynolds, R.J., and Hedrick, A.H.G. 1967. Effects of a high magnetic field at different osmotic pressures and temperatures on multiplication of *Saccharomyces cerevisiae. Applied Microbiology.* 15(3): 561–563.

15 Ultrasonic and UV Disinfection of Food

Sivakumar Manickam and Yuh Xiu Liew

CONTENTS

15.1 INTRODUCTION

15.1.1 Why Is Disinfection Essential?

Foodborne diseases have emerged as an important and growing public health problem with a significant impact on people's health in many countries. In Malaysia, the incidence of notifiable foodborne diseases, namely, cholera, typhoid, hepatitis A, and dysentery, is less than 5/100,000 population, sporadic in nature, and outbreaks are confined to certain areas only. In the case of food poisoning, cases notified have mainly been in institutions ranging from 20 to 30/100,000 population. Based on the laboratory-based surveillance conducted by the Ministry of Health, it was discovered that the foodborne diseases were due to *Salmonella spp.*, *Shigella spp.*, *Salmonella typhi*, and *Vibrio spp.* Hence, it is essential to disinfect the food products, especially fresh-cut products, that are meant to be consumed raw (Malaysia 2004). Disinfection can be defined as a method to remove or inactivate pathogens on a product or surface by physical or chemical means to a level previously specified as appropriate for its intended further handling or use (McDonnell 2009).

15.1.2 Methods Available for Disinfection

The conventional disinfection methods to inactivate microorganisms and enzymes include thermal pasteurization and sterilization (Chemat et al. 2011). Sterilization can be defined as a process, physical or chemical, that eliminates all forms of life, especially microorganisms (Block 2001). Sterilization can be achieved by applying heat, chemicals, irradiation, high pressure, and filtration, or combinations thereof. On the other hand, thermal pasteurization is a process where heat is applied to reach a specific temperature for a predefined length of time and then immediately cooled after it is removed from the heat (IDFA 2009). Its effectiveness is dependent on the treatment temperature and time. However, the magnitude of temperature is also proportional to the amount of nutrient loss, development of undesirable flavors, and deterioration of functional properties of the food products (Piyasena et al. 2003).

With growing negative public reaction about chemicals added to food, ultraviolet (UV) light irradiation has become a viable alternative to thermal pasteurization for commercial applications in food processing. The United States Food and Drug Administration (USFDA) and United States Department of Agriculture (USDA) have concluded that the use of UV irradiation is safe (Koutchma 2008). UV wavelength ranges from 10 to 400 nm. An alternative subdivision often used are UV-C with wavelength ranging from 100 to 280 nm, UV-B for 280–315 nm, and UV-A for 315–400 nm (Shama 2007). UV-C is normally used for surface disinfection as bacteria suspended in air are more sensitive to UV-C light than bacteria suspended in liquids.

15.1.3 Mechanism of UV Disinfection on Food Material

The effect of UV radiation on microorganisms of the same species may depend on the strain, growth media, stage of culture, density of microorganisms, and different characteristics, for instance type and composition of food. However, it may vary from species to species. DNA and RNA carry genetic information essential for reproduction. Hence, damage to either of these substances can effectively inactivate the organism. There are three primary types of pyrimidine molecules, namely, cytosine (found in DNA and RNA), thymine (found in DNA), and uracil (found in RNA). The UV radiation absorbed by the DNA causes a physical shifting of electrons to render

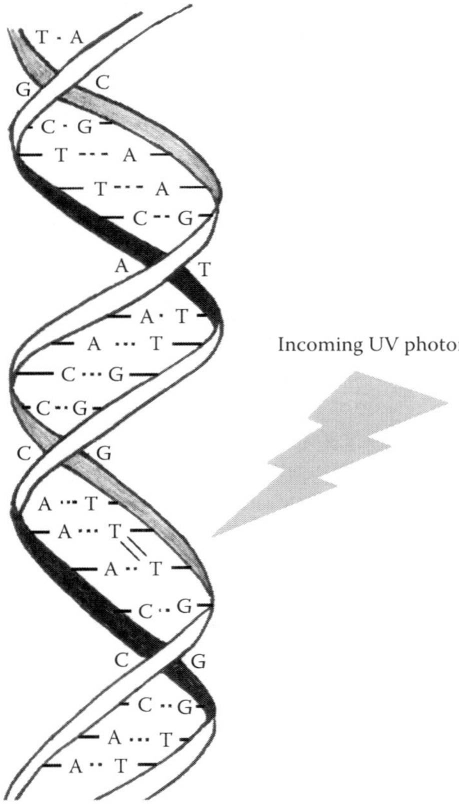

FIGURE 15.1 Dimerization of pyrimidine molecules.

splitting of the DNA bonds. A cross-linking would then occur between neighboring thymine and cytosine in the same DNA strand. This would then result in dimerization of the pyrimidine molecules. Once the pyrimidine molecules are bonded together, replication of the nucleic acid become very difficult due to the distortion of the DNA helical structure. Therefore, failure to replicate would lead to death of the microorganisms. The dimerization of the pyrimidine molecules is illustrated in Figure 15.1. The effects in the cross-linkage of DNA are proportional to the amount of UV-C light exposure (Guerrero-Beltrán and Barbosa-Cánovas 2004). The lethal effect of UV may be enhanced by combining UV treatment with the use of powerful oxidants such as hydrogen peroxide and ozone (Shama 2007).

Research investigations indicate that a constant fraction of the living population is inactivated during each progressive increment in time, when microorganisms are exposed to UV radiation. As the UV radiation exposure time increases, a constant fraction of the remaining active microorganisms will be inactivated. Hence, applying the same total energy will result in the same degree of disinfection irrespective of the duration and intensity of the dosage. Table 15.1 shows the typical dosage required to eliminate some of the microorganisms (Solsona and Méndez 2003). Currently, most UV equipments use a minimum exposure of 30,000 μWs/cm^2, which is sufficient to inactivate pathogenic bacteria and viruses in water. However, this amount of exposure may be inadequate to eliminate certain pathogenic protozoa, protozoan cysts, and nematode eggs that can require up to 100,000 μWs/cm^2 for total inactivation.

15.1.4 Advantages and Disadvantages of UV Disinfection

Advantages	Disadvantages
• Environment-friendly.	• Low dosage may not effectively inactivate some viruses, spores, and cysts.
• Chemical-free process.	• Organisms can sometimes repair and reverse the destructive effects of UV through *photoreactivation*.
• Requires no transportation, storage, or handling of toxic chemicals.	• No residual effect to indicate efficiency of UV disinfection.
• No carcinogenic disinfection of by-products which would affect the quality of the water.	• Lack of penetration in highly adsorptive media.
• Reduces pathogenic load significantly.	• Higher cost of equipment.
• Preserves flavor, color, texture, and enzyme activity.	• Exposure to UV is harmful to humans.
• No residual effect harmful to humans or aquatic life.	
• Requires less space than other methods.	
• Simplicity and ease of maintenance.	
• Easy installation.	

Sources: Guerrero-Beltrán, J.A. and Barbosa-Cánovas, G.V., *Food Sci. Technol. Int.*, 10, 137, 2004; Solsona, F. and Méndez, J.P., Water disinfection, 2003, http://whqlibdoc.who.int/paho/2003/a85637.pdf, Accessed August 5, 2013.

15.2 ULTRASOUND DISINFECTION

Another promising nonthermal technology is the application of ultrasound (US). US refers to pressure waves with a frequency of 20 kHz or more. Ultrasound equipment generally uses frequencies from 20 to 10 MHz. "Power ultrasound" refers to higher power US at lower frequencies (20–100 kHz) and has the ability to cause cavitation. The main advantage of using US is that the sound waves are generally considered safe, nontoxic, and environment-friendly (Kentish and Ashokkumar 2011). The use of US is able to prolong the shelf life of food and also maintain the sensory, nutritional quality, and functional characteristics of food especially in the case of heat-sensitive food items (Cao et al. 2010, Wang et al. 2011). The impact of US on disinfection may be classified under two categories, namely, the physical and the chemical effects, which will be discussed in the following section.

TABLE 15.1
Typical Dosage for Disinfection of Microorganisms

Bacteria	Power (μW/cm²)	Other Organisms	Power (μW/cm²)
Bacillus antharacis	8.700	**Yeasts**	
Salmonella enteritidis	7.600		
Bacillus megatherium sp. (veg)	2.500	*Saccharomyces ellipsoideus*	13.200
B. megatherium sp. (spores)	5.200	*Saccharomyces* sp.	1.600
B. peratyphosus	6.100	*Saccharomyces cerevisiae*	13.200
Bacillus subtilis	11.000	Brewer's yeast	660
B. subtilis spores	22.000	Baker's yeast	800
Clostridium tetani	22.000	Pastry yeast	13.200
Corynebacterium diphtheriae	6.500		
Eberthella typosa	4.100	**Spores**	
Escherichia coli	6.600		
Micrococcus candidus	12.300	*Penicillium roqueforti*	26.400
Mycobacterium tuberculosis	10.000	*Penicillium expansum*	22.000
Neisseria catarrhalis	8.500	*Mucor racemosus* A	35.200
Phytomonas tumefaciens	500	*M. racemosus* B	5.200
Proteus vulgaris	6.600	*Oospora lactis*	1.100
Pseudomonas aeruginosa	10.500		
Pseudomonas fluorescens	6.600	**Viruses**	
Salmonella typhimurium	15.200		
Salmonella	10.000	Bacteriophage (*E. coli*)	6.600
Sarcina lutea	26.400	Influenza (Flu) virus	6.600
Serratia marcescens	6.160	Hepatitis virus	8.000
Dysentery bacilli	4.200	Poliovirus (Poliomyelitis)	1.000
Shigella paradysenteriae	3.400	Rotavirus	24.000
Spirillum rubrum	6.160		
Staphylococcus alous	5.720	**Algae**	
Staphylococcus aureus	6.600		
Streptococcus hemolyticus	5.500	*Chlorella vulgaris*	2.000
Streptococcus lactis	8.800		
Streptococcus viridans	3.800		
Vibrio cholerae	6.500		

Source: Solsona, F. and Méndez, J.P., Water disinfection, 2003, http://whqlibdoc.who.int/paho/2003/a85637.pdf, Accessed August 5, 2013.

15.2.1 Physical Effects of US

Since US is represented as a high-frequency wave, the pressure wave would create regions of high and low pressures as it passes through the medium. Amplitude is then used to refer to the magnitude of pressure variation. Amplitude is directly proportional to the amount of energy applied to the system. When the wave passes through a viscous medium, it dissipates the energy in the form of viscous flow. For the same power intensity, higher frequency US would lead to higher energy absorption and hence generates greater acoustic flow than lower frequencies.

Due to inelastic and incompressible properties of most of the liquids, the movement of liquid is insufficient to accommodate the pressure variation. Therefore, the liquid can be torn apart if the changes in pressure are great enough, which causes the molecules to exceed the minimum molecular distance (Bilek and Turantas 2013). This would result in the formation of microbubbles of gas and vapor in order to relieve the tensile stresses. These bubbles would then begin to expand and collapse under the influence of the acoustic field. This process of formation, growth, and collapse of bubbles is known as *cavitation*.

The expansion/collapse cycle can be further divided into two different patterns. First is sinusoidal cavitation, where the cycle mimics the acoustic wave. Next is inertial cavitation, where for a certain bubble size and acoustic pressure, the bubble expansion phase is extended, followed by a violent collapse back to a smaller bubble size. Inertial cavitation may then be further divided into repetitive transient cavitation and transient cavitation. At repetitive transient cavitation, bubble oscillation can persist for many hundreds of acoustic cycles. On the other hand, transient cavitation refers to the growth and collapse of bubbles spectacularly within few acoustic cycles, and the collapsed bubbles then disintegrate into a mass of smaller bubbles if the acoustic amplitude is higher. This phenomenon is generally observed at low frequency (20–100 kHz) (Kentish and Ashokkumar 2011). Air is transferred into the bubble during expansion phase, then leaked out during collapse. This phenomenon is attributed to thinner mass transfer boundary layer and greater interfacial area during expansion than during collapse, resulting in larger bubbles growing over a very large number of acoustic cycles, which is also known as rectified diffusion (Lee et al. 2005). This is not the only reason for the formation of large bubbles by the coalescence of smaller bubbles.

The collapse of a bubble during repetitive transient or transient cavitation can generate extremely high pressures (70–100 MPa). This would cause outward propagating shockwaves, which then result in severe turbulence within the immediate surroundings. These shockwaves are strong enough to shear and break cell walls and membrane structure (Butz and Tauscher 2002, Piyasena et al. 2003, Bilek and Turantas 2013).

However, an increase in external pressure may reduce the number of bubbles formed. In contrast, it would also increase the collapse pressure during transient cavitation, hence producing more violent collapse. On the contrary, the water vapor pressure inside the cavitating bubbles increases as the external temperature increases. This would restrain the collapse event due to the "cushioning" effect of the water vapor. Hence, US is less effective at temperatures significantly above ambient levels.

15.2.2 Chemical Effects of US

Regardless of different mechanisms to dissipate acoustic energy, this energy is ultimately converted to heat. Hence, all applications of US will result in an increase in temperature. This can be explained by the enormous temperature at a localized level (>5000 K) being generated during violent collapse of bubbles throughout repetitive transient and transient cavitation (Ashokkumar and Mason 2007). The high temperature would then result in the formation of reactive hydroxyl and hydroperoxyl radicals that act as strong oxidizing agents. There are three potential sites for the sonochemical activities, which are as follows (Berberidoua et al. 2007):

1. The first potential site is the gaseous region in the bubble. At this region, volatile and hydrophobic species are easily degraded through pyrolytic reactions. Water sonolysis would also occur here, whereby hydroxyl radicals are formed (Equation 15.1).

$$H_2O \rightarrow H\bullet + OH\bullet \quad (15.1)$$

2. Another potential site is the bubble–liquid interfacial region between gas-phase cavitation bubble and the liquid-phase bulk solution, where hydroxyl radicals are localized. At this region, pyrolytic reactions may occur; however, radical reactions would predominate.
3. Lastly, the bulk liquid is another potential site. At this region, secondary sonochemical activity may take place, where free radicals escaping from the interface migrate to the bulk

liquid. The hydroxyl radicals would have the possibility to recombine yielding hydrogen peroxide, which may then react with hydrogen to regenerate hydroxyl radicals as shown in the following Equations 15.2 and 15.3:

$$OH\bullet + OH\bullet \rightarrow H_2O_2 \quad (15.2)$$

$$H_2O_2 + H\bullet \rightarrow H_2O + OH\bullet \quad (15.3)$$

When the temperature inside the collapsing bubble is at maximum, the number of radicals generated is also considered to be high. This temperature can be increased by increasing the sonication power, increasing the external pressure or decreasing the external temperature. The size of the bubble may also affect the amount of heat generated. The production of free radicals may cause inactivation of the microorganisms (Butz and Tauscher 2002).

15.2.3 Parameters Affecting the Effectiveness of US in Disinfecting the Food Materials

The parameters that would affect the effectiveness of an US treatment in disinfecting the food materials will be described in this section. The parameters include

1. US frequency
2. Amplitude of the ultrasonic waves
3. Treatment time
4. Treatment temperature
5. pH
6. Type of bacteria being tested

The effect of US frequency on microbial inactivation was investigated by Zhou et al. (2012) using 20, 40, and 75 kHz as well as by Joyce et al. (2011) using 20, 40, and 580 kHz. Similar observations have reported that lower US frequencies could lead to better microbial inactivation. The main reason for this phenomenon is due to the limited size of bubbles that would be formed at higher frequencies. On the other hand, fewer but larger bubble size can be obtained at lower frequencies. Hence, these bubbles will release higher energies during the implosion of bubble collapse (Awad 2011).

Table 15.2 shows the results obtained by Guerrero et al. (2001) with the effect of pH, temperature, and amplitude on the inactivation of *S. cerevisiae*. It can be clearly seen that at the temperature of 35°C and 45°C, the decimal reduction time, *D*-value, decreases as the amplitude increases. In other words, a shorter time is required to inactivate 90% population of *S. cerevisiae* present in the culture medium. Patil et al. (2009) have also obtained similar results as shown in Table 15.3. In both of these studies, higher amplitude levels have been corresponded to higher US intensities. Therefore, the intensity of the US effect is directly proportional to the amplitude. As US amplitude increases, more intense cavitation occur; hence, leading to more inactivation. The measurement of amplitude instead of power as an indication of the ultrasonic intensities is reported to be a reliable method for indicating the US power (Tsukamoto et al. 2004).

Treatment time is another factor that would affect the efficiency of disinfection or pasteurization. Salleh-Mack and Roberts (2007) observed an increase in microbial reduction with an increase in sonication time. Intense cavitation may be produced when the exposure time is increased, which can then reduce the higher surface tension caused by organic material and hence an increase in the activation rate (Adekunte et al. 2010b). Not only that, convective heat transfer can be accelerated, which helps to facilitate the breakup and dispersal of microorganism clumps and flocks that render the individual microorganisms more susceptible to heat and chemical

TABLE 15.2
Decimal Reduction Time, *D* (min), and Their Confidence Intervals for *S. cerevisiae* Inactivation by Ultrasonic Treatment in Sabouraud Broth (Influence of pH, Temperature, and Wave Amplitude)

		Temperature					
		35°C		45°C		55°C	
Wave Amplitude (μm)	pH	*D*	R^2_{adj}	*D*	R^2_{adj}	*D*	R^2_{adj}
71.4	5.6	29.1 ± 3.2	0.97	26.3 ± 1.7	0.988	4.3 ± 0.5	0.988
	3	30.9 ± 4.1	0.959	17.7 ± 1.1	0.987	0.6 ± 0.1	0.97
83.3	5.6	26.8 ± 1.3	0.99	18.1 ± 0.5	0.997	2.1 ± 0.2	0.977
	3	24.6 ± 2.1	0.98	16.9 ± 1.1	0.989	1.0 ± 0.1	0.989
95.2	5.6	21.2 ± 0.9	0.993	15.1 ± 0.8	0.99	1.9 ± 0.2	0.986
	3	22.9 ± 2.1	0.981	14.3 ± 0.8	0.992	1.7 ± 0.2	0.991
107.1	5.6	21.4 ± 1.6	0.988	14.1 ± 0.7	0.994	1.3 ± 0.2	0.98
	3	19.5 ± 3.4	0.981	14.7 ± 0.7	0.993	1.7 ± 0.2	0.985

Source: Guerrero, S. et al., *Innov. Food Sci. Emerg. Technol.*, 2(1), 31, 2001.

TABLE 15.3
***D*-Values and R^2 Values for US Treatment of Control and Acid-Adapted *E. coli* ATCC 25922 and *E. coli* NCTC 12900**

	Control		1 h		4 h		18 h	
Amplitude (μm)	*D*	R^2	*D*	R^2	*D*	R^2	*D*	R^2
***E. coli* ATCC 25922**								
0.4	13.73 ± 0.9	0.99	8.83 ± 0.03	0.99	12.46 ± 0.1	0.97	14.16 ± 1.0	0.97
7.5	3.44 ± 0.03	0.99	3.21 ± 0.22	0.98	3.29 ± 0.1	0.99	3.34 ± 0.03	0.99
37.5	2.23 ± 0.1	0.99	2.12 ± 0.16	0.98	2.43 ± 0.3	0.96	2.98 ± 0.17	0.98
***E. coli* NCTC 12900**								
0.4	15.26 ± 0.1	0.99	13.47 ± 0.12	0.99	15.78 ± 1.5	0.98	13.48 ± 1.1	0.97
7.5	3.05 ± 0.3	0.95	4.02 ± 0.2	0.99	4.15 ± 0.08	0.99	4.48 ± 0.09	0.99
37.5	2.75 ± 0.1	0.99	2.55 ± 0.09	0.98	2.60 ± 0.09	0.99	2.69 ± 0.09	0.99

Source: Patil, S. et al., *Innov. Food Sci. Emerg. Technol.*, 10(4), 486, 2009.

products (Nafar et al. 2013). However, a contradicting result was obtained by Zhou et al. (2012) as they strongly suggest that the logarithmic count reduction is dependent on the deposited energy.

Ugarte-Romero et al. (2006) found that with an increase in temperatures, that is, 40°C, 50°C, and 60°C, US can be more effective and increase cell destruction by 5.3, 5.0, and 0.1 log cycles. Bubbles are formed more rapidly at higher temperatures due to an increase in the vapor pressure and decrease in tensile strength. However, there should be an upper limit threshold in which no significant effect may be observed beyond this temperature. This may be attributed to higher vapor tension producing vapor-filled bubbles, which cushion the intensity of bubble collapse, thereby affecting microbial inactivation (Bermúdez-Aguirre et al. 2009).

Salleh-Mack and Roberts (2007) observed that a decrease in pH has a significant effect on increasing the effectiveness of microbial inactivation, regardless of the organic acids used as shown

TABLE 15.4
Average Reduction for Organic Acid and pH Study (No Soluble Solids, 0 g/100 mL) with Temperature Control

Sample	pH	Avg. T_f (°C)	Time (min)	Reduction[A] (CFU/mL)
Control	7.0	27.1	10	5.54 ± 0.25
Citric	2.5	32.5	9	5.42 ± 0.24[a]
Citric	4	28.4	9	5.07 ± 0.03[b]
Malic	2.5	29.9	9	5.37 ± 0.10[a]
Malic	4	29.7	9	5.06 ± 0.03[b]

Source: Salleh-Mack, S.Z. and Roberts, J.S., *Ultrason. Sonochem.*, 14(3), 323, 2007.
[a,b] Different letters indicate a significant difference in descending order (least significant difference at P = 0.05).
[A] Mean ± standard deviation for n = 3 experiments and P = 0.05.

in Table 15.4. The possible reason for this phenomenon is the lowering of pH that might reduce the resistance of microbial organisms toward sonication. In contrast, Guerrero et al. (2001) have shown an opposite result where pH has no significant effect on the microbial inactivation, as seen in Table 15.2, until the temperature increased to 55°C. This may be attributed to cell disruption due to high temperature instead of the effect of pH.

The type of bacteria being treated would also affect the efficiency of microbial inactivation. Gao et al. (2013) investigated that bacteria comprising thick and soft capsules (biopolymer layers) prevent cavitation bubbles from collapsing near the plasma membrane. Also, the soft capsules are highly hydrated and help absorb the mechanical forces exerted on the bacteria cell, thus preventing the breakup of the bacterial cell.

15.2.4 Effect of US on Food Quality (Food Regulatory)

Sonication has shown no significant changes in terms of physiochemical properties, which include pH, total soluble solids (Brix), and titratable acidity (Bull et al. 2004, Tiwari et al. 2008, Bhat et al. 2011).

However, significant differences in color were observed with the treatment of US. Santhirasegaram et al. (2013) observed that lightness (L^*) has increased, while redness ($+a^*$) and yellowness ($+b^*$) have decreased after chokanan mango juice was subjected to US treatment. Similar results have also been obtained by Tiwari et al. (2008) on orange juice. This phenomenon was due to the influence of particulate fractions (Ugarte-Romero et al. 2006). Not only that, cavitation, which governs various physical, chemical, or biological reactions such as acceleration of chemical reactions, increasing of diffusion rates, dispersing the aggregates, or breakdown of susceptible particles (microorganisms and enzymes), may also be one of the reasons for this result (Sala et al. 1995). Color degradation may also be due to the extreme physical phenomena during bubbles implosion.

Santhirasegaram et al. (2013) and Adekunte et al. (2010a) observed a lesser or slight decrease in ascorbic acid (vitamin C) content in juice processed by sonication when compared to thermal treatment. This is mainly attributed to the thermolysis of ascorbic acid in cavities, which subsequently triggered the Maillard reaction (Tiwari et al. 2009). On the other hand, Aadil et al. (2013) and Abid et al. (2013) reported an increase in ascorbic acid content. This increase could be ascribed to the absence of heat supply and the removal of entrapped oxygen due to cavitation (Cheng et al. 2007). The possible reason for this contradiction may be due to different frequency and amplitude of US used and also the treatment time.

15.2.5 Combination of US with Other Techniques to Enhance Microbial Inactivation

Sonication has been proved to effectively inactivate most of the microorganisms. However, sonication alone might not be as effective as expected if it is to be applied to practical use. The main reason is the cost, which would be high due to long contact time and US power required to achieve higher level of microorganism inactivation. Hence, using US in conjunction with other treatments might be a better way to enhance microbial inactivation.

Several researches have been done on the combination of US with heat (thermosonication, TS), pressure (manosonication, MS) or with both heat and pressure (manothermosonication, MTS) to investigate the synergistic effect between the combination.

Temperature has a large, direct effect on ultrasonic inactivation. Higher temperature treatment had a significantly less sonication time requirement than those maintained at a lower temperature (Salleh-Mack and Roberts 2007). Sonication weakens the cell wall of microorganisms, hence causing it to be more vulnerable to any external stresses such as heat. López-Malo et al. (2005) have shown that the *D*-values were lowered significantly for TS as compared to thermal treatments alone, as shown in Table 15.5. A decrease in spore resistance can be clearly seen at an US amplitude of 90 or 120 μm and at a temperature of 57.5°C for *Aspergillus flavus* and 50°C for *Penicillium digitatum*.

TABLE 15.5
Decimal Reduction Time (*D*, min) for *A. flavus* and *P. digitatum* Inactivation during Thermal and Thermosonication Treatments at Selected US Wave Amplitudes and Water Activities (a_w)

		A. flavus		*P. digitatum*	
pH	**US Amplitude (μm)**	***T* (°C)**	**a_w 0.99**	***T* (°C)**	**a_w 0.99**
5.5	120	52.5	6.2	45	29.89
	90		16.28		35.47
	60		28.18		43.92
	0		30.42		62.74
3	120		6.15		20.01
	90		6.77		22.46
	60		30.24		36.65
	0		31.7		50.35
5.5	120	55	5.06	47.5	26.74
	90		6.26		28.24
	60		7.44		35.63
	0		17.4		51.53
3	120		5.15		17.59
	90		5.34		21.15
	60		5.89		27.41
	0		11.04		33.42
5.5	120	57.5	1.59	50	9.59
	90		1.93		11.56
	60		3.72		14.7
	0		4.55		25.42
3	120		1.94		6.03
	90		2.14		8.63
	60		2.34		12.01
	0		4.36		22.01
5.5	120	60	1.2	52.5	5.33
	90		1.59		7.66
	60		2.46		9.85
	0		2.6		13.3
3	120		0.8		3.81
	90		0.82		4.8
	60		1.67		6.95
	0		1.63		9.54

Source: López-Malo, A. et al., *J. Food Eng.*, 67(1–2), 87, 2005.

However, the benefits of US application were reduced after this temperature. This may be attributed to cavitation dampening due to high vapor pressure within the bubble. Significant water vapor pressure in the bubble cushions the intensity of the collapse. Similar results have been obtained by Kiang et al. (2013), where the introduction of high-intensity US enhanced the inactivation of pathogens as compared to thermal treatment alone, as seen in Table 15.6. It was also observed that *S. enteritidis* was more sensitive to thermosonication than *E. coli* O157:H7.

High-power ultrasonic waves under pressure at nonlethal temperature MS have great potential for the inactivation of microorganisms of temperature-sensitive food as thermal treatment would impair the quality of this food. Lee et al. (2003) obtained a slight enhancement in the inactivation of *E. coli* using MS treatment when compared to both the individual treatments, as shown in Figure 15.2. Huang et al. (2006) obtained similar results, where a significant reduction of *S. enteritidis* was observed when US and high hydrostatic pressure treatment were used. However, no synergistic effect was observed, as shown in Figure 15.3. Higher hydrostatic pressure during ultrasonic treatment decreases the vapor pressure inside the bubble, which then results in a higher intensity of bubble implosion. The lower effect in their study (Lee et al. 2003) was mainly attributed to the low temperature of liquid whole egg (LWE) (5°C) since most microorganisms showed greater sensitivity to US at higher temperature, generally above 50°C. Similar to thermosonication, an upper pressure limit exists in MS. When MS exceeds its upper limit, the ultrasonic amplitude is insufficient to overcome the hydrostatic pressure and cohesive force of the liquid. Therefore, the number of bubbles undergoing cavitation will decrease leading to a decrease in lethality of MS.

As mentioned previously, MS was conducted under sublethal temperature, which led to an improvement in the inactivation rate, but no synergistic effect was observed between US and high hydrostatic pressure. Lee et al. (2009) observed that the inactivation rate of *E. coli* by MS improved as compared to sonication alone. However, the lethal effect of MS was not influenced by a temperature up to 54°C. After the lethal temperature of 61°C, a drastic increase in the inactivation rate was observed. Therefore, studies have been carried out to analyze the synergistic effect between sublethal heat and MS (manothermosonication, MTS). Arroyo et al. (2011) obtained synergistic lethal effect of MTS in *Cronobacter sakazakii* after the sublethal temperature of 60°C as damaged cells were not detected after MS treatments at 35°C but were detected after MTS treatment after 60°C. Hence, they deduced that this synergistic effect was solely due to heat and not due to a higher sensitivity of heat-damaged cells to US under pressure.

TABLE 15.6
Decimal Reduction Time (*D*-Value) of *E. coli* O157:H7 and *S. enteritidis* in Mango Juice Treated with and without Sonication at 50°C and at 60°C

			***D*-Value (min)**	
Temperature(°C)	**Treatment Condition**	**Agar**	***E. coli* O157:H7**	***S. enteritidis***
50	Without sonication	Nonselective	4.81	3.47
		Selective	3.31	1.72
	With sonication	Nonselective	3.20	3.09
		Selective	2.04	1.02
60	Without sonication	Nonselective	1.48	0.81
		Selective	1.42	0.97
	With sonication	Nonselective	1.36	0.87
		Selective	0.86	1.12

Source: Kiang, W.-S. et al., *Lett. Appl. Microbiol.*, 56(4), 251, 2013.

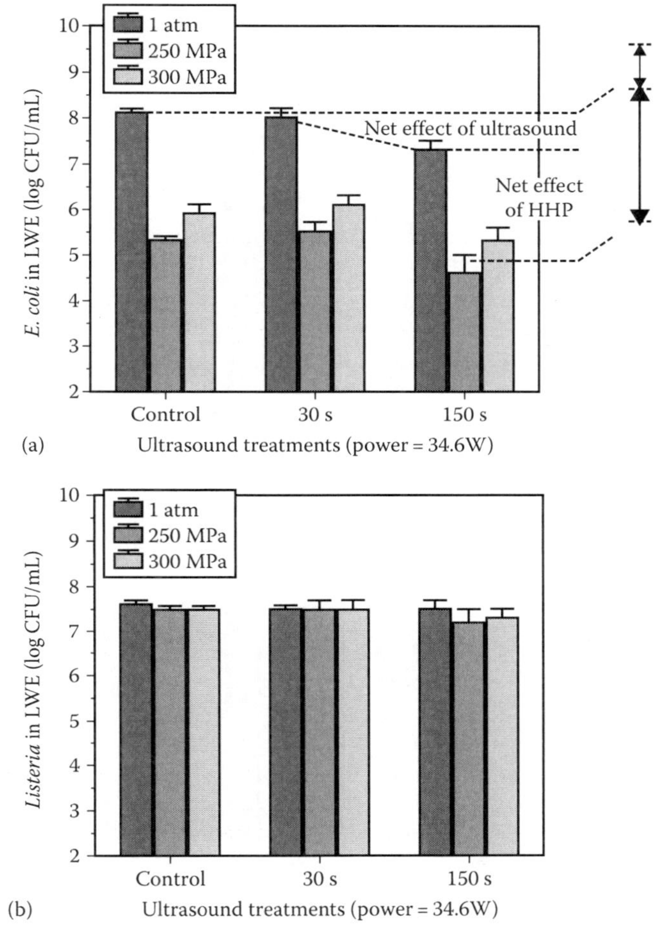

FIGURE 15.2 Combined treatments of US and HHP on (a) *E. coli* DH 5α and (b) *Listeria seeligeri* in LWE. (From Lee, D.U. et al., *Innov. Food Sci. Emerg. Technol.*, 4(1), 387, 2003.)

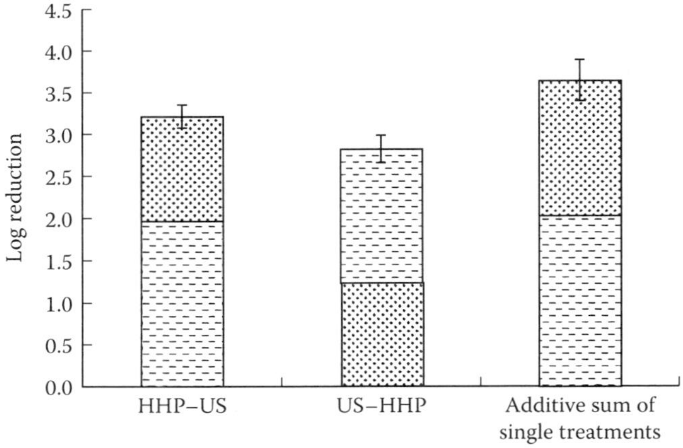

FIGURE 15.3 Combination of high hydraulic pressure with US. (From Huang, E. et al., *Biosyst. Eng.*, 94(3), 403, 2006.)

As mentioned in the previous section, UV is another promising nonthermal disinfection technology. The combined effect of UV-C and high-intensity US on *S. cerevisiae* and *Listeria monocytogenes* in buffer and clarified apple juice has been investigated by López-Malo et al. (2006). They found out that the combination of UV-C/US enhanced the individual inactivation on *L. monocytogenes* and *S. cerevisiae* in apple juice with the majority of the population dead in the first few minutes of treatment. Similar results have been reported by Char et al. (2010), where the poor effect of UV-C in orange juice was enhanced by the combination with US. The poor effect is attributed to the presence of colored compounds and pulp particles, which caused poor UV-C light transmission. In addition to that, Char et al. (2010) discovered that the combined treatment was more effective in a simultaneous arrangement rather than in a series arrangement.

15.3 CONCLUSIONS AND FUTURE DIRECTIONS

US or sonication is a promising nonthermal technique in microbial inactivation. However, sonication alone might not be feasible in practical usage. For that reason, the future use of US for microbial inactivation lies in multifactorial processes. As more investigations are under progress to identify the controlling factors to disinfect different microorganisms, this technique could be used in multiple stages to disinfect the undesired pathogens.

REFERENCES

Aadil, R. M., X.-A. Zeng, Z. Han, and D.-W. Sun. Effects of ultrasound treatments on quality of grapefruit juice. *Food Chemistry* 141(3) (2013): 3201–3206.

Abid, M. et al. Effect of ultrasound on different quality parameters of apple juice. *Ultrasonics Sonochemistry* 20(5) (2013): 1182–1187.

Adekunte, A. O., B. K. Tiwari, P. J. Cullen, A. G. M. Scannell, and C. P. O'Donnell. Effect of sonication on colour, ascorbic acid and yeast inactivation in tomato juice. *Food Chemistry* 122(3) (2010a): 500–507.

Adekunte, A., B. K. Tiwari, A. Scannell, P. J. Cullen, and C. O'Donnell. Modelling of yeast inactivation in sonicated tomato juice. *International Journal of Food Microbiology* 137(1) (2010b): 116–120.

Arroyo, C., G. Cebrián, R. Pagán, and S. Condón. Inactivation of *Cronobacter sakazakii* by ultrasonic waves under pressure in buffer and foods. *International Journal of Food Microbiology* 144(3) (2011): 446–454.

Ashokkumar, M. and T. J. Mason. *Sonochemistry. Kirk-Othmer Encyclopedia of Chemical.* New York: Wiley, 2007.

Awad, S. B. High power ultrasound in surface cleaning. In *Ultrasound Technologies for Food and Bioprocessing*, (eds.) H. Feng, G. V. Barbosa-Cánovas, and J. Weiss. New York: Springer, 2011, pp. 545–582.

Berberidoua, C., I. Pouliosa, N. P. Xekoukoulota, and D. Mantzavinosb. Sonolytic, photocatalytic and sonophotocatalytic degradation of malachite green in aqueous solutions. *Applied Catalysis B: Environmental* 74(1–2) (2007): 63–72.

Bermúdez-Aguirre, D., M. G. Corradini, R. Mawson, and G. V. Barbosa-Cánovas. Modeling the inactivation of *Listeria innocua* in raw whole milk treated under thermo-sonication. *Innovative Food Science and Emerging Technologies* 10(2) (2009): 172–178.

Bhat, R., N. S. Kamaruddin, L. Min-Tze, and A. A. Karim. Sonication improves kasturi lime (*Citrus microcarpa*) juice quality. *Ultrasonics Sonochemistry* 18(6) (2011): 1295–1300.

Bilek, S. E. and F. Turantas. Decontamination efficiency of high power ultrasound in the fruit and vegetable industry: A review. *International Journal of Food Microbiology* 166(1) (August 2013): 155–162.

Block, S. S. *Disinfection, Sterilization, and Preservation*, 5th ed. (ed.) Seymour Stanton Block. Philadelphia, PA: Lippincott Williams & Wilkins, 2001.

Bull, M., M. Zerdin, E. Howe, D. Goicoechea, P. Paramanandhan, and R. Stockman. The effect of high pressure processing on the microbial, physical and chemical properties of Valencia and Navel orange juice. *Innovative Food Science and Emerging Technologies* 5(2) (2004): 135–149.

Butz, P. and B. Tauscher. Emerging technologies: Chemical aspects. *Food Research International* 35(2/3) (2002): 279–284.

Cao, S., Z. Hu, B. Pang, H. Wang, H. Xie, and F. Wu. Effect of ultrasound treatment on fruit decay and quality maintenance in strawberry after harvest. *Food Control* 12(4) (2010): 529–532.

Char, C. D., E. Mitilinaki, S. N. Guerrero, and S. M. Alzamora. Use of high-intensity ultrasound and UV-C light to inactivate some microorganisms in fruit juices. *Food and Bioprocess Technology* 3(6) (2010): 797–803.

Chemat, F., Y.-J. Zill-e-Huma, and M. K. Khan. Applications of ultrasound in food technology: Processing, preservation and extraction. *Ultrasonics Sonochemistry* 18(4) (July 2011): 813–835.

Cheng, L. H., C. Y. Soh, S. C. Liew, and F. F. Teh. Effects of sonication and carbonation on guava juice quality. *Food Chemistry* 104(4) (2007): 1396–1401.

Gao, S. P., G. D. Lewis, M. Ashokkumar, and Y. Hemar. Inactivation of microorganisms by low-frequency high-power ultrasound: 1. Effect of growth phase and capsule properties of the bacteria. *Ultrasonics Sonochemistry* 21(1) (2013): 446–453.

Guerrero, S., A. López-Malo, and S. M. Alzamora. Effect of ultrasound on the survival of *Saccharomyces cerevisiae*: Influence of temperature, pH and amplitude. *Innovative Food Science and Emerging Technologies* 2(1) (2001): 31–39.

Guerrero-Beltrán, J. A. and G. V. Barbosa-Cánovas. Advantages and limitations on processing foods by UV light. *Food Science and Technology International* 10 (2004): 137–147.

Huang, E., G. S. Mittal, and M. W. Griffiths. Inactivation of *Salmonella enteritidis* in liquid whole egg using combination treatments of pulsed electric field, high pressure and ultrasound. *Biosystems Engineering* 94(3) (2006): 403–413.

IDFA. *Pasteurization: Definition and Methods*. International Dairy Foods Association. June 2009. http://www.idfa.org/files/249_Pasteurization%20Definition%20and%20Methods.pdf (accessed September 13, 2013).

Joyce, E., A. Al-Hashimi, and T. J. Mason. Assessing the effect of different ultrasonic frequencies on bacterial viability using flow cytometry. *Journal of Applied Microbiology* 110 (2011): 862–870.

Kentish, S. and M. Ashokkumar. The physical and chemical effects of ultrasound. In *Ultrasound Technologies for Food and Bioprocessing*, (ed.) H. Feng, G. V. Barbosa-Cánovas, and J. Weiss. New York: Springer-Verlag New York Inc., 2011, pp. 1–12.

Kiang, W.-S., R. Bhat, A. Rosma, and L.-H. Cheng. Effects of thermosonication on the fate of *Escherichia coli* O157:H7 and *Salmonella Enteritidis* in mango juice. *Letters in Applied Microbiology* 56(4) (2013): 251–257.

Koutchma, T. UV light for processing foods. *IUVA News* 10(4) (December 2008): 24–29.

Lee, D. U., V. Heinz, and D. Knorr. Effects of combination treatments of nisin and high-intensity ultrasound with high pressure on the microbial inactivation in liquid whole egg. *Innovative Food Science and Emerging Technologies* 4(1) (2003): 387–393.

Lee, H., B. Zhou, W. Liang, H. Feng, and S. E. Martin. Inactivation of *Escherichia coli* cells with sonication, manosonication, thermosonication, and manothermosonication: microbial responses and kinetics modeling. *Journal of Food Engineering* 93(3) (2009): 354–364.

Lee, J., S. Kentish, and M. Ashokkumar. Effect of surfactants on the rate of growth of an air bubble by rectified diffusion. *The Journal of Physical Chemistry* 109(30) (2005): 14595–14598.

López-Malo, A., S. Guerrero, A. Santiesteban, and S. M. Alzamora. Inactivation kinetics of Saccharomyces cerevisiae and Listeria monocytogenes in apple juice processed by novel technologies. In *Proceedings of 2nd Mercosur Congress on Chemical Engineering. 4th Mercosur Congress on Process Systems Engineering*. Paper no. 0681. 2005.

López-Malo, A., E. Palou, M. Jiménez-Fernández, S. M. Alzamora, and S. Guerrero. Multifactorial fungal inactivation combining thermosonication and antimicrobials. *Journal of Food Engineering* 67(1–2) (2005): 87–93.

Malaysia. Food-borne disease surveillance at the national level: A Malaysian perspective. In *Second FAO/WHO Global Forum of Food Safety Regulators*. Bangkok, Thailand: Agriculture and Consumer Protection, 2004.

McDonnell, G. *Sterilization and Disinfection*, Vol. I. In *Encyclopedia of Microbiology*, (ed.) M. Schaechter. Basingstoke, England: Elsevier Inc., 2009, pp. 529–548.

Nafar, M., Z. Emam-djomeh, S. Yousefi, and M. H. Ravan. An optimization study on the ultrasonic treatments for *Saccharomyces cerevisiae* inactivation in red grape juice with maintaining critical quality attributes. *Journal of Food Quality* 36(4) (2013): 269–281.

Patil, S., P. Bourke, B. Kelly, J. M. Frías, and P. J. Cullen. The effects of acid adaptation on *Escherichia coli* inactivation using power ultrasound. *Innovative Food Science and Emerging Technologies* 10(4) (2009): 486–490.

Piyasena, P, E. Mohareb, and R. C. McKellar. Inactivation of microbes using ultrasound: A review. *International Journal of Food Microbiology* 87(3) (November 2003): 207–216.

Sala, F. J., J. Burgos, S. Condon, P. Lopez, and J. Raso. Effect of heat and ultrasound on microorganisms and enzymes. In *New Methods of Food Preservation*, (ed.) G. W. Gould. London, U.K.: Blackie Academic & Professional, 1995, pp. 176–204.

Salleh-Mack, S. Z. and J. S. Roberts. Ultrasound pasteurization: The effects of temperature, soluble solids, organic acids and pH on the inactivation of *Escherichia coli* ATCC 25922. *Ultrasonics Sonochemistry* 14(3) (2007): 323–329.

Santhirasegaram, V., Z. Razali, and C. Somasundram. Effects of thermal treatment and sonication on quality attributes of Chokanan mango (*Mangifera indica* L.) juice. *Ultrasonics Sonochemistry* 20(5) (2013): 1276–1282.

Shama, G. UV disinfection in the food industry. *Controlled Environments Magazine* (Vicon Publishing Inc.) 10(4) (2007): 10–15.

Solsona, F. and J. P. Méndez. Water disinfection. 2003. http://whqlibdoc.who.int/paho/2003/a85637.pdf (accessed August 5, 2013).

Tiwari, B. K., K. Muthukumarappan, C. P. O'Donnell, and P. J. Cullen. Colour degradation and quality parameters of sonicated orange juice using response surface methodology. *LWT—Food Science and Technology* 41(10) (2008): 1876–1883.

Tiwari, B. K., C. P. O'Donnell, K. Muthukumarappan, and P. J. Cullen. Ascorbic acid degradation kinetics of sonicated orange juice during storage and comparison with thermally pasteurized juice. *LWT—Food Science Technology* 42 (2009): 700–704.

Tsukamoto, I., E. Constantinoiu, M. Furuta, R. Nishimura, and Y. Maeda. Inactivation effect of sonication and chlorination on *Saccharomyces cerevisiae*. Calorimetric Analysis. In *4th Conference on the Applications of Power Ultrasound in Physical and Chemical Processing*, (ed.) J.-Y. Hihn. Besancon, France, 2004, pp. 167–172.

Ugarte-Romero, E., H. Feng, S. E. Martin, K. R. Cadwallader, and S. J. Robinson. Inactivation of *Escherichia coli* with power ultrasound in apple cider. *Journal of Food Science* 71(2) (2006): 102–108.

Wang, Y., Y. Hu, J. Wang, Z. Liu, G. Yang, and G. Geng. Ultrasound-assisted solvent extraction of swainsonine from Oxytropis ochrocephala Bunge. *Journal of Medicinal Plants Research* 5(6) (2011): 890–894.

Zhou, B., H. Feng, and A. J. Pearlstein. Continuous-flow ultrasonic washing system for fresh produce surface decontamination. *Innovative Food Science and Emerging Technologies* 16 (October 2012): 427–435.

16 Edible Coatings and Films to Preserve Quality of Fresh Fruits and Vegetables

Constantina Tzia, Loucas Tasios, Theodora Spiliotaki, Charikleia Chranioti, and Virginia Giannou

CONTENTS

16.1 INTRODUCTION

The use of edible films and coatings in fresh fruits and vegetables preservation is not a new concept. Coating techniques to retard moisture loss in fresh produce had been in use for decades. For example, coating of fresh oranges and lemons with beeswax was practiced in China in the twelveth and thirteenth century. Yuba (soy protein edible film) was traditionally used in Asia since the fifteenth century (Gennadios et al., 1993) and "larding" (coating food with fat) was practiced in UK in the sixteenth century. In the 1930s, hot-melt paraffin waxes became commercially available for coating citrus fruits, and in the early 1950s, carnauba wax oil-in-water emulsions were developed for coating fresh fruits and vegetables (Kaplan, 1986). Development of edible coatings for use on meat products was first reported in the late 1950s.

Over the past 20 years, considerable work, reported in both the scientific and patent literature, has been done on the formation and characterization of edible films and coatings. A variety of GRAS polysaccharides, proteins, and lipids has been utilized, either alone or in mixtures, to produce composite edible films. Today, more than 60 edible films and coatings are commercially available for fresh fruits and vegetables and optimized for specific use. However, most of them are wax-based and intended for appearance improvement and prevention of water loss. Most commercial films and coatings have been tested on fresh fruits and vegetables, while recent research is referred particularly to the use of new ingredients and composite materials (Baldwin and Baker, 2002; Krochta, 2002; Olivas and Barbosa-Canovas, 2005).

The development of edible coatings technology with applications to fresh and minimally processed fruits is one of the challenges of the postharvest industry. Fruits and vegetables are perishable products with active metabolism during the postharvest period that can result in deteriorative changes. Various approaches have been used to minimize postharvest deterioration of fresh and minimally processed produce, which include low-temperature storage, special preparation techniques, use of additives, modified-atmosphere (MA) or controlled-atmosphere (CA), and application of edible coatings. The rationale for using edible coatings to extend shelf life and improve the quality of these produce is based on the formation of an artificial barrier that may result in (1) the reduction of moisture migration, (2) selective control of gas diffusion, and (3) suppression of undesirable physiological changes. Film's and coating's ingredients, method of formation, and type of application are the key factors that affect the characteristics (gas permeability, water vapor permeability) of any film or coating (Gontard et al., 1996). The kind of fruit or vegetable and the specific demands specify the above selections. Edible coatings, in particular, find applications in fresh-cut fruits which are more susceptible to alteration (enzymatic browning, texture decay, microbial contamination, and undesirable volatile production), improving their handling and functionality.

According to new trends on edible coatings, through the incorporation and/or controlled release of antioxidants, vitamins, nutraceuticals, and/or natural antimicrobial agents, the design of new products is allowed, and the production of high added-value products is possible (Baldwin, 2007; Vargas et al., 2008; Pavlath and Orts, 2009).

16.2 RATIONALE FOR USING EDIBLE FILMS AND COATINGS ON FRESH PRODUCE

Fresh fruits and vegetables are essential components of the human diet as they contain a number of important nutritive compounds, such as vitamins, which cannot be synthesized by the human body. A fruit or vegetable is a living, respiring, edible tissue that has been detached from the parent plant. Fruits and vegetables are perishable products with active metabolism during the postharvest period (Robertson, 1993). Water loss and postharvest decay account for most losses. These have been estimated to be more than 40%–50% in the tropics and subtropics (Kadam and Salunkhe, 1995). The main causes of quality and nutritional value loss of fresh produce are all interconnected. Once harvested, a fruit or vegetable no longer has any new nutrients coming into it from the root system. It must thus rely on the endogenous nutrients along with oxygen to continue its respiratory processes. Since enzymes are not inactivated as it occurs in canning, or partially inactivated as in freezing or drying, the fresh, respiring produce may improve in quality in the short term but eventually will deteriorate and decay (Labuza and Breene, 1989). The shelf life of fruits and vegetables can be extended by, in simple terms, retarding the physiological, pathological, and physical deteriorative processes (generally referred to as postharvest handling) or by inactivating the physiological processes (generally referred to as food preservation) (Robertson, 1993). One possible method of extending postharvest storage life of fresh fruits and vegetables is the use of edible coatings. Such coatings are made of edible materials that are used to enrobe fresh product, providing a semipermeable barrier to gases and water vapor, thereby reducing respiration and water loss. Some background knowledge of postharvest physiology, storage techniques, and edible coating technology is helpful to understand the effect of coatings on fresh fruits and vegetables.

16.3 FRUIT/VEGETABLE PHYSIOLOGY

16.3.1 General

Fruits and vegetables have many similarities with respect to their compositions, methods of cultivation and harvesting, storage properties, and processing. In fact, many vegetables are considered fruits in the true botanical sense (Potter and Hotchkiss, 1995). Botanically, the word "fruit" refers to the mature seed-bearing structures of flowering plants; this covers a very wide and heterogeneous group of plant products, including cereals, pulses, oilseeds, spices, and fleshy fruits. The edible fleshy fruits, however, represent a well-defined class on their own and exhibit a wide variety of plant products. They have much in common, from a culinary point of view, with the soft, edible structures developed from other parts of the plant body, commonly referred to as vegetables. Although botanically the line between fruits and vegetables cannot be clearly drawn, the products have been differentiated based on common verbal usage and the way in which they are consumed (Desai and Salunkhe, 1991). A consumer definition of fruit would be "plant products with aromatic flavors, which are either naturally sweet or normally sweetened before eating"; they are essentially dessert foods (Wills et al., 1998). The term "vegetable," in contrast, is applied to all the other soft, edible plant products that are usually eaten with meat, fish, or other savory dish, either fresh or cooked (Desai and Salunkhe, 1991).

The composition of fruits and vegetables depends not only on botanical variety, cultivation practices, and weather but also on the degree of maturity prior to harvest and the condition of ripeness, which continues after harvest and is influenced by storage conditions. Nevertheless, some generalizations can be made.

Most fresh fruits and vegetables are high in water, low in protein, and low in fat. The water content is generally greater than 70% and frequently greater than 85%. Interestingly, the water content of milk and apples is similar. Commonly, protein content is no greater than 3.5% and fat content no greater than 0.5% (Potter and Hotchkiss, 1995).

16.3.2 Postharvest Physiology

The life cycle of fruits and vegetables can be conveniently divided into three major physiological stages following germination. These are growth, maturation, and senescence (Figure 16.1). However, clear distinction between the various stages is not easily made. Growth involves cell division and subsequent cell enlargement, which accounts for the final size of the produce. Maturation usually commences before growth ceases and includes different activities in different commodities. Growth and maturation are often collectively referred to as the development phase. Senescence is defined as the period when anabolic (synthetic) biochemical processes give way to catabolic (degradative) processes, leading to ageing and finally death of the tissue. Ripening (a term reserved for fruit) is generally considered to begin during the later stages of maturation and to be the first stage of senescence. The change from growth to senescence is relatively easy to delineate. Often the maturation phase is described as the time between these two stages, without any clear definition on a biochemical or physiological basis (Robertson, 1993; Wills et al., 1998).

16.3.2.1 Respiration

Respiration (biological oxidation) is the oxidative breakdown of the more complex substrates normally present in the cells, such as starch, sugars, and organic acids, to simpler molecules (CO_2 and H_2O), with the concurrent production of energy and other molecules, which can be used by the cell for synthetic reactions (Kader, 1987). Respiration can occur in the presence of oxygen (aerobic respiration) or in the absence of oxygen (anaerobic respiration, which is sometimes called

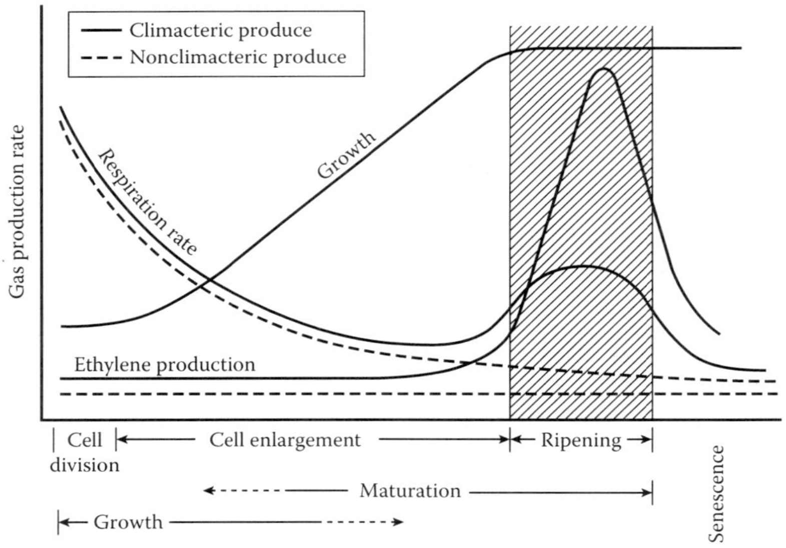

FIGURE 16.1 Growth, respiration, and ethylene production patterns of climacteric and nonclimacteric plant organs. (From Wills, R. et al., Structure and composition, in: *Postharvest, an Introduction to the Physiology and Handling of Fruit, Vegetables and Ornamentals*, 4th edn., UNSW Press, Sydney, New South Wales, Australia, pp. 15–32, 1998.)

fermentation). The greatest yield of energy is obtained when the process takes place in the presence of molecular oxygen (aerobic respiration). If hexose sugar is used as the substrate, the chemical reaction can be written as follows:

$$C_6H_{12}O_6 + 6O_2 \rightarrow 6CO_2 + 6H_2O + \text{energy} \qquad \text{Chemical Reaction 16.1}$$

This transformation actually takes place in a large number of individual stages with the participation of many different enzymes systems (Robertson, 1993). The water produced remains within the tissue but the CO_2 escapes and accounts for part of the weight loss of harvested fruits and vegetables, in the range of 3%–5%. However, the generation of heat within the fruit may lead to additional loss of weight. This heat is dissipated through direct heat transfer to the environment and through evaporation of water. The respiration heat raises the tissue temperature and, therefore, increases transpiration (Ben-Yehoshua, 1987). When one mole of hexose sugar is oxidized, 38 moles of ATP (each possessing 32 kJ of useful energy) are formed. This represents about 40% of total free energy change, the remainder being dissipated as heat. Rapid removal of this heat is usually desirable and it is important that the packaging assists rather than impedes this process (Kader, 1987; Robertson, 1993).

The respiration rate of produce is an excellent indicator of metabolic activity of the tissue and thus is a useful guide to the potential storage life of the produce: the higher the rate, the shorter the life; and the lower the rate, the longer the life. A characteristic respiratory pattern is obtained if the respiration rate of a fruit or vegetable is measured (as either oxygen consumed or carbon dioxide evolved) during the course of its development, maturation, ripening, and senescence periods. Respiration rate per unit weight is highest for the immature fruit or vegetable and then steadily declines with age (Figure 16.1). A significant group of fruits, including tomato, mango, banana, and apple, shows a variation from the described respiratory pattern in that they undergo a pronounced increase in respiration coincident with ripening. Such an increase in respiration is known as a respiratory climacteric, and this group of fruits is known as the climacteric class of fruits. The respiratory climacteric as well as the complete ripening process may proceed while the fruit is either attached to or detached from the plant. Fruits such as citrus, pineapple, and strawberry that do not exhibit a respiratory climacteric are known as the nonclimacteric class of fruits. Nonclimacteric fruits exhibit most of the ripening changes, although these usually occur more slowly than those of the climacteric fruits. All vegetables can also be considered to have a non-climacteric type of respiratory pattern (Wills et al., 1998).

From measurements of CO_2 and O_2, it is possible to evaluate the nature of the respiratory process. The ratio of the volume of CO_2 released to the volume of O_2 absorbed in respiration is termed as the respiratory quotient (RQ). Values of RQ range from 0.7 to 1.3 for aerobic respiration, depending on the substrate being oxidized; for carbohydrates $RQ = 1$, for lipids $RQ \leq 1$, and for organic acids $RQ \geq 1$ (Kader, 1987).

Anaerobic respiration involves the incomplete oxidation of compounds in the absence of O_2 and the accumulation of ethanol, acetaldehyde, and CO_2. Much lower amounts of energy (2 moles of ATP) and CO_2 are produced from 1 mole of hexose sugar than that produced under aerobic conditions. Significantly less heat energy (approximately 5%) is produced as well for a given amount of carbohydrate oxidation in anaerobic respiration compared to aerobic respiration. The oxygen concentration at which a shift from aerobic to anaerobic respiration occurs varies among tissues and is known as the extinction point. Very high RQ values (>1.3) usually indicate anaerobic respiration (Robertson, 1993).

16.3.2.1.1 Effect of Oxygen and Carbon Dioxide Concentration

A simple consideration of the chemical reaction (16.1) would suggest that if the CO_2 in the atmosphere were augmented (or the O_2 decreased), the respiration rate and storage life would be extended.

Reduction of the O_2 concentration to less than 10% provides a tool for controlling the respiration rate and slowing down senescence, although an adequate O_2 concentration must be available to maintain aerobic respiration. Vegetable crops usually require a minimum O_2 content of 1%–3% in the storage atmosphere, and at O_2 contents below 2%, most vegetables react with a sudden increase in CO_2 production.

At high CO_2 concentrations (above 20%), a significant increase in anaerobic respiration occurs, which can irreversibly damage plant tissue. Low O_2 and/or high CO_2 can reduce the incidence and severity of certain physiological disorders such as those induced by ethylene (scald of apples and pears) and chilling injury of some commodities (e.g., avocado, citrus fruits, chili pepper, and okra). On the other hand, O_2 and CO_2 levels beyond those tolerated by the commodity can induce physiological disorders, such as brown stain on lettuce, internal browning and surface pitting of pome fruits, and blackheart of potato (Robertson, 1993; Beaudry, 1999).

16.3.2.1.2 Effect of Ethylene

All fruits produce minimum quantities of ethylene during development. However, climacteric fruits produce much larger amounts of ethylene during ripening than nonclimacteric fruits. The internal ethylene concentration of climacteric fruits varies widely but that of nonclimacteric fruits changes little during development and ripening. Ethylene, applied at a concentration as low as 0.1–1.0 μL/L for 1 day, is normally sufficient to hasten full ripening of climacteric fruit, but the magnitude of the climacteric is relatively independent of the concentration of applied ethylene. In contrast, applied ethylene merely causes a transient increase in the respiration of nonclimacteric fruits, the magnitude of the increase depending on the concentration of ethylene. Moreover, the rise in respiration in response to ethylene may occur more than once in nonclimacteric fruits in contrast to the single respiration increase in climacteric fruits (Wills et al., 1998).

16.3.2.1.3 Effect of Temperature

Temperature is the most important environmental factor in the postharvest life of horticultural products because of its dramatic effect on the rates of biological reactions including respiration. Typical Q_{10} values (the ratio of the respiration rates for a 10°C interval) for vegetables are 2.5–4.0 at 0°C–10°C, 2.0–2.5 at 10°C–20°C, 1.5–2.0 at 20°C–30°C, and 1.0–1.5 at 30°C–40°C. Taking mean Q_{10} values, it can be calculated that the relative velocity of respiration would increase from 1.0 at 0°C to 3.0 at 10°C, 7.5 at 20°C, 15.0 at 30°C, and 22.5 at 40°C. These figures dramatically illustrate the need to reduce the temperature of fresh fruits and vegetables as soon as possible after harvesting in order to maximize the shelf life. The rate of increase in respiration rates declines with an increase in temperature up to 40°C, with the Q_{10} becoming less than 1 as the tissue nears its thermal death point (about 50°C–55°C), when enzyme proteins are denatured and metabolism becomes disorderly (Kader, 1987; Robertson, 1993).

16.3.2.1.4 Effect of Mechanical Damage

Physical damage also causes an increase in respiration. Mechanical stress of fresh produce that occurs during harvesting and in postharvest handling operations inevitably results in some bruising and abrasion, scuffing, and skin/periderm damage. This contributes to subsequent quality loss from induction of wound ethylene evolution and further increases respiration rate and accelerates the onset of senescence (Schlimme and Rooney, 1994). The increase of respiration, as measured by the CO_2 evolved, is the result of many biochemical events that subsequently cause changes in color, flavor, texture, and nutritional quality. The increase in the rate of respiration can be in the range of 2–10-fold (Wong et al., 1994b). Therefore, protective packaging has an important role to play in maximizing the shelf life of lightly processed fruits and vegetables.

16.3.2.2 Transpiration

All fruits and vegetables continue to lose water through transpiration after they are harvested, and this loss of water is one of the main processes that affects their commercial and physiological deterioration. If transpiration is not retarded, it induces wilting; shrinkage; and loss of firmness, crispness, and succulence, with concomitant deterioration in appearance, texture, and flavor. Most fruits and vegetables lose their freshness when their water loss exceeds 3%–10% of their initial weight. Along with loss of weight and freshness, transpiration induces water exudation, which has been shown to accelerate senescence of fruits and vegetables (Ben-Yehoshua, 1987).

Several factors, such as the surface area to volume ratio, nature of the surface coating, relative humidity, temperature, atmospheric pressure, and extent of any mechanical damage influence the transpiration process in fruits and vegetables.

The surface area to volume ratio can range from 50–100 cm^2/mL for individual edible leaves to 5–10 for small soft fruits, such as currants; 2–5 for larger soft fruits, such as strawberries; 0.5–1.5 for tubers, pome, stone and citrus fruits, bananas and onions; and 0.2–0.5 for densely packed cabbage. With other factors being equal, a leaf will lose water and weight much faster than a fruit, and a small fruit or root or tuber will lose weight faster than a larger one (Robertson, 1993).

As mentioned earlier, respiration generates heat, which is dissipated through direct heat transfer to the environment and through evaporation of water. The heat of respiration raises the tissue temperature, and therefore, increases transpiration (Ben-Yehoshua, 1987).

The ambient relative humidity is not a very reliable guide to determine the potential water loss, but it is more useful for calculating the water vapor pressure deficit (VPD) at the particular temperature and humidity. The VPD of the air is defined as the difference between the water vapor pressure of the ambient air and that of saturated air at the same temperature. Thus, the drier the air, the greater its VPD and the more rapidly any produce held in that environment will transpire. To minimize transpiration, fresh produce should be held at low temperature, high relative humidity, and as small a VPD as possible (Ben-Yehoshua, 1987; Robertson, 1993; Wills et al., 1998).

Mechanical damage can greatly accelerate the rate of water loss from produce. Bruising and abrasion damages the surface organization of the tissue, thereby allowing much greater flux of water vapor through the damaged area. Cuts are of even greater importance as they completely break the protective surface layer and directly expose underlying tissues to the atmosphere (Wills et al., 1998).

16.3.2.3 Postharvest Disorders

Physiological disorders refer to the breakdown of tissue that is not caused by either invasion by pathogens (diseases-causing organisms) or by mechanical damage. They may develop in response to an adverse environment, especially temperature, or to a nutritional deficiency during growth and development (Wills et al., 1998).

The storage life of some fruits and vegetables, primarily those of tropical or subtropical origin, can be limited by chilling injury, a disorder induced by low (0°C–12°C) but nonfreezing temperatures in whole plants or susceptible tissues. The extent of the chilling injury is influenced by the temperature, the duration of the exposure to a given temperature, and the chilling sensitivity of the particular fruit or vegetable. The symptoms of chilling injury may not be evident while the produce is held at chilling temperature but become apparent after transfer to a high temperature. Chilling injury prevents some fruits ripening and increases their susceptibility to fungal spoilage. Chilling injury is generally associated with necrosis of certain groups of cells situated either externally, leading to the formation of depressed areas, pitting, and external discoloration, or internally leading to internal browning (Robertson, 1993; Wills et al., 1998).

Also, if the storage conditions provided are not proper and adequate, certain undesirable changes may occur, including the following (Robertson, 1993):

- *Sprouting* especially in onions, ginger, garlic, potatoes, and root crops, greatly reducing their utilization value and accelerating deterioration
- *Elongation* and *curvature* of existing structures, exemplified by products such as asparagus, carrots, and beets, accompanied by increased toughness and decreased palatability
- *Rooting*, which may be initiated by high humidities, resulting in rapid decay, shriveling, and exhaustion of food reserves, especially in roots and tubers

- *Seed germination* within mature fruits such as tomatoes, papaya, and pod-bearing vegetables, which may occur during storage
- *Greening* of potatoes on exposure to light during storage, producing green tissues, which contain the toxic alkaloid solanine
- *Toughening* due to the development of spongy tissue, which may occur during the prolonged storage of green beans and sweet corn
- *Tropic response* (the bending of tissue in response to gravity and light), resulting in products of uneven shapes, which are difficult to pack and have an adverse appearance

16.3.2.4 Postharvest Decay

Many bacteria and fungi can cause postharvest decay. However, it is well established that major postharvest losses of fruit and vegetables are caused by species of the fungi *Alternaria*, *Botrytis*, *Botryosphaeria*, *Colletotrichum*, *Diplodia*, *Monilinia*, *Penicillium*, *Phomopsis*, *Rhizopus*, and *Sclerotinia* and of the bacteria *Erwinia* and *Pseudomonas*. Most of these organisms are weak pathogens in that they can only invade damaged produce. Few of them, such as *Colletotrichum*, are able to penetrate the skin of healthy produce (Wills et al., 1998). Acidic fruit tissues are generally attacked and rotted by fungi, while many vegetables having a tissue pH above 4.5 are more commonly attacked by bacteria. Initially, only one or a few pathogens may invade and break down the tissues, followed by a broad spectrum attack of several weak pathogens resulting in the complete decomposition of the produce (Robertson, 1993).

The development of postharvest decay is favored by high temperatures and high humidities, with the latter often being a result of condensation, which may affect spore germination and cause tissue anaerobiosis (Robertson, 1993).

16.4 STORAGE TECHNIQUES

16.4.1 Cold Storage

The shelf life of fruits and vegetables can be extended by storage at optimum refrigerated temperature and humidity. Storage at 0°C–15°C slows down respiration, retards ethylene production, delays ripening, and usually retards microbial growth. In some cases, low-temperature storage can result in chilling injury of some tropical and subtropical produce. Also, at very high humidities, the potential benefit of reducing water loss from plant tissues generally tends to be outweighed by the risk of rotting.

Produce may be cooled by means of cold air (room cooling, forced-air, or pressure cooling), cold water (hydrocooling), ice, and evaporation of water (evaporative cooling, vacuum cooling). Fruits are normally cooled with cold air, although stone fruits benefit from hydrocooling. Any of the above mentioned cooling methods may be used for vegetables depending upon the structure and physiology of the specific commodities and upon market requirements/expectations. Cut flowers and foliage are usually forced-air cooled, although they may also be vacuum cooled if a small amount of water loss is acceptable (Wills et al., 1998).

16.4.2 Controlled-Atmosphere (CA), Modified-Atmosphere (MA), and Subatmospheric (Hypobaric) Storage

As an adjunct to low-temperature storage or a substitute for refrigeration, the addition or removal of gases resulting in an atmosphere different from that of normal air—as in CA, MA, or subatmospheric (hypobaric or low-pressure) storage—has been widely employed to extend the storage life of fruits and vegetables. These methods aim at reducing respiration and other metabolic reactions by increasing CO_2 and decreasing O_2 concentrations. They also lower the rate of natural ethylene production (as in banana), as well as the sensitivity of fruits to ethylene (Desai and Salunkhe, 1991).

CA technology is utilized for extending the shelf life of bulk stored fruits and vegetables and widely practiced in many countries. Gas-tight warehouse stores, which can contain many hundreds of tons of product, are equipped with systems that carefully control the gas atmospheres within the stores. CA technology for commodity products was originally developed in the UK in the 1920s and 1930s, particularly by Kidd and West for storage of apples and tomatoes. More recently, CA technology has been extended to shipping containers (Brimelow and Vadehra, 1991).

There are several ways to actively control the atmospheric composition inside a storage chamber. Each has its merits and disadvantages. Furthermore, most of the active methods have higher operation and maintenance costs than the passive and semipassive (partial control) technologies. One may divide active systems into O_2-control and CO_2-control systems. The O_2 systems are external gas generators, liquid nitrogen atmospheric generators, gas separators, and hypobaric storage. The CO_2 control systems are lime, water, activated charcoal, and molecular-sieve scrubbers. Also, C_2H_4 control systems can be important in applications to fruit such as apples, which are susceptible to problems of quick ripening if C_2H_4 is not removed (Raghavan et al., 1996).

Hypobaric storage is a form of CA storage in which the produce is stored under partial vacuum. The vacuum chamber is vented continuously with water-saturated air to maintain oxygen levels and to minimize water loss. Ripening of fruit is retarded by hypobaric storage due to the reduction in the partial pressure of oxygen and for some fruits also due to the reduction in ethylene levels. A 10 kPa (0.1 atmosphere) reduction in air pressure is equivalent to reducing the oxygen concentration to about 2% at normal atmospheric pressure. Hypobaric stores are expensive to construct because of the low internal pressure required, and this high cost of application appears to limit hypobaric storage to high-value produce (Wills et al., 1998).

MA can be created inside a package either passively through product respiration or actively by replacing the atmosphere in the package with a desired gas mixture. With commodity-generated or passive MA, if product and film permeability characteristics are properly matched, the desired MA can passively evolve within a sealed package through consumption of O_2 and production of CO_2 by respiration. The gas permeability of the selected film must allow O_2 to enter the package at a rate offset by the consumption of O_2 by the commodity. Similarly, CO_2 must be vented from the package to offset the production of CO_2 by the commodity. Furthermore, the desired MA must be established within 1–2 days without creating anoxic conditions or injuriously high levels of CO_2 that may induce fermentative metabolism (Kader and Watkins, 2000).

Because of the limited ability to regulate a passively established atmosphere, actively establishing the atmosphere is becoming more preferred. This can be done by pulling a slight vacuum and replacing the package atmosphere with the desired gas mixture. This mixture can be further adjusted through the use of absorbing or adsorbing substances inside the package with the respiring commodity that scavenge O_2, CO_2, and/or ethylene (C_2H_4). Although active establishment implies some additional costs, its main advantage is that it ensures the rapid attainment of the desired atmosphere. Ethylene absorbers can help delay the climacteric rise in respiration and associated ripening for some fruit. Carbon dioxide absorbers can prevent the buildup of CO_2 to injurious levels, which can occur for some commodities during passive modification of the package atmosphere. Super-atmospheric O_2 levels (>21%) may be used in combination with fungistatic CO_2 levels (>15%) for a few commodities that do not tolerate these elevated CO_2 atmospheres when combined with air or low O_2 atmospheres. There is no evidence supporting the use of argon, helium, or other noble gases as a replacement for N_2 in MA packaging (MAP) of fresh produce (Kader and Watkins, 2000).

Polymeric films semipermeable to gases are the most popular for MAP. However, for actively respiring fruits and vegetables, commercially available polymeric films have not been very successful because these films are not sufficiently permeable to O_2 (Mujica-Paz and Gontard, 1997). Low-density polyethylene, polyvinyl chloride, and polypropylene are the main films used to package fruit and vegetables. Polystyrene has been used, but polyvinylidene and polyester have such low gas permeabilities that they would be suitable only for commodities with very low respiration rates (Kader and Watkins, 2000). Microperforation of common packaging films is generally used to obtain the required permeability of oxygen and carbon dioxide. Furthermore, the key parameter of MAP is its selectivity, which refers to the ratio of CO_2 permeability to O_2 permeability of the packaging materials and determines the possible combination of oxygen and carbon dioxide concentrations inside the package. Microperforated films are obviously not selective (CO_2/O_2 permeability ratio close to one) (Irissin-Mangata et al., 1999).

To overcome this problem, the potential use of edible films to generate a favorable MAP has been considered (Mujica-Paz and Gontard, 1997).

16.4.3 Packaging with Edible Coating Materials

Several substances deriving from biomaterials have drawn attention for their film-forming ability. Edible films and coatings produced from these substances are used as food protective materials (Gennadios and Weller, 1991).

Edible films and coatings generally can be defined as thin layers of edible material applied on (or even within) foods by wrapping, immersing, brushing, or spraying in order to offer a selective barrier against the transmission of gases, vapors, and solutes, while also offering mechanical protection. There is no clear distinction between films and coatings, and often the two terms are used interchangeably. Coatings are usually directly applied and formed on the surface of products, while films are formed separately as thin sheets and then applied on the products (Gennadios and Weller, 1990; Aydt et al., 1991).

During the last 30 years, considerable research work aiming at the development of edible packaging films and coatings has been conducted. However, few of these films have been applied commercially. This fact can be attributed mainly to the limitations of the films in comparison with traditional polymeric films. The polymer industry has been able to provide food processors with a wide variety of packaging materials characterized generally by better physical and barrier properties than edible films. Research on edible films continues and seems to have intensified over the last few years. Considering the number of advantages these films have over polymeric materials, one might anticipate that the future of food packaging belongs to edible films (Gennadios and Weller, 1990).

The advantages of edible films over other traditional nonedible polymeric packaging materials are summarized below (Gennadios and Weller, 1990, 1991; Debeaufort et al., 1998):

1. The films can be consumed with the packaged product. This is obviously of critical importance since it represents the environmentally ideal package. There is no package to dispose of.
2. Even if the films are not consumed, they could still contribute to the reduction of the environmental pollution. The films are produced exclusively from renewable, edible ingredients, and therefore, are anticipated to degrade more readily than polymeric materials.
3. The films can enhance the sensorial properties of packaged foods provided that various components (flavorings, colorings, and sweeteners) are incorporated to them.
4. The films can supplement the nutritional value of the foods. This is particularly true for films made from proteins.
5. The films can be used for individual packaging of small portions of food, particularly products that currently are not individually packaged for practical reasons such as peas, beans, nuts, and strawberries.
6. The films can function as carriers for antimicrobial and antioxidant agents. In a similar application, they can also be used at the surface of foods to control the diffusion rate of preservative substances from the surface to the interior of the food.

16.4.4 Osmotic Membrane (OSMEMB) Coatings

Reduction of water activity (α_w) increases product stability. The infusion of fruit slices with soluble solids such as fruit juices and sucrose syrups results in a reduction of surface water activity. Another process known as "dehydrofreezing" involves osmotic dehydration followed by low-temperature freezing. Dehydrofrozen products can be rehydrated to a fresh-like quality and, hence, can be regarded as minimally processed products. Coating cut pieces of fruit with edible films, followed by limited osmotic dehydration, improves the stability and also maintains the product's soft texture and fresh character. The products could be reconstituted, resulting in fresh-like quality (Wong et al., 1994b).

16.5 EDIBLE COATING TECHNOLOGY FOR FRESH AND LIGHTLY PROCESSED FRUITS AND VEGETABLES

16.5.1 General

The use of coatings for fresh fruits and vegetables is not really a new concept. Mother nature provides fruits and vegetables with a natural waxy coating called a cuticle, consisting of a layer of cutin (fatty acid–related substances such as waxes and resins with low permeability to water). The pathway for water loss varies for different commodities, but for most fruits and vegetables, the majority of water is lost because of cuticular transpiration, while some also exits through stomates and lenticels. The purpose of edible coatings for fruits and vegetables is basically to mimic or enhance this natural barrier, if already present, or to replace it in cases where light processing has partially removed or altered it (Baldwin, 1994). There are documents that describe the use of waxes to fruits during the twelveth and thirteenth century in China (Hardenburg, 1967).

The development of the so-called wax coatings (which may or may not actually include a wax) emphasized the reduction of moisture loss due to the addition of hydrophobic components such as waxes, oils, and resins (similar to cutin), and imparted sheen due to wax or resin components such as shellac or carnauba wax, respectively (Baldwin, 1994).

16.5.2 Film/Coating Components and Additives

Components of edible films and coatings can be divided into three categories: hydrocolloids, lipids, and composites. Useful hydrocolloids include proteins, cellulose derivatives, alginates, pectins, starches, and other polysaccharides. Useful lipids include waxes, acylglycerols, and fatty acids. Composites contain both lipid and hydrocolloid components. A composite film can exist as a bilayer, in which one layer is a hydrocolloid and the other a lipid, or as a conglomerate, where the lipid and hydrocolloid components are interspersed throughout the film (Donhowe and Fennema, 1994). In addition, plasticizers, such as polyhydric alcohols, waxes, and oils, are added to improve flexibility and elongation of polymeric substances. Addition of surface-active agents (surfactants) and emulsifiers reduces superficial water activity and rate of moisture loss in food products. Release agents and lubricants are added to prevent coated food products from sticking. These can include fats and oils, emulsifiers, petrolatum, polyethylene glycol, and silicone. For minimally processed fruits and vegetables, preservatives can be incorporated to retard surface growth of yeasts, molds, and bacteria during storage and distribution. Antimicrobials that may be used include benzoic acid, sodium benzoate, sorbic acid, potassium sorbate, and propionic acid. Another example of useful additives to coatings is antioxidants. These compounds are added to edible coatings to protect against oxidative rancidity, degradation, and discoloration. Certain phenolic compounds (butylated hydroxyanisole [BHA], butylated hydroxytoluene [BHT], or tertiary butylated hydroxyquinone [TBHQ]), tocopherols, or acids such as propyl gallate have antioxidant properties and inhibit oxidation of fats and oils in foods. There is also a synergistic effect between polyphenolic compounds and certain acidic substances, such as ascorbic, citric, and phosphoric acid, which are effective chelating agents.

Ethylene diamine tetraacetic acid (EDTA) and its salts are also widely used as metal chelators in food systems. Calcium chloride may be incorporated in coatings to improve the texture and color of food products (Cuppett, 1994; Baldwin et al., 1995a,b).

16.5.2.1 Hydrocolloids

Hydrocolloid films can be used in applications where control of water vapor migration is not the objective. These films possess good barrier properties to oxygen, carbon dioxide, and lipids. Most of these films also have desirable mechanical properties, making them useful for improving the structural integrity of fragile products.

Hydrocolloids used for films and coatings can be classified according to their composition, molecular charge, and water solubility. In terms of composition, hydrocolloids can be either carbohydrates or proteins. Film-forming carbohydrates include starches, plant gums (e.g., alginates, pectins, and gum Arabic), and chemically modified starches. Film-forming proteins include gelatin, casein, soy protein, whey protein, wheat gluten, and zein.

The charged state of a hydrocolloid can be useful for film formation. Alginates and pectins require the addition of a polyvalent ion, usually calcium, to facilitate film formation. They, as well as proteins, are susceptible to pH changes because of their charged state. For some applications, an advantage can be gained by combining hydrocolloids of opposite charge such as gelatin and gum arabic.

Although hydrocolloid films generally have poor resistance to water vapor because of their hydrophilic nature, those that are only moderately soluble in water, such as ethylcellulose, wheat gluten, and zein, do provide somewhat greater resistance to the passage of water vapor than do the water-soluble hydrocolloids. However, certain polysaccharides, when used in the form of high-moisture gelatinous coatings, will retard moisture loss from fresh products during short-term storage. In this application, the gel coating acts as a sacrificing agent rather than a barrier to moisture transmission. With time, the coating will eventually dry out and dehydration of the enrobed food will commence (Kester and Fennema, 1986; Donhowe and Fennema, 1993).

16.5.2.1.1 Carbohydrates

16.5.2.1.1.1 Cellulose Derivatives Cellulose is present in all land plants and is the structural material of plant cell walls. Cellulose is a polysaccharide composed of linear chains of $(1 \rightarrow 4)$-β-D-glucopyranosyl units. In its native state, the hydroxymethyl groups of anhydroglucose residues are alternately located above and below the plane of the polymer backbone. This results in very tight packing of polymer chains and a highly crystalline structure that resists solvation in aqueous media. Water solubility can be increased by treating cellulose with alkali to swell the structure, followed by reaction with chloroacetic acid, methyl chloride, or propylene oxide to yield carboxymethylcellulose (CMC), methylcellulose (MC), hydroxypropyl methylcellulose (HPMC), or hydroxypropyl cellulose (HPC). Placement of bulky substituents along the cellulose molecule, in the form of ether linkages at reactive hydroxyls, separates the polymer chains and interferes with formation of the crystalline unit cell, thereby enhancing aqueous solubility. The anionic CMC and nonionic MC, HPMC, and HPC possess excellent film-forming characteristics. HPC is unique among hydrophilic polymers in that it is a true thermoplastic resin and is, therefore, capable of being extruded into films from the molten state.

MC and HPMC display the unique property of reversible thermal gelation in aqueous systems. Both polymers are soluble in cold water, and as their aqueous solutions are heated, viscosity will decrease with an increase in temperature. However, at some critical temperature, a three-dimensional gel structure will form. This phenomenon is attributable to increased intermolecular hydrophobic interactions caused by thermal disruption of hydration shells surrounding polymer chains. Upon cooling below the critical temperature, the gel will revert back to a solution. Aqueous solutions of MC form relatively strong gels at a critical temperature of approximately 50°C. Solutions of HPMC form thermally induced gels of lower strength at 50°C–85°C, the temperature depending on the relative degree of methyl and hydroxypropyl substitution. Aqueous solutions of HPC do not form gels on heating but rather flocculate at about 40°C (Kester and Fennema, 1986; Hui, 1992; Nisperos-Carriedo, 1994; Krochta and Mulder-Johnston, 1997).

16.5.2.1.1.2 Chitosan Chitosan ([(1,4)-linked 2-amino-2-deoxy-β-D-glucan]) is a deacetylated form of chitin, a naturally occurring cationic biopolymer. After cellulose, chitin is the next most abundant polysaccharide on the planet. It is a component of the supporting material of Crustacea, insect skeletons, and fungi.

Chitosan might be an ideal preservative coating for fresh fruit and vegetables because of its film-forming and biochemical properties. Unlike other coating materials, chitosan is known to be antifungal to several fungi, including *Botrytis cinerea* Pers. ex Fr. and *Rhizopus stolonifer* Her ex Fr., to induce chitinase, a defense enzyme, and to elicit phytoalexin (pisatin) accumulation in pea (*Pisum sativum* L.) pods. Since chitosan can form a semipermeable film, coating fruits and vegetables with chitosan may modify the internal atmosphere of the tissue and consequently delay ripening (Ghaouth et al., 1992).

16.5.2.1.1.3 Pectin Pectin is a complex group of structural polysaccharides found in the middle lamella of plant cells. The major commercial sources of pectins are citrus peel and apple pomace. It is composed mainly of D-galacturonic acid polymers with varying degrees of methyl esterification. Chemical de-esterification yields low-methoxyl pectins which, when dissolved in aqueous media, are capable of forming gels in the presence of calcium ions. The function of the ionic calcium is to bridge free carboxyl groups on adjacent polymer molecules. Once the polymer chains are aligned, hydrogen bonding between neighboring chains strengthens the association (Kester and Fennema, 1986; Nisperos-Carriedo, 1994). These coatings have high water vapor permeabilities, and the only way they can prevent dehydration is by acting as sacrificial agents. Coating the film with lipids may increase the resistance to water vapor transmission rates (Nisperos-Carriedo, 1994).

16.5.2.1.1.4 Starches and Derivatives Starch, the reserve polysaccharide of most plants, occurs widely in nature and is the most commonly used food hydrocolloid. This is partly because of the wide range of functional properties it can provide in its natural and modified forms and partly because of its low cost relative to alternatives. Starches can be derived from tubers (potato, tapioca, arrowroot, and sweet potato), stem (sago), and cereals (corn, waxy maize, wheat, and rice) (Nisperos-Carriedo, 1994). Normal corn starch consists of approximately 25% amylose and 75% amylopectin. Amylose is a linear chain of D-glucose residues linked through α-1,4 glycosidic bonds. Amylopectin is a branched molecule consisting of glucose units connected by α-1,4 and α-1,6 linkages. Because linear polymers have better film-forming properties, the commercial development of high amylose corn starch in the 1950s led to the investigation of these starches as edible coatings and barriers (Kester and Fennema, 1986). Mutant varieties of corn are produced, which contain starch with up to 85% amylose. Linear polymers like amylose, cellulose, and mannan tend to crystallize and in their pure form are insoluble in water. The use of amylose in film-forming requires high temperatures and pressures or chemical modification of the amylose to form the more soluble hydroxypropyl amylose. In general, fruits and vegetables can be coated with solutions of amylose containing plasticizers such as glycerol or emulsifiers. Of special interest are the reported low-oxygen barrier properties of amylose films (Hui, 1992).

Dextrins, starch hydrolysates of low-dextrose equivalent (DE), have been suggested for use as protective coatings. Although hydrophilic in nature, starch hydrolysates do provide a limited resistance to the transport of water vapor. Starch films have very minimal resistance to water vapor transport, while films of low-DE dextrin and corn syrup are approximately two- to three-fold more resistant. They have found application on minimally processed fruits (Kester and Fennema, 1986).

16.5.2.1.1.5 Seaweed Extracts

Alginates

Alginates are salts of alginic acid. They occur naturally as the major structural polysaccharides of brown seaweeds known as Phaeophyceae. Alginic acid is considered to be a linear (1 → 4) linked polyuronic acid containing three types of block structures: poly-β-D-mannopyranosyluronic acid (M) blocks, poly-α-L-gulopyranosyluronic acid (G) blocks, and MG blocks containing both polyuronic acids (Nisperos-Carriedo, 1994).

Film formation, which may or may not involve gelation, can be achieved by evaporation, electrolyte crosslinking, or injection of a water-miscible nonsolvent for alginate. Film strength and permeability can be altered through the concentration of the polyvalent cation, the rate of its addition and time of exposure, pH, temperature, and the presence of other constituents (e.g., hydrocolloids, fillers, etc.).

Alginates react with several polyvalent cations to form gels, and this is useful in film formation. Calcium ions are the most effective gelling agents. The mechanism of calcium gelation involves cooperative association of M and G polymer segments to form aggregate structures with calcium ions coordinated in the interstices. The effect of the calcium ions is to pull the alginate chains together via ionic interactions, after which interchain hydrogen bonding occurs. Chain segments consisting of alternative mannuronic and guluronic acid residues do not interact with calcium but rather serve to link the aggregate structures, thus yielding a three-dimensional gel network (Kester and Fennema, 1986).

Alginate films are impervious to oils and fats but are poor moisture barriers. Despite this, alginate gel coatings can significantly reduce moisture loss from foods by acting sacrificially. In other words, moisture is lost from the coating before the food significantly dehydrates. Alginate coatings are good oxygen barriers, can retard lipid oxidation in foods, and can improve flavor, texture, and adhesion (Krochta and Mulder-Johnston, 1997).

Carrageenan

Carrageenan is extracted from several species of red seaweeds and is a complex mixture of several polysaccharides. The three principal carrageenan fractions, kappa (κ), iota (ι), and lambda (λ), differ in sulfate ester and 3,6-anhydro-α-D-galactopyranosyl content (Baldwin et al., 1995b).

Upon cooling a warm aqueous solution of the polymer, gelation occurs, presumably by the formation of a double-helix structure to yield a three-dimensional polymer network. In addition, since gelation is salt-specific, interchain salt bridges must be important. As is true of other gelling polysaccharides, a carrageenan gel coating will act as a sacrificing agent to retard moisture loss from an enrobed food (Kester and Fennema, 1986).

16.5.2.1.2 Proteins

16.5.2.1.2.1 Corn Zein Zein is the only corn protein that continues to be produced commercially. It is characterized by its ability to form tough, glossy, hard, grease-proof coatings after evaporation of aqueous alcoholic solutions. Zein is a mixture of proteins with an average molecular weight of 45,000 in the native state. However, during commercial extraction some disulfide bonds among polypeptide chains break, yielding a product with molecular weight of 25,000–35,000. Two zein fractions can be isolated on the basis of differential solubility: α-zein soluble in 95% ethanol and β-zein soluble in 60% ethanol. In terms of the amino acid composition, zein has a high content of nonpolar hydrophobic amino acids, such as leucine, alanine, and proline. This fact is responsible for the insolubility of zein in water, the insolubility in anhydrous alcohols and the solubility in a mixture of the two. Zein also contains a high level of glutamic acid, about 20%–22%, which exists mostly as glutamine. Glutamine contributes to the insolubility of zein in water through hydrogen bonding (Gennadios and Weller, 1990).

In the past, zein has been used in numerous nonedible applications based on its coating-forming ability. These applications include coatings for flexographic inks in the printing industry, cap liners, special coatings in photographic films, textile fibers, floor and label coatings, and can linings. Even today small quantities of zein are used for the formulation of some of the above products. Currently, the major use of zein is in the formulation of edible coatings for pharmaceutical tablets and confectionery products (Gennadios and Weller, 1990; Beck et al., 1996). Corn zein coating is a good barrier to oxygen, but its water vapor permeability is about 800 times higher than a typical shrink wrapping film (Park et al., 1994b).

16.5.2.1.2.2 Wheat Gluten Wheat gluten (a general term for the water-insoluble proteins of wheat flour) is composed of a mixture of polypeptide molecules considered to be globular proteins. Cohesiveness and elasticity of gluten give integrity to wheat dough and facilitate film formation.

Gluten, which accounts for 80%–85% of the total wheat flour proteins, can be obtained by kneading wheat flour dough and gently washing away starch and other soluble materials in a dilute acid solution or in an excess of water. The gluten can then be separated into two fractions of almost equal quantities. One fraction is soluble in 70% ethanol and is called gliadin, while the other fraction, insoluble in the same solvent, is called glutenin (Gennadios and Weller, 1990).

Wheat gluten films are reported to be good barriers to O_2 and CO_2 and their mechanical properties are comparable to polymeric films. However, these films have high water permeability due to their hydrophilic nature (Baldwin et al., 1995b). The addition of a plasticizing agent is considered necessary in order to overcome the brittleness of the film, a condition that may exist for almost all edible protein films. This brittleness is due to extensive intermolecular forces. Plasticizers reduce these forces, soften the rigidity of the film structure, increase the mobility of the biopolymer chains, and thereby improve the mechanical properties of the films. On the other hand, the resulting looser structure reduces the ability of the film to act as a barrier to the diffusion of various gases and vapors (Gennadios and Weller, 1990).

16.5.2.1.2.3 Soy Protein The protein content of soybeans (38%–44%) is much higher than the protein content of cereal grains (8%–15%). Most of the protein in soybeans can be classified as globulin. A widely used nomenclature system for soy proteins is based on relative sedimentation rates of protein ultracentrifugal fractions. Four such fractions are separable and are designated as 2S, 7S, 11S, and 15S fractions. 7S and 11S are the main fractions making up about 37% and 31%, respectively, of the total water-extractable protein (Gennadios et al., 1994).

Protein in the form of meal is one of the typical end products from soy-bean industrial processing. Protein meal can be further "concentrated" for the production of soy protein concentrates and soy protein isolates. Soy protein concentrates contain at least 70% protein on a dry basis and are produced by extracting and removing soluble carbohydrates and other minor components from the defatted protein meal. Soy protein isolates contain at least 90% protein on a dry basis. Their production from defatted protein meal involves extraction with dilute alkali and centrifugation (Gennadios et al., 1994).

The film-forming ability of soy protein has been noted along with a number of other functional properties, such as cohesiveness, adhesive mess, water and fat absorption, emulsification, dough and fiber formation, texturizing capability, and whippability. Edible films from soybeans have been traditionally produced in the Orient on the surface of heated soymilk. Protein-based films in general are very effective oxygen barriers. Their ability to act as water vapor barriers is limited due to the hydrophilic nature of proteins. Films in contact with high-moisture foods do not maintain their integrity. Also, high relative humidity gradients across the films reduce their oxygen barrier ability. This results from the plasticization induced by the high amounts of water vapor transmitted through the films (Gennadios and Weller, 1991; Gennadios et al., 1994).

16.5.2.1.2.4 Milk Protein Milk has a protein content of about 33 g/L. Whey proteins and casein are the main milk protein fractions, with casein representing 80% (27 g/L) of the total milk proteins. The casein fraction predominantly consists of phosphoproteins characterized by their precipitation at pH 4.6 and 20°C. It contains four principal components: *a*s1-, *a*s2-, β-, and κ-caseins. The primary and secondary structures of these proteins affect their functional properties. Caseins contain low levels of cysteine, with the exception of κ-casein. Whey proteins are characterized by their solubility at pH 4.6. Whey protein contains five protein types: α-lactalbumin, β-lactoglobulin, bovine serum albumin, immunoglobulins, and proteosepeptones. Liquid whey is a by-product of cheese manufacture, and the annual production of fluid whey is rising. However, the majority of it is not used. High nutritional quality, water solubility, and emulsification capability of milk proteins make their use in edible film formation very attractive (Gennadios et al., 1994; McHugh and Krochta, 1994b).

Caseins have not been as extensively investigated as other protein ingredients for their film-forming potential. Caseins form films from aqueous solutions without further treatment due to their random coil nature and ability to extensive hydrogen bonding. It is believed that electrostatic interactions also play an important role in the formation of casein-based edible films. The ability to function as a surfactant makes casein a very promising material for the formation of emulsion films. Casein films, being transparent, flavorless, and flexible, are attractive for use on food products (Gennadios et al., 1994; McHugh and Krochta, 1994a).

16.5.2.2 Lipids

Lipid films are often used as a barrier to water vapor, or as coating agents for adding gloss to confectionery products. Their use in a pure form as free-standing films is limited because most lack sufficient structural integrity and durability.

Waxes are commonly used for coating fruits and vegetables to retard respiration and lessen moisture loss. Formulation for wax coatings varies greatly and their composition is often proprietary. Acetylated mono-glycerides are frequently added to wax formulations to add pliability to the coating.

Shellac coatings, when formed on a supporting matrix, provide effective barrier properties to gases and water vapor (Hagenmaier and Shaw, 1991). Although fatty acids and fatty alcohols are effective barriers to water vapor, their fragility requires that they be used in conjunction with a supporting matrix.

Many lipids exist in crystalline form and their individual crystals are highly impervious to gases and water vapor. Since the permeate can pass between crystals, the barrier properties of crystalline lipids are highly dependent on the intercrystalline packing arrangement. Lipids consisting of tightly packed crystals offer greater resistance to diffusing gases than those consisting of loosely packed crystals. Also, crystals oriented with their major planes normal to permeate flow provide better barrier properties than crystals that are oriented differently.

Lipids existing in a liquid state or having a large proportion of liquid components offer less resistance to gas and water vapor transmission than those in a solid state, indicating that the molecular mobility of lipids detracts their barrier properties.

The barrier properties of lipids having crystalline properties can be influenced both by tempering and polymorphic form (Donhowe and Fennema, 1994).

16.5.2.2.1 Waxes and Oils

16.5.2.2.1.1 Beeswax Also known as white wax, beeswax is secreted by honey bees for comb building. The wax is harvested by certifuging the honey from the wax combs and is then melted with hot water, steam, or solar heating. Beeswax consists mostly of monofunctional alcohols C_{24}–C_{34}. This wax is very plastic at room temperature but becomes brittle at colder temperatures. It is soluble in most other waxes and oils (Hernandez, 1994).

16.5.2.2.1.2 Candelilla Wax Candelilla wax is an exudate of the Candelilla plant (*Euphorbia cerifera*, *Euphorbia antisyphilitica*, *Pedilanthus parvonis*, *P. aphyllus*), a red-like plant that grows mostly in Mexico and southern Texas. The wax is recovered by immersing the plant in boiling water, after which the wax is skimmed off the surface, refined, and bleached. This wax sets very slowly, taking several days to reach maximum hardness (Hernandez, 1994).

16.5.2.2.1.3 Carnauba Wax Carnauba wax is extracted from the leaves of the palm tree known as the "Tree of Life (*Copernica cerifera*)," found mostly in Brazil. It has the highest melting point and specific gravity compared to other commonly found natural waxes and is added to other waxes to increase melting point, hardness, toughness, and luster. Refined carnauba wax consists mostly of saturated wax acid esters with 24–32 hydrocarbons and saturated long-chain monofunctional alcohols. There are several grades of carnauba wax available (from crude to refined), but coating companies are reluctant to indicate which grades are most commonly used in edible coatings (Hernandez, 1994).

16.5.2.2.1.4 Paraffin Wax Paraffin wax is derived from the wax distillate fraction of crude petroleum. It is composed of hydrocarbon fractions of the generic formula C_nH_{2n+2} ranging from 18 to 32 carbon units. Synthetic paraffin wax is allowed for food use in the United States. It consists of a mixture of solid hydrocarbons resulting from the catalytic polymerization of ethylene. Both natural and synthetic paraffins are refined to meet FDA specifications for ultraviolet absorbance. Synthetic paraffin wax should not have an average molecular weight lower than 500 or higher than 1200 (Hernandez, 1994).

16.5.2.2.1.5 Polyethylene Wax Oxidized polyethylene is defined as the basic resin produced by the mild air oxidation of polyethylene, a petroleum by-product. It should have a minimum average molecular weight of 1200, as determined by high-temperature vapor pressure osmometry, a maximum of 5% total oxygen by weight, and an acid value of 9–19 according to the Federal Code of Regulations. Several grades of polyethylene wax are available to obtain desired emulsion properties. These waxes differ in molecular weight (affecting viscosity), density (affecting hardness), and softening point. Polyethylene waxes, although being less popular lately, are used to make emulsion coatings. Methods of emulsification include wax to water emulsification, where the wax is combined with a fatty acid (food grade oleic acid) and heated. Morpholine is then added, and the mixture is combined with hot water, agitated, and cooled (Hernandez, 1994).

16.5.2.2.1.6 Mineral and Vegetable Oils White mineral oil consists of a mixture of liquid paraffinic and naphthenic hydrocarbons and is allowed for use as a food release agent and as a protective coating agent for fruits and vegetables in an amount not exceeding good manufacturing practices. More edible coatings, derived from vegetable oils (e.g., palm oil, sunflower oil, safflower oil, coconut oil, canola oil, castor oil, peanut oil, and soybean oil), have been suggested as substitutes for the petroleum-based mineral oils. However, these coatings may suffer from flavor-stability problems. Use of commercial, partially hydrogenated vegetable oil that is resistant to rancidity gives better results (Kester and Fennema, 1986; Hernandez, 1994; Scott et al., 1995).

16.5.2.2.2 Acetoglycerides

Long-chain alcohols, such as stearyl alcohol ($C_{18}H_{38}O$), are commonly used as additives in edible coatings due to their high melting point and hydrophobic characteristics. They are usually extracted from sperm whale oil. Long-chain fatty acids, such as stearic and palmitic acid, are also commonly used in edible coatings for their higher melting points and hydrophobicity (Hagenmaier and Shaw, 1990).

Acetylation of glycerol monostearate, by its reaction with acetic anhydride, yields 1-stearodiacetin. This acetylated monoglyceride displays the unique characteristic of solidifying from the molten state into a flexible wax-like solid. Most lipids in the solid state can be stretched to only about 102% of their original length before fracturing. Acetylated glycerol monostearate, however, can be stretched up to 800% of its original length. It has been determined that highly stretchable, solid-phase acetylated monoglycerides exist in the alpha (α) polymorphic form. Stability of the α crystalline form of acetylated monoglycerides is exceptionally good at room temperature, especially if impurities are present. Flexibility of acetylated glycerol monostearate in the α polymorphic form is attributed to the unordered network of interlocking ribbon-like crystals. Upon stretching, order of the crystalline matrix is enhanced by alignment of neighboring crystallites. Barrier properties of acetylated glycerol monostearate improve as the degree of acetylation increases. This is, of course, due to the removal of free hydroxyl groups, which would otherwise interact directly with migrating water molecules (Kester and Fennema, 1986).

16.5.2.2.3 Sucrose Polyester (SPE)

SPE is a fat-like material, miscible in triglycerides and insoluble in water. SPE consists of a mixture of hexa-, hepta-, and octo-esters formed by the reaction of sucrose with long-chain fatty acids, which are not hydrolyzed by the lipolytic enzymes in the intestinal tract and thus are not absorbed. For this reason, SPE is unable to be absorbed by the body although it has the taste and consistency of conventional vegetable oil. In addition, SPE reduces the absorption and increases the excretion of bile acids containing cholesterol (Toma et al., 1988).

16.5.2.2.4 Resins and Rosins

16.5.2.2.4.1 Shellac Resin Shellac resin is a secretion by the insect *Laccifer lacca* and is mostly produced in central India. This resin is composed of a complex mixture of aliphatic alicyclic hydroxy acid polymers (e.g., aleuritic and shelloic acids). Shellac is soluble in alcohols and in alkaline solutions. It is also compatible with most waxes, resulting in improved moisture-barrier properties and increased gloss for coated products. Other natural resins used in food coatings include copal, damar, and elemi (Hernandez, 1994).

16.5.2.2.4.2 Coumarone-Indene Resin Coumarone-indene resin is a petroleum and/or coal tar by-product. It is 100% aromatic in content and exhibits excellent resistance to alkalis, dilute acids, and moisture. Several grades are available, varying in melting point, molecular weight, and viscosity (Hernandez, 1994).

16.5.2.2.4.3 Wood Rosin Rosins are obtained from the oleoresins of pine trees, either as an exudate or as tall oil, a by-product from the wood pulp industry. It is the residue left after distillation of volatiles from the crude resin. Wood rosin is composed of approximately 90% abietic acid ($C_{20}H_{32}O_2$) and its isomers and 10% dehydroabietic acid ($C_{20}H_{28}O_2$). Wood resin can be modified by hydrogenation, polymerization, isomerization, and decarboxylation to make it less susceptible to oxidation and discoloration and to improve its thermoplasticity. Drying oils, such as some vegetable oils, may be esterified with some glycol derivatives (e.g., butylenes, ethylene, polyethylene, and polypropylene) to form resinous and polymeric coating components (Hernandez, 1994).

16.5.2.3 Composites/Bilayers

Edible films and coatings formed with several compounds (composite films or bilayer films if each compound forms a separate layer) have been developed to take advantage of the complementary functional properties of these different constitutive materials and to overcome their respective drawbacks. Most composite films studied up to now combine a lipidic compound and a hydrocolloid-based structural matrix. When a barrier to water vapor is desired, the lipid component can serve this function, while the hydrocolloid component provides the necessary durability and the appropriate selective barrier to gases. These films are recommended as coatings for processed fruits and vegetables (Cuq et al., 1995; Wu et al., 2002).

16.5.3 Safety and Health Issues

Determination of the acceptability of materials for edible polymer films follows procedures identical to determining the appropriateness of such materials for food formulation:

1. An edible polymer will be generally recognized as safe (GRAS) for use in edible films if the material has previously been determined GRAS, and its use in an edible film is in accordance with current good manufacturing practices (food grade, prepared and handled as a food ingredient, and used in amounts no greater than necessary to perform its function) and within any limitations specified by the Food and Drug Administration (FDA) (Table 16.1).
2. If the edible polymer film material used is not currently GRAS, but the manufacturer can demonstrate its safety, the manufacturer may either file a GRAS affirmation petition to the FDA or proceed to market without FDA concurrence (self-determination).
3. The manufacturer may not need to establish that the use of the edible polymer in edible films is GRAS if the material received pre-1958 FDA clearance and thus has "prior sanction."
4. Finally, if the material cannot be demonstrated to be GRAS or does not have "prior-sanction," the manufacturer must submit a food additive petition to the FDA prior to its use.

Food processors considering the use of protein-based films must be aware that some consumers have wheat gluten intolerance (celiac disease), milk protein allergies, or lactose intolerance. Use of such films as coatings on foods must be declared appropriately to the consumer, no matter how small the amount used. The nutritional quality of materials used for edible films may be affected, negatively or positively, by the temperature, pH, and/or solvents used in film preparation. Aside from these considerations, no intrinsic nutritional or health problems have been identified for edible films. In fact, edible films can be carriers of nutritional supplements, and protein-based films, depending on protein quality, can provide an important nutritional enhancement to the food. Attention to the microbial safety of edible films is guided by standard considerations of water activity, pH, temperature, oxygen supply, and time. Importantly, edible films are effective carriers of antimicrobials, which improve the microbial stability of film and food alike (Krochta and Mulder-Johnston, 1997).

16.5.4 Film Formation and Application

16.5.4.1 Formation

Film-forming substances are able to form a continuous structure by settling the interactions between molecules under the action of a chemical or physical treatment. The film and coating formation involves one of the following processes (Kester and Fennema, 1986; Donhowe and Fennema, 1994; Debeaufort et al., 1998; Schmitt et al., 1998):

1. Melting and solidification of solid fats, waxes, and resins
2. Simple coacervation where a hydrocolloid dispersed in aqueous solution is precipited or gelified by the removal of solvent, the addition of a nonelectrolyte solute in which the polymer is not soluble, the addition of an electrolyte substance inducing a "salting-out" effect, or the modification of the pH of the solution
3. Complex coacervation, where two hydrocolloid solutions with opposite charges are combined, inducing interactions and the precipitation of the polymer mixture.
4. Thermal gelation or coagulation, for example, through heating of a macromolecule solution, which involves denaturation, gelification, precipitation, or by rapid cooling of a hydrocolloid solution that induces a sol–gel transition.

TABLE 16.1
Polysaccharide, Protein, and Lipid Compounds Permitted for Use as Components (Film Formers, Emulsifiers, Plasticizers, Surfactants, etc.) in Commercial and Experimental Coatings for Food Systems and Their Citation in the *Code of Federal Regulations*

Ingredient	Food Use	21 CFR Citation
Polysaccharides		
Cellulose and derivatives		
Carboxymethyl cellulose	Multipurpose GRAS substance	182.1745
Methylcellulose	Multipurpose GRAS substance	182.1480
Hydroxypropyl cellulose	Emulsifier, film former, protective colloid, stabilizer,suspending agent, thickener	172.870
Hydroxypropyl methylcellulose	Emulsifier, film former, protective colloid, stabilizer,suspending agent, thickener	172.874
Starches and derivatives		
Raw	GRAS substances	182.70, 182.90
Modified starch	Component of batter, multipurpose substance	172.892, 182.70
Pregelatinized	Multipurpose additive	172.892
Dextrin	Formulation aid, processing aid, stabilizer, thickener, surface-finishing agent	184.1277
Maltodextrin	GRAS substance	184.1444
Pectins	Emulsifier, stabilizer, thickener	184.1588
Seaweed extracts		
Alginates	Stabilizer, thickener, humectant, texturizer, formulation aid, firming agent, flavor adjuvant, flavor enhancer, processing aid, surface active agent	184.1133, 184.1187, 184.1610, 184.1724
Carrageenans	Emulsifier, stabilizer, thickener	172.620
Gums		
Gum arabic	Emulsifier, formulation aid, stabilizer, thickener, humectant, texturizer, surface-finishing agent, flavoring agent, adjuvant	184.1330
Guar gum	Emulsifier, formulation aid, stabilizer, thickener, firming agent	184.1339
Xanthan	Stabilizer, emulsifier, thickener, suspending agent, bodying agent, foam enhancer	172.695
Chitosan	Not approved	
Proteins		
Zein	GRAS	184.1984
Wheat gluten	GRAS	184.1322
Whey proteins	GRAS	184.1979c
Lipids		
Acetylated monoglyceride	Multipurpose additive	172.828, 175.230
Beeswax	GRAS	184.1973
Candelilla wax	GRAS	184.1976
Carnauba wax	GRAS	184.1978
Castor oil	Release agent, component of coatings	172.876
Coumarone-indene resin	Coating component for fresh citrus	172.215
Diacetyl tartaric acid esters of mono- and diglycerides	GRAS	184.1101
Edible oils (palm, soy, corn, coconut, etc.)	Allowed as food components not subject to FDA regulation	—
Fatty acids	Coating component for fresh citrus, lubricant, defoamer	172.860, 172.210

(Continued)

TABLE 16.1 (*Continued*)

Polysaccharide, Protein, and Lipid Compounds Permitted for Use as Components (Film Formers, Emulsifiers, Plasticizers, Surfactants, etc.) in Commercial and Experimental Coatings for Food Systems and Their Citation in the *Code of Federal Regulations*

Ingredient	Food Use	21 CFR Citation
Glyceryl monostearate	GRAS	184.1324
Lactylic esters of fatty acids	Emulsifier, plasticizer, surfactant for dried fruits and vegetables	172.848
Methyl and ethyl esters of fatty acids	Coating component for grape dehydration	172.225
Mineral oil, white	Coating component for fresh fruits and vegetables, hot melt for frozen meat, lubricant, release agent	172.878
Mono- and diglycerides	GRAS	184.1505
Morpholine salts of fatty acids	Coating component for fresh fruits and vegetables	172.235
Oleic acid	Coating component for fresh citrus fruit, lubricant, defoamer	172.210, 172.862
Oxidized polyethylene	Protective coating or component of coating for certain fruits and vegetables	172.260
Petrolatum	Protective coating for certain fresh fruits and vegetables	172.880
Petroleum naphtha	Coating component for fresh citrus fruit	172.250
Petroleum wax	Coating component for cheese, fresh fruits and vegetables, microcapsules for spice flavoring	172.886
Polyethylene glycol	Coating component for fresh citrus fruit, coating, plasticizer, lubricant in tablets	172.210, 172.820
Rice bran wax	Coating for candy, fresh fruits and vegetables, plasticizer for chewing gum	172.890
Shellac resin	Coating component	175.300
Sodium lauryl sulfate	Coating component for confectionery, fresh fruits and vegetables, emulsifier, surfactant	172.210, 172.822
Stearic acid	GRAS	184.109
Sucrose fatty acid esters	Emulsifier, coating component for certain fresh fruits	172.859
Synthetic isoparaffinic petroleum hydrocarbons	Coating component for fruits and vegetables	172.882
Synthetic paraffin and succinic derivatives	Protective coating or component of coating for certain fruits and vegetables	172.275
Synthetic petroleum wax	Coating component for cheese, fresh fruits, and vegetables, defoamer	172.888
Wood rosin	Coating component	175.300

Coatings are formed directly on the food product using either liquid film-forming solutions (or dispersions) or molten compounds (e.g., lipids). They can be applied by different methods: with a paint brush or by spraying, dipping–dripping, fluidizing, etc. Films are preformed separately from the food product. They can be produced, for instance, by drying a film-forming solution on a drum-drier, by cooling a molten compound, or through standard techniques used to form synthetic packaging, for example, thermoforming or extrusion techniques for thermoplastic materials (Kester and Fennema, 1986; Cuq et al., 1995).

16.5.4.2 Application

16.5.4.2.1 Dip Application

Dipping fruits and vegetables into a tub or tank of the coating material is adequate usually for small quantities of commodities. The produce is washed, dried, and then immersed in the dip tank.

Duration of immersion is not important, but complete wetting of the fruit or vegetable is imperative for good coverage. The commodity is then either conveyed to a drier where water is removed or allowed to dry under ambient conditions. The continuous dipping of fruits and vegetables into a largely static milieu results in unacceptable buildup of decay organism, soil, and trash in the dip tank. The dip tanks can be equipped with a porous basket, which can be lifted to strain and remove debris. In addition, fruit entering the dip tank must be completely dry to avoid dilution of the resin solution or emulsion coating. For these reasons, other coating methods are more desirable where considerable quantities of fruits and vegetables are to be coated (Grant and Burns, 1994).

16.5.4.2.2 Spray Application

Spray application is the conventional method for applying most coatings to fruits and vegetables. Films applied by spraying can be formed in a thinner, more uniform manner than those applied by dipping. Low-pressure spray applicators used in the past delivered coating in excess. Often, recovery wells were utilized, and excess coating was recirculated. As with dip application, dilution and contamination were of concern. Later, high-pressure spray applicators delivering coatings at 60–80 psi became available, which used much less coating material and gave equal or better coverage, negating the need for recovery wells and recirculation (Donhowe and Fennema, 1994; Grant and Burns, 1994).

16.5.4.2.3 Casting

This technique, useful for free-standing films, is borrowed from methods developed for nonedible films. Coating is simple and allows film thickness to be controlled accurately on smooth, flat surfaces. Casting can be accomplished by controlled-thickness spreading or by pouring (Donhowe and Fennema, 1994).

The degree of cohesiveness of the matrix is a critical parameter affecting the functional properties of edible films. It is sometimes difficult to obtain adequate adhesion of the film to the food product, for instance, when a hydrophobic film-forming material is used to protect a hydrophilic food product (e.g., minimally processed fruits and vegetables). In such cases, surface-active agents can be coated on the food or added to the film-forming solution, or a material capable of adhering to both components can be applied as an intermediate precoating (Cuq et al., 1995). Sometimes, trapping of lipid molecules within a matrix is useful. For example, an emulsion mixture of casein and acetylated monoglycerides will form a coagulum by adjusting the pH to the isolectric point (pH = 4.6). The lipid molecules are presumably trapped within the matrix of the casein coagulum. The same result can be obtained if a polymer with functional groups, such as alginate, is added at the emulsion, which along with the association of polyglucuronic acid (of the endemic pectin on the cut surface of fruit or vegetable) and calcium ions form a three-dimensional network containing casein, with acetylated monoglyceride dispersed in the interstices (Wong et al., 1994b).

16.5.5 Film Permeability

Since several of the functional properties of an edible film or coating relate to resistance to gas, vapor, or solute transport, it is appropriate to briefly review the theoretical derivation of the permeability equation as it applies to films. Permeability is defined as transmission of a penetrant through a resisting material. In the absence of cracks, pinholes, or other flaws, the primary mechanism for gas and vapor flow through a film or coating is by activated diffusion; that is, the penetrant dissolves in the film matrix at the high-concentration side, diffuses through the film driven by a concentration gradient, and evaporates from the other surface. The second step of the process, that is, diffusion, depends upon size, shape, and polarity of the penetrating molecule as well as polymer-chain segmental motion within the film matrix. Factors affecting segmental motion of polymer chains include interchain attractive forces such as hydrogen bonding and van der Waals' interactions, degree of crosslinking, and amount of crystallinity. The dissolution and evaporation steps are influenced by the solubility of the penetrant in the film (Kester and Fennema, 1986).

The permeation can be described mathematically by Fick's first law. The flux (J), which is proportional to the concentration gradient, can be defined in one direction as follows:

$$J = -D\left(\frac{\partial C}{\partial X}\right) \tag{16.1}$$

where

J is the flux, the net amount of solute that diffuses through unit area per unit time ($g/m^2 \cdot s$ or $mL/m^2 \cdot s$)
D is the diffusivity constant (m^2/s)
C is the concentration gradient of the diffusing substance
X is the thickness of the film (m)

Given the assumptions that (a) the diffusion is in steady state and (b) there is a linear gradient through the film, the flux (J) can be expressed as

$$J = \frac{D(C_2 - C_1)}{X} = \frac{Q}{A \cdot t} \tag{16.2}$$

where

Q is the amount of gas diffusing through the film (g or mL)
A is area of the film (m^2)
t is the time (s)

After application of Henry's law, the driving force is expressed in terms of partial pressure differential of gas and a rearrangement of terms yields the following equation in terms of permeability.

$$\frac{Q}{A \cdot t} = \frac{D \cdot S \cdot (p2 - p1)}{X} = \frac{P \cdot \Delta p}{X} \tag{16.3}$$

where

S is the Henry's law solubility coefficient (mole/atm)
Δp is partial pressure difference of the gas across the film (Pa)
P is the permeability ((mL or g) $m/m^2 \cdot s \cdot Pa$)

Then, the permeabilities of O_2, CO_2, and H_2O vapor can be calculated from the following equation (Park, 1999):

$$P = \frac{Q \cdot X}{A \cdot t \cdot \Delta p} \tag{16.4}$$

The diffusion and solubility of permeants are affected by temperature and by the size, shape, and polarity of the diffused molecule. Moreover, these two parameters depend on film characteristics, including the type of forces influencing molecules of the film matrix, the degree of crosslinking between molecules, the crystallinity, the presence of plasticizers or additives, etc. (Gontard and Guilbert, 1992; Cuq et al., 1995).

Permeability is only a general feature of films or coatings when the diffusion and solubility coefficients are not influenced by permeant content, that is, when Fick's and Henry's laws apply. In practice, for most edible films, the permeant interacts with the film, and the D and S coefficients are dependent on the difference in partial pressure. For instance, in relation to the water vapor permeability of hydrophilic polymer films, the water solubility and diffusion coefficients increase when the water vapor differential partial pressure increases because of the moisture affinity of the film (nonlinear sorption isotherm) and because of increased plasticization of the film due to water absorption. The film thickness can also influence permeability when using film-forming materials that do not behave ideally. The permeability of an edible film is thus defined as a property of the film-permeant complex, under specified temperature and water activity conditions (Kester and Fennema, 1986; Cuq et al., 1995).

16.5.5.1 Water Vapor Permeability

One of the most important properties of an edible film is its water vapor permeability (WVP). The WVPs of several edible films are available in the literature. When comparing these values, it is important to recognize two items. First, different conditions were employed for the testing of these films. Therefore, experimental differences must be taken into account when examining these data. For example, relative humidity (RH) effects must be considered. Also, the effect of RH on WVP of edible films is substantial. RH was shown to have an exponential effect on WVP of whey protein films (McHugh et al., 1994). Therefore, small differences in RH during testing can result in drastic changes in permeability. In any case, true experimental conditions should be considered as may result in errors (McHugh et al., 1993).

Polysaccharide films are generally rather poor water barriers due to their hydrophilic nature. However, most pure polysaccharide films do not require the addition of plasticizers and, therefore, tend to exhibit lower WVPs than most protein films. The incorporation of lipids to polysaccharide-based films resulted in films that are exceptionally good barriers to water. Permeability of a HPMC film is very dependent on the type and quantity of fatty acids. The longer chain fatty acids make better barriers (Hagenmaier and Shaw, 1990). Besides, WVPs of cellulose-derivative-based films increased as the molecular weight (MW) of the cellulose increased (Park et al., 1993). Composite films of chitosan–laurate were shown to have low permeability, while films containing other fatty acids or esters were not effective (Wong et al., 1992). Another polysaccharide-based edible film constructed from peach puree had similar water vapor permeability (McHugh et al., 1996).

Protein-based edible films have relatively high WVPs. The water vapor permeability of gluten- and milk protein-based edible films is several times larger than that of zein films. This is probably attributable to the greater hydrophobicity of zein. All protein films, however, exhibit increased WVPs at elevated RH. Sodium caseinate films, adjusted to their isoelectric points, were shown to have lower WVP than films formed at neutral pH. Furthermore, increased concentrations of plasticizers in protein films resulted in increased WVPs. The incorporation of lipid to protein films was found to decrease the WVP of these films. Calcium caseinate–beeswax emulsion films had water vapor permeability up to 90% lower than pure sodium caseinate films, according to Avena-Bustillos and Krochta (1993). Also, beeswax incorporation into sodium caseinate films was more effective in reducing WVP than stearic acid and acetylated monoglyceride. Films formed from protein emulsion systems exhibit good water barrier properties.

Pure lipid films have extremely low WVPs. This is due to their hydrophobic, crystalline nature. Candellila wax was shown to have the lowest permeability to water vapor compared to other lipid films. The WVP values of protein and polysaccharide films indicate that these films do not possess good water barrier properties compared to films like high-density polyethylene. Their overall hydrophilic nature renders them poor water barriers. However, by incorporating lipids into these films, WVPs can be reduced. Pure lipid and many composite films exhibit water barrier properties comparable to those of nonedible packaging films.

16.5.5.2 Gas Permeability

Gas barrier properties of several edible films and nonedible packaging films can also be found in the literature. As with WVPs, one must be cautious when comparing data due to experimental variations in temperature and RH.

The development of edible films with selective gas permeability (oxygen, carbon dioxide, and ethylene) allows the control of respiration exchange and microbial development and seems very promising for achieving an "MA" effect in fresh fruits and vegetables and for improving the storage potential of these products.

Films formed with hydrocolloids (proteins, polysaccharides) generally have good oxygen barrier properties, particularly under low-moisture conditions. The oxygen permeability of hydrocolloid-based films (at 0% RH) is often lower than that of common synthetic films such as polyethylene and PVCs. For example, the oxygen permeability of soy protein film was 436 times lower than that of low-density polyethylene, 124 times lower than that of high-density polyethylene, and 297 times lower than that of nonplasticized PVC.

Films formed with lipid derivatives have suitable oxygen barrier properties. For example, the oxygen permeability value for beeswax film lies between those of low-density and high-density polyethylene. Lipids with the best oxygen barrier properties are those formed with straight-chain and saturated fatty acids, according to Blank (1962, 1972). Increased unsaturation (or branching) and a reduction in the length of the carbon chain result in decreased oxygen permeability (Cuq et al., 1995). The following barrier efficiency order was observed by Kester and Fennema (1989): stearic alcohol > tristearin > beeswax > acetylated monoglycerides > stearic acid > alkanes.

As previously mentioned for water vapor permeability, formulation of composite films allows advantage to be taken of the complementary barrier properties of each component. At high RH, where hydrophilic materials are not effective as gas barriers, the addition of lipidic compounds results in a decrease in the gas permeability of the film. For example, at 91% RH, the oxygen permeability is reduced by about 30% for a composite gluten–beeswax film.

The effect of temperature on gas permeability is similar to that reported for water vapor permeability. These variations can be characterized by Arrhenius-type representations. But, as far as gas solubility decreases with temperature increase, the increase of gas permeability with temperature is lower than for water vapor permeability (Gontard et al., 1994).

In hydrophilic films, increased RH promotes both gas diffusivity (due to the increased mobility of hydrophobic macromolecule chains) and gas solubility (due to the water swelling of the matrix), leading to a sharp increase in gas permeability.

Carbon dioxide permeability in hydrocolloid-based films is often much higher than oxygen permeability. The effect of film a_w on carbon dioxide permeability is similar to that on oxygen permeability, but the sharp increase of permeability is more important. This could be explained by the differences in water solubility of these gases, that is, carbon dioxide is very soluble (carbon dioxide solubility in water = 34.5 mmol/L at 25°C and 10^5 Pa; oxygen solubility = 1.25 mmol/L at 25°C and 10^5 Pa).

At high RH, the addition of lipid components to gluten film results in a high decrease of carbon dioxide permeability. For example, at 91% RH, the carbon dioxide permeability is reduced by about 75% for a composite gluten and beeswax film. This could be related to the hydrophobic characteristics of these components, which for the same RH reduce the amount of water available for the solubilization of carbon dioxide.

The selectivity coefficient between carbon dioxide and oxygen is defined as the ratio of the respective permeabilities of both gases. In hydrophilic materials, the effect of an a_w increase on permeability is greater for carbon dioxide than for oxygen. The selectivity of these materials is thus sensitive to moisture variations (e.g., the selectivity coefficient of edible gluten films varies from 1.8 at 00% RH, to 28.4 at 95% RH), whereas the selectivity coefficient for synthetic polymers remains relatively constant at 3–6.

16.5.6 Commercial Edible Coatings

16.5.6.1 Hydrocolloid-Based Coatings

The first commercially available polysaccharide-based coating was called TAL Pro-long (Courtaulds Group, London), and later Pro-long, which was made up of sucrose polyesters of fatty acids and sodium salt of carboxymethyl cellulose. The dry ingredients were mixed with water to specific concentrations, depending on the commodity, and applied as a dip, drench, or spray to cover fruits and vegetables. Another coating, Semperfresh (United Agriproducts, Greeley, CO), appeared on the market with an apparently similar composition to TAL Pro-long. This coating claims to be an improved formulation of earlier sucrose polyester products with the major difference being the incorporation of a higher proportion of short-chain unsaturated fatty acid esters in its formulation (Drake et al., 1987). It was sold in powder or granular form which, when mixed with water to a concentration of 0.75%–2.0%, is used to coat fruits and vegetables. A polysaccharide-based coating called Nature-Seal was formulated at the U.S. Department of Agriculture, Agricultural Research Service, and was patented by the U.S. Department of Agriculture. It also uses cellulose derivatives as film formers, but does not contain sucrose fatty acid esters as does TAL Pro-long and Semperfresh. Using chitosan as a film former and natural preservative, a differentially permeable fruit and vegetable coating called Nutri-Save (Nova Chem, Halifax, NS, Canada) was developed from *N,O*-carboxymethyl chitosan. Methylation of the chitosan polymer in Nutri-Save resulted in a twofold increase in the resistance to CO_2 compared to the unmethylated polymer. These coatings, however, are not as effective in this respect as are conventional wax formulations (Baldwin, 1994). Mitsubishi International Corp., Fine Chemicals Department announced two polysaccharide films: Soageena (carrageenan-based line of coatings) and Soafil (edible polysaccharide) (Institute of Food Technology, 1991). Only one commercial protein-based coating has been produced. Its name is Cozeen (Zumbro, Inc., Hayfield, MN) and contains corn zein as its major component (Aydt et al., 1991).

16.5.6.2 Wax and Oil Coatings

Many coatings used today are similar to those used in the past. The wax-type coatings are made with natural or synthetic waxes, fatty acids (carnauba, polyethylene, oleic acid), oils (vegetable and mineral oil), wood rosin, shellac, coumarone-indene resin, emulsifiers, plasticizers, anti-foam agents, surfactants, and preservatives.

16.5.7 The Effect of Edible Coatings and Films on the Quality Characteristics of Fresh and Minimally Processed Fruits and Vegetables

16.5.7.1 General

Over the last 20 years, a great number of works on the formulation and characterization of edible films and coatings have been done in both scientific and patent literature. However, the application of edible films and coatings on fresh and minimally processed fruits and vegetables remains tricky. The research on edible films and coatings at real conditions, after their application on fresh produce, is more difficult and complex. Partly, this is due to the limited work done in this direction. Only wax-based commercial films and coatings have been tested on fresh fruits and vegetables, and no research has been done with new ingredients and composite materials.

16.5.7.2 Commercial Cellulose-Based Edible Coatings

16.5.7.2.1 TAL Pro-long

TAL Pro-long has been applied widely to apples, pears, plums, bananas, limes, oranges, mangoes, and tomatoes. When applied on fruits or vegetables, Tal Pro-long

1. Created MA by changing the internal concentrations of CO_2, O_2, and ethylene (Lowings and Cutts, 1982; Banks, 1984; Smith and Stow, 1984; Banks, 1985a,b; Lee and Kwon, 1990)
2. Retarded color development (Smith and Stow, 1984; Banks, 1985b; Dhalla and Hanson, 1988; Motlagh and Quantick, 1988)
3. Reduced weight loss (Smith and Stow, 1984; Dhalla and Hanson, 1988; Motlagh and Quantick, 1988; Lee and Kwon, 1990)

4. Retained firmness (Smith and Stow, 1984; Dhalla and Hanson, 1988; Lau and Meheriuk, 1994)
5. Retained titratable acidity, soluble solids, and ascorbic acid (Dhalla and Hanson, 1988; Lau and Meheriuk, 1994)
6. Increased flavor volatiles (Dhalla and Hanson, 1988; Nisperos-Carriedo et al., 1990).
7. Reduced core flush and scald (Lau and Meheriuk, 1994)
8. Increased resistance to fungal rots (Lowings and Cutts, 1982)

Concentration of coating materials is a critical parameter for the preservation of fresh fruits and vegetables. A 1.25% sucrose ester-carboxymethyl cellulose formulation (Tal Pro-long) failed to retard detrimental changes in fruit firmness, yellowing, and weight loss in Cox's Orange Pippin' apples during storage, but when a 1%–4% coating formulation was applied after storage, it delayed yellowing and loss of firmness, increased internal CO_2, and reduced weight loss during a 21 day marketing period (Smith and Stow, 1984). Mango fruits also exhibited retarded ripening and, therefore, increased storage life when coated with 0.75%–1.0% TAL Pro-long and stored at 25°C (Dhalla and Hanson, 1988). The authors also reported reduced weight loss in the coated fruit compared to uncoated controls and increased ethanol formation in fruit pulp after 13 days with 1% TAL Pro-long. No adverse effects on sensory quality were detected when a 0.75% formulation was used, however. Coated mangoes showed a slower decrease in titratable acidity and ascorbic acid as well as a retarded softening and carotenogenesis (loss of green color). McIntosh, Delicious, and Spartan apples were coated with several concentrations of Pro-long and stored in air at 0°C for 120–150 days or in air at 20°C for 14 days. Retention of flesh firmness and acidity were generally better in coated fruit than in control fruit during a 3-year study. Core flush and scald were lower in coated-McIntosh fruit, but core flush and breakdown were slightly higher in coated Delicious apples. The higher concentrations of coating materials caused significant skin discoloration (purplish hue) in McIntosh and Spartan, but discoloration was less severe in Delicious apples. Sensory quality was not affected by the coating materials (Lau and Meheriuk, 1994).

Besides, sucrose polyesters were shown to be effective in increasing resistance of apples, pears, and plums to some fungal rots of *Sclerontina* species and *Rhizopus nigricans* (Lowing and Cutts, 1982).

However, there were several cases where the coating did not assist preservation. For example, TAL Pro-long was not effective in decreasing the respiration rate and water loss in tomatoes (Nisperos-Carriedo and Baldwin, 1988).

16.5.7.2.2 Semperfresh

Semperfresh has been applied widely to apples, pears, citrus, cherries, melons, breadfruits, and zucchinis. When applied on fruits or vegetables, Semperfresh

1. Created MA by changing the internal concentrations of CO_2, O_2, and ethylene (Drake et al., 1987; Park et al., 1994a; Worrel et al., 2002)
2. Retarded color development (Drake et al., 1987; Santerre et al., 1989; Chai et al., 1991; Sumnu and Bayindirli, 1994, 1995; Yaman and Bayoindirli, 2002)
3. Reduced weight loss (Curtis, 1988; Drake et al., 1988; Park et al., 1994a; Sumnu and Bayindirli, 1995; Yaman and Bayoindirli, 2002; Worrel et al., 2002)
4. Retained firmness (Drake et al., 1987; Curtis, 1988; Santerre et al., 1989; Chai et al., 1991; Sumnu and Bayindirli, 1994, 1995; Yaman and Bayoindirli, 2002)
5. Retained titratable acidity, soluble solids, and ascorbic acid (Drake et al., 1987; Chai et al., 1991; Sumnu and Bayindirli, 1994, 1995; Yaman and Bayoindirli, 2002)
6. Increased flavor volatiles (Curtis, 1988)
7. Reduced core flush and scald (Bauchot et al., 1995)

Treated apples showed a delay in the change of firmness, titratable acidity, vitamin C content, soluble solid content, and weight loss as compared to the untreated ones. The concentration of the coating is a significant factor for the retention of fruits quality. Increasing the concentration increased the quality retention of apples. The Semperfresh coating concentration should exceed 5 g/L, since 5 g/L was not shown to have significantly different effect on most of the ripening parameters compared to control. Concentrations of 15 and 20 g/L affected ripening parameters, firmness, acidity, and ascorbic acid without any significant difference (Sumnu and Bayindirli, 1995). Coatings with 1% and 1.5% concentrations of Semperfresh were effective for delaying ripening of "Ankara" pears. Coated pears had higher color, firmness, acidity, and soluble solid retention. Since 1.0% and 1.5% concentrations affected most of the ripening parameters similarly and many pears treated with 1.5% concentrations showed uneven coloration, 1% is selected as the optimum concentration of Semperfresh for "Ankara" pears. When the efficiency of Semperfresh was compared to Johnfresh fruit coating, Johnfresh was determined to be the effective coating for the reduction of the weight loss. There was no significant difference between Johnfresh coating and Semperfresh in other quality factors. That is why the cost of these coatings should also be considered before deciding the best coating for "Ankara" pears (Sumnu and Bayindirli, 1994).

In mature apples, stored at 3°C–4°C for 4 months, scald incidence was reduced by postharvest dipping in a Semperfresh coating, formulated with the antioxidants, ascorbyl palmitate, and *n*-propyl gallate. The same result was not achieved, however, when the apples were stored for 10 days at room temperature (Bauchot et al., 1995).

In some cases, the results of coating was not as expected. Zucchini fruit were coated with 0.5%, 0.75%, or 1.0% aqueous solutions of Semperfresh and with different formulations of calcium caseinate-acetylated monoglyceride aqueous emulsions ranging from 2.5% to 7.0% total solids. Semperfresh did not increase water vapor resistance of zucchini. Coating did not affect internal carbon dioxide or ethylene concentrations. Hue angle and lightness values were not significantly different for coated and uncoated zucchini. Maximum water vapor resistance resulted from high sodium caseinate and low acetylated monoglyceride contents (Avena-Bustillos and Krochta, 1994). Coated apples had delayed color development (internal and external Hunter color reflectance measurements) during 4 months of storage at 5°C. Coating increased fruit firmness and did not affect pH, total acidity, and soluble solids. Sensory evaluation indicated that flavor and textural changes were not detected when apples treated with 1.2% Semperfresh were compared to untreated apples after 2 months of storage (Santerre et al., 1989).

16.5.7.2.3 Nature-Seal

Nature-Seal has been applied to oranges, mangoes, guavas, tomatoes, cut apple, and potato. When applied on fruits or vegetables, Nature-Seal

1. Created MA by changing the internal concentrations of CO_2, O_2, and ethylene (Baldwin et al., 1995c, 1999)
2. Retarded ripening (Nisperos-Carriedo et al., 1992; Baldwin et al., 1999)
3. Retarded color development (McGuire and Hallman, 1995)
4. Retained firmness (McGuire and Hallman, 1995)
5. Increased flavor volatiles (Baldwin et al., 1995c, 1999)
6. Reduced decay (Baldwin et al., 1999)
7. Improved appearance (Baldwin et al., 1999)
8. Controlled browning and microbial populations by acting as carrier of antioxidants, acidulants, and preservatives (Baldwin et al., 1996)

Nature-Seal when applied to mangoes created MAs and delayed ripening, increased concentrations of flavor volatiles, reduced decay, and improved appearance by imparting a subtle shine (Baldwin et al., 1999). In guava fruits, coatings containing 2% or 4% hydroxypropyl cellulose significantly slowed softening on average at 35% or 45%, respectively, compared to uncoated fruits (McGuire and Hallman, 1995). Use of Nature-Seal as carrier of antioxidants, acidulants, and preservatives prolonged the storage life of cut apple and potato by about 1 week when stored in overwrapped trays at 4°C. Adjustment of coating pH to 2.5 gave optimal control of browning and microbial populations. Addition of soy protein to the original cellulose-based Nature-Seal formulations reduced coating permeability to oxygen and water vapor (Baldwin et al., 1996).

16.5.7.3 Chitosan-Based Edible Coatings

16.5.7.3.1 Nutri-Save

The commercial chitosan-based Nutri-Save has been applied to apples, pears, strawberries, bread fruits, tomatoes, peppers, squash, cauliflower, brussel sprouts, and broccoli. When applied on fruits or vegetables, Nutri-Save

1. Created MA by changing the internal concentrations of CO_2, O_2, and ethylene (Elson et al., 1985; Worrel et al., 2002)
2. Retarded color development (Meheriuk, 1990; Worrel et al., 2002)
3. Reduced weight loss (Elson et al., 1985; Worrel et al., 2002)
4. Retained firmness (Elson et al., 1985; Meheriuk, 1990; Lau and Meheriuk, 1994)
5. Retained titratable acidity, soluble solids, and ascorbic acid (Elson et al., 1985; Meheriuk, 1990; Lau and Meheriuk, 1994)
6. Reduced core flush and scald (Elson et al., 1985; Lau and Meheriuk, 1994)

Certain formulations of Nutri-Save were reported to retain fruit firmness and titratable acids over 9 months of storage for "Golden" and "Red Delicious" apples and "Barlett" and "Clapps Favorite" pears, which is comparable to CA storage. Also, encouraging results were reported when Nutri-Save was applied to some vegetables, such as, peppers, squash, cauliflower, brussel sprouts, and broccoli (Elson et al., 1985). However, high concentrations of the coating material caused significant skin discoloration (purplish hue) in McIntosh and Spartan apples (Lau and Meheriuk, 1994).

16.5.7.3.2 Pure Chitosan

Pure chitosan has been applied to apples, strawberries, litchi, longan fruit, bread fruit, and tomatoes. When applied on fruits or vegetables, pure chitosan

1. Created MA by changing the internal concentrations of CO_2, O_2, and ethylene (El Ghaouth et al., 1992b; Yong Soo et al., 1998; Yueming and Yuebiao, 2001; Worrel et al., 2002)
2. Retarded color development (El Ghaouth et al., 1991, 1992b; Zhang and Quantick, 1997; Yueming and Yuebiao, 2001; Worrel et al., 2002)
3. Reduced weight loss (Zhang and Quantick, 1997; Yong Soo et al., 1998; Yueming and Yuebiao, 2001; Worrel et al., 2002)
4. Retained firmness (El Ghaouth et al., 1991, 1992b; Yong Soo et al., 1998)
5. Retained titratable acidity, soluble solids, and ascorbic acid (El Ghaouth et al., 1991, 1992b; Yong Soo et al., 1998)
6. Delayed the increase in PPO activity (Zhang and Quantick, 1997; Yueming and Yuebiao, 2001)
7. Reduced decay (El Ghaouth et al., 1991, 1992a,b; Yueming and Yuebiao, 2001)

Coating tomatoes with chitosan solutions reduced the respiration rate and ethylene production, with greater effect at 2% than 1% chitosan. Coating increased the internal CO_2 and decreased the internal O_2 levels of the tomatoes. Chitosan-coated tomatoes were firmer, higher in titratable acidity, less decayed, and exhibited less red pigmentation than the control fruit at the end of storage (El Ghaouth et al., 1992b). Litchi fruit was treated with aqueous solutions of 1.0% or 2.0% chitosan coating 1 h after dipping in 0.1% thiabendazole and then stored at 4°C and 90% relative humidity. The application of chitosan coating delayed changes in contents of anthocyanin, flavonoid, total phenolics, delayed the increase in PPO activity, reduced weight loss, and partially inhibited the increase in POD activity. All these changes corresponded to changes in browning. Also, chitosan coating partially inhibited decay of the fruit during storage. However, increasing the concentration of chitosan coating did not significantly increase the beneficial effects of chitosan on browning and decay of the fruit (Zhang and Quantick, 1997). Also, strawberries are stored at 1°C with high RH and elevated CO_2 to deter infection by *B. cinerea* and *Rhizopus* sp. Chitosan-coated berries were firmer and higher in titratable acidity and synthesized anthocyanin at a slower rate than those treated with the fungicide iprodione or untreated fruit (El Ghaouth et al., 1991, 1992a).

16.5.7.4 Protein-Based Edible Coatings

The only protein materials that have been tested on fruits or vegetables are corn zein and soybean protein isolate. When corn zein was applied to tomatoes, it created MA by changing the internal concentrations of CO_2, O_2, and ethylene; delayed color development; and prevented loss of firmness and weight (Park et al., 1994b,c). Corn zein coatings of 5 and 15 μm thickness delayed the ripening of coated tomatoes for 6 days without adverse effects. Also, a 66 μm coating markedly delayed color development, while showing weight loss and alcohol fermentation due to anaerobic fermentation (Park et al., 1994b). Likewise, a corn zein film delayed color change and loss of firmness and weight in coated tomatoes during storage at 21°C. Shelf life was extended by 6 days with film coatings as determined by sensory evaluation (Park et al., 1994c). An edible coating composed of soybean protein isolate retarded the senescence process of kiwifruit and reduced softening (Shiying et al., 2001). Also the coating created MA and prevented weight loss. Thus, the shelf life of kiwifruit coated with edible film was extended to about three times.

16.5.7.5 Commercial Wax Coatings

Commercial waxes such as shellac, carnauba, candelilla, Johnfresh, Fomesa, and Sta-Fresh have been widely applied to some fruits (apples, pears, oranges, mangoes, papayas, guavas, breadfruit, and grapefruit) but not to vegetables. When applied on fruits, commercial waxes

1. Created MA by changing the internal concentrations of CO_2, O_2, and ethylene (Paull and Chen, 1989; Drake and Nelson, 1990; Drake et al., 1991; Sumnu and Bayindirli, 1994; Baldwin et al., 1995a,b, 1999; Hagenmaier and Grohmann, 1999; Saftner, 1999; Worrel et al., 2002; Bai et al., 2003a,b)
2. Retarded color development (Drake and Nelson, 1990; Sumnu and Bayindirli, 1994, 1995; McGuire and Hallman, 1995; Worrel et al., 2002; Bai et al., 2003a,b)
3. Reduced weight loss (Paull and Chen, 1989; Drake and Nelson, 1990; Drake et al., 1991; Sumnu and Bayindirli, 1994, 1995; Baldwin et al., 1995a,b, 1999; Hagenmaier and Baker, 1995; Hagenmaier and Grohmann, 1999; Saftner, 1999; Worrel et al., 2002; Bai et al., 2003a,b)
4. Retained firmness (Drake et al., 1991; Sumnu and Bayindirli, 1994, 1995; McGuire and Hallman, 1995; Saftner, 1999; Bai et al., 2003a,b)
5. Retained titratable acidity, soluble solids, and ascorbic acid (Sumnu and Bayindirli, 1994, 1995; Bai et al., 2003a,b)
6. Increased flavor volatiles (Baldwin et al., 1995a,b)
7. Reduced decay (Baldwin et al., 1999)
8. Improved appearance (Hagenmaier and Baker, 1995; Baldwin et al., 1999; Hagenmaier and Grohmann, 1999)

In a study with "Delicious" and "Golden Delicious" apples, no quality differences were evident among the apples waxed with commercial shellac, carnauba, or resin-based waxes. The waxed apples exhibited improved color, higher internal levels of CO_2 and ethylene, and reduced weight loss, but wax treatments had no effect on firmness (Drake and Nelson, 1990). A commercial carnauba–shellac coating (Shield-Brite PR 160C) was reported to delay ripening of "d'Anjou" pears, resulting in lower external and higher internal concentrations of CO_2 than nonwaxed fruit, retained firmness, and delayed color changes (Drake et al., 1991). Waxing of papaya fruit with various resin-, polyethylene-, and carnauba-based waxes resulted in some delay in certain ripening parameters, a decrease in weight loss, and, in some cases, some off-flavor (Paull and Chen, 1989). A 5% carnauba formulation slowed softening by 10%–30% and was most effective at reducing weight loss in guava fruit (McGuire and Hallman, 1995). Shellac- and wax-based apple coatings transiently inhibited total volatile levels in "Golden Delicious" while not affecting those in "Gala" apples during 6 months of storage in air at 0°C. Holding the fruit at 20°C for up to 3 weeks following cold storage increased volatile levels with coated and nontreated fruit having similar amounts. Only shellac-coated "Golden Delicious" apples accumulated ethanol and ethyl acetate when held at 20°C (Saftner, 1999).

Cut apple pieces coated with double layers of buffered polysaccharide/lipid showed a 50%–70% reduction in the rate of CO_2 evolution and about 90% decrease in C_2H_4 as compared with uncoated controls at ~23°C. This substantial decline in the rates of gas evolution was attributed primarily to the diffusion barrier properties of the lipid layer and secondarily to the inhibitory effect of the ascorbate buffer which contained calcium (Wong et al., 1994a).

16.5.7.6 Vegetable Oils

Vegetable oils have been applied especially to apples and pears for the prevention of superficial scald. Moreover, oil-treatments delayed ripening, retained firmness and titratable acidity, and retarded color development (Scott et al., 1995; Ju et al., 2000a,b).

Vegetable oils (sunflower, canola, castor, palm, peanut) considerably reduced superficial scald in "Granny Smith" apples after several months of storage at 0°C (Scott et al., 1995). Plant oils (corn, soybean, peanut, cottonseed, and linseed) similarly inhibited ethylene production and accumulation in the first 2 weeks in "Bartlett" pears and in the first 3 months in "Golden Delicious." Compared with the untreated controls, oil-treated fruits were firmer, greener, and contained a higher level of titratable acidity after 6 weeks or 6 months at 0°C plus 7 days at 20°C. The effectiveness of oil treatment was higher in preclimacteric than in climacteric fruit (Ju et al., 2000a,b).

16.6 FUTURE—NEW TRENDS IN THE EDIBLE COATING FIELD

16.6.1 Development of New Technological Approaches

New technologies, designed to extend the shelf life of highly perishable products, such as fresh and minimally processed fruits and vegetables, and ensure their safety are in demand. Edible coating technology seems to be very promising; however, there are limitations mainly arising from the control of the film properties and the sensory acceptance of the obtained products. New trends in the edible coating field have focused on the development of new technologies that allow a more efficient control of coating properties and functionality. To this end, new methodologies have been developed, most of them focusing on composite or multilayered systems (Vargas et al., 2008). The development of composite systems is promising in improving several properties of the coating materials such as moisture barrier properties, adhesion, and/or durability. Another important factor is the sensory quality, which is essential as it affects consumer acceptance and market potential of coated products. Exogenous flavor impact by the coating materials, unattractive surface appearance of coatings, and other factors may affect consumer acceptance of the coated products. Therefore, it is important to evaluate the sensory quality of both coating materials and coated final products by measuring parameters such as appearance, color, aroma, taste, and texture (Lin and Zha, 2007).

The development of multilayered coatings by means of layer-by-layer (LbL) electrodeposition (Weiss et al., 2006) is one of these new methodologies. LbL technique is performed by alternating the immersion of substrates in solutions of oppositely charged polyelectrolytes with rinsing steps, producing, in this way, ultrathin polyelectrolyte multilayers on charged surfaces. The most common biopolymers that can be used in the formation of these multilayered structures (Marudova et al., 2005; Krzemiski et al., 2006) include chitosan, poly-L-lysine, pectin, and alginate, while other charged species, such as lipid droplets, solid particles, micelles, or surfactants, can also be used. Multilayered edible coatings receive great attention as various functional or antimicrobial agents can be incorporated into them.

Another promising technique that can be potentially used to incorporate functional ingredients and antimicrobials into edible coatings for fruits is the encapsulation technique. Encapsulation is defined as a technology in which a core material (liquid droplet, solid particle, or gas compound) is entrapped (coated or embedded) into a food grade wall material to give a product with many useful properties (Risch, 1995). This technique is especially suitable for incorporating ingredients that add value to the food product (such as enzymes and pro- and prebiotics), as well as functional ingredients, susceptible to lipid oxidation (such as omega-3-fatty acids) or to mask odors or tastes (Lopez-Rubio et al., 2006).

Finally, the most recent approach to improve coating properties is to make nano-composites by incorporating nanosized clay materials (such as silicates) into biopolymer-based matrices. However, even if this seems to be promising, the major concern of the scientific community when incorporating these nanomaterials into edible coatings is the lack of studies concerning their possible toxicity (Vargas et al., 2008).

16.6.2 Incorporation of Antimicrobials and Functional Ingredients into Edible Coatings

16.6.2.1 Antimicrobials into Edible Coatings

The ability of edible films to retard moisture, oxygen permeability, flavorings, and solute transport may be improved by incorporating additives, such as antioxidants, antimicrobials, colorants, flavors, fortifying nutrients, and spices, in film formulation (Pranoto et al., 2005). Given the fact that consumers demand less use of chemicals, the scientific research is focused on the development of antimicrobial edible coating films able to act as alternative antimicrobials and antioxidants (Ponce et al., 2008). The above films can prolong the shelf life and maintain the quality of the food products, preventing microbiological growth and deterioration. Moreover, antimicrobial substances imbedded in films can be gradually released on the food surface, therefore, requiring smaller amounts to achieve the target shelf life (Gennadios and Kurth, 1997; Min and Krochta, 2005). There are different types of antimicrobials that can be incorporated into edible films and coatings including organic acids (acetic, benzoic, lactic, propionic, sorbic), fatty acid esters (glyceryl monolaurate), polypeptides (lysozyme, peroxidase, lactoferrin, nisin), plant essential oils (cinnamon, oregano, lemongrass), and nitrites and sulfites (Franssen and Krochta, 2003). Among the above categories, plant essential oils are outstanding alternatives to chemical preservatives, and their use in foods meets consumer demands for minimally processed natural products (Burt, 2004). These compounds can also be added to edible films and coatings to modify food flavor, aroma, and odor (Cagri et al., 2004). The incorporation of oregano, cinnamon, and lemongrass oils into apple puree and alginate-apple puree edible films proved effective against *Escherichia coli* O157:H7 showing that edible films constitute a feasible approach for incorporating plant essential oils onto fresh food surfaces (Rojas-Graü et al., 2006, 2007).

Another challenging area is the use of antimicrobials into edible films and coatings, especially, in meat products. For instance, antimicrobial chitosan films, containing acetic or propionic acid, inhibited growth of *Enterobacteriaceae* and *Serratia liquefaciens* on bologna, cooked ham, and pastrami (Ouattara et al., 2000). *Listeria monocytogenes*, *E. coli* O157:H7, and *Salmonella enterica* growth on bologna and summer sausage was also inhibited through the incorporation of *p*-aminobenzoic (PABA) and sorbic acids into whey protein isolate films (Cagri et al., 2002). Milk protein-based edible films containing oregano, pimento, or oregano–pimento essential oil mix onto beef muscle slices achieved to control growth of *E. coli* and *Pseudomonas* spp., thereby increasing product shelf life during refrigerated storage (Oussalah et al., 2004). Moreover, chitosan films enriched with essential oils (anise, basil, coriander, and oregano) had strong antimicrobial effects on *L. monocytogenes* when applied to inoculated bologna meat samples (Zivanovic et al., 2005). Alginate-based edible films, containing Spanish oregano or Chinese cinnamon essential oils, were also tested onto beef muscle slices for control of *E. coli* O157:H7 and *Salmonella typhimurium* and managed to reduce bacterial growth during storage (Oussalah et al., 2006). Chitosan coatings, added individually or in combination with rosemary or a-tocopherol, effectively inhibited microbial growth, retarded lipid oxidation, and extended shelf life of fresh pork sausages of the traditional Greek-type during refrigerated storage (4°C) for 20 days (Georgantelis et al., 2007). Recently, Kanatt et al. (2008) studied the potential of using chitosan and mint extract as a preservative for meat and meat products. They found that the above mixture enhanced the shelf life of pork salami, as determined by total bacterial count, when stored at 0°C–3°C. Therefore, the use of natural antimicrobials into edible films and coatings and their application in meat products poses a valuable potential for the commercial use in order to improve preservation of these products without the use of nitrites or other additives.

16.6.2.2 Probiotic Edible Films and Coatings

Currently, the incorporation of lactic acid bacteria or probiotics to films and coatings has attracted great attention. With the use of films and coatings, the obstacles relating to probiotics lethality due to food processing can be surpassed. Concerning probiotic edible films and coatings, a number of applications including fruits, fish products, as well as bakery products have been developed.

In particular, probiotic edible films of alginate (2% w/v) and gellan (0.5% w/v) were developed as carriers for organisms, such as bifidobacteria, for coating fresh-cut fruits in an attempt to obtain functional probiotic-coated fruits (Tapia et al., 2007). Fresh-cut apples and papayas were also successfully coated with alginate or gellan film-forming solutions containing viable bifidobacteria. The potential for obtaining functional bread combining the microencapsulation of *Lactobacillus acidophilus* into starch-based coatings was also studied (Altamirano-Fortoul et al., 2012). Different probiotic coatings (dispersed or multilayer) were applied onto the surface of partially baked breads. The survival of microencapsulated *L. acidophilus* demonstrated the ability of starch solution to protect the bacteria during baking and storage time, leading to a functional bread with similar characteristics to common bread but with additional healthy benefits. Probiotic baked cereal products were also prepared through the use of air-dried probiotic edible films (Soukoulis et al., 2014). The probiotic pan bread was produced by the application of film-forming solutions, based either on individual hydrogels, for example, 1% w/w sodium alginate, or binary blends of 0.5% w/w sodium alginate and 2% whey protein concentrate containing *Lactobacillus rhamnosus* GG, followed by an air drying step at 60°C for 10 min or 180°C for 2 min. It proved that the presence of whey proteins in the film-forming solution preserved *L. rhamnosus* GG viability throughout drying and storage, while the use of film systems based exclusively on sodium alginate exhibited a very good performance under in vitro digestion.

Finally, gelatin edible coatings and films can be a suitable matrix for the incorporation of probiotic bacteria. This was confirmed through their use for the incorporation of *Bifidobacterium bifidum* in fish applications (Lopez De Lacey et al., 2012). Therefore, the development of probiotics edible films and coatings and their application in food products could confer stability during storage while enhancing their functionality.

16.7 CONCLUSIONS

Fresh fruits and vegetables contain valuable nutritious components for our diet that we cannot draw from other food sources. The preservation of fresh produce has occupied the human for a lot of centuries, since they are perishable products with small postharvest life. The combination of preservation at low temperatures and the use of edible coating materials have been proved very efficient, if of course certain requirements are met. It should be marked that each product needs particular treatment, as long as its respiration and transpiration rates are particulars. Moreover, the optimal preservation conditions (temperature, relative humidity) are different for each product. Thus, the formation of edible films and coatings should be performed based on the particular needs and characteristics of each product. Film's/coating's ingredients, method of formation, and type of application are the main factors that affect the characteristics (gas permeability, water vapor permeability, etc.) of film or coating. More research is needed to combine the desirable gas permeability and textural benefits of polysaccharide and protein coatings with the better water-barrier properties associated with lipid components. Use of additives can improve the formation of films or coatings. Since the coating may be consumed with the product, its components must be GRAS.

Most of the commercially available edible coatings for fruits and vegetables are wax-based. Although these coatings considerably reduce water loss, they also inhibit respiratory gas exchange to such an extent that it induces fermentation. Most commercial films and coatings have been tested on fresh fruits and vegetables, and no particular research has been done using new ingredients and composite materials. Therefore, only few edible films or coatings have been reported with an overall balanced performance. The development of advanced coatings and films for fresh fruits and vegetables, as well as in further food applications, can offer much more than we can imagine.

REFERENCES

Altamirano-Fortoul, R., Moreno-Terrazas, R., Quezada-Gallo, A., Rosell, C.M. 2012. Viability of some probiotic coatings in bread and its effect on the crust mechanical properties. *Food Hydrocolloids*, 29: 166–174.

Avena-Bustillos, R., Krochta, J. 1993. Water vapor permeability of caseinate–based edible films as affected by pH, calcium crosslinking and lipid content. *Journal of Food Science*, 58(4): 904–907.

Avena–Bustillos, R., Krochta, J. 1994. Optimization of edible coating formulations on zucchini to reduce water loss. *Journal of Food Engineering*, 21: 197–214.

Aydt, T.P., Weller, C.L., Testin, R.F. 1991. Mechanical and barrier properties of edible corn and wheat protein films. *Transactions of the ASAE*, 34: 207–211.

Bai, J., Alleyne, V., Hagenmaier, R., Mattheis, J., Baldwin, E. 2003a. Formulation of zein coatings for apples (*Malus domestica* Borkh). *Postharvest Biology and Technology*, 28: 259–268.

Bai, J., Hagenmaier, R., Baldwin, E. 2003b. Coating selection for Delicious and other apples. *Postharvest Biology and Technology*, 28: 381–390.

Baldwin, E.A. 1994. Edible coatings for fresh fruits and vegetables: Past, present, and future. In: *Edible Coatings and Films to Improve Food Quality*. pp. 25–64. Eds.: Krochta, J.M., Baldwin, E.A., Nisperos–Carriedo, M.O. Technomic Publishing Co. Inc., Basel, Switzerland.

Baldwin, E.A. 2007. Surface treatments and edible coatings in food preservation (Chapter 21). In: *Handbook of Food Preservation* (2nd Ed.). pp. 478–507. Taylor & Francis Group, LLC, Boca Raton, FL.

Baldwin, E.A., Baker, R.A. 2002. Use of proteins in edible coatings for whole and minimaly processed fruits and vegetables. In: *Protein-Based Films and Coatings*. pp. 501–516. Ed.: Gennadios A. CRC Press, Boca Raton, FL.

Baldwin, E.A., Burns, J.K., Kazokas, W., Brecht, J.K., Hagenmaier, R.D., Bender, R.J., Pesis, E. 1999. Effect of two edible coatings with different permeability characteristics on mango (*Mangifera indica* L.) ripening during storage. *Postharvest Biology and Technology*, 17: 215–226.

Baldwin, E.A., Nisperos-Carriedo, M.O., Baker, R.A. 1995a. Edible coating for lightly processed fruits and vegetables. *HortScience*, 30(1): 35–38.

Baldwin, E.A., Nisperos-Carriedo, M.O., Baker, R.A. 1995b. Use of edible coatings to preserve quality of lightly (and slightly) processed products. *Critical Reviews in Food Science and Nutrition*, 35(6): 509–524.

Baldwin, E.A., Nisperos-Carriedo, M.O., Chen, X., Hagenmaier, R.D. 1996. Improving storage life of cut apple and potato with edible coating. *Postharvest Biology and Technology*, 9: 151–163.

Baldwin, E.A., Nisperos-Carriedo, M.O., Shaw, P.E., Burns J.K. 1995c. Effect of coatings and prolonged storage conditions on fresh orange flavor volatiles, degrees brix, and ascorbic acid levels. *Journal of Agricultural and Food Chemistry*, 43: 1321–1331.

Banks, N.H. 1984. Studies of the banana fruit surface in relation to the effects of TAL Pro-long coating on gaseous exchange. *Scientia Horticulturae*, 24: 279–286.

Banks, N.H. 1985a. Internal atmosphere modification in pro-long coated apples. *Acta Horticulturae*, 157: 105–112.

Banks, N.H. 1985b. Responses of banana fruit to pro-long coating at different times relative to the initiation of ripening. *Scientia Horticulturae*, 26: 149–157.

Bauchot, A., John, P., Soria, Y., Recasens, I. 1995. Sucrose ester-based coatings formulated with food-compatible antioxidants in the prevention of superficial scald in stored apples. *Journal of the American Society for Horticultural Science*, 120(3): 491–496.

Beaudry, R.M. 1999. Effect of O_2 and CO_2 partial pressure on selected phenomena affecting fruit and vegetable quality. *Postharvest Biology and Technology*, 15: 293–303.

Beck, M.I., Tomka, I., Waysek, E. 1996. Physicochemical characterization of zein as a film coating polymer—A direct comparison with ethyl cellulose. *International Journal of Pharmaceutics*, 141: 137–150.

Ben-Yehoshua, S. 1987. Transpiration, water stress, and gas exchange. In: *Postharvest Physiology of Vegetables*. pp. 113–170. Ed.: Weichmann, J. Marcel Dekker, Inc., New York.

Blank, M. 1962. The permeability of monolayers to several gases. In: *Retardation of Evaporation by Monolayers: Transport Processes*. pp. 75–79. Ed.: La Mer, V.K. Academic Press, New York.

Blank, M. 1972. In: *Techniques of Surface and Colloidal Chemistry and Physics*. pp. 41–88. Eds.: Good, R.J., Stromberg, R.R., Patrick, R.L. Marcel Dekker, New York.

Brimelow, C., Vadehra, D. 1991. Chilling. In: *Vegetable Processing*. pp. 123–153. Eds.: Arthey, D., Dennis, C. Chapman & Hall, London, U.K.

Burt, S. 2004. Essential oils: Their antibacterial properties and potential applications in foods: A review. *International Journal of Food Microbiology*, 94: 223–253.

Cagri, A., Ustunol, Z., Ryser, E.T. 2002. Inhibition of three pathogens on bologna and summer sausage slices using antimicrobial edible films. *Journal of Food Science*, 67: 2317–2324.

Cagri, A., Ustunol, Z., Ryser, E.T. 2004. Antimicrobial edible films and coating. *Journal of Food Protection*, 67: 833–848.

Chai, Y.-L., Ott, D., Cash, J. 1991. Shelf-life extension of Michigan apples using sucrose polyester. *Journal of Food Processing and Preservation*, 15: 197–214.

Cuppett, S.L. 1994. Edible coatings as carriers of food additives, fungicides and natural antagonists. In: *Edible Coatings and Films to Improve Food Quality*. pp. 121–137. Eds.: Krochta, J.M., Baldwin, E.A., Nisperos-Carriedo, M.O. Technomic Publishing Co., Inc., Basel, Switzerland.

Cuq, B., Gontard, N., Guilbert, S. 1995. Edible films and coatings as active layers. In: *Active Food Packaging*. pp. 111–135. Ed.: Rooney, M.L. Blackie Academic & Professional, London, U.K.

Curtis, G.J. 1988. Some experiments with edible coatings on the long-term storage of citrus fruits. *Proceedings of the 6th International Citrus Congress*, 3: 1514–1520.

Debeaufort, F., Quezada–Gallo, J.A., Voilley, A. 1998. Edible films and coatings: Tomorrow's packaging: A review. *Critical Reviews in Food Science*, 38(4): 299–313.

Desai, B.B., Salunkhe, D.K. 1991. Fruits and vegetables. In: *Foods of Plant Origin, Production, Technology, and Human Nutrition*. pp. 301–412. Eds.: Salunkhe, D.K., Deshpande, S.S. AVI Book, New York.

Dhalla, R., Hanson, S.W. 1988. Effect of permeable coatings on the storage life of fruits. II. Pro-long treatment of mangoes (*Mangifera indica* L. cv. Julie). *International Journal of Food Science and Technology*, 23: 107–112.

Donhowe, I.G., Fennema, O. 1993. Water vapor and oxygen permeability of wax films. *JAOCS*, 70(9): 867–873.

Donhowe, I.G., Fennema, O. 1994. Edible films and coatings: characteristics, formation, definitions, and testing methods. In: *Edible Coatings and Films to Improve Food Quality*. pp. 1–24. Eds.: Krochta, J.M., Baldwin, E.A., Nisperos-Carriedo, M.O. Technomic Publishing Co., Inc., Basel, Switzerland.

Drake, S.R., Caverlieri, R., Kupferman, E.M. 1991. Quality attributes of d'Anjou pears after different wax drying temperatures and refrigerated storage. *Journal of Food Quality*, 14: 455–465.

Drake, S.R., Fellman, J.K., Nelson, J.W. 1987. Postharvest use of sucrose polyesters for extending the self-life of stored 'Golden Delicious' apples. *Journal of Food Science*, 52: 685–690.

Drake, S.R., Kupferman, E.M., Fellman, J.K. 1988. 'Bing' sweet cherry (*Prunus avium*) quality as influenced by wax coatings and storage temperature. *Journal of Food Science*, 53: 124–126.

Drake, S.R., Nelson, J.W. 1990. Storage quality of waxed and nonwaxed 'Delicious' and 'Golden Delicious' apples. *Journal of Food Quality*, 13: 331–341.

El Ghaouth, A., Arul, J., Grenier, J., Asselin, A. 1992a. Antifungal activity of chitosan on two postharvest pathogens of strawberry fruits. *Postharvest Pathology and Mycotoxins*, 82(4): 398–402.

El Ghaouth, A., Arul, J., Ponnampalam, R., Boulet, M. 1991. Chitosan effect on storability and quality of fresh strawberries. *Journal of Food Science*, 56: 1618–1631.

El Ghaouth, A., Ponnampalam, R., Castaigne, F., Arul, J. 1992b. Chitosan coating to extend the storage life of tomatoes. *HortScience*, 27(9): 1016–1018.

Elson, C.M., Hayes, E.R., Lidster, P.D. 1985. Development of the differentially permeable fruit coating 'Nutri–Save' for the modified atmosphere storage of fruit. In: *Controlled Atmosphere for Storage and Transport of Perishable Agricultural Commodities*. pp. 248–262. Ed.: Blankenship, M. North Carolina State University, Raleigh, NC.

Franssen, L.R., Krochta, J.M. 2003. Edible coatings containing natural antimicrobials for processed foods. In: *Natural Antimicrobials for Minimal Processing of Foods*. pp. 250–262. Ed.: Roller S. CRC Cambridge, Boca Raton, FL.

Gennadios, A., Kurth, L.B. 1997. Application of edible coatings on meats, poultry and seafoods: A review. *Lebensmittel-Wissenschaft and Technologie*, 30: 337–350.

Gennadios, A., McHugh, T.R., Weller, C.L., Krochta, J.M. 1994. Edible coatings and films based on proteins. In: *Edible Coatings and Films to Improve Food Quality*. pp. 201–277. Eds.: Krochta, J.M., Baldwin, E.A., and Nisperos-Carriedo, M.O. Technomic Publishing Co., Inc., Basel, Switzerland.

Gennadios, A., Weller, C.L. 1990. Edible films and coatings from wheat and corn proteins. *Food Technology*, 10: 63–69.

Gennadios, A., Weller, C.L. 1991. Edible films and coatings from soymilk and soy protein. *Cereal Food World*, 36(12): 1004–1009.

Gennadios, A., Weller, C., Testin, R. 1993. Property modification of edible Wheat Gluten-based films. *Transaction of the ASAE*, 36(2): 465–470.

Georgantelis, D., Ambrosiadis, I., Katikou, P., Blekas, G., Georgakis, S. 2007. Effect of rosemary extract, chitosan and α-tocopherol on microbiological parameters and lipid oxidation of fresh pork sausages stored at 4°C. *Meat Science*, 76:172–181.

Ghaouth, A., Ponnampalam, R., Castaigne, F., Arul, J. 1992. Chitosan coating to extend the storage life of tomatoes. *HortScience*, 27(9): 1016–1018.

Gontard, N., Duchez, C., Cuq, J.L., Guilbert, S. 1994. Edible composite films of wheat gluten and lipids: Water vapour permeability and other physical properties. *International Journal of Food Science and Technology*, 29: 39–50.

Gontard, N., Guilbert, S. 1992. Bio-packaging: Technology and properties of edible and/or biodegradable material of agricultural origin. In: *Food Packaging and Preservation*. pp. 159–181. Ed.: Mathlouthi, M. Springer Science & Business Media, New York.

Gontard, N., Thibault, R., Cuq, B., Guilbert, S. 1996. Influence of relative humidity and film composition on oxygen and carbon dioxide permeabilities of edible films. *Journal of Agricultural and Food Chemistry*, 44: 1064–1069.

Grant, L.A., Burns, J. 1994. Application of coatings. In: *Edible Coatings and Films to Improve Food Quality*. pp. 189–200. Eds.: Krochta, J.M., Baldwin, E.A., Nisperos-Carriedo, M.O. Technomic Publishing Co., Inc., Basel, Switzerland.

Hagenmaier, R., Baker, R.A. 1995. Layered coatings to control weight loss and preserve gloss of citrus fruit. *HortScience*, 30(2): 296–298.

Hagenmaier, R., Grohmann, K. 1999. Polyvinyl acetate as a high-gloss edible coating. *Journal of Food Science*, 64(6): 1064–1067.

Hagenmaier, R., Shaw, P. 1990. Moisture permeability of edible films made with fatty acid and (hydroxypropyl) methylcellulose. *Journal of Agricultural and Food Chemistry*, 38: 1799–1803.

Hagenmaier, R., Shaw, P. 1991. Permeability of coatings made with emulsified polyethylene wax. *Journal of Agricultural and Food Chemistry*, 39: 1705–1708.

Hardenburg, R.E. 1967. Wax and related coatings for horticultural products: A bibliography. *Agriculture Research Bulletin*, 57-15. U.S. Department of Agriculture, Washington, DC.

Hernandez, E. 1994. Edible coatings from lipids and resins. In: *Edible Coatings and Films to Improve Food Quality*. pp. 279–303. Eds.: Krochta, J.M., Baldwin, E.A., Nisperos-Carriedo, M.O. Technomic Publishing Co., Inc., Basel, Switzerland.

Hui, Y.H. 1992. Edible films and coatings. In: *Encyclopedia of Food Science and Technology*. pp. 659–663. Vol. 2, E-H. Ed.: Hui, Y.H., John Wiley & Sons, Inc., New York.

Institute of Food Technology. 1991. New from Mitsubishi. *Annual Meeting and Food Expo Program and Exhibit Directory*, Dallas Convention Center, June 1–5, 1991, Chicago, IL.

Irissin-Mangata, J., Boutevin, B., Bauduin, G. 1999. Bilayer films composed of wheat gluten and functionalized polyethylene: Permeability and other physical properties. *Polymer Bulletin*, 43: 441–448.

Ju, Z., Duan, Y., Ju, Z. 2000a. Plant oil emulsion modifies internal atmosphere, delays fruit ripening, and inhibits internal browning in Chinese pears. *Postharvest Biology and Technology*, 20: 243–250.

Ju, Z., Duan, Y., Ju, Z., Curry, E. 2000b. Stripped plant oils maintain fruit quality of 'Golden Delicious' apples and 'Bartlett' pears after prolonged cold storage. *The Journal of Horticultural Science and Biotechnology*, 75(4): 423–427.

Kadam, S.S., Salunkhe, D.K. 1995. Fruits in human nutrition. In: *Handbook of Fruit Science and Technology*. pp. 593–596. Eds.: Salunkhe, D.K., Kadam, S.S. Marcel Dekker, Inc., New York.

Kader, A.A. 1987. Respiration and gas exchange of vegetables. In: *Postharvest Physiology of Vegetables*. pp. 25–43. Ed.: Weichmann, J. Marcel Dekker, Inc., New York.

Kader, A.A., Watkins, C.B. 2000. Modified atmosphere packaging toward 2000 and beyond. *HortTechnology*, 10(3): 483–486.

Kanatt, S.R., Chander, R., Sharma, A. 2008. Chitosan and mint mixture: A new preservative for meat and meat products. *Food Chemistry*, 107: 845–852.

Kaplan, H.J. 1986. Washing, waxing, and color-adding. In: *Fresh Citrus Fruits*. pp. 379. Eds: Wardowski, W.F., Nagy, S., Grierson, W. AVI, Westport, CT.

Kester, J., Fennema, O. 1986. Edible films and coatings: A review. *Food technology*, 40(12): 47–59.

Kester, J., Fennema, O. 1989. An edible film of lipids and cellulose ethers: Barrier properties to moisture vapor transmission and structural evaluation. *Journal of Food Science*, 54(6): 1383–1389.

Krochta, J.M. 2002. Proteins as raw materials for films and coatings: Definitions, current status, and opportunities (Chapter 1). In: *Protein-Based Films and Coating*. Ed.: Gennadios, A. CRC Press, LLC, Boca Raton, FL.

Krochta, J.M., De Mulder-Johnston, C. 1997. Edible and biodegradable polymer films: Challenges and opportunities. *Food Technology*, 51(2): 61–74.

Krzemiski, A., Marudova, M., Moffat, J., Noel, T.R., Parker, R., Welliner, N., Ring, S.G. 2006. Deposition of pectin/poly-L-lysine multilayers with pectins of varying degrees of esterification. *Biomacromolecules*, 7: 498–506.

Labuza, T.P., Breene, W.M., 1989. Applications of "Active Packaging" for improvement of shelf-life and nutritional quality of fresh and extended shelf-life foods. *Journal of Food Processing and Preservation*, 13: 1–69.

Lau, O., Meheriuk, M. 1994. The effect of edible coatings on storage quality of McIntosh, Delicious and Spartan apples. *Canadian Journal of Plant Science*, 74: 847–852.

Lee, J.C., Kwon, O.W. 1990. Effect of Prolong on storage quality and ethylene evolution in 'Jonathan' and 'Fuji' apple fruits. *Journal of the Korean Society for Horticultural Science*, 31: 247–254.

Lin, D., Zha, Y. 2007. Innovations in the development and application of edible coatings for resh and minimally. Processed fruits and vegetables. *Comprehensive Food Science and Food Safety Reviews*, 6: 60–75.

Lopez De Lacey, A.M., Lopez-Caballero, M.E., Gomez-Estaca, J., Gomez-Guillin, M.C., Montero, P. 2012. Functionality of *Lactobacillus acidophilus* and *Bifidobacterium bifidum* incorporated to edible coatings and films. *Innovative Food Science and Emerging Technologies*, 16: 277–282.

Lopez-Rubio, A., Gavara, R., Lagaron J.M. 2006. Bioactive packaging: Turning foods into healthier foods through biomaterials. *Trends in Food Science and Technology*, 17: 567–575.

Lowings, P.H., Cutts, D.G. 1982. The preservation of fresh fruits and vegetables. *Procedings of the Institute of Food Science and Technology Annual Symposium,* July 1981, Nottingham, U.K., pp. 52.

Marudova, M., Lang, S., Brownsey, G.J., Ring, S.G. 2005. Pectin-chitosan multilayer formation. *Carbohydrate Research*, 340: 2144–2149.

McGuire, R.G., Hallman, G.J. 1995. Coating guavas with cellulose- or carnauba-based emulsions interferes with postharvest ripening. *HortScience*, 30(2): 294–295.

McHugh, T.H., Avena-Bustillos, R., Krochta, J.M. 1993. Hydrophilic Edible Films: Modified procedure for water vapor permeability and explanation of thickness effect. *Journal of Food Science*, 58(4): 899–903.

McHugh, T., Aujard, J.F., Krochta, J.M. 1994. Plasticized whey protein edible films: Water vapor permeability properties. *Journal of Food Science*, 59(2): 416–419.

McHugh, T., Huxsoll, C., Krochta, J. 1996. Permeability properties of fruit puree edible films. *Journal of Food Science*, 61(1): 88–91.

McHugh, T., Krochta, J. 1994a. Milk–protein-based edible films and coatings. *Food Technology*, 1: 97–103.

McHugh, T., Krochta, J. 1994b. Sorbitol vs glycerol–plasticized whey protein edible films: Integrated oxygen permeability and tensile property evaluation. *Journal of Agricultural and Food Chemistry*, 42(4): 841–845.

Meheriuk, M. 1990. Skin color in 'Newton' apples treated with calcium nitrate, urea, diphenylamine, and a film coating. *HortScience*, 25: 775–776.

Min, S., Krochta, J. 2005. Inhibition of *Penicillium* commune by edible whey protein films incorporating lactoferrin, lacto–ferrin hydrolysate, and lactoperoxidase systems. *Journal of Food Science*, 70(2): 87–94.

Motlagh, F.H., Quantick, P.C. 1988. Effects of permeable coatings on the storage life of fruits. I. Pro-long treatment of limes. *International Journal of Food Science and Technology*, 23: 99–105.

Mujica-Paz, H., Gontard, N. 1997. Oxygen and carbon dioxide permeability of wheat gluten film: Effect of relative humidity and temperature. *Journal of Agricultural and Food Chemistry*, 45: 4101–4105.

Nisperos-Carriedo, M.O. 1994. Edible coatings and films based on polysaccharides. In: *Edible Coatings and Films to Improve Food Quality*. pp. 305–335. Eds.: Krochta, J.M., Baldwin, E.A., Nisperos-Carriedo, M.O. Technomic Publishing Co., Inc., Basel, Switzerland.

Nisperos-Carriedo, M.O., Baldwin, E.A. 1988. Effect of two types of edible films on tomato fruit ripening. *Proceedings of the Florida State Horticultural Society*, 101: 217–220.

Nisperos-Carriedo, M.O., Baldwin, E.A., Shaw, P.E. 1992. Development of edible coating for extending post-harvest life of selected fruits and vegetables. *Proceedings of the Florida State Horticultural Society*, 104: 122–125.

Nisperos-Carriedo, M.O., Shaw, P.E., Baldwin, E.A. 1990. Changes in volatile flavor components of pineapple orange juice as influenced by the application of lipid and composite film. *Journal of Agricultural and Food Chemistry*, 38: 1382–1387.

Olivas, G.I., Barbosa-Canovas, G.V. 2005. Edible coatings for fresh-cut fruits. *Critical Reviews in Food Science and Nutrition*, 45: 657–670.

Ouattara, B., Simard, R., Piette, G., Begin, A., Holley, R. 2000. Diffusion of acetic and propionic acids from chitosan-based antimicrobial packaging films. *Journal of Food Science*, 65(5): 768–772.

Oussalah, M., Caillet, S., Salmieri, S., Saucier, L., Lacroix, M. 2004. Antimicrobial and antioxidant effects of milk protein-based film containing essential oils for the preservation of whole beef muscle. *Journal of Agricultural and Food Chemistry*, 52: 5598–5605.

Oussalah, M., Caillet, S., Salmieri, S., Saucier, L., Lacroix, M. 2006. Antimicrobial effects of alginatebased film containing essential oils for the preservation of whole beef muscle. *Journal of Food Protection*, 69: 2364–2369.

Park, H.J. 1999. Development of advanced edible coatings for fruits. *Trends in Food Science and Technology*, 10: 254–260.

Park, H.J., Bunn, J., Vergano, P., Testin, R. 1994a. Gas permeation and thickness of the sucrose polyesters, Sepmerfresh™ coatings on apples. *Journal of Food Processing and Preservation*, 18: 349–357.

Park, H.J., Chinnan, M., Shewfelt, R. 1994b. Edible corn–zein film coatings to extend storage life of tomatoes. *Journal of Food Processing and Preservation*, 18: 317–331.

Park, H.J., Chinnan, M., Shewfelt, R. 1994c. Edible coating effects on storage life and quality of tomatoes. *Journal of Food Science*, 59(3): 568–570.

Park, H.J., Weller, C., Vergano, P., Testin, R. 1993. Permeability and mechanical properties of cellulose-based edible films. *Journal of Food Science*, 58(6): 1361–1364.

Paull, R.E., Chen, N.J. 1989. Waxing and plastic wraps influence water loss from papaya fruit during storage and ripening. *Journal of the American Society for Horticultural Science*, 114: 937–942.

Pavlath, A.E., Orts, W. 2009. Edible films and coatings: Why, what, and how? (Chapter 1). In: *Edible Films and Coatings for Food Applications*. Eds.: Embuscado, M.E., Huber, K.C. Springer Science & Business Media, LLC, New York.

Ponce, A., Roura, S., Del Valle, C., Moreira, M. 2008. Antimicrobial and antioxidant activities of edible coatings enriched with natural plant extracts. *Posthatvest Biology and Technology*, 49(2): 294–300.

Potter, N.N., Hotchkiss, J.H. 1995. Vegetables and fruits. In: *Food Science* (5th Ed.) p. 409. Eds.: Potter, N.N., Hotchkiss, J.H. Chapman & Hall, London, U.K.

Pranoto, Y., Rakshit, S., Salokhe, V. 2005. Enhancing antimicrobial activity of chitosan films by incorporating garlic oil, potassium sorbate and nisin. *Lebensmittel-Wissenschaft and Technologie*, 38: 859–865.

Raghavan, G.S., Alvo, P., Gariepy, Y., Vigneault, C. 1996. Refrigerated and controlled modified atmosphere storage. In: *Biology, Principles, and Applications*. pp. 135–167. Eds.: Somogyi, L.P., Ramaswamy, H.S., and Hui, Y.H. Technomic Publishing Co., Inc., Basel, Switzerland.

Risch, S.J. 1995. Encapsulation: Overview of uses and techniques. In: *Encapsulation and Controlled Release of Food Ingredients*. p. 2. Eds.: Risch, S.J., Reineccius, G.A. ACS Symposium Series No. 590. American Chemical Society, Washington, DC.

Robertson, G.L. 1993. Packaging of horticultural products. In: *Food Packaging, Principles and Practice*. pp. 470–506. Ed.: Robertson, G.L. Marcel Dekker, Inc., New York.

Rojas-Graü, M.A., Avena-Bustillos, R., Friedman, M., Henika, P., Martín–Belloso, O., McHugh, T. 2006. Mechanical, barrier and antimicrobial properties of apple puree edible films containing plant essential oils. *Journal of Agricultural and Food Chemistry*, 54: 9262–9267.

Rojas-Graü, M.A., Olsen, C., Avena-Bustillos, R.J., Friedman, M., Henika, P.R., Martín-Belloso, O., Pan, Z., McHugh, T.H. 2007. Effects of plant essential oils and oil compounds on mechanical, barrier and antimicrobial properties of alginate–apple puree edible films. *Journal of Food Engineering*, 81: 634–641.

Saftner, R. 1999. The potential of fruit coating and film treatments for improving the storage and shelf-life qualities of 'Gala' and 'Golden Delicious' apples. *Journal of the American Society for Horticultural Science*, 124(6): 682–689.

Santerre, C., Leach, T., Cash, J. 1989. The influence of the sucrose polyester, Semperfresh™, on the storage of Michigan grown "McIntosh" and "Golden Delicious" apples. *Journal of Food Processing and Preservation*, 13: 293–305.

Schlimme, D.V., Rooney, M.L. 1994. Packaging of minimally processed fruits and vegetables. In: *Minimally Processed Refrigerated Fruits and Vegetables*. pp. 135–182. Ed.: Wiley, R.C. Chapman & Hall, Inc., London, U.K.

Schmitt, C., Sanchez, C., Desobry-Banon, S., Hardy, J. 1998. Structure and technofunctional properties of protein–polysaccharide complexes: A review. *Critical Reviews in Food Science and Nutrition*, 38(8): 692–753.

Scott, K., Yuen, C., Kim, G. 1995. Reduction of superficial scald of apples with vegetable oils. *Postharvest Biology and Technology*, 6: 219–223.

Shiying, X., Xiufang, C., Da-Wen, S. 2001. Preservation of kiwifruit coated with an edible film at ambient temperature. *Journal of Food Engineering*, 50: 211–216.

Smith, S.M., Stow, J.R. 1984. The potential of a sucrose ester coating material for improving the storage and self-life qualities of 'Cox's Orange Pippin' apples. *Annals of Applied Biology*, 104: 383–391.

Soukoulis, C., Yonekura, L., Gan, H.H., Behboudi-Jobbehdar, S., Parmenter, C., Fisk, I. 2014. Probiotic edible films as a new strategy for developing functional bakery products: The case of pan bread. *Food Hydrocolloids*, 39, 231–242.

Sumnu, G., Bayindirli, L. 1994. Effects of Semperfresh™ and Johnfresh™ fruit coatings on poststorage quality of "Ankara" pears. *Journal of Food Processing and Preservation*, 18: 189–199.

Sumnu, G., Bayindirli, L. 1995. Effects of coatings on fruit quality of 'Amasya' apples. *Lebensmittel-Wissenschaft and Technologie*, 28(5): 501–505.

Tapia, M.S., Rojas-Graó, M.A., Rodrνguez, F.J., Ramνrez, J., Carmona, A., Martin-Belloso, O. 2007. Alginate and gellan-based edible films for probiotic coatings on fresh-cut fruits. *Journal of Food Science*, 72: 190–196.

Toma, R., Curtis, D., Sobotor, C. 1988. Sucrose polyester: Its metabolic role and possible future applications. *Food Technology*, 1: 93–95.

Vargas, M., Pastor, C., Chiralt, A. 2008. Recent advances in edible coatings for fresh and minimally processed fruits, *Critical Reviews in Food Science and Nutrition*, 48: 496–511.

Weiss, J., Takhistov, P., McClements, D.J. 2006. Functional materials in food nanotechnology. *Journal of Food Science*, 71: 107–116.

Wills, R., McGlasson, B., Graham, D., Joyce, D. 1998. Structure and composition. In: *Postharvest: An Introduction to the Physiology and Handling of Fruit, Vegetables and Ornamentals* (4th Ed). pp. 15–32. Eds.: Wills, R., McGlasson, B., Graham, D., Joyce, D. UNSW Press, Sydney, New South Wales, Australia.

Wong, D., Gastineau, F., Grekoski, K., Tillin, S., Pavlath, A. 1992. Chitosan-lipid films: Microstructure and surface energy. *Journal of Agricultural and Food Chemistry*, 40: 540–544.

Wong, D., Tillin, S., Hudson, J., Pavlath, A. 1994a. Gas exchange in cut apples with bilayer coatings. *Journal of Agricultural and Food Chemistry*, 42: 2278–2285.

Wong, D.W.S., Camirand, W.M., Pavlath, A.E. 1994b. Development of edible coatings for minimally processed fruits and vegetables. In: *Edible Coatings and Films to Improve Food Quality*. pp. 65–88. Eds.: Krochta, J.M., Baldwin, E.A., Nisperos-Carriedo, M.O. Technomic Publishing Co., Inc., Basel, Switzerland.

Worrell, D., Sean Carrington, C., Huber, D. 2002. The use of low temperature and coatings to maintain storage quality of breadfruit, Artocarpus altilis (Parks.). *Postharvest Biology and Technology*, 25: 33–40.

Wu, Y., Weller, C.L., Hamouz, F., Cuppett, S.L., Schnepf, M. 2002. Development and application of multicomponent edible coatings and films: A review. *Advances in Food and Nutrition Research*, 44: 347–394.

Yaman, O., Bayoindirli, L. 2002. Effects of an edible coating and cold storage on shelf–life and quality of cherries. *Lebensmittel-Wissenschaft and Technologie*, 35(2): 146–150.

Yong Soo, H., Yo An, K., Jae Chang, L. 1998. Effect of postharvest application of chitosan and wax, and ethylene scrubbing on the quality changes in stored 'Tsugaru' apples. *Journal of the Korean Society for Horticultural Science*, 10: 579–582.

Yueming, J., Yuebiao, L. 2001. Effects of chitosan coating on postharvest life and quality of longan fruit. *Food Chemistry*, 73: 139–143.

Zhang, D., Quantick, P.C. 1997. Effects of chitosan coating on enzymatic browning and decay during postharvest storage of litchi (*Litchi chinensis* Sonn.) fruit. *Postharvest Biology and Technology*, 12: 195–202.

Zivanovic, S., Chi, S., Draughon, A.F. 2005. Antimicrobial activity of chitosan films enriched with essential oils. *Journal of Food Science*, 70: 45–52.

17 Food Packaging and Aseptic Packaging

Spyridon E. Papadakis

CONTENTS

17.1 INTRODUCTION

17.1.1 Definitions—Significance of Food Packaging

Food packaging is an integral and essential part of food processing and preservation. Developments in food packaging follow the evolutions in science and technology and are determined by developments in society such as the rise in the standards of living and the changes in the nutritional preferences of humans.

The term "packaging" is used to describe both the procedure and the means of containing goods and consequently several definitions are found in the literature. The British Standards Institution (Anonymous 1996) defines packaging (as a procedure) as: "the art of and the operations involved in the preparation of articles and commodities for carriage, storage, and delivering to the consumer." According to the European Parliament and Council Directive 94/62/EC of 20 December 1994: "the term packaging (as a means) signifies all products made of any materials of any nature to be used for the containment, protection, handling, delivery, and presentation of goods, from raw materials to processed goods, from the producer to the user or the consumer."

In developing countries, 30%–50% of the food produced is wasted and never reaches the consumer for several reasons. One of them is the lack of adequate preservation and packaging technology. Thanks also to packaging, less than 2% of all food is wasted in developed countries (Anonymous 1996). For the food industry of the United States, the cost of packaging represents on the average 15% of the exfactory cost of the foods (Brody et al. 2008), and thus there is a big incentive and challenge to reduce the cost of packaging without compromising the protection it offers to the food.

In the United States, approximately 50% (by weight) of the packaging materials consumed each year are used by the food and beverage industry (Marsh and Bugusu 2007a), while \$55–\$65 billion are spent annually on the packaging of food and beverages (Brody 2008). The distribution of these \$55–\$65 billion to the various sectors of the food and beverage packaging supply industries (Brody 2008) is 33% paper and paperboard, 22% plastic bottles and jars, 19% flexible plastic packaging, 17% metal cans, 5% glass, and 4% wood.

The basic food packaging materials are glass, metals, plastics, paper and paperboard, and laminates/coextrusions. These materials are examined in detail in separate sections of this chapter. It should be pointed out that each material presents advantages and disadvantages. The key to successful packaging is to select the material and design of the container that best satisfy many considerations. These often competing needs include product protection, marketing considerations, environmental issues, and cost. Taking into account so many factors make food packaging partly science and partly art. Marsh and Bugusu (2007b) present a table summarizing the properties of the basic categories of packaging materials related to food protection, marketing issues, environmental issues, and relative cost.

Food packaging encompasses a vast range of subjects. But, due to space limitations, the topics discussed in this chapter are the various packaging materials (glass, metals, plastics, paper, laminates/coextrusions), corrosion of metal containers, permeability of thermoplastic polymers, and aseptic packaging. Very important topics like food and packaging interactions, food packaging laws and regulations, food packaging and the environment, active and intelligent packaging, packaging requirements of various food categories are not presented here. Edible films and coatings and modified atmosphere packaging are discussed in other chapters of this book. For the food packaging topics not dealt with in this chapter and for further details on the topics presented, the interested reader is referred to textbooks on food packaging (Coles et al. 2003, Robertson 2006, Lee et al. 2008), books on food packaging (Palling 1980, Kadoya 1990, Jenkins and Harrington 1991), and books on packaging in general (Hanlon 1984, Soroka 1996, Brody and Marsh 1997, Yam 2009).

17.1.2 Functions of Food Packaging

The basic functions of food packaging (Brennan et al. 1990, Robertson 2006, Krochta 2007, Lee et al. 2008) are

- Containment of the product
- Protection of the product
- Convenience to the consumer
- Communication with the consumer

These are often interdependent. The fulfillment of the four functions must be combined with the minimum possible cost. Since the cost of packaging for food products is usually a significant proportion of the total cost, the use of the least expensive, but appropriate for the particular food, packaging is preferred. Thus, an ideal package delivers to the consumer a safe, quality food in a convenient way and at minimum cost. In addition to the four basic functions, three other functions have been introduced (Krochta 2007) placing more requirements on food packaging:

- Production efficiency (or machinability)
- Minimal environmental impact
- Package safety

17.1.2.1 Containment

This function is so obvious and as a result it is very often ignored. The first forms of packaging were probably devised to facilitate the transportation of goods. A liquid or particulate food must first be placed in a container in order to be transported. A good package minimizes the product losses and consequent environmental pollution during transportation.

17.1.2.2 Protection

This is the most significant function of food packaging. The time interval between the production of the food and its consumption may vary between a few days and a few years. During this period the food must be protected from all environmental factors so that its quality remains at the same level it had when it was first placed into the package. If the package fails to protect the food adequately, then much of the expense and energy put into the production and processing of the food will be wasted.

More specifically, the packaging must protect the contained food from mechanical damage, moisture and gases of the surrounding atmosphere, light, microbial contamination, dust, filth and foreign materials, insects, and rodents. All these requirements must be fulfilled by the packaging without altering the sensory characteristics of the food or transferring to it substances dangerous to the human health.

17.1.2.2.1 Mechanical Damage

All foods are susceptible to mechanical damage. Bruising, deformation, cracking, and breakage may happen to them during transportation and handling, from sudden impacts or shocks, vibration, and compression. These damages may lead to chemical and biological deterioration and are considered as indications of inferior quality by the consumers. Suitable packaging may decrease the probability and extend of mechanical injury. Rigid and semirigid packaging materials (metal, glass, wood, and fiberboard) protect the food from compression. The inclusion in the package of cushioning materials like soft paper and foamed plastics can reduce the effect of shock and vibration on the food.

17.1.2.2.2 Moisture and Gases of the Surrounding Atmosphere

Foods of high water activity (e.g., meat, cheese) when exposed to the normal atmosphere will lose moisture through desorption of water vapor with consequent loss of weight and deterioration in appearance, texture, and flavor. On the other hand, foods of low water activity (e.g., cookies, snacks) when exposed to high-humidity atmospheres will gain moisture, and this is also detrimental to their quality. Then dry powders may cake, cookies may lose their crispness, and dehydrated foods may spoil if their resulting water activity permits the growth of microbes. It is obvious that in both cases the packaging should block the transfer of moisture to or from the food product. There are cases where the transfer of water vapor from the product to the environment is desirable, as, for example, in the packaging of some fresh fruits and vegetables with high respiration rates. Otherwise the water vapors produced may condense inside the package affecting the appearance and the quality of the product.

For many foods, exposure to oxygen may cause several quality problems due to oxidation of lipids and of other desirable components of the foods. The shelf life of these foods may be extended by vacuum packaging or modified atmosphere packaging (different ratios of oxygen, nitrogen, and carbon dioxide than the atmospheric air). Therefore, the packaging material in these cases should provide a barrier to gases (oxygen, nitrogen, and carbon dioxide) in order for the composition of gases inside the package to remain unchanged. On the other hand, the packaging material of fresh fruits and vegetables must allow permeation of oxygen into the package and carbon dioxide out at the appropriate rates. Otherwise, because of the respiration, the oxygen within the pack will be consumed completely and replaced by the produced carbon dioxide, and off-flavors may develop and discoloration of the fruit may occur.

To prevent the loss of the characteristic aroma of some foods (e.g., coffee), it is necessary for the packaging material to have a low permeability to volatile compounds responsible for the odor of the food. Similarly, the packaging must block the transfer of foreign odors from the environment to the product.

17.1.2.2.3 Interaction of Food with Light

Light, particularly of the high ultraviolet (2900–4000 Å) and visible blue (4000–4500 Å) parts of the spectrum, can catalyze several deteriorative chemical reactions in foods. These include vitamin, color, and flavor degradation and lipid oxidation. To prevent these changes, the packaging used is either opaque or colored to block the transmission of short wavelength light.

17.1.2.2.4 Microbial Contamination

This is one of the most important functions of packaging. For sterilized foods, any postprocess contamination must be absolutely prevented. Metal cans, glass containers, and laminates (incorporating a barrier layer) are the most suitable packaging for these foods. For pasteurized products, dried, frozen or cured, etc., this role of packaging is not that crucial; however, packaging should still provide adequate protection.

17.1.2.2.5 Interactions of Packaging with the Contained Food

All food packaging materials have been found to interact in varying degrees with food. This interaction might be: migration of components of the packaging material to the food, adsorption (scalping) of food components by the packaging material, and permeation through the packaging material of foreign substances from the external environment and transfer of them to the food. Since migration is directly related to food safety, rigorous laws and regulations on the subject have been introduced in several countries. For example, according to the Regulation (EC) No. 1935/2004 of the European Parliament and of the Council of 27 October 2004, among the general requirements for the materials and articles intended to come into contact with food is the following:

> Materials and articles, including active and intelligent materials and articles, shall be manufactured in compliance with good manufacturing practice so that, under normal or foreseeable conditions of use, they do not transfer their constituents to food in quantities which could
>
> (a) Endanger human health, or
> (b) Bring about an unacceptable change in the composition of the food, or
> (c) Bring about a deterioration in the organoleptic characteristics thereof

17.1.2.3 Convenience to the Consumer

In the last 60–70 years, lifestyle in the developed countries has changed dramatically. Consumers spend less time buying their foods at the supermarkets and preparing their meals. Very often the food is consumed directly from the packaging with minimum meal preparation. Therefore, the consumers demand packaging that is convenient and user friendly. The package should be easy to open, hold, use, reseal, store in the refrigerator, cupboard, etc., and recycle.

17.1.2.4 Communication with the Consumer

Packaging is the "silent salesman" of the contained product. There is a very true old saying: "*packaging should protect what it sells and sell what it protects*" (Robertson 2006). Therefore, packaging should attract the attention of the consumer and convince him to buy the product. In addition, through the labels written on the package, information is given to the buyer concerning the name of the product and its manufacturer, the address of the responsible company, the quantity contained, the quality and composition, nutritional information, and often guidelines on how to use it. Information about the quantity, composition, and nutritional content is required by law in several countries. The contemporary methods of marketing of foods would fail completely if the packaging did not pass messages to the consumer. Supermarkets can function as self-service stores exactly because consumers can recognize the brand of the product and its expected quality from the particular characteristics of the packaging (material, shape, size, color, symbols, illustrations).

Another important communication function of the packaging is the existence of the bar code on it, which permits the recognition of the product by the scanner at the checkout counters of supermarkets, and this facilitates rapid and accurate pricing and checkout. Bar codes on secondary and tertiary packages are very useful for tracking and inventory checking at warehouses and distribution centers. During the last decade, the use of Radio Frequency Identification (RFID) technology revolutionized the logistics systems of the store supply chain. RFID tags are attached to the secondary and tertiary packages and in their microchip information about the product, the package location, etc., are stored. This information is transferred by radio waves to a reader and is finally retrieved as digital information.

17.1.3 Categories of Packaging

Food packages can be classified into various categories based on the following criteria: whether the packaging is in direct contact with the food, whether the packages are single units for retail sale or each package contains a number of units, whether the packages deform when compressed by hand, whether they are preformed or they are formed on-line, whether they are sealed hermetically or not, etc.

17.1.3.1 "Direct" and "External" Packaging

"Direct" packaging is the packaging which is in direct contact with the food, and "external" packaging is another packaging containing the "direct" one. For example, breakfast cereals are placed inside a plastic bag (the "direct" package), and the plastic bag is placed inside a paperboard box (the "external" package). The first protects the food from moisture gain, etc., while the second protects the first and the product from mechanical damage, facilitates handling, and has printed on it all the necessary information to communicate with the consumer.

17.1.3.2 Levels of Packaging

A similar distinction concerns the levels of packaging. According to the European Parliament and Council Directive 94/62/EC of 20 December 1994 "packaging" consists of

(a) Sales packaging or *primary packaging*, which constitutes a sales unit to the final user or consumer at the point of purchase
(b) Group packaging or *secondary packaging,* which at the point of purchase constitutes a grouping of a certain number of sales units. It can be sold as such to the final user or consumer, or it serves as a means to replenish the shelves at the point of sale. It can be removed from the product without affecting its characteristics
(c) Transport packaging or *tertiary packaging,* which facilitates handling and transport of a number of sales units or group packaging in order to prevent physical handling and transport damage.

17.1.3.3 Rigid, Semirigid, and Flexible Packaging

Another distinction between packages is based on whether they deform when compressed by hand. Thus, packaging may be classified into three forms: rigid, semirigid, and flexible (Lee et al. 2008). *Rigid* packages (e.g., metal cans, glass containers) have a fixed shape and do not change shape or break unless excessive external force is applied. *Semirigid* packages (e.g., plastic bottles, paperboard cartons, aluminum cans), like the rigid ones, have a fixed shape that does not significantly change when they are filled with product and sealed. However, they can be deformed with finger pressure (about 60 kPa) and return to the original shape when the application of pressure is stopped. *Flexible* packages (e.g., plastic pouches, paper bags) have a shape or contour that is determined by the product they are filled with. Depending on the external pressure, they inflate or deflate until the internal and external pressures equalize.

17.2 GLASS PACKAGING

Glass containers are the preferred package for a variety of foods and drinks. Narrow-mouth containers, with a closure diameter less than 35 mm (Cavanagh 1997), are called bottles and are used mainly for the packaging of liquids, whereas wide-mouth containers, with a closure diameter more than 35 mm, almost without neck, are called jars and are used for the packaging of both liquids and solids.

17.2.1 Glass

17.2.1.1 Composition and Structure

Glass, according to the definition of American Society for Testing and Materials (ASTM) (1999), is "*an amorphous, inorganic product of fusion that has been cooled to a rigid condition without crystallizing.*" Another (Shreve and Brink 1977) definition along the same lines is the following: Glass physically is a rigid, undercooled liquid with no definite melting point and a high viscosity (more than 10^{12} Pa s) at ambient temperature, which prevents crystallization. Chemically, common glass for packaging is an amorphous mixture of oxides of Si, Na, Ca, and Al.

The main ingredient of glass is silica (SiO_2), which forms tetrahedra with silicon at the center and oxygen at the four corners (Lee et al. 2008). Each silicon atom is connected to four oxygen atoms, and each oxygen atom is connected to two silicon atoms. In glass, the silica tetrahedra form a network with several empty spaces between them. It is possible to make pure silica glass, but silica has a very high melting point (1723°C) and molten silica has high viscosity. To reduce them both, soda (Na_2O) is added to the silica, and this is achieved by the soda filling the spaces between the silica tetrahedra. However, Na_2O tends to dissolve in water, and to prevent this, lime (CaO) and MgO are added and act as stabilizers. Alumina (Al_2O_3) is added to increase hardness and durability. Thus, the common glass for packaging applications is soda-lime glass and typically consists of 73% SiO_2, 13% Na_2O, 12% CaO, 1% Al_2O_3, and 1% other oxides. To produce colored glass, minor ingredients are added: chromium oxide for green and iron oxide, sulfides, and carbon for amber.

17.2.1.2 Advantages and Disadvantages of Glass

The attributes of glass (Osborne 1980) that make it an attractive material for the construction of containers for packaging of foods are

- Transparency, which allows viewing of the contents
- Good strength under compression
- Ability to mold containers of various shapes
- Chemical inertness
- Impermeability to gases and vapors
- Satisfactory heat resistance to permit thermal processing of foods
- High quality image associated to products packed in glass containers by the consumers
- Recyclability

Glass containers can be designed either as one-trip or returnable (refillable). The adoption of the second option depends on several economic parameters and is used mainly for beer and milk. Of course glass containers are also recycled as glass is returned to the glass manufacturer for remelting. Glass can be produced in various colors, by adding minor amounts of colorants. There are three main color categories: flint, amber, and green glass. Amber is the only one which blocks the transmission of UV light of wavelength of 300–400 nm in the critical range for the quality of light-sensitive foods and beverages.

The main disadvantages of glass are fragility, brittleness, and heavy weight. The heavy weight and safety concerns related to broken glass have reduced the use of glass in food packaging and limited its use to small and medium-size containers.

17.2.2 Glass Containers

17.2.2.1 Manufacture of Glass Containers

The main raw materials for making glass are silica sand (SiO_2), soda ash (Na_2CO_3), limestone ($CaCO_3$), and alumina (Al_2O_3). These along with minor ingredients and crushed recycled glass (called cullet) are melted together in a gas-fired furnace maintained at approximately 1500°C. The gases formed (mainly CO_2 and SO_2 and water vapor) are removed, and the molten glass is cooled to about 1150°C. Glass containers are formed from lumps of molten glass (called gobs) exiting from the furnace, and this is accomplished in two steps. There are two main types of forming machines: blow-and-blow (B&B, for making narrow-neck bottles) and press-and-blow (P&B, for making wide-mouth jars). For both process, in the first step, gobs at about 1100°C, with the appropriate shape and weight, are fed into blank molds, where they are transformed into hollow, partially formed containers with thick walls called blanks, preforms, or parisons. In the B&B process, the parison is made by blowing pressurized air into the gob, while in the P&B process it is made by pressing a metal plunger into the gob. In the second step, for both process, the semi-molten parisons are transferred into blow molds, where air blown inside the parisons pushes the glass to conform to the shape of the blow mold. Developments during the last few decades permitted the production of bottles with a third process called narrow-neck press-and-blow process, which produces containers with less glass and consequently lighter in weight.

After formation, the temperature of the glass dropped from 1100°C to about 450°C within 10–12 s. This rapid and uneven cooling results in the development of internal stresses in the formed container, making it very fragile. To relieve the stresses and stabilize the containers, they are passed through an annealing oven or lehr, where they are reheated to 540°C, held at this temperature for some time, and then cooled slowly to prevent fracture from thermal shock.

Coatings are applied to the outer surface of the glass container to strengthen it and minimize scratching. These can be hot-end and cold-end coatings. The former are applied prior to the annealing process and create a very thin film of tin or titanium oxide over the surface. The latter are applied just before the end of the annealing, by spraying the surface with stearates, waxes, silicones, or polyethylene.

Details on the manufacture of glass containers can be found in the following references: Lee et al. (2008), Cavanagh (1997), Soroka (1996), Robertson (1993), and Osborne (1980). Also, useful information about glass containers and the glass packaging industry can be obtained from the educational CD-ROM (Anonymous 2005) of the Glass Packaging Institute.

17.2.2.2 Mechanical Properties

Fracture of glass starts with small imperfections or flaws on its surface (Robertson 2006). These cracks can be so small that they may be invisible to the naked eye. When a tensile stress is applied to a glass object, stress concentration occurs at the bottom of the crack. The value of the local stress is many times greater than the average value of the stress applied to the object and the narrower the crack, the greater the local stress. Since glass is not ductile, it cannot yield, and the stress concentration remains active causing the crack to propagate, which eventually leads to fracture. Therefore, extra care is taken to prevent the formation of surface flaws during handling. In addition to surface condition, the shape of the container is very crucial for vertical load strength and resistance to impact. In general, in the regions of the container where sharp transitions of the shape occur (e.g., at the corners of a rectangular cross section), the risk of fracture is greater because of high stress concentration.

The types of fractures (Hanlon 1984) important to food packaging are fractures from: thermal shock, high internal pressure, and sudden impact. The resistance of a glass container to thermal failure depends on the glass composition itself, the shape of the container, and the thickness of its walls. When one surface of the container is at a temperature and the other surface at a different temperature, then one surface tends to expand and the other to contract. However, these changes cannot happen freely in the container and as a result stresses are set upon the surfaces. When a hot glass container is suddenly cooled (e.g., by immersion in a cold water bath), the outer surface tends to contract, and thus tensile stresses are set up on it. At the same time compensating compressional stresses of smaller magnitude are set up on the inner surface. In the opposite case (i.e., sudden heating) the bigger stresses develop again at the outer surface, but they are compressive this time. Compensating tensional stresses of smaller magnitude are set up on the inner surface. Since the mechanical strength of glass is much greater in compression than in tension, fracture of the container is more likely to occur in the case of sudden cooling when the outside surface, where the major changes are taking place anyway, is in tension. Thick-walled glass bottles and jars are more vulnerable to thermal shock than thin-walled ones. To avoid thermal shock, it is generally recommended to keep temperature differences at the two surfaces of the container smaller than 27°C by performing heating and cooling very gradually.

In bottles and jars containing carbonated beverages and vacuum-packed foods, the internal pressure is different from the external atmospheric pressure and consequently internal pressure stresses develop. These are circumferential and longitudinal. The former, in the cylindrical part of a bottle, are proportional to the diameter of the bottle and inversely proportional to the wall's thickness. Thus, to increase the resistance of a glass container to internal pressure stresses, the practical approach followed is to increase the thickness of its walls and make the container much heavier. However, such a container is more vulnerable to fracture from thermal shock.

To increase the resistance of containers to impact, coatings are applied to their outside surface. They provide lubrication and minimize abrasion from bottle against bottle and from bottle against machinery. Resistance is also improved with more uniform distribution of glass in the container.

17.2.2.3 Quality Control of Class Containers

In the glassworks, following the annealing process, all containers go through a number of inspection operations. These automatic tests (Soroka 1996) are

- Squeeze test
- Bore gauge
- Defects' detection
- Wall's thickness measurement
- Hydraulic pressure test
- Visual inspection

Containers outside specifications are automatically rejected; they are collected as cullet and recycled back to the glass furnace. Advanced photoelectric methods are now being used for the detection of critical faults, while new methods of quality inspection are constantly being proposed and investigated.

There are many defects that may be encountered in glass containers, some of them are presented by Hanlon (1984) and in posters provided by glassware manufacturers. They are classified (Hanlon 1984) as "critical" if they are hazardous to the user and make the container completely unusable, as "major" if they reduce the usability of the container or its contents, and as "minor" if they diminish the acceptability of the container by the consumers.

17.2.3 Closures for Glass Containers

Closures are made of metal or plastic. Metals used are tinplate (steel coated by tin), Electrolytic Chrome-Coated Steel (ECCS), and aluminum. They are usually coated with enamels and frequently printed. Plastic caps can be made from urea-formaldehyde or phenol-formaldehyde resins or from PP, PE, PS, and PVC thermoplastic resins. Common caps include screw caps, crown caps, and lug caps. On the inside surface of the closure, there is a gasket or a liner and that is the part which is in intimate contact with the sealing surface of the glass container finish and thus provides an effective seal. Gaskets are made from rubber or plastisol (usually suspension of finely divided PVC in a plasticizer). Liners consist of a resilient backing and a facing material and can be made from a variety of packaging materials (cork, paper, plastic films, and metal foil).

The closures of glass containers can be classified (Robertson 2006) into four categories: caps to retain high internal pressure, caps to contain and protect the product, caps to maintain vacuum inside the container, and common caps to just secure the product inside the container. Caps of the first category are used in carbonated beverages and beer and must maintain pressures from 2 to 8 atm. They are the traditional crown caps, and their improved versions are the twist-off crown caps and the roll-on tamper evident screw caps. The last are made of aluminum or plastic and are applied to both glass and plastic bottles, being popular for soft drinks in large containers where reclosure is essential. The second category includes cork stoppers and their various replacements that have appeared in the market for wine bottles. The third category caps are used in jars that typically contain heat sterilized foods. The pry-off cap, the lug-type twist cap, and the PT (Press-on Twist-off) cap belong to the third category, and they are made of tinplate or ECCS (Gavin and Weddig 1995).

Vacuum closures often have a safety button, which is a circular area around the center of the top surface of the metallic cap thinner and more flexible than the rest of the cap. When there is proper vacuum inside the jar, the button is in the down position providing a visual indication of vacuum in the headspace of the jar. When the cap is twisted open, air is admitted into the headspace and the button is pushed to the up position making a characteristic popping sound.

17.3 METAL PACKAGING

Four metals are commonly used as food packaging materials: steel, aluminum, tin, and chromium (Robertson 2006). Cans are made from tinplate or ECCS. Toward the end of the 1950s the first aluminum cans were introduced in the market.

Details on the metals and lacquers used in metal food containers and on the manufacture of three-piece and two-piece cans can be found in several references (Malin 1980, Anonymous 1986, Matsubayashi 1990, Robertson 1993, Soroka 1996, Turner 2001, Page et al. 2003, Robertson 2006, Lee et al. 2008).

Metal containers possess several advantages as food packaging means. They offer very effective protection to the food. They are sealed hermetically and thus protect the food from microbial contamination, insects, and rodents. They are total barriers to gases, moisture, vapors, and light. They have great mechanical strength and good heat resistance. Aluminum cans are completely recyclable, while for tinplate cans recycling is somewhat problematic. Their main disadvantage is their lack of inertness. They interact with the food and can be corroded. For this reason they are often coated with lacquers. Other disadvantages are their heavy weight (for steel cans) and the difficulty to be reclosed.

17.3.1 Tinplate

Tinplate (Lee et al. 2008) is a sheet of mild steel (called black plate), of thickness 0.13–0.38 mm, coated on both sides with a layer of tin (Sn) of thickness 0.2–2.5 μm. Steel is a general term referring to iron alloys with a carbon content less than 2% (w/w). The steel used for the manufacture of tinplate is low-carbon mild steel with a carbon content of around 0.13% (w/w). Next to iron, the main component of this steel is manganese (0.60%, w/w). There are several types of steel used for tinplate manufacture: type L for very corrosive foods, type MR, the most common, for moderately corrosive, type D for higher ductility, and type N for greater strength and stiffness.

Steel is produced in the steel mill in the form of slabs, and then the slabs are hot rolled and cold rolled in a series of steps into sheets with the desired thickness for the manufacture of containers. The coating of steel sheets with tin is made by an electrolytic process in an acidic bath of tin sulfate. There is the possibility to coat the two surfaces of the steel sheet with layers of tin of different thickness, the thicker layer usually being on the side of the food. After this stage, tinplate is heated at 260°C–270°C and then rapidly quenched, a process called flow melting. That leads to the formation of an iron-tin alloy ($FeSn_2$) on the surface of the steel. Next, to increase its corrosion resistance, the tinplate is passivated. Electrolysis is carried out in a sodium dichromate bath with the tinplate sheet as the cathode, and as a result a thin film of chromium, chromium oxides, and tin oxides are formed on the tinplate. Finally, an oily lubricant is applied on the tinplate sheet.

17.3.2 ECCS

ECCS, also called Tin-Free Steel (TFS), is produced from a steel sheet that is coated on both sides with metallic chromium and chromium oxide. The coating is applied electrolytically. The processes of flow melting and passivation are not carried out in this case. Like tinplate, the ECCS plate receives an oil coating at the end of manufacture. The main advantages of ECCS compared to tinplate are its lower price, good adhesion of lacquers, and resistance to sulfur staining. However, it is less resistant to corrosion than tinplate, and for this reason it is always enameled on both sides. Its main uses (Lee et al. 2008) in food packaging are for can ends, crown caps, vacuum caps for glass jars, and deep-drawn cans.

17.3.3 Aluminum

Aluminum has been used for food packaging quite recently. Aluminum cans for beer packaging were introduced to the market in 1959. Various types of aluminum alloys are used in food packaging. They are identified by four-digit numbers. In all alloys the aluminum content is very high. Types 1050 and 1100 are almost pure aluminum and are used for making foil and extruded containers. The 5000 series alloys containing 0.5%–5% Mg and 0.15%–0.5% Mn are very rigid and are used for making beverage can ends and ring-pull tabs. Aluminum is lighter and weaker than tinplate and ECCS, but it is more ductile and more easily formed into cans. Aluminum cans are always coated with lacquers to prevent interaction with the food.

17.3.4 Protective Lacquers

In several cases there is the probability of interactions of the food with the metal container and this may result in reduction of the shelf life of the food. To deal with this problem, the body and the ends of cans are coated on the inside and outside surface with lacquers or enamels. These are solutions of synthetic resins in organic solvents or water sometimes. The main function of internal lacquers is to protect the can metals from corrosion from the contained food and the food from contamination by the corrosion products. They also improve the appearance of the inside of the package, and in some cases they provide nonstick properties to the surface and facilitate product removal from the can. External lacquers protect from external corrosion and abrasion and improve the appearance of the can.

In food packaging it is imperative to use enameled cans in the following cases:

- When packaging very corrosive foods such as pickles, tomato paste, apple juice, etc.
- When the dissolved tin and iron transferred from the can to the food are detrimental to the food's sensory characteristics, such as in the case of beer and soft drinks.
- When the tin reacts with plant pigments of the food. If products like raspberries, cherries, strawberries that contain anthocyanins (red pigments) are canned in plain tinplate cans, they become either decolorized or get a bluish color.
- When packaging foods like meat and meat products, fish, and dairy products that during thermal processing liberate H_2S and S^{2-}, which react with iron and tin and form black iron sulfide stains and blue-black or brown tin sulfide stains (sulfide staining) on the internal surface of the can.

To almost all containers the wet coating is applied to the metal sheet before the formation of cans. Only the D&I cans are coated after fabrication. In the first case the application is done by roller coating and in the second by spraying. To remove the solvent and cure the lacquer, the sheets or the cans are heated or "baked" in a hot air oven at 210°C for up to 15 min. The thickness of the dry lacquer is 4–5 μm for single coating and 8–10 μm for the very corrosive foods that require double coating.

Several types of internal enamels are used depending on the canned product. These include acrylic, epoxyamine, epoxyphenolic, oleoresinous, phenolic, polybutadiene, vinyl, and vinyl organosol resins. Robertson (2006) presents a table with the types of lacquers available, their characteristics, and their uses. The sulfide stain-resistant lacquers may contain ZnO, which reacts with the H_2S and white ZnS is produced, or contain metallic aluminum powder or white pigments to mask the tin sulfide stains.

17.3.5 Manufacture of Three-Piece Cans

The classical cylindrical can is composed of three pieces: the cylindrical body, the bottom, and the lid. The last two are exactly the same, and they are called ends. In Figure 17.1 a photo of three-piece cans is presented.

17.3.5.1 Ends

The ends are specially designed to have the optimum deformation behavior during heating and cooling of the cans (Robertson 2006). During these stages, pressure differences develop between the inside and outside of the can, and the deformations of the ends must be reversible. The deformations depend on the thickness of the end, the pattern of the expansion rings, and the countersink depth.

The formation of the ends is carried out in the following stages. Initially a press stamps out a circular blank from a tinplate or ECCS sheet, which has been previously lacquered. Next the rim and the curl are formed. A lining or sealing compound is applied around the inside of the curl. The sealing compound is a solution of natural or synthetic rubber and during the double seaming operation fills the space between the body and end hooks and aids in the formation of a hermetic seal. Finally, beaded circles, called expansion rings, are made on the surface of the end with presses.

The increased demand by consumers for convenience features led to the development of cans with easy-open ends. There are two types of easy-open ends (Lee et al. 2008): pour-aperture types for beverages and full-aperture types for easy removal of solid foods. Nowadays the easy-open lids are made from lacquered tinplate, ECCS, or aluminum alloy. With presses the rivet is formed, and then the surface is scored to make the part of the lid that can be removed by pressing or pulling with the fingers. Finally, the tab is attached to the rivet. Easy-open lids are attached to the can body with double seams, exactly the same way as the common lids.

17.3.5.2 Body

At a first stage the tinplate or ECCS sheets are enameled and decorated with printing design if necessary. Then the sheets are slit into individual body blanks. Each blank is rolled into a cylinder with the two edges slightly overlapping, and the side seam is formed. There are several methods of creating the side seam. The traditional method was by soldering with lead and tin alloy. However, because of concerns on lead toxicity, this method has been abandoned completely and has been replaced by welding or cementing.

In the 1960s, two side seam welding processes were developed: the copper wire system by the Soudronic company in Switzerland and the Conoweld roll-welding system by the Continental Can Company in the United States. Both process were improved considerably and gained very wide acceptance in the 1980s.

FIGURE 17.1 Photo of three-piece cans and lid.

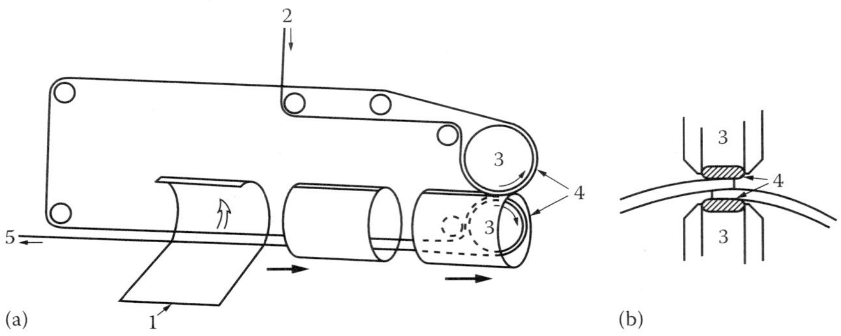

FIGURE 17.2 Schematic representation of the manufacture of a welded cylindrical can body. (a) Overview and (b) detail of welding process. 1, Blank rolled into a cylinder in the roll former; 2, copper welding-wire loom; 3, welding rolls; 4, welding wire in contact with cylinder overlap; 5, used copper wire to scrap. (From Turner, T.A., *Canmaking for Can Fillers*, Sheffield Academic Press, Sheffield, U.K., 2001. With permission.)

In the Soudronic process, shown schematically in Figure 17.2 (Turner 2001), the two overlapping edges of the body cylinder are placed between the two welding electrodes. A high intensity alternating electrical current passes through and melts and fuses together the two metal surfaces while at the same time the electrodes apply pressure to the seam. Between the two welding electrodes there is a copper wire that acts as an intermediate electrode. Its use is necessary because it removes small amounts of tin liberated from the tinplate, which otherwise could contaminate the main electrodes and lead to faulty seam. In the side seam the width of overlap between the edges of the body cylinder varies from 2–4 mm (Butterfly seam) to 0.8 mm (Wima seam) to 0.4 mm (superwima seam). During the welding process N_2 is fed in the seam area because the atmospheric oxygen may oxidize the molten metals. After welding, the side seam is coated with enamel for protection regardless of whether the rest of the body is enameled or not.

The welding of ECCS has technical difficulties and to deal with them the Conoweld technique was developed, initially for welding ECCS cans for beverages. At first, the layers of chromium and chromium oxide are removed completely and thoroughly from the steel in the area where the side seam is to be formed, and then welding is carried out. A newer technique is laser welding. Its advantages are that it can handle ECCS and the width of overlap is only 0.25–0.30 mm. However, it is a rather slow process.

A cementing technique, called Miraseam process, has been developed by the American Can Company for cans with ECCS bodies. A thermoplastic polyamide adhesive is utilized, and it is applied to one edge of the body blank before this is rolled into a cylinder. The body blank has been preheated, and the final seam obtained is very strong, capable of withstanding the high pressures generated in the cans of carbonated beverages and beer. The technique can also be applied to aluminum cans.

17.3.5.3 Double Seam

After "baking" the enamel of the side seam, the bodies are directed to a flanger where necking and flanging is usually performed for the beverage cans and beading and flanging for the food cans. Necking results in a reduction of the diameter of the top part of the cylindrical body and permits the use of smaller diameter lids. Flanging is the bending outward of the edges of the two bases of the body to create the flange of the body. To increase the resistance of the body to paneling, that is, distortion due to high internal vacuum, beads are formed in parts of the body.

The ends (bottom and lid) are joined to the body with a mechanical construction called double seam (Figure 17.3). Double seaming is carried out in two steps. In the first operation, the curl of the end is pushed radially inward and is tucked up underneath the flange of the body. In this way the end hook and the body hook are formed, the first tucked under the second. In the second operation,

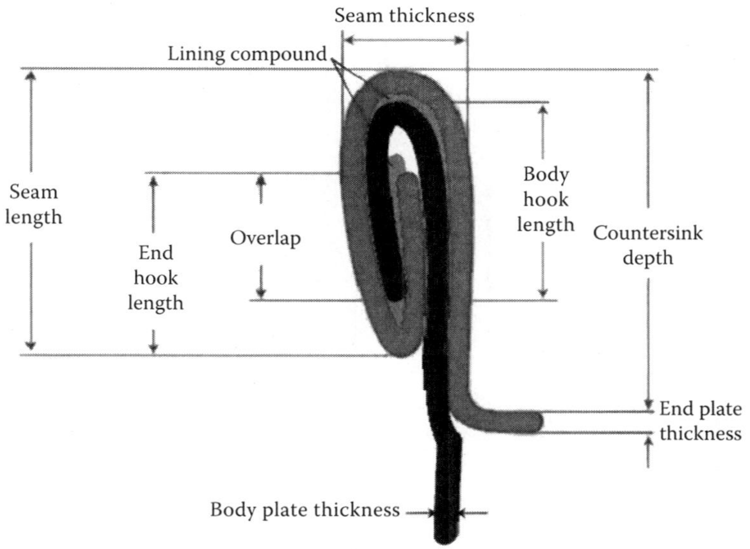

FIGURE 17.3 Cross section of a double seam and its main dimensions.

the seam is compressed and flattened ("ironed"). The double seam plays a crucial role in securing the hermetic seal of the can and thus the safety of the canned product. For this reason its quality is checked by the canmaker and the food packer according to relevant guidelines (Gavin and Weddig 1995, Canadian Food Inspection Agency 1997). There are specifications and limits for the various lengths and thicknesses, the degree of overlap of the two hooks, and the seam tightness. The double seaming machines may have up to 12 seaming heads and operate at speeds of 100–1000 cans min^{-1} (Lee et al. 2008).

17.3.6 Manufacture of Two-Piece Cans

In two-piece cans, the first piece consists of the body and the bottom end, which originate from the same piece of metal. The second independent piece is the lid, which is joined to the first with a double seam. The two-piece cans have several advantages over the three-piece ones. They offer better hermetic integrity because of the lack of the side seam, and the whole surface of the body is available for decoration. In addition, they have lower raw material cost since about 35% less metal is used for their production compared to the three-piece cans. They may be manufactured by two methods: draw and wall-iron (DWI) or draw and iron (D&I) and draw and redraw (DRD). By the first method, aluminum or tinplate cans are made with a thick bottom and thin wall and are used for packaging of beer and carbonated beverages, in which case the high internal pressure helps support the thin wall. By the DRD method tinplate, ECCS, and aluminum cans are made with a thicker wall for the packaging of heat sterilized foods. They can withstand the pressure reversals occurring during heat processing of the cans. The production rate of D&I cans is rather high reaching 2000 cans min^{-1} per production line (Lee et al. 2008).

DWI cans are made in the way shown schematically in Figure 17.4 (Malin 1980). First, from a sheet or coil of aluminum or tinplate a circular disk is stamped and from that a shallow cup is drawn. Next each cup, held on a punch, is pushed and rammed through a series of tungsten carbide rings of progressively smaller diameter. This process, while leaving the bottom unchanged, stretches the wall and thins it typically from 0.30 to 0.10 mm. At the same time the can height is increased to three times the original height. The bottom is domed to offer strength and stability to the can. The thickness of the bottom remains at 0.30 mm. Since cans are overdrawn to a bigger and irregular

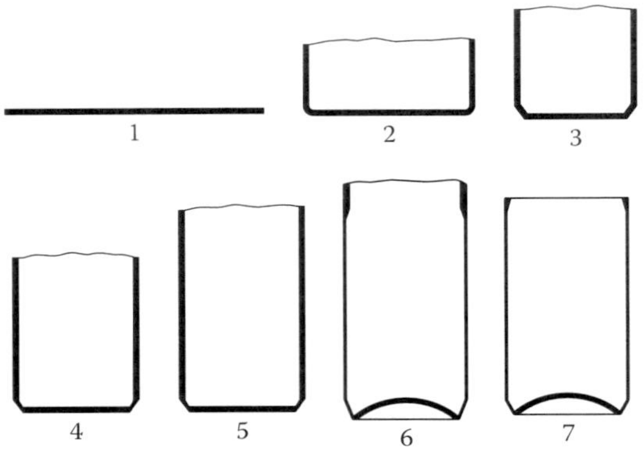

FIGURE 17.4 Sequential stages in the production of drawn and wall-ironed (DWI) cans: 1, body blank; 2 and 3, drawn and redrawn cups; 4, 5, and 6, first, second, and third stages of wall ironing and base formation; 7, finished can trimmed to required height. (From Malin, J.D., Metal containers and closures, in: S.J. Palling (ed.), *Developments in Food Packaging*-1, Applied Science Publishers Ltd., London, U.K., pp. 1–26, 1980. With permission.)

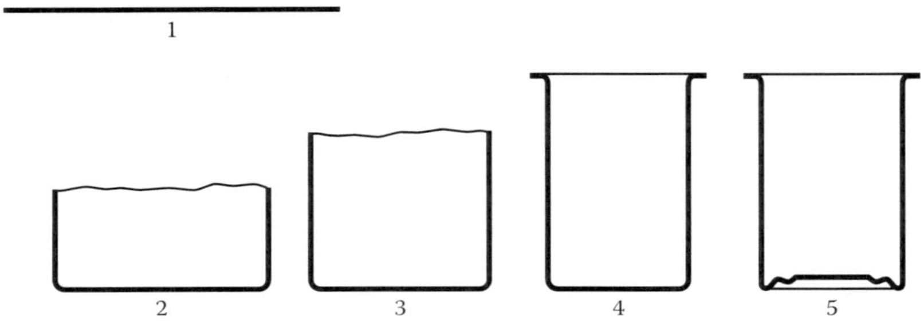

FIGURE 17.5 Sequential stages in the production of drawn and redrawn (DRD) cans: 1, body blank; 2, drawn cup; 3 and 4, first and second stage redrawn cups; 5, finished trimmed can with profiled base. (From Malin, J.D., Metal containers and closures, in: S.J. Palling (ed.), *Developments in Food Packaging*-1, Applied Science Publishers Ltd., London, U.K., pp. 1–26, 1980. With permission.)

height, the body is next trimmed to the desired height. The steps that follow are washing, printing the external surface of the body, coating both surfaces with lacquer, heating in an oven to "bake" the lacquers, necking, and flanging.

The DRD method is shown schematically in Figure 17.5 (Malin 1980). It is the same as the D&I method up to the point of formation of the drawn shallow cup. Then, using punches and dies of progressively smaller diameter, the cup is redrawn, its diameter is reduced, and its height is increased. The thickness of the bottom and of the body remains unchanged during drawing and redrawing. DRD cans are made from precoated lacquer sheets of tinplate, ECCS, or aluminum, with ECCS of thickness 0.2 mm being the most often used material.

17.3.7 Aluminum Foil

Aluminum foil is a sheet of aluminum alloy with thickness between 4 and 150 μm. A variety of alloys (e.g., the 1000, 2000, 3000, 5000, and 8000 series) can be used for foil manufacture depending on the intended application (cooking foil, laminated on plastic film or paper, semirigid trays). Aluminum foil is an excellent barrier to water vapor, gases, and aromas if its thickness is more than 25 μm. At lower thicknesses it is permeable because of formation of small pinholes.

17.4 CORROSION OF METAL CONTAINERS

A very important phenomenon which determines the shelf life of canned foods is the corrosion of their metallic containers. Corrosion is a chemical reaction between a metal and its environment to form largely useless corrosion products. Since corrosion occurs on the surface of the metal, its rate can be reduced by changing the conditions at the surface (e.g., by coating with lacquer). Corrosion may have several negative consequences on the metal container itself (swelling or perforation and leaking of the product) and on the quality of the packaged food through the transfer of metals and rust to it. Therefore, corrosion is a matter of great concern to package manufacturers and food processors and measures are taken to decrease its rate and control it. It is a subject that has been investigated by scientists since the beginning of the use of tinplate as a packaging material, and therefore extensive relevant literature exists (e.g., Board and Steele 1975, Anonymous 1986, Mannheim 1986, Robertson 1993, Robertson 2006, Lee et al. 2008).

The reaction of corrosion of a metal in an aqueous electrolytic solution containing oxidizing substances is an electrochemical one involving the transfer of electric charge, which in the metal are electrons and in the solution ions. When a metal surface corrodes the metal is oxidized; thus, its cations are formed and are dissolved into the solution leaving the freed electrons on the metal surface. For a metal, M, the chemical reaction is

$$M_{(s)} \rightarrow M^{z+}_{(aq)} + ze^-$$

The region of the surface where oxidation, that is, dissolution of the metal, takes places is called anode as the accumulation of electrons in the area establishes a negative charge on it. The electrons move through the mass of the metal to regions where they can cause reduction of the oxidizing substances contained in the electrolytic solution. The removal of electrons (reduction) is the cathodic reaction, and the region where it occurs is called cathode.

17.4.1 Corrosion of the Internal Surface of Tinplate Cans

17.4.1.1 Introduction

Corrosion of the internal surface of tinplate cans involves the oxidation of the two metals (Sn and Fe) of tinplate, that is, their conversion to cations and their transfer to the food. Each one of the two metals may be the anode and get oxidized according to the reactions

$$Sn_{(s)} \rightarrow Sn^{2+} + 2e^- \quad \text{and} \quad Sn^{2+} \rightarrow Sn^{4+} + 2e^-$$

$$Fe_{(s)} \rightarrow Fe^{2+} + 2e^- \quad \text{and} \quad Fe^{2+} \rightarrow Fe^{3+} + e^-$$

The cations of the two metals enter the food and are transformed into insoluble salts, hydroxides, or oxides and remain in the food and/or are deposited on the can walls.

Inside a tinplate can filled with food, pores and scratches in the tin coating may exist extending up to the steel layer and leaving tiny areas of steel exposed to the food. The food is usually an electrolyte and also contains oxidizing substances. In the area around such a pore the metals Sn and Fe are in electrical contact and also immersed in an electrolyte. Consequently, a short-circuited galvanic cell is created in the region between the two metals, with one metal becoming the anode and getting oxidized and the other becoming the cathode where the oxidizing substances are being reduced. This type of corrosion is called galvanic corrosion. Taking into account thermodynamic considerations only, one would normally expect that the metal with the larger negative value of the standard reduction potential E^0 would be the anode. In the present case that would be iron with $E^0 = -0.44$ V, while tin has $E^0 = -0.14$ V.

The oxidizing substances that cause corrosion of the can's metals are called corrosion accelerators. They oxidize the metals, while themselves are reduced at the cathode. The most common corrosion accelerators found in foods are the hydrogen ion and a group of other oxidizing compounds that compete with the hydrogen ion for their reduction at the cathode. These, by being reduced themselves, inhibit the emergence of cathode polarization from the hydrogen gas bubbles. They are called cathodic depolarizers, and the most important of them are oxygen, nitrates, sulfur, sulfur dioxide, some plant pigments, and trimethylamine oxide.

Corrosion of a can depends on many factors such as the corrosiveness of the contained food, the characteristics of the tinplate used to make the can, the use of lacquer (or the lack of use) on the internal surface of the can, processing variables (headspace, heating, and cooling), and storage parameters (temperature and time).

17.4.1.2 Mode of Action of Corrosion Accelerators

The action of some of the corrosion accelerators is described by the following chemical reactions:

$$\text{Hydrogen Ion:}\quad 2H^+ + 2e^- \rightarrow H_2 \uparrow$$

$$\text{Oxygen:}\quad O_2 + 2H_2O + 4e^- \rightarrow 4OH^-$$

$$\text{Nitrates:}\quad NO_3^- + 10H^+ + 8e^- \rightarrow NH_4^+ + 3H_2O$$

$$\text{Sulfur dioxide:}\quad HSO_3^- + 6e^- + 7H^+ \rightarrow H_2S + 3H_2O$$

Corrosion by oxygen gas accumulated in the can's headspace is dealt with better control of the headspace's volume, by increasing the vacuum, blanching of the product, and deaeration of liquid foods to remove the dissolved oxygen.

The source of nitrates in the canned food is either the water added to the food during its preparation or the nitrates encountered in fruits and vegetables, like tomatoes, green beans, etc., grown in heavily fertilized soils. Nitrates act as cathode depolarizers if the pH of the food is lower than 5.5. The problem could be overcome by reducing the quantities and the time of application of nitrate fertilizers. Since this is not very realistic, the best solution remains the use of very well-enameled cans. As far as the water added to the food is concerned, its nitrate content should not exceed 5 ppm.

Sulfur dioxide in acidic foods exists mostly in the form of HSO_3^-. In soft drink and wine cans, it may be reduced according to the reaction presented above to hydrogen sulfide giving an offensive smell.

Sulfide staining is another case of corrosion, for which sulfur compounds are responsible. It occurs to foods rich in proteins with pH > 5 (meat and meat products, fish, and dairy products, etc.) and manifests itself through the emergence of black stains of iron sulfide and blue–black or brown stains of tin sulfide on the surface of tinplate. This is not a health hazard; it is rather a problem of bad appearance of the inside of the can. There are three amino acids containing S in their molecule, and there may be disulfide bonds in the proteins. During or immediately after heat processing, the amino acids and the disulfide bonds are reduced, and H_2S and S^{2-} are produced. At the same time tin and iron are oxidized; their ions react with S^{2-} and stains of SnS and of FeS are formed. The measures taken to limit sulfide staining is rapid cooling after heat processing, the use of well-passivated tinplate, and the use of cans enameled with sulfide stain-resistant lacquers.

Some fruits (e.g., strawberries, cherries) containing the red pigments, anthocyanins, if canned in plain tinplate containers, are either decolorized or get an unpleasant violet color. Anthocyanins are cathodic depolarizers and are easily reduced by the tin. To avoid the problem only well-enameled cans are used for these fruits.

17.4.1.3 Mechanism and Progress of Corrosion

17.4.1.3.1 Plain Tinplate Cans

Two general cases of corrosion are observed for plain tinplate cans: (a) dissolution of tin called detinning and (b) dissolution of iron called pitting corrosion.

17.4.1.3.2 Detinning

As already mentioned, pores and scratches in the tin coating may exist extending up to the steel layer and leaving tiny areas of steel exposed to the food. In these pores, short-circuited galvanic cells are created and galvanic corrosion proceeds. According to the electrochemical series one would normally expect that iron would be the anode and get oxidized ($Fe_{(s)} \rightarrow Fe^{2+} + 2e^-$ and $Fe^{2+} \rightarrow Fe^{3+} + e^-$), while tin would be the cathode and remain intact. And indeed this is the case of external corrosion of tinplate cans. However, inside the can, depending on the food, tin can be cathodic or anodic. For salt solutions and pure water tin is cathodic. For many foods, like fruit juices, some vegetables, meat, fish, and milk, which contain organic acids or amino acids, called anodic depolarizers, reversal of polarity of the couple Sn–Fe occurs. Anodic depolarizers combine with Sn^{2+} and form stable complexes and in this way, by greatly reducing the concentration of Sn^{2+}, make the oxidation of tin ($Sn_{(s)} \rightarrow Sn^{2+} + 2e^-$) more favorable over the oxidation of iron. Tin becomes the anode; it dissolves preferentially and thus protects the iron, which remains intact. This is the most common case of corrosion, and tin is called a sacrificial anode. Another factor that contributes to the polarity reversal of the couple Sn–Fe is the higher hydrogen overpotential of tin over that of iron. Practically that means that the cathodic reaction of the reduction of hydrogen ions is taking place preferentially on the surface of steel than of tin.

There are three cases of detinning: normal, rapid, and partial detinning and pitting. In normal detinning, tin offers complete protection to the iron, and tin dissolution and hydrogen production are slow. Representative foods responsible for this type of corrosion are citrus, pineapple, peach, and apricot. In rapid detinning, tin still offers complete protection to the iron, but the rates of tin dissolution and hydrogen production are high. Representative foods for this case are tomato products, lemon juice, and berries. In partial detinning and pitting, the protection offered to iron by the tin is limited as there are areas of steel which are anodic, and steel corrodes. Relatively soon either hydrogen swelling or perforation occurs. This type of corrosion is observed with steel of lower quality or with prunes and pear nectar.

There is some concern that the tin dissolving from the can into the food might present a health risk to the consumer. In Europe, Commission Regulation (EC) 242/2004 sets limits for tin content of canned foods as follows: 200 ppm for canned foods in general; 100 ppm for canned beverages; and 50 ppm for canned foods for babies, infants, and young children. These limits are easily met. In most cases, under normal conditions of canning and storage, tin concentration in foods in plain tinplate cans is 50–90 ppm after several months of storage (Robertson 2006).

17.4.1.3.3 Pitting Corrosion

For foods like pears, carbonated beverages containing phosphoric acid, and pickles, tin is cathodic relative to iron. Anodic dissolution of iron takes place, ultimately leading to perforation of the container and leaking of the product. This type of corrosion is called pitting corrosion.

17.4.1.3.4 Enameled Cans

The internal surface of cans is very often covered with lacquers to be protected from corrosion. However, the use of coatings does not always assure protection of the surface, and there are even cases in which corrosion is accelerated. Therefore, the selection of a proper lacquer for the particular product and careful coating are very essential for avoiding corrosion problems.

The mechanism of corrosion in enameled cans is quite different and more complex than that in plain tinplate cans. Only at the pores, cracks, and scratches of the lacquer layer there is exposed surface of metal and of course exactly as in the case of plain cans there are pores and scratches in the tin coating extending up to the steel layer. Corrosion will start at the places where the defects of the lacquer layer coincide with defects in the tin layer leaving areas of steel exposed to the food.

Two types of corrosion are observed and in both initially tin is dissolved at the pores of the lacquer. Next, if tin is anodic with respect to iron, its dissolution continues under the lacquer and finally results in lacquer flaking. Since the exposed area of tin is relatively small, the sacrificial protection offered to iron is limited and very soon iron starts dissolving. Abnormally rapid corrosion of iron may occur leading to swelling or perforation of the container. The problem is dealt with by increasing the thickness of the enamel coating. Thus, while a thickness of 4–6 μm is adequate for nonaggressive products, for aggressive foods like tomato paste, a thickness of 8–12 μm is necessary (Robertson 1993).

In the second type of corrosion, iron is anodic to tin right from the beginning. The tin remains intact, and the iron is attacked. Ultimately, the can wall is perforated (pitting corrosion). Beets in vinegar and berries are to blame for pitting corrosion. The best measure to control this corrosion of lacquered cans is improved coverage of the surface (double coating and extra coating of the side seam of the can).

17.4.2 Corrosion of Aluminum

Although aluminum has a large negative standard reduction potential ($E^0 = -1.66$ V), it does not corrode easily because as soon as it is exposed to air or water a very thin film of aluminum oxide (Al_2O_3) is formed on its surface. This oxide film offers aluminum complete protection in the pH range 4–9 (Robertson 1993).

Aluminum is particularly vulnerable to pitting corrosion ($Al \rightarrow Al^{3+} + 3e^-$) if the food it is in contact with contains Cl^- (Wranglen 1985, Lee et al. 2008). The chloride ions start the creation of pores in the layer of Al_2O_3. Although their role is not completely known, it is postulated that they are adsorbed at defective spots of the layer and begin the formation of pores. The problem appears in both cans and foil and is controlled with very good lacquer coating of the surface.

17.5 PLASTIC PACKAGING

17.5.1 Introduction

Polymers are macromolecules, that is, molecules of high molecular weight, comprised of many simpler repeated units, joined together in a regular way. Monomers are the simple low molecular weight compounds from which polymers are made by the polymerization reaction. Plastics are polymers, either synthetic or modified natural polymers, that can be formed to the desired shape by heat and pressure. In the packaging industry, the terms plastic and polymer are frequently used one in the place of the other; however, the term plastic refers more to the final product, while polymer characterizes the raw material for making the product.

For polymers there is a direct relationship between method of production—structure—properties—applications. Plastics differ from all other materials in that they possess a great variety of properties. This is the reason for their widespread use in various fields of everyday life and one of them is food packaging. First of all most of the polymers are inexpensive, and the processes for converting them into food packaging are inexpensive too. They are very versatile in that they can be formed easily to rigid, semirigid, and flexible packages. Plastic containers have low weight, do not corrode, and can be transparent. Polymers are good thermal and electrical insulators, are heat-sealable, and can combine various functionalities. This combination of properties makes polymers ideally suitable for applications where other materials do not perform well. However, they have significant disadvantages too. Unlike glass and metal, they are permeable to gases and vapors. Most of them do not have high mechanical strength and high enough heat resistance. In general, they are not biodegradable, and this is the reason they are considered as having negative environmental impact. A major disadvantage of polymers is the risk of migration into the food of low molecular weight substances contained within the polymer. To minimize this possibility, many regulations exist for food packaging plastics, and extensive tests are carried out to ensure that the packaged food is safe for consumers. Another factor detrimental to food migration is the scalping (adsorption) of food aromas and flavors by the plastic packaging.

Polymer science and technology is a very vast scientific field, and there are numerous books on the subject. More information can be found in general references on polymer science and technology (e.g., Baird 1976, Rodriguez 1982, Grulke 1994), in the Macrogalleria® website of polymer science, general references devoted exclusively to plastic food packaging (e.g., Briston and Katan 1974, Jenkins and Harrington 1991, Brown 1992), and the relevant chapters in food packaging books (e.g., Briston 1980, Goddard 1980, Robertson 1993, 2006, Soroka 1996, Lee et al. 2008).

17.5.2 Classification of Polymers

Polymers can be classified according to various criteria, like their origin, type of monomer, molecular shape, behavior during heating, mechanism of polymerization reaction, and functional group.

The polymers that consist of one type of repeating building-block unit are called homopolymers, while those consisting of two or more different units are called heteropolymers. A heteropolymer produced by polymerizing together two different types of monomers is called copolymer, and if the types of monomers are three, the polymer is called terpolymer, etc. As far as the shape of the macromolecules is concerned there are linear (a main chain), branched (side chains attached to the main chain), and cross-linked (three-dimensional network of interconnected main chains) polymers.

Another distinction of food packaging polymers, related to their molecular structure and behavior when heated, is in thermoplastics and thermosets. Thermoplastics are linear or branched but with no cross-links between the chains. When heated they soften, and eventually they become polymer melt. Then they can be formed in various shapes and become solid upon cooling. The phenomenon is reversible, and the "heating-melting-forming-cooling-solidifying" cycle can be repeated many times. Almost all polymers used in food packaging are thermoplastic, and thermoplastics account for more than two thirds of all polymers used today.

Thermosets are cross-linked polymers. The creation of the network of polymer chains occurs during formation of the plastic object and is done by heating, irradiating, or adding cross-linking agents. Covalent bonds are formed between the macromolecular chains, and in essence the material becomes one gigantic molecule. The process is irreversible, and hard and stiff objects are produced from thermosets. Thermosetting polymers do not melt on heating and thus cannot be remolded for a second time. If heating continues they degrade, form blisters, and finally char. Thermosets find minimal applications in food packaging as coatings, adhesives, and bottle closures.

17.5.3 Polymer Morphology and Phase Transitions in Polymers

The properties of polymers depend on their molecular structure, molecular weight, morphology, and chemical composition (Robertson 2006). The subjects of morphology and phase transitions are briefly mentioned here. Detailed information on all these topics can be found in polymer science textbooks.

For thermoplastic polymers there are two ideal extreme cases concerning their morphology: crystalline and amorphous. In the first state the macromolecules are arranged in an orderly fashion parallel and closely packed in ideal crystals, while in the second they are completely disordered. Pure crystalline morphology is very difficult to be obtained, and hence there are only very few cases of completely crystalline polymers. On the other hand, some polymers that are commercially important in food packaging are totally amorphous. Most thermoplastic polymers are semicrystalline, consisting of amorphous and crystalline regions and are characterized by a degree of crystallinity, which expresses the percentage of crystalline material in the total mass of the polymer. The degree of crystallinity affects considerably the mechanical, optical, thermal properties of the polymer, as well as its permeability to gases and vapors. It depends on the chemical structure, the number of branches in the chain, the mean molecular weight and the molecular weight distribution of the polymer, and processing parameters, such as the cooling rate of the polymer melt, the presence of additives, and stretching.

Let us consider a thermoplastic polymer at high enough temperature for it to be in the liquid state and suppose we study the change of the specific volume of the polymer with respect to temperature as we cool it. The following behaviors are observed. If the polymer is totally crystalline, as it cools, its specific volume decreases with a constant slope and at the crystalline melting temperature T_m, a sharp drop of the specific volume occurs as the polymer undergoes a phase transition from liquid to crystalline solid. If the polymer is totally amorphous, as the temperature drops, the specific volume decreases with a constant slope, and then at a certain temperature the slope changes and remains constant at the new value thereafter. The temperature at which this change of slope is observed is called glass transition temperature T_g. If the polymer is semicrystalline both T_m and T_g are observed.

An amorphous polymer at temperatures above T_g is either liquid or soft and flexible like rubber. At the glass transition temperature T_g the polymer changes from a rubbery state to a glassy state, that is, it becomes solid like glass with physical properties similar to those of a crystalline solid but with the molecular disorder of a liquid. In the rubbery state, because of the increased molecular movement of the chains, a polymer is more permeable than in the glassy state. Other properties (e.g., coefficient of thermal expansion, specific heat capacity) that depend on the ease of molecular movement of the chains are different at the two states (rubbery–glassy).

The physical properties of thermoplastic semicrystalline polymers at a temperature of interest depend on the values of T_g and T_m in relation to the temperature of interest. If this temperature is below T_g then the polymer is comprised of crystalline and glassy regions and as such is an effective barrier to permeants. If it is above T_g permeability increases because the glassy regions have been transformed to rubbery ones.

The mobility of the polymer chains and the T_g value of the polymer are interrelated. Whatever factor increases the T_g value and vice versa. Thus, bulky pendant groups in the macromolecule and cross-linking of the chains raise T_g. On the other hand, symmetry in the macromolecule and the addition of plasticizers lower T_g.

17.5.4 Food Packaging Polymers

The most widely used materials in food packaging, thermoplastic polymers, are presented next, arranged more or less according to their functional group. These are *polyolefins* (polyethylenes, polypropylene); *poly(substituted olefins)* (polystyrene, poly(vinyl alcohol), poly(vinyl chloride), poly(vinylidene chloride)); *copolymers of ethylene* (ethylene-vinyl acetate, ethylene-vinyl alcohol, ionomers); *polyesters* (polyethylene terephthalate); *polycarbonates*; *polyamides*; *polyacrylonitrile and copolymers of acrylonitrile*; *and regenerated cellulose* (Robertson 1993).

A useful table summarizing the main properties and uses of the above polymers is presented by Krochta (2007). More information about the chemistry, properties, manufacture, applications, regulatory, safety, and environmental aspects of polyethylene (Tice 2003), polypropylene (Tice 2002a), polystyrene (Tice 2002b), poly(vinyl chloride) (Leadbitter 2003), and polyethylene terephthalate (Matthews 2000) can be found in the reports prepared under the responsibility of the ILSI Europe Packaging Material Task Force and available at the website of ILSI Europe.

17.5.4.1 Polyethylene (PE)

Polyethylene is the simplest, cheapest, and most widely used polymer. It is produced by polymerization of ethylene gas ($CH_2{=}CH_2$), and its chemical formula is $[-CH_2-CH_2-]_n$. There are several types of PE available, differing in molecular structure and properties. The distinction is based on density and the most common are

1. *Low Density Polyethylene* (*LDPE*): It has a branched structure with many and long side-chain branches. Because of the branches, it has low density (910–940 kg m^{-3}), degree of crystallinity 50%–70%, and softening temperature lower than 100°C. All types of PE,

because of their nonpolar nature, are excellent moisture barriers. However, they have high permeability to gases (O_2, CO_2, and N_2) and aromas. For packaging, a very useful property of LDPE is that it is heat-sealed easily giving tough, liquid-tight seals. LDPE has the disadvantage of environmental stress cracking, that is, when stressed while in contact with polar liquids or vapors (detergents, essential oils, vegetable oils) cracks are formed on its surface. It also tends to adsorb volatile constituents of the contained food, like D-limonene from orange juice. It is used mainly as a film for making bags to package fresh produce and bakery products and pouches for frozen foods. The LDPE film is soft, flexible, and stretchable. LDPE is combined with other packaging materials, like paperboard, aluminum, and other polymers for making multilayer packaging materials (laminates), where it serves as a moisture barrier, a heat-sealing layer, or as an adhesive.

2. *High Density Polyethylene* (*HDPE*): It is a linear polymer with few and short side-chain branches. Because of the absence of branches, it has high density (941–965 kg m^{-3}), degree of crystallinity up to 90%, and softening temperature about 121°C. Compared to LDPE, it has better resistance to chemicals and oils, greater tensile strength and hardness, and 5–6 times lower permeability to water vapor and gases, but it is much more opaque and significantly more difficult to heat-seal. Like LDPE, it has the problem of environmental stress cracking. It is used for making bottles and jugs for milk and water, containers for salt, and as a film for bags (e.g., grocery bags). The HDPE films have a white, translucent appearance and usually are very thin (10–12 μm).
3. *Linear Low Density Polyethylene* (*LLDPE*): It is a copolymer of ethylene with small amounts of a higher alkene. It is a linear polymer with many but short side-chain branches. The term linear refers to the absence of long chain branches. It combines the advantages of LDPE and HDPE. It has the clarity and good heat-sealabiliy of LDPE and the mechanical strength and chemical resistance of HDPE. For this reason it has been replacing LDPE in many applications.

17.5.4.2 Polypropylene (PP)

Polypropylene is produced by addition polymerization of propylene ($CH_2{=}CH(CH_3)$), and its chemical formula is $[-CH_2-CH(CH_3)-]_n$. It is a polymer with many similarities to LDPE. However, compared to LDPE, it has lower density (900 kg m^{-3}), higher softening temperature (140°C–150°C), higher tensile strength, stiffness and hardness, but lower resilience. As a polyolefin it has low permeability to water vapor, while its permeability to gases is medium. PP has good resistance and impermeability to fats and oils; so it is suitable for packaging fatty foods. Its relative high temperature stability makes it useful for hot filling, retorting, and microwaving applications. Finally, it has good gloss and clarity, properties essential for successful reverse printing.

PP is formed into bottles, cups, tubs, bowls, jars, glasses, trays, and closures. PP films are available in two forms: nonoriented (also called cast) and oriented. Orientation is accomplished by stretching the film while hot in one direction or two directions. In the first case, OPP (oriented PP) is produced and in the second, BOPP (biaxially oriented PP). There are food packaging applications for both forms of films. Cast PP is used as a component in multilayer packaging materials (laminates) when high temperature resistance and good heat-seal strength are required (e.g., retort pouches). OPP film is used for packaging bakery and confectionery products, various snacks, and cheeses. BOPP film has a high clarity; however, it is not heat-sealable, and it has to be coated with another polymer.

17.5.4.3 Polystyrene (PS)

PS is produced by addition polymerization of styrene ($C_6H_5-CH{=}CH_2$), and its chemical formula is $[-CH_2-CH(Ph)-]_n$, where Ph is the phenyl ring. It is totally amorphous and thus a hard, rigid, but brittle, clear, and transparent material, with a high T_g in the range 90°C–100°C. In that form it is called crystal grade PS. It has medium permeability to gases and high to water vapor. To deal with the problem of brittleness of PS, polybutadiene and a styrene–butadiene copolymer are blended with the PS, and the resulting material is called Toughened or High Impact PS (HIPS).

Crystal grade PS is used for making transparent jars, glasses, trays, and disposable cutlery. With HIPS tubs and cups are made for dairy products, margarine, ice cream, etc. PS finds also many applications as foamed or expanded PS. These include trays for fresh produce, meats, poultry, fish, cartons for eggs, cups, and plates.

17.5.4.4 Poly(vinyl chloride) (PVC or V)

PVC is produced by addition polymerization of vinyl chloride monomer (VCM) ($CH_2=CHCl$). If the addition of the monomer to the growing chain of the polymer is "head to head" PVC's chemical formula is $[-CH(Cl)-CH_2-CH_2-CH(Cl)-]_n$, and if it is "head to tail" the formula is $[-CH_2-CH(Cl)-CH_2-CH(Cl)-]_n$. In the PVC used in the various applications, this addition is done in a completely random manner and thus PVC is a clear, amorphous polymer. It is an inexpensive polymer, but its use in food packaging is rather limited. On the other hand, it finds many nonfood-packaging applications ranging from piping to surgical gloves.

PVC is a very stiff and brittle material and to make it flexible and soft, plasticizers (organic liquids of low volatility) are added to it. Also stabilizers are incorporated into the polymer. These substances tend to migrate into the food, and there have been safety concerns and controversies about this migration. Now internationally, there are regulations in the law concerning migration of additives of VCM and residual VCM content of the PVC. Therefore, extreme care should be taken when choosing additives for PVC and also compliance with the relevant migration limits set up in the law must be assured.

Unplasticized PVC, or with a low plasticizer content, is used to make jars for coffee and instant drinks in powder form and inserts for the boxes of chocolates and biscuits. Plasticized PVC films have good stretch and cling, clarity, and transparency and relatively high permeability to gases and water vapor. For these reasons, they are used for overwrapping fresh produce and fresh meat.

17.5.4.5 Poly(Vinylidene Chloride) (PVdC)

The PVdC homopolymer is produced by addition polymerization of vinylidene chloride ($CH_2=CCl_2$), and although it has many desirable properties, it is brittle, it does not adhere well to other materials, and presents problems in processing. For these reasons vinylidene chloride is copolymerized with vinyl chloride. It is a generally accepted convention to call the copolymer of vinylidene chloride (85%–90%) and vinyl chloride simply PVdC. Another name is Saran from the trade name Saran® of the film produced by Dow Chemical Company.

The unique property of PVdC is that it is an excellent barrier to both water vapor and gases and aromas. In addition, humidity does not affect its permeability to O_2 and CO_2. It is one of the two most important high barrier polymers employed in food packaging. It can withstand the relative high temperatures of hot filling and retorting, and thus it is used as an internal barrier layer in multilayer containers (e.g., retort pouches) for thermally processed foods. It is also used as a coating for bottles and other films, for example, regenerated cellulose. Saran wrap is a single-layer PVdC film used by the consumers to wrap food at home.

17.5.4.6 Ethylene-Vinyl Acetate (EVA) Copolymer

The copolymers of ethylene ($CH_2=CH_2$) and vinyl acetate ($CH_3COOCH=CH_2$), with a vinyl acetate content of 3%–12%, have many properties similar to those of plasticized PVC. Therefore, EVA copolymer films can replace the latter in various applications, for example, wrapping of fresh meat, since they have the advantage over the PVC films of not containing plasticizers that may migrate to food. EVA copolymers are also used as the heat-seal layer in laminates and lids.

17.5.4.7 Ethylene-Vinyl Alcohol Copolymer (EVOH or EVAL)

The copolymer of ethylene ($CH_2{=}CH_2$) and vinyl alcohol ($CH_2{=}CHOH$), with an ethylene content of 27%–48%, is an expensive polymer. Its properties depend on the relative proportions of the two monomers. EVOH is an excellent barrier to gases and aromas, being one of the two high barrier polymers. However, it is very sensitive to humidity and has to be protected from moisture. Its permeability to oxygen increases almost 50 times for an increase of the relative humidity from 65% to 100%. At low relative humidities, it is the material with the lowest permeability to oxygen, better than Saran HB (High Barrier). Its permeability to oxygen increases with increasing ethylene content. While PVdC is usually coated on the external surface of another material, EVOH must be between two layers of polyolefins to be protected from moisture. So in multilayer structures, it is combined with PP (retort pouches, bottles for hot filling), PET, or LDPE. It is also used in modified atmosphere packaging where, because of its gas barrier properties, it helps maintain the composition of the gases inside the package at the required level.

17.5.4.8 Ionomers

They are copolymers of ethylene and methacrylic acid ($CH_2{=}C(CH_3)COOH$), with a methacrylic acid content less than 15%. After production of the polymer, some carboxyl groups are neutralized by Na^+ or Zn^{2+} ions, and in this way ionic crosslinks are formed between the polymer chains. Ionomers are produced commercially only by Du Pont under the trade name Surlyn®. Ionomers are the best adhesives available in food packaging but also the most expensive. They are the most suitable for binding aluminum to other packaging materials. Surlyn forms excellent heat-seals even in the case that the product remains in the seal area and the seal has good strength and even in the case it didn't have enough time to cool and solidify completely (excellent hot tack). For these reasons, ionomers are used as the internal layer in multilayer films. They are combined with polyamides and polyesters for packaging meat and cheese.

17.5.4.9 Polyesters

Polyesters are produced by condensation polymerization of dicarboxylic (mostly aromatic) acids with diols (glycols). The most important polyester used in food packaging is polyethylene terephthalate (PET or PETE), made by reacting terephthalic acid with ethylene glycol:

$$n(\mathrm{HOOC{-}C_6H_4{-}COOH}) + n(\mathrm{HO{-}CH_2{-}CH_2{-}OH})$$

$$\downarrow$$

$$\mathrm{HO(-OC{-}C_6H_4{-}COO{-}CH_2{-}CH_2{-}O-)}_n\mathrm{H} + (2n-1)\mathrm{H_2O}$$

Amorphous-oriented PET is a strong polymer with relatively low permeability to gases. It has excellent clarity, transparency, and gloss and in that respect it is similar to glass. Since the end of the 1970s, PET is used for making bottles for carbonated beverages. A huge growth followed, new markets opened up, and today PET is the major plastic for making bottles for all kinds of beverages, water, juices, milk, vegetable oils, and dressings and even beer. PET jars are used for packaging a variety of foods from peanut butter and mustard to instant coffee. To improve their gas barrier properties PET bottles can be coated with PVdC, or SiO_x, or made multilayer with an internal layer of EVOH.

Almost all PET films are biaxially oriented and heat stabilized. They have excellent mechanical properties, chemical resistance, and elasticity. They have good dimensional stability from −60°C to 220°C. The crystalline melting point T_m of PET is 267°C and the T_g is between 67°C and 80°C. PET films are used for making "boil-in-bags" and "roast-in-bags." Since PET has poor heat sealability, the films are often coated with a heat-sealable layer of another polymer.

Another application of PET is for making "dual-ovenable" trays, that is, for both microwave and conventional oven. A special form of PET, crystallized PET (CPET) is used for this purpose because it can withstand up to 220°C without deformation.

Examples of other polyesters besides PET that can be used in food packaging are polybutylene terephthalate (PBT) and PETG. PETG is a copolyester, produced by reacting terephthalic acid with two glycols (ethylene glycol and cyclohexane dimethanol).

17.5.4.10 Polycarbonates (PC)

These polymers are linear polyesters of carbonic acid H_2CO_3 with aromatic diols, most commonly Bisphenol A (BPA–HO–C_6H_4–C$(CH_3)_2$–C_6H_4–OH). They are distinguished from the other polyesters in that they contain the characteristic group –O–CO–O– in their molecule. Polycarbonates are amorphous polymers with excellent clarity and transparency like glass, toughness, and high temperature resistance. Polycarbonates have high softening point temperature (T_m between 220°C and 250°C and T_g 150°C) and thus can withstand 130°C–140°C temperatures for a long time. They are used as glass replacements in various nonfood packaging applications. Plastic glasses and pitchers used in hotels and bars are usually made of polycarbonates. Their major application in food packaging is the 5-gal reusable water bottle. In the past plastic baby bottles were made of BPA polycarbonate. There were safety concerns for the babies about the BPA that remains in the plastic migrating into the hot milk usually consumed from these bottles. Bisphenol A (BPA) is a suspected endocrine disruptor. In Europe, the use of BPA for the manufacture of polycarbonate infant feeding bottles has been banned according to Regulation (EU) No. 321/2011. Nowadays all plastic baby bottles in the market are advertised as BPA free.

17.5.4.11 Polyamides (PA) or Nylons

Polyamides (nylons is a generic name) are produced by condensation polymerization of a dicarboxylic acid with a diamine and are identified by a couple of numbers, the first being the number of carbon atoms in the molecule of diamine and the second in the molecule of dibasic acid. Thus, nylon 6,6 is produced from the hexamethylene diamine H_2N–$(CH_2)_6$–NH_2 and adipic acid HOOC–$(CH_2)_4$–COOH according to the reaction:

$$n(H_2N-(CH_2)_6-NH_2)+n(HOOC-(CH_2)_4-COOH)$$

$$\downarrow$$

$$H(-HN-(CH_2)_6-NH-CO-(CH_2)_4-CO-)_nOH+(2n-1)H_2O$$

Polyamides can also be produced by the condensation of an ω-amino acid:

$$\mathrm{n}\left(H_2N-R-COOH\right)\rightarrow H(-HN-R-CO-)_\mathrm{n}OH+(\mathrm{n}-1)H_2O$$

In this case the polyamide is identified by a single number, which is the number of carbon atoms in the molecule of amino acid. Nylon 6 corresponds to condensation of ε-aminocaproic acid H_2N–$(CH_2)_5$–COOH, while it is actually produced from ε-caprolactam $(CH_2)_5$CONH, the cyclic amide of ε-aminocaproic acid.

The various polyamides have many properties in common. They are strong with excellent thermal stability and grease and oil resistance. Like EVOH, they are very good barriers to gases when dry and are highly permeable and sensitive to water vapor. Therefore, they are usually protected from moisture being placed between layers of polyolefins.

In food packaging, nylons are exclusively used as films (combined with LDPE or PP) for bags and wraps for packaging cheeses, meats, cured and cooked meats, nuts and frozen foods. Films from biaxially oriented nylon (BON) have very good mechanical properties and low permeability to gases and aromas. A unique characteristic of nylons is their high puncture resistance, and for this reason they are used in applications where this property is essential, for example, packaging of meat with bones and of nuts. In the United States, most polyamide packaging films are nylon 6, while in Europe, nylon 11.

17.5.4.12 Polyacrylonitrile (PAN) and Heteropolymers of Acrylonitrile

Polyacrylonitrile (PAN) is used as a fiber and not in packaging. The terpolymer acrylonitrile-butadiene-styrene (ABS) has limited use in food packaging for making tubs and trays. The copolymer acrylonitrile-styrene (ANS) has excellent gas and moisture barrier properties, and bottles for packaging carbonated beverages were made from this polymer. In 1977, FDA banned the use of these bottles because of concerns about potential migration of acrylonitrile into the beverage. In 1984, FDA withdrew the ban provided that the residual acrylonitrile content of the polymer is less than 0.1 ppm. Although this limit is met, by that time PET had already been established as the polymer for carbonated beverage packaging.

The acrylonitrile-methyl acrylate (ANMA) copolymer with the addition of a small quantity of butadiene-acrylonitrile rubber is known under the trade name Barex. It is a transparent material with good mechanical properties and excellent gas barrier properties. It was developed for making bottles for carbonated beverage packaging.

17.5.4.13 Regenerated Cellulose or Cellophane

Cellophane is a generic term (except in Britain) for regenerated cellulose film (RCF). Regenerated cellulose is produced from wood pulp. With the addition of suitable chemicals to the wood pulp cellulose is dissolved into solution. Then, with the addition of other chemicals, cellulose is coagulated (regenerated) and finally is produced in the form of a continuous transparent film. It can be regarded as transparent paper. In a dry environment it is a good barrier to gases; however, when exposed to high relative humidities it absorbs moisture and its permeability increases considerably. It is very permeable to water vapor and cannot be heat-sealed. With various coating these drawbacks are corrected. The RCF used in food packaging is always coated on one or both sides. The usual materials RCF is coated with are nitrocellulose, PVC, LDPE, and PVdC. Although in many applications nowadays BOPP is used instead of RCF, it is still used for twist-wrapping of sweets (Robertson 2006).

17.6 PLASTIC PACKAGE MANUFACTURE

17.6.1 Semirigid and Rigid Plastic Packaging

17.6.1.1 Methods of Production

Several different manufacturing methods exist for forming plastics into trays, tubs, cups, lids, jars, and bottles. In general, these containers have wall thickness greater than 75–150 μm.

17.6.1.1.1 Injection Molding

The pellets of a thermoplastic polymer are softened in a heated cylinder, and the molten plastic is injected through a nozzle into the cool mold with a reciprocating and rotating screw. After solidification of the plastic object, the two halves of the mold open, and the object is ejected. The method produces the most dimensionally accurate objects of any other process. It is used for making tubs, cups, lids, and spouts from PE, PP, and PS. It is also used for manufacturing the preforms for the PET bottles.

17.6.1.1.2 Compression Molding

It is used for constructing closures for bottles and jars from thermoset polymers, usually urea–formaldehyde resins. A measured quantity of the resin is placed into the open mold that has been preheated to 190°C–200°C. The mold closes and is pressed with a hydraulic press. The molten resin completely fills the mold cavity. The heating and the high pressure are necessary for thermosetting of the polymer (i.e., the formation of cross-links between the chains). The plastic object stays in the mold for 1–2 min for curing and then is removed.

17.6.1.1.3 Blow Molding

It is the main process for producing hollow objects like bottles. A cylindrical semimolten tube, called parison or preform, of the thermoplastic polymer is placed into a cooled mold of the desired shape. Pressurized air or nitrogen is fed into the parison and blows it to the final container shape. Blow molding is in general carried out in three steps: melting the resin, forming the parison, blowing the parison. The first two stages are accomplished either by extrusion and the whole process is extrusion blow molding or by injection and the process is injection blow molding.

In *extrusion blow molding* (Figure 17.6) (Briston 1980), the parison is continuously extruded between the two open halves of the mold and in the form of a heat-softened cylindrical hollow tube. At the moment the tube attains the right length, the extruder stops and the mold closes and thus seals the bottom of the parison. From a blow pin mounted inside the extruder die, blowing air is fed into the parison and inflates it. The polymers that are most commonly extrusion blow-molded are HDPE, PP, PVC, and heteropolymers of acrylonitrile. A variation of this process is *coextrusion blow molding*, in which two or more extruders, each one handling a different resin, are feeding the same die. In this way the parison produced and the final bottle are composed of layers of different polymers. For instance, to obtain a PP bottle or jar with good oxygen barrier properties, a high barrier polymer, like EVOH, can be coextruded between the two outer PP layers. Usually, in these structures, a thin layer of adhesive resin, or tie, is also used to better bind the different polymer layers. Bottles of the composition mentioned above are used for packaging ketchup and sauces. Another application of coextrusion is to make a bottle with a layer of regrind resin from industrial scraps or even a layer of recycled plastic, between layers of virgin plastic to avoid potential migration problems.

Injection blow molding is a two-step process, similar to the making of glass bottles (Figure 17.7) (Briston 1980). In the first step by injection molding, the preform is formed having a blowing stick inside it. In the second step, the semimolten preform, with the blowing stick inside, is transferred to the blow mold. Compressed air fed through the blowing stick inflates the preform to the final container shape. Common polymers injection blow-molded are HDPE, PP, PS, PVC, and PET.

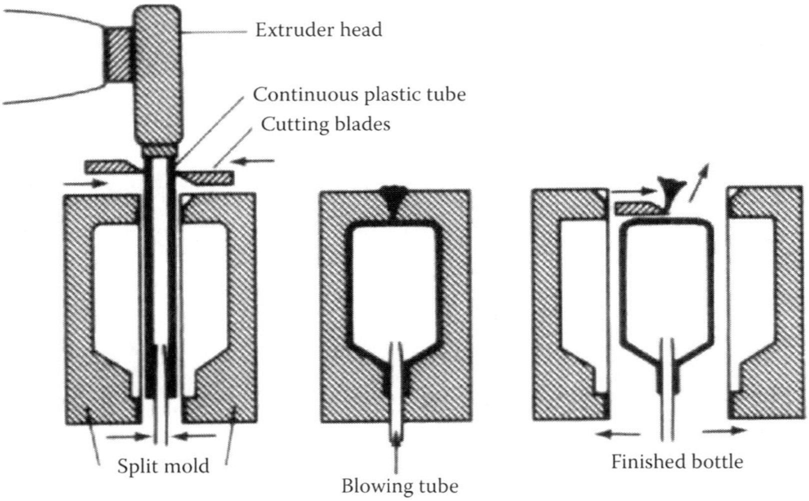

FIGURE 17.6 Extrusion blow molding of plastic bottles. (From Briston, J.H., Rigid plastics packaging, in: S.J. Palling (ed.), *Developments in Food Packaging*-1, Applied Science Publishers Ltd., London, U.K., pp. 27–53, 1980. With permission.)

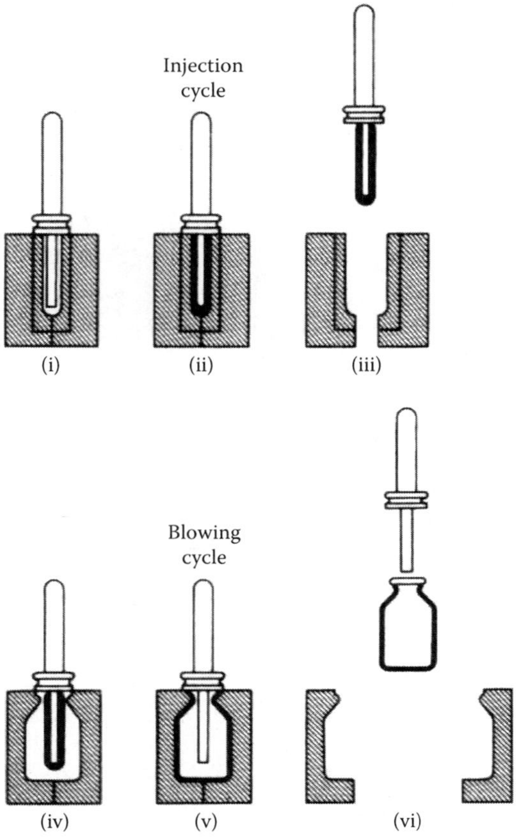

FIGURE 17.7 Diagrammatic representation of the injection blowing process. (From Briston, J.H., Rigid plastics packaging, in: S.J. Palling (ed.), *Developments in Food Packaging*-1, Applied Science Publishers Ltd., London, U.K., pp. 27–53, 1980. With permission.)

The advantages of the method over extrusion blow molding are more dimensionally accurate bottles, better quality of the neck and finish of the bottle, no scraps, and no need for trimming the bottle. As with coextrusion, with coinjection there is the possibility of producing a multilayer preform and then a multilayer bottle or jar.

An extension of the previous method is *injection stretch blow molding*. The differences are the preform is much shorter than the bottle and a rod is used to stretch the preform in the axial direction (Lee et al. 2008). With the process biaxial orientation of the polymer's macromolecules is achieved and retained. This is the reason why the process is also called orientation blow molding. The advantages of the so-formed containers are better mechanical properties, improved gas barrier properties, increased transparency and surface gloss, and reduced creep. The most common application of the method is making PET bottles and jars (Figure 17.8). PET is difficult to be blown and careful control of the temperature is required to avoid significant crystallization of the PET. Other resins like PVC, PP, and heteropolymers of acrylonitrile are also stretch blow-molded. For high volume production of the common PET bottles the following two-stage technique is implemented. In the first, the preform is injection-molded. Cooling it in the mold to less than 120°C must be very rapid to prevent PET crystallization. The rate of crystallization is high only in the temperature range of 120°C–220°C, while only amorphous PET has the required transparency and gloss. In the preform, the whole area of the finish is in its final form and won't change during blowing. The preforms are stored at ambient temperature until needed for the second stage. Then they are reheated to between 90°C and 110°C

FIGURE 17.8 PET bottles and jar and respective preforms.

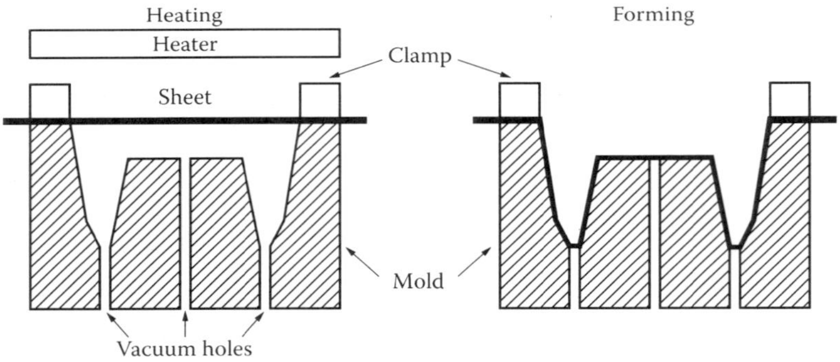

FIGURE 17.9 Simple vacuum thermoforming. (From Briston, J.H., Rigid plastics packaging, in: S.J. Palling (ed.), *Developments in Food Packaging*-1, Applied Science Publishers Ltd., London, U.K., pp. 27–53, 1980. With permission.)

and placed in the blowing mold. The stretch rod is pushed downward and stretches the preform, and the compressed air fed in inflates the preform to the final container shape.

Thermoforming (Figure 17.9) (Briston 1980) is a simple technique for making trays, cups, tubs, and cartons. A sheet of a thermoplastic material, usually 75–250 μm thick, is clamped in a frame on top of the mold. The sheet is softened by heating via an infrared heater and forced into or over the mold by vacuum, air pressure, or by mechanical means. After cooling, the plastic takes the shape of the mold. Thermoplastic polymers that are usually thermoformed are PS, PP, HDPE, nylon, and EPS (expanded or foamed PS). A major application of thermoforming is the manufacture of meat and produce trays and egg cartons from EPS sheet. Dual-ovenable trays made from thermally stable PET, also called crystallized PET (CPET), are produced with the variation of the thermoforming technique. A nucleating agent is added to the PET sheet, and the tray remains in the thermoform mold long enough for the PET to crystallize. CPET is heat stable at temperatures up to 230°C, whereas amorphous PET starts to soften at temperatures higher than 63°C.

Details on the methods of production of rigid, semirigid, and flexible plastic packaging can be found in a number of sources (Baird 1976, Briston 1980, Goddard 1980, Robertson 1993, Soroka 1996, Brody and Marsh 1997, Mauer and Ozen 2004, Robertson 2006, Krochta 2007, Lee et al. 2008, Yam 2009).

17.6.2 Flexible Plastic Film Packaging

Plastic packaging film is a flexible polymer sheet up to 0.25 mm thick. A product of thickness greater than 0.25 mm is referred to simply as a sheet, and it is usually thermoformed. Brennan et al. (1990) present a table with the general properties and applications of the plastic films most commonly used in food packaging.

17.6.2.1 Methods of Production

Extrusion. Single-layer films of any thermoplastic can be formed with this method. The components of a single-screw extruder are shown schematically in Figure 17.10 (Goddard 1980). The most important part of an extruder is the screw that rotates within a heated barrel. The resin is fed in through a hopper in pellet or powder form. The screw compresses and shears the material, which is heated by friction and through contact with the hot walls of the barrel. The plastic melts, it gets homogenized, and finally the polymer melt is forced out through the die. The die determines the final shape of the product, which is called extrudate. The extruder may also have two screws (twin-screw extruder) instead of one (single-screw extruder), corotating or counter-rotating. When the polymer melt exits through a slit die and is drawn around two or more chilled casting rolls, a flat film is produced, and the process is called *flat or cast-film extrusion*. If the die is annular, a plastic tube is extruded usually vertically upward. Air is introduced into the tube through a hole in the center of the die and inflates the tube transforming it into a large tube with thin walls. Then the tube is cooled, flattened, and wound up. The second process is called *tubular or blown-film extrusion* (Figure 17.11) (Goddard 1980). The method is employed mainly for making bags, and the polymers most commonly used are HDPE, LDPE, and LLDPE.

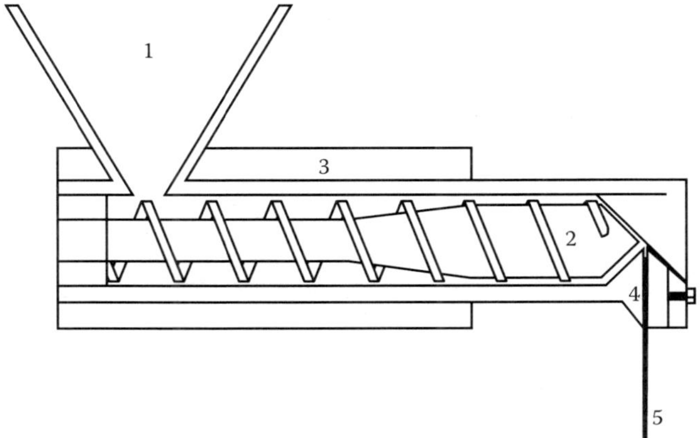

FIGURE 17.10 Single-screw extruder: 1, input hopper for resin granules; 2, feeder screw; 3, heating elements; 4, slot extrusion die; 5, extruded film. (From Goddard, R.R., Flexible plastics packaging, in: S.J. Palling (ed.), *Developments in Food Packaging*-1, Applied Science Publishers Ltd., London, U.K., pp. 55–79, 1980. With permission.)

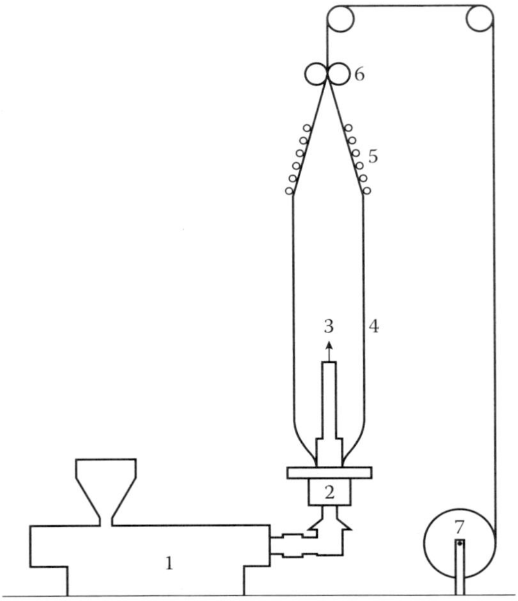

FIGURE 17.11 Blown tubular film extrusion: 1, screw extruder; 2, circular die; 3, air at constant flow; 4, tubular film; 5, guide frame; 6, nip rollers; 7, wind-up roll. (From Goddard, R.R., Flexible plastics packaging, in: S.J. Palling (ed.), *Developments in Food Packaging*-1, Applied Science Publishers Ltd., London, U.K., pp. 55–79, 1980. With permission.)

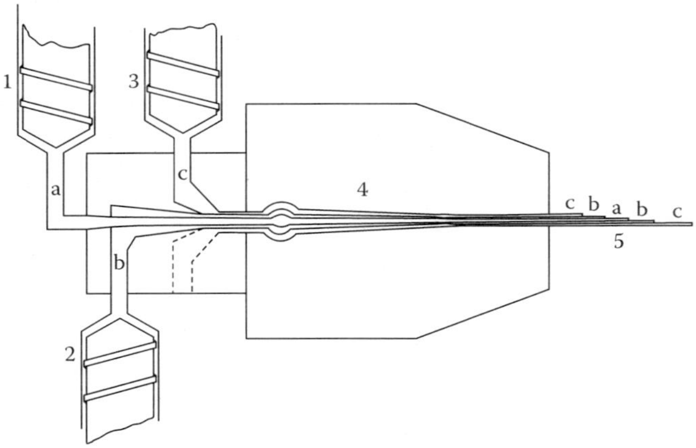

FIGURE 17.12 Five-layer and three-component extrusion combination die: 1, extruder screw (polymer a); 2, extruder screw (polymer b); 3, extruder screw (polymer c); 4, combining adaptor; 5, five-layer film. (From Goddard, R.R., Flexible plastics packaging, in: S.J. Palling (ed.), *Developments in Food Packaging*-1, Applied Science Publishers Ltd., London, U.K., pp. 55–79, 1980. With permission.)

The advantages of cast-film extrusion are better optical properties, high production rates, and better thickness uniformity of the film. The advantages of blown-film extrusion are better mechanical properties of the film and much lower cost.

Coextrusion. It is possible for two or more extruders, each one processing a different polymer, to be coupled and feed the same die (Figure 17.12) (Goddard 1980). The extrudates merge and fuse in a feed block and in this way a multilayer film is formed. The feed block is a metal block with specially designed flow channels (Lee et al. 2008). The main advantages of coextruded films are lower cost, less tendency for delamination, and greater flexibility in achieving the desired properties (Robertson 2006).

Calendering. The hot plastic material is compressed between two or more rollers. A continuous sheet of very uniform thickness is produced. PVC films and sheets are usually formed by this technique.

Solution casting. The polymer is dissolved in an appropriate solvent, and the solution is cast through a die on a moving and heated steel belt. The solvent is evaporated, and at the other side of the belt the flat plastic film is removed. Cellulose acetate, propionate, and butyrate films are produced by this method (Robertson 1993).

17.6.2.2 Orientation of Plastic Films

Usually thermoplastic films after or during their formation are oriented to acquire improved strength and durability. Orientation achieves alignment of the polymer's macromolecules in a given direction. It involves heating the film at a temperature at which it becomes soft (above T_g, but below T_m) and stretching it in one or two directions. When the film is cooled, the macromolecules remain "frozen" in this configuration and extend like springs. If an oriented film is reheated at its orientation temperature, it shrinks as the macromolecules return to their original size and random spatial orientation. Orientation in one direction (machine direction) is called uniaxial, while in two directions (machine and transverse) is called biaxial.

The advantages of biaxially oriented films are better optical and mechanical properties and increased shrinkability. Permeability to gases and water vapor may decrease by 0%–50%, depending on the polymer and the conditions of the orientation process. The disadvantages are more difficult heat sealing and easier tear propagation. If the shrinkage characteristic of oriented films is undesirable, this can be rectified by annealing, and the so-processed oriented films are called heat set.

Films made from PET, PA, PVdC, PP, and LDPE are most usually oriented, with OPP and BOPP being the ones produced in larger quantities. Both cast and tubular films can be biaxially oriented in the following way. In cast-film extrusion, if the chilled casting rolls are rotating at different and increasing speeds, the film is stretched and oriented in the machine direction. Then the film is fed into a tenter frame and is stretched and oriented in the transverse direction. In tubular film extrusion, the air inflates the plastic tube and stretches and orients it in the transverse direction, while the rolls at the end of the bubble, by rotating faster than the rolls at the beginning of the bubble, stretch and orient the film in the machine direction.

17.6.3 Multilayer Combinations of Flexible Packaging Materials

The various flexible packaging materials available have different characteristics concerning their permeability to gases and water vapor, mechanical properties, optical properties, resistance to high or low temperatures, etc. When an individual packaging material cannot meet all functional packaging needs of a food, a multilayer combination of materials is chosen because it combines the advantages of the individual materials without the disadvantages and at a low cost. For example, a two-layer film used for making vacuum packaging bags consists of PA/LDPE. PA has low permeability to gases but high to water vapor. On the other hand, LDPE has low permeability to water vapor and high to gases. Their combination has low permeability to both.

The flexible materials used in food packaging are thermoplastic polymer films, paper, aluminum foil, and metallized plastic films. Their combination is accomplished by three methods: *coextrusion*, *coating*, and *laminating*. Coextrusion, as already mentioned, is limited to combining only thermoplastic polymers. Coating is applying a thin layer of a liquid or a melt to the surface of a material. Laminating is the bonding of two or more webs. Laminate is a combination of layers of flexible packaging materials with each major web being generally thicker than 6 μm. Sometimes the term laminate is used to imply any multilayer combination regardless of the method of manufacture.

Multilayer combinations are used for packaging a wide variety of foods ranging from dried foods (dried soups, nuts, coffee) to high-moisture foods (meat, cheese) to liquids (milk, fruit juices). The selection of the most appropriate combination of materials depends on the properties of the food, the properties of the individual materials, the type of processing, packaging and storage of the food, the desired shelf life of the food, etc. In tables presented by Fellows (1990) examples of laminated and coextruded structures are given along with their food applications. A useful table presented by Goddard (1980) gives the materials that provide the various desired properties in a laminate.

Substrates like plastics, paper, and metal foil can be solution-coated or extrusion-coated with a polymer to produce a two-layer flexible material with improved properties, like strength, permeability, heat-sealability, etc. *Solution coating* involves the application of a thin layer of a solution or dispersion of a polymer onto the surface of the substrate followed by evaporation of the solvent. In *extrusion coating*, the substrate, in the form of a flat film, passes underneath the die of an extruder from where molten polymer exits and is deposited directly on the surface of the substrate. The coating is cooled and solidified as the combination of the two materials passes through two chill rolls. With this technique the most commonly used coatings are LDPE, PP, PA, and PET.

Lamination, that is, the bonding together of two or more webs, can be accomplished with adhesives (adhesive lamination) and by extrusion (extrusion lamination). *Adhesive lamination* can be *dry-bond lamination* or *wet-bond lamination*. In the former, a solution of the adhesive is coated onto one surface of one of the two substrates, then the coating is dried with hot air for the solvent to evaporate, and finally the two webs are pressed together with a heated compression nip. In the latter, the bonding of the two webs is done without drying the adhesive and in this case one of the two substrates must be permeable to the solvent of the adhesive. A development of adhesive lamination is *solventless lamination* where the adhesive layer is formed by curing (polymerization) without solvents. In *extrusion lamination* a molten polymer, usually LDPE, is continuously extruded between two other webs.

Metallization is a specialized type of coating (Robertson 2006). A very thin layer, usually between 8 and 50 nm (Lee et al. 2008), of metal (most commonly aluminum) is deposited on the surface of a plastic film or paper. The process is carried out in a vacuum chamber at pressure less than 3 kPa and is called *vacuum metallization*. The aluminum is heated at 1500°C–1800°C; it vaporizes and then condenses on the surface of the plastic film that passes past a chilled drum. Metallized films are excellent barriers to gases, water vapor, and light. Generally, a metallization of plastic films with an aluminum layer 50 nm thick results in a reduction of the permeability to gases and water vapor by 95%–99%. Metallized films have the appearance of thin metal foil, but they are more flexible, stronger, and highly reflective. Plastic films that are usually metallized include PP, unplasticized PVC, PVdC, cellophane, PS, PET, and PAs. In food packaging, metallized films are used as components of laminates and very rarely as single films. For example, a laminate consisting of 20 μm OPP/20 μm metallized OPP is widely used for packaging various kinds of snack foods. Metallized PET film laminated to paperboard is a very common susceptor used in microwave ovens for browning and crisping.

Another specialized type of coating of films and bottles is with SiO_x (x = 1.5–1.8), 150–300 nm thick. Thin glass-like coatings of SiO_x on PET, PP, and PA are produced either by physical vapor deposition (PVD) of SiO_x or by plasma-enhanced chemical vapor deposition (PECVD) of gaseous organosilane and oxygen (Lange and Wyser 2003). Films coated with SiO_x are transparent, retortable, microwavable, and have barrier properties comparable to metallized films. Their disadvantages are high production cost, limited flex resistance, and the formation of cracks on their surface.

17.7 PAPER AND PAPERBOARD PACKAGING

17.7.1 Introduction

Paper and paperboard are the most widely used packaging materials, accounting for about one-third of the total packaging market. They are used in all levels of packaging (primary, secondary, etc.). About 5% of the total paper and paperboard consumption is used for food packaging (Kirwan 2003).

Paper and paperboard are materials in the form of sheet and are made from an interlaced network of cellulose fibers. There is no clear demarcation line between the two, but generally product of thickness less than 300 μm is considered paper. ISO standards define paperboard as paper having a grammage (mass per unit area) greater than 224 g m^{-2} (Robertson 2006).

The main advantages of paper are its relatively low cost and light weight. It is a very versatile material; a variety of types are available with different properties and applications. It is easily machined and folded, easily bonded with other materials, and has good printability properties. It can be a partial or complete barrier to light. Wood, the raw material for paper production, is a renewable resource. Finally, it is recyclable and biodegradable. The main disadvantages of paper are poor barrier properties to moisture, gases, aromas and oils, limited formability, and almost complete loss of mechanical strength if it gets wet or absorbs moisture.

17.7.2 Paper and Paperboard Production

17.7.2.1 Pulping Technology

Paper pulp, the raw material for making paper and paperboard, is produced from wood. The purpose of pulping is to separate the cellulose fibers of wood without damaging them. This is accomplished either by a mechanical process (grinding of wood chips), in which case the product is groundwood pulp, or a chemical process, and the product is chemical pulp. Paper made from groundwood pulp has lower mechanical strength and brightness and is used mainly as newsprint and magazine paper. The chemical methods are basically two: the alkaline sulfate (kraft) process and the acid sulfite process. Their fundamental aim is to dissolve and remove the lignin of the middle lamella that cements together the cellulose fibers of wood. In this way the cellulose fibers are separated as pure as possible. In both methods initially the barks are removed from the logs, and the logs are passed through a chipper. The wood chips are charged into the digester with the cooking chemicals. "Digestion" of the wood is performed under pressure and at relatively high temperatures. Both methods remove the lignin and much of the hemicelluloses of the wood. The chemicals used in the sulfate process are NaOH and Na_2S, while in the sulfite process are sulfite and bisulfite salts and excess of SO_2. Combinations of chemical and mechanical pulping methods, called semichemical, are also in use. Their product is of intermediate quality, but they have high yields.

Next the pulp is bleached to improve its color or left unbleached for certain applications. For chemical pulps, bleaching involves oxidation of colored compounds and removal of the remaining lignin. In the past chlorine gas was used for bleaching chemical pulps. Byproducts of the reaction of chlorine with constituents of the pulp are the well-known toxic substances dioxins. There were concerns expressed about the trace amounts of dioxins remaining in the bleached paperboard migrating into milk packaged in aseptic cartons. Despite the extremely low risk from dioxins in this case, the manufacturers of bleached paperboard replaced chlorine gas with chlorine dioxide, oxygen, ozone, and peroxide.

17.7.2.2 Papermaking

The first step of the papermaking process is the beating/refining of the pulp, which has been diluted to a 5%–7% solids content. The fibers are beaten while suspended in water. The objective is to increase the surface area of the fibers by helping them to absorb water. By this process the fibers are flattened out, and the space between fibers is reduced. All these lead to increased interfiber bonding and eventually to stronger paper. While at the beater, various nonfibrous chemicals are added to the pulp-water slurry to improve the properties of the paper and reduce cost of manufacturing. These additives may be fillers, sizing agents, pigments, etc. Fillers like silicates, calcium carbonate, talc, etc., can improve brightness, opacity, smoothness, and ink receptivity of the paper. Sizing agents like rosin, chemically modified starches, carboxymethyl cellulose (CMC), polyurethanes, etc., fill the gaps between the fibers, so they increase paper strength, increase resistance to penetration by water and oils, and reduce ink blurring.

Next, the pulp is fed to the papermaking machine, the two main types of which are the Fourdrinier machine and the cylinder machine. In the Fourdrinier machine a very dilute suspension of fibers (solids content less than 1%) is deposited on a moving, vibrating wire-mesh belt. Over 95% of the water is drained through the screen, and the fibers interlace in a random way as they are deposited on the belt. At the last part of the Fourdrinier belt, vacuum is applied in suction boxes to remove more water. In the cylinder machine, a wire–mesh cylinder is rotated partially submerged in a fiber suspension. Vacuum is applied inside the cylinder, the water is sucked in, and the paper web is formed on the mesh. Fourdrinier machines are mainly used for paper and thinner board, while cylinder machines are for heavy multilayer boards.

The paper sheet leaves the papermaking machine at moisture content 75%–90%, depending on the type. First it is carried through press rolls that reduce its moisture to 60%–70%. Next the paper sheet is dried to final moisture 4%–10%, by passing through a series of drum driers, and is pressed between two rollers (calendering) to get a smoother surface. Other treatments of the surface follow, which depend on the required properties of the surface for the particular application of the paper. The same kinds of compounds that were added to the pulp-water slurry at the beater can be applied to the surface. For example, surface-sizing agents prevent water penetration and increase the strength of the paper.

17.7.3 Types of Paper and Board

The most common types of packaging paper are

Kraft paper, which is made from sulfate, usually unbleached pulp, has good mechanical strength, and is used for making bags and sacs. It is very rarely used in direct contact with food.

Greaseproof paper, which is a translucent paper, resistant to fat and oil penetration and for this reason is used for packaging butter and other fatty foods. Oils and fats do penetrate greaseproof paper after some time, and this is the reason greaseproof paper is also called imitation parchment because only vegetable parchment is practically impermeable to fats. The greaseproof property of this paper is obtained through very prolonged pulp beating, which results in fibrillation, extensive fiber breakage, fiber hydration, and filling of many interstitial voids.

Glassine paper, which is a glossy, almost transparent paper, more resistant to fats and oils than greaseproof paper. It is produced by treating greaseproof paper in a supercalender, where after being dampened with water, it is rolled through a battery of steam-heated rollers. Due to the compression and thermal treatment, more hydrogen bonds are formed between the cellulose fibers. Glassine paper is used for packaging bakery and confectionery products.

Vegetable parchment, which is a paper, the name of which is derived from its resemblance to animal parchment obtained from animal skin. It is made from unsized bleached chemical pulp, and at the end of its production sequence, the paper rapidly passes through a concentrated sulfuric acid bath at low temperature. This treatment makes the cellulose fibers to swell and to partially dissolve; thus, the voids between the fibers are filled and more hydrogen bonds are formed. Next the paper is neutralized, washed, and dried. The resulting paper has excellent wet strength and resistance to grease and oils.

Wet-strength papers, which are typical vegetable parchment substitutes. They have high wet strength, resistance to oils, and are produced by the addition of synthetic resins like polyamides to the pulp.

Coated papers. Papers of various types can be coated with various substances. Wax-coated papers are heat-sealable and provide a moderate barrier to water and water vapor. Polymer-coated papers are also available with various properties depending on the polymer used.

Paperboard, which is paper with thickness greater than 300 μm. There are three main types available (Brennan et al. 1990):

Chipboard, which is made from repulped waste. It is gray in color and has low mechanical strength. In food packaging it is usually used as secondary packaging, that is, with the food already contained in a bag like the breakfast cereals.

Duplex board, which is produced from a mixture of semibleached chemical pulp and mechanical pulp and it is lined on both sides with chemical pulp.

White board or *Food board*, which is made from virgin bleached Kraft pulp.

There are also available *paperboards coated* with wax, LDPE, PVdC, PAs. Paperboard coated with LDPE is called liquid packaging board (Krochta 2007), and it used for packaging pasteurized milk and juices stored in the refrigerator. Paperboard coated with PP is used for foods that are microwaved in their package. Paperboard coated with PET is used for the dual-ovenable trays, and paperboard is one of the layers of the laminate used for the aseptic cartons.

17.7.4 Paper and Paperboard Packages

The most common food packages made of paper and paperboard are

- Paper bags and wrappings
- Folding cartons and setup boxes
- Corrugated board and solid fiberboard boxes
- Composite cans
- Fiber drums
- Molded pulp containers

17.7.4.1 Paper Bags and Wrappings

Bags and sacs may be produced from a single layer of paper (e.g., grocery bags) or multiple layers (e.g., sacs for sugar and flour). They are usually made of kraft paper and because they do not have barrier properties, they serve only the functions of containing the food and protecting it from contamination and physical damage. Bags made of glassine and greaseproof papers, vegetable parchment, plastic laminated or coated papers are also available. Greaseproof paper and vegetable parchment are used for wrapping butter and margarine and as interleavers between slices of meat, pastry, etc.

17.7.4.2 Folding Cartons and Setup Boxes

Paperboard cartons and boxes are available either as folding cartons or as setup boxes. The former are delivered in a collapsed state and are assembled at the packaging point, while the latter are delivered ready for filling. Folding cartons are made from solid paperboard, consisting of one or more layers (multi-ply board) of cellulose fibers and of thickness between 300 and 1100 μm. The board can be plain, laminated, or coated with clay, LDPE, etc. Clay coatings are used to improve external appearance and printing quality. The manufacturing of folding cartons involves printing of the web or sheets of the board, cutting and creasing the board to make blanks. Then the blanks are folded to shape, the longitudinal seal is made by gluing or heat sealing, and the blanks are flattened. At the food facility the blanks are erected from the flat condition, closed or sealed at the bottom, filled with the product, and closed or sealed at the top. For setup boxes almost all the above steps are carried out at the box manufacturing facility. Setup boxes are usually made from thicker board, and therefore they are stronger than folding cartons. They have the disadvantages though of being more expensive and occupying more space. They are usually used for packaging quality candies, chocolates, and bakery products.

17.7.4.3 Corrugated Board and Solid Fiberboard Boxes

Boxes made from corrugated board or solid fiberboard are widely used as secondary, tertiary, and quaternary packaging of products to facilitate transportation and storage. They are usually the external packaging of foods already packaged, for example, in cans, jars, bottles, etc., and very seldom in direct contact with foods.

Corrugated board is a material characterized by a high compressive strength at a relatively low weight. It is composed of one or more layers of fluted or corrugated paper, called medium, and on both surfaces of it flat paperboard sheets, called liner or linerboard, have been glued to the flute tips (Robertson 1993). The flutes offer resistance to static loads in a direction parallel to them and also to dynamic stresses in a direction perpendicular to them (flat crushing). The liner is made from unbleached or bleached kraft paper of grammage 127–440 g m^{-2}, while the medium is made from mostly recycled semichemical unbleached paper of grammage 112, 127, and 150 g m^{-2} (Lee et al. 2008). The liner and the medium can be combined in many ways to form the following three main types of corrugated board: Double-faced corrugated board consists of one flute and two walls and is used for manufacturing standard boxes. Double-wall corrugated board consists of two flutes and three walls and is used for heavy duty boxes. Finally, triple-wall corrugated board consists of three flutes and four walls and is used for heavy duty boxes. In addition to the above categories of corrugated board, another parameter that may vary is the height of the flutes and their number per unit length of board. There are four basic types of flutes, designated as A, B, C, and E. For most boxes the C flute is used, characterized by 120–145 flutes m^{-1} and a flute height of 4.0 mm (Lee et al. 2008). The liner and/or the medium of corrugated board may be wax coated to improve the resistance to moisture and oils, reduce abrasion, and protect the printing.

Solid fiberboard usually consists of two to five plies of paperboard bonded together with adhesive and lined on one or both faces with kraft paper. The total thickness of the board is between 0.8 and 2.8 mm and the grammage 556–1758 g m^{-2}. Solid fiberboard containers cost two to three times more than the equivalent corrugated board containers, and for this reason they are preferred only in cases where they can be returned and reused. It is estimated that they can be reused 10–15 times (Robertson 1993).

17.7.4.4 Composite Cans and Fiber Drums

Composite cans and fiber drums are cylindrical containers consisting of layers of paper or thin paperboard or laminated layers bonded together with adhesive and used for packaging a wide variety of foods ranging from dry solids to liquids. Their cylindrical body is constructed either by spirally winding webs of paper around a stationary mandrel or by convolute or straight winding individual rectangular sheets around a rotating mandrel. The small size containers are called composite cans, and their bodies are made of paperboard or paperboard laminated with aluminum foil and a plastic film. The second option is used when a material with good barrier properties to liquids and gases is required. The top and bottom ends of the cans are made of paperboard, plastic, or metal, in which case the ends are double seamed to the body. Foods most commonly packaged in composite cans include snacks, nuts, ground coffee, refrigerated dough, and pastries (Lee et al. 2008). Fiber drums of capacity from 20 to 280 L are useful for storing and transporting dry and semiliquid foods. Usually the inside of the drum is laminated with PE, PET, or aluminum foil.

17.7.4.5 Molded Pulp Containers

Molded pulp containers are typically made from recycled paper pulp with or without the addition of virgin mechanical and chemical pulp. A pulp and water mixture is placed into a wire–mesh mold and either air under pressure and at 480°C compresses the pulp and removes water or vacuum is applied to the other side of the mold to remove water from the slurry. In both cases the formed container is dried further. The most well-known molded pulp containers are the one dozen egg cartons, bottle sleeves, and trays for fruits and vegetables (Robertson 1993, Lee et al. 2008).

More details on paper and paperboard packaging can be found in the following references: Robertson (1993), Soroka (1996), Kirwan (2003), Ottenio et al. (2004), Robertson (2006), Krochta (2007), Lee et al. (2008), and Kirwan (2008).

17.8 PERMEABILITY OF THERMOPLASTIC POLYMERS TO GASES AND VAPORS

All thermoplastic polymers permit the transport of low molecular weight substances through their mass. For foods in plastic packages this transport can greatly influence their quality and consequently their shelf life. The effective protection of packaged foods from loss to or gain from the external environment of moisture, gases, and vapors depends on package integrity and the permeability of the plastic packaging material itself to these specific substances.

There are two major mechanisms by which gases and vapors are transported through food packaging polymers:

1. Through pinholes or channel leaks, especially in the seal area of packages. The mechanism involves diffusion of gas molecules through a column of stagnant air inside the pinhole.
2. Through permeation, that is, a molecular mechanism also known as activated diffusion or solution diffusion. The mechanism involves adsorption of the gas molecules on one surface of the polymer, diffusion through the polymer's macromolecules, and desorption at the other surface of the polymer.

It should be pointed out that transport through leaks of defective packages is often more crucial than permeation in determining the total transmission rate and should not be ignored (Chung et al. 2003). The focus of this section is permeation and this mechanism will be presented exclusively next. For more details on permeation the following references may also be consulted: Karel (1975), Brown (1992), Robertson (1993), Jasse et al. (1994), Hernandez (1997), Robertson (2006), and Lee et al. (2008).

17.8.1 Theoretical Analysis of Permeation

In this section, the permeation of a gas or vapor (from now on called permeant) through a flat polymer sheet is examined. This transport is driven by a concentration gradient and naturally is in the direction from high concentration to low concentration. According to this mechanism, in the high concentration side of the sheet's surface, molecules of the permeant are adsorbed, then they diffuse through the macromolecules of the polymer because of concentration difference, and finally they are desorbed from the low concentration side of the sheet's surface. The behavior of systems in which liquid solvents or vapors of solvents at high concentrations permeate through polymeric materials is much more complex and differs considerably from that of gases and vapors at low concentrations.

Let us consider a flat polymer sheet or film, schematically shown in Figure 17.13, which has a thickness L and separates environments (1) and (2) where the partial pressures of the gas or vapor

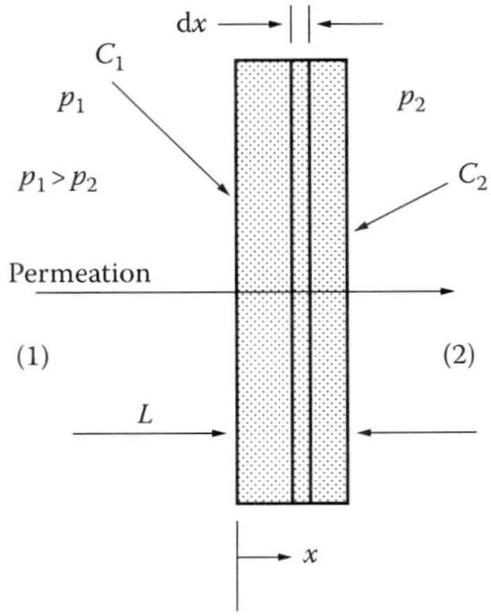

FIGURE 17.13 Permeation through a flat polymer sheet.

(permeant) under study are different. We will assume that the length and the width of the sheet are much bigger than its thickness; thus the permeation will take place in one direction, across the thickness of the sheet. Let us define A as the surface area of the sheet, perpendicular to which permeation occurs, p_1 (atm) the partial pressure of the permeant in the gas mixture of environment (1) to the left of the sheet and p_2 (atm) the equivalent partial pressure in environment (2) to the right of the sheet. If we take $p_1 > p_2$ then the transfer of the permeant is from environment (1)–(2). Let us also define C_1 and C_2 the permeant concentrations (mol cm^{-3}) at the two surfaces of the sheet in the solid polymer phase. We will also assume that the temperature is constant and the same everywhere.

For one-directional diffusion of a permeant through the polymer film, Fick's first law is applied:

$$J = -D\frac{dC}{dx} \tag{17.1}$$

where

- J is the diffusion flux (mol cm^{-2} s^{-1}) of permeant, that is, amount of permeant diffusing per unit area per unit time through a plane perpendicular to the permeation direction
- D is the diffusivity (cm^2 s^{-1}) of permeant in the polymer
- C is the concentration (mol cm^{-3}) of the permeant in the polymer
- x is the distance (cm) inside the polymer sheet in the direction of permeation

Mass balance for the permeant in combination with Equation 17.1 gives

$$\frac{\partial C}{\partial t} = -\frac{\partial}{\partial x}(J) = \frac{\partial}{\partial x}\left(D\frac{\partial C}{\partial x}\right) \tag{17.2}$$

If the diffusivity D of the particular permeant in the particular polymer is independent of concentration C Equation 17.2 gives

$$\frac{\partial C}{\partial t} = D\frac{\partial^2 C}{\partial x^2} \tag{17.3}$$

Equation 17.3 is the simplest form of Fick's second law, and solutions for various boundaries and initial conditions exist in the literature (e.g., Carslaw and Jaeger 1959). From the solution of Equation 17.3 the concentration C of the permeant in the polymer is obtained as a function of the distance x and the time t. Once the function $C(x, t)$ is known, the diffusion flux of the permeant to the left side environment of the plastic film can be calculated from

$$J\big|_{x=L} = -D\left(\frac{\partial C}{\partial x}\right)_{x=L} \tag{17.4}$$

Let us define Q (mol) as the amount of permeant transferred during time t and perpendicular to surface A to environment (2) at the right of the sheet. What we are really interested in is actually the *permeation rate* dQ/dt which is

$$\frac{dQ}{dt} = A \cdot J\big|_{x=L} \tag{17.5}$$

After a relatively short initial transient period, permeation will take place under steady-state conditions, that is, the concentration of the permeant inside the sheet will not change anymore with time. For permeation at steady state, setting $\partial C/\partial t = 0$ in Equation 17.3 one obtains:

$$\frac{d^2C}{dx^2} = 0 \tag{17.6}$$

Integration of Equation 17.6 with boundary conditions:
at $x = 0$, $C = C_1$ and
at $x = L$, $C = C_2$ gives

$$C = -[C_1 - C_2]\frac{x}{L} + C_1 \tag{17.7}$$

So the concentration C of the permeant inside the film decreases linearly with distance x.
From Equations 17.4 and 17.7:

$$J\Big|_{x=L} = D\frac{[C_1 - C_2]}{L} = \text{constant} \tag{17.8}$$

and from Equations 17.5 and 17.8 is obtained:

$$\frac{dQ}{dt} = D \cdot A \cdot \frac{[C_1 - C_2]}{L} \tag{17.9}$$

It is difficult to measure the concentrations C_1 and C_2 of the permeant at the two surfaces of the film. Fortunately they can be expressed in terms of the partial pressures p_1 and p_2 of the permeant in the gas mixtures of environments (1) and (2) to the left and the right of the plastic film, which in turn are easier to measure using gas chromatography or a gas analyzer. The simplest sorption isotherm, valid for small concentrations of permeant in the polymer, is Henry's Law which gives

$$C_1 = S \cdot p_1 \tag{17.10}$$

$$C_2 = S \cdot p_2 \tag{17.11}$$

where S (mol cm^{-3} atm^{-1}) is the solubility coefficient of the permeant in the polymer.
Substituting Equations 17.10 and 17.11 into Equation 17.9 we obtain

$$\frac{dQ}{dt} = D \cdot S \cdot A \cdot \frac{[p_1 - p_2]}{L} \tag{17.12}$$

The product of the diffusion coefficient D and the solubility coefficient S is defined as the *permeability coefficient P* (or permeability constant or simply permeability):

$$P = DS \tag{17.13}$$

Thus, the *permeation rate equation at steady state* (Equation 17.12) becomes

$$\frac{dQ}{dt} = \frac{P}{L} \cdot A \cdot [p_1 - p_2] \tag{17.14}$$

Equation 17.14 will hold provided that the assumptions that were made during its derivation are valid:

1. Permeation occurs in one direction and under steady-state conditions.
2. The diffusion coefficient D and the solubility coefficient S are constant and independent of the concentration C of the permeant. This is the case for the permanent gases (O_2, H_2, N_2, and CO_2) in polymers and at atmospheric pressure. However, it does not hold when strong interactions between the permeant and the polymer exist, for example, for water or organic vapors and polar polymers such as cellophane and nylon.

For the permeation of water vapor, Equation 17.14 can be transformed to a more useful form. The partial pressures p_1 and p_2 can be expressed in terms of the relative humidities $(\mathrm{RH})_1$ and $(\mathrm{RH})_2$ of the air in environments (1) and (2) to the left and the right of the plastic film. $(\mathrm{RH})_1$ and $(\mathrm{RH})_2$ can be measured or set at the desired values more easily than p_1 and p_2. The relative humidity (RH) of air is defined as:

$$\mathrm{RH} = \frac{p_\mathrm{w}}{p_\mathrm{w}^0} \cdot 100 \tag{17.15}$$

where

p_w is the partial pressure of water vapor in the air

p_w^0 the saturated water vapor pressure at the temperature of the air

Solving Equation 17.15 for p_w and applying it for p_1 and p_2, provided that the temperature is the same everywhere, one finally gets

$$p_1 - p_2 = p_\mathrm{w}^0 \cdot \left[\frac{(\mathrm{RH})_1}{100} - \frac{(\mathrm{RH})_2}{100}\right] \tag{17.16}$$

Thus, the permeation rate equation at steady state for water vapor (Equation 17.14) becomes

$$\frac{\mathrm{d}Q}{\mathrm{d}t} = \frac{P}{L} \cdot A \cdot p_\mathrm{w}^0 \cdot \frac{[(\mathrm{RH})_1 - (\mathrm{RH})_2]}{100} \tag{17.17}$$

In addition to the already defined terms permeation rate ($\mathrm{d}Q/\mathrm{d}t$) and permeability coefficient (P), in the relevant literature, the terms transmission rate and permeance are often used. The *transmission rate* (TR) of a permeant is defined as the permeation rate normalized for area:

$$\mathrm{TR} = \frac{1}{A} \cdot \frac{\mathrm{d}Q}{\mathrm{d}t} \tag{17.18}$$

When referring to the transmission rate of oxygen, the symbol oxygen transmission rate (OTR) is often used and when referring to water vapor the symbol water vapor transmission rate (WVTR). The common unit for OTR is [cm^3(STP) m^{-2} $(\mathrm{day})^{-1}$] and for WVTR [g m^{-2} $(\mathrm{day})^{-1}$] or [g (100 $\mathrm{in.}^2)^{-1}$ $(\mathrm{day})^{-1}$]. When reporting a TR value for a film the following must also be stated: the permeant, the polymer, the thickness L of the polymer film, the temperature, the partial pressures of the permeant gas, and the relative humidities on both sides of the film.

The *permeance* is defined as the transmission rate TR normalized for partial pressure difference:

$$\text{Permeance} = \frac{\mathrm{TR}}{(p_1 - p_2)} \tag{17.19}$$

For homogeneous films, for which the permeability coefficient P is characteristic of the bulk material and independent of L, permeance is also:

$$\text{Permeance} = \frac{P}{L} \tag{17.20}$$

It is also useful to examine the permeation during the unsteady-state period. We consider again the flat polymer sheet or film of Figure 17.13, which has a thickness L. Initially the concentration of the permeant in the polymer is uniform everywhere in the sheet and equal to C_2. At time $t = 0$ the concentration at the surface of the sheet at $x = 0$ is suddenly changed to C_1 and remains constant at this value thereafter, while the concentration at the surface of the sheet at $x = L$ remains constant at C_2. The relationship between C_1, C_2 and p_1, p_2 is again described by Equations 17.10 and 17.11. The equation describing diffusion of the permeant during the unsteady-state period is Fick's second law (Equation 17.3). Its solution (Carslaw and Jaeger 1959), for the initial and boundary conditions stated above, is

$$\frac{C(x,t)-C_2}{C_1-C_2}=1-\frac{x}{L}-\frac{2}{\pi}\sum_{n=1}^{\infty}\frac{1}{n}\sin\left(n\pi\frac{x}{L}\right)\exp\left(-\frac{n^2\pi^2 D}{L^2}t\right) \tag{17.21}$$

where $C(x, t)$ is the concentration of the permeant inside the polymer film as a function of distance x and time t. The concentration profile is a curve and progressively is transformed to a straight line. When time t attains high values, the exponential terms in the series tend to zero, the term with the series becomes zero, and Equation 17.21 is transformed to Equation 17.7, which describes the linear dependence of concentration on distance x during permeation at steady-state conditions.

We have already defined Q as the amount of permeant that has exited, from time zero to time t, the surface of the plastic film at $x = L$ and has been transferred to the environment (2) at the right of the film. From Equations 17.4 and 17.5 Q is

$$Q=-A\cdot D\cdot\int_0^t\left(\frac{\partial C}{\partial x}\right)_{x=L}\mathrm{d}t \tag{17.22}$$

If Q^* is the amount of permeant that has entered the plastic film from the surface at $x = 0$ from time zero to time t, then

$$Q^*=-A\cdot D\cdot\int_0^t\left(\frac{\partial C}{\partial x}\right)_{x=0}\mathrm{d}t \tag{17.23}$$

From Equations 17.22 and 17.23 and using Equation 17.21, after appropriate mathematical manipulation, the following expressions for the amounts of permeant Q and Q^* are obtained:

$$Q=\frac{P\cdot A\cdot(p_1-p_2)}{L}t-\frac{S\cdot A\cdot L\cdot(p_1-p_2)}{6}-\frac{2S\cdot A\cdot L\cdot(p_1-p_2)}{\pi^2}\sum_{n=1}^{\infty}\frac{1}{n^2}\cos(n\pi)\exp\left(-\frac{n^2\pi^2 D}{L^2}t\right) \tag{17.24}$$

$$Q^*=\frac{P\cdot A\cdot(p_1-p_2)}{L}t+\frac{S\cdot A\cdot L\cdot(p_1-p_2)}{3}-\frac{2S\cdot A\cdot L\cdot(p_1-p_2)}{\pi^2}\sum_{n=1}^{\infty}\frac{1}{n^2}\exp\left(-\frac{n^2\pi^2 D}{L^2}t\right) \tag{17.25}$$

The graphs of the above two equations are presented in Figure 17.14. For short times (permeation during the unsteady-state period) the plot of Q versus time is a concave curve upward, whereas

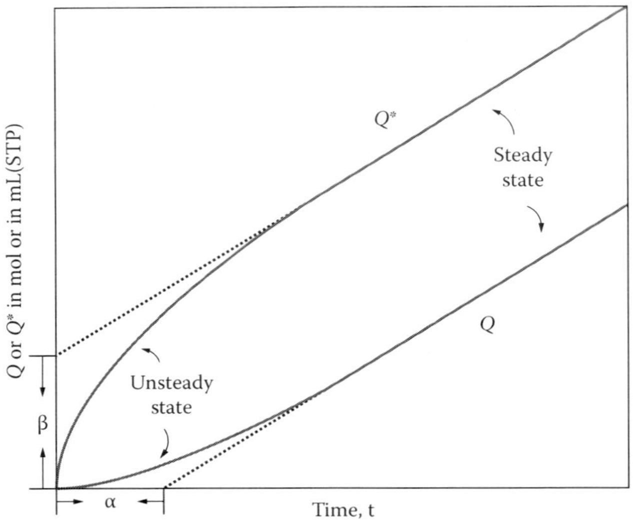

FIGURE 17.14 Q^* and Q versus time, where Q^* is the amount of permeant that has entered the plastic film from the surface at $x = 0$ from time zero to time t and Q the amount that has exited the plastic film from the surface at $x = L$ from time zero to time t.

that of Q^* is concave downward. For longer times (permeation during the steady-state period) the plots of both Q and Q^* become linear. At long times, the exponential terms in the series of Equations 17.24 and 17.25 tend to zero, therefore the series go to zero, and the equations are transformed to Equations 17.26 and 17.27. They describe only the linear parts of the two curves, which are parallel having the same slope.

$$Q = \frac{P \cdot A \cdot (p_1 - p_2)}{L} t - \frac{S \cdot A \cdot L \cdot (p_1 - p_2)}{6} \tag{17.26}$$

$$Q^* = \frac{P \cdot A \cdot (p_1 - p_2)}{L} t + \frac{S \cdot A \cdot L \cdot (p_1 - p_2)}{3} \tag{17.27}$$

In Figure 17.14 extrapolating backward the linear part of the plot of Q versus t until it intersects the horizontal axis t, the value of the lag time α can be estimated. Setting in Equation 17.26 $t = \alpha$ and $Q = 0$ we finally get

$$D = \frac{L^2}{6\alpha} \tag{17.28}$$

In the same figure extrapolating backward the linear part of the plot of Q^* versus t until it intersects the vertical axis Q^*, the value of the intercept β can be estimated. Setting in Equation 17.27 $t = 0$ and $Q^* = \beta$ we finally get

$$S = \frac{3\beta}{A \cdot L \cdot (p_1 - p_2)} \tag{17.29}$$

The practical significance of the above analysis is the following. If an experiment is carried out in which Q or Q^* is measured as a function of time and then the graph of Q or Q^* versus time is constructed from this single experiment, all parameters of the permeation, that is, P, D, and S, can be calculated in the following way:

- From the linear part of the plot of Q or Q^* versus t, the value of the slope can be calculated, and then from the slope and Equations 17.26 or 17.27 the permeability coefficient P can be estimated.
- If in the experiment Q had been measured as a function of time, from the value of the lag time α and Equation 17.28 the diffusion coefficient D can be calculated and subsequently the solubility coefficient S ($S = P/D$).
- If in the experiment Q^* had been measured as a function of time, from the value of the intercept β and Equation 17.29 the solubility coefficient S can be calculated and subsequently the diffusion coefficient D ($D = P/S$).

17.8.2 Units of Permeability Coefficient

An alternative expression for the permeability coefficient, P, is derived by solving Equation 17.14 for P

$$P = \frac{1}{A} \cdot \frac{dQ}{dt} \cdot \frac{L}{[p_1 - p_2]} \tag{17.30}$$

Therefore, the permeability coefficient P represents the amount of permeant permeating through the polymer film per unit time and per unit surface area of the film when the thickness of the film and the partial pressure difference of the permeant at the two sides of the film are unity. From Equation 17.30 the dimensions of P are

$$\begin{aligned} P = &(\text{quantity of permeant}) \times (\text{film thickness}) \times (\text{film surface area})^{-1} \\ &\times (\text{time})^{-1} \times (\text{partial pressure difference})^{-1} \end{aligned} \tag{17.31}$$

There is not a generally accepted unit for P and more than 30 different units have appeared in the literature. This is due to the fact that every quantity in Equation 17.31 can be expressed in the appropriate unit and the combination is a unit for P. Examining the quantities in Equation 17.31 one by one, we may observe the following:

1. The quantity of permeant Q is usually expressed in g, mol, or cm^3(STP). The unit cm^3(STP) expresses the volume of a gas at standard temperature and pressure (273 K and 1 atm) and is essentially a unit of mass and not of volume. 1 cm^3(STP) = 1/22,400 mol.
2. The film thickness is usually expressed in μm, mm, cm, or in mil.
3. The film surface area in cm^2, m^2, or in 100 $in.^2$
4. The time in s or days or (24 h).
5. The difference of partial pressures in mm Hg, cm Hg, atm, or in kPa.

In Table 17.1 the nine most commonly used units for the permeability coefficient are presented along with the conversion factors to convert all other units to the unit [cm^3(STP) cm cm^{-2} s^{-1} (cm Hg)$^{-1}$]. The unit barrer has been adopted by the American Society for Testing and Materials (ASTM) as a unit for the permeability coefficient, without gaining, however, very wide acceptance. 1 barrer = 1×10^{-10} [cm^3(STP) cm cm^{-2} s^{-1} (cm Hg)$^{-1}$]. For water vapor permeability a common unit is [g cm m^{-2} (day)$^{-1}$(mm Hg)$^{-1}$].

TABLE 17.1
Units of the Permeability Coefficient *P*

1 [cm^3(STP) cm cm^{-2} s^{-1} atm^{-1}] = 1.32 × 10^{-2} [cm^3(STP) cm cm^{-2} s^{-1} (cm Hg)$^{-1}$]
1 [cm^3(STP) mil m^{-2} (day)$^{-1}$ atm^{-1}] = 3.87 × 10^{-14} [cm^3(STP) cm cm^{-2} s^{-1} (cm Hg)$^{-1}$]
1 [cm^3(STP) mil (100 in^2)$^{-1}$ (day)$^{-1}$ atm^{-1}] = 5.99 × 10^{-13} [cm^3(STP) cm cm^{-2} s^{-1} (cm Hg)$^{-1}$]
1 [cm^3(STP) mm m^{-2} (day)$^{-1}$ kPa^{-1}] = 1.54 × 10^{-10} [cm^3(STP) cm cm^{-2} s^{-1} (cm Hg)$^{-1}$]
1 [cm^3(STP) cm m^{-2} (day)$^{-1}$ atm^{-1}] = 1.52 × 10^{-11} [cm^3(STP) cm cm^{-2} s^{-1} (cm Hg)$^{-1}$]
1 [cm^3(STP) cm m^{-2} (day)$^{-1}$(mm Hg)$^{-1}$] = 1.16 × 10^{-8} [cm^3(STP) cm cm^{-2} s^{-1} (cm Hg)$^{-1}$]
1 [g cm m^{-2} (day)$^{-1}$(mm Hg)$^{-1}$] = 2.59 × 10^{-4} (MW)$^{-1}$ [cm^3(STP) cm cm^{-2} s^{-1} (cm Hg)$^{-1}$]
where MW is the molecular weight of the permeant

17.8.3 Permeability to "Permanent" Gases

O_2, N_2, H_2, CO_2, and some other gases are called "fixed" or "permanent" gases (Karel 1975) because they have very low boiling points and are very difficult to condense. They all have the following "ideal" behavior when permeating through thermoplastic polymers:

1. The permeability coefficients are independent of the concentration of the permeant in the polymer.
2. The same pattern of behavior is observed when examining the permeability coefficients of any particular polymer to the various "fixed" gases. For each one of the polymers its permeability coefficient to oxygen is about four times greater than its permeability coefficient to nitrogen, while its permeability coefficient to carbon dioxide is about six times greater than its permeability coefficient to oxygen.
3. The dependence of the permeability coefficients on the temperature, for a relatively narrow temperature range, can be described by an Arrhenius type equation:

$$P = P_0 \cdot \exp\left(-\frac{E_P}{RT}\right) \quad (17.32)$$

where
P_0 is the constant, independent of the absolute temperature T
E_P is the activation energy for permeance

From Equation 17.32 it can be seen that a plot of log P versus the reciprocal of absolute temperature ($1/T$) should be a straight line. For the "permanent" gases the activation energy E_P for permeance is always positive, and therefore their permeability coefficients increase when the temperature is increased. For the vapors of organic solvents, which condense easily, their permeability coefficients may increase or decrease when the temperature is increased. In Figure 17.15, Arrhenius plots (log P vs. $1/T$) are presented for the permeability coefficients of three polymers to oxygen. The plots were constructed with the following data from Table 17.2:

LDPE	at 25°C, P = 2.93 barrer and E_P = 43 kJ mol^{-1}
PVC (unplasticized)	at 25°C, P = 0.045 barrer and E_P = 56 kJ mol^{-1}
PET (amorphous)	at 25°C, P = 0.059 barrer and E_P = 38 kJ mol^{-1}

In the plot of log P vs. 1/T of some polymers and at a certain critical temperature, a sudden change of the slope of the diagram is observed, with the material becoming much more permeable at temperatures higher than this critical value. For example, for poly(vinyl acetate) this happens at 30°C, while for polystyrene at 80°C. The change is attributed to the glass transition of the polymer.

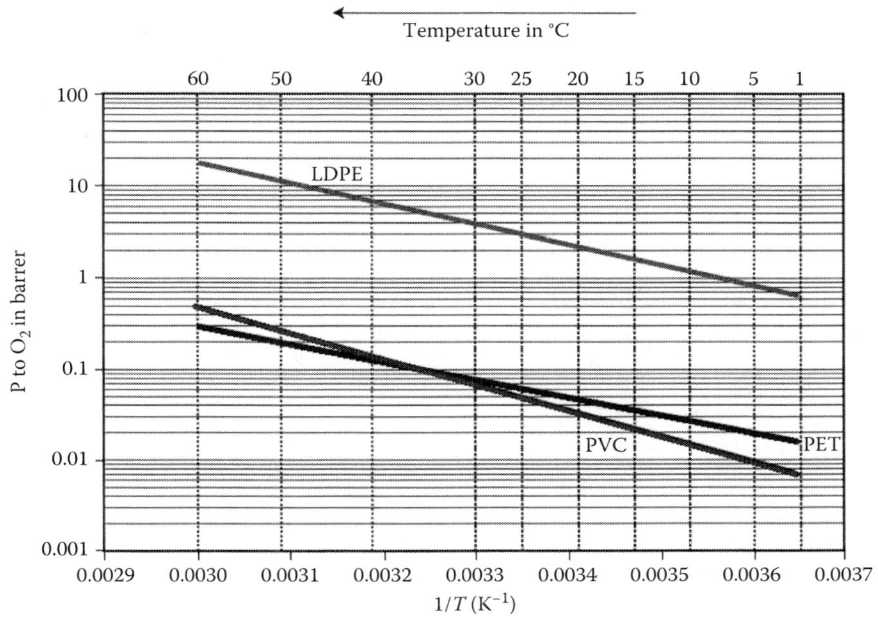

FIGURE 17.15 Permeability coefficients of LDPE, PVC, and PET to O_2 as functions of temperature (plots constructed with data from Table 17.2).

The "ideal" behavior of the "fixed" gases is not observed in the cases of permeation of water vapors in hydrophilic polymers and of permeation of vapors of volatile organic substances (e.g., solvents and fragrances) in all polymers.

In Table 17.2 (Hernandez 1997) the permeability coefficients of various polymers to O_2, CO_2, and N_2 are presented and when available the values of the solubility coefficient, the diffusion coefficient, and the activation energy for permeation are also given. Examining the values of *D* and *S* for CO_2 in relation to those for O_2, it can be easily seen that CO_2 has smaller *D* than O_2, but much bigger *S* and the final result is bigger *P*. At the same table the permeability coefficients to water vapor are shown in the same units and at the same temperature as for the "permanent" gases. For most of the polymers the permeability coefficient to water vapor is orders of magnitude bigger than the permeability coefficient to oxygen.

17.8.4 Permeability and Molecular Structure and Morphology of Polymers

As expected, the permeability coefficient of a polymer to gases and vapors also depends on the nature of the polymer. The diffusion of a dissolved gas or vapor inside the mass of a polymer is thought of as a series of activated jumps of the permeant from one roughly defined cavity to another cavity. Therefore, the following qualitative statement can be concluded: "whatever increases the number or the size of the cavities in the mass of the polymer or makes the macromolecular chains more mobile, it also increases the diffusion rate." Based on this statement and taking into account that the permeability coefficient is the product of the diffusion coefficient and the solubility coefficient, we can conclude that in order for a polymer to have low permeability to both gases and water vapor it must possess (Karel 1975, Robertson 1993) the following:

- An intermediate polarity, such as the one offered by Cl, F, CN, acrylic, and esteric functional groups. Highly polar polymers like those with many OH groups (e.g., PVOH, cellophane) have small permeability to gases under dry conditions and large to water vapor.

TABLE 17.2
Permeability, Solubility, Diffusion Coefficients, and Activation Energies for Permeation of O_2, CO_2, N_2 and Water Vapor* in Various Polymers at 25°C and 0% RH

Polymer	P in barrier	S in cm^3(STP) cm^{-3} (cm Hg)$^{-1}$	D in $cm^2\ s^{-1}$	E_P in kJ mol^{-1}
LDPE ($\rho = 0.914$ g cm^{-3})				
O_2	2.93	6.27×10^{-4}	4.60×10^{-7}	43
CO_2	12.19	3.33×10^{-3}	3.70×10^{-7}	39
N_2	0.97	3.07×10^{-4}	3.20×10^{-7}	49
H_2O	91.04			34
HDPE ($\rho = 0.964$ g cm^{-3})				
O_2	0.40	2.40×10^{-4}	1.70×10^{-7}	35
CO_2	3.55	2.93×10^{-3}	1.20×10^{-7}	30
N_2	0.15	2.00×10^{-4}	9.30×10^{-8}	40
H_2O	12.04			
Polypropylene ($\rho = 0.907$ g cm^{-3}) 50% crystallinity				
O_2	0.96			48
CO_2	3.24			38
N_2	0.12			56
H_2O (at 30°C)	67.89			42
Polystyrene, biaxially oriented				
O_2	1.70			
CO_2	5.40			
N_2	0.80			
H_2O	1126.39			
Polyacrylonitrile (Barex)				
O_2	0.005			
CO_2	0.015			
H_2O	648.060			
Poly(vinyl acetate) (at 30°C)				
O_2	0.478	8.53×10^{-4}	5.60×10^{-8}	56
EVOH (ethylene 32%)				
O_2 (0% RH)	4.63×10^{-5}			
O_2 (65% RH)	3.09×10^{-4}			
H_2O	478.33			
EVOH (ethylene 44%)				
O_2 (0% RH)	2.31×10^{-4}			
O_2 (65% RH)	4.63×10^{-4}			
H_2O	169.73			
PVC unplasticized				
O_2	0.045	3.87×10^{-4}	1.20×10^{-8}	56
CO_2	0.154	6.27×10^{-3}	2.50×10^{-9}	57
N_2	0.012	3.07×10^{-4}	3.80×10^{-9}	69
H_2O	277.74			23
PVC plasticized				
O_2	4.94			

(Continued)

TABLE 17.2 (*Continued*)
Permeability, Solubility, Diffusion Coefficients, and Activation Energies for Permeation of O_2, CO_2, N_2 and Water Vapor* in Various Polymers at 25°C and 0% RH

Polymer	*P* in barrier	S in cm^3(STP) cm^{-3} (cm Hg)$^{-1}$	*D* in $cm^2\ s^{-1}$	E_P in kJ mol^{-1}
PVdC (at 30°C)				
O_2	0.005			67
CO_2	0.029			52
N_2	0.001			70
H_2O (at 25°C)	9.260			46
PET 40% crystallinity				
O_2	0.034	9.60×10^{-4}	3.50×10^{-9}	32
CO_2	0.123	2.67×10^{-2}	6.00×10^{-10}	18
N_2	0.008	6.00×10^{-4}	1.30×10^{-9}	33
H_2O	131.160			29
PET amorphous				
O_2	0.059	1.31×10^{-3}	4.50×10^{-9}	38
CO_2	0.309	3.73×10^{-2}	8.00×10^{-10}	28
Polycarbonate (Lexan)				
O_2	1.404	6.67×10^{-3}	2.10×10^{-8}	19
CO_2	8.024	1.60×10^{-1}	4.80×10^{-9}	16
N_2	0.293			25
H_2O	1404.130			
Nylon 6				
O_2 (100% RH)	0.039			44
CO_2	0.077			41
N_2	0.005			47
Nylon 6,6 drawn				
CO_2	0.069	2.00×10^{-2}		
Cellophane				
O_2 (0% RH)	0.002			
O_2 (76% RH)	0.009			
CO_2 (0% RH)	0.005			
CO_2 (76% RH)	0.073			
N_2 (0% RH)	0.003			
N_2 (76% RH)	0.007			
H_2O	24688.0			

Source: Hernandez, R.J., Food packaging materials, barrier properties, and selection, in: *Handbook of Food Engineering Practice*, K.J. Valentas, E. Rotstein, and R.P. Singh (eds.), CRC Press, Boca Raton, FL, pp. 291–360, 1997. With permission and conversion of the units.

* $(RH)_1 = 90\%$ and $(RH)_2 = 0$.

On the contrary the nonpolar polyolefins (e.g., polyethylene) have small permeability to water vapor and large permeability to gases.

- Inertness to the permeant. Hydrophilic polymers when exposed to high humidity absorb moisture, and this has the effect of swelling or plasticizing the polymer and thus increasing the permeability.
- High degree of crystallinity. Crystalline regions are less permeable to gases and vapors than amorphous areas.

- Orientation. Orientation of amorphous polymers decreases their permeability by 10%–15%.
- Bonds or attractive forces between the macromolecular chains. Strong intermolecular interactions between the polymer chains block the passage of the permeant molecules, forcing them to diffuse via a much longer path (Lee et al. 2008), thus their diffusivity in the polymer matrix is decreased. The addition of additives, fillers, and plasticizers in the polymer matrix generally increases permeability (Lee et al. 2008).
- High glass transition temperature T_g. At temperatures below T_g less voids between the macromolecules exist and besides these voids remain fixed in shape and position. Consequently the permeant molecules have to follow a more tortuous path during their diffusion in the polymer matrix. Therefore, if a polymer has a T_g higher than ambient temperature, then it will have better barrier properties at ambient temperature.

The sorption of moisture by hydrophilic polymers decreases their T_g, causing swelling, and the water acting essentially as an internal lubricant between the macromolecules increases their mobility (Robertson 1993). The results are similar to those of other plasticizers, which by increasing the mobility of the macromolecules increase the permeability of the polymer. For example, in the polyamides when in dry condition, there are hydrogen bonds between the polymer chains that result in high T_g values. However, when even small quantities of moisture are adsorbed, these bonds break and T_g is considerably reduced. The increase of permeability to gases is particularly dramatic for polymers that have very low permeabilities when dry. On the other hand no change in the permeability to gases is observed for hydrophobic polymers (like HDPE and LDPE) when exposed to high humidities.

17.8.5 Permeability to Water Vapor

In the case of permeation of water vapor in polymers there are deviations from the "ideal" behavior of the "fixed" gases. These are more pronounced in the case of hydrophilic polymers, such as (a) the permeability coefficient cannot any longer be considered independent of the concentration of water in the polymer film and (b) the temperature dependence of the permeability coefficient often is more complex than that described by Equation 17.32. For the calculation of dQ/dt, Equations 17.14 and 17.17 are still used with the difference that the permeability coefficient depends on the partial pressures p_1, p_2 (or the relative humidities $(RH)_1$ and $(RH)_2$), and its value should be experimentally determined at the particular conditions p_1 and p_2. Thus, the values of P must be accompanied by the conditions at which P was measured, namely, temperature T (usually 38°C = 100°F or 25°C = 77°F), $(RH)_1$ (usually 95%, 90%, or 75%) and $(RH)_2$ (usually 0%). For the dependence of P on temperature the Arrhenius equation (Equation 17.32) is still used with the understanding that, for the system water vapor–polar polymers, the relationship is valid for very small temperature ranges (Karel 1975).

In the literature, very often instead of the permeability coefficient to water vapor the WVTR is specified. This is defined as:

$$\mathrm{WVTR} = \frac{1}{A} \cdot \frac{dQ}{dt} \tag{17.33}$$

and the units of WVTR are usually either [g m^{-2} (day)$^{-1}$] or [g (100 in.^{2})$^{-1}$ (day)$^{-1}$]. The value of WVTR must be accompanied by the thickness L of the polymer film (usually 25 μm), the temperature T, and the relative humidities $(RH)_1$ and $(RH)_2$ at which the WVTR was measured. From the known value of WVTR, the permeability coefficient P can be calculated via the equation:

$$P = \frac{\mathrm{WVTR} \cdot L}{p_1 - p_2} = \frac{\mathrm{WVTR} \cdot L \cdot 100}{p_w^0 \cdot [(\mathrm{RH})_1 - (\mathrm{RH})_2]} \tag{17.34}$$

TABLE 17.3
Permeability Coefficients to Water Vapor at 25°C, $(RH)_1 = 90\%$ and $(RH)_2 = 0$, and Activation Energies

Polymer	Permeability Coefficient P in [g cm m^{-2} $(day)^{-1}$ $(mm\ Hg)^{-1}$]	Activation energy E_P in kJ mol^{-1}
PVdC	6.4×10^{-5}	46
HDPE ($\rho = 0.964$ g cm^{-3})	8.4×10^{-5}	
Polypropylene ($\rho = 0.907$ g cm^{-3}) 50% crystallinity at 30°C	4.7×10^{-4}	42
LDPE ($\rho = 0.914$ g cm^{-3})	6.3×10^{-4}	34
PET 40% crystallinity	9.1×10^{-4}	29
PVC unplasticized	1.9×10^{-3}	23
EVOH (ethylene 44%)	1.2×10^{-3}	
EVOH (ethylene 32%)	3.3×10^{-3}	
Polyacrylonitrile (Barex)	4.5×10^{-3}	
Polystyrene, biaxially oriented	7.8×10^{-3}	
Polycarbonate (Lexan)	9.8×10^{-3}	
Cellophane	1.7×10^{-1}	

Source: Hernandez, R.J., Food packaging materials, barrier properties, and selection, in: *Handbook of Food Engineering Practice*, K.J. Valentas, E. Rotstein, and R.P. Singh (eds.), CRC Press, Boca Raton, FL, pp. 291–360, 1997. With permission and conversion of the units.

In Table 17.3 (Hernandez 1997) the permeability coefficients of some polymers to water vapor and the activation energies for permeation are presented.

17.8.6 Permeability of Multilayer Packaging Materials

A multilayer film consists of various films in series. Let us consider the case of three layers in series as shown in Figure 17.16. Very often the permeation rate $(dQ/dt)_{tot}$ of a permeant from environment (1) at the left side of the multilayer film to environment (4) at the right side of the multilayer film must be calculated as a function of the partial pressures p_1 and p_4 at the two sides because these two are the only known partial pressures. They may correspond to the inside and outside environment of the package. Initially, the equations for a three-layer film will be derived, and from the results the extension to any multilayer film will be obvious. We can write Equation 17.14 for $(dQ/dt)_{tot}$ as follows:

$$\left(\frac{dQ}{dt}\right)_{tot} = \frac{P_{tot}}{L_{tot}} A(p_1 - p_4) \tag{17.35}$$

where
- P_{tot} is the overall permeability coefficient of the multilayer film
- L_{tot} is the total thickness of the film

Let us define L_1, L_2, L_3 as the thicknesses of the three layers, and P_1, P_2, P_3 as the permeability coefficients of the three polymers to the permeant under study, and p_1, p_2, p_3, p_4 as the partial pressures of the permeant at the corresponding faces of the films as shown in Figure 17.16. We will also assume that permeation is taking place at steady-state conditions. The total thickness of the multilayer film is approximately equal to:

$$L_{tot} = L_1 + L_2 + L_3 \tag{17.36}$$

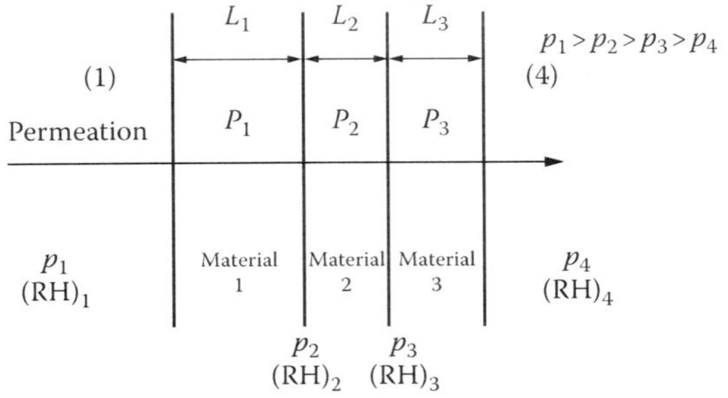

FIGURE 17.16 Permeation through a three-layer laminate.

while the surface area A, perpendicular to which permeation occurs, is the same for the three layers. Let $(dQ/dt)_{tot}$ be the permeation rate through the multilayer film and $(dQ/dt)_1$, $(dQ/dt)_2$, $(dQ/dt)_3$ the corresponding rates through films 1, 2, and 3. Since the layers are set in series one on top of the other and permeation is taking place at steady state then

$$\left(\frac{dQ}{dt}\right)_{tot} = \left(\frac{dQ}{dt}\right)_1 = \left(\frac{dQ}{dt}\right)_2 = \left(\frac{dQ}{dt}\right)_3 \tag{17.37}$$

From Equation 17.14:

$$\left(\frac{dQ}{dt}\right)_1 = \frac{P_1}{L_1} A(p_1 - p_2) \tag{17.38}$$

$$\left(\frac{dQ}{dt}\right)_2 = \frac{P_2}{L_2} A(p_2 - p_3) \tag{17.39}$$

$$\left(\frac{dQ}{dt}\right)_3 = \frac{P_3}{L_3} A(p_3 - p_4) \tag{17.40}$$

Taking into account Equation 17.37 and solving Equations 17.38, 17.39, 17.40 for the partial pressure difference we get:

$$\frac{L_1}{P_1}\frac{1}{A}\left(\frac{dQ}{dt}\right)_{tot} = p_1 - p_2 \tag{17.41}$$

$$\frac{L_2}{P_2}\frac{1}{A}\left(\frac{dQ}{dt}\right)_{tot} = p_2 - p_3 \tag{17.42}$$

$$\frac{L_3}{P_3}\frac{1}{A}\left(\frac{dQ}{dt}\right)_{tot} = p_3 - p_4 \tag{17.43}$$

Adding Equations 17.41, 17.42, 17.43 we get

$$\left[\frac{L_1}{P_1} + \frac{L_2}{P_2} + \frac{L_3}{P_3}\right]\frac{1}{A}\left(\frac{dQ}{dt}\right)_{tot} = p_1 - p_4 \tag{17.44}$$

Substituting Equation 17.35 into Equation 17.44 we finally get

$$\frac{L_{tot}}{P_{tot}} = \frac{L_1}{P_1} + \frac{L_2}{P_2} + \frac{L_3}{P_3} \tag{17.45}$$

For a multilayer film of n layers, with each layer having a permeability coefficient P_i and thickness L_i, the extension of Equation 17.45 is obviously:

$$\frac{L_{tot}}{P_{tot}} = \sum_{i=1}^{n} \frac{L_i}{P_i} \tag{17.46}$$

and the extension of Equation 17.36:

$$L_{tot} = \sum_{i=1}^{n} L_i \tag{17.47}$$

Therefore, if the thickness L_i and the permeability coefficient P_i of each of the layers of the multilayer film are known, from Equation 17.46 we can calculate L_{tot}/P_{tot}. Then taking the inverse of it and substituting into Equation 17.35, we can calculate the permeation rate $(\mathrm{d}Q/\mathrm{d}t)_{tot}$ through the multilayer film.

From the equations presented above it is possible for one to calculate the values of p_2 and p_3 inside the multilayer film. From Equation 17.41 and using Equation 17.35 we get

$$p_2 = p_1 - \frac{L_1}{P_1} \cdot \frac{P_{tot}}{L_{tot}} \cdot (p_1 - p_4) \tag{17.48}$$

and from Equation 17.43 and using Equation 17.35:

$$p_3 = p_4 + \frac{L_3}{P_3} \cdot \frac{P_{tot}}{L_{tot}} \cdot (p_1 - p_4) \tag{17.49}$$

17.8.7 Measurement of Permeability Coefficient

Most of the methods used for measuring the permeability coefficient of polymers to gases and vapors rely on the creation and maintenance of a constant partial pressure difference of the gas or vapor under study at the two sides of the polymer film. Consequently when permeation is taking place under steady-state conditions, the permeation rate $\mathrm{d}Q/\mathrm{d}t$ of the gas or vapor from one side of the film to the other is constant. Employing various methods finally an estimate of the constant value of $\mathrm{d}Q/\mathrm{d}t$ is obtained, and next the permeability coefficient P is calculated from Equation 17.30.

17.8.7.1 Permeability Coefficient of Gases

Pressure increase method. The method is described in ASTM standard D1434. The plastic film is placed in the permeability cell between the side of high pressure and the side of low pressure. The air is evacuated from both sides of the membrane. At time zero of the test, a stream of pure permeant gas at high and constant pressure is fed at the high-pressure side of the cell. The total pressure at the low-pressure side is measured as a function of time. When steady state is achieved and the permeation rate $\mathrm{d}Q/\mathrm{d}t$ of the permeant becomes constant, the total pressure at the low-pressure side increases linearly with time. The rate of pressure increase becomes constant and is calculated from the slope of a pressure versus time graph. The permeation rate is proportional to the rate of pressure increase and is calculated from that. Finally, the permeability coefficient P is calculated from Equation 17.30.

Concentration increase method. The permeability cell consists of two cylindrical compartments and between them the sample plastic film is placed. A constant partial pressure difference of the studied permeant gas is created between the two compartments. This is achieved by sweeping continuously one compartment with a stream of pure permeant gas and the other compartment with a stream of an inert carrier gas (N_2 or He) into which the permeant gas is transferred. This method is also called *isostatic* because the same total pressure is maintained at both compartments by balancing the flow rates of the two gases. In this way the plastic is not stressed. A detector is connected to the outlet of the compartment where the inert gas flows to measure the concentration of the permeant gas in the inert gas at the exit of the compartment as a function of time. The product of this concentration and the inert gas flow rate is equal to the permeation rate; thus, the data generated are permeation rate versus time. When steady state is reached, the concentration and permeation rate remain constant. Concentration of the permeant is measured with a thermal conductivity detector or a flame ionization detector or a coulometric sensor for oxygen or an infrared detector for water vapor. An extremely useful feature of the method is that it offers the capability of adjusting the humidity of the two gas streams at the desired level. This is very important when measuring films whose permeability to gases depends on the humidity as well.

Quasi-isostatic method. The method is similar to the isostatic one, the difference being that the flow rate of the inert gas is zero. The compartment with the inert gas contains a stagnant volume of inert gas, while in the other compartment the situation is as in the isostatic method. At set time intervals a small gas sample is taken from the compartment with the inert gas, and the concentration of the permeant gas in that sample is measured using gas chromatography. To compensate for the gas withdrawn, an equal amount of inert gas is injected back to the compartment. The product of the concentration of permeant at each time t and the volume of the compartment is equal to the quantity of permeant Q that has permeated the film from time zero to time t. Thus, the data generated are Q versus time, and when steady state is achieved Q increases linearly with time, with its slope being equal to the constant permeation rate dQ/dt. The experiment is terminated before the mole fraction of the permeant gas in the inert gas reaches 0.1 because the calculation for the method assumes that the partial pressure of the permeant gas in the compartment with the inert gas is practically zero during the whole experiment.

17.8.7.2 Permeability to Water Vapor

One of the most common methods of measuring WVTR through a plastic film is a gravimetric method involving an aluminum cup (ASTM E96 and BS 3177). The film is stretched on top of an aluminum cup containing a desiccant ($CaCl_2$) and is secured in place with melted wax. The cup is stored in a cabinet at constant temperature and constant relative humidity (usually 25°C or 38°C and 75% or 90% relative humidity). The mass of the cup is monitored as a function of time, and the increase of the cup's mass relative to its mass at time zero is calculated as a function of time. The desiccant maintains the relative humidity inside the cup at 0% and absorbs the water vapor that permeates through the film. The increase of the cup's mass at time t relative to its mass at time zero is the quantity Q of water vapor that permeated through the film from time zero to time t. When permeation is taking place under steady-state conditions, dQ/dt is constant; hence, Q increases linearly with time t. From the plot of Q versus t the slope dQ/dt is obtained, and then from Equations 17.33 and 17.34 the WVTR and P are calculated.

A variation of the above method is described in standards ASTM D 895 and ASTM D 1251. Instead of the aluminum cup, plastic bags are used. They are made of film, the permeability of which we wish to measure. Inside each bag a quantity of $CaCl_2$ is placed, and the bag is heat-sealed. The bags are stored in a cabinet at constant temperature and constant relative humidity. The mass of each bag is monitored as a function of time. The procedure afterward is the same as in the cup method.

In recent years instruments have been developed for rapid determination of WVTR. The sample plastic film is placed between two compartments. One is maintained at high and constant relative humidity, while the other, where the water vapor is transferred to, contains a very sensitive humidity sensor (e.g., IR detector) capable of measuring small increases of the relative humidity of the air. The time necessary to achieve a given rise of the relative humidity is recorded and from this WVTR is calculated.

17.8.8 Sample Permeability Calculations

Example 17.1

Estimate the mass of oxygen and the mass of water vapor in g, permeating through a plastic film of thickness 25 μm, of surface area 1 m², in 1 month, and at temperature 25°C, assuming that permeation is taking place under steady-state conditions. From Table 17.2 obtain the permeability coefficients of the polymers LDPE, HDPE, PS, PVC, PVdC, and PET to oxygen and water vapor. The concentrations of O_2 may be taken as 21% at one side of the film and 0 at the other, with 0% relative humidity at both sides. For water vapor the relative humidity can be taken as 90% at one side of the film and 0 at the other. Atmospheric pressure is 76 cm Hg, and the saturated water vapor pressure at 25°C is 23.756 mm Hg.

Solution

For permeation under steady-state conditions and for constant concentrations of oxygen and water vapor at the two sides of the film, the integration of Equation 17.14 gives

$$\Delta Q = \frac{P}{L} \cdot A \cdot [p_1 - p_2] \cdot \Delta t$$

For O_2:

$p_1 = 0.21 \times 76$ cm Hg $= 15.96$ cm Hg and $p_2 = 0$. $L = 25 \times 10^{-4}$ cm, $A = 1\ \text{m}^2 = 10^4\ \text{cm}^2$
$\Delta t = 1$ month $= 30 \times 24 \times 3{,}600$ s $= 2.592 \times 10^6$ s
Also 1 mol $O_2 = 32$ g $= 22{,}400\ \text{cm}^3$(STP)

Substituting in the above equation:

$$\Delta Q = P \times 10^{-10}\ \frac{\text{cm}^3(\text{STP})\ \text{cm}}{\text{cm}^2\ \text{s cm Hg}} \times \frac{1}{25 \times 10^{-4}}\ \frac{1}{\text{cm}} \times 10^4\ \text{cm}^2$$

$$\times 15.96\ \text{cm Hg} \times 2.592 \times 10^6\ \text{s} \times \frac{32}{22{,}400}\ \frac{\text{g}}{\text{cm}^3(\text{STP})}$$

So $\Delta Q = 23.639\ P$ with ΔQ in g and P in barrer.

For water vapor:

$p_w^0 = 23.756$ mm Hg $= 2.3756$ cm Hg at 25°C
$p_1 = (\text{RH})_1 \times p_w^0 = 0.90 \times 2.3756$ cm Hg $= 2.138$ cm Hg and $p_2 = 0$. $L = 25 \times 10^{-4}$ cm
$A = 10^4\ \text{cm}^2$, $\Delta t = 2.592 \times 10^6$ s
Also 1 mol $H_2O = 18$ g $= 22{,}400\ \text{cm}^3$(STP)

Substituting in the above equation:

$$\Delta Q = P \times 10^{-10}\ \frac{\text{cm}^3(\text{STP})\ \text{cm}}{\text{cm}^2\ \text{s cm Hg}} \times \frac{1}{25 \times 10^{-4}}\ \frac{1}{\text{cm}} \times 10^4\ \text{cm}^2$$

$$\times 2.138\ \text{cm Hg} \times 2.592 \times 10^6\ \text{s} \times \frac{18}{22{,}400}\ \frac{\text{g}}{\text{cm}^3(\text{STP})}$$

So $\Delta Q = 1.781\ P$ with ΔQ in g and P in barrer.

The amounts of O_2 and water vapor are estimated by substituting into the earlier two expressions for ΔQ the values of P for each polymer and are presented in the table below.

Polymer	P to Water Vapor in barrer	P to Oxygen in barrer	Mass ΔQ of Water in g	Mass ΔQ of Oxygen in g
PS	1126.39	1.70	2006.1	40.2
UPVC	277.74	0.045	494.7	1.1
PET	131.16	0.034	233.6	0.8
LDPE	91.04	2.93	162.1	69.3
HDPE	12.04	0.40	21.4	9.5
PVdC	9.26	0.005	16.5	0.1

Example 17.2

Solid food sensitive to oxygen gain is packaged under vacuum in plastic bags made from a three-layer laminate and is stored at 25°C. It can be assumed that the oxygen that permeates the bag walls is immediately absorbed by the product so that the concentration of oxygen inside the package is always practically zero. The concentration of oxygen in the atmospheric air is 21% and the atmospheric pressure 760 mm Hg. The thickness of each bag is 65 μm and its surface area 500 cm^2. The laminate is composed of the following layers: 40 μm PET/5 μm EVOH/20 μm LDPE. At 25°C the permeability coefficient of PET to oxygen is 5.0×10^{-4} [cm^3(STP) cm m^{-2} $(day)^{-1}$ $(mm\ Hg)^{-1}$], of EVOH 3.0×10^{-6} [cm^3(STP) cm m^{-2} $(day)^{-1}$ $(mm\ Hg)^{-1}$], and of LDPE 2.5×10^{-2} [cm^3(STP) cm m^{-2} $(day)^{-1}$ $(mm\ Hg)^{-1}$]. Calculate the overall permeability coefficient of the multilayer film (i.e., laminate) to oxygen and the permeation rate of oxygen to the inside of the package.

Solution

The thicknesses of the layers are

$$L_{PET} = 40\ \mu m = 40 \times 10^{-4}\ cm$$

$$L_{EVOH} = 5\ \mu m = 5 \times 10^{-4}\ cm$$

$$L_{LDPE} = 20\ \mu m = 20 \times 10^{-4}\ cm$$

and the total thickness $L_{tot} = 65\ \mu m = 65 \times 10^{-4}$ cm

Substituting into Equation 17.45 we get

$$\frac{L_{tot}}{P_{tot}} = \frac{L_{PET}}{P_{PET}} + \frac{L_{EVOH}}{P_{EVOH}} + \frac{L_{LDPE}}{P_{LDPE}} = \frac{40 \times 10^{-4}}{5.0 \times 10^{-4}} + \frac{5 \times 10^{-4}}{3.0 \times 10^{-6}} + \frac{20 \times 10^{-4}}{2.5 \times 10^{-2}}$$

$$= 8.00 + 166.66 + 0.08 = 174.74$$

$$\frac{L_{tot}}{P_{tot}} = \frac{65 \times 10^{-4}}{P_{tot}} = 174.74 \Rightarrow P_{tot} = 3.7 \times 10^{-5}\ [cm^3(STP)\ cm\ m^{-2}\ (day)^{-1}\ (mm\ Hg)^{-1}]$$

From Equation 17.35 the permeation rate of oxygen to the inside of the package is

$$\left(\frac{dQ}{dt}\right)_{tot} = \frac{P_{tot}}{L_{tot}} A(p_1 - p_4)$$

$$A = 0.0500\ m^2$$

$$p_1 = 0.21 \times 760 = 159.6 \text{ mm Hg} \quad \text{and} \quad p_4 = 0$$

$$\text{Therefore: } \left(\frac{dQ}{dt}\right)_{\text{tot}} = \frac{1}{174.74} \times 0.0500 \times 159.6 = 0.0457 \text{ cm}^3\text{(STP) (day)}^{-1}$$

Example 17.3

Let a multilayer film (laminate) composed of the following layers in series: HDPE/EVOH/PET of the following thicknesses and permeability coefficients to water vapor at 25°C:

HDPE: 40 μm, $P = 8.4 \times 10^{-5}$ [g cm m^{-2} (day)$^{-1}$ (mm Hg)$^{-1}$]

EVOH: 10 μm, $P = 3.3 \times 10^{-3}$ [g cm m^{-2} (day)$^{-1}$ (mm Hg)$^{-1}$]

PET: 50 μm, $P = 9.1 \times 10^{-4}$ [g cm m^{-2} (day)$^{-1}$ (mm Hg)$^{-1}$]

To make the calculations simpler we may assume that the above coefficients are constant and independent of RH. The multilayer film is at 25°C and at one side of it, the relative humidity of the air is RH = 90% and at the other, RH = 40%. At 25°C $p_w^0 = 23.756$ mm Hg. Calculate:

(a) The overall permeability coefficient of the multilayer film to water vapor.
(b) The WVTR from one side of the film to the other.
(c) The relative humidities at the two sides of the EVOH layer if the laminate is placed with the HDPE layer at the side of RH = 90% and the PET layer at the side of RH = 40%.
(d) Verify that the WVTR that permeates each separate layer is equal to the WVTR calculated in step (b) above using the overall permeability coefficient.
(e) Repeat the calculations of steps (c) and (d) for the case the laminate is placed with the HDPE layer at the side of RH = 40% and the PET layer at the side of RH = 90%.
(f) How should the laminate be placed so that the EVOH layer gets the maximum protection from moisture?

Solution

(a) Substituting into Equation 17.45 we get

$$\frac{L_{\text{tot}}}{P_{\text{tot}}} = \frac{L_{\text{HDPE}}}{P_{\text{HDPE}}} + \frac{L_{\text{EVOH}}}{P_{\text{EVOH}}} + \frac{L_{\text{PET}}}{P_{\text{PET}}} = \frac{40 \times 10^{-4}}{8.4 \times 10^{-5}} + \frac{10 \times 10^{-4}}{3.3 \times 10^{-3}} + \frac{50 \times 10^{-4}}{9.1 \times 10^{-4}}$$

$$= 47.62 + 0.30 + 5.49 = 53.41$$

$$\frac{L_{\text{tot}}}{P_{\text{tot}}} = \frac{100 \times 10^{-4}}{P_{\text{tot}}} = 53.41 \Rightarrow P_{\text{tot}} = 1.87 \times 10^{-4} \text{ [g cm m}^{-2} \text{ (day)}^{-1} \text{ (mm Hg)}^{-1}]$$

(b) From Equation 17.34 solving for WVTR we have

$$\text{WVTR} = \frac{P_{\text{tot}}}{L_{\text{tot}}} \cdot p_w^0 \cdot \frac{[(\text{RH})_1 - (\text{RH})_4]}{100}$$

$P_{\text{tot}} = 1.87 \times 10^{-4}$ [g cm m^{-2} (day)$^{-1}$ (mm Hg)$^{-1}$], $L_{\text{tot}} = 100 \times 10^{-4}$ cm, $(\text{RH})_1 = 90$, $(\text{RH})_4 = 40$

Therefore,

$$\text{WVTR} = \frac{1.87 \times 10^{-4}}{100 \times 10^{-4}} \times 23.756 \times \frac{[90 - 40]}{100} = 0.222 \text{ g m}^{-2} \text{ (day)}^{-1}$$

(c) For the calculation of the relative humidities of air at the two sides of the EVOH layer, that is, $(RH)_2$ and $(RH)_3$, Equations 17.48 and 17.49 can be used with p_1, p_2, p_3, and p_4 replaced by $(RH)_1$, $(RH)_2$, $(RH)_3$, and $(RH)_4$.

Also,

$$L_1 = L_{HDPE} = 40 \times 10^{-4} \text{ cm} \quad \text{and} \quad P_1 = P_{HDPE} = 8.4 \times 10^{-5} \text{ [g cm m}^{-2} \text{ (day)}^{-1} \text{ (mm Hg)}^{-1}]$$

$$L_2 = L_{EVOH} = 10 \times 10^{-4} \text{ cm} \quad \text{and} \quad P_2 = P_{EVOH} = 3.3 \times 10^{-3} \text{ [g cm m}^{-2} \text{ (day)}^{-1} \text{ (mm Hg)}^{-1}]$$

$$L_3 = L_{PET} = 50 \times 10^{-4} \text{ cm} \quad \text{and} \quad P_3 = P_{PET} = 9.1 \times 10^{-4} \text{ [g cm m}^{-2} \text{ (day)}^{-1} \text{ (mm Hg)}^{-1}]$$

$$\text{and} \quad (RH)_1 = 90 \quad \text{and} \quad (RH)_4 = 40$$

With these changes Equations 17.48 and 17.49 give

$$(RH)_2 = (RH)_1 - \frac{L_1}{P_1} \cdot \frac{P_{tot}}{L_{tot}} \cdot [(RH)_1 - (RH)_4] = 90 - \frac{40 \times 10^{-4}}{8.4 \times 10^{-5}} \cdot \frac{1}{53.41} \cdot [90 - 40] = 45.42$$

and

$$(RH)_3 = (RH)_4 + \frac{L_3}{P_3} \cdot \frac{P_{tot}}{L_{tot}} \cdot [(RH)_1 - (RH)_4] = 40 + \frac{50 \times 10^{-4}}{9.1 \times 10^{-4}} \cdot \frac{1}{53.41} \cdot [90 - 40] = 45.14$$

(d) For layer one we have

$$\text{WVTR} = \frac{P_1}{L_1} \cdot p_w^0 \cdot \frac{[(RH)_1 - (RH)_2]}{100} = \frac{8.4 \times 10^{-5}}{40 \times 10^{-4}} \times 23.756 \times \frac{[90 - 45.42]}{100} = 0.222 \text{ g m}^{-2} \text{ (day)}^{-1}$$

For layer two we have

$$\text{WVTR} = \frac{P_2}{L_2} \cdot p_w^0 \cdot \frac{[(RH)_2 - (RH)_3]}{100} = \frac{3.3 \times 10^{-3}}{10 \times 10^{-4}} \times 23.756 \times \frac{[45.42 - 45.14]}{100} = 0.220 \text{ g m}^{-2} \text{ (day)}^{-1}$$

For layer three we have

$$\text{WVTR} = \frac{P_3}{L_3} \cdot p_w^0 \cdot \frac{[(RH)_3 - (RH)_4]}{100} = \frac{9.1 \times 10^{-4}}{50 \times 10^{-4}} \times 23.756 \times \frac{[45.14 - 40]}{100} = 0.222 \text{ g m}^{-2} \text{ (day)}^{-1}$$

So the WVTR that permeates each one of the three layers is equal to the WVTR calculated in step (b) using the overall permeability coefficient.

(e) Now $(RH)_1 = 40$ and $(RH)_4 = 90$.

$$(RH)_2 = (RH)_1 - \frac{L_1}{P_1} \cdot \frac{P_{tot}}{L_{tot}} \cdot [(RH)_1 - (RH)_4] = 40 - \frac{40 \times 10^{-4}}{8.4 \times 10^{-5}} \cdot \frac{1}{53.41} \cdot [40 - 90] = 84.58$$

and

$$(RH)_3 = (RH)_4 + \frac{L_3}{P_3} \cdot \frac{P_{tot}}{L_{tot}} \cdot [(RH)_1 - (RH)_4] = 90 + \frac{50 \times 10^{-4}}{9.1 \times 10^{-4}} \cdot \frac{1}{53.41} \cdot [40 - 90] = 84.86$$

(f) From the results of steps (c) and (e) it is concluded, as expected, that for the central layer to be best protected from moisture, the polymer (HDPE in the present case) with the smallest permeability to water vapor must be placed at the side with the highest relative humidity.

17.8.9 Application of Permeability Equations to Shelf Life Calculations

17.8.9.1 Simple Shelf-Life Model for Oxygen or Moisture-Sensitive Packaged Foods

For foods sensitive to oxygen or moisture gain, a very simple method for estimating their shelf life is the following. This method may give very wrong results because of the many assumptions and simplifications used. The main assumptions are as follows:

- There exists a maximum oxygen concentration in the food or a maximum change of the moisture content beyond which the food is no longer acceptable; thus, it has reached the end of its shelf life. Representative values for these limits for various foods may be obtained from the literature (e.g., Salame 1974, Robertson 2006, Lee et al. 2008); however, for more accurate values experiments should be conducted for the specific food product.
- The total quantity of oxygen or water vapor transferred through the packaging from the outside to the inside of the package is absorbed by the food, and their concentration in the food is increased.
- In the gaseous environment inside the package the concentration of oxygen and the relative humidity remain constant. Hence, the driving force and the transfer rates of oxygen and water vapor through the packaging are constant. Most often, for oxygen-sensitive foods, the partial pressure of oxygen inside the package is considered to be practically zero, and for moisture-sensitive foods the relative humidity inside the package is equal to initial water activity of the food × 100.

Then the maximum allowable quantity of absorbed oxygen or moisture necessary for the particular quantity of food inside the package to reach the end of its shelf life is calculated. The oxygen and water vapor transmission rates are calculated from the Equations 17.50 and 17.51, which are actually Equations 17.14 and 17.17, with $p_{O_2,i} = 0$ and $(RH)_i = 100(a_w)_{initial}$.

$$\left[\frac{dQ}{dt}\right]_{O_2} = \frac{P_{O_2}}{L} A(p_{O_2,e} - p_{O_2,i}) \tag{17.50}$$

$$\left[\frac{dQ}{dt}\right]_{H_2O} = \frac{P_{H_2O}}{L} A p_w^0 \frac{[(RH)_e - (RH)_i]}{100} \tag{17.51}$$

where P_{O_2} and P_{H_2O} are the oxygen and water vapor permeability coefficients, respectively, of the polymeric packaging material, A the package surface area, L the package thickness, $p_{O_2,e}$ and $(RH)_e$ the oxygen partial pressure and relative humidity, respectively, in the external environment of the package, $p_{O_2,i}$ and $(RH)_i$ the oxygen partial pressure and relative humidity, respectively, in the internal environment of the package, and p_w^0 the saturated water vapor pressure at the constant storage temperature of the packaged food.

Thus, finally from oxygen and moisture mass balances, shelf life may be estimated from the equations:

$$\text{Shelf life } t_s \text{ of oxygen-sensitive food} = \frac{\text{maximum allowable quantity of absorbed oxygen}}{[dQ/dt]_{O_2}} \tag{17.52}$$

$$\text{Shelf life } t_s \text{ of moisture-sensitive food} = \frac{\text{maximum allowable quantity of absorbed moisture}}{[dQ/dt]_{H_2O}} \tag{17.53}$$

Sample Problem

A solid food product in powder form is to be packaged in PET jars in a pure nitrogen atmosphere. The particular product is very susceptible to oxidative rancidity, and its oxygen concentration tolerance limit is 80 ppm. Estimate the shelf life of the food at 20°C under the following conditions: Each jar contains 400 g of product. The total surface area of the jar is 350 cm^2, and the thickness of its wall is 0.0400 cm. The oxygen concentration in the atmospheric air is 21% (v/v). It can be assumed that the oxygen concentration inside the jar is practically zero. The oxygen permeability coefficient of PET at 20°C is 1.9×10^{-4} [cm^3(STP) cm m^{-2} (day)$^{-1}$ (mm Hg)$^{-1}$].

Solution

The food reaches the end of its shelf life when its oxygen concentration rises to 80 ppm = 0.080 g of O_2 contained in 1000 g of food or 0.080 × 400/1000 g O_2 contained in 400 g of food or 0.032 g O_2 contained in 400 g of food, that is, when the quantity of oxygen in the jar reaches 0.032 g. Since the molecular weight of oxygen is 32 g mol^{-1} and 1 mol corresponds to 22,400 cm^3(STP), the maximum allowable quantity of absorbed oxygen is 0.032 × 22,400/32 = 22.4 cm^3(STP).

The oxygen transmission rate from the outside environment to the inside of the jar is calculated from Equation 17.50 with $p_{O_2,i} = 0$, $p_{O_2,e} = 0.21 \times 760 = 159.6$ mm Hg, $P_{O_2} = 1.9 \times 10^{-4}$ [cm^3(STP) cm m^{-2} (day)$^{-1}$(mm Hg)$^{-1}$], $L = 0.040$ cm, and $A = 350$ cm^2.

Substituting into Equation 17.50 we get

$$\left[\frac{dQ}{dt}\right]_{O_2} = \frac{1.9 \times 10^{-4}}{0.04} \frac{\text{cm}^3\text{(STP)}}{\text{m}^2\text{ (day) (mm Hg)}} \times 350 \times 10^{-4}\ \text{m}^2 \times 159.6\ \text{(mm Hg)}$$

$$= 2.65 \times 10^{-2}\ \text{cm}^3\text{(STP) day}^{-1}$$

Therefore, from Equation 17.52 the shelf life t_s of the food is

$$t_S = \frac{22.4\ \text{cm}^3\text{(STP)}}{2.65 \times 10^{-2}\ \text{cm}^3\text{(STP)/day}} = 845\ \text{days} = 2.3\ \text{years}$$

17.8.9.2 Shelf-Life Model for Foods Sensitive to Moisture Gain

The model described in the previous paragraphs is very simplified because it assumes that the partial pressure or the relative humidity difference in Equations 17.50 and 17.51 remains constant during the whole shelf life of the food. However, this is not the case when the moisture content of the food increases considerably with time. The subject is dealt with in the model described next, which was first presented by Labuza et al. (1972).

Let us consider a moisture-sensitive food of low moisture content packaged in a plastic material with low water vapor permeability and stored at a high relative humidity environment. The model predicts the variation of the moisture content X of the food with time t. X is on a dry basis, expressed in g water per g dry solids. Assuming that the food loses its crispness and becomes inedible when its moisture content exceeds a critical value X_F, the model also predicts the time necessary for the moisture content of the food to increase from the initial value X_{IN} to the final value X_F, that is, the shelf life t_s of the moisture-sensitive food. Additional assumptions necessary for the derivation of the model are as follows:

- The water vapor permeability coefficient P of the packaging material is constant, and the temperature T and the relative humidity $(RH)_e$ of the external environment are constant too.
- The quantity of air inside the package is negligible, and the water vapors permeating the package material are all absorbed by the food increasing its moisture content.
- The main resistance to water vapor transfer is in the packaging material and not in the mass of the food. At every moment the food has uniform moisture throughout its mass and

is in equilibrium with the air inside the package, which means that the relative humidity of this air is equal to initial water activity of the food × 100. This assumption is generally valid for most plastic packaging materials, which have permeance (*P*/*L*) values smaller than 1 g m^{-2} $(day)^{-1}$ $(mm Hg)^{-1}$ (Robertson 1993).

- The moisture sorption isotherm of the food in the region of interest (for $X_{IN} \leq X \leq X_F$) can be approximated with satisfactory accuracy by a straight line segment, that is, $X = ba_w + c$.

According to the model:

For the moisture sorption isotherm between X_{IN} and X_F being approximated by:

$$X = ba_w + c \tag{17.54}$$

the moisture content *X* of the food increases with time *t* according to:

$$X = X_e - \left(X_e - X_{IN}\right)e^{-t/\tau} \tag{17.55}$$

and the shelf life t_s is given by:

$$t_s = \tau \cdot \ln\left[\frac{X_e - X_{IN}}{X_e - X_F}\right] \tag{17.56}$$

where the time parameter τ is:

$$\tau = \frac{L \cdot m_{ds} \cdot b}{P \cdot A \cdot p_w^0} \tag{17.57}$$

and

$$X_e = \frac{b(RH)_e}{100} + c \tag{17.58}$$

where

a_w is the water activity of the food
b, *c* are the slope and intercept, respectively, of the moisture sorption isotherm
$(RH)_e$ is the relative humidity of external environment
X_{IN} is the initial moisture content of food on a dry basis, in g water per g dry solids
X_F is the final critical moisture content of food on a dry basis, in g water per g dry solids, when the food reaches the end of its shelf life
L is the thickness of packaging material
m_{ds} is the mass of product dry solids contained in the package
P is the water vapor permeability coefficient of the packaging material
A is the surface area of the package
p_w^0 is the saturated water vapor pressure at the constant storage temperature of the packaged food

Therefore, the shelf life is dependent on food parameters (m_{ds}, X_{IN}, X_F, *b*, and *c*), package parameters (*P*, *A*, *L*), and environmental parameters (*T*, $(RH)_e$). Temperature *T* influences *P*, p_w^0, *b*, and *c*.

Equations 17.55 and 17.56 have been used extensively in several labs, including ours, to predict the variation of the moisture content *X* of the packaged food with time *t* and the food's shelf life. Plots of *X* predicted by Equation 17.55 versus *t* are in good agreement with experimentally determined moisture contents as functions of time for various foods. The agreement is particularly good in the region of the sorption isotherm where the linear approximation holds, and all the other assumptions made for the derivation of the model are valid. If experiments are made with the same food but packaged in various size packages and with different plastic packaging materials, all the experimental moisture content measurements can be plotted in the same graph along with the model predictions. This can be done by normalizing (i.e., dividing) the experimental time values by the equivalent for each case value of the time parameter τ. The validity of the theoretical model can also be tested by plotting the $\ln[(X_e - X_{IN})/(X_e - X)]$ (calculated with the experimental values of X) versus time t. The graph should be a straight line of slope $1/\tau$ and with zero intercept.

Combining Equations 17.56 and 17.57 and solving for P/L, one may calculate the required permeance (P/L) of the plastic packaging material in order for the food to have the desired shelf life. It is also useful for estimating the influence on shelf life of parameters like the environmental conditions (temperature and relative humidity), the size of packaging, the initial moisture content of the food, etc. As shown by Equations 17.56 and 17.57, the shelf life t_s is proportional to the ratio m_{ds}/A, equal to the volume over the surface area of the package, which in turn is approximately proportional to the thickness of the package. So t_s is approximately proportional to the thickness of the package, and this has also been experimentally verified (Taoukis et al. 1988). Therefore, for foods sold in packages of different sizes and otherwise completely identical, their shelf life will be the shortest in the smallest size package. For this reason the calculations and the experiments for shelf life must be done for the smallest size package.

For some foods, for the approximation of the moisture sorption isotherm in the region of interest, it is necessary to use two linear segments with different slopes and intercepts, instead of one. Equations for t_s, equivalent to Equation 17.56, are available for this case in the literature. Another approach is to use the experimental moisture sorption isotherm directly and not to approximate it with a mathematical expression. Then the necessary integration for the derivation of equations for t_s is done numerically instead of analytically (Azanha and Faria 2005).

Sample Problem

A single cookie of net weight 5.000 g was packaged in a polypropylene sachet and was stored in an environment of constant temperature 25°C and constant relative humidity 75%. Its initial moisture content was 0.020 g water per g dry solids, and the cookie loses its crispness and becomes inedible when its moisture content exceeds a critical value of 0.075 g water per g dry solids. The PP sachet had a surface area of 80 cm² and a thickness of 60 μm. At these conditions the water vapor permeability coefficient of PP is 3.5×10^{-4} [g cm m^{-2} (day)$^{-1}$ (mm Hg)$^{-1}$]. At 25°C the moisture sorption isotherm of the cookie in the region of 0.020–0.075 g water per g dry solids can be approximated by the straight line segment $X = 0.127\alpha_w + 0.014$. Estimate the shelf life of the cookie and derive the function predicting the variation of the moisture content X of the cookie with time t.

Solution

The mass of the cookie is 5.000 g, therefore the mass m_{ds} of the cookie's dry solids inside the package is: $m_{ds} = 5.000 \times 1/1.020 = 4.902$ g

$$b = 0.127 \text{ g water per g dry solids}$$

$$c = 0.014 \text{ g water per g dry solids}$$

$$A = 80 \text{ cm}^2 = 80 \times 10^{-4} \text{ m}^2 = 0.0080 \text{ m}^2$$

$$P = 3.5 \times 10^{-4} \text{ g cm m}^{-2} \text{ (day)}^{-1} \text{ (mm Hg)}^{-1}$$

$$L = 60\ \mu\text{m} = 60 \times 10^{-4} \text{ cm}$$

$$P/L = 3.5 \times 10^{-4}/60 \times 10^{-4} = 0.0583 \text{ g m}^{-2} \text{ (day)}^{-1} \text{ (mm Hg)}^{-1}$$

$$[(RH)_e/100] = 0.75$$

and at 25°C the saturated water vapor pressure is $p_W^0 = 23.756$ mm Hg
Substituting into Equations 17.58 and 17.57 we get

$$X_e = \frac{b(RH)_e}{100} + c = 0.127 \times 0.75 + 0.014 = 0.109 \frac{\text{g water}}{\text{g dry solids}}$$

$$\tau = \frac{L \cdot m_{ds} \cdot b}{P \cdot A \cdot p_w^0}$$

$$= \frac{1}{0.0583} \frac{\text{m}^2 \text{ (day) (mm Hg)}}{\text{g water}} \times \frac{4.902 \text{ g dry solids}}{0.0080 \text{ m}^2} \times \frac{0.127}{23.756} \frac{\text{g water}}{\text{(g dry solids) (mm Hg)}}$$

$$= \frac{4.902 \times 0.127}{0.0583 \times 0.0080 \times 23.756} \text{ days} = 56.2 \text{ days}$$

Then from Equation 17.56 the shelf life t_s is

$$t_S = \tau \cdot \ln\left[\frac{X_e - X_{IN}}{X_e - X_F}\right] = 56.2 \times \ln\left[\frac{0.109 - 0.020}{0.109 - 0.075}\right] = 56.2 \times \ln[2.618] = 54 \text{ days}$$

From Equation 17.55 the moisture content X of the cookie increases with time t according to

$$X = X_e - (X_e - X_{IN})e^{-t/\tau} = 0.109 - (0.109 - 0.020)e^{-t/56.2}$$

So, $X = 0.109 - 0.089e^{-t/56.2}$, with t in days.

17.9 ASEPTIC PACKAGING

17.9.1 Introduction

Aseptic processing and packaging is a food preservation technique that uses relatively mild sterilization treatments compared to retorting to produce high quality foods. The technique consists of two separate but integrated operations: an aseptic processing system, and, an aseptic packaging system. The aseptic processing system usually employs thermal processing to continuously sterilize a pumpable food in bulk. The aseptic packaging system consists of three operations: sterilization of the container, filling it with the presterilized and sterile food, and hermetically sealing it. The last two operations are carried out inside an aseptic zone so that reinfection is prevented. In the food industry, the terms aseptic, sterile, and commercially sterile are often used interchangeably (Stevenson 1992).

Aseptic packaging is usually applied to pasteurized or sterilized products such as milk and dairy products, puddings, desserts, juices, soups, sauces, and some products with small or large particles (e.g., potatoes or vegetable pieces) (Reuter 1989a). It is also applied to nonsterile foods, such as yogurt, with the aim of achieving extended shelf life (ESL), sometimes under refrigeration, by preventing infection with additional and other microorganisms (Reuter 1989a).

The term aseptic packaging originally meant that the food product was commercially sterile, and thus ambient temperature shelf stability was achieved. Nowadays, the situation has progressed far beyond that point. Aseptic packaging has been applied to refrigerated products to achieve ESL for these foods beyond that of traditional pasteurized products (Brody 2000, 2003, 2006). Various fluid milk products and fruit juices are currently packaged aseptically in glass or polyester bottles, and the refrigerated shelf life achieved ranges from 60 to 90 days (Brody 2000).

In 1981, the US FDA accepted the use of hydrogen peroxide for the sterilization of aseptic packages containing a low-density polyethylene layer in contact with the food (Mabee 1997), and this has led to dramatic growth of the aseptic industry. According to the Institute of Food Technologists (IFT), aseptic processing and packaging is the top invention in food science in the 50-year period from 1939 to 1989.

Compared to canning, aseptic processing and packaging has the following advantages. High-temperature short-time (HTST) sterilization processes can be used for the pumpable food product. These are more energy efficient and give products of better quality than the conventional canning process, which uses lower temperatures and longer times. Containers made of inexpensive materials like paper and plastics can be utilized. These do not have to withstand the extreme conditions of temperature and pressure encountered during retort processing (Mabee 1997). In conclusion, aseptic processing and packaging result in better product quality and cost savings due to reduced energy consumption and inexpensive packaging (Reuter 1989a).

The main problems associated with aseptic processing and packaging are the increased difficulty of securing package and seal integrity, the inadequate inactivation of some enzymes, and the complexity of the equipment and control system. Aseptic packaging is the weakest component of the whole process and thus the most likely to malfunction (Reuter 1989a). Aseptic processing of low-acid foods containing finite-size particulates (two-phase foods) presents a great challenge in terms of ensuring adequate heat treatment of the particles. In 2005, the FDA accepted a method for aseptically processing low-acid foods containing finite-size particulates, and Campbell Soup Co. began processing and packaging particulate-laden soups in cartons (Brody 2006).

17.9.2 Sterilization of Packages and Equipment

Packages and equipment surfaces get contaminated with microorganisms by air or by contact with humans or with each other. The destruction of microorganisms adhering to surfaces, by any type of sterilization method, follows, in most cases, first-order kinetics and for the mathematical description of the kinetics, the *D*-value is often used (Reuter 1993). The numerical value of *D* is the time necessary to kill the 90% of the initial number of a specific type of microorganism at constant temperature and under specific environmental conditions. The *D*-value of the sterilization process used for the most resistant type of microorganism likely to be found on the packaging material should be less than 1 s (Reuter 1989b).

17.9.2.1 Required Count Reduction of Bacterial Spores on Packaging Material

For the sterilization of the packaging material, the required number of decimal reductions (number of *D*-values) or log cycle reductions (LCR) depends on the type of product contained, its desired shelf life, and the storage temperature. For nonsterile acidic products of pH < 4.5 stored under refrigeration, for a shelf life of a few weeks, four decimal reductions are required for a determinant microorganism (e.g., *Aspergillus niger*). For sterile low-acid products of pH > 4.5 stored at ambient temperature, for a shelf life of a few months, six LCR are required for a determinant microorganism (e.g., *Bacillus subtilis*). If there is the possibility that *Clostridium botulinum* is able to grow in the product, then a full 12D process should be applied (Reuter 1989b). For heat treatment and treatment with hydrogen peroxide, six decimal reductions for *B. subtilis* are more than 12 reductions for *C. botulinum* (Reuter 1993).

The nonsterility rate F_s, that is, the proportion of nonsterile packages in a batch of sterilized packages, resulting from the sterilization process of the packaging material, can be calculated (Reuter 1989b) from the equation:

$$F_s = \frac{N \cdot A}{10^R} \qquad (17.59)$$

where

N is the number of bacterial spores per m^2 of packaging material surface
A is the surface area in contact with food in one package
R is the number of decimal reductions achieved by the package sterilization method

It is usually assumed that the maximum initial load on the package surface is 1000 microorganisms per m^2 for plastic films and paperboard laminates on reels and 3000 microorganisms per m^2 for prefabricated cups. It is also assumed that only 3% of the total number of microorganisms on the package's surface are spores (Reuter 1989b). It is obvious that it is extremely important to keep the initial load of microbes on the package surface as low as possible.

For example, in order to attain a nonsterility rate resulting from the sterilization process of the packaging material of 1 package in 1,000,000 packages ($F_s = 10^{-6}$) for prefabricated cups, with each cup having a surface area A of 400 cm^2 = 0.0400 m^2 and an assumed initial load N of 90 spores per m^2, the number R of required decimal reductions is calculated from Equation 17.59 as:

$$R = \log\left(\frac{NA}{F_s}\right) = \log\left(\frac{90 \times 0.0400}{10^{-6}}\right) = \log(3.6 \times 10^6) = 6.6$$

17.9.2.2 Methods of Sterilization of Packaging Materials

The sterilant used must fulfill (Cousin 1993) the following criteria. It must kill spores within a short contact time, be compatible with the type of packaging material, easy to apply and to remove, inexpensive, nontoxic to personnel, nondamaging to the aseptic system, and any residues left on the packaging material must be within acceptable tolerance levels. The sterilization methods generally involve heat, chemicals, irradiation, or a combination of these.

In general, heat sterilization cannot be applied to heat-sensitive plastics. Saturated steam under pressure is considered as the most reliable sterilant, despite a few problems in its use (Reuter 1989b). It is more effective in killing microorganisms than superheated steam or hot air. Saturated steam at 140°C–147°C is used to sterilize polypropylene cups for a period of 4–6 s in a pressure chamber (Reuter 1989b). Superheated steam at atmospheric pressure is used for metal cans. Dry hot air at 315°C and at atmospheric pressure is used to sterilize composite cans made from laminations of Al foil, plastics, and paper. Mixtures of hot air and saturated steam have been used to sterilize cups and lids made from PP, which are thermally stable up to 150°C (Amman 1989). The heating, occurring during the extrusion of the plastic, may also be used for the sterilization of the plastic containers. During this operation, temperatures of 180°C–230°C are reached for up to 3 min (Reuter 1989b). However, because of a nonuniform temperature distribution inside the extruder and considerable dispersion of residence time, there is no guarantee of commercial sterility. Therefore, these containers can only be used for acidic products with pH < 4.5, otherwise it is recommended that extruded containers be poststerilized chemically.

Among the chemical processes, only treatment with hydrogen peroxide, peracetic acid, and ethylene oxide have found commercial application. Although the lethal effect of hydrogen peroxide on microorganisms and spores has been known for years, the actual mechanism of death is not yet fully understood. A H_2O_2 concentration of at least 30% and a temperature higher than 80°C is necessary to achieve destruction within seconds of the most resistant spores on packaging materials (Reuter 1989b). Concentrated H_2O_2 solutions at ambient temperature do not have any destructive effect. Application of cold H_2O_2 is effective only if it is combined with hot air drying of the packaging material or with infrared irradiation or UV light. The sterilization conditions in most cases have been determined empirically. The damaging power of hydrogen peroxide comes from its transition to highly reactive hydroxyl radicals, which react with many organic compounds causing peroxidation of lipids, cross-linking and inactivation of proteins, and mutations in DNA (Nindl 2004).

A number of systems that combine peroxide treatment with heat and/or UV radiation have been developed. These are the dipping, the spraying, and the rinsing processes. In the first, the packaging material, in the form of a flat sheet, passes through a bath of 30%–33% peroxide solution and is then dried with hot air. In the spraying process, peroxide is sprayed through nozzles onto prefabricated packages, and the droplets are dried with hot air. A newer development is the direct use of a mixture of hot air and peroxide vapors. The rinsing process is applied to glass containers, metal cans, and blow-molded plastic bottles. After rinsing with peroxide or with a mixture of peroxide and peracetic acid, the container is drained and then dried with hot air. When the peroxide treatment is combined with UV radiation, they act synergistically, and the lethal effect achieved is greater than the sum of the effects of each treatment done separately (Robertson 1993). This is because the UV radiation promotes the breakdown of peroxide into hydroxyl radicals. The optimum effect is achieved when the peroxide concentrations are between 0.5% and 5%. Higher peroxide concentrations require greater UV intensities for optimum result.

Peracetic acid (CH_3–COO–OH) is effective against spores of bacteria at concentrations of 1% or less even at ambient temperature. It is produced by the oxidation of acetic acid by hydrogen peroxide, and the solution usually employed consists of peracetic acid and hydrogen peroxide (Robertson 1993).

Ethylene oxide is a toxic gas that can penetrate porous materials like paperboard. It can be used for the presterilization, in special facilities, of preformed carton blanks and bags-in-boxes (Robertson 1993). However, because of the long exposure times required for sterilization, the need to dissipate residues, and the fact that it does not have regulatory approval, it is not currently used as a packaging sterilant in aseptic packaging (Cousin 1993).

UV radiation is effective in destroying microorganisms in the wavelength range of 200–315 nm. The most effective part of the spectrum is the UV-C range between 250 and 280 nm, with an optimum effectiveness at 253.7 nm (Reuter 1989b). There are various problems, like the need for smooth surfaces and dust-free packaging materials, which limit UV irradiation's effectiveness in killing microorganisms. For these reasons, it is used commercially only in combination with hydrogen peroxide (Reuter 1989b). IR radiation is not particularly useful in sterilizing packaging materials.

Ionizing radiation in the form of gamma rays from cobalt 60 and cesium 139 is used commercially to sterilize medical instruments. Also sterilization of large volume bags made from plastic laminates for use in aseptic bag-in-box systems is carried out in this way (Robertson 1993).

One of the recent nonthermal methods of sterilization of surfaces of packaging materials, equipment, foods, and medical devices involves the use of intense and short duration pulses of broad-spectrum "white" light (Barbosa-Canovas et al. 1998; Robertson 2006). The *PureBright*™ process developed by PurePulse Technologies, Inc., uses a technique known as pulsed energy processing. Electrical energy is stored in a capacitor and is released from lamps in the form of short, high intensity pulses of light. These flashes last from 1 μs to 0.1 s and are typically applied at a rate of 1–20 flashes per second.

17.9.2.3 Verification of the Sterilization Processes

Validation of the presterilization of the aseptic packaging machine and of the sterilization of the packages requires a microbiological challenge. The method followed is presented by Elliott et al. (1992) and Cerny (1993), while the statistical considerations supporting the microbiological challenge are described by Moruzzi et al. (2000).

Challenging package sterilization. Sets of 100 containers are inoculated with 10^3, 10^4, and 10^5 spores per container of known resistance to the sterilizing agent and are left to dry. The test is performed in triplicate with resterilization of the equipment between runs. The aseptic packaging system is run as for a commercial test batch. The sterilizing critical factors, such as peroxide concentration, peroxide dosage, and drying air temperature are set at the minimum values that would be expected during regular production. Machine speed is set at the maximum. After sterilization, the containers are filled with an appropriate growth medium, incubated, and observed for growth. From the results, the LCR achieved is calculated to determine the adequacy of the sterilization process. It is very crucial to choose the proper indicator organism depending on the sterilization medium used. Bernard et al. (1993) and Cousin (1993) list recommended indicator organisms. For instance, for sterilization with H_2O_2 and UV or heat they propose *B. subtilis*.

Confirming equipment sterilization. If equipment sterilization is done with steam, sterile strips of aluminum foil are inoculated with 10^3, 10^4, and 10^5 spores of *Bacillus stearothermophilus* per strip and are placed on the surfaces to be sterilized. After the completion of the sterilization cycle, the equipment is disassembled, the strips are aseptically removed into tubes with an appropriate growth medium, and incubated at 55°C for 28 days. If all of the incubated strips are found to be sterile, then at least five decimal reductions were achieved. A similar method is used in the case of sterilization with hydrogen peroxide and heat. Spores of *B. subtilis* var. *globigii* are inoculated in this case.

Demonstrating sterile air. Air quality in the sterile zone should be demonstrated to be class 100 or better. This means that there are not more than 100 particles of size 0.5 μm or larger per cubic foot of air and not more than 100 microorganisms per m^3 of air (Elliott et al. 1992). Testing of air for particles is usually accomplished with a single-particle counter. Several kinds of air collectors are commercially available for sampling the environment for airborne microorganisms.

17.9.3 The Aseptic Zone

The aseptic zone is the area where the sterile container is filled with the sterile product and sealed. It begins at the point where the packaging material is sterilized or at the point the presterilized package enters the machine and ends after the package is sealed (Stevenson 1992). The aseptic zone must be sterilized before production begins, and sterility must be maintained during production. There should exist sterilizable physical barriers separating sterile from nonsterile areas. In addition, mechanisms should be provided to allow the entrance into the aseptic zone of sterile packaging material and the withdrawal of sealed packages, without compromising the sterility of the zone.

The aseptic zone is protected from contamination by steam or by sterile air or gases (Buchner 1993b). Steam is used in the Dole aseptic canning machine and in the filling zone of most bag-in-box systems. For the systems using sterile air or gases, either vertical laminar flow or turbulent ventilation and slight overpressure in totally closed machine cabinets are usually employed.

Before production begins, packaging machines, filling lines and fillers, and the air/gas system have to be sterilized. For the packaging machines, hydrogen peroxide solution, hot air, steam, and various disinfectant solutions are used as sterilizing agents (Buchner 1993b). Filling lines and fillers are first cleaned automatically by the CIP system and then are sterilized with saturated steam or pressurized superheated water. Sterile air is produced by filtering through high-efficiency particulate air (HEPA) filters or by incineration.

17.9.4 Aseptic Packaging Systems

Aseptic packaging systems have been classified into six categories based on the type of container used. Thus, there are can, bottle, sachet and pouch, cup, carton, and bulk packaging systems. In 1995 about 460 aseptic filling installations were operating in the Unites States; in 1999, 540; and in 2005, 560 installations (Brody 2006). In 1999, there were 23 different commercial aseptic packaging systems, while in 2005 there were 28, including 10 new ones. In 1999, the distribution of the 540 installations (Marcy 2000), according to the type of packaging used, was: 42% bag-in-box, 34% paperboard cartons, 8% metal cans, 8% plastic cups, and 8% pouches and bottles.

17.9.4.1 Can Systems

The Martin–Dole system (Lange 1989) was the first commercial aseptic packaging system and has operated successfully since 1950. It uses superheated steam to sterilize the metal cans (tinplate, ECCS, aluminum) and the can ends. For tinplate cans, the temperature must not exceed 232°C because at this temperature tin melts. Seaming is done with a conventional can double seamer, modified for aseptic operation. In the aseptic zone superheated steam is used to maintain aseptic conditions.

Dole has developed the "hot air" packaging system (Lange 1989) for packaging acid ($pH < 4.3$) noncarbonated beverages into composite cans, consisting of spirally wound body, made from laminations of Al foil, plastics and paper, and metal ends. The cans are sterilized by heating with hot air for 3 min to a temperature of 143°C. These products can also be hot-filled at 88°C–93°C, but in the case of heat-sensitive fruit juices, the aseptic filling process is preferable to preserve product quality.

17.9.4.2 Bottle Systems

There have been several attempts to commercialize aseptic packaging systems in glass bottles, but none has found widespread acceptance (Robertson 2006). In the system described by Buchner (1993a) the glass bottles are first rinsed with water, and then a mixture of hydrogen peroxide vapor and hot air is admitted into them. The peroxide condenses on all surfaces of the bottles, and after a certain exposure time it is dried off by sterile hot air.

For plastic bottles three types of aseptic packaging systems are in use (Robertson 2006). In the first system, the nonsterile bottles, after blowing, are sterilized by spraying their inside and outside surface with a H_2O_2 solution, which is subsequently evaporated as the bottles pass through a hot air tunnel. Then the bottles are rinsed with sterile water and filled. Hydrogen peroxide solution, at concentrations higher than 30%, seems to have some interaction with polyethylene terephthalate (Clark 2004). Thus, it is used only for the sterilization of HDPE bottles, whereas for PET bottles a mixture of peracetic acid and hydrogen peroxide is used.

In the second system, the bottles are blown sterile, that is, they are extruded, blown with sterile air, and then sealed to prevent contamination of their internal surface. Prior to filling, they enter a sterile chamber, where their outside surface is sterilized with a H_2O_2 spray. Then the closed tops of the bottles are cut away, and the bottles filled. Finally, they are capped with presterilized foil caps or heat-sealable closures (Robertson 1993).

In the third system, all operations, that is, bottle formation (parison extrusion and blow molding), filling, and sealing take place in sequence in a single mold (Zimmermann 1993). The thermoplastic resins usually used are HDPE, PP, and PETG. Sterility of the containers is achieved by the extrusion process at temperatures 170°C–230°C and holding time of several minutes. Sterile air is used for blowing the parison and forming the bottle. While the bottle is still inside the mold it is filled with the product, and next the top part of the bottle is formed into a cap that closes the bottle.

Newer developments of aseptic packaging in PET bottles (Brody 2009) include "dry sterilization" and "aseptic dry preform decontamination." In the first process, the PET bottles are prewarmed to 50°C–60°C with hot air. Next a mixture of H_2O_2 vapors and hot air is introduced into the bottle to sterilize it. There is no visible condensation on the surface because the bottles are hot from preheating. Finally, all peroxide is displaced from the bottle by purging with sterile air before the bottle reaches the filler. In the second process, the preforms are sterilized with gaseous H_2O_2 at the entrance of the reheat oven before blowing. Stretch blow molding is carried out with sterile air, and the empty sterile bottles are then aseptically conveyed to the aseptic filling system.

17.9.4.3 Pouch Systems

The film is drawn from a reel and is sterilized in a hot H_2O_2 bath, which also acts as a siphon lock to a sterile chamber (Buchner 1993b). Then it is drained, dried, and fed to a vertical form-fill-seal machine operating inside the sterile chamber. It is folded over a shoulder to form a tube and sealed at the longitudinal seam. The tube is closed at the bottom by the cross seal, and the pouch is filled with the product. The top transverse seal is made, and the pouch is drawn out of the sterile chamber through a flexible lock. The film used may be polyethylene or a laminate if a longer shelf life is required. Foods that are usually packaged by these machines are tomato products and sauces.

17.9.4.4 Cup Systems

The plastic cups can be either preformed or formed, filled, and sealed in a single machine. Preformed cups are usually made from high impact polystyrene (HIPS) or polypropylene. They can also be made from laminates if better barrier properties are required (Robertson 1993). The inside surface of preformed cups is sterilized either by spraying H_2O_2 and heating with hot compressed air or by dipping in a 35% peroxide bath at 85°C–90°C. The cup's closure is usually aluminum foil and has a thin coating of PE to provide heat sealability. It is typically sterilized with peroxide solution too (Robertson 1993). Foods usually packaged in aseptic cups include puddings, cream, and dairy products.

In the form-fill-seal machine, the plastic material used for thermoforming the cups is in the form of a web. Sheets made of any thermoplastic polymer can be used, the most common being polystyrene, because it is easily thermoformed. The shelf life of coffee creamer aseptically packaged in simple PS cups and at ambient conditions is at least 4 months. For longer shelf lives, improved barrier properties are required, and coextruded multilayer films are the solution. For aseptically packaged pudding, laminate of the following composition has been used: outer layer of HIPS, adhesive, barrier PVdC layer, inner layer of LDPE. The shelf life of the product in these cups and at ambient conditions is more than 12 months (Lutkemeyer 1989). Sterilization of the web is accomplished by passing it through a 35% H_2O_2 bath at room temperature. Next the web passes through a sterile tunnel where it is heated to 130°C–150°C to get sterilized, get rid of the remaining H_2O_2, and become soft enough for thermoforming. The cups are usually formed by a combination of mechanical forming and compressed air. Sterilization of the lidding material is done in a similar way to the cup web (Robertson 1993).

Another form-fill-seal system is the "Neutral Aseptic System" (NAS) supplied by ERCA and Conoffast (de Groof 1993). The system, although it did not have much commercial success, is quite ingenious. On one hand, it makes use of the high temperature reached during the extrusion process to ensure the sterility of the packaging material and on the other it protects the sterility of the food contact layer with a strippable PP layer. The film from which the cups are formed is composed of PP/PE/PVdC/PS. Within a sterile chamber first the outer layer of PP is peeled off, exposing a sterile inner surface, and then the film is thermoformed into cups. The lidding material is composed of three layers: Polybutylene/Polyethylene/Al foil. Inside the sterile chamber the polybutylene layer is peeled off, and the rest is heat-sealed on the cups.

17.9.4.5 Carton Systems

The cartons used for packaging fruit juices, long-life milk, soups, and sauces, and in general products with a minimum shelf life of 6 months at ambient temperature, are constructed from a laminate of seven layers. The layers, their typical thicknesses (Robertson 2006), and their functions (Schulte 1989; Strole 1989) are

- Outer polyethylene (15 g m^{-2}) protects the ink layer and the paperboard from moisture and permits the heat sealing of the flaps on the body of the carton; enables the package flaps to be sealed.
- Bleached paperboard serves as a carrier for the décor.
- Unbleached paperboard gives the package the required mechanical strength (typical thickness for both paperboards 186 g m^{-2}).
- Polyethylene (25 g m^{-2}) is the adhesive to bind the aluminum to the paperboard.
- Aluminum foil (6.3 μm) acts as a gas and light barrier.
- Two inner polyethylene layers (15 and 25 g m^{-2}) act as a liquid barrier and enable the formation of the transverse seals by heat sealing.

Some ingredients of fruit juices may cause, after a period of time, a reduction in the adhesion of the PE inner layer to the aluminum foil, resulting in the detachment of the PE inner layer. The problem is solved by applying two layers of PE, first a thin ethylene copolymer with good adhesion properties to aluminum and a second layer of standard PE (Strole 1989).

In contrast to the laminate described above, the cartons for pasteurized milk and pasteurized juices are constructed from laminate, which does not contain aluminum foil. It consists of polyethylene, paperboard, and polyethylene. Since the pasteurized products have a shelf life of a few days and they are stored under refrigeration, the protection offered by the aluminum foil is unnecessary; thus, a cheaper carton can be used. These cartons are usually of the gable-top type.

Through the years carton manufacturers and their customers developed cartons of new and exciting shapes, sizes, contours, and combinations of carton and plastic bottle. At the same time a whole range of add-ons, such as drinking straws, screw caps, and reclosures were applied to cartons to enhance consumer convenience (Robertson 2002). Another development is the use of barrier plastics instead of aluminum foil in the composition of the laminate. As a result there was a rapid increase in the range of foods packaged in cartons, from milk, cream, and juices at the beginning to soups, sauces, waters, wines, and teas nowadays.

A variation of the classic paperboard carton is one which features an all-plastic top on a round corner carton (Brody 2003). Another development is a retortable, square-shaped carton, for soups, ready meals, vegetables, and pet food, which was developed as a replacement for the metal can (Robertson 2002). The carton and its contents are subjected to the same high-pressure high-temperature retort process as metal cans and glass jars. The strength of the carton lies in the combination of the packaging material with a customized process to form and seal the package. The laminate has basically the same structure as the laminate of the aseptic carton but with polypropylene replacing polyethylene. Products packaged in this carton have a shelf life, under ambient conditions, of 18 months.

Stand-up aseptic pouches (100% transparent) were developed by replacing aluminum foil with silicon oxide-coated polyester film (PET/SiO_x) as the barrier layer and replacing the paperboard with plastics in the structure of the laminate of the aseptic carton. In addition, the absence of aluminum from the laminate permitted the use of these pouches in the microwave oven. This led to the introduction of microwavable stand-up aseptic pouches for packaging sauces, soups, tea, and coffee drinks.

There are two main types of aseptic packaging systems in cartons and these are described next.

17.9.4.5.1 Form-Fill-Seal Cartons

The Tetra Brik system is a typical example of a system of this type and is shown schematically in Figure 17.17 (Brennan et al. 1990). The seven-layer laminate is supplied in rolls and is already printed and creased. First, a polyethylene strip is attached to one edge of the laminate. One of the sterilization methods usually employed is the following: The laminate is fed through a deep bath containing 35% solution of H_2O_2 at 78°C. Next squeezer rollers remove surplus peroxide, and then sterile air at 125°C is blown over both surfaces of the sheet to heat and sterilize it and to evaporate the remaining peroxide. The laminate is next formed into a tube, and the longitudinal seal is made by a heat-sealer (Robertson 1993). The polyethylene strip binds together the overlapping edges of the sheet. The bottom transverse seal is made, the tube is filled with the product, and finally the top transverse seal is made. For transverse sealing the high-frequency induction heating technique (Schoefert 1993) is used. A short, high-frequency impulse of approximately 200 ms heats the aluminum layer of the laminate and melts the inner polyethylene layers, which bond together under the sealing bar pressure. It is possible to produce packages either without headspace, by making the top transverse seal below the level of the product, or with a headspace of up to 30% of the total volume by injecting sterile air or nitrogen. The sealed packages are finally pressed by molds into rectangular blocks, the top and bottom flaps are folded down and heat-sealed to the body of the carton. The sterilization of the laminate, forming of the carton, filling, and sealing are all carried

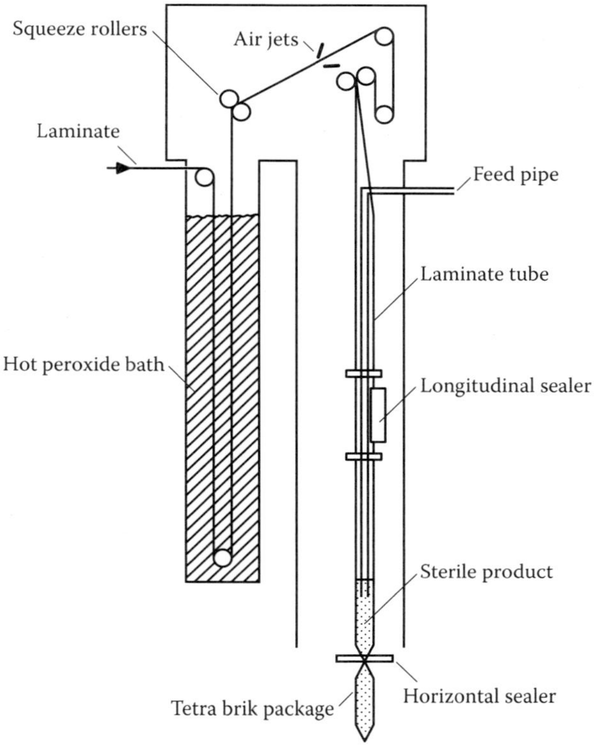

FIGURE 17.17 Schematic representation of the Tetra Brik aseptic packaging system. (From Brennan, J.G. et al., *Food Engineering Operations*, 3rd edn., Elsevier Applied Science Publishers, London, U.K., pp. 617–653, 1990. With permission.)

out inside a sterile chamber, maintained at a gauge pressure of 0.5 atm with sterile air (Robertson 1993). To meet the requirements of aseptic filling, it is essential that all machine parts that come in contact with the sterile product are sterilized before production starts. Two sterilization methods are available (Schulte 1989): treatment with hot air of 360°C or spraying with 30%–40% hydrogen peroxide solution and drying with hot air.

17.9.4.5.2 Prefabricated Cartons

The filler uses prefabricated carton sleeves or blanks, which are made at the carton manufacturer's installation according to the following procedure: First the laminate is produced in the form of a web, it is printed, and the blanks are die-cut, creased, and the longitudinal seam is made. The partly assembled lay-flat blanks are shipped in boxes to the food processor (Deimel 1989). At the processor's installation, each blank is opened up into a rectangle and pushed onto one of the mandrels of the revolving mandrel wheel. The polyethylene at the bottom end of the sleeve is softened by blasts of hot air (base activation). The base is folded, pressed, and heat-sealed against the end face of the mandrel. After that the carton's top is prefolded. All these operations take place under nonsterile conditions. The carton then passes to the aseptic zone, where sterility is maintained by a slight overpressure of sterile air. One method of sterilizing the inside surface of the carton is with 35% H_2O_2 solution. The peroxide is vaporized first and then is forced into the package by hot air under a slight overpressure. The vapor condenses on the cool carton walls. Hot air is introduced next into the carton to heat and sterilize it and to evaporate the peroxide (Deimel 1989). Alternatively, the inside surface of the carton is sprayed with a 1%–2% solution of H_2O_2, then irradiated with UV radiation, and finally the remaining peroxide is removed by jets of hot air (Robertson 1993).

Next, the carton is filled with the product, and for products that tend to foam the filler can be equipped with a defoaming unit. Most of the time a certain amount of headspace in the carton is necessary, and if required for the protection of the product the headspace can be filled with steam or nitrogen. After filling the carton, the top is folded and sealed with ultrasonics. The protruding flaps on each side are folded down and sealed to the package with hot air (Deimel 1989).

The main advantage claimed by the prefabricated carton system over the form-fill-seal carton system is that the difficult production step of longitudinal seam sealing is done by the system supplier and not the filler (Deimel 1989). The longitudinal seam of the prefabricated cartons is different from that of the form-fill-seal cartons; it is made by paring down the board layer and folding back the reduced caliper edge and sealing it by flame welding. Another advantage cited is the flexibility of the filling machines. The same machine can be used to produce different capacity cartons, provided that they are of the same cross section, and this is done simply by adjusting the height.

17.9.4.6 Bulk Packaging Systems

There are a number of bulk aseptic packaging systems used commercially, with typical container volumes of 10–1000 L (Robertson 1993). They are steel drums or bags-in-box. Large, nonrefrigerated tanks holding up to 10 m^3 of aseptic product are also used.

Steel drums have a tin coating on the inside surface. They typically have a capacity of 208 L (55 US gallons) and are used for packaging fruit juice concentrates and tomato concentrate. The ends are double seamed to the body of the drum, and filling and emptying is done through a hole. Drum sterilization is usually accomplished with saturated steam (Robertson 1993).

In *bag-in-box* packaging systems the food is filled into a plastic bag, which then is placed into a carton box, a metal drum, or a wooden crate. In the case of larger sizes, filling occurs after the bag has been placed inside the box, drum, or crate (Robertson 1993). The bags are sterilized usually by gamma irradiation in a special facility. They are delivered to the food processor premade, sterilized, and in lay-flat form. The bags are kept sealed until filling; thus, their inside surface remains sterile. They are manufactured from a variety of laminates suitable for the particular product and the shelf life required. Usually one of the laminate layers is a barrier material such as aluminum foil, PVdC, EVOH, or metallized PET. The outer layers are usually LDPE because of its good heat-sealing properties and barrier properties to liquids and moisture. A connector is fixed onto the bag. The aseptic filler consists of a filling valve and a sealed sterile chamber. Various designs of filling valves and connectors are used. Prior to filling, it is necessary to sterilize only the exposed surfaces in the filling valve and connector, which come into contact with the product. Sterilization is carried out with steam or with bactericide solutions. From the whole package only the filling valve and the connector have to be inside the sterile chamber.

17.9.5 Package Inspection and Testing

An important function of the package is to secure against the entry of microorganisms and to maintain the commercial sterility of its contents during and after processing. This function may be defeated if the package becomes defective, especially if it loses its hermetic condition. Both the FDA and USDA require that aseptical packages containing low-acid foods be inspected and tested for defects.

17.9.5.1 Definitions of Package Defects

A package defect can be classified as critical, major, and minor depending on its impact on hermetic condition (Gavin and Weddig 1995).

A *critical defect* (Gavin and Weddig 1995) provides evidence that the package has lost its hermetic condition or evidence of microbial growth in the package. A *major defect* (Gavin and Weddig 1995) does not show visible signs of the package losing its hermetic condition, but it is of such magnitude that the package is weakened and may have lost its hermetic condition. A *minor defect* has no adverse effect on the hermetic condition of the package. Although a minor defect does not cause safety problems, it can adversely affect the appearance and salability of the product.

For paperboard packages (cartons), flexible pouches, and semirigid plastic containers with heat-sealed lids, Gavin and Weddig (1995) and the NFPA Bulletin 41-L (Anonymous 1989a) present a list of the commonly observed defects and their classification as critical, major, or minor. For example, critical defects for paperboard packages are leakers (channel, corner, perforation, pull tab, and seal), cuts, and punctures that penetrate the package and swollen packages due to gas formation probably by microorganisms. The flexible package defects poster (Anonymous 1989b) has photographs of these defects to aid in their recognition. In addition, in the NFPA Bulletin 41-L good manufacturing guidelines and package integrity test methods are presented for the three types of flexible packages discussed.

17.9.5.2 Package Integrity Test Methods

Testing the integrity of a package refers either to leak detection or to structural integrity evaluation. Evaluation of seal integrity is a special case of leak detection. A leak can be defined as a physical opening on the package surface or in the seal that allows entry and/or exit of fluids, microorganisms, and/or gases (Floros and Gnanasekharan 1992). Two common types of leaks found in food packages are pinholes and channel leaks. The former are found in the body, while the latter are found in the seal area of the package. A channel leak is much longer and more difficult to detect than a pinhole. There is no agreement concerning the minimum leak size that the package integrity test should be able to detect. Obviously, this is the size that permits microbial contamination. Intuitively, one expects it to be equal to 0.5 μm, which is the size of a bacterium. In practice, this does not hold true and much research has been done on the subject and much remains to be done (Blakistone and Harper 1995). Yam (1995) believes that an inspection system can provide sufficient safety assurance if it can detect pinholes of 10 μm and channel leaks of 50 μm in diameter.

The food industry for many years has been using destructive methods for testing package integrity. Invasive tests that partially or completely destroy the package and are used on discrete samples or parts of a package are destructive (Floros and Gnanasekharan 1992). Destructive testing presumes an appropriate sampling scheme, and it is simply an indicator of packaging equipment performance. Unfortunately, it does not provide a way to isolate and reject defective packages. It is also expensive and time-consuming. On the other hand, nondestructive testing, especially if it can be on the spot, is highly desirable. Nondestructive, 100% in-line inspection is the absolute optimum.

Gavin and Weddig (1995) and the NFPA Bulletin 41-L (Anonymous 1989a) specify the frequency of testing and the minimum of tests required for paperboard packages, flexible pouches, and semirigid plastic containers with heat-sealed lids. In the first reference, a table is also presented summarizing the destructive and nondestructive methods used for testing the integrity of aseptic packages.

Destructive tests include the peel, the teardown, the bubble, the dye penetration, the electrolytic, the microbial challenge, the storage and distribution, the burst, and the seal tensile strength tests.

The *peel test* is applied to semirigid plastic containers with heat-sealed lids, while the *teardown test* is applied to both paperboard packages and flexible pouches (Anonymous 1989a, Gavin and Weddig 1995).

The *bubble test* is applied to packages containing air at atmospheric pressure in their headspace. The container is immersed in a water bath, and vacuum is drawn over the immersion vessel. Because of the pressure difference between the inside and the outside of the package, if there is a leak, air will flow from the inside to the outside. This air flow will become visible through the formation of air bubbles at the spot of the leak.

The *dye penetration test* (Anonymous 1989a, Gavin and Weddig 1995) can be implemented to all types of aseptic packages. After the product has been removed, the container is rinsed thoroughly with water and wiped dry. A dye solution is prepared by dissolving 5 g of rhodamine B powder in 1 L of isopropanol. Drops of dye are applied to the critical areas of the inside surface of the container, including the seals. Adequate time is allowed for the dye to seep into leaks, and then the seals are split open and are visually inspected for pink ink penetration. The external surface of the package is also inspected for the presence of the dye.

The *electrolytic* test can be applied to any aseptic package provided that the container has at least one layer that does not conduct electricity. The procedure described below (Anonymous 1989a, Gavin and Weddig 1995) refers to semirigid plastic containers with heat-sealed lids. The cup is cut open through the bottom, the product is drained, and the container is rinsed thoroughly with water and wiped dry. A sodium chloride solution is prepared by dissolving 10 g of sodium chloride in 1 L of water. The plastic container is half-filled with this solution and inserted upside down into a bowl containing the sodium chloride solution as well. Two electrodes are connected via wires to the poles of a 9 V battery. An ampmeter is placed in series in the circuit. One electrode is inserted into the solution in the cup and the other is placed in the solution of the bowl. Plastic packages generally do not conduct electricity; they are electrical insulators. If there is an indication in the ammeter of current flow, then there should be a hole somewhere in the cup or the seal for the circuit to close. It is possible for tight packages to give false positive results in the electrolytic test. Thus, in the case of positive result a dye test should also be performed.

Microbial challenge tests, also called *biotests*, have been designed to detect entry of microorganisms into any type of sealed container (Floros and Gnanasekharan 1992). The container is filled with a nutrient broth instead of food, sealed, and then immersed into a bacterial suspension for a specified period of time. Then the package is incubated for a certain length of time, and the extent of microbial growth in the container's contents is assessed. In theory, leaks on the container will allow the product inside to become contaminated and then to spoil. A pH indicator may also be incorporated into the nutrient broth to provide visual confirmation of microbial growth. The most commonly used microorganisms in microbial challenge tests are *Enterobacter aerogenes*, *Aerobacter aerogenes*, and *Pseudomonas aeruginosa*.

Storage and *distribution tests* include drop, vibration, and compression tests. The first two are used to simulate the abuse which packages will be subjected to during distribution (Floros and Gnanasekharan 1992). There are commercially available drop testing equipments that can simulate a variety of impact possibilities. Modern vibration testers can approximate realistically actual transportation effects. Finally, compression tests measure the ability of the package to withstand pressure during stacking in warehouses.

The *burst test* is applied to flexible pouches and provides an indication of the ability of the pouch to withstand internal pressure and of the maximum internal pressure required to cause failure of the seals (Floros and Gnanasekharan 1992). The AOAC-approved method for the burst test is described in the *AOAC International Bacteriological Analytical Manual*, 7th Edition (1992). When conducting this test, the pouch is sandwiched between two parallel restraining plates. A needle punctures the top of the pouch wall and compressed air is injected into the pouch at a controlled rate until a predetermined pressure is attained. This constant pressure is maintained for 30 s; then, it is released, and the seals are examined to ensure that they have stayed intact (Yam 1995). Burst tests have two basic controllable parameters: the pressurization rate and the distance between the restraining plates. The latter is not used in aseptic package testing, but it is applicable to retort pouches. The military specifications require that the meals-ready-to-eat (MRE) retort pouches should be able to withstand 20 psig for 30 s, when the restraining plates are 0.5 in. apart (Yam 1995).

The *seal tensile strength test* is applied to flexible pouches and is basically a tensile test performed by a universal testing machine or its equivalent. The procedure is described in ASTM D 882 method "A" or "B." A strip containing a part of the seal to be examined is cut from the package. The two ends of the strip are gripped between the jaws (one movable, the other fixed) of the universal testing machine and are pulled apart at a controlled rate. The force per unit length of the seal required to separate the two sealed surfaces is a measure of the seal's strength and should not be less than specified for the particular material and the particular application. Besides the measurement, the tear at each seal is visually inspected, and its appearance should conform to what the manufacturer considers as an adequate seal.

Nondestructive tests include the visual examination, the squeeze test, the vacuum chamber test, and many other testing methods based on pressure difference, machine imaging, infrared imaging, spectrophotometers, capacitance measurement, x-rays, ultrasound, and magnetic resonance imaging.

The *visual examination* is the simplest nondestructive test and can be applied to all types of packages. It typically involves inspecting the seals for normal appearance, absence of voids, wrinkles or pleats, correct seal alignment, presence of product in the seal area, and delamination of the packaging material. Dimensional checks of the seals may also be performed. The photographs of the various package defects contained in the flexible package defects poster (Anonymous 1989b) may be particularly useful to the operator conducting the visual examination.

The *squeeze test* for flexible pouches and semirigid plastic containers with heat-sealed lids is described by Gavin and Weddig (1995).

The *vacuum chamber test* (Gavin and Weddig 1995) is for semirigid plastic containers with heat-sealed lids. The containers are placed in carriers within a vacuum chamber. There are proximity sensors with a dial indicator just touching the lids. The chamber is evacuated to 20 in. Hg. The headspace gases will expand causing the lids to dome and the needle of the dial indicator of the proximity sensors to deflect. If the needle deflection does not remain constant and decreases over time, there might exist a microleakage in the container.

When a pressure differential exists across the wall of a package, if there is a leak, gas will flow in or out of the package. An observed gas flow is an indication of leaks. Gas flow is usually detected by measuring either pressure changes using a very sensitive pressure sensor or deflections of the package wall caused by the gas flow, using a proximity sensor, or detecting the presence in the outside environment of a trace gas which is naturally present in the package headspace or had been introduced into it. The pressure outside the package can be lower than atmospheric (vacuum method) or higher (external pressure method). This is the basic principle of all *pressure difference techniques* used for leak detection and all possible combinations of the ways to create the pressure difference and to detect the gas flow have been applied to commercial equipment. Pressure difference testing is currently the most popular nondestructive technique for flexible package evaluation. Blakistone and Harper (1995) present a list of commercially available testing equipment that employ the pressure difference technique, along with details about the way the pressure difference is created and the gas flow is detected.

In the vacuum method, the package is placed inside an enclosed chamber where vacuum is drawn. Because of the pressure difference, gas or liquid will flow out through the leaks. For the vacuum method to work, the package must have a headspace, and a certain amount of residual gas must exist in the headspace. It works best for packages containing dry products and detects gross leaks (>100 μm). In the external pressure method, the high (up to 7 atm) pressure outside the package causes the test gas to flow to the inside. Compared to the vacuum method, this method does not require residual gas in the package; it is quicker and can detect smaller leaks (Yam 1995).

A whole array of other methods for packaging defect detection have been tried and are being developed, without achieving commercial success yet. These include machine imaging, infrared imaging, spectrophotometers, capacitance measurement, x-rays, ultrasound, and magnetic resonance imaging. Blakistone and Harper (1995) present a list of these methods and a few details about the resolution, cost, speed, applications, and limitations of each method.

REFERENCES

Amman, S. 1989. Aseptic packaging in polypropylene cups and their sterilization with hot air/superheated steam mixture. In *Aseptic Packaging of Food*, H. Reuter (ed.), pp. 173–189. Lancaster, PA: Technomic Publishing Company Inc.

Anonymous. 1986. *Guidelines for Can Manufacturers and Food Canners*. Rome, Italy: FAO Food and Nutrition Paper 36.

Anonymous. 1989a. *Flexible Package Integrity Bulletin.* NFPA Bulletin 41-L. Washington, DC: Flexible Package Integrity Committee of the National Food Processors Association.

Anonymous. 1989b. *Flexible Package Defects Poster*, Flexible Package Integrity Committee of the National Food Processors Association and U.S. Food and Drug Administration. Gaithersburg, MD: AOAC International.

Anonymous. 1996. *Packaging the Facts 4.* Leicestershire, U.K.: The Institute of Packaging.

Anonymous. 2005. *Glass Packaging Essentials: A Multimedia Resource CD-ROM.* Alexandria, VA: Glass Packaging Institute.

ASTM. 1999. *Standard Terminology of Glass and Glass Products*, Annual Book of ASTM Standards C162. Philadelphia, PA: American Society for Testing and Materials.

Azanha, A. B. and J. A. F. Faria. 2005. Use of mathematical models for estimating the shelf life of cornflakes in flexible packaging. *Packaging Technology and Science* 18:171–178.

Baird, R. J. 1976. *Industrial Plastics, Basic Chemistry, Major Resins, Modern Industrial Processes.* South Holland, IL: The Goodheart-Willcox Co. Inc.

Barbosa-Canovas, G. V., U. R. Pothakamury, E. Palou, and B. G. Swanson. 1998. *Nonthermal Preservation of Foods*, pp. 139–159. New York: Marcel Dekker Inc.

Bernard, D. T., A. Gavin, V. N. Scott, D. I. Chandarana, G. Arndt, and B. Shafer. 1993. Establishing the aseptic processing and packaging operation. In *Principles of Aseptic Processing and Packaging*, J. V. Chambers and P. E. Nelson (eds.), 2nd edn., pp. 223–243. Washington, DC: The Food Processors Institute.

Blakistone, B. A. and C. L. Harper. 1995. New developments in seal integrity testing. In *Plastic Package Integrity Testing, Assuring Seal Quality*, B. A. Blakistone and C. L. Harper (eds.), pp. 1–9. Herndon, VA: Institute of Packaging Professionals.

Board, P. W. and R. J. Steele. 1975. Diagnosis of Corrosion Problems in Tinplate Food Cans. Technical Paper No. 41, CSIRO Division of Food Research, Sydney, Australia.

Brennan, J. G., J. R. Butters, N. D. Cowell, and A. E. V. Lilley. 1990. *Food Engineering Operations*, 3rd edn., pp. 617–653. London, U.K.: Elsevier Applied Science Publishers.

Briston, J. H. 1980. Rigid plastics packaging. In *Developments in Food Packaging*-1, S. J. Palling (ed.), pp. 27–53. London, U.K.: Applied Science Publishers Ltd.

Briston, J. H. and L. L. Katan. 1974. *Plastics in Contact with Food.* London, U.K.: Food Trade Press Ltd.

Brody, A. L. 2000. The when and why of aseptic packaging. *Food Technology* 54(9): 101–102.

Brody, A. L. 2003. A wonderful world for aseptic packaging. *Food Technology* 57(5): 92–94.

Brody, A. L. 2006. Aseptic and extended-shelf life packaging. *Food Technology* 02.06: 66–68.

Brody, A. L. 2008. Packaging by the numbers. *Food Technology* 02.08: 89–91.

Brody, A. L. 2009. Aseptic Packaging 2009. *Food Technology* 07.09: 70–72.

Brody, A. L., B. Bugusu, J. H. Han, C. K. Sand, and T. H. McHugh. 2008. Innovative food packaging solutions. *Journal of Food Science* 73(8): R107–R116.

Brody, A. L. and K. S. Marsh (eds.). 1997. *The Wiley Encyclopedia of Packaging Technology*, 2nd edn. New York: John Wiley & Sons Inc.

Brown, W. E. 1992. *Plastics in Food Packaging: Properties, Design and Fabrication.* New York: Marcel Dekker, Inc.

Buchner, N. 1993a. Aseptic packaging in glass and plastic bottles. In *Aseptic Processing of Foods*, H. Reuter (ed.), pp. 235–243. Lancaster, PA: Technomic Publishing Company Inc.

Buchner, N. 1993b. Aseptic processing and packaging of food particulates. In *Aseptic Processing and Packaging of Particulate Foods*, E. M. A. Willhoft (ed.), pp. 1–22. London, U.K.: Blackie Academic & Professional.

Canadian Food Inspection Agency. 1997. *Metal Can Defects Identification and Classification Manual.* Government of Canada. Available at: http://www.inspection.gc.ca/food/fish-and-seafood/manuals/metal-can-defects/eng/1348848316976/1348849127902, (accessed May 31, 2015).

Carslaw, H. S. and J. C. Jaeger. 1959. *Conduction of Heat in Solids*, 2nd edn. Oxford, U.K.: Oxford University Press.

Cavanagh, J. 1997. Glass container design and glass container manufacturing. In *The Wiley Encyclopedia of Packaging Technology*, A. L. Brody and K. S. Marsh (eds.), 2nd edn., pp. 471–484. New York: John Wiley & Sons Inc.

Cerny, G. 1993. Testing of aseptic machines for their efficiency of sterilization of packaging materials by means of hydrogen peroxide. In *Aseptic Processing of Foods*, H. Reuter (ed.), pp. 307–313. Lancaster, PA: Technomic Publishing Company Inc.

Chung, D. H., S. E. Papadakis, and K. L. Yam. 2003. Simple models for evaluating effects of small leaks on the gas barrier properties of food packages. *Packaging Technology and Science* 16:77–86.

Clark, J. P. 2004. Aseptic processing: New and old. *Food Technology* 58(11): 80, 88, 89.

Coles, R., D. McDowell, and M. J. Kirwan (eds.). 2003. *Food Packaging Technology*. Oxford, U.K.: Blackwell Publishing Ltd.

Cousin, M. A. 1993. Microbiology of aseptic processing and packaging. In *Principles of Aseptic Processing and Packaging*, J. V. Chambers and P. E. Nelson (eds.), 2nd edn., pp. 47–86. Washington, DC: The Food Processors Institute.

de Groof, B. 1993. Thermoformable barrier sheets for shelf stable container in dairy applications. In *Aseptic Processing of Foods*, H. Reuter (ed.), pp. 281–291. Lancaster, PA: Technomic Publishing Company Inc.

Deimel, G. 1989. Aseptic carton packaging on the basis of prefabricated carton sleeves. In *Aseptic Packaging of Food*, H. Reuter (ed.), pp. 126–133. Lancaster, PA: Technomic Publishing Company Inc.

Elliott, P. H., G. M. Evancho, and D. L. Zink. 1992. Microbiological evaluation of low-acid aseptic fillers. *Food Technology* 1992: 116–122.

Fellows, P. J. 1990. *Food Processing Technology: Principles and Practice*, pp. 429–439. London, U.K.: Ellis Horwood Ltd.

Floros, J. D. and V. Gnanasekharan. 1992. Principles, technology and applications of destructive and nondestructive package integrity testing. In *Advances in Aseptic Processing Technologies*, R. K. Singh and P. E. Nelson (eds.), pp. 157–188. London, U.K.: Elsevier Sci. Publ.

Gavin, A. and L. M. Weddig 1995. *Canned Food: Principles of Thermal Process, Control, Acidification and Container Closure Evaluation*, 6th edn. Washington DC: The Food Processors Institute.

Goddard, R. R. 1980. Flexible plastics packaging. In *Developments in Food Packaging*-1, S. J. Palling (ed.), pp. 55–79. London, U.K.: Applied Science Publishers Ltd.

Grulke, E. A. 1994. *Polymer Process Engineering*. Englewood Cliffs, NJ: PTR Prentice Hall.

Hanlon, J. F. 1984. *Handbook of Package Engineering*, 2nd edn. New York: McGraw-Hill Book Co.

Hernandez, R. J. 1997. Food packaging materials, barrier properties, and selection. In *Handbook of Food Engineering Practice*, K. J. Valentas, E. Rotstein, and R. P. Singh (eds.), pp. 291–360. Boca Raton, FL: CRC Press.

Jasse, B., A. M. Seuvre, and M. Mathlouthi 1994. Permeability and structure in polymeric packaging materials. In *Food Packaging and Preservation*, M. Mathlouthi (ed.), pp. 1–22. Glasgow, U.K.: Blackie Academic and Professional.

Jenkins, W. A. and J. P. Harrington. 1991. *Packaging Foods with Plastics*. Basel, Switzerland: Technomic Publishing AG.

Kadoya, T. (ed.). 1990. *Food Packaging*. San Diego, CA: Academic Press Inc.

Karel, M. 1975. Protective packaging of foods. In *Principles of Food Science, Part II, Physical Principles of Food Preservation*, M. Karel, O. R. Fennema, and D. B. Lund (eds.), pp. 399–466. New York: Marcel Dekker Inc.

Kirwan, M. J. 2003. Paper and paperboard packaging. In *Food Packaging Technology*, R. Coles, D. Mcdowell, and M. J. Kirwan (eds.), pp. 241–281. Oxford, U.K.: Blackwell Publishing Ltd.

Kirwan, M. J. 2008. *Paper and Paperboard Packaging Technology*. Oxford, U.K.: Blackwell Publishing Ltd.

Krochta, J. M. 2007. Food packaging. In *Handbook of Food Engineering*, D. R. Heldman and D. B. Lund (eds.), 2nd edn., pp. 847–927. Boca Raton, FL: CRC Press, Taylor & Francis Group.

Labuza, T. P., S. Mizrahi, and M. Karel. 1972. Mathematical models for the optimization of flexible film packaging of foods for storage. *Transactions of the ASAE* 15:150–155.

Lange, H. J. 1989. Aseptic processing and packaging (APP) of foods in cans. In *Aseptic Packaging of Food*, H. Reuter (ed.), pp. 190–206. Lancaster, PA: Technomic Publishing Company Inc.

Lange, J. and Y. Wyser. 2003. Recent innovations in barrier technologies for plastic packaging—a review. *Packaging Technology and Science*, 16:149–158.

Leadbitter, J. 2003. *Packaging Materials: 5 Polyvinyl Chloride (PVC) for Food Packaging Applications*. Brussels, Belgium: ILSI Europe.

Lee, D. S., K. L. Yam, and L. Piergiovanni 2008. *Food Packaging Science and Technology*. Boca Raton, FL: CRC Press, Taylor & Francis Group.

Lutkemeyer, B. 1989. Aseptic packaging in plastic materials from the reel. In *Aseptic Packaging of Food*, H. Reuter (ed.), pp. 134–141. Lancaster, PA: Technomic Publishing Company Inc.

Mabee, M. S. 1997. Aseptic packaging. In *Encyclopedia of Packaging Technology*, A. L. Brody and K. S. Marsh (eds.), 2nd edn., pp. 41–45. New York: John Wiley & Sons.

Macrogalleria®, http://pslc.ws/macrog/index.htm, (accessed May 31, 2015).

Malin, J. D. 1980. Metal containers and closures. In *Developments in Food Packaging*-1, S. J. Palling (ed.), pp. 1–26. London, U.K.: Applied Science Publishers Ltd.

Mannheim, C. 1986. Interaction between metal cans and food products. In *Food Product—Package Compatibility*, J. I. Gary, B. R. Harte, and J. Miltz (eds.), pp. 105–133. Lancaster, PA: Technomic Publishing Co. Inc.

Marcy, J. E. 2000. *Aseptic Processing and Packaging Course*. New Brunswick, NJ: Continuing Professional Education, Cook College, Rutgers, The State University of New Jersey, March 22–24.

Marsh, K. and B. Bugusu. 2007a. Food packaging and its environmental impact. *Food Technology* 04.07: 46–50.

Marsh, K. and B. Bugusu. 2007b. Food packaging-roles, materials, and environmental issues. *Journal of Food Science* 72(3): R39–R55.

Matsubayashi, H. 1990. Metal containers. In *Food Packaging*, T. Kadoya (ed.), pp. 85–104. San Diego, CA: Academic Press Inc.

Matthews, V. 2000. *Packaging Materials: 1. Polyethylene Terephthalate (PET) for Food Packaging Applications*. Brussels, Belgium: ILSI Europe.

Mauer, L. J. and B. F. Ozen. 2004. Food packaging. In *Food Processing: Principles and Applications*, J. S. Smith and Y. H. Hui (eds.), pp. 101–131. Oxford, U.K.: Blackwell Publishing Ltd.

Moruzzi, G., W. E. Garthright, and J. D. Floros. 2000. Aseptic packaging machine pre-sterilization: statistical aspects of microbiological validation. *Food Control* 11: 57–66.

Nindl, G. 2004. Hydrogen peroxide-from oxidative stressor to redox regulator. *Cell Science* 1(2): 1–12.

Osborne, D. G. 1980. Glass. In *Developments in Food Packaging*-1, S. J. Palling (ed.), pp. 81–115. London, U.K.: Applied Science Publishers Ltd.

Ottenio, D., J. Y. Escabasse, and B. Podd. 2004. *Packaging Materials: 6. Paper and Board for Food Packaging Applications*. Brussels, Belgium : ILSI Europe.

Page, B., M. Edwards, and N. May. 2003. Metal cans. In *Food Packaging Technology*, R. Coles, D. McDowell, and M. J. Kirwan (eds.), pp. 120–151. Oxford, U.K.: Blackwell Publishing Ltd.

Palling, S. J. (ed.). 1980. *Developments in Food Packaging*-1. London, U.K.: Applied Science Publishers Ltd.

Reuter, H. 1989a. Aseptic packaging of food. In *Aseptic Packaging of Food*, H. Reuter (ed.), pp. 3–10. Lancaster, PA: Technomic Publishing Company Inc.

Reuter, H. 1989b. Evaluation criteria for aseptic filling and packaging systems. In *Aseptic Packaging of Food*, H. Reuter (ed.), pp. 95–108. Lancaster, PA: Technomic Publishing Company Inc.

Reuter, H. 1993. Processes for packaging materials sterilization and system requirements. In *Aseptic Processing of Foods*, H. Reuter (ed.), pp. 155–165. Lancaster, PA: Technomic Publishing Company Inc.

Robertson, G. L. 1993. *Food Packaging: Principles and Practice*. New York: Marcel Dekker Inc.

Robertson, G. L. 2002. The paper beverage carton: Past and future. *Food Technology* 56(7): 46–52.

Robertson, G. L. 2006. *Food Packaging: Principles and Practice*, 2nd edn. Boca Raton, FL: CRC Press, Taylor & Francis Group.

Rodriguez, F. 1982. *Principles of Polymer Systems*, 2nd edn. New York: Hemisphere Publishing Corporation, McGraw-Hill Book Co.

Salame, M. 1974. The use of low permeation thermoplastics in food and beverage packaging. In *Permeability of Plastic Films and Coatings*, H. B. Hopfenberg (ed.), p. 275. New York: Plenum Publishing Corporation.

Schoefert, E. 1993. Roll-fed carton packaging. In *Aseptic Processing of Foods*, H. Reuter (ed.), pp. 167–178. Lancaster, PA: Technomic Publishing Company Inc.

Schulte, D. 1989. Aseptic filling of carton packages from the roll. In *Aseptic Packaging of Food*, H. Reuter (ed.), pp. 109–125. Lancaster, PA: Technomic Publishing Company Inc.

Shreve, R. N. and J. A. Jr. Brink. 1977. *Chemical Process Industries*, 4th edn., pp. 179–197. New York: McGraw-Hill Book Co.

Soroka, W. 1996. *Fundamentals of Packaging Technology*, revised UK edition. Leicestershire, U.K.: The Institute of Packaging.

Stevenson, K. 1992. Aseptic processing and packaging systems. In *Encyclopedia of Food Science and Technology*, Y. H. Hui (ed.), pp. 128–136. New York: John Wiley & Sons.

Strole, U. 1989. Carton laminates for aseptic packaging. In *Aseptic Packaging of Food*, H. Reuter (ed.), pp. 221–226. Lancaster, PA: Technomic Publishing Company Inc.

Taoukis, P. S., A. El Meskine, and T. P. Labuza. 1988. Moisture Transfer and Shelf Life of Packaged Foods. In *Food and Packaging Interactions*, ACS Symposium Series No. 365, J. H. Hotchkiss (ed.), pp. 243–261. Washington, DC: American Chemical Society.

Tice, P. 2002a. *Packaging Materials: 3. Polypropylene as a Packaging Material for Foods and Beverages*. Brussels, Belgium: ILSI Europe.

Tice, P. 2002b. *Packaging Materials: 2. Polystyrene for Food Packaging Applications*. Brussels, Belgium: ILSI Europe.

Tice, P. 2003. *Packaging Materials: 4. Polyethylene for Food Packaging Applications*. Brussels, Belgium: ILSI Europe.

Turner, T. A. 2001. *Canmaking for Can Fillers*. Sheffield, U.K.: Sheffield Academic Press.

Wranglen, G. 1985. *An Introduction to Corrosion and Protection of Metals*. London, U.K.: Chapman and Hall.

Yam, K. L. June/July 1995. On-line, non-destructive system inspects integrity of pouches. *Packaging Technology & Engineering* 46–49.

Yam, K. L. 1995. Pressure differential techniques for package integrity inspection. In *Plastic Package Integrity Testing, Assuring Seal Quality*, B. A. Blakistone and C. L. Harper (eds.), pp. 137–145. Herndon, VA: Institute of Packaging Professionals.

Zimmermann, L. 1993. Manufacturing, filling and sealing of plastic bottles in the blow mould. In *Aseptic Processing of Foods*, H. Reuter (ed.), pp. 225–234. Lancaster, PA: Technomic Publishing Company Inc.

18 Modified Atmosphere Packaging of Fruits and Vegetables

E. Manolopoulou and Theodoros Varzakas

CONTENTS

18.1 INTRODUCTION

Fruits and vegetables are important parts of the human diet as they are the major sources of essential nutrients such as vitamins, minerals, complex carbohydrates, and antioxidants (Lee et al., 1995). They also contain an immense variety of biologically active secondary metabolites that reduce the risks of cancer and heart/circulation diseases (Johnson et al., 1994). Consumption of fresh fruits and vegetables is on the rise because of the increasing awareness of their nutritional importance. Fruits and vegetables are highly perishable, as they continue their metabolic processes after harvest and must be stored to maintain the fresh quality as long as possible. The most important factor in maintaining quality and extending the shelf life of fruit and vegetables after harvest is temperature. Most of the physical, biochemical, microbiological, and physiological reactions contributing to deterioration of produce quality are largely dependent on temperature. Metabolic processes including respiration, transpiration, and ripening are particularly temperature dependent (Ryall and Lipton, 1979; De Wel et al., 1982; Mitchell, 1992). Conventional cold storage can be optimized by modification of the atmosphere surrounding the product to create a new atmosphere that usually has a lower level of O_2 and a higher level of CO_2. At these levels of O_2 and CO_2, the respiration rate of most commodities will decrease and their shelf life will be extended (Geeson, 1990). Creating and maintaining the optimal atmosphere to achieve this benefit is based on packaging with plastic films known as modified atmosphere packaging (MAP; Marcellin, 1974; Lee et al., 1995, 1996).

MAP may be defined as "the enclosure of food products in gas-barrier materials, in which the gaseous environment has been changed" (Young et al., 1988). The purpose of MAP is to inhibit spoilage agents and therefore maintain a higher quality within a perishable food during its natural life or actually extend the shelf life. The absolute or desirable levels of these environmental variables differ according to commodity and stage of development. Moreover, tissue response may vary because of interactions among these variables as influenced by variety, preharvest conditions, and climatic factors.

The fact that this technology utilizes natural gases and would not cause harmful effects to human health and to the environment made it very appealing as an alternative to chemicals commonly used in foods for the control of diseases and insects. MAP has been developed over the last decades as a technique to retain high quality of fruits and vegetables. Commercially, MAP is used in several intact and fresh-cut commodities.

In recent years, the rapid growth of fresh-cut (minimally processed) products made the use of MAP necessary (Lange, 2000). MAP is especially important for these products because of their greater susceptibility to water loss, cut-surface browning, higher respiration rates, enhanced ethylene biosynthesis and action, and microbial growth (Gorny, 1997).

Recent advances in the design and manufacturing of polymeric films with a wide range of gas-diffusion characteristics have given impetus to the implementation of MAP. Also, the increased availability of various absorbers and adsorbers of O_2, CO_2, C_2H_4, and water vapor provides possible additional tools.

18.2 MODIFIED ATMOSPHERE PACKAGING

The term MAP can be defined as an alteration in the composition of gases in and around fresh produce through respiration and transpiration when such commodities are sealed in plastic films. With proper temperature control and atmosphere modification, the shelf life of many agricultural products may be extended for long periods. Modified atmosphere is used as a supplement to low-temperature preservation of fruits and vegetables. Fruits and vegetables are enclosed in a sealed pack, typically covered with a thin, gas-permeable plastic. The equilibrium levels of O_2 and CO_2 achieved inside the package are functions of the commodity, product weight, respiration rate, package gas permeability and area, and temperature (Chinnan, 1989). An equilibrium modified atmosphere is established when the rate of O_2 and CO_2 transmission through the package equals the product's respiration rate.

The composition of the atmosphere inside the package is complicated by the fact that both respiration rates and package permeability are variables and directly proportional to temperature (Ooraikul, 1991; Tano et al., 2007). It should be noted that the temperature dependence of the respiration rate is different from that of the permeability of films (Cameron, 2003). In fact, respiration rates of fruits and vegetables increase more with temperature than gas permeability of films do (Exama et al., 1993). Owing to this fact, it is difficult to maintain an optimum atmosphere inside a package when the surrounding temperature is not constant. The maintenance of a constant optimal temperature throughout the postharvest handling chain (i.e., from the grower to the retail display) is one of the most difficult tasks. In modified atmosphere, the composition of the storage atmosphere is not closely controlled.

The atmosphere is commonly composed of N_2, O_2, and CO_2. Although ordinary air consists mainly of these three gases (78% N_2, 21% O_2, 0.04% CO_2), the beneficial effect of MAP is obtained by applying the gases in an altered ratio (low in O_2 and/or high in CO_2), which may be different for different products. The choice of gas is totally dependent on the food product being packaged. An optimal atmosphere composition influences the metabolism (reduce respiration rate, ethylene biosynthesis, and sensitivity to ethylene) of the product being packaged and the activity of decay-causing organisms and increases the storability and/or the shelf life (Kader et al., 1989; Ooraikul and Stiles, 1991; Gorrish and Peppelenbos, 1992; Church and Parsons, 1995;

Beaudry, 2000; Iqbal et al., 2009). In addition, MAP improves moisture retention, which can have a greater influence on preserving quality than O_2 and CO_2 levels in some raw products (Faber, 1991; Blakistone, 1997; Kader, 1997; Richardson and Kupferman, 1997; Saltveit, 1997; Mahajan et al., 2007; Goulas, 2008).

Oxygen concentrations greater than 21 kPa may influence postharvest physiology and maintain the quality of fresh horticultural perishables. The beneficial effects of high O_2 include preventing anaerobic fermentation and avoiding the development of off-flavor and off-odor, retarding microbial growth, and inhibiting enzymic discoloration in fruits (Day, 1996; Tian et al., 2002). Superatmospheric O_2 levels (>21%) may be used in combination with fungistatic CO_2 levels (>15%) for a few commodities that do not tolerate these elevated CO_2 atmospheres when combined with air or low O_2 atmospheres (Kader and Watkins, 2000).

The three major gases used in the MAP of foods are oxygen (O_2), nitrogen (N_2), and carbon dioxide (CO_2).

Oxygen sustains aerobic respiration, inhibits the growth of anaerobic microorganisms, promotes the growth of aerobic microbes, and is responsible for several undesirable reactions in foods, including oxidation and rancidity of fats and oils and rapid ripening and senescence of fruits and vegetables (Floros and Matsos, 2005).

Nitrogen is a relatively inert and odorless gas, with low solubility in both water and lipid. It is used to displace O_2 in packs and storage vessels so as to delay oxidative rancidity and inhibit the growth of aerobic microorganisms. Owing to its low solubility, it is used as a filler gas to prevent pack collapse (snuffing), which can be a problem in atmospheres containing high CO_2 concentrations (Church, 1994; Fellows, 2000).

Carbon dioxide is a colorless gas with a slightly pungent odor at very high concentrations. It is an asphyxiant and slightly corrosive in the presence of moisture. CO_2 has a bacteriostatic effect; it slows down the respiration of many products, inhibits ethylene action, retards fruit softening, and in certain chilling-sensitive fruits reduces chilling injury (Burg and Burg, 1967; Abeles et al., 1992; Saltveit et al., 1998; Wang, 2006). It is soluble in both water and lipids, and its solubility increases with decreasing temperatures. The dissolution of CO_2 in the product can result in package collapse (Floros and Matsos, 2005); the dissolution in water produces carbonic acid (H_2CO_3) that increases the acidity of the solution and reduces the pH.

Several other gases such as carbon monoxide, ozone, ethylene oxide, nitrous oxide, helium, neon, argon (increases shelf lives of some fruits and vegetables), propylene oxide, ethanol vapor, hydrogen, sulfur dioxide, and chlorine have been used experimentally or on a restricted commercial basis to extend the shelf life of a number of food products (Day, 1993; Barry and O'Beirne, 2000; Rocculi et al., 2005); however, few scientific reports have confirmed this (Sivertsvik et al., 2000).

MAP may be accomplished with polymeric films, rigid plastic trays, or preformed pouches closed by heat sealing (Geeson et al., 1985; Barmore, 1987; Geeson, 1988). With polymeric films, the desired interior atmosphere is obtained by selecting a film area and permeability that are appropriate for the respiration rate of the product inside the package (Stannett, 1968; Doyon et al., 1991). When impermeable plastics are used, proper gas exchanges may be achieved by using perforations or silicone rubber window (Marcellin, 1974, Geeson, 1988; Watkins et al., 1988; Emond et al., 1991; Renault et al., 1994a,b). The consequences of poor package design are significant. If a film of excessive gas permeability is used, there will be no atmosphere modification. Conversely, if a film of insufficient permeability is used, an atmosphere with an O_2 content of <2% (v/v) will develop (Priepke et al., 1976; Myers, 1989), causing quality losses due to both anaerobic respiration and specific physiological disorders (Kasmire et al., 1974; Brecht, 1980; Knee and Hadfield, 1981; Rizvi, 1981; Zagory and Kader, 1988). The exact O_2 concentration at which anaerobic respiration begins also depends on the type of produce, the storage temperature, and the CO_2 concentration (Zagory et al., 1989). The effects of low O_2 and high CO_2 concentrations on fruit and vegetables have been shown to depend on the duration of storage under these conditions. The damage caused by anoxic conditions has been found to be irreversible when caused by CO_2 concentrations >20% (Kader, 1986). The efficacy of MAP requires that the recommended steady atmosphere must be reached quickly without anoxia conditions or generating excessive CO_2 levels. For respiration product levels of approx. 5% CO_2 and O_2 are usually used with the remainder being N_2 in order to minimize the respiration rate (Day, 1993).

MAP of fresh or minimally processed fruits and vegetables can be achieved in either a passive or an active way. In the first case, modification of the atmosphere is attained through respiration of the commodity within the package and depends on the characteristics of the commodity and the packaging film (Smith et al., 1987). The main disadvantage of the passive atmosphere modification method is that the desired atmosphere is achieved very slowly. This can sometimes result in uncontrolled levels of oxygen, carbon dioxide, or ethylene, with a detrimental effect on the quality of the product; for these reasons, atmospheres within MAP may be actively established. In order to speed this process, the pack can be flushed with N_2 to reduce the O_2 rapidly, or the atmosphere can be flushed with an appropriate mixture of CO_2, O_2, and N_2. In other cases, the pack can be connected to a vacuum pump to create a slight vacuum, replacing the package atmosphere with the desired gas mixture. Compared with the passive method, active atmosphere modification is practically instantaneous and takes place at the beginning of storage. The main goal of active MAP is to shorten or avoid the transient period that could be detrimental when products are sensitive to enzymatic browning, such as fresh-cuts (Charles et al., 2003; Guillaume et al., 2011). Active packaging extends the shelf life of foods while maintaining their nutritional quality, inhibiting the growth of pathogenic and spoilage microorganisms (Labuza and Breene, 1989; Hotchkiss, 1995).

The gas concentration in the packs can be further adjusted and maintained through the use of absorbing or adsorbing substances in the package to scavenge O_2, CO_2, H_2O, and C_2H_4. This procedure is most suitable for highly perishable commodities. Typical examples of active packaging methods are as follows (Hurme et al., 2000; Brody, 2005):

- Oxygen-permeable films to obviate respiratory anaerobiosis
- Oxygen (O_2) scavengers or absorbers
- Carbon dioxide (CO_2) absorbers or generators
- Ethanol emitters
- Ethylene absorbers
- Moisture absorbers
- Odor controllers
- Flavor enhancement
- Antioxidants and/or other preservative emitters

Absorbing (scavenging) systems remove undesired compounds such as oxygen, carbon dioxide, ethylene, excessive water, and other specific compounds. Releasing systems actively add or emit compounds to the packaged food or into the headspace of the package such as carbon dioxide, antioxidant systems, and preservatives (Ahvenainen, 2003). Absorbing or adsorbing substances can potentially address several problems in MAP, for example, CO_2 absorbers can prevent the buildup of CO_2 to injurious levels, which can occur for some commodities during passive modification of the package atmosphere. O_2 absorbers can help maintain a low O_2 atmosphere when the film has been selected to produce a low O_2 atmosphere (Kader and Watkins, 2000; Mangaraj et al., 2009).

Moisture and oxygen absorbers were among the first series of active packaging to be developed and successfully applied for improving food quality and shelf-life extension. Next to these, numerous other concepts such as ethanol emitters (e.g., for bakery products), ethylene absorbers (e.g., for climacteric fruits), and carbon dioxide emitters/absorbers have been developed (Dainelli et al., 2008).

18.3 OXYGEN ABSORBERS

High levels of oxygen present in food packages may facilitate the growth of aerobic microbes, off-flavors and off-odors development, color change, and nutritional losses, thereby causing significant reductions in the shelf life of foods (Ozdemir and Floros, 2004). However, its presence in small quantities is necessary for fruits and vegetables to avoid fermentation. Oxygen absorbers maintain food product quality by decreasing food metabolism, suppressing ethylene production, reducing oxidative rancidity, inhibiting undesirable oxidation of labile pigments and vitamins, controlling enzymic discoloration, and inhibiting the growth of aerobic microorganisms (Day, 1989, 2001; Rooney, 1995). O_2 absorbers use iron powder as the main, active ingredient.

18.4 CARBON DIOXIDE ABSORBERS AND EMITTERS

High levels of CO_2 within the package retard microbial growth but may cause excessive browning, off-flavors, and increase ageing rate of the product (Pascall, 2011). Elevated CO_2 suppresses respiratory metabolism and inhibits ethylene action. Some of the absorbents used to remove excess CO_2 from CA storage rooms such as lime, activated charcoal, and magnesium oxide could be adapted for their utilization in MAP (Kader et al., 1989). For instance, where the package has a high permeability to carbon dioxide, a carbon dioxide-emitting system may be necessary to reduce the rate of respiration and suppress microbial growth (Ozdemir and Floros, 2004).

18.5 ETHYLENE ABSORBERS

Ethylene accelerates the respiration rate and subsequent senescence of horticultural products such as fruit, vegetables, and flowers and decreases their shelf life. Ethylene also accelerates the rate of chlorophyll degradation in leafy vegetables and fruits (Knee, 1990). The removal of ethylene gas from the package headspace slows senescence and prolongs shelf life. The commonly used ethylene controllers are silica gel, porous alumina, or vermiculite impregnated with potassium permanganate ($KMnO_4$) (Ozdemir and Floros, 2004); other compounds such as hydrocarbons (squalane, Apiezon) and silicones (Kader et al., 1989) activated carbon-based scavengers.

18.6 MOISTURE SCAVENGERS

Excess water development inside a food package usually occurs due to the respiration of fresh produce, temperature fluctuations, and low permeability of the film. The control of excess moisture in food packages is important to suppress microbial growth. Silica gel is the most widely used desiccant because it is nontoxic and noncorrosive (Ozdemir and Floros, 2004).

Another technique of packaging is "intelligent packaging," which monitors the condition of packaged food or the environment surrounding the food (Dainelli et al., 2008). Intelligent packaging contains an external or internal indicator for the active product history and quality determination. Typical examples of smart packaging methods are

- Time–temperature indicators intended to be fixed onto a package surface (Taoukis et al., 1991)
- O_2 indicators (Ahvenainen et al., 1995)
- CO_2 indicators (Plaut, 1995)
- Spoilage or quality indicators, which react with volatile substances from chemical, enzymatic, and/or microbial spoilage reactions released from food (Mattila and Auvinen, 1990a,b; Smolander et al., 1998)

18.7 BIOLOGICAL BASIS OF O_2 AND CO_2 EFFECTS ON POSTHARVEST LIFE OF FRUITS AND VEGETABLES

The beneficial effect of MAP on fresh produces is the reduction of the rate of respiration and the rate of ethylene production, the reduction or the inhibition of senescence and physiological disorders induced by ethylene (Kader, 1985; Herner, 1987; Lougheed, 1987), and the suppression of ripening, senescence, and growth of microorganisms. Low levels of O_2 and high levels of CO_2 in the atmosphere reduce the overall metabolic activity and preserve produce quality. Heat production generated by respiratory activities would also be reduced by lowering respiration rates.

Respiration is the most important metabolic process that takes place in any living tissue and is an indicator of the metabolic rate. During the respiration process, there is a loss of stored food reserves in the commodity, which leads to hastening of senescence because the reserves that provide energy are exhausted. The rate of deterioration of harvested commodities is generally proportional to the respiration rate. As O_2 concentration is reduced below that in air (20.9%), especially below 8%–10%, a significant reduction in respiration rate is observed (Toledo et al., 1969), implying the reduction in the rate of utilization of plant reserves (carbohydrate, acids, and moisture), thereby extending the storage life (Burton, 1974; Herner, 1987). The aerobic process is desirable as it is associated with the natural quality and flavor of fresh produce. When O_2 concentration drops below its critical value (extinction point) by about 2%, aerobic respiration is terminated, and anaerobic respiration becomes important (Boersig et al., 1988; Beaudry, 2000). At this level of O_2, pyruvic acid is not oxidized but is decarboxylated to form acetaldehyde, CO_2, and finally ethanol. Anaerobic process is harmful to tissue and will result in metabolic disorder in addition to production of substances (ethanol and acetaldehyde) that create off-odors and off-flavors (Kader et al., 1989). The oxygen level at which anaerobic respiration occurs differs from variety to variety and depends on commodity, temperature, and duration (Boersig et al., 1988; Schulz, 1989; Beaudry, 2000).

Respiration rate will decrease as CO_2 concentration is increased in the atmosphere. If the gas concentration is too high (>20%), then anaerobic respiration is induced with consequent quality problems. When the level of CO_2 rises above a critical value, the product develops physiological disorders (Lougheed, 1987; Beaudry, 1999). CO_2 sensitivity is both species- and cultivar-dependent. A more or less strong dose has varied actions: it reduces oxidation; slows down some synthesis (proteins, pigments, flavors); and slows down the loss of turgidity, firmness, acidity, and chlorophyll (Côme and Corbineau, 1999).

The stress caused by elevated CO_2 is additive and sometimes synergistic with stress caused by low O_2 levels. A 10% CO_2 added to the air influences respiratory metabolism by about the same extent as 2% O_2; a combination of 2% O_2 + 10% CO_2 has approximately twice the effect of either component (Burg and Burg, 1967).

The combination of low oxygen (O_2) levels and medium carbon dioxide (CO_2) content is used to reduce respiration of raw or fresh-cut products and delay spoilage. Nevertheless, there is no unique recommended atmosphere because optimal environmental conditions vary according to species, variety, and processing (Wang, 2006).

Exposure to O_2 concentration >21% may stimulate, have no effect, or reduce rates of respiration and ethylene production. Factors affecting these changes are commodity, maturity and ripeness stage, O_2, CO_2, and C_2H_4 concentration, storage time, and temperature (Fridovich, 1986; Lu and Toivonen, 2000). Sensitivity to O_2 toxicity varies among species and developmental stages. High O_2 concentrations enhance some of the effects of ethylene on fresh fruits and vegetables, including ripening, senescence, and ethylene-induced physiological disorders (such as bitterness of carrots and russet spotting on lettuce) (Kader and Ben-Yehoshua, 2000).

The best way to reduce respiratory metabolism and thus conserve the plant stores of carbohydrate, acids, and moisture is to reduce the temperature. Biological reactions generally increase two- to threefold for every 10°C rise in temperature. The creation and maintenance of an optimal atmosphere inside an MAP depends on the respiration rate of the product and the permeability of the films to O_2 and CO_2 (Beaudry et al., 1992), factors which are affected by temperature (Kader et al., 1989). Therefore, when the temperature increases, respiration tends to increase more than the permeation of the package, thus creating fermentative conditions. Rigorous temperature control is vital for an MAP system. Fluctuating temperatures during postharvest handling can have particularly negative impacts on the quality of products in MAP (Chambroy et al., 1993; Sanz et al., 1999; Tano et al., 1999) due to the danger of reaching injurious levels of O_2 or CO_2.

Ethylene is a simple organic molecule produced by higher plants, which affects many phases of plant growth and development. The sensitivity of fruits to ethylene varies with the stage of development and maturity, cultivar, and postharvest storage conditions such as temperature and atmospheric gas compositions. In MAP storage, ethylene is accumulated within the package or the storage environment and influences the product quality. Temperature is an important factor in ethylene production (Knee, 1990). The production and action of ethylene are influenced by O_2 and CO_2. When O_2 concentration is lowered below 8% (Abeles, 1973; Yang, 1985), the production of ethylene by fruits and vegetables is reduced by suppressing 1-amino-cyclopropane-1-carboxylic acid (ACC) synthase activity, and this effect is much more significant at lower O_2 levels, between 1% and 3% (Kader, 1980; Wang, 2006). High CO_2 has an antagonistic effect on ethylene action in addition to inhibiting the activities of ACC synthase and ACC oxidase (Wang, 2006) and suppresses plant tissue sensitivity to the effects of ethylene (Mullan and McDowell, 2003). These effects are additive to those of reduced O_2 atmospheres (Pretel et al., 2000). The effects of raised CO_2 on ethylene synthesis vary with produce. Inhibition effects have been observed with tomato (Buescher, 1979) and apple (Sisler and Goren, 1988), whereas stimulation effects have been observed with leaf plants (Rebeille et al., 1980; McRae et al., 1983). During MAP storage, a compound precursor to ethylene is accumulated in the products. Therefore, when the products are transferred to air, ethylene is rapidly produced, and the products ripen faster (Wang, 1990).

18.8 RELATIVE TOLERANCE TO REDUCED O_2 AND ELEVATED CO_2 LEVELS OF FRUITS AND VEGETABLES

Factors influencing the tolerance of fruits and vegetables to reduced O_2 or elevated CO_2 are species, cultivar, temperature and duration, concentration of O_2/CO_2, physiological age at harvest, and initial quality.

- The limit of tolerance to low O_2 would be higher as storage temperature and/or duration increases, as O_2 requirements for aerobic respiration of the tissue increase with higher temperatures (Kader, 2002).
- Production of CO_2 increases with temperature, but its solubility decreases. The physiological effect of CO_2 could be temperature dependent (Kader, 2002).
- Tolerance limits to elevated CO_2 decrease with a reduction in the O_2 level, and similarly the tolerance limits to low O_2 concentrations increase with an increase in the CO_2 level (Artés et al., 2006).
- Ripe fruits often tolerate higher levels of CO_2 than mature green fruits.
- Minimally processed (cut, sliced, or otherwise prepared) fruits and vegetables have fewer barriers to gas diffusion, and consequently they tolerate higher concentrations of CO_2 and lower O_2 levels than intact commodities (Watkins, 2000).

The tolerance of a specific crop to a low O_2 level and/or a high CO_2 level may be evaluated by the onset of fermentation. Simple indicators of fermentation are increases in the RQ (the ratio between CO_2 production and O_2 consumption rates) and the production of ethanol (Beaudry, 1993; Joles et al., 1994). RQ remains relatively constant in aerobic respiration and increases in anaerobic respiration (fermentation).

TABLE 18.1
Tolerance to Low O_2 and Elevated CO_2 of Fruits and Vegetables

Minimum O_2 Concentration Tolerated (%)	Commodities	Maximum CO_2 Concentration Tolerated (%)	Commodities
0.5	Dried fruits and vegetables		
1	Apples, pears (some cultivars), broccoli, mushrooms, onion, minimally processed fruits, vegetables	2	Apple (G. delicious), pear, apricot, grape, olive, tomato, pepper, lettuce, celery, artichoke, Chinese cabbage
2	Apples and pears (most cultivars), kiwifruit, apricot, cherry, peach, strawberry, sweet corn, green bean, lettuce, cabbage, cauliflower	5	Apple (most cultivars), peach, nectarine, orange, banana, kiwifruit, pepper, cauliflower, cabbage, carrot
3	Avocado, tomato, pepper, cucumber, artichoke, carrot	10	Lemon, lime, pineapple, cucumber, okra, asparagus, broccoli, leek, onion, potato
5	Citrus fruits, potato, asparagus, green pea	15	Strawberry, raspberry, cherry, fig, cantaloupe
7	Sweet potato	20	Spinach, sweet corn, mushrooms

Sources: Kader, A.A., Modified atmospheres during transport and storage, in: *Postharvest Technology of Horticultural Crops*, University of California, Agriculture and Natural Resources, Davis, CA, Publication No. 3311, pp. 135–144, 2002; Baccaunaud, M., Conservation et conditionnement des legumes sous gaz, in: Tirilly, Y. and Bourgeois, C.M. (eds.), *Technologie des Legumes*, Tec & Doc, Londres, NY, pp. 297–316, 1999; Côme, D. and Corbineau, F., Bases de la physiologie des légumes après récolte, in: Tirilly, Y. and Bourgeois, C.M. (eds.), *Technologie des Légumes*, Tec & Doc, Londres, NY, pp. 209–224, 1999.

The tolerance of fruits and vegetables in low O_2 and elevated CO_2 concentration is summarized in Table 18.1.

These limits are the levels beyond which physiological damage would be expected. The subjection of a fruit or vegetable to O_2 levels below or CO_2 levels above its tolerance limits at a specific temperature will result in stress to the organ with various symptoms such as irregular ripening, initiation and/or aggravation of certain physiological disorders, development of off-flavors, and increased susceptibility to decay (Lipton, 1975; Isenberg, 1979; Smock, 1979).

18.9 BENEFICIAL AND DETRIMENTAL EFFECTS OF MAP

When used properly, MAP helps refrigeration to achieve one or more beneficial effects in the storage life of fruits and vegetables. MAP suppresses metabolic activities, retards the conversion of carbohydrates and cell wall constituents, delays ripening and senescence, and extends storage life. An incorrect gas composition may change the biochemical activity of tissues, leading to the development of off-odors, off-flavors, a reduction in characteristic flavors, or anaerobic respiration.

Plastic films influence the rates of cooling and warming of the commodity and must be considered in selecting the appropriate temperature-management procedures for a packaged commodity.

The effectiveness of MAP on extending shelf live is dependent on different factors such as species, variety, growing conditions, harvesting system, stage of ripening, initial quality of the raw material, gas mixture, storage temperature and time, hygiene during handling and packaging, gas/product volume ratio, and the barrier properties of the packaging material (Artés et al., 2006). All these factors will explain the wide variability of results and recommendations for a certain produce that can be found in the literature.

The benefits of MAP are based on the fact that lowering the O_2 level and increasing the CO_2 concentration in the atmosphere reduces the overall metabolic activity and preserves produce quality. The decrease in O_2 and the increase in CO_2 suppress the rate of respiration, ripening, senescence, growth of microorganisms, and the production and action of ethylene (Wang, 2006).

MAP offers many advantages to consumers and food producers such as

- Reduction of the respiration rate
- Retardation of senescence (ripening) (Stenvers and Bruinsma, 1975; Brecht, 1980; Kader et al., 1989)
- Reduction or inhibition of ethylene biosynthesis (Kader, 1980; Kader et al., 1989; Fernández-Trujillo and Artés, 1997; Pretel et al., 2000)
- Reduction of the commodity's sensitivity to ethylene action at O_2 levels approximately below 8% and/or CO_2 levels above 1% (Lee et al., 1995)
- Limitation of the softening and preservation of firmness (Kader et al., 1989; González-Aguilar et al., 2003)
- Color preservation (Brackett, 1990; Zhuang et al., 1994; Gómez and Artés, 2004)
- Reduction of certain physiological disorders, such as chilling injury of various commodities and russet spotting in lettuce (Serrano et al., 1997; Porat et al., 2004)
- Inhibition of the growth of many pathogenic species (Yahia and Carrillo-López, 1993; Mitcham et al., 1997)
- Increase in shelf life
- Reduction and sometimes elimination of the need for chemical preservatives
- Reduction of the refrigeration load due to lower respiration rates
- Higher acceptable storage temperatures

MAP also has several disadvantages such as

- Loss of ascorbic acid with high CO_2 treatment (Agar et al., 1997, 1999).
- Exposure of fresh fruits and vegetables to O_2 levels below their tolerance limits or to CO_2 levels above their tolerance limits may increase anaerobic respiration and the consequent accumulation of ethanol and acetaldehyde causing off-flavors (Kader et al., 1989).
- O_2 and CO_2 levels beyond those tolerated by the commodity can induce physiological disorders, such as brown stain on lettuce, internal browning and surface pitting of pomes fruits, and black heart of potato (Kader et al., 1989).
- Unfavorable conditions can induce physiological breakdown and render the product more susceptible to pathogen.
- Irregular ripening of fruits (melons, tomato) can result from O_2 <2% or CO_2 >5% development of off-flavors and odors at <0.5% O_2 and/or >20% CO_2 as a result of fermentative metabolism.
- Plastic films can reduce the rate of cooling (Kader et al., 1989).
- Potential increase of water condensation within packages due to temperature fluctuations (Watada and Qi, 1999).
- Each MAP product needs a different gas formulation (Floros and Matsos, 2005).
- MAP causes larger package volumes, which leads to increased transportation and retail display space needs (Floros and Matsos, 2005).

18.10 EFFECT OF MAP ON THE QUALITY OF FRESH FRUITS AND VEGETABLES

The main objectives of fruits and vegetables packaging are shelf-life extension and maintenance of natural color, texture, flavor, and nutrients. Modified atmosphere storage is one of the food preservation methods that maintain the quality of food products and extend the storage life.

Consumer satisfaction is related to fresh product quality. This quality is generally associated with visual appearance, with color being one of the most important aspects in the consumer's purchase decision. The color of fruits and vegetables is a direct consequence of their natural pigment composition. Color changes during produce ripening and senescence are highly impacted by storage atmosphere conditions, particularly the change from green to yellow. MAP affects biochemical reactions related to pigment synthesis and degradation (Artés, 1993, 2000), although responses to MAP depend on the type of fruit or vegetable. Storage in high CO_2 and/or low O_2 results in reduced loss of chlorophyll as well as reduced accumulation of other pigments including anthocyanin, lycopene, xanthophylls, and carotenoids (Wang et al., 1971; Salunkhe and Wu, 1973; Barth et al., 1993; Zhuang et al., 1994; Barth and Hong, 1996). The inhibition of chlorophyll degradation in green vegetables by low O_2 is due to the reduction of ethylene synthesis (Makhlouf et al., 1989a,b). Knee (1980) suggested that the chlorophyll breakdown is reduced at an oxygen content of 2.5%–4%. Low O_2 inhibits browning reactions of the cut surfaces on lightly processed products (lettuce and salad mixes) catalyzed by polyphenol oxidase (PPO; Makhlouf et al., 1989a). Cutting results in the mixing of cellular contents so that the various phenolic substrates such as mono-, di-, and triphenols (Mayer and Harel, 1979) come into contact with PPO, leading to the formation of brown pigments. Another method of inhibiting enzymatic browning is an atmosphere containing 20% CO_2 and 80% N_2 combined with citric and ascorbic acids as browning inhibitors. This gas mixture and pretreatment gave the best sensory quality of sliced potatoes after 7 days of storage (Laurila et al., 1998; Sivertsvik et al., 2000).

High CO_2 concentrations often provoke cell membrane damage in outer tissues. Once membrane integrity is lost, phenolic compounds normally present in cellular compartments are exposed to O_2 and oxidized by the catalyzing activity of PPO liberated by the degrading membrane. The resulting compounds then polymerize to form brown pigments (Tano et al., 2007). If the CO_2 concentration within the package is too high, excessive browning and an increased ageing rate of the product could occur (Pascall, 2011). In some cases, at high CO_2 levels, fruit became dark red and accumulated anthocyanin due to decay caused by high level of CO_2 (Holcroft and Kader, 1999).

MAP conditions that delay ripening result in delayed synthesis of anthocyanins in some fruits, such as nectarines, peaches, and plums, but the synthesis of these pigments resumes upon transfer of the fruits to air at ripening temperatures (15°C–25°C) (Kader, 2009). Anthocyanins are very unstable pigments. Phenylalanine ammonia lyase and glucosyltransferase, two key enzymes in the synthetic pathway of anthocyanins in strawberry, were adversely affected by high CO_2 levels during cold storage (Siriphanich and Kader, 1985).

Methods that reduce product respiration immediately after harvest are often necessary to maintain firmness of harvested fruits and vegetables. Optimal concentration of O_2 and CO_2 reduce the rate of maturation and accompanying lignification. High CO_2 and/or low O_2 within the tolerance range reduce undesirable changes in texture (softening, toughness after processing) (Lougheed and Dewey, 1966; Weichmann, 1986). Outside the tolerance range, processes of softening may be accelerated (Patterson, 1982; Nanos and Mitchell, 1991). The degree of toughening of asparagus is reduced by 3%–5% O_2 and 16%–18% CO_2 at 1°C (Herregods, 1995). Knee (1974) reported that, reduction of the oxygen concentration to 2% delays the softening of apples. In the case of tomatoes, an O_2 concentration of 3% or less reduced textural changes (Kim and Hall, 1976). The effect of low O_2 concentration on firmness depends on variety and picking time (Herregods, 1995). Reduction in the levels of ethylene is important in retarding ripening and softening.

Flavor can be altered by storage in MAP. When CO_2 and O_2 are maintained in the tolerance range, flavor deterioration is slowed by the combination of reduced loss of sugar, acid, and changes in other compounds that contribute to flavor. An incorrect gas composition may change the biochemical activity of tissues, leading to development of off-odors, off-flavors, a reduction in characteristic flavors, or anaerobic respiration. In anaerobic conditions, off-flavors result from the accumulation of ethanol and acetaldehyde. Responses to low O_2 and/or high CO_2 concentrations that impact fruit aroma vary with commodity, cultivars, and storage conditions.

The production of volatile esters, which contribute to characteristic aromas of a number of fruits including apple, banana, pear, peach, strawberry, and others are affected by atmosphere modification (Shamaila et al., 1992; Song et al., 1998; Mattheis and Fellman, 2000). Low O_2 concentration suppresses the production of aroma compounds that confer characteristic odors through the ethylene mechanism. In general, most products recover from moderate to low O_2 suppression of aroma volatile production and eventually develop characteristic flavor. In the case of broccoli, low O_2 levels (0.25% and 0.1%) induced off-odors and flavors, the combination of 1% O_2 and 10% CO_2 also induced off-flavors and odors, but they are different from those induced by low O_2. Storage in 10% CO_2 resulted in some off-flavors, but these tended to dissipate after cooking (Lipton and Harris, 1974).

Fruits and vegetables play a very essential role in human nutrition and health, especially as sources of vitamins, minerals, dietary fiber, and phytonutrients (phytochemicals) (Johnson et al., 1994). The nutrient content of fruits and vegetables can be influenced by various factors such as genetic and agronomic factors, maturity and harvesting methods, and postharvest handling procedures. The effect of MAP on the nutritional quality of foods during storage is not well understood.

MAP can reduce the use of carbohydrates and titratable acids resulting in slower acid and sugar loss during storage (Salunke and Wu, 1973). Modified atmosphere with low O_2/high CO_2 preserve total sugars. Some enzymes involved in sugar degradation require O_2 and thus their efficacy is reduced in low O_2 atmospheres (Manes and Perkins-Veazie, 2003). The accumulation of reducing sugars in potato tubers at 5°C is prevented by 5% or more CO_2, however sucrose accumulation increases (Denny and Thornton, 1941).

Vitamin C is one of the most important vitamins in fruits and vegetables for human nutrition. More than 90% of the vitamin C in human diets is supplied by the intake of fresh fruits and vegetables. Ascorbic oxidase is the major enzyme responsible for enzymatic degradation of L-ascorbic acid. The rate of postharvest oxidation of ascorbic acid in plant tissues depend upon several factors such as temperature, water content, storage atmosphere, and storage time (Lee and Kader, 2000). The loss of vitamin C after harvest can be reduced by storing fruits and vegetables in an atmosphere of reduced O_2 and/or up to 10% CO_2 (Lee and Kader, 2000). The lower the O_2 content, the smaller were the losses of ascorbic acid in green bean, spinach, broccoli, and Brussels sprouts (Platenious and Jones, 1944). Enhanced losses of vitamin C in response to CO_2 higher than 10% may be due to the stimulating effects on oxidation of ascorbic acid and/or inhibition of DHA reduction to ascorbic acid (Agar et al., 1999). High CO_2 at injurious concentrations for the commodity may reduce ascorbic acid by increasing ethylene production and therefore the activity of ascorbate peroxidase (Devlieghere and Debevere, 2003). In conclusion, the effects of carbon dioxide on ascorbic acid may be positive or negative depending on the commodity, the CO_2 concentration and duration of exposure, and temperature (Weichmann, 1986; Lee and Kader, 2000).

Carotenoids form one of the more important classes of plant pigments and play a crucial role in defining the quality parameters of fruits and vegetables. Modified atmospheres with low O_2 concentrations or elevated CO_2 are able to reduce the loss of provitamin A, but also inhibit the biosynthesis of carotenoids (Kader et al., 1989). In carrots, very low O_2 concentration enhances the retention of carotene (Weichmann, 1986).

Phenolic compounds act as antioxidants by virtue of the free radical scavenging properties of their constituent hydroxyl groups (Vinson and Hontz, 1995). Atmospheres with high concentrations of CO_2 (>20%), which are usually employed to extend the postharvest life of strawberries, induce a remarkable decrease in anthocyanin content of internal tissues compared with the external ones (Gil et al., 1997). This decrease in color is related to a decrease in important enzyme activity involved in the biosynthesis of anthocyanins, phenylalanine ammonialyase, and glucosyltransferase (Holcroft and Kader, 1999). Modified atmospheres can also have a positive effect on phenolic-related quality, as in the case of the prevention of browning of minimally processed lettuce (Saltveit, 1997; Gil et al., 1998).

Low temperatures usually prolong the storage life of most commodities. Some crops cannot tolerate temperatures below ~10°C without developing severe physiological disorders that are grouped together under the term "chilling injury." MAP can reduce or eliminate the symptoms of chilling injury in some products such as citrus (Porat et al., 2004), okra (Finger et al., 2008), and melon (Flores et al., 2004). Zainon et al. (2004) concluded that suppression of the enzyme activities in fruits in MAP appeared to contribute to increased tolerance to chilling injury. Very low oxygen or too high CO_2 and the presence of excessive ethylene concentrations may exacerbate the severity of physiological disorders related to storage conditions. MAP maintains high humidity surrounding the products; according to Wang (1993), storage in high relative humidity can reduce symptoms of chilling injury as many symptoms appear to entail increased water loss. Elevated CO_2 and H_2O and reduced O_2 are beneficial in alleviating chilling injury symptoms in chilling-sensitive crops (Forney and Lipton, 1990). However, elevated CO_2 concentrations were found to increase the symptoms of chilling injury in cucumbers at 5°C (Eaks, 1956).

Scald and core flush are two disorders reduced by CA storage. The content of alfa-farnesene in scald-sensitive apple varieties is strongly inhibited under modified atmosphere (Herregods, 1995).

18.11 POLYMER PROPERTIES

The use of MAP has increased steadily, contributing significantly to extending the postharvest life and maintaining the quality of fruits and vegetables. The commercial expansion of MAP has become possible because of the recent development and proliferation of plastic polymers. Nevertheless, just a few polymers are commonly used in the manufacture of flexible films for packaging fresh fruits and vegetables.

The desirable characteristics of a polymeric film for MAP depend on the respiration rate of the produce at the transit and storage temperature to be used and on the known optimum O_2 and CO_2 concentrations for the produce. The uses of polymeric films for products with low to medium respiration rates have been successfully developed. Products such as broccoli, mushrooms, leeks, minimally processed fruits and vegetables, etc., exhibit very high rates of respiration such that conventional films can potentially overmodify the pack atmosphere and result in fermentation.

The major factors to be taken into account while selecting the packaging materials are (Kader et al., 1989; Smith, 1993; Artés et al., 2006)

- The type of package (i.e., flexible pouch or rigid or semirigid, lidded tray)
- The barrier properties needed (i.e., permeabilities to various gases and the water vapor transmission rate)
- The required selectivity (CO_2 permeability/O_2 permeability)
- The machine capability (resistant to tearing, possibility to be heat-formed)
- Integrity of closure (heat sealing)
- Antifog properties: good product visibility, prevention of the formation of water-condensed drops
- Sealing reliability (ability to seal itself and to the container)
- Resistance to chemical degradation
- Resistance to heat and to ozone
- Nontoxic and chemically inert
- Able to guarantee food safety
- Printability
- Commercial suitability with economic feasibility

The films used for creating MAP are continuous films that control movement of O_2 and CO_2 into or out of the package and perforated films with small holes or microperforations as the primary route of gas exchange. The movement of gases (O_2, CO_2) in continuous films is usually directly proportional to the gas gradient difference across the film. The steady-state condition is achieved in the package when the O_2 consumption and CO_2 production are equal to the permeating rate through the film, a state that is achieved when the respiratory rate is constant. The rate of gas movement through microperforated films is the sum of gas diffusion through microperforations and the gas permeation through the solid phase of the polymeric film. Generally, total gas flow through the perforations is much greater than gas movement through the film.

For most produce, a suitable film must be much more permeable to CO_2 than to O_2 (Exama et al., 1993; Lange, 2000; Lee et al., 2008). In continuous films, the permeability to CO_2 is usually 2–8 times that of O_2, whereas in microperforated films this proportion is 0.77. In the latter case, the sum of O_2 and CO_2 partial pressures is usually in the range of 18%–20% (Mir and Beaudry, 2002). Different films have different O_2 and CO_2 permeability, which is a function of thickness, density, presence of additives, and gradient concentration modification. The permeability depends on the chemical composition of the film, the number of layers (single or multilayers), the production technology of the film (nonstretched, uniaxial orientation, biaxial orientation, etc.), the degree of crystallinity, and the surface treatment of the film (Zanderighi, 2002).

Temperature is extremely important in package design, due to the influence of the permeability of films and the respiration rate of fruits and vegetables. The O_2 and CO_2 permeability of continuous films increases with temperature, whereas the diffusion of gases through perforations is insensitive to temperature. O_2 permeation through low-density polyethylene (LDPE) can increase 200% from 0°C to 15°C but only 11% through perforations (Mir and Beaudry, 2002). The respiration rate for most fruits and vegetables increases by 4–6 times from 0°C to 15°C (Beaudry et al., 1992; Manolopoulou and Papadopoulou, 1998; Lakakul et al., 1999). The product respiration becomes two or three times bigger than the rate of LDPE permeability and 30 times the rate of permeation with increasing temperature. These problems are developed (anaerobic conditions) when the temperature of the produce surpasses the temperature range for which the polymeric film has been designed.

Plant tissues lose moisture when the RH is below 99%–99.5%. In most commodities, the symptoms of wilting or wrinkling are visible when the loss of water is in the order of 4%–6%. Most films are relatively impermeable to water, and the RH of packages is near saturation. Controlling the internal humidity, particularly preventing the formation of moisture condensed droplets on the produce and the inner packaging film surface, retards surface mold development which in turn prolongs the shelf life of the fresh produce (Ben-Yehoshua et al., 1995).

The application of polymeric films for MAP are most often found in flexible package structures; they may also be used as a component in rigid or semirigid package structures, for example, as a liner inside a carton or as a lid on a cup or tray. The plastic film used in MAP is LDPE, linear low-density polyethylene (LLDPE), high-density polyethylene (HDPE), polypropylene (PP), polyvinyl chloride (PVC), polyester, that is, polyethylene terephthalate (PET), polyvinylidene chloride (PVDC), polyamide (Nylon), and other suitable films (Lange, 2000; Abdel-Bary, 2003; Ahvenainen, 2003; Massey, 2003; Del Nobile et al., 2007; Marsh Bugusu, 2007).

LDPE is very flexible and translucent. It is not a particularly high barrier to gas but is a good barrier to moisture and water vapor. It is generally used in film form.

LLDPE has a density range similar to that of LDPE. It is superior to LDPE in most properties such as tensile and impact strength and also is puncture-resistant.

HDPE is flexible but more rigid than LDPE, semitranslucent depending on density, a good vapor barrier but a poor gas barrier. It is commonly used for rigid and semirigid structures.

PP is rigid, solid, and durable in container or cap forms, opaque, grayish yellow in natural form, an excellent moisture barrier but a poor gas barrier.

PVC is transparent to yellowish color in natural state, flexible to rigid, good for coating, and a fair water and a good oxygen barrier. It is used with MAP for thermoformed trays to pack salads, sandwiches, etc.

PET is semirigid to rigid depending on the wall thickness, clear and transparent by nature, and a good gas and fair moisture barrier (Anon, 2001; Mark et al., 2003).

18.12 MAP RECOMMENDATIONS FOR SPECIFIC FRUITS AND VEGETABLES

Several recommendations exist on the optimal conditions for the storage or transport of fruits and vegetables. Different types of fruits, and even different cultivars of the same species, require different atmospheres for successful storage, and each therefore needs to be independently assessed.

Not all products benefit from exposure to low O_2 and high CO_2 levels. In Table 18.2 are listed the fruits and vegetables (intact and minimally processed) with high or moderate potential benefits for application of MAP.

The foremost problem with trying to establish an optimal MA atmosphere for every commodity is the problem inherent with all biological material, the variability. Due to the large variations among species, cultivars, and individual examples of the same cultivar the optimal O_2 and CO_2

TABLE 18.2
Fruits and Vegetables (Intact and Minimally Processed) with High or Moderate Potential for Application of MAP

Fruits		Vegetables	
Intact	**Fresh-Cut**	**Intact**	**Fresh-Cut**
Apple	Apple (sliced)	Artichokes	Beets (grated, cubed, peeled)
Apricot	Cantaloupe (cubed)	Asparagus	Broccoli (florets)
Avocado	Grapefruit (sliced)	Beans (processed)	Cabbage (shredded)
Banana	Honeydew (cubed)	Pepper	Carrots (shredded, sticks, sliced)
Blackberry	Kiwifruit (sliced)	Broccoli	Leek (sliced)
Blueberry	Mango (cubed)	Cabbage	Lettuce (chopped, shredded)
Cherimoya	Orange (sliced)	Cantaloupes	Onion (sliced, diced)
Cherry	Pomegranate (arils)	Lettuce	Peppers (sliced)
Durian	Strawberry (sliced)	Mushrooms	Potatoes (sliced, whole-peeled)
Fig	Watermelon (cubed)	Ripe tomatoes	Pumpkin (cubed)
Grape			Rutabaga (sliced)
Grapefruit			Spinach (cleaned)
Guava			Tomato (sliced)
Kiwifruit			Zucchini (sliced)
Lemon			
Lime			
Plum			
Mango			
Olive			
Peach			
Pear			
Persimmon			
Pineapple			
Pomegranate			
Raspberry			
Strawberry			

Sources: Adapted from Kader, A.A., *Acta Horticulturae*, 600, 737, 2003; Saltveit, M.E., *Acta Horticulturae*, 600, 723, 2003.

levels vary. The optimal O_2 concentration occurs at the boundary between the beneficial and injurious O_2 concentrations. An optimal storage environment could be defined as those storage conditions that produce the best quality product.

Before designing the optimum storage environment, we must define the quality criteria we use to evaluate the effectiveness of the storage environment. The criteria could be the retention of quality or the lowering of respiration, or the lowering of ethylene production or action, or to have better color, or better flavor and aroma, or better texture. It is inconceivable that one storage environment would produce an optimum for all criteria (Kader and Saltveit, 2003). Current recommended storage conditions of fruits and vegetables are presented in Tables 18.3 and 18.4.

A brief summary of MAP recommendations for fresh-cut fruits and vegetables is presented in Table 18.5.

TABLE 18.3
Recommended Storage Conditions of Selected Fruits

Commodity	°C	%O_2	%CO_2	Notes
Apple	0–5	1–2	0–3	60% of production is stored in CA
Apricot	0–5	2–3	2–3	Rapid cooling
Avocado	5–12	2–3	3–10	Marine transport, polybags + C_2H_4 scrubbing, individual wrapping
Banana	12–16	2–7	2–7	Marine transport, PSL improved if C_2H_4 scrubbing in polybags
Cherry, sweet	−1 to 0	3–10	10–15	Rapid cooling, pallet covers, or marine containers during transport, CO_2 improvement (≤20%)
Fig (fresh)	0–5	5–10	15–20	Limited use during transport
Grape	0–5	2–5	1–3	Incompatible with SO_2
Kiwifruit	0–5	1–2	3–5	Transport and storage, C_2H_4 below 20 ppb
Mango	7–14	3–5	5–10	Marine transport
Melon cantaloupe	2–5	3–5	10–20	Application moderate
Nectarine	−1 to 0	1–2	3–5	Limited use during marine transport
Olive	5–10	2–3	0–1	Limited use
Papaya	10–15	2–5	5–8	
Peach clingstone	−0.5	1–2	3–5	Limited use to extend canning season
Peach freestone	−1 to 0	1–2	3–5	Marine transport
Pear Asian	0–5	2–4	0–3	Limited use on some cultivars
Pear European	0–5	1–3	0–3	25% of production is stored in CA
Persimmon	0–5	3–5	5–8	Limited use of MAP
Pineapple	8–13	2–5	5–10	
Plum	−1 to 0	1–2	0–5	Limited use for some cultivars
Raspberry	0–5	5–10	15–20	Pallet covers during transport
Strawberry	0–5	5–10	15–20	Pallet covers during transport
Citrus fruits				
Grapefruit	10–15	3–10	5–10	Film individual packaging
Lemon	10–15	5–10	0–10	
Lime	10–15	5–10	0–10	Removal of C_2H_4
Mandarin	4–5	10	0	
Orange	5–10	5–10	0–5	Film individual packaging

Sources: International Institute of Refrigeration, *Recommendations for Chilled Storage of Perishable Produce*, IIF/IIR, Paris, France, 2000; Kader, A.A., Modified atmospheres during transport and storage, in: *Postharvest Technology of Horticultural Crops*, University of California, Agriculture and Natural Resources, Davis, CA, Publication No. 3311, pp. 135–144, 2002; Manolopoulou, E. et al., *Acta Horticulturae*, 2, 619, 1997.

TABLE 18.4
Recommended Storage Conditions of Selected Vegetables

Commodity	°C	%O_2	%CO_2	Notes
Artichoke	0–5	2–4	4–6	Application moderate, perforated polybags
Asparagus	0–2	Air	10–14	Application high
Beans, snap	7–8	2–3	3–5	Limited use
Broccoli	0–5	1–2	5–10	High application
Brussels sprouts	−1 to 0	2–3	4–5	Perforated polybags
Cabbage	0–5	2–3	3–6	High application
Carrot	0–1	5	3–4	Perforated polybags
Cauliflower	0–5	2–3	3–4	Slight application
Celery	0–5	1–4	3–5	Slight application
Cucumber	10–13	1–4	0	Slight application
Lettuce (crisphead)	0–1	1–3	0	Moderate application, perforated wrapping
Lettuce (leaf)	0–1	1–3	0	
Mushrooms	0–5	3–21	5–15	Microperforated films
Okra	7–10	Air	4–10	Slight application
Pepper (sweet)	5–12	2–5	2–5	Slight application
Spinach	0–5	7–10	5–10	Slight application
Tomatoes (green)	14–16	3–5	2–3	Slight application
Tomatoes (ripe)	8–10	3–5	3–5	Moderate application

Sources: International Institute of Refrigeration, *Recommendations for Chilled Storage of Perishable Produce*, IIF/IIR, Paris, France, 2000; Kader, A.A., Modified atmospheres during transport and storage, in: *Postharvest Technology of Horticultural Crops*, University of California, Agriculture and Natural Resources, Davis, CA, Publication No. 3311, pp. 135–144, 2002; Manolopoulou, E. et al., *J. Food Quality*, 30, 646, 2007; Manolopoulou, E. et al., *Biosyst. Eng.*, 106, 535, 2010b.

TABLE 18.5
MA Recommendations for Selected Fresh-Cut Fruits and Vegetables

Fresh-cut Product	°C	% O_2	%CO_2	Notes
Apple, sliced	0–5	<1	4–12	Efficacy moderate
Cantaloupe, cubed	0–5	3–5	6–15	Efficacy good
Grapefruit, slices	0–5	14–21	7–10	Efficacy moderate
Kiwifruit, sliced	0–5	2–4	5–10	Efficacy good
Mango, cubed	0–5	2–4	10	Efficacy good
Pear, sliced	0–5	0.5	<10	Efficacy poor
Strawberry, sliced	0–5	1–2	5–10	Efficacy good
Watermelon, cubed	0–5	3–5	10	Efficacy good
Vegetables				
Broccoli, florets	0–5	2–3	5–7	Efficacy good
Cabbage, shredded	0–5	5–7	15	Efficacy good
Carrots, shredded	0–5	2–5	15–20	Efficacy good
Lettuce, chopped (butterhead)	0–5	1–3	5–10	Efficacy moderate
Lettuce, chopped (green leaf)	0–5	0.5–3	5–10	Efficacy good
Lettuce, chopped (iceberg)	0–5	0.5–3	10–15	Efficacy good
Peppers, sliced	0–5	3	5–10	Efficacy moderate
Spinach, cleaned	0–5	0.8–3	8–10	Efficacy moderate

Sources: Gorny, R.J., A summary of CA and MA requirements and recommendations for fresh-cut fruits and vegetables, in: *Optimal Controlled Atmospheres for Horticultural Perishables*, Postharvest Horticullture Series No. 22A, University of California, Davis, CA, pp. 95–152, 2001; Manolopoulou, E. et al., *J. Food Quality*, 33, 317, 2010a; Manolopoulou, E. and Varzakas, T., *Food Nutr. Sci.*, 2, 956, 2011; Manolopoulou, E. et al., *J. Food Res.* 1(3), 148, 2012.

REFERENCES

Abdel-Bary EM. 2003. *Handbook of Plastic Films*. Rapra Technology Ltd., Shawbury, U.K.

Abeles FB. 1973. *Ethylene in Plant Biology.* Academic Press, New York.

Abeles FB, Morgan PW, Saltveit ME. 1992. *Ethylene in Plant Biology*, Vol. xv, 2nd ed. Academic Press, New York, p. 414.

Agar IT, Massantini R, Hess-Pierce B, Kader AA. 1999. Postharvest CO_2 and ethylene production and quality maintenance of fresh-cut kiwifruit slices. *Journal of Food Science* 64:433–640.

Agar IT, Streif J, Bangerth F. 1997. Effect of high CO_2 and controlled atmosphere (CA) on the ascorbic and dehydroascorbic acid content of some berry fruits. *Postharvest Biology and Technology* 11:47–55.

Ahvenainen R. 2003. Active and intelligent packaging: An introduction. In: Ahvenainen R (ed.), *Novel Food Packaging Techniques*. CRC Press, Boca Raton, FL, pp. 24–40.

Ahvenainen R, Hurme E, Randell K, Eilamo M. 1995. The effect of leakage on the quality of gas-packed foodstuffs and the leak detection. VTT Research Notes 1683, Espoo, Finland.

Anon. 2001. Plastics in packaging. Association of plastics manufacturers in Europe. Technical and Environmental Centre, pp. 1–18.

Artés F. 1993. Diseño y cálculo de polímeros sintéticos de interés para la conservación hortofrutícola en atmósfera modificada. In: Madrid A (ed.), *Nuevo Curso de Ingeniería del Frío*. Colegio Oficial de Ingenieros Agronómos de Murcia, Murcia, Chapter 16, pp. 427–454.

Artés F. 2000. Conservación de los productos vegetales en atmósfera modificada. In: Lamúa M (ed.), *Aplicación del Frío a Los Alimentos*. A. Madrid-Mundi-Prensa, Madrid, Spain, Chap. 4, pp. 105–125.

Artés F, Gómez A, Artés-Hernández F. 2006. Modified atmosphere packaging of fruits and vegetables. *Stewart Postharvest Review* 2(3):1–13.

Baccaunaud M. 1999. Conservation et conditionnement des legumes sous gaz. In: Tirilly Y, Bourgeois CM (eds.), *Technologie des Legumes*. Tec & Doc, Londres, NY, pp. 297–316.

Barmore CR. 1987. Packing technology for fresh and minimally processed fruits and vegetables. *Journal of Food Quality* 10:207–217.

Barry RC, O'Beirne D. 2000. Novel high oxygen and noble gas modified atmosphere packaging for extending the quality and shelf-life of fresh prepared produce. *Advances in Refrigeration Systems, Food Technologies and Cold-Chain*, Sofia, Bulgaria, September 23–26, 1998, pp. 417–424.

Barth MM, Hong ZA. 1996. Packaging design affects antioxidant vitamin retention and quality of broccoli florets during postharvest storage. *Postharvest Biology and Technology* 9:141–150.

Barth MM, Kerbel EL, Perry AK, Schmidt SJ. 1993. Modified atmosphere packaging affects ascorbic acid, enzyme activity and market quality of broccoli. *Journal of Food Science* 57:954–957.

Beaudry RM. 1993. Effect of carbon dioxide partial pressure on blueberry fruit respiration and respiratory quotient. *Postharvest Biology and Technology* 3:249–258.

Beaudry RM. 1999. Effect of O_2 and CO_2 partial pressure on selected phenomena affecting fruit and vegetable quality. *Postharvest Biology and Technology* 15:293–303.

Beaudry RM. 2000. Responses of horticultural commodities to low oxygen: Limits to the expanded use of MAP. *HortTechnology* 10(3):491–500.

Beaudry RM, Cameron AC, Shirazi A, Dostal-Lange DL. 1992. Modified atmosphere packaging of blueberry fruit: Effect of temperature on package O_2 and CO_2. *Journal of the American Society for Horticultural Science* 117:436–441.

Ben-Yehoshua S, Fang D, Rodov V, Fishman S. 1995. New developments in modified atmosphere packaging (Part II). *Plasticulture* 107(3):33–40.

Blakistone BA. 1997. *Principles and Applications of Modified Atmosphere Packaging of Foods*. Lavoisier Booksheller, Librairie, France.

Boersig MR, Kader AA, Romani RJ. 1988. Aerobic–anaerobic respiratory transition in pear fruit and cultured pear fruit cells. *Journal of American Society for Horticultural Science* 113:869–974.

Brackett RE. 1990. Influence of modified atmosphere packaging on the microflora and quality of fresh bell peppers. *Journal of Food Protection* 53(3):255–257.

Brecht PE. 1980. Use of controlled atmospheres to retard deterioration of produce. *Food Technology* 34(3):45, 46, 47–50.

Brody LA. 2005. Commercial uses of active food packaging and modified atmosphere packaging systems. In: Han J (ed.), *Innovation in Food Packaging*. Elsevier Ltd, London, U.K., pp. 457–474.

Buescher RW. 1979. Influence of carbon dioxide on postharvest ripening and deterioration of tomatoes. *Journal of American Society for Horticultural Science* 104:545–549.

Burg SP, Burg EA. 1967. Molecular requirements for the biological activity of ethylene. *Plant Physiology* 42:114–152.
Burton WG. 1974. Some biophysical principles underlying the controlled atmosphere storage of plant material. *Annals of Applied Biology* 78:149–168.
Cameron AC. 2003. Modified atmosphere packaging of perishable horticultural commodities can be risky business. *ISHS Acta Horticulturae 600: VIII International Controlled Atmosphere Research Conference 2003*, Rotterdam, the Netherlands. http://www.actahort.org/books/600/600_42.htm. March 10, 2003.
Chambroy Y, Guinebretire M-H, Jacquemin G, Reich M, Breuils L, Souty M. 1993. Effects of carbon dioxide on shelf-life and post harvest decay of strawberries fruit. *Sciences des Aliments* 13:409–423.
Charles F, Sanchez J, Gontard N. 2003. Active modified atmosphere packaging of fresh fruits and vegetables: Modelling with tomatoes and oxygen absorber. *Journal of Food Science* 68:1736–1742.
Chinnan MS. 1989. Modeling gaseous environment and physiochemical changes of fresh fruits and vegetables in modified atmospheric storage. In: Jen JJ (ed.), *Quality Factors of Fruits and Vegetables*. American Chemical Society, Washington, DC, pp. 189–202.
Church N. 1994. Developments in modified-atmosphere packaging and related technologies. *Trends in Food Science & Technology* 5:345–352.
Church IJ, Parsons AL. 1995. Modified atmosphere packaging technology: A review. *Journal of the Science of Food and Agriculture* 67:143–152.
Côme D, Corbineau F. 1999. Bases de la physiologie des légumes après récolte. In: Tirilly Y, Bourgeois CM (eds.), *Technologie des Légumes*. Tec & Doc, Londres, NY, pp. 209–224.
Dainelli D, Gontard N, Spyropoulos D, Zondervan-van den Beuken E, Tobback P. 2008. Active and intelligent food packaging: Legal aspects and safety concerns. *Trends in Food Science & Technology* 19:S103–S112.
Day BPF. 1989. Extension of shelf-life of chilled foods. *European Food and Drink Review* 4:47–56.
Day BPF. 1993. Fruit and vegetables. In: Parry RT (ed.), *Principles and Applications of Modified Atmosphere Packaging of Food*. Blackie, Glasgow, U.K., pp. 114–133.
Day BPF. 1996. High oxygen modified atmosphere packaging for fresh prepared produce. *Postharvest News and Information* 7(3):31N–34N.
Day BPF. 2001. Active packaging—A fresh approach. *The Journal of Brand Technology* 1(1):32–41.
Del Nobile ME, Licciardello F, Scrocco C, Muratore G, Zappa M. 2007. Design of plastic packages for minimally processed fruits. *Journal of Food Engineering* 79:217–224.
Denny EF, Thornton NC. 1941. Carbon dioxide prevents the rapid increase in the reducing sugar content of potato tubers stored at low temperatures. *Contributions from Boyce Thompson Institute* 12:79–84.
Devlieghere F, Debevere J. 2003. MAP, product safety and nutritional quality. In: Ahvenainen R (ed.), *Novel Food Packaging Techniques*. CRC Press, Boca Raton, FL, pp. 225–247.
De Wel R, Strubbe Y, Deforche B, Frison R, Herregods M. 1982. Commercialisation, qualite, emballage, conservation. *Revue de l'Agriculture* 3(35):2521–2538.
Doyon G, Gagnon J, Toupin C, Castaigne F. 1991. Gas transmission properties of polyvinyl chloride (PVC) films studied under subambient and ambient conditions for modified atmosphere packaging applications. *Packaging Technology and Science* 4:157–165.
Eaks IL. 1956. Effect of modified atmospheres on cucumbers at chilling and nonchilling temperatures. *Proceedings of the American Society for Horticultural Science* 67:473–478.
Emond JP, Castaigne F, Toupin CJ, Desilets D. 1991. Mathematical modeling of gas exchange in modified atmosphere packaging. *Transactions of the ASAE* 34(1):239–245.
Exama A, Arul J, Lencki RW, Lee LZ, Toupin C. 1993. Suitability of plastic films for modified atmosphere packaging of fruits and vegetables. *Journal of Food Science* 58(6):1365–1370.
Faber JM. 1991. Microbiological aspects of modified atmosphere packaging technology—A review. *Journal of Food Protection* 54:58–70.
Fellows PJ. (ed.) 2000. Controlled- or modified-atmosphere storage and packaging. In: *Food Processing Technology. Principles and Practice*, 2nd ed. CRC Press, Boca Raton, FL, pp. 406–436.
Fernández-Trujillo JP, Artés F. 1997. Quality improvement of peaches by intermittent warming and modified atmosphere packaging. *Zeitschrift fur Lebensmittel Untersuchung und Forschung* 205(1):59–63.
Finger FL, Della-Justina ME, Casali VWD, Puiatti M. 2008. Temperature and modified atmosphere affect the quality of okra. *Scientia Agricola* 65:360–364.
Flores FB, Martinez-Madrid MC, Ben-Amor M, Pech JC, Latche A, Romojaro F. 2004. Modified atmosphere packaging confers additional chilling tolerance on ethylene-inhibited cantaloupe Charentais melon fruit. *European Food Research and Technology* 219(6):614–619.
Floros DJ, Matsos KL. 2005. Introduction to modified atmosphere packaging. In: Han J (ed.), *Innovation in Food Packaging*. Elsevier Ltd, London, U.K., pp. 159–171.

Forney CF, Lipton WJ. 1990. Influence of controlled atmospheres and packaging on chilling sensitivity. In: Wang CY (ed.), *Chilling Injury of Horticultural Crops*. CRC Press, Boca Raton, FL, pp. 257–267.

Fridovich I. 1986. Biological effects of the superoxide radical. *Archives of Biochemistry and Biophysics* 247:1–11.

Geeson JD. 1988. Modified atmosphere packaging of fruits and vegetables. *Acta Horticulturae* 258:143–147.

Geeson JD. 1990. Packaging to keep produce fresh. *Nutrition & Food Science* 123(March:April):2–4.

Geeson JD, Browne KM, Maddison K, Sheperd J, Guaraldi F. 1985. Modified atmosphere packaging to extend shelf life of tomatoes. *Journal of Food Technology* 20:339–349.

Gil MI, Castañ ERM, Ferreres F, Artés F, Tomás-Barberán FA. 1998. Modified-atmosphere packaging of minimally processed Lollo Rosso (*Lactuca sativa*). *Zeitschrift für Lebensmittel-Untersuchung und -Forschung* 206:350–354.

Gil MI, Holcroft DM, Kader AA. 1997. Changes in strawberry anthocyanins in response to carbon dioxide treatments. *Journal of Agricultural and Food Chemistry* 45:1662–1667.

Gómez P, Artés F. 2004. Controlled atmospheres enhance postharvest green celery quality. *Postharvest Biology and Technology* 34:203–209.

González-Aguilar GA, Buta JG, Wang CY. 2003. Methyl jasmonate and modified atmosphere packaging (MAP) reduce decay and maintain postharvest quality of papaya 'Sunrise'. *Postharvest Biology and Technology* 28(3):361–370.

Gorny RJ. 1997. A summary of CA and MA recommendations for selected fresh-cut fruits and vegetables. Postharvest Horticulture Series No. 19. University of California, Davis, CA.

Gorny RJ. (ed.) 2001. A summary of CA and MA requirements and recommendations for fresh-cut fruits and vegetables. In: *Optimal Controlled Atmospheres for Horticultural Perishables*. Postharvest Horticullture Series No. 22A, University of California, Davis, CA, pp. 95–152.

Gorrish LGM, Peppelenbos LW. 1992. Modified atmosphere and vacuum packaging to extend the shelf-life of respiring food products. *HortTechnology* 2:303–309.

Goulas AE. 2008. Combined effect of chill storage and modified atmosphere packaging on mussels (*Mytilus galloprovincialis*) preservation. *Packaging Technology and Science* 21(5):247–255.

Guillaume C, Guillard V, Gontard N. 2011. Modified atmosphere packaging of fruits and vegetables modelling approach. In: Martín-Belloso O, Soliva-Fortuny R (eds.), *Advances in Fresh-Cut Fruits and Vegetables, Processing*. CRC Press, Boca Raton, FL, pp. 255–280.

Herner RC. 1987. High CO_2 effects on plant organs. In: Weichmann J (ed.), *Postharvest Physiology of Vegetables*. Marcel Dekker, New York, pp. 239–253.

Herregods M. 1995. Preservation of the quality and nutritional value of fruit and vegetables by CA storage. *Acta Horticulturae* 379:321–329.

Holcroft DM, Kader AA. 1999. Controlled atmosphere-induced changes in pH and organic acid metabolism may affect color of stored strawberry fruit. *Postharvest Biology and Technology* 17:419–432.

Hotchkiss JH. 1995. Safety considerations in active packaging. In: Rooney ML (ed.), *Active Food Packaging*. Blackie Academic and Professional, London, U.K., pp. 238–255.

Hurme E, Sipiläinen-Malm T, Ahvenainen R. 2000. Active and intelligent packaging. In: Ohlsson T, Bengtsson N (eds.), *Minimal Processing Technologies in the Food Industry*. CRC Press, Boca Raton, FL, pp. 87–123.

International Institute of Refrigeration. 2000. *Recommendations for Chilled Storage of Perishable Produce*. IIF/IIR, Paris, France.

Iqbal T, Rodrigues FAS, Mahajan PV, Kerry JP. 2009. Mathematical modeling of the influence of temperature and gas composition on the respiration rate of shredded carrots. *Journal of Food Engineering* 91:325–332.

Isenberg MFR. 1979. Controlled atmosphere storage of vegetables. *Horticultural Reviews* 1:337–373.

Johnson IT, Williamson GM, Musk SRR. 1994. Anticarcinogenic factors in plant foods: A new class of nutrients. *Nutrition Research Reviews* 7:175–204.

Joles DW, Cameron AC, Shirazi A, Petracek PD, Beaudry RM. 1994. Modified-atmosphere packaging of 'Heritage' red raspberry fruit: Respiratory response to reduced oxygen, enhanced carbon dioxide, and temperature. *Journal of American Society of Horticultural Science* 119:540–545.

Kader AA. 1980. Prevention of ripening in fruits by use of controlled atmospheres. *Food Technology* 34(3):51–57.

Kader AA. 1985. An overview of the physiological and biochemical basis of CA effects on fresh horticultural crops. In: *Proceedings of the Fourth National C.A. Research Conference*, July 23–26, Horticultural Report No. 126, Department of Horticultural Science, N.C.U., Raleigh, NC, pp. 1–9.

Kader AA. 1986. Biochemical and physiological basis for effects of controlled and modified atmospheres on fruits and vegetables. *Food Technology* 40:99–100, 102–104.

Kader AA. 1997. A summary of CA requirements and recommendations for fruits other than apples and pears. In: Kader A (ed.), *Fruits Other than Apples and Pears*, Vol. 2. Postharvest Horticulture Series No. 17. University of California, Davis, CA, pp. 1–36.

Kader AA. (ed.) 2002. Modified atmospheres during transport and storage. In: *Postharvest Technology of Horticultural Crops*. University of California, Agriculture and Natural Resources, Davis, CA, Publication No. 3311, pp. 135–144.

Kader AA. 2003. A summary of CA requirements and recommendations for fruits other than pome fruits. *Acta Horticulturae* 600:737–741.

Kader AA. 2009. Effects on nutritional quality. In: Yahia EM (ed.), *Modified and Controlled Atmospheres for the Storage Transportation and Packaging of Horticultural Commodities*. CRC Press, Boca Raton, FL, pp. 111–117.

Kader AA, Ben-Yehoshua S. 2000. Effects of superatmospheric oxygen levels on postharvest physiology and quality of fresh fruits and vegetables. *Postharvest Biology and Technology* 20:1–13.

Kader AA, Saltveit ME. 2003. Atmosphere modification. In: Bartz AJ, Brecht JK (eds.), *Postharvest Physiology Pathology of Vegetables*. Marcel Dekker, New York, pp. 229–247.

Kader AA, Watkins CB. 2000. Modified atmosphere packaging—Toward 2000 and beyond. *HortTechnology* 10(3):483–486.

Kader AA, Zagory D, Kerbel EL. 1989. Modified atmosphere packaging of fruits and vegetables. *Critical Reviews in Food Science and Nutrition* 28:1–30.

Kasmire RF, Kader AA, Klaustermeyer JA. 1974. Influence of aeration rate and atmospheric composition during simulated transit visual quality and off-odor production by broccoli. *HortScience* 9:228–229.

Kim BD, Hall CB. 1976. Firmness of tomato fruits subjected to low concentration of oxygen. *HortScience* 11:466–470.

Knee M. 1974. *Facteurs et Régulations de la Maturation des Fruits*. Colloques Internationaux du CNRS, Paris, France, 1–5 Juillet, pp. 1–5.

Knee M. 1980. Physiological response of apple fruits to oxygen concentrations. *Annals of Applied Biology* 96:243–247.

Knee M. 1990. Ethylene effects in controlled atmosphere storage of horticultural crops. In: Calderon M, Barkai-Golan R (eds.), *Food Preservation by Modified Atmospheres*. CRC Press, Boca Raton, FL, pp. 225–236.

Knee M, Hadfield SGS. 1981. The metabolism of alcohols by apple fruit tissue. *Journal of the Science of Food and Agriculture* 32:593–600.

Labuza TP, Breene WM. 1989. Applications of "active packaging" for improvement of shelf-life and nutritional quality of fresh and extended shelf-life foods. *Journal of Food Processing and Preservation* 13:1–69.

Lakakul R, Beaudry RM, Hernandez RJ. 1999. Modeling respiration of apple slices in modified-atmosphere packages. *Journal of Food Science* 64:105–110.

Lange DL. 2000. New film technologies for horticultural commodities. *HortTechnology* 10(3):487–490.

Laurila E, Hurme E, Ahvenainen R. 1998. The shelf life of sliced raw potatoes of various cultivar varieties: Substitution of bisulphites. *Journal of Food Protection* 61(10):1363–1371.

Lee KE, Jin Kh, Soon AD, Soon LE, Sun LD. 2008. Effectiveness of modified atmosphere packaging in preserving a prepared ready-to-eat food. *Packaging Technology and Science* 21(7):417–423.

Lee LZ, Arul J, Lencki R, Castaigne F. 1995. A review on modified atmosphere packaging and preservation of fruits and vegetables: Physiological basis and practical aspects. Part I. *Packaging Technology and Science* 8:315–331.

Lee LZ, Arul J, Lencki R, Castaigne F. 1996. A review on modified atmosphere packaging and preservation of fruits and vegetables: Physiological basis and practical aspects. Part II. *Packaging Technology and Science* 9:1–17.

Lee SK, Kader AA. 2000. Preharvest and postharvest factors influencing vitamin C content of horticultural crops. *Postharvest Biology and Technology* 20:207–220.

Lipton WJ. 1975. Controlled atmospheres for fresh vegetables and fruits—Why and when. In: Haard NF, Salunkhe DK (eds.), *Postharvest Biology and Handling of Fruits and Vegetables*. AVI Publishing, Westport, CT, pp. 130–143.

Lipton WJ, Harris CM. 1974. Controlled atmosphere effects on the market quality of stored broccoli (*Brassica oleracea* L., Italica Group). *Journal of the American Society for Horticultural Science* 99:200–205.

Lougheed EC. 1987. Interactions of oxygen, carbon dioxide, temperature, and ethylene that may induce injuries in vegetables. *HortScience* 22:791–794.

Lougheed EC, Dewey DH. 1966. Factors affecting the tenderizing effect of modified atmospheres on asparagus spears during storage. *Proceedings of the American Society for Horticultural Science* 89:336–345.

Lu C, Toivonen PMA. 2000. Effect of 1 and 100 kPa O_2 atmospheric pretreatment of whole 'Spartan' apples on subsequent quality and shelf-life of slices stored in modified atmosphere packages. *Postharvest Biology and Technology* 18:99–107.

Mahajan PV, Oliveira FAR, Montanez JC, Frias J. 2007. Development of user-friendly software for design of modified atmosphere packaging for fresh and fresh-cut produce. *Innovative Food Science and Emerging Technologies* 8:84–92.

Makhlouf J, Willemot C, Arul J, Castaigne F, Emond J-P. 1989a. Long-term storage of broccoli under controlled atmosphere. *HortScience* 24:637–639.

Makhlouf J, Willemot C, Arul J, Castaigne F, Emond J-P. 1989b. Regulation of ethylene biosynthesis in broccoli flower buds in controlled atmospheres. *Journal of the American Society for Horticultural Science* 114:955–958.

Maness N, Perkins-Veazie P. 2003. Soluble and storage carbohydrates. In: Bartz AJ, Brecht KJ (eds.), *Postharvest Physiology and Pathology of Vegetables*. Marcel Dekker Inc., New York, pp. 361–382.

Mangaraj S, Goswami T, Mahajan PV. 2009. Applications of plastic films for modified atmosphere packaging of fruits and vegetables: A review. *Food Engineering Reviews* 1:133–158.

Manolopoulou E, Lambrinos G, Assimaki H. 1997. Modified atmosphere storage of Hayward kiwifruit. *Acta Horticulturae* 2:619–624.

Manolopoulou E, Lambrinos Gr, Chatzis E, Xanthopoulos G, Aravantinos E. 2010a. Effect of temperature and modified atmosphere packaging on storage quality of fresh-cut *Romaine* lettuce. *Journal of Food Quality* 33:317–336.

Manolopoulou E, Lambrinos Gr, Xanthopoulos G. 2012. Active modified atmosphere packaging of fresh-cut bell peppers. Effect on quality indices. *Journal of Food Researches* 1(3):148–158.

Manolopoulou E, Papadopoulou P. 1998. A study of respiratory and physico-chemical changes of four kiwi fruit cultivars during cool-storage. *Food Chemistry* 63(4):529–534.

Manolopoulou E, Philippoussis A, Lambrinos G, Diamantopoulou P. 2007. Evaluation of productivity and postharvest quality during storage of five *Agaricus Bisporus* strains. *Journal of Food Quality* 30:646–663.

Manolopoulou E, Varzakas T. 2011. Effect of storage conditions on the sensory quality, colour and texture of fresh-cut minimally processed cabbage with the addition of ascorbic acid, citric acid and calcium chloride. *Food and Nutrition Sciences* 2:956–963.

Manolopoulou E, Xanthopoulos G, Douros N, Lambrinos Gr. 2010b. Modified atmosphere packaging storage of green bell peppers: Quality criteria. *Biosystems Engineering* 106:535–543.

Marcellin P. 1974. Conservation des fruits et légumes en atmosphère contrôlée à l'aide des membranes de polymères. *Revue Générale du Froid* 3:217–236.

Mark J, Kirwan M, Strawbridge J. 2003. Plastics in food packaging. In: Coles R, McDowell D, Kirwan M (eds.), *Food Packaging Technology*. CRC Press, Boca Raton, FL, pp. 174–240.

Marsh K, Bugusu B. 2007. Food packaging—Roles, materials, and environmental issues. *Journal of Food Science* 72(3):R39–R54.

Massey LK. 2003. *Permeability Properties of Plastics and Elastomers. A Guide to Packaging and Barrier Materials*. Plastic Design Laboratory/William Andrew Publishing, New York.

Mattheis JP, Fellman JK. 2000. Impacts of modified atmosphere packaging and controlled atmospheres on aroma, flavor and quality of horticultural commodities. *HortTechnology* 10(3):507–510.

Mattila T, Auvinen M. 1990a. Headspace indicators monitoring the growth of *B. cereus* and *Cl. perfringens* in aseptically packed meat soup (part I). *Lebensmittel-Wissenschaft und -Technologie* 23:7–13.

Mattila T, Auvinen M. 1990b. Indication of the growth of *Cl. perfringens* in aseptically packed sausage and meat ball gravy by headspace indicators (part II). *Lebensmittel-Wissenschaft und -Technologie* 23:14–19.

Mayer AM, Harel, E. 1979. Polyphenol oxidases in plants. *Phytochemistry* 18:193–215.

McRae DG, Coker CA, Legge RL, Thompson JE. 1983. Bicarbonate/CO_2 facilitated conversion of 1-aminocyclopropane-1-carboxylic acid to ethylene in model system and intact tissues. *Plant Physiology* 73:784–790.

Mir N, Beaudry RM. 2002. Modified atmosphere packaging. In: Kenneth CG, Wang CY, Saltveit M (eds.), *The Commercial Storage of Fruits, Vegetables and Florist and Nursery Stocks*. Agricultural Handbook Number 66. USDA, United States.

Mitcham EJ, Attia MM, Biasi W. 1997. Tolerance of Fuyu persimmons to low oxygen and high carbon dioxide atmospheres for insect disinfestation. *Postharvest Biology and Technology* 10(2):155–160.

Mitchell FG. 1992. Cooling horticultural commodities. In: Kader AA (ed.), *Postharvest Technology of Horticultural Crops*. University of California, Berkeley, CA, Spec. Publ. 3311, pp. 53–68.

Mullan M, McDowell D. 2003. Modified atmosphere packaging. In: Coles R, McDowell D, Kirwan MJ (eds.), *Food Packaging Technology*. CRC Press, Boca Raton, FL, pp. 303–339.

Myers RA. 1989. Packaging considerations for minimally processed fruits and vegetables. *Food Technology* 43(2):129–131.

Nanos GD, Mitchell FG. 1991. Carbon dioxide injury and flesh softening following high-temperature conditioning in peaches. *HortScience* 26:562–563.

Ooraikul B. 1991. Technological considerations in modified atmosphere packaging. In: Ooraikul B, Stiles ME (eds.), *Modified Atmosphere Packaging of Food*. Ellis Horwood, Chichester, U.K., pp. 26–48.

Ooraikul B, Stiles ME. 1991. *Modified Atmosphere Packaging of Food*. Ellis Horwood, Chichester, U.K.

Ozdemir M, Floros JD. 2004. Active food packaging technologies. *Critical Reviews in Food Science and Nutrition* 44:185–193.

Pascall AM. 2011. Packaging for fresh vegetables and vegetable products. In: Sinha NK (ed.), *Handbook of Vegetables and Vegetable Processing*. Wiley Blackwell, Hoboken, NJ, pp. 405–422.

Patterson ME. 1982. CA storage of cherries. In: Richardson DG, Meheriuk M (eds.), *Controlled Atmospheres for Storage and Transport of Perishable Agricultural Commodities*. Timber Press, Beaverton, OR, pp. 149–154.

Platenius H, Jones JB. 1944. Effect of modified atmosphere storage on ascorbic acid content of some vegetables. *Food Research* 9:378–382.

Plaut H. 1995. Brain boxed of simply packed? *Food Processing* 7:23–25.

Porat R, Weiss B, Cohen L, Daus A, Aharoni N. 2004. Reduction of postharvest rind disorders in citrus fruit by modified atmosphere packaging. *Postharvest Biology and Technology* 33(1):35–43.

Pretel MT, Souty M, Romojaro F. 2000. Use of passive and active modified atmosphere packaging to prolong the postharvest life of three varieties of apricot (*Prunus armeniaca* L.). *European Food Research and Technology* 211(3):191–198.

Priepke PE, Wei LS, Nelson AL. 1976. Refrigerated storage of pre-packed salad vegetables. *Journal of Food Science* 41:379–382.

Rebeille R, Bligny R, Douce R. 1980. Rôle de l'oxygène et la température sur la composition en acides gras des cellules isolées d'érable (*Acer pseudoplanus* L.). *Biochimica Biophysica Acta* 620:1–9.

Renault P, Houal L, Jacquemin G, Chambroy Y. 1994a. Gas exchange in modified atmosphere packaging. 2: Experimental results with strawberries. *International Journal of Food Science & Technology* 29(4):379–394.

Renault P, Souty M, Chambroy Y. 1994b. Gas exchange in modified atmosphere packaging. 1: A new theoretical approach for micro-perforated packs. *International Journal of Food Science & Technology* 29(4):365–378.

Richardson DG, Kupferman E. 1997. Controlled atmosphere storage of pears. In: Mitcham EJ (ed.), *Apples and Pears*, Vol. 2. Postharvest Horticulture Series No. 16. University of California, Davis, CA, pp. 31–35.

Rizvi SSH. 1981. Requirements for foods packaged in polymeric films. *Critical Reviews in Food Science and Nutrition* 14:111–134.

Rocculi P, Romani S, Dalla Rosa M. 2005. Effect of MAP with argon and nitrous oxide on quality maintenance of minimally processed kiwifruit. *Postharvest Biology and Technology* 35:319–328.

Rooney ML (ed.). 1995. *Active Food Packaging*. Blackie Academic & Professional, Glasgow, U.K.

Ryall AL, Lipton WJ. 1979. *Handling, Transportation, and Storage of Fruits and Vegetables, Vol. 1: Vegetables and Melons*. 2nd ed. AVI Publishing Company, Inc., Westport, CT.

Saltveit ME Jr. 1997. A summary of CA and MA requirements and recommendations for harvested vegetables. In Saltveit ME (ed.), *Proceedings of the 7th International Controlled Atmosphere Research Conference*, Davis, CA, Vol. 4, pp. 98–117.

Saltveit ME. 2003. A summary of CA requirements and recommendations for vegetables. *Acta Horticulturae* 600:723–729.

Saltveit ME, Yang SF, Kim WT. 1998. History of the discovery of ethylene as a plant growth substance. In: Kung SD, Yang SF (eds.), *Discoveries in Plant Biology*, Vol. 1. World Scientific, Singapore, pp. 47–70.

Salunkhe DK, Wu MT. 1973. Effects of subatmospheric pressure storage on ripening and associated chemical changes of certain deciduous fruits. *Journal of American Society of Horticultural Science* 98:12–14.

Sanz C, Pérez AG, Olías R, Olías JM. 1999. Quality of strawberries packed with perforated polyethylene. *Journal of Food Science* 64:748–752.

Schulz H. 1989. Internationaler Stand und Trend CA Lagerung von Kernobsifruchten und Schlussfolgerungen fur die Weiterentwicklung der CA-Lagerung in der D.D.R. 1. CA—Lagersymposium, Elbingerode, Germany, November 6–10, pp. 162–195.

Serrano M, Martínez-Madrid MC, Pretel MT, Riquelme F, Romojaro F. 1997. Modified atmosphere packaging minimizes increases in putrescine and abscisic acid levels caused by chilling injury in pepper fruit. *Journal of Agricultural and Food Chemistry* 45(5):1668–1672.

Shamaila M, Powire WD, Skura BJ. 1992. Analysis of compounds from strawberry fruit stored under modified atmosphere packaging (MAP). *Journal of Food Science* 5:1173–1176.

Siriphanich J, Kader AA. 1985. Effects of CO_2 on total phenolics, phenylalanine ammonia lyase, and polyphenol oxidase in lettuce tissue. *Journal the American Society for Horticulture Science* 110(2):249–253.

Sisler EC, Goren R. 1988. Ethylene binding the basis of hormone action in plants? What's new. *Plant Physiology* 12:37–42.

Sivertsvik MJ, Rosnes T, Bergslien H. 2000. Modified atmosphere packaging. In: Ohlsson T, Bengtsson N (eds.), *Minimal Processing Technologies in the Food Industry*. CRC Press, Boca Raton, FL, pp. 61–86.

Smith JP. 1993. Bakery products. In: Blackie PRT (ed.), *Principles and Applications of Modified Atmosphere Packaging of Food*. Suffolk, U.K., Springer Science+Business Media Dordrecht, pp. 135–169.

Smith S, Greeson J, Stow J. 1987. Production of modified atmospheres in deciduous fruits by the use of films and coatings. *HortScience* 22:772–778.

Smock RM. 1979. Controlled atmosphere storage of fruits. *Horticultural Reviews* 1:301–336.

Smolander M, Hurme E, Siika-Ahvo M, Ahvenainen R. 1998. Biological freshness indicator. In: Poutanen K (ed.), *Biotechnology in the Food Chain: New Tools and Applications for Future Foods*. VTT Symposium 177, Espoo, Finland, p. 256.

Song J, Deng W, Fan L, Verschoor J, Beaudry R. 1998. Aroma volatiles and quality changes in modified atmosphere packaging. In: Gorny JR (ed.), *CA'97 Proceedings, Vol. 4. Vegetables and Ornamentals*. University of California, Davis, CA, Postharvest Horticulture Series No. 18, pp. 85–95.

Stannett V. 1968. Simple gases. In: Crank J, Park GS (eds.), *Diffusion in Polymers*. Academic Press, London, U.K., pp. 41–73.

Stenvers N, Bruinsma J. 1975. Ripening of tomato fruits at reduced atmospheric partial oxygen pressures. *Nature* 253:532–533.

Tano K, Arul J, Doyon G, Castaigne F. 1999. Atmospheric composition and quality of fresh mushrooms in modified atmosphere packages as affected by storage temperature abuse. *Journal of Food Science* 64:1073–1077.

Tano K, Oule MK, Doyon G, Lencki RW, Arul J. 2007. Comparative evaluation of the effect of storage temperature fluctuation on modified atmosphere packages of selected fruit and vegetables. *Postharvest Biology and Technology* 46:212–221.

Taoukis PS, Fu B, Labuza TP. 1991. Time–temperature indicators. *Food Technology* 10(45):70–82.

Tian S, Xu Y, Jiang A, Gong Q. 2002. Physiological and quality responses of longan fruit to high O_2 or high CO_2 atmospheres in storage. *Postharvest Biology and Technology* 24:335–340.

Toledo R, Steinberg MP, Nelson AI. 1969. Heat of respiration of fresh produce as affected by controlled atmosphere. *Journal of Food Science* 34:261–265.

Vinson JA, Hontz BA. 1995. Phenol antioxidant index: Comparative antioxidant effectiveness of red and white wines. *Journal of Agricultural and Food Chemistry* 43:401–403.

Wang CY. 1990. Physiological and biochemical effects of controlled atmosphere on fruits and vegetables. In: Calderon M, Barkai-Golan R (eds.), *Food Preservation by Modified Atmospheres*. CRC Press, Boca Raton, FL, pp. 197–224.

Wang CY. 1993. Approaches to reduce chilling injury of fruits and vegetables. *Horticultural Reviews* 15:63–95.

Wang CY. 2006. Biochemical basis of the effects of modified and controlled atmospheres. *Stewart Postharvest Review* 5(8):1–6.

Wang SS, Haard NF, DiMarco GR. 1971. Chlorophyll degradation during controlled atmosphere storage of asparagus. *Journal of Food Science* 36:657–661.

Watada AE, Ling-Qi. 1999. Quality of fresh-cut produce. *Postharvest Biology and Technology* 15(3):201–205.

Watkins CB. 2000. Responses of horticultural commodities to high carbon dioxide as related to modified atmosphere packaging. *HortTechnology* 10(3):501–506.

Watkins CB, Hewett EW, Thompson CJ. 1988. Effects of microperforated polyethylene bags on the storage quality of Cox's Orange Pippin apples. *Acta Horticulturae* 258:225–235.

Weichmann J. 1986. The effect of controlled atmosphere storage on the sensory and nutritional quality of fruits and vegetables. *Horticultural Reviews* 8:101–125.

Yahia EM, Carrillo-López A. 1993. Responses of avocado fruit to insecticidal O_2 and CO_2 atmospheres. *Lebensmittel Wissenschaft und Technologie* 26(4):307–311.

Yang SF. 1985. Biosynthesis and action of ethylene. *HortScience* 20:41–45.

Young LL, Reverie RD, Cole AB. 1988. Fresh red meats: A place to apply modified atmospheres. *Food Technology* 42(9):64–66, 68–69.

Zagory D, Kader AA. 1988. Modified atmosphere packaging of fresh produce. *Food Technology* 42:70–74.

Zagory D, Kerbel EL, Kader AA. 1989. Modified atmosphere packaging of fruits and vegetables. *Critical Reviews in Food Science and Nutrition* 28:1–30.

Zainon MA, Hong, CL, Marimuthu M, Lazan H. 2004. Low temperature storage and modified atmosphere packaging of carambola fruit and their effects on ripening related texture changes, wall modification and chilling injury symptoms. *Postharvest Biology and Technology* 33:181–192.

Zanderighi L. 2002. How to design perforated polymeric films for modified atmosphere packs. *Packaging Technology and Science* 14:253–266.

Zhuang H, Barth MM, Hildebrand DF. 1994. Packaging influenced total chlorophyll, soluble protein, fatty acid composition and lipoxygenase activity in broccoli florets. *Journal of Food Science* 59(6):1171–1174.

19 Biosensors in Food Technology, Safety, and Quality Control

Theodoros Varzakas, Georgia-Paraskevi Nikoleli, and Dimitrios P. Nikolelis

CONTENTS

19.1 INTRODUCTION

A chemical sensor is a device that transforms chemical information, ranging from the concentration of a specific sample component to total composition analysis, into an analytically useful signal. Chemical sensors usually contain two basic components connected in series: a chemical recognition element ("receptor") and a physicochemical transducer. The biological recognition system translates the chemical information (i.e., concentration of the analyte) into a chemical or physical output signal. The transducer (i.e., a physical detection system) serves to transfer the signal from the output domain of the recognition element to the electrical, optical, piezoelectric, etc., domain. A biosensor is a self-contained integrated device that is capable of providing specific quantitative analytical information using a biological recognition element (e.g.., enzymes, antibodies, natural receptors, cells, etc.), which is retained in direct spatial contact with a transduction element.

Biosensors clearly offer advantages in comparison to standard analytical methods, such as minimal sample preparation and handling, real-time detection, rapid detection of the analytes of concern, use of nonskilled personnel, etc. Because of the importance of the ability of biosensors to be repeatedly calibrated, the term multiple-use biosensor is limited to devices suitable for monitoring both the increase and decrease in the analyte concentrations. Thus, single-use devices that cannot be regenerated rapidly and reproducibly be regenerated should be named single, etc.

A large class of chemical and biological sensors was based on the physical characterization of interfaces. More specifically, electronic (bio)chemical sensing is often related to the characterization of interfaces between ion-based and electron-based conductive materials by means of electrical variables such as voltage, current, and charge. Also, recent trends of integrated electronics that have started a revolution in this field, allowing the shrink of very complex electronic systems into millimeter square sizes, were a follow-up in the literature. This would allow implementing complex and sophisticated instrumentation in cheap and portable devices for fast detection of harmful and toxic agents.

The aim of this chapter is to bring into focus this important research area and the advances of biosensors, more specifically to those related to the rapid detection of food toxicants and environmental pollutants. The scope is to provide a comprehensive review of the research topics most pertinent to the advances of devices that can be used for the rapid real-time detections of food toxicants such as microbes, pathogens, toxins, nervous gases such as botulinum toxin, *Escherichia coli*, *Klebsiella pneumoniae*, sarin, VX, *Listeria monocytogenes*, *Salmonella*, marine biotoxins (such as palitoxins, spirolides, etc.), Staphylococcal enterotoxin B, saxitoxin, gonyautoxin (GTX5), francisella spore virus, *Bacillus subtilis*, ochratoxin, and even simple chemical compounds. Biosensors have found a large number of applications in the area of food analysis. Recent advances include portable devices for the rapid detection of insecticides, pesticides, food hormones, toxins, and carcinogenic compounds in the environment, such as polycyclic biphenols, etc.

These types of sensors are based on measuring responses to light emission or to illumination. Optical biosensors are based on well-founded methods including fluorescence, phosphorescence, light absorbance, photothermal techniques, chemiluminescence, surface plasmon resonance (SPR), total internal reflectance, light rotation, and polarization and could employ a number of techniques to detect the presence of a target analyte. As an example, these technical usages have been demonstrated to detect the presence of allergens, particularly peanuts, during food production [1,2].

19.1.1 Classification of Biosensors

Biosensors can be classified according to the type of recognition element (enzymatic, whole cell, or affinity-based biosensor) used. Enzymes were the first recognition elements included in biosensors. Enzymatic biosensors measure the selective inhibition or the catalysis of enzymes by a specific target. Enzymatic biosensors for the detection of contaminants in food samples have been extensively described in several reviews (Dzyadevych et al., 2003; Cock et al., 2009; Li et al., 2009; Manco et al., 2009). Another frequently used recognition element, especially for the monitoring of environmental pollutants, are whole cells such as bacteria, fungi, yeast, animal, or plant cells. These whole-cell biosensors detect responses of cells after exposure to a sample, which are related to its toxicity. These (toxic) responses can be nonspecific, such as DNA damage, heat shock, and oxidative stress or specific to a class of environmental pollutants, such as metals, organic compounds, and compounds with biological importance (such as nitrate, ammonia, and antibiotics). More information about whole-cell biosensors for different food applications can be found in several reviews (Harms et al., 2006; Ron, 2007; Yagi, 2007; Ding et al., 2008; Tecon and van der Meer, 2008; Close et al., 2009; Liu et al., 2010). A third group of recognition elements are the affinity-based recognition elements; they specifically bind to individual targets or groups of structurally related targets. Affinity-based sensors are very sensitive, selective, and versatile as affinity-based recognition elements can be generated for a wide range of targets.

Antibodies have long been the most popular affinity-based recognition elements. A wide variety of antibody biosensors reported for different food applications exist and are summarized and discussed in several reviews (Suri et al., 2002; Franek and Hruska, 2005; Gonzalez-Martinez et al., 2007; Skottrup et al., 2008; Tokarskyy and Marshall, 2008; Byrne et al., 2009; Conroy et al., 2009; Prieto-Simon and Campas, 2009; Ramirez et al., 2009). The main advantage of the use of antibodies as recognition elements is their sensitivity and selectivity; however, antibodies also have some limitations for the detection of contaminants in food. Polyclonal antibodies (PAbs) are relatively cheap, but animals have to be immunized to obtain them. Moreover, specific antibodies require isolation and purification. Besides the ethical problems of the use of animals, it is also difficult to generate antibodies for toxic compounds or small compounds that cannot elicit an immune response. Another important disadvantage of PAbs is nonspecific binding. These disadvantages can partially be minimized by using monoclonal antibodies, but hybridoma techniques are very expensive and time-consuming. Moreover, both poly- and monoclonal antibodies lose their binding properties under unfavorable environmental conditions and, therefore, cannot sustain the conditions in which environmental or food analysis are performed. Recent evolutions in biotechnology, nanotechnology, and surface chemistry have created the possibility to develop novel affinity-based recognition elements that can overcome the limitations of antibodies. Additionally, there is a growing concern about risk and safety of contaminants in food and environment from environmental, food, and safety authorities. These needs and opportunities create the perfect synergy for the recent development of advanced sensors based on alternative affinity molecules. Phages, nucleic acids, and molecular imprinted polymers are novel, innovative, affinity-based recognition elements that are becoming increasingly important for food sensors because of their exceptional characteristics, such as their high affinity and specificity for targets, their fast, cheap, and animal-friendly production avoiding batch-to-batch variations, their stability, and their ease to be modified.

19.1.1.1 Optical Biosensors

The development of optical-fiber sensors during recent years is related to two of the most important scientific advances: the laser and modern low-cost optical fibers. Recently, optical fibers have become an important part of sensor technology. Their use as a probe or as a sensing element is increasing in clinical, pharmaceutical, industrial, and military applications. Excellent light delivery, long interaction length, low cost, and ability not only to excite the target molecules but also to capture the emitted light from the targets are the main points in favor of the use of optical fibers in biosensors. Optical fibers transmit light on the basis of the principle of total internal reflection (TIR). Fiber-optic biosensors are analytical devices in which a fiber-optic device serves as a transduction element. The usual aim is to produce a signal that is proportional to the concentration of a chemical or biochemical to which the biological element reacts. Fiber-optic biosensors are based on the transmission of light along silica glass fiber, or plastic optical fiber (POF) to the site of analysis. Optical fiber biosensors can be used in combination with different types of spectroscopic technique, for example, absorption, fluorescence, phosphorescence, SPR, etc. Optical biosensors based on the use of fiber optics can be classified into two different categories: intrinsic sensors, where interaction with the analyte occurs within an element of the optical fiber, and extrinsic sensors, in which the optical fiber is used to couple light (Bosch et al., 2007).

Organophosphate (OP) neurotoxins comprise a unique class of contaminants and chemical warfare agents which have a high acute toxicity. These neurotoxins are powerful inhibitors of esterase enzymes, such as acetyl- and butyryl-cholinesterases or neurotoxic esterase. Simple methods for OP neurotoxin detection using fluorescence assays based on specific recognition of OP by organophosphate hydrolase (OPH) enzyme have been developed.

A biosensor for direct detection of OP neurotoxins such as paraoxon has been reported by Simoniana et al. (2005).

The biosensing method was based on the change in fluorescence of a competitive inhibitor of the OPH enzyme when the inhibitor is displaced by the OP substrate. The change in fluorescence intensity correlated with concentration of paraoxon presented in the solution. The sensitivity to paraoxon was obtained when enzyme inhibitor and OPH–gold nanoparticle conjugates were present at near-equimolar levels.

An optical glutamate biosensor test strip based on stacked membranes of nafion/sol–gel (bottom layer) and chitosan (uppermost layer) was fabricated on a piece of paper as a support to form a test strip by Muslim et al. (2012). The use of a stacked membrane system allows multiple immobilizations of sensing components directly without any covalent attachment via straightforward procedures. The uppermost membrane consisted of immobilized enzymes L-glutamate oxidase (GLOD) and horseradish peroxidase (HRP), which sensed the presence of L-glutamate, and the bottom membrane contained the indicator dye 3,3′,5,5′-tetramethylbenzidine (TMB). A test strip can be used to measure L-glutamate quantitatively by observing a color change from light green to dark green by increasing L-glutamate concentrations. Quantitative analysis could be performed by measuring the reflectance intensity of the color change at 550 nm. The glutamate biosensor test strip gave a linear response range of 0.01–0.30 mM to L-glutamate with a limit of detection of 5 μM. The strips were successfully applied for the estimation of L-glutamate in common food items such as sauces, soups, processed food, and flavor enhancers. Results of analysis of L-glutamate in various food samples using the glutamate test strip were comparable to a standard procedure employing high-performance liquid chromatography method.

The uniqueness of this optical biosensor design is a simultaneous immobilization of several sensing agents (one indicator dye, TMB, and two enzymes, i.e., GLOD and HRP) via a stacked membrane system without any covalent attachment of the sensing components. Avoiding covalent attachment of sensing materials on the test strip support enables a simple and straightforward fabrication procedure for the device apart from preventing deactivation of the enzymes during chemical attachment (Wong et al., 2006). In addition to that, the application of reflectance spectrophotometry in the detection of color change of the test strip in the presence of L-glutamate has the advantage of not only allowing opaque support material to be used for the test strips construction but also enable nontransparent sample matrices, especially food samples to be analyzed directly without the need of any sample pretreatments.

POF biosensors consist a viable alternative for rapid and inexpensive scheme for detection. In order to study the sensitivity of tapers for microbiological detection, geometric parameters are studied, such as the taper waist diameter as the formation of taper regions are the key sensing elements in this particular type of sensors. Beres et al. (2011) prepared a series of POF taper sensors using a specially developed tapering machine, and the dispersion of geometric dimensions is evaluated, aiming to achieve the best tapering characteristics that will provide a better sensitivity on the sensor response. The fiber tapers that presented the finest results were those constructed in U-shaped (bended) configurations, with taper waist diameters ranging from 0.40 to 0.50 mm. These fiber tapers were used as the main section of the monitoring device and when chemically treated as immunosensors for the detection of bacteria, yeast, and erythrocytes (Beres et al., 2011). A variety of studies approach the application of straight silica optical fibers in the manufacture of tapered sensors; nevertheless, the employment of U-shaped tapers in POF can afford several advantages, such as increased sensitivity, smaller taper length, economic use of reagents, an improved handling and fabrication, and a greater mechanical resistance (Frazao et al., 2008; de Nazare et al., 2011).

Fiber-optic sensors can be combined with antibodies that are able to recognize and bind to a defined antigen that induces immediate environmental changes, such as the refraction index, around the probe containing the antibody. Also, the large diameter of POFs facilitates installation and alignment, unlike their glass counterparts in which a few microns of misalignment results in heavy losses. Other well-known advantages are the efficient light coupling owing to the large numerical aperture, high ductility, low cost of production, and easy handling.

19.1.1.2 SPR Biosensors

SPR is a modern analysis technique based on the changes in the refractive index of material on the metal surface. In 1982, Liedberg first realized the biosensing potential of a prism SPR sensor with an immunoglobulin G (IgG) antibody adsorbed overlayer on the gold-sensing film, which allowed selective binding detection of IgG (Liedberg et al., 1983). Originally, the SPR technique was applied in the analysis of gases, liquids, and solids. In recent years, the SPR technique has an increasing application in biochemistry (Homola et al., 1999; Green et al., 2000), clinical diagnosis (Kanoh et al., 2006), food analysis (Spadavecchia et al., 2005), environmental monitoring (Inamori et al., 2005), and so on. The popularization of SPR is due to its properties of label-free, real-time detection, and high sensitivity. SPR can not only be used in analyte detection but also provide rich information on the specificity, affinity, and kinetics of biomolecular interactions. In the last 10 years, much attention has been focused on the detection of low-molecular-weight analyses in food and environmental fields using SPR biosensors. In this chapter, we address the basic principle of SPR, the existing detection methods, and the progress on mycotoxin detection using the SPR biosensor.

SPR is a physical optics phenomenon based on the change in the refractive index on the metal surface. Several reviews have described the basic principle and operation of SPR (Shankaran et al., 2007; Hodnik and Anderluh, 2009) the plane polarized light shines directly through the prism to the metal/solution dielectric interface over a wide range of incident angles; the evanescent wave will be generated under TIR condition. At a selected incident light wavelength or angle, the evanescent waves can resonate with surface plasmons (SP) produced by free electrons on the metal film of the sensor surface. Then, the energy of incident light is absorbed by SP, resulting in a narrow dip in the spectrum of reflected light. The angle at which the drop is maximum (minimum of reflectivity) is denoted as the "SPR angle." This "SPR angle" is extremely sensitive to the refractive index of the sample contacting with the metal surface, so that it is also highly influenced by the species and the amount of biomolecules immobilized on the gold layer. Furthermore, the kinetics information of the interaction between molecules can also be obtained (Huang et al., 2008).

As one of the relatively new analytical techniques, SPR has been proved particularly advantageous for rapid, label-free, sensitive analyte detection. Using SPR, qualitative and quantitative analyses can be performed in real time. Mycotoxins are a group of small, toxic products formed as secondary metabolites by a few fungal species. They can contaminate foodstuffs on a large scale and consequently threaten human health through the food chain. Thus, rapid, sensitive, and selective determination of mycotoxin is of great significance for food safety. This contribution addresses the basic principle of SPR, the existing detection methods, and the progress on mycotoxin detection using the SPR biosensor (Li et al., 2012).

SPR biosensing has matured into a valuable analytical technique for measurements related to biomolecules, environmental contaminants, and the food industry. Contemporary SPR instruments are mainly suitable for laboratory-based measurements. However, several point-of-measurement applications would benefit from simple, small, portable, and inexpensive sensors to assess the health condition of a patient, potential environmental contamination, or food safety issues. The trend article by Breault-Turcot and Masson (2012) explores nanostructured substrates for improving the sensitivity of classical SPR instruments and nanoparticle (NP)-based colorimetric substrates that may provide a solution to the development of point-of-measurement SPR techniques. Novel nanomaterials and methodology capable of enhancing the sensitivity of classical SPR sensors are destined to improve the limits of detection of miniature SPR instruments to the level required for most applications. In a different approach, paper or substrate-based SPR assays based on NPs are a highly promising topic of research that may facilitate the widespread use of a novel class of miniature and portable SPR instruments (Breault-Turcot and Masson, 2012).

Nanotechnology involves the characterization, fabrication, and/or manipulation of structures, devices, or materials that are between 1 and 100 nm in size. One of the major advantages of using nanomaterials for biosensing is their large surface area, allowing a greater number of biomolecules to be immobilized and this consequently increases the number of reaction sites available for interaction with a target species. This property, coupled with excellent electronic and optical properties, facilitates the use of nanomaterials in "label-free" detection and in the development of biosensors with enhanced sensitivities and improved response times (Xu et al., 2009).

Advances in the manipulation of nanomaterials has permitted the development of nanobiotechnology with enhanced sensitivities and improved response times. Low levels of infection of the major pathogens require the need for sensitive detection platforms, and the properties of nanomaterials make them suitable for the development of assays with enhanced sensitivity, improved response time, and increased portability.

Nanobiotechnologies focusing on the key requirements of signal amplification and preconcentration for the development of sensitive assays for food-borne pathogen detection in food matrices have been described and evaluated by Gilmartin and O'Kennedy (2012). The potential that exists for the use of nanomaterials as antimicrobial agents has also been examined by the same authors.

Cantilevers function by detecting differences in the stress, forces, or vibration frequency occurring when molecules bind to the surface. For example, bacteria can be detected due to the additional mass resulting from the binding of specific antibodies immobilized on the cantilever surface and the bacterial cell. There are several reports of sensitive detection of bacteria using cantilever technology such as a single *L. innocua* cell (Li et al., 2011), *E. coli* O157:H7 in spinach, spring lettuce mix, and ground beef (LOD 100 CFU/mL) (Gupta et al., 2004), and an antibody-functionalized piezoelectric-excited millimeter-sized cantilever sensor capable of identification of 1 *E. coli* O157:H7 cell/mL (Maraldo and Mutharasan, 2007). However, differences in pathogen adherence to food matrices, which affects target binding to the sensor surface, can affect sensitivity (Campbell and Mutharasan, 2007). The scaling down of this technology to the nanoscale has increased the capacity of nanocantilevers for ultrasensitive, faster detection due to higher frequencies and better mass resolution.

SPR technology has been used for the detection of food-borne pathogens using gold surfaces. Gold nanorods are elongated nanoparticles with distinct optical properties that depend on their shape. Compared with the single plasmon adsorption band of nanoparticles, excitation of surface plasmons by light in nanorods can be seen as two plasmon absorption bands, one corresponding to light absorption and scattering along the width of the particle, and the other along the length of the particle. Changes in the aspect ratio (ratio of the width of the object to its length) of a nanorod give rise to plasmon adsorption bands at different positions, hence, different-sized nanorods can be used as labels in a multiplex assay. The position and intensity of these bands can be affected by changes, for example, binding events, in the dielectric constant around the vicinity of these nanorods known as localized SPR (LSPR) or nanoSPR (Perez-Juste et al., 2005).

The sensing capacity of the detection systems is being improved lastly by using nanomaterials such as magnetic nanoparticles, carbon nanotubes, nanorods, quantum dots (QDs), nanowires, nanochannels, etc. These nanomaterials, used in electrical biosensors, have a very high capacity for charge transfer, which makes them suitable to reach lower detection limits and higher sensitivity values. Nanomaterials can contribute as label or transducer modifiers so as to improve the performance of the biosensor. Some of the reported nanomaterials are QDs. QDs are crystalline clusters with a nanosize (Murphy, 2002) that can be synthesized from semiconductor materials [e.g., cadmium sulfide (Merkoci et al., 2006), cadmium selenide (Steigerwald and Brus, 1990), cadmium telluride (Eychmuller and Rogach, 2000), indium phosphide (Guzelian et al., 1996), or gallium arsenide (Olshavsky et al., 1990)].

"Nanosized" and nanomaterial-based biosensors, also called nanobiosensors, are a modern and efficient class of detection systems (Delmulle et al., 2005; Lin et al., 2008; Sanvicens et al., 2009). The application of nanobiosensors in food industry could lead to immense improvements in quality control, food safety, and traceability.

Nanobiosensors can achieve very low detection limits (LODs; even single molecule or cell). In addition, they offer multidetection possibilities and may ensure a high stability (i.e., nanoparticles such as QDs are more stable than enzymes or fluorescence dyes). The main advantage besides the reduction of reagent volumes, detection time, and keeping the same sensitivity, is the user-friendly applicability; there is no need for professional users. The idea is to develop one-push button-like devices that can give a fast "yes–no" response or ensure a similar simple communication with the end-user.

Although some interesting nanobiosensors based on the use of nanoparticles and techniques such as optical microscopy (i.e., based in light absorption, scattering, fluorescence of nanoparticles, etc.) and electrochemistry (i.e., stripping analysis, potentiometry, etc.) have been developed and reported in several publications (Antiochia et al., 2004; Yang and Li, 2005; Merkoci et al., 2006; De la Escosura-Muniz et al., 2008; Dungchaia et al., 2008; Lin et al., 2008; Ambrosi et al., 2009; Ozdemir et al., 2010), Perez-Lopez and Mercoci (2011) gave a general overview of some of the most important nanomaterial-based biosensing systems based on various detection technologies and applied in the food field. In addition, they revised and gave opinions on the current status of detection systems, the obstacles, and some suggestions for the future development of this technology.

19.1.2 Antibody-Based Biosensors

An immunological method is one of the promising tools for the development of easy-handling biosensors. Besides, immunoassay methods are sensitive, cost-effective, easy to perform, and require a small sample volume. However, such techniques often require long reaction times and involve multiple steps. Antibodies are a common tool in analytical methods, and a number of PAbs raised to PA have been used in enzyme-linked immunosorbent assays (ELISA) protocols (Morris et al., 1987; Bertelsen et al., 1988; Finglas et al., 1988; Song et al., 1990; Gonthier et al., 1998a,b) and a radio immunoassay (RIA) method (Wyse et al., 1979) to determine PA levels in foods. In addition, a PAb to PA is the basis of a commercially available SPR biosensor assay (Qflex Kit Pantothenic Acid PI), which has previously been compared favorably to MBA and LC–MS analysis (Haughey et al., 2005, 2012). This kit has achieved AOAC PTM accreditation and has also been independently validated by Gao et al. (2008). The molecular structure of the PA molecule indicates that it has two distinct terminal moieties, carboxylic acid and hydroxyl groups. The presence of these two functional groups results in a choice of chemistries both for hapten production and for the immobilization of the antigen onto a solid support. These synthetic chemistry approaches, used in the development of ELISA and RIA systems, signify successful chemical pathways used during antibody production (Wyse et al., 1979; Morris et al., 1987; Bertelsen et al., 1988; Finglas et al., 1988; Gonthier et al., 1998a,b).

19.1.2.1 Development of Biosensors for Assaying the Contents of Starch, Glucose, Ethanol, and BOD

Biosensors for starch, glucose, and ethanol assays were developed using amylase, glucose oxidase, and alcohol oxidase enzyme preparations. Wastewaters and process waters of fermentation units contain mainly carbohydrates, alcohols, and amino acids, which is a guiding point in choosing microbial cultures for fabricating the bioreceptor element of the BOD biosensor. To form it, we used the methylotrophic yeast strain, *Pichia angusta* VKM Y-2518 (the aerobic microorganisms *P. angusta* VKM Y-2518 belong to the All-Russian Collection of Microorganisms, FSBIS G.K. Skryabin Institute of Biochemistry and Physiology of Microorganisms, Russian Academy of Sciences). In the work, these cells were also used to obtain the alcohol oxidase enzyme. Biosensors based on *Pichia* yeasts had been used for rapid assays of alcohols, so it was justified to apply them for the development of specialized receptor elements for assaying alcohol production wastewaters.

Starch, glucose, and ethanol contents were determined and BOD assessed using Clark electrode-based amperometric biosensors. This system enables recording the dependences of biochemical reaction rates on concentrations of substrates related to the consumption of oxygen during their oxidation. An IPC 2L galvanopotentiostat (Kronas, Russia) integrated with a personal computer and operated by specialized software for recording and processing electrode signals served as biosensors' electronic unit. Sodium–potassium phosphate buffer, pH 7.5, the concentration of the salts 60 mM, which provided for an optimum of bioreceptors' operation, was used for measurements. The measured parameter (biosensor response) was the maximal rate of change of the output signal, which emerged at an addition of substrates (nA/min) (Reshetilov et al., 2014).

19.1.2.2 Bioelectronic Tongues

Electronic tongue systems are novel analytical systems that develop applications in liquid samples from the use of an array of sensors plus an advanced data treatment stage. One recent progress in the design of electronic tongue systems has been the incorporation of biosensors, in order to engage in new application fields or to move existing ones. Bioelectronic tongues, as already known, are then differentiated from conventional ones in the incorporation of one or several biosensors into the sensor array, normally sharing the same transduction principle, which facilitates compatibility. After reviewing existing examples from the literature, this chapter will go more deeply into two case studies: one with potentiometric sensors and the other with devices of the voltammetric type. The first study case is related to the monitoring of the hemodialysis process, where the biosensor array will incorporate urea and creatinine biosensors constructed on the basis of the proper enzymes and ammonium ion-selective electrodes. The second case is motivated by the simultaneous determination of phenolic antioxidant compounds in beer and uses an array of voltammetric biosensors incorporating different phenol-oxidizing enzymes to generate the cross-sensitive responses. In both cases, the data processing stage used was artificial neural networks, able to provide a precise numeric response model to estimate concentrations of involved species (del Valle et al., 2014).

19.1.2.3 Electrochemical DNA Biosensors in Food Safety—Determination of Phenolic Compounds and Antioxidant Capacity in Foods and Beverages

The development and characteristics of the DNA sensors for the detection of phenolic antioxidants and antioxidant activity assessment for foodstuffs, beverages, and food additives are considered. Because most of the DNA sensors estimate protecting effect of antioxidants toward the DNA damage caused by reactive oxygen species, the mechanism of their generation in Fenton reaction and related processes as well as main approaches to the electrochemical detection of the DNA damage are also briefly discussed (Evtugyn and Porfireva, 2014).

There are many strategies for electrochemical detection of the DNA damage. Not all of them are applied for the quantification of the protecting effect of antioxidants. Thus, the analysis of the DNA unwinding and the influence of the DNA damage on the hybridization efficiency are too complicated by various experimental factors, for example, nucleic bases distribution, length of the main chain, etc. The terminal labels introduced in the DNA probe molecules are not so sensitive toward the ROS influence so that the sensitivity of such an assay is insufficient for the aims of the DNA sensors declared. Of other approaches, the following should be considered:

- Monitoring changes in direct or mediated DNA oxidation signals;
- Application of electrochemically active intercalators;
- Detecting changes in the permeability or charge-transfer properties of the surface DNA layer.

19.1.2.4 Nanotechnology and Nanofabrication Applications in Chemical Sensing

Recently, with the revolutionary advent of nanotechnology, a number of new nanomaterials have been synthesized, and their properties have been investigated. Moreover, these novel nanomaterials are being used strongly in biosensor technology. The use of nanomaterials in biosensing describes the consolidation of different sciences such as material science, molecular engineering, chemistry, and biotechnology. The nanomaterials have a great impact on the working performance of biosensors by improving the sensitivity and recognition of biomolecules at small volumes. The biosensors based on nanomaterials have a great practical usability in biomolecular recognition, pathogenic diagnosis, and environmental applications (Cui, 2007; Pan et al., 2008; Zhang et al., 2008). The most advantageous properties of nanomaterials for biosensing applications include larger surface-to-volume ratio, different physical and chemical properties from microsurface, and easy-to-detect target analytes in the small volumes. Moreover, the nanomaterials have a great diagnosing ability inside the biological cell. Use of nanomaterials inside the biological cell is making a clear difference of nanodevices from the bulk device that has been used for the biosensing purpose (Willander and Ibupoto, 2014; Otles and Yalcin, 2012).

19.1.2.5 Biosensors for Pesticides and Foodborne Pathogens

State of the art literature in the biosensor development for the detection of pesticide residues and foodborne pathogens is reviewed by Evtugyn (2014), with particular attention to real samples analysis. All the biosensors are classified in accordance with the nature of the biorecognition element and of the target analyte. The backgrounds of inhibition detection and the relationships between the assembling of the surface layer and the performance of enzyme sensors are considered. Signal transduction modes and the implementation of antibodies and DNA probes/aptamers are described for particular targets. The advantages and limitations of different approaches to the detection of pesticides and pathogenic bacteria are discussed together with the potential direction of their future progress.

Cholinesterase sensors are most intensively investigated for pesticide detection. Indeed, their progress was first affected by military purposes requiring compact robust devices for chemical weapons detection. Besides direct applications for the detection of individual organophosphate and carbamate pesticides, cholinesterase sensors have been successfully applied as models for consideration of the factors affecting the inhibition in heterogeneous conditions (e.g., the influence of enzyme immobilization on the sensitivity of inhibition, the effect of organic solvents, the optimization of the signal detection, the signal shifts during the storage, etc.).

The immunosensors involve as biorecognition elements antibodies (*Ab*) or antibody fragments able to form with analytes, also called antigen (*Ag*)-specific complexes. Contrary to enzyme sensors, especially those based on cholinesterase inhibition, the immunosensors show very high specificity of the response. For example, they can distinguish thiophosphates and their phosphoryl analogs. This is an important advantage, and meanwhile a weak point of immunosensors, because each analyte requires a specific set of immunoreagents for its quantification. Recently, the immunosensors specific to a group of chemically relative compounds containing a certain functional fragment (*Ag* determinant) have been developed (Kaufman and Clower, 1995) and are preferred compared to pharmaceuticals.

Considering the detection of pesticide residues, most of the developed immunosensors are based on a *competitive assay* (Ricci et al., 2007). Two approaches can be followed for the development of competitive immunosensors. In the first one, antibodies react with a mixture of free and labeled antigens, and the signal recorded is related to their ratio in solution. In the second approach, free antigen first reacts with an excessive amount of antibodies to form Ag–Ab complex in solution.

19.1.2.6 Biosensors in Quality of Meat Products

Different types of biosensors are mentioned by Varzakas et al. (2014) involved in meat quality such as amperometric xanthine biosensors, use of novel biological components such as mammalian cells and bacteriophages for the detection of bacterial pathogens, Biacore assay kit, potentiometric sensors based on measuring the potential of an electrochemical cell while drawing negligible or no current with common examples the glass pH electrode and ion selective electrodes for ions such as K^+, Ca^{2+}, Na^+, and Cl^-, optical fiber, and a capillary-based biosensor for calpastatin detection in heated meat samples as well as immunological capacitive biosensors for calpastatin.

Amperometric biosensors are also mentioned. The mechanism of amperometric hypoxanthine (Hx) biosensor is based on the direct oxidation of H_2O_2 formed from the enzyme reaction or on O_2 consumption.

Finally, nanobiosensors are mentioned such as a novel class of nanobiosensor developed by integrating a 27-nucleotide *Alu*I fragment of swine cytochrome b (cytb) gene to a 3-nm diameter citrate–tannate-coated gold nanoparticle (GNP).

Nanoscaffolding and nanoquenching properties of thiol-capped gold nanocrystals (GNCs), covalently linked to fluorophore-labeled oligonucleotide through metal–sulfur bond, were extensively studied for decades to detect specific sequences and single-nucleotide mismatches.

Moreover, the use of biosensors in boar taint detection along with bioluminescent bacterial sensors (development of an assay for the detection of tetracyclines in poultry based on a bioluminescent bacterial sensor reported as an alternative for "classical" microbial screening methods) is addressed.

An SPR biosensor inhibition immunoassay for the determination of ractopamine (Rac) residue in pork constructed by immobilizing Rac derivative on the SPR-2004 biosensor chip is described.

Ractopamine (Rac) belongs to the phenolic group of beta-agonists. Rac was originally used as tocolytics, bronchodilators, and heart tonics in human and veterinary medicine. Subsequently, this compound may be used illegally as growth promoters to acquire an economic interest similar to clenbuterol.

Finally, dioxin detection in meats using biosensors is reported with examples of that of a preliminary disposable electrochemical immunosensor for detection of nondioxin-like PCBs in ruminant milk, adipose tissue, and meat extracts.

19.2 MICROBIAL BIOSENSORS FOR ENVIRONMENTAL APPLICATIONS

Microbial biosensors have been developed for assaying BOD, a value related to the total content of organic materials in wastewater. BOD sensors take advantage of the high reaction rates of microorganisms interfaced to electrodes to measure the oxygen depletion rates. Standard BOD assay requires 5 days compared to 15 min for a biosensor-based analysis (Marty et al., 1997).

The biosensor should also be stable to environmental adversaries such as heavy metal toxicity, salinity, etc.

Other advances include the development of a disposable BOD sensor (Yang et al., 1996). Significant efforts have been made toward the development of a portable BOD biosensor system incorporating disposable electrodes.

Miniature Clark-type oxygen electrode arrays were fabricated using thin-film technology for mass production with assured quality (Yang et al., 1997).

Optical fiber (Preininger et al., 1994) and calorimetry (Weppen et al., 1991) based transducers have been used in BOD biosensors.

Microbial biosensors have been investigated for a variety of other environmental applications (D'Souza, 2001). Halogenated hydrocarbons used as pesticides, foaming agents, flame retardants, pharmaceuticals, and intermediates in the polymer production are one of the largest groups of environmental pollutants.

A microbial biosensor consists of a transducer in conjunction with immobilized viable or nonviable microbial cells. Nonviable cells obtained after permeabilization or whole cells containing periplasmic enzymes have mostly been used as an economical substitute for enzymes. Viable cells make use of the respiratory and metabolic functions of the cell; the analyte to be monitored being either a substrate or an inhibitor of these processes. Bioluminescence-based microbial biosensors have also been developed using genetically engineered microorganisms constructed by fusing the *lux* gene with an inducible gene promoter for toxicity and bioavailability testing. In this review, some of the recent trends in microbial biosensors with reference to the advantages and limitations are discussed. Some of the recent applications of microbial biosensors in environmental monitoring and for use in food, fermentation, and allied fields have been reviewed. Prospective future microbial biosensor designs have also been identified (D'Souza, 2001).

Biosensors for environmental analysis and monitoring are extensively reviewed by Rodriguez-Mozaz et al. (2004). Examples of biosensors for the most important families of environmental pollutants, including some commercial devices, are presented. Finally, future trends in biosensor development are discussed. In this context, bioelectronics, nanotechnology, miniaturization, and especially biotechnology seem to be growing areas that will have a marked influence on the development of new biosensing strategies in the next future.

Whole organisms are used to measure the potential biological impact (toxicity) of a water or soil sample. That is the case of the toxicity assays Microtox® (Azure, Bucks, UK), or ToxAlert®

(Merck, Darmstadt, Germany). These systems are based on the use of luminescent bacteria, *Vibrio fischeri*, to measure toxicity from environmental samples. Bacterial bioluminescence has proved to be a convenient measure of cellular metabolism and, consequently, a reliable sensor for measuring the presence of toxic chemicals in aquatic samples. Some bioassay methods are integrated now in biosensors such as the Cellsense®, which is an amperometric sensor that incorporates *Escherichia coli* bacterial cells for rapid ecotoxicity analysis. It uses ferricyanine, a soluble electron mediator, to divert electrons from the respiratory system of the immobilized bacteria of a suitable carbon electrode (Rodriguez-Mozaz et al. 2004).

Cell sense has been proposed as one of the newer rapid toxicity assessment methods within the direct toxicity assessment (DTA) demonstration program of the UK Environmental Agency (1999).

Nowadays, there is an increasing concern regarding many environmental contaminants that produce adverse effects by interfering with endogenous hormone systems, the so-called endocrine disrupting compounds (EDCs). EDCs constitute a class of substances not defined by chemical nature but by biological effect and thus, taking advantage of this feature, "endocrine effect biosensors" have been developed. Steroid hormones induce different effects in mammalian cells after binding to specific intercellular receptors, which are ligand-dependent transcription factors. Many endocrine disruptors are also believed to bind to the estrogen receptor (ER) as agonists or antagonists. Thus, the binding ability of the chemicals toward the ER would be a crucial factor for screening or testing their potential environmental toxicity. Based on estrogen receptors, several biosensors have been developed, which provide significant and useful information about estrogenic potency of the sample. The advantage of receptor assays is that they are quite simple to perform and allow the identification of all endocrine disruptors that act through the corresponding receptor (Oosterkamp et al., 1997).

Chloramphenicol (Cam), although an effective antibiotic, has lost favor due to some fatal side effects. Thus, there is an urgent need for rapid and sensitive methods to detect residues in food, feed, and environment. Mehta et al. (2011) engineered DNA aptamers that recognize Cam as their target by conducting in vitro selections.

Aptamers are nucleic acid recognition elements (DNA or RNA) that are highly specific and sensitive toward their targets and can be synthetically produced in an animal-friendly manner, making them ethical innovative alternatives to antibodies ranging from small molecules (Mann et al., 2005) to whole cells (Cerchia and de Franciscis, 2010). The name aptamer is derived from the Latin word *aptus* meaning to fit, and the Greek word *meros* meaning part or portion, referring to the folding properties of single-stranded nucleic acids, responsible for their specific three-dimensional structures. None of the isolated aptamers in this study shared sequence homology or structural similarities with each other, indicating that specific Cam recognition could be achieved by various DNA sequences under the selection conditions used. Analyzing the binding affinities of the sequences demonstrated that dissociation constants (K_d) in the extremely low micromolar range, which were lower than those previously reported for Cam-specific RNA aptamers, were achieved. The two best aptamers had G-rich (>35%) nucleotide regions, an attribute distinguishing them from the rest and apparently responsible for their high selectivity and affinity (K_d approx. 0.8 and 1 μM, respectively). These aptamers open up possibilities to allow easy detection of Cam via aptamer-based biosensors.

A new "signal-on" aptasensor for ultrasensitive detection of Ochratoxin A (OTA) in wheat starch was developed by Tong et al. (2011) based on exonuclease-catalyzed target recycling. To construct the aptasensor, a ferrocene (Fc) labeled probe DNA (S1) was immobilized on a gold electrode (GE) via Au–S bonding for the following hybridization with the complementary OTA aptamer, with the labeled Fc on S1 far from the GE surface.

In the presence of analyte OTA, the formation of aptamer–OTA complex would result in not only the dissociation of aptamer from the double-strand DNA but also the transformation of the probe DNA into a hairpin structure. Subsequently, the OTA could be liberated from the aptamer–OTA complex for analyte recycling due to the employment of exonuclease, which is a single-stranded DNA-specific exonuclease to selectively digest the appointed DNA (aptamer). Owing to the labeled Fc in close proximity to the electrode surface caused by the formation of the hairpin DNA and to the analyte recycling, differential pulse voltammetry signal could be produced with enhanced signal amplification. Based on this strategy, an ultrasensitive aptasensor for the detection of OTA could be exhibited with a wide linear range of 0.005–10.0 ng mL^{-1} with an LOD of 1.0 pg mL^{-1} OTA (at 3σ). The fabricated biosensor was then applied for the measurement of OTA in real wheat starch sample and validated by the ELISA method.

Not only enzymes but also nucleic acids can be immobilized on electrochemical transducers. DNA layers can act as biomolecular recognition elements for diagnostics of genetic or infection diseases as well as the detection of pathogens in food and environmental samples taking advantage of one of the most specific reactions known: hybridization. Likewise, the so-called aptamers (synthetic single-stranded oligonucleotides) can act as high-affinity receptors similar to antibodies for a great variety of ligands (Miranda-Castro et al., 2009). However, the usefulness of DNA layers is not restricted to these important applications. They can be the target for the antioxidant assessment by mimicking the damage caused in vivo by ROS (Prieto-Simon et al., 2006; Mello and Kubota, 2007).

REFERENCES

Ambrosi, A., M. T. Castaneda, A. De la Escosura-Muniz, and A. Merkoci. (eds.). 2009. Gold nanoparticles a powerful label for affinity electrochemical biosensors. Chapter 6. In *Biosensing Using Nanomaterials e Bionano*, 1st edn. Wiley-Interscience, United States.

Antiochia, R., I. Lavagnini, and F. Magno. 2004. Amperometric mediated carbon nanotube paste biosensor for fructose determination. *Analytical Letters* 37(8): 1657–1669.

Beres, C., F. V. B. de Nazare, N. C. C. de Souza, M. A. L. Miguel, and M. M. Werneck. 2011. Tapered plastic optical fiber-based biosensor—Tests and application. *Biosensors and Bioelectronics* 30: 328–332.

Bertelsen, G., P. M. Finglas, J. Loughridge, R. M. Faulks, and M. R. A. Morgan. 1988. Investigation into the effects of conventional cooking on levels of thiamin (determined by HPLC) and pantothenic acid (determined by ELISA) in chicken. *Food Sciences and Nutrition* 42F: 83–96.

Bosch, M. E., A. J. Ruiz Sánchez, F. Sánchez Rojas, and C. Bosch Ojeda. 2007. Recent development in optical fiber biosensors. *Sensors* 7: 797–859.

Breault-Turcot, J. and J.-F. Masson. 2012. Nanostructured substrates for portable and miniature SPR biosensors. *Analytical and Bioanalytical Chemistry* 403(6): 1477–1484.

Byrne, B., E. Stack, N. Gilmartin, and R. O'Kennedy. 2009. Antibody-based sensors: Principles, problems and potential for detection of pathogens and associated toxins. *Sensors* 9: 4407–4445.

Campbell, G.A. and R. Mutharasan. 2007. A method of measuring *E. coli* O157:H7 at 1 cell/mL in 1 liter sample using antibody functionalized piezoelectric-excited millimeter-sized cantilever sensor. *Environmental Science and Technology* 5: 1668–1674.

Cerchia, L. and V. de Franciscis. 2010. Targeting cancer cells with nucleic acid aptamers. *Trends in Biotechnology* 28: 517–525.

Close, D. M., S. Ripp, and G. S. Sayler. 2009. Reporter proteins in wholecell optical bioreporter detection systems, biosensor integrations, and biosensing applications (Review). *Sensors* 9: 9147–9174.

Cock, L. S., A. M. Z. Arenas, and A. A. Aponte. 2009. Use of enzymatic biosensors as quality indices: A synopsis of present and future trends in the food industry. *Chilean Journal of Agricultural Research* 69(2): 270–280.

Conroy, P. J., S. Hearty, P. Leonard, and R. J. O'Kennedy. 2009. Antibody production, design and use for biosensor-based applications. *Seminars in Cell and Developmental Biology* 20(1): 10–26.

Cui, D. 2007. *Journal of Nanoscience and Nanotechnology* 7: 1298–1314.

De la Escosura-Muniz, A., A. Ambrosi, and A. Merkoci. 2008. Electrochemical analysis with nanoparticle-based biosystems. *Trends in Analytical Chemistry* 27(7): 568–584.

Delmulle, B. S., S. M. De Saeger, L. Sibanda, I. Barna-Vetro, and C. H. Van-Peteghem. 2005. Development of an immunoassay based lateral flow dipstick for the rapid detection of aflatoxin B1 in pig feed. *Journal of Agricultural and Food Chemistry* 53: 3364–3368.

del Valle, M., X. Cetó, and M. Gutiérrez-Capitán. 2014. BioElectronic tongues. Chapter 13. In: *Portable Biosensing of Food Toxicants and Environmental Pollutants* (A Volume in the Series in Sensors). Nikolelis, D., Varzakas, T., Arzum, E., and Nikoleli, G.P. (eds.). CRC Press, Taylor & Francis Group, Boca Raton, FL.

de Nazare, F. V. B., C. Beres, N. C. C. de Souza, M. M. Werneck, and M. A. L. Miguel. 2011. *2011 IEEE International Instrumentation and Measurement Technology Conference Proceedings*, Hangzhou, China, pp. 959–963.

D'Souza, S. F. 2001. Microbial biosensors. *Biosensors & Bioelectronics* 16: 337–353.

Ding, L., D. Du, X. J. Zhang, and H. X. Ju. 2008. Trends in cell-based electrochemical biosensors. *Current Medicinal Chemistry* 15(30): 3160–3170.

Dungchaia, W., W. Siangprohb, W. Chaicumpac, P. Tongtawed, and O. Chailapakula. 2008. *Salmonella typhi* determination using voltammetric amplification of nanoparticles: A highly sensitive strategy for metalloimmunoassay based on a copper-enhanced gold label. *Talanta* 77: 727–732.

Dzyadevych, S. V., A. P. Soldatkin, Y. I. Korpan, V. N. Arkhypova, A. V. El'skaya, J. M. Chovelon, C. Martelet, and N. Jaffrezic-Renault. 2003. Biosensors based on enzyme field-effect transistors for determination of some substrates and inhibitors. *Analytical and Bioanalytical Chemistry* 377: 496–506.

Environmental Agency UK. 1999. Direct toxicity assessment. *Newsletter*, July 1999.

Evtugyn, G. A. 2014. Biosensors for pesticides and foodborne pathogens. Chapter 23. In: *Portable Biosensing of Food Toxicants and Environmental Pollutants* (*A Volume in the Series in Sensors*). Nikolelis, D., Varzakas, T., Arzum, E., and Nikoleli, G.P. (eds.). CRC Press, Taylor & Francis Group, Boca Raton, FL.

Evtugyn, G.A. and A. V. Porfireva. 2014. Electrochemical DNA biosensors in food safety—Determination of phenolic compounds and antioxidant capacity in foods and beverages. Chapter 25. In: *Portable Biosensing of Food Toxicants and Environmental Pollutants* (*A Volume in the Series in Sensors*). Nikolelis, D., Varzakas, T., Arzum, E., and Nikoleli, G.P. (eds.). CRC Press, Taylor & Francis Group, Boca Raton, FL.

Eychmuller, A. and A. L. Rogach. 2000. Chemistry and photophysics of thiol-stabilized II-VI semiconductor nanocrystals. *Pure and Applied Chemistry* 72: 179–188.

Finglas, P. M., M. F. Faulks, H. C. Morris, K. J. Scott, and M. R. A. Morgan. 1988. The development of an enzyme-linked immunosorbent assay (ELISA) for the analysis of pantothenic acid and analogues, Part II—Determination of pantothenic acid in foods. *Journal of Micronutrient Analysis* 4: 47–59.

Franek, M. and K. Hruska. 2005. Antibody based methods for environmental and food analysis: A review. *Veterinarni Medicina* 50: 1–10.

Frazao, O., J. M. Baptista, J. L. Santos, and P. Roy. 2008. Curvature sensor using a highly birefringent photonic crystal fiber with two asymmetric hole regions in a Sagnac interferometer. *Applied Optics* 47(13): 2520–2523.

Gao, Y., F. Guo, S. Gokavi, A. Chow, Q. Sheng, and M. Guo. 2008. Quantification of water-soluble vitamins in milk-based infant formulae using biosensor-based assays. *Food Chemistry* 110: 769–776.

Gilmartin, N. and R. O'Kennedy. 2012. Nanobiotechnologies for the detection and reduction of pathogens. *Enzyme and Microbial Technology* 50: 87–95.

Gonzalez-Martinez, M. A., R. Puchades, and A. Maquieira. 2007. Optical immunosensors for environmental monitoring: How far have we come? *Analytical and Bioanalytical Chemistry* 387(1): 205–218.

Gonthier, A., P. Boullanger, V. Fayol, and D. J. Hartmann. 1998a. Development of an ELISA for pantothenic acid (vitamin B5) for application in the nutrition and biological fields. *Journal of Immunoassay* 19(2 and 3): 167–194.

Gonthier, A., V. Fayol, J. Viollet, and D. J. Hartmann. 1998b. Determination of pantothenic acid in foods: Influence of the extraction method. *Food Chemistry* 63: 287–294.

Green, R. J., R. A. Frazier, K. M. Shakesheff, M. C. Davies, C. J. Roberts, and S. J. B. Tendler. 2000. Surface plasmon resonance analysis of dynamic biological interactions with biomaterials. *Biomaterial* 21: 1823–1835.

Gupta, A., D. Akin, and R. Bashir 2004. Detection of bacterial cells and antibodies using surface micromachined thin silicon cantilever resonators. *Journal of Vacuum Science & Technology B* 6: 2785–2791.

Guzelian, A. A., J. E. B. Katari, A. V. Kadavanich, U. Banin, K. Hamad, E. Juban et al. 1996. Synthesis of size-selected, surface-passivated InP nanocrystals. *Journal of Physical Chemistry* 100: 7212–7219.

Harms, H., M. C. Wells, and J. R. Van der Meer. 2006. Whole-cell living biosensors—Are they ready for environmental application? *Applied Microbiology and Biotechnology* 70(3): 273–280.

Haughey, S. A., C. T. Elliott, M. Oplatowska, L. D. Stewart, C. Frizzell, and L. Connolly. 2012. Production of a monoclonal antibody and its application in an optical biosensor based assay for the quantitative measurement of pantothenic acid (vitamin B_5) in foodstuffs. *Food Chemistry* 134(1): 540–545.

Haughey, S. A., A. A. O'Kane, G. A. Baxter, A. Kalman, M.-J. Trisconi, H. E. Indyk et al. 2005. Determination of pantothenic acid in foods by optical biosensor immunoassay. *Journal of AOAC International* 88(4): 1008–1014.

Hodnik, V. and G. Anderluh. 2009. Review: Toxin detection by surface plasmon resonance. *Sensors* 9: 1339–1354.

Homola, J., S. S. Yee, and G. Gauglitz. 1999. Surface plasmon resonance sensors: Review. *Sensors and Actuators B: Chemical* 54: 3–15.

Huang, Y., M. C. Bell, and I. I. Suni. 2008. Impedance biosensor for peanut protein Ara h 1. *Analytical Chemistry* 80: 9157–9162.

Inamori, K., M. Kyo, Y. Nishiya, Y. Inoue, T. Sonoda, E. Kinoshita et al. 2005. Detection and quantification of on-chip phosphorylated peptides by surface plasmon resonance imaging techniques using a phosphate capture molecule. *Analytical Chemistry* 77: 3979–3985.

Kanoh, N., M. Kyo, K. Inamori, A. Ando, A. Asami, A. Nakao et al. 2006. SPR imaging of photo-cross-linked small-molecule arrays on gold. *Analytical Chemistry* 78: 2226–2230.

Kaufman, B. M. and M. Clower Jr. 1995. Immunoassay of pesticides: An update. *Journal of AOAC International* 78: 1079–1090. Retrieved from http://cat.inist.fr/?aModele=afficheN&cpsidt=3610947 [PubMed], [Web of Science®], [CSA].

Li, B., A. D. Ellington, and X. Chen. 2011. Rational, modular adaptation of enzyme-free DNA circuits to multiple detection methods. *Nucleic Acids Research* 16: e110–e114.

Li, Y., X. Liu, and Z. Lin. 2012. Recent developments and applications of surface plasmon resonance biosensors for the detection of mycotoxins in foodstuffs. *Food Chemistry* 132(3): 1549–1554.

Li, X., Y. Zhou, Z. Zheng, X. Yue, Z. Dai, S. Liu, and Z. Tang. 2009. Glucose biosensor based on nanocomposite films of CdTe quantum dots and glucose oxidase. *Langmuir* 25(11): 6580–6586.

Liedberg, B., C. Nylander, and I. Lundstrom. 1983. A new carrier-domain magnetometer. *Sensors and Actuators* 4: 229–236.

Lin, Y.-H., S.-H. Chen, Y.-C. Chuang, Y.-C. Lu, T. Y. Shen, C. A. Chang et al. 2008. Disposable amperometric immunosensing strips fabricated by Au nanoparticles-modified screenprinted carbon electrodes for the detection of foodborne pathogen *Escherichia coli* O157:H7. *Biosensors and Bioelectronics* 23: 1832–1837.

Liu, H. Y., R. Malhotra, M. W. Peczuh, and J. F. Rusling. 2010. Electrochemical immunosensors for antibodies to peanut allergen Ara h2 using gold nanoparticle—Peptide films. *Analytical Chemistry* 82: 5865–5869.

Manco, G., R. Nucci, and F. Febbraio. 2009. Use of esterase activities for the detection of chemical neurotoxic agents. *Protein and Peptide Letters* 16(10): 1225–1234.

Mann, D., C. Reinemann, R. Stoltenburg, and B. Strehlitz. 2005. In vitro selection of DNA aptamers binding ethanolamine. *Biochemical and Biophysical Research Communication* 338: 1928–1934.

Maraldo, D. and R. Mutharasan. 2007. Preparation-free method for detecting *E. coli* O157:H7 in the presence of spinach, spring lettuce mix, and ground beef particulates. *Journal of Food Protection* 11: 2651–2655.

Marty, J. L., D. Olive, and Y. Asano. 1997. Measurement of BOD-Correlation between 5-day BOD and commercial BOD biosensor values. *Environmental Technology* 18: 333–337.

Mehta, J., B. Van Dorst, E. Rouah-Martin, W. Herrebout, M.-L. Scippo, R. Blust, and J. Robbens. 2011. In vitro selection and characterization of DNA aptamers recognizing chloramphenicol. *Journal of Biotechnology* 155: 361–369.

Mello, L. D. and L. T. Kubota. 2007. Biosensors as a tool for the antioxidant status evaluation. *Talanta* 72(2): 335–348.

Merkoci, A., S. Marın, M. T. Castaneda, M. Pumera, J. Ros, and S. Alegret. 2006. Crystal and electrochemical properties of water dispersed CdS nanocrystals obtained via reverse micelles and arrested precipitation. *Nanotechnology* 17: 2553–2559.

Miranda-Castro, R., N. de-los-Santos-Alvarez, M. J. Lobo-Castanon, A. J. Miranda-Ordieres, and P. Tunon-Blanco. 2009. *Electroanalysis* 21(19): 2077–2090.

Morris, H. C., P. M. Finglas, R. M. Faulks, and M. R. A. Morgan. 1987. The development of an enzyme-linked immunosorbent assay (ELISA) for the analysis of pantothenic acid and analogues, part I—Production of antibodies and establishment of ELISA Systems. *Journal of Micronutrient Analysis* 4: 33–45.

Murphy, C. J. 2002. Optical sensing with quantum dots. *Analytical Chemistry* 74: 520–526.

Muslim, N. Z. M., M. Ahmad, L. Y. Heng, and B. Saad. 2012. Optical biosensor test strip for the screening and direct determination of L-glutamate in food samples. *Sensors and Actuators B* 161: 493–497.

Nikolelis, D., T. Varzakas, E. Arzum, and G. P. Nikoleli. (eds). *Portable Biosensing of Food Toxicants and Environmental Pollutants* (A Volume in the Series in Sensors). CRC Press, Taylor & Francis Group, Boca Raton, FL.

Olshavsky, M. A., A. N. Goldstein, and A. P. Alivisatos. 1990. Organometallic synthesis of gallium-arsenide crystallites, exhibiting quantum confinement. *Journal of the American Chemical Society* 112: 9438–9439.

Oosterkamp, A. J., B. Hock, M. Seifert, and H. Irth. 1997. Novel monitoring strategies for xenoestrogens. *TrAC, Trends in Analytical Chemistry* 16: 544–553.

Otles, S. and B. Yalcin. 2012. Review on the application of nanobiosensors in food analysis. *ACTA Scientiarum Polonorum Technologia Alimentaria* 11(1): 7–18.

Ozdemir, C., F. Yeni, D. Odaci, and S. Timur. 2010. Electrochemical glucose biosensing by pyranose oxidase immobilized in gold nanoparticle–polyaniline/AgCl/gelatin nanocomposite matrix. *Food Chemistry* 119: 380–385.

Pan, B., D. Cui, C. S. Ozkan, M. Ozkan, P. Xu, T. Huang et al. 2008. Effects of carbon nanotubes on photoluminescence properties of quantum dots. *The Journal of Physical Chemistry C* 112: 939–944.

Perez-Juste J, I. Pastoriza-Santos, L. Liz-Marzan, and P. Mulvaney. 2005. Gold nanorods: Synthesis, characterization and applications. *Coordination Chemistry Reviews* 1(7–18): 1870–1901.

Perez-Lopez, B. and A. Merkoci. 2011. Nanomaterials based biosensors for food analysis applications. *Trends in Food Science & Technology* 22: 625–639.

Preininger, C., I. Klimant, and O. S. Wolfbeis. 1994. Optical fiber sensor for biological oxygen demand. *Analytical Chemistry* 66: 1841–1846.

Prieto-Simon, B. and M. Campas. 2009. Immunochemical tools for mycotoxin detection in food. *Monatshefte Fur Chemie* 140(8): 915–920.

Prieto-Simon, B., M. Campŏs, S. Andreescu, and J.-L. Marty. 2006. *Sensors* 6: 1161–1186.

Ramirez, N. B., A. M. Salgado, and B. Valdman. 2009. The evolution and developments of immunosensors for health and environmental monitoring: Problems and perspectives. *Brazilian Journal of Chemical Engineering* 26(2): 227–249.

Rana, J. S., J. Jindal, V. Beniwal, and V. Chhokar. 2010. Utility biosensors for applications in agriculture—A review. *Journal of American Science* 6(9): 353–375.

Reshetilov, N., V. A. Arlyapov, M. G. Zaitsev, and V. A. Alferov. 2014. Microbial cells and enzymes for assaying the fermentation processes of alcohol production: Starch, glucose, ethanol, BOD. Chapter 27. In: *Portable Biosensing of Food Toxicants and Environmental Pollutants* (A Volume in the Series in Sensors). Nikolelis, D., Varzakas, T., Arzum, E., and Nikoleli, G.P. (eds.). CRC Press, Taylor & Francis Group, Boca Raton, FL.

Ricci, F., G. Volpe, L. Micheli, and G. Palleschi. 2007. Amperometric biosensors for lactate alcohols and glycerol assays in clinical diagnostics. *Analytica Chimica Acta* 605: 111–129.

Rodriguez-Mozaz, S., M.-P. Marco, M. J. Lopez de Alda, and D. Barceló. 2004. Biosensors for environmental applications: Future development trends. IUPAC. *Pure and Applied Chemistry* 76(4): 723–752.

Ron, E. Z. 2007. Biosensing environmental pollution. *Current Opinion in Biotechnology* 18(3): 252–256.

Sanvicens, N., C. Pastells, N. Pascual, and M.-P. Marco. 2009. Nanoparticle-based biosensors for detection of pathogenic bacteria. *Trends in Analytical Chemistry* 28(11): 1243–1252.

Shankaran, R., K. V. Gobi, and N. Miura. 2007. Recent advancements in surface plasmon resonance immunosensors for detection of small molecules of biomedical, food, and environmental interest. *Sensors and Actuators B: Chemical* 121: 158–177.

Simoniana, A. L., T. A. Good, S.-S. Wang, and J. R. Wild. 2005. Nanoparticle-based optical biosensors for the direct detection of organophosphate chemical warfare agents and pesticides. *Analytica Chimica Acta* 534: 69–77.

Skottrup, P. D., M. Nicolaisen, and A. F. Justesen. 2008. Towards on-site pathogen detection using antibody-based sensors. *Biosensors and Bioelectronics* 24(3): 339–348.

Song, W. O., M. S. Smith, C. Wittwer, B. Wyse, and G. Hansen. 1990. Determination of plasma pantothenic acid by indirect enzyme linked immunosorbent assay. *Nutrition Research* 10: 439–448.

Spadavecchia, J., M. G. Manera, F. Quaranta, P. Siciliano, and R. Rella. 2005. Surface plasmon resonance imaging of DNA based biosensors for potential applications in food analysis. *Biosensors and Bioelectronics* 21: 894–900.

Steigerwald, M. L. and L. E. Brus. 1990. Semiconductor crystallites: A class of large molecules. *Accounts of Chemical Research* 23: 183–188.

Suri, C. R., M. Raje, and G. C. Varshney. 2002. Immunosensors for pesticide analysis: Antibody production and sensor development. *Critical Reviews in Biotechnology* 22: 15–32.

Tecon, R. and J. R. Van der Meer. 2008. Bacterial biosensors for measuring availability of environmental pollutants. *Sensors* 8(7): 4062–4080.

Tong, P., L. Zhang, J.-J. Xu, and H.-Y. Chen. 2011. Simply amplified electrochemical aptasensor of Ochratoxin A based on exonuclease-catalyzed target recycling. *Biosensors and Bioelectronics* 29: 97–101.

Tokarskyy, O. and D. L. Marshall. 2008. Immunosensors for rapid detection of *Escherichia coli* O157:H7—Perspectives for use in the meat processing industry. *Food Microbiology* 25(1): 1–12.

Varzakas, T., G.-P. Nikoleli, and D. P. Nikolelis. 2014. Biosensors in quality of meat products. Chapter 26. In: *Portable Biosensing of Food Toxicants and Environmental Pollutants* (A Volume in the Series in Sensors). Nikolelis, D., Varzakas, T., Arzum, E., and Nikoleli, G.P. (eds.). CRC Press, Taylor & Francis Group, Boca Raton, FL.

Weppen, P., J. Ebens, B. G. Muller, and D. Schuller. 1991. On-line estimation of biological oxygen demand using direct calorimetry on surface attached microbial cultures. *Thermochimica Acta* 193: 135–143.

Willander, M. and Z. H. Ibupoto. 2014. The nanotechnology and nanofabrication applications in the chemical sensing. Chapter 12. In: *Portable Biosensing of Food Toxicants and Environmental Pollutants* (A Volume in the Series in Sensors). Nikolelis, D., Varzakas, T., Arzum, E., and Nikoleli, G.P. (eds.). CRC Press, Taylor & Francis Group, Boca Raton, FL.

Wong, F. C. M., M. Ahmad, L. Y. Heng, and L. B. Peng. 2006. An optical biosensor for dichlovos using stacked sol–gel films containing acetylcholinesterase and a lipophilic chromoionophore. *Talanta* 69: 888–893.

Wyse, B. W., C. Wittwer, and R. G. Hansen. 1979. Radioimmunoassay for pantothenic acid in blood and other tissues. *Clinical Chemistry* 25(1): 108–111.

Xu, Q., C. Mao, N.-N. Liu, J.-J. Zhu, and J. Sheng. 2006. Direct electrochemistry of horseradish peroxidase based on biocompatible carboxymethyl chitosan–gold nanoparticle nanocomposite. *Biosensors and Bioelectronics* 22: 768–773, doi: 10.1016/j.bios.2006.02.010.

Xu, T., N. Zhang, H. L. Nichols, D. Shi, and X. Wen. 2007. Modification of nanostructured materials for biomedical applications. *Material Science and Engineering C* 27: 579–594.

Yang, L. J. and Y. B. Li. 2005. Quantum dots as fluorescent labels for quantitative detection of *Salmonella typhimurium* in chicken carcass wash water. *Journal of Food Protection* 68(6): 1241–1245.

Yang, Z., S. Sasaki, I. Karube, and H. Suzuki. 1997. Fabrication of oxygen electrode arrays and their incorporation into sensors for measuring biochemical oxygen demand. *Analytica Chimica Acta* 357: 41–49.

Yang, Z., H. Suzuki, S. Sasaki, and I. Karube. 1996. Disposable sensor for biochemical oxygen demand. *Applied Microbiology and Biotechnology* 46: 10–14.

Yagi, K. 2007. Applications of whole-cell bacterial sensors in biotechnology and environmental science. *Applied Microbiology and Biotechnology* 73: 1251–1258.

Zhang, Y., M. Yang, N. G. Portney, D. Cui, G. Budak, E. Ozbay, M. Ozkan, and C. S. Ozkan. 2008. Zeta potential: A surface electrical characteristic to probe the interaction of nanoparticles with normal and cancer human breast epithelial cells. *Biomedical Microdevices* 10: 321–328.

20 Ozone Applications in Food Processing

Daniela Bermúdez-Aguirre and Gustavo V. Barbosa-Cánovas

CONTENTS

20.1 INTRODUCTION

Ozone has been used in the food and beverage industry for a long time; in some European countries, it has been used for several decades, mainly to treat drinkable water. In the United States, the use of ozone to treat water and other food items has been recognized as a generally recognized as safe (GRAS) additive for about a decade.

The main idea of using ozone to treat food products and water is to inactivate microorganisms because of the potential oxidizing effect of this molecule, which is produced when air or oxygen passes through an electrical discharge (mainly), and then ozone is produced along with some free radicals. This oxidant creates antimicrobial activity not only in bacteria but also in other organisms such as molds, parasites, and even small infectious agents such as viruses. It is important to highlight that ozone can also degrade some toxic products such as mycotoxins in most of the cases without leaving any chemical residue in the treated product.

Vegetable and animal products have also been tested under ozone treatment, as an aqueous solution or gaseous treatment, again with the idea of disinfection. In most cases, the products retain their original properties after treatment, although in certain more specific products, undesirable chemical reactions are promoted between ozone and components such as anthocyanins or vitamins. Thus, it is important to consider the concentration of ozone and processing times before treating a product, having in mind not only the effect on microbial inactivation but also considering sensory quality and nutritional compounds.

Finally, even drinkable water has long been treated with ozone, especially in European and Asian countries. Some beverages, such as fruit juices, currently represent a challenge to food scientists to treat the product with minimal or no changes in their properties.

This chapter presents a general overview of the use of ozone in the food industry along with a discussion of the chemical properties of this chemical to better understand the chemical reactions that can take place in a specific food. The production of ozone, such as aqueous solution or gas treatment, is briefly explained. Also, the mechanism of inactivation in some microorganisms, as well as examples of target organisms in specific foods, is presented. Changes in quality and nutrition in ozone-treated foods and possible chemical explanations to these changes are also presented. Finally, the use of ozone to treat water and fruit juices is briefly discussed to provide the reader with a general overview of the use of this disinfectant agent for beverages.

20.2 OZONE

20.2.1 Physicochemical Properties

Ozone is a molecule composed of three atoms of oxygen (O_3). It is an unstable and very reactive gas with a pungent and characteristic odor. It is a colorless gas at low concentrations but is bluish at high concentrations. Ozone does not leave any residue on the product because of the quick decomposition in other species. The half-life of ozone is about 20–30 min in distilled water at 20°C (Cullen et al., 2009).

This gas is a powerful oxidant; in food processing, ozone can be used as a gas or dissolved in water. However, solubility of ozone in water follows Henry's law, meaning that its solubility at specific temperature is directly proportional to the pressure it exerts above the liquid. The reported solubility of ozone in water varies from 1130 to 307 mg/L in the range of temperature from 0°C to 60°C, respectively. This solubility can be modified for food processing by decreasing the temperature of water, increasing water purity and ozone concentration, increasing the pressure of gas and processing times, or improving the gas–water mixture, among others (Yousef et al., 2011). Ozone is more soluble in water than oxygen (by about 13 times), and it has an oxidation potential of 2.07 mV (Güzel-Seydim et al., 2004). Solubility of ozone is also affected by the pH and purity of the water; deionized and distilled water were much more efficient in dissolving ozone than tap water because tap water might contain organic matter that could cause a reaction when ozone is introduced. When the pH of water was about 5.6–5.9 (deionized and distilled water), ozone showed higher solubility than when the pH was about 8.23–8.39 (tap water); the presence of minerals in water also decreases the solubility of ozone (Khadre et al., 2001).

Toxicity of ozone in human beings is observed mainly in the respiratory tract. Some symptoms of ozone intoxication are headache, dizziness, cough, and burning sensation in the eyes and throat (Güzel-Seydim et al., 2004). It is very important to monitor workers in ozone facilities to ensure that they are not exposed to high doses of ozone.

20.2.2 Ozone Generation

Ozone is produced when oxygen (O_2) comes into contact with chemicals, electrical discharges, or ultraviolet radiation. Other methods include photochemical, thermal, chemonuclear, electrolytic, and electrochemical techniques. Because of these factors, atoms of oxygen are rearranged, and the formation of a triatomic molecule (O_3) is produced (Yousef et al., 2011).

The most common method for producing ozone at the industrial level is by the use of an electrical discharge system (corona discharge) in which oxygen or dried air is fed into the reactor, following which high voltage is applied between the two electrodes, generating an ozone flow (Güzel-Seydim et al., 2004; Yousef et al., 2011). The representation of the process for ozone production using corona

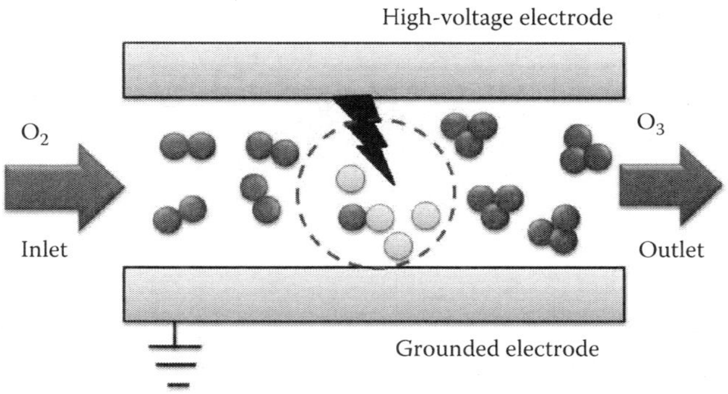

FIGURE 20.1 Representation of ozone generation using corona discharge.

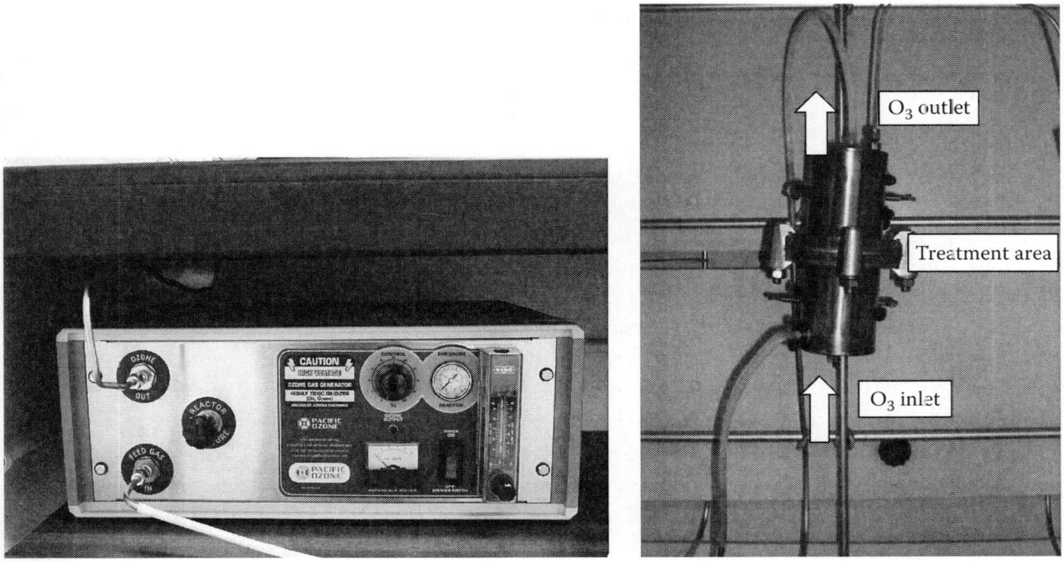

(a) (b)

FIGURE 20.2 Example of ozone generator (a) using corona discharge and (b) processing chamber to treat fresh produce.

discharge is shown in Figure 20.1. A common ozone generator and the system connected to the treatment chamber installed in a laboratory are shown in Figure 20.2. This chamber has been used to treat vegetable products such as tomatoes, carrots, and lettuce. The ozone generator uses a corona discharge to produce ozone starting with a known flow of oxygen. In the ozone generator, the pressure of the gas and the concentration of ozone are set. Several valves control the inlet and outlet of the ozone that is supplied in the treatment chamber. This system has been used under ambient temperature (21°C). Further details about the use of this system have been described in detail, as well as the results of microbial disinfection of the previously mentioned vegetables (Bermúdez-Aguirre and Barbosa-Cánovas, 2013).

20.2.3 Regulations of Ozone for Food Processing

Early use of ozone to treat food and beverages dates back to 1906 when it was used to produce drinkable water in France and 1910 when it was used in a meat-packaging facility in Germany (Sopher et al., 2002).

In the United States, ozone was used as a GRAS component in bottled water in 1982; it was reaffirmed as GRAS in 1995 for the same purpose. However, it was in August 15, 2000, when a food additive petition was filed requesting that the FDA recognize ozone as a food additive. The final ruling, approved by the FDA, was published in June 2001, and the USDA FSIS recognition was published in December 21, 2001. By 2002, approximately 400 water districts in the United States were using ozone to produce drinkable public water (Sopher et al., 2002).

20.3 EFFECTS ON QUALITY AND NUTRITION

Ozone was first investigated as a possible microbial inactivation technology in food processing with positive results. It was first used in water, and later in other foods and beverages. However, as previously mentioned, ozone is a potent oxidant that can react not only with components of the cell wall to inactivate microorganisms but also with chemical components of the food and at some point promotes undesirable chemical changes. Several studies have been conducted by specific research groups to evaluate the possible chemical reactions when a food product is treated with ozone; some of them are described in the following paragraphs.

20.3.1 Vegetable Products

In a recent study, three fruits, pineapple, banana, and guava, were treated with ozone; total phenol, flavonoid, and vitamin C contents were evaluated. The flow rate of ozone was 8 mL/s, and the maximum processing time was 30 min; results showed an increase in total phenol and flavonoid contents for pineapple and banana, but the opposite behavior was found in guava. Significant decrease was observed for vitamin C content in the three fruits. These changes can be attributed to the activation of phenylalanine ammonia lyase (PAL), which is a target in the synthesis of phenolic compounds, damage to the cell wall of the fruit tissue and the production of free radicals during ozone decomposition and possible scavenging of these radicals (Alothman et al., 2010). In another study conducted with fresh-cut asparagus treated with an aqueous ozone solution, some of the main quality enzymes such as PAL, superoxide dismutase, ascorbate peroxidase, and glutathione reductase were inactivated. Also, changes in cell wall components, such as the production of lignin, were reduced after treatment with ozone and further modified atmosphere packaging (An et al., 2007).

In a different study using spinach leaves, the main objective was to inactivate *Escherichia coli* O157:H7 inside the packaging container with ozone in a gaseous state (5 min) using different concentrations and storage temperatures to inactivate between 3 and 5 log of cells after 24 h of storage. However, after 24 h of storage, samples showed discoloration for the treatment with ozone, showing a spinach color quality between 3.83 and 1 (highest value is 5). The discoloration was observed on the leaf surface as yellowish-brown regions that could include some chemical reactions between ozone, packaging material, and spinach components (Klockow and Keener, 2009). A similar case was observed when lettuce was treated with ozone to inactivate *E. coli* cells. Samples were treated with gaseous ozone (5 ppm) for 15 min. After storage at 4°C (5 days), lettuce leaves lost the green color showing a translucent appearance. This result can be explained as the potent oxidant activity of ozone acting on the PAL of the lettuce. Lettuce has a very complex system of enzymes and pigments such as peroxidase, polyphenol oxidase, ascorbic acid, carotenoids, and chlorophyll, all of which affect the color of the product and might react with the ozone or by-products (Bermúdez-Aguirre and Barbosa-Cánovas, 2013). The same problem was reported in fresh-cut lettuce when treated with 4.5 ppm of ozonized water and processing times longer than 2.5 min, losing not only color but also its firm texture after 5 days of storage. However, vitamin C and β-carotene levels were not affected by the ozone treatment and remained almost constant during storage. The sensory quality also remained good during storage up to 7 days. In addition, ozone prevented browning of the cut edge, possibly because of the ozone action against polyphenol oxidase and peroxidase (Ölmez and Akbas, 2009).

Cilantro is popular in many dishes in American cuisine, even though it has its origins in Mexico. As an aromatic herb, it is most often added raw to dishes. Recently, cilantro has been involved in some foodborne outbreaks because of the presence of pathogenic microorganisms such as *Salmonella* spp. or *E. coli* O157:H7. The disinfection of this product is complex because the treatment should be strong enough to inactivate the microorganisms, without damaging the sensory properties of cilantro, such as color, smell, and texture. Ozone was tested with fresh-cut cilantro using an aqueous ozone solution for 5 min; in addition to studying the possible reduction of mesophiles and coliforms, sensory properties were evaluated. Results showed that ozone was able to extend the shelf life of the vegetable and also to preserve the aroma, color, and texture (Wang et al., 2004).

Carrots were also studied under ozonation and controlled atmosphere after treatment. Ozone aqueous solution (10 mg/L) was used to treat fresh-cut carrots for 10 min and results were positive in terms of reduction of lignification and decrease in respiration and ethylene production rates during storage. There was also a reduction in ascorbic acid, carotenoids, polyphenol oxidase, and peroxidase because of the oxidant effect of ozone (Chauhan et al., 2011).

Wheat flour is used to process noodles that are characteristic in Asian countries; however, when the noodles are processed to be sold "fresh," shelf life is very short because of mesophilic growth. The use of ozone to treat wheat flour and noodles and find whether there is any effect on the quality of the final product has been studied. When flour was previously treated with ozone and after the noodles were processed, shelf life was extended up to 10 days because of the delay of microbial growth and decrease in pH due to the interaction with ozone. However, sensory quality of the noodles was compromised, probably because of the strong oxidation of the fat in flour. Whiteness, dough stability, and peak viscosity of noodles were increased after treatment with ozone; however, loss of free cysteine content and aggregation of proteins were observed in the treated noodles. Ozone-treated noodles show higher texture values in firmness, springiness, and chewiness but lower values in adhesiveness (Li et al., 2012).

20.3.2 Animal Products

Ozone has also been used in animal products, mainly for microbial decontamination. However, some of these products have not shown positive results in quality characteristics after treatment. As ozone is a strong oxidant, some chemical reactions take place between the product and the gas. For example, in a study conducted with beef muscle, the tissue was treated with ozone gases at 0°C for 24 h to inactivate mesophiles and inoculated *E. coli*; however, at the end of the process, the meat exhibited undesirable color, and lipid oxidation made the meat unacceptable for consumption. Color changes are due to the oxidation of myoglobin and oxymyoglobin to metmyoglobin. If treatment time was reduced to 3 h, no changes in color or rancidity were detected, but microbial inactivation was seriously compromised and reduced to almost half (Coll-Cárdenas et al., 2011). In a contrasting study, ozone was used not only to treat salmon fillets in order to reduce microbial loads but also to ensure minimal chemical changes in the product. In this case, ozone was applied as spray aqueous solution (1 and 1.5 mg/L) using 1–3 passes through the nozzles. *Listeria innocua* was inoculated as a surrogate of the pathogenic strain (*Listeria monocytogenes*). Results showed that the highest concentration (1.5 mg/L) was effective against the inactivation of mesophiles and *L. innocua* without significantly affecting the lipid oxidation of the product during storage at 4°C (Crowe et al., 2011). In this specific case, ozone might be a good option to treat fish muscle since by quick application using the spray nozzles, the contact time is reduced, and then the possibility to develop rancidity and other chemical reactions with the food is reduced. This result is opposite to the previous example, in which beef tissue was in constant contact for 24 h with ozone producing undesirable effects.

Shucked mussels were treated with ozone not only to reduce microbial loads with the processing conditions as shown in Table 20.1 but also to study quality indicators during shelf life. Trimethylamine values remained low during storage for all treated and control samples; total volatile basic nitrogen values were lower for the ozone-treated samples; meanwhile, thiobarbituric acid values decreased

TABLE 20.1
Examples of Microbial Inactivation in Selected Foods and Surfaces Using Ozone

Product	Treatment Conditions	Target Micro-Organism	Results	Reference
Fruits and vegetables				
Dried figs	0.1–9 ppm, O_3 gas, 20°C	*E. coli*, *B. cereus*, *B. cereus* spores	3.5 log for *E. coli*, *B. cereus*; 2 log for *B. cereus* spores	Akbas and Ozdemir (2008)
Red bell peppers	0.3 ppm aqueous solution, 15°C	*L. innocua*	1.5 log	Alexandre et al. (2011b)
Fresh-cut lettuce	0.5 mg/L continuous ozonation, 30 min, 15°C	Mesophiles, coliforms, molds, and yeasts	3 log, 2.5 log, 2.1 log reduction, respectively	Alexopoulos et al. (2013)
Date fruits	5 ppm, 60 min, ozone gas	*S. aureus*, mesophiles, molds and yeasts, coliforms	3.5 log, 4 log, 3.93 log, and 3.54 log reduction, respectively	Habibi Najafi and Haddad Khodaparast (2009)
Pre-cut green peppers	Ozonated water (highest concentration 3.95 mg/L), 9°C–17.5°C, 20 s–30 min	Mesophiles	0.72 log reduction	Ketteringgham et al. (2006)
Tomatoes	5 ppm, 3 min, ozone gas	*E. coli* ATCC 11775	2.2 log reduction	Bermúdez-Aguirre and Barbosa-Cánovas (2013)
Fresh-cut lettuce	4.5 ppm, 3.5 min, ozonated water (10°C–26°C)	*L. monocytogenes*	2.5 log reduction	Ölmez and Akbas (2009)
Papaya	O_3 gas, 4 ppm, 144 h	*Colletotrichum gloeosporioides*	Inhibition of mycelia growth and spore growth	Ong et al. (2012)
Cantaloupe	O_3 gas, 10,000 ppm, 30 min + hot water (75°C, 1 min)	Mesophiles, psychrotrophic, coliforms, molds	3.8 log, 5.1 log, 2.2 log, 2.3 log reduction	Selma et al. (2008)
Shredded lettuce	Ozonated water, 2 ppm, 5 min	*Shigella sonnei*	1.7 log	Selma et al. (2007)
Meat products				
Chicken breasts	>2000 ppm, O_3 gas, 20°C	*S. infantis*, *P. aeruginosa*	97% and 95% reduction, respectively	Al-Haddad et al. (2005)
Chicken	0–9 min, O_3 gas, 33 mg/min	*L. monocytogenes*	6 log reduction	Muthukumar and Muthuchamy (2013)
Miscellaneous				
Wheat flour	Air flow rate 5 L/min, 30 min, O_3 gas	Mesophiles	50% reduction	Li et al. (2012)
Shucked mussels	Aqueous solution (1 mg/L), 90 min	Mesophiles, *Pseudomonas* spp., H_2S-producing bacteria, *Brochothrix thermosphacta*, lactic acid bacteria and *Enterobacteriaceae*	2.1 log, 1.1 log, 2.5 log, 1.4 log, 0.8 log, and 1.5 log reduction, respectively	Manousaridis et al. (2005)
Salmon trout	O_3 gas, 0.1×10^{-3} g/L, 20 min (cold smoke, 5°C)	*L. innocua*	1 log reduction	Vaz-Velho et al. (2006)
Surfaces				
Polypropylene	1 μg/g ozonized water, 2 min, 25°C	*S. aureus*	99% reduction	Cabo et al. (2009)

during storage. Sensory characteristics of the ozone-treated mussels also showed good values, and in general terms, the shelf life of treated mussels was extended by 3 days compared with the control sample (Manousaridis et al., 2005).

Whey proteins and egg white proteins were also studied under ozone treatments, along with gaseous and aqueous solutions. The treatments were carried out for 15 min at 4.5 ppm and 8°C. Results were positive regarding foam stability and foam formation, as in both cases foam stability was enhanced after ozone treatment. However, solubility of proteins was reduced, specifically for egg white proteins, and also emulsion stability and formation was affected negatively for whey proteins (Uzun et al., 2012). This example also shows how animal proteins can be affected by treatment with ozone, similar to the case of wheat flour and noodles, in which proteins were also affected by oxidation. Although for some products these changes are not considered to be positive, some of these new properties after ozone treatment could be considered for the development of new products or ingredients.

20.3.3 Proteins

Ozone has been tested in a number of vegetable and animal products and quality in terms of physicochemical properties, and nutritional components have been evaluated after treatment to ensure the quality of the product. Most studies have focused on microbial inactivation, and just a few studies have focused on the evaluation of specific nutrients treated with ozone. For example, ozone is able to oxidize the polypeptide support of proteins, to break peptide bonds and to modify amino acids. Some amino acids are more sensitive to oxidation because of the action of ozone, like aromatic amino acids such as tyrosine, tryptophan, and phenylalanine; amino acids containing sulfur, such as cysteine and methionine; and aliphatic amino acids such as arginine, lysine, proline, and histidine (Uzun et al., 2012).

20.4 ANTIMICROBIAL ACTIVITY

Ozone has been shown to be effective against a number of microorganisms, including Gram-positive and Gram-negative bacteria, viruses, spores, and fungi. Its efficacy depends on the type and population of the microorganism, temperature, pH, and purity of the medium (Alexandre et al., 2011a). Güzel-Seydim et al. (2004) mentioned a list of products that have been tested with ozone for microbial disinfection such as poultry carcasses, poultry chill water, apples and oranges, raspberries and grapes, onions, lettuce, black peppercorn, ground black pepper, and cheddar cheese. Other tested and inactivated microorganisms using ozone are shown in Table 20.1. The number of products is also shown; as can be observed, these products represent a high variety of food items from very different sources.

20.4.1 Mechanism of Inactivation in Bacteria, Viruses, and Spores

One microbial inactivation theory regarding ozone relates to the powerful action of this oxidant gas against the glycoproteins and glycolipids of the cell surface membrane, changing the cell permeability with a further cell lysis and leakage of cellular material. Ozone oxidases polyunsaturated fatty acids to acid peroxides. Also, ozone can cause action in the sulfhydryl and amino acids groups of certain vital enzymes, peptides, and proteins, thus modifying all the cellular activities (Khadre et al., 2001; Güzel-Seydim et al., 2004; Al-Haddad et al., 2005). Reports about the action of ozone on bacterial nucleic acids have also been discussed; some authors believe that ozone can act on some components such as thymine, being the most sensitive, and others such as cytosine and uracil, being less sensitive. Decrease in transcription activity of DNA and also some mutations have been observed in bacteria after treatment with ozone; thus, some authors consider these effects as possible mechanisms of inactivation (Khadre et al., 2001).

In the case of viruses, ozone acts on the RNA and modifies the polypeptide chains in protein coats (Güzel-Seydim et al., 2004). Once the coat is attacked by ozone, there is liberation and inactivation of nucleic acids. RNA is inactivated and released from the virus; similar behavior has been observed for DNA. In a bacteriophage, ozone was able to destroy the head, collar, contractile case, end plate, and tail fibers, with further release of DNA from the head. In general terms, it is believed that the main damage of ozone to viruses is related to the alteration of proteins in coats or envelopes (Khadre et al., 2001).

Spores are known for their resistance to physical and chemical agents and the hard coat that protects them from these agents. Finding specific lethal agents that can inactivate these spores is a challenge. Bacterial and fungal spores have been shown to be more resistant to ozone than bacteria vegetative cells (Khadre et al., 2001). Ozone has been shown to have some sporicidal activity in certain species of *Bacillus* spp. Based on the results of those studies, it is possible to see that ozone, as a powerful oxidant agent, degrades some of the outer components and allows the spore core to be in contact with this sanitizer. The spore coat represents 50% of its volume and is composed of about 80% of proteins that act as a barrier to some enzymes, which could affect spore viability. In a study conducted by Khadre and Yousef (2001), ozone damaged the outer spore coat, although the inner coat was not highly damaged. Most of those spores lost their viability.

20.4.2 Microbial Inactivation

As previously mentioned, ozone can inactivate microorganisms in foods. The number of tested products in the literature is extensive, ranging from fruits and vegetables to animal products and other food-related items such as packaging materials and food-processing equipment. Here, some examples will be presented based on additional characteristics that have shown interesting results. Also, more examples are shown in Table 20.1 along with the respective references for further details about equipment and specific conditions.

Ozone in combination with slurry ice was used to preserve a flatfish with high demand for European Turbot (*Psetta maxima*). The ice/water system was composed in a ratio of 40:60 and 3.3% salinity; working temperature was −1.5°C. The mix with ozone was able to delay the growth of mesophiles and psychrotrophic bacteria in skin and muscle. Not only were the microbiological aspects improved, but the chemical reactions of lipid hydrolysis and lipid oxidation were also retarded. Sensory analysis of fish storage with ozone in the slurry-ice was also superior during 14 days, duplicating the expected shelf life of fish that was not stored with ozone (Campos et al., 2006).

In another example, ozone (in a gaseous state, 4.3 mg/L, 5 min) was used to inactivate *Salmonella enterica* and *E. coli* O157:H7 in tomato, lettuce, and cantaloupe seeds. Initial counts for both microorganisms were about 6 log, and after the treatment the inactivation was about 4 log for all products regardless of the microorganisms (Trinetta et al., 2011). Even though ozone could be considered effective in the reduction of pathogenic bacteria, it is clear that under these conditions, there are some limitations to achieving total inactivation of the microorganisms. Also, in the same study, the germination rate of the seeds was examined after the ozone treatment and the results showed that there were significant differences between the control seeds and the ozonized ones.

Ozone is also able to inactivate certain viruses that are more resistant than vegetative bacteria. Some of the tested viruses against ozone treatment are bacteriophage f2, MS2, hepatitis A, poliomyelitis, poliovirus type 1, rotavirus human, and vesicular stomatitis virus, among others. Bacteriophages are the most resistant during ozone inactivation (Khadre et al., 2001).

20.4.3 Mycotoxins

Mycotoxins produce molds, mainly on grains, and they are important in the food industry because of the large economic losses that are generated due to their presence in several products, in addition to the fatal health effects that might occur if contaminated products are consumed. Techniques such as the use of pesticides have been used to control the growth of molds and to reduce the concentration of mycotoxins in some food items. However, consumers around the world are vocally opposed to the use of pesticides in crops and other food items; thus, there is a need to find alternatives to control the growth of molds and their products in foods, whereas at the same time providing consumers with safe products. Ozone has become an interesting technology that is being explored for this purpose. Some of the common mycotoxins that have been reported due to the presence of species of *Aspergillus*, *Penicillium*, and *Fusarium* are aflatoxins, cyclopiazonic acid, zearalenone, fumonisin B_1, patulin, ochratoxin, and secalonic acid D (McKenzie et al., 1997).

The specific case of aflatoxins is well known because of their high toxicity and mutagenic and carcinogenic properties, in addition to the high stability of these metabolites and their main sources *Aspergillus parasiticus* and *Aspergillus flavus* during food processing and storage. Ozone has been successfully used to degrade aflatoxin B_1 in a number of products, and in a recent study, ozone treatment was applied to identify the toxicity of those components generated after degradation of the toxin. Aqueous ozone was used to degrade aflatoxin B_1 and to considerably reduce the toxicity of by-products of the toxin after treatment (Luo et al., 2013). However, in a similar study, ozone was tested as an aqueous solution (1.71 mg/L) and as a gas treatment (13.8 mg/L) to inactivate aflatoxin B_1 in dried figs; results showed that the gaseous treatment was more effective in degradation of the toxin, and also the effectiveness was increased as processing time was increased. The total time was up to 180 min, where 95% of reduction of the concentration of aflatoxin B_1 was observed (Zorlugenç et al., 2008).

High concentration of ozone delivered quickly on grains has also been shown to be effective in degradation of aflatoxins without nutrient destruction. Aflatoxins B_1 and G were degraded quickly with 2% of ozone; meanwhile, aflatoxins B_2 and G_2 required up to 20% to be oxidized by ozone. Patulin, cyclopiazonic acid, ochratoxin A, secalonic acid D, and zearalenone required only 15 s to be degraded with ozone and without formation of by-products. Also, toxicity of these compounds after 3 s of treatment was reduced considerably. Fumonisin B_1 was degraded after 15 s of treatment, but there was still formation of some toxic by-products (McKenzie et al., 1997).

In a study related to the growth control of *Eurotium* species in naan bread, a typical Indian product, ozone was used for fumigation of the product after it was processed with or without sucrose. Two different treatments were used: ozone low concentration for long term (0.4 μmol/mol during 21 days), and ozone high concentration for short term (300 μmol/mol for 5–120 min). The long-term treatment was able to reduce the number of spores in those media without sucrose only; the short-term period was able to reduce all the spores detected by conventional methods only after 120 min of treatment. The possible mechanism of action is related to oxidative stress generated in the cells because of ozone, changing the intracellular redox potential with a subsequent modification of the basic cellular processes (Antony-Babu and Singleton, 2011).

20.5 SPECIFIC CASES

20.5.1 Food Industry Equipment, Surfaces, and Packaging Materials

In the food industry, there is a current need to find good alternatives for disinfection of surfaces and equipment that are in contact with food. After processing several items, the constant problem of biofilm and growth of microorganisms is common. Currently, some alternatives include the use of hot water and steam or chlorinated solutions. However, the first option represents an expensive method because of the amount of water that is required; the use of chlorinated solutions is avoided by some food processors because of the chemical residues that can be deposited on surfaces or the resistance of pathogenic microorganisms to chlorinated compounds. A good alternative that has been suggested recently has been the use of ozone, which is most often used as an aqueous solution. Some examples of the use of ozone are in the dairy industry, where it has been able to destroy the characteristic biofilm on stainless steel surface sand to inactivate dairy industry microorganisms. In another example, the comparison between three disinfection methods, that is, heat, ozone, and chlorinated water, were tested against a number of common food industry microorganisms. After 10 min, ozone (0.6 ppm) was able to reduce about 7.3 log of all of the bacterial species, showing the effectiveness of the technology when followed by the heat treatment (Güzel-Seydim et al., 2004). It has also been shown that the effectiveness of ozone disinfection for food-processing equipment can be reduced when organic compounds are present and, based on that finding, the cleaning process should be designed based on time and concentration. In that study, conducted by Güzel-Seydim et al. (2004), the power of ozone as a disinfectant was evaluated using typical compounds of food together with microorganisms, showing that starch does not offer any protection for microorganisms, followed by gums. However, in the presence of fat or proteins, microorganisms can be protected in these compounds because ozone cannot easily penetrate them. So, based on that, the correct dose of ozone should be applied to ensure that all of the microorganisms are reached and inactivated regardless of the presence of these food components.

The presence of spores in packaging materials and food-processing surfaces represents a challenge for the food industry because of the high resistance of these microorganisms to conventional cleaning and disinfection methods such as heat, hydrogen peroxide, and chlorine. Several species of the spores of *Bacillus* spp. were treated with concentrations of 11 μg/mL of ozonated water and reductions resulted in 1.3–6.1 log reduction, depending on the species, with *Bacillus stearothermophilus* being the most resistant micro-organism, and *Bacillus cereus* less resistant. Using hydrogen peroxide (10%, w/w), spore reductions were from 0.32 to 1.6 log (Khadre and Yousef, 2001), showing the effectiveness of ozone treatment for spore inactivation.

20.5.2 Water and Liquid Food Products

20.5.2.1 Water

Ozone has been used for several decades to pasteurize bottled water in many European countries; it is also permitted in some Asian countries. The use of ozone to treat water was limited in the United States up until a few decades ago, but in 1982, the USDA approved the use of ozone as a GRAS substance for bottled water (Kim et al., 2003). Ozone is safe to use for drinkable water because after a few seconds, ozone is decomposed to nontoxic products without leaving any chemical residues, in contrast to chlorination treatment (Selma et al., 2007). However, the process of treating drinkable water with ozone also depends on water quality, such as dissolved organic carbon concentration, alkalinity, temperature, pH, and bromide concentration. This last parameter must be controlled in the original water source because bromide oxidases to bromate when it reacts with ozone, with bromate being a potential human carcinogen. So far, bromate is the only by-product controlled in drinkable water treated with ozone. According to the European Union and the United States Environmental Protection Agency (USEPA), the standard for bromide is 10 mg/L (Meunier et al., 2006). In general terms, it is preferred that water without inorganic or organic matter be treated with ozone, since contact with this potent oxidant will promote chemical reactions and generation of by-products (Camel and Bermond, 1998), while also reducing microbial inactivation effectiveness.

However, drinkable water can be contaminated from the source with some pathogenic bacteria such as *Salmonella* spp. or *E. coli*, as well as hosting viruses and protozoa that are resistant to conventional treatments of drinkable water. For example, one of the target microorganisms found in drinkable water is the parasite *Cryptosporidium parvum* that has been related to several cryptosporidiosis outbreaks worldwide (Biswas et al., 2003). Other protozoa linked with waterborne are *Giardia lamblia* and *Cyclospora*, which have been found in several outbreaks around the world. However, ozone is able to inactivate these microorganisms, even though some parasites are highly resistant to chlorine treatment. During ozone treatment, there is also a high resistance to being inactivated in *C. parvum* compared with *Giardia* cysts (Yousef et al., 2011).

In other studies, water inoculated with *Shigella sonnei* was treated with ozone (1.6 and 2.2 ppm) for 1 min, and it was possible to inactivate 3.7 and 5.6 log reduction of the bacteria (Selma et al., 2007).

20.5.2.2 Fruit Juices

It is well known in the beverage industry that fruit juices have been related to foodborne outbreaks involving the presence of pathogenic bacteria. Several reasons for these outbreaks have been suggested such as the increase in thermal resistance of microorganisms to conventional thermal pasteurization or under-processing conditions. Also, there have been recent trends in finding alternatives to pasteurizing fruit juices with minor changes in their sensory and nutritional quality. One of these new trends is the use of ozone treatments to inactivate pathogenic and spoilage microorganisms but at the same time to retain the fresh-like characteristics of the juice to yield a product similar to fresh-squeezed juice.

Several attempts have been made with successful results. For example, in a study of orange juice processed with gaseous ozone (75–78 μg/mL) for 0–18 min, several varieties of juice were used, such as a model system, fresh unfiltered juice, juice without pulp, and juice filtered through two different sieves. The objective of the research was to inactivate *E. coli* cells. Results showed that gaseous ozone was able to inactivate the FDA-required 5-log reduction of the pathogenic microorganism; however, processing times were dependent on the juice. As expected, when the juice was free of organic matter (such as pulp), inactivation was faster, such as in the model system (60 s) or in the low pulp product (6 min). Meanwhile, in the unfiltered juice, inactivation required longer time (15–18 min) to achieve microbial death because of the interaction of ozone with the pulp (Patil et al., 2009). However, color degradation has been observed in orange juice after being treated with ozone. Because of the high oxidant power of this triatomic molecule, ozone can react and degrade some organic compounds. Ozone is able to attack conjugated double bonds in chromophores of some organic dyes and carotenoids; these last ones having aromatic rings in their chemical structure. Ozone and free radicals such as OH^- from the aqueous solution can open the rings and promote the oxidation of some organic acids, ketones, and aldehydes (Tiwari et al., 2008).

Similar behavior was observed for strawberry juice processed with ozone (1.6%–7.8% w/w, 0–10 min). The interaction of this molecule with anthocyanins of strawberry juice generated a significant reduction up to 98.2% and also the reduction of ascorbic acid (85.8%) after the longest processing time and highest ozone concentration. The color of the strawberry juice was also degraded because of the decrease in anthocyanin content, which provides the strawberry with its typical red color. The reaction between ozone and anthocyanins can occur directly as an oxidation process, or the reaction can be promoted because of the production of free radicals, again with the breakage and formation of new bonds between compounds (Tiwari et al., 2009c). Similar results in anthocyanin content and color degradation were found for blackberry juice after treatment with ozone in similar conditions (Tiwari et al., 2009a) and grape juice (Tiwari et al., 2009b).

Finally, in the last example of fruit juices, apple juice was also treated with ozone under concentration between 1% and 4.8% (w/w) and processing times from 0 to 10 min. Again, the ozone-treated product showed color degradation and considerable reduction in chlorogenic acid, caffeic acid, cinnamic acid, and total phenol content. Also, some rheological parameters, such as consistency index and flow behavior index, changed drastically after processing (Torres et al., 2011). It is evident from these few examples of fruit juices processed with ozone that this technology might represent a good alternative for microbial inactivation. However, much more research should be conducted to minimize undesirable changes in physicochemical properties and nutrient content.

20.6 CONCLUSIONS

Ozone represents a viable option for microbial decontamination in the food industry. As can be observed, the uses of ozone are variable, and in most cases the physicochemical properties of the product remain without significant changes. Ozone is easy to produce, and the treatment is not very expensive compared with other more sophisticated technologies. Also, ozone represents the option of treating foods without leaving chemical residues or by-products, such as in the case of chlorine solutions. Some minor issues have been observed in specific products that could be addressed by finding an effective combination between ozone dose and processing times. Major concerns have been observed in fruit juices treated with ozone, showing that, indeed, more research should be conducted to achieve the FDA 5-log reduction standard for pasteurization and the possibility for using ozone as an alternative in this thermal process.

REFERENCES

Akbas, M.Y. and Ozdemir, M. 2008. Application of gaseous ozone to control populations of *Escherichia coli*, *Bacillus cereus* and *Bacillus cereus* spores in dried figs. *Food Microbiology.* 25: 386–391.

Alexandre, E.M.C., Brandão, T.R.S., and Silva, C.M.S. 2011a. Modeling microbial load reduction in foods due to ozone impact. *Procedia Food Science.* 1: 836–841.

Alexandre, E.M.C., Santos-Pedro, D.M., Brandão, T.R.S., and Silva, C.M.S. 2011b. Influence of aqueous ozone, blanching and combined treatments on microbial load of red bell peppers, strawberries and watercress. *Journal of Food Engineering.* 105: 277–282.

Alexopoulos, A., Plessas, S., Ceciu, S., Lazar, V., Mantzourani, I., Voidarou, C., Stavropoulou, E., and Bezirtzoglou, E. 2013. Evaluation of ozone efficacy on the reduction of microbial population of fresh cut lettuce (*Lactuca* sativa) and green bell pepper (*Capsicum annuum*). *Food Control.* 30: 491–496.

Al-Haddad, K.S.H., Al-Qassemi, R.A.S., and Robinson, R.K. 2005. The use of gaseous ozone and gas packaging to control populations of *Salmonella infantis* and *Pseudomonas aeruginosa* on the skin of chicken portions. *Food Control.* 16: 415–420.

Alothman, M., Graham, D.M., Rice, R.G., and Strasser, J.H. 2010. Ozone-induced changes of antioxidant capacity of fresh-cut tropical fruits. *Innovative Food Science and Emerging Technologies.* 11: 666–671.

An, J., Zhang, M., and Lu, Q. 2007. Changes in some quality indexes in fresh-cut green asparagus pretreated with aqueous ozone and subsequent modified atmosphere packaging. *Journal of Food Engineering.* 78: 340–344.

Antony-Babu, S. and Singleton, I. 2011. Effects of ozone exposure on the xerophilic fungus, *Eurotium amstelodami* IS-SAB-01, isolated from naan bread. *International Journal of Food Microbiology.* 144: 331–336.

Bermúdez-Aguirre, D. and Barbosa-Cánovas, G.V. 2013. Disinfection of selected vegetables under nonthermal treatments: Chlorine, citric acid, ultraviolet light and ozone. *Food Control.* 29: 82–90.

Biswas, K., Craik, S., Smith, D.W., and Belosevic, M. 2003. Synergistic inactivation of *Cryptosporidium parvum* using ozone followed by free chlorine in natural water. *Water Research.* 37: 4737–4747.

Cabo, M.L., Herrera, J.J., Crespo, M.D., and Pastoriza, L. 2009. Comparison among the effectiveness of ozone, nisin and benzalkonium chloride for the elimination of planktonic cells and biofilms of *Staphylococcus aureus* CECT4459 on polypropylene. *Food Control.* 20: 521–525.

Camel, V. and Bermond, A. 1998. The use of ozone and associated oxidation processes in drinking water treatment. *Water Research.* 32(11): 3208–3222.

Campos, C.A., Losada, V., Rodríguez, O., Aubourg, S.P., and Barros-Velázquez, J. 2006. Evaluation of an ozone-slurry ice combined refrigeration systems for the storage of farmed turbot (*Psetta maxima*). *Food Chemistry.* 97: 223–230.

Chauhan, O.P., Raju, P.S., Ravi, N., Singh, A., and Bawa, A.S. 2011. Effectiveness of ozone in combination with controlled atmosphere on quality characteristics including lignification of carrot sticks. *Journal of Food Engineering.* 102: 43–48.

Coll-Cárdenas, F., Andrés, S., Giannuzzi, L., and Zaritzky, N. 2011. Antimicrobial action and effects on beef quality attributes of a gaseous ozone treatment at refrigeration temperatures. *Food Control.* 22: 1442–1447.

Crowe, K.M., Skonberg, D., Bushway, A., and Baxter, S. 2011. Application of ozone sprays as a strategy to improve the microbial safety and quality of salmon fillets. *Food Control.* 25: 464–468.

Cullen, P.J., Tiwari, B.K., O'Donnell, C.P., and Muthukumarappan, K. 2009. Modeling approaches to ozone processing of liquid foods. *Trends in Food Science and Technology.* 20: 125–136.

Güzel-Seydim, Z., Bever Jr., P.I., and Greene, A.K. 2004. Efficacy of ozone to reduce bacterial populations in the presence of food components. *Food Microbiology.* 21: 475–479.

Güzel-Seydim, Z.B., Greene, A.K., and Seydim, A.C. 2004. Use of ozone in the food industry. *Lebensmittel-Wissenschaft & Technologie.* 37: 453–460.

Habibi Najafi, M.B. and Haddad Khodaparast, M.H. 2009. Efficacy of ozone to reduce microbial populations in date fruits. *Food Control.* 20: 27–30.

Ketteringgham, L., Gausseres, R., James, S.J., and James, C. 2006. Application of aqueous ozone for treating pre-cut green peppers (*Capsicum annum* L.). *Journal of Food Engineering.* 76: 104–111.

Khadre, M.A. and Yousef, A.E. 2001. Sporicidal action of ozone and hydrogen peroxide: A comparative study. *International Journal of Food Microbiology.* 71: 131–138.

Khadre, M.A., Yousef, A.E., and Kim, G.J. 2001. Microbiological aspects of ozone applications in food: A review. *Journal of Food Science.* 66(9): 1242–1252.

Kim, J.G., Yousef, P.A., and Khadre, M.A. 2003. Ozone and its current and future application in the food industry. *Advances in Food and Nutrition Research.* 45: 167–218.

Klockow, P.A. and Keener, K.M. 2009. Safety and quality assessment of packaged spinach treated with a novel ozone-generation system. *LWT—Food Science and Technology.* 42: 1047–1053.

Li, M., Zhu, K.X., Wang, B.W., Guo, X.N., Peng, W., and Zhou, H.M. 2012. Evaluation of the quality characteristics of wheat flour and shelf-life of fresh noodles as affected by ozone treatment. *Food Chemistry.* 135: 2163–2169.

Luo, X., Wang, R., Wang, L., Wang, Y., and Chen, Z. 2013. Structure elucidation and toxicity of the degradation products of aflatoxin B1 by aqueous ozone. *Food Control.* 31: 331–336.

Manousaridis, G., Nerantzaki, A., Paleologos, E.K., Tsiotsias, A., Savvaidis, I.N., and Kontominas, M.G. 2005. Effect of ozone on microbial, chemical and sensory attributes of shucked mussels. *Food Microbiology.* 22: 1–9.

McKenzie, K.S., Sarr, A.B., Mayura, K., Bailey, R.H., Miller, D.R., Rogers, T.D., Norred, W.P., Voss, K.A., Plattner, R.D., Kubena, L.F., and Phillips. 1997. Oxidative degradation and detoxification of mycotoxins using a novel source of ozone. *Food and Chemical Toxicology.* 35: 807–820.

Meunier, L., Canonica, S., and von Gunten, U. 2006. Implications of sequential use of UV and ozone for drinking water quality. *Water Research.* 40: 1864–1876.

Muthukumar, A. and Muthuchamy, M. 2013. Optimization of ozone in gaseous phase to inactivate *Listeria monocytogenes* on raw chicken samples. *Food Research International.* 54(1): 1128–1130. http://dx.doi.org/10.1016/j.foodres.2012.12.016.

Ölmez, H. and Akbas, M.Y. 2009. Optimization of ozone treatment of fresh-cut green leaf lettuce. *Journal of Food Engineering.* 90: 487–494.

Ong, M.K., Kazi, F.K., Forney C.F., and Ali, A. 2012. Effect of gaseous ozone on papaya anthracnose. *Food and Bioprocess Technology.* doi: 10.1007/s11947-012-1013-4.

Patil, S., Bourke, P., Frias, J.M., Tiwari, B.K., and Cullen, P.J. 2009. Inactivation of *Escherichia coli* in orange juice using ozone. *Innovative Food Science and Emerging Technologies.* 10: 551–557.

Selma, M.V., Beltrán, D., Allende, A., Chacón-Vera, E., and Gil, M.I. 2007. Elimination by ozone of *Shigella sonnei* in shredded lettuce and water. *Food Microbiology.* 24: 492–499.

Selma, M.V., Ibañez, A.M., Allende, A., Cantwell, M., and Suslow, T. 2008. Effect of gaseous ozone and hot water on microbial and sensorial quality of cantaloupe and potential transference of *Escherichia coli* O157:H7 during cutting. *Food Microbiology.* 25: 162–168.

Sopher, C.D., Graham, D.M., Rice, R.G., and Strasser, J.H. 2002. Studies on the use of ozone in production agriculture and food processing. *Proceedings of the International Ozone Association*, Scottsdale, AZ, pp. 1–15.

Tiwari, B.K., Muthukumarappan, K., O'Donnell, C.P., and Cullen, P.J. 2008. Modeling color degradation of orange juice by ozone treatment using response surface methodology. *Journal of Food Engineering.* 88: 553–560.

Tiwari, B.K., O'Donnell, C.P., Muthukumarappan, K., and Cullen, P.J. 2009a. Anthocyanin and color degradation in ozone treated blackberry juice. *Innovative Food Science and Emerging Technologies.* 10: 70–75.

Tiwari, B.K., O'Donnell, C.P., Patras, A., Brunton, N., and Cullen, P.J. 2009b. Anthocyanins and color degradation in ozonated grape juice. *Food and Chemical Toxicology.* 47: 2824–2829.

Tiwari, B.K., O'Donnell, C.P., Patras, A., Brunton, N., and Cullen, P.J. 2009c. Effect of ozone processing on anthocyanins and ascorbic acid degradation of strawberry juice. *Food Chemistry.* 113: 1119–1126.

Torres, B., Tiwari, B.K., Patras, A., Wijngaard, H.H., Brunton, N., Cullen, P.J., and O'Donnell, C.P. 2011. Effect of ozone processing on the color, rheological properties and phenolic content of apple juice. *Food Chemistry.* 124: 721–726.

Trinetta, V., Vaidya, N., Linton, R., and Morgan, M. 2011. A comparative study on the effectiveness of chlorine dioxide gas, ozone gas and e-beam irradiation treatments for inactivation of pathogens inoculated onto tomato, cantaloupe and lettuce seeds. *International Journal of Food Microbiology.* 146: 203–206.

Uzun, H., Ibanoglu, E., Catal, H., and Ibanoglu, S. 2012. Effects of ozone on functional properties of proteins. *Food Chemistry.* 134: 647–654.

Vaz-Velho, M., Silva, M., Pessoa, J., and Gibbs, P. 2006. Inactivation by ozone of *Listeria innocua* on salmon-trout during cold-smoke processing. *Food Control.* 17: 609–616.

Wang, H., Feng, H., and Luo, Y. 2004. Microbial reduction and storage quality of fresh-cut cilantro washed with acidic electrolyzed water and aqueous ozone. *Food Research International.* 37: 949–956.

Yousef, A.E., Vurma, M., and Rodriguez-Romo, L.A. 2011. Basics of ozone sanitization and food applications. In: *Nonthermal Processing Technologies for Food*, H. Zhang, G.V. Barbosa-Cánovas, V.M. Balasubramaniam, C.P. Dunne, D.F. Farkas, and J.T.C. Yuan, eds. Ames, IA: IFT-Press, Wiley-Blackwell Publishing, pp. 291–313.

Zorlugenç, B., Zorlugenç, F.K., Öztekin, S., and Evliya, I.B. 2008. The influence of gaseous ozone and ozonated water on microbial flora and degradation of aflatoxin B_1 in dried figs. *Food and Chemical Toxicology.* 46: 3593–3597.

Index

G

H

N

O

P

U

V

W

X